Reconceptualizing Mathematics for Elementary School Teachers

Second Edition

Judith Sowder

Larry Sowder

Susan Nickerson

San Diego State University

W.H. Freeman and Company
New York

Senior Publisher: Ruth Baruth

Executive Editor: Terri Ward

Marketing Manager: Steve Thomas

Developmental Editors: Elka Block, Katrina Wilhelm

Senior Media Editor: Laura Judge

Associate Media Editors: Courtney Elezovic, Catriona Kaplan

Associate Editor: Jorge Amaral

Editorial Assistant: Liam Ferguson

Marketing Assistant: Alissa Nigro

Photo Editor: Ted Szczepanski

Cover and Text Designer: Blake Logan

Senior Project Editor: Vivien Weiss

Illustrations: MPS Ltd.

Illustration Coordinator: Bill Page

Production Coordinator: Susan Wein

Composition: MPS Ltd.

Printing and Binding: Quad Graphics

Note: The materials in this book were developed at San Diego State University in part with funding from the National Science Foundation Grant No. ESI 9354104. The content of this book is solely the responsibility of the authors and does not necessarily reflect the views of the National Science Foundation.

Library of Congress Control Number: 2012943870

ISBN-13: 978-1-4641-0335-3
ISBN-10: 1-4641-0335-6

© 2014, 2010 by W. H. Freeman and Company

Printed in the United States of America

First printing

W. H. Freeman and Company
41 Madison Avenue
New York, NY 10010
Houndmills, Basingstoke RG21 6XS, England
www.whfreeman.com

CONTENTS IN BRIEF

Note: Appendices F–K are available via the Book Companion Website.

CONTENTS

Note: Appendices F–K are available via the
Book Companion Website.

About the Authors

Judith Sowder is a Professor Emerita of Mathematics and Statistics at San Diego State University. Her research has focused on the development of number sense and on the instructional effects of teachers' mathematical knowledge at the elementary and middle school level. She served from 1996 to 2000 as editor of the *Journal for Research in Mathematics Education* and served a three-year term on the National Council of Teachers of Mathematics Board of Directors. She is an author of the middle school content chapter of the Conference Board of Mathematical Sciences document *The Mathematical Education of Teachers,* published in 2001 by the Mathematical Association of America. She has directed numerous projects funded by the National Science Foundation and the Department of Education. In 2000 she received the Lifetime Achievement Award from the National Council of Teachers of Mathematics.

Larry Sowder is Professor Emeritus of Mathematics and Statistics at San Diego State University. He taught mathematics to preservice elementary school teachers for more than 30 years. His work in a special program in San Diego elementary schools also shaped his convictions about how courses in mathematics for preservice teachers should be pitched, as did his joint research investigating how children in the usual Grades 4–8 curriculum solve "story" problems. He served on the National Research Council Committee that published *Educating Teachers of Science, Mathematics, and Technology* (NRC, 2001).

Susan Nickerson was involved in the development of these materials as a graduate student and has taught both preservice and inservice teachers using these materials. Now an Associate Professor in San Diego State University's Department of Mathematics and Statistics, her research interest is long-term professional development of elementary and middle school teachers. In particular, her focus is describing, analyzing, and understanding effective contexts that promote teachers' knowledge of mathematics and mathematics teaching.

All three authors consider themselves as having dual roles—as teacher educators and as researchers on the learning and teaching of mathematics. Most of their research took place in elementary and middle school classrooms and in professional development settings with teachers of these grades.

Message to Prospective and Practicing Teachers

Touch the future—be a teacher.[1]

Teachers matter. A variety of recently published documents support the notion that the key to increasing students' mathematical knowledge and to closing the achievement gap is to put knowledgeable teachers in every classroom.[2]

Research on the relationship between teachers' mathematical knowledge and students' achievement confirms the importance of teachers' content knowledge. It is self-evident that teachers cannot teach what they do not know.[3]

Reflect on these questions:

Do you remember how you thought and felt about mathematics during your K–8 experiences? How would an emphasis on *making sense of mathematics* have changed your thoughts and feelings about mathematics?

There is more to teaching than conveying mathematical knowledge. Although this course is about mathematics rather than about methods of teaching mathematics, you will learn a great deal that will add to your background for teaching mathematics knowledgeably and for creating positive experiences for your students. When you can converse with your students about mathematical ideas, reasons, goals, and relationships, they begin to make sense of mathematics. Sense making is a theme that permeates all aspects of this course. Students who know that mathematics makes sense will search for meaning and become successful learners.

If you gain an appreciation of the fact that the rules of mathematics have underlying reasons and that mathematical ideas and vocabulary can add precision to a teacher's talk and thought, then you will indeed *reconceptualize mathematics*. As you further study mathematics and especially as you teach mathematics to a variety of children, you will have more opportunities to learn to touch the future in positive ways. Teachers *do* matter.

How to Use This Textbook

The first step toward using a textbook productively is to understand the structure of the book and what it contains. This textbook is separated into four parts—"Part I: Reasoning About Numbers and Quantities," "Part II: Reasoning About Algebra and Change," "Part III: Reasoning About Shapes and Measurement", and "Part IV: Reasoning About Chance and Data".

Each part includes several chapters, each with the following format:

▶ A brief description of what the chapter is about.

▶ Chapter sections that contain the following elements:

- An introduction to the section.
- Prose that introduces and explains the section's content. This prose is interspersed with activities, discussions, and reflective questions called "Think Abouts."

[1]From a recruiting poster for a teacher education program

[2]Sowder, J. T. (2007). The mathematical education and development of teachers. In F. K. Lester, Jr. (Ed.), *Second handbook of research on mathematics teaching and learning*, pp. 157–223. Charlotte, NC: Information Age Publishing. Quote from p. 157.

[3]National Mathematics Advisory Panel. (2008). *Foundations for success: The final report of the National Mathematics Advisory Panel*. Washington, DC: U.S. Department of Education. Quote from p. xxi.

- *Activities* intended to be worked in small groups or pairs, providing some hands-on experiences with the content. In most instances, they can be completed and discussed in class. Discussion on activities is worthwhile because many times other groups will take a different approach.

- *Discussions* intended primarily for whole-class discussion. These discussions provide more opportunities to converse about the mathematics being learned, to listen to the reasoning of others, and to voice disagreement when appropriate. (Remember to disagree with an idea, not with the person.)

- *Think Abouts* intended to invite you to pause and reflect on what you have just read.

- *Information Boxes* that contain definitions and other important ideas. This information needs to be considered very thoughtfully to fully understand it. Examples are provided to help you make sense of them. If you don't understand the boxed information, ask your instructor or discuss it with another student.

- *Examples* providing needed clarification and opportunities to explore meaning and demonstrate procedures.

- *Take-Away Messages* summarizing the overriding messages of the section.

- *Learning Exercises* to be used for homework and sometimes for classroom discussion or activities. Note the term "learning" used here. Although the exercises provide opportunities for practice, they are intended primarily to help you think though the section content, note the relationships, and extend what you have learned. *Not all problems in the exercises can be solved quickly*. Some are challenging and may take more time than you are accustomed to spending on a problem. By knowing this, you should not become discouraged if the path to an answer is not quickly apparent. (Be sure to read the upcoming section on how to write your reasoning.) Most of these exercise sets end with *Supplementary Learning Exercises* that provide more examples and learning experiences if you need additional help to understand or consolidate the content.

▶ Most chapters have a section called *Issues for Learning*, which very often contains a discussion of some of the research about children's learning of topics associated with the content of the chapter. Reading about these issues will help you understand some of the conceptual difficulties children have in learning particular content, and will help you relate what you are learning now to teaching children in the future.

▶ *Elementary Textbook Pages* that show you how the mathematics you are learning is connected to the classroom.

▶ A *Check Yourself* section at the end of every chapter that will help you organize what you have learned (or should have learned) in that chapter. This list can serve to organize your review of the chapter for examinations.

In addition to chapters, you will find the following at the end of the book:

▶ *Appendix A: Video Clips Illustrating Children's Mathematical Thinking* (for Parts I and III only) provides a Website for video clips that show student thinking, and gives questions for reflection and discussion about what you see in these clips. Your instructor will tell you when to view these video clips.

▶ *Appendix B: Summary of Formulas* provides a summary of the formulas used in your work involving geometry in Part III.

▶ *Appendix C: About the Geometer's Sketchpad® Lessons* gives some information on using the online lessons for the *Geometer's Sketchpad®* software.

▶ *Appendices D and E* provide a table of random numbers and data sets.

▶ *Selected Answers (and Hints)* for many Learning Exercises. Space does not permit each answer to include all of the rationale in writing about your reasoning, but you should provide this information as you work through the exercises. Answers for the Supplementary Learning Exercises can be found in the *Student Solutions Manual*.

▶ A **Glossary** of important terms.

▶ **Masters** provide pages containing dot paper and templates for base materials, pattern blocks, and nets to be copied if needed.

You will see that this text includes a large margin on the outside of each page so that you can freely write notes to help you remember how a problem was worked or to clarify the text, based on what happens in class. We suggest placing the appropriate text pages in a three-hole binder so that you can intersperse papers on which you have worked assignments. These materials are produced at the lowest possible cost so that you can mark up the text, keep it, and use it to help you plan lessons for teaching.

Finally, Appendices F–K are available on the Book Companion Website. They include a review of some basic arithmetic skills you may have forgotten, a quick reminder on how to use a protractor, and instructions for using *TI-73, Fathom,* and *Excel.*

How to Write Your Reasoning

You will have opportunities to write your reasoning when you work on assignments and on exams, and frequently when you work through classroom activities. Writing your reasoning begins with describing your understanding of the situation or context that gives meaning to the present task. Do not take this aspect of writing lightly. Many difficulties arise because of initial misunderstandings of the situation and of the task being proposed to you. Another way to look at this aspect of writing is that you are making your tacit assumptions explicit because, with the exception of execution errors (such as "2 + 3 = 4"), most difficulties can be traced back to tacit assumptions or inadequate understandings of a situation.

Focus on the decisions you make, and write about those decisions and their reasons. For example, if you decide that some quantity has to be cut into three parts rather than five parts, then explain what motivated that decision. Also mention the consequences of your decisions, such as "This area, which I called '1 square inch,' is now in pieces that are each 1/3 square inch." When you make consequences explicit, you have additional information with which to work as you proceed. Additional information makes it easier to remember where you've been and how you might get to where you are trying to go.

You will (should) do a lot of writing in this course. The reason for having you write is that, in writing, you have to organize your ideas and understand them coherently. If you give incoherent explanations and shaky analyses for a situation or idea, you probably don't understand it. But if you do understand an idea or situation deeply, you should be able to speak about it and write about it coherently and conceptually.

Feel free to write about insights gained as you worked on the problem. Some people have found it useful to do their work in two columns—one for remarks to yourself about the ideas that occur to you and one for scratch. Be sure to include sketches, diagrams, etc., and write in complete, easy-to-read sentences.

Learn to read your own writing as if you had not seen it before.

▶ Is what you wrote just a sequence of "things to do"? If so, then someone reading it with the intention of understanding *why* you did what you did will be unable to replicate your reasoning. There is an important distinction between reporting what you did and explaining what you did. The first is simply an account or description of what you did, but the latter includes reasons for what you did. The latter is an explanation; the former is a report.

▶ Does your explanation make sense? Do not fall into the trap of reading your sentences merely to be reminded of what you had in your mind when you wrote them. Another person will not read them with the advantage of knowing what you were trying to say. The other person can only try to make sense of what you actually wrote. Don't expect an instructor to "know what you really meant."

▶ Are sentences grammatically correct?

Assignments will contain questions and activities. You should present your work on each question logically, clearly, and neatly. Your work should explain what you have done—it should reflect your reasoning. Your write-up should present more than an answer to a question. It should tell a story—the story of your emerging insight into the ideas behind the question. Think of your write-up as an assignment about how you present your thinking. Uninterpretable or unorganized scratch work is not acceptable.

How to Read Others' Work (the Instructor's, a Classmate's, or the Textbook)

▶ Make sure you are clear about the context or situation that gives rise to an example and its main elements. Note that this is not a straightforward process. It will often require significant reflection and inner conversation about what is going on.

▶ Interpret what you read. Be sure that you know what is being said. Paraphrasing, making a drawing, or trying a new example may help.

▶ Interpret one sentence at a time. If a sentence's meaning is not clear, then do not go further. Instead,

• Think about the sentence's role within the context of the initial situation, come up with a question about the situation, and assess the writer's overall aims and method.

• Rephrase the sentence by asking, "What might the writer be trying to say?"

• Construct examples or make a drawing that will clarify the situation.

• Try to generalize from examples.

• Ask someone what he or she thinks the sentence means. (This should not be your initial or default approach to resolving a confusion.)

The overall goal of this course is that you come to understand the mathematics deeply so that you are able to participate in meaningful conversations about this mathematics and its applications with your peers and eventually with your students.

New to This Edition

▶ Chapter 1, "Reasoning About Quantities," begins with the motivation for making sense of mathematical situations and for quantitative reasoning.

▶ Chapter 6, "Meanings for Fractions," and Chapter 7, "Computing with Fractions," were revised for richer development of fractions and more support for operations with fractions.

▶ Chapter 10, "Integers and Other Number Systems," has been rewritten to reflect new research on children's thinking with respect to signed numbers.

▶ Chapter 12, "What Is Algebra?," ties mental math, children's thinking, and *Common Core State Standards* mathematical models to the properties. This is in accord with the CCSS's increased emphasis in the elementary grades on operation properties.

▶ The order of Chapters 16 and 17 ("Polygons" and "Polyhedra") was reversed in this edition since many instructors reported having taught them in this new sequence.

They are written to be more independent of each other so that instructors may teach them in either order.

▶ Chapter 27, "Quantifying Uncertainty," has been revised to reflect emphases suggested by users (with more focus on clarifying the meaning of probability).

▶ Many chapters reflect revisions to align with the CCSS:

- Five "Issues for Learning" specifically address CCSS's influence on curriculum with respect to Number and Operations, Algebra, Geometry, Measurement, Probability and Statistics. A sixth Issue for Learning specifically addresses the *CCSS Standards for Mathematical Practice.*

- Dozens of mathematical definitions are introduced earlier or have introductions that were moved into the text exposition from the Learning Exercises.

- Additional physical and mathematical models for mathematical operations are incorporated with a particular focus on the number line, area model, and base-ten blocks. The number line and area model are particularly significant in the *Common Core State Standards.*

- Changes to some language that is more in line with the *Common Core State Standards* (e.g., relevant and irrelevant attributes changed to defining and nondefining attributes).

▶ Additional Activities, Think Abouts, and Examples throughout the book.

▶ Approximately 150 new Learning Exercises and Supplementary Learning Exercises.

▶ Appendix A includes new video clips of students engaged in mathematical tasks. These videos help teachers better understand students' reasoning about and understanding of mathematics. We have found that watching these videos motivates the need to study elementary mathematics deeply.

▶ Expanded number of MathClips available through MathPortal. An icon in the margin of the Instructor's Edition indicates areas where these animated whiteboard videos are available to further illustrate key concepts.

Media and Supplements

For Students

 **mathportal**

www.yourmathportal.com

(Access code or online purchase required.) MathPortal is the digital gateway to *Reconceptualizing Mathematics,* Second Edition, designed to enrich the course and enhance students' study skills through a collection of Web-based tools. MathPortal integrates a rich suite of diagnostic, assessment, tutorial, and enrichment features, enabling students to master mathematics at their own pace. It is organized around three main teaching and learning components:

1. **Interactive e-Book** offers a complete and customizable online version of the text, fully integrated with all of the media resources available with *Reconceptualizing Mathematics.* The e-Book allows students to quickly search the text, highlight key areas, and add notes about what they're reading. Instructors can customize the e-Book to add, hide, and reorder content; add their own material; and highlight key text for students.

2. **Resources** organizes all of the resources for *Reconceptualizing Mathematics* into one location for ease of use. Includes the following for students:

   - **Student Solutions Manual** provides solutions to all the Supplementary Learning Exercises in the text.

   - **Gradable Vocabulary Flashcards** offer additional practice to students learning key terms.

- **MathClips** present animated whiteboard videos illuminating key concepts in each chapter.
- **LEARNING**Curve is a formative quizzing system that offers immediate feedback at the question level to help students master course material.

Includes the following for instructors:

- **Instructor's Resource Manual with Full Solutions** includes detailed instructor notes that are referenced in the marginal notes throughout the instructor's edition of *Reconceptualizing Mathematics*. These notes provide additional content background on several topics in the lessons and suggest explanations and advice for teaching particular topics. This resource manual also provides full solutions to all exercises in the text.
- **Test Bank** offers more than a thousand multiple-choice, true/false, and short-answer questions.
- **Lecture PowerPoint Slides** offer a detailed lecture presentation of key concepts covered in each chapter of *Reconceptualizing Mathematics*.
- **SolutionMaster** is a Web-based version of the solutions in the *Instructor's Resource Manual with Full Solutions*. This easy-to-use tool allows instructors to generate a solution file for any set of homework exercises. Solutions can be downloaded in PDF format for convenient printing and posting.

3. **Assignments** organizes assignments and guides instructors through an easy-to-create assignment process providing access to questions from the Test Bank and Exercises from the text, including many algorithmic problems. The Assignment Center enables instructors to create their own assignments from a variety of question types for machine-gradable assignments. This powerful assignment manager allows instructors to select their preferred policies in regard to scheduling, maximum attempts, time limitations, feedback, and more.

Reconceptualizing Mathematics, Second Edition, e-Book

The complete e-Book is also available stand-alone, outside of MathPortal, at approximately one-half the cost of the printed textbook.

Companion Website at www.whfreeman.com/reconceptmath2e

For students, this site serves as a *free* 24/7 electronic study guide, and it includes such features as self-quizzes and vocabulary terms.

Printed Student Solutions Manual

ISBN: 1-4641-0899-4

This printed manual provides full solutions for all the Supplementary Learning Exercises in the text.

For Instructors

The **Instructor's Website** (www.whfreeman.com/reconceptmath2e) requires user registration as an instructor and features all of the student Web materials plus:

▶ PowerPoint Slides containing all textbook figures and tables

▶ Lecture PowerPoint Slides

Printed Instructor's Resource Manual with Full Solutions

ISBN: 1-4641-0900-1

This printed manual includes detailed instructor notes that are referenced in the marginal notes throughout the instructor's edition of *Reconceptualizing Mathematics,* Second

Edition. These notes provide additional content background on several topics in the lessons and suggested explanations and advice for teaching particular topics. This resource manual also provides full solutions to all exercises in the text.

Computerized Test Bank

ISBN: 1-4641-2382-9

The Test Bank contains more than a thousand multiple-choice, true/false, and short-answer questions to generate quizzes and tests for each chapter of the text. Available on CD-ROM (for Windows and Mac), questions can be downloaded, edited, and re-sequenced to suit each instructor's needs.

Enhanced Instructor's Resource CD-ROM

ISBN: 1-4641-2385-3

Allows instructors to search and export (by key term or chapter):

- All text images
- *Instructor's Resource Manual with Full Solutions*
- Lecture PowerPoint Slides
- Test Bank files

Course Management Systems

W. H. Freeman and Company provides courses for Blackboard, WebCT (Campus Edition and Vista), Angel, Desire2Learn, Moodle, and Sakai course management systems. These are completely integrated solutions that you can easily customize and adapt to meet your teaching goals and course objectives. Visit www.macmillanhighered.com/Catalog/other/Coursepack for more information.

i‑clicker is a two-way radio frequency classroom response solution developed by educators for educators. University of Illinois physicists Tim Stelzer, Gary Gladding, Mats Selen, and Benny Brown created the i-clicker system after using competing class-room response solutions and discovering they were neither classroom-appropriate nor student-friendly. Each step of i-clicker's development has been informed by teaching and learning. i-clicker is superior to other systems from both a pedagogical and technical standpoint. To learn more about packaging i-clicker with this textbook, please contact your local sales rep or visit www.iclicker.com.

Acknowledgments

We wish to thank the following instructors from across the United States and Canada who offered comments that assisted in the development and refinement of the second edition. We are grateful for their contributions.

Robin Ayers, *Western Kentucky University*

Kate Best, *Western Carolina University*

Darryl Corey, *Kennesaw State University*

Terry Crites, *Northern Arizona University*

Kathryn Ernie, *University of Wisconsin, River Falls*

Axelle Faughn, *Western Carolina University*

Amy Goodman, *Baylor University*

Angela Harlan, *University of Wisconsin, Whitewater*

Susan Herring, *Sonoma State University*

Judy Bowman Kidd, *James Madison University*

Danny Lau, *Gainesville State College*

Andrea S. Levy, *Seattle Central Community College*

Sandra Martinez, *University of Texas at El Paso*

Petra Menz, *Simon Fraser University*

Eva Thanheiser, *Portland State University*

Elizabeth Thompson, *East Carolina University*
Susan Tummers, *El Camino College*
Agnes Tuska, *California State University, Fresno*

Ruth S. Whitmore, *University of Wisconsin, Whitewater*
Rebecca Wong, *West Valley College*

We are also grateful to those who reviewed the previous edition:

Paul S. Ache, III, *Kutztown University of Pennsylvania*
Holly Garrett Anthony, *Tennessee Technological University*
Judith M. Arms, *University of Washington*
Edward L. Barszcz, *Buffalo State College*
Linda E. Barton, *Ball State University*
Sister Rita Basta, BVM, *California State University, Northridge*
Mary Beard, *Kapi'olani Community College*
William Blubaugh, *University of Northern Colorado*
Phyllis Barsch Bolin, *Abilene Christian University*
Jeremy Boggess, *Indiana University*
Mark Bollman, *Albion College*
Fernanda Botelho, *The University of Memphis*
Tambi C. Boyle, *Mt. Hood Community College*
Ivette Chuca, *El Paso Community College*
José N. Contreras, *The University of Southern Mississippi*
Judith Covington, *Louisiana State University in Shreveport*
Malgorzata Dubiel, *Simon Fraser University*
Laurie A. Dunlap, *The University of Akron*
Kimberly R. Elce, *California State University, Sacramento*
Aimee Ellington, *Virginia Commonwealth University*
Steven Greenstein, *The University of Texas at Austin*
Shannon M. Guerrero, *Northern Arizona University*
Karen Heinz, *Rowan University*
Joan E. Henn, *Eastern Illinois University*
Emam Hoosain, *Augusta State University*

Chris Jepsen, *San Diego State University*
Shelly M. Jones, *Central Connecticut State University*
Laurie Jordan, *Loyola University Chicago*
Bernice Kastner, *Simon Fraser University*
Margaret L. Kidd, *California State University, Fullerton*
Perry Lee, *Kutztown University of Pennsylvania*
Jane-Jane Lo, *Western Michigan University*
Jodelle S. W. Magner, *Buffalo State University*
Tracie McLemore Salinas, *Appalachian State University*
Maria T. Mitchell, *Central Connecticut State University*
Barbara Moses, *Bowling Green State University*
Diana S. Perdue, *Virginia State University*
Calvin Piston, *John Brown University*
Benjamin T. Rhodes Jr., *The University of Texas at Austin*
Amy Diekelman Rushall, *Northern Arizona University*
Ann Sitomer, *Portland Community College*
Kelli M. Slaten, *University of North Carolina Wilmington*
Diana F. Steele, *Northern Illinois University*
Allison Sutton, *Austin Community College*
Debora Tamanaha-Justeson, *Grossmont College*
Virginia McShane Warfield, *University of Washington*
Karen Waters, *Columbus State University*
Andrew T. Wilson, *Austin Peay State University*

In particular, we wish to thank Jim Lewis of the University of Nebraska for his use of these instructional materials, for his support along the way, and for telling W. H. Freeman that they should publish our text. We warmly thank Sister Rita Basta and Professors Joel Zeitlin and Jerry Gold, of California State University Northridge, for their many helpful comments on the use of this text and for allowing us to videotape their classes. We also want to thank the many graduate students at San Diego State for their enthusiasm and feedback as they taught our courses for prospective elementary teachers and in particular Ian Whitacre, who supervised many of these classes and provided us with problems and student feedback. We were fortunate to have the assistance of Lisa Lamb, Ian Whitacre, Jessica Pierson Bishop, Randy Philipp, and Bonnie Schapelle in revising Chapter 10. We appreciate Ian Whitacre's and Bridget Druken's thoughtful contributions to the media accompanying the text.[*] To faculty in our Professional Development Collaborative, Karen Payne, Steve Klass, and Mike Maxon, who used our text materials over the years in their professional development classes, thank you for your invaluable feedback and ideas for improvement. We thank Janet Bowers for her particular expertise in producing the *Geometer's Sketchpad*® lessons and the Over and Back applet, both of which add value for students.

We know that the production of a textbook so different from earlier endeavors at W. H. Freeman presented a challenge for the staff, one they undertook with skill and patience. In particular, we want to thank Terri Ward, Executive Editor, who supported us throughout this process; Katrina Wilhelm, who worked with us through the many details of preparing the text for publication; and Steve Rigolosi, Director of Market Research and Development, who also guided us through the preliminary edition and cheered us on with positive comments from users of the text. We also want to thank Vivien Weiss and Patti Brecht for their careful work on the second edition. Three others contracted by W. H. Freeman also deserve recognition. First and foremost, we thank Elka Block, our development editor, for her careful editing, thoughtful suggestions, patience, and support during the long process of producing the preliminary edition, the first edition, and the second edition. We thank both John Samons and Keith Harrington for their careful reading and for finding the errors we missed in our own proofreading.

We acknowledge the substantial contributions of Alba (now deceased) and Pat Thompson, Joanne Lobato, Janet Bowers, Nicholas Branca (now deceased), Randy Philipp, and the doctoral students who worked on this project with us when it was in its infancy. Our doctoral program will be a major benefactor of this work, as we continue to support a new generation of mathematics educators who will carry on the practice and study of the teaching and learning of mathematics.

We three wish to thank Susan's daughter Rachel for her inspiration and her unique contributions to this venture. Susan thanks her husband Steve for his love, encouragement, and counsel. May we continue to learn from each other how to foster and educate developing mathematicians.

Judith Sowder   Larry Sowder   Susan Nickerson

[*] Their contributions are based upon work supported by the National Science Foundation under grant number DRL-0918780, *Mapping Developmental Trajectories of Students' Conceptions of Integers*. Any opinions, findings, conclusions, and recommendations expressed in their material are those of the authors and do not necessarily reflect the views of NSF.

Part I

Reasoning About Numbers and Quantities

Much of what you see in these first eleven chapters will look familiar. However, you will be asked to begin understanding the mathematics of arithmetic at a more fundamental level than you may now do. Knowing *how* to compute is not sufficient for teaching; a teacher should understand the underlying reasons we compute as we do.

An important focus of these chapters is on *using* numbers. We do this by helping you develop the skill of quantitative reasoning. This reasoning is fundamental to solving problems involving quantities (such as the speed of a car or the price of chips). You may recognize such problems as *story problems* or *word problems.* In Chapters 1, 8, and 9 you will learn how to approach such problems using quantitative reasoning.

Our base-ten system of numbers, introduced in Chapter 2, is the foundation for all of the procedures we use for computing with whole numbers. A teacher must acquire a deep understanding of the base-ten system and the arithmetic procedures, which is the focus of Chapters 3, 4, and 5. We then turn to the study of fractions, decimals, and percents in Chapters 6 and 7. Many people in our society do not know how to compute with these numbers, or how these numbers are related. The emphasis on understanding these numbers and how to operate on them will give you a foundation to teach fractions, decimals, and percents well. Your students will benefit if they do not develop the belief that they cannot "do" fractions. Teachers who can teach arithmetic in a manner that students understand will better prepare their students for the mathematics yet to come.

Chapter 10 expands the type of numbers we used in previous chapters to include negative numbers, and discusses arithmetic operations on these numbers. Again, the focus is on understanding. Chapter 11, on number theory, provides different ways of thinking about and using numbers. Topics such as factors, multiples, prime numbers, and divisibility are found in this chapter.

An underlying theme throughout these chapters is an introduction to children's thinking about the mathematics they do. You will be surprised to learn the ways that children do mathematics, and we think that knowing how children reason mathematically will convince you that you need to know mathematics at a much deeper level than you likely do now. Prospective teachers are often fascinated with these glimpses into the ways children reason, particularly if they have a good foundation in their work with numbers.

Chapter 1

Reasoning About Quantities

We encounter quantities of many kinds each day. In this chapter you will be asked to think about quantities and the manner in which we use them to better understand our lives. In particular, you will encounter "story problems" and learn how to solve them using a powerful reasoning process, most likely aided by a drawing of some sort.

1.1 Ways of Thinking About Solving Story Problems

In solving story problems, how do children decide what to do? In one study[1] students were found to use the following seven different strategies. The first six strategies are based on something other than understanding the problem.

1. Find the numbers in the problem and just do something to them, usually addition because that is the easiest operation.

2. Guess at the operation to be used, perhaps based on what has been most recently studied.

3. Let the numbers "tell" you what to do. (One student said, "If it's like 78 and maybe 54, then I'd probably either add or multiply. But if the numbers are 78 and 3, it looks like a division because of the size of the numbers.")

4. Try all the operations and then choose the most reasonable answer. (This strategy often works for one-step problems, but rarely works for two-step problems.)

5. Look for "key" words to decide what operation to use. For example, "all together" means to add. (This strategy works sometimes, but not all the time. Also, words like "of " and "is" often signify multiplication and equals, but some students confuse the two.)

6. Narrow the choices, based on expected size of the answer. (For example, when a student used division on a problem involving reduction in a photocopy machine, he said he did because "it's reducing something, and that means taking it away or dividing it.")

7. Choose an operation based on understanding the problem. (Often students would make a drawing when they used this strategy. Unfortunately, though, few of the children [sixth- and eighth-graders, of average or above average ability in mathematics] used this strategy.)

Only the last strategy was considered a mature strategy. These children understood the problem because they had understood the relationships among the quantities in the situation.

DISCUSSION 1   Which Strategy?

1. Do you recognize any of these strategies as ones you have used?

2. Some teachers teach their students to use Strategy 5. Give an example of how this strategy could lead to an incorrect answer. ●

What is a quantity? Consider the following five questions:

• How long do humans live?
• How fast is the wind blowing?

- Which is more crowded, New York City or Mexico City?
- How big is this room?
- How far is it around the earth?

The answer to each of these questions involves some quantity.

> A **quantity** is anything (an object, event, or quality thereof) that can be measured or counted. The **value of a quantity** is its measure or the number of items that are counted. A value of a quantity involves a number and a unit of measure or number of units.

For example, the length of a room is a quantity. It can be measured. Suppose the measurement is 14 feet. Note that *14* is a number, and *feet* is a unit of measure: 14 feet is the value of the quantity "length of the room." The number of people on the bus is another example of a quantity. Suppose the count is 22 people. Note that *22* is a number, and the unit counted is *people*, so 22 is the value associated with the quantity "number of people on the bus."

A person's age, the speed of the wind, the population density of New York City, the area of this room, and the distance around the earth at the equator are all examples of quantities.

Think About …
What are some possible values of the five quantities mentioned above?
Notice that population alone will not be sufficient to address the question of how crowded a city or a country is. Why?

Not all qualities of objects, events, or persons can be quantified. Consider *love*. Young children sometimes attempt to quantify love by stretching their arms out wide when asked, "How much do you love me?" but love is not a quantity. Love, anger, boredom, and interest are some examples of qualities that are not quantifiable. Feelings, in general, are not quantifiable—thus, they are difficult to assess.

Think About …
Name some other "things," besides feelings, that are not quantities.

The fact that a quantity is not the same as a number should be clear to you. In fact, one can think of a quantity without knowing its value. For example, the amount of rain fallen on a given day is a quantity, regardless of whether or not someone measured the actual number of inches of rain fallen. One can speak of the amount of rain fallen without knowing how many inches fell. Likewise, one can speak of a dog's weight, a tank's capacity, the speed of the wind, or the amount of time it takes to do a chore (all quantities) without knowing their actual values.

DISCUSSION 2 Identifying Quantities and Measures

1. Identify the quantity or quantities addressed in each of the following questions:
   a. How tall is the Eiffel Tower?
   b. How fast does water come out of a faucet?
   c. Which country is wealthier, Honduras or Mozambique?
   d. How much damage did the earthquake cause?
2. Identify an appropriate unit of measure that can be used to determine the value of the quantities involved in answering the questions in Problem 1. Is the wealth of a country measured the same way as the wealth of an individual? Explain. ●

🔵 DISCUSSION 3   Easy to Quantify?

1. Many attributes or qualities of objects are easily quantifiable. Others are not so straightforward. Of the following items, which are easy to quantify and which aren't?

   a. the weight of a newborn baby   b. the gross national product
   c. student achievement   d. blood pressure
   e. livability of a city   f. infant mortality rate
   g. teaching effectiveness   h. human intelligence
   i. air quality   j. wealth of a nation

2. How is each item in Problem 1 typically quantified?

3. What sorts of events and things do you think primitive humans felt a need to quantify? Make a list. How do you think primitive societies kept track of the values of those quantities?

4. Name some attributes of objects (besides those listed in Problem 1) that are not quantifiable or that are hard to quantify.

5. Name some quantities for which units of measure have been only recently developed. 🔵

TAKE-AWAY MESSAGE . . .   In this introductory section you have learned to identify quantities and their values and to distinguish between the two. This understanding is the basis for the quantitative analyses in the next section.

1.2 Quantitative Analysis

In this section you will use what you have learned about quantities in Section 1.1 to analyze problem situations in terms of their quantitative structure. Such analyses are essential to being skillful at solving mathematical problems.

> For the purposes of this course, to *understand a problem situation* means to understand the quantities embedded in the situation and how they are related to one another.

Understanding a problem situation "drives" the solution to the problem. Without such understanding, the only recourse a person has is to guess at the calculations that need to be performed. It is important that you work through this section with care and attention. Analyzing problem situations quantitatively is central to the remainder of this course and other courses that are part of your preparation to teach school mathematics.

📝 ACTIVITY 1   The Hot Dog Problem

▶ Albert ate $2\frac{3}{4}$ hot dogs and Reba ate $1\frac{1}{2}$ hot dogs. What part of all the hot dogs they consumed did Albert eat? ◀

This fairly simple problem is given here to illustrate the process of using a quantitative analysis to solve a problem. We first analyze the situation in terms of the quantities it involves and how those quantities are related to one another: its **quantitative structure**. To do so productively, it is extremely important to be specific about what the quantities are. For example, it is not sufficient to say that "hot dogs eaten" is a quantity in this problem situation. If you indicated "hot dogs eaten" as one of the quantities and specified no more, then someone could ask: Which hot dogs? The ones Albert ate? The ones Reba ate? The total number of hot dogs eaten?

To understand the quantitative structure of this problem situation, you can do the following:

1. Name as many quantities as you can that are involved in this situation. Be aware that some quantities may not be explicitly stated although they are essential to the situation. Also, just because the value of a quantity may not be known or given, the quantity is still part of the situation's quantitative structure.

2. For each quantity, if the value is given, write it in the appropriate space. If the value is not given, indicate that the value is unknown and write the unit you would use to measure it. You may need more space than provided here. (*Reminder:* There should be no numbers appearing in the Quantity column.) Compare your list with those of other students in your class.

Quantity	Value
a. number of hot dogs eaten by Albert	$2\frac{3}{4}$
b.	
c.	etc.

3. Make a drawing that illustrates this problem. Here, the outer rectangle indicates the total number of hot dogs eaten, and the parts show the portions eaten by each person. (Note that drawings such as this one need not be drawn to scale. Here, the number of hot dogs eaten by Albert is shown only to be more than the hot dogs eaten by Reba.)

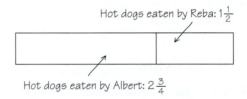

Hot dogs eaten by Reba: $1\frac{1}{2}$

Hot dogs eaten by Albert: $2\frac{3}{4}$

4. Finally, use the drawing to solve the problem. The drawing shows that the total number of hot dogs eaten is $2\frac{3}{4} + 1\frac{1}{2}$, or $4\frac{1}{4}$. What part of the total was eaten by Albert?

    Albert ate $2\frac{3}{4}$ of the total ($4\frac{1}{4}$) hot dogs, which is $\frac{11}{17}$ of the hot dogs. This value is about two-thirds of the hot dogs, a reasonable answer to this question. ●

> One must *analyze* a situation *in terms of the quantities and the relationships present* in order to gain an understanding that can lead to a meaningful solution. Such an activity is called a **quantitative analysis**.

In problems such as the one in Activity 1, you may find that you don't really need to think through all four steps. However, more difficult problems, such as those in the next two activities, are more easily solved by undertaking a careful quantitative analysis. You will improve your skill at solving problems when you engage in such analyses because you will come to a better understanding of the problem.

 ACTIVITY 2   Sisters and Brothers

Try this problem by undertaking a quantitative analysis before you read the solution. Then compare your solution with the one shown here.

▶ Two women, Alma and Beatrice, each had a brother, Alfred and Benito, respectively. The two women argued about which woman stood taller over her brother. It turned out that Alma won the argument by a 17-cm difference. Alma was 186 cm tall. Alfred was 87 cm tall. Beatrice was 193 cm tall. How tall was Benito? ◀

First, name the quantities involved in the problem.

a. Alma's height

b. Beatrice's height

c. Alfred's height

d. Benito's height

e. the difference between Alma's and Alfred's heights

f. the difference between Beatrice's and Benito's heights

g. the difference between the differences in the heights of the sister and brother pairs

Next, identify the values of the quantities.

a.	Alma's height	186 cm
b.	Beatrice's height	193 cm
c.	Alfred's height	87 cm
d.	Benito's height	not known
e.	the difference between Alma's and Alfred's heights	not known
f.	the difference between Beatrice's and Benito's heights	not known
g.	the difference between the differences in the heights of the sister and brother pairs	17 cm

[Quantity (g) is a given quantity and crucial to solving the problem.]

Draw a picture involving these quantities. Here is one possibility:

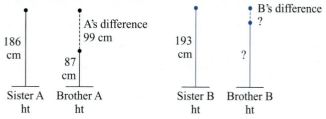

Need to relate the two differences:
A's difference is 17 cm more than B's difference.

99 cm   17 cm
         99 − 17 = 82 cm

Or, perhaps this alternative drawing would help clarify the problem. Next to each quantity, place its value if known, and use these values known to find values unknown.

Alma's height (186 cm)
Alfred's height (87 cm)   Difference A (99 cm)
                                                        Difference of differences (17 cm)
Beatrice's height (193 cm)
Benito's height (???)   Difference B (???)

Finally, use the drawing and/or diagram to solve the problem. Notice that in this case the solution is shown in the first drawing, and can easily be found in the second drawing. But note that the question asked was Benito's height. According to the drawings, it is 193 cm − 82 cm = 111 cm. ●

Often students begin a problem such as the one in Activity 2 by asking themselves, "What operations do I need to perform, with which numbers, and in what order?" Instead of these questions, you would be much better off asking yourself questions such as those given on the next page.

Notes

Quantitative Analysis Questions

- Can I imagine the situation, as though I am acting it out?
- What quantities are involved? What do I know about their nature? Do some change over time? Do some stay constant?
- What is the *quantity of interest*? That is, what quantity, and its value, am I asked to find or describe?
- How could making a diagram (or a sequence of diagrams) help reveal relationships involving the quantity of interest and other quantities?
- What quantities and values do I know that could help me find the value of the quantity of interest?
- What quantities and values can I derive to help me find the value of the quantity of interest?

And so on.

The point is that you need to be specific when doing a quantitative analysis of a given situation. How specific? It is difficult to say in general, because it will depend on the situation. Use your common sense in analyzing the situation. When you first read a problem, avoid trying to think about numbers and the operations of addition, subtraction, multiplication, and division. Instead, start out by posing the questions such as those listed above and try to answer them. Once you've done that, you will have a better understanding of the problem, and thus you will be well on your way to solving it. Keep in mind that understanding the problem is the most difficult aspect of solving it. Once you understand the problem, *what to do to solve it* often follows quite easily.

ACTIVITY 3 Down the Drain

Here is another problem situation on which to practice quantitative analysis.

▶ Water is flowing from a faucet into an empty tub at 4.5 gallons per minute. After 4 minutes, a drain in the tub is opened, and the water begins to flow out at 6.3 gallons per minute. ◀

a. Will the tub ever fill up completely without overflowing?

b. Will it ever empty completely?

c. What if the faucet is turned off after 4 minutes?

d. What if the rates of flow in and out are reversed?

e. What assumptions do we have to make in order to answer these questions? ●

The following problem is a common, though not always popular, problem type to solve in algebra classes. Although you may be able to solve the following problem using an algebraic solution, use quantitative reasoning so as not to mask any real understanding of this type of problem.

ACTIVITY 4 All Aboard!

Continental Coast Railroad provides service between Chicago, Illinois, and Birmingham, Alabama. A new passenger train leaves Chicago headed toward Birmingham at the same time that a freight train leaves Birmingham for Chicago. The trains are 583 miles apart on parallel tracks. The new passenger train travels at a constant speed of 120 miles per hour, and the freight train travels at a constant speed of 75 miles per hour.

a. Where do they meet?

b. Suppose the passenger train left one hour later than the freight train. Where do they meet? ●

When you carry out a quantitative analysis of a situation, the questions you ask yourself should be guided by your common sense. A sense-making approach to understanding a situation and then solving a problem is much more productive than trying to decide right away which computations, formulas, or mathematical techniques you need in order to solve the problem at hand. Using common sense may lead you to make some sort of a diagram. Never be embarrassed to use a diagram. Such diagrams often enhance your understanding of the situation, because they help you to think more explicitly about the quantities that are involved. Deciding what operations you need to perform often follows naturally from a good understanding of the situation, the quantities in it, and how those quantities are related to one another.

TAKE-AWAY MESSAGE . . . You should be able to determine the quantities and their relationships within a given problem situation, and you should be able to use this information, together with drawings as needed, to solve problems. These steps are often useful: (1) List the quantities that are essential to the problem. (2) List known values for these quantities. (3) Determine the relationships involved, which is frequently done more easily with a drawing. (4) Use the knowledge of these relationships to solve the problem. This approach may not seem easy at first, but it becomes a powerful tool for understanding problem situations. This type of quantitative analysis can also be used with algebra story problems, and thus, when used with elementary school students, prepares them for algebra.

Learning Exercises for Section 1.2

1. Some problems are simple enough that the quantitative structure is obvious, particularly after a drawing is made. The following problems are from a fifth-grade textbook.[2] For each problem given, make a drawing and then provide the answer to the problem.

   a. The highest elevation in North America is Mount McKinley, Alaska, which is 20,320 feet above sea level. The lowest elevation in North America is Death Valley, California, which is 282 feet below sea level. What is the change in elevation from the top of Mount McKinley to Death Valley?

   b. The most valuable violin in the world is the Kreutzer, created in Italy in 1727. It was sold at auction for $1,516,000 in England in 1998. How old was the violin when it was sold?

   c. Two sculptures are similar. The height of one sculpture is four times the height of the other sculpture. The smaller sculpture is 2.5 feet tall. How tall is the larger sculpture?

   d. Aiko had $20 to buy candles. She returned 2 candles for which she had paid $4.75 each. Then she bought 3 candles for $3.50 each and 1 candle for $5.00. How much money did Aiko have then?

   e. In Ted's class, students were asked to name their favorite sport. Football was the response of $\frac{1}{8}$ of them. If 3 students said football, how many students are in Ted's class?

   f. The first year of a dog's life equals 15 "human years." The second year equals 10 human years. Every year thereafter equals 3 human years. Use this formula to find a 6-year-old dog's age in human years.

2. These problems are from a sixth-grade textbook[3] from a different series. This time, undertake a full quantitative analysis to solve each of the problems.

   a. At Loud Sounds Music Warehouse, CDs are regularly priced at $9.95 and tapes are regularly priced at $6.95. Every day this month the store is offering a 10% discount on all CDs and tapes. Joshua and Jeremy go to Loud Sounds to buy a tape and a CD. They do not have much money, so they have pooled their funds. When they get to the store, they find that there is another discount plan just for that day—if they buy 3 or more items, they can save 20% (instead of 10%) on each item. If they buy a CD and a tape, how much money will they spend after the store adds a 6% sales tax on the discounted prices?

   b. Kelly wants to fence in a rectangular space in her yard, 9 meters by 7.5 meters. The salesperson at the supply store recommends that she put up posts every 1.5 meters. The posts cost $2.19 each. Kelly will also need to buy wire mesh to string between the posts. The wire mesh is sold by the meter from large rolls and costs $5.98 a meter. A gate to fit in one of the spaces between the posts costs $25.89. Seven staples are needed to attach the wire mesh to each post. Staples come in boxes of 50, and each box costs $3.99. How much will the materials cost before sales tax?

3. Amtrail trains provide efficient, nonstop transportation between Los Angeles and San Diego. Train A leaves Los Angeles headed toward San Diego at the same time that Train B leaves San Diego headed for Los Angeles, traveling on parallel tracks. Train A travels at a constant speed of 84 miles per hour. Train B travels at a constant speed of 92 miles per hour. The two stations are 132 miles apart. How long after they leave their respective stations do the trains meet?

4. My brother and I walk the same route to school every day. My brother takes 40 minutes to get to school and I take 30 minutes. Today, my brother left 8 minutes before I did.[4]

   a. How long will it take me to catch up with him?
   b. Part of someone's work on this problem included $\frac{1}{30} - \frac{1}{40}$. What quantities do the two fractions in $\frac{1}{30} - \frac{1}{40}$ represent?
   c. Suppose my brother's head start is 5 minutes instead of 8 minutes. Now how long would it take me to catch up with him?

5. At one point in a Girl Scout cookie sales drive, region C had sold 1500 boxes of cookies, and region D had sold 1200 boxes of cookies. If region D tries harder, they can sell 50 more boxes of cookies every day than region C.

   a. How many days will it take for region D to catch up?
   b. If sales are stopped after 8 more days, can you tell how many total boxes each region sold? Explain.

6. The last part of one triathlon is a 10K (10 kilometers or 10,000 meters) run. When runner Aña starts this last running part, she is 600 meters behind runner Bea. But Aña can run faster than Bea: Aña can run (on average) 225 meters each minute, and Bea can run (on average) 200 meters each minute. Who wins, Aña or Bea? If Aña wins, when does she catch up with Bea? If Bea wins, how far behind is Aña when Bea finishes?

7. Research on how students solve word problems contained the following incident.[5] Dana, a seventh-grader in a gifted program in mathematics, was asked to solve the following problem:

   ▶ A carpenter has a board 200 inches long and 12 inches wide. He makes 4 identical shelves and still has a piece of board 36 inches long left over. How long is each shelf? ◀

   Dana tried to solve the problem as follows: She added 36 and 4, then scratched it out, and wrote 200 × 12, but she thought that was too large so she scratched it out. Then she tried 2400 − 36, which was also too large and discarded it. Then she calculated 4 × 36 and subtracted that from 200, getting 56. She then subtracted 12, and got 44.

   Dana used a weak strategy called "Try all operations and choose." She obviously did not know what to do with this problem, although she was very good at solving one-step problems.

   Do a quantitative analysis of this problem situation, and use it to make sense of the problem in a way that Dana did not. Use your analysis to solve the problem.

8. The problems listed below, and in Exercise 9, are from a Russian Grade 3 textbook.[6] Solve the problems and compare their conditions and solutions.

   a. Two pedestrians simultaneously left two villages and walked toward each other, meeting after 3 hours. The first pedestrian walked 4 km in an hour, and the second walked 6 km in an hour. Find the distance between their villages.

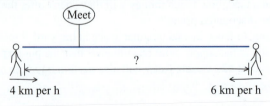

4 km per h   6 km per h

b. Two pedestrians simultaneously left two villages 27 km apart and walked toward each other. The first one walked 4 km per hour, and the second walked 6 km per hour. After how many hours did the pedestrians meet?

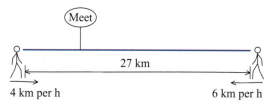

c. Two pedestrians simultaneously left two villages 27 km apart and walked toward each other, meeting after 3 hours. The first pedestrian walked at a speed of 4 km per hour. At what speed did the second pedestrian walk?

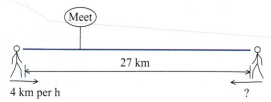

9. Two trains simultaneously left Moscow and Sverdlovsk, and traveled toward each other. The first traveled at 48 km per hour, and the second at 54 km per hour. How far apart were the two trains 12 hours after departure if it is 1822 km from Moscow to Sverdlovsk?

Supplementary Learning Exercises for Section 1.2

1. We are comfortable in saying, for example, "She is prettier than she used to be." Explain why "prettiness" is difficult to quantify.

2. Identify quantities suggested by the following situations, and make a drawing that could be useful in describing the situation.

   a. Walt is taller than Sheila, and Tammi is shorter than Sheila.
   b. Pedro makes $3.75 more an hour than Jim does, but less than Kay.
   c. I read 3 novels, with the first having 120 more pages than the second, and the second having 75 fewer pages than the third.

3. The last part of the Tri-City Triathlon is a 10,000-meter run. When Les starts the run, she is 800 meters behind Pat. Les can run (on average) 250 meters per minute, but Pat can run (on average) only 225 meters per minute. Who wins, Les or Pat? Tell how far ahead the winner is when she crosses the finish line.

4. What sort of drawing might help one solve these problems?

   a. To make 6 popcorn balls, you need 8 cups of popcorn. How much popcorn would you need to make 24 popcorn balls?
   b. A roll of film for an old camera cost $4.80. You can take 12 pictures with such a roll. Then it cost 35¢ to get each picture developed. What was the cost of film and development for 12 pictures?
   c. There were 270 students at the first game. That was 15 more students than attended the second game, and 18 more than the third game. What was the total attendance for the 3 games?

5. Give two quantities that could be in mind for "That's an impressive building."

6. Give two quantities that could be in mind for "Our basketball team played a good game."

7. Towns P and Q are 500 miles apart. Train A leaves P, heading toward Q, at 50 miles per hour. On the next track, Train B leaves Q, heading toward P, at 45 miles per hour. (*Hint:* Drawing?)

   a. How far apart are the locomotives of Trains A and B after 5 hours?
   b. How far apart are the locomotives of the two trains after 8 hours?

8. Towns P and Q are 500 miles apart. Train C leaves P, heading toward Q, at 60 miles per hour. Train D leaves Q, heading *away from* P, at 50 miles per hour. (*Hint: Drawing?*)

    a. How far apart are the locomotives of Trains C and D after 1 hour?

    b. How far apart are the locomotives of Trains C and D after 7 hours?

    c. Will Train C ever catch Train D, if they continue at the given speeds? If yes, tell in how many hours (make your reasoning clear). If not, explain your reasoning.

The following page,* from a third-grade mathematics textbook, provides some interesting story problems. Try using the Quantitative Analysis Questions on p. 8 and a drawing to solve each one.

5. There are 10 fish in all in Percy's fish tank. Four of the fish are angel fish. There are 4 more mollie fish than tetra fish. How many of each kind of fish are in the tank?

6. Norah's dog weighs 9 pounds more than her cat. Her dog weighs 6 pounds less than Jeff's dog. Norah's cat weighs 7 pounds. How much does Jeff's dog weigh?

7. The students in Mr. Cole's class voted on which kind of collection they should start as a class. The graph shows the results. How many more votes did the collection with the greatest number of votes get than the collection with the least number of votes?

8. Isadora has 15 seashells in her collection. The seashells are oyster shells, clam shells, and conch shells. There are 6 clam shells. There are 2 fewer clam shells than oyster shells. How many conch shells are in the collection?

9. Lyn, Kurt, and Steve wrote a riddle about their ages. Lyn is 7 years older than Steve. Steve is 5 years old. The sum of their ages is 25 years. How old is Kurt?

10. Stonehenge is an ancient monument in England made up of a pattern of rocks that looks like this:

Draw the shape that comes next in this pattern.

11. **Think About the Process** At the town pet show, Dina saw 48 pets. There were 6 birds and 7 cats. The remaining pets were dogs. Which number sentence shows one way to find the number of pets that were dogs?

    A $48 - 6 - 7 = $ ▨

    B $48 + 6 \div 7 = $ ▨

    C $48 - 6 \times 7 = $ ▨

    D $6 \times 7 \times 48 = $ ▨

240

1.3 Values of Quantities

The value of a quantity may involve very large or very small numbers. Furthermore, since the value is determined by counting or other ways of measuring, it can involve any type of number—whole numbers, fractions, decimals.

Think About …

Consider the following quantities. Which would you expect to have large values? Small values?

a. the distance between two stars
b. the diameter of a snowflake
c. the weight of an aircraft carrier
d. the national deficit
e. the thickness of a sheet of paper

💬 DISCUSSION 4   Units of Measure

1. What determines the appropriateness of the unit chosen to express the value of a quantity?

2. For each quantity listed in parts (a–e) above, name a unit of measure that would be appropriate to measure the quantity.

3. Explain how you would determine the thickness of a sheet of paper.

4. What determines the "precision" of the value of a quantity? ⬤

In a particular book the height of the arch in St. Louis is reported to be 630 feet. In another book the height is 192 meters. Why are the numbers different?

Think About …

What determines the magnitude of the number that denotes the value of a given quantity? Can we measure the speed of a car in miles per day? Miles per year? Miles per century? Is it convenient to do so? Explain.

Unless one has some appreciation and understanding of the magnitude of large numbers, it is impossible to make judgments about such matters as the impact of a promised $5 million in relief funds after a catastrophic flood or earthquake, the level of danger of traveling in a country that has experienced three known terrorist attacks in a single year, the personal consequences of the huge national debt, or the meaning of costly military mistakes. For example, people are shocked to hear that the Pentagon spent $38 for each simple pair of pliers bought from a certain defense contractor, yet it pays little attention to the cost of building the Stealth fighter or losing a jet fighter during testing.

✏️ ACTIVITY 5   Jet Fighter Crashes

In 2011, a West Coast newspaper carried a brief article saying that a $60 million jet fighter crashed into the ocean off the Mexican coast. How many students could go to your university tuition-free for 1 year with $60 million? ⬤

💬 DISCUSSION 5   What Is Worth a Trillion Dollars?

Suppose you hear a politician say, "A billion dollars, a trillion dollars, I don't care what it costs, we have got to solve the AIDS problem in this country." Would you agree? Is a billion dollars too much to spend on a national health crisis? A trillion dollars? How do the numbers 1 billion and 1 trillion compare? ⬤

Reminder: Values of quantities, like 16 tons or $64, involve units of measure—ton, dollar—as well as numbers.

Notes

Think About ...

Name several units of measure that you know and use. What quantities is each used to measure? Where do units come from?

Units can be arbitrary. Primary teachers have their students measure lengths and distances with pencils or shoe lengths, or measure weights using plastic cubes. The intent of these activities is to give the children experience with the measuring process so that later measurements will make sense. As you know, there are different systems of **standard units**, like the English or "ordinary" system (inches, pounds, etc.) and the **metric system** (meters, kilograms, etc.), more formally known as **SI** (from Le Système International d'Unités).

Virtually the rest of the world uses the metric system to denote values of quantities, so many are surprised that the United States, a large industrial nation, has clung to the English system for so long. Although the general public has not responded favorably to governmental efforts to mandate the metric system, international trade efforts are having the effect of forcing us to be knowledgeable about, and to use, the metric system. Some of our largest industries have been the first to convert to the metric system from the English system.

 DISCUSSION 6   Standard Units

Why are standard units desirable? For what purposes are they necessary? Why has the public resisted adopting the metric system? ●

Scientists have long worked almost exclusively in metric units. As a result, you may have used metric units in your high school or college science classes. Part of the reason for this is that the rest of the world uses the metric system because it is a sensible system that relates directly to our decimal system. A basic metric unit is carefully defined (for the sake of permanence and later reproducibility). Larger units and smaller subunits are related to each other in a consistent fashion, so it is easy to work within the system. In contrast, the English system unit, *foot,* might have been the length of a now-long-dead king's foot, and it is related to other length units in an inconsistent manner: 1 foot = 12 inches; 1 yard = 3 feet; 1 rod = 16.5 feet; 1 mile = 5280 feet; 1 furlong = $\frac{1}{8}$ mile; 1 fathom = 6 feet. Quick! How many rods are in a mile? A comparable question in the metric system is just a matter of adjusting a decimal point.

To get a better idea of how the metric system works, let's consider something that we frequently measure in metric units—length. Length is a quality of most objects. We measure the lengths of boards and pieces of rope or wire. We also measure the heights of children; height and length refer to the same quality but the different words are used in different contexts.

Think About ...

What are some other words that refer to the same quality as do "length" and "height"?

The basic SI unit for length (or its synonyms) is the **meter**. (The official SI spelling is **metre**. You occasionally see "metre" in books published in the United States.) The meter is too long to show with a line segment here, but two subunits fit easily, and illustrate a key feature of the metric system: Units larger and smaller than the basic unit are multiples or submultiples of powers of 10.

0.1 meter   _____

0.01 meter   _____

Furthermore, these subunits have names—decimeter, centimeter—that are formed by putting a prefix on the word for the basic unit. The prefix "deci-" means one-tenth, so "decimeter" means 0.1 meter. Similarly, "centi-" means one one-hundredth, so "centimeter" means 0.01 meter. You have probably heard "kilometer"; the prefix "kilo-" means 1000, so "kilometer" means 1000 meters. On reversing one's thinking, so to speak, there are 10 decimeters in 1 meter, there are 100 centimeters in 1 meter, and 1 meter is 0.001 kilometer.

Another feature of SI is that there are symbols for the basic units—m for meter—and for the prefixes—d for "deci-," c for "centi-," etc. By using the symbol for meter together with the symbol for a prefix, a length measurement can be reported quite concisely: e.g., 18 cm and 2.3 dm. The symbols cm and dm do not have periods after them, nor do the abbreviations in the English system: ft, mi, etc., except that for inch (in.).

If you are new to the metric system, your first job will be to familiarize yourself with the prefixes so you can apply them to the basic units for other qualities. The table below shows some of the other metric prefixes. Combining the symbol k for "kilo-" and the symbol m for "meter" gives km for kilometer.

Prefix	Symbol	Meaning of Prefix		Applied to Length
kilo-	k	1000	or 10^3	km
hecto-	h	100	or 10^2	hm
deka-	da	10	or 10^1	dam
no prefix		1	or 10^0	m
deci-	d	0.1	or 10^{-1}	dm
centi-	c	0.01	or 10^{-2}	cm
milli-	m	0.001	or 10^{-3}	mm

 ACTIVITY 6   It's All in the Unit

1. Measure the width of your desk or table in decimeters. Express that length in centimeters, millimeters, meters, and kilometers.

2. Measure the width of your desk or table in feet. Express that length in inches, yards, and miles.

3. A box of cereal is 10 ounces. Express that weight in pounds.

4. What is the weight of that same box of cereal expressed in grams? Kilograms? Centigrams?

5. Water from your showerhead likely flows at a rate of 2.5 gallons per minute. How many quarts flow in 1 minute? How many ounces flow in 1 minute?

6. In which system are conversions easier? Explain why. ●

TAKE-AWAY MESSAGE . . .   The value of a quantity is expressed using a number and a unit of measure. Commerce depends on having agreed-on sets of measures. Standard units accomplish this. The common system of measurement in the United States is the English system of measurement. We use the metric system, another measurement system, for science and for international trade. Most countries of the world use only the metric system. The metric system is based on powers of 10 and on common prefixes, making the system an easier one to use.

Notes

Learning Exercises for Section 1.3

1. Name an appropriate unit for measuring each given quantity.

    a. the amount of milk that a mug will hold
    b. the height of the Empire State Building
    c. the distance between San Francisco and New York
    d. the capacity of the gas tank in your car
    e. the safety capacity of an elevator
    f. the amount of rainfall in one year

2. **a.** What does it mean to say that "a car gets good mileage"?
    b. What unit is used to express gas mileage in the United States?
    c. How could you determine the mileage you get from your car?
    d. Do you always get the same mileage from your car? List some factors that influence how much mileage you get.
    e. How would you measure gas consumption? Is gas consumption related to mileage? If so, how?

3. Explain how rainfall is quantified. You may need to use resources such as an encyclopedia or the Internet.

4. **a.** Calculate an approximation of the amount of time you have spent sleeping since you were born. Explain your calculations. Express your approximation in hours, in days, and in years.
    b. What part of your life have you spent sleeping?
    c. On average, how many hours do you sleep each day? What fractional part of the day is this?
    d. How does your answer in part (b) compare to your answer in part (c)?

5. Name an item that can be used to estimate the following metric units:

    a. a centimeter   **b.** a gram   **c.** a liter
    d. a meter   **e.** a kilometer   **f.** a kilogram

6. There are some conversions from English to metric units that are commonly used, particularly for inch, mile, and quart. What are they?

Supplementary Learning Exercises for Section 1.3

1. Which of these, if any, might be 5 centimeters long: a car, a finger, a new pencil, a kite tail, an adult python?

2. Which, if any, of the objects in Problem 1 might be 5 meters long?

3. For each pair, write a sentence that tells how the first compares with the second.

    a. a meter and a centimeter   **b.** a centimeter and a meter
    c. a decimeter and a centimeter   **d.** a centimeter and a decimeter
    e. a kilometer and a meter   **f.** a meter and a kilometer

4. One might think that a *milli*meter is "a million meters." Check a dictionary for the root of "milli-" and its meaning.

5. In each part, which unit would be more appropriate?

    a. cm or m, for the length of a newly born baby
    b. mi or in., for the distance between two cities
    c. mg or kg, for the weight of a fourth-grader

6. Arrange the following in order of increasing size: 510 cm, 300 dm, 2 m.

7. Complete the following, using your knowledge of prefixes:

    a. 5000 m = _____ km   **b.** 500 m = _____ km
    c. 3.4 m = _____ cm   **d.** 3.4 cm = _____ m
    e. 200 g = _____ kg   **f.** 200 kg = _____ g

1.4 Issues for Learning: Ways of Illustrating Story Problems

Making drawings or mental pictures can play an essential role in coming to understand a problem. Here is an excerpt from one interview with a child in the study described at the beginning of the chapter:

> *Emmy:* I just pictured the post, how deep the water was. . . . Sometimes I picture the objects in my mind that I'm working with, if it's a hard problem. . . .
>
> *Interviewer:* Does that help?
>
> *Emmy:* Yeh, it helps. That's just one way of, kind of cheating, I guess you'd say.

Unfortunately, too many students seem to believe that making drawings is cheating, or is juvenile, yet making a drawing is a valuable problem-solving process.

In another study[7] two researchers compared how drawings are used in U.S. textbooks and in Japanese textbooks. They found that many elementary school students in the United States are not encouraged to make drawings that will help them understand a story problem. However, even just flipping through the pages of Japanese textbooks shows that drawings are used throughout. Teachers say to students, "If you can draw a picture, you can solve the problem."

Think About ...

In what ways have drawings helped you so far with solving the problems in this first chapter?

All too often, adults try to avoid making a drawing because they think that the need for drawings is childish. But this is *not* the case. The ability to represent a problem with a drawing is an important component of problem solving, no matter what the age of the problem solver is. Young children often draw a picture to help them understand a problem. For example, for the problem "There are 24 legs in the sheep pen where two men are shearing sheep. How many sheep are there?", one child might draw something similar to this:

Yet another child might represent this problem as

|| || |||| |||| |||| |||| ||||

There are, of course, multiple ways this problem could be represented with a drawing, but here perhaps the first is more appropriately called a picture, and the second would be more appropriately called a diagram. Students will quite naturally evolve from pictures to diagrams. "A diagram is a visual representation that displays information in a spatial layout."[8] Although diagrams are very useful in understanding the structure of a problem, students are often unable to produce an appropriate diagram without some assistance. Whether or not a picture or a diagram is appropriate depends on how well it represents the structure of the problem. A diagram is more abstract than a picture in that it often does not contain extraneous information. It takes less time to draw, but it continues to represent the problem.

In the sheep problem, the objects are discrete, i.e., they can be counted. But not all problems about discrete quantities need to be represented by individual objects because diagramming large numbers can become a chore, and pictures would take far too long to produce. Thus, we often use lines or boxes to represent the problems, even those involving discrete quantities.

🗨 DISCUSSION 7   Drawings and Diagrams

1. A fourth-grader was asked to solve some story problems you encountered in the Learning Exercises for Section 1.2. Two problems are shown on the next page. Before looking at the student's solutions, go back and see how you solved the problems.

2. Discuss the types of drawings used by this student. How did they help her visualize the problems leading to solutions?

Problem 1

[Modified from Learning Exercise 1(c) in Section 1.2]

► Two sculptures are similar. The height of one sculpture is $2\frac{1}{2}$ times the height of the other sculpture. The smaller sculpture is 3 feet tall. How tall is the larger sculpture? ◄

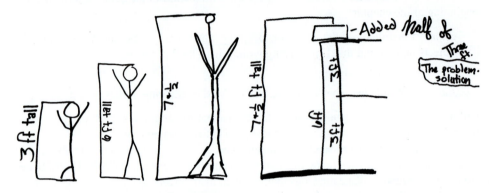

Problem 2

(From Learning Exercise 7 in Section 1.2)

► A carpenter has a board 200 inches long and 12 inches wide. He makes 4 identical shelves and still has a piece of board 36 inches long left over. How long is each shelf? ◄

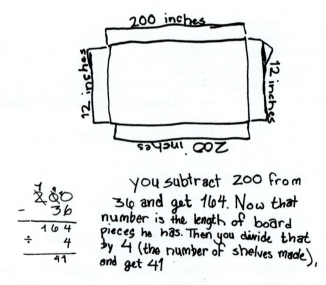

you subtract 200 from 36 and get 164. Now that number is the length of board pieces he has. Then you divide that by 4 (the number of shelves made), and get 41

Drawing diagrams to represent and then solve problem situations is common in textbooks in Singapore.[9] They use diagrams called strip diagrams. Here are some examples like those they use.

EXAMPLE 1

► The library bought 576 new books. The library loaned some of them. If 198 were left over, how many of the new books did the library lend? ◄

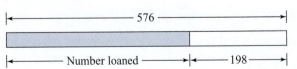

EXAMPLE 2

▶ Anna and Suzy have $28.90 altogether. Anna has $5.48 more than Suzy. How much money does Suzy have? ◀

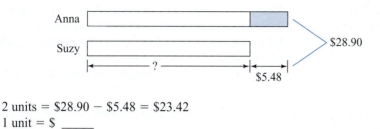

2 units = $28.90 − $5.48 = $23.42
1 unit = $ _____
Suzy has $ _____

Think About …
Can the sheep problem be represented with a strip drawing?

In this chapter we have used diagrams rather than pictures to represent problems. Analyzing situations quantitatively, which may include making drawings, can help one understand the problems being solved. "Quantitative reasoning is more than reasoning about numbers, and it is more than skilled calculating. It is about making sense of the situation to which we apply numbers and calculations."[10]

Learning Exercises for Section 1.4

Use a mature strategy and a strip diagram to work each of the following:

1. Kalia spent a quarter of her weekly allowance on a movie. The movie was $4.25. What is her weekly allowance?

2. Calle had five times as many stickers as Sara did, and twice as many as Juniper. Juniper had 25 stickers. How many stickers did they have altogether?

3. Nghiep gave his mother half of his weekly earnings, and then spent half of what was left on a new shirt. He then had $32. What were his weekly earnings?

4. Zvia's sweater cost twice as much as her hat, and a third as much as her coat. Her coat cost $114. How much did her sweater cost?

5. One number is four times as large as another number, and their sum is 5285. What are the two numbers?

6. Jinfa, upon receiving his paycheck, spent two-thirds of it on car repairs and then bought a $40 gift for his mother. He had $64 left. How much was his paycheck?

7. Rosewood Elementary School had 104 students register for the fourth grade. After placing 11 of the students in a mixed grade with fifth-graders, the remaining students were split evenly into 3 classrooms. How many students were in each of these 3 classrooms?

8. Jacqui, Karen, and Lynn all collect stamps. Jacqui has 12 more stamps than Karen, and Karen has three times as many as Lynn. Together, they have 124 stamps. How many does each person have?

9. Jo-Jo has downloaded 139 songs on his iPod. Of those songs, 36 are jazz, twice that number are R&B, and the remainder are classical. How many classical songs has Jo-Jo downloaded?

1.5 Check Yourself

In this first chapter you have learned about the role quantities play in our lives and the ways we express quantities and their values. You have learned about dealing with problem situations by analyzing the problem in terms of its quantities and their relationships to one another. Quantitative analysis helps solve a problem in a meaningful, sense-making way. The same kind of analysis can be applied to arithmetic problems and to algebra problems.

You also learned how quantities are measured in terms of numbers of units and how to express values of quantities in standard units, including metric units.

You should be able to work problems like those assigned and to meet the following objectives:

1. Identify the quantities addressed by such questions as, "How much damage did the flooding cause?"

2. Name attributes of objects that cannot be quantified.

3. Distinguish between a quantity and its value.

4. Given a problem situation, undertake a quantitative analysis of the problem and use that analysis to solve it.

5. Determine appropriate units to measure quantities.

6. Discuss reasons why the metric system is used for measurement in most countries.

7. Discuss some incorrect ways that children solve story problems.

8. Discuss the importance of appropriate drawings in problem solving.

9. Make strip diagrams to represent simple arithmetic and algebraic problems.

References for Chapter 1

[1]Sowder, L. (1988). Children's solutions of story problems. *Journal of Mathematical Behavior, 7,* 227–239.

[2]Maletsky, E. M., Andrews, A. G., Burton, G. M., Johnson, H. C., Luckie, L., Newman, V., Schultz, K. A., Scheer, J. K., & McLeod, J. C. (2002). *Harcourt math.* Orlando, FL: Harcourt.

[3]Lappan, G., Fey, J. T., Fitzgerald, W. M., Friel, S. N., & Phillips, E. D. (1998). *Connected mathematics: Bits and pieces II: Using rational numbers* (Teachers edition). Menlo Park, CA: Dale Seymour Publications.

[4]Krutetskii, V. A. (1976). *The psychology of mathematical abilities in schoolchildren* (J. Teller, Trans.). Chicago: University of Chicago Press.

[5]Sowder, L. (1995). Addressing the story problem problem. In J. Sowder & B. Schappelle (Eds.), *Providing a foundation for the teaching of mathematics in the middle grades* (pp. 121–142). Albany, NY: SUNY Press.

[6]*Russian grade 3 mathematics.* (1978). Translated by University of Chicago School Mathematics Project.

[7]Shigematsu, K., & Sowder, L. (1994). Drawings for story problems: Practices in Japan and the United States. *Arithmetic Teacher, 41* (9), 544–547.

[8]Diezmann, C., & English, L. (2001). Promoting the use of diagrams as tools for thinking. In A. C. Cuoco & F. R. Curcio (Eds.), *The roles of representation in school mathematics* (pp. 77–89). Reston, VA: NCTM.

[9]Beckmann, S. (2004). Solving algebra and other story problems in simple diagrams: A method demonstrated in grade 4–6 texts used in Singapore. *The Mathematics Educator, 14,* 42–46.

[10]Thompson, P. (1995). Notation, convention, and quantity in elementary mathematics. In J. Sowder & B. Schappelle (Eds.), *Providing a foundation for the teaching of mathematics in the middle grades* (pp. 199–221). Albany, NY: SUNY Press.

Chapter 2

Numeration Systems

Contrary to what you may believe, there are many ways of expressing numbers. Some of these ways are cultural and historical. Others are different ways of thinking about what the digits of our conventional number system mean. For example, you probably think of 23 as meaning 2 tens and 3 ones. We think this way because we use a base-ten system of counting. (How many fingers do you have?) But why not base five (using only five fingers)? What would 23 mean then? You are about to find out.

2.1 Ways of Expressing Values of Quantities

The need to quantify and express the values of quantities led humans to invent numeration systems. Throughout history, people have found ways to express values of quantities they measured in several ways. A variety of words and special symbols, called **numerals**, have been used to communicate number ideas. How one expresses numbers using these special symbols makes up a **numeration system**. Our Hindu-Arabic system uses ten **digits**: 0, 1, 2, 3, 4, 5, 6, 7, 8, 9. Virtually all present-day societies use the Hindu-Arabic numeration system. With the help of decimal points, fraction bars, and marks like square root signs, these ten digits allow us to express almost any number and therefore the value of almost any quantity.

> *Think About …*
>
> Why are numerals used so much? What are advantages of these special symbols over using just words to express numbers? What are some exceptions to representing numbers with digits?

 ACTIVITY 1   You Mean People Didn't Always Count the Way We Do?

A glimpse of the richness of the history of numeration systems lies in looking at the variety of ways in which the number twelve has been expressed. In the different representations shown below, see if you can deduce what each individual mark represents. Each representation expresses this many: ▢▢▢▢▢▢▢▢▢▢▢▢

Representations of the Number Twelve

‖	ιβ	XII	⟨‖
Old Chinese	Old Greek	Roman	Babylonian

⋮⋮	◇ ⋮	12	30
Mayan	Aztec	Today, base ten	Today, base four

> *Think About …*
>
> How would ten have been written in each of these earlier numeration systems?

Some ancient cultures did not need many number words. For example, they may have needed words only for "one," "two," and "many." When larger quantities were encountered, they could be expressed by some sort of matching with pebbles or sticks or parts of the body, but without the use of any distinct word or phrase for the number involved. For example, in a recently discovered culture in Papua New Guinea, the same word "doro" was used for 2, 3, 4, 19, 20, and 21. But by pointing also to different parts of the hands, arms, and face when counting and saying "doro," these people can tell which number is intended by the word. This method of pointing allowed Papua New Guineans to easily express numbers up through 22.[1] It was only when this culture came into contact with the outside world and began trading with other cultures that they needed to find ways of expressing larger numbers.

Think About ...

Why do you think we use ten digits in our numeration system? Would it make sense to use twenty? Why or why not?

DISCUSSION 1   Changing Complexity of Quantities over Time

What quantities, and therefore what number words, would you expect a caveman to have found useful? (Assume that the caveman had a sufficiently sophisticated language.) A person in a primitive agricultural society? A pioneer? An ordinary citizen living today? A person on Wall Street? An astronomer? A subatomic physicist? ●

TAKE-AWAY MESSAGE . . .   Mathematical symbols have changed over the years, and they may change in the future. Symbols used for numbers depend on our need to determine the value of the quantities with which we work.

Learning Exercises for Section 2.1

1. Based on what you have seen of the old counting systems such as Greek, Chinese, Roman, Babylonian, Mayan, and Aztec, which systems make the most sense to you? Explain.

2. Symbols for five and for ten often have had special prominence in geographically and chronologically remote systems. Why?

3. Numbers can be expressed in a fascinating variety of ways. Different languages, of course, use different words and different symbols to represent numbers. Some counting words are given below.

English	Spanish	German	French	Japanese	Swahili
zero	cero	null	zero	zero	sifuri
one	uno	eins	un	ichi	moja
two	dos	zwei	deux	ni	mbili
three	tres	drei	trois	san	tatu
four	cuatro	vier	quatre	shi	nne
five	cinco	fünf	cinq	go	tano
six	seis	sechs	six	roku	sita
seven	siete	sieben	sept	shichi	saba
eight	ocho	acht	huit	hachi	nane
nine	nueve	neun	neuf	kyu	tisa
ten	diez	zehn	dix	ju	kumi

Which two sets of these counting words most resemble one another? Why do you think that is true? Do you know these numbers in yet another language?

4. Roman numerals have survived to a degree, as in motion picture film credits and on cornerstones. Here are the basic symbols: I = one, V = five, X = ten, L = fifty, C = one hundred, D = five hundred, and M = one thousand. For example, CLXI is 100 + 50 + 10 + 1 = 161. What number does each of these numerals represent?

   a. MMCXIII   **b.** CLXXXV   **c.** MDVII

5. How would each of the following be written in Roman numerals? For example, one thousand one hundred thirty would be MCXXX.

   a. two thousand sixty-six   **b.** seventy-eight   **c.** six hundred five

6. Other systems we have seen all involve addition of the values of the symbols. Roman numerals use a **subtractive** principle as well; when a symbol for a smaller value comes before the symbol for a larger value, the former value is subtracted from the latter. For example, IV means 5 − 1 = 4, or four; XC means 100 − 10 = 90; and CD = 500 − 100 = 400. Note that no symbol appears more than three times together, because with four symbols we would use this subtractive property. What number does each of these represent?

   a. CMIII   **b.** XLIX   **c.** CDIX

7. Even within the same language, there are often several words for a given number idea. For example, both "two shoes" and "a pair of shoes" refer to the same quantity. What are some other words for the idea of two-ness?

8. Continue to twenty, using Roman numerals: I, II, III, . . .

9. Using Roman numerals, write the numbers that come before and after the given numbers. (*Hint:* See Learning Exercises 4 and 6.)

   a. _____, C, _____
   b. _____, MM, _____
   c. _____, CMXIV, _____

2.2 Place Value

What does each 2 in 22,222 mean? The different 2s represent different values because our *Hindu-Arabic numeration system* is a place-value system. This system depends on the powers of 10 to tell us the meaning of each digit. Once this system is understood, arithmetic operations are much easier to learn. Understanding of place value is a fundamental idea underlying elementary school mathematics.

But first, what does it mean to have a place-value system?

> In a **place-value** system, the value of a digit in a numeral is determined by its position in the numeral.

EXAMPLE 1

In 506.7, the 5 is in the hundreds place, so it represents five hundred. The 0 in 506.7 is in the tens place, so it represents 0 tens or just zero. The 6 is in the ones place, so it represents 6 ones or six. And the 7 is in the tenths place, so it represents seven-tenths. The complete 506.7 symbol then represents the sum of those values: five hundred six and seven-tenths. ●

Notice that we do not say "five hundred, zero tens, six and seven-tenths," although we could. This is symptomatic of the relatively late appearance, historically, of a symbol for zero. The advantage of having a symbol to say that nothing is there is apparently a difficult idea, but the idea is vital to a place-value system. Would 506.7 mean the same number if we omitted the 0 to get 56.7? The 0 may have evolved from some type of round mark written in clay by the Babylonians to show that there were zero groups of a particular place value needed.

Notes

Think About ...

In the Hindu-Arabic place-value system, how many different places (positions) can you name and write numerically? (Don't forget places to the right of the decimal point.)

Before the use of numerals became widespread, much calculation was done with markers on lines for different place values. The lines could be on paper, or just drawn in sand, with small stones used as markers. (Our word "calculate" comes from the Latin word for "stones.") One device that no doubt was inspired by these methods of calculating is the **abacus**, which continues to be used in some parts of the world.

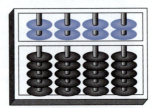

A Chinese Abacus

Notice that we have often used words to discuss the numbers instead of the usual numerals. The reason is that the symbol "12" is automatically associated with "twelve" in our minds because of our familiarity with the usual numeration system. We will find that the numeral "12" could mean five or six, however, in other systems! (If no base is indicated, assume the familiar base ten is intended.)

In our *base-ten* numeration system, the whole-number place values result from *groups of ten*—ten ones, ten tens, ten hundreds, etc. The digits 0, 1, 2, 3, 4, 5, 6, 7, 8, 9 work fine until we have ten of something. But there is no single digit that means ten. When we have ten ones, we think of them as one group of ten, without any leftover ones, and we take advantage of place value to write "10," one ten and zero leftover ones. Similarly, two place values are sufficient through nine tens and nine ones, but when we have ten tens, we then use the next place value and write "100." It is like replacing ten pennies with one dime, and trading ten dimes for a dollar.

EXAMPLE 2

If I want to find the number of ten dollar bills I could get for $365, the answer is not just 6, it is 36.

If I want to know how many dollar bills I could get from $365, the answer is 365.

If I want to know the number of dimes I could get from $365, the answer is 3650.

But if I want to know how many tens are in 365, I could say either 36 or 36.5, depending on the context. There are 36 whole tens but *exactly* 36.5 tens.

If I have 365 bars of soap and I want to know how many full boxes of 10 I could pack, the answer would be 36.

If I am buying 365 individual bars of soap priced at $6 per 10 bars, then I would have to pay 36.5 times $6. ●

With a good understanding of place value, the problems like those in Example 2 can be solved easily without undertaking long division or multiplication by 10 or powers of 10. Children who do not understand place value will often try to solve the problem of how many tens are in 365 by using long division to divide by 10, rather than observing that the answer is obvious from the number.

 **DISCUSSION 2   Money and Place Value**

Explain your answers to each of the following:

1. How many ten dollar bills does the 6 in $657 represent? The 5?

2. How many tens are in 657?

3. How many one hundred dollar bills can you get for $53,908?

4. How many one hundreds are in 53,908?

5. How many pennies can you get for $347? For $34.70? For $3.47?

6. How many ones are in 347? In 34.70? In 3.47? ●

The decimal point indicates that we are beginning to break up the unit one into tenths, hundredths, thousandths, etc. But the number *one*, not the decimal point, is the focal point of this system. So, 0.642 is 642 thousandths of one. Put another way, 0.6 is six tenths of one, while 6 is six ones, and 60 is six tens, or 60 ones. But just as 0.6 is six tenths of one, 6 is six tenths of 10, 60 is six tenths of one hundred, etc. up the line. Or starting with smaller numbers, 0.006 is six tenths of 0.01, while 0.06 is six tenths of 0.1. Likewise, 6000 is 60 hundreds, 600 is 60 tens, 60 is 60 ones, 6 is 60 tenths, 0.6 is 60 hundredths, 0.06 is 60 thousandths, etc.[2] Although this at first might seem confusing, it becomes less so with practice and thought.

 ACTIVITY 2   Many Names

A teacher challenged her class to figure out how many ways the number 154.25 could be thought about. Following are five children's answers. Indicate whether each child's reasoning is correct or incorrect. If it is incorrect, please explain.

   Anita's answer: 154.25 could be thought about as 15 tens, 4 ones, and 25 tenths.

   Bekah's answer: 154.25 could be thought about as 150 ones and 42.5 tenths.

   Cedric's answer: 154.25 could be thought about as 1542.5 tenths.

   Dario's answer: 154.25 could be thought about as 1.5425 hundreds.

   Estoban's answer: 154.25 could be thought of as 154 ones and .25 hundredths. ●

TAKE-AWAY MESSAGE . . .   Our base-ten place-value numeration system is adequate for expressing all whole numbers and decimal numbers. The value of each digit in a numeral is determined by the position of the digit in the numeral. Digits in different places have different values. Finally, the number 1, not the decimal point, serves as the focal point of decimal numbers.

Learning Exercises for Section 2.2

1. **a.** How many tens are in 357? How many whole tens?
   b. How many hundreds are in 4362? How many whole hundreds?
   c. How many tens are in 4362? How many whole tens?
   d. How many thousands are in 456,654? How many whole thousands?
   e. How many hundreds are in 456,654? How many whole hundreds?
   f. How many tens are in 456,654? How many whole tens?
   g. How many tenths are in 23.47? How many whole tenths?
   h. How many thousandths are in 23.47? How many whole thousandths?
   i. How many ones are in 23.47? How many whole ones?
   j. How many hundredths are in 23.47? How many whole hundredths?
   k. How many tenths are in 2347? How many whole tenths?
   l. How many tenths are in 234.7? How many whole tenths?

2. In 123.456, the hundre*d*s place is in the third place to the *left* of the decimal point; is the hundred*th*s place in the third place to the *right* of the decimal point? In a long numeral like 333331.333333, what separates the number into two parts that match in the way hundreds and hundredths do?

3. **a.** Is the statement "For a set of whole numbers, the longest numeral will belong to the largest number" true or false? Why?
   b. Is the statement "For a set of decimals, the longest numeral will belong to the largest number" true or false? Why?

4. Pronounce 3200 in two different ways. Do the pronunciations describe the same value?

5. Write in words the way you would pronounce each:
   a. 407.053   **b.** 30.04   **c.** 0.34   **d.** 200.067   **e.** 0.267

6. Each of the following represents the work of students who did not understand place value. Find the errors made by these students, and explain their reasoning.

   a. 15   **b.** 55   **c.**   7   **d.**   36   **e.**   36
         $+95$ $+48$ $4^1 8$ $7\overline{)441}$ $\times\ 8$
         ‾‾‾‾   ‾‾‾‾ $-2\,6$ $\underline{42}$ ‾‾‾‾
         1010   913   ‾‾‾‾   21   2448
                                      1 1 $\underline{21}$

7. In base ten, 1635 is exactly _____ *ones,* is exactly _____ *tens,* is exactly _____ *hundreds,* is exactly _____ *thousands;* it is also exactly _____ *tenths,* or exactly _____ *hundredths.*

8. In base ten, 73.5 is exactly _____ *ones,* is exactly _____ *tens,* is exactly _____ *hundreds,* is exactly _____ *thousands;* it is also exactly _____ *tenths,* or exactly _____ *hundredths.*

9. Do you change the value of a whole number by placing zeros to the right of the number? To the left of the number?

10. A child asks, "Are there a million millions in a billion?" How do you respond?

Supplementary Learning Exercises for Section 2.2

1. **a.** How many thousands make a million?
   b. How many whole hundreds are in 5386.492?
   c. How many whole hundredths are in 5386.492?
   d. How many whole tens are in 679.43?
   e. How many whole tenths are in 679.43?
   f. How many whole ten-thousands are in 82,475.32987?
   g. How many whole ten-thousandths are in 82,475.32987?

2. Write each of the following without any words:
   a. 8.5 million   **b.** 92 billion   **c.** $6.45 trillion

3. Write in words the way you would pronounce each:
   a. 600.023   **b.** 0.623   **c.** 20.08   **d.** 0.28

4. In base ten, 87.645 is exactly _____ *tens,* is exactly _____ *ones,* is exactly _____ *tenths,* is exactly _____ *hundredths,* is exactly _____ *thousandths,* is exactly _____ *ten-thousandths,* is exactly _____ *hundreds.*

5. In base ten, 0.5 is exactly _____ *ones,* _____ *tenths,* _____ *hundredths.*

6. Write in words the way you would pronounce each:
   a. 0.00407   **b.** 51.0005   **c.** 400.05   **d.** 0.405

7. Write each of the following without any words:
   a. 32 tenths   **b.** 753 hundredths
   c. 2 thousand and 17 thousandths   **d.** 2017 thousandths

2.3 Bases Other Than Ten

Too often children learn to operate on numbers without having a deep understanding of place value, the lack of which leads them to make many computational errors. The purpose of this section is to provide experiences with base numeration systems other than ten so you understand the underlying structure of the base-ten system of numeration. You are not expected to become fluent in a base other than ten. Rather, you should be able to calculate in different bases to the extent that is needed to understand the role of place value in calculations.

> *Think About ...*
>
> We use a base-ten system of counting because we have ten fingers. Other cultures have used other bases. For example, some Eskimos were found to count using base five. Why would that be? What other bases might have been used for counting?

Cartoon characters often have three fingers and a thumb on each hand, a total of eight fingers (counting thumbs) instead of ten. Suppose that we live in this cartoon land and instead of having ten digits in our counting system (0, 1, 2, 3, 4, 5, 6, 7, 8, 9), we have only eight digits (0, 1, 2, 3, 4, 5, 6, 7). Using this new counting system, we write the number eight as 10_{eight}, meaning 1 group of eight and 0 ones. Thus, we would write as we count in base eight:

$$1, 2, 3, 4, 5, 6, 7, 10, 11, 12, 13, 14, 15, 16, 17, 20, 21, \ldots$$

We read this list of numbers as one, two, three, . . . , one-zero, one-one, one-two, . . . , two-zero, ...

This many apples

can be grouped in groups of tens and ones in our base-ten numeration system: 26.

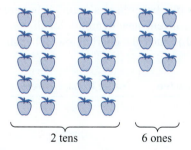

2 tens   6 ones

But in cartoon land, the objects are grouped in groups of eight.

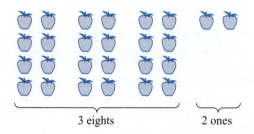

3 eights   2 ones

ACTIVITY 3   Bundling Apples in Cartoon Land

To continue the analogy further, suppose Aloolalee in cartoon land had an apple farm. He collects apples in baskets. Eight apples fit in a basket. Eight baskets fit in a bushel. Once Aloolalee has eight bushels, he fills a cartoon truck. Aloolalee has eight cartoon trucks.

Today, Aloolalee picked enough apples to load one cartoon truck, six bushels, and three baskets. Aloolalee had 3 loose apples. How many apples did he pick? ●

If Aloolalee needed a way to record the number of trucks, bushels, baskets, and loose apples, he could write the numbers in order and others in cartoon land would understand what is meant. Cartoon land characters count and record in base eight.

 ACTIVITY 4   Place Value in Cartoon Land

1. Show the value of each place in base eight by completing this pattern:

8^5	8^4	8^3	8^2	8^1	8^0
?	?	?	sixty-fours	eights	ones

2. What would follow 77 in base eight?
3. What would each digit indicate in the numeral 743 in base eight?
4. Which places correspond to the apple farm's basket, bushel, and truck? ●

Notice that the base-eight numeration system has eight digits, 0–7. Writing 6072 in base eight would require the use of the first four places to the left of the decimal point and represents 2 ones (8^0), 7 eights (8^1), 0 sets of eight squared (8^2), and 6 sets of eight cubed (8^3). The digits 6, 0, 7, and 2 would be placed in the Activity pattern in the four places to the left of the decimal point. We call this number "six zero seven two, base eight" and write it as 6072_{eight}.

Think About …

If Aloolalee had 602_{eight} apples, how many apples would Aloolalee have, written in base ten?

 DISCUSSION 3   Place Value in Base Three

What are the place values in a base-three system? What are the digits, and how many do we need? (Rather than invent new symbols for digits, let's use whichever of the standard symbols we need.) Study the chart below. What should appear in place of the question marks?

Items	Name in base ten	Name in base three	Base-three symbol
	zero	zero	0_{three}
☐	one	one	1_{three}
☐ ☐	two	two	2_{three}
☐ ☐ ☐	three	???	???

Naming *three* in base three is a key step in understanding base three. Since there are three single boxes above, they will be grouped to make *one group of three*, and the base-three symbol is 10! Notice that in base three "10" does not symbolize *ten* as we think about ten. In base three, "10" means "one group of three and zero left over." Since it does not mean ten, we should not pronounce the numeral as "ten." The recommended pronunciation is "one zero, base three," saying just the name for each digit and for the base. Notice how the following chart differs from the one above:

Items	Name in base ten	Name in base three	Base-three symbol
	zero	zero	0_{three}
▫	one	one	1_{three}
▫ ▫	two	two	2_{three}
▫▫▫	three	one-zero	10_{three}
▫▫▫ ▫	four	one-one	11_{three}

If there are four boxes, as in the last line of this table, we can make one group of three, and then there will be one leftover box, so in base three, four is written as "11." Because we have the strong link between the marks "11" and eleven from all of our base-ten experience, the notation 11_{three} is often used for clarity to show that the symbols should be interpreted in base three. Recall that 11_{three} should be pronounced "one one, base three," and not as "eleven."

ACTIVITY 5   Count in Base Three and in Base Four

Continue to draw more boxes and to write base-three symbols. What do you write for five boxes? (Now you see why the symbol 12 might mean five.) Six? Seven? Eight? And, at another dramatic point, nine? Did you write "100_{three}" for nine? What would 1000_{three} mean?

Check your counting skills by following along with counting in base four: 1, 2, 3, 10, 11, 12, 13, 20, 21, 22, 23, 30, 31, 32, 33, 100, 101, 102, 103, 110, 111, 112, 113, 120, 121, 122, 123, 130, 131, 132, 133, 200 . . . .

What does 1000_{four} mean? ●

DISCUSSION 4   Working with Different Bases

1. What are the place values in base five? What digits are needed? How would thirty-eight (in base ten) be expressed in base five? Record the first fifteen counting numbers in base five: 1, 2, ...

2. What are the place values in a base b place-value system? What digits are needed?

3. What are the place values in a base-two place-value system? How would eighteen (in base ten) be written in base two? The inner workings of computers use base two; do you see any reason for this fact?

4. Perhaps surprisingly, there is a duodecimal society that promotes the adoption of a base-twelve numeration system. What are the place values in a base-twelve system? What new digits would have to be invented? ●

With several numeration systems possible, there can be many "translations" among the symbols. For example, given a base-ten numeral (or its corresponding name), find the base-six (or four or twelve) numeral for the same number, and vice versa, given a numeral in some other base, find its base-ten numeral (or its corresponding name). In each case, the key is knowing, and probably writing down, the place values in the unfamiliar system. (Recall that any nonzero number to the 0 power is 1. Example: $5^0 = 1$.)

EXAMPLE 3

Changing from a non-ten base to base ten: What does 2103_{four} represent in base ten?

Solution

1. 2103_{four} has four digits. The first four place values in base four are written here, and the given digits put in their places:

2	1	0	3
of four sixteens, or sixty-four, or 4^3	of four fours, or sixteen, or 4^2	of four ones, or four, or 4^1	ones, or 4^0

continue

2. What does the 2 tell us? The 2 stands for two of 4^3, which is $2 \times 64 = 128$ in base ten.

3. What does the 1 tell us? The 1 stands for one 4^2, which is $1 \times 16 = 16$ in base ten.

4. What does the 0 tell us? The 0 stands for zero of 4^1, which is $0 \times 4 = 0$ in base ten.

5. What does the 3 tell us? The 3 indicates 4^0 is used three times, $3 \times 1 = 3$ in base ten.

6. Thus, $2103_{four} = (128 + 16 + 0 + 3)_{ten} = 147_{ten}$, i.e., $2103_{four} = 147_{ten}$. ●

EXAMPLE 4

Suppose instead we want to change a number written in base ten, say, 236, to a number written in another base, say, base five. We know that the places in base five are the following:

$\cdots$ _____   _____   _____   _____

one-hundred-   twenty-fives (5^2)   fives (5^1)   ones (5^0)
twenty-fives (5^3)

Solution

(You may find these steps easier to follow by dropping the ten subscript for now, for numbers in base ten.)

1. Look for the highest power of 5 in the base-ten number; here it is 5^3 because 5^4 is 625_{ten} and 625_{ten} is larger than 236_{ten}. Are there any 5^3's in 236_{ten}? Yes, just one 5^3 because $5^3 = 125$, and there is only one 125 in 236. Place a 1 in the first place above to indicate one 5^3. Now you have "used up" 125, so subtract $236_{ten} - 125_{ten} = 111_{ten}$.

2. The next place value of five is 5^2. Are there any twenty-fives in 111_{ten}? There are 4, so place a 4 above 5^2. Now four twenty-fives, or 100, have been "used," and $111_{ten} - 100_{ten} = 11_{ten}$.

3. The next place value is 5^1, which is 5. How many fives are in 11_{ten}? It has two fives, so place 2 above 5^1. There is 1 one left, so place a 1 above 5^0. Thus, $236_{ten} = 1421_{five}$. ●

    Working with different bases can be easier when you are able to physically move pieces that represent different values in a base system. Often, after doing physical manipulation, you can mentally picture the manipulation and work without physical objects. Multibase blocks are manipulatives that have proven to be extremely useful in coming to understand any base system, but primarily base ten in elementary school. Multibase blocks are wooden or plastic blocks that can be used to demonstrate operations in different bases. For base ten, a centimeter cube can be used to represent a unit or one; a long block 1 cm by 1 cm by 10 cm (often marked in ones) would then represent ten; ten longs together form a flat that is 1 cm by 10 cm by 10 cm and that represents one hundred, ten flats form a 10 cm by 10 cm by 10 cm cube that represents thousands. If the long is used for the unit, then the small cube would represent one-tenth, the flat would represent ten, etc. The multibase blocks can be used to strengthen place-value understanding.

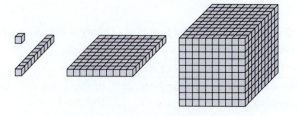

If the multibase blocks are not available, then they can easily be sketched as shown at right: The materials are often called "small cube, long, flat, big cube." Any size of the multibase blocks  can be used to represent one unit. Familiarize yourself with the multibase blocks by doing this section's Learning Exercises and making up more problems until you feel you are familiar with the blocks and their relationships.

EXAMPLE 5

The sketch below represents numbers with bases larger than five because there are five flats. If the little cube represents the unit one, the number here is 520 for any base larger than five. If the long represents one, then the number represented here is 52 in any base larger than five. If the flat represents one, then the number represented here is 5.2 in any base larger than five.

DISCUSSION 5   Representing Numbers with Multibase Drawings

1. Here is a representation of a number:

   Which bases could use this representation if it is in the final form, with no more "trades" possible? Why? What are some possible numbers that can be represented by this drawing?

2. In base eight, how many small cubes are in a long? How many in a flat? How many in a large cube? How many longs in a flat? How many flats in a large cube? Answer the same questions for base ten; for base two. ●

   You can also represent decimal numbers with base-ten blocks or drawings. You must first decide which block represents the unit. If the unit is the long, then the small block is one-tenth, the flat is ten, and the large block is 100. Thus, 2.3 in base ten could be represented as

ACTIVITY 6   Representing Numbers with Multibase Blocks

For these problems, use your cut-out blocks (from the "Masters" section) or use drawings such as those shown earlier. Note that the drawings do not show the markings of the base that appears in the picture of the blocks, and thus do not clearly indicate the base in the way that multibase blocks do.

1. Represent 2.3 in base ten using the long as one unit. Represent 2.3 using another size of block as the unit. Compare your representation with a neighbor's.

2. Use the base-five blocks to represent 2.41_{five} in two different ways. Be sure to indicate which piece represents the unit in each case. ●

TAKE-AWAY MESSAGE . . . We just as easily could have based our number system on something other than ten, but ten is a natural number to use because we have ten fingers. By working in bases other than ten, you have probably gained a new perspective on the structure and complexity of our place-value system, particularly the importance of the value of each place. This understanding underlies all the procedures we use in calculating with numbers in base ten. As teachers, you will need this knowledge to help students understand computational procedures.

Notes

Notes

Learning Exercises for Section 2.3

1. If you have access to the Internet, go to http://www.nlvm.usu.edu/en/nav/ and find Virtual Library, then Numbers and Operations, then 3–5, then Base Blocks. You cannot choose numbers to represent, but you can set the base and the number of decimal points. Practice doing this with the following:

   a. whole numbers in base ten   **b.** decimal numbers in base ten
   c. whole numbers in base five   **d.** "basimal" numbers in base five

2. Write ten (this many: ★ ★ ★ ★ ★ ★ ★ ★ ★ ★) in each given system.

   a. base four   **b.** base five   **c.** base eight

3. Write each of these numbers.

   a. four in base four   **b.** eight in base eight
   c. twenty in base twenty   **d.** b in base b
   e. b^2 in base b **f.** $b^3 + b^2$ in base b
   g. 29_{ten} in base three   **h.** 115_{ten} in base five
   i. 69_{ten} in base two   **j.** 1728_{ten} in base twelve

4. Write the numerals for counting in base two, from one through twenty.

5. How do you know that there is an error in each statement?

   a. ten $= 24_{three}$ **b.** fifty-six $= 107_{seven}$
   c. thirteen and three-fourths $= 25.3_{four}$

6. Write each number as a base-ten numeral, along with its corresponding base-ten words.

   Examples: $111_{two} = (1 \times 2^2) + (1 \times 2) + (1 \times 1) = 7_{ten}$, or seven

   $$31.2_{four} = (3 \times 4) + (1 \times 1) + \frac{2}{4} = 12 + 1 + \frac{5}{10} = 13.5_{ten}, \text{ or}$$
   thirteen and five-tenths

   a. 37_{twelve} **b.** 37_{nine} **c.** 207.0024_{ten}
   d. 1000_{two} **e.** $1,000,000_{two}$ **f.** 221.2_{three}

7. For a given number, which base—two or twelve—will usually have a numeral with more digits? What are the exceptions?

8. In what bases would 4025_b be a legitimate numeral?

9. Compare these pairs of numbers by placing <, >, or = in each box.

   a. 34_{five} ☐ 34_{six} **b.** 4_{five} ☐ 4_{six} **c.** 43_{five} ☐ 25_{six}

   d. 100_{five} ☐ 18_{nine} **e.** 111_{two} ☐ 7_{ten} **f.** 23_{six} ☐ 23_{five}

10. On one of your space voyages, you uncover an alien document in which some "one, two, . . ." counting is done: obi, fin, mus, obi-na, obi-obi, obi-fin, obi-mus. What base does this alien civilization apparently use? Continue counting through twenty in that system.

11. Hints of the influence of other bases remain in some languages. What base could have led to each of these?

    a. French for eighty is *quatre-vingt.*
    b. The Gettysburg Address, "Four score and seven years ago. . . ."
    c. A gross is a dozen dozen.
    d. A minute has 60 seconds, and an hour has 60 minutes.

12. What does 34.2_{five} mean? What is this number written in base ten?

13. In each number, write the "basimal" place values and then the usual base-ten fraction or mixed number.

    Example: $10.111_{two} = \left(2 + 0 + \frac{1}{2} + \frac{1}{4} + \frac{1}{8}\right)_{ten} = 2\frac{7}{8}$ (*Recall:* $4 = 2^2$ and $8 = 2^3$.)

    a. 21.23_{four} **b.** 34.3_{twelve}

14. Write each of these in "basimal" notation.

Example: Three-fourths in base ten is *what* in base two?

$$\left(\tfrac{3}{4}\right)_{\text{ten}} = \left(\tfrac{1}{2} + \tfrac{1}{4}\right)_{\text{ten}} = 0.11_{\text{two}}$$

 a. one-fourth, in base twelve
 b. three-fourths, in base twelve
 c. one-fourth, in base eight

15. Give the base-ten numeral for each given number.

 a. $101{,}010_{\text{two}}$ **b.** 912_{twelve} **c.** 425_{six}
 d. 41.5_{eight} **e.** 1341_{five}

16. Write this many  in each given base. (Note that there are 12_{ten} diamonds.)

 a. nine   **b.** eight   **c.** seven   **d.** six   **e.** five   **f.** four   **g.** three   **h.** two

17. Write 100_{ten} in each given base.

 a. seven   **b.** five   **c.** eleven   **d.** two   **e.** thirty-one

18. Complete with the proper digits.

 a. $57_{\text{ten}} =$ _____ five   **b.** $86_{\text{nine}} =$ _____ ten   **c.** $312_{\text{four}} =$ _____ ten
 d. $237_{\text{ten}} =$ _____ eight   **e.** $2101_{\text{three}} =$ _____ ten   **f.** $0.111_{\text{two}} =$ _____ ten

19. Represent 34 in base ten, with the small block as the unit; with the long as the unit.

20. a. Represent 234_{five} with the small cube as the unit. (Notice that 234 does not mean two hundred thirty-four here.)
 b. Represent 234_{six} with the small cube as the unit.

(If you have only base-ten blocks available, then sketch drawings for these exercises.)

21. In base six, 5413 is _____ *ones,* is _____ *sixes,* is _____ *six²'s;* is _____ *six³'s.*

22. Represent 2.34 in base ten with the flat as the unit.

23. Decide on a representation with base-ten blocks for each number.

 a. 3542   **b.** 0.741   **c.** 11.11

24. Represent 5.4 and 5.21 with base-ten blocks, using the same block as the unit. (What will you use to represent one?) Many schoolchildren say that 5.21 is larger than 5.4 because 21 is larger than 4. How would you try to correct this error using base-ten blocks?

25. Someone said, "A number can be written in many ways." Explain that statement.

26. Here is a representation of a number: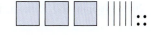

Which bases could use this representation if no more trades are possible? Why?

Supplementary Learning Exercises for Section 2.3

1. Write the corresponding base-six numerals as you count, "One, two, three, . . . ," until you need three digits. (When do you need three digits, in base six?)

2. Write the next six base-two numerals, in order:

 . . . 100_{two}, 101_{two}, 110_{two}, _____, _____, _____, _____, _____, _____

3. Why is it incorrect to write 123_{three} for eighteen?

4. Write two hundred seventy-six in each of the following:

 a. base-five numerals   **b.** base-eight numerals   **c.** base-two numerals

5. Write each of the following in base-ten numerals:

 a. $30{,}012_{\text{four}}$ **b.** $30{,}012_{\text{five}}$ **c.** 801_{nine}

6. Write two base-ten numerals for the base-ten blocks shown here. (*Hint*: What equals 1?)

7. If the blocks in the previous exercise were base seven, with the *long* as the unit, what base-seven numeral is illustrated, and what is its base-ten value?

8. How could each of the following be illustrated, with base-ten blocks?

   a. 4002   **b.** 4.325   **c.** 752.3

9. Explain this sentence: "There are only 10 types of people in the world: Those who understand binary, and those who do not."

10. Write an algebraic expression for each of the following:

   a. $10{,}000_b$ **b.** 3002_b **c.** $102{,}030_b$

11. Write the base-b numeral for each expression:

   a. $2b^3 + 3b$ **b.** $3 + 2b + b^3 + 4b^4$ **c.** $6 + b^3 + 2b + 3b^5$

Base-ten blocks are commonly used in many elementary classrooms to represent numbers, as you can see from the following page of problems* for fourth-graders. As you work through these exercises, you will be reminded how important it is to keep in mind how the unit (one) is represented.

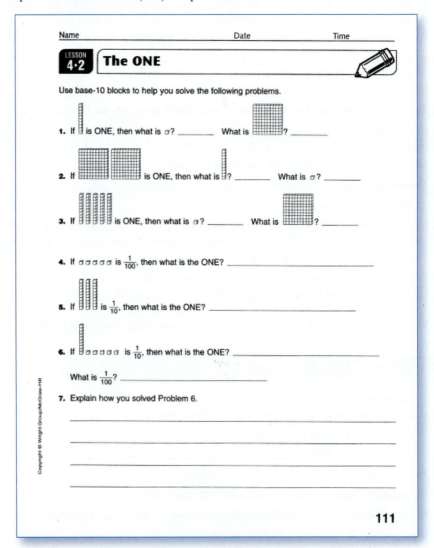

*Wright Group, *Everyday Mathematics, Math Masters*, Grade 4, p. 111. Copyright © 2008, McGraw-Hill. Reproduced with permission of The McGraw-Hill Companies.

2.4 Operations in Different Bases

Just as we can add, subtract, multiply, and divide in base ten, so can we perform these arithmetic operations in other bases. However, the intent here is to introduce you to different bases so that you have a better understanding of our own base-ten system, and you understand why children need time to learn to operate in base ten. We focus on addition and subtraction.

The standard algorithm for addition, depicted below, is commonly used and probably known to all of you. The expanded algorithms make the processes easier to understand. Once it is well understood, an expanded algorithm is easily adapted to become the standard algorithm. Not all standard algorithms in this country are used in other countries, so the word "standard" is a relative one.

In base ten, we could add 256 and 475 in these two ways, as shown here. (There are other ways, of course.) The first way is called an *expanded algorithm,* and the second, called the *standard algorithm,* is probably the one you were taught.

Notes

$$
\begin{array}{rl}
256 & \\
+\ 475 & \\
\hline
11 & \text{(thinking } 6 + 5) \\
120 & \text{(thinking } 50 + 70) \\
\underline{600} & \text{(thinking } 200 + 400) \\
731 &
\end{array}
\qquad
\begin{array}{r}
{\scriptstyle 1\ 1} \\
256 \\
+\ 475 \\
\hline
731
\end{array}
$$

The expanded algorithm is now being taught in some schools as a preparation for the standard algorithm. Note how place value is attended to in the expanded algorithm: Add the ones $6 + 5$, then add the tens $50 + 70$, then add the hundreds $200 + 400$, then add the resulting sums $11 + 120 + 600$. In the standard algorithm, each "column" is treated the same: $6 + 5$ in the column on the right, $5 + 7 + 1$ in the middle column, and $2 + 4 + 1$ in the column to the left. Although the standard algorithm leads to the correct answer, students frequently do not know why each step is taken. But when the expanded algorithm is understood, it can be condensed into the standard algorithm as shown above.

We can also use either method for adding in other bases, but the expanded algorithm is sometimes easier to follow until adding in another base is well understood.

EXAMPLE 6

Here is an example using both the standard and expanded algorithms to add the same two numbers in base ten and base eight. Make sure you can understand each way in each given base.

$$
\begin{array}{r}
{\scriptstyle 1} \\
351_{ten} \\
+\ 250_{ten} \\
\hline
601_{ten}
\end{array}
\qquad
\begin{array}{rl}
351_{ten} & \\
+\ 250_{ten} & \\
\hline
1_{ten} & \text{thinking } (1 + 0) \\
100_{ten} & \text{thinking } (50 + 50) \\
\underline{500}_{ten} & \text{thinking } (300 + 200) \\
601_{ten} & \text{thinking } (1 + 100 + 500)
\end{array}
$$

$$
\begin{array}{r}
{\scriptstyle 1} \\
351_{eight} \\
+\ 250_{eight} \\
\hline
621_{eight}
\end{array}
\qquad
\begin{array}{rl}
351_{eight} & \\
+\ 250_{eight} & \\
\hline
1_{eight} & \text{thinking } (1 + 0)_{eight} \\
120_{eight} & \text{thinking } (50 + 50)_{eight} \\
\underline{500}_{eight} & \text{thinking } (300 + 200)_{eight} \\
621_{eight} & \text{thinking } (1 + 20 + 100 + 500)_{eight}
\end{array}
$$

It may be helpful to represent addition and subtraction problems using base blocks.

EXAMPLE 7

Add 38 and 75 using base-ten blocks, with the small block as the unit.

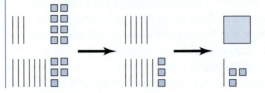

We have too many ones, so regroup ten ones into a long. Now we have too many longs. Group ten longs into a flat. The answer is 113. ●

 ACTIVITY 7 Adding in Base Four

Add these two numbers in base four in both expanded and standard algorithms: 311_{four} and 231_{four}. (Drawings of base-four pieces may be helpful.) ●

If we can add in different bases, we should be able to subtract in different bases. Here is an example of how to do this.

EXAMPLE 8

Find $321_{\text{five}} - 132_{\text{five}}$.
    One way to think about this problem is to regroup in base five just as we do in base ten, then use the standard way of subtracting in base ten.

Solution
$$\begin{array}{r} 321 \\ -132 \\ \hline \end{array}$$

Step 1: We cannot remove 2 ones from 1 one, so we need to take one of the fives from 321_{five} and trade it for 5 ones:

$$321_{\text{five}} \rightarrow 300_{\text{five}} + 20_{\text{five}} + 1_{\text{five}} \rightarrow 300_{\text{five}} + 10_{\text{five}} + 11_{\text{five}}$$

Step 2: We can now take 2 ones from 11 ones (in base five), leaving 4 ones. (Notice how 321 has changed with 3 five squareds, then 1 five, then 11 ones, from Step 1.)

$$\begin{array}{r} 1 \\ 3 \ \ 2 \ \ ^1 1_{\text{five}} \\ -1 \ \ 3 \ \ 2_{\text{five}} \\ \hline 4_{\text{five}} \end{array}$$
means 3 (five squareds) + 1 five + 11 ones, as in Step 1.

Step 3: In the fives place: We cannot subtract 3 fives from 1 five, so we must change 1 five squared to five sets of five. This, together with the 1 five already in place, gives us 11 fives (or six fives). That is,

$$300_{\text{five}} + 10_{\text{five}} \rightarrow 200_{\text{five}} + 110_{\text{five}}, \text{ so}$$

$$\begin{array}{r} 2 \ \ 11 \\ 3 \ \ 2 \ \ ^1 1_{\text{five}} \\ 1 \ \ \underline{3 \ \ 2_{\text{five}}} \\ 1 \ \ 3 \ \ 4_{\text{five}} \end{array}$$
means 11 fives, not 11 ones, so the 11 stands for 110 and 11 fives minus 3 fives is 3 fives, or $110 - 30$ is 30.

We now have 2 (five squareds) from which 1 (five squared) is subtracted, leaving 1 (five squared). The answer is 1 five squareds plus 3 fives plus 4 ones, which is 134_{five}. ●

 ACTIVITY 8 Subtracting in Base Four

Subtract 231_{four} from 311_{four} in base four. ●

Subtracting in base four is similar to adding in base four. However, for both operations we can use base materials to help visualize adding and subtracting in other bases. We will do that next. You can cut out and use materials from the "Masters." As you use the base materials, notice how they support the symbolic work you did earlier in this section.

EXAMPLE 9

Suppose we want to add 231_{four} and 311_{four} using base-four blocks, with the small block as the unit. We could first express the problem as

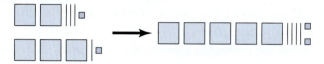

We have too many longs (in base four), so trade four longs for a flat. Now we have too many flats (each representing four squared). Trade four flats for a large cube (which represents four cubed):

Represents the answer, which is 1202_{four}. The blocks represent 1 four cubed, 2 fours squared, and 2 ones.

EXAMPLE 10

Suppose we want to subtract 23_{four} from 32_{four}.
This time let us use the *long* as the unit:
32_{four} is represented as

We cannot remove three longs (ones) until we change a flat to four longs (which means change one four into four ones):

Remove two flats and three longs.

To take away 23_{four}, we must remove three longs (3 ones) and two flats (2 fours), and we are left with 3_{four} as the difference.

Think About …

If we had used the small block as the unit in the above subtraction example, would the numerical answer be different? Try it.

ACTIVITY 9   Subtracting in Base Four

Once again, subtract 231_{four} from 311_{four} in base four, this time using drawings.

There are no examples or exercises provided here for multiplication and division in different bases, although it is certainly possible to carry out multiplication and division. This omission allows more time for other topics. Your experience with addition and subtraction in other bases should allow you to appreciate children's difficulties in base ten.

TAKE-AWAY MESSAGE . . .   Arithmetic operations in other bases are undertaken in the same way as in base ten. However, because we have less familiarity with other bases, arithmetic operations in those bases take us longer than operations in base ten. For children not yet entirely familiar with base ten, the time needed to complete arithmetic operations is longer than for us.

Learning Exercises for Section 2.4

1. Add 1111_{three} and 2102_{three} without drawings and then with drawings in the ways illustrated above. Which way did you find it easier?

2. Do these exercises in the designated bases, using the cardboard cut-outs in the "Masters," or with drawings.

   a. 341_{five} **b.** 101_{two} **c.** 321_{four} **d.** 296_{ten}
   $+ 220_{five}$ $+ 110_{two}$ $- 123_{four}$ $-\ \ 28_{ten}$

3. Go to http://nlvm.usu.edu/en/nav/ on the Internet and find Virtual Library, then Numbers and Operations, then 3–5. Go to Base Blocks Decimals. You cannot choose numbers to add and subtract, but you can set the base and the number of decimal points. Do the following:

   a. Practice adding and subtracting numbers in base ten using whole numbers.
   b. Practice adding and subtracting numbers using one decimal place.
   c. Practice adding and subtracting numbers in base four using whole numbers.
   d. Practice adding and subtracting numbers in base four using one "decimal" place.

4. Add the following in the appropriate bases, without blocks unless you need them.

   a. 2431_{five} **b.** 351_{nine} **c.** 643_{seven} **d.** 99_{eleven}
   $+ 223_{five}$ $+ 250_{nine}$ $+ 134_{seven}$ $+ 88_{eleven}$

5. Subtract in different bases, without blocks unless you need them.

   a. 351_{nine} **b.** 643_{seven} **c.** 2431_{five} **d.** 772_{eleven}
   $-\ 250_{nine}$ $-\ 134_{seven}$ $-\ 223_{five}$ $-\ 249_{eleven}$

6. Do you think multiplying and dividing in different bases would be difficult? Why or why not?

7. Use the cut-outs from the "Masters" for the different bases to act out the following. As you act out each, record what would take place in the corresponding numerical work.

   a. 232_{four} **b.** 232_{five} **c.** 232_{eight} **d.** 101_{two}
   13_{four} 13_{five} 13_{eight} 11_{two}
   $+ 113_{four}$ $+ 113_{five}$ $+ 113_{eight}$ $+ 111_{two}$

8. Use the cut-outs from the "Masters" for the different bases to act out the following. As you act out each, record what would take place in the corresponding numerical work.

   a. 200_{four} **b.** 200_{five} **c.** 200_{eight} **d.** 100_{two}
   -13_{four} -13_{five} -13_{eight} -11_{two}

9. Describe how cut-outs for base six would look. For base twelve.

Supplementary Learning Exercises for Section 2.4

1. Use the cut-outs from the "Masters" for the different bases to act out the following. As you act out each, record what would take place in the corresponding numerical work.

   a. 212_{five} **b.** 212_{four} **c.** 212_{seven} **d.** 1101_{two}
   33_{five} 33_{four} 33_{seven} 11_{two}
   $+ 113_{five}$ $+ 113_{four}$ $+ 113_{seven}$ $+ 101_{two}$

2. Use the cut-outs from the "Masters" for the different bases to act out the following. As you act out each, record what would take place in the corresponding numerical work.

   a. 300_{four} **b.** 300_{five} **c.** 300_{seven} **d.** 1000_{two}
   -21_{four} -21_{five} -21_{seven} -11_{two}

3. Describe how cut-outs for base seven would look. For base twenty.

4. Make up and solve an addition calculation and a subtraction calculation in base seven.

5. **a.** Complete the given table for the one-digit-plus-one-digit results (the "basic addition facts") for base two. Write the entries in base two. How many entries are there?

+	0	1
0		
1		

  b. Similarly, complete the table of basic addition facts for base three. Write the entries in base three. How many entries are there?

+	0	1	2
0			
1			
2			

  c. How many entries would there be in a table of basic addition facts for base seven?

  d. How many basic addition facts are there for base ten?

  e. How many basic addition facts would there be for base n?

2.5 Issues for Learning: Understanding Place Value

Understanding place value is considered to be foundational to elementary school mathematics. Yet place-value instruction in schools is often superficial and limited to studying only the placement of digits. Thus, children are taught that the 7 in 7200 is in the thousands place, the 2 is in the hundreds place, a 0 is in the tens place, and a 0 is in the ones place. But when asked how many hundred dollar bills could be obtained from a bank account with $7200 in it, or how many boxes of ten golf balls could be packed from a container with 7200 balls, children almost always do long division, dividing by 100 or by 10. They do not read the number as 7200 ones, 720 tens, or 72 hundreds, and certainly not as 7.2 thousands. But why not? These are all names for the same number, and the ability to rename in this way provides a great deal of flexibility and insight when working with the number.

One activity-centered primary program incorporates many activities involving grouping by twos, by threes, etc., even before extensive work with base-ten groupings, to accustom the children to counting not just one object at a time, but groups each made up of several objects. Ungrouping needs to be included also. That is, 132 could be regarded as 1 one hundred, 3 tens, and 2 ones. Or, it could be regarded as 1 one hundred and 32 ones. Here, the 3 tens are "unbundled" to make 30 ones. Regarding a group made up of several objects as one thing is a major step that needs instructional attention.

Over the years many different methods have been used to teach place value. An abacus with nine beads on each string is one type of device used to represent place value. The base-ten blocks pictured in Section 2.3 have been extensively used to introduce place value and operations on whole numbers and decimal numbers. One problem with these representations, however, is that students do not always make the connections between what is shown with the manipulative devices and what they write on paper.

The manner in which we vocalize numbers can sometimes cause problems for students. For example, some young U.S. children will write 81 for eighteen, whereas scarcely any Hispanic children (*diez y ocho* = eighteen) or Japanese children (*ju hachi* = eighteen) do so. (Some wishfully think we should say "onetyeight" for eighteen in English.) What other numbers can cause the same sort of problem that eighteen does?

Our place-value system of numeration also extends to numbers less than 1. The naming of decimal numbers needs special attention. The place-value name for 0.642 is six hundred forty-two thousandths. Compare this to reading 642, where we simply say six hundred forty-two, not 642 ones. This is a source of confusion that is compounded by the use of the ten*ths* or hundred*ths* with decimal numbers, the use of ten or hundred with whole numbers, and the additional digits in the whole number with a similar name. The number 0.642 is read 642 thousand*ths*, meaning 642 thousand*ths* of one, while 642,000 is read 642 thousand, meaning 642 thousand ones. That tens and ten*ths*, hundreds and hundred*ths*, etc., sound so much alike no doubt causes some children to lose sense-making when it comes to decimals. Some teachers resort to a digit-by-digit pronunciation—"two point one five" for 2.15—but that removes any sense for the number; it just describes the numeral. Plan to give an artificial emphasis to the *-th* sound when you are discussing decimals with children. (You can also say "decimal numeral two and three-tenths" and "mixed numeral two and three-tenths" to distinguish 2.3 and $2\frac{3}{10}$.)

To compare 0.45 and 0.6, students are often told to "add a zero so the numbers are the same size." The strategy works, in the sense that the student can then (usually) choose the larger number, but since it requires no knowledge of the size of the decimal numbers, it does not develop understanding of number size. Instead of annexing zeros, couldn't we expect students to recognize that six-tenths is more than forty-five hundredths because 45 hundredths has only 4 tenths and what is left is less than another tenth? But for students to do this naturally, they must have been provided with numerous opportunities to explore—and think about—place value. Comparing and operating on decimals, if presented in a non-rule-oriented fashion, can provide these opportunities. If teachers postpone work with operations on decimals until students conceptually understand these numbers, students will be much more successful than if teachers attempt to teach computation too early. Some researchers[3] have shown that once students have learned rote rules for calculating with decimals, it is extremely difficult for them to relearn how to calculate with decimals meaningfully.

2.6 Check Yourself

Understanding place value and its role in the elementary school mathematics curriculum is crucial. Too many teachers think that teaching place value is simply a matter of noting which digit is in the ones place, which is in the tens place, etc. But it is only when students have a deep understanding of place value that they can make sense of numbers larger than 10 and smaller than 1, and understand how to operate on these numbers. Most arithmetic errors (beyond careless errors) are due to a lack of understanding of place value. Unfortunately, the algorithms we teach usually treat digits in columns without attending to their values, and students who learn these algorithms without understanding the place value of each digit are far more likely to make computational errors.

You should be able to work problems like those assigned and to meet the following objectives:

1. Discuss the advantages of a place-value system over other ancient numeration systems.

2. Explain how the placement of digits determines the value of a number in base ten, on both sides of the decimal point.

3. Explain how the placement of digits determines the value of a number in any base, such as base five or base twelve, and answer questions such as "What does 346.3 mean in base twelve?" Convert that number to base ten.

4. Given a particular base, write numbers in that system beginning with one.

5. Make a drawing with base materials that demonstrates a particular addition or subtraction problem, e.g., 35.7 + 24.7 or 35.7 − 24.7 in base ten.

6. Write base-ten numbers in another base, such as 9 in base nine, or 33 in base two.

7. Add and subtract in different bases.

8. Understand the role of the unit one in reading and understanding decimal numbers.

9. Discuss problems that children who do not have a good understanding of place value might have when they do computation problems.

References for Chapter 2

[1]Saxe, G. B. (1981). Body parts as numerals: A developmental analysis of numeration among the Oksapmin in Papau New Guinea. *Child Development, 52,* 306–316.

[2]Sowder, J. T. (1997). Place value as the key to teaching decimal operations. *Teaching Children Mathematics, 3* (8), 448–453.

[3]Hiebert, J., & Wearne, D. (1986). Procedures over concepts: The acquisition of decimal number knowledge. In J. Hiebert (Ed.), *Conceptual and procedural knowledge: The case of mathematics* (pp. 199–223). Hillsdale, NJ: Erlbaum.

Chapter 3

Understanding Whole Number Operations

There are two major ideas interspersed in this chapter. The first is that an arithmetic operation such as subtraction can be modeled by many different situations. Teachers need to know what these situations are in order to understand and extend children's use of the operations. The second is that the kind of procedures that children can develop for computing, when they have not yet been taught the usual standard algorithms, can demonstrate a deep conceptual understanding that might be lost if standard algorithms are introduced too soon. The examples of student work given in this section are all taken from published research, although in some cases the numbers have been changed. Most of the examples of nonstandard algorithms come from students who have been in classrooms where there is a strong emphasis on building on intuitive knowledge and on place-value understanding. The students demonstrate that they are able to compute with ease. Some of the examples are from classrooms where calculators are always available to students. However, they tend not to use calculators if they can do the calculation easily themselves using paper and pencil and/ or mental computation. The use of calculators introduced these students to new ways of using numbers, including, in some cases, using negative numbers to help them in their computation.

3.1 Additive Combinations and Comparisons

When does a problem situation call for adding? When does it call for subtracting? Both types of situations are considered in this section. At times, knowing when and what to add or subtract is not at all easy. To solve more difficult problems, we turn once again to undertaking quantitative analyses of problems.

 **ACTIVITY 1**   Applefest

Consider the following problem situation:

▶ Tom, Fred, and Rhoda combined their apples for a fruit stand. Fred and Rhoda together had 97 more apples than Tom. Rhoda had 17 apples. Tom had 25 apples.[1] ◀

Perform a quantitative analysis of this problem situation with these four steps.

1. Identify as many quantities as you can in this situation, including those for which you are not given a value. Can you make a drawing?
2. What does the 97 stand for in this situation?
3. How many apples did Fred and Rhoda together have? How many apples did all three of them combined have?
4. How many apples did Fred have? ●

Note in the Activity 1 problem that when Fred and Rhoda combined their apples, they had 97 *more* apples than Tom. The 97 apples does not refer to Fred and Rhoda's combined total; rather, it refers to the *difference* between their combined total and Tom's number of apples.

Notes

Consider the following drawing that represents the applefest problem:

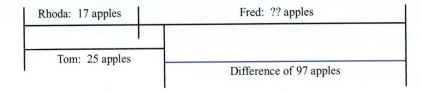

| Rhoda: 17 apples | Fred: ?? apples |
| Tom: 25 apples | Difference of 97 apples |

Quantities are often ***combined additively*** (put together) and the new quantity has the value represented by the sum of the values of the quantities being combined. The sum of the number of Rhoda's apples and the number of Fred's apples is the value of the number of apples belonging to the two. Quantities can also be ***compared additively***. In the applefest problem, the quantities consisting of the value of the quantity represented by Rhoda's and Fred's combined apples are compared to the value of the quantity represented by Tom's apples.

When quantities are combined additively, they are joined together, so the appropriate arithmetic operation on their values is usually addition. This operation is called an **additive combination**. The result of an additive combination is a **sum** of the values combined. Any time we compare two quantities to determine how much greater or less one is than the other, we make an **additive comparison**. The **difference** of two quantities is the quantity by which one of them exceeds or falls short of the other. The appropriate arithmetic operation on their values is usually subtraction.

The reason this comparison in this problem is called additive is that two quantities can also be compared multiplicatively. We will study multiplicative comparisons in a later section.

Typically, subtraction is the mathematical operation used to find the difference between the known values of two quantities. We can think of the difference as the amount that has to be added to the lesser of the two values to make it equal in value to the greater of the two values. Thus, rather than use *additive/subtractive comparison,* we use the shorter *additive comparison.* The following diagrams illustrate how one might think of a situation as either a difference of two quantities or as a combination of two quantities (even when, in the latter case, the value of one of the quantities being combined is unknown).

EXAMPLE 1

▶ Julian wants to buy a bicycle. The bike costs $143.95. Julian has a total of $83.48 in cash and savings. How much more does he need? ◀

Two ways to conceive of the situation

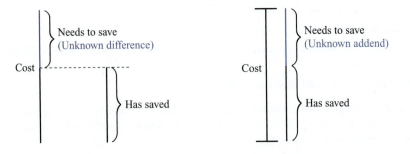

The drawings help in seeing how the quantities are related in this situation. ●

Although typically subtraction is the operation used to find a difference, this is not always the case. Consider the following example.

EXAMPLE 2

▶ Jim is 15 cm taller than Sam. This difference is five times as great as the difference between Abe's and Sam's heights. What is the difference between Abe's and Sam's heights? ◀

Solution

The difference between Abe's and Sam's heights is $15 \div 5$ cm, or 3 cm. ●

> ### *Think About ...*
> What would the diagram for Example 2 look like?

Example 2 illustrates that the additive comparison of two quantities does not automatically signal that one must subtract. How the quantities in a situation are related to each other determines the mathematical operations that make sense. Understanding the quantitative relationships is essential to being able to answer questions reliably involving the values of quantities in the given situation. Without such an understanding, one has no recourse but to guess what mathematical operations are needed, as was illustrated in the last chapter. In the absence of real understanding, it is the unreliability of the guessing games that students often play which makes story problems difficult for many of them.

 ACTIVITY 2　It's Just a Game

Practice doing a quantitative analysis to solve this problem.

▶ Team A played a basketball game against Opponent A. Team B played a basketball game against Opponent B. The captains of Team A and Team B argued about which team beat its opponent by more. Team B won by 8 more points than Team A won by. Team A scored 79 points. Opponent A scored 48 points. Team B scored 73 points. How many points did Opponent B score? ◀

1. What quantities are involved in this problem? (*Hint:* There are more than four.)
2. What does the 8 points refer to?
3. There are several differences (results of additive comparisons) in this situation. Sketch a diagram to show the relevant differences in the problem.
4. Solve the problem.
5. What arithmetic operations did you use to solve the problem?
6. For many students this is a difficult problem. Why do you suppose this is the case? Is the computation difficult? What is difficult about this problem?
7. Suppose you are the teacher in a fifth-grade class. Do you think telling your students that all they need to solve this problem is subtraction and addition will help them solve it? Explain. ●

TAKE-AWAY MESSAGE ...　The problems undertaken in this section involved additive combinations (which can be expressed with an addition equation) and additive comparisons (which can be expressed with a subtraction equation). What makes these problems difficult is the complexity of the quantitative structures of the problems, not the arithmetic. The problems involve only addition and subtraction. Understanding the quantitative relationships in a problem is what's crucial. Once again, these problems illustrate how one can understand a complex problem by undertaking a quantitative analysis—by listing quantities and using diagrams to explore the relationships of the quantities.

Learning Exercises for Section 3.1

To develop skill in analyzing the quantitative structure of situations, work the following tasks by first identifying the quantities and their relationships in each situation. Feel free to draw diagrams to represent the relationships between relevant quantities.

1. **a.** Bob is taller than Laura. Suppose you were told Bob's height and Laura's height. How would you calculate the difference between their heights?
    b. Bob is taller than Laura. Suppose you were told the difference between Bob's height and Laura's height. Suppose you were also told Laura's height. How would you calculate Bob's height?

2. Kelly's mom timed her as she swam a 3-lap race in 1 minute 43 seconds. Her swimming coach timed Kelly only on her last two laps. Describe how Kelly might calculate how long it took her to swim the first lap using the information from her mom and the coach.

3. Metcalf School has two third-grade rooms (A and B) and two fourth-grade rooms (C and D). Together, rooms C and D have 46 students. Room A has 6 more students than room D. Room B has 2 fewer students than room C. Room D has 22 students. How many students are there altogether in rooms A and B?

4. In Exercise 3, what quantities are combined? Which quantities are compared?

5. Following are two variations of the activity "It's Just a Game," from the last Activity.
    a. Variation 1: Team A scored 79 points. Opponent A scored 53 points. Fill in the blanks with scores for Team B and Opponent B so that Team B won by 8 more points than Team A won by.
        Team B _____ Opponent B _____
    b. Variation 2: Team B scored 75 points. Opponent B scored 69 points. Give possible scores for Team A and Opponent A so that Team B won by 8 more points than Team A won by.
        Team A _____ Opponent A _____
    c. Compare your response to Variation 1 with those of classmates. Are they correct? How many possible responses are there?
    d. What did you find out after thinking through Variation 2? State your conclusion and explain why it is the case.

6. Connie bought several types of candy for Halloween: Milky Ways, Tootsie Rolls, Reese's Cups, and Hershey Bars. Milky Ways and Tootsie Rolls together were 15 more than the Reese's Cups. There were 4 fewer Reese's Cups than Hershey Bars. There were 12 Milky Ways and 14 Hershey Bars. How many Tootsie Rolls did Connie buy?

7. **a.** One day Annie weighed 24 ounces more than Benjie, and Benjie weighed $3\frac{1}{4}$ pounds less than Carmen. How did Annie's and Carmen's weight compare on that day?
    b. Why can't you tell how much each person weighed?

8. You have two recipes that together use a pound of butter. One recipe takes $\frac{1}{4}$ pound more than the other one. How much butter does each recipe use?

9. A hospital needs a supply of an expensive medicine. Company A has the most, 1.3 milligrams, which is twice the difference between the weight of Company B's supply and Company C's, and 0.9 milligram more than Company C's supply. How many milligrams can the hospital get from these three companies?

10. A city spent about $\frac{1}{4}$ of its budget on buildings and maintenance, which was about $\frac{1}{8}$ of the total budget less than it spent on administrative personnel. The amount spent on administrative personnel was about $\frac{1}{8}$ of the budget more than was spent on public safety. About what part of the budget was left for other expenses?

11. All the problems in the previous exercises involve the quantitative operations of combining and comparing quantities. Choose two of them and for each state one additive combination or one additive comparison that is involved.

Supplementary Learning Exercises for Section 3.1

1. What quantities do the three values in the following story represent?

▶ I read three novels, with the first having 120 more pages than the second, and the second having 75 fewer pages than the third. ◀

2. Analyze the quantitative structure of each of the following problems to help in solving them. Drawings are recommended.

    a. The winner of the skiing race won by 0.064 seconds. The second-place skier beat the third-place skier by 0.92 seconds. The second-place skier took 23.4 seconds for the race. How long did the winner and the third-place skiers take?

    b. On two trips to the grocery store, Judy spent a total of $60.45. She spent $11.85 more on the second trip than she did on the first trip. How much did she spend on each trip?

    c. Von, Will, and Xavier are looking at their baseball cards. They notice that Will and Xavier together have 9 more cards than Von does, that Xavier has 23 cards, and that Von has 40 cards. How many cards does Will have?

3. For each part, what quantities are compared, and what quantities are combined?

    a. In playing Monopoly, Nia and her sister together collected more money than did her two brothers, Dan and Ian, combined. The four players collected $17,500 in all.

    b. Bank X has $137 million in assets less than Bank Y does. But together, the two banks have $25 million more in assets than Bank Z does.

4. In Supplementary Learning Exercise 3, in each of parts (a) and (b), give values for the quantities that fit the descriptions. Will your values be the same as those given by others?

5. Write an *additive comparison* story problem that uses 90¢ and 49¢.

6. Make a diagram to help solve the following problem: Sue's GPA is 0.3 honor points greater than Tia's. If Tia can improve her GPA by 0.5 honor points, her GPA will be 3.2.

    a. What is Sue's current GPA?

    b. Identify any additive combinations that are involved in the story problem in part (a).

    c. Identify any additive comparisons that are in the story problem in part (a).

3.2 Ways of Thinking About Addition and Subtraction

Although the idea of combining two or more quantities additively seems quite simple, situations involving additive combinations can vary in difficulty for young children. Situations involving subtraction are quite varied, and when only one type of situation is taught (e.g., take away), children can have difficulty with other situations (e.g., comparing) that also call for subtraction. In this chapter we consider two types of situations calling for addition and three calling for subtraction.

    A problem situation that calls for addition often describes one quantity being *physically*, or *actively*, put with another quantity, as in Example 3. The term *joining* is often used for this type of situation.

EXAMPLE 3

▶ Four girls were in the car. Two more got in. How many girls were in the car then? ◀ ●

Another type of problem situation that calls for addition involves *conceptually* (rather than physically or literally) placing quantities together, thus leading to another view of what addition means. *Part* and *whole* can also be used here, as in *part + part = whole*.

EXAMPLE 4

▶ Only 4 cars and 2 trucks were in the lot. How many vehicles were there altogether? ◀ ●

Even though both problems involve the additive combination of two quantities, and they both call for the mathematical operation of addition, many research studies have shown that the second problem is more difficult for young children. They may not yet understand concept relations as "cars and trucks are simply two different kinds of vehicles."

> Addition can describe situations that involve an **additive combination** of quantities, either literally (as in the first example) or conceptually (as in the second example). Numbers added together are called **addends**. The number that is the result of an addition is called the **sum**.

In either of the problems given in Examples 3 and 4, the numbers being added (2 and 4) are addends, and the result is called the sum (6).

An important distinction calling for awareness by the teachers is this: The calculation one does to solve a problem may be different from that suggested by the action described in the problem or by the problem's underlying structure. The following illustrates this important point.

EXAMPLE 5

▶ Josie needs to make 15 tacos for lunch. She has made 7 already. How many more tacos does she have to make? ◀ ●

You would probably regard this as a subtraction problem, because one can calculate the solution from the number sentence $15 - 7 = n$. Yet, when young children have solved similar problems, perhaps with the aid of concrete materials, they often count out 7 blocks, and then put out additional blocks, one at a time, until they have 15. Their actions are a reenactment of the problem's story. The action of the problem suggests addition of an unknown number of tacos to the 7 tacos, to give a total of 15. Mathematically, this problem can also be described by the sentence $7 + n = 15$, where the sum and one addend are known, but the other addend is missing. If the numbers were larger, say, $1678 + n = 4132$, you would probably *want* to subtract to find n. Such **missing-addend (or missing-part) situations**, then, can be solved by subtraction, and missing-addend story problems may be in a subtraction section of an elementary textbook. Note the potential confusion for students due to the mismatch of the joining action suggesting addition and the calculation that uses subtraction. The classification of such problems as "missing-addend" problems serves to point out this connection.

> A problem situation that can be represented as $a + ? = b$ is called a **missing-addend problem**. Although the action suggests addition, the missing value is $b - a$. It is therefore classified as a subtraction problem.

Think About …

▶ Ana has to practice the piano for 15 minutes every day. Today she has practiced 7 minutes. How many more minutes does she have to practice today? ◀

Why is this problem a missing-addend problem calling for subtraction?

In the typical elementary school curriculum, missing-addend situations are not the first ones students encounter when dealing with subtraction. Usually, *take-away* situations are the first introduced under subtraction. In fact, the minus sign ($-$) is often read as "take away" instead of as "minus," a reading that can limit one's understanding of subtraction.

Contrast the following story problem with the missing addend given in the above Think About problem:

▶ Josie made 15 tacos for her friends at lunch. They ate 7 of them. How many tacos did they still have? ◀

Think About …

How do you suppose children would act out this story problem using counters?

In situations like the problem above, children typically start with 15 counters, and then they *take away* or *separate* 7 counters. The remaining counters stand for the uneaten tacos. This enables them to answer the question. In a take-away situation, it is natural to call the quantity that is left after the taking away is done the remainder.

A situation in which one quantity is removed or separated from a larger quantity is called a **take-away subtraction situation**. What is left of the larger quantity is called the **remainder**.

As in the case of addition, one can distinguish between two types of situations: one in which there is a physical taking away, and one in which there is no "taking-away" action as such, but there is a missing part, given a whole.

Think About …

How would you classify the following problem situation?

▶ Of Josie's 15 tacos, 7 contained chicken and the rest beef. How many of the tacos contained beef? ◀

Because the discussion of missing addends is fresh in your mind, you may be saying to yourself, "I think I can see both of the last two taco problems as missing-addend problems."

[number of tacos eaten (7)] + [number still to eat (?)] = total number (15)

[number of chicken tacos (7)] + [number of beef tacos (?)] = total number (15)

This similarity of structure is one reason that some textbooks emphasize the missing-addend view when dealing with subtraction.

EXAMPLE 6

One can think about $7 - 3 = ?$ as $3 + ? = 7$. ●

Familiarity with $3 + 4 = 7$ enables one to see $7 - 3 = 4$. Some curricula build considerably on this view and may at some point teach *families of facts*. The family of (addition and subtraction) facts for 3, 4, and 7 would include all of these: $3 + 4 = 7, 4 + 3 = 7, 7 - 3 = 4$, and $7 - 4 = 3$. Do you see what might be some advantages to knowing families of facts?

A third type of subtraction comes from the additive comparison idea. This type of situation is also found in elementary school curriculum. Consider the following situation:

▶ Josie made 15 tacos and 7 enchiladas. How many more tacos than enchiladas did Josie make? ◀

Two *separate* quantities, the number of tacos and the number of enchiladas, are being compared in a how-many-more (or how-many-less) sense.

> A problem situation involving an additive comparison is referred to as a **comparison subtraction situation**.

Subtraction can tell how many more, or less, there are of one than of the other. The result of the comparison is the *difference* between the values of the two quantities being compared. Comparison subtraction usually appears in the curriculum after take-away subtraction.

> When one number is subtracted from another, the result is called the **difference** or **remainder** of the two numbers. The number from which the other is subtracted is called the **minuend** and the number being subtracted is called the **subtrahend**.

EXAMPLE 7

In $50 - 15 = 35$, 50 is the minuend, 15 is the subtrahend, and 35 is the difference (or remainder). ●

The operation needed to answer the question asked in a comparison subtraction situation, however, can be addition, as in this example.

▶ Josie made 7 more beef tacos than chicken tacos. If she made 15 chicken tacos, how many beef tacos did she make? ◀

Similarly, some take-away subtraction situations can actually require addition to answer the questions.

▶ Jan spent $29.85 shopping and found that she had $11.65 left. How much money did she have before she went shopping? ◀

In all the previous examples, the quantities under consideration were separate, disconnected objects, like tacos. But there are other situations in which we combine or compare quantities such as distances, areas, volumes, etc.

> Quantities being considered are called **discrete** if they are separate, nontouching objects that can be counted. Quantities are **continuous** when they can be measured only by length, area, etc. Continuous quantities are measured, not counted.

EXAMPLE 8

The following problem situation is also a comparison subtraction problem. Height is a *continuous* quantity. The heights given are only approximations of the real height of each. One cannot *count* an *exact* number of inches.

▶ Johann is 66 inches tall. Jacqui is 57 inches tall. How much taller is Johann than Jacqui? ◀ ●

EXAMPLE 9

The following problem situation is a comparison problem calling for subtraction of discrete quantities. One can count siblings; one cannot have parts of siblings.

▶ Johann has 5 siblings and Jacqui has 4 siblings. How many more siblings does Johann have than Jacqui? ◀ ●

 ACTIVITY 3 Which Is Which?

Pair up with someone in your class and together classify each of the following problem situations (e.g., comparison subtraction):

1. Velma received 4 new sweaters for her birthday. Two of them were duplicates, so she took 1 back. How many sweaters does she have now?
2. Velma also received cash for her birthday, $36 in all. She has an eye on some software, which costs $49.95. Her dad offered to pay her to wax and polish his car. How much does she need to earn to buy the software?
3. Velma also received 2 mystery novels and 3 romance novels for her birthday. How many novels did she receive?
4. There were 6 boys and 8 girls at her birthday party. How many more girls than boys from her class were at the party? How many friends were at the party? ●

 ACTIVITY 4 Writing Story Problems

1. Pair up with someone in your class. Write problem situations (not necessarily in the order given) that illustrate these different views of addition and subtraction:

    a. addition that involves putting together two quantities
    b. addition that involves thinking about two quantities as one quantity
    c. take-away subtraction
    d. comparison subtraction
    e. missing-addend subtraction
    f. addition arising from a comparison situation

2. Share your problems with another group. Each group should identify the types of problems illustrated by the other group, and discuss whether each group can correctly identify the problem situations with the description. In each problem, identify whether the quantities are discrete or continuous. ●

 DISCUSSION 1 Identifying Types of Problem Situations

If there are any disagreements in the story problems written for the last activity, discuss these as a class. ●

TAKE-AWAY MESSAGE . . . Addition is called for in problems of two different generic types: combining literally or conceptually. Subtraction is called for in three different generic types: taking away, comparing, and finding the missing addend. Missing-addend problems are sometimes called additive comparison problems, but they are solved by subtracting. If a teacher illustrates subtraction only with take-away situations, then that teacher should not be surprised if his or her students cannot recognize other types of subtraction situations. Also, if students are presented only with discrete objects when they add or subtract, they will not necessarily be able to transfer this knowledge to situations in which the quantities are continuous.

One might ask how children cope with the diversity of situations that call for addition or subtraction. The manner in which they act out the story usually shows how they have conceived the situation. The physical situations can be quite different even though the symbolic mathematics may be the same.

Notes

Learning Exercises for Section 3.2

1. Illustrate the computation 8 − 5 with both continuous and discrete drawings for each of the three generic situations for subtraction.

   a. take-away   **b.** comparison   **c.** missing-addend

   Notice how the drawings differ.

2. **a.** The narrative in this section has neglected story problems like this one: "Jay had 60 pieces of clean paper when she started her homework. When she finished, she had 14 clean pieces. How many pieces of paper did she use?" What mathematical sentence would one write for this problem? Notice that the action here is take-away, but the correct sentence *for the action* would be $60 − n = 14$, not $60 − 14 = n$ (one could calculate $60 − 14$ to answer the question though). Such complications are usually avoided in current K–6 treatments in the United States, but several other countries give considerable attention to these types. Are we "babying" our children intellectually?

   b. Make up a story problem for this "missing-minuend" number sentence: $n − 17 = 24$.

3. Teachers sometimes emphasize "key words" to help children with story problems. For example, "altogether" suggests addition, "left" suggests subtraction. The intent is good: Have the children think about the situation. But unfortunately children often abuse the key words. They skim, looking just for the key words, or they trust them too much, taking what is intended as a rough guide as a rule. In the following, tell how the italicized key words could mislead a child who does not read the whole problem or who does not think about the situation.

   a. Dale *spent* $1.25. Then Dale had 55 cents. How much did Dale have at the start?

   b. Each classroom at one school has 32 children. The school has 12 classrooms. How many children are at the school *altogether*?

   c. Ben *divided* up his pieces of candy evenly with Jose and Cleveland. Each of the 3 boys got 15 pieces of candy. How many pieces did Ben start with?

   d. Flo has 3 *times* as much money as Lacy does. Flo has 84 cents. How much does Lacy have?

   e. Manny's mother bought some things at the grocery store. She gave the clerk $10 and got $1.27 in change. *In all*, how much did she spend at the store?

   f. Each package of stickers has 6 pages, and each page has 12 stickers on it. What is the *total* number of stickers in 4 packages?

4. Consider this comparison subtraction situation: "Ann has $12.85 and Bea has $6.43. How much less money does Bea have than Ann?" Note that the question is awkward. This is believed to be one explanation of why young children have more trouble with comparison story problems than with take-away problems. When story problems are phrased less awkwardly, children tend to do much better. For example, "There are 12 bowls and 8 spoons. How many bowls do not have a spoon?" is easier for children than, "There are 12 bowls and 8 spoons. How many more bowls are there than spoons?" This may be due to the fact that the phrasing in the former case is suggestive of an action (matching bowls with spoons) that enables students to obtain an answer.

   Rephrase the questions to make these problems easier.

   a. The baseball team has 12 players but only 9 gloves. How many more players are there than gloves?

   b. There are 28 children in the teacher's class. The teacher has 20 suckers. How many fewer suckers are there than children?

5. Give the families of facts for the following:

   a. 2, 6, 8

   b. 12, 49, 61

   c. x, m, and p, where $m − p = x$

6. Choose a family of facts in Exercise 5 above; write one addition word problem and three subtraction word problems, one of each type, for that family of facts.

7. In a first-grade class the teacher gave the following problem:[2]

   ▶ There are 12 boys and 8 girls. How many more boys than girls? ◀

   A majority of the children answered "four more boys" correctly, but 5 children chose to add and gave "20 children" as their answer. One of these 5 children insisted that subtraction could not be used because it is impossible to **take away** 8 girls from 12 boys. None of the other children could respond to this child's claim. In light of what you have learned in this section, describe how you would handle this situation if you were the teacher, in order to help this child understand that subtraction is the appropriate operation in this situation. Note that simply telling a student "the right way" does not necessarily help him or her understand.

8. Tell which type of subtraction is indicated in each of these story problems, with an explanation of your choice.

   a. Villi is running a 10-kilometer race. He has run 4.6 kilometers. How far does he have yet to go?

   b. Diego is saving up for a car. He needs $2000 to buy one from his uncle. Thus far he has $862. How much more does he need?

   c. Laresa just got paid for the week. She received $200. She owes her mother $185 for some clothes. Once she pays it, what will she have left?

   d. Laresa's friend works at a different place and is paid $230 a week. How much more does she make per week than Laresa?

   e. Bo just bought gas for $3.79 per gallon. Last week he bought gas for $3.72. How much more did he pay for gas this week than he did last week?

9. Make up a word problem for each of the two different types of addition situations; make one for the three different types of subtraction situations.

10. Mr. Lewis teaches second grade. He has been using "take-away" problems to illustrate subtraction. A district text contained the following problem:

    ▶ Vanessa's mother lost 23 pounds on a diet last year. Vanessa herself lost 9 pounds on the diet. How many more pounds did Vanessa's mother lose? ◀

    Most of Mr. Lewis's students were unable to work the problem, and he was very discouraged because he had spent so much time teaching subtraction. Do you have any advice for him?

11. For each problem on the next page, name the quantities and their values, note the relationships among the values, make a drawing for the problem, then compute the answer.

    Example: JJ earned $2.50 an hour for helping a neighbor. JJ worked 5 hours. Then JJ bought a T-shirt for $7.35. How much money did JJ have left?

    Solution: Quantities with values are: Amount JJ earns per hour: $2.50. Number of hours JJ works: 5. Cost of T-shirt: $7.35. Amount of money left after purchase of T-shirt: unknown.

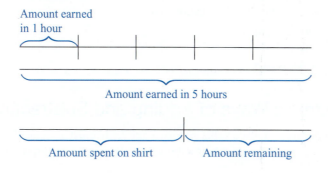

To find the amount earned, I would multiply her wage per hour by the number of hours she worked or add five $2.50s. Then I would subtract the amount she spent on the shirt because that cost is taken away from the total earned: $2.50 \times 5 - \$7.35 = \$5.15$.

  a. A post 12 feet long is pounded into the bottom of a river. 2.25 feet of the post are in the ground under the river. 1.5 feet stick out of the water. How deep is the river at that point?

  b. At one school $\frac{3}{5}$ of all the eighth-graders went to one game. $\frac{2}{3}$ of those who went to the game traveled by car. What part of all the eighth-graders traveled by car to the game?

  c. A small computer piece is shaped like a rectangle that is 2.5 centimeters long. Its area is 15 square centimeters. How wide is the piece?

  d. Maria spends two-thirds of her allowance on school lunches and one-sixth for other food. What fractional part of her allowance is left?

  e. On one necklace, $\frac{5}{8}$ of the beads are wooden. There are 40 beads in all on the necklace. How many beads are wooden?

  f. A painter mixes a color by using 3.2 times as much red as yellow. How much red should he use with 4.8 pints of yellow?

Supplementary Learning Exercises for Section 3.2

1. Without looking them up, illustrate the three ways of thinking about $5 - 2$, either with drawings or with story problems. (Be sure to label them.)

2. In $165.8 - 82.3 = 83.5$, tell which is the difference (or remainder), the minuend, and the subtrahend.

3. Use the labels for ways of thinking about subtraction to classify each of the following story problems. Use the action of the problem to help you decide. Be certain that you can solve the problems.

   a. Monday morning, the farmer had $12\frac{1}{2}$ acres to plow. She managed to plow $8\frac{3}{4}$ acres on Monday. How many acres did she still have to plow?

   b. The farmer planned to plant corn in the $12\frac{1}{2}$ acres and then to rent some more land so that she would have at least 35 acres planted in corn. How many acres must she rent?

   c. In addition to her 35 acres of corn, the farmer has 52 acres of soybeans. How many more acres of soybeans does she have than of corn?

4. Write an equation in which x is the difference, y is the minuend, and z is the subtrahend.

5. Make up story problems answerable by calculating $28 - 7$ and meeting the given criteria. (Use marbles as a theme if you cannot think of one.)

   a. comparison subtraction
   b. missing-addend subtraction
   c. take-away subtraction

6. For each part, finish this story to illustrate the different uses of subtraction: "The gardener had 48 feet of white irrigation pipe. . . ."

   a. comparison subtraction
   b. missing-addend subtraction
   c. take-away subtraction

3.3 Children's Ways of Adding and Subtracting

Children who understand place value and have not yet learned a standard way to subtract often have unique ways of undertaking subtraction.[3] Prepare to be surprised!

ACTIVITY 5 Children's Ways

1. Consider the work of nine second-graders, all solving $364 - 79$ (in written form, without calculators or base-ten blocks). Identify:

   a. which students clearly understand what they are doing;
   b. which students might understand what they are doing; and
   c. which students do not understand what they are doing.

(1)

```
  364
-  79
─────
   -5
  -10
  300
─────
  285
```

(2)

```
   2 5⁴
  3̶4̶'4
 -  7 9
─────
  2 3 5
```

(3)

```
   2 '5
  3̶6'4
 -  7 9
─────
  2 8 5
```

(4)

```
  3'6̶4
 -  8̶9
─────
  2 8 5
```

(5)

```
  3'6̶4¹
 -   7 9
─────
  3̶9 5
```

(6)

FIRST I TAKE THE 70 FROM 360
AND THATS 290 THEN I PUT THE 4
BACK AND ITS 294 THEN I TAKE 9
AWAY 4 FIRST TO 290 THEN 5
SO ITS 285.

(7)

WELL I KNOW its the same as 365-80
AND THATS the same as 385-100
So 285

(8)

```
  364
-  79
─────
  300
  290·
  285
```

(9)

```
  364
-  79
─────
  300
-  10
  290
-   5
─────
  285
```

2. Pick out places where errors were made, and try to explain why you think the errors occurred. ●

In Activity 5 the solutions of Students 2, 3, and 5 are all based on the *standard algorithms* that are taught in school. Student 3 correctly uses the "regrouping" algorithm taught in most U.S. schools.

Describe in writing the steps followed when one uses the standard regrouping algorithm for subtraction.

 ACTIVITY 6 Making Sense of Students' Reasoning

1. Justify, using your knowledge of place value and the meanings of addition and subtraction, the procedure used in Solution 3 of the activity "Children's Ways." Make up some other problems, and talk them through, using the language of "ones, tens, hundreds," etc.

2. Discuss what you think happened in the solutions of Students 2 and 5.

continue

Notes

3. The procedure used by the fourth student is called the "equal additions" method. Figure out how this algorithm works, and the mathematical basis for it.
Hint: Because 9 cannot be taken from 4, ten was added to both numbers, but in different places. ●

Sometimes teachers dismiss solutions that do not follow a standard procedure. Yet sometimes these solutions show a great deal of insight and understanding of our base-ten numeration system and how numbers can be decomposed in many helpful ways. But there are times that a solution which is "different," such as the equal additions method of subtraction, may not demonstrate any insight. Rather, any procedure can be learned as a sequence of rules. In the United States, subtraction is usually taught by a "regrouping method" (see Student 3's method in Activity 5) but in some other countries children are taught the "equal additions" method (see the work of Student 4). Either can be taught rotely or meaningfully.

 ACTIVITY 7   Give It a Try

Go back over the solutions of Students 1, 6, 7, 8, and 9. Make sure you understand the thinking of each student. You can do this by thinking through each method as you try 438 − 159. ●

The other students appear to have used "nonstandard algorithms" or "invented algorithms," i.e., procedures they have developed on their own, or perhaps as a class, for solving subtraction problems. In Activity 7, you should have thought about place-value understanding for each of the methods here.

> *Think About ...*
> If you were (or are) a parent, what would you say to your child if he or she subtracted like Student 1? Or Student 9? What about Student 4? Have you ever seen this subtraction method before? If so, tell your instructor so that your method might be shared with others.

 DISCUSSION 2   Is One Way Better?

Is any one method of subtraction (from those illustrated at the beginning of this lesson) better than others? Why or why not? Which methods do you think could be more easily understood? Are there any methods that should not be taught (rather than invented by a student)? Why or why not? ●

TAKE-AWAY MESSAGE ... Children do not think about mathematics in the same way that adults do. Part of the reason for this is that adults are more mature and have a wider variety of experiences with numbers. But also part of the reason is that we have had limited opportunities to explore numbers and the meaning of operations on numbers. When children enter school, they are often inquisitive and willing to explore in ways that they often lose in later grades, unfortunately. But when children are given numerous opportunities to think about numbers and come to understand the place-value system we use, and then apply that understanding in a variety of situations, they often come up with novel ways of approaching problems. The ways they invent for undertaking addition and subtraction are useful to them, and often can be generalized. This knowledge can also help them understand and remember any traditional algorithms they are taught. This has the advantage that they are unlikely to make the common errors that result when place value is not well understood.

Learning Exercises for Section 3.3

1. Make up some new subtraction problems and try solving them by each of the methods illustrated in the activity "Children's Ways." Try inventing some other ways you think a child might approach your subtraction problems.

2. Here are several cases of addition and subtraction problems solved by other first- and second-graders in classrooms where the standard algorithms had not been taught. Some procedures are more sophisticated than others, based on the individual student's understanding of numeration. Some were done mentally; others were written down. Study each method until you understand the thinking of the student, and then do the problem given using the same strategy.

   Case A. Teacher: What is 39 + 37?

   Student 1: 30 and 30 is 60, then 9 more is 69, then 7 more is 70, 71, 72, 73, 74, 75, 76.
   You do: 48 + 59.

   Case B. Teacher: What is 39 + 37?

   Student 2: 40 and 40 is 80, but then you need to take away. First 1, and get 79, then 3, and get 76.
   You do: 48 + 59.

   Case C. Teacher: What is 39 + 37?

   Student 3: 40 and 37 is, let's see, 40 and 30 is 70, and 7 more is 77, but you need to take away 1, so it's 76.
   You do: 48 + 59.

   Case D. A student talks aloud as he solves 246 + 178:

   Student 4: Well, 2 plus 1 is 3, so I know it's 200 and 100, so now it's somewhere in the three hundreds. And then you have to add the tens on. And the tens are 4 and 7. . . . Well, um. If you started at 70; 80, 90, 100. Right? And that's four hundreds. So now you're already in the three hundreds because of this [100 + 200], but now you're in the four hundreds because of that [40 + 70]. But you've still got one more ten. So if you're doing it 300 plus 40 plus 70, you'd have 410. But you're not doing that. So what you need to do then is add 6 more onto 10, which is 16. And then 8 more: 17, 18, 19, 20, 21, 22, 23, 24. And that's 124. I mean 424.[4]
   You do: 254 + 367.

   Case E. Teacher: What is 65 − 7?

   Student 5: 65 take away 5 is 60, and take away 2 more is 58.
   You do: 58 − 9.
   Student 6: 65 − 7. 7 and 8 is 15, and 10 is 25, and 10 is 35 and 10 is 45 and 10 is 55 and 10 is 65. Five tens is 50, and 8 is 58. (Uses fingers to count tens.)
   You do: 58 − 9.

   Case F. Written solution for 654 − 339.

   Student 7: 6 take away 3 is 3 and you make it into hundreds so 300. Then you add 50 → 350 − 30 and it comes to 320, +4 is 324, −9 is first −4 is 320 then −5 is 315.
   You do: 368 − 132.

   Case G. Oral solution for 500 − 268.

   Student 8: 2 to get to 270, then 30 more to get to 300, then 200 more, so 232.
   You do: 800 − 452.

3. Discuss the procedures used in Exercise 2 in terms of how well you think the student understands the procedure; how long you think it took for the student to develop and understand the process being used; and how robust the procedure is, i.e., can it be used on other problems?

4. Discuss the procedures used in Exercise 2 in terms of whether the procedure would be difficult for students to remember. Do you think the student, when faced with another subtraction problem, will be able to use his or her procedure? Why or why not?

5. In many countries, addition and subtraction are taught using an "empty" number line that is not marked with 0 and 1. Some U.S. texts are now introducing this method. Here are two examples for addition, with 48 + 39:

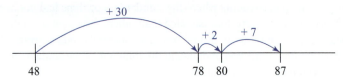

or

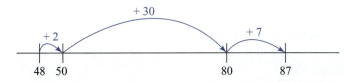

Subtraction is done similarly. Consider 364 − 79, a problem you saw earlier:

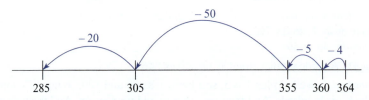

or

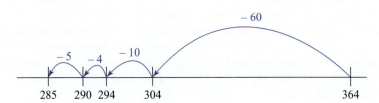

Try the following computations using the empty number line method.

a. 62 + 49   **b.** 304 − 284   **c.** 72 − 38   **d.** 253 − 140

6. A child's work is typed up below. What is the child doing? Does the procedure make sense? If not, explain why.

$$4^15^13$$
$$\underline{-6\ 5}$$
$$\cancel{4}\ \cancel{9}$$
$$3\ 8\ 8$$

7. A reviewer contributed the work in Learning Exercise 6 and pointed out that the procedure reflects work that children often do with base-ten materials. What does she mean?

Supplementary Learning Exercises for Section 3.3

1. A child is learning addition of one-digit numbers. But he writes results like 1 + 6 = 16 and 1 + 2 = 12. What might the child be thinking?

2. Use "empty" number lines (Learning Exercise 5 in Section 3.3) to calculate these sums and differences.

a. 189 + 43   **b.** 936 + 89   **c.** 1000 − 732   **d.** 381 − 62

3. Write down the calculations this child is working.

   a. "400, then 140 is 540, then ten more is 550. 150."
   b. "25 more to get to 100 and 300 more to get to 400, so 325."

4. Make sketches to show how the numerical work by Students 3 and 6 in Activity 5 can reflect work with base-ten materials.

5. Devise a nonstandard algorithm for addition of multidigit numbers, like those below. (Notice that the second example is trickier.) Does your algorithm make sense, with base-ten materials in mind?

   $$527 \qquad 527$$
   $$+432 \qquad +673$$

3.4 Ways of Thinking About Multiplication

Just as there are different ways of thinking about addition and subtraction, depending on the quantities involved and the type of situation involved, there are different ways of thinking about multiplication and division. Many children learn to think of multiplication *only* as repeated addition, which limits the type of situations in which they know to multiply.

 DISCUSSION 3   When Do We Multiply?

How would you solve these two problems?
1. One kind of cheese costs $2.19 a pound. How much will a package weighing 3 pounds cost?
2. One kind of cheese costs $2.19 a pound. How much will a package weighing 0.73 pound cost? ●

If you even hesitated in deciding to multiply to solve the second problem, you are not alone. Researchers across the world have noticed that success on the second problem is usually 35 to 40% *less* than success on the first problem, even among adults! Many solvers think they should divide or subtract on the second problem. Why does this happen? This section will offer a possible reason: Each of multiplication and division, like addition and subtraction, can represent quite different situations, but solvers may not be aware of this fact.

Just as there are different ways of thinking about subtraction, there are several ways to think about multiplication and division. We begin with different meanings of multiplication.

> When a whole number *n* of quantities, each with value *q*, are combined, the resulting quantity has a value of $q + q + q + \cdots$ (*n* addends), or $n \times q$. This is called the **repeated-addition view of multiplication**.

| **EXAMPLE 10**

▶ Karla invited 4 friends to her birthday party. Instead of receiving gifts, she gave each friend a dozen roses. How many roses did she give away? $4 \times 12 = 12 + 12 + 12 + 12.$ ◀ ●

Some critics say that repeated addition receives attention for so long (Grade 2 or 3 on) that it restricts children's attention when other situations that use multiplication appear. Nonetheless, this view does build on the children's experience with addition, and it does fit many situations.

When we use the notion of repeated addition, 3×4 means the sum of 3 addends, each of which is 4: $3 \times 4 = 4 + 4 + 4$. Notice the order; 3×4 means 3 fours, not 4 threes. At least that is the case in every U.S. text series; the reverse is the case in some other countries (British or British-influenced countries, e.g.), which can cause confusion. Of course, you know from commutativity of multiplication that $3 \times 4 = 4 \times 3$, but as a teacher, you would want to model the standard meaning, even after the children have had experience with commutativity. In contrast with commutativity of addition, commutativity of multiplication is much less intuitive: It is not obvious ahead of time that 3 eights (3×8) will give the same sum as 8 threes (8×3).

> The numbers being multiplied are called **factors**, and the result is called the **product** (and sometimes a **multiple** of any factor). The first factor is sometimes referred to as the **multiplier**, and the second the **multiplicand**. In some contexts, **divisor** is used as a synonym for factor, except divisors are not allowed to be 0 (but factors may be).

Note that under the repeated-addition view of multiplication, the first factor must be a whole number. Under this view, 2.3×6 cannot be interpreted; it is meaningless to speak of the sum of 2.3 addends. However, the commutative property of multiplication (which children should know by the time they are solving problems with decimal numbers) tells us that 2.3×6 is the same as 6×2.3. Overall, any story situation in which a quantity of any sort is repeatedly combined could be described by multiplication. 6×2.3 could be used, e.g., for "How much would 6 boxes of that dog treat weigh? Each box weighs 2.3 pounds."

Think About ...
Read again the cheese problems in Discussion 3. Does the idea of multiplication as repeated addition fit both problems?

An emphasis on multiplication only as repeated addition to the exclusion of other interpretations of multiplication appears to lead many children into the unstated and dangerous over-generalization that the product is always larger than the factors, expressed as follows: *Multiplication always makes bigger*.

Think About ...
When does multiplication NOT "make bigger"?

A second type of situation calling for multiplication is a rectangular array with n items across and m items down, for a total of $m \times n$ items. Or we could generate a rectangle of n squares across and m squares down for a total of $m \times n$ squares. (This is, of course, what we use to find the area of a rectangle.)

> The **array (or area) model of multiplication** occurs in cases that can be modeled as a rectangle n units across and m units down. The **product** is $m \times n$.

Note: When the factors are represented by letters, we often use a dot or drop the multiplication symbol: $m \times n = m \cdot n = mn$.

If n and m are whole numbers, the array model could be considered a special case of repeated addition. One attraction of this model is that for continuous quantities the n and the m do not have to be whole numbers. Another feature of this model is that commutativity of multiplication is easily seen: Length times width (or number of rows times number in each column) is just reversed if the array or rectangle is turned sideways.

Think About ...
Draw a rectangle that is $3\frac{1}{2}$ inches across and $2\frac{1}{4}$ inches high. What is the total number of square inches shown? Find the answer first by counting square inches and parts of square inches, and then by multiplying. Did you get the same answer?

There is a third way to think about multiplication.

> The **fractional part-of-a-quantity model of multiplication** occurs when we need to find a fractional part of one of the two quantities. This is sometimes referred to as the **operator view of multiplication**.

With this model we can attach a meaning to products such as $\frac{2}{3} \times 17$ (pounds, say), meaning two-thirds *of* 17 pounds, or to a product of 0.35×8.2 (kilograms), meaning thirty-five hundredths *of* eight and two-tenths kilograms.

Think About …
Suppose you are dealing with cookies. How could you "act out" $6 \times \frac{1}{2}$? How could you "act out" $\frac{1}{2} \times 6$? How are these different? Do you get the same answer for both? What is the "unit" or the "whole" for the $\frac{1}{2}$ in each case?

You should have said that the unit or whole for the $\frac{a}{b}$ in the second situation, $\frac{1}{2} \times 6$, is the amount represented by the second factor, 6 cookies. We want one-half of *six cookies*. The reason this is referred to as the operator view is because the $\frac{a}{b}$ acts on, or operates on, the amount represented by the second factor. The operator view will be treated in more detail later, but you now have the background to consider the following question.

Think About …
Using the part-of-an-amount way of thinking about multiplication by a fraction, tell what 0.4×8 or 0.7×0.9 might mean. Does that help you locate the decimal point in the $0.4 \times 8 (= 3.2)$ and $0.7 \times 0.9 (= 0.63)$ products?

Think About …
If you buy 5.3 pounds of cheese costing \$4.20 per pound, could this be represented as a repeated-addition problem?

Think About …
What is meant by 5.3×4.20, as in finding the cost of buying 5.3 pounds of meat at \$4.20 a pound?

There are other situations in which ordered "combinations" of objects, rather than sums of objects themselves, are being counted and which can be described by multiplication. This fourth type of situation uses what is commonly called the *fundamental counting principle*. These situations show that, mathematically at least, multiplication need not refer to addition at all. The next activity leads to a statement of that principle.

 ACTIVITY 8   Finding All Orders with No Repeats Allowed

You hear that Ed, Fred, Guy, Ham, Ira, and Jose ran a race. You know there were no ties, but you do not know who was first, second, and third. In how many ways could the first three places in the race have turned out? ●

If you were successful in identifying all 120 possible orders for the first three finishers, you no doubt used some sort of systematic method to keep track of the different possible outcomes. One method that you may encounter in elementary school books is the use of a tree diagram. Here is an example, for a setting that is often used in elementary books:

► You have 3 blouses and 2 pairs of pants in your suitcase, all color-compatible. How many different blouse-pants outfits do you have? ◄

At the start, you have two choices to make—a choice of blouse, and a choice of pants. A **tree diagram** records these choices this way:

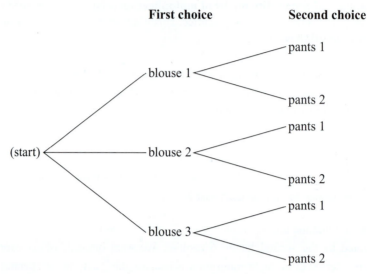

Notice that after a blouse is chosen, both possibilities for the choice of pants make the next "branches" of the tree. You make a choice by going through the tree. For example, blouse 1 and pants 2 would be one choice. You make all possible choices by going through the tree in all possible ways. Do you find 6 different outfits? What would the tree diagram look like if you chose pants first? Would there be 6 choices then also?

    With a situation as complicated as the 6-racers problem in Activity 8, making a tree diagram is quite laborious (try it). Although it is good experience to make tree diagrams, there is an alternative. Notice that in the outfits example, the first act, choosing a blouse, could be done in 3 ways and then the second act, choosing pants, could be done in 2 ways no matter what the choice of blouse was. And, $3 \times 2 = 6$, the number of possible outfits. Situations that can be thought of as a sequence of acts, with a number of ways for each act to occur, can be counted efficiently by this principle.

> In a case where two acts can be performed, if Act 1 can be performed in m ways, and Act 2 can be performed in n ways no matter how Act 1 turns out, then the sequence Act 1–Act 2 can be performed in $m \cdot n$ ways. This is the **fundamental counting principle** view of multiplication.

The principle can be extended to any number of acts. For example, the racers' problem has three "acts": finishing first (6 possibilities), finishing second (5 possibilities after someone finished first), and finishing third (now 4 possibilities). The number of possible first-second-third outcomes to the race can then be calculated by the fundamental counting principle: $6 \times 5 \times 4$, or 120, without having to make a tree diagram. Notice that here the unit for the end result is different from any unit represented in the factors. (As an aside, notice also that the tree diagram here is not the same as the "factor tree" that you may remember from elementary school.)

Think About …

At an ice cream shop, a double-decker ice cream cone can be ordered from 2 choices of cones, 12 choices for the first dip, 12 for the second dip, and 8 kinds of toppings. How many choices of double-decker ice cream cones are possible? Does your answer include 2 dips of the same flavor?

    A later section on multiplicative comparisons (not additive comparisons) will give a fifth important way of thinking about multiplication. Until then, when you use multiplication for a problem, identify which way is involved: repeated addition, array, part-of-a-quantity, or fundamental counting principle.

 ACTIVITY 9 Merrily We Multiply

Devise a multiplication problem for each of the types described thus far: repeated addition, array, part-of-a-quantity, or fundamental counting principle. If you do this in pairs or groups, compare your problem situations with those of a neighboring group. ●

Earlier we referred to the **commutative property** of multiplication, i.e., for any numbers m and n, $mn = nm$. This property tells us we can treat either m or n as the multiplier. You should also remember the **associative property** of multiplication, i.e., for numbers p, q, and r, $(pq)r = p(qr)$. This property tells us that when multiplying, we do not need parentheses, because it does not matter which product we find first.

EXAMPLE 11

$$(3 \times 2) \times 8 = 3 \times (2 \times 8) \quad \text{or} \quad 6 \times 8 = 3 \times 16$$

Thus, we could write the product as $3 \times 2 \times 8$. Parentheses tell us which operation to do first, but the associative property tells us that when multiplying, it does not matter which multiplication we do first. ●

The distributive property of multiplication over addition (often just shortened to the **distributive property**) is a particularly useful property: For numbers c, d, and e, $c(d + e) = cd + ce$.

EXAMPLE 12

Here is an illustration of this property, showing that $2 \times (4 + 3) = (2 \times 4) + (2 \times 3)$. (The right side could have been written without parentheses because in a string of operations, all multiplication must be done before addition.)

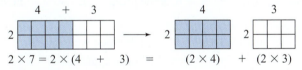

$$2 \times 7 = 2 \times (4 + 3) = (2 \times 4) + (2 \times 3)$$

Learners often wonder, in the symbolic form, where the other 2 came from. The drawing shows it was there all the time. ●

TAKE-AWAY MESSAGE . . . Problem situations that call for multiplication can be categorized in different ways, including the following:

1. Finding the sum when a whole number of like quantities is combined is called the repeated-addition view of multiplication. In the United States, finding mn means $n + n + n + \cdots$ with n as an addend m times. Using this form consistently makes understanding easier at first. Later, when commutativity is understood, students recognize that the addend and the number of addends could be reversed.

2. Finding the area of a rectangle or the number associated with a rectangular array is called the array (or area) model of multiplication. This view highlights the commutativity of multiplication and the distributive property of multiplication over addition.

3. Finding the fractional part of a quantity is sometimes called the operator view of multiplication because it appears that the multiplier is "operating" on the multiplicand. This third type of situation can involve a product that is smaller than one or both of the factors, which causes difficulty for students who have only thought about multiplication as repeated addition. The misconception that "multiplication makes bigger" is common and needs to be countered during instruction.

continue

4. In situations where finding the number of ways in which one act follows another, the fundamental counting principle view of multiplication applies. Tree diagrams help to illustrate this principle.

Learning Exercises for Section 3.4

1. **a.** Make sketches for 6×2 and for 2×6 and contrast them.

   b. Make sketches for $5 \times \frac{1}{2}$ and for $\frac{1}{2} \times 5$ and contrast them.

   c. Make a systematic list of possible choices, such as vanilla-coconut (or v-c), for this problem:

   ▶ Janice is ordering ice cream. There are 5 kinds available: vanilla, strawberry, chocolate, mocha, and butter pecan. There are also 5 kinds of sprinkles to put on top: M&Ms, coconut, Heath bar pieces, chocolate chips, and walnut bits. If she orders 1 scoop of ice cream and 1 kind of sprinkles, how many choices does she have? ◀

2. A teacher quite often has to make up a story problem on the spot. For *each* view of multiplication, make up *TWO* story problems that you think children would find interesting. Label them for later reference (e.g., array model). Use a variety of sizes of numbers. You might want to share yours with others.

3. Will multiplying a whole number (>1) by *any* fraction always result in a product smaller than the whole number we started with? When *does* multiplication make it bigger?

4. Does repeated addition make sense for $4 \times {}^-2$? For ${}^-2 \times 4$? For $8 \times \frac{2}{3}$? For $\frac{2}{3} \times 8$? Explain your thinking for each case.

5. In terms of repeated addition and/or part-of-an-amount, what does $6\frac{1}{2} \times 12$ mean? (Here is a context that may be helpful: $6\frac{1}{2}$ dozen eggs.)

6. For each, make up a story problem that could be solved by the given calculation.

   a. 32×29 **b.** 4×6.98 **c.** 0.07×19.95

   d. $12 \times \frac{3}{8}$, about pizza   **e.** $\frac{2}{3} \times 6$, about pizza   **f.** $40\% \times 29.95$

7. Here are the ingredients in a recipe that serves 6:

   6 skinned chicken breast halves   1 tablespoon margarine

   1 cup sliced onion $1\frac{1}{2}$ cups apple juice

   1 tablespoon olive oil $2\frac{1}{2}$ tablespoons honey

   2 cups sliced tart apples $\frac{1}{2}$ teaspoon salt

   a. You want enough for 8 people. What amounts should you use?

   b. You want enough for 4 people (and no leftovers). What amounts should you use?

8. The fact that $m \times n$ and $n \times m$ are equal is a very useful idea for, say, learning the basic multiplication facts. For example, if you know that $6 \times 9 = 54$, then you automatically know that $9 \times 6 = 54 \ldots$ if you believe that $m \times n = n \times m$. Text series may use the array model to illustrate this idea. An array for 3×5 would have 3 rows with 5 in each row (rows go sideways in almost every text series, and in advanced mathematics).

A 3×5 array   With a quarter-turn,
of square regions   becomes a 5×3 array

Since the two arrays are made up of the same squares, even without counting you know that $3 \times 5 = 5 \times 3$.

   a. Make an array for 4×7 and turn the paper it is on 90°. What does the new array show?

   b. Use a piece of graph paper to show that $12 \times 15 = 15 \times 12$.

   c. Show how this idea works in verifying that $6 \times \frac{1}{2} = \frac{1}{2} \times 6$.

9. **a.** Show with a drawing that $5 \times (6 + 2) = 5 \times 6 + 5 \times 2$. What is the name of this property?

   b. How could you illustrate, without a drawing, the associative property of multiplication? [*Hint:* $(a \times b) \times c = a \times (b \times c)$.]

   c. Show with a drawing that $3 \times 6.5 = 6.5 \times 3$.

10. Make a tree diagram or a systematic list for three of these, and check that the fundamental counting principle gives the same answer. Tell how many you found.

   a. You flip a spinner that has 4 differently colored regions (red, white, blue, green) all equal in area, and toss 1 die and count the dots on top (1 through 6 possible). How many color-dot outcomes are possible?

   b. A couple is thinking of a name for their baby girl. They have thought of 3 acceptable first names and 4 acceptable middle names (all different). How many baby girls could they have without repeating the whole name!?

   c. In a sixth-grade election, Raoul, Silvia, Tien, Vena, and Wally are running for president; Angela and Ben are running for vice president; and Cara and Don are running for treasurer. In how many ways could the election come out?

   d. In a game you toss a red die and a green die, and count the number of dots on top of each one, e.g., R2, G3. (The numbers of dots are *not* added in this game.) In how many different ways can a toss of the dice turn out?

11. A hamburger chain advertised that it made 256 different kinds of hamburgers. Explain how this claim is possible. (*Hint:* One can choose 1 patty or 2, mustard or not, etc.)

12. An ice cream store has 31 kinds of ice cream and 2 kinds of cones.

   a. How many different kinds of single-scoop ice cream cones can be ordered at the store?

   b. How many different kinds of double-scoop ice cream cones are there? (*Decision:* Is vanilla on top of chocolate the same as chocolate on top of vanilla?)

13. You have become a car dealer! One kind of car you will sell comes in 3 body styles, 4 colors, and 3 interior-accessory "packages," and costs you, on average, $12,486. If you wanted to keep on your lot an example of *each* type of car (style with color with package) that a person could buy, how much money would this part of your inventory represent?

14. License plate numbers come in a variety of styles.

   a. Why might a style using 2 letters followed by 4 digits be better than a 6-digit style?

   b. How many more are there with a style using 2 letters followed by 4 digits than the 6-digit style? How many times as large is the number of license plates possible with a style using 2 letters followed by 4 digits as is the number possible with a 6-digit style? (Ignore the fact that some choices of letters could not be used because they would suggest inappropriate words or phrases.)

   c. Design a style for license plate numbering for your state. Explain why you chose that style.

15. Locate the decimal points in the products by thinking about the part-of-an-amount meaning for multiplication. Insert zeros in the product, if needed.

   a. $0.6 \times 128.5 = 7\,7\,1$ (Where does the decimal point go?)

   b. $0.8 \times 0.95 = 7\,6$

   c. $0.75 \times 23.8 = 1\,7\,8\,5$

   d. $0.04 \times 36.5 = 1\,4\,6$

   e. $0.328 \times 0.455 = 1\,4\,9\,2\,4$

   f. $0.65 \times 0.1388 = 9\,0\,2\,2$

16. Spacecraft *Voyager 2* was launched in 1977. In late 2011 a signal from Earth took 13.5 hours to reach *Voyager 2*. If such a signal travels 186,000 miles each second, how far away from Earth was *Voyager 2* at that time?

17. How many seconds are in a day?

18. How many seconds are in a 365-day year?

19. A *light year* is actually a distance—the distance light can travel in one year. If light travels 186,000 miles in one second, how many miles are in a light year? (*Hint:* Exercise 18.)

20. Write a single equation that could be used to describe each story problem. Be careful about the order of any factors that you use. You do not have to give a final numerical answer unless your instructor says to do so.

> **Example:** Chuck has spent $\frac{5}{8}$ of his paycheck. He plans to pay bills with half of the rest of his paycheck. What part of his whole paycheck will he spend on bills?
>
> **Solution:** $\frac{1}{2} \times \left(1 - \frac{5}{8}\right) = n$

    a. You have used $1\frac{3}{4}$ pounds of flour from a 4-pound bag. You plan to use $\frac{2}{3}$ of the remaining flour to make cupcakes. How many pounds of flour are you planning to use for cupcakes?

    b. The recipe for biscuits from one mix calls for $2\frac{1}{4}$ cups of the mix and makes 9 biscuits. You plan to make only half the recipe. How many cups of mix should you use?

    c. Another time, you want to make biscuits, allowing 3 biscuits for each of 5 people. How much mix should you use? (Two equations may be more natural.)

21. A restaurant offers spaghetti, rotini, penne, and bow ties to be topped with plain sauce, meat sauce, or clam sauce. How many different orders of (one) pasta plus sauce are possible?

22. One auditorium has 20 rows, each with 25 seats, and an additional 12 rows, each with 18 seats. How many can be seated in those rows?

23. What kind of multiplication—repeated addition, array, fractional part of a quantity, or fundamental counting principle—is involved for each occurrence of multiplication in Learning Exercises 16–23?

24. For 5×3, people sometimes say, "Add 3 five times." Why is it inaccurate to imply that there are five additions for 5×3?

Supplementary Learning Exercises for Section 3.4

1. Without looking them up, give four ways of thinking about multiplication, and give the common label for each.

2. Make sketches for 4×3 and 3×4 and contrast them.

3. Both products in Exercise 2 are 12. What property does this illustrate?

4. For each, make up a story problem that could be solved by the given calculation.

    a. 5×1.99 **b.** 35×12

    c. $6 \times \frac{3}{4}$, about fruit   **d.** $\frac{3}{4} \times 6$, about fruit

5. Write a numerical equation for (a) the associative property of multiplication and (b) the distributive property. For each, verify that the two sides of the equation are indeed equal.

6. Which one of the four views of multiplication in this chapter does each of the following involve?

    a. Dee is designing the cover for her project report. She has in mind 3 possible colors for the cover page and 6 possible illustrations to put on it. How many cover designs are possible?

    b. A snack size of peanuts is about $\frac{1}{4}$ cup. About how many cups of peanuts would be used in making 12 servings?

    c. Diem made 12 servings of dessert for a party, but $\frac{1}{4}$ of them were not eaten. How many servings were not eaten?

    d. Each row in the auditorium will seat 24 people. How many people can be seated if there are 12 rows?

7. Make a sketch for each of the story problems in Exercise 6.

8. Without calculating, answer the following for the product $\frac{92}{113} \times \frac{578}{234}$. Explain your reasoning.

   a. Is the product greater than, or less than, $\frac{578}{234}$?

   b. Is the product greater than, or less than, $\frac{92}{113}$?

9. **a.** Put the words "product," "multiplier," and "multiplicand" in the blanks. [Sometimes a teacher's manual for text series uses these words; "factor" is more common, as in part (b).]

   _____ × _____ = _____

   b. Put the words "factor" and "product" in the blanks.

   _____ × _____ = _____

10. Give an example in which multiplication does *not* make bigger.

11. Sketch a number line and show each of the following.

    a. 5×2 **b.** 2×5

12. What multiplication expression would the following number line sketch illustrate?

13. A student was asked to solve this story problem: "Ann, Bea, and Carl are running for class president. Dee and Eli are running for vice president. In how many ways could the election turn out?" The student replied, "I don't know. There's not any numbers." What was apparently missing in the student's background?

14. Here is an old riddle: "As I was going to St. Ives, I met a man with 7 wives. Each wife had 7 sacks. Each bag had 7 cats. Each cat had 7 kittens. Kittens, cats, sacks, wives. How many are headed to St. Ives?" What is your answer?

3.5 Ways of Thinking About Division

Next we turn to division. There are two common types of situations that call for division, and they both appear in virtually every curriculum. A third view, which you may recognize from your earlier coursework, is more encompassing and mathematically includes the first two views. Under any of these views, this vocabulary applies.

> In a division situation that can be described by $a \div b = q$, the a is called the **dividend**, b is called the **divisor**, and q is called the **quotient**. In a division situation in which b is not a factor of a, the situation can be described as $a \div b = q + \frac{r}{b}$, the quotient is $q + \frac{r}{b}$, and the quantity r is called the **remainder**. This situation is also written as $a \div b = q$ remainder r. Note that the divisor can never be 0.

EXAMPLE 13

In $28 \div 4 = 7$, 28 is the dividend, 4 is the divisor, and 7 is the quotient. In the usual U.S. calculation form, $divisor \overline{)dividend}$, the quotient is written on top of the dividend. There may also be a remainder, as in $13 \div 4 = 3$ R 1, or $3\frac{1}{4}$. Division calculations that have a remainder of 0 are informally said to "come out even." ●

> One basic view of the division $a \div b$ is called the **measurement view**. This follows from the question, "How many measures of b are in a?" Because the answer can often be found by repeatedly subtracting b from a, this view of division is also sometimes referred to as the **repeated-subtraction view of division**. This view is also called the **quotitive view of division**.

You will find that different textbooks use different terms.

Notes

Think About ...

Suppose a child had this problem situation:

▶ Eli and his friends baked 15 cookies. If shared fairly, there are no cookies left over. How many children were there? ◀

How might a child who does not yet understand division try to model this situation?

This page is from a Grade 3 textbook.* The preceding textbook page treated division as equal sharing.

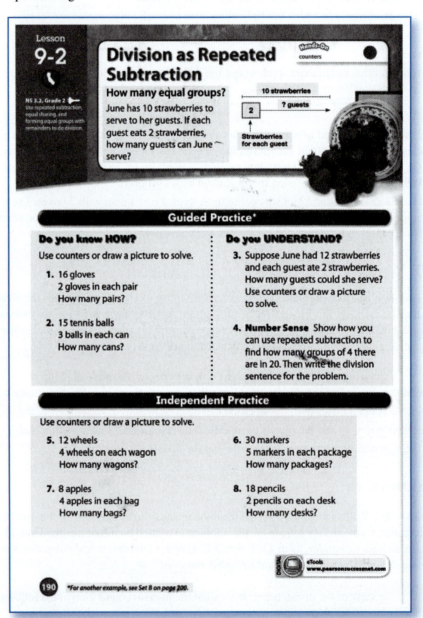

*From Randall I. Charles et al., *enVision*, California Math Edition, Grade 3, p. 190. Copyright © 2009 Pearson Education, Inc. or its affiliate(s). Used by permission. All rights reserved.

EXAMPLE 14

When asked what $12 \div 4$ means, one would say it means "How many 4's are there in 12" or "How many measures or counts of 4 are in 12" and the answer can be obtained by repeatedly subtracting 4 until the 12 are gone, resulting in a quotient of 3. When asked what $13 \div 4$ means, we can repeatedly subtract 4 three times again, resulting in a partial quotient of 3, but this time there is a remainder, 1, because it was not possible to again subtract 4. ●

> *Think About ...*
> What is meant by $\frac{3}{4} \div \frac{1}{2}$, using the measurement view? How can one think of using repeated subtraction to find the quotient? What is meant by $0.6 \div 0.05$?

Using the measurement view of division, one could ask, "How many measures of $\frac{1}{2}$ are in $\frac{3}{4}$?"

> *Think About ...*
> Suppose a child has this problem situation: "Danesse has 3 friends. She has 12 cookies to share with them. How many cookies does each of the 4 people get?" How do you suppose a child who does not yet understand division might model this situation?

Another basic view of division is **sharing equally**. When one quantity (the dividend) is shared equally by a number of objects (the divisor), the quantity associated with each object is the quotient. This sharing notion of division is also called **partitive division**.

EXAMPLE 15

Danesse probably shared her cookies by saying one for you, one for you, etc. Thus, the 12 cookies are shared equally by 4 people, and each gets 3 cookies. Or, you might say that the 12 cookies have been partitioned into 4 sets of the same size. ●

 ACTIVITY 10   Highway Repair

Are these two problems the same? Which view of division is represented in each? Would you make the same drawing for each?

1. A highway crew has several miles of highway to repair. Past experience suggests that they can repair 2.5 miles in a day. How many days will it take them to repair the stretch of highway?

2. A highway crew has the same stretch of highway to repair. But they must do it in 2.5 days. How many miles will they have to repair each day to finish the job on time? ●

 DISCUSSION 4   Practicing the Meanings

Tell what the following could mean with each view of division:

1. $85 \div 7$
2. $22.5 \div 2$
3. $\frac{7}{8} \div \frac{1}{8}$ (repeated subtraction view only)
4. $1.08 \div 0.4$ (repeated subtraction view only) ●

A more general way of thinking about division is the **missing-factor view of division** in which a question is asked about a missing factor that, when multiplied by the divisor, would result in the dividend. That factor is the quotient.

EXAMPLE 16

$12 \div 4 = n$ would be associated with $n \times 4 = 12$ (with repeated subtraction in mind) or $4 \times n = 12$ (with sharing in mind). Indeed, the common way to check a division calculation is to calculate divisor $\times$ quotient, and see whether the product is the dividend. ●

A missing-factor view could replace the earlier two views of division, once they are understood. Because our emphasis is on mathematics for the elementary school, our attention will center on the repeated subtraction and sharing views. Keep them in mind.

 DISCUSSION 5 Dividing by Zero

"You can't divide by 0." An elementary school teacher should know why not. Perhaps the easiest way to think about the issue is to think of checking a division calculation: The related multiplication, divisor $\times$ quotient, should equal the dividend. So $12 \div 4 = 3$ is true because $3 \times 4 = 12$. Do the following:

1. "Check" $5 \div 0 = n$. What number, if any, checks?
2. "Check" $0 \div 0 = n$. What number, if any, checks?
3. A child says, "Wait! Last year my teacher said, 'Any number divided by itself is 1.' So isn't $0 \div 0 = 1$?" What do you reply?

Thus, either *no* number checks (as in $5 \div 0$) or *every* number checks (as in $0 \div 0$) when dividing by zero. Either is clearly a good reason to say that division by zero is undefined, or "You can't divide by zero." ●

TAKE-AWAY MESSAGE . . . Just as with other operations, there are different contexts that can be described by division:

1. Finding the number of times a particular value can be repeatedly subtracted from another value. This view of division has different labels: repeated-subtraction division, measurement meaning of division, and quotitive meaning of division.
2. Finding the number in each share if a quantity is shared equally by a number of objects. This view of division is called sharing-equally division, or partitive division.
3. Finding the missing factor that, when multiplied by another known factor, would result in a known product. This view of division is called the missing-factor view of division. Mathematically, this view can encompass the first two views, but we depend on the first two views when helping children to understand division.
4. When one value (the dividend) is divided by another (the divisor, which cannot be 0), the result is a value called the quotient. In situations where the quotient must be expressed as a whole number, there is sometimes a value remaining, called the remainder.
5. For good reasons, division by zero is undefined. That is, one cannot divide by zero.

Learning Exercises for Section 3.5

1. For the repeated-subtraction (also called measurement or quotitive) view of division, make up *TWO* story problems that you think children would find interesting. Do the same for the sharing (partitive) view of division.

2. Illustrate $12 \div 3$ with drawings of counters using the repeated-subtraction view and then the sharing view. Notice how the drawings differ.

3. Write story problems that could lead to the following equations:

    a. $150 \div 12 = n$ (partitive)   **b.** $150 \div 12 = n$ (repeated subtraction)
    c. $3 \div 4 = n$ (partitive)   **d.** $3 \div 4 = n$ (repeated subtraction)

4. After years of working with division involving whole numbers, children often form the erroneous generalization, *division makes smaller.* Give a division problem that shows that division does not always make smaller.

5. Consider the following situation:

   ▶ Mount Azteca is 1.3 km high. There are 3 different trails from the starting place at the bottom to the top of the mountain. One trail is 4.5 km long, another is 3.5 km long, and the most difficult is 2.4 km long. ◀

   For each computation indicated below, finish the story problem started above so that your problem could lead to the computation described. As always, it may be helpful to make a drawing of the situation and think about the quantities involved.

   a. 3×2.4
   b. $(4.5 + 3.5) - (2 \times 2.4)$
   c. $4.5 \div 3$ (partitive)
   d. $4.5 \div 3$ (repeated subtraction)
   e. $10 \div 2.4$

6. a. A national test once used a question like "The bus company knows that after a ball game, 1500 people will want to catch a bus back to the main bus terminal. Each bus can hold 36 people. How many buses should the company have ready?" What is your answer to the question?

   b. There will be 300 children singing in a district songfest. Each singer gets a sash. Sashes come in boxes of 8. Calculate $300 \div 8$, and then answer the following from your calculation:
      i. How many boxes of sashes should the district order?
      ii. If the district orders only 37 boxes, how many children would not get a sash?

7. Analyze the quantities in this setting to generate some questions you can answer:

   ▶ One company plans to open 505 acres around an old gold-mining site and to process the ground chemically. The company hopes to extract 636,000 ounces of gold from 48 million tons of ore and waste rock, over a $6\frac{1}{2}$ year period. The company expects to get about $200 million for the gold. ◀

8. The six problems below are like those that appear in elementary school mathematics textbooks. Discuss the meaning or interpretation of division in each problem and explain how you would help students visualize the action in the problem and connect it to a meaning of division. Identify the quantities in each and tell how to use them to draw a diagram for each problem.

   a. The school district bought 900 kg of cheese. The cheese comes in boxes of 12 kg each. How many boxes did the school district buy? (Fourth grade)
   b. The school district used 525 kg of cheese in 3 weeks. What is the average number of kilograms of cheese used each week? (Fourth grade)
   c. A deliveryman drove about 5760 mi in 12 weeks. What was the average distance the deliveryman drove each week? (Fifth grade)
   d. The teacher has 256 sheets of paper for the copy machine. One copy of a special article takes 16 sheets. How many copies can the teacher make? (Fifth grade)
   e. Each of 4 bags of potatoes weighs the same. They weigh 27 lb in all. How much does each bag of potatoes weigh? (Sixth grade)
   f. A box holds 10 balls. How many boxes does it take to hold 100 balls? (Third grade)

9. What is $24 \div 3$? $240 \div 30$? $2400 \div 300$? $24,000 \div 3000$? $2.4 \div 0.3$? Formulate a rule for finding quotients such as these.

10. *Reflect on this:* If you were not taught by a teacher who clarified the meanings of operations, how did this affect your ability to solve, and develop an attitude toward, story problems?

Notes

Supplementary Learning Exercises for Section 3.5

1. Make drawings to contrast the repeated-subtraction (or measurement or quotitive) view and the sharing-equally (or partitive) view of $6 \div 2$.

2. Label the divisor, the dividend, and the quotient in each:

   a. $1620 \div 36 = 45$ **b.** $36\overline{)1620}$ with 45 above

3. In $96 \div 10 = 9$ R 6, what do the 9 and the 6 tell you
   a. if the division is for a repeated-subtraction situation?
   b. if the division is for a sharing-equally situation?

4. **a.** Explain why $6 \div 0$ is not defined (use the repeated-subtraction view).
   b. Explain why $0 \div 0$ is not defined (use the repeated-subtraction view).
   c. Explain why $6 \div 0$ is not defined (use the missing-factor view).
   d. Explain why $0 \div 0$ is not defined (use the missing-factor view).

5. Make up three story problems that could be answered by calculating $150 \div 5$.
   a. One should involve repeated-subtraction division.
   b. Another should involve sharing-equally division.
   c. Another should involve a missing-factor situation.

6. Explain what $322 \div 14 = 23$ tells you for each of the following:
   a. repeated-subtraction division
   b. sharing-equally division
   c. missing-factor division

7. Each of the following equations can be solved by calculating $168 \div 12$. Use the repeated-addition meaning for multiplication and indicate which way of thinking about division (repeated subtraction, sharing equally) corresponds to each equation. Explain your thinking.
   a. $12 \times n = 168$
   b. $n \times 12 = 168$

8. The sun is about 93 million miles from Earth. Light travels 186,000 miles in one second. About how long does it take light from the sun to reach Earth
   a. in seconds?
   b. in minutes?
   c. Were your divisions for parts (a) and (b) repeated-addition or sharing-equally divisions?

9. Illustrate each of the following in finishing this story: "The local food bank has 200 pounds of rice. . . ."
   a. repeated-subtraction division
   b. sharing-equally division

10. Using teacher-lounge language, explain how the drawing below can show each of the following:
    a. $15 \div 6$ **b.** $15 \div 2.5$

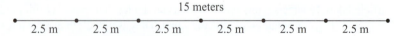

15 meters

2.5 m   2.5 m   2.5 m   2.5 m   2.5 m   2.5 m

11. Sketch how you could illustrate $12 \div 3$ on a number line.

12. Child: "I think $0 \div 0$ should be 0. 'Cause if you've got nothing, that's what the answer should be." What is your response?

13. Child: "You said that $0 \div 0$ is not defined, because it could be anything. I don't see how it could be, like 100." What is your response?

14. Rehearse how you would explain that division by 0 is not defined. Your explanation should cover the case of $0 \div 0$ as well as cases such as $9 \div 0$.

15. One office has 12 reams of paper (1 ream contains 500 sheets). A special mailing to their customers uses 350 sheets of paper. How many mailings can they make with their supply of paper?

16. Under certain conditions, sound can travel 768 miles in an hour. Under those conditions, you hear thunder 3.5 minutes after seeing lightning, and a friend hears the same thunder in 2 minutes. How close might you and your friend be, in miles?

3.6 Children Find Products and Quotients

When students can analyze situations that involve finding a product or a quotient, they might use novel methods for computing, perhaps not even realizing that they are multiplying or dividing. In this section you will see a variety of ways that children think about multiplication and division problems.

DISCUSSION 6   Indian River Oranges

In a third-grade classroom the focus had been on understanding problem situations and the children knew that multiplication can be thought of as repeated addition. They were working on this problem.[5]

► Every Christmas my father gets big boxes of fruit, like Indian River oranges, from his company. We get so many pieces of fruit that we have to give some away. This year we ended up making 24 grocery bags of fruit, with 16 pieces in each bag. How many pieces did we bag altogether? ◄

Discuss what you think each child was thinking about in each case below:

Student 1

$20 \times 10 = 200$
$20 \times 6 = \underline{120}$
320
$4 \times 10 = 40$
$4 \times 6 = \underline{24}$
64
$\underline{320}$
384

Student 2

(continued
24 times)

Student 3

96
$\underline{96}$
192
$\underline{192}$
384

Student 4

$16 \times 12 = 10 \times 12 = 120$
$6 \times 12 = \underline{72}$
192
$\underline{192}$
384

(*Note*: Student 2 actually wrote the number 16 twenty-four times.) ●

DISCUSSION 7   What Were They Thinking?

1. Justify, using your knowledge of place value and multiplication, *each* of the methods used in Discussion 6.

2. Discuss the relative clarity and the relative efficiency of the four methods.

continue

3. Another child (Student 5) was asked to find 54×62 and wrote: "First I did 50×60 wich (*sic*) came to 3000. Then I did 60×4 and then I added it on to 3000 wich (*sic*) came to 3240. Then I did 50×2 which came to 100 and this I added it on to the 3240 which came to 3340 then I did 4×2 which came to eight and that's how it came to 3348." Work through this thinking to see whether it is correct. ●

DISCUSSION 8   Thinking About Division

Here is some children's work[6] on what we typically regard as division problems. None of these children had learned the standard division algorithm but coped quite well without it. You are to study each method and tell what you think each child is thinking. The first child was about 7 years old and had not learned about division, but he knew how to multiply using the calculator. The second child, also about 7, knew simple division facts and used them to his advantage. Note the way he deals with remainders.

Student 1
How many cups containing 160 mL are in 2200 mL?

$$10 \times 160 = 1600$$
$$13 \times 160 = 2080$$
$$14 \times 160 = 2240$$

So, 13 cups, 120 mL left over.

Student 2
What is 78 divided by 3?

$$20 + 20 + 20 = 60$$
$$8 \div 3 = 2 \text{ R } 2$$
$$10 \div 3 = 3 \text{ R } 1$$
$$2 + 1 = 3$$
$$3 \div 3 = 1$$

So, $78 \div 3 = 26$.

The next two solutions were from children in the fourth grade who had been solving problems requiring division. The teacher had not yet taught them a particular method for dividing. However, they knew that division could be thought of as repeated subtraction. $12 \div 3$ could be found by asking how many times 3 could be subtracted from 12; the answer is 4 times.

Student 3

$$280 \div 35$$

$$
\begin{array}{r}
280 \\
-70 \\
\hline
210 \\
-70 \\
\hline
140 \\
-70 \\
\hline
70 \\
-70 \\
\hline
0
\end{array}
$$

So, four 70s is eight 35s.

Student 4

$$
\begin{array}{r}
27\overline{)3247} \\
\underline{2700} \quad 100 \\
547 \\
\underline{270} \quad 10 \\
277 \\
\underline{270} \quad \underline{10} \\
7 \quad\; 120
\end{array}
$$

So, $3247 \div 27 = 120 \text{ R } 7$.

The fourth way is the Greenwood or "scaffold" algorithm, dating back to the seventeenth century.

The next child was in first grade and knew a few simple division and multiplication facts. He also knew what division meant, and in his head he made up a story that he explained after he had given the method that appears here.[7]

Student 5
What's $42 \div 7$? Well, 40 divided by 10 is 4, and 3 times 4 is 12, and 12 and 2 is 14, and 14 divided by 7 is 2, and 2 plus 4 is 6, so it's 6.

Do you understand the thinking of each student? (The last one may be difficult.) What knowledge of place value did these students have? ●

TAKE-AWAY MESSAGE . . . When young children can work problems such as "Amelia has 6 vases and 24 roses. She wants to put the same number of roses in each vase. How many roses will she put in each vase?" or "Amelia has 24 roses and some vases. If she wants to place 4 roses in each vase, how many vases will she need?", they are doing division. They may find the quotient in each case by acting it out. As the numbers in each situation become larger, the methods they use may seem more complex because we find it more difficult to follow their reasoning. In actuality, they are using one of the three different views of division to solve each problem. They deal with the situation at hand, solving the problem in a sense-making manner. Student-devised methods may help them to later develop and understand algorithms for finding quotients.

Learning Exercises for Section 3.6

Exercises 1 through 5 refer to the children's work in Discussion 8.

1. How is the first student's method related to estimating?

2. The second student shows remarkable facility in dealing with remainders. Try this method yourself on 56 divided by 4.

3. Student 3 solved the division problem by repeatedly subtracting. What are advantages and disadvantages of this method?

4. Student 4's method of dividing is sometimes called the *scaffolding* method, and in some schools it is taught as a first algorithm for doing division. Compare it to the standard division algorithm in terms of advantages and disadvantages.

5. Student 5's method for dividing is perhaps the most difficult one to understand. Suppose you start by telling yourself you have 42 candies to share among 7 people, and since you don't know how many each person should get, you begin with 10 piles. When you have figured out this method, try it on $63 \div 9$.

6. The problem given next was written by a third-grader who was challenged by her father to make up some story problems for him to solve. How much of the information given is used to solve the problem? What would you ask this child about this problem (in addition to: "Where did you learn the word 'heedless'?")?

   ▶ Alicia was a heedless breaker, and she broke 24 lamps in one week. Her parents paid $7.00 for every lightbulb she cracked to replace it. Her parents only paid for 5 bulbs in a week, like Monday–Friday. How much money did her parents pay for one week, not counting the weekend? ◀

Supplementary Learning Exercises for Section 3.6

1. Why might a child make the errors below? What meaning is missing in the child's thinking?

$$\overset{1\ 7}{4\overline{)428}} \qquad 37 \div 5 = 7\,R\tfrac{2}{5}$$

2. A young child knows a little bit about multiplication but has not yet encountered two-digit factors. Why might such a child make the errors below?

$$\begin{array}{r} 42 \\ \times 23 \\ \hline 86 \end{array} \qquad \begin{array}{r} 34 \\ \times 12 \\ \hline 38 \end{array}$$

3. What calculations are the students doing?
    a. "Two 432's is 864, then double that to get 1600 plus 128, which is 1728."
    b. "I can take away ten 15's, which is 150, and can do this three times to get 450, which is thirty 15's. I have 70 left and that is four 15's with 10 left over. So the answer is 34 remainder 10."

4. "When I see a division like $560 \div 20$, I change it to $56 \div 2$ if I can do it in my head. Or, if it's like $120 \div 14$, I change it to $60 \div 7$. Or $175 \div 15$, I try to change to . . . $35 \div 3$, I guess. Then sometimes I can do them in my head." This child's (unusual) idea could be generalized as $(na) \div (nb) = a \div b$. Are the technique and the generalization correct $(n, a, b \neq 0)$?

3.7 Issues for Learning: Developing Number Sense

The idea of developing number sense in the elementary grades is a crucial one. Too often, when focusing just on answers rather than on both reasoning and answers, children do not try to make sense of mathematics, as shown by contrasting the two elementary classrooms described by Howden.[8] She asked students in one first-grade classroom in a very transient neighborhood to tell what came to mind when she said twenty-four. The children immediately gave a variety of answers: two dimes and four pennies, two dozen eggs, four nickels and four pennies, take a penny away from a quarter, the day before Christmas, my mother was 24 last year, when the hand (on a grocery scale hanging in the room) is almost in the middle of twenty and thirty. When she asked this question in a third-grade classroom in a professional community, 24 was just a number that is written in a certain way, that appears on a calendar or on a digital watch. Howden claimed the students in the first-grade class had more *number sense*. She said:

> Number sense can be described as a good intuition about numbers and their relationships. It develops gradually as a result of exploring numbers, visualizing them in a variety of contexts, and relating them in ways that are not limited by traditional algorithms. Since textbooks are limited to paper-and-pencil orientation, they can only suggest ideas to be investigated, they cannot replace the "doing of mathematics" that is essential for the development of number sense. No substitute exists for a skillful teacher and an environment that fosters curiosity and exploration at all grade levels. (p. 11)

There has been a great deal written about number sense in the past two decades. Educators have come to realize that developing good number sense is an important goal in the elementary school. In fact, an important document from the National Research Council,[9] called *Everybody Counts,* contains the statement: "The major objective of elementary school mathematics should be to develop number sense. Like common sense, number sense produces good and useful results with the least amount of effort" (p. 46).

Research has shown that children come to school with a good deal of intuitive understanding, but that for many this understanding is eroded by instruction that focuses on symbolism and does not build on intuitive knowledge. Many educators now recognize that we can introduce symbols too soon and that students need to build an understanding of a mathematical phenomenon before we attempt to symbolize it.

Unfortunately, many teachers believe that although an emphasis on understanding is good for some students, others, particularly those in inner-city schools and in remedial programs, need a more rigid approach to mathematics. Yet a very large research study[10] of academic instruction for disadvantaged elementary school students showed that mathematics instruction that focused on understanding was highly beneficial to the students. "By comparison with conventional programs, instruction that emphasizes meaning and understanding is more effective at inculcating advanced skills, is at least as effective at teaching basic skills, and engages children more extensively in academic learning" (p. i).

Number sense is a way of thinking about numbers and their uses, and it has to permeate all of mathematics teaching if mathematics is to make sense to students. It develops gradually, over time. All the instruction in these pages focuses on helping you develop number sense and recognize it when it occurs. For example, mental computation and computational estimation both build on number sense and continue to develop it. The examples of children's methods of operating on numbers demonstrate, in most cases, a good deal of number sense on the part of the students.

TAKE-AWAY MESSAGE ... Students exhibit good number sense when they can find ways to tackle problems, such as subtraction, multiplication, and division, even though they have not been taught procedures to solve these types of problems. Teachers must have good number sense if they are going to be successful in helping their students develop it. Do *you* have good number sense? Do you think you are developing better number sense?

Learning Exercises for Section 3.7

The first five Learning Exercises will, like the earlier children's work, give you a better idea of what number sense means.

1. In each pair, choose the larger. Explain your reasoning (and use number sense rather than calculating).

   a. $135 + 98$, or $114 + 92$ **b.** $46 - 19$, or $46 - 17$
   c. $\frac{1}{2} + \frac{3}{4}$, or $1\frac{1}{2}$ **d.** 0.0358, or $0.0016 + 0.313$

2. Is 46×91 more, or less, than 5000? Is it more, or less, than 3600? Explain.

3. Suppose you may round only one of the numbers in 32×83. Which one would you choose, 32 or 83, to get closer to the exact answer? Explain with a drawing.

4. Without computing exact answers, explain why each of the following is incorrect:

a.	**b.**
310	119
520	46
630	137
150	940
+470	+300
2081	602

   c. $27 \times 3 = 621$ **d.** $36 \div 0.5 = 18$

5. Suppose that the sum of 5 two-digit whole numbers is less than 100. Decide whether each of the following must be true, false, or may be true. Explain your answers.

   a. Each number is less than 20.
   b. One number is greater than 60.
   c. Four of the numbers are greater than 20, and one is less than 20.
   d. If two are less than 20, at least one is greater than 20.
   e. If all five are different, then their sum is greater than or equal to 60.

6. Look for *compatible numbers* to estimate these:
   a. $36 + 47 + 52 + 18 + 69$ **b.** 39×42
   c. $1268 - 927$ **d.** $34{,}678 \div 49$
   e. $19 + 26 + 79 + 12 + 74$ **f.** $4367 \div 73$

7. **a.** A quick estimate of $56 \div 9.35$ is 5.6, from $56 \div 10$. A better estimate would adjust the 5.6 up. Explain why, using a meaning for division.

   b. A quick estimate of $715 \div 10.2146$ is 71.5. A better estimate would adjust the 71.5 down. Explain why, using a meaning for division.

8. Consider 150.68×5.34. In estimating the product by calculating 150×5, one has ignored the decimal part of each number. In refining the estimate, which decimal part (the 0.68 or the 0.34) should be the focus? Explain.

9. In multiplying decimals by the usual algorithm, the decimal points do not need to be aligned (in contrast to the usual ways of adding and subtracting decimals). Why not?

3.8 Check Yourself

This very full chapter focused on how children can come to understand the four arithmetic operations on whole numbers. As teachers, either now or in the future, you need to be able to reason through a child's solution that is different from one you may have seen before. Of course, this cannot always be done quickly, but children should not be told their work is incorrect when it is not. Sometimes a student has a unique and wonderful way of using knowledge of number structure to find a solution—such a child has good number sense. At other times the answers are wrong because of a lack of understanding of the underlying place-value structure of the numbers.

You should be able to work problems like those assigned and to meet the following objectives:

1. Identify the operations and the view of that operation that fits a problem situation involving any of the four operations.

2. Write story problems that correctly illustrate any specified view of an operation; e.g., write a story problem with the missing-addend view of subtraction.

3. Provide reasons why some views of operations are more difficult than others.

4. Explain why using "key words" is not a good strategy for answering story problems (see the Learning Exercises in Section 3.2).

5. Study students' arithmetic procedures and explain how the students are reasoning.

6. Describe why, in many countries, the empty number line is a popular way of beginning addition and subtraction of multiple-digit numbers.

7. Describe how you would explain to someone that "multiplication makes bigger" is not always true.

8. Describe what is meant by "number sense" and how you would recognize it.

References for Chapter 3

[1]Thompson, P. W. (1996). Imagery and the development of mathematical reasoning. In L. P. Steffe, B. Greer, P. Nesher, & G. Goldin (Eds.), *Theories of learning mathematics* (pp. 267−283). Hillsdale, NJ: Erlbaum.

[2]Hatano, G., & Inagaki, K. (1996). *Cultural contexts of schooling revisited: A review of* The Learning Gap *from a cultural psychology perspective.* Paper presented at the Conference on Global Prospects for Education: Development, Culture and Schooling, East Lansing, MI.

[3]Shuard, H., Walsh, A., Goodwin, J., & Worcester, V. (1991). *Primary initiatives in mathematics education: Calculators, children and mathematics.* London: Simon and Schuster.

[4]Carpenter, T. P. (1989, August). *Number sense and other nonsense.* Paper presented at the Establishing Foundations for Research on Number Sense and Related Topics Conference, San Diego, CA.

[5]Kamii, C., & Livingston, S. J. (1994). Classroom activities. In L. Williams (Ed.), *Young children continue to reinvent arithmetic in 3rd grade: Implications of Piaget's theory* (pp. 81−146). New York: Teachers College Press.

[6]Shuard, H., Walsh, A., Goodwin, J., & Worcester, V. (1991). *Primary initiatives in mathematics education: Calculators, children and mathematics.* London: Simon and Schuster.

[7]Harel, Gershon. Personal conversation with a child he interviewed.

[8]Howden, H. (1989). Teaching number sense. *Arithmetic Teacher, 36* (6), 6−11.

[9]National Research Council. (1989*). Everybody counts: A report to the nation on the future of mathematics education.* Washington, DC: National Academy Press.

[10]Knapp, M. S., Shields, P. M., & Turnbull, B. J. (1992). *Academic challenge for the children of poverty* (Summary LC88054001). Washington, DC: U.S. Department of Education.

Chapter 4

Some Conventional Ways of Computing

The previous chapter introduced you to some of the ways children invent for carrying out arithmetic operations. The procedures you yourself use are probably the standard ones you learned in elementary school. Contrary to what some believe, these algorithms, or procedures, took centuries to evolve, and other algorithms are standard elsewhere. In this section we will look at those procedures for carrying out arithmetic operations more carefully, and come to understand how and why they work.

4.1 Operating on Whole Numbers and Decimal Numbers

The arithmetic operations on whole numbers or on decimal numbers are similar since the same base-ten structure underlies both of these forms. The computations are highly dependent on this base-ten structure. Developing sensible step-by-step procedures, or **algorithms**, to carry out the arithmetic operations of addition, subtraction, multiplication, and division demands understanding our base-ten system. When students have difficulty with computational algorithms, many errors they make can be traced back to a lack of understanding of our base-ten system.

> *Think About ...*
> Are the algorithms you use for the four arithmetic operations based on right-to-left or left-to-right procedures? Can all be undertaken working from left to right?

Some children understand algorithms better when they are illustrated with base-ten materials or drawing.

EXAMPLE 1

Use base-ten drawings to illustrate $174 + 36$.

Solution

Step 1: Represent the problem, using the small cube as 1.

$$\begin{array}{r} 174 \\ +36 \end{array}$$

Step 2: Place all cubes together; replace 10 small cubes with one long.

$$\begin{array}{r} {\scriptstyle 1} \\ 174 \\ +36 \\ \hline 0 \end{array}$$

Step 3: Place all longs together; replace 10 longs with one flat.

11
174
+36
 10

Step 4: Put like blocks together to represent the sum.

11
174
+36
210

 ACTIVITY 1   Acting It Out, Writing It Out

Act out the following calculations using base-ten blocks or drawings, whichever is appropriate. Then redo each and record numerically at each step. It is important that one write down the steps as the procedures are undertaken, with the numerical work linked to the work with the blocks, software, or drawings. When you use base-ten blocks or drawings, be sure you specify which block or drawing is being used to represent one whole. For the subtraction problems, illustrate as (a) take-away and (b) comparison.

1. $312 - 124$ 2. $31.2 - 1.24$ 3. $123 + 88$
4. $12.3 + 88$ 5. $12.3 + 8.8$ 6. $12.3 + 0.88$
7. Describe what new blocks would be needed to compute $235.42 + 6.345$.

Multiplication and division algorithms are more difficult to understand than addition and subtraction algorithms. In the last section you saw some multiplication algorithms that could easily lead to the standard algorithm that you use. If children understand place value and use that to multiply, they will understand the process. The first example below on the left shows all six **partial products**. The second example, shown on the right, is a shortened way to find the product using just two partial products. If children first learn to write all partial products, they will understand where the partial products come from in the condensed algorithm. Note the use of decomposing 348 as $8 + 40 + 300$ and then using the distributive property.

$$
\begin{array}{rl}
348 & \\
\times\ 26 & \\
48 &= 6 \times 8 \\
240 &= 6 \times 40 \\
1800 &= 6 \times 300 \\
160 &= 20 \times 8 \\
800 &= 20 \times 40 \\
\underline{6000} &= 20 \times 300 \\
9048 &
\end{array}
\qquad
\begin{array}{rl}
348 & \\
\times\ 26 & \\
2088 &= 6 \times 348 \\
\underline{6960} &= 20 \times 348 \\
9048 &
\end{array}
$$

The illustration below shows how each of the six partial products corresponds to the area of a piece of the rectangular array.

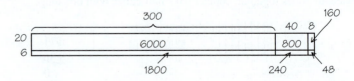

We can also use all partial products to make sense of multiplying by decimal numbers.

$$
\begin{array}{r}
3.48 \\
\times\ 2.6 \\
\hline
.048 = .6 \times .08 \\
.240 = .6 \times .4 \\
1.800 = .6 \times 3 \\
.160 = 2 \times .08 \\
.800 = 2 \times .4 \\
\underline{6.000} = 2 \times 3 \\
9.048
\end{array}
\qquad
\begin{array}{r}
3.48 \\
\times\ 2.6 \\
\hline
2.088 = .6 \times 3.48 \\
\underline{6.960} = 2 \times 3.48 \\
9.048
\end{array}
$$

 ACTIVITY 2   Your Turn

Multiply 1.3×2.4, showing all partial products. Illustrate with a rectangular array model showing the partial products. ●

Here is an example of a series of division algorithms for $472 \div 37$ that lead to the standard algorithm that you probably use. In the first example, 37s are subtracted 10 at a time or 1 at a time. In the second, we subtract other multiples of 37 (in this case 2). In the third, we simply move the 10 and 2 to the top. In the last, we do not write 10, but only the first digit of 10, aligned with the tens place in the dividend, and thus leaving room for a digit in the ones place so that the 2 can be filled in. Notice that the standard algorithm you learned is just a condensed version of subtracting 37s.

$$
\begin{array}{r}
37\overline{)472} \\
\underline{370} \quad 10 \\
102 \\
\underline{37} \quad 1 \\
65 \\
\underline{37} \quad \underline{1} \\
28 \quad 12
\end{array}
\qquad\qquad
\begin{array}{r}
37\overline{)472} \\
\underline{370} \quad 10 \\
102 \\
\underline{74} \quad \underline{2} \\
28 \quad 12
\end{array}
$$

$$
\begin{array}{r}
\underline{12} \\
2 \\
10 \\
37\overline{)472} \\
\underline{370} \\
102 \\
\underline{74} \\
28
\end{array}
\qquad\qquad
\begin{array}{r}
12 \\
37\overline{)472} \\
\underline{37} \\
102 \\
\underline{74} \\
28
\end{array}
$$

Thus, 472 divided by 37 is 12 with a remainder of 28. Notice that each successive time the division problem is worked, it becomes a little more condensed, so that the final illustration matches the way most of us learned to divide. However, the previous three illustrations show that the problem can be thought of as how many 37s are in 472, and subtracting multiples of 37 until no more can be subtracted. While this is happening, we keep count at the side or on top of the number of 37s that have been subtracted and record them. Studying this series of algorithms will help you understand the last one, which is the one most widely used in the United States but often never understood.

This textbook page* for fifth-grade students uses the same methods as illustrated in this chapter for 472 ÷ 37. The third way, however, is different. Can you calculate 472 ÷ 37 this third way? Is this way more difficult using a two-digit number as the divisor?

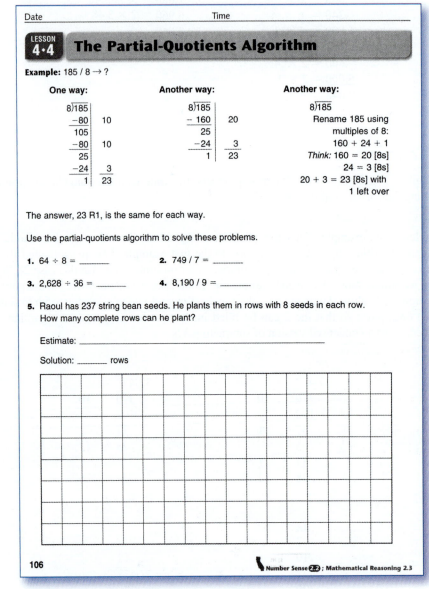

Date   Time

LESSON 4·4   The Partial-Quotients Algorithm

Example: 185 / 8 → ?

One way:

$$8\overline{)185}$$
$$\underline{-80} \quad 10$$
$$105$$
$$\underline{-80} \quad 10$$
$$25$$
$$\underline{-24} \quad 3$$
$$1 \quad 23$$

Another way:

$$8\overline{)185}$$
$$\underline{-160} \quad 20$$
$$25$$
$$\underline{-24} \quad 3$$
$$1 \quad 23$$

Another way:

$$8\overline{)185}$$

Rename 185 using multiples of 8:

160 + 24 + 1

Think: 160 = 20 [8s]

24 = 3 [8s]

20 + 3 = 23 [8s] with 1 left over

The answer, 23 R1, is the same for each way.

Use the partial-quotients algorithm to solve these problems.

1. 64 ÷ 8 = _____   **2.** 749 / 7 = _____

3. 2,628 ÷ 36 = _____   **4.** 8,190 / 9 = _____

5. Raoul has 237 string bean seeds. He plants them in rows with 8 seeds in each row. How many complete rows can he plant?

Estimate: _____

Solution: _____ rows

106   Number Sense 2.2 ; Mathematical Reasoning 2.3

*Wright Group, *Everyday Mathematics*, Student Math Journal, Grade 5, Vol. 1, p. 106. Copyright © 2008, McGraw-Hill. Reproduced with permission of The McGraw-Hill Companies.

ACTIVITY 3   Why Move the Decimal Point?

As you know, the first step in using the usual algorithm for calculating, say, 56.906 ÷ 3.7, is to move the decimal point in the divisor to make a whole number (37) and to move the decimal point in the dividend that number of places (giving 569.06).

1. For each of these, compute each number string one at a time, then make a statement about relationships within and among the equations. For example, "3 × 5 = 15,

6 × 5 = 30. When one factor doubled and the other factor remained the same, the product doubled."

   a. $56 \div 7 =$
   b. $56 \div 0.7 =$
   c. $5.6 \div 0.7 =$
   d. $144 \div 12 =$
   e. $14.4 \div 1.2 =$
   f. $1440 \div 120 =$
   g. $1.728 \div 1.44 =$

2. What insights into the "move the decimal point in the divisor" rule does this activity suggest? ●

 ACTIVITY 4   Now I Know How to Divide Decimal Numbers

Try dividing 0.18 by 1.5 using each of the techniques demonstrated for $472 \div 37$. Use knowledge gained from the previous activity to account for "moving the decimal point." ●

TAKE-AWAY MESSAGE . . .   Undertaking the fundamental arithmetic operations using the standard procedures is easier when you can first see them "unpacked," with each step along the way clarified by using properties and thinking about the meaning of the operation involved. This work can lead to an understanding of where all the numbers come from and how the algorithm can then be condensed into the procedures we all know and probably all use.

Learning Exercises for Section 4.1

1. Work through the following problems, thinking about each step you perform, and why. If you have used base-ten blocks or sketches of base-ten blocks, you can use them to help you with these problems. *In each case, specify what you decide to use as your unit.* For example, if the flat is used to represent 1, then 34.52 can be represented as follows:

   a. $56.2 + 34.52$
   b. 4×0.39
   c. $45.6 - 21.21$
   d. $2912 \div 8$ (Think of this as a sharing problem.)

2. Work through the following problems in the base indicated. Write down your procedure, and *specify what you use as your unit in each case.*

   a. $231 + 342$ in base five
   b. $1000 - 555$ in base six

   If you feel at all uncertain about these problems, make up and try some of your own.

3. Show how you could use a series of algorithms, such as those in this section, to teach another person to understand the standard algorithms for multiplication and division for 35×426; $14,910 \div 426$.

4. There have been, in the past, many different algorithms developed for carrying out computation. Here are two such algorithms for computing 36×342. The first is called the **lattice method for multiplication**, used by the Arabs in the 1600s and carried to Europe. The method depends on knowing the multiplication facts, but not much on place value. The factors are written across the top and right, and the answer, 12,312, is read off going down on the left and around the bottom. Compare this method with showing all partial products.

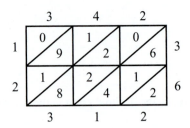

The second is called the **Russian peasant algorithm**. One number is successively halved until 1 is reached (if it is odd, 1 is subtracted before halving) and the other number is doubled the same number of times the first is halved. Numbers in the second column are crossed out when the corresponding number in the first column is even, and the remaining numbers are added. Reread the preceding steps as you study the following calculation of 36×342:

36	~~342~~
18	~~684~~
9	1368
4	~~2736~~
2	~~5472~~
1	10,944
	12,312

Try these algorithms on 57×623. They are not magic! Each can be justified mathematically.

5. Use your understanding of division to complete each equation.

If $4000 \div 16 = 250$, then

a. $8000 \div 16 =$ _____   b. $16,000 \div 16 =$ _____

c. $2000 \div 16 =$ _____   d. $4000 \div 32 =$ _____

e. $4000 \div 64 =$ _____   f. $4000 \div 8 =$ _____

g. $4000 \div 4 =$ _____   h. $4000 \div 0.4 =$ _____

6. In calculating with decimals, why do you "line up" the decimal points for addition and subtraction, but not necessarily for multiplication, such as with 1.92×0.3?

7. Mark the array below to show the four partial products in calculating 12×23.

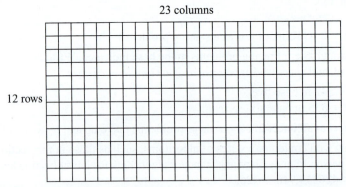

23 columns

12 rows

8. Make sketches of work with base-ten materials to support the usual algorithm for calculating each of the following. (*Hint*: Is repeated subtraction or sharing equally more practical here?)

a. $396 \div 3$ b. $252 \div 4$ c. $187 \div 4$

9. As in Learning Exercise 8, make sketches of work with base-ten materials to support the usual algorithm for calculating each of the following. (*Hint*: Is repeated subtraction or sharing equally more practical for these three?)

    a. $284 \div 142$ **b.** $960 \div 320$ **c.** $1000 \div 275$

10. Show the lattice method (Learning Exercise 4) for calculating 127×25.

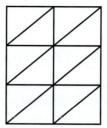

Supplementary Learning Exercises for Section 4.1

1. Which of the following computations might have led to the block work pictured below? Explain your thinking.

    a. $2353 + 134 = n$ **b.** $2353 + 1034$ **c.** $235.3 + 103.4$
    d. $2.353 + 1.034$ **e.** $2353_{\text{seven}} + 1034_{\text{seven}}$

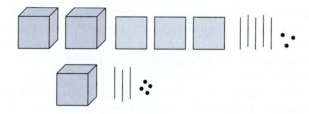

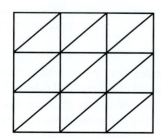

2. What base-ten block work would be involved in acting out $136 - 89$, using the *take-away* view and working to support the usual algorithm?

3. What base-ten block work would be involved in acting out $136 - 89$, using the *comparison* view and working to support the usual algorithm?

4. What base-ten block work would be involved in acting out $136 - 89$, using the *missing-addend* view and working to support the usual algorithm?

5. What base-ten block work would enable you to show $52.3 - 34.6$?

6. What block work would enable you to illustrate 3×24 in a way that would support the usual algorithm?

7. What modification in your work for Exercise 6 would enable you to show 3×0.24? To show 3×0.024?

8. Show the lattice method for calculating 286×492.

9. $32 \times 95 = 3040$. Use that information to find the following (mentally, without calculator or paper-and-pencil calculating).

    a. 64×95 **b.** 96×95 **c.** 128×95

    d. $3040 \div 32$ **e.** $3040 \div 64$ **f.** $3040 \div 16$

10. Without calculating products, tell which of these will give the same product.

    a. 16×83 **b.** 8×166 **c.** 4×332 **d.** 2×664

4.2 Issues for Learning: The Role of Algorithms

In Section 4.1 we focused on the methods used in carrying out the arithmetic operations on whole numbers and decimal numbers, methods that are likely to be already familiar to you. By learning well one algorithm for each operation, e.g., traditional long division, that skill becomes automatized and allows one to think about other things, e.g., the problem at hand that led to the division.

 DISCUSSION 1   What Do You Remember?

What do you remember about learning algorithmic procedures in school? Did you understand these procedures? Were they easy or painful to learn? Were they easy or difficult to remember? Do you use them often now? How important is it to be able to carry out operations rapidly with paper and pencil? ●

The standard algorithms are called "standard" because children are taught to do them the same way. The evolution of these algorithms began in the 1600s, with the demise of the use of the Roman numeral system due to the hard-won acceptance of the Hindu-Arabic decimal system of numeration (our place-value system). The algorithms came into existence as people attempted to become as efficient and speedy as possible in their calculations. Even so, some of the algorithms we use here are not standard in other parts of the world. For example, our subtraction algorithm was selected for teaching in U.S. schools because a research study[1] in 1941 showed it to be "slightly better" than the equal-additions method of subtraction you encountered in Chapter 3, a method that is taught in many other countries. Those who learned the equal-additions algorithm find the algorithm used in the United States to be strange.

    Many mathematics educators say that there should be less emphasis on teaching paper-and-pencil algorithms because students can use calculators to compute. But they do not argue that we should therefore *not* teach methods of calculating that are independent of calculator use. One reason for teaching calculation methods is because calculation is a useful tool. Another reason is that it may lead to a deeper understanding of numbers and number operations. Students make better sense of standard algorithms when they have had opportunities to explore arithmetic operations and find their own ways of operating on numbers, exemplified by those children whose work you have studied in the last chapter. Students can then learn standard algorithms as required, because in most states and school districts they are required. If they learn standard algorithms with an understanding of why the procedures lead to correct answers, they are more likely to remember them and use them correctly.

> ### *Think About …*
> Do you have a better understanding of the multiplication and division algorithms after seeing, in the previous section, how longer algorithms can be used to show why standard algorithms work?

    What, then, is the role of calculators in elementary school? Calculators are ubiquitous, and are much faster than any human calculator. Will calculators someday completely replace paper-and-pencil calculations? Not likely. But efficiency and automaticity should be reconsidered as valid reasons for teaching arithmetic skills. This does not mean that people should not have these skills, but rather that finding an answer using paper and pencil may no longer need to be rapid (except in some testing situations). There are times when paper-and-pencil computing should be preferred, and times when calculators should be used.

 DISCUSSION 2   What Is the Role of Calculators?

What is the role of calculators in school mathematics? A newspaper cartoon showed parents and grandparents sitting at desks using calculators for computing, while the 12-year-old son was sitting at his desk doing worksheets of long-division practice. What do you think was the message? ●

Contrary to what is believed by many, research has shown that availability of calculators does not hinder the learning of basic skills and can, in fact, enhance learning and skill development when used appropriately in the classroom.[2–4] The National Assessment of Educational Progress (NAEP) from years 1990 to 2003 shows no difference in scores for fourth-grade students in classrooms where calculators were frequently used and where calculators were not used.[5] (The NAEP test in mathematics is administered nationally at Grades 4, 8, and 12 every 2–4 years.) In eighth grade, students who frequently used calculators scored significantly higher than students who did not. Keep in mind that students often do not understand enough about arithmetic operations to be able to choose appropriate operations to solve real problems, and in such cases a calculator is useless. If understanding algorithms leads to less need for practice, more time can be devoted to learning which operations are appropriate and when.

In the next chapter you will be given opportunities to consider a third option to using standard algorithms or calculators. In many cases an exact answer is not needed. What should be done then?

4.3 Check Yourself

By now you should be able to explain the multiplication and division algorithms to someone else who does not understand the processes. You should know the pros and cons of the standard algorithms and student-invented algorithms, and provide a cogent argument defending your view of what is important for children to learn about the procedures for carrying out arithmetic procedures.

You should also be able to do problems like those assigned and to meet the following objectives:

1. Explain each step of the addition (e.g., for 378 + 49) and subtraction algorithms (e.g., for 378 – 49) and show how renaming is used.

2. Explain each step of whole number multiplication (e.g., 375 × 213) using partial products.

3. Explain division of whole numbers (e.g., 764 ÷ 46) using a series of algorithms that leads to the standard algorithm.

4. Recognize the role of place value in the algorithms.

5. Discuss the role of both nonstandard and standard algorithms.

References for Chapter 4

[1]Brownell, W. A. (1941). *Arithmetic in grades I and II: A critical summary of new and previously reported research.* Durham, NC: Duke University Press.

[2]Shuard, H., Walsh, A., Goodwin, J., & Worcester, V. (1991). *Primary initiatives in mathematics education: Calculators, children and mathematics.* London: Simon and Schuster.

[3]Kloosterman, P., & Lester, F. K., Jr. (Eds.). (2004). *Results and interpretations of the 1990 through 2000 mathematics assessments of the National Assessment of Educational Progress.* Reston, VA: National Council of Teachers of Mathematics.

[4]Kilpatrick, J., Swafford, J., & Findell, B. (Eds). (2001). *Adding it up.* Washington, DC: National Research Council, National Academy Press.

[5]Kloosterman, P., Sowder, J., & Kehle, P. (2004, April). *Clearing the air about school mathematics achievement: What do NAEP data tell us?* Presentation at the American Educational Research Association, San Diego, CA.

Chapter 5

Using Numbers in Sensible Ways

People who are comfortable working with numbers have many ways to think about numbers and number operations. For example, they use "benchmarks," i.e., numbers close to particular numbers that are easier to use in calculating. Thinking about numbers in various ways allows them to estimate answers, do mental arithmetic, and recognize when answers are wrong. This chapter will help you acquire better *number sense*, i.e., the ability to recognize when and how to use numbers in sensible ways.

5.1 Mental Computation

When people hear the word "computation," they usually think of paper-and-pencil computation using the methods they learned in school. But computation can take many forms, and an individual with a good understanding of numbers will choose the most appropriate form of computation for the problem at hand. Today, this often means using a calculator because it is fast and efficient. Often, though, one can mentally compute an answer faster than one can compute with paper and pencil or even a calculator. At other times, one might wish to compute mentally simply because it is more convenient than finding pencil and paper or a calculator. To become skilled at mental computation takes some practice. Number properties must be understood and used, although you may not even be aware of the properties because they are so natural for you to use. One important advantage of practicing mental computation, for children and for adults, is that it helps develop flexibility of use of numbers and properties of operations and gives them a sense of control over numbers.

Consider some of the following ways of mentally computing 12×50:

1. 12×50 can be thought of as $12 \times 5 \times 10 = 60 \times 10 = 600$. This required decomposing 50 into 5×10, and then using the associative property of multiplication: $12 \times (5 \times 10) = (12 \times 5) \times 10 = 60 \times 10$.

2. 12×50 can be thought of as $(10 + 2) \times 50 = (10 \times 50) + (2 \times 50)$ (using the distributive property of multiplication over addition), which is $500 + 100 = 600$.

3. $12 \times 50 = (6 \times 2) \times 50 = 6 \times (2 \times 50) = 6 \times 100 = 600$ (using the associative property).

Remember that addition is also commutative and associative, giving a great deal of flexibility in adding numbers.

There are also several ways of thinking about how to mentally compute a difference. For example:

1. $147 - 66$ could be found by adding 4 to each number: $151 - 70$ is $151 - 50 - 20 = 101 - 20 = 81$.

2. $147 - 66$: Count up 4 to 70, then 30 to 100, then 40 to 140, then 7 to 147. This is $4 + 30 + 40 + 7$, which is 81.

3. $147 - 66$ is $140 - 60$ that is 80, plus 7 is 87, minus 6, is 81.

And, of course, there are other ways. . . .

Notes

◻ ACTIVITY 1   I Can Do It in My Head!

Before reading ahead, do the following calculations in as many ways as possible:

1. $152 - 47$ 2. $1000 - 729$ 3. 25×24

4. $38.6 + 27.2 + 4.8 - 38.6$ 5. $12 \div \frac{1}{4}$ 6. $6 \times 43 + 6 \times 7$ ●

Many times the ease of mental computation depends on the numbers used, and on familiarity with properties of operations. The computation should not begin before looking over the numbers and thinking of different ways of reorganizing or rewriting the numbers. Consider Problem 4 in Activity 1. Many people would begin by trying to add 38.6 and 27.2 mentally, a challenge but certainly possible. But as one becomes more adept at mental computation, the knowledge that the numbers could be reordered, allowing the $38.6 - 38.6$ to be the first operation, makes this a much easier problem to compute mentally: $27.2 + 4.8$ is $31 + 1$ is 32.

Now consider Problem 3, which is 25×24. One way to think about this problem is as $(5 \times 5) \times (4 \times 6) = (5 \times 4) \times (5 \times 6) = 20 \times 30 = 600$. Notice that you have really done the following, but without writing it all:

$$25 \times 24 = (5 \times 5) \times (4 \times 6)$$
$$= 5 \times (5 \times 4) \times 6 \text{ using the associative property of multiplication}$$
$$= 5 \times (4 \times 5) \times 6 \text{ using the commutative property of multiplication}$$
$$= (5 \times 4) \times (5 \times 6) \text{ using the associative property again}$$

When 25 is a factor, it could also be thought of as $\frac{100}{4}$. Now 25×24 becomes

$$\frac{100}{4} \times 24 = \frac{100}{4} \times 4 \times 6 = 100 \times 6 = 600$$

For Problem 6, the arithmetic is easy if we use the distributive property:

$$(6 \times 43) + (6 \times 7) = 6 \times (43 + 7) = 6 \times 50 = 300$$

For Problem 2, which is $1000 - 729$, the mental computation by counting up is actually easier than using the standard algorithm, which, because of the zeros, often leads to errors. Add 1 to 730, 70 to 800, then 200 to 1000, $1 + 70 + 200$ is 271. Notice that this is based on the missing-addend view of subtraction.

In Problem 5, think of how many quarters are in $12. Too often, people give the answer 3. Why do you think this happens?

◻ DISCUSSION 1   Becoming Good at Mental Computation

Try these individually, and then together discuss different ways of undertaking these mental computations.

1. Mentally compute $50 \times 35 \times 2 \div 5$.

2. Mentally compute 19×21.   (Is this the same as 20×20?)

3. Mentally compute $5,400,000 \div 600,000$.

4. Mentally compute $3476 + 2456 - 1476$.

5. Mentally compute 50×340.

6. Mentally compute $24 \times 38 + 24 \times 12$.

7. Mentally compute $8 \times 32 \times 4$.   (*Hint*: You can use powers of 2.)

8. Mentally compute $35 \times 14 + 35 \times 6$. ●

Mental calculation with some percents is not difficult. Here are some hints about calculating with percents.

1. 10% of a number is $\frac{1}{10}$ or 0.1 times the number: 10% of 50 is 5. 10% of 34.5 is 3.45.

2. Multiples of 10% can be found using the 10% calculation: 20% of 50 is twice 5, or 10. 20% of 34.5 is 3.45 twice, or 6.90. 70% of 20 is 10% seven times, or 14. (Or think: 10% of 20 is 2, and $2 \times 7 = 14$ is 70% of 20.)

3. 15% is 10% plus half that again. 15% of 640 is 64 + 32 or 96.

4. 25% is a quarter of an amount, and 50% is half an amount. Thus, 25% of $360 is $90 because one-quarter of 360 is 90. Or, you could say 10% is 36, 10% more is 36 more, then 5% is half of 10%, and half of 36 is 18. Finally, 36 + 36 + 18 is 90.

 ACTIVITY 2   I Can Do Percents in My Head

Determine the following mentally. Does a mental drawing like the one shown here help in some cases?

a. What is 25% of 40?

b. Twelve is 20% of what number?

c. Thirteen is 50% of what number?

d. Ten percent of what number is 10?

e. What is 30% of 55?

f. What percent of 66 is 22? ●

TAKE-AWAY MESSAGE . . .   There are many ways to mentally compute if one knows the properties of operations on numbers. As people become more comfortable using numbers, they are more likely to mentally compute answers when this is easy to do. Computing mentally leads to greater facility with numbers and more confidence in one's ability to work with numbers.

Learning Exercises for Section 5.1

1. How can the following computations be done mentally using the distributive property? Write out your solutions:

    a. 43×9 b. $23 \times 98 + 23 \times 2$ c. 72×30

2. Tell how you could mentally compute each of the following:

    a. $365 + 40 + 35$ b. $756 - 28$ c. $2391 + 431 - 1391$

    d. $499 - 49$ e. 124×25 f. $42 - 29$

    g. 44×25 h. 75×88 i. 8×32

3. Compute each of the following mentally:

    a. 25% of 60   b. 10% of 78   c. 30% of 15   d. 80% of 710   e. 15% of $40

    f. 100% of 57   g. 125% of 40   h. 5% of 64   i. 20 is 20% of ?

4. Compute each of the following mentally:

    a. 24 is 20% of what number?   b. 38 is 10% of what number?

    c. 89 is 100% of what number?   d. 50 is 125% of what number?

5. The figure shown here is 120% of a smaller figure. Can you show 20% of the smaller figure? Shade 100% of the smaller figure.

6. Shade a grid to help you find the fractional equivalent in simplest form for each of the following percents:

    50% =_____   33⅓% =_____   75% =_____

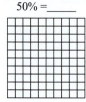

    66⅔% =_____   12.5% =_____   20% =_____

Supplementary Learning Exercises for Section 5.1

1. Each of the following can be calculated mentally and exactly. Do the mental calculation first, and then write down how you thought.

   a. $988 + 567 + 12$ **b.** $67 + 98 + 33 + 13$
   c. $845 - 210$ **d.** $3129.6 - 18$
   e. 36×100 **f.** 36×50
   g. 36×25 **h.** 36×75
   i. $19 \times 85 + 19 \times 15$ **j.** $96 \times 23 + 4 \times 23$

2. Give the exact answers mentally. Do the mental calculation first, and then write down how you thought.

   a. 10% of 8500   **b.** 20% of 8500   **c.** 25% of 128   **d.** 25% of 1280
   e. 80% of 320   **f.** 110% of \$12   **g.** 15% of \$32   **h.** 15% of \$50

5.2 Computational Estimation

Many times an estimate rather than an exact answer to a calculation may be sufficient. Estimates are used not only when an exact answer could be found but is not needed, but also when an exact answer is not possible, such as when you make a budget and estimate how much your total utilities bills will be each month. Mental computation is used in computational estimation, but when you estimate, you are looking only for an approximate answer. However, estimation is sometimes considered more difficult because it involves both rounding *and* mental computation.

> ### Think About …
> Think back on the times you have needed a numerical answer to a computation in the past week. How many times did you need an exact answer? How many times did you need just an estimate?

💬 DISCUSSION 2   Is One Way Better Than Another?

1. Estimate 36×55 alone before discussing this problem.

2. Carefully read these students' solutions to the first exercise. Be sure you understand the thinking behind each one. Then discuss them in terms of whether each way is a good way to estimate. Did you use one of these ways?

   Shawn: Round to 40 and 60. $40 \times 60 = 2400$.
   Jack: First round down: $30 \times 50 = 1500$. Then round up: $40 \times 60 = 2400$. So it's about in the middle, maybe a little past. So I'd say 2000.
   Maria: Rounding both up would make it too big, so I'll round 36 to 40 and 55 to 50. $40 \times 50 = 2000$.
   Jimmy: A little more than 36×50, which is $36 \times 100 \div 2$ and that's $18 \times 100 = 1800$. It's about 5×36 more, or about 180 more, so I'll say 1980.
   Deb: Rounding both up gives $40 \times 60 = 2400$. Since that's too big, I'll say it's about 2200.
   Sam: A little more than $6 \times 6 \times 50$, which is $6 \times 300 = 1800$. So I'll say 1900. ●

Each of the above ways provides good estimates, but some are better than others because they are efficient and/or because they are more accurate. For example, Shawn rounded both numbers up to the closest multiple of 10. Rounding up gives a high estimate. Jack rounded both up and down, then takes something in the middle. His way takes slightly more work but gives a closer estimate. But depending on the context, you might want a high estimate.

Suppose you are at the hardware store buying drawer pulls that cost 55¢ each. You need 36 of them and wonder if you have enough cash with you. According to Shawn, you need $24, and according to Jack, you need $20. Before you reach for your wallet, which estimate would you use? Why?

 ACTIVITY 3   Now I'll Try Some

1. Try each of the strategies from the above discussion (if appropriate) to estimate 16 × 48. Which estimates do you think are better? Why?

2. If a city's property tax is $29.87 per $1000 of assessed value, what would be an estimate of the tax on property assessed at $38,600?

3. Estimate 789 × 0.52.   4. Estimate 148.52 + 49.341. ●

> ### *Think About …*
>
> In a research study[1] many middle school students were asked to estimate 789 × 0.52. Some students rounded 0.52 to 1, rounded 789 to 800, and said their answer was 800 × 1 = 800. Is 1 a good substitute for 0.52? Why or why not? Suppose something cost $0.52, and you bought 789 of these objects. Would $800 be a good estimate of the price? Why do you think students rounded 0.52 to 1?

> ### *Think About …*
>
> In the same study the students were asked to estimate 148.52 + 49.341. Some students said, "148 is almost 150 and 49 is almost 50. And 0.52 is almost 0.6, and 0.341 is almost 0.3. And 0.6 + 0.3 is 0.9. So the answer is 200.9." Find the flaw in this argument. (*Hint*: Is the 148 being rounded up to 150, or is the 148.52 being rounded up to 150?)

Research has shown that good estimators use a variety of strategies and demonstrate a deep understanding of numbers and of operations. They are flexible in their thinking and are disposed to make sense of numbers. A good estimator uses the rounding of numbers a great deal, but seldom uses the formal rounding procedures often taught in schools.[2] Proficiency in flexible rounding requires that an individual has an intuitive notion of the magnitude of the number in question, and this intuition is acquired through practice in comparing and ordering numbers and using benchmarks, especially for fractions and decimals. For example, to estimate 257 + 394 + 2 + 49, a good strategy (but not the only one) would be to round 257 to 250, 394 to 400, 49 to 50, and drop the 2 since it is insignificant here (or think of rounding it to 0). School students who are inflexible in their rounding will insist that 257 must be rounded to 300, or perhaps 260 if rounding to the nearest 10, but not to 250. In another example, if you want to estimate the quotient of 6217 ÷ 87, you might find it more convenient to round to 6300 and 90 rather than 6000 and 90, since 70 × 90 is 6300. The activity below is intended to provide you with practice in estimating.

 ACTIVITY 4   Finding "Just About"

Be sure to check your solutions and methods with others. You may be surprised to find many good ways of doing these estimations.

1. Estimate $\dfrac{71 \times 89}{8}$.   2. Estimate 42 × 34.   3. Estimate 5.8 ÷ 12.

4. If 34 × 86 is estimated as 30 × 86, then the estimate is 2580. The exact answer is 2924. The difference between these two numbers is 344. If 496 × 86 is estimated as 500 × 86, then the exact answer is 42,656 and the estimate is 43,000. Again, the difference is 344. Which of these would you choose: (a) The first estimate is better; (b) they are the same; (c) the second estimate is better. Why?

continue

5. 18×86 can be estimated as (a) 20×90, (b) 20×86, or (c) 18×90. Which of these three ways gives the closest estimate?

6. If you estimate 53×27 by saying 50×30 is 1500, is the exact answer less than, equal to, or more than your estimate? ●

By now you know that when multiplying two numbers, rounding both up gives a high estimate, and rounding both down gives a low estimate. Thus, a more accurate estimate of a product could be found by rounding one number up and one number down.

DISCUSSION 3   Rounding Up and Down

If you are estimating the sum of two numbers, is it usually best to round both numbers up? If not, what is the better thing to do? What about subtraction? What about division? ●

Estimating with percents is very common. When you shop and see a sale at "30% off," you need to know what you have to pay. Here are some hints about estimating with percents, building on the earlier mental computation with percents.

1. Estimating 10% of 34.5 could be 10% of 35, which is 3.5.

2. If you estimate 20% of 34.5, it is 3.5 twice or 7. An estimate of 30% of 50.3 is 10% of 50 three times: $3 \times 5 = 15$.

3. 15% is 10% plus half that again. 15% of 642 is $64.2 + 32.1$ or 96.3. But if you are estimating, such as for a tip, you could say 15% of 640 is $64 + 32$, 96, or even 100.

4. 25% is a quarter of an amount. 25% off $350 is about 25% of $360, which is $90. (Note that $\frac{1}{4}$ of 360 is 90.) Or, you could say 10% is 35, 10% more is 70, half of 35 is about 18, and $35 + 35 + 18$ is 88.

5. When estimating a discount on a price, it may be easier to estimate the sale value. If a vase is regularly priced at $140 and is 20% off, then the actual sale value is 80% of $140: 50% of 140 is 70; 10% of 140 is 14, and so 30% of 140 is 3×14, or 42. So, the actual sale value is $70 + $42 = 112. To estimate, 80% of $150 would be $8 \times 15, $4 \times 30, or $120. In this case, finding the actual sale value was as easy as estimating it.

ACTIVITY 5   What Does It Cost? Estimate.

1. A coat is 30% off. It was originally $150. How much is it now?

2. A dress, marked as $90 before the sale, is 40% off. What percent of the original price must one pay to buy it? What is the discounted cost?

3. A jacket cost $45 after the 50% discount. How much was it originally? ●

An eighth-grade teacher of a low-ability class told one of the authors that she spent the last 8 weeks of math classes on mental computation and estimation because her students were so weak in this area, and she wanted them to better understand how to operate with numbers. Her students did extremely well. She said that they came to feel very confident, mathematically, and were no longer afraid of numbers.

TAKE-AWAY MESSAGE . . . Computational estimation builds a greater facility with numbers but at the same time reinforces facility. Flexibility and good number sense are essential when estimating. Estimation plays a major role in our daily lives. Learning how to make better estimates can be a valuable skill, one we want to pass on to students. You should be estimating calculations throughout this course.

Learning Exercises for Section 5.2

1. Estimate 0.76×62.

2. You buy 62 tablets at 76¢ each. You want to estimate the price before you get to the cash register to make certain you have enough money with you. Do you estimate differently than in Exercise 1? Discuss times when an estimate should be an overestimate, and times when an underestimate is preferable.

3. Which is greater (without calculating): 0.21×84.63 or $84.63 \div 0.21$? Explain your answer.

4. Your restaurant bill is $27.89. What should you leave as a tip if you want to tip 15%?

5. Reflect in writing on the six ways of estimating the product of 36 and 55 used in the first discussion in this section. Can all be done mentally? Are some better than others? Why or why not? Are some easier than others? Which ones show a better "feel" for numbers?

6. Estimate the following products using whatever strategies you prefer:
   a. 49×890 b. 25×76 c. 16×650
   d. 341×6121 e. 3×532

7. Explain how you would estimate the cost of the following items with the discounts indicated:
   a. a dress marked $49.99 at a 25% discount
   b. a CD marked $16.99 at 15% off
   c. a blouse at 75% off the $18.99 clearance price
   d. a suit at 60% off the already marked-down price of $109.99

8. Use your "close to" knowledge to *estimate* the following. The first one is done for you.
   a. 0.49×102 is about half of 100, or 50
   b. 32% of 12   c. 94% of 500   d. 0.52×789
   e. 23% of 81   f. 35% of 22   g. 76% of $210

9. Evaluate each student's reasoning. If the reasoning can easily be improved, tell how.
   a. "For $20 \times \$19.95$, I said $20 \times \$20$ and $20 \times \$1$, because $19 is about $20 and point 95 is about 1. So, about $400 + \$20 = \$420$."
   b. "25.3% of 119.2 would be about 0.3×120, or 36."
   c. "0.514×8.06 is about $1 \times 8 = 8$."

10. Return to the activities in this section and once again work through each part on your own.

Supplementary Learning Exercises for Section 5.2

Exercises of this sort are rather easy to make up; check your estimates with a calculator.

1. Give an estimate for each, calculating mentally. Then write down how you thought.
   a. the 10.5% tax on a hotel room that rents for $148 before taxes
   b. the 5.5% sales tax on a sale of $48.95
   c. the 7.75% sales tax on a sale of $74.95
   d. a 15% tip on a restaurant bill of $16.85

2. Give an estimate for each, calculating mentally. Then write down how you thought.
   a. 51.3×695.72 b. 32.8×41.09
   c. 8.1×602.89 d. 65% of $584.38
   e. 23% of $79.95   f. 33% of $35.95

This page* from a fourth-grade textbook is an example of how estimation problems occur in elementary school mathematics. Try answering the questions on this page yourself.

Date _____   Time _____

LESSON 4·4   A Bicycle Trip

Diego and Alex often take all-day bicycle trips together. During the summer, they took a 3-day bicycle tour. They carried camping gear in their saddlebags for the two nights they would be away from home.

Alex had a **trip meter** that showed miles traveled in tenths of miles. He kept a log of the distances they traveled each day before and after lunch.

Travel Log		
	Distance Traveled	
Timetable	**Before lunch**	**After lunch**
Day 1	27.0 mi	31.3 mi
Day 2	36.6 mi	20.9 mi
Day 3	25.8 mi	27.0 mi

Use estimation to answer the following questions. Do not work the problems out on paper or with a calculator.

1. On which day did they travel the most miles?   _____

2. On which day did they travel the fewest miles?   _____

3. During the whole trip, did they travel more miles before or after lunch? _____

4. Estimate the total distance they traveled. Choose the best answer.

⊂⊃ less than 150 miles   ⊂⊃ between 150 and 180 miles

⊂⊃ between 180 and 200 miles   ⊂⊃ more than 200 miles

5. Explain how you solved Problem 4.

_____

_____

_____

6. On Day 1, about how many more miles did they travel after lunch than before lunch?

_____

7. Diego said that they traveled 1.2 more miles before lunch on Day 1 than on Day 3. Alex disagreed. He said they traveled 2.2 more miles. Who is right? Explain your answer.

_____

_____

Number Sense 2.0; Number Sense 2.1; Number Sense 2.2   **85**

*Wright Group, *Everyday Mathematics*, Student Math Journal, Grade 4, Vol. 1, p. 85. Copyright © 2008, McGraw-Hill. Reproduced with permission of The McGraw-Hill Companies.

5.3 Estimating Values of Quantities

Many times when people talk about estimation, they mean estimating numerosity rather than estimating calculations, e.g., the crowd at some big event. Perhaps you have at some time entered a contest where you had to guess the number of jelly beans in a large jar and found it very difficult to do. Numerosity refers not just to guessing but to making intelligent guesses. If you went home and found a similar jar that you filled with jelly beans, then counted them, you might make a better guess than had you not done so. For a long time you will be able to make fairly good judgments concerning the number of jelly beans in jars.

Although that may not be a skill worth developing, there are other contexts where estimating skills can be useful. Skill in estimating the values of quantities, whether discrete counts (like jelly beans) or continuous measures (like the height of a tree), requires practice.

Money is certainly one area in which understanding estimates can be useful. A recent news report recommended that those speaking about Social Security legislation speak only about numbers in the millions, because people don't know what a trillion means. Perhaps large numbers will make more sense to you after this section.

ACTIVITY 6 Developing a "Feel" for the Size of Quantities

1. How big is a crowd of 100 people? Of 1000 people? Of 10,000 people? Do you have a sense of how large each of these groups would be?

    a. Would 100 people fit in a typical classroom? If so, how crowded would the room be?

    b. Would 1000 people fit in a typical classroom? Inside your home? Where would it be natural to find a group this size? Are there enough seats in a standard movie theater for 1000 people?

    c. Where would you expect to see 10,000 people? Do you have a feel for how large a group this would be? If exactly this many people were in attendance at your local stadium, would it be completely full? Half full? How full?

    d. How many stadiums of this size would be needed for 6 billion people?

2. How long is 10 ft? 100 ft? 1000 ft? 10,000 ft? Do you have a sense of how long each of these lengths is?

    a. Name at least five objects whose typical length or height is approximately 10 ft. (Think of the length of your bedroom. Is it more or less than 10 ft?)

    b. Do the same for 100 ft. Think about a point or object in the environment surrounding you that is approximately 100 ft away from you, 100 ft tall, or 100 ft long. If necessary, go outside and find something that will help you make sense of 100 ft.

    c. Name some things that you could use as a reference for a length, height, or distance of 1000 ft. (One way—a mile is 5280 ft, so 1000 ft is about. . . . ?) ●

Having good number sense requires having some sense not only of the relative size of a number (e.g., the relative size of 3 and 30, or of $\frac{1}{3}$ and $\frac{1}{2}$) but also of the absolute size of a number. (For example, how much is a million? Could you reasonably ask a third-grade class to bring in their pennies until they reached a million?) To develop a "feel" for the size of numerical amounts, it is helpful to compare them to some reference amount with which you are familiar. Such a reference amount is often called a **benchmark**. A benchmark is a personally meaningful and recognizable amount that can be used to make size estimates. Benchmarks are particularly useful when we want to develop a feel for very large or very small numerical amounts. Benchmarks can involve money, population, time, distance, height, weight, a collection of objects, or any other physical or nonphysical attributes.

EXAMPLE 1

▶ It is 250,000 miles from the earth to the moon. How far is that? ◀

If you know that it is approximately 25,000 miles around the earth along the equator, you could think that traveling 10 times around the earth would be comparable in distance to traveling to the moon.

Or,

it is about 2500 miles by air between New York City and Los Angeles, so you can imagine flying between the two cities 100 times.

Or,

you could think of a place that is roughly 10 miles away from where you are and imagine covering that distance 25,000 times! ●

Think About ...

Does the example above give you a "sense" of how far away the moon is? Is it farther away than you thought? Or closer? How long would it take you to walk to the moon?

Having one's own benchmarks can be very helpful. If these referents are to be useful to you, they must be personalized. This means that you are the one who needs to be familiar with the amount. For example, if you attend games at the local stadium and you know its capacity, you can use that as a benchmark for that many people. In one city, the football stadium holds more than 60,000 people. Such a reference can give you a feel for that many people. Having a feel for 60,000 people, however, does not necessarily help you develop a feel for how far 60,000 miles is, or how much $60,000 can buy. Therefore, the nature of the quantity, not just its numerical value, is important in deciding the usefulness of a reference amount.

TAKE-AWAY MESSAGE . . . Many people have little "feel" for the size of numbers unless they practice, within some context. For example, a builder can easily estimate the height of a building and its square footage. Having some benchmarks for numbers as they are used in different contexts is an important aspect of living with numbers every day.

Learning Exercises for Section 5.3

1. Think of a personal referent that will give you a feel for each of the following distances:

   a. 3 miles
   b. 30 miles
   c. 300 miles
   d. 3000 miles
   e. 30,000 miles

2. How long would it take you to travel each of the distances in Exercise 1 if you were to drive your car?

3. Find personal benchmarks for at least five large numbers (groups) of people. For example, you might say that 1000 people is about the number of students that attended your high school; your university has about 25,000 students; the city you live in has a population of about half a million people, etc. (These are estimates of numerosity.)

   Do the same thing for amounts of money. For example, my car is worth about $10,000; the salary of a beginning teacher in my city is about $35,000; the average price of a home in my city is $250,000, etc. But find your own!

4. A congressman once said, "I don't care how much it costs—a billion, a trillion. We need to solve the AIDS problem." Can the United States afford to spend 1 million dollars to fund AIDS research? One billion dollars? One trillion dollars? Why or why not? The population of the United States is now about 310 million. What would each of these amounts mean for each person in this country, in terms of a person's share of the national debt?

5. Write a children's story (not a school lesson) that uses at least five quantities with large values in ways that will help children understand how big the values really are. Include references to places, things, and events that will make sense to them. The story should have between 500 and 1000 words. (Do you have a feel for how many pages that is?)

As you develop your own personal referents, make a list of them so that you can refer to them at any time. In a later study of measurement, you will again be asked to prepare a list of personal referents.

Supplementary Learning Exercises for Section 5.3

1. About how many school classrooms could a 12,000-seat arena hold?

2. About how wide is the house or building you live in?

3. If your small dog weighs about 25 pounds, how much does a raccoon weigh?

5.4 Using Scientific Notation for Estimating Values of Very Large and Very Small Quantities

Most very large and very small numbers are actually estimates. For a 12-digit whole number, the digit in the ones place is not going to matter. Indeed, for a large number, only a few digits to the far left really matter.

Many people do not even know the names of numbers past 1 trillion, even though uses for very large numbers exist. Very small numbers also need to be expressed in some standard way that can be shared. We can express very large and very small numbers using scientific notation.

> A number is written in **scientific notation** when it has the form $a \times 10^b$, where a is a number between 1 and 10, including 1, and b is an integer.

EXAMPLE 2

a. 50 million is 50,000,000, and in scientific notation this is 5×10^7.

b. For small numbers, scientific notation requires the use of negative integers. 10^{-3} means $\frac{1}{10^3}$. We formalize this by saying that for any nonzero c and any d, $c^{-d} = \frac{1}{c^d}$.

c. 6 ten-thousandths is .0006 or $\frac{6}{10,000}$, which in scientific notation is 6×10^{-4}. Sometimes we write .0006 as 0.0006 so that the decimal point is more obvious.

d. $350,000,000 = 3.5 \times 100,000,000$, which in scientific notation is 3.5×10^8.

e. $0.00052 = 5.2 \times 0.0001$, which in scientific notation is 5.2×10^{-4}. ●

Note that both the 3.5 and the 5.2 are between 1 and 10, and that the second part of the numerals have ten to a power.

Operating on numbers in scientific notation requires knowing how to use exponents to your advantage. Recall that $5^3 \times 5^4$ means $(5 \times 5 \times 5) \times (5 \times 5 \times 5 \times 5)$, which is $5 \times 5 \times 5 \times 5 \times 5 \times 5 \times 5$ or 5^7, and that $5^3 \div 5^4$ means $\frac{5 \times 5 \times 5}{5 \times 5 \times 5 \times 5}$, which is $\frac{1}{5}$ or 5^{-1}. In general, we have the following rules.

> **Two laws of exponents** useful in scientific notation are
>
> $$a^m \times a^n = a^{m+n} \quad \text{and} \quad a^m \div a^n = a^{m-n}$$

EXAMPLE 3

$(3.5 \times 10^8) \times (5.2 \times 10^4) = (3.5 \times 5.2) \times (10^8 \times 10^4) = 18.2 \times 10^{12}$. But this is not in scientific notation because 18.2 is larger than 10. So, $18.2 \times 10^{12} = 1.82 \times 10^{13}$. ●

EXAMPLE 4

$(3.5 \times 10^8) \div (5.2 \times 10^4) = (3.5 \div 5.2) \times (10^8 \div 10^4) \approx 0.67 \times 10^4$. But this is not in scientific notation because 0.67 is smaller than 1. So, $0.67 \times 10^4 = 6.7 \times 10^3$. ●

Notes

EXAMPLE 5

▶ Light travels 186,000 miles in a second. A light-year is the distance that light travels in one year, even though it might sound like a time unit. About how many miles are in a light-year? ◀

Solution

$$1.86 \times 10^5 \, \frac{\text{mi}}{\text{s}} \times 60 \, \frac{\text{s}}{\text{min}} \times 60 \, \frac{\text{min}}{\text{h}} \times 24 \, \frac{\text{h}}{\text{day}} \times 365 \, \frac{\text{days}}{\text{year}} \approx 5.87 \times 10^{12} \, \frac{\text{mi}}{\text{year}}$$

So, a light-year is about 5.87×10^{12} miles. (Is this a meaningful number?) ●

 ACTIVITY 7   Do You Agree?

Write out two very large numbers and two very small numbers and give them to a partner to write in scientific notation. Your partner should do the same with you. Check the numbers to see if you agree with the ways in which these numbers are written in scientific notation. Make certain that you are using the notation. Then write other numbers in scientific notation and give them to one another to write in full notation. If you have questions, seek clarification. ●

TAKE-AWAY MESSAGE . . . Scientific notation provides an efficient way to express and operate on very large and very small numbers, using powers of 10 and laws of exponents.

Learning Exercises for Section 5.4

1. **a.** What is $(6.12 \times 10^4) \times (3 \times 10^2)$?
    b. What is $(6.12 \times 10^4) \div (3 \times 10^2)$?
    c. What is $(6.12 \times 10^2) \div (3 \times 10^4)$?

2. Why is $(3 \times 10^4) \times (4 \times 10^6)$ *not* 12×10^{10} in scientific notation?

3. **a.** Write $45,000,000 \times 220,000,000,000$ in scientific notation; then compute and be sure the answer is in scientific notation.
    b. Do the same for $6,900,000 \div 23,000,000,000$.
    c. Do the same for $0.0000000000056 \div 70,000$.
    d. Do the same for $0.0084 \div 0.000004$.

4. **a.** Write 1.5 billion in scientific notation.
    b. Write 4.27 trillion in scientific notation.

5. Find $(3.14 \times 10^6) + (2.315 \times 10^4)$. Write a sentence about the role of scientific notation in addition and subtraction.

6. Experiment with scientific notation on a scientific calculator. Where does the power of 10 appear on the calculator? Do the computations in Exercises 1, 3, and 5 using a scientific calculator. Think about how to enter and read the numbers.

7. Choose some number to be n, and then find $2n$, 2^n, n^2, 10^n. As n gets larger, describe how the results change in each case. Does the result grow faster when n is an exponent, or when it is a base?

8. How long is 1 billion seconds in hours? In days? In years?

9. A googol is 10^{100}. Have you ever heard this term used? If so, where?

10. Not all large numbers are commonly expressed in scientific notation. Computer memory is expressed in powers of 2. A kilobyte is 2^{10} (kilo means 1000, and $2^{10} = 1024$); a megabyte is 2^{20}; a gigabyte is 2^{30}. Which of these is close to 1 billion?

Supplementary Learning Exercises for Section 5.4

1. Why is 21.7×10^5 not in scientific notation?

2. Write each of these in regular notation. (It is interesting to see whether you can pronounce the ones in regular notation!)

   a. 5.804×10^6 **b.** 8×10^9 **c.** 1.86×10^5 **d.** 3.9×10^{12}
   e. 1.75×10^{-2} **f.** 2.703×10^{-5} **g.** 7.4×10^{-8} **h.** 5.2368×10^{-3}

3. Write each of the following in scientific notation:

   a. 26,829,000,000,000   **b.** 8236   **c.** 945,000,000
   d. 0.418923722   **e.** 0.23   **f.** 0.000485

4. What power of 10 is a thousand? A million? A billion? A trillion?

5. Write these in scientific notation.

   a. 26 billion   **b.** 0.5 million   **c.** 26 trillion

6. Use scientific notation to help with these calculations. Answers should be in scientific notation.

   a. $(4.2 \times 10^6) \times 10\ 000$ (The metric system uses a space where we ordinarily use a comma: $10\ 000 = 10{,}000$.)
   b. $(3.4 \times 10^5) \times 200\ 000\ 000$ **c.** $460\ 000 \times 500\ 000$ **d.** 5×4000 million

7. If you have a calculator, what is the largest number you can enter without using scientific notation?

8. Write each of the following in words and decimals, without scientific notation.

   a. 5×10^{-8} **b.** 166×10^{-4} **c.** 19×10^{-6}

5.5 Issues for Learning: Mental Computation

In a research study[3] fourth- and sixth-grade students received almost daily instruction on mental computation that always allowed students to discuss a variety of strategies for mental computation problems. At the beginning, these students wanted to perform the mental analogue of the paper-and-pencil procedure, i.e., they would say (when asked to add 38 and 45 mentally), "I put the 45 under the 38. Then I add 5 and 8 and that's 13 and I write down a 3 and carry the 1. Then I add 1 and 3 and 4 and I get 8, so the answer is 83." When asked to add 345 and 738, these students could not mentally remember the steps and made frequent errors. By the end of the research study, students would have other methods of mental computation that did not depend on trying to remember all the numbers from the standard algorithms. For example, they would add 38 and 45 by saying, "Thirty and 40 is 70, and 8 and 5 is 13, and 70 and 13 is 83." Now they were using place value. Instead of adding 3 and 4, they were adding 30 and 40. They could now mentally compute with skill and accuracy.

Mental computation is used in many cultures by people who need to do frequent calculations. For example, in studies[4] of the Diola people of the Ivory Coast, unschooled children who worked with their parents in the marketplace developed excellent mental computation skills that showed deep insight into the structure and properties of the whole number system.

Mental computation can play an important role in developing number sense through explorations that force students to use numbers and number relations in novel ways that are likely to increase awareness of the structure of the number system. Unfortunately, some teachers view mental computation as a skill to be practiced very rapidly. In such cases instruction tends to focus on drill using chain calculations (e.g., $4 + 15; -9; \times 2; \times 3; -15; \div 5$ is 9) and on learning "tricks" such as those for multiplying by 9 or by 11. Although such skills are valuable and could have a place in the curriculum, it is questionable whether they should be emphasized at the expense of instruction that could lead children to better number sense.

5.6 Check Yourself

The power of mental computation and computational estimation skills is that they both cause and result from number flexibility. As you develop these skills, you will feel more comfortable with numbers. This is true for elementary school children as well. By providing them with opportunities to work with numbers in this personal way, you will allow them to develop a comfort with numbers that can permeate their study of numbers and operations.

You should be able to work problems like those assigned and to meet the following objectives:

1. Perform mental calculations when appropriate, depending on the numbers involved, and do so in a variety of different ways.

2. Identify the associative and commutative properties of addition and multiplication, and the distributive property of multiplication over addition when they are used in a mental computation.

3. Estimate calculations in a variety of ways, for a variety of forms of numbers.

4. Evaluate computational estimations in terms of what they reveal about the number understanding of the person doing the calculations.

5. Use personal benchmarks to estimate, at the least, numerosity and distance.

6. Determine when a number is in scientific notation, and what it means.

7. Discuss why scientific notation makes it easier to express large and small numbers.

8. Express given numbers or do calculations involving scientific notation.

References for Chapter 5

[1]Threadgill-Sowder, J. (1984). Computational estimation procedures of school children. *Journal of Educational Research, 77,* 332–336.

[2]Sowder, J. T. (1992). Making sense of numbers in school mathematics. In G. Leinhardt, R. Putnam, & R. Hattrup (Eds.), *Analysis of arithmetic for mathematics education.* Hillsdale, NJ: Erlbaum.

[3]Markovits, Z., & Sowder, J. (1988). Mental computation and number sense. In M. J. Behr, C. B. Lacampagne, & M. M. Wheeler (Eds.), *Proceedings of the Tenth Annual Meeting of the North American Chapter of the International Group for the Psychology of Mathematics Education.* DeKalb, IL.

[4]Ginsburg, H. P., Posner, J. K., & Russell, R. L. (1981). The development of mental addition as a function of schooling and culture. *Journal of Cross-Cultural Psychology, 12*(2), 163–178.

Meanings for Fractions

The symbol $\frac{2}{5}$ can be interpreted in several ways. In this chapter we discuss two such ways, and we make the connections between the different meanings of the fraction symbol.

 DISCUSSION 1 Which Is Greater?

1. In which of the following descriptions is the teacher referring to the largest part of his or her class? Be prepared to defend your answer.

   a. $\frac{3}{5}$ of the class   **b.** $\frac{18}{30}$ of the class   **c.** 0.6 of the class   **d.** 60% of the class

2. Which represents the greatest quantity of cheese? Be prepared to defend your answer.

   a. $1\frac{3}{4}$ pounds   **b.** $\frac{7}{4}$ pounds   **c.** 1.75 pounds   **d.** 28 ounces ●

These questions illustrate an especially important idea: The value of a quantity may be expressed in many ways. When a part of a quantity is involved, there is an especially rich variety of forms that the numerical value can take: fractions, decimals, mixed numbers, and percents.

   In general, the term "fractional expression" includes all of the following:

$$\frac{\sqrt{3}}{4} \qquad \frac{3 + \sqrt[3]{7}}{5} \qquad \frac{9}{10} \qquad \frac{2}{3} \qquad \frac{1.382}{0.94} \qquad \frac{12\frac{3}{4}}{1\frac{5}{6}} \qquad \frac{2.3 \times 10^3}{1.7 \times 10^{-2}} \qquad \frac{a^2}{b}$$

For now, however, the term **fraction** will usually be limited to the form most prominent in elementary school: $\dfrac{\text{whole number}}{\text{nonzero whole number}}$. You may recall the terms **numerator** and **denominator**, which can be used with any fraction: $\dfrac{\text{numerator}}{\text{denominator}}$.

6.1 Understanding the Meanings of $\dfrac{a}{b}$

There are different meanings associated with a fraction in the usual elementary school curriculum. The first one introduced to the children is the *part-whole* meaning of a fraction, then the *division* meaning (although sometimes not made explicit), and later the *ratio* meaning of the symbol $\frac{a}{b}$. The part-whole and division meanings are the focus of this section. Ratio is considered in Chapter 8.

> The **part-whole** meaning for the fraction $\frac{a}{b}$ has these three elements:
>
> $\rightarrow \frac{a}{b} \leftarrow$
>
> 1. The unit, or whole, is clearly in mind. (What = 1?)
> 2. The denominator tells how many pieces of equal size the unit is cut into (or thought of as being cut into).
> 3. The numerator tells how many such pieces are being considered.

$\frac{a}{b}$ is the quantity formed by a parts of size $\frac{1}{b}$. There are two basic kinds of wholes, depending on whether the unit is *discrete* or *continuous*. The whole can be a single thing or a group of things. Recall from Chapter 3 that quantities that are separate, countable objects are **discrete**. If one child of a team of six is a fourth-grader, then $\frac{1}{6}$ of the team is in fourth grade. The whole is a group of discrete (countable) things that are considered as a unit (the team). Or we might say, "One half of the team is boys," when each of two subsets of the team, one subset being girls and the other boys, has the same number of children. In this case, we consider one of two equal-sized parts, each consisting of countable objects. If there are 6 players on the team, then $\frac{1}{2}$ of the team is 3 boys. But not all fractions make sense in a given discrete setting. It would not make much sense to talk about $\frac{1}{7}$ of a team of 6 players, e.g.

Quantities are **continuous** when they can be *measured* by length, area, mass, etc. Continuous quantities cannot be counted; rather, they are measured. Even though we might say something is 5 inches long, this is only approximate. With lengths, it *is* possible to talk about $\frac{1}{7}$ of a quantity, no matter what that quantity is. This second kind of whole, like a pizza, can be thought of as cut up or marked off into *any* number of equal-sized pieces. (Notice that we cannot do this with a quantity such as number of people.) A pizza (the whole) can be cut up into (approximately) 6 equal-sized pieces, each of which is $\frac{1}{6}$ of the whole pizza, or 13 equal-sized pieces, each being $\frac{1}{13}$ of the whole pizza. We say that such a whole is continuous. The size of a continuous whole and its equally sized parts can be measured, e.g., by length, area, volume, weight, or other noncount measures, depending on its nature. The equally sized pieces of pizza have the same area on top.

When we model fractions by drawing rectangles, the *area* of the rectangle is the whole. When we model fractions on a number line, the whole is the *length* of the unit segment from 0 to 1.

Whether the whole is continuous or discrete, it should be clear that the denominator can never be 0. There will always be *at least one piece* in a continuous unit, or whole, or *at least one object* with a discrete unit, or whole, so the denominator must be at least 1. Note that "equally sized" is an ideal. In real life, the pieces can be only approximately equal.

In the early grades, the pieces of continuous quantities are most often the same size and shape, but having the same shape is not a requirement. It is only necessary that the unit be in pieces with the same length, area, volume, or count, e.g., or that the unit can be thought of in terms of such pieces.

DISCUSSION 2   When Are the Parts Equal?

Explain how you know that the pieces are or are not the same size.

Careful use of drawings, making certain the pieces have the same area, can help develop an understanding of fractions.

ACTIVITY 1   What Does $\frac{3}{4}$ Mean?

Explain how each of the following diagrams could be used to illustrate the part-whole meaning for $\frac{3}{4}$, making clear what the unit, or whole, is in each case. In which diagrams does $\frac{3}{4}$ refer to a part of a continuous length or region? In which diagrams does it refer to a discrete unit?

a.   b.

c.   d.

Depending on the context, what the word "size" refers to may vary. If the whole is continuous, say, area, or length, or volume, then the part-whole interpretation requires that the pieces have equal sizes: For example, the phrase "3 out of 4" is not good all-purpose language; instead, use the phrase "3 pieces when the whole has 4 *equal* pieces." In a discrete situation the parts do not have to be equal in size, but all the pieces must belong to some set of objects for which it makes sense to talk about parts of the whole, such as in the case of "3 of the 4 pieces of fruit are apples." An important concept, illustrated in part (d), is that $\frac{3}{4}$ is 3 one-fourths.

Under the **part-whole** view, $\frac{5}{6}$ means 5 one-sixths, where 6 one-sixths equals one whole. However, there are other meanings for a fraction. An important meaning is that a fraction can denote **division**: $\frac{5}{6}$ can tell how many 6s are in 5, e.g.

> The fraction $\frac{a}{b}$ can be interpreted to mean **sharing-equally division** (or **partitive division**), denoted by $a \div b$, in which case a wholes would be partitioned into b equal parts, each part being an equal "share."

> The fraction $\frac{a}{b}$ can also be interpreted to mean **repeated-subtraction division** (**measurement** or **quotitive division**): How many of b are in a? If the fraction can be represented as $\frac{1}{n}$, or $1 \div n$, it is called a **unit fraction**.

In contrast to the part-whole view, when $\frac{5}{6}$ is interpreted in terms of division, $5 \div 6$, we mean that 5 wholes could be partitioned into 6 equal parts, each of which is an equal "share." This interpretation requires a quite different display from that under the part-whole view.

ACTIVITY 2   What Does $\frac{5}{6}$ Mean?

Show that the part-whole $\frac{5}{6}$ is indeed the same amount as $5 \div 6$ (using a partitive, i.e., sharing interpretation). Use the square region as the unit in each case.

a. Show $\frac{5}{6}$.

b. Show $\frac{5}{6}$ with the number line model.

c. Show $5 \div 6$ (partitive).

d. Show $5 \div 6$ with the number line model.

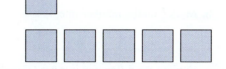

[For $\frac{5}{6}$, think: "One person gets $\frac{5}{6}$ of a brownie." How can this be shown? For $5 \div 6$, think (e.g.): "Five brownies are being shared by 6 people." How much does each person get?] ●

ACTIVITY 3   More Meanings?

a. Show $5 \div 6$ (quotitive).
For $\frac{5}{6}$ (quotitive), think: Suppose brownies are placed 6 to a plate. How many plates (or how much of a plate) of brownies are needed if you have 5 brownies?

b. Show $12 \div 8$ (quotitive) using a carton of 12 eggs. ●

> ### Think About ...
> Suppose one wants to show what 0.6 means by using drawings. One could begin with one unit rectangle as the unit whole, or with six unit rectangles. In each case, how would 0.6 be shown? How do the representations differ?

Notes

ACTIVITY 4   What Do the Rules Say? What Do They Mean?

One rule often taught for changing a mixed number like $2\frac{1}{3}$ to a fraction is to multiply the 2 and 3, add the 1, and then write this result over 3, i.e., $2\frac{1}{3} = \frac{7}{3}$. Explain how to make that rule sensible, starting with these drawings.

a. [drawings of squares]   **b.** [number line with points marked at 0, 1, 2, 3]

c. Can you also justify the reverse rule: Change $\frac{7}{3}$ into a mixed number by dividing the 7 by 3? [*Hint*: Use the drawings in parts (a) and (b).] ●

Note that in the first two parts of Activity 4, the whole number part of $2\frac{1}{3}$, the 2, needs to be considered in terms of thirds. There are three thirds in 1 whole and therefore six thirds in 2 wholes. If you begin with an "improper fraction" such as $\frac{7}{3}$ to be replaced by a mixed number, then you first need to find out how many wholes there are. In this case, because each 3 of the 7 pieces makes a whole, there are 2 wholes. What is left over?

Likewise, $\frac{7}{3}$ can be seen as $\frac{3}{3} + \frac{3}{3} + \frac{1}{3} = 2\frac{1}{3}$.

TAKE-AWAY MESSAGE . . . The fraction symbol can mean different things, but if it is interpreted within a context, one usually knows what the symbol means. The two most common uses in the elementary grades are as part of a whole, and as a way to indicate division. Any time you use fractions in the part-of-a-whole sense, you should be clear about *what unit or whole a given fraction is a fraction of*. Not doing so is frequently a source of many errors in work with fractions. Even in a given situation involving several fractions, not all the fractions have to refer to the same whole. The same pertains to decimals.

Learning Exercises for Section 6.1

1. In how many ways can you "cut" a square region into two equal pieces?

2. Using a rectangular or circular region as the whole, make sketches to show each of these:

   a. $\frac{2}{3}$ **b.** $\frac{7}{4}$ **c.** $2\frac{3}{8}$ **d.** $\frac{0}{4}$ **e.** $\frac{3}{1}$

3. **a.** Mark this representation of 4 feet of a licorice whip to show $\frac{3}{4}$ of the licorice whip.

   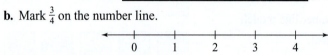

   b. Mark $\frac{3}{4}$ on the number line.

   [number line marked 0, 1, 2, 3, 4]

   c. Explain the difference in parts (a) and (b) by referring to the unit.

4. It is important to keep in mind what the unit or whole is for a particular fraction. Make a drawing for this situation: John ate $\frac{1}{4}$ of a cake. Maria ate $\frac{1}{4}$ of the remaining cake. What was the unit or whole in John's case? In Maria's case? How much of a full pan was there in each case?

5. With a square region as unit, show $\frac{3}{2}$ under

   a. the part-whole view   **b.** the division view

6. **Pattern Blocks** Using pattern blocks, take the yellow hexagon, the red trapezoid, the blue rhombus, and the green triangle. (Cut-outs of the pattern blocks are included in the "Masters" section. Depending on the colors used, they may or may not be the colors indicated here.)

   a. Let the hexagon = 1. Give the value for each of the other three pieces. (*Note:* = means *represent*.)

   b. Let the trapezoid = 1. Give the value for each of the other three pieces.

   c. Let a pile of two hexagons = 1. Give the values for the hexagon and each of the other three pieces.

7. **Pattern Blocks** In how many different ways can you cover the hexagon? Write an addition equation for each way.

8. Using drawings of circular regions, justify that each of these is true.

   a. $\frac{5}{8} > \frac{1}{2}$ b. $\frac{5}{8} < \frac{3}{4}$ c. $2\frac{5}{6} > \frac{5}{2}$

9. Use drawings of rectangular regions to find the larger of

   a. $\frac{2}{3}$ and $\frac{3}{4}$ b. $\frac{1}{2}$ and $\frac{4}{7}$ c. $\frac{1}{8}$ and $\frac{1}{9}$

   Using the part-whole meaning of a fraction, find the larger of each. Write down your explanation.

   d. $\frac{12}{17}$ and $\frac{12}{19}$ e. $\frac{10}{9}$ and $\frac{11}{9}$ f. $\frac{62}{101}$ and $\frac{62}{121}$

10. You are in charge of the Student Education Association cake sale. As you leave to buy some paper plates, you ask someone you do not know to cut all the cakes. When you return and look at the cakes (see the diagrams below), you gasp! The plan had been to charge $5 for a whole cake, and you had thought that each piece could be priced by dividing $5 by the number of pieces if necessary.

    a. For each cake, is it fair to charge the same amount for each piece? (Do not worry about how the icing is spread, or whether a piece close to an edge might be a little smaller in height.)

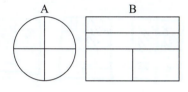
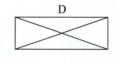

Cake C has cuts
evenly spaced.

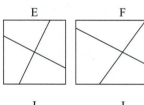

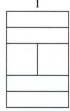

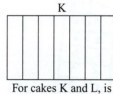
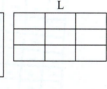

For cakes K and L, is
$1.25 for 2 slices fair?

    b. Cut your own cake *creatively*, but so that every piece is worth the same amount.
    c. Which of your fair/not fair decisions in diagrams A–L would change if the price of a whole cake changed from $5 to some other amount?
    d. What does part (a) (A–L) have to do with fractions?

11. Can you say what fraction of the second rectangular region is shaded? Explain.

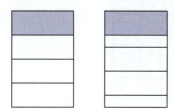

Notes

12. a. If you were offered a piece of cake from one of the following, which cake would give you a piece with more cake?[1]

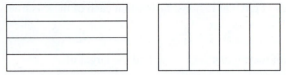

b. It is interesting that until about Grade 3 or 4, children often believe that one cake or the other will give a larger piece. If you have access to a young child, see what he or she says. (First make sure that the child agrees that the cakes are the same size.)

13. $\frac{5}{6}$ of an amount is represented here. How many marks would there be in $\frac{2}{3}$ of the same amount? Explain how you got your answer.

14. Children were asked in one class to show what $\frac{3}{4}$ means in as many ways as possible. Here is the work of three children. How would you evaluate the understanding in each case? (Make certain that you first understand what the child is thinking.)[2]

Sally:

Sam:

1. $\frac{3}{4}$ is bigger than $\frac{5}{8}$.   2. $\frac{3}{4}$ is smaller than 1 whole.   3. $\frac{4}{4}$ is bigger than $\frac{3}{4}$.

4. $\frac{13}{16}$ is bigger than $\frac{3}{4}$.   5. $\frac{32}{16}$ is $\frac{20}{16}$ bigger than $\frac{3}{4}$.

Sandy:

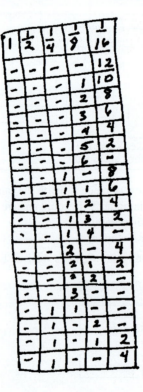

continue

Sandy claimed to have "found them all." What do you think he meant by that? (Note that this is a table, and that the first row contains headings for each column.)

15. The given region shows $\frac{3}{4}$ of a candy bar.

    a. Show how large the original candy bar was. Write down your reasoning.

    b. Show how large $2\frac{1}{6}$ of the original candy bar would be.

16. The piece of carpet runner below is $2\frac{2}{3}$ units long (the units are usually yards or feet). Drawing as accurately as you can, show each of these. (Trace the piece below to make a separate drawing for each part.)

    a. $\frac{3}{4}$ unit

    b. 4 units

    c. $1\frac{1}{3}$ units

    d. If the $2\frac{2}{3}$ unit piece was for sale for \$4.80, what would be a fair price for 1 unit? Write down your reasoning.

17. Pretend that the piece of expensive wire below is 1.25 decimeters long.

    a. Show how long 1 decimeter of wire would be (without using a ruler).

    b. If the 1.25-decimeter piece cost \$1.60, what would you expect the 1-decimeter piece to cost?

18. Here is part of a number line, with only two points labeled.

$$\longleftarrow \quad \overset{\displaystyle +}{1} \qquad \overset{\displaystyle +}{1.5} \quad \longrightarrow$$

Drawing as accurately as you can, locate the points for 2 and for $\frac{-3}{4}$. Explain your reasoning.

19. Use the part-whole notion to explain why $\frac{1}{7} > \frac{1}{8}$; why $\frac{7}{10} > \frac{7}{11}$.

20. A second-grader announced to his father that $\frac{2}{7}$ was smaller than $\frac{1}{3}$ but bigger than $\frac{1}{4}$. When asked how he knew, he said "You take 7 sevenths, and you know that 7 divided by 3 is bigger than 2, but 7 divided by 4 is smaller than 2." Can you explain his thinking?

21. Which label, discrete or continuous, applies to each of these "wholes"?

    a. the number of fans at a ball game

    b. the mileage traveled by a wagon train

    c. the amount of Kool-Aid drunk at a third-graders' party

    d. the number of cartons of milk ordered for snack time in kindergarten

22. Make a sketch to show each unit. Then use shading or some sort of marks to indicate the fractional part. Be sure that the thirds are clear.

    a. $\frac{2}{3}$ of the 6 children in the play were girls.

    b. $\frac{2}{3}$ of the 6 yards of cloth were needed for the play.

    c. $\frac{2}{3}$ of the 6 cartons of ice cream were chocolate.

    d. $\frac{2}{3}$ of each of the 6 cartons of ice cream were eaten.

    e. $\frac{2}{3}$ of the 6 moving vans were painted red.

    f. $\frac{2}{3}$ of each of the 6 moving vans were painted red.

    g. The store had 6 cartons of paper towels and used $\frac{2}{3}$ of them.

    h. The store had 6 cartons of paper goods and used $\frac{2}{3}$ of each of them.

23. Given a number line and only the points for the indicated fractions, accurately locate and label the point for the number in each part. If your markings do not make your reasoning clear, add an explanation. Extend the given line as needed.

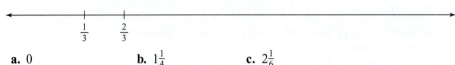

$\frac{1}{3}$ $\frac{2}{3}$

   a. 0   **b.** $1\frac{1}{4}$ **c.** $2\frac{1}{6}$

24. Locate and label the points for the given numbers, as in Learning Exercise 23.

1                $1\frac{2}{3}$

   a. 0   **b.** $\frac{3}{4}$ **c.** $\frac{7}{3}$

25. Mentally solve the equation in each of parts (a–c).

   a. $4 = \frac{x}{1}$ **b.** $7 = \frac{y}{3}$ **c.** $115 = \frac{z}{2}$

   d. Is it surprising that whole numbers can be regarded as naming fractions?

Supplementary Learning Exercises for Section 6.1

1. What does the part-whole fraction $\frac{5}{9}$ mean?

2. Which of the following wholes is/are cut so that sixths can be shown?

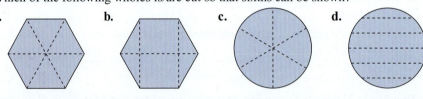

   a.   **b.**   **c.**   **d.**

3. Change the mixed number $3\frac{2}{5}$ to a fraction, and explain why your answer is correct by using a drawing.

4. Change the fraction $\frac{19}{4}$ to a mixed number, and explain why your answer is correct by using a drawing.

5. Rewrite the following as mixed numbers, without a calculator:
   a. $\frac{22}{7}$ **b.** $\frac{193}{15}$ **c.** $\frac{57}{4}$ **d.** $\frac{684}{35}$ **e.** $\frac{385}{384}$ **f.** $\frac{851}{219}$

6. What is the "sharing-equally division" or the "repeated-subtraction division" meaning for $\frac{x}{y}$?

7. What fractions can be illustrated rather easily with a regular pentagon, like the one to the right, as the whole? Why is the pentagon not good for showing thirds?

8. What is the unit or whole for each fraction in this monologue?
   "I spent about $\frac{1}{3}$ of my paycheck on groceries for the party, with about $\frac{3}{4}$ of that going for drinks, and I only filled my grocery cart $\frac{1}{2}$ full. And my rent will take $\frac{4}{5}$ of what I have left from my paycheck."

9. In each pair, tell which fraction is larger and explain how you decided this.
   a. $\frac{2}{3}$ and $\frac{2}{9}$ **b.** $\frac{15}{23}$ and $\frac{15}{22}$ **c.** $\frac{2 \times 27}{53 \times 62}$ and $\frac{3 \times 20}{53 \times 62}$ **d.** $\frac{197}{192}$ and $\frac{349}{360}$

10. The region given at right shows $\frac{2}{3}$ of a cake.
   a. Show how large the original cake was, giving your reasoning.
   b. Then show how large $2\frac{3}{4}$ cakes of the same size as the original would be.
   c. If the piece of cake shown is worth $4.50, how much is a whole cake worth?

11. Locate and label the points for the given numbers, as in Learning Exercise 23.

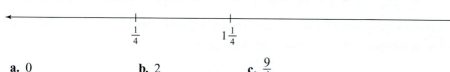

$\frac{1}{4}$ $1\frac{1}{4}$

a. 0   **b.** 2   **c.** $\frac{9}{4}$

12. Locate and label the points for the given numbers, as in Learning Exercise 23.

$109\frac{7}{8}$

109.5

a. 110   **b.** 109   **c.** $109\frac{1}{4}$

6.2 Equivalent (Equal) Fractions

At times fractions that look different can refer to the value of the same quantity. For example, the numerals $\frac{2}{3}$ and $\frac{100}{150}$ certainly look different to children, but as you know, these two fractions are **equivalent** or **equal fractions** (elementary school textbooks use both terms). For example, $\frac{3}{4} = \frac{6}{8} = \frac{9}{12}$ is reasonable to conclude from these drawings because the same amount of the unit is shaded in each drawing:

Starting with $\frac{3}{4}$, each of the original four pieces is cut into two equal pieces, giving $4 \times 2 = 8$ equal pieces all together in the second drawing, of which 6 are shaded: Thus, $\frac{6}{8}$ is shaded. If each of the original four pieces is cut into three equal-sized pieces, then 9 of the 12 pieces are shaded: $\frac{9}{12}$ is shaded.

 ACTIVITY 5   More About $\frac{3}{4}$

a. Give a similar analysis to show that $\frac{3}{4} = \frac{3 \times 4}{4 \times 4}$, and make drawings using the area model to show $\frac{3}{4} = \frac{3 \times 5}{4 \times 5}$.

b. Make a diagram to show $\frac{3}{4}$ is equivalent to $\frac{12}{16}$ because their position on the number line is the same. ●

> As a general principle, for any n not equal to 0, $\frac{a}{b} = \frac{a \times n}{b \times n}$. We call $\frac{a \times n}{b \times n}$ an equivalent fraction of $\frac{a}{b}$.

Notice that the above equation can be read in reverse: $\frac{a \times n}{b \times n} = \frac{a}{b}$. This understanding is also important because it shows that a **common factor** of both the numerator and the denominator, like n, can be "ignored" to give a simpler-looking fraction. (You may recall that a **factor** of a number divides the number, leaving no remainder. For example, 3 is a factor of 6, 2 is a factor of 6, but 4 is not a factor of 6.)

Thus, $\frac{12}{30} = \frac{6 \times 2}{15 \times 2} = \frac{6}{15}$. Also, $\frac{6}{15} = \frac{2 \times 3}{5 \times 3}$, so $\frac{6}{15} = \frac{2}{5}$. Because 2 and 5 have no common factors (except 1), we say that $\frac{2}{5}$ is the **simplest form** of $\frac{12}{30}$. Alternatively, we say $\frac{12}{30}$ in **lowest terms** is $\frac{2}{5}$. Note that we could have done this in one step:

$$\frac{12}{30} = \frac{2 \times 6}{5 \times 6} = \frac{2}{5}$$

Finding a common factor in a fraction is often difficult for children because the common factor is not visible; the common factor must be arrived at by thinking about factors of the numerator and denominator. The common phrase "reduced fraction" is potentially misleading. *Reduce* usually means to decrease in size, but a *reduced* fraction still has the same value as the original. *Simplest form* does not carry any incorrect ideas.

As you know, it can be useful to write equivalent fractions so that two fractions have a common denominator. $\frac{2}{3} = \frac{8}{12}$ and $\frac{3}{4} = \frac{9}{12}$. The fractions $\frac{8}{12}$ and $\frac{9}{12}$ have a common denominator.

> The **least common multiple (LCM)** of two (or more) whole numbers is the smallest number that is a multiple of the two (or more) numbers.

For example, 12 is the least common multiple of 3 and 4 because if you list the multiples of 3 and the multiples of 4, it is the smallest number they have in common.

$$3, 6, 9, 12, 15, 18, 21, 24, 27, 30, \ldots$$
$$4, 8, 12, 16, 20, 24, 28, 32, 36, 40, \ldots$$

> The **greatest common factor (GCF or gcf)** of two (or more) whole numbers is the largest number that is a factor of the two (or more) numbers. For example, 2, 3, and 6 are all common factors of 12 and 18, and 6 is the *greatest* common factor of 12 and 18. The GCF is sometimes also called the **greatest common divisor (GCD or gcd).**

In a discrete context, students often have difficulty in seeing that two fractions are equivalent. For example, how might you illustrate that $\frac{3}{12} = \frac{1}{4}$ within a discrete context? The same diagram representing one whole should demonstrate 3 parts out of 12 equal parts while *at the same time* demonstrating 1 part out of 4 equal parts. An illustration such as the tray of cupcakes here can be used. The given drawing shows 12 cupcakes, with 3 cupcakes having chocolate icing. The same drawing also shows 4 groups (the columns), with 1 group having chocolate icing. So, $\frac{3}{12}$ does equal $\frac{1}{4}$.

You often can visualize a collection of discrete objects in multiple ways. A collection of 12 objects such as the cupcakes (or perhaps with another arrangement) can be seen as made up of 2, 3, 4, or 6 equal groups. Notice that each of 2, 3, 4, and 6 is a factor of 12.

An important number concept is that of prime number. A **prime number** is a whole number greater than 1 that has only 1 and itself as factors. For example, the numbers 3, 29, and 101 are prime numbers. A related idea is that of **relatively prime numbers**. Numbers are relatively prime when they have no common factors except 1. For example, 3 and 5 are relatively prime, as are 3 and 4, even though 4 is not itself prime. Thus, we can say that a fraction is in its lowest terms when the numerator and denominator are relatively prime.

DISCUSSION 3 Do They Have Common Factors?

Which of these pairs of numbers is relatively prime? How did you decide?

a. 16 and 27   **b.** 75 and 102   **c.** 625 and 835 ●

TAKE-AWAY MESSAGE . . .   The ability to recognize equivalent fractions is fundamental to operating with them. The focus in this section was to find a meaning for the ways we used to find equivalent fractions. Of next importance is being able to compare fractions: Which of two fractions is larger or smaller?

There are several ways of comparing fractions shown on this sixth-grade textbook page.* Why do you think so many different ways are illustrated?

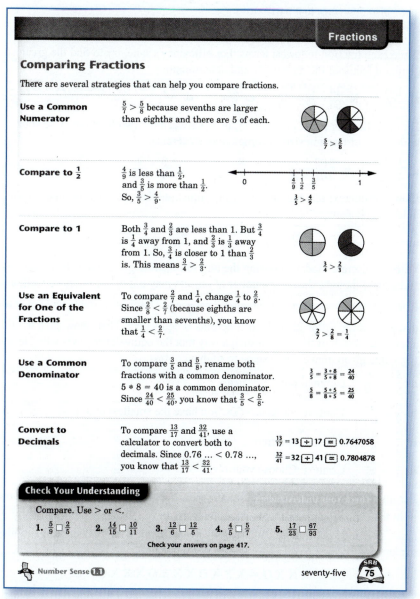

*Wright Group, *Everyday Mathematics*, Student Reference Book, Grade 6, p. 75. Copyright © 2008, McGraw-Hill. Reproduced with permission of The McGraw-Hill Companies.

Notes

Learning Exercises for Section 6.2

1. **a.** Write 10 fractions that are equivalent to $\frac{2}{3}$ and 10 fractions that are equivalent to $\frac{5}{8}$.
    b. Do any of your 20 fractions have **common denominators**?
    c. When would having common denominators be useful?

2. Use sketches of regions or number lines to show each of the following:

    a. $\frac{2}{3} = \frac{10}{15}$ **b.** $1\frac{4}{10} = 1\frac{2}{5}$ **c.** $\frac{5}{4} = \frac{15}{12}$

3. Find at least one equivalent fraction for each of the following and demonstrate on a number line that the two fractions are equivalent.

    a. $\frac{3}{5}$ **b.** $\frac{3}{4}$ **c.** $\frac{0}{2}$

4. Suppose a school committee is made up of 6 girls and 4 boys. What fraction describes the part of the committee that is girls? How would you show, through a diagram, that $\frac{3}{5}$ also describes that part?

    (*Hint:* GGGGGGBBBB can be organized as GG GG GG. . . .)

5. Write each fraction in lowest terms. For efficiency's sake, try to find the greatest common factor of the numerator and denominator.

    a. $\frac{450}{720}$ **b.** $\frac{24 \cdot 45 \cdot 17}{64 \cdot 12 \cdot 17}$ **c.** $\frac{225}{144}$ **d.** $\frac{x^2 y^3 z^6}{xy^4 z^3}$ **e.** $\frac{(4.2 \times 10^7) \times (1.5 \times 10^8)}{4.9 \times 10^4}$

6. For the following pairs, place $<$, $>$, or $=$ between the two fractions. Use your knowledge of fraction size to complete this exercise.

    a. $\frac{3}{4}$ $\frac{3}{5}$ **b.** $\frac{102}{101}$ $\frac{75}{76}$ **c.** $\frac{8}{9}$ $\frac{40}{45}$ **d.** $\frac{8}{9}$ $\frac{9}{10}$ **e.** $\frac{13}{18}$ $\frac{29}{62}$

7. Using the context of a box of 24 crayons, illustrate that the fractions in each part are indeed equivalent.

    a. $\frac{4}{24} = \frac{1}{6}$ **b.** $\frac{18}{24} = \frac{3}{4}$ **c.** $\frac{4}{12} = \frac{2}{6}$

8. With the discrete model, choosing the unit to illustrate given equivalent fractions takes some thought. Just any number of objects may not fit the fractions well; a set of 4 objects would not work for showing $\frac{2}{4} = \frac{3}{6}$, e.g. Describe a usable unit for showing that each of the following pairs of fractions is equivalent. Then show each fraction on a separate sketch of the unit, in such a way that the equivalence is visually clear. If a context is given, make your sketch to fit that context.

    a. $\frac{4}{5} = \frac{8}{10}$ Context: popsicle sticks

    b. $\frac{6}{9} = \frac{4}{6}$ Context: marbles   (Do you have the smallest unit?)

    c. $\frac{2}{3} = \frac{4}{6}$ Context: children

    d. $\frac{2}{3} = \frac{12}{18}$ Context: Choose your own.

9. For illustrating that $\frac{3}{4} = \frac{9}{12}$, a student repeats a pattern that represents $\frac{3}{4}$ to show $\frac{9}{12}$, as in the following diagram. He explains that you can multiply the fraction by 3 to make an equivalent fraction. What problems are there with this diagram and his explanation?

    $$X X X O = X X X O X X X O X X X O$$

10. Sometimes a textbook or a teacher will refer to *reducing* a fraction like $\frac{9}{12}$. What possible danger do you see in that terminology?

11. Each letter on this number line represents a number. Match the numbers to each operation shown *without doing any calculation*. Use good number sense.

```
 A  B     C   D     E              F          G
 ├──┼─────┼───┼─────┼──────────────┼──────────┼──
 0         1         2
```

a. $\frac{3}{4} \times \frac{4}{3}$ b. $2 \times \frac{2}{3}$ c. $\frac{3}{4} \times \frac{1}{2}$

d. $2 \div \frac{2}{5}$ e. $\frac{1}{3} \times \frac{2}{5}$

12. In each case, use number sense rather than equivalent fractions to order these fractions and decimals. Use only these symbols: $<$ and $=$. Explain your reasoning.

a. $\frac{2}{5}$ $\frac{2}{3}$ $\frac{40}{80}$ 0.60 $\frac{7}{10}$ b. $\frac{3}{15}$ $\frac{1}{10}$ 0.4   0.1 $\frac{1}{5}$ 0.2

13. Go to http://illuminations.nctm.org/, then to Activities. Place checkmarks in the box before 3–5 and another in the box 6–8 and click on Search. Scroll down to Equivalent Fractions and click to open. Click on the box to the left of Instructions. Do several sets of problems, as directed in the instructions, sometimes using circles, sometimes squares.

14. In the same activity given in Exercise 13, use the circle choice. Ignore the red circle. Use only the blue and green circles to find which is the larger. Use the $<$, $>$, and $=$ symbols as appropriate. You may need to use the number line below the circles.

a. $\frac{2}{3}$ $\frac{1}{2}$ b. $\frac{3}{5}$ $\frac{7}{11}$ c. $\frac{3}{10}$ $\frac{5}{12}$

d. $\frac{2}{5}$ $\frac{5}{16}$ e. $\frac{3}{4}$ $\frac{7}{11}$ f. $\frac{5}{14}$ $\frac{5}{16}$

Supplementary Learning Exercises for Section 6.2

1. What does it mean to say that three fractions are "equivalent"?

2. a. First show $\frac{2}{5}$ of the pentagonal region to the right.
   b. Then show that $\frac{2}{5}$ is equivalent to $\frac{4}{10}$ by adding marks to the work for part (a).
   c. How might one show that $\frac{6}{5} = \frac{12}{10}$, with the pentagonal region as the whole?

3. a. Write six fractions that are equivalent to $\frac{5}{6}$.
   b. Which, if any, of your fractions in part (a) is largest?

4. a. Show twelfths on the circular region to the right.
      (*Hint*: Think of a clock, with marks for 12, 3, 6, and 9 first.)
   b. How would you illustrate that $\frac{9}{12} = \frac{3}{4}$ using the circular region?

5. Write the following fractions in simplest form (or lowest terms):

a. $\frac{3 \times 43}{4 \times 43}$ b. $\frac{42}{64}$ c. $\frac{1540}{8800}$ d. $\frac{105}{140}$ e. $\frac{288}{1000}$

f. $\frac{xy}{xz}$ g. $\frac{2.4 \times 10^5}{3.6 \times 10^7}$ h. $\frac{a^2 b^6}{a^3 b^{12}}$ i. $\frac{75x}{90}$ j. $\frac{x^4(x + y^2)}{x^8(x + y^2)}$

6. a. Show and label the points for $\frac{5}{6}$ and $\frac{13}{6}$ on the number line.

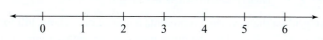

```
 ├───┼───┼───┼───┼───┼───┼───┼──
     0   1   2   3   4   5   6
```

b. Show $\frac{5}{6}$ of this 6-unit-long stick.

_____

c. Why are the results different in parts (a) and (b)?

7. Use your fraction sense or common denominators to find the larger in each pair. (Try fraction sense first.)

   a. $\frac{29}{65}$ and $\frac{163}{190}$ **b.** $\frac{29}{65}$ and $\frac{86}{195}$ **c.** $\frac{37}{72}$ and $\frac{44}{81}$ **d.** $\frac{55}{72}$ and $\frac{83}{108}$

8. A box contains only 12 black marbles and 3 white marbles. What fraction describes the part of the marbles that are black? How would you show, *using a diagram of marbles*, that $\frac{4}{5}$ also describes the part made up of black marbles?

9. Given a number line and only the points for the indicated fractions, accurately locate and label the point for the number in each part. If your markings do not make your reasoning clear, add an explanation.

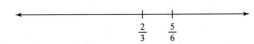

   a. $\frac{0}{6}$ **b.** $1\frac{1}{6}$ **c.** $\frac{1}{2}$

10. Locate and label the points for the given numbers, as in Supplementary Learning Exercise 9.

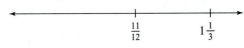

   a. $\frac{1}{2}$ **b.** $\frac{3}{4}$ **c.** $\frac{2}{3}$

6.3 Relating Fractions, Decimals, and Percents

When children in a research study[3] were asked to find the sum $5 + 0.5 + \frac{1}{2}$, many said it could not be done because they thought fractions and decimals could not be combined. But as you know, fractions and decimals (and percents) are very closely related even though they look different. All fractions can be expressed as decimals, using a division interpretation of a fraction, and many decimals can be expressed as fractions in a $\frac{\text{whole number}}{\text{nonzero whole number}}$ form. And both fractions and decimals can be represented as percents.

> **Think About …**
>
> In the last chapter you practiced estimating with percents. How are percents related to fractions and to decimals?

 ACTIVITY 6   Reviewing Something I Learned Long Ago

Write the following as percents:

   a. $\frac{3}{4}$ **b.** $\frac{1}{2}$ **c.** 0.37   **d.** 1.45   **e.** 0.028   **f.** 43.21

Write the following as decimals and as fractions:

   g. 32%   **h.** 123%   **i.** 0.01%   **j.** 43.2% ●

 ACTIVITY 7   Many Different Meanings

1. Each square grid on the next page represents a unit of one whole. Fill in the amount of each square grid indicated by the given number. For each, *explain and justify* how you know how much to fill in. Be sure you can say what the 4 represents.

a. 0.4

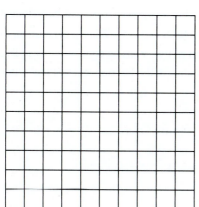

b. $\frac{1}{4}$

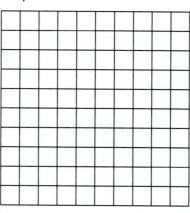

c. 0.45

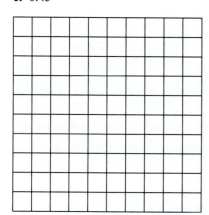

d. 4%

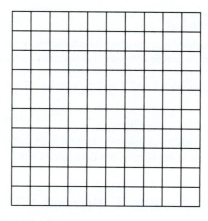

2. Shade 6 of the small squares in the rectangle shown below.

Using the diagram, explain how to determine each of the following:

a. the percent of the area that is shaded
b. the fractional part of the area that is shaded
c. the fractional part expressed as a decimal ●

 ACTIVITY 8   Can Every Fraction Be Represented as a Decimal?

Write an exact decimal equivalent for each fraction in (a–h). Be sure to discuss any that cause difficulty, and to reach consensus.

a. $\frac{2}{5}$ **b.** $\frac{2}{3}$ **c.** $\frac{2}{40}$ **d.** $\frac{1}{7}$

e. $\frac{23}{800}$ **f.** $\frac{16}{250}$ **g.** $\frac{3}{12}$ **h.** $\frac{24}{35}$

i. Which of these decimals terminate? What can be said about the denominators in each of the fractions? (*Hint*: Rewrite the denominators as products of 2s and 5s. If this cannot be done, see if you can first simplify the fraction.)

j. Which decimals did not terminate? Can you find a part of each of these decimals that repeats? ●

When the denominator of a fraction is a power of 10, changing the fraction to a decimal representation is easy: e.g., $\frac{3}{100} = 0.03$. Note that any power of 10 can be written as a product of 2s and 5s. For example, $100 = 2^2 \cdot 5^2$ and $1000 = 2^3 \cdot 5^3$. When a denominator has factors of only 2s and 5s, we can rewrite the fraction so that it has an equal number of 2s and 5s in the denominator. Doing so allows us to easily change the fraction to a decimal number.

EXAMPLE 1

$$\frac{7}{40} = \frac{7}{2^3 5} \times \frac{5^2}{5^2} = \frac{175}{1000} = 0.175 \; \bullet$$

EXAMPLE 2

$$\frac{42}{175} = \frac{7 \times 6}{7 \times 25} = \frac{6}{25} = \frac{6}{5^2} = \frac{6 \times 2^2}{5^2 \times 2^2} = \frac{24}{100} = 0.24 \; \bullet$$

Think About ...

$\frac{3}{12}$ yielded a terminating decimal representation, but 12 cannot be factored as a product of only 2s and 5s. Why did this happen? Why did $\frac{3}{12}$ give a terminating decimal?

From Examples 1 and 2 and the previous Think About, we can conclude the following:

> A fraction $\frac{a}{b}$ in simplest form can be represented with a terminating decimal when the denominator has only 2s and 5s as factors, because we can always find an equivalent fraction with a denominator that is a power of 10.

DISCUSSION 4   Remainders That Repeat

1. What are the possible remainders when you divide a whole number by 7? How many different possible remainders are there? Test your conclusion by calculating 1 divided by 7. When a remainder repeats, what happens when you continue the division calculation?

2. How many possible remainders are there when you divide a whole number by 11? Test your conclusion on $\frac{3}{11}$ and revise your conclusion if necessary. ●

Nonterminating, repeating decimals, like those you get for $\frac{1}{7}$ and $\frac{3}{11}$, are often abbreviated by putting a bar over the repeating part.

EXAMPLE 3

$4.333333. \ldots = 4.\overline{3}$ and $1.7245245245245. \ldots = 1.7\overline{245}$ ●

Think About ...

How many repeating digits *might* there be in the decimal equivalent for $\frac{n}{73}$, for different whole number values for n?

We now can conclude one more fact about a fraction in simplest form.

> In general, a fraction $\frac{a}{b}$ in simplest form can be represented with a nonterminating, repeating decimal if the denominator has factors other than 2s and 5s.

The reverse question can now be considered: *Can every decimal be represented as a fraction?*

✏️ ACTIVITY 9   From Decimals to Fractions

Problem 1 gives some practice in finding a fraction for a terminating decimal, and Problem 2 starts an examination of nonterminating, repeating decimals.

1. For each of the following terminating decimals, write the number as a fraction:

    a. 0.62   **b.** 1.25   **c.** 0.125   **d.** 0.3333

2. From the past (or division), you know that $\frac{1}{3} = 0.33333\ldots = 0.\overline{3}$.

    a. Use $0.33333\ldots = \frac{1}{3}$ to find a fraction for $0.11111\ldots$.

    b. Use part (a) to find fractions for $0.77777\ldots$ and $0.44444\ldots$.

    c. Now find fractions for $9.\overline{7}$ and $0.044444\ldots$. ●

Problem 2 in Activity 9 suggests how to find the fraction for any nonterminating, repeating decimal with a repeating block *one* digit long. But what if the repeating block has two or more digits, as in $0.\overline{85}$ or $6.3\overline{927}$? Fortunately, the idea from Activity 9 can be extended. To work with a decimal with a repeating block of *two* digits, notice the following:

$$0.111111\ldots = \frac{1}{9} = \frac{11}{99} = \frac{1}{99} + \frac{1}{99} + \frac{1}{99} + \frac{1}{99} + \frac{1}{99} + \frac{1}{99} + \frac{1}{99} + \frac{1}{99} + \frac{1}{99} + \frac{1}{99} + \frac{1}{99}$$

But this means that each $\frac{1}{99}$ must equal $0.01010101\ldots = 0.\overline{01}$. So, e.g., $0.\overline{85} = 85 \times 0.\overline{01} = 85 \times \frac{1}{99} = \frac{85}{99}$, and $0.\overline{13} = 13 \times 0.\overline{01} = \frac{13}{99}$. For a repeating block of *three* digits, the same sort of reasoning (or the division $1 \div 999$) shows that $0.\overline{001} = \frac{1}{999}$. This fact enables us to see, e.g., that $0.\overline{384} = 384 \times 0.\overline{001} = \frac{384}{999}$, and that $0.0\overline{492} = \frac{1}{10} \times 0.\overline{492} = \frac{1}{10} \times \frac{492}{999} = \frac{492}{9990}$.

> ### *Think About ...*
> $0.\overline{1} = \frac{1}{9}$, $0.\overline{01} = \frac{1}{99}$, and $0.\overline{001} = \frac{1}{999}$. If the pattern continues, what basic relationship would enable you to change a decimal such as $0.\overline{7265}$ to a fraction? What fraction would be equal to $0.\overline{7265}$?

✏️ ACTIVITY 10   Your Turn

Find fraction equivalents for

a. $0.\overline{8}$ **b.** $3.\overline{08}$ **c.** $3.0\overline{8}$ **d.** $0.7\overline{5}$ **e.** $0.\overline{328}$ ●

> ### *Think About ...*
> 1. What is $0.\overline{9}$? What is $0.\overline{3} + 0.\overline{3} + 0.\overline{3}$?
> 2. Write 0.24 as a fraction, and 0.240 as a fraction. Are the fractions equal? What does this result imply about placing zeros after a decimal number?

The decimals we have considered thus far all terminate or repeat. All can also be represented as fractions.

> Any fraction or its equivalent terminating or repeating decimal names a **rational number**. (Negative rational numbers are introduced in Chapter 10.)

The word "rational" comes from the word "ratio," which is one way of interpreting a fraction symbol.

Nonterminating, nonrepeating decimals cannot be represented as fractions with whole numbers as the numerator and as the denominator, so they are not rational numbers. They are called **irrational numbers**. For example, $\sqrt{3}$ is an irrational number. It cannot be written as a fraction in which the numerator and denominator are whole numbers, and its decimal is nonterminating and nonrepeating. The set of rational numbers together with the set of irrational numbers form the set of **real numbers**. All numbers used in this course are real numbers, and they are sufficient for all of elementary and middle school mathematics. At right is a Venn diagram showing how some sets of numbers are related.

Think About ...

$\sqrt{3}$ on a calculator may give 1.7320508. But 1.7320508 = $\frac{1732508}{10000000}$, so why is $\sqrt{3}$ said to be irrational?

TAKE-AWAY MESSAGE ...   All fractions of the form $\frac{a}{b}$ with whole numbers a and b where $b \neq 0$ can be written in decimal form; some terminate and some repeat. Vice versa, decimals that terminate or repeat can be written as fractions. These numbers are known as rational numbers.

    If a nonterminating decimal does not repeat, it names an irrational number. Together, the rational numbers and the irrational numbers make up the set of numbers we call real numbers.

Learning Exercises for Section 6.3

1. Determine which of the following fractions have repeating decimals and tell why. Then write each fraction as a terminating or as a repeating decimal.

   a. $\frac{3}{8}$ **b.** $2\frac{3}{10}$ **c.** $\frac{3}{7}$ **d.** $\frac{3}{11}$ **e.** $\frac{9}{24}$ **f.** $\frac{5}{6}$ **g.** Two answers are the same. Why?

2. Write the following decimals as fractions, if possible. Change to simplest form or lowest terms.

   a. 0.625   **b.** 0.49   **c.** 91.333...   **d.** 1.7

   e. $0.\overline{9}$ **f.** $0.9\overline{3}$ **g.** $0.\overline{53}$ **h.** $0.0\overline{53}$

   i. $4.\overline{76}$ **j.** $8.\overline{094}$ **k.** 0.12131415161718191101111121131114115...

   [There is a pattern in part (k). Assume that the pattern you see continues. Is there a repeating block of digits?]

3. **a.** Are there any fractions between $\frac{7}{15}$ and $\frac{8}{15}$? If so, how many are there?

   b. Are there any decimal numbers between $\frac{7}{15}$ and $\frac{8}{15}$?

4. Find decimal numbers between each pair of numbers:

   a. 0.4 and 0.5   **b.** 0.444 and 0.445

   c. 1.3567 and 1.35677777...

   d. 0.00000111111... and 0.000001100110011...

5. Place the following in order from smallest to largest without using a calculator:

   $$\frac{2}{3} \quad \frac{5}{6} \quad 0.\overline{56} \quad 1.23 \quad \frac{3}{17} \quad \frac{12}{15} \quad \frac{11}{29} \quad \frac{11}{24} \quad 0.\overline{26} \quad \frac{1}{4} \quad 0.21$$

6. Explain how you know there is an error, without calculating:

   $$\frac{\text{some whole number}}{7} = 4.29\overline{23456789}$$

7. One type of calculator can show only 8 digits. Suppose the calculator answer to a calculation is 8.1249732. Explain how the exact answer could be

   a. a terminating decimal   **b.** a repeating decimal   **c.** an irrational number

8. Draw lines to show what value these decimal numbers and percents are close to. (Don't use a calculator.)

   0.29   0.52   94%   32%   0.78   0.49   24%   5%   70%

9. Complete the following table:

Decimal form	Fraction	Percent
0.48	$\frac{48}{100}$	
	$\frac{4}{5}$	
		457%
		0.1%

10. Chris says, "0.3 is bigger than 0.65 because tenths are bigger than hundredths."

    Sam says, "0.28 is bigger than 0.3 because 28 is bigger than 3."

    Lars says, "0.72 is bigger than 0.6 because 0.72 has seven tenths and 0.6 has only six tenths and the hundredths don't matter here."

    a. Discuss these answers. Which is correct? Would you have made any of these errors?
    b. Order these from smallest to largest: 0.7   0.645   0.64.
    c. Here are the ways that Chris, Sam, and Lars ordered these numbers from smallest to largest. Match the name with the order. Use the reasoning above to figure out which name goes with which ordering.

    _____ 0.7   0.64   0.645

    _____ 0.645   0.64   0.7

    _____ 0.64   0.645   0.7

11. A student says, "My calculator shows 0.0769231 for $\frac{1}{13}$ when I divide 1 by 13." Does this mean that $\frac{1}{13}$ is equivalent to a terminating decimal? Explain.

12. A student says, "My calculator shows that $\sqrt{2}$ is 1.4142136. And 1.4142136 = $\frac{14,142,136}{10,000,000}$, which is a rational number. But you said $\sqrt{2}$ is an irrational number.

    Which is it, rational or irrational?" How would you respond to the student?

13. Go to http://illuminations.nctm.org/, then to Activities. Check 3–5 and 6–8, then Search. Scroll down to Fraction Models and click to open. Read Instructions. Use this applet to complete the following table:

Fractions	Decimal	Percent
	0.25	
$\frac{11}{10}$		
	3.1	
$\frac{13}{20}$		
		250%
$\frac{17}{4}$		
	1.8	
$\frac{14}{5}$		
	3.4	
		30%

14. Complete the following table:

Fraction	Decimal	Percent
		500%
$\frac{50}{54}$		
$\frac{95}{67}$		
	1.5	
	0.04	
		73%
		235%
$\frac{1}{3}$		

Supplementary Learning Exercises for Section 6.3

1. Which of the following numbers will have terminating decimals? Tell how you know without extensive calculation.

    a. $\frac{699}{300}$ **b.** $\frac{1197}{5^2 \times 2^5}$ **c.** $\frac{231}{625}$ **d.** $\frac{985}{1024}$ **e.** $\frac{531 \times 293}{5 \times 10^8}$

2. Write these decimals as (exact) fractions of the form $\frac{\text{whole number}}{\text{nonzero whole number}}$.

    a. 0.3333   **b.** 8.374   **c.** $0.\overline{7}$ **d.** $0.\overline{63}$

    e. $0.6\overline{32}$ **f.** $4.3\overline{57}$ **g.** $17.19\overline{2}$ **h.** $2.3\overline{015}$

3. In terms of decimals, what is a rational number? An irrational number? A real number?

4. Are whole numbers rational or irrational? Explain.

5. "$\sqrt{5}$ is irrational." What does that (true) statement tell you?

6. "π is $\frac{22}{7}$. And $\frac{22}{7} = 3.\overline{142857}$. Why do you say that π is irrational?" What would be your response?

7. Give two decimals between each pair, if it is possible. If it is not possible, explain why.

    a. 0.98 and 0.99   **b.** $1.\overline{04}$ and 1.04   **c.** 0.4999 and 0.5

8. The decimal 3.1416 is sometimes used for the number π, which is an irrational number. What does that tell you about $\pi = 3.1416$?

9. Sometimes memory fails you. But often you can regenerate facts, using modest mental work to build on facts you do remember (and without a calculator). For example, writing $\frac{5}{8}$ as a decimal can be determined by noting that $\frac{1}{8}$ is half of $\frac{1}{4}$ and $\frac{1}{4} = 0.25$. So $\frac{1}{8}$ = half of 0.25, or 0.125. Then $\frac{5}{8} = 5 \times 0.125 = 0.625$. Show how the decimals and percents for these could be determined.

    a. $\frac{1}{16}$ **b.** $\frac{3}{8}$ **c.** $\frac{1}{6}$ **d.** $\frac{1}{12}$ **e.** $\frac{11}{12}$

6.4 Estimating Fractional Values

In Chapter 5 you practiced estimating with whole numbers and percents. An understanding of fractions makes estimating with fractions also quite easy. Using numerical benchmarks with fractions can simplify estimating computational results. Recognizing errors when computing with fractions is also a by-product of knowing benchmarks for fractional values.

ACTIVITY 11 Happy Homeowner?

Consider the following situation:

▶ You have heard that housing costs should be about $\frac{1}{3}$ of your income. Your income is $3500 per month, and you currently spend $965 per month in housing payments. ◀

a. Name as many quantities as you can that are involved in this situation.

b. Which quantity in this situation has a value of $\frac{1}{3}$?

c. Which quantity in this situation has a value of $\frac{965}{3500}$? Is this fraction close to a familiar fraction?

d. Can you tell whether you are paying more or less for housing than the guidelines suggest without a calculator or lengthy computations? ●

Just as benchmarks can be useful in getting a feel for large values of quantities, they can also be helpful in dealing with small numbers. Whenever we have a quantity whose value is expressed as a fraction, it may be useful to compare this fractional value to a more familiar value, such as 0, $\frac{1}{3}$, $\frac{1}{2}$, $\frac{2}{3}$, or 1. In Activity 11, for example, being able to tell quickly whether the fraction is more or less than $\frac{1}{3}$ allows us to gain an understanding of the situation without extensive calculations. We were able to make this decision based on the fact that $\frac{965}{3500}$ is close to $\frac{1000}{3500}$, which is compared to $\frac{1}{3}$. (Note that the denominator would need to be 3000 for the fraction to be equivalent to $\frac{1}{3}$.) Making these types of comparisons can give us a feel for the relative size or magnitude of less familiar fractions. Developing the ability to estimate with fractions is the focus of this section.

 DISCUSSION 5 Formalizing "Close To"

1. Describe ways of telling (by simple inspection and without using a calculator) when a fraction has a value close to 0. Is $\frac{2}{3}$ close to 0? How about $\frac{2}{5}$? How about $\frac{2}{9}$? How about $\frac{2}{18}$?

2. Describe ways of telling when a fraction has a value close to 1. Is $\frac{7}{16}$ close to 1? How about $\frac{9}{18}$? How about $\frac{17}{18}$? How about $\frac{97}{108}$?

3. Describe ways of telling when a fraction has a value close to $\frac{1}{2}$. Is $\frac{4}{8}$ close to $\frac{1}{2}$? Is $\frac{5}{8}$ close to $\frac{1}{2}$? How about $\frac{9}{16}$? How about $\frac{7}{16}$?

4. Describe ways of telling when a fraction is close to $\frac{1}{3}$.

5. Describe ways of telling when a number is larger than 1; smaller than 1.

6. Describe ways of telling when a number is a little more than $\frac{1}{2}$; a little less than $\frac{1}{2}$. ●

 ACTIVITY 12 Number Neighbors

For each of the following fractions, indicate whether it is closest to 0, $\frac{1}{2}$, 1, or whether it really is not close to any of these three numbers. Think of each fraction as a point on the number line.

1. $\frac{3}{8}$ **2.** $\frac{5}{4}$ **3.** $\frac{2}{9}$ **4.** $\frac{4}{7}$ **5.** $\frac{1}{3}$ **6.** $\frac{1}{6}$ **7.** $\frac{8}{9}$ **8.** $\frac{9}{8}$

9. $\frac{4}{9}$ **10.** $\frac{5}{6}$ **11.** $\frac{21}{45}$ **12.** $\frac{13}{15}$ **13.** $\frac{34}{64}$ **14.** $\frac{13}{84}$ **15.** $\frac{3}{7}$ ●

In Activity 12, a fraction was close to 1 when its numerator and denominator were close. For example, 13 is close to 15, so $\frac{13}{15}$ is close to 1. This "rule" works when the numbers in the numerator and denominator are fairly large. For example, 13 and 15 are only two apart, as are 2 and 4. But to say that $\frac{2}{4}$ is close to 1 is not accurate because $\frac{2}{4}$ is the same as $\frac{1}{2}$.

Which number is closer to 0, $\frac{1}{8}$ or $\frac{1}{9}$? $\frac{1}{9}$ is the smaller of the two (i.e., if something is cut into 9 equal pieces, one of them is smaller than one of 8 equal pieces of the same quantity). So, we know that $\frac{1}{9}$ is closer to 0.

Notes

✏️ ACTIVITY 13   Reasoning with Fractions

1. You now know that $\frac{1}{8}$ is larger than $\frac{1}{9}$. How can you use this information to tell which fractional value is closer to 1: $\frac{7}{8}$ or $\frac{8}{9}$? Explain your reasoning (without using decimals).

2. For each given pair of numbers, tell which fractional value is closer to $\frac{1}{2}$. Explain your reasoning.

    a. $\frac{11}{25}$ or $\frac{12}{25}$ **b.** $\frac{5}{8}$ or $\frac{9}{20}$ **c.** $\frac{12}{25}$ or $\frac{7}{15}$

    d. $\frac{11}{24}$ or $\frac{9}{20}$ **e.** $\frac{7}{15}$ or $\frac{13}{25}$ ●

The knowledge gained in Activities 12 and 13 can be useful when estimating with fractions. For example, $\frac{15}{16} + \frac{1}{9}$ would be close to $1 + 0$, or simply 1.

> **Think About ...**
> Find an estimate for $3\frac{4}{9} - 2\frac{15}{16}$.

TAKE-AWAY MESSAGE ... Having rules of thumb about fractions close to, larger than, and smaller than 0, $\frac{1}{2}$, or 1 can be very useful in making estimates and checking whether an answer is reasonable. Thinking about rules of thumb for $\frac{1}{3}$, $\frac{2}{3}$, and $\frac{1}{4}$ comes up in the exercises.

Learning Exercises for Section 6.4

1. A sign says that all merchandise is being sold for a fraction of its original price. Your friend cynically says, "Yeah, $\frac{5}{4}$ of the original price." Why is this a cynical remark?

2. A student worked the problem: "The sewing pattern called for $\frac{7}{12}$ yard of silk and $\frac{5}{8}$ yard of linen. How much fabric was in the blouse?" He wrote $\frac{7}{12} + \frac{5}{8} = \frac{23}{24}$. Is the result of the calculation sensible? How can you tell by simple inspection, without actually performing the computation?

3. Is it possible for a fraction to have a value less than 0? Explain.

4. Draw lines from the fractions to the baskets. Compare them with a classmate's when you have the opportunity. Discuss any differences. What did you do when a number could go in more than one basket?

    $\frac{4}{7}$ $\frac{2}{9}$ $\frac{11}{12}$ $\frac{99}{152}$ $\frac{17}{35}$ $\frac{15}{34}$ $\frac{11}{108}$ $\frac{3}{12}$ $\frac{9}{8}$

    | Close to 0 | Close to $\frac{1}{2}$ | Close to 1 |

5. When is a fraction close to $\frac{1}{3}$? When is a fraction close to $\frac{2}{3}$? After you have answered these questions, draw lines from fractions to the appropriate baskets.

    $\frac{4}{9}$ $\frac{21}{29}$ $\frac{49}{99}$ $\frac{15}{47}$ $\frac{27}{52}$ $\frac{42}{63}$ $\frac{31}{45}$ $\frac{22}{47}$ $\frac{99}{152}$ $\frac{2}{7}$

    | Close to $\frac{1}{3}$ | Close to, but less than $\frac{1}{2}$ | Close to, but more than $\frac{1}{2}$ | Close to $\frac{2}{3}$ |

6. In each case, tell which quantity is larger by thinking about how close or far each fraction is from 0, 1, or $\frac{1}{2}$.

    a. $\frac{3}{4}$ or $\frac{4}{9}$ of the class?   **b.** $\frac{1}{3}$ or $\frac{1}{4}$ of a cup?

    c. $\frac{3}{4}$ or $\frac{9}{10}$ of a meter?   **d.** $\frac{2}{5}$ or $\frac{3}{6}$ of those voting?

    e. $\frac{13}{24}$ or $\frac{11}{20}$ of the money budgeted?   **f.** $\frac{11}{25}$ or $\frac{17}{15}$ of the distance?

7. How can you tell when a fraction has a value close to $\frac{1}{4}$?

8. Find a fraction *between* each pair of fractional values. Assume that the two fractions refer to the same unit. Justify your reasoning. (Do not change to decimal form.)

    a. $\frac{2}{5}$ and $\frac{3}{5}$ b. $\frac{8}{15}$ and $\frac{9}{12}$ c. $\frac{7}{8}$ and $\frac{9}{10}$

    d. $\frac{17}{18}$ and $\frac{17}{16}$ e. $\frac{1}{4}$ and $\frac{1}{3}$

9. A child claims that $\frac{4}{5}$ and $\frac{5}{6}$ are the same because, "They are both missing one piece." What would you do to help this child come to understand these fractions better?

10. There are some commonly used fractions for which you should know the decimal equivalent without needing to divide to find it. And, of course, if you know decimal equivalents, it is quite easy to find percent equivalents. For each of the following, give the decimal and percent equivalents to all the ones you know. Find the others by dividing, and then memorize these, too.

    a. $\frac{1}{8}$ b. $\frac{1}{5}$ c. $\frac{1}{4}$ d. $\frac{1}{3}$

    e. $\frac{2}{5}$ f. $\frac{3}{8}$ g. $\frac{5}{4}$ h. $\frac{1}{2}$

    i. $\frac{3}{5}$ j. $\frac{5}{8}$ k. $\frac{2}{3}$ l. $\frac{3}{4}$

    m. $\frac{4}{5}$ n. $\frac{7}{8}$ o. 1   p. $\frac{7}{4}$

11. Name a fraction or whole number that is *close to* being equivalent to each percent.

    a. 24%   b. 95%   c. 102%

    d. 35%   e. 465%   f. 12%

12. In Chapter 5 you learned ways of finding fractional equivalents for percents. For example, $10\% = \frac{1}{10}$, $25\% = \frac{1}{4}$, and $50\% = \frac{1}{2}$. Use this knowledge to *estimate* each of the following amounts:

    a. 15% of 798   b. 90% of 152   c. 128% of 56

    d. 65% of 24   e. 59% of 720   f. 9% of 59

    g. 148% of 32   h. 0.1% of 24   i. 32% of 69

13. Draw a diagram that shows why $\frac{4}{5}$ of a given amount is the same as $\frac{8}{10}$ of the same amount.

14. Draw a diagram that shows the sum $\frac{4}{5} + \frac{3}{10}$.

15. An anonymous donor has offered to pledge $1 to a local charity for every $3 that is pledged by the general public. Draw a picture that represents the public's donation, the matching pledge by the anonymous donor, and the total amount donated to the charity. What fractional part of the total amount donated is pledged by the general public?

16. A child claims that $\frac{1}{4}$ can be greater than $\frac{1}{2}$, and draws the picture shown to support her argument. How would you respond to this child's claim? Can you adapt this picture to explain why $\frac{1}{2}$ is larger than $\frac{1}{4}$?

Cake

Cookie

17. Estimate answers for each of the following fraction computations. Can you say whether your estimate is larger than or smaller than the exact answer would be? Remember what you learned earlier about number neighbors.

    a. $\frac{5}{8} + \frac{9}{10}$ b. $1\frac{7}{16} + 4\frac{3}{27} + 7$ c. $14\frac{9}{10} - 5\frac{1}{16}$

    d. $4\frac{11}{12} - 2\frac{15}{16}$ e. $\frac{20}{31} - \frac{1}{3}$ f. $4\frac{7}{16} + 5\frac{11}{12} - 2\frac{1}{8}$

18. a. *About* what *percent* of the circle at the right is each piece?

A _____ B _____ C _____ D _____ E _____

What percent is the sum of your estimates?

Is this reasonable?

b. *About* what *fraction* of the circle is each piece?

A _____ B _____ C _____ D _____ E _____

What should the sum of these fractional parts be?

Supplementary Learning Exercises for Section 6.4

1. How do you know that something is incorrect in $\frac{8}{15} + \frac{6}{11} + \frac{9}{17} = 1\frac{179}{2805}$?

2. From memory, give familiar fractions that are roughly equal to the given decimals.

   a. 0.48176   **b.** 0.3411   **c.** 0.74622   **d.** 0.655   **e.** 0.67

3. Estimate each of the following. Indicate whether you think your estimate is too large or too small.

   a. $2\frac{7}{16} \times \$51.53$ **b.** $3\frac{1}{10} + 7\frac{8}{9} + 2\frac{2}{3} + 4\frac{5}{16}$ **c.** $13\frac{7}{16} \div \frac{8}{15}$

4. Your waitress was a friend of yours, so you left about a 20% tip ($3.00). What was your approximate bill before the tip?

5. With the denominator given, write one fraction that is close to, but larger than, the given fraction, and one fraction that is close to, but smaller than, the given fraction.

   a. $\frac{1}{2}$, denominator 70   **b.** $\frac{1}{3}$, denominator 60   **c.** $\frac{3}{4}$, denominator 160

6. **a.** The accompanying diagram is split into several regions. Each region is about what percent of the entire figure?

   A _____ B _____ C _____ D _____ E _____ F _____

   b. Each region is about what fraction of the entire figure?

   A _____ B _____ C _____ D _____ E _____ F _____

7. **News Report** "The water level in the river is 3.8 feet above flood stage. We expect the water level to decrease about a foot by tomorrow." The anticipated decrease will be about what percent of flood water?

8. Sketch a line segment about 68% longer than this one:

   _____

9. In an election with only two candidates, A and B, A won by 18%. What percent of the whole vote did each A and B receive?

6.5 Issues for Learning: Understanding Fractions and Decimals

Even young children in the early grades have some understanding of parts and wholes, but they do not always relate this to the fraction symbol. There are several critical ideas[4] that children need to understand before they can successfully operate on fractions. They include the following:

1. When a whole (in this case a continuous whole) is broken into equal parts, then the more parts there are, the smaller each part will be. Thus, $\frac{1}{6}$ is smaller than $\frac{1}{5}$. Even though children might know this in a particular setting such as pieces of pizza, they still need to learn this symbolically.

2. Fractions represent specific numbers. That is, $\frac{3}{5}$ itself is the value of a quantity; $\frac{3}{5}$ is *one* number, not *two* numbers separated by a bar. When students do not understand this relationship between the numerator and the denominator, they operate on them separately. They might say, e.g., that $\frac{3}{5} + \frac{2}{3} = \frac{5}{8}$.

3. Equivalence is one of the most crucial ideas students must understand before operating on fractions and decimals. Children will not be able to add or subtract fractions until they can identify and generate equivalent fractions, because they often need to find an equivalent form of a fraction before they can add or subtract.

4. Children must understand why *like denominators* are necessary before one can add or subtract fractions. Only when there is a common denominator for two fractions are they adding or subtracting like pieces, because now both fractions refer to a common size for the pieces.

There are similar problems working with decimal notation. Students may not understand what the decimal point indicates. If they estimate the value of 48.85 as, e.g., 50.9, then they are treating the 48 separately from the .85.

The work of Chris, Sam, and Lars shown in the Learning Exercises for Section 6.3 illustrates some common misunderstandings of decimals that have been found in a number of countries. Some children think that because hundredths are smaller than tenths, 2.34, which has hundredths, is smaller than 2.3, which has tenths. Others believe that the number of digits indicates relative size, as is true for whole numbers. Thus, 2.34 is considered larger than 2.4 because 234 is larger than 24.

Teachers often have students compare numbers such as 2.34 and 2.3 by "adding zeros until each number has the same number of places." Now the comparison is between 2.34 and 2.30. Although this technique works, often it is not meaningful and becomes just another trick of the trade, unless children are also taught that 2.30 and 2.3 are equal; that if 2.30 and 2.3 are represented by base-ten blocks, they would be represented identically. Another approach is to have students express 0.3 and 0.30 as fractions in simplest form to show that they are equivalent.

6.6 Check Yourself

This chapter focused on understanding the meaning of the fraction symbol, particularly the part-whole meaning and the division meaning. You should be able to clearly distinguish between these two meanings by drawings that would indicate how they would be illustrated differently, but still represent the same amount. Some attention is also given to decimals and percents.

You should be able to work problems like those assigned and to meet the following objectives:

1. Distinguish between, and identify, discrete and continuous units used for a fraction representation.

2. Generate sketches that illustrate a given fraction or mixed number, and identify fractions or mixed numbers shown in any drawing, whether the number is less than 1 or greater than 1.

3. Given a part of a whole and the fraction this represents, find the whole.

4. Generate drawings that show why two equivalent fractions represent the same amount.

5. Change fractions to decimal form and percent form and change repeating and terminating decimals to fractions and to percents.

6. Define *rational number* and *irrational number* and be able to recognize each.

7. Describe the numbers that are real numbers.

continue

Notes

8. Be able to order a set of fractions, decimal numbers, and percents from smallest to largest, without (in most cases) resorting to changing all the numbers to decimal numbers or using equivalent fractions.

9. Identify fractions that are close to 0; fractions that are close to, smaller than, or larger than $\frac{1}{2}$; and fractions that are close to, smaller than, or larger than 1. You should be able to explain your reasons for these identifications.

10. Discuss fundamental ideas about fractions that are necessary for addition and subtraction of fractions.

References for Chapter 6

[1]Armstrong, B. E., & Larson, C. N. (1995). Students' use of part-whole and direct-comparison strategies for comparing partitioned rectangles. *Journal for Research in Mathematics Education, 26*, 2–19.

[2]Kieren, T. Problems shared with the authors.

[3]Sowder, J., & Markovits, Z. (1991). Students' understanding of the relationship between fractions and decimals. *Focus on Learning Problems in Mathematics, 13*, 3–11.

[4]Mack, N. K. (1995). Critical ideas, informal knowledge, and understanding fractions. In J. T. Sowder & B. P. Schappelle (Eds.), *Providing a foundation for teaching mathematics in the middle grades* (pp. 67–84). Albany, NY: SUNY Press.

Chapter 7

Computing with Fractions

Why do you need a common denominator to add or subtract fractions, but not for multiplying and dividing them? Why do you end up *multiplying* to divide by a fraction? If mathematics is to make sense, questions like these must have answers. This chapter reviews the step-by-step processes—the algorithms—used in computing with fractions, with a focus on making sense of these algorithms.

7.1 Adding and Subtracting Fractions

In the last chapter you had the opportunity to develop a good understanding of the nature of fractions. In this chapter you will study the ways in which we operate with fractions and why these ways work. We begin here with addition and subtraction of fractions.

> ### Think About ...
> Which of these two calculations is easier? Why?
>
> $$\frac{79}{144} + \frac{35}{144} - \frac{13}{144} \quad \text{or} \quad \frac{3}{8} + \frac{1}{6} + \frac{2}{15}$$

No doubt you recognized that the first set of calculations, despite the "ugly" fractions, is easier than the second because the denominators are all the same. Why does that make the calculations easier? Under a part-whole view of fractions, having the same denominator (or *common denominator*) means that if the unit is cut into that number of equally sized pieces, then it is easy to add or subtract the numerators for every fraction in order to count the number of equally sized pieces in the result. In effect, the first calculation involves only $79 + 35 - 13$, with the result telling the number of $\frac{1}{144}$ pieces.

   In the second set of calculations, however, the different denominators mean that the pieces involved are (probably) different sizes, so it is *not* just a matter of counting pieces. In the diagram shown, the pieces are of different sizes, and thus they cannot just be counted to give a sum. Even though it is easy to *see* the total amount for the sum in the drawing, naming that quantity is a puzzle. That is why having a common denominator is a first step in calculating sums and differences of fractions with paper and pencil. Each fraction is replaced with an equal fraction having the common denominator, which gives all underlying pieces the same size.

   The procedure for finding a common denominator is: Find the product of all the given denominators (this product gives a **multiple** of each denominator). For $\frac{3}{8} + \frac{1}{6} + \frac{2}{15}$, $8 \times 6 \times 15 = 720$, so 720 could be a common denominator. We would get

$$\frac{3}{8} + \frac{1}{6} + \frac{2}{15} = \frac{270}{720} + \frac{120}{720} + \frac{96}{720} = \frac{270 + 120 + 96}{720} = \frac{486}{720} = \cdots = \frac{27}{40}$$

The drawback of this method for finding a common denominator is that you usually do more calculation than necessary because smaller denominators would also work here: 120, 240, 360, 480, 600. The smallest denominator that works here is 120.

Notes

A number is the **least common multiple (LCM or lcm)** of a set of numbers if it is a common multiple of each of the numbers (i.e., each number in the set is a factor) and it is the smallest number greater than zero with this property. If the set of numbers are all denominators of fractions, then the LCM is often called the **least common denominator (LCD or lcd)**.

EXAMPLE

0, 30, 60, 90, 120 . . . are all common multiples of 6 and 15. But the smallest nonzero multiple is 30, so 30 is the LCM. ●

Try computing $\frac{3}{8} + \frac{1}{6} + \frac{2}{15}$ with the least common denominator, 120, and compare the difficulty of calculation to that with the 720 denominator.

Finding the least common denominator is not a trivial task; one way is mentioned in the exercises if you do not remember any method. More attention is paid to finding the LCD in Chapter 11.

 DISCUSSION 1   Algorithms for Adding and Subtracting Fractions

Algorithms for adding and subtracting fractions are sometimes expressed symbolically as follows: $\frac{a}{b} + \frac{c}{d} = \frac{ad + bc}{bd}$, and $\frac{a}{b} - \frac{c}{d} = \frac{ad - bc}{bd}$. Do they make sense? Apply the first rule to $\frac{3}{8} + \frac{1}{4}$. Is there an easier way to add these fractions than using the rule? ●

Once all fractions in a set have common denominators, you can perform the operations of addition and subtraction on the numerators to find the answer. As you may recall from the previous chapter, these fractions now describe pieces of the same size, which means you are adding and subtracting like parts.

ACTIVITY 1   Can You Picture This?

1. Describe how the following drawing can be used to show $\frac{1}{2} + \frac{1}{3}$:

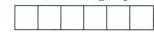

2. Describe how the following drawing can be used to show $\frac{1}{2} - \frac{1}{3}$:

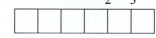

3. Use the number line model to show

$\frac{1}{2} + \frac{1}{3}$:

4. Use the number line model to show

$\frac{1}{2} - \frac{1}{3}$: ●

 DISCUSSION 2   Properties of Addition with Fractions

1. Does commutativity of addition hold for fractions? Explain.
2. Does associativity of addition hold for fractions? Explain. ●

DISCUSSION 3   Comparing Fractions Additively

Comparing fractional values additively calls for subtraction. That is, to find how much more $\frac{7}{8}$ is than $\frac{1}{2}$, you can subtract. What other situations involving fractional values would call for subtraction? Give some examples of story problems that involve fractions or mixed numbers and could be answered by subtracting. ●

TAKE-AWAY MESSAGE . . .   Addition and subtraction of fractions require understanding that a fraction names a number, a value for a quantity; that fractions can easily be added and subtracted when they refer to a common subunit, which requires that they have a common denominator; and that changing the representation of a fraction to one with the common denominator requires understanding of equivalent fractions.

Learning Exercises for Section 7.1

1. Explain your method for finding a common denominator for $\frac{3}{8} + \frac{1}{6} + \frac{2}{15}$.

2. Embedding calculations in contexts like story problems is often helpful in keeping track of the unit. Compare these two problems. How do you think the misunderstanding in part (a) arises?

   a. A student thinks that $\frac{3}{4} + \frac{2}{4} = \frac{5}{8}$ because of his drawing:

   b. In a cake-eating contest, Ann ate $\frac{3}{4}$ of a cake and Bea ate $\frac{2}{4}$ of a cake. How much cake did they consume in all?

3. Using a rectangular region as the unit, illustrate each of the following:

   a. $\frac{2}{5} + \frac{1}{2}$ b. $\frac{16}{20} - \frac{2}{5}$ c. $\frac{2}{3} - \frac{1}{2}$

4. (Use pattern blocks for this exercise. Pattern blocks, if not available in wood, can be cut out from the "Masters" section.) Choose a unit and do each of the following. Write numerical expressions for your work with the pattern blocks. Some can be done in more than one way. Record your work in sketches.

   a. Combine $\frac{1}{2}$ and $\frac{1}{6}$.   c. How much more is $\frac{5}{6}$ than $\frac{1}{2}$?

   b. Combine $2\frac{1}{2}$ and $\frac{2}{3}$.   d. Take $1\frac{5}{6}$ from $2\frac{2}{3}$.

   e. Take $2\frac{1}{4}$ from 4. (*Hint:* Let 2 yellow hexagons = 1 unit.)

   f. What needs to be combined with $\frac{5}{6}$ to get $2\frac{2}{3}$?

5. Eight loaves of bread are to be shared equally among 10 men. How might this be done? (This problem is from the Rhind papyrus, 1700 BC.[1]) The following diagram suggests one way:

   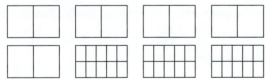

   Each man gets $\frac{1}{2} + \frac{3}{10}$.

   a. What part of a loaf does each man get in all?
   b. How else might the loaves be shared?
   c. Does each person get the same total amount in each situation? Justify that each person gets the same amount in each of your cuttings.

6. Solve this problem (written by a fifth-grader):

   ▶ Four guys, Chico, Roberto, Bob, and Phil, eat 9 bags of corn chips. They each ate their equal shares. Later Johnny came with 11 bags and helped eat them. How many bags did Johnny eat? How many bags did Chico eat altogether? ◀

7. Jim ate $\frac{1}{3}$ of a whole pizza at lunch and $\frac{1}{6}$ of the pizza later for a snack. How much of the pizza did Jim eat? Show this with a drawing.

8. In his will Thomas gave $\frac{1}{3}$ of his estate to his daughter Jenny, $\frac{2}{5}$ of his estate to his son Will, $\frac{1}{6}$ of his estate to his granddaughter Karen, and the remaining amount to the American Heart Association. What part of Thomas's estate did the American Heart Association receive?

9. A wealthy woman's will calls for $\frac{2}{3}$ of her estate to go to her favorite charity and for each of her three great-nieces to receive $\frac{1}{16}$ of her estate. What fraction of her estate is left for other beneficiaries?

10. A cable installer working in an isolated area has 200 feet of cable left. One of his remaining jobs will take $29\frac{1}{2}$ feet of cable, a second will take $42\frac{3}{4}$ feet, and a third, 118 feet. Will he be able to do the three jobs without going to headquarters to get more cable? If he can, will he have enough cable left for a job that will require $24\frac{5}{8}$ feet? If he cannot, how many feet will he be short?

11. You have a recipe that gives $1\frac{3}{4}$ quarts of punch, and an insulated jug that holds $1\frac{1}{4}$ gallons. If you make 3 recipes of the punch, will it all fit in the jug? If not, how much punch will be left over? (1 gallon = 4 quarts.)

12. A youngster mows his neighbors' yards. His lawnmower tank holds $2\frac{1}{3}$ gallons of gasoline mixture. Two of the yards take $\frac{5}{8}$ of a gallon of gasoline mixture each, a third takes $\frac{1}{2}$ gallon, and a fourth takes $\frac{2}{3}$ of a gallon. Can he mow all 4 yards on one tank of gasoline mixture? If he can, how much gasoline mixture will be left? If not, how much more will he need?

13. Arnie is $5\frac{1}{4}$ inches taller than Ben, Ben is $3\frac{7}{8}$ inches shorter than Carlo, and Carlo is $2\frac{1}{4}$ inches shorter than Donell. Who is taller, Arnie or Donell, and by how much?

14. Make up a story problem that might lead to each of these.

    a. $\frac{2}{3} - \frac{1}{2}$ **b.** $12 - 7\frac{3}{4}$ **c.** $2\frac{5}{8} + n = 6$

15. Describe or show a strategy you could use to compute each of these mentally.

    a. $\frac{3}{4} + \frac{5}{6} + \frac{1}{4}$ **b.** $\frac{3}{8} + \frac{1}{2}$ **c.** $6\frac{3}{4} - 3\frac{1}{2} - 1\frac{1}{2}$ **d.** $2\frac{2}{3} + 5\frac{1}{6} + 3\frac{1}{6} - 4$

    e. $7 - 2\frac{3}{16}$ **f.** $\frac{7}{16} + \frac{0}{72}$ **g.** $\frac{0}{42} + \frac{20}{63}$

16. Here is one technique for finding the least common denominator for a given addition/subtraction calculation. Determine whether the largest denominator would work as a common denominator; if it does not, then try multiples of the largest denominator (2 ×, 3 ×, etc.). For the calculation in Exercise 1, $\frac{3}{8} + \frac{1}{6} + \frac{2}{15}$, try 15 (won't work because 8 is not a factor of 15), then 30 (won't), 45 (won't), 60 (won't), . . . , 120 (will! $120 = 8 \times 15$, and $120 = 6 \times 20$, so fractions equal to the given ones can be found easily.) Practice this technique:

    a. $\frac{17}{40} + \frac{19}{8} - \frac{1}{5}$ **b.** $\frac{25}{36} + \frac{62}{135}$ **c.** $\frac{3}{4} + 0.7 - \frac{17}{25}$

17. Give an example that illustrates each statement.

    a. Subtraction of fractions is not commutative.
    b. Subtraction of fractions is not associative.

18. Jenny and Leslie made chocolate milk. Jenny used $\frac{1}{3}$ glass of chocolate syrup. Leslie, whose glass is twice as large as Jenny's, used $\frac{1}{4}$ glass of syrup. They decide to combine their drinks into a larger pitcher. What part of the combined mixture would be syrup?[2] (*Hint*: Notice that there are three different wholes involved.)

19. Given a number line and only the points for the indicated fractions, accurately locate and label the point for the number in each part. If your markings do not make your reasoning clear, add an explanation.

    a. 1   **b.** $\frac{1}{2}$ **c.** $\frac{11}{6}$

20. Locate and label the points for the given numbers, as in Learning Exercise 19.

    a. 12   **b.** 13   **c.** $10\frac{1}{2}$

21. Locate and label the points for the given numbers, as in Learning Exercise 19.

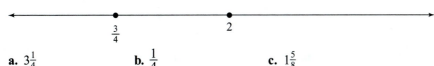

  a. $3\frac{1}{4}$ **b.** $\frac{1}{4}$ **c.** $1\frac{5}{8}$

22. Locate and label the points for the given numbers, as in Learning Exercise 19.

  a. $1\frac{3}{4}$ **b.** $2\frac{1}{4}$ **c.** $1\frac{5}{8}$

Supplementary Learning Exercises for Section 7.1

1. If you have an 8-foot-long piece of string and cut off and use a piece $4\frac{3}{4}$ inches long and a piece $2\frac{3}{8}$ inches long, how much string do you still have?

2. **a.** What is the perimeter (the distance around) of the rectangular field to the right?

  b. If the lengths of the sides of the field are tripled, what is the new perimeter?

    $5\frac{1}{8}$ miles

    $3\frac{3}{4}$ miles

3. You have a full 2-gallon container of a fertilizer mix for your flowers. You use half a gallon for one watering, $\frac{2}{3}$ gallon for another, and $\frac{1}{4}$ gallon for a third. How much mix do you have left?

4. Write a story problem that can be answered by $\frac{7}{8} - \frac{3}{4} = n$, for each view of subtraction. Use milk if you cannot think of another context.

  a. take-away   **b.** comparison   **c.** missing-addend

5. Calculate the exact answers mentally. Be able to describe your thinking.

  a. $73 - \frac{3}{4}$ **b.** $129 - \frac{7}{8}$ **c.** $248 - 2\frac{7}{9}$ **d.** $1000 - \frac{17}{18}$

6. Here is an example of how 12-year-old Killie subtracted some mixed numbers:

$$8\frac{2}{5} - 5\frac{3}{5} = 3 - \frac{1}{5} = 2 + \frac{5}{5} - \frac{1}{5} = 2\frac{4}{5}$$

Use Killie's method on these two calculations.

  a. $7\frac{5}{8} - 5\frac{7}{8}$ **b.** $10\frac{1}{3} - 6\frac{4}{9}$

  c. Would Killie have used the method on $9\frac{13}{16} - 4\frac{5}{16}$?

   Killie's method was reported in the *Air Force Print News Today* (November 18, 2003). Killie's teacher said that he had never seen such a solution and had to prove to himself that evening that it was all right. The principal said, "I know lots of math teachers who would have looked at Killie's work and just said it was wrong."

  d. How would *you* have reacted to Killie's work?

7. (This is an old puzzle problem.) In his will, a rancher decides to give his horses to his children this way: $\frac{1}{3}$ to Curly, $\frac{4}{9}$ to Buck, and $\frac{1}{6}$ to Cindy Lou. After the rancher dies, his lawyer realizes that the rancher made an error, but the lawyer rides out to the ranch and makes everyone satisfied. How did the lawyer realize that the rancher had made an error? And how did he satisfy everyone?

8. Calculate the following. If the answer is greater than 1, write it as a mixed number.

  a. $\frac{7}{8} + \frac{5}{6} + \frac{11}{12} - \frac{2}{3}$ **b.** $\frac{13}{16} + 2\frac{1}{2} - \left(\frac{3}{4} - \frac{1}{6}\right)$ **c.** $10 - 7\frac{7}{12} + \frac{1}{3} + \frac{7}{18} - \frac{4}{9}$

  (See also the varied Supplementary Learning Exercises for Chapters 7–9, following those for Section 9.3.)

9. As in Learning Exercises 19–22, locate and label the points for the given numbers.

$2\frac{3}{4}$ $5\frac{1}{4}$

a. $6\frac{1}{2}$ **b.** $\frac{1}{4}$ **c.** $4\frac{1}{4}$

10. How can properties of addition help students do the following calculation *mentally*?

$$\left(\frac{40}{99} + \frac{7}{3}\right) + \left(\frac{5}{39} + \frac{2}{3}\right) + \frac{59}{99}$$

7.2 Multiplying by a Fraction

> ### Think About …
>
> Is there a rationale for the way we multiply fractions:
>
> $$\frac{a}{b} \times \frac{c}{d} = \frac{a \times c}{b \times d}?$$

Here is one argument for the above Think About. Suppose you want to know $\frac{2}{3}$ of $\frac{4}{5}$, or $\frac{2}{3} \times \frac{4}{5}$, perhaps to solve a story problem like this:

▶ Juanita had mowed $\frac{4}{5}$ of the lawn, and her brother Jaime had raked $\frac{2}{3}$ of the mowed part. What part of the lawn had been raked? ◀

A drawing might start like this:

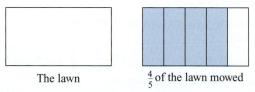

The lawn $\frac{4}{5}$ of the lawn mowed

Now comes the key step. You could take $\frac{2}{3}$ of the mowed part by first cutting each fifth in the mowed part into three equal parts, as shown below:

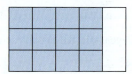

Each fifth cut into three equal parts.

Now it is a matter of taking two of the small pieces in each fifth to get $\frac{2}{3}$ of $\frac{4}{5}$. We will assume a very tidy raker.

The checked part is $\frac{2}{3}$ of $\frac{4}{5}$.   The checked part is $\frac{8}{15}$ of the lawn.

The dilemma of how to describe numerically the checked part is settled by realizing that cutting the unmowed fifth into three equal parts as well will cut the whole lawn (the unit) into 15 equal pieces—fifteenths! As far as the $\frac{a}{b} \times \frac{c}{d} = \frac{a \times c}{b \times d}$ algorithm is concerned, the drawing gives a bonus: The whole lawn is cut into 3 rows of 5 identical pieces, or $3 \times 5 = 15$ identical pieces, and the mowed and raked part is made up of 2 rows of 4 of those pieces, or $2 \times 4 = 8$. Hence, $\frac{2}{3} \times \frac{4}{5}$ does equal $\frac{2 \times 4}{3 \times 5} = \frac{8}{15}$.

Notice again that in the first problem, the unit for $\frac{4}{5}$ is the entire lawn. That is, the $\frac{4}{5}$ *refers to* the *whole lawn*. But the $\frac{2}{3}$ refers to the *mowed part*, i.e., $\frac{2}{3}$ has $\frac{4}{5}$ of the lawn as its whole (or more briefly, we can say the unit for $\frac{2}{3}$ is the $\frac{4}{5}$). Thus, the $\frac{2}{3}$ and the $\frac{4}{5}$ refer to different units. And what about the unit for $\frac{2}{3} \times \frac{4}{5}$, the $\frac{8}{15}$? This is the part of the lawn that is mowed and raked. Thus, the referent unit for the $\frac{8}{15}$ is again the entire lawn. The fact that the three fractions in $\frac{2}{3} \times \frac{4}{5} = \frac{8}{15}$, or in $\frac{a}{b} \times \frac{c}{d} = \frac{a \times c}{b \times d}$ do *not* all refer to the same unit is fundamental to understanding multiplication of fractions, and a lack of understanding of this principle leads to a great deal of confusion when children are first learning to multiply fractions. One way to keep these referent units in mind is to think of $\frac{2}{3} \times \frac{4}{5} = \frac{8}{15}$ as $\frac{2}{3}$ *of* $\frac{4}{5}$ *of* 1 is $\frac{8}{15}$ *of* 1, giving a name for the unit: one whole lawn.

To show a number line drawing for $\frac{2}{3} \times \frac{4}{5}$, we begin by taking a segment of length 1 and dividing it into 5 equal parts. If we take 4 of these 5 parts (represented by the blue segment in the diagram), this is our referent unit for the $\frac{2}{3}$. Each of these four $\frac{1}{5}$ segments can be divided into 3 equal parts, so the $\frac{4}{5}$ referent unit is further divided into 12 equal, smaller segments. Four of these 12 smaller segments represent $\frac{1}{3}$ (the gray segment) of the $\frac{4}{5}$ (the blue segment). Now, it is a matter of taking two $\frac{1}{3}$ gray segments to get $\frac{2}{3}$ of $\frac{4}{5}$.

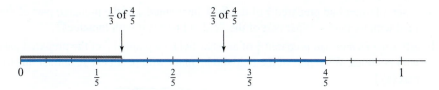

Think About …

1. Some teachers say that multiplication of fractions is easier than addition of fractions. Why might they say that? What are they overlooking?

2. Which of these multiplications, $\frac{1}{3} \times \frac{1}{2}$ or $\frac{1}{2} \times \frac{1}{3}$, means that you begin with $\frac{1}{3}$ of a whole?

ACTIVITY 2   Juanita and Jaime Again

1. How would the drawing look if the problem had been the following:

   ▶ Juanita had mowed $\frac{2}{3}$ of the lawn, and her brother Jaime had raked $\frac{4}{5}$ of the mowed part. What part of the lawn had been raked? ◀

   Make the drawing, and name the referent unit for each of the units in the drawing.

2. Make a drawing for each of these two stories:

   ▶ Pam had $\frac{2}{3}$ of a cake. She ate half of it. How much cake did she eat? ◀

   ▶ Pam ate $\frac{2}{3}$ of a cake, and Jojo ate half of a cake. How much cake did they eat? ◀

   Tell what the referent units are for the fractions in each story. ●

The given argument glosses over another point that should be justified: How does the *of* in $\frac{2}{3}$ *of* $\frac{4}{5}$ become $\times$ in $\frac{2}{3} \times \frac{4}{5}$? One justification is to mention that the total of 3 groups *of* 6 units can be translated into 3×6 and the total of 5 amounts *of* $\frac{3}{4}$ unit can be translated into $5 \times \frac{3}{4}$. So, it is plausible to translate $\frac{2}{3}$ *of* $\frac{4}{5}$ unit into $\frac{2}{3} \times \frac{4}{5}$. Another way is to link the multiplication to finding the area of a rectangular region that has dimensions $\frac{2}{3}$ unit by $\frac{4}{5}$ unit. (Try this to see whether you get $\frac{8}{15}$ of a square unit.)

Multiplying numerators to find the numerator of the product, and multiplying denominators to find the denominator of the product, can sometimes be simplified. Consider the following example of the algorithm for multiplying fractions:

$$\frac{15}{32} \times \frac{8}{25} = \frac{15 \times 8}{32 \times 25}$$

By using what we know about equivalent fractions and first identifying common factors in the numerator and the denominator, we can simplify the product to be

$$\frac{15}{32} \times \frac{8}{25} = \frac{15 \times 8}{32 \times 25} = \frac{3 \times 5 \times 2 \times 2 \times 2}{2 \times 2 \times 2 \times 2 \times 2 \times 5 \times 5} = \frac{3}{2 \times 2 \times 5} = \frac{3}{20}$$

(See Exercise 14 for a shortcut.)

TAKE-AWAY MESSAGE . . . Understanding multiplication of fractions requires knowing more than just a rule. Multiplying by a fraction $\frac{m}{n}$ means considering m parts when the whole is divided into n equal parts. The product $\frac{m}{n} \times \frac{p}{q}$ means $\frac{p}{q}$ is the referent whole to be divided. To make multiplication of fractions meaningful, one must understand the referent units for the fractions and what the product actually stands for.

Learning Exercises for Section 7.2

1. Write out an explanation, with sketches, to answer each of the following. Your explanations should make clear the referent unit for each fraction.

   a. After a farmer had prepared $\frac{5}{6}$ of the field, he planted $\frac{3}{4}$ of the prepared part of the field with tomatoes. What part of the field did he plant with tomatoes?

   b. After the farmer had prepared $\frac{3}{4}$ of another field, he planted $\frac{5}{6}$ of the prepared part with potatoes. What part of the field did he plant with potatoes? [Contrast this to part (a).]

2. A large candy bar is sectioned off:

   If Pat gets $\frac{1}{3}$ of the blue part, then Pat gets _____ of the candy bar.

   So, $\frac{1}{3} \times \frac{3}{4}$ is _____. (Notice that the algorithm involving twelfths is not needed.)

3. Shade in $\frac{5}{8}$ of four circles in at least two different ways: How do your methods differ? What is $\frac{5}{8}$ of 4?
   What is $4 \times \frac{5}{8}$? What do these questions have to do with the drawings?

4. **a.** Shade in $\frac{3}{4}$ of 2:

   b. Using the same-size rectangle as your unit, show two $\frac{3}{4}$ pieces.
   Are the amounts the same or different for parts (a) and (b)?

5. Do the following problems using common sense instead of algorithms. For part (a), think: Take eight halves, that's the same as _____ wholes.

   a. $8 \times \frac{1}{2} =$ **b.** $4 \times \frac{1}{2} =$ **c.** $2 \times \frac{1}{2} =$ **d.** $1 \times \frac{1}{2} =$

   e. $12 \times \frac{1}{2} =$ **f.** $14 \times \frac{1}{2} =$ **g.** $\frac{98}{98} \times \frac{49}{50} =$ **h.** $\frac{17}{24} \times \frac{0}{36} =$

6. Represent these two situations with equations. Are the equations the same or different?

   a. Darryl had 12 apricots. He ate $\frac{1}{6}$ of them for lunch. How many apricots did he eat?

   b. Darryl had some pizzas, each cut into sixths. If he ate 12 pieces of the pizza, how many (whole) pizzas did he eat?

7. **Pattern Blocks** With the hexagon as the unit (or a circular region as the unit if you are not using pattern blocks), show $4 \times \frac{2}{3}$, and show $\frac{2}{3} \times 4$. Sketch, even if pattern blocks are available, for future reference.

8. Shade in $\frac{2}{3}$ of $\frac{3}{4}$ of this rectangle:

   a. Locate the $\frac{2}{3}$. It is $\frac{2}{3}$ of what quantity?

   b. Locate the $\frac{3}{4}$. It is $\frac{3}{4}$ of what quantity?

   c. The final shaded part is *what fraction* of *what amount?*

9. Shade in $\frac{3}{2}$ of $\frac{1}{4}$ of this rectangle:

   a. Locate the $\frac{1}{4}$. It is $\frac{1}{4}$ of what quantity?

   b. Locate the $\frac{3}{2}$. It is $\frac{3}{2}$ of what quantity?

   c. The shaded part is *what fraction* of *what amount?*

10. a. The two rectangles shown are identical. Shade in $\frac{3}{4}$ of $\frac{4}{3}$ of the first rectangle, and $\frac{4}{3}$ of $\frac{3}{4}$ of the second rectangle.

    Why are $\frac{3}{4}$ and $\frac{4}{3}$ called **multiplicative inverses**? (They are also called **reciprocals**.)

    b. Exercises 4, 6, and 7 illustrate that *computationally* $\frac{a}{b} \times \frac{c}{d} = \frac{c}{d} \times \frac{a}{b}$ even though $\frac{a}{b} \times \frac{c}{d}$ and $\frac{c}{d} \times \frac{a}{b}$ are *conceptually* different. What is this property called?

    c. Why are these calculations easy? $\frac{72}{72} \times \frac{99}{181}$ and $\frac{571}{623} \times \frac{103}{103}$? What property did you use?

11. **Pattern Blocks** Show each of these with pattern blocks or drawings. Make sketches of your pattern block work for later reference.

    a. $\frac{2}{3} \times \frac{1}{2}$ b. $\frac{2}{3} \times 6$ c. $\frac{4}{3} \times 6$ d. $\frac{4}{3} \times 1\frac{1}{2}$ e. $\frac{2}{3} \times \left(\frac{3}{2} \times 4 \right)$

12. Draw a diagram that represents each of the following situations:

    a. There was $\frac{3}{4}$ of a pie left in the refrigerator. John ate $\frac{2}{3}$ of what was left. How much pie did he eat? Also tell what the whole (the unit) is for each given fraction and for the answer.

    b. Three-fourths of the class were girls. Two-thirds of the girls have dark hair. What fraction of the class is female and dark-haired? (*Note:* What was the *whole* for each fraction?)

    c. Jackie napped for $\frac{3}{4}$ of an hour. Corrina napped $\frac{7}{6}$ as long. How long did Corrina nap? (What is the whole for each given fraction and for the answer?)

    d. The following diagram and explanation are one student's solution to Exercise 12(c). Compare it to yours.

    The $\frac{3}{4}$ of an hour is shaded in. I broke it into 6 parts. Now each part is $\frac{1}{8}$ of an hour. I added another sixth of the $\frac{3}{4}$ hour, which is another $\frac{1}{8}$ of a total hour. So now I know Corrina slept $\frac{7}{8}$ of an hour.

13. Make up a story problem that might lead to each calculation. (If you have trouble thinking of a situation, think about quantities of cheese.)

    a. $\frac{1}{4} \times 2\frac{3}{4}$ b. $3 \times \left(\frac{1}{16} + \frac{1}{8} \right)$ c. $2 \times \left(3 - 1\frac{1}{3} \right)$

14. a. Students are often taught to "cancel," without justification for why it works. Explain why canceling works, as here, with a common factor of 3 ignored in the *first* denominator and the *second* numerator:

$$\frac{19}{\underset{7}{\cancel{21}}} \times \frac{\overset{2}{\cancel{6}}}{35} = \frac{19 \times 2}{7 \times 35} = \frac{38}{245}$$

    (*Hint for one explanation:* Here are all the steps in calculating $\frac{19}{21} \times \frac{6}{35}$, including steps that one often just thinks rather than writes:

continued

$$\frac{19}{21} \times \frac{6}{35} = \frac{19 \times 6}{21 \times 35} = \frac{114}{735} = \frac{3 \times 38}{3 \times 245} = \frac{3}{3} \times \frac{38}{245} = 1 \times \frac{38}{245} = \frac{38}{245}.\Big)$$

    b. Canceling does result in less demanding calculation, so it is very attractive to students—so attractive, in fact, that they will use it on calculations like $\frac{19}{21} + \frac{6}{35}$ or in working with proportions like $\frac{19}{21} = \frac{6}{x}$. Does canceling work with these problems?

15. Make sketches to illustrate the following situations. Notice that the settings involve discrete things.

    a. One fifth-grade class has 24 students. One-half of them are girls. Three-fourths of the boys in the class like to play football. What part of the whole class are the boys who like to play football? Would your answer change if you did not know the number of students in the class?

    b. Another fifth-grade class has 32 students. One-half of them are girls. Three-fourths of the boys in the class like to play football. What part of the whole class are the boys who like to play football? Would your answer change if you did not know the number of students in the class?

    c. Perhaps because there is a tendency to focus on the actual *number* of boys who like to play football, fraction multiplication seems to be difficult conceptually in settings such as the following, when the number of discrete things is unknown— one-half of a fifth-grade class is girls. Three-fourths of the boys in the class like to play football. What part of the whole class are the boys who like to play football? How do you "see" this last version?

16. a. The distributive property is very important with whole numbers: $k \times (m + n) = (k \times m) + (k \times n)$ for every choice of whole numbers k, m, and n. Is the distributive property also true if the numbers are fractions rather than whole numbers?

    b. Calculate each in *two* ways. [*Hint:* See part (a).]

        i. $\frac{3}{4} \times \left(\frac{4}{5} + \frac{8}{15}\right)$ **ii.** $\left(\frac{5}{6} \times 11\frac{7}{10}\right) + \left(\frac{5}{6} \times \frac{3}{10}\right)$

17. Calculate each in *two* ways.

    a. $8 \times 6\frac{2}{7}$ $\left(\textit{Hint:} \text{ Remember that } 6\frac{2}{7} = 6 + \frac{2}{7}.\right)$ **b.** $7\frac{1}{2} \times 8\frac{4}{9}$

18. Order each set of fractions from least to greatest.

    a. $\frac{3}{4}$ $\frac{9}{19}$ $\frac{23}{25}$ $\frac{12}{78}$ **b.** $\frac{3}{9}$ $\frac{3}{2}$ $\frac{3}{14}$ $\frac{3}{7}$ $\frac{3}{4}$

19. Jim, Ken, Len, and Max have a bag of miniature candy bars. Jim takes $\frac{1}{4}$ of all the bars, and Ken and Len each take $\frac{1}{3}$ of all the bars. Max gets the remaining 4 bars.

    a. How many bars were in the bag originally?

    b. How many bars did Jim, Ken, and Len each get?

20. Jim, Ken, Len, and Max have a bag of miniature candy bars. Jim takes $\frac{1}{4}$ of the bars. Then Ken takes $\frac{1}{3}$ of the remaining bars. Next, Len takes $\frac{1}{3}$ of the remaining bars. Max takes the remaining 8 bars.

    a. How many bars were in the bag originally?

    b. How many bars did Jim, Ken, and Len each get?

    c. How are the fractions in this exercise different from the fractions in Learning Exercise 19?

Supplementary Learning Exercises for Section 7.2

1. In $\frac{a}{b} \times \frac{c}{d} = \frac{e}{f}$, which fractions refer to the same unit?

2. a. With a drawing of a rectangular region, show $\frac{2}{5} \times \frac{3}{4}$.

    b. Tell where in your drawing the 2×3 and the 5×4 of the usual algorithm show up.

3. Three adults in the same family gave their parents an ocean cruise as a present. Because they had different financial resources, Ally paid $\frac{4}{9}$ of the cost, Bella paid $\frac{3}{4}$ as much as Ally paid, and Ronnie paid the rest. Ronnie paid $800.

   a. How much did the cruise cost?

   b. How much did Ally and Bella pay?

4. "Cancel" [see Learning Exercise 14(a) in Section 7.2] to find the simplest forms of the fractions in parts (a–d). If the product is greater than 1, write it as a mixed number.

   a. $\frac{5}{12} \times \frac{7}{10}$ **b.** $\frac{9}{16} \times \frac{8}{3}$ **c.** $\frac{45}{64} \times \frac{8}{27} \times \frac{8}{15}$ **d.** $\frac{56}{15} \times \frac{36}{35} \times \frac{25}{24}$

   e. Why can one cancel in parts (a–d), but not in $\frac{4}{3} + \frac{1}{8}$?

5. Jay repaid an interest-free loan from an uncle in three payments. Jay's first payment was for $\frac{2}{5}$ of the loan. Jay's second payment was $\frac{3}{4}$ times as large as the first payment. Jay's final payment was for $60. How much was the loan?

6. **a.** At a sale, if an item is 30% off, what percent will you pay?

   b. At a sale, if an item is x% off, what percent will you pay?

   c. In two ways, calculate the sale price for an item that is regularly priced at $79.95 but is 25% off. [*Hint*: Part (b).]

   See also the varied Supplementary Learning Exercises for Chapters 7–9 following those for Section 9.3.

7. Make up a story problem that could lead to each calculation. Answer the question you ask. (If you have trouble thinking of a context, use quantities of sugar.)

   a. $5 \times \frac{1}{4}$ **b.** $\frac{2}{3} \times 1\frac{1}{2}$ **c.** $\frac{3}{4} \times \left(4 - 2\frac{1}{2}\right)$

8. Children may make the following error: $2\frac{1}{3} \times 3\frac{1}{4} = 6\frac{1}{12}$ (from $2 \times 3 = 6$ and $\frac{1}{3} \times \frac{1}{4} = \frac{1}{12}$).

   a. Why is that an error?

   Give the correct answer for each of these erroneous ones.

   b. $5 \times 3\frac{1}{4} = 15\frac{1}{4}$ **c.** $3\frac{1}{3} \times 7\frac{2}{5} = 21\frac{2}{15}$ **d.** $14\frac{1}{9} \times 12\frac{7}{8} = 168\frac{7}{72}$

9. The president of a company says, "This year we hired 60 new people, so the number of our employees is now 120% of what it was last year."

   a. Make a drawing for the situation.

   b. How many employees did the company have last year?

   c. How many employees does the company have this year?

10. In a campus election, 20% of the students did not vote. Of those who voted, $\frac{1}{4}$ voted for candidate A, 35% voted for candidate B, and the remaining 800 students voted for candidate C.

    a. Name the unit for each of these given values: 20%, $\frac{1}{4}$, and 35%.

    b. Make a drawing for this problem.

    c. Who won the election? Tell how you know.

    d. How many students are in the whole student body?

7.3 Dividing by a Fraction

Earlier you learned about the meanings of division. In this section we consider three questions: (1) Why do we invert and multiply when we divide by a fraction? (2) Do we also need to attend to the referent units for division the way we did with multiplication? (3) *When* do we divide?

We begin with the first question: *Why do we invert and multiply when we divide by a fraction?* The repeated-subtraction way of thinking about division is to ask how many of the divisors can be subtracted from the dividend. For example, $12 \div 3 = 4$ because you can subtract 3 from 12 four times: $12 - 3 - 3 - 3 - 3 = 0$. Thus, $12 \div 3$ answers the question, "How many 3s are in 12?" We can use this meaning of subtraction to justify the *invert and multiply rule*.

Possibility 1: Dividing by a unit fraction $\frac{1}{n}$.

We use the repeated-subtraction meaning of division to find $1 \div \frac{1}{4}$.

$1 \div \frac{1}{4}$ can mean, "How many one-fourths are in 1?" (*Think:* How many times can I subtract $\frac{1}{4}$ from one?) There are 4 one-fourths in 1. Therefore, $1 \div \frac{1}{4} = 4 = \frac{4}{1}$.

Think About ...

Can you complete the following division in the same way?

$1 \div \frac{1}{3}$ can mean, "How many _____ are in 1?" There are _____ in 1.

Therefore, $1 \div \frac{1}{3} = ? = \frac{?}{1}$.

We now can generalize and say that $1 \div \frac{1}{b} = b$ or $\frac{b}{1}$.

Recall that the reciprocal of a fraction $\frac{a}{b}$ is $\frac{b}{a}$. When dividing 1 by $\frac{1}{b}$, we see that the quotient is $\frac{b}{1}$, the reciprocal of $\frac{1}{b}$. And we know that $\frac{b}{1} = b$.

Return again to $1 \div \frac{1}{3} = 3$. This says that there are 3 one-thirds in 1. But $3 = \frac{3}{1}$, so to say that there are 3 one-thirds in 1 is the same as saying there are $\frac{3}{1}$ one-thirds in 1. Then it would make sense to say there are *twice as many* one-thirds in 2 as in 1. $2 \div \frac{1}{3} = 2 \times \left(1 \div \frac{1}{3}\right) = 2 \times \frac{3}{1} = \frac{2}{1} \times \frac{3}{1} = 6$. Similarly, $7 \div \frac{1}{3}$, telling how many one-thirds are in 7, should have 7 times as many one-thirds as 1 does, or $7 \div \frac{1}{3} = 7 \times \frac{3}{1} = 21$.

ACTIVITY 3   Dividing by Unit Fractions

Use the reasoning above to complete the following:

a. $11 \div \frac{1}{3}$ **b.** $4 \div \frac{1}{6}$ **c.** $7 \div \frac{1}{4}$ **d.** $2\frac{1}{2} \div \frac{1}{3}$ ●

The same reasoning should apply to $2\frac{1}{2} \div \frac{1}{3}$. Two and a half should have $2\frac{1}{2}$ as many one-thirds as 1 does. $2\frac{1}{2} \div \frac{1}{3} = 2\frac{1}{2} \times \frac{3}{1} = \frac{5}{2} \times \frac{3}{1} = \frac{15}{2} = 7\frac{1}{2}$.

To test the reasoning with a diagram, consider $2\frac{1}{2} \div \frac{1}{3}$ again. If a large rectangle represents one unit, then the first row of the diagram has $2\frac{1}{2}$ units (or wholes, or ones). In the second row, each unit is cut into 3 pieces, each representing $\frac{1}{3}$. But how many one-thirds are in the $2\frac{1}{2}$? Count. There are $7\frac{1}{2}$ thirds in the $2\frac{1}{2}$ whole rectangles.

How many thirds?

1 2 3   4 5 6 $7\frac{1}{2}$ thirds

$2\frac{1}{2} \div \frac{1}{3} = 2\frac{1}{2} \times \frac{3}{1} = 7\frac{1}{2}$

Thus, we can generalize and say that

$$\frac{a}{b} \div \frac{1}{c} = \frac{a}{b} \times \frac{c}{1} = \frac{ac}{b}$$

Possibility 2: Dividing by non-unit fractions.

Thus far we have considered dividing by unit fractions. (Remember that unit fractions have the form $\frac{1}{n}$.) We learned that the quotient can be found by multiplying by the reciprocal of the unit fraction. Does this shortcut apply to division by fractions that are not unit fractions? Consider $1 \div \frac{2}{3}$ in the drawing shown:

$1 \div \frac{2}{3}$ can mean, "How many two-thirds are in one whole unit?" Visually, we can see that there is *one* $\frac{2}{3}$ and *half* of another $\frac{2}{3}$ in the circle. Thus, $1 \div \frac{2}{3} = 1\frac{1}{2}$, which is the same as $\frac{3}{2}$. So, $1 \div \frac{2}{3} = \frac{3}{2} \left(= 1\frac{1}{2} \right)$.

Once again, we see that dividing 1 by a fraction is the same as multiplying 1 by the reciprocal of the fraction: $1 \div \frac{2}{3} = 1 \times \frac{3}{2}$. So $1 \times \frac{3}{2}$ does tell how many two-thirds are in 1. In $4 \div \frac{2}{3}$ there should be 4 times as many two-thirds as for 1, so $4 \div \frac{2}{3} = 4 \times \frac{3}{2}$.

> ### *Think About ...*
>
> How could you argue that $\frac{4}{5} \div \frac{2}{3} = \frac{4}{5} \times \frac{3}{2}$?

> Dividing by a fraction gives the same result as multiplying by the reciprocal of the fraction. Symbolically, when the divisor is a fraction, $n \div \frac{c}{d} = n \times \frac{d}{c}$, or, if n is itself a fraction,
>
> $$\frac{a}{b} \div \frac{c}{d} = \frac{a}{b} \times \frac{d}{c} \quad \left(\frac{c}{d} \text{ cannot equal } 0. \right)$$

The reciprocal of a nonzero fraction is obtained by inverting the fraction. Thus, the *invert and multiply rule* is simply a different phrasing for the *multiply by the reciprocal rule*.

Drawings can be used to find quotients, such as $\frac{5}{6} \div \frac{2}{3}$. Shade $\frac{5}{6}$. How many $\frac{2}{3}$ in $\frac{5}{6}$? There is one $\frac{2}{3}$ and a bit more. How much more? The remaining segment is $\frac{1}{6}$ of the whole but $\frac{1}{4}$ of the $\frac{2}{3}$. Thus, there are 1 and $\frac{1}{4}$ two-thirds in $\frac{5}{6}$.

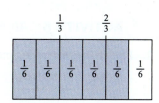

 ACTIVITY 4   Can We Draw Pictures?

Make a drawing that shows the quotient of

a. $\frac{4}{5} \div \frac{1}{2} = 1\frac{3}{5}$.   **b.** $\frac{1}{2} \div \frac{4}{5} = \frac{5}{8}$.   **c.** $1\frac{1}{2} \div \frac{4}{5} = \frac{15}{8} = 1\frac{7}{8}$. ●

The second question asked at the beginning of this section is: *Do we also need to attend to the referent units for division the way we did with multiplication?* The answer is *yes*. Consider this problem:

▶ Kathleen had $\frac{3}{4}$ of a gallon of milk. She gave each of her cats $\frac{1}{12}$ of a gallon of milk. How many cats got milk? ◀

We are asking: How many one-twelfths of a gallon are there in $\frac{3}{4}$ of a gallon? $\frac{3}{4} \div \frac{1}{12} = ?$ We could subtract $\frac{1}{12}$ from $\frac{3}{4}$ nine times, because there are three one-twelfths in every $\frac{1}{4}$. We can, we now know officially, write this as $\frac{3}{4} \div \frac{1}{12} = \frac{3}{4} \times 12 = 9$.

In this problem, both the $\frac{3}{4}$ and the $\frac{1}{12}$ refer to a gallon, whereas the 9 refers to the number of cats receiving milk. Actually, the $\frac{1}{12}$ can be thought of as $\frac{1}{12}$ *of a gallon per cat*. The point is that the fractions in this division have different referent units. To understand this problem, one must know the referent unit for each of the following numbers: the dividend, the divisor, and the quotient.

 DISCUSSION 4   Highway Repair

▶ A stretch of highway is $3\frac{1}{2}$ miles long. Each day, $\frac{2}{3}$ of a mile is repaved. How many days are needed to repave the entire stretch? ◀

continued

How can you think of this as a repeated-subtraction problem? Do you get the same answer through repeated subtraction as you do by the *invert and multiply rule*? ●

Here, the dividend and the divisor in the discussion problem both refer to miles. For repeated subtraction to fit a problem, both the dividend and the divisor must refer to the same unit. As an aside, if we look in a more sophisticated way at the units, the quotient refers not to that unit (miles) but to the divisor value, the $\frac{2}{3}$ of a mile *per day*. That is, $5\frac{1}{4}$ days refers to the number of days for which $\frac{2}{3}$ of a mile is repaved per day. Continuing a closer look at the units, $3\frac{1}{2}$ *miles* at $\frac{2}{3}$ of a *mile per day* = $5\frac{1}{4}$ *days* can also be thought of as $3\frac{1}{2}$ *miles* at $\frac{3}{2}$ *days per mile* = $5\frac{1}{4}$ *days*.

The third question asked at the beginning is: *When do we divide?* Just as with whole number problems, division is used in repeated-subtraction (measurement or quotitive) situations, as above, but also in sharing-equally (partitive) situations. Contrast the previous cats-and-milk problem with this problem:

▶ John had $\frac{3}{4}$ gallon of milk to feed his 9 cats. How much milk does each cat get? ◀

Here is a solution: $\frac{3}{4} \div 9$ because the $\frac{3}{4}$ of a gallon is shared by the 9 cats, presumably equally. Each cat gets $\frac{1}{12}$ of a gallon of milk.

> ### Think About ...
> How do the two cats-and-milk problems differ in terms of the division involved? Can you make a drawing to clarify what is happening for each problem? Which one calls for a partitive view of division? Which one calls for a quotitive view of division?

 ACTIVITY 5   An Alternative Approach—Draw

1. Find the answers using pictures or diagrams rather than the *multiply across rule*.

$$\frac{2}{3} \times \frac{1}{6} = \qquad \frac{3}{8} \times \frac{3}{4} = \qquad \frac{5}{6} \times \frac{3}{3} = \qquad \frac{1}{2} \times \frac{3}{8} = \qquad \frac{3}{4} \times 1\frac{1}{2} =$$

2. Find the answers using pictures or diagrams rather than the *invert and multiply rule*.

$$\frac{2}{3} \div \frac{6}{1} = \qquad \frac{3}{8} \div \frac{4}{3} = \qquad \frac{5}{6} \div \frac{3}{3} = \qquad \frac{1}{2} \div \frac{8}{3} = \qquad \frac{3}{4} \div \frac{2}{3} =$$

3. Compare your answers in Problems 1 and 2. Can you generalize what is happening?

4. In the given calculations, what referent unit does each fraction refer to? It may be helpful to put each calculation in a context. ●

$$\frac{2}{3} \times \frac{3}{4} = \frac{1}{2} \qquad \frac{1}{2} \div \frac{3}{4} = \frac{2}{3}$$

 ACTIVITY 6   Illustrating

For each problem first choose an interpretation for the operation; second, act out the problem with pattern blocks (or drawings) and record the solution; and third, write a story problem for your interpretation.

1. $\frac{2}{3} \div \frac{1}{2}$ 2. $\frac{2}{3} \times \frac{2}{1}$ 3. $\frac{1}{2} \times \frac{2}{3}$ 4. $\frac{1}{2} \div \frac{3}{2}$ 5. $3 \times \frac{2}{3}$ 6. $3 \div \frac{3}{2}$ ●

 DISCUSSION 5   Can You See What I See?[3]

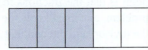

1. Can you see $\frac{3}{5}$ of something in this picture? Where? Be explicit. ($\frac{3}{5}$ of *what*?)

2. Can you see $\frac{5}{3}$ of something in the given picture? Where? How did you change the way you looked at the picture in order to see $\frac{5}{3}$?

3. How can you see $\frac{6}{10}$ of something in the given picture? How did you change the way you looked at the picture in order to see $\frac{6}{10}$?

4. Can you see $\frac{5}{3}$ of $\frac{3}{5}$ in the given picture? How did you have to change the way you looked at the picture in order to see $\frac{5}{3}$ of $\frac{3}{5}$?

5. Can you see $\frac{2}{3}$ of $\frac{3}{5}$? What is the whole (the unit)? What part is $\frac{3}{5}$ of the whole? What part is $\frac{2}{3}$ of $\frac{3}{5}$ of the whole?

6. Can you see $1 \div \frac{3}{5}$? ●

TAKE-AWAY MESSAGE . . . Division of fractions is often associated with *invert and multiply*, but most people do not understand why this rule works and even sometimes forget which fraction to invert. By making sense of this rule, such errors should not be made. Understanding referent units for the fractions is also important when dividing so that one can make sense of the quotient. Finally, elementary school students (and adults) also need to know *when* to divide; without this knowledge, knowing how cannot be of much use.

Learning Exercises for Section 7.3

1. Do these problems using common sense instead of using rules. For part (a), think, "How many halves are in eight wholes?"

   a. $8 \div \frac{1}{2}$ b. $4 \div \frac{1}{2}$ c. $2 \div \frac{1}{2}$ d. $1 \div \frac{1}{2}$ e. $\frac{1}{2} \div \frac{1}{2}$ f. $\frac{1}{4} \div \frac{1}{2}$ (Did this one surprise you?)

2. A company is promoting its pizza by giving away slices. In the first hour, they gave away 5 pizzas, each cut into sixths. How many pieces did they give away? One way to answer is to say that each of the 5 pizzas had 6 pieces, so they gave away $5 \times 6 = 30$ pieces. Another way to answer is to ask how many sixths are in 5, i.e., $5 \div \frac{1}{6}$. *Show* with the given drawing that the answer is 30:

3. Use your understanding of repeated-subtraction division to complete each equation, building on the given $1 \div \frac{2}{3} = 1\frac{1}{2}$ or $\frac{3}{2}$, as in part (a).

   If $1 \div \frac{2}{3} = 1\frac{1}{2}$ or $\frac{3}{2}$, then . . .

   a. $2 \div \frac{2}{3} = 2 \times 1\frac{1}{2} \left(\text{or } 2 \times \frac{3}{2} \right) = 3.$ b. $5 \div \frac{2}{3} =$ _____ (or _____)

   c. $24 \div \frac{2}{3} =$ _____ (or _____)   d. $7\frac{1}{2} \div \frac{2}{3} =$ _____ (or _____)

   e. $\frac{4}{5} \div \frac{2}{3} =$ _____ (or _____)

4. The company also gave away $\frac{1}{2}$ of a cup of soda with each slice of pizza. Use thinking rather than calculation to answer each question. How many could be served with

   a. 8 cups of soda? $8 \div \frac{1}{2} =$ _____   b. 4 cups of soda? $4 \div \frac{1}{2} =$ _____

   c. 2 cups of soda? $2 \div \frac{1}{2} =$ _____   d. 1 cup of soda? $1 \div \frac{1}{2} =$ _____

   e. $\frac{1}{2}$ cup of soda? $\frac{1}{2} \div \frac{1}{2} =$ _____

5. A recipe calls for $\frac{1}{2}$ of a cup of sugar. You have $\frac{3}{4}$ of a cup of sugar. How many recipes can you make (assuming that you have the other ingredients on hand)? How many one-halves are in $\frac{3}{4}$? $\left(\text{Or } \frac{3}{4} \div \frac{1}{2} = ? \right)$ What does the $\frac{1}{2}$ refer to? What does the $\frac{3}{4}$ refer to? What does the solution $\frac{3}{4} \div \frac{1}{2}$ refer to? ($1\frac{1}{2}$ of *what*?)

Notes

6. How would you illustrate (with a drawing) the division $\frac{2}{3} \div \frac{1}{6}$?

   Your answer to "How much is $\frac{2}{3} \div \frac{1}{6}$?" tells you *what* about $\frac{2}{3}$ and about $\frac{1}{6}$? That is, what does this answer refer to?

7. How would you illustrate (with a drawing) the division $\frac{1}{2} \div \frac{3}{4}$?

   Your answer to "How much is $\frac{1}{2} \div \frac{3}{4}$?" tells you *what* about $\frac{1}{2}$ and about $\frac{3}{4}$? That is, what does this answer refer to?

8. **Pattern Blocks or Drawings** "Act out" each of the following, and be able to explain your reasoning. (Be sure to specify the unit before you begin.)

   a. $\frac{5}{6} \times \frac{2}{1}$ **b.** $\frac{5}{6} \div \frac{1}{2}$ **c.** $4\frac{2}{3} \times \frac{1}{2}$ **d.** $4\frac{2}{3} \div 2$ **e.** $\frac{3}{2} \times \frac{3}{1}$ **f.** $\frac{3}{2} \div \frac{1}{3}$

9. Here is a drawing for $\frac{3}{4} \div \frac{2}{3}$. How many two-thirds of a whole are in $\frac{3}{4}$ of a whole? There is obviously 1 and a little more. Describe what is represented by the light gray part and justify your answer without calculating.

10. Make a drawing and give an explanation for this story problem: Todd can paint $\frac{3}{4}$ of a wall with 1 gallon of paint. How much of the wall can he cover with $\frac{3}{5}$ of a gallon?

11. Make a drawing and give an explanation for this story problem: A piece of fabric is $\frac{3}{4}$ of a yard long. Costume belts for a choir each requires $\frac{3}{16}$ of a yard. How many belts can be made from the piece of fabric?

12. Sketch $\frac{7}{8} \div \frac{4}{16}$ and explain your thinking.

13. Answer the questions below for part (a) and then for part (b).

    a. $\frac{2}{3} \times \frac{3}{4} = \frac{1}{2}$ What does the $\frac{2}{3}$ refer to? The $\frac{3}{4}$? The $\frac{1}{2}$?

    b. $\frac{1}{2} \div \frac{3}{4} = \frac{2}{3}$ What does the $\frac{1}{2}$ refer to? The $\frac{3}{4}$? The $\frac{2}{3}$?

14. The following are typical word problems found in elementary school textbooks. Make a drawing for each problem, and solve.

    a. At a bake sale, one pan of brownies was $\frac{1}{3}$ full. Mr. Fuller bought $\frac{3}{4}$ of what was left. What part of the pan of brownies did Mr. Fuller buy?

    b. Latisha wanted some cake to share with her boyfriend. She bought $\frac{3}{4}$ of a carrot cake at the sale, and then gave $\frac{2}{3}$ of the cake she bought to her boyfriend. What part of the whole cake did her boyfriend receive?

    c. Bronson made cupcakes for the sale. The recipe called for $1\frac{1}{3}$ cups of sugar. He used the 5 cups of sugar that he had. How many recipes of cupcakes did he make if he had all the other ingredients on hand?

    d. The recipe called for $3\frac{1}{2}$ cups of flour. He could make only $3\frac{3}{4}$ recipes. How much flour did he have?

    e. Each band banner requires $\frac{2}{5}$ of a yard of fabric. How many banners could be made with $5\frac{1}{2}$ yards of the fabric?

    f. Nellie is planting a flower garden. She wants $\frac{1}{2}$ of the garden to be planted with pansies, $\frac{2}{5}$ of the garden to be daisies, and the remaining part to be petunias. What part of the garden will be planted in petunias?

15. Write a story problem for each part that could lead to the calculation given.

    a. $6 \div 1\frac{2}{3}$ **b.** $7\frac{1}{2} \div 2\frac{3}{4}$ **c.** $1\frac{7}{8} \div 3$

16. Here is one popular activity that might result in some greater understanding of operations with fractions and that involves computational practice. The teacher gives a "target" number and four numbers. The student is to use each of the four numbers *once*, and whatever combination of operations it takes to get the target number.

Example: Target = 24; given 1, 2, 3, and 5.

Solution: $(5 + 3 = 8, 2 + 1 = 3,$ and $8 \times 3 = 24,$
or at your level, $24 = (5 + 3) \times (2 + 1).$

Here are two target numbers that involve fractions and decimals.

a. Target = 24; given $\frac{1}{2}, \frac{2}{3}, 2,$ and 9. **b.** Target = 2.8; given 0.5, 0.8, 1, and 6.

c. Make one up.

17. **a.** Is division of fractions commutative? Associative? Explain.

b. Are $\frac{a}{b} \div 1 = \frac{a}{b}, \frac{a}{b} \div \frac{n}{n} = \frac{a}{b},$ and $1 \div \frac{a}{b} = \frac{a}{b}$ correct equations?

18. Expressions like

$$\frac{\frac{1}{4}}{\frac{2}{3}} \quad \text{or} \quad \frac{7}{\frac{5}{6}}$$

are called *complex fractions*, even though as parts of an amount, they are difficult to
think about. But extending the $\frac{a}{b} = a \div b$ link to include complex fractions allows
a different, symbolic justification for the invert-and-multiply algorithm for dividing
fractions:

$$\frac{1}{4} \div \frac{2}{3} = \frac{\frac{1}{4}}{\frac{2}{3}} = \frac{\frac{1}{4} \times \frac{3}{2}}{\frac{2}{3} \times \frac{3}{2}} = \frac{\frac{1}{4} \times \frac{3}{2}}{1} = \frac{1}{4} \times \frac{3}{2}, \quad \text{showing that } \frac{1}{4} \div \frac{2}{3} = \frac{1}{4} \times \frac{3}{2}.$$

Use this complex fractions method to confirm the following:

a. $\frac{8}{9} \div \frac{5}{6} = \frac{8}{9} \times \frac{6}{5}$ **b.** $2\frac{1}{2} \div \frac{7}{8} = 2\frac{1}{2} \times \frac{8}{7}$

Supplementary Learning Exercises for Section 7.3

1. Give the *repeated-subtraction* meaning for each of the following, and then use a
rectangular region as the unit to confirm each of the quotients.

a. $1\frac{1}{2} \div \frac{1}{4} = 6$ **b.** $1\frac{1}{2} \div \frac{2}{3} = 2\frac{1}{4}$ **c.** $1\frac{1}{2} \div \frac{7}{8} = 1\frac{5}{7}$

d. $1 \div 1\frac{1}{2} = \frac{2}{3}$ **e.** $2 \div 1\frac{1}{2} = 1\frac{1}{3}$ **f.** $3 \div 1\frac{1}{2} = 2$

2. Try each of the confirmations in Exercise 1, with a circular region as the unit.

3. Give the sharing-equally meaning for each of the following, and then use a
rectangular region as the unit to confirm each quotient.

a. $1\frac{1}{2} \div 2 = \frac{3}{4}$ **b.** $1\frac{1}{2} \div 3 = \frac{1}{2}$ **c.** $1\frac{1}{2} \div 4 = \frac{3}{8}$

4. Awareness of the unit or whole for every fraction is important for the teacher (and the
student). In each part, identify which fractions refer to the same unit.

a. $\frac{a}{b} + \frac{c}{d} = \frac{e}{f}$ **b.** $\frac{a}{b} - \frac{c}{d} = \frac{e}{f}$ **c.** $\frac{a}{b} \times \frac{c}{d} = \frac{e}{f}$ **d.** $\frac{a}{b} \div \frac{c}{d} = \frac{e}{f}$

5. Are the following problems the same? Explain your decision and answer the questions.

a. You measure out $6\frac{3}{4}$ cups of sugar and use $\frac{1}{3}$ cup. How much sugar is left?

b. You measure out $6\frac{3}{4}$ cups of sugar and use $\frac{1}{3}$ of it. How much sugar is left?

6. You use $\frac{3}{4}$ cup of hummingbird food each time you fill the feeder. How many times
can you fill the feeder, if you mix the following amounts of food? [Can you use
part (a) to help answer the other parts?]

a. 1 cup **b.** 2 cups **c.** 3 cups **d.** $4\frac{1}{2}$ cups **e.** $\frac{7}{8}$ cup

7. You copy a drawing in a photocopy machine at the 75% setting. You lose the original! If you copy the smaller version, what photocopy setting will give the original size?

8. Child (or adult!): "I don't get it. . . . When we do it your way, we get, like, $\frac{3}{4} \div \frac{1}{4} = 3$. But 3 is larger than both $\frac{3}{4}$ and $\frac{1}{4}$! How can division give a larger number?" How would you respond?

9. Distributivity with multiplication and addition is so handy, it is natural to ask about the division analogies.

   a. Is $(a + b) \div c = (a \div c) + (b \div c)$? Argue that it is, using a meaning for division. Check the equality with numbers, even fractions.

   b. Is $a \div (b + c) = (a \div b) + (a \div c)$? Argue that it is *not*, using a meaning for division. Check that the equality is not true, using numbers (avoid a, b, and c being 0).

We use unit fractions in this chapter to explain the rule for dividing fractions. But unit fractions can also help solve problems, such as in this fifth-grade textbook page.* Use unit fractions to solve the Check Your Understanding problems.

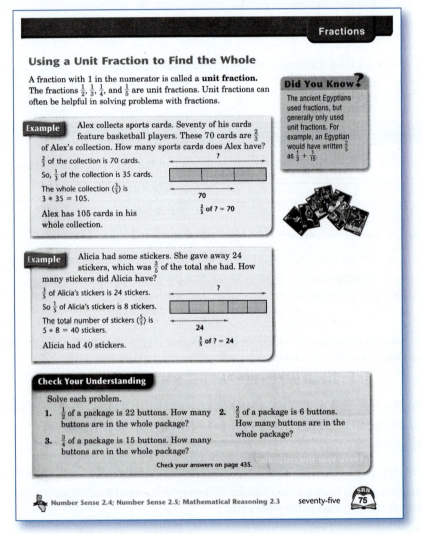

Fractions

Using a Unit Fraction to Find the Whole

A fraction with 1 in the numerator is called a **unit fraction**. The fractions $\frac{1}{2}$, $\frac{1}{3}$, $\frac{1}{4}$, and $\frac{1}{5}$ are unit fractions. Unit fractions can often be helpful in solving problems with fractions.

Did You Know?

The ancient Egyptians used fractions, but generally only used unit fractions. For example, an Egyptian would have written $\frac{2}{5}$ as $\frac{1}{3} + \frac{1}{15}$.

Example Alex collects sports cards. Seventy of his cards feature basketball players. These 70 cards are $\frac{2}{3}$ of Alex's collection. How many sports cards does Alex have?

$\frac{2}{3}$ of the collection is 70 cards.

So, $\frac{1}{3}$ of the collection is 35 cards.

The whole collection ($\frac{3}{3}$) is
$3 * 35 = 105$.

Alex has 105 cards in his whole collection.

70

$\frac{2}{3}$ of ? = 70

Example Alicia had some stickers. She gave away 24 stickers, which was $\frac{3}{5}$ of the total she had. How many stickers did Alicia have?

$\frac{3}{5}$ of Alicia's stickers is 24 stickers.

So $\frac{1}{5}$ of Alicia's stickers is 8 stickers.

The total number of stickers ($\frac{5}{5}$) is
$5 * 8 = 40$ stickers.

Alicia had 40 stickers.

24

$\frac{3}{5}$ of ? = 24

Check Your Understanding

Solve each problem.

1. $\frac{1}{2}$ of a package is 22 buttons. How many buttons are in the whole package?

2. $\frac{2}{3}$ of a package is 6 buttons. How many buttons are in the whole package?

3. $\frac{3}{4}$ of a package is 15 buttons. How many buttons are in the whole package?

Check your answers on page 435.

Number Sense 2.4; Number Sense 2.5; Mathematical Reasoning 2.3   seventy-five   75

7.4 Issues for Learning: Teaching Calculation with Fractions

Calculating with fractions is something perceived to be very difficult, in part because it is so poorly understood.

> ### *Think About ...*
> Would a teacher treat these story problems in the same way pictorially? Numerically?
>
> ▶ For a pizza party, you expect that each of the 12 attendees will eat $\frac{1}{6}$ of a pizza. How many pizzas should you order? ◀
>
> ▶ For a large pizza party, you plan to order 12 pizzas. One-sixth of them should be vegetarian. How many vegetarian pizzas should you order? ◀

Although more latitude would be allowed with children, the teacher should write $12 \times \frac{1}{6}$ for the first story problem, but $\frac{1}{6} \times 12$ for the second one. Do you see why? Be attentive to the order in which you write mathematical expressions, since you want your expressions to support the concepts involved rather than just give a correct numerical solution. The two problems in the Think About above deal with the commutative property of multiplication, which is not at all obvious to elementary grade children, and teachers will need to illustrate it. Did you show 12 one-sixths, which can be thought of in terms of repeated addition, and then $\frac{1}{6}$ of 12, or a fractional part of a number of wholes? In the second case, we cannot use repeated addition. For elementary grade children who think of multiplication only in terms of repeated addition, this is a stumbling block that needs discussion.

Here are other points to be aware of, as you deal with children in the intermediate grades. Eighth-graders in one fairly large-scale testing were given this story problem to solve:

▶ At one school $\frac{3}{4}$ of all the eighth-graders went to one game. Two-thirds of those who went to the game traveled by car. What part of all the eighth-graders traveled by car to the game? ◀

Only about 12% of the children chose to multiply, while about 55% decided to subtract and about 8% to divide! Some interviews with children suggested this thinking, "I want less than $\frac{3}{4}$. So I've got to do something to get a smaller number. . . ," and to them only subtraction and division give smaller numbers. Their idea of multiplication was that "multiplication makes bigger," so they did not consider multiplication.[4] Research shows that many elementary students, and even some adults, think that "multiplication makes bigger and division makes smaller" is always true. This common misconception leads to a great deal of confusion when solving problems. Notice how important it is to realize that when you take only a part of a quantity (with a positive value), you will get less than the quantity. This operation on quantities is reflected in multiplying by a fraction less than 1: Multiplication *can* make smaller. Notice also that the *whole* for the two-thirds in the story above is different from the *whole* for the $\frac{3}{4}$; again, a key for understanding many problems involving fractions is to keep track of the whole for each fraction mentioned.

Learning Exercises for Section 7.4

We include here a set of problems that help you review the content of this chapter. Drawings are recommended.

1. Jerry and Joel bought a large pizza. Jerry ate $\frac{1}{3}$ of it, and Joel ate $\frac{2}{5}$ of it. Then James came over and ate $\frac{3}{4}$ of what was left. What part of the whole pizza did James eat?

2. How many germs 0.002 cm long would make a line 0.9 cm long?

3. To prepare for a large gathering at church, you use a recipe that calls for $\frac{3}{4}$ of a cup of milk. You have 1 quart of milk (4 cups). How many full recipes can you make?

4. For your vegetable garden, you plan to plant $\frac{5}{12}$ of it in corn, $\frac{1}{6}$ in different kinds of beans, and $\frac{1}{12}$ in tomatoes. You will plant the remainder equally in eggplant, onions, and squash. What part of the garden will be planted in onions and squash?

5. One medical pill uses 0.2 gram of a certain chemical and 0.6 gram of filler. You have 75 grams of the chemical and 200 grams of filler. How many pills can you make?

6. Your diet allows you to have $\frac{3}{4}$ of a cup of cottage cheese every day. You buy 4 containers of cottage cheese on sale, each holding 2 cups. How many full days will this cottage cheese last you?

7. You buy a large piece of chocolate, planning to eat $\frac{1}{5}$ of it every day, Monday−Friday. But you eat $\frac{1}{3}$ of it on Monday! By what percent did your Monday consumption exceed the planned amount?

8. A recipe calls for 2 cups of sugar.
   a. You used $2\frac{1}{2}$ cups instead. By what percent did you exceed the recipe amount?
   b. Another time (diet time) you used only $1\frac{1}{4}$ cups. What percent less than the recipe amount did you use?

7.5 Check Yourself

This chapter focused on understanding how to perform arithmetic operations on fractions, and the reasons behind the rules that all of us learned long ago. With care, these operations do make sense, once you understand them. You should be able to work problems and answer questions like those assigned and that deal with the following:

1. Fractions are numbers that can be operated on just like other numbers, but the algorithms are different from those with whole numbers.

2. Adding and subtracting require that students understand *that* they need a common denominator to get pieces that are all the same size, *how* to find the least common denominator, and *why* it is needed. Facility with adding and subtracting fractions is expected.

3. Multiplying fractions is easy to do, but more difficult to understand. One must understand the referent units for the multiplicand, the multiplier, and the product, and be able to interpret the product in terms of the original problem context.

4. Dividing fractions calls for an understanding of reciprocals, the *invert and multiply rule* and the ability to describe the referent units for the divisor, dividend, and quotient. You should be able to give a rationale for the *invert and multiply algorithm*.

5. The ability to write story problems that illustrate the different operations (and views of the operations) is essential to show your understanding, as is recognizing which view a particular problem fits.

References for Chapter 7

[1]Streefland, L. (1991). *Fractions in realistic mathematics education*. Boston: Kluwer Academic.

[2]Adapted from Newton, K. J. (2008). An extensive analysis of preservice elementary teachers' knowledge of fractions. *American Educational Research Journal, 45*(4): 1080–1110.

[3]Thompson, P. W. (1995). Notation, convention, and the quantity in elementary mathematics. In J. T. Sowder & B. P. Schappelle (Eds.), *Providing a foundation for teaching mathematics in the middle school* (pp. 199−221). Albany, NY: SUNY Press.

[4]Greer, B. (1992). Multiplication and division as models of situations. In D. A. Grouws (Ed.), *Handbook of research on mathematics teaching and learning* (pp. 276–295). New York: Macmillan.

Chapter 8

Multiplicative Comparisons and Multiplicative Reasoning

In Chapter 1 you learned how to go about undertaking a quantitative analysis of a problem situation. In Chapter 3 you learned about additive comparisons: Any time you want to compare two quantities to determine how much greater or less one is than another, you additively compare the quantities by finding the difference of their values. There are times, however, when you want to compare two quantities by not *how much larger* one is than another, but rather *how many times as large* one is than another. The second kind of comparison is called a *multiplicative comparison* and involves ratios. Analyzing multiplicative comparisons is the focus of this chapter.

Notes

8.1 Quantitative Analysis of Multiplicative Situations

Suppose you plant two trees and compare their growth over a year's time. During one measuring, you find that the first tree has grown from 2 feet to 4 feet, and that the second tree has grown from 3 feet to 5 feet. Which one grew the most? There are different ways of answering this problem. One way is to look at the difference. In each case, a tree grew 2 feet. Another way to consider the growth question is to note that one tree doubled its height and the other did not. This second way of comparing two quantities is explored in this section.

> Given two quantities, whenever we want to determine *how many times as large one of them is than the other one,* we do a **multiplicative comparison** of the two quantities.

EXAMPLE 1

If length A has a value of 12 meters and length B has a value of 3 meters, the *multiplicative comparison* of A to B is 4. The multiplicative comparison of B to A is $\frac{1}{4}$ because length B is $\frac{1}{4}$ of length A. An *additive comparison* of the two lengths might say that A is 9 meters longer than B. ●

Given two quantities, we can compare them either additively or multiplicatively.

EXAMPLE 2

If a city is expected to grow from 100,000 people to 130,000 people, we could claim that either the population will grow by 30,000 people (an additive comparison) or the population will become 130% of its current value (from the multiplicative comparison of the expected 130,000 persons with the 100,000 current population). ●

Which type of comparison we use depends on the question being asked and on the needs of the person asking the question. In Example 2 the person planning the water supply for the town will be interested in the additive comparison of the population before and the population after. But this person might also use a multiplicative comparison to estimate a new budget or to judge the adequacy of a reservoir. A government agency concerned with population growth patterns in several localities of different sizes might be more interested in multiplicative comparisons.

 DISCUSSION 1   Population Growth

Does a population growth from 1000 to 2000 people have the same implications (social, economic, etc.) for a community as a growth from 100,000 to 101,000 people? Explain. ●

As you may have noted in Discussion 1, the number of schools in a town that experiences a population growth of 100% will likely have to be doubled, whereas a growth of 1% (the second case) can be absorbed by the existing schools. Thus, even though "how many more people" the community has to deal with is the same in each case, what matters for planning purposes is the growth *relative* to the original size—i.e., the multiplicative relation between the current and the original populations, not the difference (additive comparison) between the current and original populations.

ACTIVITY 1   Where Should I Put My Money?

A fifth-grade teacher posed this problem to his students:

▶ Jackie invested $20 in her bank, and 3 months later she received $40 back. Jolanda invested $10 in her bank, and 3 months later she received $30 back. Which bank would you use?[1] ◀

The students said it didn't matter. Why do you think they said that? What would be a good follow-up problem to pose? ●

ACTIVITY 2   As Time Goes By

▶ Today is Sally's birthday. She is 7 years old. At some time in the future, John will have his 39th birthday. At that time, he will be 3 times as old as Sally. How old is John now?[2] ◀

a. Identify all the quantities in the situation, including those whose values are not known.
b. What does the 3 in the problem refer to? What quantities are being compared?
c. How would you as a teacher respond to a student who says that since John is 3 times as old as Sally, and Sally is 7 years old, John must be 21 years old now?
d. How much time will elapse between now and the time when John is 39 years old?
e. How old is John now?
f. What is the difference between Sally's and John's ages when John is 39? Twenty-five years from now? Now?
g. What can you conclude about the difference between John's and Sally's ages as time goes by?
h. As long as John is alive, will he always be 3 times as old as Sally? Explain. ●

DISCUSSION 2   Analyzing Sally's and John's Ages

Consider the following diagram of the quantitative structure of the problem in Activity 2:

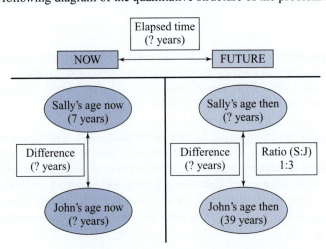

1. Based on this diagram, describe two different ways of determining John's age now.

2. In the diagram, the quantities involved in the problem appear in rectangles and in ovals. Some of these quantities are explicitly stated in the problem, while others are not mentioned; rather, they are implicit in the problem. Which quantities are explicit and which ones are implicit?

3. Sketch an alternative diagram that captures the quantities and quantitative relationships in the given problem. ●

Notes

The phrase "he will be 3 times as old as Sally" is a multiplicative comparison and leads to the idea of *ratio*.

> The result of comparing two quantities multiplicatively is called a **ratio**. If x is the value of quantity A and y is the value of another quantity B, then the ratio $x{:}y$, or $\frac{x}{y}$, tells us how many times as large A is as B. The ratio $x{:}y$ is often pronounced "x to y" or "x is to y."

The context usually makes clear whether to use the $x{:}y$ or $\frac{x}{y}$ notation.

EXAMPLE 3

In the problem in Activity 2, the ratio of John's age to Sally's age on John's 39th birthday is 3 to 1 $\left(\text{expressed as } 3{:}1 \text{ or } \frac{3}{1}\right)$.

> ### Think About ...
> What is the ratio of Sally's age to John's age on John's 39th birthday? How many times as old as John will Sally be then?

EXAMPLE 4

If time period A is 192 hours and time period B is 12 hours, then A is 16 times as long as B. The 16 comes from $\frac{192}{12}$, or its link, $192 \div 12$. For this reason, an elementary school book might say that a ratio is a comparison of two amounts by division. ●

ACTIVITY 3   Unreal Estate

Here is another problem for you to analyze in terms of its quantitative structure, perhaps making a drawing. Be sure to include all quantities, explicit and implicit ones. The intent of this activity is to analyze the problem. Once you have analyzed it, solving it is easy.

▶ A $140,000 estate was sold, and the money was split among two children and two grandchildren. The two grandchildren get the same amount of money, and each child gets two times as much as each grandchild. How much did each child get? ◀ ●

Observe from Activity 3 that there are many ways of expressing multiplicative relationships between quantities. For example, we can say, "Each child gets twice as much as each grandchild," or "The ratio of a child's amount to a grandchild's amount is 2 to 1." Also, notice that many (although not all) questions about quantitative relationships can be answered when one simply knows the ratio of the two quantities, child's amount and grandchild's amount, even without actually knowing the values of the individual quantities.

TAKE-AWAY MESSAGE . . .   This section builds on earlier work in which a quantitative analysis was required for a problem situation. This time, however, the situations were multiplicative in nature; multiplication and/or division is needed to solve at least part of the problem. Multiplicative reasoning is distinguished from additive reasoning, which calls for addition and/or subtraction.

Notes

Learning Exercises for Section 8.1

Drawings will help in these exercises.

1. Four of every 5 dentists interviewed recommend Yukky Gum.

   a. Among those interviewed, what is the ratio of those who recommend Yukky Gum to those who do not?

   b. Among those interviewed, there are _____ times as many dentists who do not recommend Yukky Gum as dentists who do.

   c. What fraction of the dentists interviewed recommend Yukky Gum?

2. In Mensville, for every 3 women, there are 4 men.

   a. What is the ratio of men to women in Mensville?

   b. The number of women is ____ times the number of men.

   c. The number of men is ____ times the number of women.

   d. Women make up what fraction of the total population of Mensville?

   e. What fraction of the total population of Mensville are men?

3. Two of every 3 seniors at Lewis High apply for college. Three of every 5 seniors at Lewis High who apply for college are female students.

   a. How does the number of seniors who apply for college compare to the number of seniors at Lewis High?

   b. How does the number of seniors who apply for college compare to those who don't?

   c. What fraction of the graduating seniors at Lewis High do not apply to college?

   d. What fraction of the seniors are males who apply for college? How does the number of males who apply for college compare to the number of graduating seniors?

   e. The number of females who apply for college is ____ times as large as the number of students in the senior class. What part of the senior class consists of females who apply for college?

   f. How does the number of males who apply for college compare to the number of females who apply?

   g. For every ____ females who apply for college, there are ____ males who apply.

   h. The number of males who apply for college is ____ times as large as the number of females who apply for college.

   i. The number of females who apply for college is ____ times as large as the number of males who apply.

   j. What is the ratio of female seniors who do not apply for college to males who do not apply for college?

4. Did you make a drawing for Learning Exercise 3? If so, was it helpful? If not, use a rectangular region for all of the seniors to see whether the drawing could help with parts of the exercise.

5. At Poly High, 5 out of every 6 seniors drive to school. Three of every 4 seniors at Poly High who drive to school are male students.

   a. Make a drawing for this situation, using a rectangular region for all the seniors at Poly High.

   b. What fraction of the senior class are females who drive to school?

   c. The number of males who drive to school is ____ times as large as the number of students in the senior class. What part of the senior class consists of males who drive to school?

   d. The number of senior females who drive to school is ____ times as large as the number of senior males who drive to school.

   e. What is the ratio of the number of male seniors who do not drive to school to the number of females who do not drive to school?

   f. Do you know how many seniors are at Poly High?

   See the Supplementary Learning Exercises following Section 8.2.

8.2 Fractions in Multiplicative Comparisons

The problems in the preceding section, e.g., Learning Exercises 2(a) and (c), illustrate the relationship between ratios and fractions. A ratio is not the same as a fraction even though the same symbol $\frac{a}{b}$ is often used for both. Recall that a ratio is the result of a multiplicative comparison of two quantities. When we compare two quantities and know the value of the corresponding ratio, it is usually possible to deduce what fractional part one quantity is of the other quantity from that comparison. The tasks in Activity 4 will help you see the relationship between fractions and ratios more clearly.

ACTIVITY 4   Candy Bars

Below are diagrams of regions. Consider each region as a candy bar that is being shared between two people. For each candy bar, cut the bar into two pieces, A and B, so that it represents the given description of Part A and Part B.

1.
   a. Part A is $\frac{1}{2}$ as large as Part B.
   b. Part B is ___ times as large as Part A.
   c. Part A is how much of the bar?
   d. What is the ratio of Part A to Part B?

2.
   a. Part A is $\frac{1}{4}$ as large as Part B.
   b. Part B is ___ times as large as Part A.
   c. Part A is how much of the bar?
   d. What is the ratio of Part B to Part A?

3.
   a. Part A is $\frac{2}{3}$ as large as Part B.
   b. Part B is ___ times as large as Part A.
   c. Part A is how much of the bar?
   d. What is the ratio of Part A to Part B?

4.
   a. Part A is $\frac{4}{5}$ as large as Part B.
   b. Part B is ___ times as large as Part A.
   c. Part A is how much of the bar?
   d. What is the ratio of Part B to Part A?

5.
   a. The ratio of Part A to Part B is 3:2.
   b. Part A is ___ times as large as Part B.
   c. Part B is how much of the bar?
   d. Part B is how many times as large as Part A?

6.
   a. The ratio of Part A to Part B is 3:4.
   b. Part B is ___ times as large as Part A.
   c. Part A is how much of the bar?
   d. Part A is how many times as large as Part B? ●

When using ratios to express the comparison between two quantities (e.g., Part A is $\frac{1}{2}$ as large as Part B), it is extremely important to be clear about the quantities being compared (in this case, Part A to Part B, not Part A to the whole bar or Part B to the whole bar). The ratio 1:2 can also be interpreted as saying that Part A is $\frac{1}{2}$ *of* Part B, although strictly speaking, this is not really the case. Part A is not *part of* Part B in the sense that we do not get Part A from Part B. However, *the measure of* Part A is $\frac{1}{2}$ the measure of Part B. In this sense, ratios and fractions are closely related.

   In comparing Part A to Part B, it is important to understand that a ratio of, say, 4:5, means that Part A consists of 4 of whatever unit Part B has 5 of. Therefore, the measure of Part A is $\frac{4}{5}$ the measure of Part B.

   The same ideas used with the candy bar problems in Activity 4 can be applied to similar but more complicated situations, such as those involving more than two sharers.

EXAMPLE 5

▶ Annie, Bob, and Cass ate a whole pizza. Annie ate twice as much as Bob did, and Bob ate twice as much as Cass. What part of the pizza did each person eat? If the pizza cost $14, how much should each person pay? ◀

continue

Solution

Using ratios and natural labels for the quantities eaten, we have A:B = 2:1 and B:C = 2:1. Bob is involved in both ratios, so we can link the two. For every 2 shares Annie had, Bob had 1, or for every 4 shares Annie had, Bob had 2. Symbolically, A:B = 2:1 = 4:2 and B:C = 2:1. Although pizzas are generally circular, using rectangles to help in thinking (as in the drawing) is all right. Then, for every 4 shares Annie had, Bob had 2, and Cass had 1. Or for every 7 shares, Annie had 4, Bob 2, and Cass 1. Thus, Annie ate $\frac{4}{7}$ of the pizza, Bob ate $\frac{2}{7}$, and Cass ate $\frac{1}{7}$. From those fractions and the $14 cost for the pizza, Annie should pay $8, Bob $4, and Cass $2. ●

Think About …

Suppose that the ratios in Example 5 had been linked this way: A:B = 2:1 and B:C = 1:$\frac{1}{2}$.

Reason through to see that you would get the same answers as those in the solution given.

ACTIVITY 5   The Orient Express

You may wish to use diagrams like those used in Activity 4 to solve the following problem:

▶ While riding the train from Chicago to New York, Pat fell asleep after traveling half of the trip. When she woke up, she still had to travel half the distance she had traveled while sleeping.[3] ◀

1. Draw a diagram showing the entire trip with the portion that Pat slept darkened, and then compare your diagram to the following diagrams. For each of the diagrams, *explain* whether or not it fits the description of the situation. Assume the shaded part is the portion Pat traveled while sleeping.

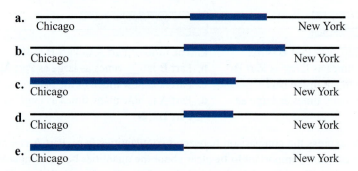

2. For what part of the trip was Pat asleep? ●

DISCUSSION 3   Elapsed Time

At some time before the end of the day, there remains $\frac{4}{5}$ of what has elapsed since the day began at midnight.[4]

a. Is it before noon or after noon? Explain how you can tell.

b. The number of hours that have elapsed since the day began is _____ times the number of hours that remain until the end of the day.

c. What time is it? ●

Notes

Notice that some of these problems from a Grade 5 textbook* require a drawing. For each problem, make a drawing and find the answer. Which of the problems is similar to the candy bars activity in Section 8.2?

Independent Practice

For **6**, use the chart at the right.

6. The school cafeteria manager needs to know how many food trays are needed during a week. All of the students eat lunch each school day, and half of all the students eat breakfast. How many trays will be needed in one week?

Grade	Number of Students	Grade	Number of Students
K	95	3	107
1	112	4	100
2	104	5	114

7. Juan used first-class mail to send two baseballs to his grandson. Each baseball weighed 5 ounces. The postage was $0.39 for the first ounce and $0.24 for each additional ounce. How much was the postage?

8. The Meadows Farm has 160 acres. Three times as many acres are used to plant crops as are used for pasture. Draw a picture and write an equation to find how many acres are used for pasture.

9. A youth group charged $6 per car at their car wash to raise money. They raised $858. Of that amount, $175 was given as donations and the rest of the money came from washing cars. Stella estimated that they washed more than 100 cars. Is her estimate reasonable? Explain your reasoning.

10. A hardware store has 5 employees. Each employee works the same number of hours every week, and each one earns $10.50 per hour. Last week they worked a total of 167.5 hours. Draw a picture and write an equation to find how many hours each employee worked.

Think About the Process

11. Matt is saving to buy a skateboard and a helmet. The skateboard costs $57 and the helmet costs $45. Matt has saved $19 so far. Which hidden question needs to be answered before you can find how much more he needs to save?

   A Is the skateboard on sale?

   B How much more does the skateboard cost than the helmet?

   C What is the price of the skateboard minus the price of the helmet?

   D What is the total price of the helmet and the skateboard?

12. Two restaurant waiters share $\frac{1}{4}$ of their tips with the dishwasher. On Saturday, one waiter earned $122 in tips, and the other waiter earned $136 in tips. Which expression shows how to find the solution to the hidden question?

   A $122 + 136$

   B $\frac{1}{4} \times 136$

   C $\frac{1}{4} \times 122$

   D $136 - 122$

*From, *enVision*, California Math Edition, Grade 5, p. 166. Copyright © 2009 Pearson Education, Inc., or its affiliate(s). Used by permission. All rights reserved.

TAKE-AWAY MESSAGE . . . The activities and discussions in this section have focused on distinguishing between a ratio and a fraction. For the ratio 1:2 or $\frac{1}{2}$, when we compare two quantities A and B, Quantity A need not be *part of* Quantity B in the sense that we do not necessarily get A from B. However, *the measure of* A is $\frac{1}{2}$ *the measure of* B. In the part-whole notion of a fraction, Quantity A *is part of* Quantity B. This section provides opportunities to think about the types of comparisons being made.

Learning Exercises for Section 8.2

For Exercises 1 through 3, explain which questions address a multiplicative comparison and which questions address an additive comparison, in addition to finding the answers.

1. Anh uses $\frac{3}{8}$ times as many spools of thread in one month as Jack. Anh uses 2 spools of thread.
   a. How many spools does Jack use?
   b. Who uses more thread?
   c. How much more?

2. Claudia ran $9\frac{2}{7}$ laps and Juan ran 5 laps. Claudia ran how many times as many laps as Juan? How much farther did Claudia run?

3. Les is using 10 sticks of butter to make cookies. The actual recipe uses only $2\frac{3}{8}$ sticks of butter. Les is using how many times as many sticks of butter as the actual recipe? How much more butter than called for did Les use?

4. Mary used 14 meters of ribbon to make bows and strips. She used $\frac{3}{4}$ as much ribbon for bows as she did for strips. How many meters of ribbon did she use for bows and how many for strips?

5. The big dog weighs 5 times as much as the little dog. The little dog weighs $\frac{2}{3}$ as much as the medium-sized dog. The medium-sized dog weighs 12 pounds more than the little dog. How much does the big dog weigh?

6. The following problem[5] is from an SAT exam, and very few students solved it. Can you solve this problem?

   ▶ A flock of geese on a pond were being observed continuously.
   At 1:00 PM, $\frac{1}{5}$ of the geese flew away.
   At 2:00 PM, $\frac{1}{8}$ of the geese that remained flew away.
   At 3:00 PM, 3 times as many geese as had flown away at 1:00 PM flew away, leaving 28 geese on the pond. At no other time did any geese arrive, fly away, or die. How many geese were in the original flock? ◀

7. In each statement, the people named shared a large candy bar. Make a drawing and give your reasoning in finding the fraction of the bar each person ate.
   a. Al ate $\frac{4}{5}$ as much as Babs ate.
   b. Cameron ate $\frac{4}{5}$ of the bar, and Don ate the rest.
   c. Emily ate three times as much as Fran ate.
   d. Gay ate one-half as much as Haille, but three times as much as Ida.
   e. Judy ate $\frac{2}{3}$ of the bar, and Keisha ate twice as much as Lannie.
   f. Mick and Nick each ate a quarter of the bar, and Ollie ate one-half as much as Pete did.

8. If each candy bar in Exercise 7 cost $1.80, how much should each person pay?

9. A cook used all of a 2-pound chunk of cheese in making three recipes. Recipe I used one-third as much cheese as Recipe II did, and Recipe II used one-half as much cheese as Recipe III.
   a. What part of the whole chunk of cheese, and how many pounds, did each recipe use?
   b. If the whole chunk of cheese cost $6.60, how much was the cheese in each recipe worth?

10. Arica planned to donate a total of $100 to four charities. She wanted to give twice as much to Charity A as she did to Charity B, but only one-fourth as much to Charity A as to Charity C. It worked out best if she gave $12 to Charity D. How much did she give to each of Charities A, B, and C?

11. Quinn, Rhonda, and Sue shared $\frac{3}{4}$ of a whole pizza. Quinn was hungrier than Rhonda and ate twice as much as Rhonda did. Sue ate $\frac{3}{4}$ as much as Rhonda. What part of the whole pizza did each eat?

Supplementary Learning Exercises for Section 8.1 and 8.2

1. Four people share a pan of brownies, as suggested by the drawing to the right.

A	A	C	C
A	A	B	C
A	B	B	D

   a. What is the ratio of A's share to B's share?
   b. What part of the pan did A get?
   c. A's share is _____ times as large as B's.
   d. B's share is _____ as large as A's.
   e. A and B together got _____ times as much as D did.
   f. A and B together got _____ times as much as C and D together.
   g. The ratio of the total of A's and B's shares to the total of C's and D's shares is _____.

Use drawings rather than algebra in the following:

2. Mama Bear ate 80% as much porridge as Papa Bear did. Baby Bear ate $\frac{1}{3}$ as much as Mama Bear did. Papa Bear ate 1.2 liters more porridge than Mama Bear did. How much porridge did the three bears eat, in all?

3. During a recent year, Company A's revenues were $\frac{3}{5}$ as large as Company B's revenues, but twice as large as Company C's. Company B had $3.4 million in revenues more than Company A. What were Company C's revenues?

4. a. Team A and Team B together won 50% more games than Team C did. Team A won 50% as many games as Team B did. The three teams won 60 games in all. How many games did each team win? (*Hint*: What is the ratio of the number of A's wins to the number of B's wins?)

   b. Team D and Team E together won 25% more games than Team F did. Team D won $\frac{2}{3}$ as many games as Team E did. The three teams won 72 games in all. How many games did each team win?

5. This problem is adapted from the column "Ask Marilyn," by Marilyn vos Savant, *Sunday Parade*, October 30, 2005, p. 13. Make a diagram for the problem.

   A husband and wife are driving to her mother's house. They have an appointment with a marriage counselor en route. After some time passes, the wife asks, "How far are we from our house?" The husband replies, "Half as far as from here to Pollyanna's office." A while after the appointment, they are 200 miles from where the wife asked her first question. She now asks a second question: "How far are we from Mother's house?" The husband takes advantage of this golden opportunity and responds, "I told you already: half as far as from here to Pollyanna's office." How far apart are the two houses?

6. Darla, Ellie, and Fran ate a whole container of ice cream. Darla ate half as much as Ellie ate, and Fran ate 5 times as much as Darla ate. If the container of ice cream cost $6.00, how much should each person pay?

7. In each part, the people named shared all of a large candy bar. Make a drawing and give your reasoning in finding what fraction of the bar each person ate.

   a. Tom ate $\frac{2}{3}$ as much as Ulysses ate.
   b. Vicky ate $\frac{3}{4}$ as much as Willie, who ate half as much as Xavier.
   c. Yolanda ate $\frac{1}{3}$ of the bar, Zeb ate $\frac{3}{4}$ of what Yolanda left, and then Arnie ate the rest.

8. If the candy bar in each part of Supplementary Learning Exercise 7 cost $3.00, how much should each person pay?

9. Monday evening: The jar of M&M's is full.

   Tuesday: $\frac{1}{6}$ of the M&M's are eaten.

   Wednesday: $\frac{1}{3}$ of the M&M's that remained disappeared.

   Thursday morning, first thing: The jar had 45 M&M's remaining.
   How many M&M's were in the jar on Monday evening?

8.3 Issues for Learning: Standards for Learning

Over the past several years, mathematics achievement in the United States has been shown to be lower than that of many other countries, particularly many in Asia and in Europe. This fact does not bode well for the future of America. For this reason, the president of the United States, in 2006, created a National Mathematics Advisory Panel to make recommendations for improving students' knowledge and performance in mathematics. The final report[6] of this panel was published in March 2008.

The panel began with evidence that algebra is the place in the curriculum where we see large failure rates because students are not properly prepared for algebra. This fact led to the primary recommendation that the mathematics curriculum in Grades pre-K–8 should be streamlined to emphasize focused, coherent progression of mathematics learning, with an emphasis on proficiency with key topics, primarily fractions, so that students are prepared for algebra (and all future courses in mathematics). Furthermore, "proficiency with whole numbers is a necessary precursor for the study of fractions, as are aspects of measurement and geometry. These three areas—whole numbers, fractions, and particular aspects of geometry and measurement—are the Critical Foundations of Algebra" (p. xviii). Proficiency is more than computational skill—the panel members consider conceptual understanding and problem-solving skills as equally important and see these capabilities as mutually supportive.

But how can district and state framework committees design curricula for these grades to assure that when students reach algebra, they will have the background needed to be successful?

Equally important, how can we ensure that our students are well-prepared to be a part of the global workforce? Studies of mathematics education in high-performing countries suggest that we would benefit from having a curriculum that is better focused and more coherent. Unlike many other countries that have a national curriculum, our educational standards have been developed and implemented by each state. Our curriculum has been described as "a mile wide and an inch deep."

Implementing the recommendations of the panel report depends on national consensus about the mathematics curriculum. Work on a national framework of content and skills that should be learned at each grade level began in 2009. In mid-2010 the National Governors Association and the Council of Chief State School Officers released a set of education standards, the *Common Core State Standards* (CCSS).[7] The CCSS reflect a state-led initiative to establish a shared set of educational standards that each state can adopt. As of this writing, all but five states have adopted the CCSS (with each state making a small percentage of modifications and additions).

The CCSS for Mathematics consist of two intertwined parts: Standards for Content and Standards for Mathematical Practice. The Content Standards define what students should know and be able to do, and the Standards for Mathematical Practices describe the processes and proficiencies that students should develop. The practices standards have foundations in National Council of Teachers of Mathematics' (NCTM) (2000)[8] process standards as well as in the strands of mathematical proficiency described in the National Research Council's (NRC) (2001) report *Adding It Up*.[9] The NCTM process standards and NRC report address aspects such as problem solving, reasoning and proof, communication, connections, conceptual understanding, procedural fluency, and productive disposition (briefly, an inclination to see mathematics as sensible, coupled with self-efficacy). The Standards for Mathematical Practice describe mathematically proficient students as having the ability to make sense of problems, to persevere in solving them, and to look for and make use of structure. These are discussed more fully in a later Issues for Learning section.

The Standards for Mathematical Practice are connected to the Standards for Mathematical Content, especially with those standards that explicitly state *understanding* as an expectation. When students lack understanding, it prevents them from engaging in the mathematical practices that would support their understanding of the mathematics content. The content standards define grade-specific standards for a sequence of topics and performances. The authors of the CCSS endeavored to design standards with focus and coherence through organizing structure, such as place value and the properties of operations.

The authors of the Standards for Mathematical Content describe a progression of a topic across grade levels, an approach informed by the logical structure of mathematics and by research on children's cognitive development. What follows is a brief summary of the K–5 standards regarding Numbers and Operations, Whole Number and Fractions (the focus of Chapters 1–11 of this text). Issues for Learning sections in the other three parts of the text contain a brief description of the content standards regarding expressions and equations, geometry and measurement, and probability and statistics. The Standards for Mathematical Practice, described in more detail elsewhere, are embedded throughout the course materials.

K–5 Numbers and Operations. Students should learn to count with meaning, develop a basic understanding of our base-ten system, and be able to work comfortably with whole numbers including their composition. For example, they should recognize that 23 could be decomposed into 2 tens and 3 ones, but also into 1 ten and 13 ones. They should develop an understanding of unit fractions and all other fractions based on unit fractions. Various meanings of addition and subtraction of whole numbers should be understood, together with the relationship between these two operations. Situations calling for multiplication and division should be recognized, such as equal groupings and equal sharing. Numbers and operations with numbers should be appropriately modeled with the number line diagrams, the area model, and story problems. Students should be fluent when working with different number combinations for addition and subtraction (often called the *basic facts*) such as sums for numbers through 10. They should be able to use a variety of methods and tools for computing with numbers, including counting with objects, mental computation, estimation, paper and pencil, and calculators.

Grades 3–5. Students' understanding of place value should expand to include large numbers and decimal numbers, and they should be able to compare these numbers and recognize equivalent ways of expressing them. A sound understanding of fractions will include the ability to compare fractions and operate on them. They should be able to recognize and generate equivalent forms of fractions, decimals, and percents. Numbers less than 0 are introduced. The various meanings of arithmetical operations and the ways that they relate to one another should be understood, as well as the effects of those operations. Properties of operations must be learned and used correctly.

Students must be fluent with operations on whole numbers, decimal numbers, and fractions. They should develop strong number sense as evidenced by their ability to mentally compute with numbers and estimate appropriately with whole numbers, fractions, and decimal numbers, and to judge the reasonableness of results of these operations. They should be able to use various types of representations and benchmarks, and they should be able to select appropriate methods and tools, including mental computation, estimation, paper and pencil, and calculators to carry out computations of various types.

One way to gain more familiarity with the national vision of mathematics education and the Common Core State Standards is to utilize the information offered by the National Council of Teachers of Mathematics (NCTM). NCTM is an organization of approximately 100,000 mathematics teachers that has been providing information and assistance to mathematics teachers for nearly a century, primarily through conferences and publications. Your university library and most elementary schools have access to one or more of the NCTM practitioner journals: *Teaching Children Mathematics* (aimed at elementary grades), *Mathematics Teaching in the Middle School*, and *The Mathematics Teacher* (aimed at secondary school mathematics). The NCTM website www.nctm.org has information about effectively teaching mathematics and implementing the Common Core State Standards.

NCTM, the national panel, and the CCSS initiative recognize that teachers have a central role in mathematics education and to this end recommend that the government support initiatives for attracting and appropriately preparing, evaluating, and retraining effective teachers of mathematics. *Teaching that demands students learn, at any given grade level, a small number of topics well will require that a teacher must know these topics exceedingly well.*

Notes

> ### *Think About ...*
> How well do you understand the mathematics you will be teaching? In this course are you going beyond procedural learning to include conceptual understanding and problem-solving skill?

8.4 Check Yourself

This chapter returned to quantitative analyses of problems, but now the problems require multiplicative reasoning. A distinction was made between additive reasoning and multiplicative reasoning. An additive comparison requires using the difference between values of two quantities, whereas a multiplicative comparison requires using the ratio between values of two quantities. A distinction was also made between fractions and ratios, which often use the same symbols.

You should be able to work problems and answer questions like those assigned and that deal with the following:

1. Differentiate between additive and multiplicative reasoning both by contrasting the two and by recognizing when each is used in a problem solution.

2. Identify problems that require a multiplicative approach to solve.

3. Undertake quantitative analyses of problems that require multiplicative reasoning. (There will be more chances to practice in Chapter 9.)

4. Use ratios to compare quantities.

5. Distinguish between ratios and fractions and the manner in which they are used.

6. Identify resources for determining appropriate curricula for elementary school grades.

References for Chapter 8

[1]Sowder, J., & Philipp, R. (1999). Promoting learning in middle-grade mathematics. In E. Fennema & T. A. Romberg (Eds.), *Mathematics classrooms that promote understanding* (pp. 89–108). Mahwah, NJ: Erlbaum.

[2]Adapted from Thompson, A. G., Philipp, R. A., Thompson, P. W., & Boyd, B. A. (1994). Calculational and conceptual orientations in teaching mathematics. In A. Coxford (Ed.), *Professional development for teachers of mathematics: 1994 Yearbook of the NCTM* (pp. 79–92). Reston, VA: National Council of Teachers of Mathematics.

[3,4]Adapted from Krutetskii, V. A. (1976). *The psychology of mathematical abilities in schoolchildren* (J. Teller, Trans.). Chicago: University of Chicago Press.

[5]Dancis, J. (no date given). *Reading instruction for arithmetic word problems: If Johnny can't read and follow directions, then he can't do math.* http://www.math.umd.edu/~jnd/subhome/Reading_Instruction.htm, April 4, 2012.

[6]U.S. Department of Education. (2008). Final Report of the National Mathematics Advisory Panel. http://www.ed.gov/about/bdscomm/list/mathpanel/index.html

[7]National Governors Association Center for Best Practices, Council of Chief State School Officers. (2010). *Common Core State Standards Mathematics*. Washington, DC: National Governors Association Center for Best Practices, Council of Chief State School Officers.

[8]National Council of Teachers of Mathematics. (2000). *Principles and standards for school mathematics.* Reston, VA: Author.

[9]National Research Council. (2001). *Adding it up: Helping children learn mathematics.* J. Kilpatrick, J. Swafford, & B. Findell (Eds.). Mathematics Learning Study Committee, Center for Education, Division of Behavioral and Social Sciences and Education. Washington, DC: National Academies Press.

Chapter 9

Ratios, Rates, Proportions, and Percents

In this chapter we continue our work on multiplicative reasoning. We have talked in the past about how we measure certain quantities. (Recall from Section 1.1 that a quantity is any attribute of an object that can be measured or counted.) For example, we can measure length with linear units such as inches or centimeters; we can measure area in terms of square feet or square meters; and we can measure speed in terms of miles per hour or meters per second. There are, of course, many other ways to measure these quantities. Measuring via ratios and rates is the focus of this chapter, along with proportional thinking and the special case of percent.

9.1 Ratio as a Measure

Ratio was introduced in Chapter 8 as a way of comparing quantities. In this section we build on that idea in new contexts.

 DISCUSSION 1 Lot Sizes

Consider the following problem[1]:

▶ A new housing subdivision offers rectangular lots of three different sizes:

a. 75 feet by 114 feet   **b.** 455 feet by 508 feet   **c.** 185 feet by 245 feet

If you were able to view these lots from above, which would appear most nearly square? Which would appear least square? Explain your answers. ◀

Be sure to think about what attribute or characteristic of the lots we are interested in and in what ways this attribute can be quantified. ●

Notice that because we are interested in comparing the three lots in terms of their *squareness,* it would help to assign each lot a measure of its squareness to facilitate the comparison. Without a numerical value for squareness, it would be difficult to decide which lot is the *most square.* How would you quantify *squareness* in this case?

 DISCUSSION 2 Downhill

In Japan, indoor skiing is very popular. Ski slopes are built in very large indoor arenas and covered with a plastic fiber that simulates packed snow. In one arena there are three ski slopes. Suppose you have measurements on each of the three slopes that tell you the length of the base,

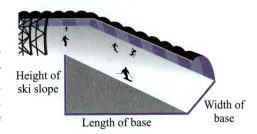

Height of ski slope

Length of base

Width of base

the width of the base, and the height of the ski slope. How could you decide which of the three slopes is the most steep or the least steep?[2] ●

TAKE-AWAY MESSAGE . . . Multiplicative comparisons involve the use of ratios. Ratios can be used to indicate some measurements, such as steepness. One way to find out which of two slopes is steeper is to compare ratios. Recognizing when to use ratios is an indicator of the ability to reason multiplicatively.

Learning Exercises for Section 9.1

1. You are running a contest to see if people can tell which of three different-sized batches of coffee (from the same type of coffee bean) is strongest. As the manager of the contest, how would *you* measure the strength of each batch?

2. Devise a way of quantifying population growth to enable comparisons among the different countries.

Country	Population	No. of Live Births*	No. of Deaths*
China	1,336,922,550	14,116,242	7,238,311
Mexico	111,254,336	1,705,811	406,874
South Africa	43,616,972	600,305	769,448
United Kingdom	60,972,219	502,455	474,148
United States	305,214,694	3,335,170	1,945,124

*July 1, 2008, to April 9, 2009 (282 days).

3. Discuss how each of the following attributes might be quantified:
    a. the clarity of a computer screen
    b. the shades of gray from white to black that a printer is able to make
    c. the steepness of a line on a coordinate plane
    d. the distance from one city to another
    e. population density
    f. the likelihood of drawing a red ball from a bag containing a mixture of red and blue balls
    g. quality of performance of an undergraduate in all her coursework
    h. the sweetness of a drink
    i. the strength of iced tea

4. By what process could you enlarge polygons while maintaining their shape?

5. Suppose you have two similar triangles (same shape, perhaps different sizes). Discuss the different ways the two shapes could be compared. Do any of these ways involve ratios? Do any not involve ratios?

Supplementary Learning Exercises for Section 9.1

1. A student says, "You claim that a rectangle 105 feet by 100 feet is more nearly square than one that is 10 feet by 5 feet. But they're both off by the same amount, 5 feet. Why do you say the larger one is more nearly square?" What do you say?

2. How would you decide which of these two boxes is more nearly a cube (all the flat pieces are squares of the same size)?

    Box 1: length 9 meters, width 6 meters, height 4 meters
    Box 2: length 50 meters, width 45 meters, height 40 meters

3. How might each of the following be quantified?
    a. the strength of a weight lifter
    b. the strength of lemonade
    c. the strength of a bar of steel
    d. the strength of a salt-water mixture

4. A fun-house mirror might make you look taller but leave your width the same. Use the idea of ratio to explain why you look funny in such a mirror, whereas with a regular mirror, you look all right.

5. Tell why or why not these two parts of a recipe will give the same taste.

    The recipe calls for $\frac{1}{4}$ teaspoon (tsp) of chili powder and 2 tsp of garlic powder. Someone is making a larger amount and, adding 1 tsp each, uses $1\frac{1}{4}$ tsp of chili powder and 3 tsp of garlic powder.

6. Father 1 says, "My daughter got 2 hits in 5 tries!" Father 2 thinks, "Well, my daughter has 4 hits in 12 tries, so it looks as though my daughter is twice as good as his." Tell why Father 2 should not say that out loud.

9.2 Comparing Ratios

Thus far we have used ratios to compare qualities such as squareness, steepness, or coffee strength. In this section we continue such comparisons and consider situations when two ratios are proportional (equal) and when they are not proportional. Unit ratios are introduced, and the $x:y$ and $\frac{x}{y}$ link is again illustrated.

ACTIVITY 1   Orange Juicy

The orange drink in Pitcher A is made by mixing 1 can of orange concentrate with 3 cans of water. The mixture in Pitcher B is made by mixing 2 cans of orange concentrate with 6 cans of water. Which will taste more "orangey": the mixture in Pitcher A or the mixture in Pitcher B or are they the same?[3]

a. Give an argument to support your answer to the above question.

b. How would you, as a teacher, respond to a student who says that the mixture in Pitcher A is more orangey because less water went into making it?

c. In a fifth-grade class some students reasoned like the student in part (b). Other students in the same class argued that the mixture in Pitcher B is more orangey because it has more orange concentrate. How would you settle the argument? What is wrong with the reasoning of the students in each group?

d. There was another student who argued as follows:

    Pour 1 can of orange concentrate and 1 can of water into Pitcher A. Take 2 cans of orange concentrate and 2 cans of water and pour them into Pitcher B. The two mixtures are equally orangey because they are made with equal parts orange and water, that is, 50-50. Now there are only 2 cans of water left to go into Pitcher A and 4 cans of water left to go into Pitcher B. Because the mixtures are the same strength, when you add 4 cans of water to the mixture in Pitcher B, it will be more watery than the mixture in Pitcher A, which gets only 2 more cans of water. Therefore, the mixture in Pitcher A is more orangey.

    How would you deal with this student's thinking?

e. How can "oranginess" be quantified in this situation to facilitate the comparison? ●

Once we recognize that a ratio is, itself, a quantity, then we can compare ratios just as we compare quantities. In the problem above, we can ask questions such as these: 1 can orange concentrate and 3 cans of water, 1 to 3; 2 cans orange and 6 cans water, 2 to 6; which mixture is more orangey? Or equivalently, which is the larger ratio?

The students in Activity 1 were reasoning additively when they gave their reasons. Focusing on just one of the two quantities being combined to make the mixture, orange or water, but not both, will lead to faulty reasoning. In the Orange Juicy situation, one has to focus on the two quantities, amount of orange *and* amount of water. That is why a ratio of the two quantities is an appropriate measure of the strength of the mixture.

 DISCUSSION 3   Other Ratio Situations

What are some other situations in which a ratio would be needed in order to make a comparison? ●

When we compare two ratios, we are trying to compare the *sizes* of the ratios. If two ratios are equal (equivalent), we say that they form a *proportion*, or that the two quantities involved are *proportional*. So, the question about the orange juice mixtures could be phrased as: Are the amounts of concentrate and water proportional in the two recipes?

> A **proportion** is a statement that two ratios are equal to one another. The quantities are said to be proportional.

EXAMPLE 1

In Activity 1, the ratio of orange concentrate to water was $1:3$. Any other orange juice of the same strength would have some n such that $1:3 = 1n:3n$, viewing it as n repetitions of the $1:3$ recipe. In the case where $n = 2$, we have the proportion $1:3 = 2:6$. This also can be written as $\frac{1}{3} = \frac{2}{6}$, so the earlier work on equal fractions is applicable here. ●

The close relationship between ratios and fractions, and the flexible use of $\frac{a}{b}$ for $a:b$, is strengthened by the important similarity in calculating equal fractions and equal ratios, shown in the following equations:

$$\frac{x}{y} = \frac{nx}{ny}\ (n \neq 0) \qquad \text{vs.} \qquad x:y = nx:ny\ (n \neq 0)$$

This similarity allows us to use the $\frac{x}{y}$ form for the $x:y$ form in many calculations.

    If another batch of juice is made up with 3 cans of concentrate to 7 cans of water, the ratio of concentrate to water could be expressed as $3:7$. This ratio is not equal to $1:3$. The new mixture is not proportional to the other mixtures. Thus, a question can be asked: Which juice is stronger, the juice with 1 can of concentrate per 3 cans of water, represented as $\frac{1}{3}$ (for the $\frac{1}{3}:1$ unit ratio), or the juice with 3 cans of concentrate to 7 cans of water, represented as $\frac{3}{7}$ (for the $\frac{3}{7}:1$ unit ratio)?

> *Think About ...*
>
> Which do you think would taste juicier, 2 cans of concentrate to 5 cans of water or 3 cans of concentrate to 6 cans of water?

Comparing these ratios is much like comparing fractions. Which is larger? We know $\frac{1}{3} = \frac{3}{9}$, and we can reason that $\frac{3}{9} < \frac{3}{7}$ because ninths are smaller than sevenths. Thus, the first juice is weaker. Alternatively, you could compare $\frac{1}{3}$ and $\frac{3}{7}$ by finding a common denominator: $\frac{7}{21}$ and $\frac{9}{21}$, but you must take care that this procedure does not simply become a rule that is not understood. The first way of comparing fractions, which depends on number sense, assures that you understand the problem. In other words, the first way is more transparent.

    Another kind of comparison problem occurs when asked, "*How much water should be used with 12 cans of concentrate, given that the strength is to be equal to the 1 can of concentrate to 3 cans of water?*" This question is often called a **missing-value problem** and is represented as $\frac{1}{3} = \frac{12}{x}$. There are several ways to think about this proportion when asked to find the value of x. Here, for every can of concentrate you would need 3 cans of water. So for 12 cans of concentrate, you would need 12×3 cans of water, or 36 cans of water.

But suppose, instead, *I have 12 cans of water, and I want to know how much concentrate to use.* The $\frac{1}{3}$ can also be thought of as the concentrate-to-water ratio; there is $\frac{1}{3}$ can of concentrate per can of water, i.e., every $\frac{1}{3}$ can of concentrate calls for 1 can of water. So for 12 cans of water, there would be $12 \times \frac{1}{3}$ cans of concentrate, or 4 cans of concentrate.

These two equivalent expressions for ratios can be written as

$$1:3 \textit{ is the same as } \tfrac{1}{3}:1, \textit{ or } \tfrac{\frac{1}{3}}{1}.$$

Notice how these expressions were used in the previous two paragraphs. They all represent the ratio of concentrate to water. Just as the ratio $1:3$ can be read as "1 can concentrate for every 3 cans water," the ratio $\frac{1}{3}:1$ can be read "$\frac{1}{3}$ can concentrate for every can of water." Having equivalent expressions for ratios often can be useful in solving ratio problems.

Both of the comparison problem solutions involving 1 can of concentrate per 3 cans of water took advantage of the unit ratio.

A **unit ratio** is a ratio for which the first quantity is represented by some nonzero number and the second quantity is represented by 1.

 DISCUSSION 4   Using Unit Ratios

Suppose the missing-value problem had asked for orange juice of the same strength as orange juice with 3 cans of concentrate per 7 cans of water. For this problem, use unit ratios to answer the following questions:

1. What is the unit ratio for concentrate to water? How much concentrate should be used to make the same-strength orange juice if you have 12 cans of water?

2. What is the unit ratio for water to concentrate? How many cans of water would be needed to make orange juice of the same strength if you use 12 cans of concentrate? ●

Many textbooks would have the student set up a proportion to solve the first discussion problem: $\frac{3}{7} = \frac{x}{12}$, where x is called the "missing value." There are many ways of solving such missing-value proportion problems. They vary not only by efficiency but also by the transparency of the method, i.e., whether the procedure makes sense. In the following three examples, consider the different ways of solving this problem in which 3 cans of concentrate and 7 cans of water make orange juice the same strength as x cans of concentrate with 12 cans of water.

EXAMPLE 2

Three cans of concentrate for 7 cans of water means that for $\frac{3}{7}$ can of concentrate, 1 can of water is used. Thus,

$$\text{the unit ratio} = \frac{\frac{3}{7} \text{ can concentrate}}{1 \text{ can water}}$$

So for 12 cans of water, $12 \times \frac{3}{7} = \frac{36}{7} = 5\frac{1}{7}$ cans of concentrate are needed. ●

The method in Example 2 uses a unit ratio. It is commonly employed by children even if the cross-multiplication method in Example 4 has been shown to them as the "correct" way to solve proportion problems.[4]

EXAMPLE 3

In $\frac{3}{7} = \frac{x}{12}$, the ratios can be replaced by equal ratios using the same second entries (a *common denominator* in fraction terms):

$$\frac{3 \cdot 12}{7 \cdot 12} = \frac{7 \cdot x}{7 \cdot 12}$$

Clearly, the ratios will be equal when $3 \cdot 12 = 7x$, or $x = \frac{36}{7}$. ●

EXAMPLE 4

For $\frac{3}{7} = \frac{x}{12}$, we "cross-multiply" to get $7x = 36$, so $x = \frac{36}{7}$. ●

> ### *Think About …*
> How are the methods in Examples 3 and 4 related?

 DISCUSSION 5   Making Sense of Proportions

Discuss each of the three solutions in Examples 2–4 in terms of whether or not they make sense of the original problem. Why might a teacher want to delay presenting the solution in Example 4? ●

The idea of *rate* has been used above, without actually using the term. The ideas and techniques of ratios will also apply to rates. There is not a uniform usage of the term "rate" across curricula, but we will use the term as defined below.

> A **rate** is a ratio of quantities that change without changing the value of the ratio.

EXAMPLE 5

In Activity 1, the rate is the same for the two recipes, because the ratios in the two mixtures are equal. From rate of speed expressions, such as 50 miles per hour, meaning 50 miles for every 1 hour, you know that units with "per" are describing rates. **An elementary school textbook might explain "per" by saying it means "for every" or "for each," both of which are useful ways of thinking about "per" (and rates in general).** If we consider the two ratios in Activity 1 as rates, we say the rates are 2 cans of concentrate per 6 cans of water and 1 can of concentrate per 3 cans water. In a common form, the rate might be given as $\frac{2}{6}$, or $\frac{1}{3}$, can of concentrate per 1 can of water, again showing the close relationship between a ratio and a fraction. Even a ratio for a fixed situation might be considered a rate: 12 boys for 4 girls $(12:4)$ could be viewed as the rate 3 boys for every girl $(3:1)$. ●

Students first learning about ratios, proportions, and rates can be assisted by setting up a table. Thus, for the problem with the ratio of $\frac{1}{3}$ representing the ratio of the orange juice concentrate to water and 12 cans of concentrate are used, a fifth-grade teacher might show a table such as the following to find the amount of water needed:

No. of cans of OJ concentrate	1	2	12
No. of cans of water	3	6	36

ACTIVITY 2   Cupcakes and Perfume

1. A researcher[5] studying the ways in which students think about and solve proportion problems taught them to make tables to show their thinking. She added a third column to explain how each number in the second column was obtained. A typical problem with a corresponding table is given as an illustration.

   ▶ If 15 cupcakes cost $3.36, find the cost of 38 cupcakes. ◀

Cupcakes	Cost in dollars	Notes
15	3.36	Given
30	6.72	× 2
5	1.12	÷ 6
3	0.672	30 ÷ 10
38	8.512	30 + 5 + 3

The cost of 38 cupcakes was $8.51.

2. Solve the following problem by designing a three-column table similar to the table in Problem 1.

▶ A famous designer gave away free perfume samples in the mall, and in 15 minutes, 600 people picked up his free gifts at the cosmetic counter. If this give-away is going to be repeated at this rate, but this time for 1.5 hours, how many gift packages will be needed? ◀ ●

 ACTIVITY 3 Pepperoni to Go![6]

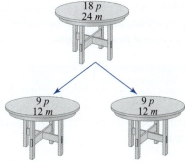

Luigi's is a pizza parlor that caters to the local college crowd. The 24 members of the chess club come in to celebrate. Eighteen pizzas had been ordered in advance (all the same size). None of the tables will hold 24 persons, so they sit at two tables, each with 9 pizzas and 12 members. We designate the change from one to two tables like this:

We say the distribution now is $\frac{9p}{12m}$ (9 pizzas for 12 members) at each of the two new tables.

1. a. Construct diagrams of alternative ways the 24 members can be seated and show how they might share equitably the 18 pizzas.

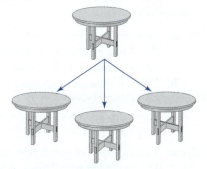

   b. Here is a seating arrangement where pizzas and members were moved from one table to three different tables in such a way that everyone got a fair share of pizza: In how many different ways could 18 pizzas and 24 members have been arranged at each table so everyone got a fair share of pizza? How much pizza does each person get? Is there more than one way to distribute pizzas among members?

   c. Return to part (a). Discuss your work in terms of equivalent ratios.

2. Consider some new situations, keeping in mind that pizzas are shared fairly at each table.

   a. Would the distribution $\frac{10p}{14m}$ and $\frac{8p}{10m}$ be fair for each member?

   b. Who would get more pizza, a person sitting at a table of $\frac{3p}{5m}$ or a person sitting at the table of $\frac{7p}{9m}$?

   c. Another person has $1\frac{2}{3}$ pizzas. At what table might she be sitting?

   d. Can someone who is served $\frac{1}{2} + \frac{1}{3}$ pizza have sat at the table $\frac{2p}{5m}$?

   e. If someone is given a serving of $\frac{1}{2}$ pizza, at which tables might he be sitting?

   f. Someone is sitting at a table designated as $\frac{4p}{5m}$. At which table would she get only half as much? ●

Notes

 DISCUSSION 6   When Not to Use Proportions

Explain why proportional reasoning might not be appropriate to use in the following problems:

1. Jake drove 72 miles during the first hour of his trip. How long will it take to drive the entire 144 miles of his trip?

2. In a pie-eating contest, Juarez ate two pies in the first 5 minutes. How many pies can he eat in 1 hour?

3. It took Denise 20 minutes to complete 10 out of the 20 problems that were assigned. How long will it take her to complete all 20 problems?

4. Jim can mow the lawn in 45 minutes. Today Janyce is helping him. How long will it take for the two of them to mow the lawn? ●

Proportional reasoning, i.e., reasoning about proportions, is the major type of multiplicative thinking introduced in upper elementary and middle schools. Many elementary and middle-school textbooks devote only a few pages to ratio and proportion, and simply state what a ratio is, what a proportion is, and then ask students to solve problems by setting up proportions and cross-multiplying. There is often little to no understanding of proportions developed in such lessons.

> **TAKE-AWAY MESSAGE . . .**   Ratios are quantities that can be compared; when equal, the quantities are proportional. When students are asked to find the missing value of a proportion, they often do so mechanically. Proportional reasoning is a vitally important type of reasoning that undergirds much of the mathematics that students will encounter in secondary school and college.

Learning Exercises for Section 9.2

Treat Exercises 1–15 as thinking exercises, rather than relying on computational procedures that you may remember. That is, your explanations should not reside solely on elaborate calculations.

1. **a.** Three scoops of coffee are used with 4 cups of water in a coffee machine; 4 scoops of coffee are used with 6 cups of water in another coffee machine. Which brew will be stronger? How do you know?
    b. Which is stronger, 4 scoops for 8 cups of water or 7 scoops for 15 cups of water? How do you know?
    c. How many scoops of coffee were there per 1 cup of water in each case? How do you know?

2. **a.** A certain kind of punch mix uses a 1:2 ratio, i.e., each cup of punch concentrate is to be diluted with 2 cups of water. Sonia was mixing a large bowl of punch for a party. She miscounted and put in 8 cups of concentrate to 15 cups of water. Was the mix stronger or weaker than intended?
    b. On the second batch she used 9 cups of concentrate and 19 cups of water. Was the second batch stronger or weaker than intended by the instructions? Was it stronger or weaker than the first batch?
    c. List some mixes that are slightly stronger than intended.
    d. List some mixes that are slightly weaker than intended.
    e. List some mixes that are much stronger than intended.
    f. List some mixes that are much weaker than intended.

3. Repeat Exercise 2, parts (c–f), with a punch that uses a 1:1 ratio: 1 cup of concentrate to 1 cup of water.

4. **a.** Use unit ratios on this problem:

   ▶ A manufacturer recommends 10 tablespoons of cocoa be mixed with 4 cups of milk to make hot cocoa. A school cafeteria is fixing hot cocoa for 50 first-graders. How many tablespoons of cocoa should they use if they are to mix enough to give every first-grader a cup? Given that 2 tablespoons of cocoa are equivalent to $\frac{1}{8}$ cup of cocoa, how many cups of cocoa should be used? ◀

   b. Name one rate in the problem given in part (a).

5. Adam and Matt are using different maps. On Adam's map, a line 6 inches long represents a road that is really 9 miles long. On Matt's map, a line 8 inches long represents a road that is really 12 miles long. If both boys were to measure the distance from City A to City B, who would have a longer line (in terms of inches)?

6. Car A can travel a greater distance in 3 hours than Car B can travel in 2 hours. If possible, find which car will travel a greater distance: Car A in 5 hours or Car B in 6 hours.

7. Read the following two problems:

   A. Which is stronger: 3 cups of orange concentrate mixed with 4 cups of water or 5 cups of orange concentrate mixed with 8 cups of water?
   B. Which is stronger: 3 cups of orange concentrate mixed with 5 cups of water or 5 cups of orange concentrate mixed with 8 cups of water?

   More students answer Problem A correctly than Problem B. Explain why you think this is so.

8. Horse A can travel 40 km in 3 hours. Horse B can travel 67 km in 5 hours. If possible, find which horse can travel faster.

9. Workers A and B, working 9 hours, made 243 parts. Worker A makes 13 parts in 1 hour. If the workers work at a steady rate throughout the day, who is more productive?

10. Tuna can be bought in 12-oz cans that sell for $1.09 each or in 10.75-oz cans that sell for 98¢ each. Which is the better buy? Use unit ratios to answer this problem.

11. Jane and Scott were given identical boxes of crayons. Two weeks later, Jane had $\frac{4}{9}$ of the box left and Scott had $\frac{3}{7}$ of his box left. Who has more crayons left?

12. In Mrs. Heath's class there are 13 girls and 11 boys. In Mrs. Lauri's class 15 of 28 children are girls. In which class are the girls better represented?

13. Car A started out with a full tank of gas (12 gallons) and traveled 250 miles before it ran out of gas. Car B, whose tank capacity is 15 gallons, started out with half a tank and traveled 145 miles. Which car is more economical in terms of gasoline consumption?

14. In Boogleville, two-thirds of the men are married to three-fourths of the women. What is the ratio of men to women? (*Hint*: Draw a diagram.)

15. How can you tell whether the two ratios, $133:161$ and $95:115$, are equal?

16. **a.** Rewrite $a:b$ and $c:d$ as equal ratios with the same (algebraic) second entries to give a criterion for when $a:b$ is equal $c:d$. (*Hint*: First write the ratios in fraction form: $\frac{a}{b}$ and $\frac{c}{d}$, and then replace those with equal fractions having the same denominator.)

   b. **Cross-multiplication algorithm** The idea in part (a) can be used in solving proportions with an unknown entry—e.g., $57:32 = 171:x$. Writing the ratios as fractions $\left(\frac{57}{32} = \frac{171}{x}\right)$ and then replacing the fractions with equal fractions having the same denominator $\left(\frac{57x}{32x} = \frac{32 \cdot 171}{32x}\right)$ give $57x = 32 \cdot 171$, an equation that can be solved by dividing both sides by 57, giving $x = 96$. Notice that the $57x$ and $32 \cdot 171$ could be obtained from the original $\frac{57}{32} = \frac{171}{x}$ by multiplying diagonally.

   continue

This procedure illustrates a useful shortcut called *cross-multiplication*. Solve

$$\frac{51}{x} = \frac{17}{48} \quad \text{and} \quad \frac{11.7}{2.16} = \frac{x}{7.2}$$

using cross-multiplication. We don't focus on it here because it is often used mindlessly, without understanding.

   c. Once children learn cross-multiplication, they often use it on addition and multiplication of fractions. Does cross-multiplication give correct answers then?

17. There is a minor controversy about whether or not ratios and fractions are exactly the same. As you have seen, *computationally* they behave alike in some important situations. But consider this setting: "A good batter made 3 hits in 4 at-bats in one game, and 2 hits in 4 at-bats in the next game. How did the batter do in the two games together?" Does the natural response $\frac{3}{4} + \frac{2}{4} = n$ give the correct answer?

18. Complete the table for the following problem:[7]

   ▶ Cheese costs $4.25/pound. Nancy selects several chunks for a large party and when they are weighed, she has 12.13 pounds of cheese. How much will it cost her? ◀

Pounds	Cost	Notes
1	$4.25	Given
10		
2		
0.1		
12.1		
0.05		
0.01		
0.03		
12.13		

19. A donut machine produces 60 donuts every 5 minutes. How many donuts does it produce in an hour?

   a. Is there a *rate* in this problem? If so, what is it and why is it a rate? If not, why not?
   b. What unit fraction is associated with this rate?
   c. Answer the question in part (a) using the unit rate.
   d. Set up a proportion and solve the problem without using the unit rate.

20. A certain car gets 23 miles to the gallon on open highway. How far can it travel on 12 gallons of gas on open highway?

   a. Is there a *rate* in this problem? If so, what is it and why is it a rate? If not, why not?
   b. What unit fraction is associated with this rate?
   c. Answer the question in part (a) using the unit rate.
   d. Set up a proportion and solve the problem.

21. Name University tries to maintain an average of seven undergraduate students for every two graduate students. For every 100 graduate students admitted, how many undergraduate students would you expect to be admitted?

   a. Is there a *rate* in this problem? If so, what is it and why is it a rate? If not, why not?
   b. What unit fraction is associated with this rate?
   c. Answer the question in part (a) using the unit rate.
   d. Set up a proportion and solve the problem without using the unit rate.

22. In Mrs. Heath's class there are 12 girls and 11 boys.

    a. What is the ratio of girls to boys?

    b. Why is there not a rate associated with this problem?

23. Find two examples of the use of unit ratios in a grocery store.

24. "Numberless" problems like the ones below allow a focus on the quantities and their relationships rather than on computation. Explain your thinking for your answers for each.

    a. Andy drove more miles than Bob did, and Andy drove fewer hours than Bob did. Who drove faster—Andy, Bob, the same, or you can't tell?

    b. Carla swam fewer laps than Donna did. Carla took more time than Donna did. Who swam faster—Carla, Donna, the same, or you can't tell?

    c. Evan made more cookies than Fawn did. Evan took longer than Fawn did. Who baked cookies faster—Evan, Fawn, the same, or you can't tell?

Supplementary Learning Exercises for Section 9.2

1. a. Without looking, tell what a proportion is.

    b. Without looking, tell what a rate is.

2. Why isn't the proportional reasoning used appropriate?

    a. One person can sing the song in 2 minutes, 30 seconds, so a trio can sing the song in 7 minutes, 30 seconds.

    b. If a 5-person crew can do the job in 8 hours, then a 15-person crew should take 24 hours.

3. The scale on one map is $1:2\frac{1}{2}$, meaning 1 inch on the map represents $2\frac{1}{2}$ miles.

    a. $\frac{3}{8}$ inch on the map represents how many miles?

    b. $3\frac{1}{4}$ inches on the map represents how many miles?

    c. $6\frac{1}{4}$ miles would be how many inches on the map?

    d. 60 miles would be how many inches on the map?

    e. $\frac{1}{2}$ mile would be how many inches on the map?

4. A recipe calls for $2\frac{1}{2}$ cups of flour and $\frac{2}{3}$ cup of sugar. If you adjust the recipe, how much flour or sugar should you use for each of the following?

    a. 4 cups of flour   **b.** $5\frac{1}{3}$ cups of flour   **c.** 1 cup of sugar   **d.** 3 cups of sugar

5. Within certain limits, a manufacturer of a vehicle knows that for every 50 pounds of added (or lost) weight, the vehicle will lose (or gain) 1.5% in fuel efficiency.

    a. What is the effect of 80 pounds *more* weight on the fuel efficiency?

    b. What is the effect of 80 pounds *less* weight on the fuel efficiency?

    c. What is the effect of 35 pounds more weight on the fuel efficiency?

    d. What change in weight will give a 5% increase in fuel efficiency?

    e. What change in weight will give a 6% decrease in fuel efficiency?

6. One basketball statistic is the "assist-to-turnover" ratio (assists are good, turnovers bad). During part of the season, Andy had 4 assists and 10 turnovers, Bob had 13 assists and 17 turnovers, and Cam had 9 assists and 13 turnovers.

    a. A student reasons that Bob is the best player of the three because he made the most assists. Why is this reasoning faulty?

    b. Another student thinks Andy is the best player, because he had the fewest turnovers. Why is this thinking faulty?

    c. Who of Andy, Bob, and Cam has the best assist-to-turnover ratio?

7. If a team has a wins-to-losses ratio of $4:3$, does that mean they won exactly 4 games and lost 3 games? Give your thinking.

8. Write each of the following ratios as a unit ratio.

    a. $8:3$ **b.** $3:8$ **c.** $14:5$

9. What does it mean to say, "The city budget has grown at a rate of 5% per year for the last decade"?

10. Dale charged $30 for shoveling snow for $2\frac{1}{2}$ hours. At that rate, how much can Dale expect to earn by shoveling snow for 6 hours?

11. The birthrate gives the number of births per 1000 population.
    a. In 1910 the population was about 92,228,000 and the birthrate was about 30.1. About how many births were there in 1910?
    b. By 2000, 4,058,814 births gave a birthrate of 14.7. What was the approximate population in 2000?

12. The total fertility rate (TFR) is the average number of children that would be born to an average woman over her lifetime. Here are some 2011 estimates of TFRs (http://www.cia/gov/library):

    Canada 1.58   Mexico 2.29   Russia 1.42   United States 2.06

    a. If roughly half a country's population is female, what does the TFR predict for the U.S. population of about 313,232,000?
    b. What does a TFR less than 2, such as Canada's and Russia's, predict about population growth?
    c. Which, birthrate (Exercise 11) or TFR, would a young woman find more interesting? A public health planner?

13. A woman's pulse rate might be 72 beats per minute.
    a. What does that mean?
    b. Is it possible to find the number of beats in 10 seconds? If so, do it. If not, explain why.

9.3 Percents in Comparisons and Changes

You have worked on some problems involving percents in previous sections, and of course you have encountered them many times in your daily life. Percents provide us with a handy way of comparing fractions when the values are "messy." Percents also are used as a way to discuss change, but care needs to be taken that percent problems are understood. *Percent* comes from words meaning "per hundred," so the hidden ratio or rate meaning is clear.

> A **percent** is a ratio for which the value of the second quantity is understood to be 100.

So, 8.5% means 8.5 per 100. Calculating percents often uses the $x:y$, $\frac{x}{y}$, and $x \div y$ links.

 DISCUSSION 7   When Are Percents Handy?

Consider each of these two problems. What is the advantage of using percents in situations such as these?

1. One discussion class has 28 females and 17 males. A large lecture class has 106 females and 62 males. In which class is the female population more dominant?

2. Suppose you scored 15.5 points out of a possible total of 20 points on the first quiz and 59 points out of a possible total of 75 points on the second quiz. How would you figure out on which quiz you did better? ●

One can think of percent as a standardized way to express ratios or fractions in order to facilitate comparisons. For example, when we find an equivalent ratio for which the value of the second quantity is 100, we are finding a percent. If we compare test grades by finding ratios that are equivalent to $\frac{15.5}{20}$ and $\frac{59}{75}$ but with 100 as the denominator, we are finding a percent. We can write $\frac{59}{75}$ as 78.7% and $\frac{15.5}{20}$ as 77.5%. Now the two quiz performances given in Discussion 7 are easy to compare.

> ⋮ *Think About ...*
> ⋮ In Activity 1 in the last section, you were asked which of two mixtures tasted more
> ⋮ "orangey." One could (but need not) think in terms of percents. What percent of the
> ⋮ mixture in Pitcher A (1 orange to 3 water) is orange? What percent is water? How about in
> ⋮ Pitcher B (2 orange to 6 water)?

Solving percent problems is really no different from solving ratio and proportion problems.

Some textbooks teach percents by introducing three different types of percent problems. All have the form $a\%$ of $b = c$, and in each case, only one of a, b, or c is unknown. These three types of problems are sometimes taught as three different types, with a rule for each. The rules are usually forgotten. There is really just one rule for solving these problems.

 ACTIVITY 4   Three Kinds of Percent Problems? Or Just One?

Each of these problems can easily be solved mentally. Compare your mental strategies that depend on good number sense. Then for each problem, set up a proportion for which one ratio has 100 as the value of its second quantity. Compare these methods.

1. 30 is what percent of 45?

2. What number is 50% of 60?

3. 16 is 25% of what number?

Did these seem like three different kinds of problems to you, or basically the same kind of problem? Could you solve each without thinking about which "type" it was? ●

Just as a fraction is a fraction *of some quantity,* a percent is a percent *of some quantity.* The next activity will help to make that clear.

 ACTIVITY 5   Fair or Not?

1. The boss says, "You remember when business was bad last year, I had to cut everyone's pay by 10%? Well, business is better, so I can raise your pay by 10% now. That will put you back to where you were before the cut." Is the boss correct?

2. You buy an article on sale for 20% off in a locale where there is an 8% sales tax. You expected the clerk to figure the discount first and then the sales tax on the reduced price, but the clerk figured the sales tax first and then the discount on that price. Were you cheated? If so, by what percent? If not, explain. ●

Examples like those in Activity 5 show how important it is to know what a percent is a percent *of,* i.e., the *base* for the percent. For Problem 2 in Activity 5, the base could be the price of the purchase, or it could be the price of the purchase plus tax. Your answers should have referred to the base (the referent) for each percent occurring in the problems.

For a problem in which there is a percent increase or a percent decrease, knowing that the base is the starting quantity helps to determine how to solve them.

EXAMPLE 6

News Item 1: *A company announced that it is increasing its workforce by 15%, or 30 workers.* Can you tell how many workers will be in the workforce after the increase?

Solution

The **15%** refers to the starting workforce, which is the base for the percent. If 15% of the starting workforce is 30 workers, then 1% of the starting workforce would be 2 workers, and 100% of the starting workforce would be 200 workers. After the hiring, the company will have 230 workers. ●

Notes

EXAMPLE 7

News Item 2: *The town's budget is now $3 million. That is 25% more than it was last year.* What dollar amount was last year's budget?

Solution

The **25% more** is 25% more than last year's budget, which is the base. The new budget, $3 million, is 125% of last year's budget. If we divide both by 5, $0.6 million is 25% of last year's budget, and so 100% of last year's budget is $2.4 million, i.e., last year's budget was $2.4 million. ●

EXAMPLE 8

One school had a decline in enrollment of 12%, and now has 352 students. How many students did the school have before the decline?

Solution

The **12%** is 12% of the original enrollment, so the 352 students would be 88% of the original enrollment. One percent of the original enrollment would then be $352 \div 88 = 4$ students, so 100% of the original enrollment would have been 400 students. ●

 ACTIVITY 6   What's the Base?

1. In a 1-hour TV slot, one week's program started with a 3-minute review of last week as a lead-in. Commercials took 30% of the total time for the slot. What percent of the total show time was left for new developments?

2. A mayor notices these facts: The city budget last year was 25% more than the $15,000,000 budget of the previous year, and this year's budget is 12% more than last year's budget.

    a. What is this year's budget?
    b. Why is this year's budget not 37% of the budget two years ago?

3. In preparing the county budget for next year, planners intended to keep costs the same as this year's, except for increases in the total budget of 5% for salaries, 3% for benefits, 3% for energy, and 4% for maintenance and supplies. The new budget calls for $32.2 million dollars. What is the dollar amount for this year's budget? ●

These examples should help to make clear how important it is to know what a given percent is a percent *of*. To repeat, in problems that involve a percent increase or decrease, the base for the percent is the original amount.

   Percent can also be thought about as a way of expressing parts of wholes. Percent can be a way of introducing children to decimals and fractions, although the sequence in elementary schools is usually the reverse: fractions, then decimals, then percents. In a research study[8] in Canada with fourth-graders, students were introduced to percents first because the researchers felt that students had a better intuitive knowledge of percents than of fractions. The students became quite adept at solving percent problems mentally (and later were able to transfer that knowledge to decimals and fractions). For example, two typical students in the study were asked for 65% of 160. Their two answers were

> "Okay, 50% of 160 is 80. Half of 80 is 40 so that is 25%. So if you add 80 and 40, you get 120. But that's too much because that's 75%. So you need to minus 10% (of 160) and that's 16. So, 120 take away 16 is 104."

> "The answer is 104. First I did 50% which was 80. Then I did 10% of 160 which is 16. Then I did 5% which was 8. I added them $(16 + 8)$ to get 24, and added that to 80 to get 104."

 ACTIVITY 7   Your Turn

Use either student's procedure described above to find 45% of 180. ●

DISCUSSION 8 Do You Talk the Talk?

We often misinterpret common phrases using percents. For example, 25% *larger than* $1200 is correctly interpreted to mean $1500, and 25% *as large as* 1200 is correctly interpreted to mean $300. However, 300% *larger than* 1200 and 300% *as large as* 1200 are often interpreted to mean the same thing, although 300% *larger than* $1200 is $4800 and 300% *as large as* $1200 is 3600.

Once again, drawings can help make the needed distinction.

$1200

25% larger than $1200

25% as large as $1200

You make drawings of 300% larger than $1200 and 300% as large as $1200. ●

TAKE-AWAY MESSAGE . . . Percent is simply a special ratio and, in fact, a very useful one that can be used in a variety of everyday situations. Percent is often used for comparisons because the denominator is always 100, so only the *numerators* have to be compared. Problems can occur when people are not paying attention to the base of the percent: percent *of what?* With practice, one can often find percents mentally. In percent-change situations, the base for the percent is the starting value.

Learning Exercises for Section 9.3

Reminder: Drawings are often very useful.

1. One coat was originally $120 but is on sale for $90; another coat was $150 but is on sale for $120. What is the percent of discount in each case? Which is the better buy? (This last question is very ambiguous. What are the different ways it could be answered?)

2. On one test you received 21 out of 28 points and on the second test you received 38 out of 50 points. On which test did you do better, in terms of percent correct? How should an instructor average these grades?

3. Consider again the problem:

   ▶ Jane and Scott were given identical boxes of crayons. Two weeks later, Jane had $\frac{4}{9}$ of the box left and Scott had $\frac{3}{7}$ of his box left. Who has more crayons left? ◀

   Solve this problem again, but this time standardize each fraction by changing it to a percent, and then answer the question.

4. Jaqi owes Katie $60 and pays $35 toward her debt.
   a. What percent of the debt has she paid?
   b. What percent is still owed?
   c. What is her debt now?
   d. If Jaqi pays Katie another $10 toward her new debt, what percent of her new debt is paid off?
   e. What percent of her old debt is paid off?

5. If the box shown represents 75% of an amount, show a box that represents 125% of the same amount. What percent of that box is the given one?

6. If the box shown represents 150% of an amount, show a box that represents $33\frac{1}{3}$% of the same amount.

7. The school budget just passed for this year is $4.2 million, which is 10% less than the budget last year. How much was the budget last year?

8. Find the percent change for each of the following cities:
   a. The population of City P increased to 125,000 from 100,000.
   b. City Q's population decreased to 100,000 from 125,000.
   c. City R's population grew to 87,450 from 72,625.
   d. City S's population went from 20,125 to 17,750.

9. City A's population is 15,000; City B's is 12,000.
   a. City A's population is what percent larger than City B's?
   b. City B's population is what percent less than City A's?
   c. City A's population is what percent of City B's?
   d. City B's population is what percent of City A's?

10. Street Scene went to 3 days instead of the usual 2 days. What percent change was that?

11. The decrease in price for notebooks was $1.60, a 40% decrease. What was the original price and what is the new price?

12. One day the Dow Jones was off by 2%. It closed at 10,567. About what was the Dow Jones the previous day?

13. Dan has a novel to read. He reads $\frac{1}{4}$ of the pages on each of Monday and Tuesday, 65 pages on Wednesday, another 20% of the pages on Thursday, and the final 61 pages on Friday. How many pages were in the novel?

14. "I'll double your pay," said the eager employer. What percent increase over your current pay would that be?

15. In changing jobs, Pat increased her hourly take-home pay by 24%, or by $3.48 an hour. What is her new hourly pay, and what was her old one?

16. "Accidents were down in our town, with about 75% as many this year as last year. This year there were about 840 accidents." If the speaker was correct, about how many accidents were there last year?

17. One department head is chatting with another. "Today, our two budgets total $1.2 million. But that's only 3% of the company's whole budget. And you spend $2 for every $1 I do." What is the company's whole budget, and what is each of the two department's budgets?

18. In one basketball game, Angie scored $\frac{1}{3}$ of her team's points, Beth scored 15% of them, and Carlita scored $\frac{1}{4}$ of the points. The rest of the team scored 16 points. The team made 60% of its points in the second half by hitting 55% of its shots.
   a. How many points did the team score in the first half?
   b. For what other quantities do you know values?

19. Dee has a part-time job and wants to buy a car. She figures that she spends about half her income on her apartment rent and utilities and about $12\frac{1}{2}$% on food, for a total of $450 per month. She wants to allow $10 per week for entertainment and incidentals. About how many dollars are left for car payments and car expenses?

20. Review: Estimate each of the following:
   a. 35% of 121 people   b. 52% of 12 pounds   c. 65% of 67 kilometers
   d. 15% of $39.15   e. 23% of 102 miles   f. 35% of 66 minutes
   g. 76% of $399   h. ___% of $402 is $66 i. 74 miles is ___% of 298 miles
   j. 24 pounds is 49% of what?

21. Review: Find each of the following mentally:
   a. 25% of 80 people   b. 40% of 160 pounds   c. 65% of 80 kilometers
   d. 15% of $64   e. 12.5% of 80 miles   f. $33\frac{1}{3}$% of 66 minutes
   g. 75% of $700   h. ___% of $350 is $70 i. 75 miles is ___% of 750 miles
   j. 24 pounds is 200% of what?

22. What else can you tell from the following news clips?

    a. Four percent of the high school graduates will qualify. Nearly two-thirds of the students in the 4% group already qualify. . . . About 3,600 students across the state will become newly eligible when the policy takes effect. . . . A third of these newly eligible students are expected to come from urban schools; a quarter will be drawn from urban areas.

    b. The company will lay off 11% of its workforce, about 1500 workers worldwide.

23. A newspaper reported the estimate that 30%, or $48 billion, of food in the United States is wasted each year in restaurants, cafeterias, and grocery stores. About how much is all the food in such places worth in a year?

24. One election report stated, "Candidate Smith has 445,015 votes, or about 54% of the total vote."

    a. About how many votes make up 1% of the total vote?
    b. About how many total votes are there?
    c. About how many votes did the other candidates get?

25. In each case, give the *percent* of the new hourly wage needed to return to the original hourly wage.

    a. Your hourly wage of $10.00 is cut by 10%.
    b. Your hourly wage of $18.50 is cut by 10%.
    c. Your hourly wage is cut by 10%.

26. Rosie weighs 160 pounds and gains 10%. What *percent* of her new weight must she lose to get back to 160 pounds?

27. One radio station reports the daily stock-market activity in terms of the percent change from the start of the day. What is the *exact* percent change from the beginning of Monday to the end of Wednesday, if the daily reports are as follows: Monday—plus 5%, Tuesday—plus 2%, Wednesday—minus 3%? (*Hint:* The answer is not plus 4%.)

Supplementary Learning Exercises for Section 9.3

1. Why is $\frac{697}{1000}$ easier to rewrite as a percent than $\frac{7}{13}$ is, by hand?

2. **a.** If a sale item is 25% off, what percent of the original price does one pay?
   b. If a sale item is 40% off, what percent of the original price does one pay?
   c. Losing 15% of one's weight means that the person still has what percent of his or her original weight?
   d. If X is 150% as large as Y, then X is _____ % more than Y. (*Hint:* Y is 100% as large as Y.)
   e. If John eats 175% as many calories as Jane does, then John eats _____% more calories than Jane does.
   f. If Kaylene spends 200% as much as Lana spends, then Kaylene spends _____% more than Lana does.

3. **Percent increase or decrease** Teachers—and citizens in general—often see phrases like "a 15% increase" or "a decrease of 25%," so people should be able to interpret them. Recall that the key is that actual amount of change, either the amount of increase or the amount of decrease, is compared to the *starting* value.

   For example, going from 2 people to 5 people represents an increase of 150% (the actual increase of 3 people is compared to the starting value of 2 people). The 5 people are 250% as many as the original 2 people. Or, going from 5 people to 2 people is a decrease of 60% (the decrease of 3 people is compared to the starting value of 5 people). The 2 people are 40% as many as the original 5 people.

   a. Average class size in one school district went to 28.6 students from 26.8 students. What is the percent increase in average class size?

continue

   b. End-of-year enrollment at one school was 552 after being 576 in September. What was the percent decrease?

   c. The average rent for a 2-bedroom apartment in one community was $800/month last year, but $850/month this year. What is the percent increase per month? Per year?

   d. The starting salary increased to $32,400, an increase of 8%. What was the starting salary before the increase? (*Hint*: 8% of the earlier salary + 100% of the earlier salary = 108% of the earlier salary.)

   e. The new math textbook in your school cost $22.95; the old one cost $19.95. What is the approximate percent increase in the price of a math textbook?

   f. Employment in one industry was down by 6%, or about 1200 people. How many did the industry employ, before and after the change? (In your drawing, the 6% is 6% of what?)

   g. Is raising a tax from 10% to 11% a 1% change, or a 10% change? Justify each possibility.

4. Calculate the missing information for these sale items.

	Usual price	Sale price	% savings
a.	99¢	$0.69	?
b.	$1.29	99¢	?
c.	$11.98	?	25%
d.	?	$1.87	25%
e.	59¢	?	$33\frac{1}{3}$%
f.	?	$15.15	20%

5. In a pre-election poll, one candidate received 37% of her party's vote. If the 37% is assumed to be a rate, about how many votes can the candidate expect on election day for the following?

   a. In a county with 7400 voters of her party?

   b. In a state with 158,000 voters of her party?

   c. Why might the 37% *not* be a rate?

6. A champion hot dog (and bun) eater ate 62 hot dogs in 10 minutes in one competition.

   a. What rate is that, in hot dogs (and buns), per minute?

   b. Why is it not reasonable to predict that the champion would eat 372 hot dogs in an hour?

   c. The second-place finisher ate 53 hot dogs in 10 minutes. If he and the champion ate at consistent rates, how far behind was the second-place finisher after 4 minutes?

   d. Is it reasonable to assume that the eaters' rates would be consistent, from start to finish? Explain your thinking.

7. For a particular brand of snack crackers, a serving of 9 crackers has 110 calories and 3.5 grams of fat.

   a. How many calories and how many grams of fat are in 14 of those crackers?

   b. A 2000-calorie diet allows 65 grams of fat. How many crackers would total 65 grams of fat?

   c. What percent of the 65 grams of fat would a 9-cracker serving have?

Supplementary Learning Exercises for Chapters 7–9[9]

1. Lisa and Rachel each mixed a can of paint. The colors they mixed were exactly the same shade of green. Lisa mixed 6 ounces of white paint with 4 drops of green tint. How much white paint did Rachel mix with 6 drops of green tint?

2. Todd can paint $\frac{3}{5}$ of a square wall with 1 gallon of paint. How much of that same wall could he paint with $\frac{3}{4}$ of 1 gallon? How many gallons of paint will Todd need to paint the entire wall?

3. Two workers working 8 hours made 512 parts. One of the workers makes 36 parts in 1 hour. If the workers work at a steady rate throughout the day, how many parts does the second worker make in 1 hour?

4. Sixteen liters of water were poured into a tank, filling it to $\frac{2}{5}$ of its volume. What is the volume of the tank?

5. For each present John needs $\frac{3}{8}$ meters of ribbon. John wishes to wrap 15 presents. How many meters of ribbon does he need?

6. If  represents $\frac{3}{2}$ of the total number of trees in a park, how many trees are there in the park?

7. The parking lot was full, and $\frac{1}{3}$ of the cars were red. This amounted to 18 cars. How many cars were in the parking lot?

8. If the group of dogs shown represents $\frac{3}{7}$ of the dogs at the pound, how many dogs altogether are at the pound?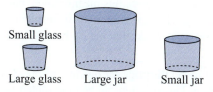

9. A group of 50 people go to a holiday camp for 28 days. They need to buy enough sugar. They read in a book that the average consumption of sugar for 10 persons is 3.5 kg per week. How much sugar do they need?

10. The drawing shows 2 glasses, a small one and a large one. It also shows 2 jars, a small one and a large one. It takes 15 small glasses of water or 9 large glasses of water to fill the large jar. It takes 10 small glasses of water to fill the small jar. How many large glasses of water does it take to fill the same small jar?

Small glass

Large glass   Large jar   Small jar

11. In 3 weeks a horse eats 10 lb of hay. How much hay will the horse eat in 5 weeks?

12. The chips shown represent $\frac{5}{6}$ of a unit. How many chips would there be in $\frac{2}{3}$ of that same unit? How many chips would there be in $1\frac{1}{2}$ units?

13. A train runs at a constant speed. Two brothers, Peter and Tom, are on the train and try to calculate the distances between the different stations by using their watches. Between A and B, the time it takes the train is 16 minutes, and they know the distance is 40 kilometers. Between two other stations, C and D, the time it takes the train is 36 minutes. What is the distance between C and D?

14. Officials in Antville, Beeburg, and Cowtown make these observations: The Antville budget went up 25% to $15 million; the Beeburg budget went up 20% to $15 million; and the Cowtown budget went up 15% from its former $12 million. Of Antville, Bugburg, and Cowtown, which had the largest budget last year?

15. Bigtown has 4 times the population that Littletown does. Littletown has $\frac{5}{8}$ as many people as Middletown does. Middletown has 40,000 people. What is the population of Bigtown?

16. Six children are sharing a large bag of chocolates. Abe eats an eighth of the chocolates, as do Bob and Cass. Dannie eats a third of the chocolates, and Ella eats a fourth of them. Fran's share finished the bag; Fran ate two chocolates. How many chocolates were in the bag originally?

17. Joel and Kendrick paint at the same rate. Joel gets a head start and has painted 120 square yards of a wall when Kendrick has painted 40 square yards. When Joel has painted a total of 600 square yards, how many square yards has Kendrick painted? (Caution!)

18. Lewis is more experienced than Max and can paint 120 square yards in the time that Max can paint 40 square yards. When Lewis has painted a total of 600 square yards, how many square yards has Max painted?

19. Samantha, Tyler, and Ullie are comparing notes on their wages. Samantha makes 50% more per hour than Tyler does, and Tyler makes $1.50 less than Ullie does. Samantha makes $10.50 per hour. How much do Tyler and Ullie make, per hour?

20. Velma, Winnie, and Zoe want to rent a 3-bedroom apartment. Velma wants the corner bedroom, so she will pay 25% more of the rent than Winnie will. Zoe wants the biggest bedroom and agrees to pay 50% more than Winnie. What share of the rent will each woman pay?

9.4 Issues for Learning: Developing Proportional Reasoning

Most adults use a cross-multiplication method to solve for a missing value in a proportion. However, research[10] shows that even when this strategy is explicitly taught to 11- and 12-year-olds, they often become confused and are unsuccessful with this method. If left on their own to solve such a problem, they commonly used the *unit method*, such as with this problem:

▶ John purchased 24 loaves of bread to sell in his grocery store last week, at a total price of $26. If he wishes to buy 30 loaves next week, how much will he have to spend? ◀

A typical answer was

> If John is paid $26.00 for 24 loaves, this means that he must have been paying a little over $1.00 for every loaf. The exact amount is 26 divided by 24, or $1.08\overline{3}$. If John wants to buy 30 loaves next week, that means he will have to pay 30 times $1.08\overline{3}$, or $32.50.

Note that the student first found the price for *one* loaf of bread, then multiplied by 30. The result is $30 \times 1.08\overline{3} = 32.5$. This solution can be represented as

$$\frac{26 \text{ dollars}}{24 \text{ loaves}} = \frac{1.08\overline{3} \text{ dollars}}{1 \text{ loaf}} \qquad \text{or} \qquad 1.08\overline{3} \text{ per loaf}$$

So, 30 loaves cost $32.50.

Someone using a cross-multiplication approach to solving a proportion would write

$\frac{26}{24} = \frac{x}{30}$ gives $24x = 26 \times 30 = 780$. So, $x = 780 \div 24$, 32.5, or $32.50.

Susan Lamon,[11] who has been investigating proportional thinking for many years, has listed the following characteristics of proportional thinkers. Use them to evaluate whether or not you are a proportional thinker. Here are some of the characteristics she described:

1. Proportional reasoners can think both in terms of unit rates, such as 25 miles per gallon of gas, and in terms of multiple units, such as $5.15 per 6 bottles of water.

2. Proportional reasoners are more efficient problem solvers. If water is $5.15 for 6 bottles of water and they wanted to know the cost of 24 bottles of water, they would think: 4 groups of 6 is $4 \times \$5.15$. Finding the unit price per bottle and then multiplying by 24 is an earlier way of finding the price of 24 bottles, but it is less efficient and more prone to errors.

3. They can use partitioning to help them. For example, if 3 people share 5 pizzas, then each person gets $\frac{5}{3}$ pizzas. That is, each person gets $\frac{1}{3}$ of the total 5 pizzas, which is $\frac{1}{3} \cdot 5$ or $\frac{5}{3}$ pizzas.

4. They can think flexibly about quantities and find unit quantities. If cans of spinach are 3 for 99¢, then 1 can is 33¢.

5. They are not afraid of fractions and decimals and can think flexibly with these numbers.

6. They can mentally compute with fractions, decimals, and percents. For example, they know that they can find $\frac{3}{5}$ of a quantity if they know $\frac{1}{5}$ of a quantity, and 70% if they know 10%.

7. They can identify everyday situations where proportions are not useful. For example, if told that Tommy ran a mile in 5 minutes, and asked how long it would take Tommy to run 10 miles, they would not simply multiply by 10, because Tommy cannot run that fast over several miles.

8. They can solve both missing-value problems and comparison problems by reasoning about them, not just rotely using the cross-multiplication strategy.

Research has shown that many children prefer using the *unit method* rather than the more common cross-multiplication method when solving proportions. Now some textbooks include the unit method, as this page in a Grade 6 textbook* does. Work the problems on this page.

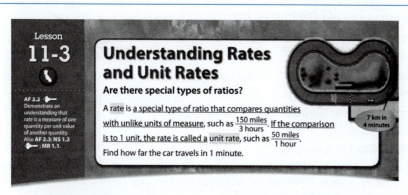

Lesson 11-3

AF 2.2
Demonstrate an understanding that rate is a measure of one quantity per unit value of another quantity. Also AF 2.3; NS 1.2 ; MR 1.1.

Understanding Rates and Unit Rates

Are there special types of ratios?

A rate is a special type of ratio that compares quantities with unlike units of measure, such as $\frac{150 \text{ miles}}{3 \text{ hours}}$. If the comparison is to 1 unit, the rate is called a unit rate, such as $\frac{50 \text{ miles}}{1 \text{ hour}}$.

Find how far the car travels in 1 minute.

7 km in 4 minutes

Guided Practice*

Do you know HOW?

Write each as a rate and as a unit rate.

1. 60 km in 12 hours

2. 26 cm in 13 s

3. 230 miles on 10 gallons

4. $12.50 for 5 lb

Do you UNDERSTAND?

5. What makes a unit rate different from another rate?

6. Explain the difference in meaning between these two rates: $\frac{5 \text{ trees}}{1 \text{ chimpanzee}}$ and $\frac{1 \text{ tree}}{5 \text{ chimpanzees}}$.

Independent Practice

In **7** through **18**, write each as a rate and a unit rate.

7. 38 minutes to run 5 laps

8. 36 butterflies on 12 flowers

9. 252 days for 9 full moons

10. 18 eggs laid in 3 days

11. 56 points scored in 8 games

12. 216 apples growing on 9 trees

13. 125 giraffes on 50 hectares

14. 84 mm in 4 seconds

15. 123 miles driven in 3 hours

16. 210 miles in 7 hours

17. 250 calories in 10 crackers

18. 15 countries visited in 12 days

Animated Glossary
www.pearsonsuccessnet.com

270 *For another example, see Set C on page 282.*

9.5 Check Yourself

This chapter focused on learning to understand and use ratios, proportions, and percents to solve problems. You should be able to work problems like those assigned and to meet the following objectives:

1. Identify problems where a solution can be found by comparing ratios, and explain why.

2. Compare and contrast an additive comparison with a multiplicative comparison.

3. Give examples of problems where it is convenient to use ratios as measures, and explain why.

4. Compare ratios.

5. Solve proportion problems in ways other than cross-multiplying, and explain why other ways can be more meaningful than using the cross-multiplication strategy.

6. Illustrate how tables can be used to solve proportion problems. Describe why using tables might be a better way to begin teaching proportions.

7. Solve a variety of proportion problems.

8. Solve percent problems of all types.

9. Estimate with percents.

10. Perform a quantitative analysis of a variety of types of problems in order to solve them in a meaningful way.

11. Understand all of the definitions given in this chapter.

References for Chapter 9

[1,2]Simon, M. A., & Blume, G. W. (1994). Mathematical modeling as a component of understanding ratio-as-measure: A study of prospective elementary teachers. *Journal of Mathematical Behavior, 13*(2), 183–197.

[3]Noelting, G. (1980). The development of proportional reasoning and the ratio concept: Differentiation of stages (Part I). *Educational Studies in Mathematics, 11*(2), 217–253. (Problems are adapted from Noelting's well-known experiments with these types of ratios and proportions.)

[4]Smith, J. P. III. (2002). The development of students' knowledge of fractions and ratios. In B. Litwiller & G. Bright (Eds.), *Making sense of fractions, ratios, and proportions*, 2002 Yearbook (pp. 3–17). Reston, VA: National Council of Teachers of Mathematics. (See, in particular, p. 16.)

[5,7,11]Lamon, S. J. (1999). *Teaching fractions and ratios for understanding.* Mahwah, NJ: Erlbaum. (See, in particular, pp. 232–233.)

[6]Streefland, L. (1991). *Fractions in realistic mathematics education.* Boston: Kluwer Academic.

[8]Moss, J., & Case, R. (1999). Developing children's understanding of the rational numbers: A new model and an experimental curriculum. *Journal for Research in Mathematics Education, 30,* 122–147.

[9]Vernaud, G. (1983). Multiplicative structures. In R. Lesh & M. Landau (Eds.), *Acquisition of mathematics concepts and processes* (pp. 127–174). New York: Academic Press. Many of the Supplementary Learning Exercises for Chapters 7–9 were adapted from Vernaud's problems.

[10]Case, R. (1985). *Intellectual development: Birth to adulthood.* Orlando, FL: Academic Press. (See, in particular, p. 398.)

Chapter 10

Integers and Other Number Systems

This chapter deals with extensions of our number system. We focus on the integers, which represent our introduction to the notion of signed numbers—positive and negative. This is a critical point in children's learning of mathematics. As with the rational numbers, the advent of integers broadens the domain of children's mathematical worlds and presents conceptual difficulties. In their early experiences with numbers, zero is thought of as representing nothing. Suddenly, there are numbers less than zero and addition can make them smaller. However, it may surprise you to know that mathematicians historically struggled with some of the same counterintuitive ideas. For example, Augustus De Morgan, British mathematician of the 1800s, found the "existence of witches" to be more believable than the idea of a "quantity less than nothing." When children have difficulty making sense of ideas related to positive and negative numbers, we should remember that they are in good company.

At the same time, however, children are capable of mathematical intuitions that will enable them to make sense of integers and integer arithmetic. Researchers have found that, when given the opportunity, children can reason about integers in productive and relatively sophisticated ways, even in the lower elementary grades. These ways of thinking about signed numbers will be discussed in Section 10.2. In subsequent sections, we look at models for calculating with integers, review the properties of operations on numbers, and once again discuss rational and real numbers.

Notes

10.1 Big Ideas About Signed Numbers

Positive and negative numbers are sometimes called **signed numbers** because of the $+$ sign (for a positive number) or the $-$ sign (for a negative number) that may introduce the symbol, as in $+2$ or $^+2$ or -2 or $^-2$. Except for emphasis, the $+$ sign is often omitted, suggesting that the numbers we have used thus far can be regarded as special signed numbers. It is not uncommon to place the negative sign higher so that it is not confused with subtraction, until addition and subtraction of signed numbers are well understood: -3 may be written $^-3$. Later, when there is no fear of confusion, we may use the $-$ symbol for negative numbers and for subtraction. When signed numbers are first introduced to students, it is important to pay attention to the multiple ways the $-$ sign is used. It is conceptually very different to think about a negative number than to think of the operation of subtraction.

The use of the number line to illustrate signed numbers is likely familiar to you and builds on work with whole numbers and fractions. If we use the usual number line as a model for the numbers, then we can answer the question, "What's to the left of 0?" by describing the numbers to the left of zero as negative numbers. For the time being and with the usual elementary school curriculum in mind, we will first restrict ourselves to those numbers that are the opposites of the whole numbers, $^-1$ is the opposite of 1, 1 is the opposite of $^-1$, $^-2$ is the opposite of 2, etc.

When we combine the set of whole numbers with their opposites, including zero, we obtain a set of numbers we call **integers**. That is, $I = \{\ldots, ^-3, ^-2, ^-1, 0, 1, 2, 3, \ldots\}$. These numbers can be represented on the number line as follows:

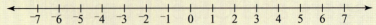

We say that 6 and $^-6$ are *opposites* or **additive inverses** of one another because their sum is 0. Generally speaking, the opposite of a is also denoted by ^-a, whether a is positive or negative, and $a + {}^-a = 0$.

Think About ...

What is the opposite of 2? What is the opposite of $^-2$? What is the opposite of the opposite of 2? What is the opposite of the opposite of $^-2$? What is the opposite of 0? Is ^-a always a negative number? What are the different meanings of the $-$ sign in $5 - {}^-(^-2)$?

Although $^-6$ can be read as either "negative six," "the additive inverse of six," or "the opposite of six," ^-a should be read as "the additive inverse of a" or "the opposite of a," but not as "negative a," because ^-a can be positive. The context usually makes clear which of the three interpretations of the $-$ sign is intended: subtraction (as in $67 - 29$), negative number (as in $^-5$ or -5), or additive inverse (as in ^-a or $-a$). All three interpretations can be seen in the expression $5 - {}^-(^-2)$.

Keep in mind that the labeling of the points for numbers on a number line comes about because of their distances from 0, a starting point. But $^+2$, say, could be thought of as any jump of length 2 units toward the right but starting anywhere (not necessarily at the 0 point), just as $^-3$ could be thought of as any jump of length 3 units, but toward the left and starting anywhere. Thinking of a signed number as a description of a jump size rather than only as a point on the number line has value in working with the addition and subtraction of signed numbers.

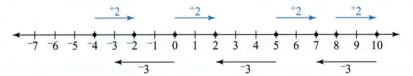

Zero can be represented with a number line in a variety of ways, such as the following:

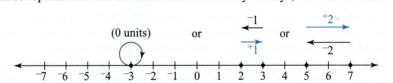

Consistent with the points for whole numbers and fractions, on the usual number line the point for a smaller number is to the left of the point for a larger number. Here, order and magnitude can be visible, with numbers to the left being smaller. Hence, e.g., $^-100$ is less than $^-5$, or $^-100 < {}^-5$. Thinking of debts and "worse off" for $<$ helps to make such inequalities believable.

We have focused primarily on the integers because they are the first of the signed numbers to appear in the usual elementary school curriculum. But the familiar fractions and decimal numbers can also have opposites (also called additive inverses). For example, $^-\left(\frac{3}{4}\right)$ is a negative number between $^-1$ and 0 on the number line. Its additive inverse is $\frac{3}{4}$. (We have used parentheses here simply to show that the negative sign is for the entire fraction, not just the numerator.) We will discuss this further in Section 10.6.

Another big idea is order. Numbers have an order going left ($^-1$, $^-2$, $^-3$, $^-4$, . . .) even as their distance from zero increases. One source of difficulty with negative numbers is the need to reconcile the order relation with magnitude (absolute value). For example, $^-4$ is smaller than $^-2$, even though there is more of a quantity in $^-4$.[1]

ACTIVITY 1   There Are Numbers to the Left of Zero

Locate the following on a number line and explain your thinking.

1. $^-2$ 2. Additive inverse of $^-3$ 3. Opposite of 5
4. $-(^-7)$ 5. $^-1.5$ 6. Opposite of .75
7. $\left|^-3\frac{4}{5}\right|$ (its distance from 0)   8. ^-x (on what does this depend?) ●

TAKE-AWAY MESSAGE . . .   There are big conceptual ideas in thinking about integers and other signed numbers. One is the opposing effects of a number and its additive inverse or opposite. The key feature of the additive inverse of a number a is that $a + {^-a} = 0$. Another big idea involves distinguishing between the sign of a number and an operation. A third big idea is one of order.

Learning Exercises for Section 10.1

1. Reorder each group of numbers from smallest to largest.

    a. $^+50$, $^-3$, $^-22$, $\frac{3}{4}$, $^-75.2$, $^-\left(\frac{2}{3}\right)$, $^-1$, 1

    b. 3.1, $^-5$, $2\frac{9}{10}$, $^-0.9$, $\frac{4}{9}$, 0.5, $\frac{13}{10}$, $^-\left(\frac{4}{9}\right)$, $^-0.1$, $^-1.2$

2. **a.** What is $^-(^-8)$? Explain why you think your answer is correct.
    b. Zero is regarded as neither positive nor negative, but one occasionally runs into $^-0$ in calculations. What is $^-0$? Explain.
    c. Is ^-a always negative? Explain.

3. In each part, what number is being described?

    a. the additive inverse of the additive inverse of the additive inverse of 9
    b. the additive inverse of the additive inverse of negative 9
    c. the additive inverse of the additive inverse of 6
    d. $^-(^-(^-(^-(^-10.3))))$

4. A jump on a number line may be followed by another jump that starts where the first jump ends. What single integer describes the net result in each drawing?

    a.   **b.**    **c.**

    d.   **e.**   **f.**

5. For $3 + 5 = n$, a child might draw a number line like the following. What might the child be thinking?

    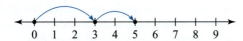

6. Sometimes the negative sign is not raised, so instead of writing $^-3$, we write -3. Why do you think this is done? What are the pros and cons of writing negative numbers this way?

7. Locate the following on a number line and explain your thinking.

    a. 6.25   **b.** $-\frac{3}{2}$ **c.** opposite of $\frac{2}{3}$

    d. additive inverse of $-5/4$ **e.** $-(0.5)$

8. Show the following as a jump on the number line, but not starting at zero.

    a. $^-2$ **b.** $^-5$ **c.** additive inverse of 8

Supplementary Learning Exercises for Section 10.1

1. Put these numbers in order, from smallest to largest, without using decimals.

$$\frac{53}{6} \qquad ^-\left(\frac{3}{4}\right) \qquad ^-\left(\frac{5}{4}\right) \qquad \frac{53}{7} \qquad ^-\left(\frac{53}{7}\right) \qquad \frac{1}{1000} \qquad ^-\left(\frac{1}{10,000}\right) \qquad ^-\left(\frac{53}{6}\right)$$

2. What single integer is represented by each arrow below?

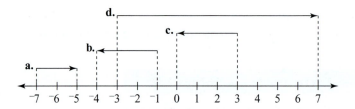

10.2 Children's Ways of Reasoning About Signed Numbers

We know that middle school students sometimes struggle to make sense of and solve problems that involve negative numbers. You may have a memory of your own struggle to make sense of operations with negative numbers. However, researchers[2] were surprised to find that many children (some as young as the first grade!) have productive ideas about negative numbers that enable them to solve a variety of problems. Below we share three powerful ways of reasoning that children use: motion on the number line, treating negative numbers like positive numbers, and determining the sign of the answer before calculating a final answer.

    The first is **motion on the number line**. Many young children are familiar with the number line and can use it to reason about problems with negative numbers. Consider Violet, a second-grade student, who was interviewed by a researcher. Violet was given a number line with the numbers 0, 1, 2, 3, 4, 5, 6, and 7 already drawn on it. When asked to solve $^-5 + 2$, she wrote in the number 8 and the negative integers shown as below:

    Notice that Violet correctly ordered the negative integers on the number line. She also used her ideas about addition and subtraction to support her thinking. For example, she thought about addition as "making more" and subtraction as "making less," which, on the number line, signifies that *addition* means to "move right" and *subtraction* means to "move left."

    Violet used her ideas about which direction to move on the number line to solve many problems with negative numbers. For example, she solved the problem $^-5 + 2 = \square$ by marking $^-5$ on the number line. Because addition means to "move right," she moved right 2 units and landed at $^-3$. Violet similarly solved the problem $5 - 8 = \square$ by marking 5 on the number line and then moving left (for subtraction) 8 units, to land at her ending point of $^-3$. Violet used what she already knew about operations with whole numbers to help her solve problems with negative integers. We can think of her reasoning in the following way:

$$\boxed{\text{Starting point}} + \boxed{\text{Change}} = \boxed{\text{Ending point}} \qquad \text{OR} \qquad \boxed{\text{Starting point}} - \boxed{\text{Change}} = \boxed{\text{Ending point}}$$

🗨 DISCUSSION 1

How do you think Violet would solve $^-5 - 4 = \boxed{}$ and $3 - \boxed{} = {}^-2$? How would she illustrate her thinking? ●

The problem $\boxed{} + 5 = 3$ was more difficult for Violet because the starting point was not given to her. Instead, the starting point is the number she needs to find. To solve this problem, she used trial and error on the number line. Violet makes a "guess" that the number that goes in the blank is $^-5$ (her start), so she marks $^-5$ on the number line and moves right 5 units (because addition means to move right), one at a time, to end at 0. She did not end at 3 (the ending point in the problem), so she tries $^-4$ and $^-3$ as her starting points and finally $^-2$. She moved right 5 units, ended at 3, and realized that the number that goes in the blank is $^-2$.

Violet used the number line in a way that is similar to the empty number line described in Chapter 3. This strategy allows students to begin the problem by treating the first number as the "start," and then (as one strategy) to move left or right according to the operation and the number of units of the "change" given in the problem.

📝 ACTIVITY 2   How Would Violet Reason About It?

Use Violet's way of reasoning for the following five problems. Describe and illustrate the thinking on the number line.

1. $4 - 9 = \boxed{}$ **2.** $^-8 + \boxed{} = {}^-2$ **3.** $^-3 - \boxed{} = {}^-10$

4. $^-3 + 7 = \boxed{}$ **5.** $\boxed{} - 5 = {}^-2$ ●

Notice that Violet's ways of reasoning are appropriate for solving problems for which the change is a positive number. When given a problem like $4 + \boxed{} = {}^-3$, she at first thought that it could not be solved, because addition means to move right—so how could a person begin at 4, move right (for addition), and land at $^-3$?

Violet was eventually able to make sense of these problems by conjecturing that the negative sign means to "do the opposite." She shared her thinking while solving the problem $4 + \boxed{} = {}^-3$. She counted 7 to get to $^-3$. She wrote $^-7$. Violet then compared the sum $4 + {}^-7$ to see what happens when she solves $4 + 7$. "Because positive seven gets you to eleven . . . that would be adding positive seven. If you add negative seven, to me, it's kind of like you're going backwards."

In her explanation, Violet informally uses a critical aspect of negative number understanding—what we call *negation*. Although Violet did not use this language, she treated the negative sign as negating, or doing the opposite, of the positive sign. Because she already had a meaning for adding (and subtracting) a positive number, she was able to use the meaning of a negative as an opposite to make sense of adding (and subtracting) a negative number.

Violet was able to extend her ideas of addition and subtraction toward the more nuanced ideas that she could move right when adding a positive number. So adding a negative number *had* to mean moving *left*, and she could move left for subtracting a positive number, and so subtracting a negative number *had* to mean moving *right*.

> *Think About …*
>
> How would Violet solve $2 - \boxed{} = 6$?

What makes Violet's ideas especially interesting for teachers is that she was *not* taught rules for how to add and subtract negative numbers when using the number line. Instead, Violet used her existing ideas about adding and subtracting and was able to think about how those ideas could be adjusted to add and subtract negative numbers.

ACTIVITY 3   And We Think Like Violet Again

Use Violet's way of reasoning for the following four problems. Explain each problem to a partner, using the reasoning that Violet used.

1. $6 + {}^-9 = \square$ **2.** $2 - {}^-8 = \square$

3. ${}^-3 - \square = 5$ **4.** $7 + \square = 2$ ●

It is important to understand that other children interviewed in the study thought about the problem as Violet did. Motion on the number line was a useful strategy for other children and, as you will see, connects nicely to a formal model for teaching operations with integers.

In their early school years, young children spend most of their mathematics lessons learning about natural numbers, also called the positive integers $\{1, 2, 3, \ldots\}$. As it turns out, some of these children use what they know about natural numbers (what young children often call "real" or "regular" numbers) to solve problems about negative numbers. Children who use this type of reasoning **treat negative numbers like positive (or regular) numbers**. Computing with negative numbers is often explicitly compared to computing with positive numbers. First-grader Jacob illustrates another form of reasoning that children use when he says, "${}^-5 + {}^-3$ is *just like* $5 + 3$. I know that $5 + 3$ is 8 so ${}^-5 + {}^-3$ is ${}^-8$."

When shown the problem ${}^-6 + \square = {}^-9$, Jacob counted 6 cubes and answered, "Negative 3." He explained his answer, saying, "If you're doing it with real numbers—it's like 6 plus 3 equals 9. So I added the real numbers . . . and the negative numbers are like them [the real numbers] because you just add a minus sign." Here, we see Jacob treat negative numbers like positive numbers to productively reason about and solve problems involving negative integers. Similarly, he answered ${}^-2$ for the problem ${}^-7 - \square = {}^-5$. "Well, for this one I need little cubes. It would be like real numbers but you just add the minus sign. You just do . . . seven minus two equals five. That's the answer for real numbers, so I just added a negative to all of them and there is my answer."

ACTIVITY 4   Jacob's Ways

Use Jacob's way of reasoning to solve the problems below. Identify the one problem for which "treating negative numbers like positive numbers" does not make sense.

1. ${}^-9 - {}^-2 = \square$ **2.** $\square + {}^-6 = {}^-10$ **3.** ${}^-3 + {}^-8 = \square$

4. ${}^-8 - \square = {}^-3$ **5.** $3 + {}^-5 = \square$ ●

As it turns out, problems such as Problem 4 are particularly challenging for seventh-grade students to solve. Yet for Jacob and others who use this type of reasoning, solving this kind of problem presents no special challenge.

There is a third powerful way of reasoning when solving problems with negative numbers. Middle and high school students will sometimes first take into account the operation and relative size of the numbers to help them **determine the sign of the unknown** (i.e., whether the answer is positive or negative) **before calculating the final answer**. For example, when solving $6 + \square = 4$, Angel wrote ${}^-2$ in the blank and remarked, "I thought, '6 plus a positive number cannot be 4,' so it took a negative number." Similarly, others who first considered the sign of the unknown before calculating its final answer described thinking about the problem $5 - \square = 8$: "Well, since I'm subtracting but the answer, 8, is larger than what I'm subtracting from, 5, I must be subtracting a negative number." When students use this way of reasoning, they are first using their *number sense* (see Chapter 5) to reason about the sign of the number prior to calculating a final answer.

 **DISCUSSION 2**

How would students use this way of reasoning to solve the following problems?

1. $8 - \boxed{} = 13$ **2.** $4 - 9 = \boxed{}$

3. $10 + \boxed{} = 7$ **4.** $^-8 - \boxed{} = 4$ ●

    It is important to understand that children have productive ways of reasoning, so you should listen for reasoning on which to build instruction.

> ### Think About …
> What connections can you see among these children's ways of reasoning (motion on the number line, treating negative numbers like positive numbers, and determining the sign of the answer before calculating) and the big ideas of integers mentioned in Section 10.1?

Learning Exercises for Section 10.2

1. Describe and illustrate how a child can use *motion on the number line* to solve the following problems.

    a. $1 - \boxed{} = ^-3$ **b.** $^-6 + 3 = \boxed{}$ **c.** $\boxed{} + 5 = ^-1$

2. What addition problem might a child be solving with the following sketch?

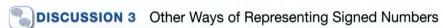

3. Make up three problems for which a young child could productively reason about the sign of the unknown before calculating the final answer.

4. Make up a problem for which *treating negative numbers like positive numbers* does not make sense.

10.3 Other Models for Signed Numbers

One reason that signed numbers are important is that they have many applications, and so there are many ways of thinking about them. Here are some.

 **DISCUSSION 3**   Other Ways of Representing Signed Numbers

How could each of the following be used to think about signed numbers? Describe what positive and negative numbers, and zero, would mean.

1. Financial matters like bank balances, profit/loss, paycheck/bill, income/debt, credit cards, etc.

2. Temperature changes

3. Sea levels

4. Sports settings like football and golf

5. Diets

6. Atomic charges (although atomic charges may not be part of the K−6 curriculum)

7. Games in which you can "go in the hole" ●

    Teachers often use story problems set in contexts (e.g., money, temperature, sports) to help students reason about operations with signed numbers. These contexts abound. No one model can help students understand all the important aspects of integers. The focus as you work with the materials should be on the big ideas, such as additive inverses, without the expectation that the models will carry meaning. Nevertheless, with proper development, the models are useful for developing understanding. Our emphasis in this chapter will be

Notes

on the chip model and the number line, which are prevalent in textbooks; the number line can also be built on for the formal mathematics to follow.

We will focus now on another way to represent signed numbers: chips of two colors. Chips of two colors are an adaptation of the ancient Chinese method of 200 BC. We will use blue for positive and gray for negative. For example, three blue chips can represent $^{+}3$ (or 3), and 4 gray chips can represent $^{-}4$. Of course, any two colors can be used, so long as it is clear which represents positive numbers and which negative numbers.

 represents $^{+}3$ represents $^{-}4$

Just as a gain of $2 can be cancelled by a loss of $2, giving a zero change in finances, a chip representation for 0 can occur in many forms.

Each of these three drawings is
a way of showing 0:    or   or

Notice that there is a degree of abstraction here, requiring an understanding of the representation. Two blue and two gray chips represent zero, even though there are four chips involved. Two blue chips have, in a sense, a canceling effect on two gray chips (or two gray chips have the opposite effect of two blue chips). It is natural to think of addition as describing two blue chips put with two gray ones, and the canceling effect gives the important numerical result, $^{+}2 + {^{-}2} = 0$. With the canceling idea in mind, chip drawings like each of the following can be interpreted as having the value $^{-}3$. Do you see why?

 but also   or

These more complicated ways of showing an integer can be handy in subtraction, as we will see.

> ### Think About …
> What are several ways of showing $^{+}2$ with the chips, besides with two blue chips? Why are these ways more abstract than showing just two blue chips?

DISCUSSION 4 Pulling It Together

How is it possible to illustrate the big ideas of additive inverses and order with the number line, colored chips, and a sports setting, such as football? Are some models easier than others for a particular big idea? ●

Learning Exercises for Section 10.3

1. Make drawings of chips to show the following. Use blue for positive and gray for negative, for consistency.
   a. 5   **b.** $^{-}6$

2. Make drawings of chips to show zero in at least four ways different from those in the text. Again, use blue for positive and gray for negative, for consistency.

3. Give the single integer that each of the following chip drawings can represent. Be ready to explain your thinking. (Blue—positive, gray—negative)

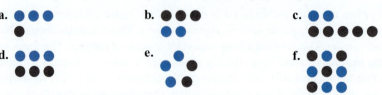

4. Give, if possible, an example of each type of number. If it is not possible, explain why.
   a. a negative real number that is not rational
   b. a negative real number that is not irrational
   c. a negative integer that is not real
   d. a negative real number that is not an integer

5. Interpret $^{+}10$ and $^{-}4$ in these settings.
   a. a financial situation of some sort
   b. a sport
   c. a temperature change
   d. a temperature

Supplementary Learning Exercises for Section 10.3

1. What single integer is being represented by each chip display? (Use blue for positive and gray for negative, for consistency.)
   a.
   b.
   c.

2. "I don't get it! There are 8 chips, but you say that it shows 2." To what chip display is the student reacting, and what is missing in the student's understanding?

3. Interpret this record of plays in a football game, in terms of yards gained or lost:
   $$^{+}4,\ ^{+}8,\ ^{-}3,\ ^{-}15,\ ^{+}11$$

4. On a number line, a child may think that 0 and $^{-}0$ correspond to different points. How might you try to convince the child that 0 and $^{-}0$ correspond to the same point?

10.4 Operations with Signed Numbers

Just as there are several ways in which to think about signed numbers, there are several ways to think about adding and subtracting them. Here, we will focus first on adding and subtracting integers via colored chips, the number line, and a money argument. Then we summarize the rules symbolically, using absolute-value language. The rules that arise with integers are applicable to all signed numbers.

We will use blue chips to indicate positive integers, and gray ones to indicate negative integers. We can think of the chips as positive and negative "charges" that "cancel out" one another, if there is the same number of blues as grays. If we begin with three blue chips and add three gray chips, the chips cancel each other out, and we have zero: $3 + (^{-}3) = 0$, incorporating the important additive inverse property into this model. Again, notice that although there are 6 chips visible, the meaning attached to them allows one to say "0" for this arrangement, much like having a check for $3 and a bill for $3 gives, in effect, $0.

Using an optional box to surround the work is occasionally useful, especially when the sum is zero.

With the chips, adding integers with the same sign is straightforward and involves showing each addend with chips and then counting the total.

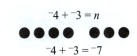

If the signs of the addends differ, we can use the additive inverse feature for integers. For example, for $3 + {}^-5 = n$, we begin with three blue chips and add five gray chips, but three of the gray chips cancel out the three blue chips, and we are left with two gray chips: $3 + {}^-5 = {}^-2$.

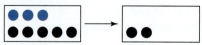

For ${}^-2 + 6 = n$, two gray chips cancel two of the six blue chips, leaving four blue chips, so ${}^-2 + 6 = 4$.

Notice that in effect, when the signs differ, just finding the difference in the numbers of chips for the addends, and then giving that difference the sign of the larger number of chips yield the sum. You may have learned something like that as a rule for adding numbers with different signs.

Let us turn to subtraction of integers with the chips.

💬 DISCUSSION 5   Subtracting with the Chip Model

How can you use the take-away interpretation for subtraction to show each of the following with chips?

1. $4 - 3$ **2.** ${}^-5 - ({}^-2)$ **3.** $4 - 7$

As you probably noticed, some subtractions are very easy and can be shown in ways similar to those used for whole numbers.

For example, for $4 - 3 = 1$, we could show the following:

Or, for ${}^-5 - {}^-2 = {}^-3$,

In Discussion 5, the last problem raises the question, "What if there are not enough chips to remove, as with $4 - 7 = n$?" Two ways may have arisen, each adding equal numbers of both negative and positive chips—in effect, adding 0—and then subtracting.

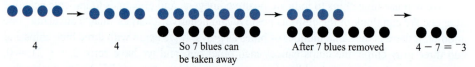

| 4 | 4 | So 7 blues can be taken away | After 7 blues removed | $4 - 7 = {}^-3$ |

You may also have thought, "Why not just put in 3 more blues and 3 grays; then you could take away 7 blues?" The answer is, "You could." The way illustrated, however, suggests at the third step, $4 - 7 = 4 + {}^-7$, and you may recognize in that equation the basic rule for subtracting signed numbers: Change the sign of the subtrahend (the number being subtracted), and add.

It may be instructive to see the symbolic form for the steps in the work for $2 - ({}^-3) = n$.

Replace 0 with $3 + ({}^-3)$.

$$2 - ({}^-3) = [2 + 0] - ({}^-3) = \underline{[2 + 3 + ({}^-3)]} - ({}^-3) = \underline{[5 + ({}^-3)]} - ({}^-3) = 5$$

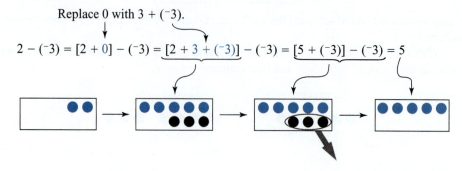

Notice that $2 - (^-3)$ gives the same result as $2 + 3$, supporting the usual rule for subtraction. You can even see the $2 + 3$ in the last drawing.

Let us turn to the number line. Addition and subtraction of signed numbers on a number line may already be familiar to you.

ACTIVITY 5   Hopping on the Number Line

Using the number line, find the following in a way that makes sense.

1. $6 + 5$ 2. $3 + 4$

3. $^-6 + 5$ 4. $6 + {}^-9$

5. $6 + {}^-5$ 6. $6 - 5$

7. $3 + {}^-4$

Then check your answers by reading the next paragraphs. ●

If you have forgotten how to use the number line for these sums, here is some help. (Draw a number line, and read the instructions slowly.) To find $3 + 4$, start at 0, move 3 units to the right, then move 4 more units to the right, to 7. (Starting at 0 allows one to read the answer from the number-line markings.) Think of a positive addend as moving to the right and a negative addend as moving to the left.

To subtract $6 - 5$ using the number line, begin at 0, move to the right 6 units, and then "take away" 5 units to the left, to 1. Continuing on the number line you drew (or a copy), if you want to find $6 + (^-5)$, begin at 0 and move 6 units to the right, then move 5 units to the left. Note that $6 + (^-5)$ takes us to the same place on the number line as does $6 - 5$, i.e., $6 - 5 = 6 + (^-5)$. Similarly, if you want to find $6 + (^-9)$, you go from 0 to 6 and move 9 units to the left, ending at $^-3$. The $6 - 5 = 6 + (^-5)$ equation suggests that $6 + (^-9)$ should be $6 - 9$. And from the number line, $6 - 9 = ^-3$. Again, notice that the equations $6 - 5 = 6 + {}^-5$ and $6 - 9 = 6 + {}^-9$ suggest the eventual rule for subtraction, even though that may seem irrelevant at this point.

Think About ...
Do you see connections to the ways in which children reason about signed numbers?

How would one show $6 - (^-9)$ on the number line? Rather than introduce a new interpretation of subtraction (e.g., *do the opposite*) as is often done, we can introduce 0 in a clever way, as was done with the chips, and continue with *take-away* as the meaning for subtraction. First, show 6 as $6 + 0$, in the form $6 + 9 + {}^-9$, so that the $^-9$ can be "taken away."

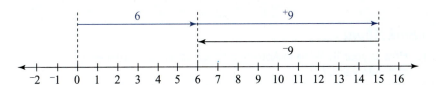

Once the $^-9$ segment is taken away, we are left with $6 + (^+9)$. So, our illustration means that $6 - (^-9) = 6 + (^+9)$, an equation again supporting the eventual rule.

As a final way of thinking about the addition and subtraction of signed numbers, let us consider money, first looking at addition and the symbolic rules governing addition, and then similarly at subtraction. The symbolic rules use the idea of *absolute value*, so we will first review that topic.

There are times when we are interested in a number's direction—i.e., whether its place on the number line is to the left or to the right of 0. At other times we are interested only in a number's distance away from 0 on the number line, and do not care in which direction we must go to arrive at the number.

Notes

Notes

A number's distance from 0 on the number line is called the **absolute value** of the number, and we consider this value to be positive (or zero in the case of zero). We denote the absolute value of a number b as $|b|$.

EXAMPLE 1

We can say that $|6|$ is 6, and similarly, $|{}^-6|$ is 6. Both 6 and ${}^-6$ are 6 units away from zero. Opposite numbers (always) have the same absolute value. ●

In terms of chips, absolute value can be interpreted as just how many uncanceled chips there are. For example, $|{}^-3| = 3$ because there are 3 gray chips (or a surplus of 3 gray chips), and $|{}^+9| = 9$ because there are 9 blue chips (or a surplus of 9 blue chips). The following diagrams show two ways to represent $|{}^-3|$ using chips.

$|{}^-3| = 3$ $|{}^-3| = 3$

With absolute value in mind, we now return to the addition of signed numbers.

 ACTIVITY 6   Leading to the Rules for Addition

Use examples from a money context to consider these situations.

1. Adding two signed numbers that have the same sign (e.g., ${}^+53 + {}^+79.95$, and ${}^-19 + {}^-56$).
2. Adding two signed numbers with different signs (e.g., $253 + {}^-79$, and $57 + {}^-84$). ●

Your thinking in Activity 6 could be formalized using absolute value, as is often done in algebra classes. It can be difficult to understand the formal statements. The children's thinking described earlier is closely related.

Addition of signed numbers when both numbers have the same sign:

If both numbers are positive, then $a + b = |a + b|$. If both numbers are negative, then $a + b = {}^-(|a| + |b|)$.

Addition of signed numbers when one is positive and the other is negative:

Consider a to be positive and b to be negative.
If $|a| > |b|$, then $a + b = |a| - |b|$.
If $|a| < |b|$, then $a + b = {}^-(|b| - |a|)$.

Think About …

1. How is your thinking in Activity 6 reflected in the formal addition rules using absolute value?
2. Is $|a - b| = |b - a|$ always? Is $|a| + |b| = |a + b|$ always? Is $|a - b| = |a| - |b|$ always?

Two special cases for addition, with the relevant vocabulary, should be highlighted.

Special cases:

If $a = 0$, then $a + b = b$ and $b + a = b$. We call 0 the **additive identity**.
If $a = {}^-b$, then $a + b = {}^-b + b = 0$ (the additive identity). We call each of b and ${}^-b$ the **additive inverse** of the other because their sum is 0.

Notice that work with chips or the number line or some other representation of signed numbers can suggest the formal statements. Having a mental image of chips or a money situation, e.g., can allow you to understand where the rules come from.

 DISCUSSION 6   Connecting Children's Thinking with the Formal

Recall from Section 10.2 that children's thinking about signed numbers is productive and useful; i.e., their ways of thinking could be connected to the more formal ways of mathematicians. How is the thinking of children (motion on a number line, treating negatives like positives, or determining the sign of the unknown before calculating) closely related to the formally expressed rules for adding and subtracting? ●

This page* from a Grade 6 textbook gives two ways to add integers with different signs. Do you think this page could be followed by a similar page on subtraction of integers? Explain your answer.

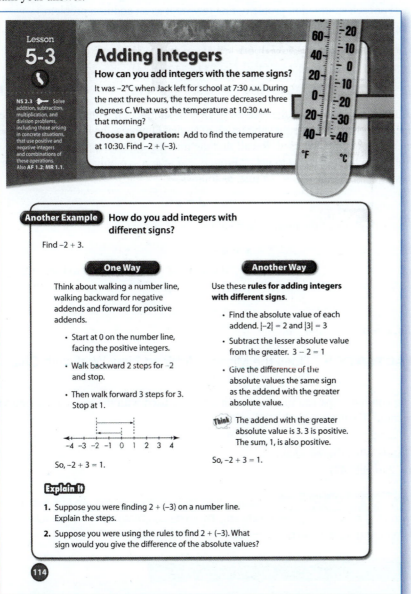

Notes

Notes

Fortunately, we can use the rules for addition when we subtract, by changing from subtraction to addition according to the rule below. We noted this relationship in several of the previous calculations with the chips and the number line.

> **Subtraction of signed numbers:**
>
> If c and d are signed numbers, then $c - d = c + (^-d)$.

EXAMPLE 2

a. $^-5 - 2 = ^-5 + ^-2 = ^-7$ b. $11 - ^-5 = 11 + ^+5 = 16$

 ACTIVITY 7   Can You Add and Subtract?

Calculate, referring to the rules for addition and subtraction given on page 192 and directly above. (For each subtraction problem, first rewrite it as an addition problem.)

1. $^-141 + 141$	**2.** $^-1 + ^-1$	**3** $^-18.2 + ^-4.83$	**4.** $^-75 + 413$
5. $413 + ^-75$	**6.** $^-4.37 + 6.1$	**7.** $^-483 + 217$	**8.** $483 + 217$
9. $^-1 - 1$	**10.** $^-1 - ^-1$	**11.** $^-12 - ^-41$	**12.** $^-12 - 41$
13. $41 - ^-12$	**14.** $^-38 - 654$	**15.** $^-431 - 22$	**16.** $6.8 - 212$

Although we have used the take-away view of subtraction with the chips and the number line to motivate the rule for subtraction of signed numbers, the missing-addend view of subtraction could also be used. Recall that the missing-addend approach for $c - d$ would ask: What can be added to d to get c? From $c - d = x$, a symbolic line of reasoning could also result in the $c - d = c + ^-d$ rule, as follows:

$$c - d = x$$

Then, thinking of the missing addend for $c - d$, we get

$$x + d = c$$
$$x + d + ^-d = c + ^-d \quad \text{(by adding } ^-d \text{ to both sides)}$$
$$x + 0 = c + ^-d$$
$$x = c + ^-d$$

So, $c - d = x = c + ^-d$.

 ACTIVITY 8   Does the Missing-Addend View Work with Chips and the Number Line?

Can the colored chips and the number line also be used with the missing-addend view of subtraction? Try them with the following:

1. $^-4 - ^-2$ **2.** $5 - ^-1$ **3.** $^-6 - 2$
    (*Think*: What can be added
    to $^-2$ to get $^-4$?)   ●

TAKE-AWAY MESSAGE ... Colored chips and the number line are just two methods of illustrating the addition of signed numbers. Their use can motivate the usual rules for adding and subtracting signed numbers.

Learning Exercises for Section 10.4

1. Using drawings of two colors of chips, find the following:

a. $^-4 + ^-2$ b. $^-4 + ^+2$ c. $^-4 - ^-2$ d. $^-4 - ^+2$
e. $(4 + ^-3) + ^-1$ f. $4 + ^-6$ g. $4 - ^-3$ h. $4 - ^-6$

2. Make number-line drawings to find the following
   a. $^+5 + ^-7$
   b. $^-3 + ^-4$
   c. $^-2 + ^+5$
   d. $^+2 + ^-6$
   e. $^+6 - ^+2$
   f. $^+6 - ^-2$
   g. $^-5 - ^-2$
   h. $^-5 - ^+2$

3. Add and subtract these numbers using drawings of chips, the number line, or a money situation.
   a. $^-5 + ^-5$
   b. $^-(^-5) + ^-5$
   c. $^-5 + ^-(^-5)$
   d. $^-5 - ^-5$
   e. $^-5 - ^-(^-5)$
   f. $5 - ^-5$

4. Calculate the following and illustrate on a number line:
   a. $\frac{14}{15} + ^-\left(\frac{3}{5}\right)$
   b. $^-\left(\frac{4}{9}\right) - 4$
   c. $\left(\frac{7}{13}\right) + ^-\left(\frac{7}{13}\right)$
   d. $^-\left(\frac{5}{9}\right) - \left[^-\left(\frac{2}{3}\right)\right]$
   e. $2.34 - 5.612$
   f. $^-3.4 + ^-7$
   g. $^-5.567 - 2.33$
   h. $3.5 - (^-8)$

5. In Learning Exercise 5 in Section 3.2, we considered families of addition and subtraction facts such as the one shown at right:

$3 + 2 = 5$	$5 - 2 = 3$
$2 + 3 = 5$	$5 - 3 = 2$

   Complete these fact families for integers:

   a. $^-3 + 5 = 2$
      $5 + ^-3 = 2$

   b. $^-2 - ^-5 = 3$
      $^-5 + 3 = ^-2$ | $^-2 - 3 = ^-5$

   c. $^-32 + ^-29 = ^-61$

6. Solve and show its location on the number line:
   a. $|^-3| = ?$
   b. $^-|^-3| = ?$
   c. $\left|17\frac{4}{7}\right| = ?$
   d. $|^-6| + |^-6| = ?$
   e. $|^-6| + |6| = ?$
   f. $|^-6| - |6| = ?$
   g. $|^-6| - |^-6| = ?$

7. Temperature change is often used as a setting for adding and subtracting integers. Design some problems that you could use to teach someone else how to add and subtract integers.

8. Is it possible to use the comparison view of subtraction with the chip model with signed numbers? With the number line? With money? (The comparison for $7 - 2$, e.g., would tell how much greater 7 is than 2, or how much less 2 is than 7.)

9. For each story and with signed numbers, write an equation that describes the situation, and answer the question.
   a. An official from Company A said, "Here's how we did last year. The first quarter we earned $57,000, and the second quarter, $35,000. But during each of the third and fourth quarters, we lost $16,000." How did Company A fare, for the whole year? (Remember to use signed numbers in your equation.)
   b. Company B reports, "During the second quarter, we earned $92,000, so for the first two quarters, we have earned a total of $15,000." How did Company B do during the first quarter?
   c. Company B later reports, "We lost $125,000 in the fourth quarter, so now we have earned only $11,000 in all, for the last two quarters." How much did the company gain or lose during the third quarter?
   d. How did Company B [see parts (b) and (c)] do overall, for the first and fourth quarters only?

10. Write a story problem involving financial matters for an individual (paycheck/bill, income/debt, credit cards, etc.) that could be described by each of the following. Give the answers to your questions.
    a. $105.89 + ^-75 + 92.73 + 11.68 + ^-99.15 = n$
    b. $87.58 + 100 - ^-22.75 - 69 = n$

11. For each equation below, write a story problem involving football or golf or diets that could be described by the equation.

a. $2 + {}^-5 + 14 + 2 + {}^-6 + 3 = n$ **b.** ${}^-6 + {}^-2 + 3 + {}^-1 + {}^-2 + 1 = n$

12. The newspaper *Honolulu Advertiser* gives the heights of the tides for different locations relative to those in Honolulu. What are the heights of the tides in the following locations when high tide in Honolulu is 0.8 ft and low tide is ${}^-0.1$ ft?

a. Lahaina (relative to Honolulu): high, ${}^+0.2$ ft; low, ${}^+0.1$ ft
b. Hilo (relative to Honolulu): high, ${}^+0.3$ ft; low ${}^-0.2$ ft

13. What is the difference in the highest and lowest temperatures for each of the following continents?

a. Africa   135.9°F (highest), ${}^-11$°F (lowest)
b. Asia   129°F (highest), ${}^-96.16$°F (lowest)
c. Australia   128°F (highest), ${}^-10.4$°F (lowest)
d. Europe   122°F (highest), ${}^-67$°F (lowest)
e. North America   134°F (highest), ${}^-81.4$°F (lowest)
f. South America   120°F (highest), ${}^-27.4$°F (lowest)

14. Use an empty number line to illustrate the differences in Learning Exercise 13.

Supplementary Learning Exercises for Section 10.4

1. Illustrate $5 + {}^-2$ with each of the following:

a. colored chips   **b.** a number line   **c.** a financial situation

2. a. Illustrate $5 - {}^-2$ with colored chips.
   b. Illustrate ${}^-5 - {}^-2$ with colored chips.
   c. Illustrate ${}^-5 - 2$ with colored chips.

3. Calculate the following and illustrate on a number line:

a. $112 + {}^-78 + {}^-47 - 65$ **b.** ${}^-563 - {}^-386 + {}^-291$

c. ${}^-\left(\frac{3}{4}\right) + \frac{5}{12} - \frac{3}{8}$ **d.** ${}^-4 + \frac{2}{3} - {}^-1\frac{1}{3} + 2\frac{1}{2}$

e. ${}^-1.19 + 2.3 - {}^-0.6 + 0.003$

4. Some accountants show losses by enclosing the number in parentheses. For example, a loss of $1500 could be shown by ($1500).

a. How would the accountant record the amounts in Learning Exercise 9(a)?
b. How would the account record a gain of $750.80? Losses of $1000 and $110.50?

5. Using a variety of positive and negative numbers, test with three different choices for a and for b to see how ${}^-(a - b)$ and ${}^-a + b$ compare.

10.5 Multiplying and Dividing Signed Numbers

Addition and subtraction of signed numbers usually first appear in Grades 5–6, perhaps just with the integers. Multiplication and division of signed numbers are then treated in Grades 6–7. You may have heard the rhyme: "Minus times minus is plus, the reason for this we need not discuss." But here we do discuss it. If you have your own doubts about why multiplying two negative numbers gives a positive number, you are in good company. Until the mid-1800s there was a great deal of resistance to that result even by many mathematicians.

   We will offer different arguments as to what the sign of a product involving negative numbers should be. The first argument will assume commutativity of multiplication and consider a pattern. The second will show that the desirable properties of multiplication necessarily lead to "negative times negative is positive." A final argument, using chips, is offered. The chips argument is given last.

Multiplications involving positive numbers or zero are already familiar. The product of two positive numbers is positive, and if zero is a factor, the product is zero. The other cases involve multiplying numbers (1) of opposite signs and (2) when both are negative. We will focus on integers, although the same results will apply to all real numbers.

Earlier we found that one way of thinking about multiplication is as repeated addition. Applying this view of multiplication to $4 \times {}^-2$ gives

$$4 \times {}^-2 = {}^-2 + {}^-2 + {}^-2 + {}^-2 = {}^-8$$

suggesting that (positive) $\times$ (negative) = (negative). But it is not so easy to think about what ${}^-2 \times 4$ could mean as repeated addition. *However, if multiplication of integers is to be commutative, then ${}^-2 \times 4$ must equal $4 \times {}^-2$, which we have just shown is ${}^-8$.* In other words, ${}^-2 \times 4 = {}^-8$, suggesting that (negative) $\times$ (positive) = (negative).

EXAMPLE 3

a. $3 \times {}^-4 = {}^-12$ **b.** ${}^-6 \times 14 = 14 \times {}^-6 = {}^-84$ ●

The only case left is that of having two factors that are both negative. The pattern in the next activity is suggestive (and is common in the elementary school curriculum).

ACTIVITY 9   A Strange Rule?

Using the results for multiplying a negative number and a positive number, complete the patterns in these two columns. (Even if you know the answers already, look for the patterns as you go down a column.)

$$
\begin{array}{ll}
4 \times 2 = 8 & {}^-4 \times 4 = {}^-16 \\
3 \times 2 = 6 & {}^-4 \times 3 = {}^-12 \\
2 \times 2 = 4 & {}^-4 \times 2 = {}^-8 \\
1 \times 2 = 2 & {}^-4 \times 1 = ? \\
0 \times 2 = 0 & {}^-4 \times 0 = ? \\
{}^-1 \times 2 = ? & {}^-4 \times {}^-1 = ? \\
{}^-2 \times 2 = ? & {}^-4 \times {}^-2 = ? \\
{}^-3 \times 2 = ? & {}^-4 \times {}^-3 = ? \\
\text{etc.} & \text{etc.}
\end{array}
$$
●

Although the result may seem counterintuitive, the pattern suggests that the product of two negative numbers must equal a positive number. Here is a summary of all the results from this first line of reasoning. (The results actually apply to all real numbers.)

Multiplying two signed numbers:

If the signs of the two numbers are the same, the product will be positive. If the signs of the two numbers are different, the product will be negative.

EXAMPLE 4

a. ${}^-3 \times {}^-4 = 12$ **b.** ${}^-0.2 \times {}^-0.4 = 0.08$ ●

DISCUSSION 7   Convinced?

Did you find the pattern argument for the product of two negative numbers convincing? Do you think young students would find it convincing? (Mathematicians like patterns, but they do not trust them completely because sometimes the patterns can break down.) ●

Notes

The second line of reasoning for the product of two negatives rests solely on properties of multiplication that we would want to continue to be true for signed numbers, so we will look at those properties. We have already used commutativity of multiplication, but there are other important properties as well, to be discussed in Section 10.6.

A second way to show that defining the product of two negative numbers to be positive makes sense mathematically is illustrated in this example, which depends heavily on the distributive property: Suppose the product of concern is $^-3 \cdot {}^-2$.

Start with $^-3 \cdot 0 = 0$. Substitute $^+2 + {}^-2$ for the first 0.

$^-3 \cdot ({}^+2 + {}^-2) = 0$ after the substitution.

$({}^-3 \cdot {}^+2) + ({}^-3 \cdot {}^-2) = 0$ using the distributive property.

$^-6 + ({}^-3 \cdot {}^-2) = 0$ using the known $^-3 \cdot {}^+2 = {}^-6$.

So, $^-3 \cdot {}^-2$ must be equal to $^+6$ to make the equation true.

Basically, all this is saying is that *if* the product of two negative numbers were *not* defined to be a positive number, then at least some of the rules of numbers we so far know to be true would fail when negative numbers are included.

The rules for multiplication of integers automatically provide us with ways of dividing signed numbers, both positive and negative. The missing-factor way of thinking about division is useful. Recall that for $^-16 \div 8$, e.g., this view of division says to think, "What times 8 gives $^-16$?" Because $^-2 \cdot 8 = {}^-16$, $^-16 \div 8 = {}^-2$. In general, division can be defined as follows:

> If a, b, and c are real numbers and b is not 0, then $c \div b = a$ if $a \cdot b = c$.

EXAMPLE 5

a.	$12 \div 4 = 3$	because	$3 \cdot 4 = 12$
b.	$^-12 \div 4 = {}^-3$	because	$^-3 \cdot 4 = {}^-12$
c.	$12 \div {}^-4 = {}^-3$	because	$^-3 \cdot {}^-4 = 12$
d.	$^-12 \div {}^-4 = 3$	because	$3 \cdot {}^-4 = {}^-12$ ●

Think About ...

Why can't 0 be the divisor in the definition of division ("... and b is not 0 ...")? How does the missing-factor view lead to (negative) ÷ (negative) = (positive)?

For completeness, we summarize a rule for multiplication and division of signed numbers.

> **Multiplication and division of signed numbers:**
>
> The product or quotient of two numbers with the same sign is positive. The product or quotient of two numbers with opposite signs is negative.

The equality link between $\frac{a}{b}$ and $a \div b$ extends to signed numbers. Hence, $\frac{^-17}{5}$, $\frac{17}{^-5}$, and $^-\left(\frac{17}{5}\right)$ are all equal, because $\frac{^-17}{5} = ({}^-17) \div 5 = {}^-\left(\frac{17}{5}\right)$ and $\frac{17}{^-5} = 17 \div {}^-5 = {}^-\left(\frac{17}{5}\right)$. This equality between $\frac{a}{b}$ and $a \div b$ leads to a common way of defining the rational numbers in advanced work.

> A **rational number** is any number that can be expressed in the form
>
> $$\frac{\text{integer}}{\text{nonzero integer}}.$$

You might wonder whether the chips of two colors can also be used to demonstrate multiplication of integers. Whenever possible, make connections with whole numbers. Consider the repeated addition model of multiplication. We know that $^+2 \times {}^+4 = 2 \times 4$ can be thought of as $4 + 4$. So, we can show 2×4 as an array of repeated addition:

Likewise, we can show $2 \times {}^-4$ as an array of repeated addition:

However, *repeated addition cannot serve as a model for multiplication when the first factor is negative.* Thinking about $^-2 \times 4$ as repeated addition does not make sense; how would one *add* 4 negative 2 times? Rather than appealing to commutativity of multiplication, the key is in recognizing that since repeated addition does not make sense for $^-2 \times 4$, then when the first factor is negative, we need a *new* interpretation. Since in $^+2 \times 4$ we can think of adding 4 two times, *it is not unnatural to think of* $^-2 \times 4$ *as subtracting 4, two times.* But how do we get started? As with repeated addition, the answer is to start with a "neutral" amount (0), but here cleverly chosen so that the subtractions are possible.

For $^-2 \times 4$, the work might proceed as follows:

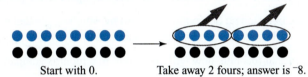

Start with 0. Take away 2 fours; answer is $^-8$.

For $^-2 \times {}^-4$, we would take away two sets each with 4 gray chips, so the answer is the 8 blue chips left, or $^-2 \times {}^-4 = {}^+8$.

This process would lead us to the same rules for multiplying with signed numbers. Again, if division is considered in missing-factor terms, the rules for division of signed numbers would also continue to hold.

DISCUSSION 8 Motion on a Number Line

How can you use the thinking of multiplying whole numbers times whole numbers or whole numbers times fractions to illustrate the multiplication of integers on a number line? ●

TAKE-AWAY MESSAGE . . . Multiplication and division of signed numbers were considered in this section. New are cases where one or both of two numbers are negative. In multiplying or dividing two positive numbers or two negative numbers, the answer is a positive number. If the two numbers have opposite signs, the answer is a negative number. You should be able to see connections to the meanings for operations of multiplication and division.

Learning Exercises for Section 10.5

1. Does a fraction with a negative numerator and positive denominator have the same value as a similar fraction but this time with a positive numerator and negative denominator? That is, is $\frac{^-2}{5}$ equal to $\frac{2}{^-5}$? Is either or both of these equal to $^-\left(\frac{2}{5}\right)$? Explain in terms of a fraction representing a division.

2. Practice operations on signed numbers by completing the following computations:

   a. $^-12 \div 6$ **b.** $^-13 - {}^-21$ **c.** $^-7 \cdot (3 + {}^-5)$ **d.** $^-121 \div {}^-11$

   e. $\frac{14}{15} \times \left(\frac{^-5}{7}\right)$ **f.** $\frac{^-2}{5} \div 4$ **g.** $\frac{7}{13} \div \frac{^-7}{13}$ **h.** $\frac{^-7}{13} \times \frac{^-13}{7}$

   i. $^-2 \times {}^-7 \times {}^-9$ **j.** $\frac{^-144}{16}$ **k.** $\frac{28}{^-1120}$ **l.** $({}^-8 + {}^-2) \cdot {}^-5$
   (two ways?)

   m. $({}^-1)^{100}$ **n.** $({}^-1)^{999}$

3. Demonstrate, with drawings, how obtaining the following products could be demonstrated using two colors of chips:

   a. $3 \times {}^-5$ **b.** ${}^-3 \times 5$ **c.** ${}^-2 \times {}^-3$ **d.** ${}^-3 \times {}^-2$

4. Give the line of reasoning similar to the one given for ${}^-3 \times {}^-2 = {}^+6$ to show that ${}^-7 \times {}^-5$ must be ${}^+35$.

5. What is missing in each student's understanding?

   Ann: "I had ${}^-3 \times {}^-2 \times {}^-1 = {}^+6$, and you marked it wrong. But you said when you multiply negatives, you get a positive."

   Bobo: "You said two negatives make a positive, but when I did ${}^-2 + {}^-3 = {}^+5$, my Mom said it wasn't correct."

6. Using signed numbers, write an equation that describes each of these story problems.

   ▶ Little Bo-Peep loses 4 sheep every week from her very large flock, and they never come home! ◀

   a. In 5 weeks, how will the number in her flock compare to the present number? (Remember to use signed numbers.)
   b. Six weeks ago, how did the number in her flock compare to the present number?

   ▶ Godzilla has been losing 30 pounds a month by watching his diet and by exercising. ◀ (Remember to use signed numbers.)

   c. If he continues at this rate, how will his weight in 6 months compare to his present weight?
   d. Three months ago, how did his weight compare to his present weight?

7. Is there an identity for multiplication of integers? If so, what is it?

8. **a.** Suppose you take any integer a. Can you always find another integer b such that $a \times b = 1$? (That is, does a have a multiplicative inverse in the set of integers?)

   b. Which two integers are their own multiplicative inverses?

9. Identify which of the eleven properties of addition and multiplication is exhibited in each of the following. Or, if a statement is not true, fix it so that it is, and tell which property or properties you used.

   a. ${}^-7 \cdot (3 + {}^-5) = (3 + {}^-5) \cdot {}^-7$ **b.** ${}^-7 \cdot (3 + {}^-5) = {}^-7 \cdot ({}^-5 + 3)$

   c. ${}^-7 \cdot (3 + {}^-5) = {}^-7 \cdot 3 + {}^-7 \cdot {}^-5$

   d. ${}^-7 \cdot [(3 + {}^-5) + 4] = {}^-7 \cdot [3 + ({}^-5 + 4)]$

   e. ${}^-7 \cdot [(3 + {}^-5) + 4] = {}^-7 \cdot (3 + {}^-5) + {}^-7 \cdot 4$

   f. ${}^-7 \cdot [(3 + {}^-5) + 4] = [(3 + {}^-5) + 4] \cdot {}^-7$

   g. $\dfrac{{}^-4}{5} \cdot \dfrac{5}{{}^-4} \cdot {}^-7 = \dfrac{{}^-4}{5} \cdot {}^-7 \cdot \dfrac{5}{{}^-4}$ **h.** $\dfrac{{}^-4}{5} \cdot \dfrac{5}{{}^-4} \cdot {}^-7 = 1 \cdot {}^-7 = {}^-7$

   i. $(3 + 0) + 4 = 3 + 4$ **j.** $(3 + {}^-5) \cdot 6 = 3 \cdot 6 + {}^-5 \cdot 6$

Supplementary Learning Exercises for Section 10.5

1. Give a pattern argument to suggest that ${}^-2 \times {}^-5 = +10$, assuming that (negative) $\times$ (positive) = (negative) has been established.

2. Give the following products and quotients. (Can you do these without a calculator? You should be able to do so.)

   a. ${}^-57.3 \times {}^-19$ **b.** ${}^-\left(\dfrac{7}{24}\right) \div {}^-\left(\dfrac{49}{15}\right) \times \dfrac{14}{25}$ **c.** $7296 \div {}^-57$

   d. ${}^-235.752 \div {}^-0.57$

3. Can your calculator multiply integers, or do you have to be aware of the sign for the product?

4. Use the chip model and repeated subtraction to make drawings to calculate the following:

    a. $^-10 \div {}^-2$ b. $^-16 \div {}^-2$

5. What is your favorite way of making $^-2 \times {}^-3 = 6$ seem reasonable to someone who cannot see how multiplying two negative numbers can give a positive number?

6. a. How would you read $^-(a \cdot b)$, $^-a \cdot b$, and $a \cdot {}^-b$? They each have different meanings.

   b. Using a variety of positive and negative numbers, test with three different choices for a and for b to see how the three expressions compare.

10.6 Number Systems

Signed numbers include all positive and negative integers, positive and negative fractions, and positive and negative repeating or terminating decimal numbers, i.e., all **rational numbers,** extending the term introduced in Chapter 6 for nonnegative rational numbers. Similarly, as with $^-\sqrt{3}$, there are negative **irrational numbers** (numbers that have nonterminating and nonrepeating decimals). Because the rational numbers and irrational numbers together are called the **real numbers,** every real number has an additive inverse. Just as every integer can be matched to a point on the number line, so can *every* real number be matched to a point on the number line. And every point on a number line corresponds to some real number. With two or more number lines arranged to give the familiar *x-y* coordinate system that you studied in algebra, this match of numbers and geometry allows many geometric shapes to be studied with algebra and many algebraic topics to be represented geometrically.

 DISCUSSION 9   Between Any Two Rational Numbers

1. Think of any two rational numbers, such as $\frac{7}{12}$ and $\frac{13}{15}$. Find another rational number between the two numbers.

2. Find another rational number between $\frac{7}{12}$ and the number you found in part (a). This will give a second number between $\frac{7}{12}$ and $\frac{13}{15}$.

3. How many rational numbers are there between $\frac{7}{12}$ and $\frac{13}{15}$ in all? ●

   Your answer to Question 3 in Discussion 9 is perhaps a surprising consequence of the line of reasoning for Question 1. The mathematical term for this phenomenon is called the **density** property of rational numbers. The rational numbers are said to be *dense*.

> A set of numbers is **dense** if, for every choice of two different numbers from the set, there is always another number from the set that is between them (the ***density property***).

Think About …

How does the density property assure that there are infinitely many rational numbers, not just one, between every two different rational numbers?

   Hence, with the number line in mind, one might think that the points for the rational numbers completely fill up the line. But, as you know, the irrational numbers also have points on the number line. Even though the set of rational numbers is dense, there are still "empty" spaces for the irrational numbers.

   We have used some properties of addition without comment. The properties reviewed earlier for whole numbers and rational numbers do continue to be true when negative and irrational numbers are involved, i.e., when any real numbers come up. The commutative property of addition allows us to commute the order of the addends, e.g., $^-13 + 4 = 4 + {}^-13$. The associative property of addition allows us to change the order in which we add. For example, $(3 + {}^-7) + {}^-5 = 3 + ({}^-7 + {}^-5)$. In checking that these two sums are equal, we have $^-4 + {}^-5 = {}^-9$ and $3 + {}^-12 = {}^-9$ for both sides of the equation.

Think About ...

Does the associative property of addition allow you to ignore the parentheses when only addition is involved? Why or why not?

Rational numbers include all whole numbers, fractions, and repeating decimal numbers. Negative integers and negative fractions (or their decimal forms) are also rational numbers. In fact, when we add *any* two rational numbers, the sum is also a rational number. This is an example of what is called the **closure property,** a property that is not usually emphasized in grades K−6 but is useful in more advanced work.

> A set of numbers is **closed** under an operation if, when operating on every two numbers in the system, the result is also in the set of numbers.

EXAMPLE 6

When we add any two positive rational numbers, such as $9\frac{3}{4}$ and 5, the sum, $14\frac{3}{4}$, is also a rational number. This property extends to include negative and positive numbers. $^-9\frac{3}{4} + 5 = {}^-4\frac{3}{4}$, a rational number. In both cases, the sum of two rational numbers was another rational number, so we could say that this example illustrates that *the set of rational numbers is closed under addition.* ●

Even though the full set of real numbers is not commonly encountered in grades K−6, it is also true that the set of real numbers is closed under addition. For example, $\sqrt{2} + 5\sqrt{3}$ and $\sqrt{7} + {}^-6$ are real numbers.

Finally, addition of signed numbers has two additional properties, noted earlier. The first is that 0 is the *additive identity*. The second is that every real number has an *additive inverse*.

EXAMPLE 7

a. $3 + 0 = 3$ and $^-3 + 0 = {}^-3$, because 0 is the additive identity.
b. $3 + {}^-3 = 0$, because 3 and $^-3$ are additive inverses of one another.
c. What is the additive inverse of $2\frac{1}{8}$? It is $^-2\frac{1}{8}$ because $2\frac{1}{8} + {}^-2\frac{1}{8} = 0$. The number $2\frac{1}{8}$ is also the additive inverse of $^-2\frac{1}{8}$.
d. Similarly, $\sqrt{7}$ and $^-\sqrt{7}$ are additive inverses of each other because their sum equals 0. ●

Think About ...

Suppose you are restricted to the set of integers. Do all five of the properties for addition—closure, commutativity, associativity, additive identity, and existence of additive inverses—hold true for all integers?

The properties often (but not always) occur in mathematical situations, perhaps situations not even involving numbers. Mathematicians look for situations in which the properties do occur because the properties may lead to still other results that may be useful in the situations.

In the following activity and discussion, notice the parallels to the corresponding properties of addition. Many will be stated in terms of rational numbers, but they are also true for real numbers.

 DISCUSSION 10   More Properties

1. Provide several examples to test whether multiplication of signed numbers is associative. That is, when three rational numbers are multiplied, does it matter which

multiplication is done first: $a(bc) = (ab)c$, for every choice of rational numbers a, b, and c? For example, will $^-3 \cdot (^-2 \cdot ^-4)$ give the same result as $(^-3 \cdot ^-2) \cdot ^-4$? (Recall that the multiplication symbol $\times$ is often replaced with $\cdot$. In fact, when one or both factors are represented with letters, there often is no symbol between the letters if multiplication is intended: $2 \times b = 2 \cdot b = 2b$, or $a \times b = a \cdot b = ab$.)

2. Is the set of rational numbers closed under multiplication? That is, when every choice of two rational numbers is multiplied, is the product always a rational number?

3. Is there an identity for multiplication of rational numbers? That is, for each rational number a, is there a rational number x for which $a \cdot x = x \cdot a = a$? If so, what is it?

4. Does every rational number have a multiplicative inverse? That is, if c is any rational number, then does there exist a rational number d such that $c \cdot d = d \cdot c = 1$ (where 1 is the identity for multiplication)?

5. For rational numbers, is multiplication distributive over addition? That is, if a, b, and c are any rational numbers, is it true that $a \cdot (b + c) = a \cdot b + a \cdot c$? Substitute numbers for a, b, and c and test whether or not this property appears to be true for all rational numbers. The commutative property of multiplication also gives $(x + y) \cdot z = x \cdot z + y \cdot z$ as a form of this distributive property. ●

> The *multiplicative identity* for the set of rational numbers is 1 because for every rational number a, $1 \cdot a = a$, and $a \cdot 1 = a$.

> If the product of two numbers is 1, each number is the **multiplicative inverse** of the other number. If a is not 0, its multiplicative inverse is often written $\frac{1}{a}$ or even a^{-1}. The multiplicative inverse of a (nonzero) fraction is sometimes called its **reciprocal**.

Think About …

What is the reciprocal of $\frac{m}{n}$? How does your answer satisfy the description above? Is there a multiplicative identity for the set of integers? Do integers have multiplicative inverses? Why doesn't 0 have a multiplicative inverse?

~~Five properties were listed as~~ being true for addition, five for multiplication, and one property connects multiplication and addition. Can you list them? When all eleven of these properties hold for addition and multiplication on any set of numbers, mathematicians call this set with its two operations a *field*. (This perhaps surprising term was given by a mathematician who had a wide view of numbers.)

DISCUSSION 11   What Makes a Mathematical Field?

1. Is the set of even integers (e.g., . . ., $^-6$, $^-4$, $^-2$, 0, 2, 4, 6, . . .), with addition and multiplication defined as usual, a field? If not, which of the eleven properties fails?

2. Is the set of positive rational numbers a field? If not, which of the eleven properties fails? ●

In an earlier chapter we also talked about irrational numbers, i.e., numbers that cannot be expressed as a fraction or as a repeating decimal. Numbers such as π and $\sqrt{2}$ are irrational numbers. You learned that the set of rational numbers combined with the set of irrational numbers is called the set of **real numbers.**

Although we will not spend more time here on real numbers, it suffices to say that (1) the real numbers, with operations of addition and multiplication, form a field, and (2) every real number corresponds to a point on the number line, and every point on the number line corresponds to a real number.

Students will encounter real numbers primarily when they reach algebra, and it will be necessary for them to use the field properties to operate with real numbers in algebra and beyond.

Yet, some mathematically important number systems do not have infinitely many numbers. One such system is sometimes called *clock arithmetic.* Suppose you have a five-hour clock, i.e., the numbers 0, 1, 2, 3, and 4 are evenly spaced around the clock, such as this:

Sometimes 5 is used instead of 0, but you will soon see why 0 is used here.

When one adds in this arithmetic, it is like going around the clock, the number of hours indicated by the addends, starting at 0. Thus, 2 + 4 begins at 0, goes two spaces to 2, and then goes clockwise four spaces, ending at 1. So, 2 + 4 = 1, a result that looks quite strange but makes sense in this number system. Also, 4 + 2 begins at 0, moves to 4, then moves two spaces clockwise, landing on 1, so 4 + 2 = 1. Likewise, 4 + 0 would mean beginning at 0, moving to 4, and then traveling 0 hours, so 4 + 0 = 4. But 0 + 4 would mean beginning at 0, going zero spaces, and then going four spaces, landing on 4, so 0 + 4 = 4.

ACTIVITY 10   Which Properties Hold?

1. Complete the following, using clock arithmetic with five numbers: 0, 1, 2, 3, and 4.

   a. 3 + 2   **b.** 4 + 4 + 4   **c.** 1 + 4   **d.** 3 + 3   **e.** 2 + (3 + 4)

   f. (2 + 3) + 4   **g.** (1 + 2) + 4   **h.** 1 + (2 + 4)   **i.** 2 + 3   **j.** 4 + 1

2. Have you illustrated, in Problem 1, any instances of the commutative property of addition? If so, which one(s)? Try some others.

3. Have you illustrated, in Problem 1, any instances of the associative property of addition? If so, which one(s)? Try some others.

4. Is this set of numbers closed under addition? That is, for any two clock numbers in the set, would the sum be in the set?

5. Is there an additive identity in this system? If so, what is it? (Now you see why 0 rather than 5 was chosen.)

6. Does each number have an additive inverse? That is, for any given clock number c, is there a number d such that $c + d = 0$? (Of course, this assumes that 0 is the additive identity.) ●

We can also define multiplication in this system, using repeated addition. That is, $4 \times 2 = 2 + 2 + 2 + 2 = 3$ because 2 + 2 (starting at 2 and moving clockwise two places) is 4. Then 4 + 2 is 1, and finally 1 + 2 = 3. Also, $2 \times 4 = 4 + 4 = 3$.

ACTIVITY 11   Do the Field Properties Hold in Clock Arithmetic?

1. Fill in these two tables. Some results have been entered for you.

+	0	1	2	3	4
0					
1					
2		3		0	
3					
4					

×	0	1	2	3	4
0	0				
1					
2	0			1	
3					
4					

2. Try several examples to illustrate that multiplication in the five-hour clock system is commutative and associative.

3. Is the set closed under multiplication?

4. Is there a multiplicative identity? If so, what is it?

5. Does each number in the system have a multiplicative inverse?

6. Finally, is multiplication distributive over addition? If all eleven properties hold, then clock arithmetic for 5 is a field. Is this a field? ●

TAKE-AWAY MESSAGE . . . All the rational and irrational numbers make up the real numbers, with every real number corresponding to a point on a number line, and vice versa. The rational numbers are dense, meaning that there is always another rational number between two given rational numbers; indeed, there are infinitely many. The rational numbers, together with the operations of addition and multiplication, form what is called a field, as do the real numbers with addition and multiplication. A mathematical field is defined as a set of numbers with two operations for which the eleven properties we have discussed all hold. Some sets of numbers we have worked with, such as the set of whole numbers, do not form a field with addition and multiplication because one or more properties fail.

The rational numbers and real numbers are both infinite number systems. Are there any finite number systems that form a field? You have found one: clock arithmetic for a clock with numbers 0, 1, 2, 3, and 4, with operations defined as in the tables. Are there other finite fields? That's an interesting question.

Learning Exercises for Section 10.6

1. What is $^-(^-3)$?

2. What is the additive inverse of $^-\left(\frac{3}{5}\right)$? Of 0? Of 14? Of 3.67?

3. **a.** What is the multiplicative inverse of $^-\left(\frac{3}{5}\right)$?
   b. Of 14?
   c. Of 3.67?
   d. Why does 0 not have a multiplicative inverse?

4. **a.** Is the set of odd whole numbers (with addition and multiplication) a field? Why or why not?
   b. Is the set of even whole numbers (with addition and multiplication) a field? Why or why not?
   c. Is the set of integers (with addition and multiplication) a field? Why or why not? In each case, tell which properties hold, which ones do not, and provide an example in each case.

5. Consider clock arithmetic using a clock with just four numbers: 0, 1, 2, and 3, along with addition and multiplication defined similarly to the example of a clock with five numbers. Do all eleven field properties hold here? If not, which ones do not?

6. Do the same with clock arithmetic using six numbers; then do the same using seven numbers. Can you make any conjectures about when all eleven properties needed for a field will hold for a clock arithmetic? You may wish to examine other clock systems to test your conjecture further.

7. The set of real numbers is dense. What does that mean?

8. In each part, tell whether the set of numbers is dense. Justify your answers.

   a. the set of integers
   b. the set of positive rational numbers
   c. the set of negative rational numbers

9. Give eight fractions between each pair.

   a. $\frac{7}{4}$ and $\frac{9}{5}$ **b.** $^-\left(\frac{2}{3}\right)$ and $^-\left(\frac{31}{50}\right)$

Notes

10. a. Is the set of even integers closed under addition? Why or why not? (The even integers are . . . , $^-4, ^-2, 0, 2, 4, \ldots$.)

   b. Is the set of multiples of 3 closed under addition? Why or why not?

   c. Is the set of odd numbers closed under addition? Why or why not? (The odd integers are . . . $^-3, ^-1, 1, 3, 5, \ldots$.)

   d. Is the set of whole numbers closed under subtraction? Why or why not?

   e. Is the set of all integers closed under addition? Why or why not?

   f. Is the set of all integers closed under subtraction? Why or why not?

   g. Is the set of all positive rational numbers closed under subtraction? Why or why not?

11. For each of the following, say whether or not the statement is true. If it is true, state the property that makes it true.

   a. $(3 + {}^-4) + 6 = ({}^-4 + 3) + 6$ **b.** $(3 + {}^-4) + 6 = 3 + ({}^-4 + 6)$

   c. $2 + 0 = 2$ **d.** $4 + {}^-4 = 0$

   e. $17 + (4 + {}^-4) = 17 + 0$ **f.** $17 + (4 + {}^-4) = 21 + {}^-4$

   g. $\frac{306}{18} + {}^-4$ is a rational number. In fact, it is an integer.

   h. $(3 + {}^-4) + 6 = 6 + (3 + {}^-4)$ **i.** $289\frac{1}{2} + 0 = 289\frac{1}{2}$

   j. $17.638 + {}^-17.638 = 0$

12. Provide the additive inverse for each of the following:

   a. 13 **b.** $^-4\frac{3}{10}$ **c.** $^-\left(\frac{4}{9}\right)$ **d.** $\sqrt{11}$

13. Use examples to test which property or properties of the five properties of addition, the five properties of multiplication, and the distributive property of multiplication over addition do or do not hold for just the set of integers.

14. Use examples (including negative rational numbers) to illustrate that the eleven properties all hold for the set of rational numbers.

Supplementary Learning Exercises for Section 10.6

1. Why is the set of whole numbers *not* dense?

2. a. Give five rational numbers between $^-\left(\frac{3}{4}\right)$ and $^-\left(\frac{1}{4}\right)$.

   b. How many rational numbers are there between $^-\left(\frac{3}{4}\right)$ and $^-\left(\frac{1}{4}\right)$?

3. Use *only* the property given to change the following expressions:

   a. $\left(\frac{5}{8} \times {}^-20\right) \times \left({}^-4\frac{5}{6} \times \frac{8}{35}\right)$, associativity of multiplication

   b. $\left(\frac{5}{8} \times {}^-20\right) \times \left({}^-4\frac{5}{6} \times \frac{8}{35}\right)$, commutativity of multiplication

   c. $\frac{5}{6} \times (24 + {}^-600)$, distributive property

   d. $\left(\frac{2}{3} \times {}^-298\right) + \left(\frac{2}{3} \times {}^-2\right)$, distributive property

   e. $2x + {}^-5x$, distributive property

   f. $\left(5 \times \frac{1}{5}\right) \times 1$, multiplicative inverses

   g. $\left(5 \times \frac{1}{5}\right) \times 1$, multiplicative identity

4. Use *only* the property given to change the following expressions:

   a. $(7 + {-}3) + ({-}15 + 8)$, commutative property of addition

   b. $(5 + {-}5) + 0$, additive inverse property

   c. $(5 + {-}5) + 0$, additive identity

   d. $[{}^-2 + ({}^-4 + 5)] + {}^-5$, associative property of addition

5. **a.** Is the set of numbers $-1, 0, 1$ closed under multiplication? Explain.
    b. Is the set of numbers $-1, 0, 1$ closed under addition? Explain.
    c. Is the set of numbers $-2, 0, 2$ closed under multiplication? Explain.

10.7 Issues for Learning: Open Number Sentences

In this chapter, you have had opportunities to think about the integers and other number systems in a variety of ways. You may have gained a new appreciation for the challenges that teachers and students face in integer instruction. At the same time, you have seen examples of productive mathematical intuitions that children can bring to learning about integers. How can we teach children about integer arithmetic in ways that invite them to build on their intuitions?

Some textbooks use the number line extensively in integer instruction; others use a chip model. Many focus on the signs rules for integer arithmetic. Regardless of which of these approaches is emphasized, textbook instruction concerning integer arithmetic concentrates overwhelmingly on *result-unknown* problems, i.e., problems of the form a + b = ☐ or a − b = ☐. These are also more amenable to the application of rules for computing with integers (see Section 10.4).

Researchers have found that changing the position of the unknown can invite students to reason in different ways.[3] Open number sentences, which include unknown starting points and unknown change problems, can encourage students to reason flexibly about integer arithmetic. These include opportunities for counterintuitive problems, including subtraction problems in which two whole numbers are given but the unknown is negative, or addition problems in which the result is smaller than the start. For example, many young children will say that 6 + ☐ = 4 is not possible because addition makes bigger, and 4 is less than 6. Older children often build productively on this same reasoning: They recognize that 6 is greater than 4, and they know that addition cannot make smaller *unless* the number added is negative. In this way, they infer that the unknown must be a negative number.

Examples of open number sentences:

Start Unknown:	☐ + ⁻8 = ⁻10	☐ − 5 = ⁻1
Change Unknown:	6 + ☐ = 4	5 − ☐ = 8
Result Unknown:	3 − 5 = ☐	⁻3 + 6 = ☐

Posing open number sentences affords opportunities to discuss how students' strategies relate to the big ideas of positive and negative numbers. For example, some children may solve ☐ − 5 = ⁻1, reasoning in terms of order: ⁻1 is 5 units less than what number? Or using an empty number line, ⁻1 is one unit to the left of 0. What number after subtracting 4 will get to zero? In selecting open number sentences, it is useful to think about these opportunities to connect to the big ideas.

One big idea relates to the different meanings of the minus sign. One meaning of the minus sign, opposite of, is often neglected in integer instruction. Look for opportunities to discuss the *opposite-of* meaning. For example, what is the relationship between 4 and ⁻4? Many textbooks may refer to these as opposites, but in what sense are they opposite? Formally, ⁻4 is the additive inverse of 4. This means that 4 + ⁻4 = 0. We can think about the opposite of any number in this way: What do we have to add to the given number to get 0? The opposite-of meaning becomes more clear when we consider examples like −(⁻4). In this case, the minus sign outside of the parentheses does not mean negative. We can make sense of the expression −(⁻4) by reading it as the opposite of ⁻4, which is 4. How can we use the opposite-of meaning to interpret the expression ⁻x?

Subtraction and negative are familiar meanings of the minus sign. For example, many students say that the problem ⁻8 − 3 = ☐ involves two negative numbers. They solve by adding ⁻8 and ⁻3. In doing this, they implicitly treat subtracting 3 as equivalent to adding ⁻3. People who are flexible in reasoning about integers often make these kinds of moves without giving them much thought, and this level of procedural fluency is desirable.

Notes

At the same time, the question of justifying mathematically why we can transform $^-8 - 3$ into $^-8 + {}^-3$ deserves attention. Thus, it is important to pay attention to the language you use when talking about integer arithmetic. For example, the language "minus eight minus three" does not distinguish the two meanings of the minus sign, whereas "negative eight minus three" does. Invite students to reason in new ways with open number sentences and find ways to make connections to the big ideas of signed numbers.

10.8 Check Yourself

In this chapter, we have extended our number system to include all of the signed numbers. We can now add, subtract, multiply, and divide signed numbers, and in particular we considered the properties of addition and of multiplication. This number system, together with the eleven properties we found to be true of addition and multiplication, form a *mathematical field*.

You should be able to work problems like those assigned and to meet the following objectives:

1. Add, subtract, multiply, and divide rational numbers, including negative numbers, according to the rules described in this chapter.

2. Understand absolute value, apply it, and explain why it is useful in the context of working with negative and positive numbers.

3. Check whether a clearly defined set of numbers is closed with respect to a particular operation.

4. Provide examples of the closure, commutative, associative, identity, and inverse properties for addition and multiplication of rational numbers, and of the distributive property of multiplication over addition.

5. Recognize and name the properties when they are used.

6. Examine other number systems (e.g., even integer arithmetic; clock arithmetic for numbers such as 4, 5, and 6) and tell whether the eleven field properties hold in these systems.

7. Explain why operations on negative numbers, particularly multiplication, are sometimes difficult to teach.

References for Chapter 10

[1]Thomaidis, Y., & Czanakis, C. (2007). The notion of historical "parallelism" revisited: Historical evolution and students' conception of the order relation on the number line. *Educational Studies in Mathematics, 66*(2), 165–183.

[2]Pierson, J., Lamb, L., Philipp, R., Schappelle, B., Whitacre, I., & Lewis, M. (2012). *Ways of reasoning about integers: Order, magnitude, and formalisms*. Paper prepared for the 2012 annual meeting of the American Educational Research Association, New Orleans, LA.

[3]Fennema, E., Carpenter, T., Levi, L., Franke, M. L., & Empson, S. B. (1999). *Children's mathematics: Cognitively guided instruction*. Portsmouth, NH: Heinemann.

Chapter 11

Number Theory

Number theory is one of the oldest branches of mathematics. For many years people who studied number theory delighted in its "pure" nature because there were few practical applications of number theory. It is therefore somewhat ironic that number theory now plays important roles in keeping military and diplomatic messages secret and in making certain that people are authorized to withdraw money in electronic financial transactions.[1] (These naturally are more complicated than your secret PIN for an automatic teller machine.) Our attention will be restricted largely to the ideas from number theory that come up in the elementary school curriculum. *Although number theory ideas can be applied to negative integers as well as positive ones, we will have in mind only the whole numbers, 0, 1, 2, 3, . . . , in this chapter.* Fractions use number theory ideas, but only in a context where number theory is applied to whole number numerators and denominators.

11.1 Factors and Multiples, Primes and Composites

Numbers are related to one another in many ways. In this section we examine the fundamental ways that whole numbers exist when they are expressed multiplicatively.

Recall the discussion of factors in Chapter 3. Since $5 \times 6 = 30$, and $15 \times 2 = 30$, each of the numbers 5, 6, 15, and 2 is a factor of 30. (There are more.) Some numbers have many more factors. Indeed, 240 has 20 factors in all! Even a small number can have several factors: $2 \times 3 = 6$ and $1 \times 6 = 6$, so the numbers 2, 3, 1, and 6 are factors of 6. Some numbers have exactly two different factors—e.g., 13 has only 1 and 13 as factors. Such numbers play an important role in number theory and are called **prime numbers**. The number 29 is another example of a number that has exactly two factors: 1 and 29, so 29 is a prime number.

It may be surprising to you that there are infinitely many prime numbers, a fact known to the ancient Greeks. There are, e.g., 455,052,512 prime numbers less than 10^{10}. Indeed, with the advancing capabilities of computers (and knowledge of number theory), larger and larger primes are occasionally found. In 1978, e.g., the largest known prime required 6533 digits to write. By 1985 other new primes had been found, the largest one requiring 65,050 digits to write. (How many pages would that require?) By 1992 mathematicians had found a prime number requiring 227,832 digits to write. In 1997 they found a prime requiring 895,932 digits, which would fill 450 pages of a paperback book. As of this writing, the largest known prime number has 12,978,189 digits—there is good reason for not printing it here! In fact, if you were able to write 10 digits per second (a feat in and of itself), it would take you 15 days to write this number.

A **prime number** is a whole number that has exactly two different whole number factors. A **composite number** is a whole number greater than 1 that has more than two factors.

⋮ *Think About ...*

⋮ Does the number 1 fit the description of a prime number? Of a composite
⋮ number? What about the number 0?

Notes

 DISCUSSION 1   Representing Primes and Composites

Suppose you have *n* tiles or counters as "chairs." If $n = 6$, in how many different ways can you arrange the chairs in complete rows with the same number in each row? If $n = 13$? If $n = 14$? Discuss how the possible arrangements of *n* chairs in a rectangular array are different for *n* as a prime vs. *n* as a composite number. ●

Eratosthenes, a Greek who lived more than 2200 years ago, devised the following method for identifying primes.

 ACTIVITY 1   Eratosthenes' Sieve for Finding Primes

1. **a.** Cross out 1 in the array of numbers shown below.

   b. The number 2 is prime. Circle 2 in the array. Cross out all the larger multiples of 2 in the array ($2 \times 2 = 4, 3 \times 2 = 6, 4 \times 2 = 8, \dots$).

   c. The number 3 is prime. Circle 3 in the array. Cross out all the larger multiples of 3 in the array.

   d. The number 5 is prime. Circle 5 in the array. Cross out all the larger multiples of 5 in the array.

   e. The number 7 is prime. Circle 7 in the array. Cross out all the larger multiples of 7 in the array.

   f. What is circled next? Does this procedure ever end? Explain.

   g. Circle 11 in the array. Cross out all the larger multiples of 11 in the array.

   h. Circle all the numbers not yet crossed out. Are the numbers circled all primes?

1	2	3	4	5	6	
7	8	9	10	11	12	
13	14	15	16	17	18	
19	20	21	22	23	24	
25	26	27	28	29	30	
31	32	33	34	35	36	
37	38	39	40	41	42	
43	44	45	46	47	48	
49	50	51	52	53	54	
55	56	57	58	59	60	
61	62	63	64	65	66	
67	68	69	70	71	72	
73	74	75	76	77	78	
79	80	81	82	83	84	
85	86	87	88	89	90	
91	92	93	94	95	96	
97	98	99	100	101	102	
103	104	105	106	107	108	
109	110	111	112	113	114	
115	116	117	118	119	120	
121	122	123	124	125	126	etc.

2. **a.** If this array were extended, which column would 1000 be in? Which column would 1,000,000 be in?

   b. $2^{10} = 1024$. Which column would this number be in?

   c. $2^{11} = 2048$. Which column would this number be in?

3. Find a column in the array for which the following is true: If two numbers in the column are multiplied, the product is also in that column. Each of four original columns in the array (if continued infinitely) has this property. Which four? What is the name of this property? ●

Think About ...

If this array were written in four columns rather than six columns, which column would the number 1000 be in? How did you determine that?

You probably observed that all the numbers in the last column of the array are multiples of 6. You can represent each of these numbers as rectangular arrays with exactly 6 tiles in each row. Rectangular arrays can be used to illustrate some relationships between numbers. You can easily draw a rectangular array of 18 tiles with 6 in each row.

This rectangular array can illustrate the following:

$3 \times 6 = 18$.
18 is a multiple of 6.
18 is the product of 6 and 3, which are factors of 18.

Think About ...

Can you draw a rectangular array with 15 tiles that has 6 tiles in each row? Why or why not?

You should recall the following definitions from earlier work. Pay particular attention to these vocabulary words. They are often misused.

If $mn = p$, then m and n are called **factors** of p and p is called a **multiple** of m (and of n). If $mn = p$ and m is not 0, then m is called a **divisor** of p. We say that p is *divisible* by m. Recall also that p is the **product** of m and n.

EXAMPLE 1

$12 \times 15 = 180$, so 12 and 15 are factors of 180; they are also divisors of 180. We can also say 180 is divisible by 12 (and by 15). 180 is the product of 12 and 15 or, in number theory lingo, 180 is a multiple of 12 (and of 15). $0 \times 5 = 0$, so 0 and 5 are factors of 0, and 5 is a divisor of 0 but 0 is not a divisor of 0. ●

Think About

Why is 0 never a divisor? Think back to Chapter 3 in which dividing by 0 was discussed.

ACTIVITY 2   Vocabulary Practice

1. Use 6, 8, and 48 in sentences that involve "factor," "divisor," "product," and "multiple." Use rectangular arrays, equations, and words to describe the relationship between 6, 8, and 48.

2. Use these vocabulary words to describe the variables in $mn = p$.

3. If m is a factor of n, is n a multiple of m? If p is a multiple of q, is q a factor of p?

4. If $m = 2n$, what can you say about m? ●

The following activity will help us think about these relationships in different ways.

ACTIVITY 3   Which Lockers Are Open?

In a certain school there are 100 lockers lining a long hallway. All are closed. Suppose 100 students walk down the hall, in file, and the first student opens every locker. The second student comes behind the first and closes every second locker, beginning with locker no. 2. The third student changes the position of every third locker; if it is open, this student closes it; if it is closed, this third student opens it. The fourth student changes the position of every fourth locker, etc., until the 100th student changes only the position of the 100th locker. After this procession, which lockers are open? Why are they open? At the end of this procession, how many times did locker 9 get changed? How many times did locker 10 get changed? ●

TAKE-AWAY MESSAGE . . . Understanding distinctions between prime and composite numbers, and between multiples and factors (or divisors) is essential before continuing on to the remaining sections of this chapter. The sieve of Eratosthenes provides one way to find prime numbers, but is not efficient for large numbers. Prime numbers are used in cryptography and in businesses such as banking.

Learning Exercises for Section 11.1

1. **a.** Give three factors of 25. Can you find more? If so, how? If not, why not?
   b. Give three multiples of 25. Can you find more? If so, how? If not, why not?

2. **a.** Write an equation that asserts 25 is a factor of k. How could a rectangular array show this?
   b. Write an equation that asserts m is a factor of w.
   c. Write an equation that asserts v is a multiple of t.

3. **a.** If 216 is a factor of 2376, what equation must have a whole number solution?
   b. How does one find out whether 144 is a factor of 3456?

4. Use the notion of rectangular arrays to assert that 21 is not divisible by 5.

5. Explain why these assertions are not quite correct:
   a. "A factor of a number is always less than the number."
   b. "A multiple of a number is always greater than the number."

6. **a.** Give two factors of 506.
   b. Give two multiples of 506.

7. True or false? If false, correct the statement.
   a. 13 is a factor of 39.
   b. 12 is a factor of 36.
   c. 24 is a factor of 36.
   d. 36 is a multiple of 12.
   e. 36 is a multiple of 48.
   f. 16 is a factor of 512.
   g. 2 is a multiple of 1.

8. **a.** Write an equation that asserts that 15 is a multiple of a whole number k.
   b. Write an equation that asserts that a whole number m is a factor of a whole number x.

9. **a.** Suppose that k is a factor of m and m is a factor of n. Is k a factor of n? Is n a multiple of k? Justify your decisions.
   b. Suppose that k is a factor of both m and n. Is k a factor of $m + n$ also? Justify your decision.
   c. Suppose k is a factor of m but k is not a factor of n. Is k a factor of $m + n$ also? Justify your decisions. (You may want to try this with numbers first. For example, 5 is a factor of 15, but is not a factor of _____.)

10. You know that the even (whole) numbers are the elements of the set of numbers 0, 2, 4, 6, 8, . . . , and that the odd (whole) numbers are the elements of the set of numbers 1, 3, 5, 7, 9, ...

    a. Write a description of the even numbers that uses "2" and the word "factor."
    b. Write a description of the even numbers that uses "2" and the word "multiple."
    c. Write a description of odd numbers that uses "2."

11. Complete the following addition and multiplication tables for even and odd numbers:

+	even	odd
even		
odd		

×	even	odd
even		
odd		

Can you then make any definite assertions about

    a. the sum of any number of even numbers?
    b. the sum of any number of odd numbers?
    c. the product of any number of even numbers?
    d. the product of any number of odd numbers?
    e. whether it is possible for an odd number to have an even factor?
    f. whether it is possible for an even number to have an odd factor?
    g. Is the set of even numbers closed under addition?
    h. Is the set of odd numbers closed under addition?
    i. Is the set of even numbers closed under multiplication?
    j. Is the set of odd numbers closed under multiplication?

12. Explain why each of these is a prime number: 2, 3, 29, 97.

13. List all the primes (prime numbers) less than 100. (You can use the sieve of Eratosthenes described in Activity 1.)

14. Explain why each of these is a composite number: 15, 27, 49, 119.

15. **a.** Why is 0 neither a prime nor a composite number?
    b. Why is 1 neither a prime nor a composite number?
    c. What is the drawback to the following "definition" of prime number: a whole number with only 1 and itself as factors?

16. Give two factors of each number (there may be more than two):

    a. 829   **b.** 5771   **c.** 506   **d.** n (if $n > 1$)

17. Explain why 2 is the only even prime number. (Can you always find a third factor for larger even numbers?)

18. Conjecture: Given two whole numbers, the larger one will have more factors than the smaller one will. Gather more evidence on this conjecture by working with several (4 or 5) pairs of numbers.

19. **a.** Just above a number line (at least to 50), mark each factor of 24 with a heavy dot and mark each multiple of 6 with a square.
    b. Just below the same number line, mark each factor of 18 with a triangle and mark each multiple of 18 with a circle.
    c. What are common factors of 18 and 24? What are common multiples of 6 and 18?

20. Explain without much calculation how you know that 2, 3, 5, 7, 11, 13, and 17 are not factors of $n = 2 \cdot 3 \cdot 5 \cdot 7 \cdot 11 \cdot 13 \cdot 17 + 1$.

21. 6 is called a **perfect number** because the sum of its factors (other than itself) equals the number: $1 + 2 + 3 = 6$. What is the next perfect number?

Notes

Supplementary Learning Exercises for Section 11.1

1. Without looking up the phrases, tell what a prime number is, and what a composite number is.

2. Is it possible for some composite numbers to also be prime numbers?

3. An array for a prime number can have exactly one row or exactly one column. True or false? Explain your decision.

4. Except for 1, only composite numbers are crossed out in the sieve of Eratosthenes. True or false?

5. In the sieve of Eratosthenes, if 19 is the next number not crossed out, which one of the following will be the next new number to be crossed out? Explain.

   a. 23   **b.** 361 ($= 19 \times 19$)   **c.** 323 ($= 19 \times 17$)   **d.** 20

6. If k is a factor of m, is m always a factor of k? Explain.

7. Use an example to show how the words "multiple" and "factor" can be used for the same multiplication equation.

8. For each number below, write a sentence that refers to the number, for $87 \times 24 = 2088$.

   a. 87   **b.** 24   **c.** 2088

9. For each (whole number) variable below, write a sentence that refers to the variable, for $g \times s = w$.

   a. g **b.** s **c.** w

10. **a.** Give three factors of 49. Can you find more? If so, how? If not, why not?
    b. Is 49 a composite number? Explain why or why not.
    c. Give three multiples of 49. Can you find more? If so, how? If not, why not?

11. Tell whether each of the following is a prime or composite number. Explain how you know.

    a. 31   **b.** $13 \times 23 = 299$ **c.** 27   **d.** 999   **e.** $\frac{7}{11}$

12. Is the set of all prime numbers closed under multiplication? Explain.

13. **a.** List all the multiples of 0.
    b. What whole numbers are factors of 0?

14. Find at least one example that shows the following conjecture is false:

    The sum of two prime numbers is an even number.

15. $97 \times 103 = 9991$. How do you know that 9991 is not a prime?

16. Why does the sieve of Erathosthenes give only prime numbers? That is, how do you know that some composite number does not "sneak" through? (*Hint:* Exercise 15)

17. The number 49 has exactly three factors: 1, 7, and 49. Find three other numbers that have exactly three factors. What seems to be true about such numbers?

11.2 Prime Factorization

How do you know whether a number is prime or composite? The number 6 can be written as a product of prime numbers: 2×3. The number 18 can be written as the product of three primes: $2 \times 3 \times 3$. Can other composite numbers be written as a product of primes? These questions and others are explored in the next activity.

ACTIVITY 4   I'm in My Prime (Part 1)

Write the numbers from 2 to 48 using only prime numbers.

		13	25	37
2	2	14	26	38
3	3	15	27	39
$2 \times 2 = 2^2$	4	16	28	40
5	5	17	29	41
2×3	6	18	30	42
7	7	19	31	43
$2 \times 2 \times 2 = 2^3$	8	20	32	44
	9	21	33	45
	10	22	34	46
	11	23	35	47
	12	24	36	48

Every whole number except 1 can be written as a prime number or as a product of prime numbers. (Why is that?)

> A number written as a product of prime numbers is in **prime factorization** form.

 ACTIVITY 5   I'm in My Prime (Part 2)

a. Compare your table with others.

b. Write down as many patterns as you can find in this table.

c. If the list continued, what would be the prime factorization of 1008?

d. Did you have any prime factorizations different from those of other students?

e. What can you say about numbers that are divisible by 2? By 5? By 10?

All of you should have exactly the same factorizations in Activity 4, except possibly for the order of the factors (and notational shortcuts like 3^2 for $3 \cdot 3$). This result is true in general.

> The fact that every whole number greater than 1 is either a prime or can be expressed as the product of prime numbers uniquely (except possibly for order) is called the **Unique Factorization Theorem**, or sometimes, the **Fundamental Theorem of Arithmetic**.

The Fundamental Theorem of Arithmetic means that, in some sense, the prime numbers can be regarded as the building blocks for all the whole numbers other than 0 and 1. Other whole numbers are primes or can be expressed in exactly one way as the product of primes. Thus, a number theorist often finds it most useful to think of 288 as $2 \cdot 2 \cdot 2 \cdot 2 \cdot 2 \cdot 3 \cdot 3$, or $2^5 \times 3^2$.

  By asking that the prime factors be given in increasing order and that exponents be used if a prime factor is repeated, one gets the standard prime factorization. For example, $180 = 2^2 \times 3^2 \times 5$ and $288 = 2^5 \times 3^2$. This form is not essential, but it makes quick comparisons of two prime factorizations easier.

Notes

Notes

ACTIVITY 6   Detective Work on Factorizations

1. Francisco thinks he's found a contradiction to Unique Factorization Theorem. How would you respond?

$$2^8 \cdot 7^5 \cdot 89^2 = 2^7 \cdot 7^4 \cdot 14 \cdot 89^2$$

2. These people are finding the prime factorization of the same number x. No one is finished. Answer (and explain) the questions below without doing any computation:

Aña: $x = 3^8 \cdot 7^4 \cdot 4797134197203$ Ben: $x = 3^7 \cdot 7^4 \cdot$ an odd number

Carlos: $x = 21^2 \cdot 7^9 \cdot 3^6 \cdot$ an even number   Dee: $x = 3 \cdot 3 \cdot 3 \cdot 49 \cdot$ an odd number

   a. Who might agree when they finish?
   b. Who definitely will disagree?
   c. Might all four respondents be correct, if they work forward from where they are now? ●

One consequence of unique factorization into primes is that any factor (except 1) of a number greater than 1 can involve only primes that appear in the number's prime factorization. For example, suppose that $n = 2100$. Then 42 is a factor of 2100 because $42 \times 50 = 2100$. If we continue to factor the 42 and 50, eventually we will get prime factors that *must* appear in the prime factorization of 2100 because the prime factorization is unique. Any prime number that does not appear in the prime factorization of 2100 cannot be "hidden" in some factor of the number.

   One way to visually organize the work when finding the prime factorization of a number is to make a **factor tree.**

Thus,

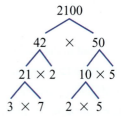

Looking at the ends of each branch of the previous factor tree, we have factors 3, 7, 2, 2, 5, 5. Thus, $2100 = 2^2 \cdot 3 \cdot 5^2 \cdot 7$. Note that we could have thought of the composite numbers in other ways, e.g., 42 as 6×7, or 2100 as 2×1050.

> ### Think About …
> If different factorizations for composite numbers in the factor tree are used, will the prime factorization be the same?

   The general argument for finding the prime factorization of a number follows the same pattern. Suppose that m is a factor of n. Then there is some whole number k such that $m \cdot k = n$. This process starts a factorization of n, and the factorization *must* lead eventually to a unique set of prime factors, according to the Unique Factorization Theorem. This result means that some of the *same* primes in the prime factorization of n must be factors of m (and of k) and that no other primes can be involved.

ACTIVITY 7   More on Primes

1. If $n = 2 \times 3^5 \times 11^2 \times 19^3$, what are four prime factors of n? What are 10 composite factors of n?

2. Find the *smallest* number with factors . . .

   a. 2, 6, 8, 22, 30, and 45   b. 50, 126, and 490 ●

How many factors does a number have? For example, consider the number 72. We could write 72 as 6×12, 2×36, 3×24, 4×18 or 8×9, or 1×72. We could organize these by listing each pair of numbers. In other words, we can find pairs of "buddy" factors. 1, 2, 3, 4, 6, 8, match with 72, 36, 24, 18, 12, and 9, respectively (i.e., 1 and 72 are buddy factors, 2 and 36 are buddy factors, etc.). So, 72 has 12 factors.

We can find the prime factorization of 72 by using a factor tree or by simply factoring 72 as 6×12, and then continuing by writing 6 and 12 as a product of prime numbers. The prime factorization of 72 is $(2 \times 3) \times (2 \times 2 \times 3)$, or $2^3 \times 3^2$.

How can we use the prime factorization of a number to list all the factors of a number? We could use the knowledge of the prime factorization of $72 = 2^3 \times 3^2$ to list the factors systematically as follows:

$2^0 \cdot 3^0$;	$2^1 \cdot 3^0$;	$2^2 \cdot 3^0$;	$2^3 \cdot 3^0$

gives us factors 1, 2, 4, and 8.

$2^0 \cdot 3^1$;	$2^1 \cdot 3^1$;	$2^2 \cdot 3^1$;	$2^3 \cdot 3^1$

gives us factors 3, 6, 12, and 24.

$2^0 \cdot 3^2$;	$2^1 \cdot 3^2$;	$2^2 \cdot 3^2$;	$2^3 \cdot 3^2$

gives us factors 9, 18, 36, and 72.

> ### Think About …
>
> What is the same about each row of factors? What is different between the rows of factors?

Counting is one of the "big ideas" of mathematics that can be found throughout mathematics. The questions "Do we have them all?" and "Are any repeats?" can usually only be answered if the counting was systematic. Be *sure* to note how the system worked to assure finding all factors.

 ACTIVITY 8 Use My Rule

Systematically list all the factors of $5000 = 2^3 \cdot 5^4$. You should have 20 numbers listed. ●

Understanding the system will be critical in generalizing a rule that will help you effectively predict the number of factors a number has.

 DISCUSSION 2 How Many Factors?

Return to your table in Activity 4 and write each number using exponents where possible. How can you determine the number of factors of a composite number by knowing the exponents in its prime factorization? Determine a rule that will tell you the number of factors for any whole number except 0. *Hint*: Recall the Fundamental Counting Principle from Chapter 3. For any prime number a to the *n*th power, a can appear in a factor in $n + 1$ ways where the exponents are . . . (you finish the sentence—see the rows of factors of 72 above). Take note of the difference between finding *all* factors and finding the *prime* factors. ●

TAKE-AWAY MESSAGE . . . Every whole number greater than 1 can be uniquely factored into primes (disregarding order). This fact is referred to as the Unique Factorization Theorem (sometimes as the Fundamental Theorem of Arithmetic). Another way to think about factors of a number is to list *all* factors of the number. A "buddy" system for finding all factors of a number was described, but a better way to ensure you have listed them all is to systematically list all factors using the prime factorization, and taking every possible selection from the prime factors.

Learning Exercises for Section 11.2

1. State the Unique Factorization Theorem. What does it assert about 239,417?

2. Find the prime factorization of each of the following, using a factor tree for each.
   a. 102   **b.** 1827   **c.** 1584   **d.** 1540   **e.** 121   **f.** 1485

3. Find the prime factorization of each of these numbers, using a factor tree for at least two of them.
   a. 5850   **b.** 256   **c.** 2835   **d.** 10^4 **e.** 17,280
   f. Does a complete factor tree for a number show all the factors of the number? All the prime factors of the number?

4. Name three prime factors of each of the following products:
   a. $3 \times 7^3 \times 22$ **b.** 27×22 **c.** $29^4 \times 11^6 \times 2^5$

5. What is the difference between *prime factor* and *prime factorization*?

6. Is it possible to find nonzero whole numbers m and n such that $11^m = 13^n$? Explain.

7. Which cannot be true, for whole numbers m and n? Explain why not. For the ones that can be true, give values for m and n that make the equation true.
   a. $2^9 \cdot 17^3 \cdot 67^2 = 2^7 \cdot 17^2 \cdot 34 \cdot 67 \cdot m$ **b.** $2^9 \cdot 17^3 \cdot 67^2 = 2^9 \cdot 17^4 \cdot m$
   c. $2^9 \cdot 17^3 \cdot 67^2 = 2^8 \cdot 17^2 \cdot n$ **d.** $2^9 \cdot 17^3 \cdot 67^2 = 2^9 \cdot 17^3 \cdot 134 \cdot m$
   e. $4^m = 8^n$ **f.** $6^m = 18^n$

8. Consider $m = 2^9 \cdot 17^3 \cdot 67^2$. Without elaborate calculation, tell which of the following could *NOT* be factors of m. Explain how you know.
   a. $2^8 \cdot 7$ **b.** $2^{10} \cdot 17^2 \cdot 67$ **c.** $2^8 \cdot 17^2 \cdot 67^2$ **d.** 34^3 **e.** 134^2

9. If 35 is a factor of n, give two other factors of n (besides 1 and n).

10. How many factors does each have?
    a. 2^5 **b.** $2^2 \cdot 3^3 = 108$ **c.** 45,000   **d.** $2^7 \cdot 3^5 \cdot 11 \cdot 13^2$
    e. 10^6 **f.** 11^6 **g.** 12^6
    Explain your reasoning for two of the parts (a–g).

11. Consider $19^4 \times 11^4 \times 2^5$. Which of the following products of given numbers are factors of this number for some whole number n? If so, provide a value of n that makes it true. If not, tell why not.
    a. $19^4 \times 11^3 \times 2^5 \times n$ **b.** $19^4 \times 22 \times 2^5 \times n$
    c. $19^4 \times 11^4 \times 64 \times n$ **d.** $19 \times 11 \times 2 \times n$

12. Consider $q = 19^4 \times 11^4 \times 2^5$. Which of the following are multiples of this number? If so, what would you need to multiply this q by to get the number?
    a. $19^4 \times 11^8 \times 2^5$ **b.** $19^4 \times 22^4 \times 2^5 \times 17$
    c. $19^4 \times 11^4 \times 64$ **d.** $(19 \times 11 \times 2)^5$

13. **a.** How many factors does 64 have? List them.
    b. How many factors does 48 have? List them.
    c. How many factors does $19^4 \times 11^4 \times 2^5$ have?

14. If p, q, and r are different primes, how many factors does each of the following have?
    a. p^{10} **b.** p^m **c.** q^n **d.** $p^m \cdot q^n$ **e.** $p^m \cdot q^n \cdot r^s$

15. Give two numbers that have exactly 60 factors. (The numbers do not have to be in calculated form.)

16. Give one number that has the number 121 as a factor and that also has exactly 24 factors. Is there just one possibility?

Supplementary Learning Exercises for Section 11.2

1. Does the 10 in $1024 = 2^{10}$ keep 2^{10} from being the prime factorization of 1024? Explain.

2. What does the Unique Factorization Theorem say about $1024 = 2^{10}$?

3. Make a factor tree for each of the following, and give the prime factorization for each:

   a. 960   **b.** 9600   **c.** 1125   **d.** 8100   **e.** 29

4. Kim and Lee start their factor trees for 5000 differently. Kim has branches to 5 and 1000, and Lee has branches to 50 and 100. Does this violate the Fundamental Theorem of Arithmetic? Explain.

5. What is contradictory in this statement: If k is an odd number and $k = 2m$, then k is a factor of m?

6. Why aren't the given results a contradiction of the Unique Factorization Theorem?

   Abbie has $32{,}832 = 2^4 \times 2^2 \times 3^2 \times 57$.
   Bonita has $32{,}832 = 2^6 \times 3 \times 171$.

7. Give (a) four prime factors and (b) six composite factors of

   $$3{,}972{,}672 \ (= 2^6 \times 3 \times 19 \times 33^2)$$

8. Is there a whole number m such that $7^{15} = 9^m$? Explain.

9. For each part, find the smallest number with the given factors.

   a. 14, 105, and 63   **b.** 33, 45, and 90   **c.** 98, 147, 35, and 420

10. **a.** Every nonzero multiple of 75 will have 3×5^2 in its prime factorization. True or False?

    b. Every nonzero multiple of 24 will have $2^3 \times 3$ in its prime factorization. True or False?

    c. Use parts (a) and (b) to find the smallest number that is a multiple of 75 and also a multiple of 24 (the least "common multiple").

11. Find, if possible, nonzero values for m and n such that $25^m = 125^n$. If it is not possible, explain why.

12. **a.** How many factors does 94,325 have? (*Hint*: $94{,}325 = 5^2 \times 7^3 \times 11$.)

    b. If p, q, and r are different primes, what expression tells how many factors $p^x \times q^y \times r^z$ has?

13. Give the prime factorization of 1 million, and tell how many factors 1 million has.

14. How many factors does 1 billion have? How many of these are primes? How many are composites? (*Hint*: Don't forget 1.)

15. Give one number that has 21 as a factor and that has exactly 15 factors. Is there just one possibility?

16. Is it possible for some number to have two factor trees that are not exactly alike? If so, give an example. If not, tell why.

11.3 Divisibility Tests to Determine Whether a Number Is Prime

The secret military and diplomatic codes mentioned earlier usually involve knowing whether a large number is a prime, or finding the prime factors of a large number. It is a challenge to tackle a large number like 431,687 to see whether it is prime. (431,687 is not a large number for a computer, however. It was newsworthy in 1995 that it was possible to find out whether a number 129 digits long was a prime, using a network of 600 volunteers with computers. It took them eight months and about 1.5×10^{17} calculations.) Because a large number automatically has two different factors, 1 and itself, we need find only one other factor to settle the question of whether or not the number is prime. If it has a third factor, we know that the number is not a prime. If primeness is the only concern, we do not even have to look for other factors.

ACTIVITY 9   Back to Patterns in the Table

In Section 11.2 of this chapter, you were asked to find patterns in the Activity 4 table. Did you find patterns that would tell you when a number is divisible by 2? By 3? By 4? By 5? By 6? See if you can find any now. ●

Is 495,687,115 a prime? Is 1,298,543,316 a prime? It is likely that you saw immediately that 5 is a factor of the first number, and that 2 is a factor of the second number, so you quickly knew that each number had at least 3 factors and hence was not a prime. There are other divisibility tests beyond those for 2, 5, and 10. Perhaps you found some in the activity. Divisibility tests are useful in investigating whether a given number is a prime (and for a teacher who wishes to make up division problems that *come out even*—i.e., give a remainder of 0). They can be helpful when considering whether pairs of numbers have factors in common when writing equivalent fractions and, later, in factoring algebraic expressions.

A **divisibility test** tells whether a number is a factor or divisor of a given number, but without having to divide the given number by the possible factor. So they could be called *factor tests* but usually are not. Nor are they called *multiples tests,* even though, e.g., 112 is a multiple of 2 (but not of 5).

Here are possible statements for the divisibility tests for 2 and for 5.

> **Divisibility test for 2:**
>
> A number is divisible by 2 if, and only if, 2 is a factor of the ones digit (i.e., the final digit is 0, 2, 4, 6, or 8).
>
> **Divisibility test for 5:**
>
> A number is divisible by 5 if, and only if, 5 is a factor of the ones digit (i.e., the final digit is 0 or 5).

Think About …

The divisibility tests for 2, 5, and 10 ignore most of the digits in a large number! Why do the tests work?

There are some easy-to-use but difficult-to-explain divisibility tests. They can be demonstrated using properties of operations and using the two conjectures that you examined in Learning Exercise 9 of Section 11.1: (1) If k is a factor of both m and n, then k is also a factor of $m + n$, and (2) if k is a factor of m but k is not a factor of n, then k is not a factor of $m + n$.

Think About …

Test these conjectures by putting in different numbers for each of the conjectures to be sure you understand them. (For example, suppose k is 3, m is 12, and n is 15. Does the first conjecture hold true? Does k also divide $m + n$?)

The test for divisibility by 3 is quite different from the test for divisibility by 2. The last digit of a number does not reveal whether a number is divisible by 3. For example, 26 and 36 both end in a digit that represents a number divisible by 3, but 26 is *not* divisible by 3, whereas 36 is. Furthermore, we cannot simply say that every third number in the table in Section 11.2 is divisible by 3 and have an efficient test for a large number.

Notes

Of course, if we know the prime factorization, we can tell immediately whether a number is divisible by 3, but one reason we need divisibility tests is to *find* the prime factorization of a number. Thus, we know that $2 \cdot 2 \cdot 2 \cdot 3 \cdot 7 = 168$, so 168 is divisible by 3 because it has 3 as a factor. How could you tell that 3 is a factor of 168, except by calculating $168 \div 3$ or knowing its prime factorization? That is, what is a divisibility test for 3?

Looking at the expanded place-value expression for 168 tells the secret. (Note which properties are used in the equations below.)

$$168 = 1 \cdot 100 + 6 \cdot 10 + 8$$
$$= 1 \cdot (99 + 1) + 6 \cdot (9 + 1) + 8$$
$$= 1 \cdot 99 + 1 \cdot 1 + 6 \cdot 9 + 6 \cdot 1 + 8 \quad \text{(using the distributive property of}$$
$$\text{multiplication over addition)}$$
$$= 1 \cdot 99 + 6 \cdot 9 + 1 \cdot 1 + 6 \cdot 1 + 8 \quad \text{(using the commutative property}$$
$$\text{of addition)}$$
$$= (1 \cdot 99 + 6 \cdot 9) + (1 \cdot 1 + 6 \cdot 1 + 8) \quad \text{(using the associative property}$$
$$\text{of addition)}$$

The numbers 99 and 9 are always divisible by 3, so $(1 \cdot 99 + 6 \cdot 9)$ is divisible by 3 by the first conjecture that you tested in the Think About. And by the second conjecture, 3 would need to divide the rest of the number $(1 \cdot 1 + 6 \cdot 1 + 8)$ if 3 indeed divides 168.

But $1 \cdot 1 + 6 \cdot 1 + 8$ is just $1 + 6 + 8$, which is the sum of the digits of 168. So if 3 divides $1 + 6 + 8$, then 3 divides 168. Three does divide $1 + 6 + 8 = 15$, so 168 must be divisible by 3. (If we actually did the $168 \div 3$ calculation to check whether 3 is a factor, we would find that $168 = 3 \cdot 56$, so indeed 3 is a factor of 168.)

Think About ...

Does the following approach work for 3528?

$$3528 = 3 \cdot 1000 + 5 \cdot 100 + 2 \cdot 10 + 8$$
$$= 3 \cdot (999 + 1) + 5 \cdot (99 + 1) + 2 \cdot (9 + 1) + 8$$
$$= (3 \cdot 999 + 5 \cdot 99 + 2 \cdot 9) + (3 + 5 + 2 + 8)$$

You finish checking for divisibility by 3 using the process used for 168. Can you test for divisibility for 3 for any number, using this process?

However, if the number above had been 3527, all would be the same except the $3 + 5 + 2 + 8$ would now be $3 + 5 + 2 + 7 = 17$. The number 17 is not divisible by 3, and so 3 is not a factor of 3527. (Check it out using a calculator or long division.)

Divisibility test for 3:

A whole number is divisible by 3 if, and only if, the sum of the digits of the whole number is divisible by 3.

An interesting and useful fact about dividing a number by 9 is that the remainder for the division is always the sum of the digits of the number, if the digits continue to be added until the number is less than 9. For example, $215 \div 9 = 23$ remainder 8 and $2 + 1 + 5 = 8$, the remainder. Another way of writing this is $215 = 9 \times 23 + 8$.

Why does this work?

$$215 = 200 + 10 + 5$$
$$= 2(99 + 1) + 1(9 + 1) + 5$$
$$= (2 \times 99) + (2 \times 1) + (1 \times 9) + (1 \times 1) + 5$$
$$= [(2 \times 9 \times 11) + (1 \times 9)] + (2 + 1 + 5)$$
$$= 9[2 \times 11 + 1] + (2 + 1 + 5)$$
$$= 9 \times 23 + 8 \quad \text{(Note that the } 8 = 2 + 1 + 5.)$$

 **ACTIVITY 10**   What About 9?

Use the reasoning for the divisibility test for 3 and the reasoning about the remainder when dividing by 9 to devise a divisibility test for 9. ●

Sometimes children are taught a method for checking arithmetic calculations called *casting out nines*. Perhaps you know it. This method is easy enough for an upper elementary student to use, but understanding why it works involves some of the notions we've been discussing. Here's how it works if you wish to check for errors in addition:

$$
\begin{array}{r}
326 \\
479 \\
+ \ \ 84 \\
\hline
889
\end{array}
$$

Step 1: For each number, cross out ("cast out") digits that are 9 or whose sum is 9.
Step 2: Add the remaining digits until you have a number 0–8. This will be your "reduced number."
Step 3: Do the operation indicated on the reduced numbers.
Step 4: Check to see if the sum, difference, or product, as appropriate, of the reduced numbers matches the reduced number of the sum, difference, or product. If it doesn't, check further for an error.

For the above addition problem, here is an illustration of the method:

$$
\begin{array}{l}
/ 2 / \ \leftarrow 2 \quad \text{(Also, when 326 is divided by 9, the remainder is 2.)} \\
47 / \ \leftarrow 4 + 7 = 11; 1 + 1 = 2 \quad \text{(Also, when 479 is divided by 9, the remainder is 2.)} \\
+ \ \ \ \ 84 \ \leftarrow 8 + 4 = 12; 1 + 2 = 3 \quad \text{(Also, when 84 is divided by 9, the remainder is 3.)} \\
88 / \ \leftarrow 8 + 8 = 16; 1 + 6 = 7 \quad \text{(Also when 889 is divided by 9, the remainder is 7.)}
\end{array}
$$

If the sum is correct, then the sum of 2, 2, and 3 should equal the reduced number of the original sum. It does! $2 + 2 + 3 = 7$.

Why does this work? It is based on the fact that the sum of the digits of a number is the remainder when dividing by 9. Consider:

$$
\begin{array}{ll}
326 = 9 \times 36 + 2 & \text{(Note that } 3 + 2 + 6 = 11 \text{ and } 1 + 1 = 2.) \\
479 = 9 \times 53 + 2 & \text{(Note that } 4 + 7 + 9 = 20 \text{ and } 2 + 0 = 2.) \\
+ \ \ 84 = 9 \times \ \ 9 + 3 & \text{(Note that } 8 + 4 = 12 \text{ and } 1 + 2 = 3.) \\
\hline
889
\end{array}
$$

The number 889 should equal $9(36 + 53 + 9) + (2 + 2 + 3) = 9(98) + 7$, and it does. A similar technique works for checking products.

 **ACTIVITY 11**   Casting Out Nines

Try this technique with another sum of three large numbers using a new set of numbers. Include a number whose digits sum to 9 and a number with 9 as a digit or two digits that sum to 9. Did the sum of the reduced numbers equal the reduced number of the sum? Note that if one digit in the sum is changed (as when an error is made), the reduced number of the sum is different.

Generate a multiplication problem and use the analogous set of steps for multiplication. Now, the most important question that begs to be asked is, "Why does this work?" *Hint*: What are the reduced numbers? Another important mathematical question needs to be answered: Are there circumstances when using the method would fail to catch an error? ●

As we have seen thus far, divisibility tests can involve looking at the last digit or summing all the digits. What about a divisibility test for 4? In the Activity 4 table in

Section 11.2, numbers divisible by 4 had 2×2 as a factor. Did you notice anything about the last two digits of each of these numbers? Consider again the number 3528:

$$3528 = 3 \cdot 1000 + 5 \cdot 100 + 2 \cdot 10 + 8 = (3 \cdot 1000 + 5 \cdot 100) + (2 \cdot 10 + 8)$$

Notice that 1000 is divisible by 4 because $1000 = 4 \cdot 250$, so $3 \cdot 1000$ must be divisible by 4. Similarly, $5 \cdot 100$ is divisible by 4 because $100 = 4 \cdot 25$. We are left with $2 \cdot 10 + 8$ or 28. If this is divisible by 4, then the entire number 2528 must be divisible by 4. With similar reasoning, 2527 is not divisible by 4.

> **Divisibility test for 4:**
>
> A number n is divisible by 4 if, and only if, 4 is a factor of the number formed by the final two digits of n.

 ACTIVITY 12   Divisible by 8?

Use the reasoning from the divisibility test for 4 to construct a divisibility test for 8. (Consider the last three digits of the number.) ●

Return again to the Activity 4 table in Section 11.2 and note which numbers are divisible by 6. Note that they all have 2 and 3 as factors. Thus, applying the divisibility rules for 2 and for 3 will show whether a number is divisible by 6.

What about a divisibility test for 12? Again from the table, all of 12, 24, 36, 48, and 60 have $2 \cdot 3$ as a factor, but so do 18 and 30, which are not divisible by 12. So just testing for 2 and 3 will *not* suffice as a divisibility test for 12. All numbers divisible by 12, however, must have 4 (*two* factors that are 2s) and 3 as factors. Applying these two tests will work as a divisibility test for 12 because 4 and 3 have no prime factor in common. Thus, divisibility tests that you know can help you develop new tests.

> We say two numbers, n and m, are **relatively prime** if they have no prime factor in common or, equivalently, their only common factor is 1.

 DISCUSSION 3   Are They Relatively Prime?

Are 12 and 6 relatively prime? Are 15 and 6 relatively prime? Are 25 and 6 relatively prime? Are 7 and 11 relatively prime? Are *any* two prime numbers relatively prime? ●

> **General divisibility tests to test for composite factors $m \cdot n$:**
>
> If a number p is divisible by m and also by n, and if m and n are relatively prime, then the number p is divisible by the number $m \cdot n$. This general rule can be used to construct new divisibility tests similar to what we discussed for 6 and for 12.

Divisibility tests can help us find the prime factorizations of larger numbers. Consider $n = 12,320$. What is its (unique) prime factorization?

We know that 2 divides the number because it ends in 0. We know 5 divides the number because it ends in 0. So we know that $2 \cdot 5 = 10$ is a factor of 12,320 (and you may have noticed that immediately). Thus, we know that $12,320 = 2 \cdot 5 \cdot 1232$.

We also know that 4 divides the number 1232 because the final two digits form the number 32, which is divisible by 4. A little division then shows $1232 = 4 \cdot 308$, so we now have $12,320 = 2 \cdot 5 \cdot 4 \cdot 308$, or $2 \cdot 5 \cdot 2 \cdot 2 \cdot 308$.

But 4 is also a factor of 308 (since 8, from 08, is divisible by 4). So, $12,320 = 2 \cdot 5 \cdot 2 \cdot 2 \cdot 4 \cdot 77$. Because the sum of the digits of 77 is 14, which is not divisible by 3, we know

77 and n are not divisible by 3. Nor are 2, 4, 5, or 10 factors of 77. But 77 is $7 \cdot 11$, so we finally have

$$12{,}320 = 2 \cdot 5 \cdot 2 \cdot 2 \cdot 4 \cdot 7 \cdot 11 = 2 \cdot 2 \cdot 2 \cdot 2 \cdot 2 \cdot 5 \cdot 7 \cdot 11$$

which could be written more compactly using exponents as $12{,}320 = 2^5 \cdot 5 \cdot 7 \cdot 11$.

ACTIVITY 13   A New Way of Finding Prime Factorizations

Use the method explained here to find the prime factorizations of 1224. Of 4620. ●

The divisibility tests for the prime numbers 2, 3, 5 are not difficult, and less simple ones for 7 and 11 exist. However, for some primes it is easier to simply divide by the prime and notice whether the quotient is a whole number than to use complicated and hard-to-remember tests.

Suppose the divisibility test for 2 tells you that 2 is not a factor of some number n. Could 4 nonetheless be a factor of n? One way to consider this is as follows: If 2 is not a factor of n, 2 cannot appear in the prime factorization of n. But if 2 cannot appear in the prime factorization, 4 could not be a factor of n, because then that 4 could give 2 as a factor (twice) in the prime factorization resulting from having 4 as a factor. (Recall that there can be only one prime factorization of n.) So if 2 is not a factor, then 4 cannot be a factor either.

> ### *Think About …*
> Give a similar argument to convince yourself that if 3 is not a factor of n, then 6 cannot be a factor of n. Give an argument that if p is not a factor of n, then $k \cdot p$ is not a factor of n.

DISCUSSION 4   True or False?

Discuss whether each of the following is true. Explain your answers, giving counterexamples for false statements.

a. If 7 is not a factor of n, then 14 is not a factor of n.

b. If 7 is a factor of n, then 14 is a factor of n.

c. If 14 is not a factor of n, then 7 is not a factor of n.

d. If 20 is not a factor of m, then 60 is not a factor of m.

e. If 20 is a factor of m, then 60 is a factor of m.

f. If 60 is not a factor of m, then 20 is not a factor of m.

g. If a number is a factor of n, then the number is a factor of any multiple of n.

h. If a number is a factor of a multiple of n, then the number is a factor of n.

i. If a number is not a factor of n, then the number is not a factor of any (nonzero) multiple of n.

j. If a number is not a factor of a multiple of n, then the number is not a factor of n. ●

What the above means for testing for primeness is this important fact: ***You need test only for divisibility by primes when deciding whether or not a number is prime***. If 7, say, is not a factor, then 14, 21, or 28, etc., will not be factors either. You would be wasting your time in testing whether 14, 21, or 28 were factors, once you found out that 7 was not a factor. In trying to determine whether 187 is a prime then, you would need to find out whether any one of 2, 3, 5, 7, etc., is a factor—if one is, you have found a third factor (besides 1 and 187), and so 187 would be composite. But if a prime such as 2, 3, 5, or 7 is not a factor, then you do not need to think about their multiples being factors.

The following discussion allows us to refine a rule in testing a number for primeness. The issue is, How many primes do you have to test in deciding whether a number like 661 is a prime? If you find that a prime like 2, 3, 5, or 7 is a factor of 661, you are done, of course. The number 661 would not be a prime.

🔵 DISCUSSION 5   Testing for Primes

Check to see whether 661 is divisible by 2, by 3, by 5, by 7, by 11. To test whether or not 661 is prime, how many more primes do you think you need to test? Do you think you would need to test for divisibility by 91? Why or why not? ●

You may have found that you needed to test for divisibility by more primes than 2, 3, 5, 7, or 11 to find the prime factorization of 661. But as a matter of fact, you did not have to test for any prime factors greater than 23. This fact derives from the following: ***When testing whether or not n is prime, you need test only the primes $\leq \sqrt{n}$***. This is because if both $p > \sqrt{n}$ and $q > \sqrt{n}$, then $pq > n$. Thus, either p or q must be $\leq \sqrt{n}$ if $pq = n$.

Hence, in trying to find out whether 661 is a prime, one would at worst have to try only the primes less than or equal to $\sqrt{661}$, which is about 26. Primes less than 26 are 2, 3, 5, 7, 11, 13, 17, 19, and 23. If none of these is a factor, then 661 is a prime. Even if you have a calculator handy, it is good practice to zero in on the square root of a number with educated efforts at trial and error. For example, for $\sqrt{661}$ think 20^2 is 400 so 20 is too small, and 30^2 is 900 so 30 is too large. Check 25^2; 625 is slightly less than 661, and 29^2 will be much larger than 661, so checking for primes less than 25 is sufficient.

> ### *Think About ...*
> What is the largest prime you need to worry about to find out whether 119 is prime? What about 247?

TAKE-AWAY MESSAGE . . .   You now have divisibility tests for 2, 3, 4, 5, 6, 8, 9, 10, 12, and others that have two relatively prime factors such as 15 and 18. For primes other than those listed, dividing by the prime can be used to see if the prime is a factor. Using these tests can simplify the work of finding out whether or not a number, *n,* is prime. You need find only one divisor for *n* (other than 1 and *n*) to show that a number is not prime. Moreover, you need not check for any primes larger than the square root of *n* to determine whether or not a number is prime.

Learning Exercises for Section 11.3

1. Practice the divisibility tests for 2, 3, 4, 5, 6, 8, 9, and 10 on these numbers:
   a. 43056   **b.** 700010154   **c.** 9460000000023   **d.** 71005165

2. Give a six-digit number such that
   a. 2 and 3 are factors of the number, but 4 and 9 are not.
   b. 3 and 5 are factors of the number, but 10 is not.
   c. 8 and 9 are factors of the number.

3. What could ■ be in $n = 4187$ ■ 432, if 3 is a factor of n? If 9 is a factor of n? Give all the single-digit possibilities for ■.

4. Notice that 2 and 4 are factors of 12, but $2 \cdot 4$, or 8, is not. So a divisibility test for 8 that is *NOT* safe is to use the 2 test and the 4 test. Find counterexamples for these plausible-looking, but unreliable, divisibility tests.
   a. 12 is a factor of n if and only if 2 is a factor of n and 6 is a factor of n.
   b. 18 is a factor of n if and only if 3 is a factor of n and 6 is a factor of n.
   c. 24 is a factor of n if and only if 4 is a factor of n and 6 is a factor of n.

Notes

5. Try these conjectured divisibility tests with 3 or 4 examples each.

   a. 10 is a factor of n if and only if 2 is a factor of n and 5 is a factor of n.
   b. 12 is a factor of n if and only if 3 is a factor of n and 4 is a factor of n.
   c. 18 is a factor of n if and only if 2 is a factor of n and 9 is a factor of n.
   d. 24 is a factor of n if and only if 3 is a factor of n and 8 is a factor of n.
   e. Examine Exercises 4 and 5(a–d) to see whether you can predict when such test-two-factors approaches will work, and when they will not.

6. Using Exercise 5, determine whether:

   a. 24 is a factor of 200000000000000000000000000112.
   b. 24 is a factor of 200000000000000000000000001012.
   c. 18 is a factor of 400000000000000000000000000221.
   d. 18 is a factor of 400000000000000000000000000212.
   e. 45 is a factor of 111000000000000000000000022200.

7. Using Exercise 5, find a 15-digit number that is a multiple of 36. A 15-digit number that is not a multiple of 36.

8. The divisibility tests given here depend on the number being expressed in base ten. The tests are properties of the numeration system rather than of the numbers. Find examples with numbers written in base five to show that, say, the (base-ten) divisibility test for 2 does not work in base five. You will want to find a number for which 2 is (or is not) a factor but whose base-five representation does (or does not) satisfy the divisibility test for 2 that you know for base ten.

9. Explain why finding only one factor of n besides 1 and n is enough to show that n is composite.

10. Suppose that $n = 2^3 \times 5^2 \times 7 \times 17^3$. Give the prime factorization of n^2 and n^3. (*Hint:* Do not work too hard.)

11. Determine whether each of these is a prime.

   a. 667   **b.** 289   **c.** 3501   **d.** 47×61 **e.** 4319   **f.** 29^3

12. The numbers 2 and 3 are consecutive whole numbers, each of which is a prime. Is there another pair of consecutive whole numbers, each of which is a prime? If there is, find such a pair; if not, explain why.

13. Test each of these numbers for divisibility by 2, 3, 4, 5, 6, 8, 9, 10, 12, 15, and 18.

   a. 540   **b.** 150   **c.** 145   **d.** 369   **e.** 840

14. Which of these numbers are prime? For those not prime, give the prime factorization.

   a. 29×23 **b.** 5992   **c.** 127   **d.** 121   **e.** 31^2 **f.** 1247   **g.** 3816

15. Here is an interesting conjecture that mathematicians are uncertain about even though it has been studied for more than 100 years: Every even number greater than 2 can be written as the sum of two primes (Goldbach's conjecture). Test the conjecture for the even numbers through 36.

16. Devise a way of checking to see whether or not a number is divisible by 24. Test your method on 36. On 120.

17. Are every two different primes relatively prime? Explain.

18. Which pairs of numbers are relatively prime?

   a. 2, 5   **b.** 2, 4   **c.** 2, 6   **d.** 2, 7   **e.** 2, 8   **f.** 2, 9
   g. 8, 9   **h.** 8, 12   **i.** 3, 8   **j.** 40, 42   **k.** 121, 22   **l.** 39, 169

19. Give an example of two 3-digit nonprime numbers that are relatively prime.

20. If possible, give a composite number that is relatively prime to 22.

21. a. Find $128 + 494 + 381$ and check your answer by casting out nines.
   b. Compute 23×45 and check your answer by casting out nines.

22. a. A result in number theory states that the product of any n consecutive positive integers is divisible by the product of the first n positive integers. For example, for the product $5 \times 6 \times 7 \times 8 = 1680$, this theorem asserts that 1680 is divisible by $1 \times 2 \times 3 \times 4$, or 24. Verify that 1680 is divisible by 24, using divisibility rules.

b. Compute the product of another 4 consecutive positive integers and check to see if the product is divisible by 24.

c. Demonstrate this result for some example you choose for $n = 5$ (any five consecutive positive integers).

23. A result in number theory states that if p is a prime number and n is a positive integer, then $n^p - n$ is divisible by p. Demonstrate this for 3 cases where you choose n and p.

24. a. "I am a 3-digit number.
I am not a multiple of 2.
7 is not one of my factors, but 5 is.
I am less than 125.
Who am I?"

b. Make up a "Who am I?" involving number theory vocabulary and ideas.

Supplementary Learning Exercises for Section **11.3**

1. Which pair(s) of numbers below are relatively prime? Explain.

    a. 27, 29   **b.** 25, 35   **c.** 25, 32   **d.** 100, 81   **e.** 125, 36

2. A divisibility test could be called a factor test instead. True or False?

3. Practice the divisibility tests for 2, 3, 4, 5, 6, 8, 9, and 10 on these numbers:

    a. 97,236   **b.** 1,000,000   **c.** 714,612   **d.** 7,000,005
    e. 2299   **f.** 494   **g.** 80×54 **h.** 400

4. In verifying that each of the following is a prime, what is the largest prime that must be checked as a possible factor?

    a. 401   **b.** 509   **c.** 887   **d.** 1607

5. Which of the following are primes? For those not prime, give the prime factorization.

    a. 207   **b.** 121   **c.** 5893 ($= 83 \times 71$)   **d.** 6859 ($= 19^3$)   **e.** 247
    f. 119   **g.** 97   **h.** 197   **i.** 297

6. Give one 12-digit number that has 3 as a factor, but not 9, and also 4 as a factor, but not 8.

7. Why is it not possible to give a 10-digit number that has 9 as a factor but does not have 3 as a factor?

8. Why does writing 1527 as

$$(999 + 1) + 5 \cdot (99 + 1) + 2 \cdot (9 + 1) + 7 = (999 + 5 \cdot 99 + 2 \cdot 9) + 1 + 5 + 2 + 7$$

help to make clear that the divisibility tests for 3 and 9 depend on the sum of the digits $1 + 5 + 2 + 7$?

11.4 Greatest Common Factor, Least Common Multiple

As you would suspect about an old area of mathematics like number theory, there are entire books on the subject, exploring many different and advanced areas. So the work with number theory in the elementary school curriculum touches on only a small part of number theory. Elementary school number theory usually comes up right before work with fractions. Simplifying fractions and finding common denominators for adding and subtracting fractions use number theory ideas. The same ideas carry over to algebraic fractions, so even though fraction calculators might be available, the reasoning behind the work with regular fractions will continue to be important.

Recall from Chapter 6 that a basic result about fractions is that $1 \times \frac{a}{b} = \frac{a}{b}$. The number 1 can be written as $\frac{n}{n}$, where n is any nonzero number. So, $\frac{n}{n} \cdot \frac{a}{b} = \frac{na}{nb}$. Starting with $\frac{a}{b}$, one can generate any number of fractions equal to $\frac{a}{b}$ simply by making different choices for n.

Read "backwards," however, $\frac{na}{nb} = \frac{a}{b}$, and the result shows that one can write a simpler fraction for a given fraction by finding a common factor of the numerator and denominator, and "canceling it out," as you might have said in earlier work.

For example, because 3 is a common factor of 84 and 162, $\frac{84}{162} = \frac{3 \cdot 28}{3 \cdot 54}$, so $\frac{84}{162} = \frac{28}{54}$, an equivalent but simpler fraction. But you should notice that $\frac{28}{54}$ can be simplified further, because 2 is a common factor of 28 and 54: $\frac{28}{54} = \frac{14}{27}$. Because 14 and 27 have only 1 as a common factor (they are "relatively prime"), $\frac{14}{27}$ is the simplest form for $\frac{84}{162}$. If you can find the **greatest common factor** (GCF, also called the greatest common divisor) of the numerator and denominator (which is 6), then you can get the simplest fraction in one step: $\frac{84}{162} = \frac{6 \cdot 14}{6 \cdot 27} = \frac{14}{27}$.

> The **greatest common factor** (**GCF** or **gcf**) of two (or more) whole numbers is the largest number that is a factor of the two (or more) numbers. The GCF is sometimes also called the **greatest common divisor** (**GCD** or **gcd**).

For reasonably small pairs of numbers, you can often "see" the greatest common factor by inspection. A systematic way would involve listing all the factors of each number and then picking out the greatest common factor.

Set of factors of 84 = {1, 2, 3, 4, 6, 7, 12, 14, 21, 28, 42, 84}

Set of factors of 162 = {1, 2, 3, 6, 9, 18, 27, 54, 81, 162}

A Venn diagram is one way of illustrating the *common* factors, 1, 2, 3, and 6.

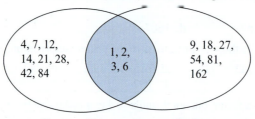

The region they share provides common factors: 1, 2, 3, and 6, and so 6 is the GCF of 84 and 162.

Factors of 84   Factors of 162

Sometimes a more efficient method would be to examine the factors of the smaller number to see which are also factors of the larger number. Usually, you start with the larger factors because you are looking for the greatest common factor. For 84, you would try 84, 42, etc., working either from the complete list or hoping not to overlook a larger common factor than any you find.

📋 ACTIVITY 14   GCFs

Find the prime factorizations of 84 and 162. What do they share in common? ●

Knowing how to find the least common multiple of fraction denominators can make the adding or subtracting of fractions easier. As you may recall, to find the sum of fractions like $\frac{9}{16}$ and $\frac{7}{12}$, the usual algorithm or method calls for replacing the given fractions with fractions that have the same denominators but are equal to the original fractions ("equivalent fractions with a common denominator" might be the language used). One can always multiply the numerator and denominator of each fraction by the other denominator $\frac{9 \cdot 12}{16 \cdot 12}$ and $\frac{16 \cdot 7}{16 \cdot 12}$, or $\frac{108}{192}$ and $\frac{112}{192}$ but usually those new fractions do not lead to the simplest arithmetic. Finding the least common denominator is the usual approach, just to keep the numbers smaller. This least common denominator is just the least common positive multiple of the denominators of the fractions involved.

As with the greatest common factor, finding the least common multiple can be approached in several ways. One way, perhaps best when the idea is new, is to list all the common multiples of each number until a common one (not 0) is found. (0 is literally the least common multiple of every two whole numbers, but it is not useful in situations where least common multiples arise, as with fractions.)

Set of multiples of 16 = {0, 16, 32, 48, 64, 80, 96, 112, 128, . . .}
Set of multiples of 12 = {0, 12, 24, 36, 48, 60, 72, 84, 96, 108, . . .}

A Venn diagram can also be used to illustrate the multiples of 16 and 12:

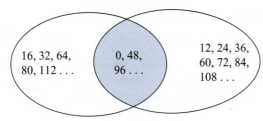

Multiples of 16   Multiples of 12

Because 48 is the least common nonzero multiple of 16 and 12, 48 would be the smallest number that could serve as a common denominator in adding, e.g.,

$$\frac{9}{16} + \frac{7}{12} = \frac{9 \cdot 3}{16 \cdot 3} + \frac{7 \cdot 4}{12 \cdot 4} \quad \text{or} \quad \frac{27}{48} + \frac{28}{48} = \frac{55}{48}$$

The sum is not always in simplest form, but if you are calculating by hand, the fractions with the **least common multiple (LCM)** as denominator offer the simpler arithmetic with the smaller numbers.

ACTIVITY 15 Using LCMs and GCFs

1. What is the GCF of 68 and 102?

2. Simplify $\frac{68}{102}$.

3. What is the LCM of 68 and 102?

4. What is $\frac{5}{68} + \frac{13}{102}$? ●

There are other methods for finding least common multiples and greatest common factors. A useful one to use with larger numbers involves their prime factorizations. Suppose that $m = 2^3 \cdot 5^2 \cdot 29 \cdot 31^2$ and $n = 2 \cdot 5^2 \cdot 31 \cdot 83$, and we need the least common multiple of m and n. First, we know this new number must be a multiple of m and of n. Any (nonzero) multiple of m will involve all of $2^3 \cdot 5^2 \cdot 29 \cdot 31^2$; similarly, any (nonzero) multiple of n will involve all of $2 \cdot 5^2 \cdot 31 \cdot 83$. Any **common multiple** of m and n, then, must involve enough 2s, 5s, 29s, 31s, and 83s to be a multiple of each of m and n. For the least common multiple, what is the fewest 2-factors, 5-factors, etc., so that the new number takes into account these dual needs? With $m = 2^3 \cdot 5^2 \cdot 29 \cdot 31^2$ and $n = 2 \cdot 5^2 \cdot 31 \cdot 83$, for starters the fewest possible 2-factors is 3, giving 2^3; the fewest 5-factors is 2, giving 5^2. Continuing, we get the least common multiple of m and n to be $2^3 \cdot 5^2 \cdot 29 \cdot 31^2 \cdot 83$. In finding the least common multiple this way, think "multiple" first, then "common multiple," and finally "least common multiple."

With this reasoning, the LCM of 72 and 108 is the LCM of $2^3 \times 3^2$ and $2^2 \times 3^3$. All common multiples of $2^3 \times 3^2$ and $2^2 \times 3^3$ must have *at least* three 2s and three 3s. Thus, the least common multiple (LCM) is $2^3 \times 3^3 = 8 \times 27 = 216$.

> ### Think About ...
> Why must all common multiples of $2^3 \times 3^2$ and $2^2 \times 3^3$ have *at least* three 2s and three 3s? Name some common multiples that satisfy this condition. Why must all common factors of $2^3 \times 3^2$ and $2^2 \times 3^3$ have *at most* two 2s and two 3s? Name some common factors that satisfy this condition.

EXAMPLE 2

Let $n = 72$ and $m = 63$. Now, $n = 2^3 \times 3^2$ and $m = 3^2 \times 7$. We could also say $n = 2^3 \times 3^2 \times 7^0$ and $m = 2^0 \times 3^2 \times 7^1$ because $7^0 = 1$ and $7 = 7^1$. Common multiples of n and m must have *at least* three 2s, two 3s, and one 7. One common multiple would be $2^3 \times 3^4 \times 7^1$; another would be $2^5 \times 7^2 \times 3^4 \times 5^2 \times 13^2$. There are an infinite number of common multiples of 72 and 63. The *least* one is the number that has each of three 2s, two 3s, one 7, and no more. That is, $2^3 \times 3^2 \times 7 = 504$. ●

 ACTIVITY 16   GCF Analysis

Do an analysis similar to the one for finding the LCM of

$$m = 2^3 \cdot 5^2 \cdot 29 \cdot 31^2 \qquad \text{and} \qquad n = 2 \cdot 5^2 \cdot 31 \cdot 83$$

to find the greatest common factor of m and n. That is, what numbers are common factors of both m and n? You know that, e.g., 2 is a common factor. So is 5. So is 5^2. Are there other common factors? What is the *greatest common factor*? ●

This analysis shows how prime factorizations can be used to find both LCMs and GCFs of two (or more) whole numbers. Consider another example, again using 72 and 63.

EXAMPLE 3

Use $72 = 2^3 \times 3^2 \times 7^0$ and $63 = 2^0 \times 3^2 \times 7^1$ again. Now some common factors would be 3, or 3^2. There are a finite number of common factors. To find the *greatest* common factor, consider the 2, the 3, and the 7, and the *least* power for which each appears: 2^0, 3^2 and 7^0 to allow the number to be a factor of both 72 and 84. So the GCF is $2^0 \times 3^2 \times 7^0$, but 2^0 and 7^0 are both just 1. Thus, the GCF is $2^0 \times 3^2 \times 7^0$, or just 3^2, so 9 is the GCF. ●

> *Think About …*
>
> What is the GCF of 72 and 108? Of 260 and 650?

Thus, using prime factorizations of numbers leads to easy ways of finding both the LCM and the GCF.

 ACTIVITY 17   Once Again . . . Using LCMs and GCFs

1. What is the GCF of 260 and 650? Of 260 and 186? Of 186 and 650?
2. Simplify $\dfrac{260}{650}$, $\dfrac{260}{186}$, and $\dfrac{186}{650}$.
3. What is the LCM of 260 and 650? Of 260, 650, and 130?
4. What is $\dfrac{5}{650} + \dfrac{13}{260}$? ●

The same method of finding LCM and GCF can be applied when renaming algebraic fractions. Given the factorizations of algebraic expressions, you can calculate the LCM and GCF.

 ACTIVITY 18   GCF and LCM in Algebra

1. Find the GCF and LCM of x^2y and y^2x.
2. Find the GCF and the LCM of $(x - 2)(x + 2)$ and $(x - 2)^2$.
3. Make up an algebraic problem of your own for others to solve. ●

The following problem involves many ideas from all four sections of this chapter. It is a puzzle credited to Brahmagupta (born 598 AD).

✏️ ACTIVITY 19   An Ancient Puzzle

▶ An old woman goes to market and a horse steps on her basket and breaks all her eggs. The rider offered to pay her for the damages and asks her how many eggs she had bought. She does not remember the exact number, but when she had taken them out two at a time, there was one egg left. The same happened when she picked out three, four, five, and six at a time, but when she took seven at a time, they came out even. What is the smallest number of eggs she could have had? ◄

(*Hints:* What are common multiples of 2, 3, 4, 5, and 6? Would any of those numbers plus 1 satisfy all the conditions?) ●

This activity from a Grade 4 textbook* focuses on common multiples. Complete them and share your results with a neighbor. Make up a Venn diagram problem similar to those for these problems and have a neighbor complete it.

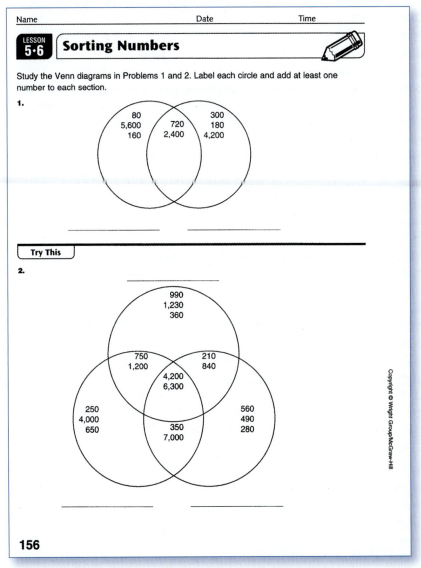

Name _____   Date _____   Time _____

LESSON 5·6 Sorting Numbers

Study the Venn diagrams in Problems 1 and 2. Label each circle and add at least one number to each section.

1.

80
5,600
160
720
2,400
300
180
4,200

_____   _____

Try This

2.

990
1,230
360

750
1,200
210
840

4,200
6,300

250
4,000
650
350
7,000
560
490
280

_____   _____

156

*Wright Group, *Everyday Mathematics*, Math Masters, Grade 4, p. 156. Copyright © 2008, McGraw-Hill. Reproduced with permission of The McGraw-Hill Companies.

TAKE-AWAY MESSAGE . . . In this section we returned to factors and multiples to examine ways of finding the greatest common factor (GCF) and the least common multiple (LCM) of two or more whole numbers or of two or more algebraic expressions. One way of finding the GCF was by listing common factors and taking the greatest. One way to find the LCM was to list common multiples and to take the least. Another was to use the prime factorizations of the numbers. The LCM of two (or more) numbers would be expressed by the product of each of the prime factors to the greatest power in which they exist in the original numbers. The GCF would be expressed by the product of each of the common prime factors to the least power in which they exist in the original numbers.

Learning Exercises for Section 11.4

1. List four factors of each of the following numbers:

   a. 2^3 **b.** 27×49 **c.** 12   **d.** 15
   e. 108   **f.** 125   **g.** 72

2. List four nonzero multiples of each of the following. You may leave them in factored form.

   a. 8   **b.** 27×49 **c.** $2^2 \times 3$ **d.** 15
   e. 108   **f.** 125   **g.** 72

3. As you know, Earth takes 365 "earth" days, i.e., 1 year, to make one revolution around our sun. It takes Saturn about 30 earth years to make one revolution around the sun. Jupiter takes 12 earth years to do so, and Mars about 2 earth years. If Earth, Saturn, Jupiter, and Mars are aligned when they begin, in approximately how many years would all four planets be aligned again?

4. You have probably noticed that when you buy hot dogs and hot dog buns, the packages do not contain the same quantities. Make up a story problem that gives quantities of hot dogs in a package and quantities of hot dog buns in a package and asks, "How many packages of each do you buy so that there are the same number of hot dogs as hot dog buns?" Would the solution involve LCM or GCF? Why?

5. For each of the following groups of numbers, find the least common (nonzero) multiple. You may leave your answers in factored form or write them out.

   a. 72 and 108   **b.** 144 and 150   **c.** 72 and 90
   d. 72 and 144   **e.** 144 and 567   **f.** 90 and 567
   g. 108 and 90   **h.** 150 and 350   **i.** 150, 35, and 270

6. For each of the following groups of numbers, find the greatest common factor. You may leave your answers in factored form or write them out.

   a. 72 and 108   **b.** 144 and 150   **c.** 72 and 90
   d. 72 and 144   **e.** 144 and 567   **f.** 90 and 567
   g. 108 and 90   **h.** 150 and 350   **i.** 150, 350, and 270

7. Give three (nonzero) multiples of each of the following (leave them in factored form):

   a. $2^3 \cdot 3$ **b.** $2^3 \cdot 7^3$ **c.** $3^2 \cdot 7^3 \cdot 11^5 \cdot 19^2$

8. In each part, find the least common (nonzero) multiple of the numbers given. Then find the greatest common factor for each part.

   a. $m = 5^2 \cdot 7^3, n = 5 \cdot 13^2$
   b. $m = 37^4 \cdot 47^5 \cdot 67^6, n = 37^6 \cdot 47^5 \cdot 71$
   c. $m = 7^2 \cdot 26, n = 2^2 \cdot 11 \cdot 17$
   d. $m = 10125, n = 26730$
   e. $m = 2^3 \cdot 7^2, n = 2 \cdot 5^3 \cdot 7, p = 3^2 \cdot 5 \cdot 7$
   f. $m = 6x^2y^5z^{12}, n = 10xy^6z^4$

9. 0 is a common multiple of 16 and 12. Why is the least common multiple defined as the smallest nonzero value?

10. Jogger A can run laps at the rate of 90 seconds per lap. Jogger B can run laps on the same track at the rate of 2 minutes per lap. If they start at the same place and time, and run in the same direction, how long (in time) will it be before they are at the starting place again, at the same time?

11. You cut a yellow cake into 6 pieces, and a chocolate cake of the same size into 8 pieces. Then you find out that you were supposed to cut both cakes into the same (unspecified) number of pieces! As a number theory student, what do you do?

12. A machine has two meshing gears. One gear has 12 teeth and another gear has 30 teeth. After how many rotations do both gears revert back to their original position?

13. Paper plates are sold in packages of 25. Paper bowls are sold in packages of 40. Plastic spoons are sold in packages of 20. How many packages of each do you need to buy to have the same number of plates, bowls, and spoons?

14. A paint manufacturer produces base in 25-gallon drums and color in 10-gallon drums. A company wants to order stock for a dark blue paint mixture for which 1 gallon of color requires 1 gallon of base. How many of each should they order so that the mix comes out without any unused base or color left unmixed?

15. As in Exercise 14, a paint manufacturer produces base in 25-gallon drums and color in 10-gallon drums. A company wants to order stock for a dark blue paint mixture for which 1 gallon of color requires 4 gallons of base. How many of each should they order so that the mix comes out without any unused base or color left unmixed?

16. The principal says that the sixth-graders are raising money for a field trip by selling caps with the school logo. He said that last week they raised $414 and the week before they raised $543. You were going to ask what price they were selling for but realized you could figure out the price from what he had just told you. What was the likely selling price of the caps?

17. The principal says the seventh-graders raised money for a field trip to Washington, D.C., by washing cars at a fixed price the past three weekends. He said that last weekend they raised $198 and two weekends ago they raised $252. Three weekends ago they raised $495. What was the price of the car wash?

18. You want to explore the concept of scale factor with your students in an activity in which they will create a small-scale earth and sun to show their relative sizes. The sun is approximately 1,400,000 km in diameter and the earth is approximately 12,800 km in diameter. Because you want them to initially use only whole numbers, find the GCF to determine what the scale factor should be.

19. You want to explore the concept of scale factor with your students. You will work with your students to draw a scale drawing of your classroom. Your classroom has dimensions of 216 inches by 282 inches. If you want them to use the smallest whole numbers possible, what should be the scale factor?

20. Find several values of m and n such that $27^m = 9^n$.

21. Write the simplest form for each fraction, using the greatest common factor:

   a. $\dfrac{135}{150}$ b. $\dfrac{36}{48}$ c. $\dfrac{84}{100}$ d. $\dfrac{180}{160}$ e. $\dfrac{2^3 \cdot 3^5}{2^4 \cdot 3^3}$ f. $\dfrac{x^2 y^3}{x^3 y}$

22. Use the phrase "relatively prime" to describe your final answers to parts (a–d) in Exercise 21.

23. Use the least common multiple in calculating these sums and differences:

   a. $\dfrac{7}{24} + \dfrac{11}{18}$ b. $\dfrac{13}{8} - \dfrac{5}{4}$ c. $\dfrac{14}{15} + \dfrac{7}{10} - \dfrac{1}{4}$

continue

Notes

d. $\dfrac{x}{y^2} + \dfrac{2x^2}{x^2 y}$ **e.** $\dfrac{6.3 \times 10^4}{2 \times 10^2} + \dfrac{7.2 \times 10^5}{3 \times 10^3}$

24. For 12 and 16, the greatest common factor is 4 and the least common multiple is 48. Compare 12×16 and 4×48. Try other pairs of numbers to see whether their greatest common factor and least common multiple are related in the same way.

25. If your calculator allows you to enter and simplify fractions, use it to find the most reduced form of the fractions for the following. (If you do not have a calculator, try them with paper and pencil.)

   a. $\dfrac{1280}{1440}$ **b.** $\dfrac{8530}{47,250}$ **c.** $\dfrac{2720}{10,000}$

26. Notice that the number 3 is a common factor of 84 and 72 and 3 is also a factor of $84 - 72 = 12$.

   a. Conjecture: If x is a common factor of m and n, then x is also a factor of ___.
   b. Test the conjecture by considering at least four examples of m and n.
   c. Find simpler names for $\frac{147}{150}, \frac{210}{216}$, and $\frac{111}{126}$, with help from part (a).
   d. Determine whether this conjecture is relevant to $\frac{81}{86}$, or $\frac{48}{55}$, or $\frac{112}{127}$. Explain.

27. **Practice** Write the following fractions in the most reduced form:

   a. $\dfrac{15}{35}$ **b.** $\dfrac{28}{54}$ **c.** $\dfrac{150}{350}$ **d.** $\dfrac{12}{144}$ **e.** $\dfrac{150}{567}$

   f. $\dfrac{15}{40}$ **g.** $\dfrac{64}{512}$ **h.** $\dfrac{21}{49}$ **i.** $\dfrac{2223}{4536}$

28. **Practice** Use the least common multiple of the denominators to calculate these sums and differences. For extra practice, also write the answer in simplified form.

   a. $\dfrac{39}{144} + \dfrac{35}{108}$ **b.** $\dfrac{25}{72} + \dfrac{81}{567}$ **c.** $\dfrac{36}{108} + \dfrac{41}{72}$

   d. $\dfrac{15}{39} + \dfrac{110}{169}$ **e.** $\dfrac{169}{500} + \dfrac{169}{650}$ **f.** $\dfrac{126}{504} + \dfrac{98}{770}$

29. As a third-grade teacher, you are designing a measurement lesson. You decide to start with a measuring lesson using nonstandard units. You create a measuring tool (a new "ruler") that will allow your students to measure classroom items and have the measure in whole units. The desks in your classroom are 27 inches wide and 36 inches long. What is the largest unit you can create that will allow this desk to be measured in whole units?

30. Go to the National Library of Virtual Manipulatives on the Internet by accessing http://nlvm.usu.edu/en/nav/index.html, then click on Virtual Library at the top of the page, then click on Numbers and Operations, and scroll down to Factor Tree and open it. Click on One Tree at the bottom of the page. Practice finding factors of numbers (chosen by the program). Then click on Two Trees. At least five times find the factors of two numbers provided by the program, then drag the factors into the appropriate area of the Venn diagram provided. Use this information to find the GCF and the LCM of the two numbers.

Supplementary Learning Exercises for Section 11.4

1. **a.** What is the GCF of 36 and 48?   **b.** What is the LCM of 36 and 48?
   c. Simplify $\dfrac{36}{48}$ in one step.   **d.** Add $\dfrac{17}{36} + \dfrac{5}{48}$.

2. **a.** What is the GCF of 150 and 270?   **b.** What is the LCM of 150 and 270?
   c. Simplify $\dfrac{150}{270}$ in one step.   **d.** Add $\dfrac{91}{150} + \dfrac{7}{270}$.

3. For each of the following pairs of numbers, find the LCM and the GCF. You may leave your answers in factored form or multiply them out.

   a. 96 and 132   **b.** 42 and 126   **c.** 36 and 64   **d.** x^4y^6 and x^2y^9

   e. $2^3 \times 7^2 \times 11$ and $2^2 \times 7^3 \times 13$

4. Here is a method that is taught in another culture (Hong Kong) for finding the least common multiple of three numbers, such as 12, 18, and 60, in the following example:

First	Then	Finally

3)12 18 60
   4   6 20
(3 is a common
factor of all three
numbers.)
	3)12 18 60
	2)4 6 20
	  2 3 10
		3)12 18 60
		2)4 6 20
		2)2 3 10
		  1 3 5
		(Notice the 3 in 2 3 10 is
		just repeated.)

The divisions stop when no pair of numbers have a common factor besides 1. The LCM of 12, 18, and 60 is then obtained by multiplying the numbers on the outside:

$$3 \times 2 \times 2 \times 3 \times 5 = 180 \quad \text{and} \quad \text{LCM (12, 18, 60)} = 180$$

   a. Try the algorithm to find the LCM of 24, 45, and 50.

   b. Why does this algorithm work?

5. Suppose there is a weird basketball game in which one team scores only 2-point field goals and the other team scores only 3-point field goals. (No free-throws were made.)

   a. Can the game ever be tied (after 0 to 0)?

   b. What number theory idea applies here?

11.6 Issues for Learning: Understanding the Unique Factorization Theorem

Whole numbers can be represented in a variety of ways. Different forms of a number can provide different information about the number. For example, suppose one is asked: What are the factors of 1334? Obviously, 2 is a factor, and if 1334 is divided by 2, one finds that $1334 = 2 \times 667$, so both 2 and 667 are factors. Are there more? Suppose one also knows that 23 is a divisor of 667. Can all the factors be listed now?

Suppose, again, that a number is represented as $5^2 \times 31 \times 43$. Now what are the factors? Is 11 a factor? How about 3?

Answers to these questions can tell a great deal about your understanding of prime numbers as building blocks of whole numbers. Once a number is represented as a product of prime numbers, it is quite easy to find factors and multiples of the number. You can, from this information, find all the prime factors, and in fact, all the factors of the number. This method is possible because the prime factorization of any number is unique. As you know, this is sometimes called the Fundamental Theorem of Arithmetic and other times referred to as the Unique Factorization Theorem. The theorem seems simple enough to understand, but research shows that many students (even at the college level) have difficulty applying this theorem. Consider this example[2, 3] from one research study:

Patty was asked whether $3^3 \times 5^2 \times 7$ is divisible by 7 or 5. Patty said yes, because these factors were clearly visible. When asked whether 11 was a factor, she multiplied the given factors and obtained 4725. She said she would divide 4725 by 11 to find out whether or not 11 is a factor. But by the prime factorization theorem, 11 cannot be a factor because it is prime and does not appear in the prime factorization. The researchers concluded that Patty did not really understand or believe that the prime factorization was unique because she thought that there might be another factorization with 11 as a factor.

Why did Patty have this problem? Perhaps she saw different factorizations with composite factors and generalized that the prime factorization is just one possible factorization and in another, 11 might be a factor. For example, $3^3 \times 5^2 \times 7$ could be written as $3 \times 5^2 \times 7 \times 9$ or as $3^2 \times 5 \times 7 \times 15$. But in these cases, the factorizations are not prime factorizations.

TAKE-AWAY MESSAGE . . . Research has shown that understanding the Unique Factorization Theorem is more complex than it might first seem.

Learning Exercises for Section 11.5

Here are some other questions that were asked of students in this study. Each of them deals with prime factorization. Try them yourself, and discuss what kinds of difficulties students might have if they do not understand that a prime factorization of a number is unique.

1. $k = 16{,}199 = 97 \times 167$ where 97 and 167 are both primes. Decide whether k can be divided by 5, by 11, by 17.

2. $a = 153 = 3^2 \times 17$. Is 51 a factor of a? How do you know?

3. $a = 153 = 3^2 \times 17$ and $b = 3^2 \times 19$. Which number do you think has more factors, a or b? Why?

11.6 Check Yourself

Many of the ideas in this chapter center on the important result appropriately called the Fundamental Theorem of Arithmetic—that the prime factorization of a number is unique. Factoring numbers, finding and using greatest common factors and least common multiples, and using divisibility rules, all relate to the Unique Factorization Theorem.

You should be able to work problems like those assigned and to meet the following objectives:

1. Use the terms even, odd, factor, multiple, prime, and composite correctly, and recognize statements using these terms as true or false, with reasons for your decisions.

2. Provide a reason for why the number 1 is considered to be neither prime nor composite.

3. Use a sieve method, such as the sieve of Eratosthenes, to find prime numbers less than a given number n.

4. State the Unique Factorization Theorem, also called the Fundamental Theorem of Arithmetic.

5. Use this theorem to determine whether or not a number is a factor of another number when both numbers are in factored form.

6. State and use divisibility tests for 2, 3, 4, 5, 6, 9, 10, 12, 15, 18.

7. Find the prime factorization of a number, perhaps using a factor tree.

8. Explain why the square root test provides a way to know which prime numbers (at most) need to be tested when determining whether or not a number is prime.

9. Determine whether a number (of a reasonable size) is prime or composite.

10. Determine the number of factors of a number from its prime factorization form.

11. Find common factors of sets of numbers; find common multiples of sets of numbers.

12. Find the greatest common factor and least common multiple of sets of numbers.

13. Discuss common errors made in using the Unique Factorization Theorem.

References for Chapter 11

[1]Clay Mathematics Institute website. http://www.claymath.org/posters/primes/April 12, 2005.

[2]Zazkis, R., & Campbell, S. (1996). Prime decomposition: Understanding uniqueness. *Journal of Mathematical Behavior, 15,* 207–218.

[3]Zazkis, R., & Liljedahl, P. (2004). Understanding primes: The role of representation. *Journal for Research in Mathematics Education, 35,* 164–186.

Part II

Reasoning About Algebra and Change

Your experience with algebra in school likely was in one or two courses devoted to the subject. So you may be surprised to learn that algebra is increasingly recognized as a continuing topic throughout the curriculum, with ideas from algebra occurring as early as the first grade and with an increased use of symbols and equations in the K–6 mathematics curriculum, often under the guise of *algebraic reasoning*.

Algebra has acquired somewhat of a negative image, and many cartoons play on the belief that algebra is difficult to understand and not too useful. Such is far from the truth. When algebra is taught with sense-making always a goal, its power in solving many types of problems becomes apparent. Of course, this is made easier when the teacher has the background to help students make sense of algebra, including algebra topics now found in the elementary grades' curriculum.

The goal of the following chapters is to provide you with the ability and confidence to teach the ideas and skills that underlie algebra. This is not an algebra course devoted to building skills by manipulating numbers and variables. We do in fact assume that you have been exposed to such skills in earlier algebra classes, but, nonetheless, we have built in some review.

Chapter 12 in this part of *Reconceptualizing Mathematics* first centers on exploring different facets of algebra. This chapter provides information about many things you may have forgotten and lays the groundwork for the remaining chapters in Part II. Chapter 13 focuses on using graphs to study quantitative relationships. Chapter 14 naturally follows this work by describing mathematical change and emphasizing what graphs tell us. Chapter 15 uses algebra to approach additional types of problems. Various issues for learning aspects of algebra are treated throughout.

Chapter 12

What Is Algebra?

There are many ways of thinking about algebra—as a symbolic language, as generalized arithmetic, as a study of patterns and functions, as reasoning about quantities, and as a powerful tool for solving problems.[1] This chapter introduces you to these different ways of thinking about algebra and provides a foundation for the next three chapters.

12.1 Algebra as a Symbolic Language

Algebra provides a symbolic language that can be used to represent quantitative relationships, such as those in story problems. We use symbols in algebra in three primary ways—to stand for a particular value in an expression or sentence, to state formulas, and to express arithmetic and algebraic properties. The next three examples show how symbols can be used to stand for particular values. The symbols most often used are letters of the alphabet, but in the primary grades other symbols, such as boxes, are sometimes used.

EXAMPLE 1

Jaime has 16 baseball cards but lost 4 of them. How many does he have left? We can represent this problem as $16 - 4 = \square$. Although you can substitute different numbers for the box, only one number will make this a true statement. ●

EXAMPLE 2

Karen is running a 5-mile race. If we let x represent the number of miles she has run at any time during the race, we can represent the number of miles remaining as $5 - x$. In this case, x can take on a variety of values between and including 0 and 5, depending on the distance Karen has run at any point in time. ●

EXAMPLE 3

If a bottle recycling center pays 5¢ for every plastic bottle recycled and 2¢ for every glass bottle recycled, we can use symbols to tell how much money an individual might receive for recycling bottles: $5p + 2g$, where p represents the number of plastic bottles and g represents the number of glass bottles. ●

In these three examples, symbols are used to represent the values of certain quantities. These symbols are called *variables*.

> A **variable** is a symbol used to stand for a value from a particular set of values.

A second way we use symbols as variables is in formulas. For example, $A = lw$ is used to represent the area of a rectangle as the product of the length l and the width w of the rectangle.

EXAMPLE 4

A rectangle is 6 inches long and 4 inches wide. Using the formula for the area of a rectangle, $A = lw$, we find that the area A is 6 inches $\times$ 4 inches, or 24 square inches. The formula for the perimeter of a rectangle is $P = 2 \times (l + w)$. Thus, the perimeter of this rectangle is $2 \times (6 \text{ inches} + 4 \text{ inches}) = 20$ inches. ●

A third way we use symbols as variables is in stating properties of arithmetic operations. These important properties are utilized when finding an unknown value that makes a statement true, such as in Example 1. As we will see, these properties are used to evaluate and simplify mathematical expressions. Number properties must be understood and used in mental computation, though you may not be aware of using them.

Children have some understanding of these properties, though they do not yet know their formal names. Consider this fourth-grader's logic for mentally obtaining the product of 4×20: "Four times twenty is double four times ten because twenty is double ten."[2] We can use symbols and properties to record this child's thinking. The property states that, in general, if we switch the order of the numbers being multiplied, the product is unaffected.

$$4 \times 20 = 2 \times (4 \times 10) \quad \text{because } 20 = 2 \times 10$$

$$4 \times (2 \times 10) = 2 \times (4 \times 10) \quad \text{associativity and commutativity of multiplication}$$

Children sometimes understand and state their observations more generally, as in the following explanation: "When you add zero to any number, you get the number you started with." This assertion is a partial statement of zero as an additive identity. It can be recorded with variables: $a + 0 = a$, where a is any number.

These properties are so important in both arithmetic and algebra that we restate them here for future reference. Let a, b, and c be any choice of numbers.

Properties of addition

1. Commutativity of addition: $a + b = b + a$. For example, $3x + 5x = 5x + 3x$.

2. Associativity of addition: $a + (b + c) = (a + b) + c$. For example, $362d + (8d + 27) = (362d + 8d) + 27$.

3. Zero is the identity for addition: $0 + a = a$ and $a + 0 = a$. For example, $0 + 505 = 505$ and $9t + 0 = 9t$.

4. Existence of additive inverses: For each number a, there is another number, denoted ^-a, such that $a + {}^-a = 0$ and $^-a + a = 0$. Each of a and ^-a is the additive inverse of the other. For example, $7x$ and ^-7x are additive inverses of each other, because $7x + {}^-7x = 0$.

Properties of multiplication

5. Commutativity of multiplication: $a \times b = b \times a$. For example, $2r \times 3t = 3t \times 2r$.

6. Associativity of multiplication: $a \times (b \times c) = (a \times b) \times c$. For example, $4z \times (25z \times 3) = (4z \times 25z) \times 3$.

7. One (1) is the identity for multiplication: $1 \times a = a$ and $a \times 1 = a$ for any number a. For example, $1 \times \frac{x}{t} = \frac{x}{t}$ and $3 \times 1 = 3$.

8. Existence of multiplicative inverses: For each nonzero number a, there is another number b such that $a \times b = 1$ and $b \times a = 1$. For example, $\frac{3}{s}$ and $\frac{s}{3}$ are multiplicative inverses of each other because $\frac{3}{s} \times \frac{s}{3} = 1$. We also say that $\frac{3}{s}$ and $\frac{s}{3}$ are reciprocals, or multiplicative inverses, of each other. In the same manner, we can say k and $\frac{1}{k}$ are reciprocals or multiplicative inverses of each other.

A property involving both addition and multiplication

9. Distributivity of multiplication over addition (often referred to simply as the distributive property): $a \times (b + c) = (a \times b) + (a \times c)$ for any numbers a, b, and c. The form $(b + c) \times a = (b \times a) + (c \times a)$ is also useful. For example, $6 \times (3a + 2) = (6 \times 3a) + (6 \times 2)$ and $(4.2 + 7) \times r = 4.2r + 7r$.

The $\times$ sign is easily confused with the variable x, so both the raised dot (as in $a \cdot b$ or $2 \cdot b$) and juxtaposition (as in ab or $2b$) are commonly used in algebra to indicate multiplication when doing so is not confusing. We also do not need a multiplication symbol when parentheses are used, i.e., $6 \times (4a + 3)$ could be written simply as $6(4a + 3)$.

Commutativity of multiplication is then compactly expressed by the statement $ab = ba$ for every choice of numbers or algebraic expressions a and b. Any algebraic expressions can play the roles of a and b. For example, commutativity of multiplication assures that $(3x + 8)(2x + 5)$ and $(2x + 5)(3x + 8)$ will be equal for any choice of the variable x. Other properties involving multiplication can be restated similarly.

Children should be encouraged to notice and describe patterns, structure, and regularities in arithmetic operations. These number properties are the foundation of algebra.

 ACTIVITY 1 Listen Carefully Now

Here are some children's mental and written computations. If the child is using correct logic, discuss what properties he or she may be recognizing and describing. If the child is making an error, describe what property he or she appears to not understand.

a. $45 + 15$ is the same as $50 + 10$ because I borrow 5 from the 15 to get to 50 and that leaves 10 more to add.

b. 5×6 is 26 because 5×5 is 25 and 6 is one more than 5. Twenty-six is 1 more than 25.

c. $\frac{2}{3} = \frac{8}{12}$ because $2 \times 4 = 8$ and $3 \times 4 = 12$. I have to multiply by the same thing in the numerator as the denominator.

d. 23×9 is 23×10 minus 23.

e. $-3 + \square = 5$. I think the answer must be 8 because I add 3 to get to zero and 5 more. That's eight. ●

ACTIVITY 2 Finding Properties

Tell what property or properties have been used:

a. $a + (n + 49)$ is rewritten as $(a + n) + 49$ **b.** $6\frac{4}{5} \times \frac{1}{2} = \frac{1}{2} \times 6\frac{4}{5}$

c. $(2b)c = 2(bc)$ (or $2bc$)   **d.** $z + {}^-z = 0$

e. $\left(29 \times \frac{7}{8} \times \frac{8}{7}\right) = 29 \times \left(\frac{7}{8} \times \frac{8}{7}\right) = 29 \times 1 = 29$

f. $\frac{1}{2} \cdot \left(6 + \frac{4}{5}\right)$ can be calculated by $\frac{1}{2} \cdot 6 + \frac{1}{2} \cdot \frac{4}{5}$

g. $xy^2 + 0 = xy^2$

h. $1 \cdot x = x$ (*Note*: x is commonly used in place of $1 \cdot x$.)

Use the properties to change the following expressions to make them easy to do mentally and tell what properties you used:

i. $(7 + 40) + 3$ **j.** $25 \times (4 \times 72.7)$

k. $24 \times 38 + 24 \times 12$ ●

Some representations (e.g., rectangular area, number lines, unit cubes) can help us see the structure and regularities of the properties. The distributive property of multiplication over addition can be represented with a rectangular area model; the area model is sometimes coupled with a number line. Commutativity and associativity of addition can be modeled on a number line or with unit cubes.

EXAMPLE 5

$$5 \times (10 + 7) = 5 \times 10 + 5 \times 7$$

	10	7
5	50	35

●

Notes

EXAMPLE 6 $13 + 17 = 17 + 13$

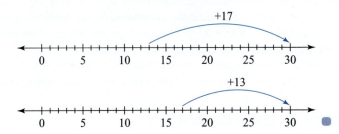

Think About ...

How can we model the commutative property of multiplication?

These same properties justify some of our routine algebraic manipulations, which helps make sense of these manipulations.

EXAMPLE 7

We can replace $5x + 3x$ with $8x$ because of the distributive property: $5x + 3x = (5 + 3)x$. ●

EXAMPLE 8

We can think of $7x + (3x + 19)$ as $(7x + 3x) + 19$ using associativity of addition, and then we can use the distributive property to write $(7 + 3)x + 19$ and arrive at $10x + 19$.

Indeed, if only addition is involved, grouping symbols are often omitted, except perhaps to add clarity or emphasis. Thus, $9x + (39 + x^2)$ can be replaced by $x^2 + 9x + 39$. The sequence of steps using addition properties that make this replacement possible is as follows:

$$9x + (39 + x^2) = 9x + (x^2 + 39) \quad \text{using the commutative property}$$
$$= (9x + x^2) + 39 \quad \text{using the associative property}$$
$$= (x^2 + 9x) + 39 \quad \text{using the commutative property}$$
$$= x^2 + 9x + 39 \quad \text{because the associative property tells us we do not need parentheses when adding three numbers} \ ●$$

EXAMPLE 9

We can say that $y(4y^2) = (4y)y^2 = 4(y \cdot y^2) = 4y^3$ due to the commutative and associative properties of multiplication. We usually would say simply that $y(4y^2) = 4y^3$, but it is important to recognize which properties allow us to do this. ●

By combining these properties with what we know about order of operations, we can perform many algebraic manipulations and solve algebraic equations. (As a reminder, first perform operations within grouping symbols, then attend to exponents, then perform all multiplications and divisions moving from left to right, and finally, perform all additions and subtractions moving from left to right.)

EXAMPLE 10

Simplifying expressions.

a. $3 + 11(15 - 3) - 12 \div 4 = 3 + 11 \cdot 12 - 3 = 11 \cdot 12 = 132$

b. $2x(5x - 13) + 3x^2 - x(12x - x - 13)$
$= 10x^2 - 26x + 3x^2 - x(11x - 13)$
$= 13x^2 - 26x - 11x^2 + 13x = 2x^2 - 13x$ [*Recall:* $^-a(^-b) = ab$.] ●

ACTIVITY 3   Time to Practice

a. Evaluate $16(12 - 8) - \frac{3}{4}(12 - 8) + 9$ in two different ways, using the distributive property in one way.

b. Evaluate $3t + r(t - 7) - 4r(t + r + 5) - 2r^2$ when $r = 7$ and $t = 14$.

c. Simplify $3t + r(t - 7) - 4r(t + r + 5) - 2r^2$ first and then evaluate your new expression when $r = 7$ and $t = 14$. How does your answer compare with that for part (b)? ●

Symbols are also used to express many other relationships that are true for any numbers substituted properly. Here are examples of more properties that are frequently used in algebra but do not have formal names:

1. For any number a, $a \cdot 0 = 0 \cdot a = 0$.

2. If $a > b$ and $c > d$, then $a + c > b + d$ for any values of a, b, c, and d.

3. $(pq)^2 = p^2 q^2$ for any values p and q.

4. $a + c - c = a$ for any values a and c.

5. $(x + y)^2 = x^2 + 2xy + y^2$ for any values x and y.

> **Think About …**
>
> Are you convinced that the properties described in the previous list hold for all possible values of a, b, c, d, p, q, x, and y? Choose some values and test these claims. Can you write down another property using s and t that is true for all values of s and t?

Algebra problems are often expressed as *equations* for which we seek values for variables that will lead to true statements. A simple example would be $3 + n = 28$. When 25 is substituted for n, the statement is true. We say that we have *solved the equation* $3 + n = 28$, and we call 25 the *solution* to this equation. If any other value is substituted for n in the equation $3 + n = 28$, the statement is false. Some equations may have more than one value that will give a true statement. For example, $x^2 = 4$ has two solutions, 2 and ⁻2, because when we substitute either value for x, we get a true statement. We say that 2 and ⁻2 are solutions to the equation $x^2 = 4$.

EXAMPLE 11

Which of these values, ⁻4, ⁻3, ⁻2, ⁻1, 0, 1, 2, 3, 4, 5, 6, can be substituted for the variable in each equation to make true statements?

a. $j + 8 = 13$ Only the value 5 makes this statement true.

b. $k + 8 = 8$ Only the value 0 makes this statement true.

c. $n^2 = 9$ Only the values 3 and ⁻3 make this statement true. ●

Obviously, substituting values for variables in an equation, which is a trial-and-error method, is not an efficient way of solving an equation. Instead, we rely on four more properties of equalities to solve equations.

> **Four important properties of equalities for any given values a, b, and c:**
>
> 1. If $a = b$, then $a + c = b + c$.
> 2. If $a = b$, then $a - c = b - c$.
> 3. If $a = b$, then $a \cdot c = b \cdot c$.
> 4. If $a = b$ and $c \neq 0$, then $a \div c = b \div c$.

Basically, these properties tell us that whatever operation we perform on the left side of the equals sign we must also perform on the right side of the equals sign (and vice versa) to assure equality is maintained.

EXAMPLE 12

Solve each of the following equations. Tell which properties you used.

  a. $10n - 14 = 22 - 8n$ **b.** $1.4x + 5 = 16.2$ **c.** $3g - 7 + 2g = 13$

  d. $\frac{2}{3}x = 18$ **e.** $2t + 125 = 9t + 20$

Solution

a. The steps for (a) are provided in more detail than they are in (b–e).

  i. Using the first property of equality listed, we can add 14 to both sides of the equation: $10n - 14 + 14 = 22 - 8n + 14$.

  ii. On the left side, subtracting 14 from $10n$ is like adding $^-14$. Because $^-14 + 14 = 0$ by the additive inverse property, the left side becomes just $10n$. On the right side, we can combine 22 and 14. The result is $10n = 36 - 8n$.

  iii. Next, we again use the first property of equality and add $8n$ to both sides: $10n + 8n = 36 - 8n + 8n$.

  iv. Using the distributive property on the left and the additive identity property on the right leaves us with $18n = 36$.

  v. Finally, using the fourth property of equality, we can divide both sides by 18 and we then arrive at $n = 2$.

b. Subtract 5 from *both* sides to get $1.4x = 11.2$. Then divide both sides by 1.4: $\frac{1.4x}{1.4} = \frac{11.2}{1.4}$. So $x = 8$ is the solution if there were no calculation mistakes. You can check by substituting 8 for x in the original equation to see whether the equation statement is true. Is $1.4(8) + 5 = 16.2$?

c. Working just on the left-hand side, we can change the original equation to $5g - 7 = 13$. Then we add 7 to each side, getting $5g = 20$. Dividing both sides by 5 gives $g = 4$.

d. Divide both sides by $\frac{2}{3}$ (or multiply both sides by $\frac{3}{2}$), getting $x = 27$.

e. You can get the answer in the "some number $= t$" form. With $9t$ on the right-hand side and $2t$ on the left-hand side, we will isolate the t on the right-hand side. Subtract $2t$ and 20 from each side, giving $105 = 7t$. Dividing both sides by 7 gives $15 = t$ as the solution. ●

> *Think About …*
>
> For $10n - 14 = 22 - 8n$, substitute 2 in place of n in every step along the way. What happens? For $5x - 3 = 2x$, is it also true (but not so useful for solving the original equation) that $5x - 3 + 500 = 2x + 500$?

Some students find it helpful to think of the process of solving an equation as "uncovering" to find the value of the unknown. We uncover when we use properties to "isolate" the variable on one side of the equation.

 ACTIVITY 4 Seeking the Truth

1. Solve these equations and tell what properties you use to do so.

  a. $5z + 14 = 7z - 30$

  b. $19 + 13t = 28t - 56$

2. Write an equation with a variable for each of the following elementary school problems, and then solve the equation:

  a. Ana has to practice the piano for 15 minutes every day. Today she has practiced 7 minutes. How many more minutes does she need to practice today?

  b. One kind of cheese costs \$2.19 a pound. How much will a package weighing 3 pounds cost?

  c. One kind of cheese costs \$2.19 a pound. How much will a package weighing 0.73 pound cost?

continue

d. A rectangle has a height of 70 mm and a base of 120 mm. What is the area of the rectangle? (Note that the height is the same as the length, and that the base is the same as the width, allowing the use of $A = lw$.)

e. A triangle with an area of 44 ft^2 has a base of 8 ft. What is its height? (Recall that the formula for the area of a triangle is $A = \frac{1}{2}bh$.) ●

💬 DISCUSSION 1   Algebra in Some Elementary Classrooms

Can you solve the following two problems? Are you surprised that these types of problems are being solved in elementary classrooms?

▶ How many hexagonal tables would it take to seat 25 guests? ◀

Assume that 6 guests could be seated at one hexagonal table, and if there are two or more hexagonal tables, they are arranged as shown in the diagram above. Guests can be seated at each side of a hexagonal table except where they are joined.

▶ What are some mathematically complex ways to arrive at 9 (e.g., $350 - 341$)? ◀

Read this excerpt from a report on mathematics in classrooms in the Lebanon, Oregon, school district, published December 29, 2008, in the Portland newspaper *The Oregonian*.

LEBANON—Lori Haley and Mya Corbett hunch over a pile of yellow hexagons, trying to figure out how many hexagonal tables it would take to seat 25 guests. The pair want to get the answer, but what they're really itching to do is to come up with a formula that will tell them how many people they could seat for any given number of tables.

Suddenly, the girls detect a pattern, and one shouts: "$(t \times 4) + 2 = s$!" They try it on one table, two tables, eight tables—it works. They beam, flashing smiles. . . . Lori and Mya just started third grade. . . .

Visit a Lebanon elementary math class, and you will see:

- First-graders set up and solve formulas such as $9 - x = 5$, as they did when Raylene Sell talked with her class about "some teddy bears" walking away from the classroom rug, leaving five behind.
- Third-graders suggest mathematically complex ways to arrive at 9: $-219 + 228$ or $(10 \times 5) - 40 - 1$, or even $(3 \times 3) + (8 \times 8) - ((4 \times 4) + (4 \times 4)) - 32$. ●

TAKE-AWAY MESSAGE . . .

- Algebra provides a symbolic language that can be used to represent quantitative relationships, i.e., word problems can be solved by first expressing them as equations.
- A variable is a symbol used to stand for a value from a particular set of values. Variables can be used (1) in expressions such as $5 - x$, (2) in formulas such as $A = bh$, and (3) to describe properties such as $a + b = b + a$ for any values a and b.
- The properties of operations together with knowledge of the order of operations allow one to evaluate algebraic expressions and solve algebraic equations.
- Solving an algebraic equation is the process of using properties of operations and properties of equalities to find the value of the variable(s) that makes the equation true.

Learning Exercises for Section 12.1

1. Evaluate and then express each of the following as a general property, using variables. Give the name of the property.

  a. $(18 \times 93) + (18 \times 7)$ can be calculated mentally and exactly by $18 \times (93 + 7)$.

  b. 12 nickels plus 8 nickels has the same penny value as 20 nickels.

continue

c. $(231 + 198) + 2$ can be calculated exactly by $231 + (198 + 2)$.

d. $(17 \times 25) \times 4$ can be calculated mentally and exactly by $17 \times (25 \times 4)$.

e. $\frac{117}{298} \times \frac{39}{39}$ is an easy mental calculation.

f. $\frac{0}{24} + \frac{13}{35}$ is an easy mental calculation.

g. Each of 4 pockets has a dime and 7 pennies. Calculate the total value in two ways.

2. Consider the related problems in parts (a–c) that lead up to the symbolic equation $a + d - d = a$.

    a. Is $65 + 108 - 108$ equal to, larger than, or smaller than 65?

    b. Is $65 + 4\frac{1}{4} - 4\frac{1}{4}$ equal to, larger than, or smaller than 65?

    c. Is $65 + d - d$ equal to, larger than, or smaller than 65 or is it not possible to tell?

    d. Write a set of sentences that leads up to the sentence "$g - h + h = g$."

3. Consider each statement below. Replace the symbol ∇ with the relation symbol $=$, $<$, or $>$ in turn, and say which replacements give true sentences. The first one is done for you.

    Example: $x \nabla x$ is *true* if $=$ is substituted for ∇ (i.e., $x = x$), but not true if either $<$ or $>$ is substituted for ∇.

    a. If $x \nabla y$, then $y \nabla x$.

    b. If $x \nabla y$ and $y \nabla z$, then $x \nabla z$.

4. The following problems are typical of those asked in grade school. Express each of the following using an equation with one variable:

    a. Malia had \$2.30 but wanted to buy a movie ticket for \$4.00. How much more does she need?

    b. Leesha had 8 stuffed animals and received 3 more. How many does she have?

    c. Jeremy downloaded 6 songs from iTunes onto his iPod. He already had 360 songs on his iPod. How many songs does he have now on his iPod?

    d. Regis is a teacher. He bought 10 boxes of colored markers for his students. Each box has 12 markers. He packages them into bags for each of his 30 students. If each student gets the same number of markers, how many markers are in each bag?

5. **a.** Choose a number, and keep it secret from others. Add 8 to your number, then multiply by 4, subtract 3, add 7, divide by 4, subtract 9. You should now have the number you started with. Why does this work?

    b. Make up a similar problem with at least five steps.

6. In each part, write the property or properties being used.

    a. replacing $6x + 4x$ with $10x$

    b. rewriting $(7x + 2) + ax$ as $(7x + ax) + 2$ (two properties)

    c. thinking of $(15x)(17y)$ as $(15 \cdot 17)(xy)$ (two properties)

    d. replacing $3x + {}^{-}7x + {}^{-}5x$ with $(3 + {}^{-}7 + {}^{-}5)x$

    e. replacing $xy + xz$ with $x(y + z)$

    f. replacing $(x + 2)(x + 7)$ with $x(x + 7) + 2(x + 7)$

7. Use the given property or properties to rewrite each expression.

    a. $(x + 5)x$, commutativity of multiplication

    b. $x(4x)$, associativity of multiplication

    c. $10n + 10m$, distributive property

    d. $y + (3y + 2)$, associativity of addition

    e. $(4 - 3)x^3$, identity for multiplication

    f. $\frac{4}{3}n^2 + \frac{0}{3}$, identity for addition

continue

g. $\left(\dfrac{3}{4} \cdot \dfrac{x}{y}\right) \cdot \dfrac{y}{x}$, associativity of multiplication, multiplicative inverse property, and identity for multiplication

8. Is 5 a solution of $2x^3 - 3x = 135$? Tell how you know.

9. Solve each of these equations. Show all your work.

a. $3x - 7 = 21.5$ **b.** $\dfrac{3}{4}y - 2 = 10$ **c.** $58 = 4 + (n - 1)3$

d. $\dfrac{3}{4}(y - 2) = 10$ **e.** $9x - 12 + 2x = 4x + 16$ **f.** $20 - 1.95x = 8.3$

g. $4x - 10 = {}^-7$ **h.** ${}^-2x - 7 = 11$ **i.** $\dfrac{5}{12}t + 3 + \dfrac{1}{3}t = t - 21$

10. Four people have these amounts of money: x, $2x$, $x + 12$, and $x - 3$.

a. Write an algebraic expression that gives the total amount of money they have.
b. Write an expression that gives each person's share if they share the money equally.
c. Write an expression that gives the total remaining if each person spends $15.
d. If the four people originally had a total of $119, how much did each person have?
e. Describe a situation that leads to the equation $5x + 9 = \$36$, and then solve the equation and use it to explain the situation.

11. Joan is x years old now. Write an algebraic expression for each part.

a. Joan's age in 10 years **b.** Joan's age in y years
c. Joan's age 3 years ago **d.** Joan's age n years ago
e. Joan's age when she is twice as old

12. Kay has k songs on her iPod. Write an algebraic expression for each situation described.

a. Lenore has 12 fewer tunes on her iPod than Kay has.
b. Kay gets n new tunes for her iPod.
c. On her iPod, Minnie has a third as many tunes as Kay has after part (b).
d. Nan has twice as many tunes as Lenore has [from part (a)].

13. An item is priced at d dollars. Write an algebraic expression for each situation described.

a. the cost of 6 of the items
b. the cost of n of the items
c. the cost of 2 of the items, along with $17.95 of other things
d. the reduced price if the item is sold at 30% off
e. your change if you pay for the item with two $20 bills
f. the amount of tax on the item if the sales-tax rate is 7%

14. Write an algebraic expression for the perimeter (distance around) each shape shown. Simplify the expression, if possible. Assume that the measurements are all in inches.

a. rectangle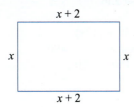

b. triangle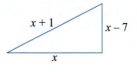

c. square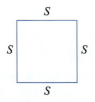

d. quadrilateral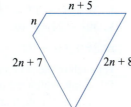

Notes

15. A researcher posed each of the following problems to Brazilian street children.[3] They could solve one but not the other. Why do you suppose this happened? Which one could they solve?

▶ A boy wants to buy chocolates. Each chocolate costs 50 cents. He wants to buy 3 chocolates. How much money does he need? ◀

▶ A boy wants to buy chocolates. Each chocolate costs 3 cents. He wants to buy 50 chocolates. How much money does he need? ◀

16. Use an array, rectangular area, number line, or base-ten blocks to model each of these properties for a particular instance.

   a. associative property of multiplication: $(3 \times 5) \times 2 = 3 \times (5 \times 2)$
   b. associative property of addition: $1.7 + 3.3 = 2.0 + 3.0$ because $1.7 + (0.3 + 3.0) = (1.7 + 0.3) + 3.0$
   c. existence of multiplicative inverses: $\frac{2}{3} \times \frac{3}{2} = 1$ and $\frac{3}{2} \times \frac{2}{3} = 1$
   d. commutativity of multiplication (show on a number line *and* with a rectangular area model): $3 \times 5 = 5 \times 3$

17. Write equations for the following "balance" displays. A black dot represents 1.

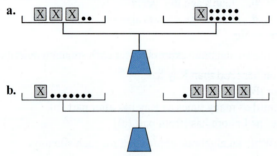

18. How do you solve each of these equations? Compare your algebraic work with actions on a balance representation.

   a. $3x + 2 = x + 10$ **b.** $x + 7 = 1 + 4x$

Supplementary Learning Exercises for Section 12.1

1. Tell whether each is true or false. If true, tell which property is demonstrated.

   a. $(x + 3) + 5(x + 3) = 6(x + 3)$
   b. $32 \times 8 = 30 \times 8 + 2 \times 8 = 240 + 16 = 256$
   c. $(2x + 17) + 13 = 2x + (17 + 13)$
   d. $3x + 5x + 17 \times 3x = 3x + 5x + (17 \times 3) + (17x)$
   e. $16 \times 25 = (4 \times 4) \times 25 = 4(4 \times 25) = 4 \times 100 = 400$
   f. $16 \times 25 = 10 \times 25 + 6 \times 25 = 250 + 150 = 400$
   g. $41 \times 28 = 40 \times 28 + 8$
   h. $3x + 5x + 51x = 59x^3$
   i. $2a + 2b = 2(a + b)$
   j. $3x + 5x + 17 \times 3x = 3x + 5x + (17 \times 3)x$
   k. $39x + 21y + 40z \times 0 = 0$ **l.** $(39x + 21y + 40z) \times 0 = 0$
   m. $2k + 13v + 24z = 2k + 24z + 13v$ **n.** $\frac{24}{7} \times \frac{13}{13} = \frac{24}{7}$

2. For each of the statements below, let @ symbolize $=$, $\leq$, or $\geq$. For each of the following, which are true for values b, m, n, r, s, and t? (For example, if $<$ is substituted for @, then for any value c, we know that c @ c is not true because it is not possible for $c < c$.)

   a. b @ b **b.** If m @ n, then n @ m.
   c. If r @ s and s @ t, then r @ t.

3. Which of these formulas are true? If not, why not?

   a. If n is the measure of each side of a pentagon (a five-sided polygon), then the perimeter P of the pentagon can be found by the formula $P = 5n$.
   b. The area A of the same pentagon can be found by the formula $A = n^2$.
   c. If the side of a square is $2s$, then the area of the square is $A = 4s^2$.
   d. If a polygon has sides f, g, h, k, m, n, and q, then the perimeter P of the polygon is $P = q + n + m + k + h + g + f$.
   e. If the radius of a circle is 5 cm, then the circumference of the circle is 25 cm.

4. A group of friends collect comic books. Write an algebraic equation or inequality for each of the following in terms of the number c of comic books that Casey owns. You may introduce new variables (but make clear what they represent).

   a. Kaye has three times as many comic books as Casey.
   b. Nancy has more than twice as many comic books as Casey.
   c. Jinfa has only half as many comic books as Casey.
   d. Fortuna has four more comic books than Kaye has.
   e. Milissa has two fewer comic books than Fortuna has.

5. Suppose n is the number of sweaters owned by Alicia, $n + 2$ is the number of sweaters owned by Kim, and $2n$ is the number of sweaters owned by Samme. Write an algebraic expression for each of the following. Simplify the expression if it can be simplified.

   a. The total number of sweaters they own altogether.
   b. Samme gives 6 of her sweaters to Goodwill and Alicia receives 2 more sweaters for her birthday. How many sweaters do they now own altogether?
   c. Later, they learn that Carolyn lost all her clothing when her home was destroyed by a hurricane. Each of the others gave her 2 sweaters. How many sweaters do the three now own in all?

6. Which of the following are solutions to $x^3 - 3x^2 = 10x$?

   a. 0   b. 2   c. ⁻2   d. 4   e. 5   f. ⁻4

7. Write an equation for each of the following using a variable, and then solve the equation:

   a. Fanny's company has a mortgage debt of $120 million and Freddie's company has a mortgage debt of $$z$ million. Together, their debt is $560 million. What is the mortgage debt of Freddie's company?
   b. Sidney woke up one morning feeling ill. Her temperature was 104°. She took aspirin and stayed in bed that day. The following morning her temperature was 99.6°. What was the change in her temperature?
   c. The price of a ticket to Animal World is $32 per day. Tanya bought one-day tickets to Animal World and to Air Park for herself and four of her friends for $300. What was the one-day price per person at Air Park?
   d. The price of a ticket to Animal World is $32 per day or $81 for a three-day pass. If Tanya expected to go for all three days, which type of pass should she buy and why?
   e. At Amco supermarket the special on mixed nuts from a bin was $3.68 per pound. Jenny bought 6 ounces of the nuts. How much did she pay?
   f. The supermarket in part (e) also had a special on jellybeans. Vinny bought 8 ounces of mixed nuts and 8 ounces of jellybeans, and his total was $3.28. What was the per-ounce cost of the jellybeans?

8. Give the answer to each part with mental calculation. Write which property enabled you to do the calculation mentally.

   a. $(59 \times 25) \times 4$

   b. $(4 \times 3.98) + (6 \times 3.98)$

   c. $\dfrac{82}{71} \times \dfrac{71}{82} \times \dfrac{97}{106}$

   d. $(24 \times 1.50) + (24 \times 4.50)$

9. Sydney sold b boxes of Girl Scout cookies. Write an algebraic expression for each situation described.

   a. Samme sold 7 boxes more than Sydney.
   b. Tyler sold 2 more than the total number of boxes that Sydney and Samme sold.
   c. Sydney sold b plus c more boxes of Girl Scout cookies.
   d. Rachel sold two fewer boxes than twice as many boxes as Samme sold.

10. Solve each of these equations. Show all your work.

   a. $5x + 17 = 33 + x$ b. $32x - 24 = 12x + 24$ c. $0.5 + c = {}^-13.5 + 0.5c$
   d. $\frac{2}{3}z = \frac{5}{6}(2 + z)$ e. ${}^-0.25n = 136.75$

11. Begin with your age and follow these steps. The final result will be 25.

   a. Add 14 to your age. Multiply the result by 2. Subtract 4. Divide by 2. Add 11. Subtract your age.
   b. Why does this work? Modify the above steps so that the result will be your age.

12.2 Algebra as Generalized Arithmetic

This section treats one prominent view of algebra: algebra as generalized arithmetic. That is, what we did in arithmetic with numbers, we can now do in algebra with variables. For example, consider the properties of operations reviewed in Section 12.1, using symbols. These are first learned as properties of arithmetic operations and then are generalized and used in algebraic operations.

Our place-value numeration system gives another illustration of algebra as a generalization of arithmetic. You know that one expanded form of 452 is $400 + 50 + 2$, or $4 \cdot 100 + 5 \cdot 10 + 2$. A similar expanded form uses exponents: $452 = 4 \cdot 10^2 + 5 \cdot 10^1 + 2 \cdot 10^0$. Because $a^0 = 1$ for any nonzero value of a, we know that $10^0 = 1$, so we can use $4 \cdot 10^2 + 5 \cdot 10 + 2$ for the expanded form.

If we use the variable x instead of 10 in the last expression, we get the expression $4 \cdot x^2 + 5 \cdot x^1 + 2$, or $4x^2 + 5x + 2$, which is a *polynomial in x*.

> A **polynomial in some variable** is any sum of number multiples of whole number powers of the variable. The expressions that are added (or subtracted) are called **terms** of the polynomial, so $4x^2$, $5x$, and 2 are the terms of the polynomial $4x^2 + 5x + 2$.

Arithmetic that has been learned conceptually can lead naturally to polynomial arithmetic. Addition of polynomials is very similar to addition of multidigit whole numbers.

EXAMPLE 13

	Consider	and contrast that with
	452	$4x^2 + 5x + 2$
	+324	$+ 3x^2 + 2x + 4$
The parallel is clearer in the expanded form:	$4 \cdot 10^2 + 5 \cdot 10 + 2$ $+ \ 3 \cdot 10^2 + 2 \cdot 10 + 4$	$4x^2 + 5x + 2$ $+ \ 3x^2 + 2x + 4$
The usual algorithm for adding whole number transfers to the polynomial form:	$4 \cdot 10^2 + 5 \cdot 10 + 2$ $+ \ 3 \cdot 10^2 + 2 \cdot 10 + 4$ $\overline{7 \cdot 10^2 + 7 \cdot 10 + 6}$	$4x^2 + 5x + 2$ $+ \ 3x^2 + 2x + 4$ $\overline{7x^2 + 7x + 6}$ ●

The usual algorithms are efficient in part because they ignore the many steps that are necessary if one had to be explicit about the properties of operations involved. The uses of the properties are clearer when the calculation is written in horizontal form.

DISCUSSION 2   Properties Working for Us

$452 + 324$ is the same as $(4 \cdot 10^2 + 5 \cdot 10 + 2) + (3 \cdot 10^2 + 2 \cdot 10 + 4)$.

When placed in vertical form as shown on the right, the algorithm usually is computed from right to left: $(2 + 4) + (5 + 2) \cdot 10 + (4 + 3) \cdot 10^2$, in that order. (In fact, the place values are often ignored, and the calculations done are $2 + 4$, $5 + 2$, and $4 + 3$. When this happens, children lose awareness of the place value implicit in the numbers.)

$$\begin{array}{r} 452 \\ +324 \\ \hline \end{array}$$

In the horizontal form, however, it is more natural to work from left to right: $(4 + 3) \cdot 10^2 + (5 + 2) \cdot 10 + (2 + 4)$.

What properties assure that. . .

$$(4 \cdot 10^2 + 5 \cdot 10 + 2) + (3 \cdot 10^2 + 2 \cdot 10 + 4)$$

is indeed the same as the following?

$$(4 + 3) \cdot 10^2 + (5 + 2) \cdot 10 + (2 + 4) \; \bullet$$

Isn't it fortunate that the usual algorithm for adding multidigit whole numbers allows us to bypass being explicit about each use of a property? The usual algorithm for multiplying multidigit whole numbers also disguises the fact that properties can explain why the algorithm gives correct answers. The properties also apply to the multiplication of polynomials.

 DISCUSSION 3   Multiplying Polynomials Is Like Multiplying
                                         Whole Numbers

1. How does
$$\begin{array}{r} 32 \\ \times 4 \\ \hline \end{array}$$
transfer to
$$\begin{array}{r} 3n + 2 \\ \times 4 \\ \hline \end{array}$$

What property or properties are involved?

2. How does
$$\begin{array}{r} 32 \\ \times 14 \\ \hline \end{array}$$
transfer to
$$\begin{array}{r} 3n + 2 \\ \times n + 4 \\ \hline \end{array}$$

What property or properties are involved? $\bullet$

Writing the calculations in Discussion 3 in horizontal form helps you see what properties are being used.

For the numerical calculation:

$$4(30 + 2) = (4 \times 30) + (4 \times 2) \quad \text{(distributivity)}$$

And for the corresponding algebraic calculation:

$$4(3n + 2) = (4 \times 3n) + (4 \times 2) \quad \text{(distributivity)}$$

Thus, distributivity (of multiplication over addition) is involved in the numerical calculation as well as in the algebraic calculation. Furthermore, $(10 + 4) \times 32 = (10 \times 32) + (4 \times 32)$ does not give all the steps in the usual algorithm until we use distributivity again:

$$(10 \times 32) + (4 \times 32) = (10[30 + 2]) + (4[30 + 2])$$

This last equation gives the following, but not in the usual algorithm order because that algorithm starts at the right:

$$(10[30 + 2]) + (4[30 + 2]) = (10 \times 30) + (10 \times 2) + (4 \times 30) + (4 \times 2)$$

The four multiplications shown on the right-hand side of the previous equation are those that we compute when multiplying 14 and 32.

Notes

Notes

Similarly, $(n + 4)(3n + 2)$ involves the sum of the four multiplications, $4 \cdot 2$, $4 \cdot 3n$, $n \cdot 2$, and $n \cdot 3n$. (Compare the vertical form.) Notice that each term in the $n + 4$ expression is multiplied by each term in the $3n + 2$ expression. In either the numerical or the algebraic case, distributivity is used more than once.

 DISCUSSION 4   Algebraic and Numerical Division Algorithms

How does $12\overline{)276}$ transfer to $x + 2\overline{)2x^2 + 7x + 6}$? ●

Thus, operations on polynomials are similar to operations on numbers. Operations on algebraic fractions are also similar to operations on arithmetic fractions.

EXAMPLE 14

With numbers	With algebra
Suppose we want to add two numeric fractions: $\frac{5}{16} + \frac{7}{72}$.	Suppose we want to add two algebraic fractions: $\frac{x}{y^4} + \frac{z}{y^3t^2}$.
We first need to find a common denominator, preferably the least common denominator. Because 16 is 2^4 and 72 is $2^3 \cdot 3^2$, the LCD is $2^4 \cdot 3^2$. $$\frac{5}{2^4} + \frac{7}{2^3 \cdot 3^2} = \frac{5 \cdot 3^2}{2^4 3^2} + \frac{7 \cdot 2}{2^4 3^2}$$ $$= \frac{45}{144} + \frac{14}{144} = \frac{59}{144}$$	We first need to find a common denominator, preferably the least common denominator. The denominators are already factored, and the LCD is y^4t^2. $$\frac{x}{y^4} + \frac{z}{y^3t^2} = \frac{xt^2}{y^4t^2} + \frac{zy}{y^4t^2}$$ $$= \frac{xt^2 + zy}{y^4t^2}$$ ●

EXAMPLE 15

With numbers	With algebra
Suppose we want to multiply two numeric fractions: $\frac{5}{21} \times \frac{33}{45}$.	Suppose we want to multiply two algebraic fractions: $\frac{a}{bc} \times \frac{bd}{ab^2}$.
We first multiply across, then cancel common factors in the numerator and denominator, which can be done by first factoring: $$\frac{5}{21} \times \frac{33}{45} = \frac{5 \cdot 33}{21 \cdot 45} = \frac{\cancel{5} \cdot \cancel{3} \cdot 11}{\cancel{3} \cdot 7 \cdot \cancel{5} \cdot 9} = \frac{11}{63}$$	We first multiply across, then cancel common factors in the numerator and denominator, which can be done by first factoring: $$\frac{a}{bc} \times \frac{bd}{ab^2} = \frac{abd}{bcab^2} = \frac{\cancel{a}\cancel{b}d}{\cancel{b}c\cancel{a}b^2} = \frac{d}{cb^2}$$ ●

 ACTIVITY 5   Your Turn

How is finding $\frac{42}{70} - \frac{14}{40}$ like finding $\frac{mnp}{mrp} - \frac{mp}{m^3r}$?

How is finding $\frac{42}{70} \div \frac{14}{40}$ like finding $\frac{mnp}{mrp} \div \frac{mp}{m^3r}$? ●

TAKE-AWAY MESSAGE ... If you understand the underlying reasons for properties and procedures with numbers, then you can generalize those reasons to corresponding situations with polynomial expressions. If you understand the underlying reasons for how we operate with arithmetic fractions, then you can generalize those reasons to corresponding situations with algebraic fractions.

Learning Exercises for Section 12.2

1. Complete the following additions side by side as shown in the examples, and explain how they are alike:

   a. Add numbers 4026 and 43.
   b. Add polynomials $4n^3 + 2n + 6$ and $4n + 3$.

2. Complete the following subtractions side by side and explain how they are alike:

   a. $598 - 347$
   b. Subtract $3x^2 + 4x + 7$ from $5x^2 + 9x + 8$.

3. Complete the following additions side by side as shown in the examples and explain how they are alike:

   a. $642 + 188$
   b. Add polynomials $6x^2 + 4x + 2$ and $x^2 + 8x + 8$.

4. a. Simplify $\dfrac{12}{18}$. b. Simplify $\dfrac{4x^3}{7x^2}$.

5. Complete the following additions and tell how they are alike:

   a. $\dfrac{5}{16} + \dfrac{7}{16}$ b. $\dfrac{3x}{y} + \dfrac{2x+1}{y}$ c. $\dfrac{5}{x+2} + \dfrac{x}{x+2}$

6. Complete the following additions and tell how they are alike:

   a. $\dfrac{5}{8} + \dfrac{3}{4}$ b. $\dfrac{3x}{(x+2)(x+3)} + \dfrac{2}{x+2}$

7. Complete the following subtractions and tell how they are alike:

   a. $\dfrac{7}{9} - \dfrac{2}{9}$ b. $\dfrac{2x}{7y} - \dfrac{4}{7y}$

8. Complete the following subtractions and tell how they are alike:

   a. $\dfrac{3}{4} - \dfrac{1}{7}$ b. $\dfrac{3}{xy} - \dfrac{4x}{9n}$

9. Complete the following multiplications and tell how they are alike:

   a. $\dfrac{2}{3} \times \dfrac{5}{7}$ b. $\dfrac{2a}{b} \times \dfrac{c}{3d}$ c. $\dfrac{4}{x+2} \cdot \dfrac{2y}{x+3}$

10. Complete the following divisions and tell how they are alike:

    a. $\dfrac{5}{6} \div \dfrac{2}{3}$ b. $\dfrac{x^2}{2y} \div \dfrac{xy}{3}$

11. Areas and volumes can give insight into some algebraically equivalent expressions. Using sketches, find or verify equivalent expressions in the following:

    a. $(x + y)^2 = ?$ b. $(x + y)^3 = x^3 + 3x^2y + 3xy^2 + y^3$

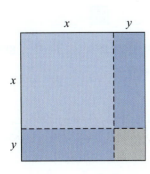
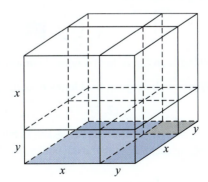

    c. $x(x + y) = ?$ (Make your own drawing.)

    d. $(x + 2)(x + 3) = ?$ (Make your own drawing.)

Supplementary Learning Exercises for Section 12.2

1. Give the polynomial suggested by the numeral.

   a. 41,025, with $x = 10$ **b.** 234,125 with $x = 10$

2. Use the numerical calculation to inform the algebraic one. Give the answer to the algebraic one.

   a.
   $$\begin{array}{r} 513 \\ +\,452 \end{array} \qquad \begin{array}{r} 5x^2 + x + 3 \\ 4x^2 + 5x + 2 \end{array}$$
   b.
   $$\begin{array}{r} 943 \\ \times\,2 \end{array} \qquad \begin{array}{r} 9x^2 + 4x + 3 \\ \times\,2 \end{array}$$

   c.
   $$\begin{array}{r} 34 \\ \times\,21 \end{array} \qquad \begin{array}{r} 3x + 4 \\ \times\,2x + 1 \end{array}$$

3. Calculate the sums and products of the polynomials in each part.

   a. $5x^2 + 3x + 9$ and $7x^2 + 2x + 10$ **b.** $\frac{3}{4}x + 2$ and $\frac{1}{3}x + \frac{2}{5}$

   c. $5x + {}^{-}3$ and $3x + {}^{-}2$ **d.** $19x + 25$ and $22x + 31$

   e. $x + 3$, $6x + 1$, and $7x + {}^{-}1$

4. Use the arithmetic calculation to inform the algebraic calculation for each of the following, and give answers for the algebraic calculation:

   a. $\frac{2}{5} + \frac{1}{5}$ and $\frac{2}{x} + \frac{1}{x}$ **b.** $\frac{2}{5} + \frac{7}{10}$ and $\frac{2}{x} + \frac{7}{2x}$

   c. $\frac{14}{25} - \frac{3}{10}$ and $\frac{14}{(y+7)^2} - \frac{3}{2(y+7)}$ **d.** $\frac{1}{25} \cdot \frac{10}{35}$ and $\frac{x}{y^2} \cdot \frac{2y}{xy}$

   e. $\frac{3}{8} \div \frac{9}{16}$ and $\frac{z}{n^3} \div \frac{z^2}{n^4}$

12.3 Numerical Patterns and Algebra

Mathematics is sometimes described as *the science of patterns* because it involves the study of patterns in numbers, in shapes or objects, and even in systems. This brief description highlights the importance of patterns in mathematics. Exercises in recognizing, describing, and extending patterns occur in the curriculum even in the first grade. Numerical patterns often involve some calculation practice within the larger goal of seeing what the pattern is, so patterns can disguise some drill. In this section our focus is on patterns and how to represent them algebraically.

    The primary focus of this section will be numerical patterns. However, you should be aware that repeating patterns of blocks or shapes, as in the following diagram, can appear in first-grade textbooks:

Following is an example of a typical numerical pattern exercise given in the early grades.

▶ What numbers go in the blanks to continue the pattern?

$$7, 17, 27, 37, \underline{\quad}, \underline{\quad}. ◀$$

In later grades, children might be asked to tell what the 51st number in the pattern is, a more challenging exercise, with the intent that the child *not* fill in all the intervening blanks.

> **Think About …**
>
> Jan decided that the 51st number would be 507. How do you think Jan reasoned?

The ability to predict ahead is an important by-product of recognizing a pattern.

EXAMPLE 16

Predict the next four numbers in each sequence of numbers. (The entries are often called **terms**, especially if a variable is involved.) The dots (. . .) are used to indicate that the pattern continues forever.

1. 6, 9, 12, 15, . . .

2. 18, 12, 22, 16, 26, 20, . . .

3. 20, 19.75, 19.5, 19.25, . . .

4. 2, 6, 12, 20, 30, . . .

Solution

1. The numbers seem to increase by 3 each time, so a good prediction is that the next four numbers would be 18, 21, 24, and 27.

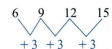

2. This pattern is more complicated.

$$18 \quad 12 \quad 22 \quad 16 \quad 26 \quad 20$$
$$-6 \quad +10 \quad -6 \quad +10 \quad -6$$

But it looks as though $20 + 10$, or 30, would be the next number, and then $30 - 6 = 24$, $24 + 10 = 34$, and $34 - 6 = 28$ would follow.

3. Subtracting 0.25 (or adding ⁻0.25) gets from one term to the next, so the next four numbers could be $19.25 + {}^{-}0.25 = 19$, $19 + {}^{-}0.25 = 18.75$, 18.5, and 18.25—i.e., the next four terms are 19, 18.75, 18.5, and 18.25.

4. This sequence of numbers also has a pattern.

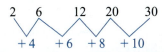

It is reasonable to predict $30 + 12 = 42$, $42 + 14 = 56$, $56 + 16 = 72$, and $72 + 18 = 90$ to be the next four numbers: 42, 56, 72, and 90. ●

The type of pattern in the first and third sequences in Example 16 is so common that it has a name. As a teacher, you should know about it even though the terms may not come up in elementary or middle school. A pattern can be formed by adding the same number, called the **common difference**, to any number in a list to get the next number. For example, in the first pattern in Example 16, the common difference is 3. In the third pattern, we could say that the common difference is ⁻0.25. Such patterns are called **arithmetic sequences**.

ACTIVITY 6 I Beg to Differ?

1. Which, if any, of the following sequences of numbers is an arithmetic sequence? Identify the common difference for any arithmetic sequence you find and give the next three numbers in the pattern.

   a. 18, 23, 28, 33, . . .   **b.** 7.2, 7.9, 8.6, 9.3, 10, . . .   **c.** 112, 109, 106, 103, 100, . . .

2. Write the first five numbers in an arithmetic sequence that starts with 18 and has a common difference of 4.

3. Identify the 20th number in each of the arithmetic sequences in Problem 1, without putting in all the intervening numbers. (*Hint*: How many times must the common difference be added to the first number to get the 20th number?) ●

Notes

The kind of thinking that you likely did for Problem 3 in Activity 6 allows you to predict *any* number—even the 100th or the 2000th—in an arithmetic sequence *without having to find the intervening terms*. To get the *n*th term, you must add to the first number in the sequence the common difference, $n - 1$ times:

The *n*th number = the first number + $(n - 1) \times$ (the common difference)

We can express this general idea more briefly by using algebraic symbols. Suppose that in an arithmetic sequence the first number is *a* and the common difference is *d*. Then the *n*th number in the arithmetic sequence is given by this equation:

The *n*th number = $a + (n - 1)d$

Use the formula to check that the 51st number in each of the arithmetic sequences in Activity 6, Problem 1 is 268, 42.2, and ⁻38, respectively.

> **Think About …**
>
> In the expression $a + (n - 1)d$ for finding the *n*th number in an arithmetic sequence, why is it $n - 1$ rather than *n*?

⊞ ACTIVITY 7 Sally Moves to the Rockies

Water does not always boil at 212° Fahrenheit. The boiling point changes depending on the altitude.

1. Find a pattern in this table of altitudes in feet above sea level and temperatures in degrees Fahrenheit. Find the numbers that are missing. Then write an expression for the boiling point at an altitude of *n* feet.

Altitude above sea level (feet)	0	500	1000	1500	2000	2500	?	n
Boiling point of water (°F)	212	211	210	?	208	?	204	?

2. Sally moves to the Rockies and now lives at 10,000 feet above sea level. What is the boiling point of water for Sally?

3. Why does the boiling point of water affect baking a cake? ●

Notice in Activity 7 that the ability to predict numbers in the boiling-point column from numbers in the altitude column is quite useful. With regular steps (500 feet here) in the first column, the decrease of 1° at each step in the second column could be used together to predict the boiling point at other altitudes.

Here is another type of numerical pattern that has a name: A pattern in which each number in a list is obtained by *multiplying* the number that precedes it by the *same factor* is called a **geometric sequence**. For example, 4, 8, 16, 32, 64, . . . is a geometric sequence, with the factor being 2:

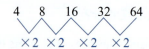

> **Think About …**
>
> The factor, which is 2 in the previous sequence, is often called the **common ratio**. Do you see why?

Geometric sequences usually get less attention than arithmetic sequences in the elementary school curriculum.

ACTIVITY 8 Go Forth and Multiply

Consider a geometric sequence that starts with 5 and has a common ratio 3.

1. Write the next four numbers in the geometric sequence.

2. Write an expression that gives the 100th number in the sequence. [*Hint*: In part 1, how many times did you multiply by 3?]

3. The table in Activity 7 organized the data so that we could predict entries in the second column from entries in the first column. Give an algebraic expression for the *n*th number in a geometric sequence that starts with *a* and has common ratio *r*.

Which number in the sequence?	1st	2nd	3rd	4th	5th	...	*n*th
The term in that position	a	ar	ar^2	ar^3	ar^4	...	?

Think About ...

Someone says, "The number of bacteria increases geometrically day by day." What does this mean?

Many sequences are neither arithmetic nor geometric. The sequence 2, 6, 12, 20, 30, . . . from Problem 4 in Example 16 is neither an arithmetic nor a geometric sequence, e.g. Even though we can predict the next numbers in that sequence, it is not easy to see how we could predict, say, the 100th number without putting in all the in-between numbers. However, notice that the **first** number $2 = 1 \cdot 2$, the **second** number $6 = 2 \cdot 3$, the **third** number $12 = 3 \cdot 4$, the **fourth** number $20 = 4 \cdot 5$, and the **fifth** number $30 = 5 \cdot 6$. Do you see a pattern in the products that would enable you to see that the 99th number in the sequence would be 9900?

Sometimes, a problem can be solved by looking at simpler or specific cases, organizing the data from those cases, and looking for a pattern in those data. Activity 9 illustrates these useful techniques.

ACTIVITY 9 How Do You Get Started?

Patterns can be useful in solving problems that have several specific cases, such as this one:

Figure out a shortcut for squaring any number ending in 5.

For example, you will be able to give the product 75^2 easily, without paper and pencil (or a calculator). Here is one way to approach a problem like this one. Let us *look at some specific cases*, possibly $25^2 = 625$, $65^2 = 4225$, $15^2 = 225$, $45^2 = 2025$, and $35^2 = 1225$. At this point, it looks as though each product ends in 25 (ending in 5 is not a surprise). But how could we predict the other digits? Let us *organize the data*. A table is a convenient way to organize numerical data. Here we might have a table like the following:

Number ending in 5	15	25	35	45	55	65
Its square	225	625	1225	2025		4225

In each squared number, the digits preceding the "25" apparently depend on the number of tens. We can focus on them in the table and *look for a pattern*.

For the number 55 and other numbers ending in 5 not given in this table, do you see a way to predict the other digits to the left of 25 in the squared number from the number of tens? Does your shortcut enable you to see instantly that $95^2 = 9025$?

EXAMPLE 17

Consider the fraction $\frac{5}{27} = 0.185185185...$ (the digits 185 repeat forever). What digit will be in the 2000th decimal place in the decimal for $\frac{5}{27}$?

Solution

The block of digits repeats, so there is a pattern. But how can we predict the 2000th decimal place without putting in a lot of 185s? Here is one way to think: The repeating block is 3 digits long, so we need to find out how many full blocks of 3 digits will get us to the 2000th decimal, or close to it. A moment's thought might suggest that what we want to know is, how many 3s are in 2000? That thinking suggests division. Because $2000 \div 3 = 666$ R 2, to get to the 2000th decimal place will take 666 full 1-8-5 blocks, with two additional digits from the next 1-8-5 block. Do you see that the digit in the 2000th decimal place must be 8? ●

Patterns will also be featured in the next section, where we relate patterns to what are called functions.

> **TAKE-AWAY MESSAGE . . .** The study of patterns underlies a good part of mathematics. Many numerical patterns can be described in ways that enable us to determine the actual number at any location in the pattern. For example, the nth term in an arithmetic sequence starting with a and having common difference d is given by $a + (n - 1)d$. The nth term in a geometric sequence that starts with a and has common ratio r is given by ar^{n-1}. In general, an unfamiliar problem might be solvable by looking at specific cases, organizing the data, and looking for a pattern.

Learning Exercises for Section 12.3

1. For each pattern, give the next four entries and any particular entry requested, as suggested by the pattern. Assume that the patterns continue indefinitely.

   a. ABABAB ___ ___ ___ ___ ; the 100th entry is ____.

   b. ABBAABBA ___ ___ ___ ___ ; the 63rd entry is ____.

   c. 6.5, 7.3, 8.1, 8.9, ___, ___, ___, ___; the 20th entry is ____.

   d. 100, 95, 90, 85, ___, ___, ___, ___; the 30th entry is ____.

   e. 2, 6, 18, 54, ___, ___, ___, ___.

   f. 5, 2.5, 1.25, 0.625, ___, ___, ___, ___.

   g. $\frac{1}{1}, \frac{1}{2}, \frac{1}{3}, \frac{1}{4}, \frac{1}{5}$, ___, ___, ___, ___; the 100th entry is ____.

   h. $\frac{1}{2}, \frac{2}{3}, \frac{3}{4}, \frac{4}{5}$, ___, ___, ___, ___; the 100th entry is ____.

   i. 2, 4, 6, 8, 2, 4, 6, 8, ___, ___, ___, ___; the 30th entry is ____.

   j. 1, 4, 2, 8, 5, 7, 1, 4, 2, 8, 5, 7, 1, 4, ___, ___, ___, ___; the 40th entry is ____.

   k. Which of the sequences in parts (a–j) are arithmetic sequences?

   l. Are there any geometric sequences in parts (a–j)? If so, tell which parts.

2. a. Examine these (correct) calculations for a pattern and look for an easy rule for multiplying by a power of 10. Write the rule.

$12.3457 \times 10 = 123.457$	$23 \times 10 = 230$
$12.3457 \times 100 = 1234.57$	$23 \times 100 = 2300$
$12.3457 \times 1000 = 12{,}345.7$	$23 \times 1000 = 23{,}000$
$12.3457 \times 10{,}000 = 123{,}457$	$23 \times 10{,}000 = 230{,}000$
$12.3457 \times 100{,}000 = 1{,}234{,}570$	$23 \times 100{,}000 = 2{,}300{,}000$

   b. Why does your rule work?

3. a. Examine these (correct) calculations for a pattern and look for an easy rule for dividing by a power of 10. Write the rule.

$512.345 \div 10 = 51.2345$ $8.41 \div 10 = 0.841$
$512.345 \div 100 = 5.12345$ $8.41 \div 100 = 0.0841$
$512.345 \div 1000 = 0.512345$ $8.41 \div 1000 = 0.00841$
$512.345 \div 10,000 = 0.0512345$ $8.41 \div 10,000 = 0.000841$
$512.345 \div 100,000 = 0.00512345$

b. Why does your rule work?

c. How can this idea be used to find 1% of a known amount? 10% of a known amount?

4. Use the equation (nth term) $= a + (n - 1)d$ for an arithmetic sequence starting with a and having a common difference d to find each of the following:

a. the 500th number in the sequence, 10, 14, 18, 22, 26, . . .

b. the 25th number in the sequence, $6\frac{3}{4}, 7\frac{1}{2}, 8\frac{1}{4}, 9, \ldots$

c. the first number in an arithmetic sequence that has a common difference of 5 and has 257 as its 51st term

d. the first number in an arithmetic sequence that has a common difference of 4.5 and has 466 as its 101st term

5. The numbers 1, 1, 2, 3, 5, 8, . . . give an example of a **Fibonacci** (pronounced "fee-ba-NAH-chee") **sequence**, which is a pattern that appears in nature, art, and geometry.

a. What are the next four numbers in that Fibonacci sequence? (*Hint*: Look at two consecutive numbers in the sequence and then the next one.)

b. Amazingly, the *n*th number in that Fibonacci sequence is

$$\frac{(1 + \sqrt{5})^n - (1 - \sqrt{5})^n}{2^n \sqrt{5}}$$

Verify that for $n = 1$ and $n = 2$, the expression does give the first two numbers in the pattern, 1 and 1.

c. Give decimals for the first ten ratios of consecutive Fibonacci numbers:

$$\frac{1}{1}, \frac{2}{1}, \frac{3}{2}, \frac{5}{3}, \frac{8}{5}, \ldots$$

The ratios get closer and closer to a special value that is called the **golden ratio**. The golden ratio is often used in art to make pleasing proportions.

6. a. What is the *n*th fraction in the following sequence?

$$\frac{1}{2}, \frac{1}{4}, \frac{1}{8}, \frac{1}{16}, \frac{1}{32}, \ldots$$

b. What is the *sum* of the first *n* of those fractions? To what number is the sum getting closer and closer?

7. What digit is in the 99th decimal place in the decimal for $\frac{1}{7}$? Explain your reasoning.

8. a. What digit is in the ones' place in the calculated form of 3^{250}?

b. What digit is in the ones' place in the calculated form of 7^{350}?

c. What digit is in the ones' place in the calculated form of $3^{250} \cdot 7^{350}$?

d. Write down how you would proceed to find the digit in the ones' place in the calculated form of 7^n for any whole number *n*.

9. The numbers $1^2 = 1, 2^2 = 4, 3^2 = 9, 4^2 = 16$, etc. are called the square numbers. Tell how the drawings below explain why they are called the *square* numbers.

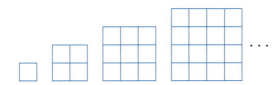

10. The *triangular* numbers 1, 3, 6, 10, etc. are suggested by the numbers of dots in the following triangles of dots (the first one is not really a triangle):

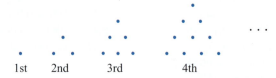

1st   2nd   3rd   4th

a. Make a table, such as the following one, for the first 8 triangular numbers and look for a pattern:

Which triangular number?	1st	2nd	etc.
Number of dots	1	3	etc.

b. Do the *drawings* support an argument that the pattern will continue?

11. A famous mathematician, Gauss ("gows"), was assigned the problem of adding the whole numbers from 1 through 100. Here is his clever approach:

$$1 + \quad 2 + \quad 3 + \cdots + \quad 98 + \quad 99 + 100$$
$$\underline{100 + \quad 99 + \quad 98 + \cdots + \quad 3 + \quad 2 + \quad 1}$$
$$101 + 101 + 101 + \cdots + 101 + 101 + 101$$

Then he reasoned: There are 100 of the 101s, or 10,100. But I added them twice, so $1 + 2 + 3 + \cdots + 98 + 99 + 100$ is half of 10,100, or 5050.

a. Apply this reasoning to the following sum:

$$1 + 2 + 3 + \cdots + (n - 2) + (n - 1) + n$$

b. How is part (a) relevant to Learning Exercise 10?

12. *Jump-It* is a label for this puzzle. There are 2 colors of round markers, 3 of each color. They are arranged on a 7-square row, as below.

a. Can you switch the colors, so that each color is on the other end with an empty square in the middle, subject to these two rules?

Rule 1. A marker can be moved to an adjacent empty square.

Rule 2. A marker can "jump" a marker in the next square, if there is an empty square on the other side of the jumped marker.

b. What is the minimum number of moves for the markers to change ends? (*Hint*: It is not 17.)

c. What is the minimum number of moves, if there are *n* markers on each end, with an empty square in the middle? (*Hints*: A good problem-solving approach is to look at simpler problems and look for a pattern. Here, there is a pattern relating the number on each end and the minimum number of jumps.)

13. On the first day of the month, a rich lady offers you this choice:

Choice I. $1000.00 at the end of the month.

Choice II. You get 1¢ if you stop after the first day, 2¢ if you stop after the second day, 4¢ if you stop after the third day, 8¢ if you stop after the fourth day, etc. until the end of the month.

To get more money, which choice should you make? (*Hints*: $2^0 = 1$; $2^{10} = 1024$.)

Supplementary Learning Exercises for Section 12.3

Notes

1. Give for each list (i) the likely 500th number and (ii) an expression for the likely *n*th number. If the sequence is arithmetic, identify *a* and *d*. If the sequence is geometric, identify *a* and *r*.

   a. 8, 12, 16, 20, 24, . . .

   b. 2.3, 3.6, 4.9, 6.2, 7.5, . . .

   c. ⁻1, 1, ⁻1, 1, ⁻1, . . .

   d. $3\frac{1}{4}$, 5, $6\frac{3}{4}$, $8\frac{1}{2}$, $10\frac{1}{4}$, . . .

   e. ⁻100, ⁻98, ⁻96, ⁻94, ⁻92, . . .

   f. 4, 7, 9, 4, 7, 9, 4, 7, 9, . . .

   g. 1, 5, 25, 125, 625, 3125, . . .

   h. 2000, 1998.3, 1996.6, 1994.9, 1993.2, . . .

   i. 3, 9, 27, 81, . . .

   j. $\frac{1}{1}, \frac{1}{2}, \frac{1}{4}, \frac{1}{8}, \frac{1}{16}, \cdots$

   k. 16, 12, 8, 4, 0, . . .

   l. 0.5, 0.75, 1, 1.25, 1.5, . . .

2. a. Give the first 5 numbers in the arithmetic sequence that has a first number 22.3 and a difference of 0.35.

   b. Give the first 5 numbers of a geometric sequence that has a first number 1.5 and a ratio of 2.

3. What would the 36th digit be in the repeating pattern 3, 6, 9, 3, 6, 9, 3, 6, 9, . . .? What would the 100th digit be?

4. What digit is in the 88th place (after the decimal point) of the decimal equivalent of $\frac{2}{11}$?

5. What digit is in the ones' place in the calculated form of 2^{120}?

6. Use the shortcut method from Activity 9 to find 85^2 and 19.5^2.

7. a. What digit is in the 38th place in the calculated form of $\frac{1}{11}$?

   b. What digit is in the 29th place in the calculated form of $\frac{3}{7}$?

   c. What digit is in the 43rd place in the calculated form of $\frac{1}{9}$?

8. a. What is the *n*th term in the following sequence?

   $$1, \frac{1}{3}, \frac{1}{9}, \frac{1}{27} \cdots$$

   b. What is the sum of the first *n* numbers in this sequence? That is, to what number is the sum getting closer and closer?

12.4 Functions and Algebra

The work with patterns can lead naturally to the idea of *function*, a topic that permeates mathematics even if not made explicit. Informal work with functions in some guise can appear even in grades K–3. We shall use your work with patterns from the previous section as our lead-in.

You may remember this elementary school problem from Section 12.3 and its more challenging extension in the Think About:

▶ What numbers go in the blanks to continue the pattern?

7, 17, 27, 37, ___, ___. ◀

Think About …
Jan decided that the 51st number would be 507. How do you think Jan reasoned?

Asking about the 51st number suggests that the list can be associated with the counting numbers by referring to a number's location in the list: 7 is the **1**(st) number in the pattern, 17 is the **2**(nd), 27 is the **3**(rd), etc. This association leads to tables, as used in Example 18.

EXAMPLE 18

Make a table for this pattern of numbers: 3, 6, 9, 12, 15, . . .

continue

Solution

Location in the list	1(st)	2(nd)	3(rd)	4(th)	5(th)
Actual number	3	6	9	12	15

(*Note*: The horizontal form for the table is common in elementary school books because it fits on a textbook page a little better than the following vertical form.)

We can write the table vertically:

Location in the list	Actual number
1	3
2	6 + 3
3	9 + 3
4	12 + 3
5	15 + 3

In the last section, we noted that each actual number increased by 3 to get the next number. Each successive number in the sequence was generated from the preceding one. In contrast, you can think of the numbers in column 2 as generated by multiplying the location in the list by 3, as the arrows in these tables suggest.

Location in the list	Actual number		n	s
1	3		1	3
2	6		2	6
3	9		3	9
4	12		4	12
5	15		5	15

This last table, with the column headings the variables n and s, should help you see that work with patterns can lead to an algebraic equation, such as $s = 3n$ for the pattern in Example 18. Such an equation allows us to predict the number located in the 100th place in the sequence just by multiplying by 3.

Tables like those for Example 18, in which each value of n corresponds to exactly one value of s, incorporate the basic idea of a mathematical *function*. The dictionary usually gives the more common meanings for the word "function" before giving the mathematical meaning. Here is one definition as the word is used in mathematics.

> A **function** from one set to another (which may be the same set) is a correspondence in which each element of the first set is assigned to exactly one element of the second set.

We can call the first set *inputs* and the second set *outputs*. Based on what is given in Example 18, the input set (for n) consists of the whole numbers 1 through 5, and the output set (for s) would be any set containing 3, 6, 9, 12, and 15. In most cases in mathematics, there is a **function rule** that describes how the correspondence works. In the previous Example 18, the function rule could be represented by the equation $s = 3n$, or by the words, multiply a number from the input set by 3 to get the number from the output set.

EXAMPLE 19

Write an equation for the perimeter (distance around) of a regular hexagon with sides of length x inches. Use the variable p for the perimeter.

Solution $p = 6x,$

or even the longer equation $p = x + x + x + x + x + x$

You may be familiar with notations such as $f(x)$, which denotes the value associated with x for some function f. The value associated with 2, e.g., might be designated $f(2)$. The equation $f(x) = 6x$ could have been used in the last example, rather than $p = 6x$, to emphasize the functional relationship. Letters other than f can be used to signal a function, as in $g(x) = 2x + 3$ for some other situation. The context usually makes clear that $f(x)$ is not describing a product, as well as what numbers are acceptable for x and for $f(x)$. Although variables x and y are commonly used with functions, you may see equations such as $s = 16t^2$ or $f(t) = 16t^2$ in function contexts.

Hence, from a more advanced mathematical viewpoint, you were also working with functions when you worked with patterns in the last section.

We have other ways to represent functions besides tables and function rules. *Dot diagrams* will be illustrated in Activity 10. Sometimes, especially when no function rule is obvious, an ordered pair such as (18, 25) is used to communicate the correspondence $18 \rightarrow 25$ (read as "18 is assigned to 25"). The first entry in the ordered pair is from the input set, and the second entry is from the output set. *A collection of ordered pairs* might then represent a function. In the next chapter, we work with coordinate systems to show functions.

Some elementary school programs may also include not only work with patterns but also exposure to functions in the guise of *function machines* like the one pictured. The words "input" and "output" might be used instead of x and y (or other variables) in tables. For example, the function rule $y = 2x + 5$ [or $f(x) = 2x + 5$] might not appear but could be expressed as "output equals two times the input, plus 5." Children might be asked to complete a table for such a machine.

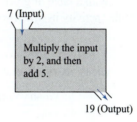

In the definition for *function*, it is important to note that *each* element in the first set corresponds to *exactly one* element in the second set. The table below does *not* describe a function, because 17 in the first set corresponds to two elements in the second set, 19 and 11.

Number from first (input) set	17	15	17	21
Number from second (output) set	19	23	11	13

ACTIVITY 10 Which Correspondences Give Functions?

Tell whether each diagram describes a (mathematical) function. Tell how you know. Notice the variety of ways of showing functions.

1.

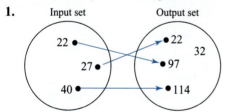

2.

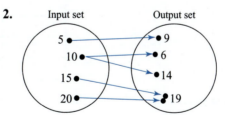

3.

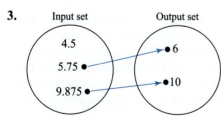

4.

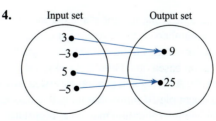

5. To each postage stamp, assign the weight of the first-class letter the stamp will pay for. For example, a 45¢ stamp will pay for weights up to and including 1 ounce.

6. The collection of ordered pairs, (0, 1), (1, 2), (2, 3), (3, 4), etc.

7. The collection of ordered pairs, (2, 5), (3, 8), (2, 9), (5, 10).

continue

8. Every number is paired with its square, so 1 is paired with 1, $\frac{1}{2}$ is paired with $\frac{1}{4}$, $1\frac{2}{3}$ is paired with $\frac{25}{9} = 2\frac{7}{9}$, $\sqrt{2}$ is paired with 2, etc.

9. In any classroom, associate with each first name a student's full name. Here is part of the class roster for a particular classroom.

First name	Full name
Emily	Emily Dickinson
Memo	Memo Flores
Rachel	Rachel Hobbs
Emily	Emily Bronte
Jon	Jon Sinclair
⋮	⋮

The restrictive *exactly one* in the definition of function assures that there is no ambiguity in determining what corresponds to a specific element of the input set.

With a table of given values and their corresponding values, you can do a What's My Rule? activity in which you look for a pattern and express its function rule.

Think About ...

What's My Rule? for the following data?

When you say. . .	25	16	40	3
I say. . .	75	66	90	53

Using a table to find a function rule is a good practice, but doing so has risks. One risk is that, even though you think that you have found the function rule, it might not be the correct one. For example, the table in the previous Think About "obviously" suggests the equation $y = x + 50$, but the function rule *might* actually be something else.

EXAMPLE 20

Consider this table for a What's My Rule? problem:

x	y
0	1
1	2
2	4

Do you see that both $y = 2^x$ and $y = \frac{1}{2}x^2 + \frac{1}{2}x + 1$ describe the table entries? ●

If the table is about some real situation, then you can check other values for x. But with just a few numbers to work with, you cannot be certain that *your* rule is the only possible one. At some stage in their mathematics education, children should know that trusting a generalization based on a pattern is risky, but most work in schools proceeds as though *a* rule is *the* rule. Arriving at a general assertion by looking at examples is called **inductive reasoning**. Example 20 shows that inductive reasoning might give good ideas but is risky, as more than one interpretation may be appropriate.

Rather than trust completely a function rule or pattern based on a table of values, the mathematically sound way to proceed is to look for a justification that the rule must be correct. We will use the problem given in Activity 11 to illustrate what a possible justification might look like in the later Discussion 5. Note that the use of a specific large number, such as 100 in Activity 11, means that a child does not have to be algebraically sophisticated to *reason* about the situation.

Try these "What's My Rule?" exercises from a Grade 4 textbook.

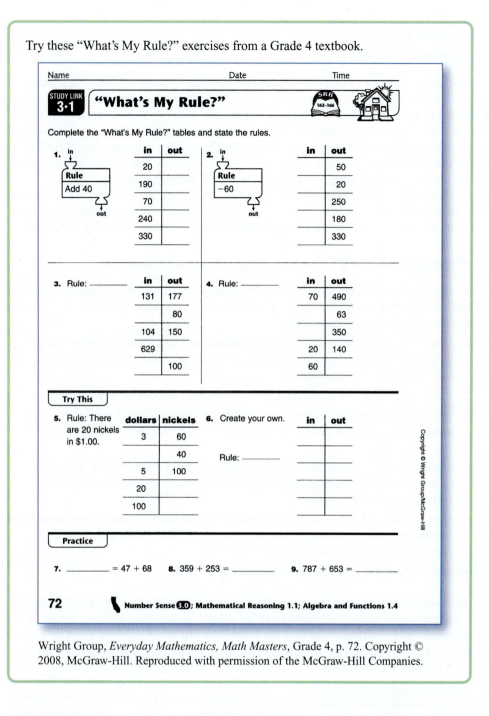

Wright Group, *Everyday Mathematics, Math Masters,* Grade 4, p. 72. Copyright © 2008, McGraw-Hill. Reproduced with permission of the McGraw-Hill Companies.

ACTIVITY 11   Making a Row of Squares from Toothpicks

Pat made a pattern of squares from toothpicks, as in this drawing:

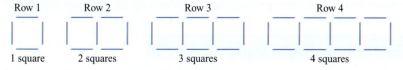

Row 1	Row 2	Row 3	Row 4
1 square	2 squares	3 squares	4 squares

How many toothpicks would Pat need to make a row with 100 squares? More generally, how many toothpicks would it take to make a row with n squares? (A good approach to solving this problem is to make a table, perhaps like the one on the following page.)

Number of squares	1	2	3	4	...	100	...	n
Number of toothpicks					...		...	

Your table in Activity 11 likely showed that the row with 1 square requires 4 toothpicks, the row with 2 squares requires 7 toothpicks, etc. Did you see a pattern that enabled you to predict that it would take 301 toothpicks for the row with 100 squares and $3n + 1$ for the row with n squares? The results for 100 and n certainly seem to fit the pattern. But is there a risk in endorsing these results enthusiastically? Some other rule might apply, perhaps as the number of squares gets much larger than 100. Is there any way to argue that the results will always be correct for all values of n? The answer happens to be yes.

When dealing with patterns in mathematics, we try to justify the results in some way to show that the result *must be* true in *all* cases, not just the cases examined. Giving a justification confirms that a particular generalization from inductive reasoning is all right.

 DISCUSSION 5   Are These Justifications Convincing?

Which of these arguments, used to justify the result for the row with 100 squares, could be modified to justify the general $3n + 1$ result for a row of n squares, from Activity 11? If an argument is not strong, explain why.

1. *Mike*: "I needed 4 toothpicks for the first square, and then 3 more for each of the 99 squares after that. I got 301." In general, Mike's argument suggests the rule $f(n) = 4 + (n - 1)3$ for n squares.

2. *Nadia*: "I checked two more cases, and they worked. So it has to be all right."

3. *Oscar*: "Well, I put down 1 toothpick, and then every time I put down 3 more, I got another square. So $1 + 100 \times 3$." In general, Oscar's reasoning suggests the rule $f(n) = 1 + 3n$.

4. *Paloma*: "Across the top, there would be 100 toothpicks. In the middle, 101. And on the bottom, 100. So, 301." In general, Paloma's reasoning suggests the rule $f(n) = n + (n + 1) + n$.

5. *Quan*: "100 squares would take 400 toothpicks by themselves. But when you put them together, you don't need the extra ones inside so subtract 99." In general, Quan's reasoning suggests the rule $f(n) = 4n - (n - 1)$.

6. *Ricky*: "Everyone I asked said the same thing, 301." ●

Because the toothpick problem in Activity 11 is based in reality, there is a good chance that a mathematically sound justification can be offered, as was illustrated by several of the arguments given in Discussion 5. If the table were just about numbers (with no context given), there would be no way to know whether the rule that you find (or an algebraic equivalent) is true in all cases.

TAKE-AWAY MESSAGE... The study of patterns is the basis for the more advanced study of mathematical functions. Many numerical patterns can be described by general function rules that enable us to determine the actual number given any location in the pattern. Functions may also be represented in tables or with dot diagrams. Although a pattern based only on examples (inductive reasoning) may have more than one interpretation, many times it is possible to find a justification showing that the pattern will be true given a context.

Learning Exercises for Section 12.4

1. In a classroom of children, do these assignments give functions? Tell why or why not.

   a. To each child, assign his or her first name.
   b. From a list of the last names of children in the class, assign the child.

2. Does each of the following dot diagrams represent a function? Tell how you know.

a.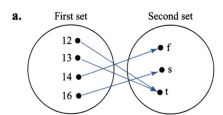

b.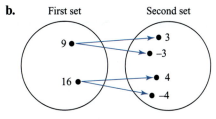

c. these ordered pairs:
   (8, 56), (11, 77), (9, 63), (0, 0)

d. these ordered pairs:
   (1.7, 2), (1.8, 2), (3.2, 3), (4.3, 4)

In Learning Exercises 3–10, find a possible function rule for each table. Then use your rule to find the numerical value for *n*, if *n* is indicated. (*Hint*: Do any of them give arithmetic sequences?)

3.

x	1	2	3	4	. . .	n
y	9	30	51	72	. . .	1038

4.

x	1	2	3	4	. . .	n
$f(x)$	78	56	34	12	. . .	⁻120

5.

input	1	2	3	4	. . .	n
output	0	8	16	24	. . .	1608

6.

input	50	51	52	53	. . .	n
output	8	8.3	8.6	8.9	. . .	15.8

7.

x	3	1	4	2	. . .	n
y	12	2	17	7	. . .	547

8.

x	2	1	3	0	4	. . .	n
$f(x)$	35	19	51	3	67	. . .	403

9.

x	0	1	2	3	4	. . .	n
$f(x)$	$\frac{1}{2}$	$1\frac{1}{6}$	$1\frac{5}{6}$	$2\frac{1}{2}$	$3\frac{1}{6}$	. . .	$8\frac{1}{2}$

10.

x	0	3	4	5	. . .	n
y	4	8	$9\frac{1}{3}$	$10\frac{2}{3}$	. . .	84

11. Find a function rule for each table.

a.

x	1	2	3	4	. . .	n
$g(x)$	1	4	9	16	. . .	900

b.

x	1	2	3	4	. . .	n
$g(x)$	1	8	27	64	. . .	729

12. Find a function rule for each of the following patterns, and justify that your rule will be true in general.

    a. the number of toothpicks to make Shape *n* in the pattern:

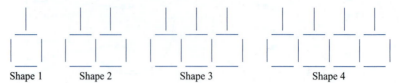

Shape 1   Shape 2   Shape 3   Shape 4

    b. the number of toothpicks to make Double-decker *n*:

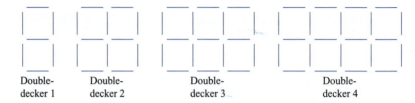

Double-   Double-   Double-   Double-
decker 1   decker 2   decker 3   decker 4

    c. the number of toothpicks to make Shape *n*:

Shape 1   Shape 2   Shape 3   Shape 4

    d. the number of toothpicks to make Shape *n*:

Shape 1   Shape 2   Shape 3

    e. the number of toothpicks to make Rowhouse *n*:

Rowhouse 1   Rowhouse 2   Rowhouse 3

    f. Make up a toothpick pattern of your own, and challenge others to find a rule for it. Can you justify the rule?

13. Some *sums* of patterns can be predicted.

    a.

How many evens (starting with 2)?	Sum of those evens
1	**2**
2	$2 + 4 = $ **6**
3	$2 + 4 + 6 = $ **12**
4	$2 + 4 + 6 + 8 = $ **20**
5	$2 + 4 + 6 + 8 + 10 = $ **30**
n	$2 + 4 + \cdots + (2n) = $?

    b. Use part (a) and algebra to show that

$$1 + 2 + 3 + 4 + \cdots + n = \frac{n(n + 1)}{2}$$

continue

c. Use part (b) and algebra to show that

$$6 + 12 + 18 + 24 + \cdots + (6n) = 3n(n + 1)$$

d. What is the sum of the first *n odd* numbers? [*Hint*: Make a table, or use your results from parts (b) and (a).]

14. Rather than just finding the *n*th number in an arithmetic sequence, many times we want to find the *sum* of the first *n* numbers. Look for a pattern in the following table to see whether the sum of the first *n* terms in an arithmetic sequence is predictable. [*Hint*: See Learning Exercise 13(b).]

How many?	Arithmetic sequence	Sum
1	a	a
2	$a + d$	$a + (a + d) = 2a + d$
3	$a + 2d$	$a + (a + d) + (a + 2d) = 3a + 3d$
4	$a + 3d$	$\ldots = 4a + 6d$
5	$a + 4d$	$\ldots = 5a + 10d$
n	$a + (n - 1)d$	$\ldots = ?$

15. How many small squares will be in an *n*-step stairway like those shown in the following diagram? [*Hint*: See Learning Exercise 13(b).]

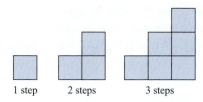

1 step   2 steps   3 steps

16. Formulas can be viewed as defining functions. For example, the formula $C = \frac{5}{9}(F - 32)$ tells what Celsius temperature, C, corresponds to a Fahrenheit temperature, F.

   a. What Celsius temperatures correspond to the boiling point of water (212°F at sea level), the freezing point of water (32°F), normal body temperature (98.6°F), and a comfortable temperature of 70°F?

   b. Use algebra to write the Fahrenheit temperature F as a function of the Celsius temperature: $F = $ ___.

17. What digit is in the 1000th decimal place in the decimal for $\frac{13}{303}$?

$$\frac{13}{303} = 0.042904290429\ldots$$

18. **a.** What digit is in the ones' place in the calculated form of 8^{999}?

   b. What digit is in the ones' place in the calculated form of 16^{423}?

   c. What digit is in the ones' place in the calculated form of $8^{999} + 16^{423}$?

19. A particular bologna machine gives 8 packages of bologna for every 7 pounds of raw food.

   a. How many packages of bologna could be produced by 1 pound of raw food?

   b. How many pounds of raw food make 1 package of bologna?

   c. If 10 pounds of raw food are put into the machine, how many packages of bologna would the machine give?

20. Each of the three students below is arguing that his or her function rule is the correct one. What do you say to them?

   Abe: $f(x) = 2(x + x)$
   Beth: $f(x) = 4x$
   Candra: $f(x) = x + x + x + x$

21. Many function rules that appear in elementary school will be simple, but polynomial rules like $f(n) = 2n^3 + 3n + 8$ can occur. The variable can also appear in exponents, as in geometric sequences with ar^{n-1}. Verify for $n = 1, 2, 3, 4, 5,$ and 6 that $f(n) = (^-1)^n$ gives the nth number in the following list:

$$^-1, 1, ^-1, 1, ^-1, \ldots$$

22. Go to http://nlvm.usu.edu/, find the applet called Function Machine, and try it out! Click in the Algebra, 6–8 box.

Supplementary Learning Exercises for Section 12.4

1. Which of the following rules are examples of functions?

a. Passengers on a plane in the first set, matched with seats on the airplane as the second set. Each passenger gets only one seat.

b. Seats on an airplane in the first set, matched with passengers on the airplane in the second set.

c.
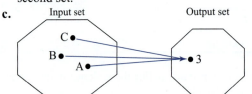

d.
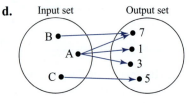

e. Every square number is matched with its square roots.

f. Every whole number is matched with its square.

g. These ordered pairs: (8, 3), (7, 4), (6, 3), (5, 2), (4, 1), (3, 0), (2, 1), (0, 2).

h. These ordered pairs: (3, 8), (4, 7), (3, 6), (2, 5), (1, 4), (0, 3), (1, 2), (2, 0).

2. Find a likely function rule for each of the following, and find the missing values for m and n:

a.

x	y
1	12
2	17
3	22
4	27
⋮	⋮
100	m
n	2007

b.

x	$f(x)$
1	153
2	149
3	145
4	141
⋮	⋮
36	m
n	$^-43$

c.

$input$	$output$
1	5
2	7
3	11
4	19
⋮	⋮
10	m
n	2051

d.

$input$	$output$
10	35
11	38
12	41
13	44
⋮	⋮
100	m
n	641

e.

x	$g(x)$
1	2
2	9
3	28
4	65
⋮	⋮
10	m
n	8001

f.

x	$h(x)$
3	18
1	4
4	25
2	11
⋮	⋮
25	m
n	2454

3. Find a function rule for each of the following, and justify that your rule will be true in general:

   a. the number of toothpicks to make Shape *n* in the pattern:

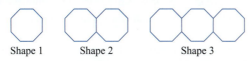

   b. the number of toothpicks to make Shape *n* in the pattern:

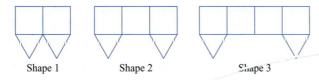

   c. the number of toothpicks to make Shape *n* in the pattern:

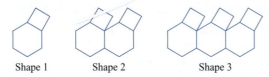

Notes

4. How are an arithmetic sequence and a geometric sequence alike? How are they different?

5. Continue each sequence for six more terms, using the given information.

   a. arithmetic; first number 12, common difference 27
   b. arithmetic; first number 3, common difference $6\frac{3}{4}$
   c. geometric; first number 3, common ratio 5
   d. geometric; first number 8, common ratio $\frac{2}{3}$

6. What would be the number of cricket chirps per minute when the temperature outside is 90°F if the pattern in the following table stays the same? Find a likely function rule.

Number of Chirps	144	152	160	168	176
Temperature (°F)	76	78	80	82	84

12.5 Algebraic Reasoning About Quantities

Many children (and adults) have difficulty solving mathematical story problems. One reason is that many times they are focusing on the numbers in the problem statement, rather than what the numbers represent. They then try something with the numbers, hoping to get an acceptable answer (and quickly move on to the next problem). Here is misguided advice from a Grade 6 student[4]:

> *Interviewer*: What can we tell students to help them solve story problems?
> *Ann*: If you see a big number and a little number, go for the division. If that doesn't work, then you can try the other ones . . . .

Notice Ann's focus on the *numbers*, not *what the numbers are about*. We distinguish between the numbers and what the numbers are about by using the terms *value* and *quantity*.

> A **quantity** is anything (an object, event, or quality thereof) that can be measured or counted. The **value** of a quantity is its measure or the number of items that are counted. A value of a quantity involves a number and a unit of measure or number of units.

For example, your height is a quantity. It can be measured. Suppose the measurement is 64 inches. Note that 64 is a number, and inches is a unit of measure: 64 inches is the value of the quantity, your height. The number of people in a room is another example of a quantity. Suppose the count is 32 people. Note that 32 is a number and the unit counted is people, so 32 (people) is the value associated with the quantity, number of people in the room.

A quantity is not the same as a number. In fact, one can think of a quantity without knowing its value. For example, the amount of snow during a month is a quantity, regardless of whether or not someone measured the actual number of inches of snow fallen. One can speak of the amount of snow fallen without knowing how many inches fell. Similarly, one can speak of a person's annual income, a school's enrollment figure, the speed of a race car, or the amount of time it takes to get to school (all are quantities) without knowing their actual values.

 DISCUSSION 6   Identifying Quantities and Measures

1. Identify the quantity or quantities addressed in each of the following questions.
   a. How tall is Mount Everest?
   b. How fast does gasoline come out of a pump?
   c. How much damage did the storm cause?

2. Identify an appropriate unit of measure that can be used to determine the value of the quantities involved in answering the questions in Problem 1. ●

The difficulty with story problems becomes greater with algebra story problems, because some relevant values may not even be apparent. To write an equation for the problem, it becomes essential to focus on the quantities involved and how the quantities are related.

In this section, you will analyze problem situations in terms of the quantities and their relationships. Such analyses are essential to being skillful at solving mathematical problems.

> For the purposes of this course, to ***understand a problem situation*** means to recognize the quantities embedded in the situation and understand how the quantities are related to one another.

Understanding a problem situation "drives" the solution to the problem. Without such understanding, the only recourse a person has is to guess at the calculations that need to be performed or to proceed as the Grade 6 student Ann does. It is important that you work through the following problems with care and attention. Analyzing problem situations quantitatively is a powerful tool that will help you become a better problem solver.

 ACTIVITY 12   Best Friends

Try this problem before you read the solution. Then compare your solution to the one shown here.

▶ Three boys spent the afternoon playing a video game and then compared their best scores. Al says, "My best today is 900 points more than yours, Bob." Carlo says, "I was having a good day. My 3600 points is $\frac{2}{3}$ as much as your two scores combined, Al and Bob." How many points did each boy get in his best game that afternoon? ◀

Possible Solution

1. Name the quantities involved in the problem.
   a. Al's best score
   b. Bob's best score
   c. Carlo's best score

continue

    d. the difference between Al's and Bob's best scores
    e. the total of Al's and Bob's best scores
    f. the comparison of Carlo's best score to Al and Bob's total

2. Identify the values of the quantities.

    a. Al's best score   unknown
    b. Bob's best score   unknown
    c. Carlo's best score   3600
    d. the difference between Al's and Bob's best scores   900
    e. the total of Al's and Bob's best scores   unknown
    f. the comparison of Carlo's best score to Al and Bob's total $\frac{2}{3}$

3. Look for relationships. This step often involves rereading the problem and making a drawing for the problem.

The second sentence relates quantities (a), (b), and (d). Al's best score is 900 more than Bob's. The third sentence relates quantities (c), (e), and (f):

$$3600 = \frac{2}{3}(\text{Al's best score} + \text{Bob's best score})$$

The following is a possible drawing:

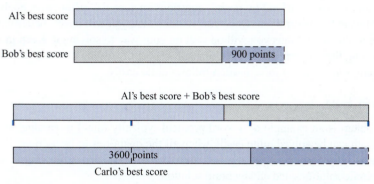

4. If we use algebra to solve the problem, introduce a variable and try to write an equation. (As you may know, some problems may require more than one variable, but these are not usually encountered until well into a study of algebra.)

It is not clear in this problem what score to have as the unknown. Let us try x as the number of points in Al's best score. Then Bob's must be $x - 900$. Carlo's information then leads to the equation $3600 = \frac{2}{3}(x + x - 900)$, which leads to $5400 = 2x - 900$. Solving this equation, we arrive at $6300 = 2x$, which in turn leads to $x = 3150$. So, Al's best score was 3150 points, Bob's was 2250 points (from $3150 - 900$), and Carlo's was 3600 points (known).

How did *your* solution compare with the elaborated one? ●

Of course, usually we do not write so much detail, by doing some mental work. But if we are stuck, we might try writing down more and referring to our diagram for help in seeing relationships. In Activity 12 you may even have noticed that you could solve the problem without algebra.

> **Think About ...**
>
> Suppose that in the Best Friends problem you had let x be the number of points in Bob's best game. What equation would you get then? Would you get the same answers for the given problem?

Often, students begin a problem such as the one in Activity 12 by asking themselves, "What operations do I need to perform, with which numbers, and in what order?" Instead of these questions, you would be much better off asking yourself general questions such as those shown on the next page:

Quantitative Analysis Questions

- Can I imagine the situation, as though I am acting it out?

- What quantities are involved? What do I know about their nature? Do some change over time? Do some stay constant?

- What is the quantity of interest? That is, what quantity, and its value, am I asked to find or describe?

- How could making a diagram (or a sequence of diagrams) help reveal relationships among the quantity of interest and other quantities?

- What quantities and values do I know that could help me find the value of the quantity of interest?

- What quantities and values can I derive to help me find the value of the quantity of interest?

The point is that you need to be specific when doing a quantitative analysis of a given situation. How specific? It is difficult to say in general, because it will depend on the situation. Use your common sense in analyzing the situation. When you first read a problem, try not to only think about the numbers and the operations of addition, subtraction, multiplication, and division. Instead, start out by posing questions such as the quantitative analysis questions and trying to answer them. Once you've done that, you will have a better understanding of the problem, and thus you will be well on your way to solving it. Keep in mind that understanding the problem is the most difficult aspect of solving it. Once you understand the problem, *what to do to solve it* often follows quite easily.

EXAMPLE 21

This problem is an example of a word problem typically found in an algebra textbook. Often, such can be solved quite easily when all the quantities to be used are listed and a drawing is made. But here we use a quantitative analysis and a drawing to guide two nonalgebraic solutions and an algebraic solution.

▶ Amtrail super-express mag-lev trains provide efficient, nonstop transportation between New York City and Washington, D.C. Suppose Train A leaves New York City headed toward Washington at the same time that Train B leaves Washington headed for New York City, traveling on parallel tracks. Train A travels at a constant speed of 65 miles per hour. Train B travels at a constant speed of 75 miles per hour. The two Amtrail stations are 210 miles apart. How long after they leave their respective stations do the trains meet? ◀

Quantities	Values
Speed of Train A	65 mph
Speed of Train B	75 mph
Distance traveled by Train A to meeting point	? miles
Distance traveled by Train B to meeting point	? miles
Total distance traveled by trains	210 miles
Time traveled by both trains, departure to meeting point	? hours

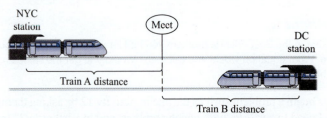

Here is a traditional algebraic solution that also follows from a quantitative analysis and a drawing. In this solution, note that the time traveled by both trains is the same, which is

t hours, and that students use the $d = rt$ formula, with Distance$_A$ and Distance$_B$ representing the distances (in miles) traveled by Trains A and B, respectively, when the two trains meet.

$$\text{Distance}_A + \text{Distance}_B = 210 \text{ miles}$$

$$65t + 75t = 210, \quad \text{so } 140t = 210 \quad \text{and} \quad t = 210 \div 140 = 1.5 \text{ hours}$$

You may have noticed that algebra is not essential for this particular problem and would not be needed in elementary school.

Here is one nonalgebraic solution: In 1 hour, the trains together will go $65 + 75 = 140$ miles. Since they need to go 210 miles, they will meet after $210 \div 140 = 1.5$ hours.

A second nonalgebraic solution might look at how far each train goes in 1 minute: $\frac{65}{60}$ miles and $\frac{75}{60}$ miles, or $\frac{65}{60} + \frac{75}{60} = \frac{140}{60}$ miles together. Again, since they need to go 210 miles, the calculation $210 \div \frac{140}{60}$ will give the number of minutes needed. ●

Here is a final example, for which quantitative reasoning is awkward for some solvers.

EXAMPLE 22

▶ Three brothers, Tom (16 years old), Dick (14 years old), and Harry (9 years old), were home alone and hungry. They decided to buy a whole apple pie and eat it. They chipped in their money—Tom, \$4; Dick, \$2; and Harry, \$3—and Tom bought a large pie. They succeeded in eating it all. Tom ate twice as much as Harry, and Dick ate $1\frac{1}{2}$ times as much as Harry. What part of the pie did each brother eat? ◀

Solution

Some of the quantities—the brothers' ages (except to justify the unequal shares), the amounts of money chipped in, the total cost of the pie—seem irrelevant to the question. The relevant quantities are the brothers' shares of the pie. These are based on Harry's share (and not the money chipped in), so let x be the fraction of the pie eaten by Harry. Then Tom's fraction is $2x$, and Dick's is $1\frac{1}{2}x$. But how are these related?

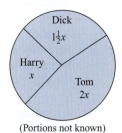

(Portions not known)

The sketch shows it: The three brothers ate the whole pie! So, $x + 2x + 1\frac{1}{2}x = 1$, and hence $4\frac{1}{2}x = 1$, or $x = \frac{2}{9}$.

Harry ate $\frac{2}{9}$ of the pie, Tom ate $\frac{4}{9}$ of the pie, and Dick ate $\frac{3}{9} = \frac{1}{3}$ of the pie. ●

In the Learning Exercises for this section, you will find many problems that can be solved either arithmetically or algebraically. A drawing is usually useful in either case. But for many problems, algebra is required or at least makes the problem easier to solve.

TAKE-AWAY MESSAGE ... Using the Quantitative Analysis Questions, you should be able to determine the quantities and their relationships within a given problem situation, and you should be able to use this information, together with drawings as needed, to solve problems. The following steps are often useful:

1. List the quantities that are essential to the problem.

2. List known values for these quantities.

3. Determine the relationships involved, which is frequently done with the help of a drawing.

4. Use the knowledge of these relationships to write an equation to solve the problem.

This approach may not seem easy at first, but it becomes a powerful tool for understanding problem situations. When used with elementary school students, this focus on quantities prepares them for algebra story problems.

Learning Exercises for Section 12.5

Solve each problem. Drawings are recommended, and the use of algebra is encouraged. If you use algebra, make clear what quantity your variable represents.

1. Dan, Lincoln, and Miguel are pooling their cash to see whether they have enough to buy a game. They have a total of $46.80. They noticed that Lincoln has twice as much as Dan and that Miguel has three times as much as Dan. How much does each boy have?

2. Your school's student council held a dance after a basketball game. Tickets were $2 each for students and $3 each for nonstudents. You know that there were 6 nonstudents at the last dance and that, overall, the Student Council made $148. How many students attended the dance?

3. The Jones family takes a car trip every December. Lee wanted to record the miles driven each day for a report. The family drove for 5 days: 316 miles during the second driving day, 280 miles during the third, and 156 during the fourth. But Lee forgot to record the mileage during the first and fifth days. Lee's dad knew that Lee had been studying algebra, so he said, "I kept track, too. We drove a total of 1112 miles. And we drove twice as many miles the fifth day as we did the first day because we got a late start the first day. So now you can figure out the missing mileages." What might Lee's algebraic solution look like?

4. Puzzle problems about coins often appear in algebra books to give students practice in representing situations algebraically.

   a. Beth notices that she has several coins in her dresser drawer, $1.41 in all. She has 3 more nickels than dimes, 16 pennies, and 2 quarters. How many nickels and how many dimes are there? (*Hint*: Notice that you must distinguish between two quantities: the *number* of each type of coin, and the *monetary value* of those coins. If there are n dimes, their value is $10n$ cents.)

   b. Serena notices that she has a lot of change in her purse. She has twice as many dimes as quarters, 6 more nickels than dimes, and 18 pennies. In all, she has $2.13. How many of each type of coin does she have?

   c. Shawna has been collecting her change for some time and has 101 coins in all. She has 6 half-dollars, half as many quarters as dimes, 85¢ in nickels, and 3 times as many pennies as nickels. How much is her collection worth? (*Hint*: First find out how many coins of each kind she has.)

   d. Make up your own coin problem.

5. You are measuring a rectangular room for painting. You notice that the room is exactly twice as long as it is wide and that the total distance around the room is 87 feet. What are the dimensions (the length and width) of the room?

6. A town has been using a recycling plan for three years. The second year of the plan saw an increase in recycled tonnage of 25% over the first year, and the third year saw an increase of 20% over the second year. The mayor proudly claims, "During those three years, we have saved 41,250 tons of trash from being put into our landfill." How many tons were recycled each of the three years? (*Hint*: For the third year, you will probably want an expression for 120% of the expression for the second year.)

7. One type of problem often found in algebra texts is called a "work problem." Here is one:

   ▶ If Thuy can paint her family room in 6 hours and her husband can paint it in 8 hours, give a good estimate of how long it should take to paint the room if they worked together (under ideal circumstances). ◀

   Give a good estimate based on a diagram. Give an exact answer using algebra.

8. Callie left home at 7:30 AM to bike 8 miles to the school where she taught. Luckily, she could ride on a bike trail where she averaged 12 miles per hour. But at 7:40 her husband Jeff noticed that she had forgotten to take her graded exam papers, which he knew she wanted to return that day. So he hopped on his bike to catch up with her. If he averaged 16 miles per hour, did he catch her before she arrived at school? If so, what time did he catch up with her? (*Hint*: How far could each ride in 1 minute?)

9. Officials from two cities, Allswell and Bestburg, are talking about their anticipated tax revenue for the coming year. The Allswell official says, "Our tax revenue will be up 20%, or $15,000,000, over this year's." The Bestburg official says, "Our revenue will be up by 25%, and even then we will have $10,000,000 less than you do." What is this year's tax revenue for Bestburg?

10. Dollie decides to drink just water after finding out that she was daily consuming too much caffeine, more than 330 milligrams a day. On a typical day, she drank a cup of coffee (120 milligrams of caffeine) and 2 cans of diet cola in the morning (each containing 42 milligrams of caffeine) and more cans of the same diet cola after noon. How many cans of diet cola did she usually drink after noon?

11. Your car repairs cost $548.95, for parts and labor. The parts cost $338.95 and labor was $35 an hour. How many hours of labor did the repairs take?

12. You joined a neighborhood health club. The one-time initiation fee is $50, and the monthly fee is $30. You also wanted one package of personal fitness consultations at $100 for four 25-minute sessions. You wrote a check for $240. How many months have you paid for?

13. These problems are from a 1924 book[5] of arithmetic story problems. (Capitalization was used more frequently in 1924, as this original text shows.) They are much easier using algebra ideas, like arithmetic sequences.

   a. "Ignatius Trott had, at the age of 7 years, a Conscience able to support a weight of 22 lb., 4 oz. The weight which it could carry increased every year thereafter 16 lb., 7 oz. What weight can his Conscience support at his present age, $47\frac{1}{2}$ years?" (p. 1). (If you have forgotten, 1 pound or 1 lb = 16 ounces or 16 oz. *Advice*: Work with fractions.)

   b. "A Boy and his Sister, ages 7 and 8, are eating Watermelons. Working jointly, it takes them 5 minutes to consume the first Melon, 10 minutes the second, 15 minutes the third, etc., etc. What will be the number of the Melon which will require 1 hour and 35 minutes for its consumption?" (p. 6).

14. Gabby and Helen try to stay in shape by working out on an exercise bicycle. Each woman sets a target for a Monday through Friday week.

   a. One week Gabby biked the following times:

   > Tuesday: 5 minutes more than on Monday
   > Wednesday: twice as long as on Tuesday
   > Thursday: twice as long as on Monday
   > Friday: same time as on Monday

   Gabby met her target of 2 hours for that week. How many minutes did Gabby bike each day?

   b. Helen sets her goal in terms of *miles* for the week, 50 miles.

   > On Monday, she biked 4 miles.
   > On Wednesday, she biked 1 less mile than on Tuesday.
   > On Thursday, she biked three times as far as on Tuesday.
   > On Friday, she biked twice as far as on Wednesday.

   Helen also met her goal. How many miles did she bike each day?

Notes

15. "Our town's budget of $19.8 million is up 20% over last year's budget, and last year's budget was 10% more than the previous year's! We taxpayers are angry." What were the town's dollar budgets for last year and the year before that?

Supplementary Learning Exercises for Section 12.5

The parts in each numbered problem follow roughly the same theme but may not give completely similar equations.

1. a. Paula and Ruell receive the same weekly allowance. One Saturday after receiving their allowances, Paula spent $5.00 and Ruell spent $2.00. Paula then had only half as much left as Ruell had left. How much was each allowance?

b. Ray and Amir each bought a used car for the same price. Each needed some repairs. The repairs on Ray's car cost an additional $800 and on Amir's, $200. Amir noted, "Ray, my total cost was really only $\frac{5}{6}$ of your total cost." How much did each pay for his used car?

c. Silvia and Mai bought complete outfits for the same price. Silvia returns the fancy $20 belt, and Mai returns her $45 shoes. Mai said, "Silvia, I actually spent only $\frac{3}{4}$ as much as you did." How much did each original outfit cost?

2. a. A (four-member) relay team ran the mile relay. The second runner was 2 seconds slower than the first runner, but the third runner took 3 seconds more than the second runner. The fourth runner ran 4 seconds faster than the second runner. Their total time was 3:59 (3 minutes, 59 seconds). What was the time for each of the four runners?

b. A (four-member) relay team swam the 400-meter relay in a time of 6:53 (6 minutes, 53 seconds). The second swimmer took 2.5 seconds longer than the first swimmer, and coincidentally, the third and fourth swimmers took the same total amount of time as the first two did. What was the time for each of the first two swimmers?

c. In one grand slalom ski race, Bjorn won with a total time of 1:40.68 (1 minute, 40.68 seconds) for two runs, by skiing 1.06 seconds faster on his second run. What was his time for each run?

3. a. Three friends went shopping. Danetta spent $25 more than Elaine spent, but Jan spent $35 less than Danetta spent. If they spent $165 in all, how much did each of the friends spend?

b. The three friends went shopping again. This time Danetta spent $12 less than Jan spent, but Elaine spent twice as much as Danetta spent. They spent $86 in all. How much did each friend spend this time?

c. Make up a shopping problem and solve it algebraically (even though you perhaps made up the problem by thinking at the start of amounts spent by each person).

4. a. You went shopping at three stores and spent $167.80 in all. You spent three times as much at the first store as you spent in all at the other two stores. You spent $8.45 more at the second store than you spent at the third store. How much did you spend at each store?

b. Sarita shopped at four stores and spent a total of $49.65. She spent $15.95 more at the third store than she spent at the first store, and $8.75 less at the fourth store than at the first store. Sarita did not buy anything at the second store. How much did she spend at each store?

c. Make up a shopping problem and solve it algebraically (even though you perhaps made up the problem by thinking at the start of the amounts spent at the different stores).

5. a. Alima went for a $2\frac{1}{2}$-mile walk with her sister Noor, Noor's husband, and their baby. The baby was heavy, so they took turns carrying him. Alima carried the baby only one-third as far as her sister, and Noor's husband carried the baby twice as far as Noor did. How far did each person carry the baby?

   b. Dien, Gia, and Minh took a vacation together, sharing the driving. One day they drove 8 hours and 35 minutes. Minh drove twice as long as Gia did, and 10 minutes longer than Dien did. How long did each man drive that day?

   c. On another day, Dien, Gia, and Minh drove 475 miles. Gia drove twice as far as Dien did, and Minh drove 115 miles. How far did each man drive that day?

6. **a.** A store manager notices that sales for last year were 25% over the previous year. She knows that sales for last year were $840,000. What were the sales in dollars the previous year?

   b. A clothing shop had sales of $180,000 over a two-year period. Sales during the second year were 40% higher than during the first year. What were the sales figures for each of the two years?

   c. Tabatha has owned a small store for three years. The store has grossed a total of $1,200,000 over those years. The gross sales for the second year were 20% more than those for the first year, and sales for the third year were 80% more than those for the first year. What were the gross sales for those three years?

7. **a.** Susan, Rachel, and Steve made 6 dozen cookies. Over a period of days, they ate them all. Susan ate 25% more cookies than Rachel ate, and Steve ate 2 more cookies than Susan ate. How many cookies did each of the three eat?

   b. Rayann made 2 dozen deviled eggs for a family gathering, but only she and her two cousins, Nell and Bell, ate any. Together, they ate $\frac{2}{3}$ of the eggs. Rayann ate one-third as many as Nell ate. Bell ate one more than Rayann ate. How many deviled eggs did each eat?

   c. Anh, Rama, and Amy together ate all of a 1-yard-long licorice whip. Anh ate part of it in 2 minutes at the rate of 4 inches a minute, Rama ate for $1\frac{1}{3}$ minutes at a rate of 1 foot per minute, and Amy finished the whip in 3 minutes. What was Amy's rate of eating, in inches per minute? (*Note:* 1 yard = 3 feet; 1 foot = 12 inches.)

   d. Steve and Susan went on a hike on a hot day, with 2 quarts of water. During the hike, Susan drank a half pint of water more than Steve did. They drank all of the water. How much water did each of them drink? (*Note:* 1 quart = 2 pints.)

8. Rushmore Park has a walking/running track. Carita and Jorge begin walking at the same place at the same time. Carita consistently walks 6 miles per hour and Jorge at 4 miles per hour. After a while, Carita shouts to Jorge, "Jorge, I'm a quarter-mile ahead of you." How many minutes had they been walking when Carita said that?

12.6 Issues for Learning: The National Assessment of Educational Progress and Achievement in Algebra

The National Assessment of Educational Progress (NAEP, usually pronounced "nape") has been recording achievement at the fourth, eighth, and twelfth grades for several decades. Measures of mathematics achievement over the years provide a snapshot of how well the youths of our country are performing. Only a few of the test items are released because, once released, items cannot be used again. The released items, and performance on these items, can be seen at http://nces.ed.gov/nationsreportcard/. All the test items discussed in this section were used in the 2011 testing.

   On the Grade 4 algebra assessment, algebra questions assess students' understanding of many of the topics covered in this chapter. Students must recognize and extend patterns

Notes

and use symbols to represent unknown quantities, e.g. The following test items were used in the Grade 4 algebra assessment.

Test Item 1. Sam folds a piece of paper in half once. There are two sections.

Sam folds the piece of paper in half again. There are four sections.

Sam folds the piece of paper in half again. There are eight sections.

Sam folds the piece of paper in half two more times. Which list shows the number of sections there are each time Sam folds the paper?

(A) 2, 4, 8, 10, 12   **(B)** 2, 4, 8, 12, 24   **(C)** 2, 4, 8, 16, 24   **(D)** 2, 4, 8, 16, 32

Recognizing and extending this pattern were very difficult for fourth-grade students. Only 23% were able to choose D as the correct answer, though this is a doubling pattern.

Test Item 2. Every 30 minutes, Dr. Kim recorded the number of bacteria in a test tube.

Time	1:00 PM	1:30 PM	2:00 PM	2:30 PM
Number of Bacteria	600	1,190	2,390	4,800

Which explanation best describes what happened to the number of bacteria every 30 minutes?

(A) The number of bacteria increased by 500.

(B) The number of bacteria increased by 1,000.

(C) The number of bacteria doubled.

(D) The number of bacteria tripled.

Did you use a calculator on this question?

The doubling pattern in Test Item 2 was still difficult to recognize in a table, but 34% of fourth-graders correctly chose answer C.

Test Item 3. Each of the 18 students in Mr. Hall's class has p pencils. Which expression represents the total number of pencils that Mr. Hall's class has?

(A) $18 + p$ **(B)** $18 - p$ **(C)** $18 \times p$ **(D)** $18 \div p$

Students in Grade 4 are asked to identify an expression that models a context. This was also a difficult question; 35% of the students answered it correctly. Knowing C is the correct answer demonstrates an understanding of the operation of multiplication as repeated addition. A student must also understand how the symbol p can stand for a variety of values.

Students in Grade 8 were generally more capable of identifying which equation was appropriate for a given context. More than half (53%) of the eighth-graders correctly identified C as the answer to Test Item 3.

Test Item 4. Robert has x books. Marie has twice as many books as Robert has. Together, they have 18 books. Which of the following equations can be used to find the number of books that Robert has?

(A) $x + 2 = 18$ **(B)** $x + x + 2 = 18$

(C) $x + 2x = 18$ **(D)** $2x = 18$

(E) $2x + 2x = 18$

Test items pertaining to other areas of algebra are also available. The Grade 8 test covers content usually included in Grades 5–7. Not all algebra test items involve variables. Interpreting graphs is a very important topic. We will explore this in subsequent chapters. The following NAEP item is one such problem. Approximately 70% of the eighth-graders could give the correct answer for this item.

Test Item 5. According to the accompanying graph, between which of the following pairs of interest rates will the *increase* in the number of months to pay off a loan be greatest?

(A) 7% and 9%   **(B)** 9% and 11%   **(C)** 11% and 13%

(D) 13% and 15%   **(E)** 15% and 17%

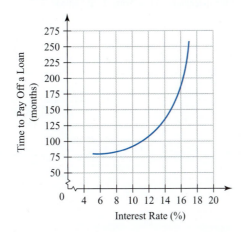

There are, of course, many other items on the NAEP test. NAEP results have become more important in the last decade because they provide not only national results but also individual state results. Hence, they serve as a way of comparing progress at the national and state levels, from one test administration to the next. Thus, it is good to be acquainted with the types of items appearing on the NAEP test and perhaps to try some of them in your own classrooms.

12.7 Check Yourself

After studying this chapter, you should be able to do problems like those assigned and to meet the following objectives:

1. Name and use the properties involved in given numerical or algebraic work.

2. Evaluate algebraic expressions and solve simple equations.

3. Give parallel numerical and algebraic calculations, and point out how they are alike.

4. Calculate the sum and product of two polynomials.

5. Illustrate, identify, and work with arithmetic sequences and geometric sequences.

6. Find a specific number, or an expression for the *n*th number (or a function rule), in a given pattern or table. Generate your own data for some situations.

7. Explain what a function is and why functions are important in mathematics.

8. Explain why a function rule based only on examples might not be 100% reliable.

9. For selected situations, find a general function rule and give a justification that it is 100% reliable.

10. Solve a story problem using algebra (and a drawing).

References for Chapter 12

[1]Conference Board of Mathematical Sciences. (2000). *The mathematical education of teachers.* Providence, RI: American Mathematical Society, p. 108.

[2]Russell, S. J., Schifter, D., & Bastable, V. (2011). *Connecting arithmetic to algebra: Strategies for building algebraic thinking in the elementary grades.* Portsmouth, NH: Heinemann.

[3]Schliemann, A. D., Araujo, C., Cassundé, M. A., Macedo, S., & Nicéas, L. (1998). Use of multiplicative commutativity by school children and street sellers. *Journal for Research in Mathematics Education, 29*, 422–435.

[4]Sowder, L. (1988). Children's solutions of story problems. *Journal of Mathematical Behavior, 7*, 227–238.

[5]Weeks, R. (1924). *Boys' own arithmetic.* New York: E. P. Dutton.

Chapter 13

A Quantitative Approach to Algebra and Graphing

In Section 12.5 you learned that words like "quantity" and "value" are a part of everyday language, but giving careful meanings to these words yields important advantages. This chapter focuses on how quantities are related and on generating representations (algebraic, graphical, and numerical) of relationships. These representations can help us understand how quantities are changing.

13.1 Using Graphs and Algebra to Show Quantitative Relationships

A quantitative approach to algebra starts by examining how quantities are related in situations. Algebra is then used as *language* to describe quantitative *relationships*. Furthermore, we have different ways to represent quantitative relationships: *Tables* of *ordered pairs* can be useful in identifying patterns in these relationships; *story problems* describe the context; *algebraic equations* can help us express these relationships generally and are sometimes used to make predictions; and *graphs* are used to represent or describe quantitative relationships in a pictorial manner to examine changes in the quantities. Thus, a quantitative approach underlies the topics of algebra, representations, and change.

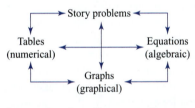

Two quantities are related when each value of one quantity corresponds to a value or values of another quantity. For example, the number of calories consumed while eating M&M's is related to the number of M&M's eaten. These two quantities are related predictably—as the number of M&M's eaten varies, the number of calories consumed varies in a systematic way. In contrast, although the money you save shopping by using coupons increases as the number of coupons increases, the quantities are not related predictably because coupons have different values.

Activity 1 illustrates how it is possible to begin algebraic instruction by examining quantitative relationships *before* any symbols, such as *x*'s, *y*'s, and *z*'s, are introduced. After you gain experience identifying quantitative relationships in this section, you will use tables, algebraic expressions, and graphs in particular contexts to study and represent changes in quantities.

ACTIVITY 1 Identifying Quantitative Relationships

1. Determine whether the following pairs of quantities are related to each other. If so, explain whether the value of the quantity given on the right increases or decreases as the value of the corresponding quantity on the left increases.

Notes

a.	Number of minutes that have passed while driving a car at 65 mi/h	Amount of fuel left in the gas tank
b.	Number of minutes that have passed while driving a car at 65 mi/h	Amount of fuel you have used
c.	Length of one side of an equilateral triangle	Perimeter of the triangle
d.	Perimeter of a rectangle	Area of the rectangle
e.	Number of potato chips eaten	Number of calories consumed
f.	Amount of money inserted into a candy vending machine	Number of candy bars that you receive

2. For each quantity, name another quantity that is related to it.

   a. the amount of money you make from selling lemonade

   b. the amount of fuel you use on a car trip

   c. the total amount of sales tax that you have to pay for a purchase

3. The following story was written by an elementary school student to illustrate a situation in which quantities are related:

▶ A girl has a bag of flour. Her dog bites a hole in it. The more she walks, the more flour comes out. ◀

   a. Name two quantities in this situation that are related. How are they related? That is, do both quantities increase, or does one quantity decrease while the other increases?

   b. Now name a different pair of related quantities in the situation. How are they related? ●

Some people might think that a quantitative approach to algebra differs from a traditional numerical approach primarily because the quantitative approach uses real-world situations. However, this distinction is not the major difference. Both approaches can use real-world situations, but they treat the situations differently. Consider a situation in which a car is traveling at a speed of 55 mi/h. In a numerical approach to algebra, the formula $d = rt$ can be given (where d = distance traveled in miles; r = rate or speed in miles per hour; and t = time traveled in hours), and students are asked to find either the distance traveled if a car travels for 2.5 hours at 55 mi/h or the time the car takes to travel 300 miles. A numerical approach emphasizes solving for an unknown value by substituting known values for some of the algebraic symbols.

   Although solving for an unknown may be part of a quantitative approach to algebra, a quantitative approach also explores how quantities like distance and time are related. For example, changes in the amount of time that has passed can produce changes in the distance traveled. Furthermore, this relationship is systematic (i.e., exhibits a pattern): If you travel 1 hour at an average speed of 55 mph, you go 55 miles. If you travel 5 hours, which is 5 times as long, you will go 5 times as far, or 275 miles (assuming an average speed of 55 mi/h). If you travel 20 minutes, which is one-third the original time, you will travel one-third the distance, or approximately 18 miles.

   Teaching arithmetic or algebra with a quantitative approach means helping your students get a feeling for how quantities are related. You might explore quantitative relationships without even referring to numbers. For example, Learning Exercise 4 in this section illustrates a rich mathematical activity (one appropriate for upper elementary students) in which the challenge is to figure out which quantities are related (without determining a specific numerical relationship).

Think About …

What do you remember about algebraic equations and graphs on coordinate systems? How are equations and graphs connected? What does something like (4, ⁻2) mean?

To understand a quantitative situation, we often represent it with a graph. Graphs in algebra are constructed according to certain conventions. For example, if there are two variables, each is represented on a separate number line scale, called an **axis**, and labeled to make clear the quantity and the scale unit.

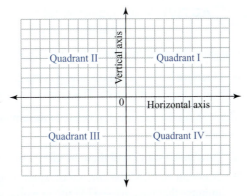

The two number lines, or **axes**, are most often placed at a right angle with their zero points coinciding; the scale units on each axis are uniformly spaced. Such axes give a **coordinate system**. There are four quadrants in a coordinate system. In the first quadrant, the values of both variables are positive; in the second quadrant, the values on the horizontal axis are negative and on the vertical axis they are positive. In the third quadrant, the values on both axes are negative. In the fourth quadrant, the values on the horizontal axis are positive and on the vertical axis they are negative.

Sometimes, we think of a graph as a picture. However, a graph as a picture is a very specialized type of picture. It shows how quantities are related. Graphs reveal particular information that a verbal description of a relationship may not. At the same time, a graph may hide or downplay other information that is highlighted by other representations, such as words or algebraic equations. In elementary school, the situations described by graphs are most often in the first quadrant because the variables involved in the situation can only have positive values.

ACTIVITY 2   A Picture Is Worth a Thousand Words

Consider the following situation as described by an elementary school student.

▶ A block of ice sits in a tub in the sun. The block of ice consists of 16 centiliters of water that have been frozen. The block of ice melts at a rate of 3 centiliters a minute. ◀

1. Describe in words how these two quantities are related: the size of the block of ice and time.

2. Construct a set of axes and plot points to show the relationship between the amount of ice in the block of ice and the number of minutes that have passed. Be sure to label each axis with the appropriate quantity. ●

DISCUSSION 1   Whose Graph, If Any, Is Best?

1. On the board, put every graph from Activity 2 that differs from others in some way. Then compare them. Are there any graphs that do not make sense? If so, why not? Are there any pairs of graphs that look different from each other but that both make sense? Does there have to be only one correct graph? Explain.

2. Does it matter which quantity is represented on the horizontal axis? (*Hint*: Try graphing the relationship twice, switching the quantities that are represented on the horizontal axis, and then comparing the two graphs. What are the advantages and disadvantages of each graph?)

3. How does each graph indicate that one quantity is increasing while the other is decreasing?

4. Why does it make sense to connect the points in this graph? (*Hint*: Does it make sense to ask how much ice is left after 2.5 minutes?) ●

Although different choices for what to show on each axis can generate different-looking graphs, they *literally* contain the same information. But *visually*, there are advantages in following some conventions. For example, having the scales increase as one goes up or to the right is extremely common. When time is involved, placing time on the horizontal axis allows you to see whether the other variable quantity increases or decreases as time goes on.

ACTIVITY 3   Using Algebraic Symbols to Represent Quantitative Relationships

Consider this modification of the melting ice block situation:

▶ A block of ice sits on the sidewalk in the sun. The block of ice consists of 100 centiliters of water that have been frozen. The block of ice melts at a rate of 2.5 centiliters a minute. ◀

Think about the relationship between the amount of time and the amount of water accumulating on the sidewalk.

a. How much water will there be after 4 minutes? 8 minutes? 20 minutes?

b. How can you find the amount of water on the sidewalk if you know the number of minutes that have passed? Write in words a description of your strategy.

c. If T represents the number of minutes that have passed and W represents the amount of water (in centiliters) that has accumulated on the sidewalk, write an equation that expresses the relationship between the two quantities.

d. How can you check your equation from part (c)? ●

If we now think about the relationship between the *amount of time* and the *amount of ice*, how much ice will be left in the block after 4 minutes? 8 minutes? 20 minutes? How can you find the amount of ice given the number of minutes that have passed? It can be helpful to complete a table that indicates the arithmetic you are performing.

Time (min)	Amount of ice (cl)
0	100
1	$100 - 2.5 = 97.5$
2	$(100 - 2.5) - 2.5 = 95$
3	$(100 - 2.5 - 2.5) - 2.5 = 92.5$
4	
5	
6	

What pattern do you notice? One could describe this pattern in words similar to these: "To find the amount of ice, start with 100 centiliters and subtract 2.5 centiliters for each minute that has passed." This can also be expressed using the language of algebra. Using algebraic symbols to describe a pattern is a very powerful tool for solving problems.

 ACTIVITY 4   Using Algebraic Symbols

a. If T represents the number of minutes that have passed and I represents the amount of ice (in centiliters) left in the block, write an equation relating the two quantities.

b. How can you check your equation from part (a)?

c. When will the block of ice be gone? How do you know?

d. For part (a), someone wrote $I + 2.5T = 100$. What was the person thinking? Is this equation all right? ●

 DISCUSSION 2   Understanding What Graphs Convey

Draw a graph of the relationship between the number of minutes that have passed and the amount of ice left in the ice block.

a. What quantity did you represent on the horizontal axis and what quantity did you represent on the vertical axis? How can this choice reveal important relationships?

continue

b. How did you decide on the scale unit? How might the picture of the situation change if the scale unit were different?

c. Where on the graph does it indicate when the block of ice will be gone? ●

As we said at the beginning of the section, relationships among quantities can be represented in many different ways. Graphs can provide a specialized picture of how quantities are changing with respect to each other. Tables can help us see patterns so that algebraic symbols can provide a powerful mechanism for solving problems. In the next section, we will look more deeply into the relationships between different representations, e.g., between algebraic equations and graphs.

TAKE-AWAY MESSAGE . . . Rather than focusing on letters like x and y, a quantitative approach focuses on specific quantities involved and how they are related. An understanding of the quantities in a given graph can give a "picture" of how one quantity changes as another quantity changes. For constructing a coordinate-system graph, graphing conventions (such as uniform spacing in a scale on an axis or having the horizontal scale increase to the right and the vertical scale increase to the top) make it easier to grasp the relationship between the quantities involved. Understanding how the quantities are related may enable you to write tables, find patterns, and write an algebraic equation for the situation.

Learning Exercises for Section 13.1

1. Elementary school students wrote the following stories. For each situation, write a sentence that shows how two quantities in the situation are related. It is possible that you can describe several relationships in each situation.

   Example: A large block of ice is in a tub in the sun. The ice melts 3 centiliters a minute.

   Solution: Here are two possible sentences: The amount of ice decreases as the number of minutes that pass increases. As the number of minutes increases, the amount of water in the tub increases.

   a. An airplane is flying miles high over Siberia, hundreds of miles from any help. Suddenly, the fuel tank ruptures and fuel starts steadily leaking out.
   b. A motor home has a water tank that holds a large amount of water. Each shower takes 5 gallons of water.
   c. You have $5 and you buy candy bars at 50¢ each.
   d. A class starts off with 16 students. Two students stop coming every week.

2. **a.** Write your own story similar to those that the students wrote in Learning Exercise 1.

   b. Write a sentence that shows how two quantities in the situation in part (a) are related. There may be several quantitative relationships that you can describe.

3. Determine whether the following pairs of quantities are related to one another. If they are, explain whether the value of the second quantity increases or decreases as the value of the first quantity increases. (*Problem-Solving Tip*: Try some examples and see what happens.)

   a. the perimeter of a square and the area of the square
   b. the base of a triangle and the area of the triangle
   c. the value of the numerator in a fraction (when the denominator stays the same) and the value of the fraction
   d. the value of the denominator in a fraction (when the numerator stays the same) and the value of the fraction

4. A pendulum gives a good place to look for relationships. You will conduct three experiments to determine what quantities affect the number of times a pendulum swings in a certain amount of time. You will need the following materials: string; several heavy

objects like washers, nuts, bolts, screws, or fishing sinkers; tape; and a watch with a second hand.

a. Make a pendulum by tying a heavy object (such as a washer) at the end of a string (Figure 1). Measure the number of swings a pendulum makes in 10 seconds. One swing is the distance the washer travels from the far right to the far left.

Figure 1 Making a pendulum from string and a weight.

b. Conduct an experiment to see if the weight of the pendulum bob (the object at the end of the string) affects the number of times the pendulum swings in 10 seconds. You will need to decide how to conduct the experiment. Then collect and record your data in a table. Finally, write a sentence expressing your conclusion.

c. Conduct an experiment to see if the length of the string affects the number of times the pendulum swings in 10 seconds. Decide how to conduct the experiment. Then collect and record your data in a table. Finally, write a sentence expressing your conclusion.

d. Conduct an experiment to see if letting the pendulum go from different positions affects the number of times a pendulum swings in 10 seconds (see Figure 2). Again, decide how to conduct the experiment, collect and record data, and write a sentence expressing your conclusion.

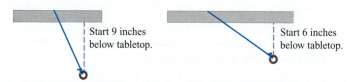

Start 9 inches below tabletop.

Start 6 inches below tabletop.

Figure 2 Letting a pendulum start from different positions.

5. Determine a pair of quantities and write a sentence that shows how the quantities are related to one another. Your sentence should explain whether the value of the second quantity increases, decreases, or is unaffected as the value of the first quantity increases.

a. Joella is going to participate in a run to raise money for cancer research. The company Joella works for has agreed to contribute $100.

b. Joella runs the same distance for the cancer fund-raiser as she did last year but faster than she did last year.

c. Bianca is going to participate in a run to raise money for cancer research. Bianca's track coach has agreed to deduct 5 minutes off Bianca's practice time for every mile she runs.

d. A group of college students arranges to rent a house at the beach for the weekend. The house costs $600 to rent and the students agree to split the cost evenly.

e. Corporations spend $30 million for a 30-second advertisement during the Super Bowl.

6. *Help! Something's Leaking, Part 1* Consider the following scenario, written by a student:

▶ An airplane is flying high over Siberia, hundreds of miles from any help. Suddenly, the fuel tank starts leaking. The tank contained 80 gallons of fuel when the tank started leaking and is leaking at a rate of 10 gallons each minute. ◀

a. How does the amount of fuel in the tank change as time passes?

b. Complete the following table of data that describe the relationship between the amount of fuel in the tank and the time spent flying since the rupture:

Time (min)	0	1	2	3	4		
Amount of fuel in tank (gal)	80	70					

c. How can you find the amount of fuel in the tank if you know the number of minutes that have passed?

d. Let T represent the number of minutes that have passed and G represent the amount of fuel in the tank. Write an equation that represents the relationship between the two quantities.

e. How can you check to see whether your equation from part (d) represents the data?

f. When will the fuel tank be empty? How do you know?

g. Graph the relationship between the amount of fuel in the tank and the time spent flying since the fuel leak started. (Save this graph; it will be referred to in Learning Exercise 2 in Section 13.2.)

h. One person wrote the equation $G + 10T = 80$ for part (d). What was this person thinking? Does this equation work?

7. *Help! Something's Leaking, Part 2* Consider the scenario in Learning Exercise 6. However, this time think about the relationship between the amount of fuel that has leaked (rather than the amount of fuel in the tank) and the time that has passed since the rupture in the tank.

a. Describe in words how the amount of fuel that has leaked changes over time.

b. Create a table of data to describe the relationship between the amount of fuel that has leaked and time.

c. Use algebraic symbols to describe the relationship between the amount of fuel that has leaked and time. Be sure to label your variables.

d. Graph the relationship between the amount of fuel that has leaked and the time spent flying. Be sure to label the quantities on each axis. (Save this graph; it is referred to in Learning Exercise 2 in Section 13.2.)

e. How do the graphs from Learning Exercises 6(g) and 7(d) differ? What does this difference say about the quantitative relationships?

f. At what rate is the fuel leaking? How can you tell using each graph? [See Learning Exercises 6(g) and 7(d).]

8. Geometric relationships can result in interesting graphs.

a. Graph the relationship between the length of the side of a square and the perimeter of the square. You may first want to generate a table of data, using several squares.

b. Graph the relationship between the length of the side of a square and the area of the square.

c. Compare the two graphs. How are they alike? How are they different?

d. Let s represent the number of length units in a side of a square, p the number of units in the perimeter of the square, and A the number of units in the area of the square. Write an equation to represent the relationship between the length of the side of a square and the perimeter of the square. How can you check to make sure your equation is correct? Now write an equation to represent the relationship between the length of the side of a square and the area of the square. Compare the two equations. How are they alike? How are they different? How are the differences related to differences between the two graphs?

9. *Real-World Relationships* Describe a quantitative relationship that occurs in daily life and then apply this situation to parts (a–d).

a. Identify two quantities in the situation you created and describe how the quantities are related.

b. Represent the relationship using a table of data (make up reasonable numbers if you do not have exact ones).

continue

    c. Represent the relationship by creating a graph. Be sure to label the quantity represented by each axis.

    d. Try to write an algebraic equation for the relationship, even though doing this might not be easy. Be sure to state the number that each algebraic symbol represents.

10. Sharing an object with others generates an interesting graph.

    a. Suppose you are sharing a pizza fairly among several people. Create a table of data that shows what fraction each person will receive for different numbers of people.

    b. Graph the relationship between the number of people sharing the pizza and the size of the piece each receives.

    c. Does it make sense to connect the points in your graph? Why or why not?

    d. Does the size of each pizza piece increase or decrease as the number of people sharing the pizza increases?

    e. Why is the shape of your graph a curve instead of a line?

    f. What, if anything, would change if the people were sharing 2 pizzas? 6 pizzas? $\frac{3}{4}$ pizza?

11. An experiment can suggest an important relationship in a circle.

    a. Locate 3 round objects of different sizes (e.g., a coffee can or a teacup) and trace around the base of each object. Alternately, use a compass to draw 3 circles of different sizes.

    b. Look at your 3 circles. As the diameters increase, what happens to the circumferences?

    c. Measure the diameter of each circle using a ruler. Measure the circumference of each circle by first placing string carefully around the rim of the circle and then measuring the string with a ruler. Record your data in a table.

    d. Graph the relationship between the diameters and the circumferences.

    e. About how many times as large is the circumference as the diameter for each circle?

    f. How can you get the information requested in part (e) above from the graph?

    g. Does it make sense to connect the points on your graph? Why or why not?

    h. Is your graph a line or a curve? Why do you think so?

    i. Let c represent the length of the circumference (in inches or centimeters) and d the length of the diameter (in inches or centimeters). Write an equation to show the relationship between the circumference and the diameter of a circle.

12. a. PharmImpact provides consulting on health plans. The company began with 30 employees but is growing rapidly. If the company hires 5 employees per month for 1 year, then 10 per month after the first year, how long will it take to increase the total number of personnel to 300?

    b. Draw a graph that describes this information.

Supplementary Learning Exercises for Section 13.1

1. For each situation, write a sentence that shows how two quantities in the situation are related. It is possible that you can describe several relationships in each situation.

    a. Joella is going to participate in a run to raise money for cancer research. Her sister pledges 50¢ for every mile that Joella runs.

    b. For each mile Joella runs, she has total pledges of $12 per mile, all to be matched by an anonymous donor.

    c. You are making lemonade for a faculty picnic. The juice of a dozen Meyer lemons, $\frac{3}{4}$ cup of sugar, and 14 cups of water are needed to make 1 gallon of lemonade.

2. Determine whether the quantities in each pair are related. (Use common sense and your own knowledge.) If the quantities are related, describe whether the value of the first quantity increases or decreases as the value of the second quantity increases.

    a. the number of miles traveled and the cost of an airline ticket

    b. the number of cans of soda bought for a party and the number of party guests

    c. the number of pounds of ground beef bought and the cost per pound of ground beef

    d. the radius of a circle and the diameter of a circle

continue

    e. the base and height of a rectangle

    f. the height and width of a set of rectangles that all have an area of 100 square inches

    g. the amount of gasoline used and the number of miles driven on the highway

3. Light travels at 300,000,000 m/s but sound travels only 343 m/s. Although light and sound originate at the same time, there is a time lag between the time you see a lightning bolt and the time you hear the thunder because of the difference in the speeds.

    a. If there is a 5-second time lag between the time you see lightning and the time you hear the accompanying thunder, how far away is the lightning?

    b. Write an equation that gives the distance the lightning is from you at the time you hear the thunder.

4. a. The Sparrow family lost their home to a fire, and the town where they live is sponsoring a series of fund-raisers to help them out. One fund-raising activity is a performance by a local band, the Pirates, with all proceeds going to the Sparrows. Tickets cost $40. At the concert, there is a raffle for three Visa debit cards, each worth $100. Raffle tickets cost $10 each. Describe two quantitative relationships suggested in this situation and write an algebraic expression for each. Is there a direct relationship between the number of people at the concert and the total amount of money raised?

    b. Another fund-raising activity is a pancake breakfast for a cost of $10 per adult and $5 per child. Two local grocers are providing all the pancake mix and the eggs. Another grocer is providing coffee, juice, and milk. A third store is providing free paper goods. The senior class at the local high school is preparing and serving the food. Describe as many quantitative relationships as you can that are suggested in this situation and write an algebraic expression for each.

5. a. A large new condo complex is being constructed. It will eventually have 750 units. The owner is seeking buyers while he constructs. His plan is to build 20 units the first month, to be sold during the second month, then 30 new units over each of the next three months, with the units sold in each of the following months, and 40 new units per month after that, each to be sold within the following month. How long will it take the owner to complete the complex and sell the units, if his plan is successful?

    b. Draw a graph that describes the total sold per month. Is the graph a straight line? If not, why not?

13.2 Understanding Slope: Making Connections Across Quantitative Situations, Graphs, and Algebraic Equations

Two quantities are related when each value of one quantity corresponds to a value or values of another quantity. When one quantity increases, another can increase, decrease, or stay the same. Sometimes, these quantities change in predictable ways. We can talk about the rate at which these quantities change. When the *rate of change* is constant, the pattern describes a line, and we call this rate of change the **slope** of the line. In this section we determine the slope of a line from ordered pairs in a table, or we read it from a graph of the line.

    Recall the following scenario from Activities 3 and 4 in Section 13.1:

▶ A block of ice sits on the sidewalk in the sun. The block of ice consists of 100 centiliters of water that have been frozen. The block of ice melts at a rate of 2.5 centiliters a minute. ◀

When we focused on the quantitative relationship between the number of minutes that have passed and the amount of water left on the frozen block, we described a pattern that changed in predictable ways. This pattern indicated that the rate of change in the amount of water frozen in the block with respect to time is a *decrease* of 2.5 centiliters each minute, as shown in the table on the next page.

Notes

Time (min)	0	1	2	3	4	5	6
Amount of water frozen in block (cl)	100	97.5	95	92.5	90	87.5	85

$^{-}2.5 \quad ^{-}2.5 \quad ^{-}2.5$

The rate of change in the amount of water frozen in the block is constant—the change in the amount of water between any two consecutive minutes is the same. The graph that you drew in Discussion 2 is a straight line, as is any quantitative relationship in which the rate of change is constant.

The algebraic equation you wrote to express this quantitative relationship is

$$I = 100 - 2.5T$$

where T represents the number of minutes that have passed and I represents the amount of ice that is left in the block. When the equation is written in this familiar format, you can see the initial value is 100 and the rate of change is -2.5. We call this rate of change the *slope* of the line in the graph.

We sometimes designate algebraic equations whose graphs are lines with the general equation: $y = mx + b$. The letter x stands for the values of the quantity on the horizontal axis. We say that it is the *independent* variable. The letter y stands for the values of the quantity on the vertical axis. We say this is the *dependent* variable because the quantity on the y-axis is said to depend on the quantity on the x-axis. In the melting ice example,

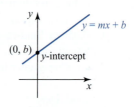

the amount of water left frozen depends on the number of minutes. In the general equation, we designate the slope using the variable m. You may also recall that the point at which the graph of a line crosses the y-axis is called the *y-intercept* and is designated by b. The *y-intercept* has the value of the dependent variable when the value of the independent variable is zero. Any equation of the form $y = mx + b$ has a line associated with it.

In what follows, we'll review computing the slope of the line, attach meaning to the formula, and investigate what the slope means.

 ACTIVITY 5 The Slippery Slope

The following work is one prospective elementary teacher's response to Learning Exercise 9 from Section 13.1. The amount of fuel your car uses depends on how far you have driven.

miles	gallons
0	0
20	1
40	2
60	3
80	4
100	5

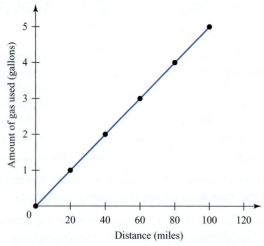

1. In high school, many of you learned to compute the slope of a line with a formula. For two known points (x_1, y_1) and (x_2, y_2),

$$\text{slope} = \frac{\text{change in } y \text{ values}}{\text{change in } x \text{ values}} = \frac{(y_2 - y_1)}{(x_2 - x_1)}$$

continue

    a. Compute the slope of the line shown in the given graph using (0, 0) and (20, 1).

    b. What does the slope that you calculated in part (a) mean in terms of the situation with gas?

2. Make a new graph of the same data. This time let the horizontal axis represent the amount of gas and let the vertical axis represent the distance.

    Suppose your friend used the points (1, 20) and (3, 60) and then correctly used the slope formula to calculate the slope as

$$\frac{60 - 20}{3 - 1} = \frac{40}{2} = 20$$

    a. What does the slope of 20 mean in this context?

    b. Explain why subtracting $60 - 20$ makes sense in this situation. What does this difference tell you and why would you want to find this difference?

    c. Explain why subtracting $3 - 1$ makes sense in this situation. What does the difference tell you and why do you want to find this difference?

    d. Explain why division $(40 \div 2)$ makes sense in this situation.

    e. Are the slopes from Problems 1(a) and 2(a) the same or different? Why? ●

 DISCUSSION 3   Slope by Any Other Name . . .

1. In Activity 5, Problem 2, would it have mattered if your friend had chosen two different points to calculate the slope, such as (0, 0) and (5, 100) or (1, 20) and (2, 40)? Why or why not?

2. Why is the slope sometimes referred to as "rise over run"? ●

Activity 5 suggests an interpretation of slope. The slope gives the change in the value of the quantity on the vertical axis for every one-unit change in the value of the quantity graphed on the horizontal axis. We see the connections among graphs, equations, and the stories that describe the relationships.

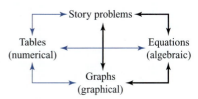

 ACTIVITY 6   Phone Call Plan

The cost of a phone call through this phone company depends predictably on the number of minutes you talk. The graph shown, which represents this relationship, is a line.

1. What is the slope of this line?

2. What does the slope mean in this context?

3. Make a table of points from the graph.

4. Let t represent the number of minutes you are engaged in phone calls. Let c represent the cost of the calls. Write an algebraic equation that describes the relationship between the time in minutes and the cost of the calls. Remember that there is an up-front charge of $2.50.

5. How is the rate of change reflected in the equation? ●

Though the cost of the phone calls is affected by the initial cost ($2.50 in this case), the rate of change is not. If the initial cost were $3.50 but the rate of change in the cost of the phone call remained the same, the resulting graph would be a line parallel to the one pictured in Activity 6. You should convince yourself that this is true by constructing a table and drawing a graph of a phone call plan with an initial cost of $3.50 at zero minutes and the same rate of change as the plan in Activity 6.

 **ACTIVITY 7**   Is the Cost of Phone Calls Escalating?

Redraw the graph of the phone call cost given in Activity 6 by allowing each unit on the horizontal axis to equal 1 minute instead of $\frac{1}{2}$ minute.

1. What is the slope of this line? Compare the slope computed here with the slope computed in Activity 6.

2. Does changing the scale alter the slope? Why or why not?

3. Does changing the scale alter the *look* of the graph? Is the fact that one line is steeper-looking reflected in the slope measurement? Explain.

4. Can you tell from the algebraic equation that one line looks steeper? ●

This introduction to slope from a Grade 6 resource focuses on how to calculate the slope, rather than the meaning of slope as the rate of change.

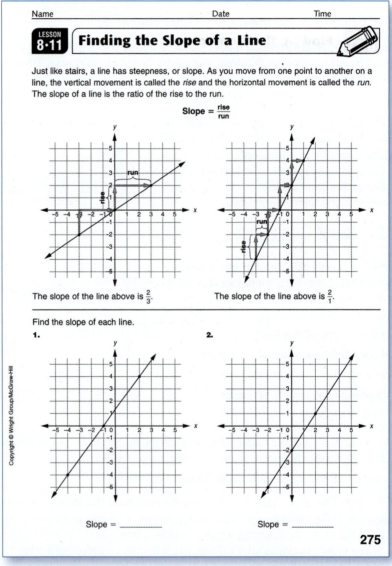

Name _____   Date _____   Time _____

LESSON 8·11 **Finding the Slope of a Line**

Just like stairs, a line has steepness, or slope. As you move from one point to another on a line, the vertical movement is called the *rise* and the horizontal movement is called the *run*. The slope of a line is the ratio of the rise to the run.

Slope = $\frac{\text{rise}}{\text{run}}$

The slope of the line above is $\frac{2}{3}$.

The slope of the line above is $\frac{2}{1}$.

Find the slope of each line.

1.

2.

Slope = _____

Slope = _____

275

Wright Group, *Everyday Mathematics, Math Masters*, Grade 6, p. 275. Copyright © 2008, McGraw-Hill. Reproduced with permission of the McGraw-Hill Companies.

You probably have heard and have used language that suggests that the steeper the line, the greater the slope. Although this is common in conversations about graphs on the same co-ordinate system, as a teacher, you will want to think carefully about what this language might suggest to students.

Think About ...

Think of a situation for which the rate of change is zero. What does the graph of this line look like?

TAKE-AWAY MESSAGE . . . With a little practice, slope is easy to calculate by finding the *rise over run*, but understanding what the resulting value means is equally important. The slope refers to the rate of change of the graphed quantities. The value of the slope can appear in some forms of an equation for a line; e.g., in the form $y = mx + b$, we know that m is the slope and $(0, b)$ is the point where the line crosses the y-axis. The visual steepness of a line can be influenced by changing the scales on one or both axes, but the slope and its meaning stay the same. For two points on the graph of a line, the slope gives the change in the values on the vertical axis for every change of one unit in the values on the horizontal axis.

Learning Exercises for Section 13.2

1. **a.** The graph shows a relationship in a bowling situation. Calculate the slope of the line.
   b. What does the slope mean in terms of the bowling situation?
   c. Draw another line on the graph that has the same slope. What does this line represent in terms of the bowling situation?
   d. What is the cost before any games are bowled? What is the y-intercept of the graph? What does the y-intercept mean in terms of the bowling situation?

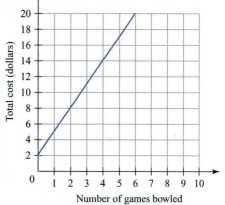

   e. Although such graphs are commonly used to show the linear pattern of data, why is a solid line literally not appropriate here?

2. **Help! Something's Leaking, Part 3** Read again Learning Exercises 6 and 7 from Section 13.1. Both exercises dealt with the situation of an airplane flying over Siberia with a leaky fuel tank.

   a. The graph that you created in Learning Exercise 6(g) showed the amount of fuel in the tank over time. Calculate the slope of this graph.
   b. What does this slope tell you about the situation with the airplane?
   c. The graph that you created in Learning Exercise 7(d) showed the amount of fuel that had leaked over time. Calculate the slope of this graph.
   d. What does the slope in part (c) tell you about the situation with the airplane?
   e. Compare the two slopes in parts (a) and (c). How are they alike and why? How are they different and why? What does a *negative* slope tell you?

3. **Burning Candles, Part 1** The following data were collected as a candle burned. The data show how the height of the candle changed over time.

Number of minutes a candle has burned	0	2	8	12	20
Height of the candle (cm)	15	14	11	9	5

continue

    a. Graph the data. Put the height values of the candle on the vertical axis and the time values on the horizontal axis. Why is it appropriate to join the points and make a solid line in this situation?

    b. Calculate the slope of the line.

    c. What does the slope tell you about the candle?

4. ***Burning Candles, Part 2*** Consider the following graphs for a group of three burning candles. Each line represents a different candle.

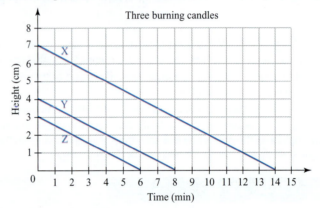

    a. What was the starting height of each candle?

    b. At what rate did each candle burn? How do you know?

    c. What is the slope of each line?

    d. Which of the groups of candles at right could be represented by the graph and why? Which candle goes with which line?

    e. Write an equation to represent each candle's burning. (Make a table of data for each candle first, if it will help.)

    f. How are the three equations in part (e) alike? How are they different?

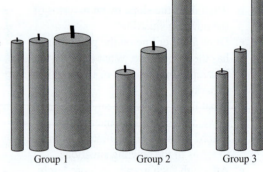

Group 1   Group 2   Group 3

    g. Write an equation for a fourth candle, Candle N, whose graph is parallel to the graphs of Candles X, Y, and Z.

    h. Sketch the graph of your equation for Candle N on the same grid with the graphs for Candles X, Y, and Z.

    i. Describe the candle (Candle N) that you've represented by the graph and equation in parts (g) and (h).

5. ***Burning Candles, Part 3*** Consider the following graphs for another group of three burning candles. Each line represents a different candle.

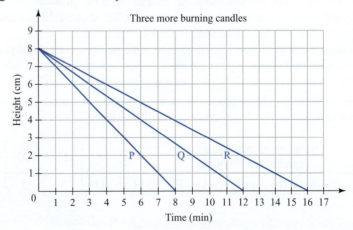

continue

Notes

a. What was the starting height of each candle?

b. At what rate did each candle burn? How do you know?

c. What is the slope of each line?

d. Which of the groups of candles at right could be represented by the graph and why? Which candle is Candle P? Candle Q? Candle R?

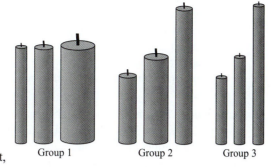

Group 1   Group 2   Group 3

e. Write an equation to represent each candle's burning. (Make a table of data for each candle first, if it will help.)

f. How are the three equations alike? How are they different?

g. Write an equation for a fourth candle, Candle S, whose graph will intersect the point (0, 8) like the other three lines.

h. Sketch the graph of your equation for Candle S on the same grid with the graphs for Candles P, Q, and R.

i. Describe or draw Candle S that you've represented by the graph and equation in parts (g) and (h).

6. The graph at right shows the number of quarts of strawberries picked by three people one morning:

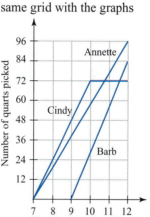

a. One person was scared by a snake and quit picking. Who was that? How do you know?

b. How many quarts does Annette pick per hour?

c. At what time did Barb start picking? How many quarts does she pick per hour?

d. Which person picks strawberries most quickly? How do you know?

7. The next morning Annette (see Learning Exercise 6) was ill. She started picking strawberries at 8:00 AM but could pick only 10 quarts of strawberries per hour. She quit picking at 11:00 AM. Graph both days of Annette's picking on the same set of axes.

8. One Monday the manager of a drug store counted the number of shoppers at various times during the day, which are represented by the solid dots on her graph.

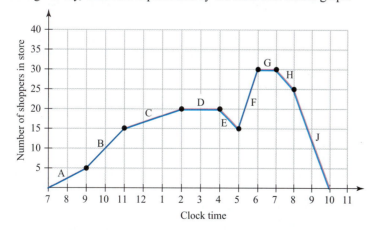

a. Why does the graph begin at 7 on the time axis rather than 0?

b. What does the manager's graph assume about what happens between the times when the manager counted the shoppers?

c. What period saw the greatest increase in customers?

d. How could segment D be interpreted?

e. How could segment B be interpreted?

f. During what hours might the manager need more cashiers?

continue

9. On Valentine's Day, a florist charges $5 per rose, plus a $20 holiday surcharge.

   a. Create a table of values to represent the relationship between the number of roses purchased and the total cost.

   b. In this situation, which quantity would you say depends on which? Why?

   c. Write an equation to express the relationship between the total cost and the number of roses purchased.

   d. Construct a graph of the relationship between total cost and number of roses purchased.

10. Jackie paid $175 to rent a booth at the local craft fair to sell "Jackie's Jams." Each jar of jam sells for $7.50. Write an algebraic equation that shows the relationship between the number of jars of jam sold and her gross income.

11. When you travel outside of the United States, it is likely that the country you are visiting reports the temperature in Celsius, rather than Fahrenheit, degrees. When you hear the temperature outside may reach 30°C, you may not know how warm or cool it is without first converting to degrees Fahrenheit.

    a. At sea level, water freezes at 32°F (which is 0°C) and boils at 212°F (which is 100°C). Write the algebraic equation that helps you compute the degrees Fahrenheit given the degrees Celsius.

    b. If the temperature is 30°C, what is the corresponding temperature in degrees Fahrenheit?

Supplementary Learning Exercises for Section 13.2

1. Because crickets are sensitive to air temperature, they can be used to determine the temperature. To calculate the temperature, count the number of chirps in 15 seconds and then add 39 (degrees in Fahrenheit).

   a. Complete this table to find the temperature when hearing a cricket.

Seconds	15	15	15
Number of chirps	30	39	
Temperature	69°F		50°F

   b. Write an equation relating the temperature and the number of cricket chirps.

   c. Make a graph that relates the temperature and the number of cricket chirps.

   d. What is the slope of the line you drew?

2. For a particular set of rectangles, the length of each rectangle is twice the width.

   a. Write an equation relating the length and width of these rectangles.

   b. Write an equation relating the width and area of these rectangles.

   c. Make a graph showing the relationship between the length and width of these rectangles.

   d. What is the slope of the line in part (c)?

3. Here are some statistics on a large passenger airplane: The length is 239 feet, the wingspan is 262 feet, the fuel capacity is 81,890 gallons, and its range is 8,000 nautical miles.

   a. Compare the length and wingspan of the airplane additively, in a sentence. Then compare them multiplicatively, in a sentence.

   b. Approximately how many gallons of fuel does the airplane use per nautical mile?

   c. Make a graph of the number of miles traveled and the number of gallons of fuel used. Plan the scales for your axes ahead of time.

   d. What is the slope of the line in part (c)?

4. The graph at right describes how one candle burns. Sketch a likely shape for the candle, being as careful as you can. Notice that some slopes of the line segments appear to be equal.

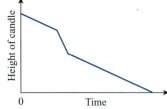

5. When you travel outside the United States, you most likely will see signs telling you the speed limit in kilometers per hour rather than miles per hour. A kilometer is about 0.62 mile.

   a. The distance from Point A to Point B is 1 mile. How many kilometers are between Points A and B?

   b. A highway sign says that the speed limit is 100 km/h, which seems incredibly fast to you. What is this speed limit in miles per hour?

   c. How long is a 10-kilometer race in miles?

   d. Draw a graph that relates kilometers and miles, with miles on the horizontal axis.

   e. Would a graph with kilometers on the horizontal axis be steeper or not as steep as the graph with miles on the horizontal axis?

13.3 Linear Functions and Proportional Relationships

The quantitative relationships introduced in the last section (those having a constant rate of change and a straight-line graph) are functions, specifically *linear functions*. The algebraic equations you have been writing are function rules. In this section we focus on very special linear functions that describe proportional relationships.

Think About ...

How could you explain to a friend that all relationships described by the general algebraic equation $y = mx + b$ describe mathematical functions?

When we earlier discussed sequences, both arithmetic and geometric, we paired the location of a number with the number in the sequence. We then found function rules for the *n*th number in the sequence. Consider the following arithmetic sequence: $\frac{2}{5}, \frac{4}{5}, 1\frac{1}{5}, 1\frac{3}{5}, \ldots$

Location		Number
1	$\longrightarrow$	$\frac{2}{5}$
2	$\longrightarrow$	$\frac{4}{5}$
3	$\longrightarrow$	$1\frac{1}{5}$
4	$\longrightarrow$	$1\frac{3}{5}$
$\vdots$		$\vdots$
n	$\longrightarrow$	$S = \frac{2}{5}n$

The constant difference here is the constant rate of change and its graph consists of points that lie on a line. In general, function rules for arithmetic sequences are linear functions.

In real-world contexts, several quantities can be significant. Some functions can involve inputs of two or more quantities, as Activity 8 suggests.

 **ACTIVITY 8**   The Soda Machine

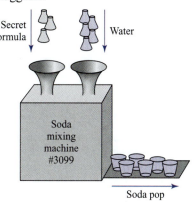

PopsiCo has a machine for restaurants into which you pour containers of Secret Formula and identical containers of water. Out of the machine come cups of soda. All the fluid that goes into the machine comes out of the machine. So, if 1 gallon of fluid (Secret Formula and water combined) goes into the machine, then 1 gallon of soda pop comes out of the machine.

Machine #3099 works this way: When you pour 3 containers of Secret Formula and 5 containers of water into the machine, you get 6 cups of soda pop out of the machine.

1. How many *cups* does 1 *container* of fluid hold? Explain.

2. How many *containers* of fluid does 1 *cup* hold? Explain.

3. What fraction of each cup of soda is made up of Secret Formula? How do you know?

4. What fraction of each cup of soda is made up of water? Explain.

5. Complete this table:

Number of containers of Secret Formula	3		4		7		n	
Number of containers of water	5	8				18		k
Number of cups of soda pop	6			9				h

Activity 8 provides an opportunity to review proportional thinking, introduced in Chapter 9 where you learned that a *rate* is a ratio of quantities that change without changing the value of the ratio. In Activity 8, Machine #3099 used 3 containers of Secret Formula for every 5 containers of water, so other mixtures would be equivalent to this mixture only when the ratio of the amounts of Secret Formula to water is $\frac{3}{5}$. That is, $f : w = 3:5$, or $\frac{f}{w} = \frac{3}{5}$, where f indicates the number of containers of Secret Formula and w the number of containers of water. Notice that $\frac{f}{w} = \frac{3}{5}$ leads to the expression $f = \frac{3}{5} w$. (Multiply both sides of $\frac{f}{w} = \frac{3}{5}$ by w.) Thus, given the number of containers of water, we could find the number of containers of the Secret Formula by multiplying the number of containers of water by $\frac{3}{5}$.

Think About …

How many containers of Secret Formula should be used with 15 containers of water to obtain soda pop with the same taste as a mixture using 3 containers of Secret Formula to 5 containers of water? Why?

A graph of the function rule $f = \frac{3}{5} w$ would look like the following:

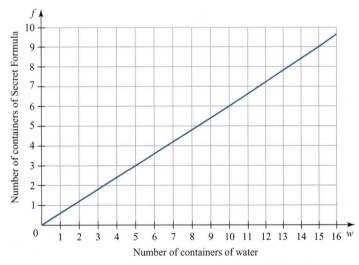

Think About …

Use the graph to answer again the question: How many containers of Secret Formula should be used with 15 containers of water to obtain soda pop of the same taste as a mixture using 3 containers of Secret Formula to 5 containers of water?

DISCUSSION 4   Is There More than One Rate Here?

Activity 8 involves another quantity: the number of cups of soda pop. Is this quantity related as a rate to either the number of containers of water or the number of containers of Secret Formula? ●

Many problems from Chapter 9 dealt with proportional relationships between quantities.

ACTIVITY 9   Donuts and Graphs

Consider the following problem.

▶ A donut machine produces 60 donuts every 5 minutes. How many donuts does it produce in an hour? ◀

a. Identify the rate in this problem.

b. What unit ratio is associated with this rate?

c. Write a proportion that could be used to find the number of donuts produced in an hour.

d. Make a graph of the number of donuts produced every minute. What is the slope of the line? (Remember to put *time* on the horizontal axis.) ●

The graphs in the soda pop and donut situations represent *proportional relationships*. In each case, the equation that describes the graph has the form $y = mx$, where m is some fixed value, and so the line begins at (0, 0).

If we can express a relationship in the form $y = mx$, the relationship between y and x is **proportional** because then the ratio $y:x$ always has the same value, m.

EXAMPLE 1

In the relationship $f = \frac{3}{5} w$, the number $\frac{3}{5}$ is the fixed value, and the relationship has the form $y = mx$, where f corresponds to y and w corresponds to x. This relationship is a linear function whose graph is a straight line beginning at (0, 0). ●

EXAMPLE 2

In the donut problem in Activity 9, 12 donuts are produced every minute. This relationship can be expressed as $d = 12t$, where d is the number of donuts and t is the number of minutes. Again, the equation has the form $y = mx$, with 12 as the constant m. ●

⋮ *Think About …*
⋮ What is the slope of the line in Example 1? In Example 2? What does each slope mean?

DISCUSSION 5   Proportionally Related?

1. Suppose that Clairisa can run 5 meters per second and Karlene can run 4 m/s. Make a coordinate graph to show the relationship between time and distance from the start. Draw a graph for each runner on the same coordinate system. How are the graphs different? How are they the same?

2. What is the equation in each case?

3. Are the distance and time proportionally related in each case?

4. What is the slope of each line graph?

5. What is the unit rate for each runner?

6. Suppose that another runner, Tabbie, begins the race at the 4-m marker and runs 4 m/s. Make a graph of Tabbie's run. Will Clairisa overtake Tabbie? If so, when?

7. Write an equation to describe Tabbie's run.

8. Are Tabbie's distance and time proportionally related? Why or why not? ●

In a proportional relationship such as $y = mx$, how are the other values affected if one of the values changes? For example, what happens to the value of y if the value of m is halved? We'll first explore this question in the particular situation of motion.

ACTIVITY 10   Road Trip

The following problem is from a seventh-grade unit on rate and proportionality. The unit was used in a large-scale study of SimCalc, a program intended to provide all students with access to the understanding of fundamental mathematics, using technology.[1]

▶ Every year a soccer team makes the trip from Abilene to Dallas for a special challenge match. They take both a bus and a van on the trip to accommodate all players and boosters. ◀

The following is a graph of the bus's and van's travel on that trip:

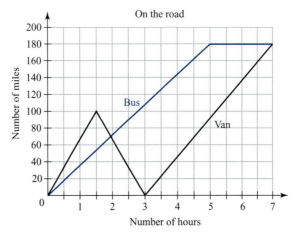

a. What did the van do after traveling for one and a half hours?

b. What happened here? Tell the story of this trip.

c. At what rate did the van travel in the first $1\frac{1}{2}$ hours?

d. At what rate did the bus travel in the first $1\frac{1}{2}$ hours?

e. At what rate did the van travel between 3 hours and 7 hours?

f. For which of the three travel segments described in parts (c–e) are time and distance proportionally related?

g. If in the first $1\frac{1}{2}$ hours the van traveled half as fast, how far would the van have traveled? How would this be reflected in the van's graph? ●

Proportional relationships are very important, and you will likely encounter them in your teaching of elementary or middle school science topics.

TAKE-AWAY MESSAGE ... A linear function takes the form $y = mx + b$ and gives a straight-line graph. As a special case, when $b = 0$, the equation takes the form $y = mx$, indicating that the variables x and y are related proportionally.

Learning Exercises for Section 13.3

1. A mile is roughly equivalent to 1.6 kilometers. What linear function relates the number of miles to the number of kilometers? Graph the function. What quantities are proportionally related? Explain.

2. As an elementary teacher, you may teach science as well. Upper elementary students will likely learn about how pressure is related to distance above sea level.

   a. Although air pressure fluctuates due to weather conditions, standard normal or average air pressure at sea level is 14.7 pounds per square inch. Standard air pressure in Denver at approximately 5280 feet (a mile high) is 12.1 pounds per square inch. A linear function describes this relationship. Use the given two points to write an algebraic equation that relates the distance above sea level with the air pressure. Is this a proportional relationship?

b. At 66 feet below sea level, the pressure equals 44.1 pounds per square inch. At 99 feet below sea level, the pressure equals 58.8 pounds per square inch. How are the quantities of depth and pressure related? Is this a proportional relationship? (*Hint*: The change is constant.)

3. On a recent trip abroad, Sandra learned that 5 dinars was equivalent to 23.25 florins. Describe how she can convert any number of dinars to florins.

4. Environmental biologists use a capture/recapture method to estimate the size of a population of animals in a given geographical area. Suppose they capture 20 birds of interest, tag them, and release them into the wild. A few weeks later in the same area, they capture 30 birds of interest and 5 of these birds have tags. They use the following proportional relationship to estimate the population size:

$$\frac{N}{C} = \frac{M}{R}$$

M is the number of birds captured and tagged on the first visit. C is the number of birds captured on the second visit. R is the number of tagged birds recaptured. N is the estimate of the population size.

a. Given the data in this example, what is the estimate of the population size?
b. What ratio is used to compute the population estimate for this bird of interest?

5. (See Activity 8 in this section.) A new PopsiCo Machine #4138 works this way: When you pour 4 containers of Secret Formula and 2 identical containers of water into the machine, you get 11 cups of soda pop out of the machine.

a. How many *cups* does 1 *container* of fluid hold? Explain.
b. How many *containers* of fluid does 1 *cup* hold? Explain.
c. What fraction of each cup is made up of Secret Formula? Explain.
d. What fraction of each cup is made up of water? Explain.
e. Show that the relationship between containers of Secret Formula and containers of water is a proportional one.

6. Chris and Sam biked from Childress to Navalo, a distance of 82 miles. Chris's average speed was 12 mph, and Sam's average speed was 10 mph.

a. Chris and Sam begin at the same time. On the same coordinate system, make distance-time graphs showing Chris's and Sam's trips from Childress to Navalo.
b. Write equations for each of the bikers.
c. What is the slope of each line?
d. Abby also biked from Childress to Navalo at 10 mph, but she began an hour earlier than Chris and Sam. Graph Abby's trip on the same coordinate system as that for Chris's and Sam's trips. Write the equation for Abby's trip, as shown on the graph. What is the slope for Abby's graph?
e. Which graphs show proportional relationships? Why?

7. Learning Exercise 9 in Section 9.2 gives the following problem:

▶ Two workers working 9 hours made 243 parts. Worker A makes 13 parts in 1 hour. If the workers work at a steady rate throughout the day, who is more productive, Worker A or Worker B? ◀

Assume the workers begin work at the same time and both work for 9 hours.

a. Write an equation for each worker.
b. Are the variables proportionally related? Why or why not?
c. Graph the two lines. What does the slope of each line tell you?

8. Which of these five graphs show a proportional relationship?

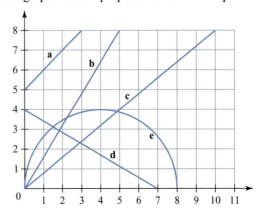

9. Which of the following situations provides a proportional relationship between the two variables suggested? Make clear what the variables are.

   a. The distance d traveled by a rock dropped from a tall building can be described by the equation $d = 16t^2$, where d is in feet and t is the number of seconds after the rock is dropped.

   b. The speed s of a falling body t seconds after it is released is given by $s = 32t$.

   c. Every 7 crackers have 120 calories.

   d. Tickets for a balcony section cost $45 each.

Supplementary Learning Exercises for Section 13.3

1. Find a likely function rule for each of the following, and find the missing values for m and n. (You may have found some of these in Chapter 12.) Which, if any, of the situations involve proportional relationships?

   a.

x	0	1	2	3	$\ldots$	100	n
y	0	3	6	9	$\ldots$	m	360

   b.

x	1	2	3	4	$\ldots$	36	n
$f(x)$	153	149	145	141	$\ldots$	m	$^-43$

   c.

input	1	2	3	4	$\ldots$	10	n
output	5	7	11	19	$\ldots$	m	2051

   d.

input	2	3	6	8	$\ldots$	12	n
output	$^-12$	$^-18$	$^-36$	$^-48$	$\ldots$	m	$^-600$

   e.

x	1	2	3	4	$\ldots$	10	n
$g(x)$	2	9	28	65	$\ldots$	m	8001

   f.

x	3	1	4	2	$\ldots$	25	n
$h(x)$	12	4	16	8	$\ldots$	m	804

2. Which of these equations gives a proportional relationship between x and y? (You may need to use a little algebra.)

   a. $y = 75x$ b. $y = \frac{7}{8}x$ c. $\frac{x}{y} = 9$

   d. $y - 3x = 0$ e. $y - 3 = x$ f. $y^2 = 16x$

3. Janet and Sylvia each drove a carload of sixth-grade soccer players from Lincoln Middle School to a tournament at a middle school in the neighboring city of Nestor. The Nestor school is 210 miles away from Lincoln Middle School. Janet drove at an average speed of 50 miles per hour. Sylvia drove an average of 60 miles per hour. When Sylvia's group reached Jaspar, 15 miles away from the middle school, one of the students realized he had left his wallet on the locker room bench. Sylvia turned around and drove back for the wallet, which took 15 minutes to retrieve from the lost-and-found office. She once again drove to the Nestor school, traveling an average of 60 miles per hour.

a. Calculate who arrived first, and by how many minutes. Then answer the following questions, based on the accompanying graph.

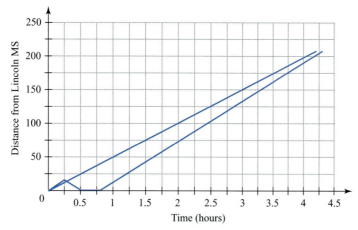

b. Which line portrays Janet's trip? Which portrays Sylvia's trip?

c. Which part of Sylvia's graph portrays the trip to Jasper and back? Which part portrays the time used to find the wallet?

d. Draw a horizontal line on the graph that indicates where each trip ended. Do the intersections of this line and the two trips match what you found in part (a)?

e. Suppose Carlene took a third carload on the same trip, leaving one hour later and driving an average of 65 miles per hour. Draw a line on the graph portraying this trip. Does she arrive before or after the first two carloads?

13.4 Nonlinear Functions

In the last section we focused on linear functions. There are other important *nonlinear* functions with predictable patterns. Though elementary students encounter mostly linear functions in their studies and may not encounter other types of functions until later grades, it is important for you to understand the distinction between linear functions and nonlinear functions. In this section we discuss a few important classes of functions.

 ACTIVITY 11 Something Really Different for a Change

1. For each of the following functions, complete a table for integer values of x from -3 to 3. Draw a graph of the data using an appropriate scale. To get an accurate picture of the graph, you'll need to plot a few additional noninteger values for x, such as $-2.5, -1.5, 1.5,$ and 2.5.

$$y = 2x \qquad\qquad y = x^2 \qquad\qquad y = 2|x|$$

x	y		x	y		x	y
-3			-3			-3	
0			0			0	
3			3			3	

continue

2. As x increases, how does y change?

3. Describe the rate of change for each function. Remember we describe a rate of change for every increase of 1 unit on the y-axis. Is the change predictable? Constant?

4. How are these graphs alike and how are they different? ●

As you know, the rate of change is constant for the linear function in Activity 11. The rate of change for each of the two nonlinear functions, though not constant, is predictable. The graphs give us some insight into how the quantities change with respect to each other.

We have seen how linear functions can be used to model many everyday situations. For example, the cost of hamburger is a function of the amount bought. The temperature in degrees Celsius is a linear function of the temperature in degrees Fahrenheit. The number of chirps a cricket makes is related to the temperature and is modeled by a linear function. You can think of many other situations.

Nonlinear functions, such as $y = x^2$ and $y = 2|x|$, describe many situations with which you are familiar as well. The algebraic equation $y = x^2$ relates the side of a square to its area (if $x > 0$). It is part of a class of functions called **quadratic functions**. Another example of a quadratic function is $y = \pi x^2$. Do you recognize that this function relates the radius of a circle x (again, radii would be greater than zero) with the circle's area y? Other relationships are modeled with quadratic functions. A quadratic function can be used to describe the height of an object as it falls with respect to time elapsed. The quadratic function $y = 30 - 16x^2$ could describe the height of a clamshell (in feet) when dropped from a height of 30 feet by a seagull trying to break it on the rocks.

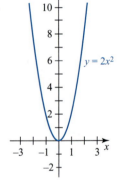

The graph of a quadratic function is called a **parabola**. For example, the graph to the right is a graph of the function $y = 2x^2$.

> **Think About ...**
> How would the graph $y = {}^-2x$ differ from the graph of $y = {}^-2x^2$?

> **Think About ...**
> What else distinguishes a quadratic function from other functions?

✏️ ACTIVITY 12   All Together Now

1. For each given function, make a table of values and graph the corresponding ordered pairs. For purposes of comparison, include several x values less than zero. Make sure you choose some nonintegral x values.

   a. $y = \pi x^2$ **b.** $y = 30 - 16x^2$

2. Compare the tables and graphs of the two new equations in Problem 1 to $y = x^2$ and to each other. Describe what is similar about the graphs, the rates of change, and the algebraic equations.

3. Write an algebraic equation for a quadratic function that you make up. Discuss with a partner your thoughts on how to form a quadratic function. ●

As you looked at an input–output table for a quadratic function such as $y = x^2$ below, you probably noticed that the differences in the table were predictable but not constant. You may or may not have noticed that the differences themselves had a pattern of constant change.

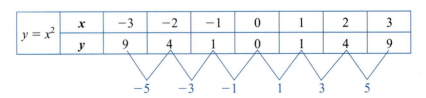

$y = x^2$	x	-3	-2	-1	0	1	2	3
	y	9	4	1	0	1	4	9

Another type of nonlinear function is the **absolute value function**. As you know, a number's distance from zero on the number line is its absolute value. Functions involving absolute value are easy to spot given the familiar absolute value sign: $|x|$. The figure below on the left is the graph of $y = {}^-2|x|$. The figure below on the right is the graph of $y = |x| - 3$.

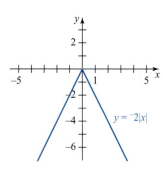
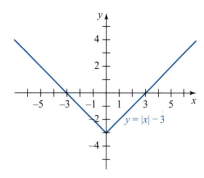

Absolute value functions can be used to describe tolerance or the amount of acceptable error that can be tolerated in a situation such as manufacturing or medicine. The graphs of absolute value functions such as $y = |x| - 3$ or $y = {}^-2|x|$ illustrate why it is important not to generalize and say "absolute value graphs are always positive."

An **exponential function** generally has the form $y = b^x$, where $b > 0$. For example, below is the graph of $y = 2^x$. Notice how it increases rapidly as x increases in size. This gives some meaning to the expression that something "grows exponentially."

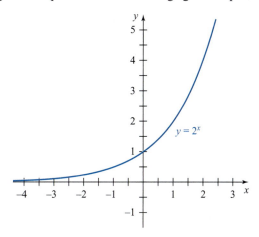

Think About ...
Can b be equal to 1? Why or why not?

Exponential functions also have quite familiar applications, as we see in Activity 13.

 ACTIVITY 13   Change Compounded

The balance A of a savings account that earns compound interest is expressed with the following $A = P(1 + r)^{nt}$. P is the principal (amount invested), r is the rate *per compounding period*, t is the time in years, and n is the number of compound periods each year.

1. Use the formula for compound interest to find the balance if $1000 is invested for one year at the annual interest rate of 5% compounded quarterly.

2. Graph this function. Plot enough x values to allow you to describe the curve.

3. Graph the simple exponential function $y = 2^x$.

4. What can you infer about the rate of change of an exponential function from an input–output table or from a graph? ●

Think About ...

Graphs of geometric sequences are nonlinear. Do geometric sequences easily fit into one of the classes of nonlinear functions we discussed?

Though most of your teaching may involve linear functions, it is important to be able to know what distinguishes linear functions from nonlinear functions by looking at an algebraic equation, an input–output table, or a graph. Nonlinear functions model important aspects of our lives. A few classes of nonlinear functions have been discussed in this section, but many other nonlinear functions do not fall into one of these categories.

TAKE-AWAY MESSAGE ...   There are important classes of nonlinear functions. Nonlinear functions do not have a constant rate of change. A few familiar nonlinear functions are quadratic functions, absolute value functions, and exponential functions. A quadratic function graphs as a parabola and has a predictable but not constant rate of change. An absolute value function is piecewise linear. An exponential function has a rate of change that changes predictably.

Learning Exercises for Section 13.4

1. Are the following functions linear or nonlinear? Explain. (You may find it useful to create an input–output table and/or draw a graph.)

    a. $y = 3x^2 + 5x - 1$ **b.** $y = x(x + 5)$

    c. $P = 3^t$ **d.** $y = -3|x|$

2. Are the following functions linear or nonlinear? Explain. (You may find it useful to create an input–output table and/or draw a graph.)

    a. $y = 5x(x - 2)$ **b.** $3y + 5 = y - 2$

    c. $A = 2\pi r$ **d.** $\frac{3}{2}y + 3x = 26$

3. For each of the following situations, describe how the quantities are related. You may be able to write an algebraic equation. Say whether the algebraic equation describes a linear or nonlinear function. Explain.

    a. The number of gallons of water remaining in a tub x minutes after the drain is opened to drain the tub.
    b. The area of a rectangle if the width is $\frac{1}{2}$ the measure of the length.
    c. You put $2500 in your checking account that paid 1.25% annually. At the end of one year, how much money do you have?

4. For each of the following situations, describe how the quantities are related. You may be able to write an algebraic equation. Say whether the algebraic equation describes a linear or nonlinear function. Explain.

    a. Suppose your pay increase as a teacher was negotiated to be 3% a year for 3 years. What is the value of your salary after 3 years?
    b. On a map drawn to scale, $\frac{1}{4}$ inch = 500 feet. How is distance on the map related to actual distance?
    c. The area of the shaded region shown is a function of the length of the side of the square.

5. Refer to the function rules generated for toothpick shapes in Learning Exercise 12 in Section 12.4. For each function rule, tell whether it is linear or nonlinear.

6. Several students share an extra-large pizza. Write a rule that relates the number of students to the fraction of the pizza they each receive. What does the graph of this function look like?

Supplementary Learning Exercises for Section 13.4

1. For the following functions, complete the input–output tables. Use that information to decide whether or not the function is linear, and tell how you know. If you want to graph the points, you can. In cases where the function is linear, tell what its rate of change is and graph the function.

a. $y = x$ **b.** $y = 3$ **c.** $y = x^2$ **d.** $y = 2x^2 + 2x$

x	y
0	
2	
5	
8	
9	

x	y
0	
1	
2	
6	
10	

x	y
0	
1	
2	
3	
4	

x	y
0	
1	
2	
3	
6	

2. Make input–output tables for the following functions and tell whether they are linear or not. Provide reasons for your decisions.

a. $x = 3y$ **b.** $h = 4v^3$ **c.** $2y = 3x$ **d.** $P = 5s$

3. Each of these situations describes a relationship between or among quantities. Describe each one algebraically and tell why it describes a linear function.

   a. A bakery sells cupcakes 5 for $1.00.
   b. A certain kind of punch mix uses a 1:2 ratio, i.e., each cup of punch concentrate is to be diluted with 2 cups of water.

4. Barbara works for a financial management firm where it is customary that each worker receive a 10% raise every January. If Barbara's base salary when she begins working is $1200 per month, what will her salary be at the beginning of the fifth year?

5. A king gives his court jester 1 grain of rice on the first day of the year, 2 grains on the second day, 4 grains on the third day, 8 grains on the fourth day, i.e., each day doubling the number of grains received the day before. How many grains will the jester have after the tenth day?

13.5 Issues for Learning: Algebra in the Elementary Grades

Although a formal treatment of algebra is usually saved for the eighth or ninth grade, most national and state recommendations and textbooks now incorporate algebraic ideas into the elementary grades. The *Principles and Standards for School Mathematics*[2] from the National Council of Teachers of Mathematics or NCTM (http://www.nctm.org) states that "by viewing algebra as a strand in the curriculum from pre-kindergarten on, teachers can help students build a solid foundation of understanding and experience as a preparation for more-sophisticated work in algebra in the middle grades and high school" (p. 370).

The *Common Core State Standards* (CCSS),[3] released in mid-2010, reflect a state-led initiative to establish a shared set of educational standards. Likewise, these Standards list *Operations and Algebraic Thinking* as a domain, or category of related standards, for kindergarten through fifth grade. Within the two documents, many of the expectations are the same.

The CCSS for Mathematics consist of two intertwined parts: Standards for Content and Standards for Mathematical Practice. The Content Standards define what students should know and be able to do, and the Standards for Mathematical Practices describe the processes and proficiencies that students should develop. The content standards define grade-specific standards for a sequence of topics and performances. The authors of the CCSS endeavored to design standards with focus and coherence. They sought coherence through organizing structure, such as place value and the properties of operations.

In the category or domain of Operations and Algebraic Thinking, the authors describe mathematically proficient K–5 students as understanding the meaning and properties of operations, generating and analyzing patterns, representing word problems with equations, and solving for unknowns. These are to be built progressively and coherently in preparation for working with the algebraic expressions, equations, and methods of middle school and beyond. Particular state standards vary, but all are influenced by the Common Core State Content Standards described below.

By Grade 2, students are expected to add and subtract numbers up to 20 with fluency and to use the properties of operations, decomposition of numbers, and the relationship between addition and subtraction as strategies to add and subtract. For example, $13 - 4 = 13 - 3 - 1 = 10 - 1 = 9$.

Students are also expected to generalize and use symbols to represent mathematical ideas. After working with the addition of two one-digit numbers for a period of time, they may come to realize that $2 + 5$ gives the same sum as $5 + 2$ and to generalize this (commutative) property for adding any two numbers. This knowledge becomes useful when it allows them to count on from 5 (6, 7) rather than from 2 (3, 4, 5, 6, 7) when finding the sum of 2 and 5. Students need not use the properties explicitly, but recognizing, identifying, and recording the properties in student reasoning will be helpful. Models and representations such as the rectangular area and number line can support their reasoning (see Section 12.1).

They also learn to represent word problems with symbols. For example, "Cathy has 4 pencils. How many more are needed to have 7 pencils?" can be represented symbolically as $4 + \square = 7$. The CCSS explicitly call for third-grade students solving two-step word problems to use the four operations and, furthermore, to represent these problems with a letter standing in for an unknown in these equations. In the fifth grade, students are expected to work more formally with writing and evaluating numerical expressions.

Students also should learn to model situations. For example, when given the problem "There are six chairs and stools. The chairs have four legs and the stools have three legs. Altogether there are 20 legs. How many chairs and how many stools are there?" (NCTM, p. 95); students can use drawings or numbers and adjust them until they reach a total of 20 legs.

Students at this level also should learn the meaning of equality and be able to express equality in different ways. Many children at this age erroneously think of the equal sign as directional, i.e., that $3 + 2 = 5$ is correct, but that $5 = 3 + 2$ is incorrect. Understanding equality means that they can write the equality in either way and they will find acceptable expressions such as $4 + 2 = 3 + 3$.

In these early grades, students also begin to learn about change. They can, e.g., talk about growing taller, or about the weather becoming colder.

Students in Grades 3–5 also investigate patterns, both numerical and geometric. Page 159 of the NCTM Standards uses a pattern of "growing squares" to describe this work. Students are expected to describe the pattern, tell what the next square would look like, and say how many small squares would be needed to build it.

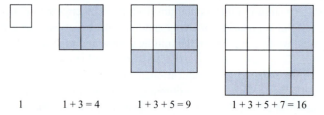

1          $1 + 3 = 4$ $1 + 3 + 5 = 9$ $1 + 3 + 5 + 7 = 16$

Modeling situations and relationships can be extended from what was presented to children in Grades Pre-K–2. For example, students might be asked to find the total surface area of consecutive towers built with cubes, placing one cube on top of another. They can use a table and sequences of ordered pairs (and, at a later grade, a graph in the first quadrant of the coordinate plane) to describe what is happening.

Cubes	Surface area
1	6
2	10
3	14
4	18
5	22

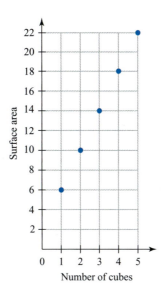

One student may describe what is happening by saying that each cube has 6 faces but when one is placed on top of another, 2 faces are lost, and then for n cubes write this as $6n - 2(n - 1)$, with sufficient algebra sophistication. Another might reason that each time a cube is added to the tower, there are 4 new faces. So, there are 6 faces to begin with and then 4 new ones each time: $6 + 4(n - 1)$. The students could then discuss whether these two expressions are equivalent.

Students also study change in more depth in these intermediate grades. They can study growth patterns in plants, rates of change such as the distance traveled in a certain amount of time, or the change in cost as the number of objects bought increases. These changes can be represented in tables and graphs.

The Operations and Algebraic Thinking domain across the elementary grades should be preparation for formal work with expressions and equations in middle school. Although younger students can represent and solve word problems with equations involving one unknown quantity, students in Grades 6–8 solve equations and inequalities using properties of operations explicitly. Mathematically proficient students solve equations with algebraic methods. They should be able to "represent, analyze, and generalize a variety of patterns with tables, graphs, words, and, when possible, symbolic rules; relate and compare different forms of representation, and identify functions as linear or nonlinear and contrast their properties from tables, graphs, and equations" (NCTM, p. 222). In these grades, students begin to develop an understanding of the different uses for variables and use symbolic algebra to represent situations and solve problems. They should be able to "model and solve contextualized [story] problems using various representations" (p. 222) and understand graphing of linear equations and the meaning of slope. The study of change becomes more formal in these grades.

One way to gain more familiarity with the national vision of mathematics education as described by NCTM and the *Common Core State Standards* is to utilize the information offered by the NCTM. It has been providing information and assistance to mathematics teachers for nearly a century, primarily through conferences and publications. The NCTM website offers information about effectively teaching mathematics and implementing the *Common Core State Standards*.

13.6 Check Yourself

After this chapter, you should be able to do problems like those assigned, and to meet the following objectives:

1. Identify increasing or decreasing quantitative relationships in a given situation.
2. Represent a given quantitative relationship with a graph (perhaps a qualitative one, without numbers) or an algebraic equation.
3. Given a graph (or perhaps an algebraic equation) for a situation, describe the quantitative relationship(s) involved.
4. Calculate the slope for a given straight-line graph, and explain the meaning of slope in a given context.
5. Relate the slope of a straight-line graph to an equation for the graph.
6. Give meaning to the equation of a line in the form $y = mx + b$.
7. Discuss how graphs portray change.
8. Write equations for lines with known slope and y-intercept.
9. Identity nonlinear functions.

References for Chapter 13

[1]http://math.sri.com/. This research project investigated the integration of technology and curriculum to improve math instruction in the middle grades.

[2]National Council of Teachers of Mathematics. (2000). *Principles and standards for school mathematics.* Reston, VA: Author.

[3]National Governors Association Center for Best Practices, Council of Chief State School Officers. (2010). *Common core state standards mathematics.* Washington, DC: Author.

Chapter 14

Understanding Change: Relationships Among Time, Distance, and Rate

In Chapter 13 you were asked to draw graphs to illustrate different types of situations in which quantities change. In this chapter the focus is on a specific type of change—changes involving time, distance, and speed relationships. You are familiar with all three of these quantities. In the many contexts involving motion, these three quantities often change but in a related way. Speed is a rate and is often given in miles per hour (mi/h, or sometimes mph) or feet per second (ft/s). These units suggest that the ratio of a change in distance to a change in time is involved. In this chapter we build on these relationships graphically, algebraically, and occasionally numerically.

14.1 Distance-Time and Position-Time Graphs

Some of you may be thinking, "This discussion will be about $d = rt$, that is, distance equals rate multiplied by time." In a way, you would be correct, but the focus here will be on understanding and interpreting situations in which the $d = rt$ formula might be relevant. In particular, you will create graphs without being concerned about the specific values of the variables, but rather by thinking about how these quantities are related.

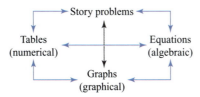

 **ACTIVITY 1**   Identifying Quantitative Relationships in a Motion Situation

Consider the following situation:

► Wile E. Coyote[1] leaves his cave, walking at a slow yet constant speed. Then he stops to build a trap for Road Runner. After several minutes, he turns around and runs back to his cave at a constant speed. ◄

(Because Wile E. and Road Runner are cartoon characters, you can assume that they can start walking or running at a constant rate instantly without first accelerating to that rate.)

a. Act out this situation with one volunteer, using a desk or some other object in the room as the cave.

b. What quantitative relationships do you see in this situation? Identify every pair of quantities that you can.

c. For each pair of quantities that you identified in part (b), describe how one quantity is changing (increasing, decreasing, or staying the same) as the other quantity increases. ●

One quantitative relationship in Activity 1 is the relationship between the total distance traveled by Wile E. Coyote and the time elapsed. In Example 1, you will see the graph of the relationship between Wile E. Coyote's distance traveled and time. Because no specific numerical information has been given regarding Wile E. Coyote's walking speed, his running speed, or the time he stops to build a trap, the graph will not have numbers. This type of graph is sometimes referred to as a **qualitative graph** or a "sketch" graph. It allows you to capture the relationship between the quantities in the situation without worrying about specific numerical relationships.[2]

EXAMPLE 1

Activity 1 describes a situation that involves distance, time, and speed. A distance-time graph can be drawn to represent these relationships as follows:

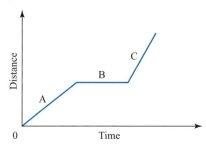

In this graph, time is displayed on the horizontal axis and distance on the vertical axis. It is important to note that here, distance means the total distance run, not how far Wile E. Coyote is from his cave. The graph has three distinct parts, labeled A, B, and C. Segment A represents Wile E. Coyote walking away from the cave at a slow, constant speed. When he is at the point at the end of the portion of the trip labeled A, both time on the horizontal axis and distance on the vertical axis have increased. We could find the slope of segment A if we knew the time values on the horizontal axis and distance values on the vertical axis. For the portion of the trip labeled B, time continues to increase but distance does not, so segment B is parallel to the horizontal axis. Finally, segment C shows that as time increases, distance increases more rapidly than it does in segment A. ●

> ### Think About …
> Why does the slope in a distance-time graph tell you what the speed is? (*Hint:* What is the *meaning* of slope?)

There are a number of choices to be made in creating a graph. In making a graph of the relationship between distance traveled and time elapsed, you might choose to make a qualitative graph (without numbers) and another person might choose to use numbers. Different decisions could be made about what quantities should be represented on each axis. These decisions can affect what is conveyed about the situation.

> ### Think About …
> If you yourself are running, can you start, stop, or change your speed immediately? Why do you suppose situations involving cartoon figures are used here, beyond hoping to amuse you?

 ACTIVITY 2   Creating a Position-Time Graph

Refer to the situation with Wile E. Coyote described in Activity 1 and Example 1. This time, graph the relationship between Wile E. Coyote's position (i.e., his distance from the cave) and time. Be sure to label each axis with a quantity. You can make a sketch without using any numbers, or you can make up reasonable data and graph that. ●

Although the story was the same, the graph of the relationship between Wile E.'s distance from the cave and time looks different from the graph of the relationship between the distance traveled and time. Yet many of the same elements of the story are conveyed. In Discussion 1, look specifically at how a position-time graph and a distance-time graph are different and what information is the same for both graphs.

DISCUSSION 1   Why Is It Back to Zero?

1. Compare your position-time graph with your distance-time graph, especially for the segment in which Wile E. Coyote runs back to the cave. How are the graphs different and why?

2. What can you tell about Wile E. Coyote's speed during each segment of the journey by using the graph? ●

TAKE-AWAY MESSAGE . . .   Even without numbers, a qualitative graph can give information about a situation. Drawn with care, both distance-time graphs and position-time graphs can reflect information about speed. Distance-time graphs show the total distance run. Position-time graphs tell how far, over time, the object is from the starting point.

Learning Exercises for Section 14.1

1. Consider the following situation:

   ▶ Wile E. Coyote leaves his cave walking at a slow, constant rate. He walks to the edge of a cliff. He is planning to build a catapult that will hit Road Runner by launching a large boulder. Just as Wile E. Coyote reaches the cliff, he remembers that he left the large rubber band for the catapult back in his cave. He turns around and walks back to his cave at the same pace as before. It takes him several minutes in the cave to find the large rubber band for the catapult. He realizes that he had better hurry so he starts running (at a constant speed). About halfway to the cliff, he remembers that he also left the boulder at home. He turns around and runs home at an even faster (yet constant) speed. ◀

   a. Graph the relationship between Wile E. Coyote's *total distance* traveled and time.
   b. Create a second graph, this time graphing Wile E.'s *position* (i.e., his distance from the cave) and time. [Using the same scales as in part (a), and placing this second graph right under the graph from part (a), make it easy to compare these graphs in part (c). Save this graph for later use in Learning Exercise 2 in Section 14.3.]
   c. Compare the two graphs. How are they alike? How are they different?

2. The graph at right shows a new journey taken by Wile E. Coyote.

   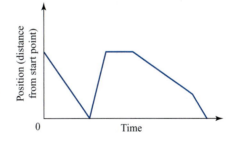

   a. Write a story that could be represented by this graph. (Remember that this graph shows Wile E.'s distance from a starting point, such as his cave. It is not a graph of the total distance traveled.)
   b. Construct a new graph of the relationship represented in the given graph. This time, graph the total distance traveled over time.

3. Graphs can involve numbers, of course. Suppose Wile E. Coyote leaves his cave walking at a constant speed of 8 ft/s for 10 seconds. Then he runs in the same direction at a constant speed of 12 ft/s for 5 more seconds. He realizes that he forgot something at home, turns around, and walks back to his cave at a constant speed of 5 ft/s.

   a. How long does it take Wile E. Coyote to get back to the cave once he turns around and starts walking back home?
   b. Graph Wile E.'s distance from the cave over time (i.e., make a position-time graph). You may want to create a table of values first.
   c. Sketch a second graph, this time graphing Wile E.'s total distance traveled over time. You may want to create a table of values first.
   d. Compare the two graphs. How are they alike? Different?
   e. How can you determine Wile E. Coyote's speed from each graph?

Notes

4. Jiminy Cricket is sitting on a tree leaf looking for a girlfriend. He catches sight of a possible one 2 feet away, and decides to go courting. He approaches her cautiously at a speed of 1 ft/s. But just when he reaches her, she notices him and jumps away at a speed of 4 ft/s. He follows her at her speed for 10 seconds but tires and turns around and returns to his leaf. He was away from his leaf for a total time of 33 seconds.

   a. Make a total distance versus time graph of Jiminy's trip away from and returning to his leaf. Mark the segments.
   b. Find the slope of each segment.
   c. Make a position versus time graph of Jiminy's trip away from and returning to his leaf. Mark the segments.
   d. Find the slope of each segment.

5. Many times, motion graphs (such as position-time graphs) can give an algebraic equation.

   a. Suppose Wile E. Coyote walks at a constant speed of 4 ft/s away from his cave. Graph the relationship between his distance from the cave over time (i.e., as time passes). Then write an equation representing this relationship.
   b. Suppose Wile E. Coyote is at a boulder about 20 feet from his cave. He starts walking in a direction that is away from both the boulder and the cave at a constant speed of 4 ft/s. Graph the relationship between his distance from the cave and time. Then write an equation representing this relationship.
   c. Suppose Wile E. Coyote walks toward his cave at a rate of 4 ft/s from a cliff that is 56 feet from the cave. Graph the relationship between his distance from the cave and time. Then write an equation representing this relationship. How are the graphs from parts (a–c) alike and different? Why?
   d. How are the equations from parts (a–c) alike and different? Why?
   e. What is the slope of the graph in part (a)? What does this slope tell you about Wile E.'s motion? Can you find the slope from your equation?
   f. What is the slope of the graph in part (b)? What does this slope tell you about Wile E.'s motion? Can you find the slope from your equation?
   g. What is the slope of the graph in part (c)? What does this slope tell you about Wile E.'s motion? Can you find the slope from your equation?

6. Suppose Wile E. Coyote is standing on a hill some distance from his cave. He spends some time looking out into the distance to see if he can spot a rabbit. He sees one very close by and runs after it, running right by his cave along the way. He does not catch the rabbit, and dejected, he mopes on his slow trip back to his cave.

   a. Make a total-distance-traveled versus time graph of his trip, labeling the different segments of your graph using A, B, etc.
   b. What is the slope of the first segment of your graph?
   c. Suppose that the hill is 64 feet from his cave and he sits there for 15 seconds. Then he chases the rabbit at 8 ft/s back past his cave and another 72 feet beyond. What is the slope of this portion of the graph?
   d. If his return to his cave takes 100 seconds, what is the slope of this portion of the graph?
   e. Make a position-time graph of his trip, using different labeling than you did in part (a).
   f. Find the slope of each segment of the position-time graph.
   g. What is the same and what is different about the slopes of the segments in the two graphs?

7. Pioneer 10 is the first man-made object to leave the solar system. Pioneer 10 will take 2 million years to get to a red star 68 light-years away. During that trip, what will Pioneer 10's average speed be, in miles per hour? (Light travels at 186,000 miles per second.)

8. Consider the situation given on the next page:[3]

continue

▶ An explorer went to visit a jungle village. She drove the first 25 miles of the trip along a rough road at a steady speed of 20 mi/h. She had to go the next 3 miles by canoe, at a speed of 5 mi/h. Then she had to cut her way through 12 miles of dense jungle and could manage a speed of only $\frac{3}{4}$ mi/h. Finally, she came to a path and walked the last 13 miles at a speed of 4 mi/h. How long did it take her in all to reach the village? ◀

Answer the question, explaining your reasoning. (Drawings may be useful.)

9. Rates are useful in situations other than those dealing with speed. Consider the following situations:

  a. You buy 3 pounds of apples at $1.49 per pound, 0.8 pound of coffee beans at $12 per pound, and 0.8 pound of cheese at $4.80 per pound. Calculate the total spent on this shopping trip.

  b. A moped went 40 miles on $\frac{5}{8}$ gallon of fuel. How far can the moped go on a gallon of fuel?

  c. A person's heart may beat at 70 beats per minute. How many beats would that translate into in a day? How long would it take for a billion beats, at this rate?

Supplementary Learning Exercises for Section 14.1

1. Consider the following situation:

  ▶ Wile E. wakes up and, after eating breakfast, leaves his cave and trots out to the spring for a drink of water. As he slowly saunters back toward the cave, about halfway he thinks he sees Road Runner on the other side of the spring, so he runs fast in that direction. When he gets to the spring, however, he realizes that he was incorrect, so he slowly saunters back to the cave. ◀

  a. Graph the relationship between Wile E.'s *total distance* traveled and time. Start the graph at the time when Wile E. wakes up.

  b. Make a second graph, this time graphing Wile E.'s position (i.e., his distance from the cave) and time. Use the same scale as in part (a) and put part (b) right under part (a) to allow an easy comparison for part (c).

  c. Compare the two graphs. How are they alike? How are they different?

2. The given graph shows a new journey taken by Wile E. Coyote.

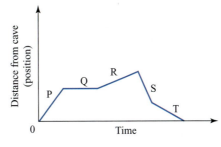

  a. Write a story that could be represented by this graph. (Remember that this graph shows Wile E.'s distance from his cave. It is not a graph of the total distance traveled. The letters in the graph allow a discussion, but you do not need to mention them in your story.)

  b. Construct a new graph of the relationship represented in the graph on the previous page. But this time, graph the total distance traveled over time.

3. In each part, graph the relationship between the distance from the cave and time, and then write an equation that expresses the relationship.

  a. Suppose Wile E. walks at a steady speed of 5 ft/s away from his cave.

  b. Wile E. is at the spring, about 15 feet from the cave. He walks away from the spring, in a line with the spring and the cave, at a steady speed of 5 ft/s.

  c. Wile E. is at his lookout point, 30 feet from his cave. He walks toward the cave at a speed of 5 ft/s.

  d. How are the graphs in parts (a–c) alike and different?

  e. How are the equations in parts (a–c) alike and different?

4. Answer the following questions about your work in Exercise 3:

  a. What is the slope of the graph in part (a) in Exercise 3? What does the slope tell you about Wile E.? Can you find the slope from your equation?

continue

   b. What is the slope of the graph in part (b) in Exercise 3? What does the slope tell you about Wile E.? Can you find the slope from your equation?

   c. What is the slope of the graph in part (c) in Exercise 3? What does the slope tell you about Wile E.? Can you find the slope from your equation?

5. Wile E. takes a trip from his cave that can be described by the equation $d = 10t$, where d is the distance from the cave in feet and t is the time in seconds. Write a story about Wile E. that fits the equation.

6. Wile E. returns to his cave in a way that can be described by the equation $d = 50 - 5t$, where d is the distance from the cave in feet and t is the time in seconds. Write a story about Wile E. that fits the equation.

7. Here is a distance-time graph for Sam's bike ride. Describe Sam's bike ride.

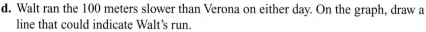

8. Verona ran 100 meters today in 20 seconds.

   a. Make a distance-time graph of Verona's run, assuming that she ran at a steady rate from the start.

   b. What is the slope of the line you graphed?

   c. The following day, she shaves 2 seconds off her previous day's run. Graph this run on the same set of axes as in part (a). What is the slope of the line showing this run?

   d. Walt ran the 100 meters slower than Verona on either day. On the graph, draw a line that could indicate Walt's run.

   e. Why is the assumption of steady speeds not realistic for most races?

14.2 Using Motion Detectors

The activities[4] in this section involve a motion detector that is connected to a computer or a calculator. The goal of working with a motion detector is to create graphs using real motion. By creating graphs of our movement, we often become more involved in thinking about the relation between position and time than we might be if we are just imagining movement from an observer's point of view.

   The motion detector is a physical device that measures how far you are from the detector over a given time interval. It works by emitting ultrasonic pulses and then recording the length of time it takes for the reflected pulses to return. Using this length of time and the known speed of sound, it calculates a distance approximately 20 times per second (although this measure can be changed using the software). The ultrasonic sounds are completely safe, so you do not have to worry about any radiation or other dangers. The range of the motion detector is between 0.45 and 6 meters.

© Courtesy of Kathleen Wise

The graph shown in the picture resulted from a student walking away from a motion detector, turning, and walking back at a steady pace in both directions. The motion detector measured the distances from itself. There is a little "noise" at the end of the graph that students were able to ignore.

Using the Motion Detector Software

The software allows you to create graphs of your distance from the motion detector over time. If your motion detector is attached correctly, you should be able to click once on the start button and begin collecting data.

ACTIVITY 3   Take a Hike

1. Create and record a graph that represents each of the following descriptions. Discuss with your group what the walker had to do to obtain the graph. Write a careful and complete explanation of what the walker did.

   a. line with a positive slope
   b. steeper line with a positive slope
   c. line with a negative slope
   d. less steep line with a negative slope
   e. horizontal line (i.e., a line with a slope of zero)

2. Start several feet from the motion detector and then walk away from the motion detector at a constant speed. Then use the data from the graph that the computer created to find the speed at which you walked. Be sure to include the unit of measurement on the graph. (*Note:* If you click on a point, you can get the point's coordinates by looking at the bottom of the screen. This may help you determine your speed.)

3. Create a graph that is a curve. Describe what the walker did to obtain the graph. Then use the graph to determine how fast you were traveling during different parts of the trip. Why is it more difficult to find the average speed of the walker here than in Problem 2 of this activity?

4. Try to recreate each of the following graphs using the motion detector. (*Note:* The vertical axis of each graph measures the walker's distance from the motion detector. The horizontal axis measures the time.) For each graph, do the following:

   • Carefully describe what the walker(s) had to do to obtain the original graph or your modified graph.

   • Describe any difficulty you may have in obtaining any part or all of the graph (some graphs may not be possible to recreate).

   • If the graph requires only some modifications, resketch it with your changes.

a. 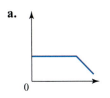 **b.** 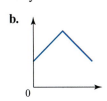 **c.** 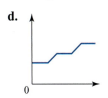 **d.**

e. 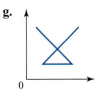 **f.** 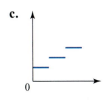 **g.**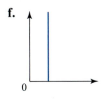

Notes

Notes

TAKE-AWAY MESSAGE . . .   Real-time movement graphs can help us visualize the relationship between position and time and can help us notice errors in graphing.

Learning Exercises for Section 14.2

Exercises 1–3 can be done without the motion detector.

1. The following graphs were obtained using the motion detector system. Describe as fully as possible the walks represented by these graphs and explain the differences between them.

a.
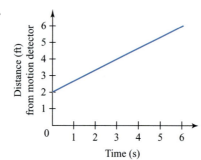

b.
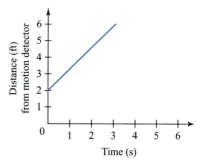

c.
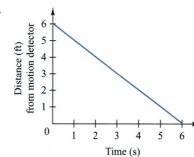

d.
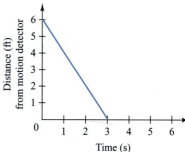

e.
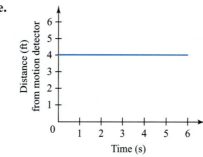

2. What was the walker's speed in each situation given in Learning Exercise 1?

3. **a.** Refer to graph (a) in Learning Exercise 1. Create a new graph (on paper) that represents a person walking at the same rate as the person represented by graph 1(a), but starting at a different distance from the motion detector.

   b. Compare your new graph with graph 1(a). How are they alike? How are they different?

   c. Create another new graph that represents a person starting in the same position from the motion detector as in graph 1(a), but walking at a slower rate.

   d. Compare your graph created in part (c) with graph 1(a). How are they alike? How are they different?

Exercise 4 requires using the motion detector.

4. Motion detector graphs involving curves require careful thought. Try to recreate each graph given on the next page using the motion detector. For each graph, do the following:

   • Carefully describe what the walker(s) had to do to obtain the original graph or your modified graph.

continue

- Describe any difficulty you may have in obtaining any part or all of the graph (some graphs may not be possible to recreate).
- If the graph requires only some modifications, resketch it with your changes.

a.

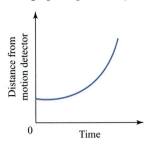

b.

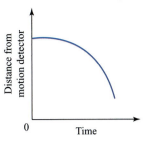

c.

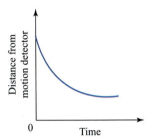

d.

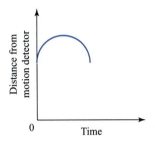

Supplementary Learning Exercises for Section 14.2

These additional exercises do not require the use of a motion detector.

1. The graphs below were based on results from a motion detector. Describe how each walk happened, and write an equation for each graph.

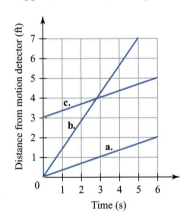

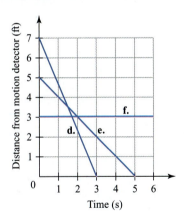

2. In Exercise 1, the graphs for walkers **b** and **c** cross. What does that mean, if the walkers started at the same time (with different motion detectors)?

3. **a.** From the graphs in Exercise 1, how can you tell which walker walked fastest away from the motion detector? Toward the motion detector?

   b. From the equations in Exercise 1, how can you tell which walker walked fastest toward the motion detector? Away from the motion detector?

4. Make graphs and write equations that describe these walks.

   a. Start 4 feet away from the motion detector and walk farther away from the detector at 2 ft/s.

   b. Start 4 feet away from the motion detector and walk toward the detector at 1 ft/s.

   c. Start 6 feet away from the detector and walk toward the detector at 2 ft/s.

14.3 Graphs of Speed Against Time

As we mentioned earlier, speed is a relationship between distance and time. For example, if you can travel 30 feet in 5 seconds at a steady speed, you are going at a much faster speed than someone who travels 20 feet in 5 seconds at a steady speed. Your speed is 6 ft/s, and the other person's speed is 4 ft/s. You can graph the relationship between distance and time in a situation involving motion and then obtain information indirectly about the speed (e.g., through the slope of the graph). Alternatively, you can represent speed directly in a graph by using speed and time as the quantities represented by the axes.

> *Think About ...*
>
> If each of two cars is going 50 mi/h but in opposite directions, why does it make sense to think of one of the speeds as a *negative* 50 mi/h?

ACTIVITY 4   Graphing Motion

Reconsider the following situation from Section 14.1.

▶ Wile E. Coyote leaves his cave walking at a slow, constant speed. Then he stops to build a trap for Road Runner. After several minutes, he turns around and runs back to his cave at a constant speed. ◀

1. Act out this situation. As you act it out, describe what is happening to Wile E. Coyote's speed over time. Is his speed increasing, decreasing, or not changing? During which segment of the journey is his speed the greatest?

2. Graph Wile E. Coyote's speed over time. (*Note:* The vertical axis should represent speed, not distance or position.)

3. Compare your graph in Problem 2 to the distance-time graph and the position-time graphs that you constructed in Section 14.1 to describe this same situation.

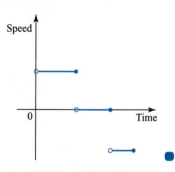

4. Could the given graph represent Wile E.'s situation? Why or why not? (For clarity, at a particular value for time, an open dot means that the point is not included; a solid dot means the point is included. Notice that no solid dot is joined to an open dot by a vertical line segment.)

Although negative speeds can be interpreted and make sense when thinking of slopes in distance-time graphs, they may not be important in some contexts. Your instructor will let you know how alert you should be to the possibility of using negative speeds in making your speed-time graphs.

> **TAKE-AWAY MESSAGE . . .** Because speed is a quantity that can change over time, a graph showing speed versus time makes sense. When a speed involves a decrease in distance, it is reasonable to consider that a negative speed, but for many purposes doing so is not essential.

Learning Exercises for Section 14.3

1. Graph each of the following situations, using the same speed-time graph. Then compare the graphs: How are they alike, and how are they different?
   a. Wile E. Coyote walks at a constant speed of 4 mi/h.
   b. Wile E. Coyote runs at a constant speed of 10 mi/h.
   c. Wile E. Coyote gets in a go-cart and travels at a constant speed of 15 mi/h.

2. In Exercise 1(b) from Section 14.1, you constructed a position-time graph for the following situation. Now construct a speed-time graph, and then compare the two graphs.

▶ Wile E. Coyote leaves his cave walking at a slow yet constant rate. He walks to the edge of a cliff. He is planning to build a catapult that will hit Road Runner by launching a large boulder. Just as Wile E. Coyote reaches the cliff, he remembers that he left the large rubber band for the catapult back in his cave. He turns around and walks back to his cave at the same pace as before. It takes him several minutes in the cave to find the large rubber band for the catapult. He realizes that he had better hurry so he starts out from his cave running (at a constant speed). About halfway to the cliff, he remembers that he also left the boulder at home. He turns around and runs home, at an even faster (yet constant) speed. ◀

3. The graph at right is a distance-time graph for a cartoon object's trip. Graph the *speed* at which the object is traveling over time.

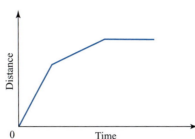

4. Consider the following situation. Since this is the cartoon world, assume it is possible to attain a constant speed immediately.

▶ Bart Simpson leaves his home and runs to a gas station at a constant rate. He stops suddenly for several minutes. Then he walks back home at a constant rate. ◀

Represent Bart's motion by sketching the following three different graphs:

a. a position-time graph (i.e., a graph of Bart's distance from home over time)
b. a distance-time graph (i.e., a graph of Bart's total distance traveled over time)
c. a speed-time graph

5. On each set of axes given below, sketch a graph to represent the following situation:

▶ A 5-inch flower is planted in a pot. It starts growing at a slow but constant rate. When it is moved into the sun, the plant starts growing at a faster (yet constant rate) until it is 3 times the original height. Then the plant is moved into a shady corner and it stops growing. ◀

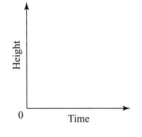 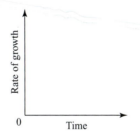

6. For each graph given, write a story about a journey that Bart Simpson took that could be described by the graph. If the graph represents an impossible situation, explain why. Add open and solid dots to the graphs (as in Activity 4, Problem 4) if your instructor wants you to.

a.

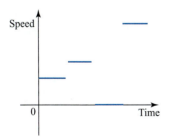

b.

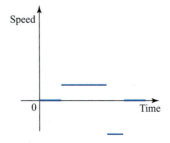

continue

c.

d.

7. For each of the following situations, draw two graphs, each of which fits the situation by showing two quantities varying simultaneously. The graphs should differ from each other in the quantitative relationship that they describe. For example, you may want to draw a distance-time graph and then a speed-time graph. Be sure to label the axes. Even though these describe real situations, it is all right to assume cartoon-like changes in speeds.

   a. Felecia walked from her house to the store. About halfway to the store, she realized that she had forgotten to bring any money. So she turned around and walked back home, this time walking a little faster. It took her several minutes at home to find her money. Then she left the house and ran all the way to the store.
   b. Pat drove to work on the city streets, stopping at three red lights along the way.
   c. A long-distance truck driver has an all-day trip of highway driving. He stops for a rest or meal break every 2 hours.
   d. A loaded truck drives up a hill and then down the hill.
   e. A space rocket was launched. It sped up when its booster rockets were fired. Then the rocket traveled at a constant speed, in a circular orbit around Earth for several days, collecting data. It reentered the atmosphere and eventually came down safely by parachute.

8. Each of the following containers pictured (side views) is being filled with water at a steady rate. Sketch qualitative graphs that show the height of the water in the container as time goes on. Assume that the water keeps running for a while after the container is full.

   a.   **b.**   **c.**

9. Below is a side view of a roller coaster.[5]

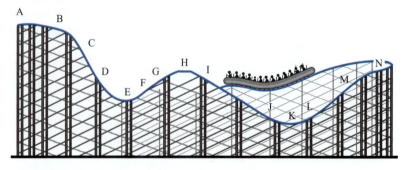

   a. Is the speed greater at B or at C? Why?
   b. Is the speed greater at G or at H? Why?
   c. Sketch a qualitative graph for the *speed* of the roller coaster, versus its position on the track.

Supplementary Learning Exercises for Section 14.3

1. Make a distance-time graph and a speed-time graph for the following story. Align them so that the pieces can be compared.

    ▶ Wile E. leaves his cave and trots steadily out to the spring. He continues on more slowly but at a steady pace, sniffing the ground as he goes. One smell was particularly interesting, so he stopped for a long sniff. He recognizes the Road Runner's smell and continues on, running fast. When he gets to the edge of a cliff, he stops and looks around. But the Road Runner is not in sight. ◀

2. For each, make a distance-time graph and a speed-time graph. Align them so the corresponding pieces can be compared. (Assume cartoon-like changes in speed.)

    a. Susan drove to the expressway without having to stop and, after reaching her exit, drove in city traffic to her office, hitting only one red light.

    b. A child rode down a hill in a wagon, dragging a foot to keep the speed the same, and turned over at the bottom of the hill. Then the child pulled the wagon back up the hill and rode down again.

    c. A plant grew steadily but slowly until it was fertilized and watered. Then it grew steadily but faster until it reached its maximum growth.

3. a. In a distance-time graph of Wile E., how can you tell when Wile E. has stopped?

    b. In a speed-time graph of Wile E., how can you tell when Wile E. has stopped?

    c. In a position-time graph of Wile E.'s distance from his cave, how can you tell when Wile E. has stopped?

4. For each of the following distance-time graphs for a trip for a cartoon character, draw the speed-time graph:

    a. 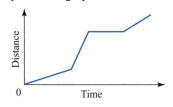   b.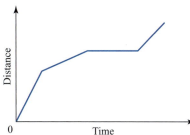

5. Below are sketches of three race courses. The race cars go clockwise on each course. Sketch a speed-time graph for a car making one lap on each course, giving an explanation of your reasoning. (*Suggestion*: Make some estimates of speeds at various points.)

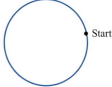  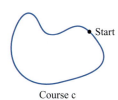

    Course a   Course b   Course c

14.4 Interpreting Graphs

In most of the work in this chapter so far, we have started with a situation, and you were asked to construct a graph that represents a relationship between quantities in that situation. That process will be reversed in this section. You will be given a graph and asked to create a situation that might be represented by the graph and to interpret the graph in terms of the quantities in that situation.

Notes

 ACTIVITY 5 Tell Me a Story

Write a story that could be represented by the given graph. In your story, identify the quantitative relationship being graphed.

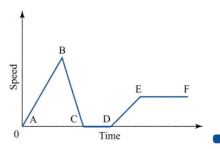

Wasn't it difficult to see that as the speed increased, it is unlikely that someone was going *up* a hill, and that as the speed decreased, it is unlikely that someone was going *down* a hill? Children have the same difficulty with distance-time graphs. (See Section 14.5.)

ACTIVITY 6 Climbing to the Top!

Consider the following situation:

▶ A child climbs up a slide at a constant speed, stops at the top, and then slides down. ◀

Examine the following graphs and for each graph answer the following questions:

1. Which graph best models the situation? Give your reasoning.
2. Which graph looks most like a playground slide? Could this graph represent the situation? Why or why not?
3. For each of the other graphs, explain how the graph fails to model the situation.

a.

b.

c.

d.

e.

f.

Many questions arise naturally from the types of graphs you have constructed and analyzed so far—e.g., during what part of a person's journey was he or she going the fastest? While it is possible to answer a question like this by examining visually the steepness of different parts of the speed-time graph, it will be important for you, as a teacher, to also create explanations that include a discussion of the quantities involved, quantities such as time versus distance or speed versus time. Giving explanations is what you will be doing when you teach, so giving explanations here is good practice, even though some of the ideas are more advanced than those you will be teaching. Good explanations tell *why* something might have happened and are not just reports of *what* happened. Good explanations should use technical language correctly and not mix up different ideas (for instance, using *height* when *rate* is intended).

ACTIVITY 7 Explanations Involving Time and Distance

The given graph represents Susan's distance from home, over time, for her trip from her house to the grocery store.

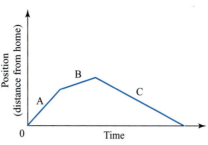

1. During which segment of her journey did Susan walk the farthest? How do you know?

2. During which segment of her journey did Susan take the longest? How do you know?

3. During which segment of her journey did Susan travel the fastest? Explain why. Be sure to use distance and time in your argument. (Do not rely simply on a visual estimation of steepness.) ●

Relating a graph and a table to a problem situation now appears in elementary school textbooks, as the excerpt below shows. This activity is from a fourth-grade textbook. Give it a try!

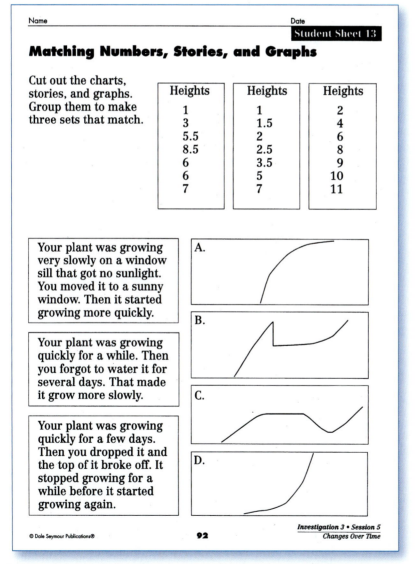

Name _____ Date _____

Student Sheet 13

Matching Numbers, Stories, and Graphs

Cut out the charts, stories, and graphs. Group them to make three sets that match.

Heights	Heights	Heights
1	1	2
3	1.5	4
5.5	2	6
8.5	2.5	8
6	3.5	9
6	5	10
7	7	11

Your plant was growing very slowly on a window sill that got no sunlight. You moved it to a sunny window. Then it started growing more quickly.

A.

Your plant was growing quickly for a while. Then you forgot to water it for several days. That made it grow more slowly.

B.

Your plant was growing quickly for a few days. Then you dropped it and the top of it broke off. It stopped growing for a while before it started growing again.

C.

D.

© Dale Seymour Publications®   92

Investigation 3 • Session 5
Changes Over Time

*From Cornelia Tierney, Ricardo Nemirovsky, and Amy Shulman Weinberg, *Changes Over Time Graphs*, Grade 4, p. 92. Copyright © 1998 Pearson Education, Inc. or its affiliate(s). Used by permission. All rights reserved.

TAKE-AWAY MESSAGE . . . Interpreting speed-time graphs requires careful thought, with a focus on how the speed (not the distance) is changing. Part of a teacher's job is to explain. Giving good explanations takes practice and, in part, careful attention to the language used.

Learning Exercises for Section 14.4

1. For each of the situations below and on the following page, circle *one* graph that best represents the situation. Explain why you selected that graph.

   a. Tanya starts to run up Mount Soledad on Mountain Road. She begins at a standstill and quickly gets up to a comfortable yet slow pace. She runs at this constant rate until she gets to the top of Mount Soledad. She slows to a stop and enjoys the view for about 15 minutes. Then she runs down the steep side of the mountain on Capri Road, going faster and faster until she reaches her top speed.

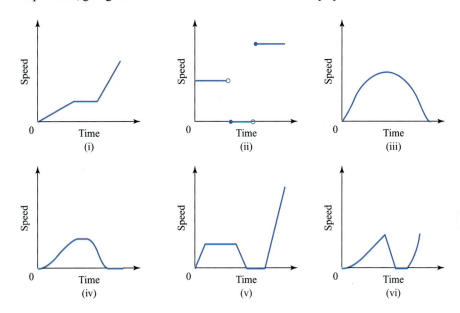

   b. A train pulls into a station and then stops to let off its passengers.

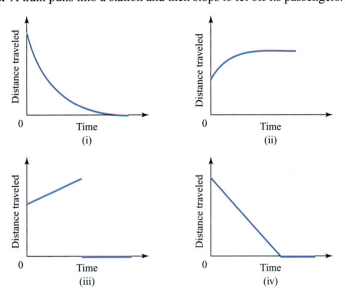

continue

c. Armando scales a rock wall at a very slow and erratic pace, stopping many times. When he reaches the top, he walks along the plateau at a brisk yet constant rate.

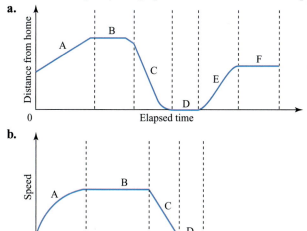

2. Tell a story about a journey that could be represented by each graph. Tell what happened in each lettered part of the graph. Be sure to talk about the speed represented by each part.

a.

b.

3. Suppose the speed-time graph in Learning Exercise 2 represents Janet's bike ride. For part of her ride, Janet was going up a steep hill and became very tired. Which segment of the graph might reflect that part of her ride? Explain.

4. The following graphs show the motion of two different boats:

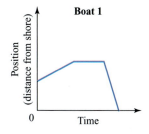

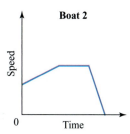

a. Tell a story about the journey of the two boats that could be described by the graphs above.

b. Notice that the graphs look identical except for the labels on the axes. Could Boat 1 and Boat 2 be the same boat—i.e., could the two graphs be describing the same motion? Why or why not?

continue

Notes

   c. Is there more than one way to interpret each graph? If so, give an alternate story for each graph. If not, explain why not.

5. The work from two students is shown below the given graph. Both were asked to write a story for a journey that could be represented by the graph. For each student, describe what you think the student understands mathematically and what he or she doesn't understand.

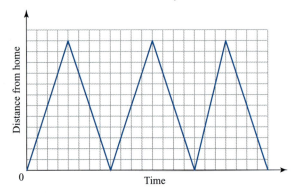

Student No. 1 (typed verbatim from the student's report):

This is a graph of my speed up a hill. It took me 4 hours to get up the hill, but I reduced the time by going down hill. I reached the bottom in half time it took me to reach the top. I continued doing this for 16 hours straight. The last few hours I got tired and didn't ride up the steep hill but a less steep hill.

Student No. 2 (typed verbatim from the student's report):

A Math Story

The Sea Wog (Militaries latest lard powered submarine) is prepared to go. The Admiral who will navigate the vessel is explaining to the crew why lard is a better fuel in times of defense cuts.

The Sea Wog sets off. By the time it is 12 miles out to sea The Admiral realizes that he had carelessly left the weapons back at the port. He sets back at the same speed he set out at. About 3 mph (lard is not a very good submarine fuel as far as speed is concerned).

The admiral had called ahead and asked for the weapons to be ready on port by the time he returned to avoid delay. When he had reached port he didn't need to stop or even slow down. They just threw the weapons in.

The Sea Wog sets out again at about 3 mph until again about 12 miles from shore. Then it was to the admirals surprise that this was not really his crew. His was back at the Dock.

When the Admiral had returned the submarine needn't even stop or slow down. The wrong crew jumped off, the right crew jumped on.

After 3 hours it was 12 miles out that it was discovered that there wasn't enough lard to make the trip. (I cant tell you to where the trip was. TOP SECRET you know.) They head back and decided to try again tomorrow.

6. The following graph represents Jordan's bike trip:

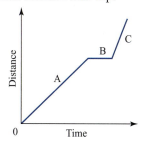

   a. Which segment of Jordan's trip took the most time? How do you know?

   b. During which segment of his trip did Jordan go the farthest? How do you know?

   c. During which segment of his trip did Jordan travel the fastest? How do you know? Use distance and time in your argument.

   d. Explain how you know that Jordan stopped during segment B. Use time and distance in your argument.

7. The following graph represents the change of the height of two candles (Candle A and Candle B) over time, as they burn:

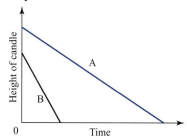

   a. Which candle was taller before they were lit (at $t = 0$)? How do you know?

   b. Which candle burned out sooner? How do you know?

   c. Which candle burned at a faster rate? How do you know? Use height and time in your argument.

8. Interpret the following graph[6] that gives information about an oven. Why is the last part of the graph wiggly?

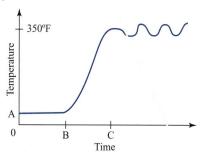

Supplementary Exercises for Section 14.4

1. Write a story that fits the distance-traveled versus time graph given.

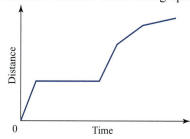

Notes

2. What does a speed-time graph like the one below mean?

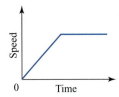

3. Below is a speed-time graph for Diane's bike ride. Describe Diane's bike ride.

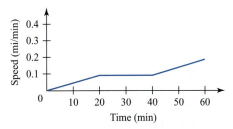

4. This graph represents Pat's walk from home to a snack shop, then to the library, and then back home.

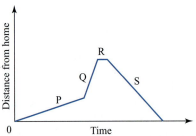

   a. In the graph, what would suggest that Pat forgot her money?
   b. During which part of the walk did Pat walk the farthest? How do you know?
   c. Which part of the trip took Pat the longest time? How do you know?
   d. In which part of the trip did Pat walk the fastest? Explain how you know, not in terms of slope, but in terms of distance and time considerations.
   e. Can you tell whether Pat's walk involved any steep hills? Explain.

5. One floral shop uses four types of vases. Below are side-view drawings of the types and four graphs that indicate the water level height, as the empty vases are filled with water. Match the vases and graphs and explain your choices.

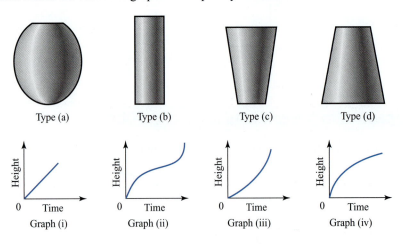

6. A special vase is used for the bouquet of flowers at the head table. Below is a graph indicating the height of the water in the vase as it is being filled with water at a steady rate. Make a sketch of what the vase should look like.

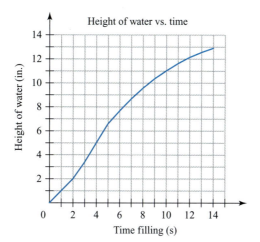

Height of water vs. time

14.5 Issues for Learning: Common Graphing Errors

In the past, algebra has often focused on problems that can be solved by symbolic representation alone.[7] As a result, students often consider visual representations as unnecessary. Our previous work (in "Part I: Reasoning About Numbers and Quantities") illustrated the efficacy of being able to represent a quantitative situation with a picture or diagram. A graph is a form of a picture that illustrates a situation and can lead to the solution of the problem. The ability to construct a visual representation such as a graph is evidence that a student understands the problem situation being presented. Yet it is well known that students have difficulties when constructing graphs.

Graphing errors result from either not knowing the *conventions* of drawing graphs or not understanding a problem well enough to represent it with a graph. Conventions of graphing have developed over time and have become agreed-on ways of representing information graphically. Following are some of these conventions:

1. A two-dimensional graph (used when two variables are being considered) has horizontal and vertical axes perpendicular to each other, with the intersection representing 0 on each axis.

2. Numbers along an axis must be equally spaced. That is, the distance between 3 and 4 must be the same as the distance between 1 and 2; the distance between 5 and 10 must be the same as the distance between 10 and 15. (There are exceptions, such as graphing involving special scales, but those are well beyond the scope of elementary and middle school mathematics.)

3. The axes typically show 0. However, if there are great distances between 0 and the points of interest, we can indicate a break in the otherwise uniform scale within the graph, e.g., by a crooked line such as the following:

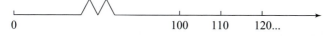

4. In analyzing relationships between quantities, we usually look at how one variable, called the dependent variable, depends on the other variable, called the independent variable. The independent variable is located along the horizontal axis, often called the *x*-axis, and the dependent variable is located along the vertical axis, often called

Notes

the y-axis. For example, the distance traveled is dependent on time traveled, not the reverse. Thus, time would be represented on the horizontal axis, distance on the vertical axis.

Some common graphing errors are demonstrated in the following student-drawn graphs based on the given table that tells the number of minutes a candle has burned.

Number of minutes a candle has burned	Height of the candle (cm)
0	15
2	14
5	11
12	9
20	5

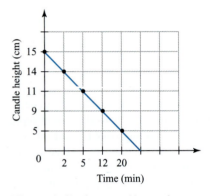

Figure 1 Student no. 1's graph.

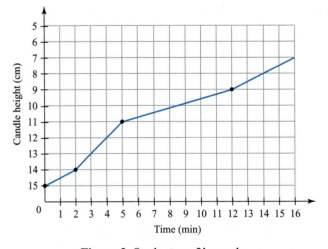

Figure 2 Student no. 2's graph.

✎ ACTIVITY 8   What Are the Graphing Errors?

1. **a.** What error(s) did Student No. 1 make?

    b. What do you think this student doesn't understand?

2. **a.** What errors(s) did Student No. 2 make?

    b. What do you think this student doesn't understand?

3. **a.** Make a correct graph of the data.

   b. What information about the quantitative relationship is captured in your graph that was lost in the students' graphs? ●

Research has also shown that when students in both elementary school and high school are asked to interpret graphs, they often exhibit what is known as the *"graph as picture" misconception*.[8] They ignore the quantities on the axes, interpreting the graph as a realistic image or picture of an object or event. For example, a group of students were shown the graph below. Then they were asked to describe a journey that they took that could be described by the graph. Two typical incorrect responses are given below:

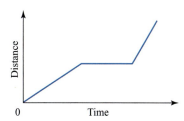

Student 1: "First I started walking up a hill. Then the hill flattened out. Then there was another, steeper hill that I walked up."

Student 2: "I rode my bike down a street. Then I turned a little to the right and rode for a while. Then I turned left and rode a long way."

DISCUSSION 2 What Does the Graph Say?

1. Explain what each student might have been thinking.

2. Why does it make sense to say that both of these students have the "graph as picture" misconception?

3. What is a correct interpretation of the graph? ●

A graph is only one way of representing a situation, but it is a powerful way that in many cases makes a problem situation easier to understand. Some situations can be represented with a table, an equation, and a graph, each showing the information given in the problem. Students should be able to move among these representations with ease, although equation representations are usually learned later than tables and graphs.

Graphing calculators are now being used in algebra classes in many institutions. These calculators offer new ways of coming to understand quantitative situations by producing graphs quickly and efficiently. They also provide the capability of considering and comparing a variety of quantitative situations, and of zooming in and out to understand better how a graph represents a situation. If used properly, graphing calculators can also help students develop a fundamental understanding of graphs—how they are made and what they tell us.

14.6 Check Yourself

After this chapter, you should be able to do problems like those assigned and to meet the following objectives:

1. Draw qualitative graphs for a situation (such as a Wile E. Coyote story), both a distance-time graph and a position-time graph.

2. Write a story for a given qualitative graph.

3. Relate speed to slope in a distance-time graph.

4. Sketch a speed-time graph or distance-time graph for a given situation.

5. Write a story for a given speed-time graph.

6. Interpret the changes illustrated by various parts of a distance-time or position-time or speed-time graph.

7. Explain and illustrate what is meant by the "graph as picture" misconception (Section 14.5).

References for Chapter 14

[1] Wile E. Coyote and Road Runner are favorite cartoon characters created by animator Chuck Jones in 1949 for a cartoon series produced by Warner Brothers.

[2] The use of qualitative graphs in this chapter builds on the work of Malcolm Swan and his colleagues at the Shell Centre, Nottingham, UK: *The language of functions and graphs,* 1985.

[3] Adapted from Greer, B. (1987). Nonconservation of multiplication and division involving decimals. *Journal for Research in Mathematics Education, 18*(1), 37–45.

[4] All activities and exercises in Section 14.2 were written by Helen Doerr from Syracuse University and revised by Janet Bowers and Joanne Lobato from San Diego State University.

[5] The sketch and problem are from the National Council of Teachers of Mathematics, *Curriculum and evaluation standards for school mathematics*, 1989, p. 83, as an example of reasoning about graphs.

[6] The graph is from the National Council of Teachers of Mathematics, *Curriculum and evaluation standards for school mathematics*, 1989, p. 155.

[7] Yerushalmy, M., & Schwartz, J. L. (1993). Seizing the opportunity to make algebra mathematically and pedagogically interesting. In T. A. Romberg, E. Fennema, & T. P. Carpenter (Eds.), *Integrating research on the graphical representation of functions* (pp. 42–68). Hillsdale, NJ: Lawrence Erlbaum.

[8] Kaput, J. J. (1987). Representation systems and mathematics. In C. Janvier (Ed.), *Problems of representation in the teaching and learning of mathematics* (pp. 19–26). Hillsdale, NJ: Lawrence Erlbaum.

Chapter 15

Further Topics in Algebra and Change

Sometimes, the algebraic description of a story can involve more than one equation. In this chapter we first consider how to get the equation for a line from information from a graph. We then examine some methods for solving story problems that can lead to more than one equation (Sections 15.2 and 15.3). We next look at rate situations that involve questions about the average rate (Section 15.4). Section 15.5 notes the wide applicability of functions and gives an approach to predicting more complicated function rules. Section 15.6 reviews some of the difficulties that students have in algebra.

Notes

15.1 Finding Linear Equations

You know how to find the equation of a line, given its slope (or rate of change) m and its y-intercept $(0, b)$: $y = mx + b$. But other graphical information can also enable you to write a linear equation. After all, isn't "Two points determine a line" a common statement in geometry? Consider this example.

EXAMPLE 1

▶ Wile E. Coyote is at the top of a ski slope when he spots Road Runner sleeping in the valley below. Wile E. skis toward Road Runner at a steady speed even though he is not a skilled skier. After 2 minutes of skiing, Wile E. is 5450 feet from the valley, and after 4 minutes of skiing, he is still 3650 feet from the valley. How high was the top of the ski slope, and when will Wile E. reach the valley? ◀ ●

A rough sketch of a graph of height above the valley versus time shows that the story in Example 1 gives information about two points: (2, 5450) and (4, 3650). If we could find the equation of the line, we could answer both questions in the problem. In other situations, you might have information about the slope and some point *not* on the y-axis. In both these cases, we will see how to find the equation of the line.

In each case the idea is simple: *Write an expression for the slope using a variable point (x, y) on the line, and fit that result to the other information.* This plan works because on a line the rate of change for every two points—i.e., the slope—is the same. So any two expressions or values for the slope of a given line will be equal.

Two-point case. Suppose, as in Example 1, the line goes through the points (2, 5450) and (4, 3650). Let h be the height above the valley (in feet) and t be the time Wile E. has been skiing (in minutes).

Step 1. Using a variable point (t, h) and either point, write an expression for the slope:

$$\frac{h - 5450}{t - 2} \quad \text{or} \quad \frac{h - 3650}{t - 4}$$

Step 2. Calculate the slope using the coordinates of the known points:

$$\frac{5450 - 3650}{2 - 4} = \frac{1800}{^-2} = {}^-900 \quad \text{or} \quad \frac{3650 - 5450}{4 - 2} = \frac{^-1800}{2} = {}^-900$$

continue

Step 3. Equate the results from Steps 1 and 2 and simplify as desired: $\dfrac{h - 5450}{t - 2} = {}^-900$.

So, $h - 5450 = {}^-900(t - 2)$, leading eventually to $h = {}^-900t + 7250$.

From $h = {}^-900t + 7250$, we see that when Wile E. was at the top of the slope ($t = 0$), the height was 7250 feet. Furthermore, he will reach the valley when $h = 0$, and from $0 = {}^-900t + 7250$, we find that will occur after slightly more than 8 minutes of skiing. (We can also tell how fast Wile E. was skiing from the slope: 900 ft/min downward.)

> *Think About …*
>
> In the two-point case illustrated, would you get the same final equation if you used the $\dfrac{h - 3650}{t - 4}$ expression for slope instead of $\dfrac{h - 5450}{t - 2}$ at Step 3?

Let us examine the situation in which we know the slope and some point (not the y-intercept). For example, we might know Wile E.'s speed and his location at some point after $t = 0$.

Point-slope case. Suppose the line goes through the point (5, 3) and has slope 7. We can get a $\dfrac{\text{change in } y \text{ values}}{\text{change in } x \text{ values}}$ expression for the slope using the coordinates of a *variable* point (x, y) and the known point (5, 3); we know that must equal 7:

$$\frac{y - 3}{x - 5} = 7$$

Multiplying both sides by $x - 5$ and simplifying give $y = 7x - 32$, the familiar form.

> *Think About …*
>
> Does the point-slope case also give $y = mx + b$ for a line with slope m and the special case when the point is the y-intercept (0, b)?

The Learning Exercises give you a chance to practice finding linear equations for the two cases.

TAKE-AWAY MESSAGE . . .   A graph of a linear equation can lead to the $y = mx + b$ form. But information about (1) any two points on the line or (2) any point on the line and the slope of the line can also give an equation for the line.

Learning Exercises for Section 15.1

1. Suppose that the points (4, 7) and (9, 17) lie on the same line.
   a. Calculate the slope of the line.
   b. Write an expression for the slope of the line, using a variable point (x, y) on the line and either of the given points.
   c. Write an equation using your answers to parts (a) and (b). Simplify the equation to the $y = mx + b$ form.

2. Suppose that a given line passes through the point (5, 9) and has slope 6.
   a. Write an expression for the slope of the line, using a variable point (x, y) on the line.
   b. Write an equation using your answer to part (a) and the given information that the slope is 6. Simplify the equation to the $y = mx + b$ form.
   c. What characteristic of lines allows you to know that the expression from part (a) is indeed equal to 6?

In Learning Exercises 3–14, write an equation for the line that is described. Simplify to the $y = mx + b$ form.

3. Has slope 4 and passes through the point (12, 6).

4. Has slope $^-2$ and passes through the point (1, 3).

5. Has slope $^-3$ and passes through the point (2, 3.5).

6. Has slope 7 and passes through the point (100, 105).

7. Has slope $\frac{3}{4}$ and passes through (24, 8).   **8.** Has slope $\frac{8}{9}$ and passes through (16, 4).

9. Passes through points (4, 1) and (7, 3).   **10.** Passes through points (2, 5) and (9, 6).

11. Passes through points ($^-2$, 3) and (5, 9).   **12.** Passes through points (4, $^-3$) and (11, 11).

13. Passes through points ($1\frac{1}{2}$, 5) and (3, $2\frac{1}{2}$).

14. Passes through points (6, $2\frac{1}{3}$) and ($3\frac{1}{3}$, 4).

15. A triangle has its vertices (corners) at the points (2, 3), (6, 3), and (5, 2). What three equations describe the lines that the three sides of the triangle are on?

16. A spring stretches when weights are hung on it (up to a certain limit): The greater the weight, the greater the amount of stretch. Write an equation that relates the weight w to the amount s of stretch, if the spring stretches 2 in. when 8 lb are hung from it, and 3 in. when 12 lb are hung from it.

17. Your upper elementary students will likely be learning about how temperature changes with increasing altitude. Within the troposphere (up to an altitude of 12 km), a linear function describes the relationship between altitude and temperature. Suppose at an altitude of 3 km, the temperature is 2.7°C. At an altitude of 7 km, the temperature is $^-23.3$°C. Given the temperature at these two different altitudes, write a linear equation that could be used to predict the temperature at any given altitude up to 12 km.

18. a. Find the slopes of line p and of line q in the graph to the right.

b. What is the product of the two slopes? (Notice that the two lines make a right angle.)

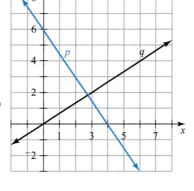

19. The result suggested in Learning Exercise 18 is true in general: If the product of the slopes of two lines equals $^-1$, the two lines are perpendicular (make a right angle of 90°). In each part, tell whether the graphs of the two equations are perpendicular, without graphing the equations.

a. $\begin{cases} y = 2x + 59 \\ y = \frac{^-1}{2}x + 42 \end{cases}$ **b.** $\begin{cases} y = \frac{^-3}{4}x + 12 \\ 3y = 4x + 7 \end{cases}$ **c.** $\begin{cases} 2x = 6y + 3 \\ 4y + 12x = 11 \end{cases}$

20. In the following, write an equation that meets each set of conditions:

a. Its graph is parallel to the graph of $y = 7x - 4$.

b. Its graph is parallel to the graph of $3y + 5x - 7 = 0$.

c. Its graph is perpendicular to the graph of $y = 4x + 8$.

d. Its graph is perpendicular to the graph of $2y - 7x = 4$.

21. What is the slope of the x-axis? What is the slope of the y-axis? The two axes are perpendicular. Is the product of their slopes equal to $^-1$?

22. In the following, write an equation that meets each set of conditions:

a. Its graph is perpendicular to the graph of $y = {}^-5x + 2$ and passes through the point (3, 7). (*Hint*: See Learning Exercise 19.)

b. Its graph is parallel to the graph of $y = {}^-5x + 2$ and passes through the point (3, 7).

c. Its graph makes right angles with the graph of $y - 3x = 9$ and passes through the point (10, 16).

d. Its graph is perpendicular to the graph of $2x - 6y = 11$ and passes through the point (5, $^-2$).

Supplementary Learning Exercises for Section 15.1

In Supplementary Exercises 1–10, write an equation of the line described. Simplify to the $y = mx + b$ form.

1. Passes through points (2, 5) and (4, 9).

2. Passes through points (1, 7) and (6, 22).

3. Passes through points (1, 9) and ($^-$4, 14).

4. Passes through points (6, $2\frac{1}{2}$) and ($4\frac{1}{2}$, 2).

5. Has slope 8 and passes through the point (1, 3).

6. Has slope $\frac{3}{4}$ and passes through the point (4, 5).

7. Has slope $^-$4 and passes through the point (1, $^-$11).

8. Has slope $^-$2 and passes through the point ($^-$1, $^-$7).

9. Is parallel to the line of $y = 6x + 1$ and passes through the point (1, 3).

10. Is parallel to the line of $y + 2x = 5$ and passes through the point (1, 7).

11. Use slope to explain how you know, without graphing, that the points (0, 0), (2, 4), and (3, 7) do *not* lie on the same line.

12. Tell how you know that the graph of $y - 5x = 7$ and the line through the points (2, 12) and (5, 27) do not have any points in common.

13. Wile E. parachuted from an airplane at a steady speed to capture Road Runner. Three minutes after jumping, Wile E. was 3200 ft up, and 7 min after jumping, he was 400 ft up.

    a. What was Wile E.'s speed on the way down?
    b. How high was Wile E. when he jumped?
    c. When did Wile E. land?

14. Road Runner saw Wile E. coming out of his cave, so Road Runner started running in a direction away from the cave at a steady rate of 45 ft/s. After 6 s, Road Runner was 500 ft from Wile E.'s cave. How far from the cave was Road Runner when he started to flee?

15.2 Solving Two Linear Equations in Two Variables

Sometimes, a problem can lead to two equations in two variables, rather than just one equation with one variable. Example 2 gives a story about elevators to illustrate a situation that can lead to two equations in two variables.

EXAMPLE 2

The elevators in the following story move steadily at a constant rate, and do not stop at particular floors, unless otherwise noted.

▶ A building has a freight elevator as well as a passenger elevator. Both elevator shafts are 600 feet long. Both elevators start at the same time, $t = 0$. We have this information about the passenger elevator: 4 minutes after starting, it was at the top of its shaft, 600 feet up; 1 minute after starting, it was 150 feet up. This information is compactly described by the ordered pairs, (4, 600) and (1, 150). With a similar notation, we have this information about the freight elevator: (2, 360) and (3, 240). Exactly when (if ever) are both elevators at the same height, and exactly what is that height? ◀

The ordered pair notation suggests a graphical approach. The following diagram shows the graphs for parts of the trips of the two elevators:

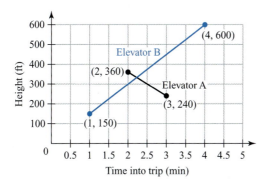

DISCUSSION 1   The Graphs and the Questions

Analyze the graphs in Example 2.

1. Which graph belongs to the passenger elevator and which to the freight elevator?

2. What background about the elevators assures that the graphs will be along straight lines?

3. What are the coordinates of the endpoints of the two segments?

4. How can the graph be useful in answering the questions given in Example 2? Try to answer the questions. ●

You have no doubt found that many times reading a graph, even a well-made one, is difficult and may not give exact answers. It is here that using algebra may help because algebra can give exact answers.

Using the information in Example 2 and the two-point approach to finding a linear equation, we can write two equations, one for each elevator, relating the height h of the elevator (in feet) to the time t elapsed (in minutes).

$$\text{For the passenger elevator:} \quad \frac{h - 600}{t - 4} = \frac{600 - 150}{4 - 1}$$

$$\text{For the freight elevator:} \quad \frac{h - 360}{t - 2} = \frac{360 - 240}{2 - 3}$$

Verify that these equations simplify to the following ones:

$$\text{Passenger:} \quad h = 150t$$

$$\text{Freight:} \quad h = 600 - 120t$$

If the two elevators are at exactly the same height, then we can *substitute* the $150t$ expression from the passenger equation for the h in the freight equation: $150t = 600 - 120t$. This resulting equation is one you know how to solve:

$$150t = 600 - 120t$$
$$270t = 600$$
$$t = \frac{600}{270} = \frac{20}{9} = 2\tfrac{2}{9}$$

That is, the two elevators are at the same height after $2\tfrac{2}{9}$ minutes have elapsed. So, we know *when* the two elevators are at the same height, but *what height* is it? Now that we know the number of minutes, we can use *either* of the *original* equations to find the exact height when the elevators are at the same level:

$$\text{From } h = 150t, \text{ we have } h = 150 \cdot 2\tfrac{2}{9} = 333\tfrac{1}{3} \text{ feet.}$$

$$\text{From } h = 600 - 120t, \text{ we have } h = 600 - 120 \cdot 2\tfrac{2}{9} = 333\tfrac{1}{3} \text{ feet.}$$

Notes

So, we now know the answers to the questions in Example 2: $2\frac{2}{9}$ minutes after both elevators started, they will be exactly at the same height of $333\frac{1}{3}$ feet. Reading the graph alone would not likely have given these exact answers.

> In answering the questions in Example 2, we have demonstrated the **substitution method**. This method can be applied whenever you are seeking a common solution to a set of equations.

ACTIVITY 1   Using the Substitution Method

The graphs of each of the following pairs of equations have a point in common. Use the substitution method to find the coordinates of that point.

1. $\begin{cases} y = 2x - 5 \\ y = 5x - 14 \end{cases}$ 2. $\begin{cases} y = x + 50 \\ 2x + y = 110 \end{cases}$ 3. $\begin{cases} s = t + 1\frac{1}{2} \\ s = 2t - 1 \end{cases}$

The substitution method is sometimes cumbersome to use, as it is with this set of two equations:

$$\begin{cases} 2y + 3x = 9 \\ 8y - 5x = 12 \end{cases}$$

After some preliminaries, we will show a second method of solving such equations, called the *addition/subtraction method*.

If someone were to ask you to graph $x = 4$, you might wonder what to do because this equation has just one variable. But in the context of two variables, $x = 4$ means $x = 4$ no matter what values y takes on. Graphically, all the points $(4, 0)$, $(4, 0.75)$, $(4, 2.3)$, etc., have $x = 4$. Hence, the graph for $x = 4$ is a vertical line through the point $(4, 0)$.

Similarly, the graph of $y = 3$ is a horizontal line, because *any* point $(x, 3)$ has $y = 3$.

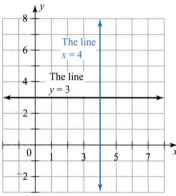

> **Think About ...**
>
> Why is the slope of the line for $x = 4$ said to be undefined? On the other hand, what is the slope of the line for $y = 3$?

The next two activities lead to a second method for finding the common solution to a set or system of equations.

ACTIVITY 2   Surprise No. 1

a. Graph $y = 2x + 1$.

b. Triple each side of $y = 2x + 1$ to get $3y = 6x + 3$, and graph that equation on the same coordinate system.

c. Multiply both sides of $y = 2x + 1$ by some other (nonzero) number and graph that equation on the same coordinate system. ●

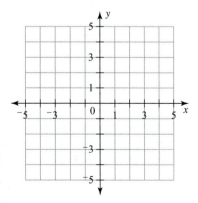

Perhaps you were not surprised that although multiplying both sides of an equation by the same (nonzero) number gives a new-looking equation, nonetheless this new equation has the *same* graph as that of the original equation. But Activity 3 surprises most people.

 **ACTIVITY 3**   Surprise No. 2

Notes

Consider the system

$$\begin{cases} 2y + 3x = 9 \\ 3y - 1.5x = 1.5 \end{cases}$$

Partial graphs of these equations are given on the following coordinate system. [In this case, the graphs *suggest* that (2, 1.5) is the common solution, but we are using this example to demonstrate eventually a second algebraic method for solving the system exactly.]

a. We first add the two equations, as follows:

$$\begin{array}{r} 2y + 3x = 9 \\ \underline{3y - 1.5x = 1.5} \\ 5y + 1.5x = 10.5 \end{array}$$

Graph the new equation on the given coordinate system.

b. How are the three lines related?

c. Suppose instead we subtract the equations, as follows:

$$\begin{array}{r} 2y + 3x = 9 \\ \underline{-(3y - 1.5x = 1.5)} \\ {}^{-}1y + 4.5x = 7.5 \end{array}$$

Graph the new equation, $^{-}1y + 4.5x = 7.5$, on the given coordinate system.

d. How is the new line related to the previous three lines? ●

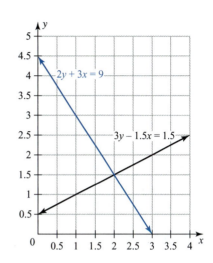

Activities 2 and 3 suggest the following two ideas, which are true in general:

> **1.** Multiplying both sides of an equation by the same nonzero number gives an equation with the same graph as the original.
> **2.** Adding or subtracting the sides of two equations gives a new equation whose graph passes through the common point for the first two equations.

The payoff comes in (usually) combining these two ideas to *eliminate* one of the variables in the new equation. Suppose we take the system of Activity 3, double the second equation, and then add that new equation to the first original equation:

$$\begin{cases} 2y + 3x = 9 \\ 3y - 1.5x = 1.5 \end{cases} \quad \text{doubling the second equation gives} \quad \begin{cases} 2y + 3x = 9 \\ 6y - 3x = 3 \end{cases}$$

Now add these last two equations:

$$\begin{array}{r} 2y + 3x = 9 \\ \underline{6y - 3x = 3} \\ 8y = 12 \end{array}$$

From $8y = 12$, we get $y = 1.5$, whose graph (a horizontal line) passes through the common solution to the original system. Substituting $y = 1.5$ into either of the original equations, $2y + 3x = 9$ or $3y - 1.5x = 1.5$, gives $x = 2$. So, the common solution to the system is $x = 2$, $y = 1.5$, which means that the graphs of the system intersect at the point (2, 1.5). This solution is illustrated in the graph on the next page.

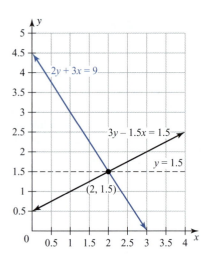

This solution method is often called the **addition/subtraction method**, in contrast with the *substitution method*. In solving the previous system, we eliminated the x terms. However, we could have chosen to eliminate the y terms instead, getting an equation such as $x = 2$. (Graphically, the line for $x = 2$ passes through the common point of intersection.) Substituting $x = 2$ into one of the original equations gives the y value of the common solution, namely $y = 1.5$.

 ACTIVITY 4   Using the Addition/Subtraction Method

Find the common solution of each system using the addition/subtraction method. Some hints follow the systems, in case you get stuck.

1. $\begin{cases} 2y + 3x = 19 \\ 2y - x = 7 \end{cases}$ **2.** $\begin{cases} 4x + 3y = 17 \\ 2x + 5y = 12 \end{cases}$

3. $\begin{cases} x + y = 85 \\ 2x + 3y = 195 \end{cases}$ **4.** $\begin{cases} 3r + 4s = 43 \\ 6r - s = 5 \end{cases}$

Hints:

1. Subtract as is, or triple the second equation and add.

2. Double the second equation and subtract.

3. Double or triple the first equation.

4. Double the first equation or quadruple the second equation. ●

The two methods for solving a system of linear equations assume that the graphs of the equation are different and do cross. What happens if they are not different or do not cross?

 ACTIVITY 5   Some Anomalies

1. Try to solve this system by substitution. What happens?

$$\begin{cases} y = 2x + 1 \\ 2y - 4x = 2 \end{cases}$$

2. Try to solve this system by substitution. What happens?

$$\begin{cases} y = 3x \\ 2y = 6x + 7 \end{cases}$$ ●

In Problem 1 of Activity 5, you may have ended with an equation like $2 = 2$, giving no information about x or y. In such cases, *all* the solutions to one equation are also solutions to the other equation. If we end with a statement that is always true, such as $2 = 2$, then the graphs of the two equations are exactly the same. Such equations are called **dependent equations**.

In Problem 2 of Activity 5, you may have reached an impossibility such as $0 = 7$. The graphs of the two equations give parallel lines (check the two slopes for the equations in Problem 2). Because parallel lines never intersect, there is no common solution. Such equations are called **inconsistent equations**.

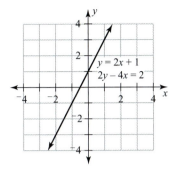

Dependent equations

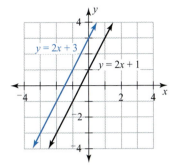

Inconsistent equations

TAKE-AWAY MESSAGE ... Although graphing to find the common solution to a system of two linear equations in two unknowns is theoretically possible, the method may not give exact results. If there is a single common solution, algebraic methods like the substitution method and the addition/subtraction method can give exact results. Dependent equations do not have a *single* common solution. Inconsistent equations have *no* common solution.

Learning Exercises for Section 15.2

1. The distance-time graph for Rolf intersects the distance-time graph for Annette at the point (2.5, 7). (Each graph gives the distance in miles and the time in hours.) What does the common point (2.5, 7) tell you?

Use the *substitution method* in Learning Exercises 2–7 to find the common solution, if there is one. If there is no common solution, tell how you know that.

2. $\begin{cases} y = 2x + 1 \\ y - x = 4 \end{cases}$ 3. $\begin{cases} y = 5 + x \\ 3x = y + 5 \end{cases}$ 4. $\begin{cases} y = x + \frac{1}{2} \\ 4y - x = 8 \end{cases}$

5. $\begin{cases} y = 1 - x \\ 3y + 2x = 5 \end{cases}$ 6. $\begin{cases} x = 6 + y \\ 6x - y = 11 \end{cases}$ 7. $\begin{cases} y = 3x - 8 \\ 2y - 6x = 21 \end{cases}$

Use the *addition/subtraction method* in Learning Exercises 8–13 to find the common solution, if there is one. If there is no common solution, tell how you know that.

8. $\begin{cases} 3y + 2x = 15 \\ 2y - x = 3 \end{cases}$ 9. $\begin{cases} 5x + 2y = 17 \\ 3x + 2y = 15 \end{cases}$ 10. $\begin{cases} x + 2y = 250 \\ x - y = 25 \end{cases}$

11. $\begin{cases} x + y = 16 \\ 3x - 2y = 13 \end{cases}$ 12. $\begin{cases} y + 4x = 9 \\ 7x = y + 2 \end{cases}$ 13. $\begin{cases} 2x + y = 9 \\ 4x = 4y - 9 \end{cases}$

Use your choice of methods to find the common solution to each system in Learning Exercises 14–19.

14. $\begin{cases} 2x - y = 13 \\ y = x - 3 \end{cases}$ 15. $\begin{cases} 17 - y = x \\ x = 2y + 2 \end{cases}$ 16. $\begin{cases} x - y = 12 \\ 2y = x + 19 \end{cases}$

17. $\begin{cases} 8x + 2y = 13 \\ y = 25 - 4x \end{cases}$ 18. $\begin{cases} 6x = y + 40 \\ \frac{3}{2}y + 3x = 26 \end{cases}$ 19. $\begin{cases} x + y = 2 \\ 3y - x = 14 \end{cases}$

20. a. Without graphing the equations, tell how you know that the graphs of $y = 4x - 7$ and $y = 4x + 5$ do not have any points in common.
   b. Joe graphs $y + 2x = 1.5$ and $2y + 4x = 3$ and is puzzled, saying, "I can't find the answer." Tell Joe what has happened.

21. a. Give an equation that is inconsistent with $y = 3x - 10$.
   b. Give an equation that is dependent with $y = 3x - 10$.

22. a. Write two linear equations whose graphs intersect only at the point (5, 7).
   b. Write a linear equation whose graph is parallel to the graph of $y + 4x = 6$.

Learning Exercises 23–29 describe elevator trips in parallel shafts. Assume that the passenger and freight elevators move at steady speeds and can start and stop instantly. The trips are nonstop unless otherwise indicated. Both elevators start at the same time. A notation like (4, 600) means that 4 minutes into a trip, the elevator is 600 feet above the bottom of the elevator shaft. Rough sketches on a coordinate system may be helpful.

23. In the Baldwin Building,

   Passenger: (0, 200) to (4, 600)

   Freight: (1, 350) to (5, 50)

   a. When, if ever, are the elevators at the same height?
   b. How fast does each elevator travel?
   c. How high was the freight elevator when the trips started?

24. In the Century Building,

   Passenger: (1, 450) to (4.6, 0)

   Freight: (2, 100) to (4, 200)

   a. When, if ever, are the two elevators at the same height during this trip? What height is that?
   b. How fast does each elevator travel?
   c. How high was each elevator when the trips started?

25. In the Dromedary Building,

   Passenger: (3, 0) to (6, 360)

   Freight: (0, 75) to (6, 225)

   a. When, if ever, are the two elevators at the same height during this trip? What height is that?
   b. How fast does each elevator travel?

26. In the Ellis Building,

   Passenger: Starts 400 ft up, travels to (2, 40)

   Freight: (0.5, 100) to (3, 250)

   a. When, if ever, are the two elevators at the same height during this trip? What height is that?
   b. How fast does each elevator travel?

27. In the Frankenstein Building,

   Passenger: (1, 220) to (3, 500)

   Freight: (0, 50), and goes up at a speed of 25 ft/min

   a. How do you know that the two elevators are not at the same height at any time during this trip?
   b. Someone says, "I got a negative number for my answer." What would that mean?

28. In the Excelsior Building, the passenger elevator started at the bottom of the parallel elevator shafts and made two stops during the first 9 min (see the following graph), with the first stop at 240 ft and the second at 360 ft. The freight elevator started at the same time as the passenger elevator, and after 1 min was 500 ft from the bottom of the elevator shaft. After 5 more min of nonstop travel, the freight elevator was at 200 ft. When, if ever, were the two elevators at the same height? (Graph the freight elevator's trip on the same coordinate system.)

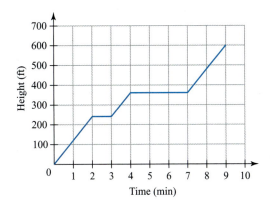

29. The following times and locations were recorded for two elevators in the Founders Building:

Passenger: 12:02 PM, at 150 ft headed up.
At 12:04:30 PM, stopped at 300 ft up.
Started again at 12:05 PM, headed down.
12:10 PM, at bottom of elevator shaft.

Freight: 12:02 PM, at 280 ft up.
Made a nonstop trip to 56 ft, arriving there at 12:10 PM.

At what *two* times are the two elevators at the same height? (*Hint*: A rough graph is very helpful.)

30. Two saleswomen talked about their pay. Gayle said, "I get $30,000 a year plus 2.4% commission on my sales." Helena said, "I get $25,000 a year plus 4% commission."

    a. Write an equation for each woman's income I per year versus her sales S in dollars.
    b. Is there a sales figure that gives the same annual income for the two women? If so, tell what it is and indicate what annual income that would yield.

Supplementary Learning Exercises for Section 15.2

1. **a.** What does it mean to say that (5, 4) is a common solution to the following system?
$$\begin{cases} 3y + 2x = 22 \\ y = x - 1 \end{cases}$$
   b. *Is* (5, 4) a common solution to the system in part (a)?
   c. If (2, 7) is a common solution of some other system of equations, what does that mean graphically for the graphs of that system?

2. **a.** What does it mean to say, "This system of equations is inconsistent"?
   b. What does it mean to say, "Those equations are dependent"?

   Find a common solution for each system in Supplementary Exercises 3–12. It is a good idea to practice both the substitution method and the addition/subtraction method.

3. $\begin{cases} y = x + 3 \\ 2y = 3x + 1 \end{cases}$ 4. $\begin{cases} 2x - 5y = 5 \\ 2x + y = 23 \end{cases}$ 5. $\begin{cases} x - y = 3 \\ 6x + 2y = 2 \end{cases}$

6. $\begin{cases} y = 5 - 2x \\ y = {}^-2x + 10 \end{cases}$ 7. $\begin{cases} s + 2t = 9 \\ 3s - 4t = 2 \end{cases}$ 8. $\begin{cases} x + 2y - 5 = 0 \\ 2x = y + 5 \end{cases}$

Notes

9. $\begin{cases} y = {}^-x - 1 \\ 3y = 4x - 1 \end{cases}$

10. $\begin{cases} x - y = 4 \\ 3y = 2x + 13 \end{cases}$

11. $\begin{cases} 2x + 4y = 17 \\ 2y - 8\frac{1}{2} = {}^-x \end{cases}$

12. $\begin{cases} 3x - 2y + 3 = 0 \\ 3y - 2x + 8 = 0 \end{cases}$

13. **a.** Write an equation that is inconsistent with $y = 5x$.

   b. Write an equation that is dependent with $y - 5x = 19$.

14. **a.** Write a linear equation that has a graph parallel to the graph of $20x - 42y = 163$.

   b. Write two linear equations that have graphs crossing at the point (2, 3).

15. The Jones, Smith, and Quan families each donated money to the same charity. The Joneses and Smiths gave a total of $325. The Smiths and Quans gave a total of $475. The Quans gave 3 times as much as the Joneses. How much did each family give to the charity? (*Hint*: There appear to be three unknowns needed here, but you can manage with two.)

16. In one basketball game, Misha and Sue scored a total of 24 points. Misha and Lavonne scored a total of 32 points. Lavonne scored twice as many points as Sue scored. How many points did each woman score?

17. In another basketball game, Misha and Sue scored a total of 26 points. Misha and Lavonne scored a total of 21 points, but Lavonne scored only half as many points as Sue. How many points did each woman score?

18. On a shopping trip, Lucia and Marcy spent a total of $44. With Angela, the three together spent $57.20, and Marcy spent one-third as much as Angela. How much did each of the three spend?

19. One building with elevator shafts 800 ft long has a passenger elevator and a freight elevator. The passenger and freight elevators left their positions at the same time. The passenger elevator was 100 ft above the bottom of the shaft and headed up; the freight elevator was 250 ft above the bottom of the shaft and headed down. After nonstop trips, the passenger elevator was 460 ft from the bottom after 4 min, and the freight elevator was 100 ft from the bottom after 5 min. When, if ever, were the two elevators at the same height?

20. (This problem is more involved; a rough graph is recommended.) In a skyscraper, two passenger elevators start trips at the same time. Passenger elevator A leaves from the bottom of the elevator shaft and stops 150 ft up after 2 min. After stopping for 1 min, A continues up at the same speed as before. Passenger elevator B starts the trip 10 ft from the bottom of the shaft and goes up nonstop at a speed of 60 ft/s. Find the *three* times that A and B are at the same heights.

15.3 Different Approaches to Problems

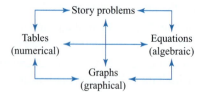

This section gives you practice with three representations of a problem—numerical, graphical, and algebraic—that might be used in solving a problem. As you work with the problems in this section, judge the relative efficiency and accuracy of these methods but also keep in mind the limited algebra background that elementary school students would likely have.

EXAMPLE 3

Solve the following problem in three different ways. (As illustrations, we will give four ways, the fourth to emphasize again the value of quantitative reasoning.)

▶ The last part of the triathlon is a 10K (10 kilometers, or 10,000 meters) run. When competitor Aña starts running this last part, she is 600 m behind competitor Bea. But Aña can run faster than Bea can. Aña can run (on average) 225 m/min, and Bea can run (on average) 200 m/min. Who wins, Aña or Bea? If Aña wins, tell when she catches up with Bea. If Bea wins, tell how far behind Aña is when Bea finishes. ◀

Approach 1

(Table) Make a table that shows Aña's and Bea's positions every minute. If the positions are ever equal before the race is over, then Aña has caught up. Here are two versions of such a table:

Version 1

Time (min)	Aña's position (m)	Bea's position (m)
0	0	600
1	225	800
2	450	1000
3	675	1200
4	900	1400
5	1125	1600
⋮	⋮	⋮

Version 2

Time (min)	Aña's position (m)	Bea's position (m)
0	0	600
5	1125	1600
10	2250	2600
15	3375	3600
20	4500	4600
25	5625	5600

Version 1, with positions recorded at each minute, might even make you think that Aña would never catch up! But if we add a fourth column that records Aña's distance behind Bea, we would see that Aña *is* catching up, but only by 25 m/min. Version 2 recognizes that Aña is going to take several minutes to catch up and records positions every 5 min. Again, adding a fourth column showing how far Aña is behind Bea might be helpful. The Version 2 table does show that Aña is ahead after 25 min, so she has caught up. A little more calculation (or a lot more, for the Version 1 table) shows that Aña catches up after 24 min, when both have run 5400 m.

Approach 2

(Graphical) The idea is to show each woman's distance-time graph. Where the graphs meet (if they do meet) indicates that the distances are the same and Aña has caught up. If Aña catches up before the 10,000 m have been run, we have an answer. If the women have run more than 10,000 m, then Bea wins, and the graph can show how far Aña is behind when Bea finishes.

Immediately, you will notice that to make the graphs, you need at least a few rows of data in a table, and more important, that it may be difficult to read the exact solution from the graph. But the graphs make clear that at some time just short of 25 min, Aña catches up.

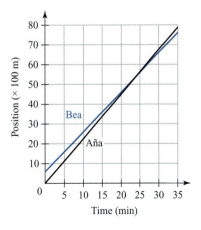

continue

Approach 3

(Equations) Either a table or a knowledge of the slopes and the starting points in the race leads to equations for the two women's distances traveled (*d* in meters versus time *t* in minutes):

$$\text{Aña: } d = 225t \qquad \text{Bea: } d = 200t + 600$$

If Aña is to catch Bea, their distances must be exactly the same. So, $225t$ must equal $200t + 600$, leading to $25t = 600$, or $t = 24$. So Aña catches Bea after 24 min. Is the race already over? No, because Aña's distance is $225 \times 24 = 5400$ m, indicating that they both are a little more than halfway into the 10,000-m race.

Approach 4

(Quantitative reasoning) Aña catches up by 25 m every minute. She needs to catch up a total of 600 m, so the question becomes: How many 25s make 600? This question suggests a repeated subtraction division, and $600 \div 25 = 24$. Next, check to see whether the race is over: $24 \times 225 = 5400$, and so the race is still on. ●

 DISCUSSION 2 Pros and Cons (for Whom?)

What are pros and cons for each of the approaches shown in Example 3? Which method(s) might be accessible to elementary school children? Why? Which method seems most insightful? Which method(s) requires more prior experience in mathematics? ●

EXAMPLE 4

Solve the following problem in three ways:

▶ My brother and I walk the same route to school every day. My brother takes 40 minutes to get to school, and I take 30 minutes. Today, my brother left 8 minutes before I did. How long will it take me to catch up with him? ◀

Approach 1

(Table) The problem resembles the triathlon problem, so a similar approach should work. But we do not have information on the actual distances traveled. Notice that the quantity, what *fraction* of the trip has been traveled, might give a breakthrough. Let's create a table, keeping in mind that we need not go minute by minute if that looks too cumbersome.

Table 1

My time (min)	Part of whole trip covered	
	Bro	Me
0	$\frac{8}{40}$	0
1	$\frac{9}{40}$	$\frac{1}{30}$
2	$\frac{10}{40}$	$\frac{2}{30}$
3	$\frac{11}{40}$	$\frac{3}{30}$
⋮	⋮	⋮

Table 2

My time (min)	Part of whole trip covered	
	Bro	Me
0	$\frac{8}{40}$	0
5	$\frac{13}{40}$	$\frac{5}{30}$
10	$\frac{18}{40}$	$\frac{10}{30}$
15	$\frac{23}{40}$	$\frac{15}{30}$
20	$\frac{28}{40}$	$\frac{20}{30}$
25	$\frac{33}{40}$	$\frac{25}{30}$

Notes

As in the triathlon problem, the first table looks as though it will take too long to create, and it involves the comparison of fractions (perhaps by using a common denominator or decimals). In the second table, using common denominators or decimals shows that I catch my brother some time before 25 min. (*Note*: We can also use excellent number sense: $\frac{33}{40}$ is $\frac{3}{40}$ more than $\frac{3}{4}$, and $\frac{3}{40}$ is less than the $\frac{1}{12}$ difference between $\frac{3}{4}$ and $\frac{5}{6}$.) So we would have more calculations to do, working backward, to get equality at $t = 24$ min.

Approach 2

(Graphical) The idea of using parts of the whole trip makes the graph easy to draw.

> **Think About …**
>
> Why was the graph in Approach 2 easy to draw?

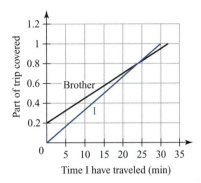

Again, it is difficult to read the exact answer from the graph drawn here, but a good guess would be about 24 min for the catch-up time.

Approach 3

(Equations) In the triathlon problem, we knew the speeds. We do not know them here, except that the total distance is traveled in a certain number of minutes. Also, in this problem we do not know the total distance, but we can try 1000 m to see how the total distance enters in.

Here is an important note: Because usually the distances are with respect to the same point and the 0 times are the same for the actors in a problem, d and t have the same interpretations. Here, however, my brother left 8 minutes earlier than I, so if t is the number of min I travel, my brother travels $t + 8$ minutes. This distinction is important in writing equations or making graphs—be especially clear about exactly what the variables represent. In the brother and I problem, d can represent the distance that each person is from home (in meters), and t can represent the number of minutes that I have traveled (so my brother's time is $t + 8$ min). We now have two equations:

$$\text{Brother:} \quad d = \frac{1000}{40}(t + 8)$$

$$\text{I:} \quad d = \frac{1000}{30}t$$

So to find if and when I catch my brother, we equate the distances:

$$\frac{1000}{40}(t + 8) = \frac{1000}{30}t$$

Notice that the 1000 on both sides of the equation divides out, which surprisingly shows that the total distance does not matter. We then have

$$\frac{1}{40}(t + 8) = \frac{1}{30}t$$

Solving this equation gives $t = 24$ (min) for the time it takes me to catch up.

Approach 4

(Quantitative reasoning) My brother has a head start of $\frac{8}{40} = \frac{1}{5}$ of the way to school. Each minute, I go $\frac{1}{30}$ of the way to school, and my brother goes $\frac{1}{40}$ of the way. So each minute, I catch up by $\frac{1}{30} - \frac{1}{40} = \frac{1}{120}$ of the way. How many $\frac{1}{120}$s are there in $\frac{1}{5}$? Because $\frac{1}{5} \div \frac{1}{120} = 24$, I catch up with my brother in 24 minutes. ●

DISCUSSION 3 Pros and Cons Again

Again, what are the pros and cons for each of the approaches used in Example 4. Why? Which method(s) might be accessible to elementary school children? Which method seems most insightful? ●

Often, there are multiple ways to solve a problem. Each way can provide a different insight into the problem, and different ways call on different abilities.

Notice the degree of symbolic work in this Grade 5 page, based on What's My Rule? problems. Try them.

Name   Date   Time

STUDY LINK 10·3 **Writing Algebraic Expressions**

Complete each statement below with an algebraic expression, using the suggested variable.

SRB
218 231
232

1. Lamont, Augusto, and Mario grow carrots in three garden plots. Augusto harvests two times as many carrots as the total number of carrots that Lamont and Mario harvest. So Augusto harvests

Augusto   Lamont and Mario harvested $L + M$ carrots.

_____ carrots.

2. Rhasheema and Alexis have a lemonade stand at their school fair. They promise to donate one-fourth of the remaining money (m) after they repay the school for lemons (l) and sugar (s). So the girls donate

_____ dollars.

3. a. State in words the rule for the "What's My Rule?" table at the right.

   _____
   _____

   b. Circle the number sentence that describes the rule.

   $Q = (3 + N) * 5$ $Q = 3 * (N + 5)$ $Q = 3N + 5$

N	Q
2	11
4	17
6	23
8	29
10	35

4. a. State in words the rule for the "What's My Rule?" table at the right.

   _____
   _____

   b. Circle the number sentence that describes the rule.

   $R = E * 6 * 15$ $R = (E * 6) + 15$ $R = E * 15 + 6$

E	R
7	57
10	75
31	201
3	33
108	663

Practice

5. $384 * 1.5 =$ _____

6. $50.3 * 89 =$ _____

7. $\frac{843}{7} =$ _____

8. $70.4 / 8 =$ _____

Algebra and Functions **1.2**, **1.5**   **299**

Wright Group, *Everyday Mathematics, Math Masters*, Grade 5, p. 299. Copyright © 2008, McGraw-Hill. Reproduced with permission of the McGraw-Hill Companies.

ACTIVITY 6   Solving a Problem Using Tables, Graphs, and Equations

Consider the following problem:

▶ A new electronic gaming company, called Pretendo, charges $180 for each gaming system and $40 for each game. The main competitor, Sega-Nemesis, charges $120 for a gaming system and $50 for each game. Sega is cheaper if a customer buys only one game. However, Pretendo eventually will be cheaper to own because each game costs less. How many games does a customer have to buy for the total expenditure for the two brands to be the same? (Remember that a customer must buy one gaming system before he or she can purchase games.) ◀

Attack the problem in the following ways:

1. *Solve using arithmetic.* You may want to use a chart or table.

2. *Solve by making a graph.* Graph the total cost of a Sega-Nemesis system for different numbers of games purchased. On the same grid, graph the total cost of a Pretendo system for different numbers of games purchased. Solve the problem by interpreting your graph.

3. *Solve using algebraic equations.* Write an equation that tells you the total cost of a Sega-Nemesis system for different numbers of games purchased. Write an equation that tells you the total cost of a Pretendo system for different numbers of games purchased. Use these two equations to find the solution.

4. Can you think of a quantitative reasoning solution? If so, describe your solution. ●

In Activity 6, you used different representations to answer the question posed in the problem. By solving the problem in different ways, you can focus more fully on the relationships among the different representations. The chart, or table, is created from the story narrative. The graph is a plot of the points in your chart or table. The algebraic equation is created from the relationships you identified when reading the narrative, from looking at entries in the table, or perhaps from the graph. An unstated quantity such as the difference in price per game may enable you to see a quantitative solution.

TAKE-AWAY MESSAGE . . .   A problem can be approached by making a table of values, by drawing a graph, or by using algebra. Because these methods can be somewhat mechanical, it is easy to overlook the fact that reasoning about the quantities involved can lead to a solution based on insight into how the quantities are related. Thus, it is always a good idea to think about a problem before starting to write. Each method of approaching a problem has its advantages and disadvantages, depending on whether understandability is an aim, on who the solver is, on how quickly a solution must be obtained, or on how accurate the solution must be.

Learning Exercises for Section 15.3

1. Suppose Tony and Rita are racing. Tony starts 50 ft ahead of the starting line because he is slower than Rita. Tony's speed is 10 ft/s. Rita's speed is 15 ft/s.

   Let t represent the number of seconds that have elapsed since the start of the race.

   a. What does the expression $15t$ tell you in terms of the race?
   b. Write an equation to show the relationship between the time that has elapsed in the race and Tony's distance from the starting line.
   c. Write an equation to show the relationship between the time that has elapsed in the race and Rita's distance from the starting line.
   d. If you solved the equation $15t = 50 + 10t$ for t, what would you have found (in terms of the situation of the race)?
   e. Use algebra to find the time at which Rita will catch up to Tony.
   f. How far will Tony have run when Rita catches up to him? How far will Rita have traveled?

continue

   g. How far apart are Rita and Tony after 35 s?

   h. Show the original situation with distance-time graphs by plotting points describing Rita's distance and Tony's distance on the same grid.

2. Going into the 10,000-m part of a triathlon, Kien is 675 m behind Leo. But Kien can run a little faster than Leo: 265 m/min versus 250 m/min. Can Kien catch Leo before the race is over? If so, tell when. If not, how far behind is Kien when Leo crosses the finish line?

3. Two snails, Pierre and Marie, are captured by a Frenchman who plans to have them for dinner. They are placed on his kitchen counter, 5 in. from one another. Then the Frenchman goes to gather some other ingredients. Pierre and Marie are doomed. All they want is to embrace one last time before they are cooked. Pierre squirms toward Marie at $\frac{1}{2}$ in./min. Marie desperately crawls toward Pierre at $\frac{3}{4}$ in./min. How long will it take the two snails to reach one another?

4. Bobby and Jaime are very competitive. They want to impress their classmate Lily by showing her how fast they can run. They agree to race 30 yd across the playground. Bobby runs at an average speed of 7 ft/s for the whole race. Jaime runs at 10 ft/s, but Bobby had a 3-s head start.

   a. Write equations to describe Bobby's and Jaime's position over time.

   b. Who wins the race (and Lily's heart)?

5. Del Mar and San Pedro are 352 mi apart. Two trains on parallel tracks are traveling toward each other. Madonna is on train A leaving from Del Mar, and the paparazzi are on train B leaving from San Pedro. Train A goes 45 mph, and train B goes 65 mph. When should the paparazzi lean out the window to take photos of Madonna—i.e., when will the two trains meet?

   a. Write an equation for each train to describe its position over time.

   b. How long does it take for the two trains to meet? How many miles had they traveled? (Solve algebraically.)

   c. What fraction of the 352 mi had each train covered, at the time they meet?

6. Sister and Brother go to the same school and by the same route. Brother takes 40 min for the trip, and Sister takes 50 min. One day, Sister gets a 15-min head start on Brother.

   a. Can Brother catch Sister before they get to school?

   b. If so, how many minutes from school are they when he catches her? If not, where is Brother when Sister gets to school?

For Learning Exercises 7–12, solve each problem in at least three ways:

   a. First solve by generating a table of data.

   b. Use graph paper to graph each relationship in the problem on the same set of axes. Interpret your graph to solve the problem.

   c. Write an equation to describe each relationship in the problem. Use the two equations to solve the problem algebraically.

   d. Use quantitative reasoning to solve the problem.

7. Suppose Turtle runs at 55 ft/s. Rabbit runs at 80 ft/s, but gives Turtle a 5-s head start. How many seconds will Turtle have run when Rabbit catches up with him?

8. Suppose Turtle runs at 42 ft/s. Rabbit runs at 53 ft/s, but gives Turtle a 3-s head start. How far will Turtle have run when Rabbit catches up with him? (Answer exactly—do not give decimal approximations.)

9. A fishing boat has been anchored for several hours 200 mi from shore. It pulls up its anchor and cruises away from shore at a rate of 33 mi/h. At the same time that the fishing boat starts moving, you leave shore in a speedboat, traveling at a speed of 50 mi/h. How long (in hours and minutes) will it take you to catch up to the fishing boat? (Give your answer to the nearest minute.)

10. Suppose you are a famous musician about to sign a recording contract. You are offered two choices:

> Option A: $2.25 profit for every CD sold
> Option B: $300,000 up front, plus $0.75 for every CD sold

How many CDs would you have to sell for the two options to be of equal value? Which option would you select and why?

11. You have just moved to a new city and have called the phone company to set up an account. The phone company tells you that it has two new plans, as follows:

> Plan A: $15.50 per month plus $0.05 per call
> Plan B: $5 per month, plus $0.40 per call

a. What is the rate of change (or slope) of each of the lines describing the plans?
b. When is Plan A the better deal?
c. When is Plan B the better deal?
d. Which plan would you select and why?

12. You are offered two different jobs selling cell phones. One has an annual salary of $24,000 plus a year-end bonus of 5% of your total sales. The other has a salary of $10,000 plus a year-end bonus of 12% of your total sales. How much would you have to sell to earn the same amount in each job?

13. Reflect on the different methods you used to solve the problems assigned from Learning Exercises 7–12. List at least one advantage *and* one disadvantage for each given method.

a. table (numerical)   **b.** graph (graphical)   **c.** equations (algebraic)

Supplementary Learning Exercises for Section 15.3

See also the Learning Exercises and Supplementary Exercises for Section 15.2.

For Exercises 1–3, solve each problem in at least three different ways:

a. First solve by generating a table of data.
b. Draw graphs and interpret them.
c. Write equations to describe the situation, and use the equations to solve the problem algebraically.
d. Use quantitative reasoning to solve the problem.

1. In one city, two taxicab companies charge as follows:

> Speedy Taxi: $1.50 plus 12¢ for every $\frac{1}{18}$ mile
> A-One Taxi: $1.20 plus 15¢ for every $\frac{1}{18}$ mile

At what distance will the fares for the two companies be the same? What is that fare?

2. You need a plumber. So, you call two companies and learn the following prices:

> Perfecto Plumbing: $30 for the visit, plus $20 for every 15 minutes of work
> Pronto Plumbers: $40 for the visit, plus $18 for every 15 minutes of work

You are not certain how long the job will take. For what amount of time for the work will Perfecto Plumbing be less expensive?

3. You read in a study of college students of the same age-height as you are that one female uses 1520 calories per day and a better-conditioned male friend uses 1730 calories per day, without special exercise. In walking 3 miles per hour, the female uses an additional 350 calories per hour, and the better-conditioned male uses an additional 320 calories an hour. How many miles would each have to walk so that they use the same total number of calories in a day? (1000 calories = 1 kilocalorie.)

Notes

Notes

4. From your work in Exercises 1–3, which method seems easiest? Most understandable? Most elementary? Most advanced? If the numbers were not as compatible as they are in the exercises, would that change your reactions? If so, why?

5. Fahrenheit (F) and Celsius (C) temperatures are related by the equation $F = \frac{9}{5}C + 32$. Graph this equation and the equation $F = C$ on the same set of axes to see whether there is any temperature reading that is the same (numerically) for the two scales.

15.4 Average Speed and Weighted Averages

The famous fable of the turtle and the rabbit involves a race between these two animals. On the one hand, we have the capricious rabbit, whose confidence in his ability to travel very fast causes him to travel erratically, at best. On the other hand, the slow but steady turtle plods along at a constant pace. As we know from the moral of the story, the turtle wins the race! In this section, we explore situations such as the Rabbit Turtle race to develop a better understanding of average speed and weighted averages in general.

 ACTIVITY 7   GPA

Consider the following story:

▶ Janice went to Viewpoint Community College for 3 semesters. She earned 42 credits and had a grade-point average (GPA) of 3.4. She then moved to State U for 5 semesters where she earned 80 credit hours before graduating and had a GPA of 2.8. What was her overall GPA when she graduated? She had to have an average GPA of 3 or better to enter a graduate program. Did she have that? ◀

a. What does "grade-point average" mean?

b. How many grade points did Janice earn at VCC?

c. How many grade points did Janice earn at SU?

d. What is the total number of grade points Janice accumulated?

e. What is the total number of credits Janice accumulated?

f. What can you do with the information from parts (d) and (e)?

g. Is the result in part (f) the GPA that you predicted? ●

 ACTIVITY 8   To Angie's House and Back

Consider the following story:

▶ Wile E. Coyote decided to visit his friend Angie Coyote, who lives 400 m away in another cave. He walked the 400 m at a steady pace of 4 m/s. When he returned, he was anxious to get home, so he ran the 400 m at 8 m/s. What was his average speed for the entire trip? ◀

You probably answered the question by saying that the average speed for the trip was 6 m/s—that is what most people would say. But consider these questions:

a. How long did it take Wile E. to get to Angie's cave?

b. How long did it take Wile E. to get home?

c. What was the total time traveling?

d. What was the total distance traveled?

e. From the formula $d = rt$, what was the rate (speed) for the entire trip? (*Hint*: It is not 6 m/s.) What is going on? ●

The two problems in Activities 7 and 8 illustrate average rates and weighted averages. Just averaging averages is not sufficient. Why? What must be considered in each problem to get the correct answer?

The key in Activities 7 and 8 is this question: What does the phrase "grade-point average" mean and what does the phrase "average speed" actually mean? *Find the average* usually signals a computational procedure, but the resulting (and important) number can be interpreted in a useful way. For example, a GPA of 3.0 for 122 credits can be interpreted to mean the grade value that Janice would have earned for each credit *if her total grade points were distributed equally over all of her credits*. A class average of 81.3% on a test can be thought of as the percent each student would have gotten on the test *if the total of the percents for the class were distributed equally among all the students taking the test*. If the average number of children per family in a community is 2.6, that number can be interpreted as how many children each family would have *if the total number of children were spread equally among all the families*.

Fortunately, the average can be determined graphically, not just numerically, as suggested by Activity 9.

 ACTIVITY 9   The Wile E.–Angie Situation with a Graph

Make a total distance versus elapsed time graph for Activity 8 on the following coordinate system:

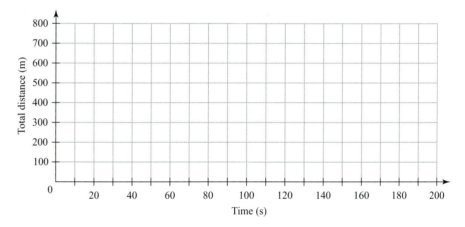

Now join the starting point (0, 0) to the final point (150, 800) with a line segment. What is the slope of this line? Why does the slope give the average speed? (*Hint*: Think of what *average* can mean.) ●

> *Think About …*
>
> How could a graph showing total distance versus elapsed time give the average speed if Wile E.'s trip involved three segments? Why does that work?

The notion of average speed can be sharpened by considering a race situation, as in the next activity.

 ACTIVITY 10   Turtle and Rabbit Go Over and Back

The software program Over & Back is available at http://crmse.sdsu.edu/nickerson. Briefly, the program acts out this situation: Turtle and Rabbit run a race to a certain place, and then run back to the starting point. Rabbit's speed over and speed back can be different, but Turtle's speed both ways is the same. The speeds and *total* distances can be changed by clicking on them. See the illustration given on the next page.

Race 1. Rabbit's speed over is 4 m/s and back is 2 m/s. Turtle's speed both ways is 3 m/s. The total distance is 20 m. Who do you think will win, or do you think they will tie? If they do not tie, how could you adjust Turtle's speed so that they do tie? Should you make Turtle's speed *more than* or *less than* 3 m/s?

Notes

Race 2. Rabbit's speed over is 8 m/s and back is 2 m/s. Total distance is still 20 m. What speed should Turtle go (the same both ways) to tie?

Race 3. What would happen in Race 1 if the total distance was 40 m instead and speeds are set so that Rabbit and Turtle tie for the 40-m race? In Race 2? If the total distance was 100 m? Explain your thinking.

Race 4. Solve the problem using your own choices for a different Rabbit-Turtle race.

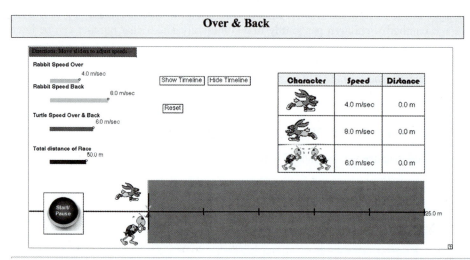

This page uses **JavaSketchpad**, a World-Wide-Web component of *The Geometer's Sketchpad*. Copyright © 1990-2001 by KCP Technologies, Inc. Used with permission. © Patrick Thompson, 1992; JavaSketchpad version developed by Janet Bowers, *Interactive Educators*.

You may have noticed that knowing how much time is spent at each of Rabbit's speeds is important in calculating Turtle's speed to tie. That is, to calculate a tying speed, Rabbit's individual speeds must be weighted by the time spent at each of them.

DISCUSSION 4   Thinking It Through

1. If Turtle and Rabbit tie, how does Turtle's speed compare to the (weighted) average of Rabbit's two speeds? Why?

2. Do you think that the outcome of the race depends on the total distance they run if they run a longer race? What if they run a shorter race? Explain your answers.

TAKE-AWAY MESSAGE . . . The average of some items can show the size each item would be if the total were spread equally among the items. When an average is requested, it is common just to add the values and then divide by how many values are used. But this technique does not give a correct *overall average* for averages and rates, in general. Unless each average is based on the same number of items, you must use a *weighted* average. In the case of distance, the weighted average reflects how much time each rate, or speed, was in effect. A distance-time graph can help make clear that the average rate may be thought of as the total distance spread evenly over the total time.

Learning Exercises for Section 15.4

1. One day, Duke earns $8.00/h for the first 4 h that he works, and then he earns $16.00/h for the next 4 h.

    a. How much money has he earned at the end of the day?
    b. What hourly rate must he work at for 8 h to earn the same daily amount?
    c. If Duke worked at $8.00/h for 2 h and at $16.00/h for 6 h, what would be his average hourly rate for the 8 h?

2. In Tokyo, a taxicab ride can commonly cost as much as $100.00 or more. The breakdown on how much you might be charged by the mile is as follows, converting from units used in Tokyo:

$8.00/mi for the first 7 mi
$4.00/mi for the next 9 mi
$2.00/mi for every mile after that

   a. How much would a 30-mi cab ride cost?
   b. What is the average dollar/mile rate for the 30-mi ride?

3. John is participating in the triathlon. He swims at a rate of 2 mi/h, and he can run at a rate of 6 mi/h. The course is broken down in the following manner:

1st leg—swimming—6 mi
2nd leg—running—15 mi
3rd leg—bicycling—18 mi

   How fast will John have to bicycle if he wants to finish the race in 7 h flat?

4. Make graphs to help find the following:

   a. the average cost per mile for the taxi ride in Learning Exercise 2(b)
   b. John's bicycle speed in Learning Exercise 3 to finish in 7 h

5. Jasmine transferred to State University from City Community College. At CCC she earned 36 credits and had a GPA of 3.2. She has been at State now for 2 semesters and has earned 24 credits with a GPA of 2.6. What is her overall GPA?

6. Kayla typed for 2 h at 70 words/min, then took a coffee break for 15 min, and then typed for another hour at 55 words/min. Homer, on the other hand, typed at one speed for the total $3\frac{1}{4}$ h. How fast did he type if he typed the same number of words as Kayla did over the $3\frac{1}{4}$ h?

7. Rabbit's pride has suffered, and he wants to run more races. His friends are all betting on which race will be closest and which will not be close. Arrange in order the races below so that your Race 1 is the closest (and most exciting), your Race 2 the next closest, and your Race 3 not very close at all. The total distance is 100 m. (If Over & Back is available, you can check your answers, but make predictions first!)

R Over (m/s)	R Back (m/s)	T Both (m/s)	Predicted winner?	Rank of how close the race will be	Speed at which T would have to run to tie
8	2	5			
0.5	9.5	5			
4.5	5.5	5			

8. In an Over & Back race of 40-m total distance, Rabbit's speeds are 5 m/s over and 10 m/s back. Turtle's speed both ways is 6 m/s.

   a. Who will be ahead 1.5 s after the race starts and by how much?
   b. Who will be ahead 5.5 s after the race starts and by how much? Can you tell when and where Rabbit catches up with Turtle?
   c. Who wins the race and by how much time? How many meters is the loser behind when the race is over?
   d. Would your answers to parts (b) and (c) change if Rabbit's speeds were reversed: 10 m/s over and 5 m/s back? (Turtle's speed stays at 6 m/s each way.)

9. Understanding average speed is often difficult because of the influence of calculating averages. But notice Xenia's reasoning given on the next page. Xenia was 7 years old when her mother shared this story!

Notes

Mother, while driving: "If we drove for two hours on the freeway, at 60 mi/h, and then got off the freeway and drove for an hour through the city at 20 mi/h, what was our average speed?"

Xenia thought for a while and said, "I think it would be something like 50." Mother had asked Xenia to explain her reasoning so often that it was a habit by now, so Xenia continued, "I think that because we went 60 a lot longer than we went 20, so the 20 will only pull the 60 down a little bit."

    a. Is Xenia's reasoning correct?   **b.** What would the exact average speed be?
    c. What does the answer to part (b) mean?

10. In a race of 120-m total distance, Rabbit gave Turtle a 4-s head start. Then Rabbit ran 2 m/s over, and ran an unknown speed back. Turtle ran 3 m/s both ways, after the head start.

    a. What fractional part of the way over was Turtle's head start?
    b. If Rabbit and Turtle tied, how fast did Rabbit run on the way back? (Do not overlook Turtle's head start.)
    c. If Turtle had run 3.173 m/s (with the same head start) and Rabbit had run as given and as in part (b), who would have won the race?

11. In a race of 60-m total distance, Rabbit ran 5 m/s over, stopped to eat a carrot for 4 s, and then ran back in 5 s. Turtle ran at the same speed, over and back.

    a. What was Rabbit's speed back?
    b. If Turtle and Rabbit tied, what was Turtle's speed over and back?
    c. Twelve seconds into the race, where were Rabbit and Turtle located?

12. Suppose Tabatha rode her bike 3 mi in half an hour, and then 12 mi in 1 h.

    a. Graph this information. What is Tabatha's average speed for each of the 2 legs of her journey?
    b. Carly rode her bike 15 mi in 1.5 h. What is her average speed for this trip? Graph this information onto the same graph as in part (a).
    c. What was Tabatha's average speed?   **d.** What was Carly's average speed?
    e. If your answers are the same for parts (c) and (d), why? If your answers are different for parts (c) and (d), why?

13. Jerold took a taxi from his hotel to the airport. For the first 15 min the taxi was slowed down by traffic and traveled just 4 mi. The taxi then entered a tunnel and drove 2 mi in 5 min. Once out of the tunnel, the taxi drove another 7 mi in 15 min. The taxi then entered the freeway and drove 15 mi in 15 min. Finally, the taxi exited the freeway, drove 4 mi in 10 min, and then dropped Jerold off for his flight.

    a. Make a graph of Jerold's trip.
    b. What was his average speed during each of the 5 legs of his trip?
    c. Find the average rate for Jerold's trip.
    d. Cassie was at the same hotel as Jerold, and she took a later taxi to the airport, 32 mi away. The trip took an hour at a steady speed. Graph her journey on the same graph as Jerold's trip.
    e. What does this new line on the graph tell you?

Supplementary Learning Exercises for Section 15.4

1. Danielle is a senior and has a GPA of 2.8 for 108 semester hours. She is taking 12 semester hours this semester and would like to graduate with at least a 3.0 GPA. If she gets a 4.0 average this semester, will she meet her goal? If not, what is the highest GPA she can get?

2. **a.** What is your average speed for a trip that is 200 mi each way, if you go 50 mi/h going and 60 mi/h returning?
    b. Suppose that an accident delayed traffic so that you averaged only 35 mi/h on the return trip. What would your average speed for the trip have been then?

continue

c. If you go at 50 mi/h, what speed returning would give an average for the whole trip of 65 mi/h?

3. The average class size is 30 students at Washington School and 24 students at Jefferson School. Why can't you safely say that the average class size at the two schools is exactly 27 without more information? What other information is needed to arrive accurately at the average class size for the two schools?

4. The average price for the tickets to a concert sold in one city was $48 and in a larger city $60. Was the average ticket price for the two cities $54? If so, how do you know? If not, explain why.

5. "I lost an average of 2 pounds a week on my diet," says one person. Another person says, "Well, I lost an average of 3 pounds a week!"

   a. Why can't you tell who lost more weight?
   b. Is the average weight loss for the two people 2.5 pounds a week? Explain.

6. In some Over & Back races of 60-m total distance, Rabbit ran at the speeds listed below. Turtle runs at the same speed over and back. What speed must Turtle run to tie Rabbit for the race?

   a. Rabbit over: 20 m/s; Rabbit back: 5 m/s
   b. Rabbit over: 3 m/s; Rabbit back: 30 m/s
   c. Rabbit over: 40 m/s; Rabbit back: 10 m/s
   d. Rabbit over: 60 m/s; Rabbit back: 1 m/s

7. **a.** In Exercise 6, what is Rabbit's average speed in each part?
   b. If the races in Exercise 6 had been 120-m total distance, which (if any) answers would change?

8. In an Over & Back race of 80-m total distance, Rabbit gives Turtle a 20-m head start. Rabbit then runs 10 m/s over and 16 m/s back. What steady speed must Turtle run (once the race starts) to tie Rabbit?

15.5 More About Functions

Many patterns and functions deal with numbers, or numbers inside problems, as with the toothpick arrangements shown in Section 12.4. But there are other situations in which numbers play only a part of the role or even no role at all. For example, in tossing a coin and noting whether heads or tails comes out on top, you may have heard the phrase "the probability of heads is $\frac{1}{2}$." In this case, each outcome of the coin toss is either heads or tails, an object rather than a number, but each outcome corresponds to a number. Hence, a notation like $P(heads) = \frac{1}{2}$ is quite reasonable and does resemble the typical $f(x)$ notation. Activity 11 gives an example of a function that does not involve numbers at all.

ACTIVITY 11   Through the Looking Glass

Suppose that the vertical line in the following drawing is a mirror. If the cartoon figure looks at itself in the mirror, which point (A, B, or C) do you think is $f(P)$? Use your answer to write $f(P) =$ ___. If S is the entire head on the left of the mirror, sketch what you think $f(S)$ is.

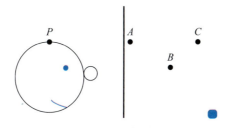

Notes

Notes

If you sketched a cartoon head of the same size looking at itself, you probably have a good feel for this sort of function, one in which points correspond to points in a particular way. Numbers need not appear at all with some functions, yet a notation like $f(P) = C$ makes sense.

But most patterns and functions do involve numbers. As with the toothpick patterns in Section 12.4, there can be a shape involved, as in Activity 12.

 ACTIVITY 12   Walk the Walk

What is the perimeter p (distance around the outside) of the nth shape in the pattern below? Measure in toothpicks. Ignore the dashed lines that show the squares and the small empty spaces between toothpicks. Write your finding in function form:

$$p(\text{Shape } n) = \underline{\hspace{1cm}}.$$

Can you justify that your function rule is indeed trustworthy?

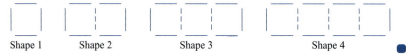

Shape 1   Shape 2   Shape 3   Shape 4

You know that two numbers can be combined in a variety of ways, e.g., using addition or multiplication. You may be surprised that functions can often be *combined* as well. The function machine representation from Section 12.4 can be used to illustrate the idea behind combining functions: *The output from the first function becomes the input to the second function.* The combination of the two machines can be thought of as a function itself. Let us consider an example.

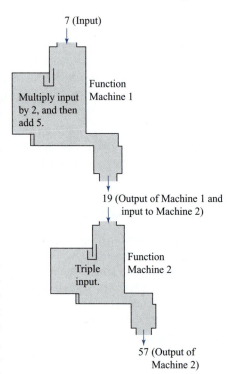

7 (Input)

Multiply input by 2, and then add 5.

Function Machine 1

19 (Output of Machine 1 and input to Machine 2)

Triple input.

Function Machine 2

57 (Output of Machine 2)

EXAMPLE 5

Refer to the previous function machine diagram. The output 19 from the first function is the input for the second function, which triples its input. The final output 57 is the result of the two-function combination. If we consider just the beginning input and the final output, we see that for an initial input of 7, the combination then gives a final output of 57. The rule for the combination can even be predicted from the two individual function rules:

$$output = 2 \times input + 5 \qquad \text{and} \qquad output = 3 \times input$$

Notice that the input to the second machine is the $2 \times input + 5$ for the original input. That is, for the combination,

$$final\ output = 3(2 \times original\ input + 5) \qquad \text{or} \qquad output = 6 \times original\ input + 15$$

The usual algebraic shorthand certainly takes less space: From $y = 2x + 5$ first and $y = 3x$ second, the combination is given by $y = 3(2x + 5) = 6x + 15$. ●

 ACTIVITY 13   Does Order Matter?

Because this combination of functions is analogous to, say, multiplication of numbers, a natural mathematical question is whether this combination of functions is commutative or associative. Investigate commutativity with the function machines pictured above. ●

The next activity involves a well-known problem. This problem leads us to the *differencing* technique, which can give information about function rules in some situations.

 ACTIVITY 14   The Painted Cube[1]

Cubes of different sizes are made from unit cubes and then dipped in paint. We start with a 2 by 2 by 2 cube (made from 8 unit cubes), then we have a 3 by 3 by 3 cube (made from 27 unit cubes), etc. The diagram below shows a 4 by 4 by 4 cube made from 64 unit cubes and then dipped in paint.

unit cube

After the paint is dried, suppose that such a cube is taken apart into its original unit cubes. Gather data on the following questions and record your results in the table. (The unit cube is omitted because all 6 of its faces would be painted when dipped.)

1. How many unit cubes will have paint on 3 squares (faces)? Can you predict the number for the general n by n by n cube?

2. How many unit cubes will have paint on only 2 squares?

3. How many unit cubes will have paint on only 1 square?

4. How many unit cubes will have paint on 0 squares?

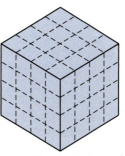

4 by 4 by 4 cube

Size of cube	Paint on 3 squares	Paint on 2 squares	Paint on 1 square	Paint on 0 squares
2 by 2 by 2				
3 by 3 by 3				
4 by 4 by 4				
5 by 5 by 5				

5. In which columns can you now predict later entries? ●

> ### Think About …
> From your table in Activity 14, predicting the number of unit cubes with 3 squares painted looks easy. Can you *reason* that the answer will indeed be what you predict?

Predicting the next entries for the number of unit cubes with paint on only 2 squares looks reasonable as well, because they are all multiples of 12: 0×12, 1×12, 2×12, 3×12. We can even predict later entries by noting the relationship between the size of the cube and the multiple of 12: 2 by 2 by 2 $\rightarrow 0 \times 12$, 3 by 3 by 3 $\rightarrow 1 \times 12$, 4 by 4 by 4 $\rightarrow 2 \times 12$, e.g., suggesting that for the n by n by n cube, there would be $(n - 2) \times 12$ unit cubes with only 2 faces painted.

In the table in Activity 14, the entries in the column labeled Paint on 1 square do suggest a pattern, so you likely would predict $54 + 42 = 96$ for the next cube, which is 6 by 6 by 6. But is there a function rule that would allow us to skip intervening cases and predict the number for, say, a 20 by 20 by 20 cube? The answer is yes. We can repeat the differencing technique on the first differences! We illustrate this technique for the number of cubes with paint on only 1 square on the next page.

Size	Paint on 1 square
2 by 2 by 2	0
	+6
3 by 3 by 3	6
	+18
4 by 4 by 4	24
	+30
5 by 5 by 5	54
	+?
6 by 6 by 6	

First differences

Size	Paint on 1 square
2 by 2 by 2	0
	+6
3 by 3 by 3	6
	+18
4 by 4 by 4	24
	+30
5 by 5 by 5	54
	+?
6 by 6 by 6	

+12

+12

Second differences

When the second differences are the same, that means there is a function rule involving the *second* power of the variable that will describe the entries in the table. We cannot spend time on the full details about how to determine this exact expression, but we give the function rule $f(n) = 6n^2 - 24n + 24$ that finds the number of unit cubes having only 1 square face painted, for the n by n by n cube.

> **Think About …**
>
> $f(n) = 6n^2 - 24n + 24$ can be written as $f(n) = 6(n - 2)^2$.
> Does the drawing help you see where the $n - 2$ comes from?
> The 6?

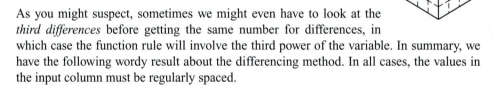

As you might suspect, sometimes we might even have to look at the *third differences* before getting the same number for differences, in which case the function rule will involve the third power of the variable. In summary, we have the following wordy result about the differencing method. In all cases, the values in the input column must be regularly spaced.

> If the first differences all equal the same number, then the function rule will involve the first power of the variable. If the first differences are different but the second differences are all the same, then the function rule will involve the second power of the variable. If the first time the differencing method gives the same number all the time occurs at the third differences, then the function rule will involve the third power of the variable. And so on.

 ACTIVITY 15   **What Do the Differences Make?**

Use the differencing method to help find a function rule for each table in Problems 1–3.

1. x	$F(x)$		2. n	$G(n)$		3. x	$f(x)$
7	50		2	8		3	11
8	65		3	27		5	27
9	82		4	64		2	6
10	101		5	125		4	18

4. Which power (as your work in Activity 14, Problem 4, suggests) would be involved in a function rule for the number of unit cubes with 0 squares painted? ●

The Painted Cube problems illustrate that function rules for real situations, not just made-up tables, can involve second and third power of numbers. Even higher powers are possible in other problems.

TAKE-AWAY MESSAGE... Functions need not involve numbers, although many do. This widespread applicability makes function ideas particularly attractive in mathematics and in scientific uses of mathematics. Hence, it is easy to see why the groundwork for functions is likely to appear in the K–8 curricula from which you teach. The differencing technique can give some information in finding function rules.

Learning Exercises for Section 15.5

1. Explain how these everyday sentences make sense mathematically (if the characteristics are quantifiable).

   a. "Record sales are a function of the singer's popularity."
   b. "Success is a function of how much you work."
   c. "Reaction time is a function of blood alcohol content."

2. Find a function rule that gives the perimeter of the *n*th shape in each of the following patterns. Justify that your rule will always work. (Use the shortest segment as the measuring unit.)

   a.

   Shape 1   Shape 2   Shape 3

   b.

   Shape 1   Shape 2   Shape 3

   c.

   Shape 1   Shape 2   Shape 3

3. For each input in parts (a–c), give the output for the combination: first Machine 1, then Machine 2.

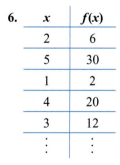

Machine 1 — Square the input

Machine 2 — Add 3, then double that

   a. Input = 5   **b.** Input = 10
   c. Input = x

   In parts (d–f), use the combination: first Machine 2, then Machine 1.

   d. Input = 5   **e.** Input = 10
   f. Input = x
   g. In combining machines, does order matter? How do you know?

Use the differencing technique as an aid in finding a function rule in Learning Exercises 4–9.

4.

x	y
2	3
3	8
4	15
5	24
6	35
⋮	⋮

5.

x	y
2	2
4	8
6	18
8	32
10	50
⋮	⋮

6.

x	f(x)
2	6
5	30
1	2
4	20
3	12
⋮	⋮

7.

x	y
5	125
1	1
3	27
7	343
9	729
⋮	⋮

8.

x	y
0	0
1	2
2	16
3	54
4	128
5	250
⋮	⋮

9.

x	f(x)
51	252
52	257
53	262
54	267
55	272
⋮	⋮

10. ***Extension of the Activity 14 Problem*** Suppose a rectangular solid is made up of 1-cm cubes, dipped in paint, and then dried and taken apart into the individual cubes. If the rectangular solid has dimensions m cm, n cm, and p cm, how many cubes have 3 squares painted, 2 squares painted, 1 square painted, and 0 squares painted? (*Hint:* Because there are three variables, finding a pattern is difficult. Instead, try the reasoning you may have used to justify the results for the Painted Cube problem. Here is a drawing of a 3 cm by 4 cm by 6 cm rectangular solid on which to practice your reasoning.)

Supplementary Learning Exercises for Section 15.5

1. Explain how these statements can make sense, mathematically (if the characteristics are quantifiable).

   a. "Profit is in large part a function of sales."
   b. "Stopping distance is a function of the square of speed."
   c. "Agricultural productivity in the area is a function of the amount of rainfall."
   d. "The life of the product is a function of how it is cared for."

2. Find a function rule that gives each of the following. Justify that your rule will always work. (Use the shortest segment as the measuring unit if measurements are involved.)

   a. the perimeter of the nth shape in the following pattern:

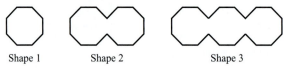

       Shape 1   Shape 2   Shape 3

   b. the cost of n boxes of nails at $2.98 per box, and with a sales tax rate of 6%
   c. the perimeter of the nth shape in the following pattern:

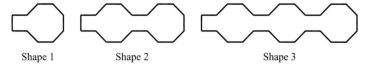

       Shape 1   Shape 2   Shape 3

   d. the perimeter of the nth shape in the following pattern:

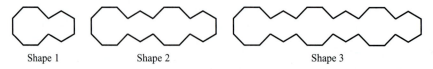

       Shape 1   Shape 2   Shape 3

   e. the cost of n twelve-packs of soda on sale at $3.95 a pack, plus a 5¢ deposit per can

3. Give the following outputs, for the functions to the right:

a. F(10)
b. G after F, for an input of 10
c. F(0)
d. G after F, for an input of 0

4. If the function rule is given by *output* = $(3 \cdot input)^2$, find the outputs for each of the following inputs:

a. 15   b. ⁻4   c. $2\frac{1}{2}$ d. 7.8

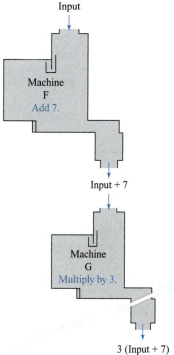

Input

Machine
F
Add 7.

Input + 7

Machine
G
Multiply by 3.

3 (Input + 7)

5. Suppose that two function machines work this way:

Machine 1 doubles the input and adds 8 to that result.

Machine 2 adds 8 to the input and doubles that result.

a. Do the two functions have the same effect on the input? Explain.
b. Give the final output for the combination, first Machine 1, then Machine 2, if the initial input is 27.7.
c. Give the final output for the combination, first Machine 2, then Machine 1, if the initial input is 27.7.

Use the differencing technique as a possible aid in finding a function rule for each of Supplementary Exercises 6–9.

6.

x	f(x)
0	4
1	5
2	8
3	13
4	20

7.

x	g(x)
0	⁻1
1	0
2	7
3	26
4	63

8.

x	y
0	⁻4
3	41
6	86
9	131
12	176

9.

x	y
0	0
3	18
1	2
4	32
2	8

15.6 Issues for Learning: Topics in Algebra

There has been a great deal of research on the learning and teaching of algebra. Now that the elementary school curriculum is including more ideas of algebra, there is naturally more research done with elementary school children. This section will draw primarily from compilations of the research[2,3] and present some of the findings, with an eye toward alerting you to the expectations that now exist and to some of the obstacles that children seem to encounter in learning algebra.

In Section 14.5 we mentioned several types of common errors made by students with co-ordinate graphs. In this section we consider four of the difficulties that children face in dealing with algebraic ideas and notations, according to the research. These include (1) difficulties in dealing with differences in algebraic notation, including differences in algebraic notation from arithmetic notation; (2) confusion about the meaning of variables; (3) conventions like the order of operations; and (4) dealing with elementary but conceptually more difficult equations.

Difficulties with algebraic notation. Algebraic notation can cause confusion with children who are quite comfortable with numbers. Some students believe that if the letter for a variable appears later in the alphabet, then the later letter represents a larger value. So, they believe, if $a = 5$, then n or x would have a value greater than 5.

Many students think that different variables must have different values. For example, only about three-fourths of a large group of sixth-graders responded correctly when asked, "Is $h + m + n = h + p + n$ always, sometimes, or never true?" To prompt students to acknowledge the possibility that two different variables, in this case m and p, could have the same value, teachers used the following problem and asked students to write an equation for it.

EXAMPLE 6

▶ Ricardo has 8 pet mice. He keeps them in two cages that are connected so that the mice can go back and forth between the two cages. One of the cages is blue, and the other is green. Show all the ways that 8 mice can be in two cages.[4] ◀ ●

Think About …

How can Example 6 lead to an acceptance that two variables can have the same value? (Use b for the number of mice in the blue cage and g for the number in the green cage.)

Children are so accustomed to replacing an indicated calculation, say, $152 + 389$, with the answer, 541, that they often have trouble regarding an expression like $x + 2$ as a single value. It is as though there is a compulsion to write a single expression, without operation signs, as the answer. So, e.g., children might write $7n$ for $3n + 4$, or write $5xy$ for $2x + 3y$. The children have what researchers call an "apparent lack of closure" for expressions like $3n + 4$ or $2x + 3y$, not seeing them as representing a single value.

Think About …

What equation would you write to describe the following situation? At one university, there are 6 times as many students as professors.

Confusion about the meaning of variables. In the Think About above, you should have said "$6P = S$" if you used P to stand for the number of professors and S to stand for the number of students. However, if you are like two out of every five college students going into engineering (who presumably should be very mathematically able), you wrote something like the equation $6S = P$. This phenomenon has proved resistant to change, even when students make a correct drawing, so teachers, mathematics educators, and psychologists have looked for explanations. One idea is that the phrasing "6 times as many students as professors" invites a translation into $6S = P$. Another hypothesis notes that, just as 12 inches = 1 foot, $6S = P$ gives a sort of measurement relationship: 6 students "make" 1 professor. Some regard the mistake $6S = P$ as further evidence of students' weak understanding of the symbol =. Many feel that students are using S as a label for "students" rather than for the number of students.

Your algebra teachers may have been insistent that you be explicit in your work, as in "x = the number of gallons of gasoline" and not just "x = gasoline," for the letters in algebra do represent numbers. You, like your algebra teachers, will have to emphasize the number nature of a variable, not its mnemonic (memory-aiding) role.

Order of operations. Some conventions in algebra (often not emphasized in arithmetic) appear to be difficult for students. Conventions are just how we do things, like write a 5 or a letter of the alphabet, but not for any intrinsic reason. If someone had decided at one time that we should make a symbol for five that looks different from the one we use, then that could have happened without loss . . . so long as the convention is widely followed. Writing "$n + n + n$" instead of "nnn" is an example, important because under our conventions, nnn means $n \cdot n \cdot n$.

A useful but arbitrary convention is the order of operations, useful because it allows us to avoid a lot of parentheses: First do calculations in parentheses (or other grouping symbols), then exponents, then multiplications and divisions as you encounter them going from left to right, and finally additions and subtractions as you encounter them going from left to right. "Please excuse my dear Aunt Sally" is a mnemonic you may have used (for *p*arentheses, *e*xponents, *m*ultiplications/*d*ivisions, *a*dditions/*s*ubtractions).

> *Think About ...*
>
> What is the answer to the calculation $3 + 15 \div 3 - 4 \times 2$?
> **a.** ⁻9   **b.** ⁻2   **c.** 0   **d.** 4   **e.** 5

Further research and development. The increased attention to algebra in elementary school has resulted in several research projects. One group of researchers[5] has been investigating how children can arrive at some of the properties of operations, like commutativity of addition or distributivity, through careful teacher planning and questioning. For example, in connection with a "How many of each?" activity focusing on the different ways that one could have 7 peas and carrots, first-grade children noted that 5 peas and 2 carrots has an "opposite," 2 peas and 5 carrots. Second-graders, in making up a list of ways to make 10, referred to "turnarounds" in noting that $7 + 3$ also gives $3 + 7$. Students were convinced that such would always be true and verified it with problems like $17 + 4$ and $4 + 17$.

Despite this apparent mastery of commutativity of addition, when the children were asked whether, say, $13 + 12 = 12 + 13$ was true, the children refused to endorse it (even after verifying that each sum was 25). The children had formed the "$=$" means "Write the answer" belief mentioned above, and for $12 + 13 = 13 + 12$, "There's no answer here." Another group of children expressed commutativity of addition as the "switch-around rule." But as an example of overgeneralizing, many of the children thought that the switch-around rule also applied to subtraction, so that $7 - 4 = 3$ and $4 - 7 = 3$.

The point here is that the basis for algebraic ideas can be laid at quite early ages, even without being formalized with conventional language or notation. Curricular materials that you use may provide opportunities to learn important ideas such as commutativity, without focusing on the term "commutativity."

Technology can be used to help young children work with algebraic ideas. There are, of course, drill and practice computer "games" for integer arithmetic. But more important, there is also computer software (like Over & Back) that involves quantities that can vary, like distance and speed, thus allowing instructors to introduce graphs, numerical tables, and the associated algebra, often with activities to be discussed among a small group of students. Spreadsheets, invented for business, can also be tailored to help to introduce algebraic variation and algebraic labeling.

Thus, with the increased attention to algebra in the elementary school curriculum, there has been a corresponding increase in novel curricular approaches to help bypass some difficulties and to give meaningful bases for algebraic ideas. With various states and districts now requiring algebra at the eighth grade, the importance of preparation for algebra in the elementary grades is clear. The topics covered in Part II (Chapters 12–15) will support your ability to deal with introductions to algebra.

Learning Exercises for Section 15.6

1. Explain why, in a Turtle-Rabbit race, writing just "T = turtle" and "R = rabbit" is not good practice. What should be written instead?

2. Practice the order of operations conventions in evaluating these expressions.

   a. $19 - 5 \times 2 + 6 \div 3 - 2^3 + 3(^-1)^3 + (8 + \ ^-8)^7$

   b. $4\frac{1}{2} + 9\frac{1}{8} + \frac{7}{8} \cdot (2^2)^4 - 3 \cdot (6 - \ ^-2)^2$

3. For each given value of x, find the value of $2x^2 - 7x - 5$.

   a. 3   **b.** $^-4$ **c.** $\frac{2}{3}$ **d.** $\frac{^-1}{2}$

4. For each of the following equations, make up a story problem that could be described by the equation:

   a. $n + 7 = 13$ **b.** $7 + n = 13$ **c.** $n - 6 = 9$

   d. $14 - n = 6$ **e.** $2n + 1.95 = 4.15$ **f.** $\frac{n}{25} = 76\%$

5. What property is suggested by viewing the given drawing in two ways, as in the curriculum adapted from Russia? (See Section 13.5.)

6. How does a drawing for the students-professors problem presumably make clear that the equation $6S = P$ is inappropriate?

15.7 Check Yourself

After studying this chapter, you should be able to do problems like those assigned and to meet the following objectives:

1. Solve a system of two linear equations, using substitution or addition/subtraction.

2. Use and explain the phrases "inconsistent equations" and "dependent equations." Describe the graphical implications of the phrases.

3. Solve algebraically story problems that can lead to two linear equations.

4. Show your grasp of different approaches to solving a story problem: table, graph, equation(s). Give the advantages and disadvantages of each approach from the elementary school perspective.

5. Deal with weighted averages, as in finding an average speed or a grade-point average.

6. Give a (noncomputational) explanation of the word "average."

7. Give an example of a function that does not involve numbers.

8. Illustrate how two functions can be combined, perhaps with function machine representations. Test this combination idea for commutativity.

9. Use the differencing technique to gain information in finding a function rule.

References for Chapter 15

[1]Phillips, E. (1991). *Patterns and functions*, from the *Addendum series, Grades 5–8*. Reston, VA: National Council of Teachers of Mathematics (NCTM).

[2]Kieran, C. (1992). The learning and teaching of school algebra. In D. Grouws (Ed.), *Handbook of research on mathematics teaching and learning* (pp. 390–419). New York: Macmillan. (A similar handbook, edited by Frank Lester, Jr., became available in 2007 from NCTM.)

[3]National Research Council. (2001). *Adding it up: Helping children learn mathematics*. J. Kilpatrick, J. Swafford, & B. Findell (Eds.). Mathematics Learning Study Committee, Center for Education, Division of Behavioral and Social Sciences and Education. Washington, DC: National Academy Press.

[4]Stephens, A. C. (2005). Developing students' understandings of variable. *Mathematics Teaching in Middle School, 11*(2), 96–100.

[5]Schifter, D., Monk, S., Russell, S. J., Bastable, V., & Earnest, D. (2003, April). Early algebra: What does understanding the laws of arithmetic mean in the elementary grades? Paper presented at the annual meeting of the NCTM, San Antonio. [This work is now in Carraher, D., Kaput, J., & Blanton, M. (2007). *Algebra in the early grades*. New York: Routledge.]

Reasoning About Shapes and Measurement

The real world is made up of objects that have shapes. It is no surprise then that the mathematics curriculum gives attention to geometric shapes, particularly those commonly seen in manufactured goods. Manufacturing naturally requires many sorts of measurements, including those of geometric shapes. So, it also is no surprise that measurement is treated in the elementary school mathematics curriculum.

Although Part III of *Reconceptualizing Mathematics* deals with several three-dimensional shapes, Part III also treats two-dimensional ideas with a teacher-view in mind (Chapters 16, 17, 21). Symmetry and tessellations (Chapters 18 and 19) show some esthetic possibilities with geometric shapes, as well as being topics of mathematical interest. Scale models and photographic enlargements are two examples of the general topic of similarity (Chapter 20). Transformation geometry (Chapter 22) allows these three topics (symmetry, tessellation, and similarity) to be placed under one umbrella.

The important topic of measurement is the focus in the remaining chapters of Part III. Measurement in the early chapters is largely restricted to length and angle size, and length and angle size are revisited more deeply with measurement basics in Chapter 23. Chapter 24 treats the measurement of area and volume, with a focus on related formulas and their origins in Chapter 25. Chapter 26 deals with the historical and practical Pythagorean theorem, along with a look at some nongeometric measurements.

Unfortunately, in some elementary school classrooms geometry and measurement are given little attention to allow more time for work with numbers. But the measurement of geometric quantities gives many opportunities to work with numbers involved in measurements, many of which students can make themselves. Your work in Part III should help you to see many ways to make shapes and measurement an important part of the mathematics you teach. For a teacher, geometry and measurement are more than vocabulary and formulas.

Chapter 16

Polygons

In geometry, flat, or planar, shapes are common models for many real-world items in Western societies. This chapter features a common type of planar shape, polygons, and several ideas and facts associated with them, including the notion of hierarchical relationships among the various types of polygons.

16.1 Review of Polygon Vocabulary

You may need to brush up on your two-dimensional geometry vocabulary. Words that you may have forgotten often come back with a little review.

> ### *Think About …*
> How many vertices (corner points) does a 10-sided shape
> (a **decagon**) have? A 100-sided shape? Does every such type
> of shape have the same number of vertices as sides? How
> many angles does a 6-sided shape have? An *n*-sided shape?
> Does every such type of shape have the same number of
> angles as sides?

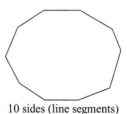

10 sides (line segments)

This section quickly reviews many of the geometric terms associated with polygons (*poly-* means "many"; *-gon* means "angle"). You should be adept at two *translations* between names and shapes: (1) Give the best name for a given shape, and (2) draw or pick a shape, given its name.

All the shapes in Figure 1 on the next page are polygons. As the arrangement suggests, one common way of classifying polygons is by the number of angles (or vertices or sides) that the polygon has. Each name involves a prefix indicating the number of angles, vertices, or sides, followed by either *-gon*, *-angle*, or *-lateral* (sides). The Glossary includes some of the prefixes; for polygons of 6 or 7 sides, the Greek prefixes are the common ones. The term ***n*-gon** allows us to be general about the number *n* of angles, vertices, or sides.

How would you define polygons? Here is a common description.

> A **polygon** is a closed planar figure made up of line segments joined end to end, with no crossings or reuse of endpoints.

> ### *Think About …*
> Explain why each of these shapes is not a polygon.
>
> a.   b.   c.   d.   e.

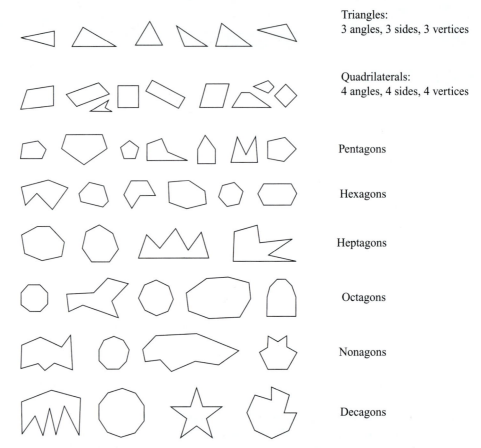

Triangles:
3 angles, 3 sides, 3 vertices

Quadrilaterals:
4 angles, 4 sides, 4 vertices

Pentagons

Hexagons

Heptagons

Octagons

Nonagons

Decagons

Figure 1

A polygon consists of just the points *on* its line segments. The points *inside* a polygon make up its **polygonal region** (pronounced "pa LIG a nal"). If the context is clear, however, even experts often describe a polygonal region by naming the polygon that bounds it. For example, a face of a cube is often called a *square* instead of the more precise *square region*.

Polygons   Polygonal regions

A **diagonal** of a polygon is a line segment joining two vertices of the polygon that are not joined by a side. The dashed segments show three of the nine diagonals in this hexagon.

A few important adjectives can be applied to all types of polygons. An **equiangular polygon** is a polygon whose angles all have the same size. An **equilateral polygon** is a polygon whose sides all have the same length. Because these adjectives refer to different parts of a polygon, they represent separate ideas, in general. However, the two ideas go hand in hand for triangles. *If a triangle is equiangular, it is automatically equilateral, and vice versa.* Other polygons also can be both equiangular and equilateral; in this case, the polygons are called **regular polygons**. For example, because a square is both equiangular and equilateral, a square is a *regular* quadrilateral. A rectangle that is not also a square is equiangular but is not regular because it is not equilateral.

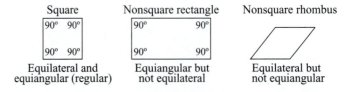

Square
Equilateral and
equiangular (regular)

Nonsquare rectangle
Equiangular but
not equilateral

Nonsquare rhombus
Equilateral but
not equiangular

Think About ...

Which polygons in Figure 1 appear to be regular polygons?

Triangles can be further classified in two ways. One way focuses on the size of the largest angle. An **acute triangle** has each angle measuring less than 90° in size; a **right triangle** has an angle that is 90° in size; and an **obtuse triangle** has an angle that is more than 90° in size. (Practice using these adjectives with the triangles in Figure 1.) The second way to classify triangles deals with the relative lengths of the sides of the triangle. The sides of a **scalene triangle** all have different lengths. If at least two sides of a triangle have the same length, the triangle is an **isosceles triangle**. If all three sides have the same length, the triangle is an **equilateral triangle**. (Practice using these adjectives with the triangles in Figure 1.)

Think About ...

Why is an equilateral triangle a special isosceles triangle?

Quadrilaterals come in a great many varieties. Several types are shown here.

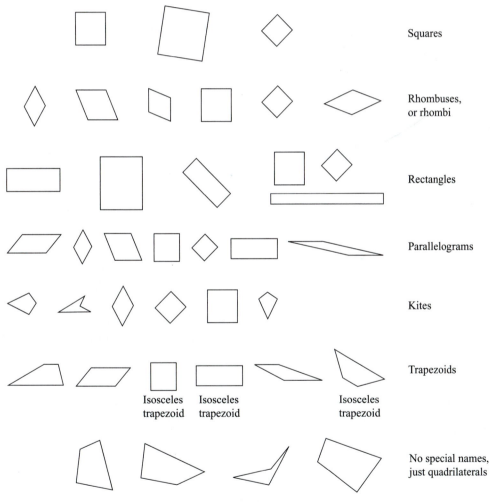

Squares

Rhombuses, or rhombi

Rectangles

Parallelograms

Kites

Trapezoids

Isosceles trapezoid   Isosceles trapezoid   Isosceles trapezoid

No special names, just quadrilaterals

Figure 2

ACTIVITY 1   From Examples to Definitions

Without looking up definitions, write definitions for these types of quadrilaterals: square, rhombus, rectangle, parallelogram, kite, and trapezoid. Compare your definitions with those of someone else in your class. Figure 2 on the previous page should help you determine the important elements for your definitions. ●

Notice the vocabulary level in this Grade 1 textbook.

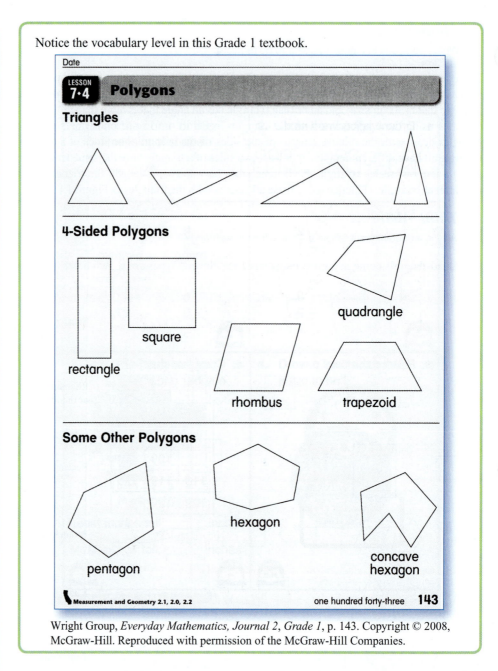

Wright Group, *Everyday Mathematics, Journal 2, Grade 1*, p. 143. Copyright © 2008, McGraw-Hill. Reproduced with permission of the McGraw-Hill Companies.

This book is written without extensive use of many of the **notations** common in advanced mathematics, partly because their use in elementary school is unpredictable. The following notations are reminders; feel free to use them (or your instructor may wish you to use them).

- Point: Capital letter (A, B, . . .)

- Line segment with endpoints at P and Q: $\overline{PQ}$

- Ray starting at C and going through D: $\overrightarrow{CD}$
- Line through D and E: $\overleftrightarrow{DE}$

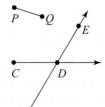

- Polygon with vertices *F*, *G*, *H*, and *I*: *FGHI*

- Line segments (and angles) may also be indicated with small letters, which may also mean their lengths (and angle sizes). The endpoints of a line segment may be highlighted by dots. We shall examine length units more closely later, but we shall use a few metric units for length now. A centimeter (cm) is smaller than an inch (about 0.4 in.); a meter (m) is slightly longer than a yard (39.37 in.); and a kilometer (km) is about 0.62 mile.

- Angles are named either by naming just the vertex, or if needed for clarity, as for angle *D* in the drawing on the previous page, naming a point on one side, the vertex, and then a point on the other side: $\angle D$, or $\angle CDE$.

Vocabulary for Angles

A review of some of the angle vocabulary that can arise in connection with polygons can be useful (we have already used some terms, as in classifying triangles). We will introduce other angle-related terms later. *Note*: Although we leave the main "theory" of measurement to later chapters (Chapter 23), our discussion assumes some familiarity with angle-size (and length) measurements. Angle size can be measured with the **protractor** and is usually measured in degrees (°) in elementary school. (Appendix G treats how to use the protractor if you have never used one.)

   A **right angle** has size 90° (like a corner of a sheet of paper); lines that make right angles are called **perpendicular**. An **acute angle** has size less than 90°; an **obtuse angle** has size greater than 90° but less than 180°; and a **straight angle** has size 180°. Strictly speaking, the sides of an angle are rays, but line segments are often used to indicate the angle. A little box in a corner may be used to make clear that an angle is a right angle.

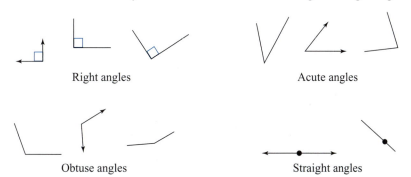

Right angles   Acute angles

Obtuse angles   Straight angles

Adjacent angles are angles with the same vertex and a common side between them. **Supplementary angles** are two angles whose sizes add up to 180°; supplementary angles do not have to be adjacent but may be. **Complementary angles** (note the spelling) are two angles whose sizes add to 90°. Examples of these types of angles are shown below.

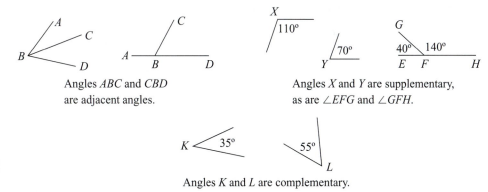

Angles *ABC* and *CBD*   Angles *X* and *Y* are supplementary,
are adjacent angles.   as are $\angle EFG$ and $\angle GFH$.

Angles *K* and *L* are complementary.

Notes

An **exterior angle** of a polygon is formed by the extension of one side of the polygon through a vertex of the polygon and the other side of the polygon through that vertex. We commonly refer to an angle "inside" a polygon as an angle of the polygon, or an **interior angle** of the polygon, for clarity. Angles *AED* and *EDC* are interior angles in pentagon *ABCDE*.

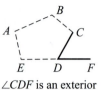

∠*CDF* is an exterior angle of *ABCDE*.

∠*KJH* is an exterior angle of *GHJ*.

ACTIVITY 2 A Possible Pattern!

Work with the given polygons to make entries into the table below. Predict what would be the sum of the sizes of all the angles of a 20-gon. Do the same for an *n*-gon.

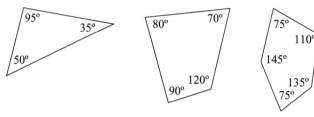

Number of Sides	Sum of All Angles
3	$180° = 1 \times 180°$
4	____° = ____ $\times 180°$
5	____° = ____ $\times 180°$
⋮	⋮
20	____° = ____ $\times 180°$
n	(_____) $\times 180°$

This result for the sum of all the angles of a polygon is useful and worth memorizing. ●

TAKE-AWAY MESSAGE . . . Dealing with polygons involves a lot of nouns and adjectives, especially with quadrilaterals. The vocabulary is simplified a little by some use of prefixes and root suffixes such as -*gon*. We also use several adjectives to describe angles, and we noticed that the sum of all the angles of an *n*-gon appears to be $(n - 2) \times 180°$.

Learning Exercises for Section 16.1

1. Name each of the following polygons. If an adjective applies, use it also.

a. b. c. d. e.

f. g. h. i. j.

2. Sketch an example, if it is possible, of each shape described. If any are not possible to sketch, explain why.

   a. a trapezoid with at least one right angle
   b. a hexagon with two sides perpendicular (make right angles, or angles of 90°)
   c. a pentagon with two sides parallel
   d. a regular quadrilateral (What is this shape usually called?)
   e. an equiangular quadrilateral that is not equilateral (What is this shape usually called?)
   f. an equilateral quadrilateral that is not equiangular (What is this shape usually called?)
   g. a parallelogram that has exactly one right angle
   h. a trapezoid that has equal-sized angles next to one of the parallel sides
   i. an isosceles right triangle
   j. a scalene obtuse triangle
   k. an isosceles obtuse triangle
   l. an equilateral right triangle
   m. a rhombus that is also a rectangle
   n. an isosceles triangle with all sides the same length
   o. a kite with exactly one right angle
   p. an equilateral hexagon that is not a regular hexagon
   q. a regular pentagon that is not an equilateral pentagon

3. The reasonings below give incorrect conclusions. Explain why.

   a. "I don't get it. A hexagon has six sides, and each side has two endpoints, which are vertices of the hexagon. Six times two is twelve. Why doesn't a hexagon have twelve vertices?"
   b. "Hmm. An octagon has eight sides. It takes two sides to make an angle of the octagon. There are *four* two's in eight, so it seems that an octagon should have *four* angles, not eight!"

4. Look around the room you are in (or a room in your imagination). What polygons and polygonal regions do you see?

5. Many classrooms have sets of **tangrams**, cardboard or plastic pieces that can be fit together to make a variety of shapes. The tangram pieces can be cut from a square region, as in the drawing shown at right.

   a. What shape is each of the seven tangram pieces?
   b. Using the segments in the drawing, find as many of each type of polygon as you can: isosceles right triangles; isosceles trapezoids; trapezoids that are not isosceles; parallelograms that are not rectangles. (*Hint*: Some may involve more than one tangram piece.)
   c. Let the entire square region be 1. Give the fractional value for each of the seven tangram pieces.
   d. Trace the tangram pieces, cut them out, and, without looking at the diagram, put them together again to form a square.
   e. Tangrams are often used in classrooms to make different shapes. Can you use the pieces to make something—a cat, e.g.?

6. Sometimes ideas are presented to elementary school children by *concept cards*. Here is a pair of concept cards for the ideas of **convex** polygon and **concave** polygon:

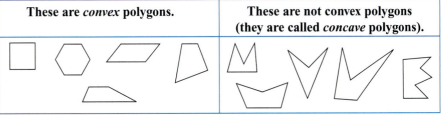

These are *convex* polygons.	These are not convex polygons (they are called *concave* polygons).

continue

a. Draw another example of a convex polygon and of a concave polygon.
b. How would you describe convex polygons, in words? Concave polygons? (Your definitions may be more intuitive than the mathematical definitions in the Glossary.)
c. Make up a pair of concept cards for the idea of parallelogram.
d. Give a word description of *kite* from the information in the following concept cards:

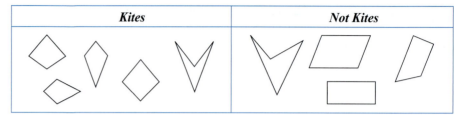

Kites	Not Kites

7. For each part (a–f), copy the regular hexagonal region. (Answers for the first three parts are in the standard set of pattern blocks in the "Masters.")

   a. Cut the hexagonal region into three identical rhombi.
   b. Cut the hexagonal region into six equilateral triangles.
   c. Cut the hexagonal region into two isosceles trapezoids.
   d. Cut the hexagonal region into 12 identical right triangles.
   e. Cut the hexagonal region into six identical kites.
   f. Cut the hexagonal region into four identical trapezoids.

8. What are some quantities associated with a polygon? With a polygonal region?

9. Make and continue a table like the one indicated, and try to predict the total number of diagonals in a 12-gon, a 20-gon, and an *n*-gon.

Number of vertices	3	4	5	(etc.)
Number of diagonals	0	2		

10. You know that the sizes of all the angles in a triangle add up to 180°. Justify that the sum of the sizes of all the angles in every quadrilateral is 360°. *Hint*: Draw one diagonal. Does your justification work on quadrilaterals such as the one shown at the right?

11. Find the missing sizes of the angles marked for each quadrilateral. (*Hint*: See Learning Exercise 10.)

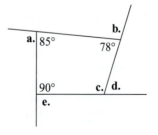

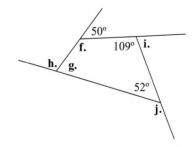

k. Make an educated guess (a conjecture) about the sum of the sizes of all the exterior angles of every quadrilateral, using just one exterior angle at each vertex.
l. Give a justification that your conjecture in part (k) will always be correct.

12. a. From Activity 2, you likely found that the sum of the sizes of all the angles of a 20-gon is probably 3240°. If the angles of the 20-gon are all the same size, how many degrees are in each angle?

b. If all the angles of a 32-gon are the same size, how many degrees are in each angle?

c. If all the angles of a 102-gon are the same size, how many degrees are in each angle?

13. Use Activity 2 to help find the sizes of each of the following angles:

a. each (interior) angle of an equiangular pentagon

b. each exterior angle of an equiangular pentagon

c. each (interior) angle of an equiangular octagon

d. each exterior angle of an equiangular octagon

e. each (interior) angle of an equiangular n-gon

f. each exterior angle of an equiangular n-gon

14. Find the missing sizes of the angles marked for each pentagon.

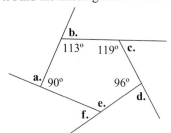

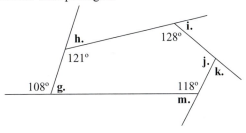

n. Make an educated guess (a conjecture) about the sum of the sizes of the exterior angles of every pentagon, using just one exterior angle at each vertex.

o. Give a justification that your conjecture in part (n) will always be correct.

15. Are the following statements true? Give your reasoning.

a. The two acute angles in a right triangle are complementary.

b. A triangle cannot have more than one right angle.

c. The supplement of an angle of size $x°$ will have size $180° - x°$.

16. The Greeks tied much of their work with numbers to geometry. For example, they might think of the first, second, third, fourth, etc., square numbers by building increasingly larger squares of dots:

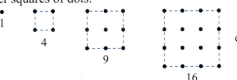

a. Draw a square to verify that the fifth square number is indeed 25.

b. Give an algebraic expression for the nth square number. Do the same for the $(x + 1)$st square number.

c. The Greeks identified the **triangular numbers** in a similar way.

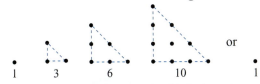

According to this pattern, the third triangular number is 6, and the fourth triangular number is 10. Make a table like the following one. Extend the results to help predict the number of dots in the 10th triangular number, the 20th triangular number, and the nth triangular number.

Triangular number	1st	2nd	3rd	4th
Number of dots involved	1	3	6	10

Notes

17. Some middle-grades teachers have their students illustrate particular geometric ideas by folding small pieces of paper. (The straight edges and right angles of the paper are not to be used. Pretend the paper is irregularly shaped.) For example, folding a piece of paper and creasing it give a straight-line segment, which can then be marked with a pencil. Use quarter-sheets of scrap paper to show each of the following geometric ideas:

   a. the line segment through two points marked on the paper

   b. the bisector of an angle drawn on the paper (Can you make the angle without using a ruler?)

   c. a right angle

   d. one line that is perpendicular to a line segment marked on the paper, at the midpoint of the marked line segment (the **perpendicular bisector** of the line segment)

   e. the perpendicular to a marked line, at a point marked on that line

   f. two parallel lines

   g. a square

18. Practice making the polygon versus polygonal region distinction by giving the best names for the following models:

   a. this sheet of paper (Does it represent a rectangle, or a rectangular region?)   **b.** a triangle made of coat-hanger wire

   c. a triangular-shaped piece of cloth   **d.** the bottom of a shoebox

   e. a "Yield" traffic sign   **f.** the edge of a picture frame

   g. a hexagonal window frame   **h.** a hexagonal window

19. Experiment to test whether this plausible-sounding conjecture (educated guess) appears to be true: In a triangle, the greater side is opposite the greater angle.

20. Trace and label the points shown to the right, and then show each geometric object on a single drawing.

   a. $\overrightarrow{IA}$ **b.** $\angle IBA$ **c.** region *BCDEFGH*   **d.** $\overline{AH}$

   e. $\overrightarrow{HI}$ **f.** $\overrightarrow{BA}$ **g.** $\overleftrightarrow{HB}$

$C_\bullet$ $E_\bullet$ $_\bullet G$

$D_\bullet$ $F_\bullet$

$B_\bullet$ $_\bullet H$

$A_\bullet$ $_\bullet I$

21. Make sketches to show each geometric object.

   a. polygon *PQRS*   **b.** $\overrightarrow{NM}$ **c.** $\overleftrightarrow{FG}$

   d. $\overline{AB}$ **e.** $\angle FDE$ **f.** region *TUV*

Supplementary Learning Exercises for Section 16.1

1. Name each of the following shapes. If an adjective applies, use it also.

a.   **b.**    **c.**   **d.**   **e.**

f.   **g.**    **h.**

2. Sketch an example, if it is possible, of each shape described. If any are not possible to sketch, explain why.

continue

a. a pentagon that is not equilateral
b. a 12-gon
c. a square that is not regular
d. an equiangular parallelogram
e. a rectangle that is not equiangular
f. an obtuse right triangle
g. a regular triangle
h. a kite with exactly two right angles

3. Find the missing sizes of the angles marked in these shapes.

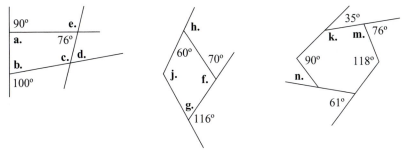

4. Give these sums.
   a. the sum of the angles of a 9-gon ("nonagon," occasionally "enneagon")
   b. the sum of the interior angles of a 12-gon
   c. the sum of the angles of an *m*-gon

5. Make a drawing to show each of the following shapes:
   a. a quadrilateral with a diagonal that is horizontal (Notice how the geometric *diagonal* differs from a common use of *diagonal*.)
   b. a quadrilateral with a diagonal that is vertical (up-and-down)
   c. a quadrilateral that has only two opposite sides equal and is not a trapezoid

6. Make sketches to show each geometric object.
   a. polygon *PQRST*
   b. $\overrightarrow{BA}$
   c. $\overleftrightarrow{CD}$
   d. obtuse ∠*EFG*
   e. $\overline{MN}$
   f. region *HIJK*
   g. straight ∠*Z*

16.2 Organizing Shapes

This section treats one way to categorize some common geometric shapes. Categorizing things and then naming the categories seem to be natural human activities.

A person not familiar with categorizing animals might focus on differences and regard all of the following as different: apes, beagles, cobras, collies, humans, sharks, rattlesnakes, salmon, and whales. Zoologists, however, might focus on *shared* characteristics rather than differences, organizing the categories as follows (with new categories in parentheses):

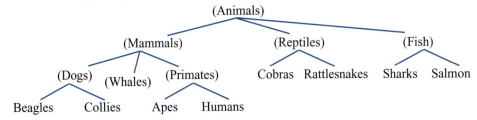

Such a **hierarchical classification** shows that lower entries can be regarded as special cases of any higher entries to which they are connected by line segments. For examples, beagles are special dogs as well as mammals and animals; apes are special mammals and special primates; salmon are special fish as well as special animals.

It is likely that at some point in your life, you were surprised to learn that humans are considered to be animals. Yet you now realize that people share many characteristics with all animals: living beings and mobility, e.g. Any characteristic that is true for all animals is true for every human, even though people have many characteristics that other kinds of animals do not have. All mammals are animals with special characteristics, such as feeding their young with milk, being warm-blooded, and having hair.

Classifying Geometric Shapes

The same focus on shared characteristics, rather than differences, enables a sophisticated clas- sification of geometric shapes. You may be quite comfortable with regarding a rectangle as a quadrilateral, but you may be less comfortable with calling a rectangle a parallelogram, which is possible by focusing on particular shared characteristics of rectangles and parallelograms. In mathematics just a few (not all) of the characteristics of a class of shapes are needed to define the class, because other characteristics can be deduced logically. In future sections, you will see examples of this type of reasoning, where some facts are deduced from other established facts.

DISCUSSION 1   A Square by Any Other Name. . .

Organize the following geometric shapes in a hierarchical classification diagram. Be ready to explain your method of organization.

circles, equilateral triangles, hexagons, isosceles trapezoids, isosceles triangles, kites, octagons, parallelograms, pentagons, polygons, quadrilaterals, rectangles, rhombuses, scalene triangles, squares, trapezoids, triangles ●

TAKE-AWAY MESSAGE . . . A classification into categories often lends itself to a further organization, one in which items in a subcategory have all the characteristics of the major cat- egory, as well as other characteristics. Hence, items in the subcategory can go by multiple names—those names of the major category as well as those of the subcategory.

Learning Exercises for Section 16.2

1. Why is the following classification system (shown using a Venn diagram) limiting to the user, for the categories named?

2. Decide whether each statement is always true, sometimes true, or never true. If a statement is sometimes true, sketch an example of when it is true and an example of when it is not.

   a. A square is a rectangle.   **b.** A rectangle is a square.
   c. A parallelogram is a rectangle.   **d.** A rectangle is a parallelogram.
   e. A trapezoid is a kite.   **f.** A kite is a trapezoid.
   g. A kite is a rhombus.   **h.** A rhombus is a kite.
   i. A square is a kite.   **j.** A kite is a square.
   k. A right triangle is an isosceles triangle.   **l.** An isosceles triangle is a right triangle.
   m. An acute triangle is a scalene triangle.   **n.** A scalene triangle is an acute triangle.

3. In each given group of shapes, what characteristics are shared by the shapes? Characteristics might include lengths of sides, sizes of angles, parallelism, length of diagonals, and so on. What characteristics can be different?

continue

a. parallelograms, rectangles, rhombi, squares
b. kites, rhombi, squares   c. quadrilaterals, trapezoids
d. trapezoids, parallelograms   e. parallelograms, rhombi

4. Give an example of each shape, if possible. If it is not possible, explain why.

a. a kite that is also a rhombus
b. an obtuse triangle that is also isosceles
c. a square that is not a parallelogram
d. a parallelogram that is not a square
e. a rhombus that is not a kite
f. a rectangle that is not a parallelogram

5. Organize these terms: angles, circles, ellipses, ovals, parabolas, squiggles, zigzags.

6. If you know a four- or five-year-old, show him or her the drawing of a square in the usual orientation (vertical and horizontal sides), and ask, "What shape is this?" Many will know what it is. If the child says the shape is a square, rotate the drawing so that the sides are not horizontal and vertical, and ask, "Is this still a square?" Many will say it is not! They have focused on an irrelevant characteristic—the horizontalness or verticality of the sides—of the squares they have seen, and then have absorbed that as a requirement to be a square.

a. What irrelevant characteristic of trapezoids is involved in the instruction, "Draw an upside-down trapezoid"?
b. What irrelevant characteristic(s) in each scenario given might these students be assuming to be important?

i. a student who does not regard a rectangle as also being a parallelogram
ii. a student who does not regard a square as a special rectangle
iii. a student who does not regard a square as a special rhombus
iv. a student who does not regard a rhombus as a special kite

Supplementary Learning Exercises for Section 16.2

1. Use the partial hierarchy of animals given below to answer the questions.

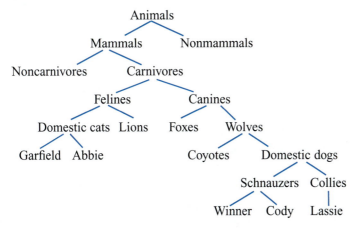

a. Is Winner a carnivore?
b. Is every coyote a carnivore?
c. Is every canine a domestic dog?
d. Is every fox a mammal?
e. Is it possible to be a carnivore, a feline, and also a schnauzer?
f. Is every wolf a special coyote?
g. Is every coyote a special wolf?
h. Is every coyote a special fox?
i. Are foxes considered to be special wolves?

Notes

2. Consider the (noncontroversial) hierarchy to the right. Use the hierarchy to answer the questions below.

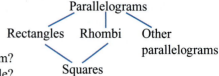

   a. Is every rectangle a special parallelogram?
   b. Is every parallelogram a special rectangle?
   c. Is every square a special parallelogram?
   d. Is every rhombus a special rectangle?
   e. Is it possible for some rhombi to be special rectangles?

16.3 Triangles and Quadrilaterals

The simplest polygons (triangles and quadrilaterals) have many properties that are not included in the statements of their definitions. For example, the opposite sides of a parallelogram are parallel by definition, but they also seem to be equal in length when you look at a drawing. However, there are risks in using drawings as proof (or in trusting patterns too much). Nonetheless, ideas have to come from somewhere. Drawings and patterns give ideas that might be justified in a general way. The basic descriptions of quadrilaterals in this section allow a sophisticated classification scheme like those given in the last section.

ACTIVITY 3   You Can't Go by Looks

In parts (a) and (b), which segment looks longer, $\overline{PQ}$ or $\overline{RS}$? Measure each segment to check your perception. What do you notice about the object in part (c)? Is it completely safe to make conjectures based on looking at drawings?

a.

b.

c.

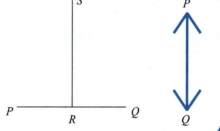

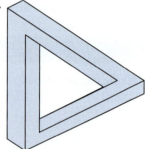

   The message from Activity 3 is, of course, that our perceptions may trick us. So, in arriving at conclusions based on drawings, we should be very cautious and tentative in asserting their truth. Drawings can be the source of ideas, however, and give us some basis for educated guesses, or **conjectures**. Activity 4 calls for making conjectures, recognizing that some conjectures may be incorrect.

ACTIVITY 4   You Can't Go by Looks, But I Think. . .

The aim of this activity is to form conjectures about properties of quadrilaterals or special quadrilaterals. One method of getting ideas is to examine drawings of several examples. Your conjectures might be about sides, angles, or diagonals. Your conjecture might also be about how the sides, angles, or diagonals appear to be related. Your group should consider these polygons: squares, rectangles, parallelograms, rhombi, trapezoids, isosceles trapezoids, kites, and quadrilaterals in general. Organize your conjectures for a whole-class discussion. ●

Your group probably found that rectangles, e.g., appear to have several properties. There are two concerns: (1) Are these conjectures indeed *always* true for *every* rectangle, and (2) which of the properties should be used to define *rectangle*? We will consider the latter concern first.

Creating a Definition

Novices often describe a shape by telling everything they know about the shape. Adopting *all* of the properties of a shape as a definition, however, would have the drawback of requiring us to verify a lot of properties unnecessarily, when checking whether a particular shape fits the definition. The unnecessary work would result because some properties are *consequences* of others, so they will happen automatically without being mentioned in the definition. Mathematicians know that there are often different ways to choose properties for an official definition. Of course, the properties chosen must indeed be adequate to identify all, and only, those shapes intended. To make clear the connections in a classification system, definitions may use properties that seem inadequate, oddly chosen, or even counterintuitive. (You should probably reread this paragraph after you have examined the definitions given below.)

As a taste of bare-bones descriptions, here are some fairly standard definitions for special triangles and quadrilaterals. A few are repeated from an earlier section for convenience. Make sketches to see that each definition does indeed describe the specific polygon. (Some drawings given in Section 16.1 may be helpful.)

An **isosceles triangle** is a triangle that has at least two sides with the same length.

An **equilateral triangle** is a triangle whose sides all have the same length.

A **kite** is a quadrilateral in which two consecutive sides have equal lengths and the other two sides also have equal lengths.

A **trapezoid** is a quadrilateral with at least one pair of opposite sides parallel.

An **isosceles trapezoid** is a trapezoid in which both the angles next to one of the parallel sides have the same size.

A **parallelogram** is a trapezoid with both pairs of opposite sides parallel.

A **rectangle** is a parallelogram with a right angle.

A **rhombus** is a kite that is also a parallelogram.

A **square** is a rectangle that is also a rhombus.

Some of these given definitions no doubt sound strange. The phrasings certainly make clear the connections between shapes. For example, the definitions explicitly say that a parallelogram is a special trapezoid and that a rectangle is a special parallelogram. Yet you know that a rectangle has *four* right angles, but the official definition says *a* right angle. Legalistically—and mathematically—a right angle does not rule out *four* right angles, in the same way as answering "Do you have a quarter?" with "Yes" does not mean you have only one quarter. Why aren't the other three right angles mentioned? You may realize that they must be a logical consequence of the parallelogram-with-right-angle requirement for a rectangle. Here, the hierarchical arrangement in a classification system comes strongly into play: Each fact about every parallelogram also applies to every rectangle! Your group may have noticed that in a parallelogram the opposite angles are the same size and the pairs of consecutive angles have sizes that add to 180° (i.e., they are supplementary). These two facts logically show that the other three angles of a rectangle must also be right angles. (Can you give a justification that uses *only* the second fact?)

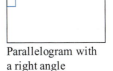
Parallelogram with a right angle

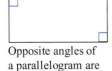
Opposite angles of a parallelogram are the same size.

Consecutive angles of a parallelogram have sizes that sum to 180°.

Hence, a hierarchical arrangement in a classification system allows us to take advantage of facts about one shape by applying them to special cases of that shape, as we just did in applying facts about parallelograms to rectangles by viewing rectangles as special parallelograms.

Notes

 **DISCUSSION 2**   A Rhombus Has. . .

The definition of *rhombus* does not mention explicitly that a rhombus has four sides with the same length. How is that information hidden in the definition? (*Hint:* Some of your conjectures from Activity 4 may be useful.) ●

Hierarchical Arrangements of Polygons

The previous definitions also give the following categorization scheme, or hierarchy, for polygons. The connection between *trapezoid* and *parallelogram* would not exist if a trapezoid is defined as having exactly one pair of parallel sides, nor would the connection between isosceles trapezoids and rectangles exist.

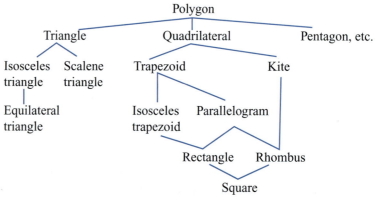

Figure 3

The hierarchy in Figure 3 shows, e.g., that a rhombus is a special parallelogram and a special kite, as well as being a special trapezoid, quadrilateral, and polygon. The hierarchy of Figure 3 is the one we will use, but your instructor may prefer the hierarchy in Activity 5.

 **ACTIVITY 5**   If the Definition of Trapezoid Changed

Many elementary school mathematics books do define a trapezoid as having exactly one pair of opposite sides parallel. In the diagram below and with that definition in mind, draw lines to show how the hierarchy for quadrilaterals would change from the previous one.

Quadrilateral

Trapezoid   Kite

Isosceles   Parallelogram
trapezoid

Rectangle   Rhombus

Square

Figure 4   ●

In establishing that certain facts are true about a shape, sometimes we cannot take advantage of any hierarchical relationships, however. For example, your group may have conjectured (or you may remember from school geometry) that in an isosceles triangle, the angles opposite the sides of equal length have equal sizes. How could you justify that conclusion? It is not true for all triangles, so there is no help from the hierarchy. Later on, we will appeal to the concept of *symmetry* to justify some of the facts about shapes. You may remember other methods, such as the use of congruent triangles, from secondary school.

   One method that we will encourage, but again with a cautionary note, is to use a pattern suggested by several examples.

ACTIVITY 6 Can Patterns Be Completely Trusted?

Look for patterns to answer the questions.

a. Notice the following pattern:

$$1^2 = 1$$
$$11^2 = 121$$
$$111^2 = 12321$$
$$\text{etc.}$$
$$1,111,111,111^2 = ?$$

b. The data in the table below are fairly accurate for humans:

Month of pregnancy	2	3	4	5
Length of fetus (cm)	4	9	16	25

How long will the baby be at birth (9 months)?

c. Can patterns be completely trusted?

Basing a conclusion solely on one or more examples is called **inductive reasoning**. Mathematicians know that conjectures from drawings and patterns cannot be completely trusted. Inductive reasoning is risky. For example, the correct value for $1,111,111,111^2$ is a **counterexample** (an example that shows that the conjecture is not true in general) to the pattern suggested by the earlier results. So, mathematicians seek a general argument that will apply to *all* cases at once, not just those represented by the drawing or examples examined. In short, be pleased with the conjectures you make using inductive reasoning from drawings or patterns, but be very careful in asserting that the conjectures are definitely true.

TAKE-AWAY MESSAGE... Many facts about types of polygons are not always evident in their mathematical definitions. Inductive reasoning through careful drawings or patterns may suggest some good but tentative conjectures, but general reasoning, perhaps aided by a hierarchy, can establish that they are true.

Learning Exercises for Section 16.3

1. Correct the classification system of the Venn diagram below:

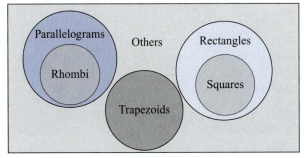

2. Organize all the conjectures about special quadrilaterals that you have. Check those that have some sort of justification besides examples. Mark out any conjectures for which someone has found a **counterexample** (an example that shows that the conjecture is not true in general). You could organize your work as shown below. The following table includes a few conjectures as samples to help get you started.

continue

Notes

Conjecture	Opposite sides are equal.	Opposite angles are equal.	Diagonals are equal.	Diagonals bisect the angles.	Diagonals bisect each other.
Quadrilaterals					
Trapezoids					
Isosceles trapezoids					
Parallelograms					
Rectangles					
Squares					
Kites					
Rhombuses					

3. Check your conjectures in Learning Exercise 2 with the hierarchy given in Figure 3 on page 388.

 a. Do all your conjectures about trapezoids appear to apply to parallelograms, rhombuses, rectangles, and squares?

 b. Do all your conjectures about kites appear to apply to rhombuses and squares?

 c. Do all your conjectures about parallelograms appear to apply to rectangles, rhombuses, and squares?

 d. Do all your conjectures about rhombuses appear to apply to squares?

 e. Do all your conjectures about rectangles appear to apply to squares?

4. a. Examine examples and make one or more conjectures about this situation: In a triangle, the midpoints of two sides are joined, as in Figure A below.

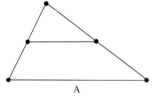 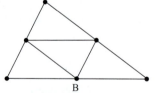

A   B

 b. Examine examples and make one or more conjectures about this situation: In a triangle, the midpoints of all three sides are joined, as in Figure B above.

5. Here are two informal methods for suggesting a fact you already know.

 a. Cut out a triangular region. Then, tear off the three angles of the triangle and rearrange them to show what their sum is. Try this with an unusually shaped triangle as well.

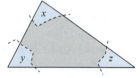

 b. Use a copy of triangle *PQR* on the next page and fold it to justify (experimentally) a fact you know about the sum of the sizes of the angles, $x + y + z$. Does the method work on other triangles?

continue

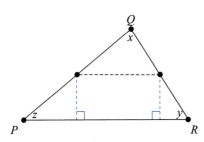

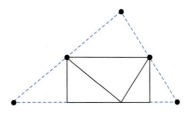

6. Find the number of degrees in each lettered angle.

a.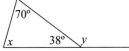

b.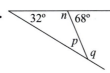

c.

7. a. Here is a What's My Shape? puzzle from a research article.[1] Uncover the clues one at a time, and see what shapes are possible after each clue.

> *Clue 1.* It is a closed figure with 4 straight sides.
> *Clue 2.* It has 2 long sides and 2 short sides.
> *Clue 3.* The 2 long sides are the same length.
> *Clue 4.* The 2 short sides are the same length.
> *Clue 5.* One of the angles is larger than one of the other angles.
> *Clue 6.* Two of the angles are the same size.
> *Clue 7.* The other two angles are the same size.
> *Clue 8.* The 2 long sides are parallel.
> *Clue 9.* The 2 short sides are parallel.

b. Make up a What's My Shape?

8. Pat says, "I was supposed to calculate $4 - \frac{4}{5}$, but I got mixed up and figured out $4 \times \frac{4}{5} = \frac{16}{5} = 3\frac{1}{5}$ instead. But when I did do $4 - \frac{4}{5}$, I noticed that I got the same answer, $3\frac{1}{5}$. So I think that when you subtract a fraction from a whole number, you get the same answer as you would if you did the whole number times the fraction."
a. Complete Pat's generalization in algebraic form: $a - \ldots$.
b. Is Pat's reasoning correct?

Supplementary Learning Exercises for Section 16.3

1. Use your results from Learning Exercise 2 on page 389 and perhaps definitions to add other measurements to these polygons.

a.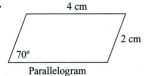

Parallelogram

b.

Kite

c.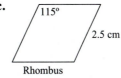

Rhombus

2. Sketch a counterexample to show that these conjectures are incorrect in general.

a. "The slanty sides in a trapezoid have the same length."
b. "The diagonals of a rectangle cut each angle of the rectangle into two angles that have the same size."
c. "The two diagonals of a trapezoid have the same length."
d. "Two angles in a parallelogram must be obtuse, and two must be acute."

3. Measure this rectangle to test the following conjectures:

   a. The two diagonals of a rectangle have the same length.

   b. Each diagonal of a rectangle cuts the other diagonal into two segments of the same length.

   c. Why isn't your evidence in parts (a) and (b) considered a mathematical proof?

16.4 Issues for Learning: Some Research on Two-Dimensional Shapes

When you were in elementary school, the coverage of geometry was not so well defined nationwide as was the coverage of numbers and operations. Some research even found that teachers in Grades 4 and 5 in some entire school districts devoted essentially *no* time to geometry![2] But with most states accepting the *Common Core State Standards in Mathematics*[3] (CCSS), the coverage should be more uniform, at least in textbooks. (Sections 18.3 and 25.3 will look at the CCSS.) Even though many teachers have in the past spent a fair amount of time on shapes, it is a bit surprising to find that children can entertain ideas such as those given below.[4] These erroneous beliefs may continue, so you will want to be alert to them when you are teaching.

Why is each of these children's ideas incorrect, in general?

- An angle must have one horizontal ray.
- A right angle is an angle that points to the right.
- A segment is not a diagonal if it is vertical or horizontal.
- A square is not a square if its base is not horizontal.
- The only way a figure can be a triangle is if it is equilateral.
- The angle sum of a quadrilateral is the same as its area.
- If a shape has four sides, then it is a square.

Where do such incorrect ideas come from? One possible source might be in the limited number of examples that children may see, both in their textbooks and in their teachers' drawings. Some of the ideas listed are natural, especially when the children have seen only examples incorporating the ideas, with angles always having a horizontal side, right angles always oriented toward the right, or equilateral triangles and squares being the only examples of triangles and quadrilaterals. Thus, a first message from the research is that a teacher should be sure that the students see a great variety of examples, with some omitting irrelevant attributes like horizontalness or a particular orientation. Look again at the sketches of polygons and quadrilaterals in Section 16.1 to see how we have attempted to give a variety of examples. Research also suggests that including a variety of nonexamples is also a good idea. The nonexamples allow a discussion to go beyond "No, it is not an example" to a focus on why the nonexample is not an example. (See also the nonexamples of polygons in Section 16.1.) Exercises like What's My Shape? (Learning Exercise 7 in Section 16.3) allow a consideration of what might be important features of a polygon.

When an early primary student was presented with the drawings of triangles shown at the top of the next page, she relied on what "looked like" a triangle and the attribute of three points. Notice that when a number of nonexamples were included, she included figures that were not constructed with straight lines. She did not include Figures 3 and 11.

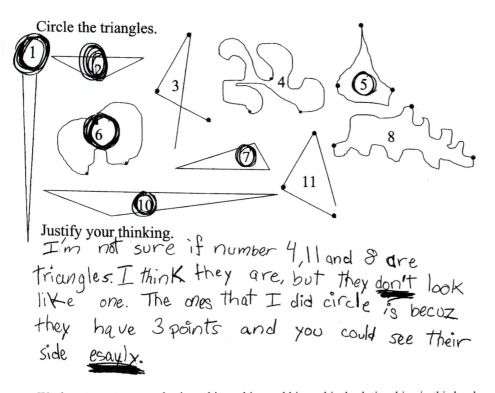

Circle the triangles.

Justify your thinking.

I'm not sure if number 4,11 and 8 are triangles. I think they are, but they don't look like one. The ones that I did circle is becuz they have 3 points and you could see their side esauly.

We do not want our emphasis on hierarchies and hierarchical relationships in this book to imply that young children can learn these ideas easily. Indeed, some studies show that some children have difficulty with multiple classifications, even with familiar objects.[5] Here are examples from another study[1] that illustrate the misconceptions and the difficulties for some children with multiple classifications. The children were presented with the drawings in Figure 5. When the children were asked to identify the squares, rectangles, and (if the child was familiar with the words) parallelograms and rhombuses in a similar collection of drawings, one third-grader chose shapes 2, 6, 7, 9, and 12 as squares. A fifth-grader chose 2, 4, 5, 7, 8, and 13 as squares, and 3, 6, 9, 10, and 12 as rectangles. The difficulty with the multiple classifications implied by a hierarchy is also illustrated by an eighth-grader who chose 2 and 7 as squares; 9 and 12 as rectangles; 3, 5, 6, and 10 as parallelograms; and 8 and 13 as rhombuses. The eighth-grader, when asked for definitions, gave definitions that made the types of quadrilaterals separate categories—e.g., for parallelogram, "Two parallel lines the same length are connected by two slanting lines the same length. The slanting lines are a different length than the parallel lines," thus ruling out rectangles, rhombuses, and squares as parallelograms.

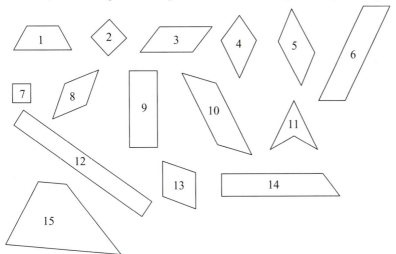

Figure 5

Notes

Notes

One viewpoint, with some research support, is that a person's understanding of a shape might develop in levels. For example, given a drawing of a rectangle, at the lowest level a child might say it is a rectangle *solely* because the shape *looks like* a rectangle (perhaps because it incorporates some irrelevant aspect such as different lengths for the two dimensions). A next level might be represented by a child's calling a shape a rectangle because of the angle sizes and the pairs of parallel sides of equal lengths. A still more advanced level might be shown by students who apply the definition of rectangle to the particular shape and find that the shape meets the conditions of the definition. The researchers think that starting at an advanced level of understanding, such as a definition, might be "over the head" of someone who is at only an earlier level, such as recognizing a shape by how it looks.

Some of the "wrong" answers can be explained by the common school practice of expecting only the most informative name for a shape. But this practice hides the hierarchical relationships that can be important in understanding shapes and in later work in geometry.

TAKE-AWAY MESSAGE . . .   Dealing with, or even avoiding, some of the misunderstandings children can have regarding conventional geometric ideas might entail (1) giving a great variety of examples that perceptually vary common but irrelevant characteristics, and (2) including nonexamples to allow a focus on the important relevant features. Better learning might result when many examples and nonexamples are used, along with a definition that might be appropriate at some grade level. Teachers should keep in mind that there are probably levels of understanding of any given geometric term and thus not assume that correct recognition of shape names is adequate evidence of the desired level of understanding. Understanding that a shape can have several names may be related to children's cognitive development and hence may require careful planning for a genuine appreciation of hierarchies.

Learning Exercises for Section 16.4

1. Based on the examples she has seen, a child might think that a trapezoid is "a shape with 4 sides, top and bottom are parallel, bottom longer than top, no sides the same size, fairly large, no right angles, and cut off at the top." Which of these characteristics are relevant? Draw some examples that vary the irrelevant characteristics that she associates with trapezoids.

2. A child once rejected a long, thin triangle as being a triangle because the shape was "too sharp." What might have led to the child's idea?

3. Suppose a child thinks that for the two angles shown below, angle *A* is larger than angle *B*. What irrelevant characteristics for angle size does the child seem to have?

4. Give a collection of examples and nonexamples that you think would be good for the idea of rhombus.

5. How might the following children have been thinking?

   a. the third-grader who chose 2, 6, 7, 9, and 12 as squares in Figure 5 on page 393

   b. the fifth-grader who chose 2, 4, 5, 7, 8, and 13 as squares and 3, 6, 9, 10, and 12 as rectangles in Figure 5

6. On the next page, the work of two primary grade children is shown. Assess the children's work for their understanding of *triangle*.

Circle the triangles.

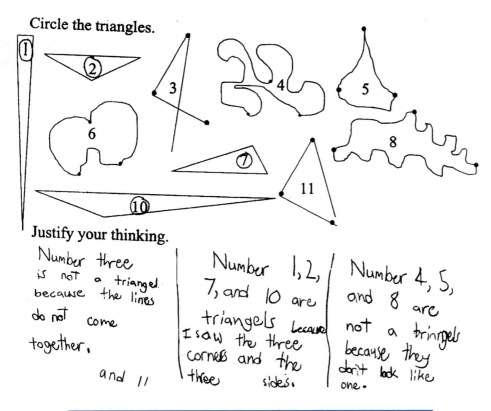

Justify your thinking.

Number three is not a triangel. because the lines do not come together.

and 11

Number 1, 2, 7, and 10 are triangels because I saw the three corners and the three sides.

Number 4, 5, and 8 are not a trinngels because they don't look like one.

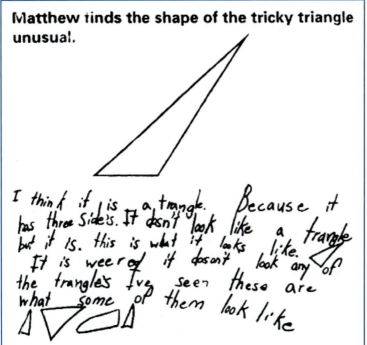

Matthew finds the shape of the tricky triangle unusual.

I think if 1 is a trange. Because it has three Sides. It dosn't look like a trange but it is. this is what it looks like. It is weered it dosn't look any of the trangles Ive seen these are what some of them look like

16.5 Check Yourself

The Check Yourself sections in the chapters give you a guide to many of the content expectations in the course. Your instructor may include others, of course, but the Check Yourself sections will give you some guidance in seeing whether you feel on firm ground in your understanding for the long term and in preparing for examinations. Writing your own "I should be able to . . ." list as you go through the course, and then comparing your list with the following guidelines, is a valuable way to see whether you are getting the point. As with most learning, the more actively involved you are, the better. Passively reading through this type of section can be just a reading exercise rather than a review, if you do not think about yourself vis-à-vis the items. For example, one way to review is to make up sample questions for the items in a Check Yourself section and then answer them. In reviewing past homework, you will likely take note of particular items that may have been difficult for you earlier, but think also about the items that were easy for you then as well (the human mind does forget!).

Chapter 16 has included a review of many terms about two-dimensional shapes and also has taken them in what may be new directions for you: writing precise definitions (especially for the types of quadrilaterals) and the hierarchical arrangement of categories and subcategories of polygons. Along the way, you should have developed and tested several conjectures about triangles and quadrilaterals, possibly with some of those conjectures being incorrect in general. Such false conjectures illustrate the dangers of expecting completely reliable conclusions from drawings or examples alone (inductive reasoning). Establishing a reliable conjecture requires a general argument that covers *all* cases.

You should be able to work problems like those assigned and to meet the following objectives:

1. Use, recognize, illustrate, spell(!), and perhaps define the various terms associated with polygons given in this chapter, including *n*-gon. You should be comfortable in those ways for the various types of quadrilaterals (trapezoid, isosceles trapezoid, kite, parallelogram, rectangle, rhombus, and square), pentagons, hexagons, heptagons, octagons, nonagons, and decagons, as well as the types of angles (acute, right, obtuse, adjacent, complementary, supplementary, exterior, and interior) and triangles (isosceles, equilateral, right, acute, obtuse, and scalene). The Glossary may be helpful in refreshing your memory.
2. Describe the difference between *polygon* and *polygonal region*.
3. Use appropriately (and spell) the adjectives *equiangular*, *equilateral*, and *regular*.
4. Give the sum of the angles of any polygon, knowing just the number of sides or angles or the name of the polygon (Activity 2 in Section 16.1).
5. Give the accepted hierarchy for the terms involved in this chapter.
6. Appreciate the role of a hierarchy in organizing relationships and information.
7. Use hierarchical relationships to answer such questions as "Is a square a rectangle?"
8. State and use the definitions in Section 16.3, or the definitions approved by your instructor. For example, you should be able to recognize and create examples of a rhombus and to recognize nonexamples of rhombuses.
9. Apply whatever conjectures turned out to be true, particularly conjectures about the types of quadrilaterals, and state the advantages from using hierarchical relationships.
10. Use terms such as *conjecture* and *counterexample* knowledgeably.
11. Give an argument to convince someone that the sum of the angles of a triangle is indeed 180° and the sum of the angles of a quadrilateral must be 360° (Learning Exercises 5 in Section 16.3 and 10 in Section 16.1, respectively).
12. Tell why even a good drawing or a pattern from several examples (inductive reasoning) cannot be trusted in creating conjectures.

References for Chapter 16

[1]Burger, W. F., & Shaughnessy, J. M. (1986). Characterizing the van Hiele levels of development in geometry. *Journal for Research in Mathematics Education, 17,* 31–48.

[2]Porter, A. (1989). A curriculum out of balance: The case of elementary school mathematics. *Educational Researcher, 18,* 9–15.

[3]National Governors Association Center for Best Practices, Council of Chief State School Officers. (2010). *Common core state standards in mathematics*. Washington, DC: Authors.

[4]Clements, D. H., & Battista, M. T. (1992). Geometry and spatial reasoning. In D. Grouws (Ed.), *Handbook of research on mathematics teaching and learning* (pp. 420–464). New York: Macmillan.

[5]Inhelder, B., & Piaget, J. (1964). *The early growth of logic in the child* (Trans. by E. Lunzer & D. Papert). New York: W. W. Norton.

Chapter 17

Polyhedra

Our physical environment, from ants to atoms to buildings to flowers to planets to zippers to zoos, is three-dimensional (or 3D). In many elementary school curricula the children's first work with geometry involves 3D shapes, often hands-on, in which the children work with wooden or plastic blocks of different shapes. So, it is appropriate to treat elementary 3D shapes, introducing more sophisticated vocabulary and ideas as we go.

In Section 17.1 we introduce some of the 3D vocabulary in a familiar context, and then in subsequent sections we discuss several special 3D shapes, showing how to draw them and giving some special relationships associated with them.

Notes

17.1 Shoeboxes Have Faces and Nets!

An ordinary shoebox has a 3D shape that is commonly encountered in a variety of sizes: e.g., children are likely familiar with a box holding an appliance, a game box, a cereal box, or a marker or crayon box, to name a few.

ACTIVITY 1 Drawing a Shoebox

You are quite familiar with the shape of a shoebox. Try to draw one. ●

There are different ways to draw a shoebox shape, but one way follows the steps shown in Figure 1 below. (Children might even call the starting shape a box, because that is a term they commonly use for a rectangle.)

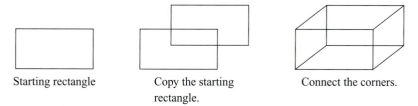

Starting rectangle | Copy the starting rectangle. | Connect the corners.

Figure 1

If you did not know how to make such a drawing, you should practice with different starting rectangles and with different depths at the second of the three steps.

The flat parts of a shoebox are examples of two-dimensional (or 2D) or **planar regions**, because they require only a flat surface. These flat parts or planar regions are called the **faces** of the shoebox (not the collo-quial term *sides*, which is reserved for line segments and rays in shapes). The corner points of the shoebox are more precisely called **vertices** (singular: **vertex**), and the line segments where two faces meet are called **edges** (and the edges *are* sides of the faces). Vertices are sometimes

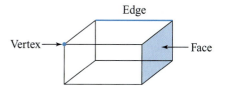

Notes

shown with a large dot for emphasis. In a drawing of a shoebox, do you see 6 faces, 8 vertices, and 12 edges?

If you start the drawing with a rectangle with all of its sides the same length (a square) and adjust the depth so that all the edges have the same length, you could get a **cube**.

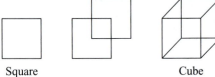

Square   Cube

Think About ...
How many faces, vertices, and edges does a cube have?

A special dot paper, called **isometric dot paper**, gives an alternative way to draw shapes made of cubes. (A sample of isometric dot paper is in the "Masters." Use it to make copies for future assignments.) A shape drawn on paper does not always show the back view, so some edges of solid cubes cannot be seen and so they may not be drawn. Figure 2 shows an example of a cube and a shape made of cubes on isometric dot paper. Do you see six cubes in the second shape? Can you identify parallel lines and what would be right angles in actual 3D cubes?

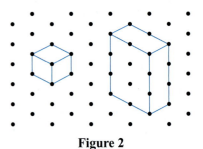

Figure 2

ACTIVITY 2   A Pile-Up

Draw a stack of cubes, 4 cubes tall, on isometric dot paper. ●

For manufacturing purposes, boxes may be made from patterns on flat material, and then folded up and taped. Vice versa, some edges of a box can be cut so that the box can be unfolded to lie flat but in one connected piece. Unfolded versions of a 3D shape are called **nets**.

DISCUSSION 1   Straighten Up

Which of the nets in Figure 3 gives a shoebox shape without its lid, if it is folded up and taped?

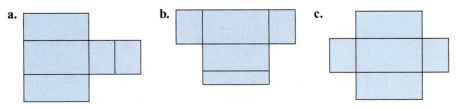

a.   b.   c.

Figure 3   ●

You will get more practice with nets in the Learning Exercises in order to prepare for the next section.

TAKE-AWAY MESSAGE . . . An ordinary shoebox allows you to talk about its faces, vertices, and edges and to illustrate the idea of a net for a 3D shape. Drawing a shoebox is relatively easy, and the same method can be used to draw a cube.

Learning Exercises for Section 17.1

1. Draw a box that would be shaped like a box holding a large refrigerator.

2. Draw a stack of four cubes free-hand (not with isometric dot paper).

3. How many faces, vertices, and edges does each of these have?

   a. an Egyptian pyramid   **b.** a prism from science class

4. How do you know that the net to the right will *not* give a complete cube?

5. For visualization practice, some K–8 curricula use actual 3D models of arrangements of cubes. These curricula have the students show the front view (as you look at the shape *directly* from the front), the view from the right, and the view from the top (again as you approach the shape from the front). A slightly more difficult version of this visualization would be to draw those views, but starting from a *drawing* rather than from a model. Here is an example:

Example   **Solution**

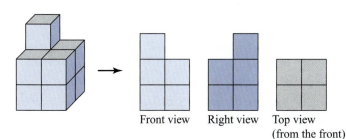

Front view   Right view   Top view
(from the front)

Sketch the front, right-side, and top views for each cube arrangement.

a.   **b.**

          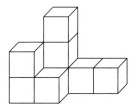

   c. Sketch the view from the *left* of the shapes in parts (a) and (b). How are they related to the views from the right?

6. a. Draw an isometric version of the cube arrangement given in the example in Learning Exercise 5. The isometric version for the example is started for you in the following diagram:

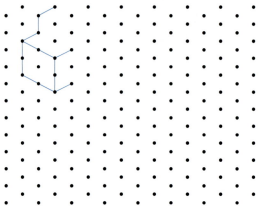

b. Draw an isometric version of the cube arrangement in Learning Exercise 5(a).

7. (*Website*) The National Council of Teachers of Mathematics' website has several sample lessons on geometry and measurement. Go to **http://illuminations.nctm.org/ Lessons.aspx** and look for "Isometric Drawing Tool" and "Dynamic Paper" under Activities. Briefly describe the software in these lessons.

Supplementary Learning Exercises for Section 17.1

1. How many faces, vertices, and edges does each of the following shapes have?

a. 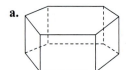   **b.** 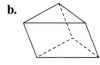   **c.**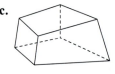

d. How many lateral edges does the shape in part (a) have? In part (b)?

2. a. Draw free-hand (not with isometric dot paper) a stack of cubes that has 2 cubes on the bottom and is 3 cubes high.)

b. Verify that your stack has 8 vertices, _____ faces, and _____ edges.

3. Use isometric dot paper to draw the stack of cubes that has 2 cubes on the bottom and is 3 cubes high. (*Hints*: Use a pencil with an eraser. Start at the top.)

4. Sketch the front view, the view from the right, and the top view for the shape below.

17.2 Introduction to Polyhedra

 ACTIVITY 3   Fold Them Up (preliminary homework activity)

Write your name inside the outlines for shapes A–M (their nets), found in the "Masters." Before cutting out the shape, try to imagine what the final, folded-up shape will look like. Cut around the outside of each shape, fold on the other line segments (a ruler helps make sharp folds; fold so the shape letter and your name will show), and then tape the shape together. Find a small container so that you can bring the shapes to class without damaging them. We will occasionally refer to the collection as your *kit*. Add to your collection any small containers or boxes or unusual shapes that you have on hand. On separate paper, jot down the names for the shapes that you know. ●

This section covers some important categories of 3D shapes and illustrates that defining them precisely is not trivial. Throughout this section (and later ones), you will be asked questions about the folded-up shapes from Activity 3.

 DISCUSSION 2   Sorting Shapes

Each of the shapes you made in Activity 3 is an example of a **polyhedron** (plural: **polyhedra**, sometimes **polyhedrons**), a closed 3D surface made up of planar regions (flat pieces). How would you sort shapes A–K into different sets? Write down your word descriptions for the sets. Do you know any vocabulary for the sets? Look at any 3D shapes other than A–K that students brought to class. Do they fit into the sets you have? Are any of the shapes not polyhedra? What else do you notice about any of the shapes? ●

The planar regions in a polyhedron are the **faces** of the polyhedron, the line segments in which faces meet are the **edges**, and the points at which edges meet are the **vertices** of the polyhedron. The word "polyhedron" comes from the Greek words for "many" (*poly-*) and "face" (*-hedron*).

 DISCUSSION 3   Defining Types of Polyhedra

1. As you may know, shapes A, D, G, and I are called **pyramids**. Come to a consensus about a definition for pyramid. (Do not look up a definition. The intent of this discussion is to have you focus on defining and nondefining characteristics of types of shapes.) One face is called the **base** of the pyramid. Which face do you think is the base? What is the shape of the base called?

2. Shapes B, C, E, and F are called **prisms**. Come to a consensus about a definition for prism, as you did in Problem 1. Two faces in each prism are called the **bases** of the prism. Which faces are the bases? How are the bases of a prism related to each other? ●

Some edges of prisms and pyramids have convenient labels. The edges of a prism or pyramid that are *not* on a base are called **lateral edges** (e.g., the thicker edges in the sketches) and the faces that are *not* bases are call **lateral faces**. If the lateral edges of a prism make right angles with the edges that they meet at the bases, the prism is called a **right prism** (and an **oblique prism**, otherwise).

A right hexagonal
prism (6 lateral edges)

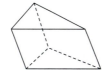

An oblique triangular
prism (3 lateral edges)

A **regular pyramid** has as its base a regular polygon, and all its lateral edges are the same length. Shape G in your kit is a regular pentagonal pyramid. If the lateral edges of a pyramid are not all the same length, it is an **oblique pyramid**. Sometimes the vertex that is not on the base is called the **vertex** of the pyramid, or the **apex** of the pyramid.

Adding pertinent information to the entries in the Glossary toward the end of this book, or making your own glossary, either on 3×5 cards or with a word processor, is a good idea. You probably have your favorite way to practice vocabulary.

 ACTIVITY 4   Quantities Associated with Polyhedra

Recall that a quantity is anything that can be measured or counted.

1. Pick one of the shapes A–K and tell what quantities you can associate with it.

2. Pick another one of the shapes A–K. Are the same quantities you selected in Problem 1 relevant to this shape as well? Explain. ●

Notice the vocabulary used in this Grade 1 textbook. (Section 17.5 treats *regular polyhedrons*.)

Date _____

LESSON 10·5   Reviewing 3-Dimensional Shapes

Word Bank

sphere	rectangular prism	pyramid
cube	cone	cylinder

Write the name of each 3-dimensional shape.

1. _____   2. _____   3. _____

4. _____   5. _____   6. _____

Five Regular Polyhedrons

The faces that make each shape are identical.

tetrahedron	cube	octahedron	dodecahedron	icosahedron
4 faces	6 faces	8 faces	12 faces	20 faces

Measurement and Geometry 2.1, 2.0, 2.2   two hundred three **203**

ACTIVITY 5 The Quantities *V*, *F*, and *E* for Pyramids

1. Find all the pyramids there are in shapes A–K. Count the number *F* of faces, the number *V* of vertices, and the number *E* of edges for each pyramid. Organize the data in a table and look for patterns or relationships.

2. Find similar counts for a hexagonal pyramid, a 100-gonal pyramid, an *n*-gonal pyramid, and an (*x* + *y*)-gonal pyramid, and add these counts to your data. Do the patterns or relationships hold for these pyramids also? Can you justify these counts without referring to the patterns?

3. One relationship that you may have overlooked relates all three quantities—*V*, *F*, and *E*—along with possibly a specific number. If you did not see such a relationship earlier, look for one now. ●

ACTIVITY 6 The Quantities *V*, *F*, and *E* for Prisms

1. Similarly to what you did in Activity 5, find all the prisms there are in shapes A–K. Count the number *F* of faces, the number *V* of vertices, and the number *E* of edges for each one. Organize the data in a table and look for patterns or relationships.

2. Find similar counts for a hexagonal prism, a 100-gonal prism, an *n*-gonal prism, and an (*x* + *y*)-gonal prism, and add these counts to your data. Do the patterns or relationships hold for these prisms also? Can you justify these counts without referring to the patterns?

3. One relationship that you may have overlooked relates all three quantities, *V*, *F*, and *E*. If you did not see such a relationship earlier, look for one now. (*Hint, if you want one:* Insert a column or row to your data that gives the value *V* + *F*.) ●

If you are studying on your own, at this point you may be stuck finding the relationship hinted at in Problem 3 of the last two activities. Patterns are not always easy to notice. If you are stuck, look up **Euler's formula** in the Glossary. (Euler is pronounced "oiler.") Euler's formula is important enough to memorize.

TAKE-AWAY MESSAGE . . . There are different types of polyhedra, with many vocabulary words associated with them. Two particularly important types of polyhedra are pyramids and prisms. Writing careful definitions for pyramids and prisms is not easy, as you found out. Euler's formula for polyhedra gives a result applicable to most of the polyhedra you are likely to encounter.

Learning Exercises for Section 17.2

(Unfamiliar words in these exercises can be found in the Glossary.)

1. Which of the shapes A–K meet the criterion given in each statement?

   a. All the faces are parallelograms (including special parallelograms, such as rectangles and squares).
   b. A base is a pentagonal region.
   c. All the triangular faces are equilateral.
   d. None of the triangular faces is equilateral.
   e. Two faces are parallel and congruent, but the shape is not a prism.
   f. More than one face is an isosceles trapezoid region.
   g. All the edges are congruent.
   h. No two edges are parallel.

2. **a.** For each prism, how do the lateral edges appear to be related? What shape are all the lateral faces?

   b. How many lateral edges does a 50-gonal pyramid have? How many lateral faces does this 50-gonal pyramid have? What shape are they?

3. a. *Who am I?* or *What am I?* exercises are common in elementary programs. Ideally, the clues are revealed one at a time, and then what is known from the revealed clues is discussed. Answer "Who am I?" for the following example.

   Example:
   I am a polyhedron.
   I have 7 faces.
   Six of my lateral edges are equal in length.
   One of my faces has 6 sides.
   Who am I?

   b. Make up a Who am I? exercise.

4. Is there a polyhedron that has the fewest vertices? The most vertices? Explain.

5. a. Test the relationship you found in Activities 5 and 6 (Euler's formula), using shapes H, J, and K, and any other polyhedron you have that is neither a pyramid nor a prism.

   b. Suppose a polyhedron has 10 vertices and 15 edges. How many faces does it have?

   c. Suppose another polyhedron has 20 faces and 30 edges. How many vertices does it have?

   d. Can the number of vertices, the number of faces, and the number of edges of a polyhedron all be odd numbers? Explain.

6. What geometric name (using adjectives, as appropriate) applies to each of these objects?

   a. a shoebox
   b. an unsharpened pencil with flat sides (and no eraser)
   c. a filing cabinet
   d. an unused eraser (the separate kind, not on a pencil)

7. React to the following scenarios:

   a. "I was thinking. A square pyramid has 4 triangular faces and a square base. Each triangular face has 3 sides, which are edges of the pyramid. So there should be 4 times 3, or 12, lateral edges, plus the 4 on the square base, 16 edges in all. But I counted and found only 8 edges. What is wrong with my reasoning?"

   b. "What is the matter with my thinking here? I know a cube has 8 vertices. But a cube has 6 square faces, and a square has 4 vertices. So it seems there should be 6 times 4, or 24, vertices, not 8."

8. For the following types of polyhedra, investigate to see whether the total number of *angles* on all the faces is predictable. For example, the cube, a special prism, has 4 angles on each of its 6 faces, so it has 24 angles.

   a. pyramids   **b.** prisms

9. a. Which, if any, of the shapes in your kit are right prisms?
   b. What shape are all the lateral faces of a right prism?
   c. Are there any regular pyramids in your kit?

10. How many of the angles on *all* the faces of a prism could be *right* angles in each case given?

   a. The prism is a triangular prism.
   b. The prism is a quadrilateral prism.

11. Geometry is all around you. Go for a ten-minute walk, with paper and pen or pencil in hand. (The walk can be inside if the weather is bad.) Sketch any shapes, 2D and 3D, that you recognize as being common or noteworthy. Next to the sketches, write the names for shapes that you can identify.

12. With a piece of string or a rubber band, show how a straight cut across a right rectangular prism (e.g., shape E or F in your kit) could give each of the following 2D figures:

    a. an equilateral triangle
    b. an isosceles triangle that is not equilateral
    c. a rectangle
    d. a parallelogram that is not a rectangle

13. Make sure that you are comfortable with these vocabulary terms: polyhedron, pyramid, prism, face, lateral face, edge, lateral edge, vertex, vertices, oblique prism, right prism, and regular pyramid. What does it mean to be *comfortable* with a word or phrase? How can you organize a lot of vocabulary to keep it straight and remember it?

14. Why are shapes J and K in your kit *not* prisms?

Supplementary Learning Exercises for Section 17.2

1. In each part, which of shapes A–K meet the criterion?

    a. It is a right prism.
    b. It has exactly three lateral edges.
    c. It has exactly three lateral faces.
    d. It is a regular pyramid.
    e. All its faces are square regions.

2. Give the following for a 50-gonal pyramid:

    a. the number of edges
    b. the number of lateral edges
    c. the number of vertices
    d. the number of faces

3. Give the following for a prism that has 60 edges (in all):

    a. the number of lateral edges
    b. the number of edges on each base
    c. the number of vertices
    d. the number of faces

4. Would your answers to Exercise 3 change if the prism were a *right* prism? If so, explain.

5. Give the following for a pyramid that has 90 edges (in all):

    a. the number of lateral edges
    b. the number of edges on the base
    c. the number of vertices
    d. the number of faces

6. Would your answers to Exercise 5 change if the pyramid were an *oblique* pyramid? If so, explain.

7. In each part, find the number of vertices of the polyhedron.

    a. 18 edges and 8 faces
    b. 90 edges and 32 faces
    c. 62 faces and 180 edges

8. In each part, find the number of faces of the polyhedron.

    a. 24 vertices and 36 edges
    b. 48 vertices and 72 edges
    c. 24 edges and 12 vertices

9. In each part, find the number of edges of the polyhedron.

    a. 60 vertices and 92 faces
    b. 30 vertices and 32 faces
    c. 26 faces and 24 vertices

17.3 Representing and Visualizing Polyhedra

Even when a 3D shape is not present, the idea of the shape may be communicated by a word or a drawing. The conventions of 2D drawings for 3D shapes must be learned, of course, if such drawings are to be interpreted as intended. This section treats the variety of ways for communicating the ideas of particular shapes.

Think About ...
What do you see when you look at this drawing?

Do you see a regular hexagon with some segments drawn to its center? Or do you see a cube? (Do you see both shapes now?) Often 3D shapes are represented by 2D drawings. Without a context suggesting three dimensions, these drawings are ambiguous, and children often do not interpret them in the way that was intended. *Reading* such drawings is important for understanding the ideas they represent, and being able to draw a variety of 2D and 3D shapes is a useful skill, especially for a teacher.

Representations of a Polyhedron

A drawing of a polyhedron is one **representation** of the polyhedron. Strictly speaking, like all geometric objects, a polyhedron is an ideal object that exists only in one's mind—an idea to which we have assigned the verbal label *polyhedron*. Physical-world representations of polyhedra might include boxes, but boxes have rounded vertices and their faces are rarely perfectly flat, so boxes are only approximate representations of polyhedra. We usually ignore these flaws and become comfortable with approximate representations of ideal geometric objects, e.g., a cardboard box for a prism, a stretched piece of string or a pencil for a line segment, or a piece of paper for a rectangular region. And in the last section, you have made very good representations of several polyhedra, using paper models. Indeed, the flat patterns or nets that you cut out to make the polyhedra give another type of representation for polyhedra. Hence, we have at least three nonverbal representations for polyhedra: physical models, drawings, and nets.

Our overall task is to be able to *translate* among the different representations. You have already practiced some of these translations. For example, you started with a 3D model of a cube, and then you gave the word "cube" for this object. Another translation you possibly have done is when shown the drawing of a cube in the last Think About, you might have then given the word "cube" for the drawing.

In the diagram to the left in Figure 4, we show in general the different representations of a polyhedron. Next, in the diagram on the right, we show the different translations among these representations of a polyhedron.

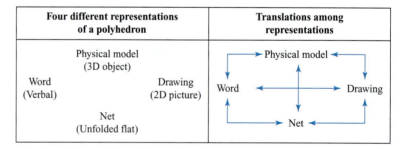

Figure 4

Some Examples of Translations

Two specific translations shown in Figure 4 are the translation going from a physical model or a word to a net. These translations are the reverse of what you did in Section 17.2: You started with a net and obtained a physical model and then eventually a word name. In

Activity 7 below, you will be asked to do a reverse task: You start with a physical model or a word, and then obtain a net for the shape.

 ACTIVITY 7 Unfolding a Cube

Using a cube or just thinking about a cube, draw a net for it. ●

Starting with the physical model is the easiest way to do the type of translation in Activity 7. Place the model on a piece of paper, trace around the face touching the paper, and then in steps "roll" the model on its edges so you can trace around the other faces. It is a good idea to have some system in mind so that some faces are not repeated or omitted. Starting with just a word, like "cube," and then trying to sketch the net involve more demanding visualization practice. This type of task forces you to visualize the shape and to do the "rolling" or unfolding mentally. Compare Activity 7 with the next one for difficulty.

 ACTIVITY 8 Reading a Net

How might a person have obtained the given net for a cube, where the marked face was drawn first? (Use a cube if you have trouble visualizing what might have happened.) ●

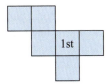

 ACTIVITY 9 Attributes of a Cereal Box

Following are three nets created by primary children who were asked to make a net of a cereal box (from an unpublished teaching experiment by Linda Jaslow). First-graders created the first two nets and a second-grader created the third.

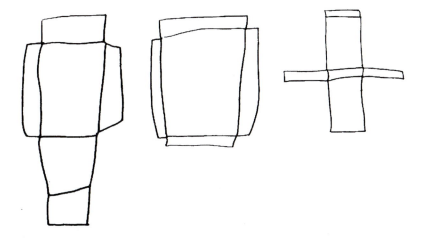

1. Consider each of the nets. Are they or are they not nets of a cereal box? Explain how you know.

2. What attributes of a cereal box do the children accurately represent or fail to accurately represent? As you discuss the children's nets, make a special effort to use the vocabulary terms "face," "edge," "base," "lateral edge," "lateral face," and others as appropriate.

3. What is the most accurate geometric name for the shape of the cereal box? ●

A second type of translation is going from a drawing to words (i.e., "reading" a drawing). There is no single standard drawing for a 3D shape, unfortunately. One kind of drawing you have probably seen is a *perspective* drawing, such as Figure 5 on the next page, which attempts to represent how your eye might see an object. In a perspective drawing, some segments that are actually parallel in the shape itself may appear to converge in the drawing, and distant segments will appear shorter in the drawing even though they are the same length in the actual shape. For example, in Figure 5, the parallel lines *AB* and *CD* appear to converge at the point *X*, called the *vanishing point* by artists. Consequently, many measurements on perspective drawings will not be to scale. For this reason, perspective drawings, although common in art, are not popular in drafting and mathematics.

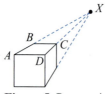

Figure 5 Perspective drawing of a cube.

Figure 6 Drawing of a cube in mathematics (no perspective).

The common type of drawing in mathematics does attempt to display actual parallel segments as being parallel in the drawing by pretending that the drawer's eye is slightly in front of, above, and to the right of the object (see Figure 6 above). The right angles in the face of the cube closest to the viewer do look almost like right angles. But most of the right angles in the actual cube are not right angles in the drawing, and so some faces look like parallelogram regions rather than the square regions they actually are. Hence, reading this representation requires understanding this distortion; a parallelogram in the drawing of a polyhedron *may* indeed represent a square or rectangle in the 3D object. (Look at an actual cube with your eye in the position described above, and with your eyes "squinty.") As Figure 7 below illustrates, showing all the edges removes some possible ambiguity. Making the **hidden edges**—the ones you can't actually see in a paper model—lighter or dashed also helps to remove some of the visual ambiguity even when all of the edges are shown. Shading may also help the eye to "see" what is intended.

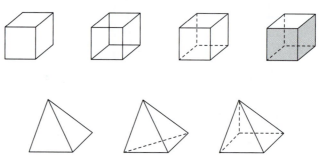

Figure 7 Alternate representations: cubes with no hidden edges shown; with all edges shown, with hidden edges dashed, and with two faces shaded; pyramid with no hidden edges shown; and pyramids with ambiguity removed by showing hidden edges

Elementary school teachers and students sometimes use everyday objects to create frameworks for different polyhedra. For example, they might use toothpicks or plastic coffee stirrers for edges, with miniature marshmallows or raisins for vertices. Using such objects can help students to see both the visible edges and those edges hidden in 2D drawings.

A third type of translation is from a word or physical model to a drawing. Drawing 3D shapes can be done in different ways, as the drawings in Figures 5–7 suggest.

Our standard drawing in this text will be the type showing parallel segments in the object as parallel in the drawing and viewing the object from slightly in front, above, and to the right. Equal lengths in the shape should be equal in the drawing, insofar as possible. To remove some ambiguity, we will show hidden edges with lighter or dashed segments and add shading as we see fit.

For drawing the face in front, it is often possible to draw some edges to be horizontal. One way to make good drawings of prisms in general is to concentrate on drawing the top base first (because all its edges are visible), recognizing that your viewpoint is in front of, above, and to the right of the prism. Hence, there is usually some distortion of the actual shape of the base, because you are looking at it from an angle. Because the lateral edges are all the same length and all parallel, next draw the lateral edges—first the visible ones and then the hidden edges, more lightly or dashed. Sometimes these other edges are too close to each other so you have to adjust the original base. With experience, when you sketch the top base, you can avoid such difficulties by *not* locating a vertex right above or right below another vertex on the other base. Finally, sketch the bottom base, first the visible edges and then the hidden ones. These edges should be parallel to ones in the top base, so you have a chance to adjust any lengths of lateral edges that might be off. An example is in order.

EXAMPLE 1

Draw a right rectangular prism.

Solution

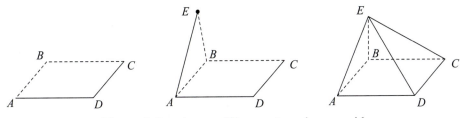

Top base first (Equal) parallel lateral edges, with one hidden Base edges, with two hidden

Similarly, one way to draw pyramids is to start with the base, trying to anticipate which edges will be hidden. Then locate the other vertex of the pyramid and join it to each vertex of the base, keeping in mind which edges will be hidden. Figure 8 below shows an example of how to draw an oblique rectangular pyramid. Notice that from our viewing angle, the rectangular base will appear to be a parallelogram.

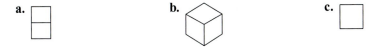

Figure 8 Drawing an oblique rectangular pyramid.

TAKE-AWAY MESSAGE . . . There are several ways of representing the idea of a 3D shape: word, net, model, or drawing. Starting with any of these representations, you should be able to get to the other representations. The Learning Exercises invite you to practice making sketches and many of the translations among representations.

Learning Exercises for Section 17.3

1. Describe where your eye might be if you are looking at a cube and see each of the following shapes:

   a.   b.   c.

2. a. What is the maximum number of faces you can see when looking at a paper model of a cube from outside the cube? The minimum?
   b. What is the maximum number of vertices you can see?
   c. What is the maximum number of edges you can see?

3. Draw an enlarged copy of the drawing at the right and put in hidden edges to show that it is actually a drawing of a pentagonal pyramid.

4. Make a drawing of each of the following shapes. Be sure to show the hidden edges.

   a. a right rectangular prism
   b. a triangular right prism
   c. a square pyramid
   d. a right octagonal prism
   e. a pentagonal pyramid
   f. a cube
   g. a triangular prism in which the edges make 14 right angles with each other
   h. a polyhedron with 8 edges and 5 faces

5. If the following nets were folded up to give cubes, which pairs of faces would be opposite each other?

   a.

1		
2	3	4	5
6		

   b.

5	6		
	1	2	3
	4		

   c.

1	
2	3
	4
	5	6

   d.

		1
4	3	2
6	5	

6. Using the 3D shapes from the nets for shapes A–K in the "Masters," draw nets that are different from the given original nets for the shapes specified.

   a. shape A   **b.** shape C   **c.** shape F   **d.** a different net for shape F

7. Which, if any, of the following nets will give a triangular pyramid?

   a.   **b.**   **c.** 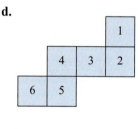   **d.**

8. Consider the shape shown below that has been drawn on isometric dot paper. (*Note*: Observe how the dots are arranged. Where is one's eye, in viewing an isometric drawing? Do isometric drawings appear to retain parallelism for segments that are parallel in the original? Are segments of equal lengths in the original also equal in the isometric drawing? Are right angles in the original also right angles in the drawing?)

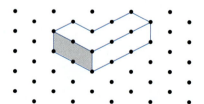

   a. Draw an isometric version of the given shape, but from a different viewpoint.

   b. Does Euler's formula apply to this polyhedron?

   c. What is the best name for the polyhedron at the bottom of the previous page?

   d. Sketch a net for the polyhedron.

   e. What are the surface area and the volume for this polyhedron? (Use the natural units—the smallest squares and smallest cubes.)

9. Sketch a cube, label the bottom face B, and then check (√) the edges that could be cut to yield each given net. (The letter B in each net marks the bottom face of the cube.)

   a. 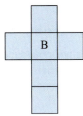   **b.** 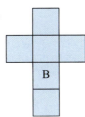   **c.**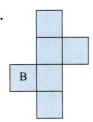

10. The following cubes have been cut on the *dashed* edges. Draw the nets that would result if the cube is then unfolded. (If the cube is not cut so it will unfold, then correct the cuts.) Try this first as an exercise in visualizing; look at a cube if you are uncertain after visualizing.

   a.   **b.**   **c.**

11. Make up an exercise like Learning Exercise 9 or 10 for shape A.

12. The game of dominoes has playing pieces that can be represented as two squares joined by a side. Connecting four squares similarly would give a 4-omino, or a **tetromino** (*tetra*- means "4").

Show all the differently shaped tetrominoes. Two tetrominoes have the same shape if one can be moved so that it matches the other. The first three tetrominoes below are all the same:

The two shapes to the right are not "legal" because squares must be joined by matching sides.

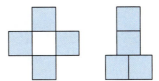

13. a. How many differently shaped **pentominoes** (*penta*- means "5") are there?

   b. Give the area and perimeter of each pentomino.

   c. Is the following statement true? *Shapes that have the same area also have the same perimeter.* Explain, using the pentominoes as a basis.

   d. Is the following statement true? *Shapes that have the same perimeter also have the same area.* Explain, using rectangles as a basis.

   e. Which pentominoes, viewed as nets, can be folded to give *open* cubes—i.e., cubes with one face missing?

14. The sketch in each drawing shows a cross section of a cube. Identify the type of shape in each cross section.

a.

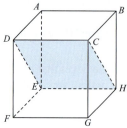

b.

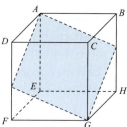

15. A teacher might wish to add tabs to a net so that the students can use glue or paste instead of tape. The strongest polyhedron will have every pair of faces sharing an edge also sharing a tab. Copy the nets below and indicate where tabs should be put to give a strong polyhedron.

a.
b.

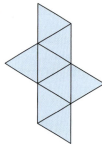

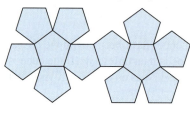

c.

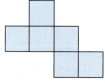

16. Give the most informative name for the polyhedron that each net would make.

a.
b.

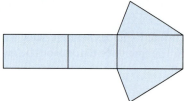

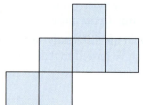

17. Draw an isometric version of 3D shapes that have the following views:

a.

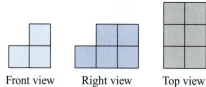

Front view   Right view   Top view

b.

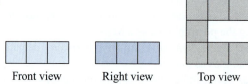

Front view   Right view   Top view

18. Consider the following right rectangular prism:

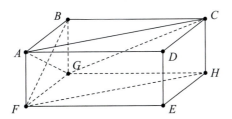

a. Edge AF is perpendicular to (makes right angles with) which of these? (The bar notation is common for line segments.)

edge FE, or $\overline{FE}$ edge AD, or $\overline{AD}$ edge AB, or $\overline{AB}$

segment AG, or $\overline{AG}$ segment BF, or $\overline{BF}$ segment FH, or $\overline{FH}$

b. Name the endpoints of three edges that are parallel to edge AD.

19. Some visualization tests include tasks such as the following: Imagine a piece of paper that is successively folded, and has a hole or holes punched into the last folded position. The task is to identify how the holes would appear when the paper is unfolded. Try the following tasks:

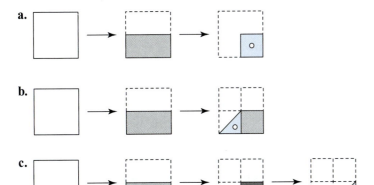

a.

b.

c.

d. Design another task for folding a piece of paper.

Supplementary Learning Exercises for Section 17.3

1. Draw a prism that has 15 edges. Show any hidden edges with dotted lines.
2. Draw each of the following polyhedra. Be sure to indicate hidden edges.
   a. an oblique pyramid with 7 *faces*
   b. an oblique pyramid with 8 *edges*
   c. an oblique pyramid with 7 *vertices*
3. Sketch a net for each polyhedron you drew in Exercises 1 and 2.
4. Draw an isometric version of 3D shapes that have the following views:

a.

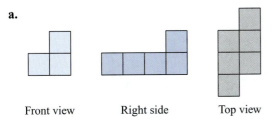

Front view   Right side   Top view

continue

b.

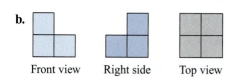

Front view   Right side   Top view

5. Which, if any, of shapes A, B, C, and D can be obtained by turning the starting shape (square and shading). Try to do the turning mentally, or analyze the figures to get your answer.

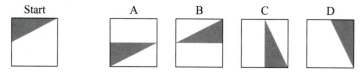

Start   A   B   C   D

6. Give the best names for all the faces in these shapes.
   a. a regular hexagonal pyramid
   b. a right rectangular prism
   c. shape D in your kit of polyhedra
   d. shape K in your kit of polyhedra

7. Organize these shapes in a classification diagram: kite prisms, isosceles-trapezoid prisms, parallelogram prisms, quadrilateral prisms, rectangular prisms, rhomboidal prisms, square prisms, and trapezoid prisms.

8. **Fact** The sum of the measures of all the angles in all of the faces (the face angles) of any trapezoidal prism is 2160°.

   a. Without calculation, give the sum of the measures of all the face angles of a rhomboidal prism. Explain your reasoning. (*Hint:* See Supplementary Learning Exercise 7.)
   b. Without calculation, give the sum of all the face angles of a rectangular prism.

17.4 Congruent Polyhedra

Replacement parts for a car or a computer, cookies cut from a cookie cutter, or blouses made from the same pattern—every object in one of these groups mentioned is expected to be *exactly the same size and shape*. They can be regarded as copies of one another. If these shapes were holograms, they could be made to coincide completely. Hence, matched, or *corresponding*, parts (such as segments, angles, curves, or faces) have the same measurements.

ACTIVITY 10   Exactly Alike

Which of the following 3D objects are exactly the same size and shape?

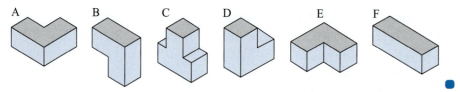

A   B   C   D   E   F

Think About …

Are 3D objects G and H exactly the same size and shape? It may be helpful to make the shapes from cubes, if you have any. *Hint:* Think of reflecting in a mirror.

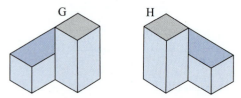

G   H

Matching shapes often is done by mentally or physically moving one shape so that it *looks exactly like* the other shape. The movement may be a turn or a slide, e.g. The movement may even be a reflection in a plane, as is the case with G and H in the Think About. The movement might even be a combination of turns, slides, and reflections.

The movements involved are important mathematically, because they are fundamental to decisions about whether two shapes can be matched exactly in size and shape.

DISCUSSION 4   How Do You Decide?

What movements might be involved in deciding whether the shapes A–F in Activity 10 are exactly the same size and shape?   ●

As you may recall, the technical word for this "exactly the same size and shape" idea is **congruence**, and shapes related by congruence are called **congruent shapes**. Shape A in Activity 10 is *congruent* to shape B, e.g., but *not congruent* to shape F: There is no way to move either shape A or shape F to match the other shape exactly, even with reflections in planes. That is, there is no way to match the points in shapes A and F so that measurements of all corresponding parts are equal. So, shapes A and F are not congruent.

TAKE-AWAY MESSAGE . . .   Three-dimensional shapes may be congruent, even if they are not in the same position. You may have to turn, slide, or reflect a shape to see that it is exactly like a congruent shape.

Learning Exercises for Section 17.4

1. **a.** Sketch shapes congruent to shapes I and J shown below by using a reflection in a plane for each shape.
   b. Sketch shapes congruent to shapes I and J by using a turn for each shape.

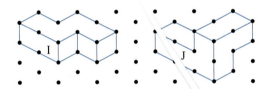

2. Are any of the polyhedra sketched below congruent? Explain how you know.

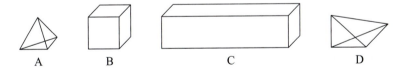

   A   B   C   D

3. A hidden shape is known to be congruent to a 3 cm by 4 cm by 6 cm right rectangular prism. What is the total of the areas of all the faces of the hidden shape? (*Hint:* Use your shape F for ideas.)

4. **a.** Is a shape congruent to itself? Explain your thinking.
   b. If shape P is congruent to shape Q, is shape Q congruent to shape P? Explain your thinking.
   c. If shape R is congruent to shape S, and shape S is congruent to shape T, is shape R congruent to shape T? Explain your thinking.

5. Sketch all possible noncongruent polyhedra in which each polyhedron is made up of four congruent cubes. When two cubes are connected, they must share a face. (*Hint:* There are more than five noncongruent ones.) Make up a name for all such polyhedra, using your knowledge of prefixes. (See **prefixes** in the Glossary.)

Notes

Notes

6. a. Suppose that you and a friend are going to talk by telephone, and that each of you has a prism. Prepare a list of all the questions you *might* need to ask your friend, so that you will be able to tell whether your friend's prism is congruent to your prism.

   b. Prepare another list of questions that would enable you to determine whether two pyramids are congruent.

7. a. On isometric paper, sketch a shape congruent to the shape having the following three views. Because shapes on isometric paper do not have a clear *front* face, interpret *front* below to be *front left*, and *right* to be *front right*.

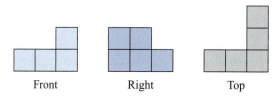

   Front   Right   Top

   b. On isometric paper, sketch a shape congruent to your shape in part (a), but from a different viewpoint.

   c. Sketch the front, right, and top views of your shape in part (b). (Call the left front the *front* view.)

   d. What movement would show that the original shape from part (a) and your shape are congruent?

8. Chemists know that molecules of some substances exist in mirror-image versions (*chiral* molecules). Drugs made with them can have quite different effects. One version of thalidomide is a sedative but the chiral version causes birth defects. One version of limonen has a lemon smell but the chiral version smells orangey. Sketch chiral versions of these "molecules." (The drawings represent molecules in either triangular pyramid or cubic arrangements, with the dots showing atoms and the dashed segments representing chemical bonds. They are not actual molecules of any substance, and X, Y, Z, and W are made-up elements.)

a.

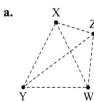

b.

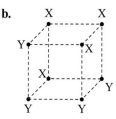

9. This section has focused on congruent *polyhedra*. How would you define congruent *polygons* in a similar way? Sketch an example of congruent quadrilaterals, and tell how you know they are congruent.

Supplementary Learning Exercises for Section 17.4

1. Is it possible for a given pyramid to be congruent to some prism? Explain your decision.

2. Prisms P and Q (not shown) are congruent. Tell why they must have the same number of faces.

3. One edge in prism X (not shown) is 4 in. long. Must there be at least one edge in hidden prism Y, which is congruent to X, that is 4 in. long? Tell how you know.

4. A 2 cm by 6 cm by 3 cm right rectangular prism is congruent to hidden prism Z. What do you know about measurements in prism Z?

5. Sketch a prism that is congruent to the following prism, but show the hidden edges in your drawing:

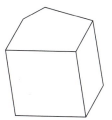

17.5 Some Special Polyhedra

The types of polyhedra that have been highlighted in earlier sections lead to a classification that bypasses some polyhedra of historical (and aesthetic) interest: the regular polyhedra.

> A **regular polyhedron,** or **Platonic solid,** is a polyhedron all of whose faces are congruent regular polygonal regions of one particular type, with the same arrangement of polygonal regions at each vertex.

In this section we will focus on the *convex* regular polyhedra. By convex, we mean those shapes that are not "dented in" anywhere.

Think About …

You may not realize that you already have at least three types of regular polyhedra in your kit of polyhedra. Can you find them? Recall that all of the faces must be identical, and the arrangement of the regions at each vertex must be the same.

A regular polyhedron is named on the basis of how many faces it has (from the Greek word for "face": *-hedron*, plural *-hedra*). A *regular tetrahedron* has four faces, each an equilateral triangular region. A *regular hexahedron* has six faces, each a square region. (A regular hexahedron is most often called a cube, if you were wondering.) A *regular octahedron* has eight faces, each an equilateral triangular region. Did you find a regular tetrahedron, a regular hexahedron, and a regular octahedron in your kit?

A natural question would be, "Is there a regular polyhedron for every number of faces, or at least for every *even* number of faces?" Perhaps surprisingly, the answer is "No." There are only five types of regular polyhedra. The other two types are the *regular dodecahedron* (*dodeca-* means "12") and the *regular icosahedron* (*icosa-* means "20"). Figure 9 shows nets for these two types; verify that they have the proper number of faces (see also your shapes L and M).

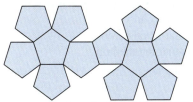   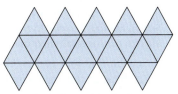

A net for a dodecahedron   A net for an icosahedron

Figure 9

The regular polyhedra were well known to the ancient Greeks, and they assumed an almost mystical air during some periods of history. For example, Kepler (1571–1630) described a model of the planets that involved (erroneously) the regular polyhedra to go along with Kepler's important laws of planetary motion. At that time, the regular polyhedra were also linked to the view that the universe was made up of four basic elements: air, fire, earth, and water. The regular octahedron represented air; the tetrahedron represented fire; the cube represented earth; and the icosahedron represented water. To complete this theory, the dodecahedron represented the universe.

You may have seen the regular polyhedra only in novelties, such as a calendar with one month per face of a dodecahedron. So it may surprise you to find that the regular polyhedra do indeed have real significance and appear in nature. Crystals of some minerals are shaped like the regular polyhedra, as are some simple organisms. For example, crystals of ordinary table salt are shaped like cubes, and many viruses are shaped like regular icosahedra.

TAKE-AWAY MESSAGE . . . Shapes that were perhaps just intellectual playthings to the ancient Greeks turn out to represent objects in the real world that were largely unknown to the Greeks. The power of mathematics to represent real-world phenomena is one reason that mathematics is so valuable and so attractive to many people.

Learning Exercises for Section 17.5

1. Decide whether each statement is always true, sometimes true, or never true.
   a. A cube is a polyhedron.
   b. A polyhedron is a cube.
   c. A right rectangular prism is a cube.
   d. A cube is a right rectangular prism.
   e. A regular polyhedron is a prism.
   f. A prism is a regular polyhedron.
   g. The diagonals of every face of a cube are the same length.
   h. The diagonals of every face of a right rectangular prism are the same length.
   i. A pyramid is a regular polyhedron.
   j. A regular polyhedron is a pyramid.
   k. A hexahedron is a cube.
   l. A cube is a hexahedron.

2. Wondering about the number of types of regular polyhedra is natural to a mathematician. Why is there an advantage in knowing how many types there are of something?

3. From their biological names, what is the general shape of these organisms?
   a. circogonia icosahedra
   b. circorrhegma dodecahedra
   c. circoporus octahedrus

4. a. If you take two regular tetrahedra (see shape A in your kit) and put them face to face, you get a polyhedron with six faces, each of which is shaped like an equilateral triangle. Why is this polyhedron *not* counted as a regular polyhedron? (Read the description of a regular polyhedron carefully.)
   b. Sketch another hexahedron that is not a regular hexahedron.
   c. Make a net for a tetrahedron that is not a regular tetrahedron.

5. a. On a cube, find four vertices that give a regular tetrahedron.
   b. What shape do the remaining four vertices give?

6. Check to see whether Euler's formula applies to each type of regular polyhedron. Record your counts in a table; do you notice anything? (*Hint*: Is Figure 9 helpful?)

7. A polyhedron may be made up of two or more types of regular polygonal regions, with the same arrangement of regions at each vertex (like your shape J). These shapes are called *semiregular polyhedra* (or sometimes *Archimedean polyhedra*), and they were also studied extensively even in ancient times (e.g., by Archimedes, 287–212 BC).

a. The sketches below represent two semiregular polyhedra. What regular polygons are involved in each polyhedron? (Hidden edges are not shown in the second sketch.)

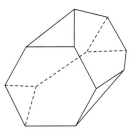 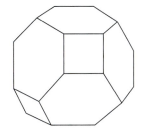

b. Check the first semiregular polyhedron in part (a) to see whether Euler's formula applies. Why is it difficult to check the second polyhedron in part (a)? Can you remedy that difficulty?

c. Make up a natural question about semiregular polyhedra. (See Learning Exercise 2.)

8. Which of the five types of regular polyhedra is your favorite? Why?

9. a. Must all the edges of a regular polyhedron be the same length? Explain.

    b. Must each angle in every face of a regular polyhedron have the same size? Explain.

    c. Must all the edges of a semiregular polyhedron be the same length? Explain.

    d. Must each angle in every face of a semiregular polyhedron have the same size? Explain.

10. Use isometric dot paper to show two different nets for a regular octahedron.

Supplementary Learning Exercises for Section 17.5

1. a. After using a net for the icosahedron (as in Figure 9), a student says, "I see the 20 faces, but I counted 41 edges and 26 vertices." Why were the student's counts incorrect?

    b. What advice is important in using a net to count edges and vertices?

2. Why is the Great Pyramid in Egypt *not* a regular polyhedron?

3. Try the *Geometric Solids* activity at http://illuminations.nctm.org.

4. Challenge Find the 11 different nets that give regular hexahedra (cubes).

17.6 Issues for Learning: Dealing with 3D Shapes

The lack of an opportunity to learn about geometry can account for some of the difficulties that children (and adults) have with 3D geometric shapes, including the conventional vocabulary. For example, only about one-fourth of the 13-year-olds in one national testing program[1] could give the term *cube* when shown a model of a cube; many children used *square* for the cube, not distinguishing between the 2D nature of the square faces and the 3D nature of the shape. With most elementary school mathematics textbooks now covering the basic vocabulary for 3D shapes, children today usually have the opportunity to learn the names of the shapes . . . unless teachers skip those pages!

    Two special aspects of dealing with drawings of 3D shapes lead to difficulties for learners: (1) making such drawings and (2) interpreting them. You perhaps have seen children's drawings of houses, such as those shown below for a house shaped like a cube, with a door in the front and windows on the right and left sides of the house. The increasing sophistication in the drawings may be due to natural development or increased exposure to, and experience with, drawings.

 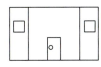

continue

A similar development occurs for drawings of differently shaped but plain blocks, as in the following illustrations that are based on work[2] with students in Grades 1, 3, 5, 7, and 9. For other shapes, most of the children drew at different levels of sophistication, suggesting that drawing some shapes is more familiar than others. For example, the cylinder was more easily drawn than the prism, the pyramid, and the cube (drawings not shown here).

Drawings of a right rectangular prism.

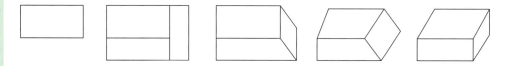

Drawings of a right square pyramid.

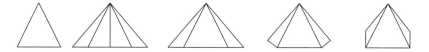

Some elementary school textbooks include special representations, such as those with isometric dot paper or the front-side-top views in your Learning Exercises, recognizing that there is a degree of learning in interpreting and making them.

But exposure helps. For example, in one unpublished report (by L. Jaslow), a teacher asked a mixed Grades 1–2 class to make nets for a cube. The illustrations below show how some of the initial ideas looked. But after discussing the nets and deciding that a turned version of, say, Harrison's would not be different, the children were able to find most of the different nets of a cube.

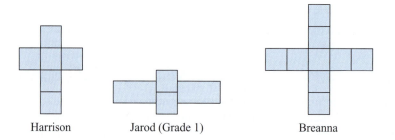

Harrison   Jarod (Grade 1)   Breanna

Textbooks, of course, rely on 2D drawings for solid shapes. But "reading" such drawings does not happen automatically for many children. For example, in one international testing,[3] given a drawing of a cube without the hidden edges indicated, only 40% of the fourth-graders and 34% of the third-graders picked the correct number of edges for the cube, from the multiple choices.

Work with physical models of 3D shapes would seem to be a natural precursor to making a definition. For example, in a teaching experiment[4] with third-graders, having the children build polyhedra (often pyramids) meeting certain criteria, but of their own designs, helped the children focus on important attributes of polyhedra (again, often pyramids). For example, the children's shapes shown on the next page helped them focus on what makes a pyramid a pyramid. (Why are these not pyramids?)

© Courtesy of Kathleen Wise

TAKE-AWAY MESSAGE . . . Although there is some evidence of developmental constraints in making drawings or in attending to all of the dimensions involved, instruction that builds carefully from actual models seems to be helpful. But in particular, the opportunity to learn is crucial.

17.7 Check Yourself

Chapter 17 has involved several vocabulary words for 3D shapes. Some of these may have been new to you, so you will need to devote some care to remember them and keep them straight. Some relationships between shapes (e.g., congruence) and within a shape (e.g., that the lateral edges of a prism are parallel and have the same length), as well as relationships about all polyhedra (e.g., Euler's formula), deserve special attention.

Here are some more specific guidelines, with more detail than will usually be the case. You should be able to work problems like those assigned and to meet the following objectives:

1. Use the vocabulary. You should be adept at giving adjectives as well as naming the basic shape: e.g., *right hexagonal prism* instead of just *prism*. There is a lot of vocabulary, and several of the words start with the letter *p* (e.g., polyhedron, pyramid, prism, parallel, etc.), so the words can easily be confused. Some vocabulary words are introduced through Learning Exercises, and you do not want to overlook them. *Learning Exercise 13 in Section 17.2 gives many of the words for 3D shapes that you should have mastered.* The approach to congruent shapes (Section 17.4) may have been new to you, as well as the names for the regular polyhedra (Section 17.5).

   Note: Measurement ideas of angle size, perimeter, area, and volume may have come up during class or in some exercises. The degree of familiarity you should have with such terms at this time ought to be clear from what your instructor has said in class. Both area and volume receive extensive attention later in this book.

2. Give, preferably by your knowledge of the type of shape, the number of faces, vertices, or edges for a given shape. For example, how many edges does a 200-gonal prism have? How many are lateral edges?

3. State and use Euler's formula for polyhedra.

4. Translate among the different representations: word, physical model, drawing, or net. For example, starting with, say, a name like *right square prism*, give the other three representations. Some of these translations involve having a good mental picture (or, if you are not a good visualizer, a good word description) of what a shape is, as in "Make a sketch of a right pentagonal prism." Your instructor may also expect you to be able to use isometric dot paper in making some drawings.

5. Show in your drawings an awareness of hidden line segments and show that you know to retain parallels and equal lengths whenever the view allows.

6. Recognize and draw congruent polyhedra.

7. Make with cubes, or sketch on isometric paper, a polyhedron that fits a given front-view, side-view, top-view information (from the Learning Exercises).

Notes

8. Tell what a *regular polyhedron* is and name the types or name the type from a given drawing.

9. Describe the difficulties that learners often have in drawing 3D shapes or interpreting drawings of 3D shapes (Section 17.6).

References for Chapter 17

[1]Carpenter, T., Coburn, T. G., Reys, R. E., & Wilson, J. W. (1978). *Results from the first mathematics assessment of the National Assessment of Educational Progress*. Reston, VA: National Council of Teachers of Mathematics.

[2]Mitchelmore, M. C. (1978). Developmental stages in children's representation of regular solid figures. *Journal of Genetic Psychology*, *133*, 229–239.

[3]Trends in International Mathematics and Science Study. http://nces.ed.gov/timss/

[4]Ambrose, R., & Kenehan, G. (2009). Children's evolving understanding of polyhedra in the classroom. *Mathematical Thinking and Learning*, *11*(3), 158–176.

Chapter 18

Symmetry

Symmetry is of interest in many areas, e.g., art, design in general, and even the study of molecules. This chapter begins with a look at two types of symmetry of two-dimensional shapes, and then moves on to introduce symmetry of polyhedra (and of three-dimensional objects in general).

18.1 Symmetry of Shapes in a Plane

Symmetry of plane figures can appear as early as Grade 1, where symmetry is restricted to **reflection symmetry**, or **line symmetry**, for a figure, as illustrated at the right. The **reflection line**, the dashed line in the figure, cuts the figure into two parts, each of which would fit exactly onto the other part if the figure were folded on the reflection line. Many flat surfaces in nature have reflection symmetry, and many human-made designs incorporate reflection symmetry.

You may have made symmetric designs (e.g., snowflakes or Valentine's Day hearts) by first folding a piece of paper, next cutting something from the folded edge, and then unfolding. The line of the folded edge is the reflection line for the resulting figure.

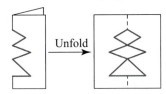

A given shape may have more than one reflection symmetry. For example, for a square there are four reflection lines, each of which gives a reflection symmetry for the square. Hence, a square has four reflection symmetries.

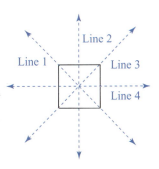

A second kind of symmetry for some shapes in a plane is **rotational symmetry**. A shape has rotational symmetry if it can be rotated around a fixed point until it fits exactly on the space it originally occupied. The fixed turning point is called the **center of the rotational symmetry**. For example, suppose square *ABCD* below is rotated counterclockwise about the blue point as the center. The blue line segment to vertex *C* is used here to help keep track of the number of degrees turned. A prime mark is often used as a reminder that the point is associated with the original location. For instance, we denote *B′* as the point to which *B* would move after the rotation.

 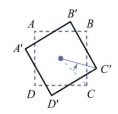 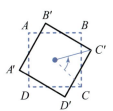 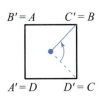

423

Eventually, after the square has rotated through 90°, it occupies the same set of points as it did originally. The square has a rotational symmetry of 90° with its center at the high-lighted point. Convince yourself that the square also has rotational symmetries of 180° and 270°.

Every shape has a 360° rotational symmetry. However, in counting the number of sym-metries, the mathematical convention is to count the 360° rotational symmetry only if there are other rotational symmetries for a figure. Hence, a square has four rotational symme-tries. Along with the four reflection symmetries, the square has eight symmetries in all. Because every shape has a 360° rotational symmetry, the 360° rotational symmetry is sometimes called the *trivial* symmetry.

Think About ...

The rotations in the preceding figure were all counterclockwise. Explain why 90°, 180°, 270°, and 360° *clockwise* rotations do not give any new rotational symmetries.

The symmetries make it apparent that they involve a movement of some sort. We can give the following general definition.

> A **symmetry of a shape** is any movement that fits the shape onto the same set of points it started with.

 ACTIVITY 1   Symmetries of an Equilateral Triangle

1. What are the reflection symmetries and the rotational symmetries of the equilateral triangle *KLM*? Be sure to identify the lines of reflection and the number of degrees in the rotations.

2. How could you have predicted that the angles of rotation here should be 120°, 240°, and 360°? Why doesn't 90° work? ●

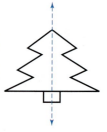

Notice that when we use the word "symmetry" with an object in geometry, we have a particular figure in mind, such as the tree shown at the right. And we must imagine some movement, such as the reflection in the dashed line, that gives the original figure as the end result. Many points have "moved," but the figure as a whole occupies the exact same set of points after the movement as it did before the movement. If you blinked during the move-ment, you would not realize that a motion had taken place.

Rather than just trusting how a figure looks, we can appeal to symmetries in some figures to justify some conjectures for those figures. For example, an isosceles triangle has a reflection symmetry. In an isosceles triangle, if we bisect the angle formed by the two sides of equal length, those two sides "trade places" when we use the bisecting line as the reflection line. Then the two angles opposite the sides of equal length (angles *B* and *C* in the figure) also trade places, with each angle fitting exactly where the other angle was.

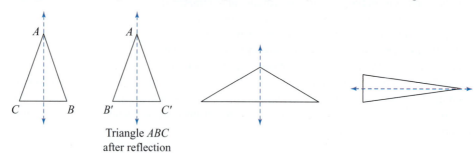

Triangle *ABC*
after reflection

So, *in an isosceles triangle the two angles opposite the sides of equal length must have equal sizes.* Notice that the same reasoning applies to every isosceles triangle, so there is no worry that somewhere there may be an isosceles triangle with those two angles having different sizes. Rather than looking at just one example and relying on what appears to be true in the example, this reasoning applies to all isosceles triangles and, as a result, gives a strong justification for not having *equal angles* as part of a definition of isosceles triangle, even though every isosceles triangle will have two angles equal in size.

 DISCUSSION 1 Why Equilateral Triangles Have to Be Equiangular

Use the previous fact about isosceles triangles to deduce that all three angles of an equilateral triangle must be of the same size.

In Discussion 1 you used the established fact about angles in an isosceles triangle to justify in a general way the fact about angles in an equilateral triangle. Contrast this method with just looking at an equilateral triangle and trusting your eyesight.

 ACTIVITY 2 Does a Parallelogram Have Any Symmetries?

Trace a general parallelogram and look for symmetries. Does a parallelogram have any reflection symmetries? Does it have any rotational symmetries besides the trivial 360° symmetry? ●

Here is another illustration of justifying a conjecture by using symmetry. Previously you may have made these conjectures about parallelograms: *The opposite sides of a parallelogram are equal in length, and the opposite angles are the same size.* The justification takes advantage of the 180° rotational symmetry of a parallelogram, as suggested in the following sketches. Notice that the usual way of naming a particular polygon by labeling its vertices provides a good means of talking (or writing) about the polygon, its sides, and its angles.

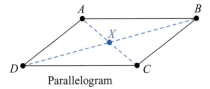
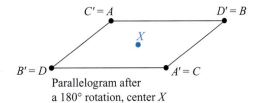

Parallelogram

Parallelogram after a 180° rotation, center X

 **ACTIVITY 3** Symmetries in Some Other Shapes

How many reflection symmetries and how many rotational symmetries does each of the following polygons have? In each case, describe the lines of reflection and the degrees of rotation.

a. regular pentagon *PQRST*

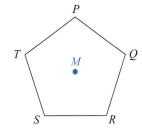

b. regular hexagon *ABCDEF*

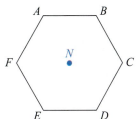

c. a regular *n*-gon ●

TAKE-AWAY MESSAGE . . . Symmetry of shapes is a rich topic. Not only do symmetric shapes have a visual appeal, they make the design and construction of many manufactured objects easier. Symmetry is often found in nature. Mathematically, symmetries can provide methods for justifying conjectures that might have come from drawings or examples.

In this page from a Grade 6 textbook, the students are asked to find reflection and rotational symmetries. Try them. ("Order of rotation symmetry" can be interpreted to mean "how many rotational symmetries.")

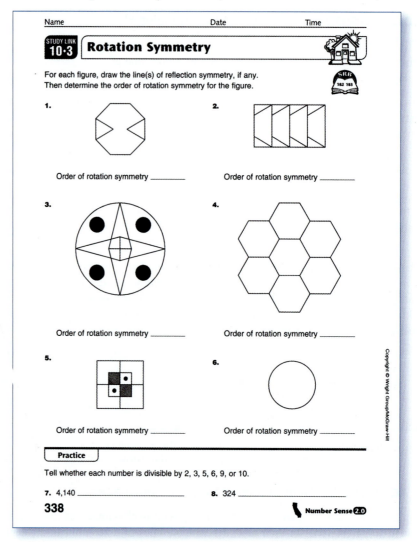

Wright Group, *Everyday Math Masters, Grade 6*, p. 338. Copyright © 2008, McGraw-Hill. Reproduced with permission of the McGraw-Hill Companies.

Learning Exercises for Section 18.1

1. Which capital letters, in a block printing style (e.g., A, B, C, D, E, F, G, H, I, J, K, L, M, N, O, P, Q, R, S, T, U, V, W, X, Y, Z), have reflection symmetry(ies)? Rotational symmetry(ies) besides the trivial one, 360°?

2. Identify some flat surface in nature that has reflection symmetry and one that has rotational symmetry.

3. Find some human-made flat surface that has reflection symmetry and one that has rotational symmetry. (One source might be company logos.)

4. What are the reflection symmetries and the rotational symmetries for each polygon given on the next page? Describe the lines of reflection and give the number of degrees of rotation. Do not count a 360° rotation unless there are other rotational symmetries.

continue

Notes

a. an isosceles triangle with only two sides the same length

b. a rectangle that does not have all its sides equal in length (Explain why the diagonals are *not* lines of symmetry.)

c. a parallelogram that does not have any right angles

d. an isosceles trapezoid

e. an ordinary, nonisosceles trapezoid

f. a rhombus that does not have any right angles

g. a kite

5. Shapes other than polygons can have symmetries.

   a. Draw a line of symmetry for an angle.

   b. Draw four lines of symmetry for two given lines that are perpendicular (i.e., that make right angles). Describe four rotational symmetries also.

   c. Draw three lines of symmetry for two given parallel lines.

   d. Draw several lines of symmetry for a circle. (How many lines of symmetry are there?)

   e. How many lines of symmetry does an ellipse have?

6. Explain why this statement is incorrect: "You can get a rotational symmetry for a circle by rotating it 1°, 2°, 3°, etc., about the center of the circle. So, a circle has exactly 360 rotational symmetries."

7. Copy each design and add to it so that the result gives the required symmetry described in parts (a–e).

   a. Design I, rotational symmetry

   b. Design I, reflection symmetry

   c. Design I, reflection symmetry with a line different from the one in part (b)

   d. Design II, rotational symmetry

   e. Design II, reflection symmetry

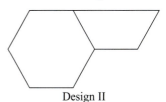

Design I

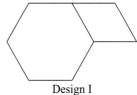

Design II

8. Pictures of real-world objects and designs often have symmetries. Identify all the reflection symmetries and rotational symmetries in the following pictures:

a.

b.

c.

d.

e.

f.
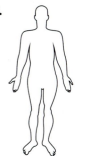

9. Pattern Blocks Make an attractive design using pattern blocks (or the paper ones found in the "Masters"). Is reflection symmetry or rotational symmetry involved in your design?

10. Suppose triangle *ABC* has a line of symmetry *k*. What does that tell you, if anything, about the following objects?

   a. segments *AB* and *AC* (What sort of triangle must *ABC* be?)
   b. angles *B* and *C*
   c. point *M* and segment *BC*
   d. angles *x* and *y*

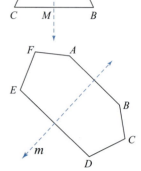

11. Suppose that *m* is a line of symmetry for hexagon *ABCDEF*. What does that tell you, if anything, about the following objects?
   a. segments *BC* and *AF* (Explain.)
   b. segments *CD* and *EF*
   c. the lengths of segments *AB* and *ED*
   d. angles *F* and *C* (Explain.)
   e. other segments or angles

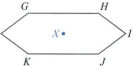

12. Suppose that hexagon *GHIJKL* has a rotational symmetry of 180°, with center *X*. What does a 180° rotation tell you about specific relationships between segments and between angles in the hexagon?

13. a. Using symmetry, give a justification that the diagonals of an isosceles trapezoid have the same length.
   b. Is the result stated in part (a) also true for rectangles? For parallelograms? Explain.

14. a. Using symmetry, give a justification that the diagonals of a parallelogram bisect each other.
   b. Is the result stated in part (a) also true for special parallelograms? For kites? Explain.

15. Examine the following conjectures about some quadrilaterals to see whether you can justify any of them by using symmetry:

   a. The longer diagonal of a kite cuts the shorter diagonal into segments that have the same length.
   b. In a kite like the one shown at the right, angles 1 and 2 have the same size.
   c. All the sides of a rhombus have the same length.
   d. The diagonals of a rectangle cut each other into four segments that have the same length.

Supplementary Learning Exercises for Section 18.1

1. What are the reflection symmetries and the rotational symmetries for each of the following shapes? Draw the lines of reflection and give the number of degrees of rotation. Do not count a 360° rotation unless there are other rotational symmetries.

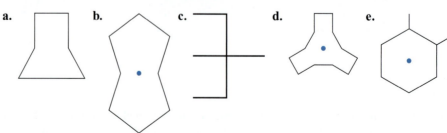

   a.   b.   c.   d.   e.

2. Add to each given shape so that the result will have the designated symmetries.

a.
square, with additions . . .
only 1 line of symmetry
0 nontrivial rotational symmetries

b.
square, with additions . . .
2 line symmetries
2 rotational symmetries (including 360°)

c.
ellipse, with additions . . .
2 line symmetries
2 rotational symmetries (including 360°)

3. Use symmetry to argue that the three diagonals of a regular hexagon have the same length.

4. Look at your face in a mirror. Does your image in the mirror have reflection symmetry, or close to it? (Most people's faces are *not* completely symmetric.)

5. Why is it incorrect to say, "Line symmetry means to cut in half"?

18.2 Symmetry of Polyhedra

In Section 17.4 we informally linked congruence of polyhedra to motions. Because we linked symmetry of 2D shapes to motions also, it is no surprise to find that symmetry of 3D shapes can also be described by motions. This section introduces symmetry of 3D shapes by looking at polyhedra and illustrating two types of 3D symmetry. Have your kit of shapes handy!

Reflection Symmetries

Clap your hands together and keep them together. Imagine a plane (or an infinite two-sided mirror) between your fingertips. If you think of each hand being reflected in that plane or mirror, the reflection of each hand would fit the other hand exactly. The left hand would reflect onto the right hand, and the right hand would reflect onto the left hand. The plane cuts the two-hands figure into two parts that are mirror images of each other; reflecting the figure—the pair of hands—in the plane yields the original figure. The figure made by your two hands has **reflection symmetry with respect to a plane**. Symmetry with respect to a plane is sometimes called **mirror-image symmetry**, or just **reflection symmetry**, if the context is clear.

   The plane can be called the *plane of symmetry*. The figure that the plane of symmetry makes with the shape gives a *cross section* of the shape.

 ACTIVITY 4 Finding a Plane of Symmetry and the Cross Section

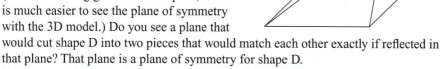

a. Look at shape D from your kit of 3D shapes. (The 2D drawing given is for shape D, but it is much easier to see the plane of symmetry with the 3D model.) Do you see a plane that would cut shape D into two pieces that would match each other exactly if reflected in that plane? That plane is a plane of symmetry for shape D.

b. Is there any other plane that would serve as a plane of symmetry?

c. What cross section of shape D would the plane of symmetry in part (a) give?

d. Repeat parts (a–c) with shape I from your kit. ●

Unless you are quite experienced at reading 2D drawings, you will agree that working with a 3D model is a great aid in seeing planes of symmetry and their cross sections. Keep that in mind in Activity 5, with your shape E or some other cube that is handy.

 ACTIVITY 5 Splitting the Cube

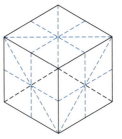

1. Can you find 9 planes of symmetry for a cube? Describe the cross section for any reflection symmetry that you find. (This drawing of a solid cube may help you.)

2. Does a right rectangular prism have any reflection symmetries? Describe the cross section for each reflection symmetry that you find. ●

Rotational Symmetries

A 3D figure has **rotational symmetry with respect to a line** if, by rotating the figure a certain number of degrees using the line as an axis, the rotated version *coincides* with the original figure. Points may now be in different places after the rotation, but the figure *as a whole* will occupy the same set of points after the rotation as before. The line is sometimes called the **axis of the rotational symmetry**, or **axis**. A figure may have more than one axis of rotational symmetry. As with the cube shown below, it may be possible to have different rotational symmetries with the *same* axis, by rotating different numbers of degrees. Because the two rotations shown— a 90° rotation and a 180° rotation—affect at least one point differently, they are considered to be two different rotational symmetries. The cube occupies the same set of points *in toto* after either rotation as it did before the rotation, so the two rotations are indeed symmetries.

After 90° rotation, clockwise (viewed from the top)   After 180° rotation, clockwise (viewed from the top)

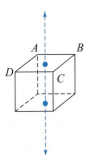 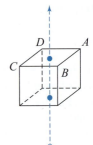 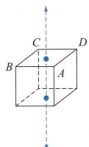

Similarly, a 270° and a 360° rotation with this same axis give a third and a fourth rotational symmetry. For this one axis, then, there are four different rotational symmetries: 90°, 180°, 270°, and 360° (or 0°).

 ACTIVITY 6 Rounding the Cube

1. Find all the axes of rotational symmetry for a cube. (There are more than three.) For each axis, find every rotational symmetry possible, giving the number of degrees for each rotational symmetry.

2. Repeat Problem 1 for an equilateral-triangular right prism (see shape C in your kit of shapes). ●

A 3D shape may not have *any* axes of rotational symmetry. For example, neither shape D nor shape I from your kit of shapes has any (nontrivial) rotational symmetry.

TAKE-AWAY MESSAGE . . . Some three-dimensional shapes have many symmetries, but the same ideas used with symmetries of two-dimensional shapes apply. Except for remarkably able or experienced visualizers, most people find a model of a shape helpful in counting all the symmetries of a 3D shape.

Learning Exercises for Section 18.2

1. Can you hold your two hands in *any* fashion so that there is a rotational symmetry for them (besides a 360° one)? Each hand should end up exactly where the other hand started and fingernails should be in the same places.

2. How many different planes give symmetries for these shapes from your kit? Record a few planes of symmetry in sketches, for practice.

   a. shape A   **b.** shape F   **c.** shape G   **d.** shape K

3. How many rotational symmetries does each shape in Learning Exercise 2 have? Show a few of the axes of symmetry in sketches, for practice.

4. You are a scientist studying crystals shaped like shape H from your kit. Count the symmetries of shape H, both reflection and rotational. (Count the 360° rotational symmetry just once.)

5. Describe the symmetries, if there are any, of each of the following shapes made of cubes:

   a.   **b.** 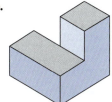   **c.**   Top view for c.

6. Copy and finish the following incomplete "buildings" so that they have reflection symmetry. Finish each one in two ways, counting the additional number of cubes each way needs. [The building in part (b) already has one plane of symmetry. Do you see it? Is it still a plane of symmetry *after* your additional cubes?]

   a.    **b.**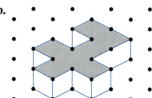

7. Design a net for a regular square pyramid that has exactly four rotational symmetries (including only one involving 360°).

8. Imagine a right octagonal prism with bases like 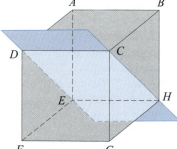. How many reflection symmetries will the prism have? How many rotational symmetries?

9. The cube shown below is cut by the symmetry plane indicated. For the reflection in the plane, to what point does each vertex correspond?

   $A \rightarrow$ ____ $B \rightarrow$ ____

   $C \rightarrow$ ____ $D \rightarrow$ ____

   $E \rightarrow$ ____ $F \rightarrow$ ____

   $G \rightarrow$ ____ $H \rightarrow$ ____

10. Explain why each pair is considered to describe only one symmetry for a figure.

    a. a 180° clockwise rotation and a 180° counterclockwise rotation (same axis)
    b. a 360° rotation with one axis and a 360° rotation with a different axis

Notes

11. (*Suggestion:* Work with a classmate.) You may have counted the reflection symmetries and rotational symmetries of the regular tetrahedron (shape A in your kit) and the cube. Pick one of the other types of regular polyhedra and count its reflection symmetries and axes of rotational symmetries.

Supplementary Learning Exercises for Section 18.2

1. How many rotational symmetries has each of the following shapes (each made of three cubes)? Do not count the 360° rotational symmetry unless there are others.

a. 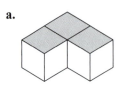   **b.**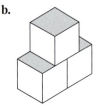

2. How many reflection symmetries does each shape in Exercise 1 have? Describe them, perhaps with a sketch of cross sections.

3. Consider a right hexagonal prism with bases that are regular hexagonal regions. How many reflection symmetries and how many rotational symmetries does this prism have?

4. Find at least two planes of symmetry for shape J from your kit of shapes.

5. Explain why shape B from your kit of shapes does not have any (nontrivial) symmetries.

18.3 Issues for Learning: What Geometry Is in the Pre-K–8 Curriculum?

Unlike the work with numbers, the coverage of geometry in Grades K–8 has not been uniform in the United States, particularly with respect to work with three-dimensional figures. Chapters on geometry in elementary school mathematics textbooks are often toward the end of the book and are not reached by the end of the school year. On an international test, students in the United States score relatively worse in geometry than in other content domains, compared to students from many other nations.[1]

The nationwide tests used by the National Assessment of Educational Progress give an indication of what attention the test writers believe should be given to geometry.[2] At Grade 4, roughly 15% of the items are on geometry (and spatial sense), and at Grade 8, roughly 20% of the items are on geometry (and spatial sense), suggesting the importance of these topics in the curriculum.

This test does not yet reflect the influence of *The Common Core State Standards* (CCSS)[3] released in mid-2010, a state-led initiative to establish a shared set of educational standards. The CCSS for mathematics consist of two intertwined parts: Standards for Content and Standards for Mathematical Practice. The content standards define grade-specific standards, in particular, the processes and proficiencies that students should develop for a sequence of topics and performances. These standards list Geometry as a domain, or category of related standards, for Grades K–8. In Grades K–5, Measurement and Data exists as a separate domain; geometric measurement ideas are a part of the geometry standards in Grades 6–8. Section 25.3 is entitled "Issues for Learning: What Measurement Is in the Curriculum?" Though each state has its own standards for mathematics, nearly all have adopted the CCSS with a small percentage of modifications.

What follows is a brief summary of the Grades K–5 standards regarding geometry, the focus of Chapters 16–22 of this text. The Issues for Learning section in four other parts of the text contains a brief description of the content standards regarding number and operations, algebraic thinking, measurement, and probability and statistics. The Standards for Mathematical Practices are described in detail in Section 21.3, "Issues for Learning."

Grades Pre-K–2. Children should be able to recognize, name, build, draw, and sort shapes, both two-dimensional and three-dimensional, and recognize them in their surroundings regardless of their orientation. They should be able to describe their physical world using language for directions, distance, and location, using words like "above," "beside," "in front of," and "between." Our students should analyze and compare similarities and differences in shapes. They should be able to describe shapes by examining sides and angles, including hexagons, trapezoids, and cubes. In preparation for creating and relating categories of figures based on a figure's properties, a Grade 1 standard describes students distinguishing between defining attributes and nondefining attributes. By Grade 2, the focus on attributes supports an emphasis on definitions. Likely, by second grade, some instructional time will focus on using definitions to determine whether or not a particular figure is in the category described. Justification for whether a particular figure is or is not as described in the definition will support the development of the mathematical practice of constructing viable arguments. Furthermore, our students should be capable of drawing shapes with specified attributes (e.g., three "corners" and sides of equal length). Finally, they should compose larger shapes from simpler shapes and partition circles and rectangles into halves, thirds, and fourths.

Three kindergarteners' illustrations of sharing with three friends (a) and with four friends (b, c).

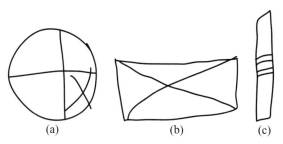

(a)   (b)   (c)

Grades 3–5. Students should focus more on the properties of two- and three-dimensional shapes. Words like "parallel," "perpendicular," "vertex," "obtuse angle," etc., should become part of their vocabulary. They should be able to classify a two-dimensional figure based on the presence (or absence) of attributes such as parallel lines or right angles. Coordinate systems should be introduced. Students graph points in the first quadrant of the coordinate plane as part of representing problems. They should be able to relate coordinate points to a given context. Students should make and test conjectures about properties of geometric shapes and should be able to justify their conclusions. Students identify line-symmetric figures and, furthermore, place the line of symmetry in the figure.

Grades 6–8. As mentioned, measurement and geometry are not separate domains in Grades 6–8. Geometry ideas support the measurement of length, area, and volume. In Grade 6, there is more focus on students' ability to compose and decompose shapes to solve real-world and mathematical problems. Another expectation for students in these grades is to be able to draw geometric shapes with given conditions. Their study of coordinate geometry continues as students find the length of a line segment in special cases where one coordinate is the same, as in the figure at right. Drawing polygons in the coordinate plane prepares them to work on scale drawings and constructions.

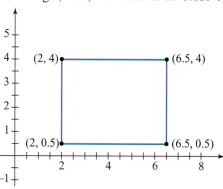

Serious study of transformation geometry begins with rigid motions of reflections, rotations, translations, and then dilations. Students use this as a basis for describing congruence (a two-dimensional figure is congruent to another when it can be obtained through a sequence of rigid motions) and similarity (includes dilations). Real-world application connections are made to scale drawings. Not all work is with two-dimensional figures. They should be able to imagine the result of slicing a three-dimensional figure, such as right rectangular prisms and pyramids.

The authors of the CCSS suggest that given three angles or sides, students should be able to determine whether a unique triangle can be drawn, more than one triangle can be drawn, or whether there is no triangle that can be drawn with the given criteria. In their deductive geometry course that follows in secondary school, the students will make formal arguments for the figures that can be created with given criteria. It will be important for you as an elementary teacher to support the students' mathematical practices (such as making sense of problems and constructing viable arguments) as they learn and apply these geometry concepts to art, science, and mathematical problems.

It may surprise you to learn that in early primary grades, you will be instructing on dividing rectangles and circles in halves, thirds, and fourths. The important geometry idea that children need to recognize is that equal shares of identical wholes need not have the same shape. They also learn that decomposing a whole into more equal shares creates smaller shares (unit fractions). Teachers of elementary school sometimes engage students in paper-folding activities to support understanding of partitioning before children have ways to express formal and sophisticated ideas of fractional relationships and area. Kieran (1995) describes activities used by middle school students that are easily modified for 7- to 8-year-olds.[4] Though some are difficult, students can generate successful folding strategies on their own. One student suggested the following to generate thirds:

"When I see halves, then I know I have thirds."

ACTIVITY 7   One-Fifth-ings and Other Fraction Folds

Start with an ordinary 8½ × 11-inch sheet of paper.

1. Using half folds and third folds, using three folds, how many different unit fractions can you make? Can you do any differently?

2. How can the folded paper be used to help children recognize that equal shares of identical wholes need not have the same shape?

3. How can you fold into fifths? (*Hint:* Use a "When I see thirds . . ." strategy like the child's "When I see halves. . . .")

4. Using half folds, third folds, and fifth folds, how many different unit fractions can you make with three folds? Can you imagine them without actually folding them? ●

18.4 Check Yourself

You should be able to work problems like those assigned and to meet the following objectives:

1. Define symmetry of a shape.

2. Sketch a figure that has a given symmetry.

3. Identify all the reflection symmetries and the rotational symmetries of a given two-dimensional figure, if there are any. Your identifications should include the line of reflection or the number of degrees of rotation.

4. Use symmetry to argue for particular conjectures. Some are given in the text and others are called for in the Learning Exercises, but an argument for some other fact might be called for.

5. Identify and enumerate all the reflection symmetries (in a plane) and the rotational symmetries (about a line) of a given three-dimensional figure.

References for Chapter 18

[1]Gonzales, P., Williams, T., Jocelyn, L., Roey, S., Kastberg, D., & Brenwald, S. (2008). *Highlights from TIMSS 2007: Mathematics and science achievement of U.S. fourth- and eighth-grade students in an international context* (NCES 2009–001 Revised). Washington, DC: National Center for Education Statistics, Institute of Education Sciences, U.S. Department of Education.

[2]Silver, E. A., & Kenney, P. A. (2000). *Results from the seventh mathematics assessment of the National Assessment of Educational Progress.* Reston, VA: National Council of Teachers of Mathematics.

[3]National Governors Association Center for Best Practices, Council of Chief State School Officers. (2010). *Common core state standards for mathematics.* Washington, DC: Author.

[4]Kieran, T. E. (1995). Creating spaces for learning fractions. In J. T. Sowder & B. P. Schappelle (Eds.), *Providing a foundation for teaching mathematics in the middle grades* (pp. 31–66). Albany: State University of New York Press.

Chapter 19

Tessellations

Covering an endless flat surface (a plane) with a given shape or shapes gives a **tessellation**. Tessellations, a topic that may appear in many elementary curricula, offer an opportunity for explorations, some surprises, connections to topics such as area, and a relation to art-work. The attention to tessellations in elementary school focuses on the plane, but the same ideas can be applied to space.

Notes

19.1 Tessellating the Plane

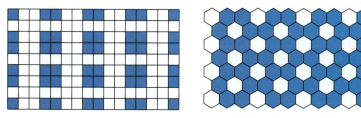

Figure 1

You have seen the two patterns shown above in tiled floors. These tilings are examples of **tessellations**, which are coverings of the plane made up of repetitions of the same region (or regions) that could completely cover the plane without overlapping or leaving any gaps. (The word "tessellation" comes from a Latin word meaning *tile*.) The first tiling in Figure 1 is a tessellation with squares (of course, it involves square *regions*), and the second tiling is a tessellation with regular hexagons. As in these two examples, enough of the covering is usually shown to make clear that it would cover the entire plane, if extended indefinitely. In the examples, we show how shading can add visual interest; colors add even more. Furthermore, a tessellation with squares can lay the foundation for children's work with area later on.

 **ACTIVITY 1** Regular Cover-Ups

Each tessellation pictured in Figure 1 involves one type of regular polygonal regions. A natural question is, "What other regular polygons give tessellations of the plane?" Test the following regions to see whether each type will give a tessellation of the plane:

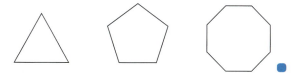

It may have been a surprise to find that some regular polygons do not tessellate the plane. Are there *any* other shapes that will tessellate?

 **ACTIVITY 2**   Stranger Cover-Ups

Another natural question is whether regions from nonregular shapes can tessellate the plane. Test the following regions to see whether any will tessellate the plane. Show enough of any tessellation to make clear that the whole plane could be covered.

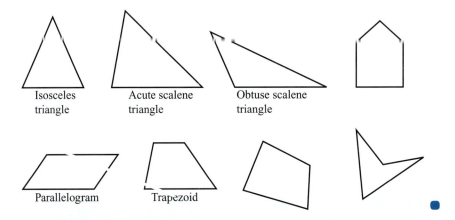

Isosceles   Acute scalene   Obtuse scalene
triangle   triangle   triangle

Parallelogram   Trapezoid

Activity 2 may have suggested that the subject of tessellations is quite rich. If you color some of the tessellations, using a couple of colors or even just shading, you can also find a degree of aesthetic appeal in the result. Indeed, much Islamic art involves intricate tessellations (Islam forbids the use of pictures of humans in its religious artwork). Islamic tessellations inspired the artist M. C. Escher (1898–1972) in many of his creations, some of which you may have seen. The drawing in Figure 2 shows two amateurish *Escher-type* tessellations, with additional features drawn; surprisingly, each tessellation starts with a simpler polygon than the final version might suggest. Elementary school students sometimes make these types of drawings as a part of their artwork, coloring the shapes in two or more ways.

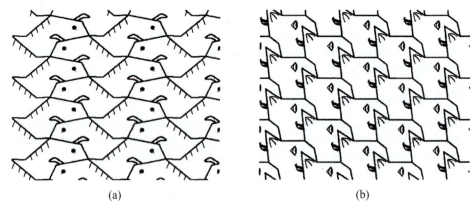

(a)   (b)

Figure 2

Notes

Compare this page from a Grade 6 textbook with our Activity 1!

Name _____ Date _____ Time _____

LESSON 10·1 | **Same-Tile Tessellations**

Decide whether each polygon can be used to create a same-tile tessellation.
Write the name of the polygon. Then record your answers in Column A. In Column B,
use your Geometry Template to draw examples illustrating your answers in Column A.

Polygon	A. Tessellation? (Yes or No)	B. Draw an example.
△ _____		
⬜ _____		
⬠ _____		
⬡ _____		
⯃ _____		

332

Copyright © Wright Group/McGraw-Hill

Wright Group, *Everyday Math Masters, Grade 6*, p. 332. Copyright © 2008,
McGraw-Hill. Reproduced with permission of the McGraw-Hill Companies.

Creating a Tessellation

How does one get polygons that will tessellate? One way is to start with a polygon that you
know will tessellate, and then modify it in one or more of several ways. For example, in the
illustration on the next page, we start with a regular hexagonal region, then cut out a piece
on one side, and finally tape that piece in a corresponding place on the opposite side.

The resulting shape tessellates. The final shape can then be decorated in whatever way the shape suggests to you—perhaps a piranha fish for this shape.

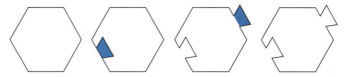

This same technique—cut out a piece and slide it to the opposite side—can be applied to another pair of parallel sides in the same region above. Again, notice that you can add extra features to the inside of any shape you know will tessellate.

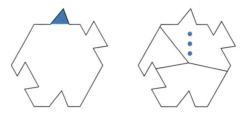

Yet another way to alter a given tessellating polygon so that the result still tessellates is to cut out a piece along one side, and then turn the cut-out piece around the midpoint of that side, as the following illustration shows:

When you draw the tessellations with such shapes, you find that some must be turned. Doing so gives a different effect when features are added to the basic shape. As always, coloring with two or more colors adds interest.

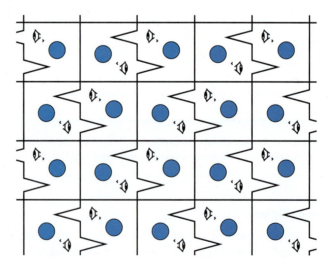

TAKE-AWAY MESSAGE ... The fact that tessellations, or coverings, of the plane are possible with equilateral triangles, squares, or regular hexagons is probably not surprising. More surprising is the fact that these are the only regular polygons that can give tessellations. Even more surprising is the fact that every triangle or every quadrilateral can tessellate the plane. Some clever techniques allow you to design unusual shapes that will tessellate the plane, often with an artistic effect.

Learning Exercises for Section 19.1

1. Test whether the regular heptagon (7-gon) or the regular dodecagon (12-gon) gives a tessellation.

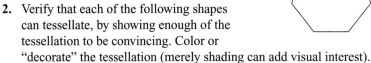

2. Verify that each of the following shapes can tessellate, by showing enough of the tessellation to be convincing. Color or "decorate" the tessellation (merely shading can add visual interest).

a.
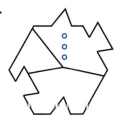
b.

c.
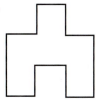

3. **a.** Start with a square region and modify it to create a region that will tessellate. Modify the shape in two ways; add features and shading or coloring as you see fit.
   b. Start with a regular hexagonal region and modify it to create a region that will tessellate. Modify the shape in two ways; add features and shading or coloring as you see fit.

4. Show that the following quadrilateral will tessellate. (If you can, use the grid as an aid, rather than tracing the quadrilateral and cutting the tracing out.) Add features and shading as you see fit. (*Hint:* Using the midpoint of a side, turn the shape 180°.)

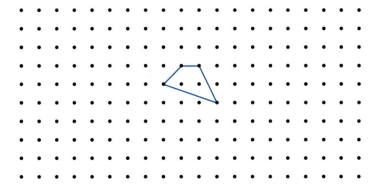

5. Which of the pattern blocks in the "Masters" give tessellations?

6. More than one type of region can be used in a tessellation, as with the regular octagons and squares shown below. Notice that the same arrangement of polygons occurs at each vertex.

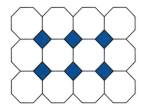

   Show that the following combinations can give tessellations:

   a. regular hexagons and equilateral triangles
   b. regular hexagons, squares, and equilateral triangles

7. Will each type of pentomino tessellate? [See Learning Exercise 13(a) in Section 17.3.]

8. Tessellations can provide justifications for some results.

    a. Label the angles with sizes x, y, and z in other triangles in the partial tessellation below to see if it is apparent that $x + y + z = 180°$. (*Hint*: Look for a straight angle.)

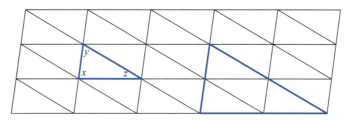

    b. Use the larger blue triangle to justify this fact: The length of the segment joining the midpoints of two sides of a triangle is equal to half the length of the third side.

    c. How does the area of the larger blue triangle compare to the area of the smaller blue one? (*Hint*: No formulas are needed; study the sketch.)

9. Which sorts of movements are symmetries for the tessellations given by the following? What basic shape gives each tessellation?

    a.

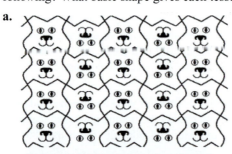

    b.

    c.

    d.

Supplementary Learning Exercises for Section 19.1

1. Show that each shape below can tessellate the plane, by showing enough of the tessellation to be convincing.

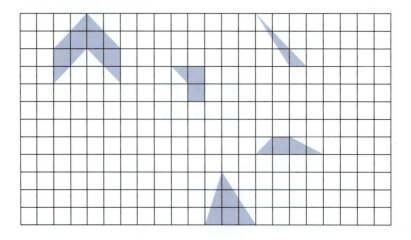

2. Are there any symmetries (reflection or rotational) in any of your tessellations for the shapes in Exercise 1?

3. The size of each angle in a regular pentagon is 108°. Use that information to explain that a tessellation with regular pentagons is not possible.

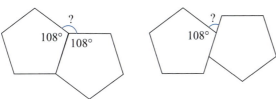

4. Use a tessellation of the plane with regular hexagons to show that the size of each angle of a regular hexagon is 120°. See the diagram at the right.

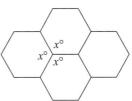

5. Will each type of tetromino (Learning Exercise 12 in Section 17.3) tessellate the plane?

6. If it is possible, draw a shape that has at least one curved part yet will tessellate the plane. If it is not possible, tell why.

19.2 Tessellating Space

The idea of tessellating the plane with a particular 2D region can be generalized to the idea of filling space with a 3D region.

> ### Think About …
>
> Why are the following objects shaped the way they are?
>
> | most boxes | bricks |
> |---|---|
> | honeycomb cells | commercial blocks of ice |
> | lockers | mailbox slots in a business |

When space is completely filled by copies of a shape (or shapes), without overlapping or leaving any gaps, space has been *tessellated*, and the arrangement of the shapes is called a **tessellation of space**. You can imagine either arrangement below as extending in *all* directions to fill space with solid regions formed by right rectangular prisms. These "walls" could easily be extended right, left, up, and down, giving an infinite layer that could be repeatedly copied behind and in front of the first infinite wall to fill space. Hence, we can say that a right rectangular prism will tessellate space and will do so in at least the two ways shown below.

or

Although it is rarely, if ever, an explicit part of the elementary school curriculum, tessellating space with cubical regions is the essence of the standard measurement of volume.

ACTIVITY 3 Fill 'Er Up

Which of shapes A–H from your kit will tessellate space? ●

After your experience with designing unusual shapes that will tessellate the plane, you perhaps can imagine ways of altering a 3D shape that tessellates space and still have a shape that will tessellate.

TAKE-AWAY MESSAGE . . . The idea of covering the plane with a 2D region can be extended to the idea of filling space with a 3D region.

Learning Exercises for Section 19.2

1. Are there arrangements of right rectangular prisms, other than those suggested in Section 19.2, that will give tessellations of space? (*Hint*: You may have seen decorative arrangements of bricks in sidewalks, where the bricks are twice as long as they are wide.)

2. Which of the following shapes could tessellate space (theoretically)?

   a. cola cans   **b.** sets of encyclopedias   **c.** round pencils, unsharpened
   d. hexagonal pencils, unsharpened and without erasers
   e. oranges

3. Which, if any, of the following could tessellate space? Explain your decisions.

   a.    **b.**    **c.**

   d. each type of the (3D) pattern block pieces
   e. each of the base *b* pieces (units, longs, flats)

4. How are tessellation of space and volume related?

5. (*Group*) Show that shape I in your kit can tessellate space.

6. (*Group*) Will either shape J or shape K tessellate space?

Supplementary Learning Exercises for Section 19.2

In Exercises 1–8, which, if any, of the shapes could (theoretically) tessellate space. (Assume that they are filled to make solid shapes.)

1. 14-ounce boxes of one kind of corn flakes   2. 42-ounce oatmeal boxes (one brand)

3. basketballs   4. door stops, all exactly like this one:

5. nickels   6. dollar bills

7. sugar cubes   8. pumpkins that are all the same size

19.3 Check Yourself

This short chapter about tessellations of a plane or of space opens up an aesthetic side of mathematics, illustrating that intricate designs can be derived from basic mathematical shapes.

   You should be able to work problems like those assigned and to meet the following objectives:

1. Tell in words what a tessellation of a plane is.

2. Determine whether a given shape can or cannot tessellate a plane. You should know that some particular types of shapes can tessellate, *without* having to experiment.

3. Create an "artistic" tessellation.

4. Tell in words what a tessellation of space is.

5. Determine whether a given shape can or cannot tessellate space.

Similarity

One important feature of congruent figures is that corresponding parts of the figures have exactly the same geometric measurements. In many applications of mathematics, however, two figures may have the exact same shape but not the same size. For example, photographic enlargements or reductions should look the same, even though corresponding lengths are different. This chapter gives a precise meaning to the idea of *same shape*, first with two-dimensional shapes and then with three-dimensional shapes.

20.1 Similarity and Dilations in Planar Figures

A moment's thought about, say, a photograph of a building and an enlargement of that photograph makes clear that for the buildings to look alike, corresponding angles have to be the same size. Activity 1 below will help us confront the less visible aspect of *exact same shape*. That is, how are corresponding lengths related?

Special tools needed: ruler with metric (protractor optional)

ACTIVITY 1   A Puzzle About a Puzzle[1]

In the diagram below, the pieces may be cut out and then, as a puzzle, reassembled to make the square. You are to make a puzzle shaped just like the one given, but larger, using this rule: The segment that measures 4 cm in the original diagram should measure 7 cm in your new version. If you work as a group, each person should make at least one piece. When your group finishes, you should be able to put together these pieces to make a square.

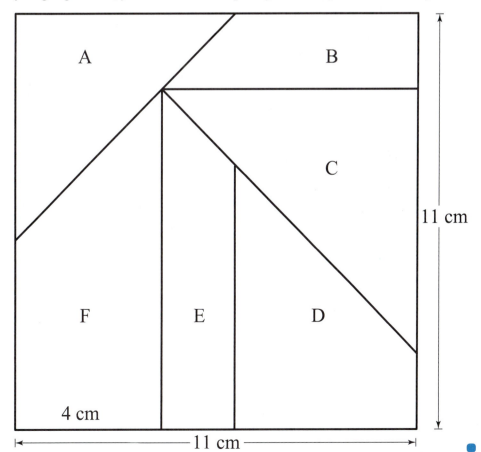

 **DISCUSSION 1**   Students Discuss Their Methods

Here are some students' descriptions of their thinking for Activity 1. They have agreed that the angles have to be the same, and they plan to use the many right angles to make the larger puzzle. Discuss each student's thinking.

Lee: "7 is 3 more than 4. So you just add 3 to each length. 3 centimeters. Add 3 centimeters to 5 centimeters, and the new side should be 8 centimeters."

Maria: "7 is $1\frac{3}{4}$ times as much as 4, so a new length should be $1\frac{3}{4}$ times the old length."

Nerida: "To be the same-shaped puzzle, it's got to be proportional to look the same. But I'm not sure how to make it proportional. Do you use ratios?"

Olivia: "From 4 centimeters to 7 centimeters is 75% more, so I would add 75% to each length. For example, take a 5-centimeter length; 75% of 5 centimeters is 3.75 centimeters, so the new length would be 5 + 3.75, or 8.75, centimeters."

Pat: "If 4 centimeters grow to 7 centimeters, each centimeter must grow to $1\frac{3}{4}$ centimeters. So 5 centimeters should grow to 5 times as much, that is, 5 times $1\frac{3}{4}$ centimeters, and 6 cm should grow to 6 times $1\frac{3}{4}$ centimeters." ●

 **ACTIVITY 2**   Super-Sizing It More

Make a sketch, share the work, and indicate all the measurements needed to get a bigger puzzle, where a 5-cm segment in the original square in Activity 1 measures 8 cm in the enlarged version. Then discuss your thinking with others. ●

ACTIVITY 3   Reducing

1. Make a sketch and indicate the measurements needed to get a smaller puzzle, where a 6-cm segment in the original square in Activity 1 should measure 4 cm in the smaller version.

2. In making an enlargement or reduction of a shape, as was done with the puzzle, how do angle sizes in the new shape compare with the corresponding ones in the original shape? How do lengths in the new shape compare with the corresponding ones in the original shape?

3. Write a set of instructions for enlarging/reducing such puzzles. Give a warning about any method that does not work, and explain why it does not work. ●

> ### *Think About …*
>
> Which of the following images would be acceptable as a reduced size of the given original drawing?
>
>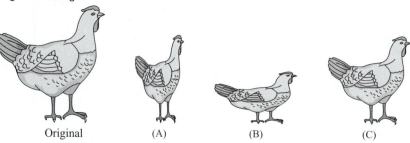
>
> Original   (A)   (B)   (C)

Similarity

Enlargements or miniatures must have the exact same shape as the original. Two shapes related this way are called **similar shapes** in mathematics, in the technical sense of the word "similar" and not just because they are the same general shape.

Two shapes are **similar** if the points in the two shapes can be matched so that (1) every pair of corresponding angles have the same size, and (2) the ratios from every pair of corresponding lengths all equal the same value, called the **scale factor**.

The second point makes clear that it is the multiplicative comparisons of corresponding lengths, not the additive comparisons, that are crucial. This point can also be expressed in different but equivalent ways:

- (*new* length) : (corresponding *original* length) = **scale factor**

- $\dfrac{\textit{new} \text{ length}}{\textbf{corresponding } \textit{original} \textbf{ length}}$ = **scale factor**

- (*new* length) = (scale factor) × (corresponding *original* length)

The last version makes explicit the multiplicative effect of the scale factor. So, if you are careful about keeping corresponding angles the same size and corresponding lengths related by the same scale factor, you can make a polygon similar to a given one.

ACTIVITY 4   Creating a Similar Polygon with Ruler and Protractor

On a piece of paper, draw a (nonspecial) quadrilateral with your ruler. Choose a scale factor (of a size so that your result will fit on the same sheet of paper). With your protractor and ruler, create a new quadrilateral similar to your quadrilateral. ●

Because the ratios of lengths in similar figures are all equal to the same value, two such ratios are equal to each other, leading to a proportion of the form $\frac{a}{b} = \frac{c}{d}$. You are probably familiar with a proportion like $\frac{6}{9} = \frac{8}{x}$ in connection with similar figures. Hence, multiplicative reasoning and proportional reasoning are often closely linked in writing.

Do you see that the scale factor for the original puzzle enlargement in Activity 1 is $1\frac{3}{4}$? In most situations, which shape is the new one and which is the original (or *old*, if you prefer) is arbitrary. *So long as the scale factor and ratios are interpreted consistently, the choice of new and original can be made either way.* For example, if the lengths of the sides of a figure A are 4 times as long as those of another figure B, then the lengths of the sides of figure B are $\frac{1}{4}$ as long as those of figure A.

Let us consider some examples that illustrate dealing with shapes known to be similar.

EXAMPLE 1

Make a rough sketch of triangles similar to the given original triangle (a) using a scale factor of 2.5 and (b) using a scale factor of $\frac{4}{5}$. Also, find the sizes of the angles and sides of the triangles of each new triangle.

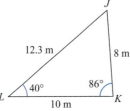

Solution

The size of the angle at J can be calculated by finding $180 - (40 + 86)$, which gives 54°. For either part (a) or (b), the angles in the similar triangles will be the same sizes as those in the original triangle: 40°, 86°, and 54°.

a. Your sketch should show a larger triangle, with sides about 2.5 times as long as those in the original. The lengths of the sides of the similar triangle will be 2.5 × 12.3 for the new $\overline{JL}$, 2.5 × 8 for the new $\overline{JK}$, and 2.5 × 10 for the new $\overline{LK}$, all in meters. That is, the lengths will be (about) 30.8 m, 20 m, and 25 m.

b. In the same way, the lengths will be 9.8 m, 6.4 m, and 8 m. Your sketch of the image triangle should be smaller, because the scale factor is less than 1.

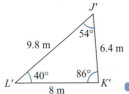

EXAMPLE 2

Suppose you are told that the following two quadrilaterals are similar. Find the missing angle sizes and lengths of sides.

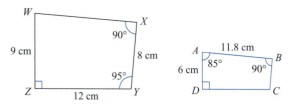

Solution

We are told that the two shapes are similar, so the first thing to do is to determine the correspondence. How the shapes are drawn helps, although the given angle sizes allow just one possibility. Because corresponding angles are the same size, the angle at W has size 85°, and the angle at C has size 95°.

To determine the missing lengths, we need the scale factor. The only pair of corresponding sides for which we know measurements are $\overline{WZ}$ and $\overline{AD}$. Thinking of $ABCD$ as the original (it is usually easier to deal with scale factors greater than 1), the scale factor is 9:6, or $\frac{9}{6} = 1.5$. So then the length of $\overline{WX}$ will be 1.5 × 11.8, or 17.7 cm. To find the missing lengths in $ABCD$, we can solve 8 = 1.5 × BC, and 12 = 1.5 × DC, or we can reverse the viewpoint to make $WXYZ$ the original, and work with a revised scale factor of 6:9, or $\frac{2}{3}$. In either case, we find that $\overline{BC}$ has length about 5.3 cm, and $\overline{DC}$ is 8 cm long.

EXAMPLE 3

You are told that the following two triangles are similar, but they are deliberately not drawn to scale. Using the given measurements, find the missing lengths and angle sizes.

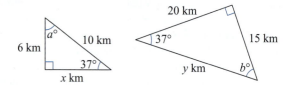

Solution

The missing angle sizes present no problem because angles a and b are corresponding and the angle sum for a triangle is 180°, so a and b are both 180 − (90 + 37), or 53°.

The complication here for finding the scale factor is that the correspondence is not obvious: Does the 6 km correspond to the 20-km or to the 15-km side? Because the 6-km side is opposite the 37° angle, its correspondent in the other triangle should be opposite the 37° angle there. That would make 6 km and 15 km corresponding. (Alternatively, the 6-km side is common to the right angle and the 53° angle, so find those angles in the other triangle and use their common side.)

Similarly, the 10-km and y-km sides correspond, as do the x-km and 20-km sides. Using the scale factor $\frac{15}{6} = 2.5$, we find that

$$y = 2.5 \times 10 = 25 \text{ km} \qquad \text{and} \qquad 20 = 2.5 \times x, \qquad \text{or} \qquad x = \frac{20}{2.5} = 8 \text{ km}$$

Dilations

The basis for a similarity is a transformation called a **dilation.** (After the size is changed, the result can be moved around.) You will notice that the following method measures only lengths, yet the sizes of corresponding angles are automatically equal. The *ruler method* for obtaining a similar polygon is given below and illustrates a dilation. Try using this method on separate paper. You will need to choose your own point for a center, your own scale factor, and your own original polygon (any triangle or quadrilateral will do). The importance of the scale factor is apparent in this method.

Steps in the Ruler Method for a Dilation

1. Pick a point (which becomes the **center** of the dilation). Draw a ray from the center through a point on the original shape. Measure the segment from the center to the point on the original shape.

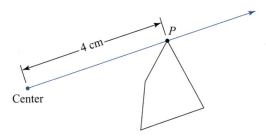

2. Multiply that measurement by your chosen scale factor (we'll use 1.7 here, so $1.7 \times 4 = 6.8$ cm).

3. Measure that distance (6.8 cm) from the *center* (that is important), along the ray starting at the center and going through the selected point. This distance gives what is called the **image** of the point.

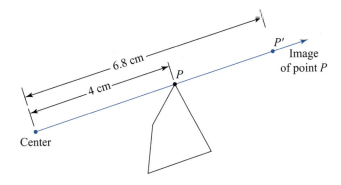

4. Repeat Steps 1–3 with the other vertices of the original polygon. Connect the images of all the vertices with the ruler. The resulting polygon is the **image** of the original polygon.

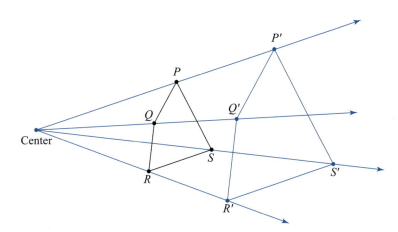

Notes

For a curved figure, the ruler method is not efficient because you have to go through the steps for too many points. But the ruler method works well for a figure made up of line segments. Starting with a more elaborate figure and then coloring the figure and its enlargement or reduction can make an attractive display. The completed drawing often carries a three-dimensional effect.

 ACTIVITY 5   Oh, I See!

a. Measure the lengths of the pairs of corresponding sides of the two quadrilaterals *PQRS* and *P′Q′R′S′* shown above to see how they are related.

b. How else do the corresponding sides appear to be related?

c. How are the pairs of corresponding angles of the polygons related in size?

d. Do other line segments, such as diagonals *PR* and *P′R′* have related lengths? ●

This page shows a Grade 6 version of our "ruler method."

Name _____   Date _____   Time _____

LESSON 5·5   **Scaling Transformations**

Some scaling transformations produce a figure that is the same shape as the original figure but not necessarily the same size. Enlargements and reductions are types of scaling transformations.

Enlargement: Follow the steps to draw a triangle *D′E′F′* with angles that are congruent to triangle *DEF* and sides that are twice as long as triangle *DEF*.

Step 1   Draw rays from *P* through each vertex. The first ray $\overrightarrow{PD}$ has been drawn for you.

Step 2   Measure the distance from point *P* to vertex *D*. Then locate the point on $\overrightarrow{PD}$ that is 2 times that distance. Label it *D′*.

Step 3   Use the same method from Step 2 to locate point *F′* on $\overrightarrow{PF}$ and point *E′* on $\overrightarrow{PE}$.

Step 4   Connect points *D′*, *E′*, and *F′*.

Reduction: Change Steps 2 and 3 to draw a triangle *D″E″F″* with angles that are congruent to triangle *DEF* and sides that are half as long as triangle *DEF*.

162

Dilations, like the one shown in the ruler method steps, are a basic way of getting similar polygons. After the size has changed, the image can be moved around by rotating or reflecting it; e.g., Figure 1 illustrates reflecting the larger image of a triangle. Triangle *A′B′C′* is similar to triangle *ABC* because it is the image of triangle *ABC* from the dilation. Triangle *A″B″C″* is a reflection of triangle *A′B′C′* about the line of reflection shown, and it is still similar to the original triangle. (Note the use of the *A*, *A′* and *A″* to make corresponding points clear.) If a rotation or reflection is involved, finding corresponding vertices and sides may require some attention.

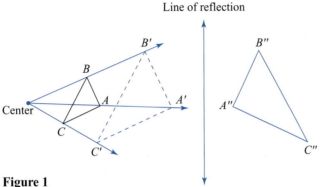

Figure 1

It is slightly digressive, but let us review some language for which everyday usage is often incorrect. For example, consider two segments, one 2 cm long and the other 3.4 cm long. The comparison of the 2-cm and the 3.4-cm values can be correctly stated in several ways:

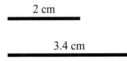

- "The ratio, 3.4 : 2, is 1.7."   (a multiplicative comparison)
- "3.4 cm is 1.7 *times as long as* 2 cm."   (a multiplicative comparison)
- "3.4 cm is 170% *as long as* 2 cm."   (a multiplicative comparison)
- "3.4 cm is 1.4 cm *longer than* 2 cm."   (an additive comparison and a true statement, but not the important one for similarity; note the *-er* ending on "longer")

EXAMPLE 4

What numbers in the blanks make the statements true?

a. 8 cm is ———— times as long as 2 cm.   **b.** 8 cm is ———— times longer than 2 cm.

c. 8 cm is ———— % as long as 2 cm.   **d.** 6 is ———— % as big as 4.

e. 6 is ———— % bigger than 4.

Solution

a. 4   **b.** 3   **c.** 400   **d.** 150   **e.** 50 ⬤

DISCUSSION 2   Saying It Correctly

Who is correct, Arnie or Bea? Explain. (The preceding sketch might help, identifying the *longer than* part.)

Arnie: "3.4 cm is 1.7 times longer than 2 cm."

Bea: "3.4 cm is 1.7 times as long as 2 cm, but 3.4 cm is only 0.7 times longer than 2 cm, or 70% longer than 2 cm." ⬤

Incorrect **comparison** language can be heard especially when both additive and multiplicative languages are used in the same sentence. Most people, however, do correctly fill in the blanks in statements such as "___ is 50% as big as 10" and "___ is 50% bigger than 10," so these examples might be helpful as checks in other sentences.

Notes

When you want to show that two shapes are indeed similar, you need to confirm these two conditions: (1) Corresponding angles must have the same size, and (2) the lengths of every pair of corresponding segments must have the same ratio, i.e., the scale factor. Vice versa, knowing that two figures are similar tells you that both these conditions have been met, which allows you to determine many missing measurements in similar figures. Dilations give the changes in size for similar shapes. Using the correct language in comparing lengths in similar shapes requires some care.

Learning Exercises for Section 20.1

Have your ruler and protractor handy for some of these exercises.

1. **a.** Summarize how the following are related for two polygons that are similar: corresponding lengths and corresponding angle sizes.

   b. Using the results from Examples 1, 2, and 3 in this section, conjecture how the perimeters of (i.e., distances around) similar polygons compare.

2. Tell whether the two shapes given in each part are similar. How do you know?

   a. a 6 cm by 7 cm rectangle, and a 12 cm by 13 cm rectangle

   b.

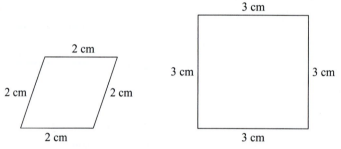

3. Copy and finish the incomplete second triangle to give a similar triangle, so that the 2 cm and 5 cm sides correspond. Does your triangle *look* similar to the original one?

4. Draw a triangle and then a similar triangle, with scale factor 4, using the following methods. Plan ahead so that the triangles will fit on the same page.

   a. with the ruler method, using your ruler and a center of your choice
   b. with a ruler and protractor

5. Draw a trapezoid and then a similar trapezoid, with scale factor 2.4, using the following methods. Plan ahead so that the trapezoids will fit on the same page.

   a. with the ruler method, using your ruler and a center of your choice
   b. with a ruler and protractor

6. Draw a triangle and then a similar triangle, with scale factor $\frac{3}{4}$, using the following methods. Plan ahead so that the triangles will fit on the same page.

   a. with the ruler method, using your ruler and a center of your choice
   b. with a ruler and protractor

7. **a.** Besides the ratio of lengths, how do a side of a polygon and its image appear to be related in the ruler method for dilations?

   b. Check the key relationships and your ideas about angles and sides from part (a) on the two similar triangles given on the next page.

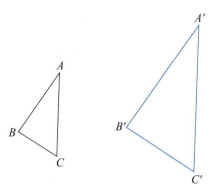

8. Find the sizes of all the angles and sides of shapes that are similar to the given parallelogram with the scale factors:

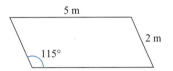

   a. 6.1   **b.** $\frac{2}{3}$

   c. What shape are the images in parts (a) and (b)?

9. Find the scale factor and the missing measurements in the similar triangles in each part. (The sketches are not drawn to scale.)

   a.

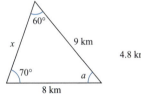

   b.

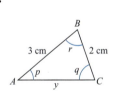

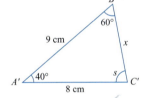

   c.

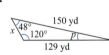

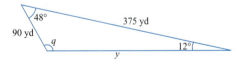

   d. (Be careful!)

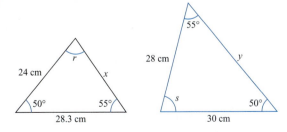

10. Can the center of a dilation be

    a. inside a figure?   **b.** *on* the figure?

11. For a given scale factor and a given figure, what changes if you use a different point for the center of a dilation?

Notes

12. For a given center and a given figure, what changes if you use a different scale factor for a dilation?

13. Suppose the scale factor is 1 for a dilation. What do you notice about the image?

14. Scale factors often are restricted to positive numbers.

    a. What would the image of a figure be if a scale factor were allowed to be 0?

    b. How could one make sense of a negative scale factor, say, ⁻2?

15. Is each of the following sentences phrased correctly? Correct any that are not by changing a number.

    a. 60 is 200% more than 30.   **b.** 12 cm is 150% longer than 8 cm.

    c. $75 is 50% more than $50.   **d.** The 10K run is 100% longer than the 5K.

16. a. Janeetha said, "I increased all the lengths by 60%." If Janeetha is talking about a size change, what scale factor did she use?

    b. Juan used a scale factor of 225% on a 6 cm by 18 cm rectangle. How many centimeters longer than the original dimensions are the dimensions of the image? How many percent longer are they than in the original?

17. Consider this original segment: ▬▬▬▬▬

Draw another segment that fits each description.

    a. 3 times as long as the original segment

    b. 1.5 times longer than the original segment

    c. 300% longer than the original segment

Consider this original region:

Draw another region that fits each description.

    d. 3 times the area of the original region (*Note:* Area = number of the small square regions inside.)

    e. 1.5 times the area of the original region

    f. 300% more than the area of the original region

18. In each part, which statements express the same relationship? Support your decisions with numerical examples or sketches.

    a. "This edge is 50% as long as that one," versus "This edge is 50% longer than that one," versus "This edge is half that one."

    b. "This quantity is twice as much as that one," versus "This quantity is 200% more than that one," versus "This quantity is 100% more than that one."

    c. "This value is 75% more than that one," versus "This value is $\frac{3}{4}$ as big as that one," versus "This value is $1\frac{3}{4}$ times as big as that one," versus "This value is 75% as much as that one."

19. In each part, give a value that fits each description.

    a. $2\frac{1}{3}$ times as long as 24 cm   **b.** $2\frac{1}{3}$ times longer than 24 cm

    c. 75% as long as 24 cm   **d.** 75% longer than 24 cm

    e. 125% more than 24 cm   **f.** 125% as much as 24 cm

    g. 250% as large as 60 cm   **h.** 250% larger than 60 cm

20. How would you make a shape similar to the given parallelogram with the scale factor 4? Give two ways.

21. Explain how size transformations are involved in each of the following situations:

    a. photographs

    b. different maps of the same location

    c. model cars or architectural plans

    d. banking interest (This won't involve shapes!)

22. What is the scale of a map if two locations 3 inches apart on the map are actually 84 miles apart in reality?

23. The diagram below shows two maps with the same two cities, River City and San Carlos. Even though the second map does not have a scale, determine the straight-line distance from San Carlos to Beantown. (*Hint*: Measure.)

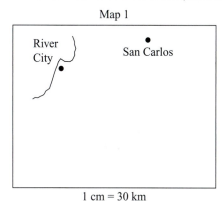

Map 1

1 cm = 30 km

Map 2

24. Time lines are representations that also use scales. Make a time line 20 cm long, starting at year 0, and mark the given dates. Add any other dates you wish.

Magna Carta 1215	Wright brothers' flight 1903
Columbus 1492	World War II 1941–1945
Declaration of Independence 1776	First atomic bomb 1945
French Revolution 1789	Commercial television 1950s
Civil War 1861–1865	Personal computers late 1970s
	Your birth

25. a. Make a time line 20 cm long to represent the following geologic times:
Cambrian, 600 million years ago (first fossils of animals with skeletons)
Carboniferous, 280 million years ago (insects appear)
Triassic, 200 million years ago (first dinosaurs)
Cretaceous, 65 million years ago (dinosaurs gone)
Oligocene, 30 million years ago (modern horses, pigs, elephants, etc., appear)
Pleistocene (first humans, about 100,000 years ago)

b. If you were to add the Precambrian, 2 billion years ago (first recognizable fossils), and use the same scale as in part (a), how long would your time line have to be?

26. Suppose a rectangle undergoes a dilation with scale factor 3, and then that image undergoes a second dilation, with scale factor 4. Are the final image and the original rectangle similar? If so, what is the scale factor?

27. a. Some teachers like to use two sizes of grids and have students make a larger or smaller version of a drawing in one of the grids. Try this method.

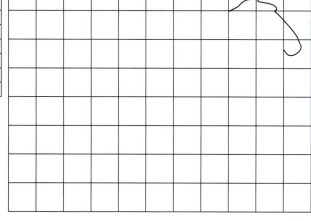

continue

Notes

b. What scale factor is involved in part (a)? (*Hint*: Measure.)

c. Can this method be used to make a *smaller* image? Explain how or why not.

28. a. Equilateral triangle X has sides 7 cm long, and equilateral triangle Y has sides 12 cm long. Are X and Y similar? Tell how you know.

b. Are two arbitrarily chosen equilateral triangles similar? Tell how you know.

c. Are every two right triangles similar? Tell how you know.

d. Are every two squares similar? Tell how you know.

e. Are every two rectangles similar? Tell how you know.

f. Are every two hexagons similar? Tell how you know.

g. Are every two regular n-gons (with the same n) similar? Tell how you know.

29. Some reference books show pictures of creatures and give the scale involved. Find the actual sizes of these creatures. (*Suggestion*: Use metric units.)

Scale factor is 1:170.   Scale factor is 7.3:1.

30. In the drawing at right, x' and x'' are the images of x for dilations with center C and the respective scale factors r and s. (These relationships are fundamental in trigonometry.)

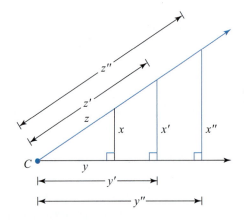

a. Find these ratios: $\dfrac{x'}{y'}, \dfrac{x''}{y''}, \dfrac{y'}{z'},$ and $\dfrac{y''}{z''}$.

b. How do these ratios compare:

$$\frac{x}{y}, \frac{x'}{y'}, \text{ and } \frac{x''}{y''}?$$

c. How do these ratios compare:

$$\frac{y}{z}, \frac{y'}{z'}, \text{ and } \frac{y''}{z''}?$$

Supplementary Learning Exercises for Section 20.1

1. Summarize the main relationships involving the sizes of corresponding angles and the lengths of corresponding sides in two similar polygons.

2. Copy parts (a) and (b) onto the same piece of paper (start your copy at the left side of the paper). Find the image of the line segment in each part, for the dilation given.

a. center P, scale factor 2.5   **b.** center P from part (a), same scale factor 2.5

P

3. In Exercise 2, what would be different if the center of the dilation were halfway between the two segments?

4. a. For the dilation indicated, with center C, describe the image of a 28 cm by 56 cm rectangle.

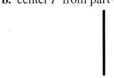

$\vdash$——4 cm——$\dashv\vdash$——3 cm——$\dashv$
C A A'

continue

b. What, if anything, would change if in part (a) points A and A' were interchanged?

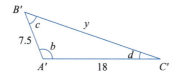

5. Find the missing measurements in the following similar triangles. Length measurements are in centimeters. (Sketches are not drawn to scale.)

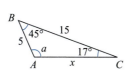

6. Find the missing measurements in the following similar triangles. Lengths are in meters. (Sketches are not drawn to scale.)

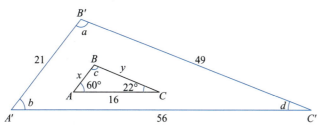

7. Find the missing measurements in the similar triangles at the right. Lengths are in feet. (Sketches are not drawn to scale.)

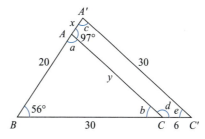

8. The following parallelograms are similar. Find the missing measurements. Lengths are in centimeters. (Sketches are not drawn to scale.)

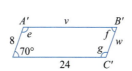

9. In each part, give a value that fits the description.
   a. $1\frac{1}{2}$ times as much as $12   **b.** $1\frac{1}{2}$ times more than $12   **c.** 150% as much as $12
   d. 150% more than $12   **e.** 4 times heavier than 8 lb   **f.** 4 times as heavy as 8 lb
   g. 400% heavier than 8 lb   **h.** 400% as heavy as 8 lb

10. Go to http://nlvm.usu.edu and click on the Geometry 3–5 (or 6–8) box. Scroll to and click on Transformation–Dilation. Try the different activities given in the right margin (use the > button).

20.2 More About Similar Figures

You know two methods for creating similar figures: (1) Apply the two criteria (make corresponding angles the same size, and use the same scale factor in changing the lengths), and (2) use the ruler method for performing a dilation. This section discusses other interesting results that arise once you have similar figures and also a very easy way of knowing that two triangles are similar.

Notes

ACTIVITY 6   Finding Missing Measurements

1. Suppose that the given original triangle *PQR* is similar to triangle *P'Q'R'* (not shown) with scale factor 5. What are the sizes of the angles and sides of triangle *P'Q'R'*? The units for lengths are kilometers (km).

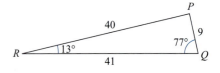

2. Find the perimeters (distances around) of triangles *PQR* and *P'Q'R'* and compare them. ●

> **Think About ...**
>
> Why are the perimeters of similar polygons also related by the scale factor?

Although the reason may be difficult to put in words, a form of the distributive property—e.g., $5p + 5q + 5r = 5(p + q + r)$—gives a mathematically pleasing justification. Do you see the two perimeters in the equation?

With lengths and perimeters of similar shapes related by the scale factor, a natural question is: How is the area of a polygon related to the area of its image, for a dilation? With a dilation drawing, you can conjecture the answer *without having to figure out the values of the two areas*! Examine the following figure, which uses 2 as the scale factor; you can see the relationship without finding the area of either triangle.

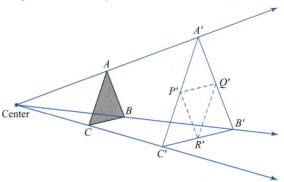

In the image the four copies of the original triangular region suggest that with a scale factor 2, the area of the image is $4 = 2^2$ times the area of the original.

> **Think About ...**
>
> For a dilation with scale factor 3, the area of the image of a shape is ____ times as large as the area of the original shape. (Make a rough sketch, using a triangle.) Make a conjecture for dilations with other scale factors.

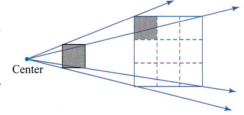

In summary, the perimeters of similar shapes are related by the scale factor (*k*, say), but their areas are related by the square of the scale factor (k^2).

Determining Similarity of Triangles

So far we have examined creating similar shapes (Section 20.1) and the relationships that exist when two shapes are known to be similar (Section 20.1 and the previous Think Abouts). But how would you know whether two triangles are similar, especially if both could not be in your field of vision at the same time? After thinking about it, your first response would likely be, "Measure all the angles and sides in both triangles. See whether they can be matched so that corresponding angles are the same size and the ratios of corresponding lengths are all the same (which would give the scale factor)." And you would be correct. Indeed, you would have described a method that could be applied to polygons of any number of sides, not just triangles. But for triangles we are especially lucky; we need to find only two pairs of angles that are the same size.

Notes

> Two **triangles** are similar if their vertices can be matched so that two pairs of corresponding angles have the same size.

The assertion is that the third pair of angles and the ratios of corresponding lengths take care of themselves. For example, if one triangle has angles of 65° and 38° and the other triangle does too, then even without any knowledge about the other angle and the sides, the two triangles must be similar. One triangle is the image of the other by a dilation, along with possibly some sort of movement like a reflection or a rotation. But will you also know the scale factor? The answer is "No." Finding the scale factor involves knowing the lengths of at least one pair of corresponding sides.

EXAMPLE 5

Using the information given in the following triangles, (a) tell how you know that they are similar, (b) find all the missing measurements, and (c) give the ratio of their areas. The triangles are not drawn to scale.

Solution

a. The 45° pair and the 17° pair assure that the triangles are similar.

b. It is easy to find the sizes of the third angles from the angle sum in a triangle: 118°. To find the missing lengths, we need the scale factor. The drawing here makes it easy to find corresponding sides. The known lengths 15 cm and 24 cm are

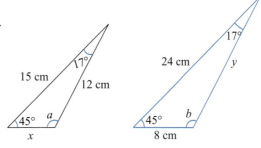

corresponding lengths. Using the left-hand triangle as the original, we get the scale factor = $\frac{24}{15}$ = 1.6. Then y is 19.2 cm, and x is 5 cm (from $1.6x = 8$).

c. The ratio of the areas is the square of the scale factor: $(1.6)^2 = 2.56$. The larger triangle has an area that is 2.56 times as large as that of the smaller triangle. ●

EXAMPLE 6

Given the information in the following drawings, find the missing measurements and give the ratio of the areas of the triangles. The triangles are not drawn to scale.

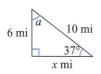
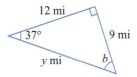

Solution

Angles a and b must have 53° because the right angles have 90° and the other given angles have 37°. The right angles and either the 53° angles or the 37° angles tell us that the two triangles are similar. Finding corresponding parts takes some care, and we need to know two corresponding lengths. Perhaps after other trials, we notice that both the 6-mi and 9-mi segments are opposite the 37° angles. So, the scale factor is 1.5. Using the scale factor, we find that y is 15, x is 8, and the ratio of the areas is $(1.5)^2$, or 2.25. ●

TAKE-AWAY MESSAGE . . .   The ratio of the perimeters of similar figures is the same as the scale factor, but the ratio of their areas is the square of the scale factor. Justifying that two triangles are similar is easy because you need to find only two pairs of angles that are the same size. Finding the correspondence in two similar figures can require some care.

Learning Exercises for Section 20.2

1. Summarize how the following items are related for a polygon and its image under a dilation: corresponding lengths, corresponding angle sizes, perimeters, and areas.

2. How would you convince someone that the ratio of the areas of two similar triangles is the *square* of the scale factor?

3. In each part, are the triangles similar? Explain how you know.

    a.

    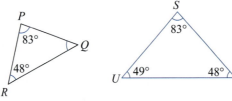

    b. Triangle 1: 50°, 25°   Triangle 2: 25°, 105°
    c. Triangle 1: 70°, 42°   Triangle 2: 48°, 70°
    d. Right triangle 1: 37°   Right triangle 2: 53°

4. In parts (a–d), find the missing lengths. Explain how you know the figures are similar and how you know which segments correspond to each other. (The sketches are not to scale.)

    a.

    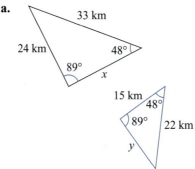

    b.

    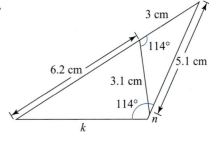

    c.

    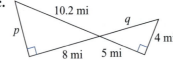

    d.

    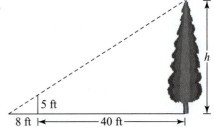

    e. Give the ratio of the areas of the triangles in parts (a) and (c).
    f. Devise a method for determining the width of a pond. [*Suggestion*: See part (d).]

5. A rectangle 9 cm wide and 15 cm long is the image for some size change of an original shape having a width of 4 cm.

    a. What is the scale factor of the size transformation?
    b. What type of figure is the original shape?
    c. What are the dimensions of the original shape?
    d. What are the areas of the original shape and the image? How are they related?

continue

e. If the description, *width* of 4 cm, in the original description was replaced by *one dimension* 4 cm, which of parts (a–d) could be answered differently? (*Hint*: What are the names for the dimensions of a rectangle?)

6. If the two large triangles in the diagram below are similar, find the scale factor. How many centimeters is *x*? (This sort of diagram is common in the study of light and lenses.)

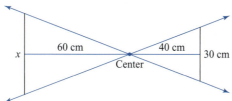

7. **a.** Your archaeological exploration has found a huge stone monument with the largest face being a triangular region. It appears to have the same proportions as a monument at another location. You telephone someone at the other location. What questions would you ask, at minimum, to determine whether the two triangles are similar?
   b. Suppose the situation in part (a) involves quadrilaterals. What questions would you ask, at minimum, to determine whether the two quadrilaterals are similar?

8. Go to http://illuminations.nctm.org, click on Activities, Grades 6–8, and scroll to Scale Factor. Try the activity.

Supplementary Learning Exercises for Section 20.2

1. Angle sizes in two triangles are given. Which pairs are similar?
   a. Triangle 1: 62°, 39°   Triangle 2: 39°, 79°
   b. Triangle 3: 92°, 38°   Triangle 4: 92°, 60°
   c. Triangle 5: 22° 35′, 74° 45′   Triangle 6: 74° 45′, 82° 40′
   d. Triangle 7: 39° 52′, 42° 8′   Triangle 8: 98°, 39° 52′

2. If the quadrilaterals are similar, find the missing measurements. Lengths are in meters. (*Hint:* What is the scale factor?)

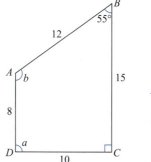

3. The following quadrilaterals are similar. Find the missing measurements.

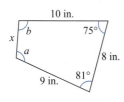
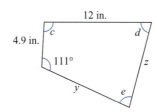

Notes

4. What is the ratio of the areas of the similar polygons in Exercises 2 and 3? What is the ratio of their perimeters? (*Remember:* You do not have to find the areas or perimeters to answer these questions.)

5. One side of a regular pentagon is 12 in. long. The perimeter of a similar polygon is 45 in.

    a. What is the best name for the second polygon?
    b. How long are the sides of the second polygon?
    c. What is the scale factor of the dilation for the two polygons?

20.3 Similarity in Space Figures

Similarity for three-dimensional figures is much like similarity for two-dimensional shapes. This section deals with the ideas and terminology as though they were new topics and should strengthen your understandings from earlier work.

 DISCUSSION 3   Related Shapes

Which of the shapes in the following figure would you say are related in some way? Is there another collection of the shapes that are related in some way?

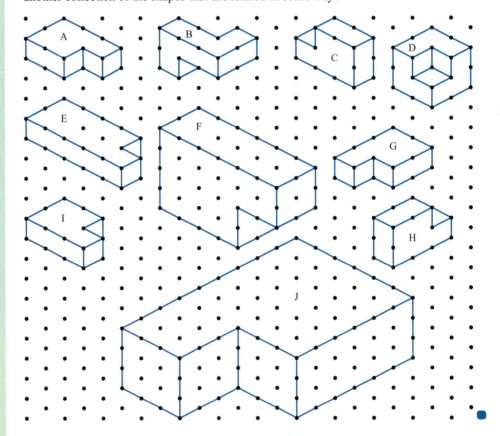

 ACTIVITY 7   Here's Mine

Make or sketch other shapes that you think would be related to shape B and to shape F above. Explain how they are related. ●

The rest of this section focuses on one of these relationships, **similarity.**

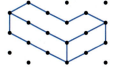

DISCUSSION 4   Larger and Smaller

An eccentric billionaire owns an L-shaped building like the one shown at the right.

a. She wants another building designed, "shaped exactly like the old one, but twice as large in all dimensions." Make or draw a model shaped like this second building that she wants.

b. Below are some diagrams. Which, if any, of the following drawings will meet the criterion? Explain.

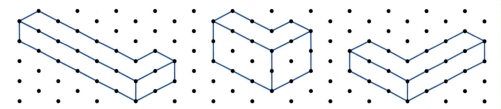

c. On isometric dot paper, show your version of a building (call it Building 2) that will meet the billionaire's criterion, and compare your drawing with those of others. Discuss any differences you notice.

d. Now the billionaire wants a third building (call it Building 3) designed to be shaped like the original, but "three-fourths as large in all dimensions." Make a drawing or describe this Building 3. ●

Buildings acceptable to the billionaire in Discussion 4 are examples of **similar shape**. The word "similar" has a technical meaning and is used when one shape is an exact enlargement (or reduction) of another. What "exact" means will come out of the next Think About.

> ### Think About ...
> Identify some quantities in the billionaire's original building. How are the values of these quantities related to the values of the corresponding quantities in Building 2?
>
> **a.** In particular, how are *new* and *original* lengths related?
> **b.** How are *new* and *original* angles related?
> **c.** How are *new* and *original* surface areas (the number of square regions required to cover the building, including the bottom) related?
> **d.** How are *new* and *original* volumes (the number of cubical regions required to fill the building) related?

For a given pair of buildings, you may have noticed that the angles at the corresponding faces are all the same size. You also may have noticed that the ratio of a *new* length to the corresponding *original* length was the same, for every choice of lengths. That is,

$$\frac{\text{length in one shape}}{\text{corresponding length in other shape}}$$

is the same ratio for all corresponding segments (if we assume the ratios are formed in a consistent fashion, each ratio starting with the same building). This ratio is called the **scale factor** for the enlargement (or reduction).

In short form, we can write

$$\frac{\text{new length}}{\text{original length}} = \text{scale factor}$$

An algebraically equivalent and useful form for similar figures is

$$(\text{new length}) = (\text{scale factor}) \times (\text{original length})$$

With either form, if you know two of the values, you can find the third one. Notice that if the scale factor is k, the last equation says that a new length is k times as long as the corresponding original length. The wealthy woman could have used the term *scale factor* in her requests: "Building 2 should be built with scale factor 2, and Building 3 should be built with scale factor $\frac{3}{4}$." When two polyhedra are similar, every ratio of corresponding lengths must have the same value, and every pair of corresponding angles must be the same size. Because lengths are affected by the scale factor, it is perhaps surprising that angle sizes do *not* change; i.e., corresponding angles in similar shapes will have the same sizes.

> Two three-dimensional shapes are **similar** if the points in the two shapes can be matched so that (1) every pair of corresponding angles have the same size, and (2) the ratios from every pair of corresponding lengths all equal the same value, called the **scale factor**.

So, corresponding lengths and angle sizes in similar figures are related. The relationships between surface areas and between volumes for similar three-dimensional shapes are important as well. From your results in the Think About, what conjectures are reasonable?

(new surface area) = _____ × (original surface area)

(new volume) = _____ × (original volume)

We end this section with this important point: For any scale factor, say, $\frac{\text{new}}{\text{original}} = \frac{6}{7}$, the ratio $\frac{6}{7}$ does *not* necessarily mean that new = 6 and original = 7. For example, *new* could be 60 and *original* could be 70, but the ratio $\frac{\text{new}}{\text{original}}$ would still equal $\frac{6}{7}$.

TAKE-AWAY MESSAGE . . . The following descriptions are all important relationships to know:

Quantities	How 2D similar shapes are related	How 3D similar shapes are related
Sizes of corresponding angles	equal	equal
Lengths of corresponding segments	ratios = scale factor	ratios = scale factor
Areas/surface areas	ratio = (scale factor)^2	ratio = (scale factor)^2
Volumes	(not applicable)	ratio = (scale factor)^3

Learning Exercises for Section 20.3

1. Summarize the relationships among length and angle measurements in similar polyhedra. In particular, how is the scale factor involved? How are the areas of similar polyhedra related? The volumes?

2. **a.** Now the billionaire from Discussion 4 wants two more buildings sketched, each similar to the original one. Building 4 should have scale factor $1\frac{2}{3}$, and Building 5 should have scale factor 2.5. Make sketches to show the dimensions of Buildings 4 and 5.
   b. What is the scale factor between Building 4 and Building 5?

3. Are any of the following shapes similar? Explain your decisions, and if two shapes are similar, give the scale factor. (Make sure that it checks for every dimension.)

 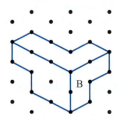 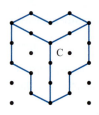

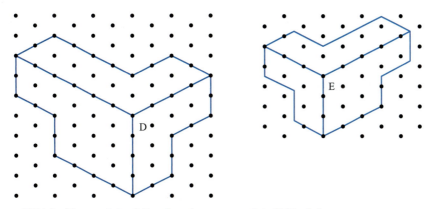

4. a. Which, if any, of the following shapes are related? Explain.

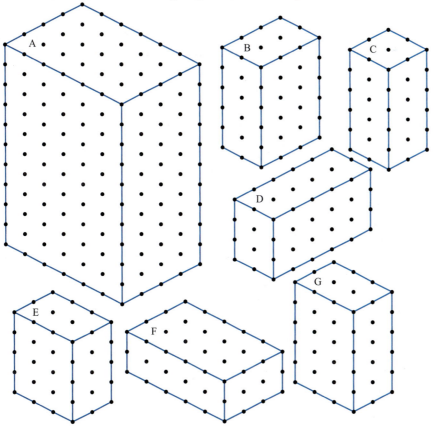

b. How many shapes like shape (B) would it take to make shape (A)?

5. Are any of the following right rectangular prisms similar to a right rectangular prism with dimensions 3 cm, 7 cm, and 8 cm (in other words, a 3 cm by 7 cm by 8 cm or a 3 cm × 7 cm × 8 cm one)? If they are similar, what scale factor is involved (i.e., is *every* ratio of corresponding lengths the same)? Explain your decisions, including a reference to corresponding angles of the two prisms.

a. 5 cm × 9 cm × 10 cm
b. 96 cm by 36 cm by 84 cm
c. 8.7 cm × 20.3 cm × 23.2 cm
d. 6 cm by 14 cm by 14 cm
e. 3 in. × 7 in. × 8 in.
f. 5 cm by 11.67 cm by 13.33 cm
g. 9 cm × 49 cm × 64 cm
h. 7 cm × 14 cm × 15 cm
i. 15 mm × 35 mm × 4 cm
j. Are the prisms in parts (b) and (c) similar?

6. a. A detailed model of a car is 8 in. long. The car is actually 12 ft long. If the model and the car are similar, what is the scale factor?

b. A natural history museum has prepared a 12-ft-long model of one kind of locust. They say the model is 70 times life size. What is the life size of this locust?

Notes

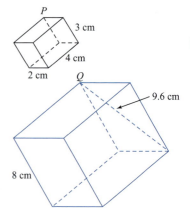

7. **a.** What other measurements of lengths and angles do you know about the given right rectangular prisms P and Q, if they are similar?

   b. What scale factor is involved if P is the original shape and Q the new one?

   c. What scale factor is involved if shape Q is the original and shape P the new one?

   d. In shape P how long is the segment that corresponds to the marked diagonal of the face in shape Q?

8. Olaf lives in a dorm in a tiny room that he shares with three others. He wants to live off campus next year with his friends, but he needs more money from his parents to finance the move. He decides to build a scale model of his dorm room so that when he goes home for break, he can show his parents the cramped conditions he lives in. He decides to let 1 inch represent 30 inches of the actual lengths in the room. His desk is a right rectangular prism 40 inches high, 36 inches long, and 20 inches wide. He decides that his scaled desk should be $1\frac{1}{3}$ inches high, but a roommate says it should be 10 inches high. Who is right and why? What are the other dimensions of the scaled desk?

9. Make sketches to guide your thinking in answering the following questions:

   a. How many centimeter cubes does it take to make a 2 cm by 2 cm by 2 cm cube?

   b. How many centimeter cubes does it take to make a 3 cm by 3 cm by 3 cm cube?

   c. Are the two large cubes in parts (a) and (b) similar? Explain.

   d. How do the surface areas of the two cubes compare? There are two ways to answer: (1) direct counting of squares to cover, and (2) the theory of how the scale factor is involved.

   e. How do the volumes of the two cubes compare?

10. **a.** Is a polyhedron similar to itself (or to a copy of itself)? If not, explain why. If so, give the scale factor.

    b. If polyhedron X is similar to polyhedron Y, is Y similar to X? If not, explain why. If so, how are the scale factors related?

    c. Suppose polyhedron X is similar to polyhedron Y and polyhedron Y is similar to polyhedron Z. Are polyhedra X and Z similar? If not, explain why. If so, how are the scale factors related?

11. **a.** Give the dimensions of a right square prism that would be similar to one with dimensions 20 cm, 20 cm, and 25 cm.

    b. What scale factor did you use?

    c. How does the total area of all the faces of the original prism compare with that of your prism? A rough sketch may be useful.

    d. How do the volumes of the two prisms compare?

12. Repeat Learning Exercise 11, letting the scale factor be r.

13. Why is it ambiguous to say, "This polyhedron is twice as large as that one"?

14. Give the dimensions of a right rectangular prism that would be similar to one that is 4 cm by 6 cm by 10 cm with

    a. scale factor 2.2   **b.** scale factor 75%   **c.** scale factor $\frac{4}{7}$

    d. scale factor $3\frac{2}{3}$ **e.** scale factor 100%   **f.** scale factor 220%

15. Suppose a 2 cm by 3 cm by 5 cm right rectangular prism is similar to another shape with scale factor 360%. What are the surface area and volume of the other shape? What are its dimensions?

16. You have made an unusual three-dimensional shape from 8 cubic centimeters and want to make another one five times as big for a classroom demonstration. If *five times as big* refers to lengths, how many cubic centimeters will you need for the bigger shape? Explain your reasoning.

17. Can a scale factor be 0? Explain.

18. Are the following shapes similar? If not, explain why. If so, tell how you would find the scale factor.

    a. a cube with 5-cm edges and a cube with 8-cm edges
    b. every two cubes, with their respective edges m cm and n cm long
    c. every two triangular pyramids
    d. every two right rectangular prisms
    e. a rhomboidal prism with all edges x cm long and a cube with edges $3x$ cm long

19. Why can't a cube be similar to any pyramid?

20. Given a net for a polyhedron, how would you make a net that will give a larger (or smaller) version of that polyhedron?

21. One student explained the relationship between the volumes of two shapes this way: "Think of each cubic centimeter in the original shape. It grows to a k by k by k cube in the enlargement. So, each original cubic centimeter is now k^3 cubic centimeters." Retrace her thinking for two similar shapes related by the scale factor 4, using a drawing to verify that her thinking is correct.

22. Suppose the scale factor relating two similar polyhedra is 8.

    a. If the surface area of the smaller polyhedron is 400 cm^2, what is the surface area of the larger polyhedron?
    b. If the volume of the smaller polyhedron is 400 cm^3, what is the volume of the larger polyhedron?

23. Polyhedron 1, which is made up of 810 identical cubes, is similar to polyhedron 2, which is made up of 30 cubes of that same size.

    a. What is the scale factor relating the two polyhedra?
    b. What is the ratio of the surface areas of the two polyhedra?

24. Legend: Once upon a time there was a powerful but crabby magician who was feared by her people. One year she demanded a cube of gold, 1 m on an edge, and the people gave it to her. The next year she demanded, "Give me twice as much gold as last year." When they gave her a cube of gold, 2 m on an edge, she was furious—"You disobedient people!"—and she cast a spell over all of the people. Why? And why should she have been pleased?

25. Make two nets for a cube so that the nets are similar as 2D shapes but also so that one net will have an area four times as large as that of the first net. When the nets are folded to give cubes, how will the volumes of the two cubes compare? (*Hint*: How will the lengths of the edges compare, in the two nets?)

26. The index finger of the Statue of Liberty is 8 ft long. Measure the length of your index finger, the length of your nose, and the width of your mouth. Use the information to predict the length of the Statue of Liberty's nose and the width of her mouth. What are you assuming?

Supplementary Learning Exercises for Section 20.3

1. "These three pyramids are similar." From this statement what do you know about the pyramids?

2. Are every two triangular prisms similar? Explain your decision.

3.	Each part gives the dimensions for two right rectangular prisms. Tell whether the two prisms are similar, and explain your decisions.

   a.	4 cm by 3 cm by 3 cm, and 6 cm by 5 cm by 5 cm
   b.	12 cm by 20 cm by 8 cm, and 9 cm by 15 cm by 6 cm
   c.	10 cm by 12 cm by 14 cm, and 20 cm by 24 cm by 21 cm
   d.	6 cm by 8 cm by 10 cm, and 12 cm by 15 cm by 9 cm

4.	One scale model of a space shuttle is 22 in. tall. The scale is 1:100.

   a.	About how tall is the space shuttle, in feet? (*Recall:* 1 foot = 12 inches.)
   b.	How does the volume of the model compare to that of the space shuttle?

5.	Give the dimensions of a right rectangular prism that would be similar to one that is 10 cm by 15 cm by 20 cm with scale factor

   a.	$2\frac{1}{5}$	b.	3.7	c.	125%

6.	What are the ratios of the areas and volumes of the right rectangular prisms to the original prism in each part of Exercise 5?

7.	Sketch and indicate the lengths in a triangular right prism similar to the one to the right, if the scale factor is

8 cm	7 cm

9 cm	3 cm

   a.	7.2	b.	0.85	c.	160%

8.	What are the ratios of the areas and volumes of the triangular right prisms to the original prism in each part of Exercise 7?

9.	Polyhedron X is made up of 5 identical cubes, and polyhedron Y is made up of 320 cubes of that same size. X and Y are similar.

   a.	What is the scale factor relating X and Y?
   b.	What is the ratio of the surface areas of X and Y?

20.4 Issues for Learning: Similarity and Proportional Reasoning

Similarity often comes up in the intermediate grades in the elementary school mathematics curriculum, but perhaps just as a visual exercise. Students are asked "Which have exactly the same shape?" for a collection of drawings. Occasionally, there is a little numerical work, usually with scale drawings and maps (and the latter may not be associated with similarity in the children's minds). Although the overall situation may be changing, there is much less research on children's thinking in geometry than on number work, with only a scattering of studies dealing with similar figures. Here is a task that has been used in interviews of children of various ages.[2]

(A drawing like the one shown here is given to the student, along with a chain of paper clips.) Mr. Short is 4 large buttons in height. Mr. Tall (deliberately not shown to the student) is similar to Mr. Short but is 6 large buttons in height. Measure Mr. Short's height in paper clips (he is 6 paper clips tall in the drawing actually used in the interviews) and predict the height of Mr. Tall if you could measure him in paper clips. Explain your prediction.

Mr. SHORT: 4 Buttons

Mr. TALL: 6 Buttons (not shown)

Would you be surprised to learn that more than half the fourth-graders (and nearly 30% of the eighth-graders) would respond something like this? "Mr. Tall is 8 paper clips high. He is 2 buttons higher than Mr. Short, so I figured he is two paper clips higher." Plainly, the students noticed the *additive* comparison of 6 buttons with 4 buttons, but they did not realize that it is the *multiplicative* comparison, the ratio, that is important for similar shapes. The younger children, of course, may not have dealt with similarity, and there may be developmental reasons why numerical work with similarity does not come up earlier in the curriculum. But the older students most likely *had* experienced instruction on proportions, yet they had not fully understood the idea of similar figures and/or the relevance of the ratio relationship in the Mr. Short–Mr. Tall task.

You, or someone else in your class, may have focused on the additive comparison in working with Activity 1 ("A Puzzle About a Puzzle") so the lack of recognition of the importance of multiplicative comparisons and proportional reasoning for similar shapes clearly can continue beyond Grade 8.

On the other hand, once students do learn about proportions, they tend to overuse them. For example, does each of the following story problems involve multiplicative reasoning and proportions? (The problems are adapted cases from a Grades 4–6 research study,[3] p. 196.)

1. ▶ Ellen and Kim are running around a track. They run equally fast, but Ellen started later. When Ellen has run 16 laps, Kim has run 32 laps. When Ellen has run 48 laps, how many has Kim run? ◀

2. ▶ A group of 25 musicians plays a piece of music in 75 min. Another group of 35 musicians will play the same piece of music. How long will it take this group to play it? ◀

3. ▶ The locomotive of a train is 12 m long. If there are 4 rail cars connected to the locomotive, the train is 44 m long in total. If there were 10 rail cars connected to the locomotive, how long would the train be? ◀

Applying (inappropriately) a proportion to each of the problems gives respective incorrect answers of 96 laps, 105 min, and 110 m.

Think About ...

What *are* the correct answers to the given three story problems? How does Problem 3 involve multiplicative reasoning, in part?

Fourth-graders have had little, if any, work with proportions, whereas sixth-graders have. The researchers found that on problems such as Problems 1 and 2, fourth-graders did somewhat *better* than sixth-graders, with the sixth-graders relying much more, and incorrectly, on proportions. On all three problems above, the sixth-graders incorrectly tried proportions more than the fourth-graders. Perhaps the sixth-graders have learned an unintended and incorrect lesson: If the problem has three numbers in it, use a proportion. Perhaps sixth-graders are just more adept at noticing how numbers are related—e.g., 75 is 3 times as many as 25, and 10 is 2.5 times as much as 4. In any case, the curriculum and teachers must avoid overselling the value of proportions by giving practice in identifying whether particular situations with three numbers do involve multiplicative relationships exclusively.

TAKE-AWAY MESSAGE ... Additive comparison seems to be natural, perhaps from many occurrences outside of school, but multiplicative relationships and proportional reasoning, even beyond those in similarity of shapes, may require schooling. Once proportions are introduced, children tend to overuse them, so the schooling must be carefully planned.

Learning Exercises for Section 20.4

1. Researchers[3] have noted the apparent influence of the numbers in a story problem. Why do you think the second problem below is more difficult than the first one, at Grades 4–6? Both are adapted from the research study.

   a. ▶ In the shop, 9 boxes of pencils cost $27. The school wants to buy 18 boxes. How much does the school have to pay? ◀

   b. ▶ In the shop, 9 boxes of pencils cost $24. The school wants to buy 12 boxes. How much does the school have to pay? ◀

2. Interestingly, numbers that do not have multiplicative relationships that are whole numbers can, in some situations, perhaps slow down the solvers enough that they have to stop and think. Without knowing that, which do you think would be more difficult at Grades 4–6, Problem 1 or Problem 2?

▶ Ellen and Kim are running around a track. They run equally fast but Ellen started later.

Problem 1   When Ellen has run 16 laps, Kim has run 32 laps. When Ellen has run 48 laps, how many has Kim run?

Problem 2   When Ellen has run 16 laps, Kim has run 24 laps. When Ellen has run 36 laps, how many has Kim run? ◀

20.5 Check Yourself

You should be able to work problems like those assigned and to meet the following objectives:

1. Appreciate that creating an enlargement or reduction of a shape involves a particular relationship among any pair of corresponding lengths.

2. More completely, know the criteria for two figures to be similar (the two angles in every pair of corresponding angles have the same size, the lengths of every pair of corresponding segments are related by the same scale factor).

3. Use the two criteria for similarity in determining missing angle sizes and lengths in similar shapes. Applications involving similarity (e.g., photographs, maps, scale drawings, and time lines) might come up.

4. Create similar polygons with the ruler method for dilations and with ruler and protractor (Section 20.1).

5. Distinguish between, and use correctly, such phrases as "times as long as" versus "times longer than," or "85% as big as" versus "85% bigger than."

6. State and use the relationships between the perimeters and the areas of two similar 2D figures.

7. Use the work-saving way of telling whether two triangles are similar.

8. Extend the ideas of similarity with 2D figures to 3D shapes. That is, be able to tell whether two given 3D shapes are similar, and for 3D shapes that are known to be similar, find missing angle sizes, lengths, surface areas, and volumes. State the relationships between the surface areas and volumes of similar 3D shapes.

9. Illustrate a difficulty that children may have with multiplicative thinking (Section 20.4).

References for Chapter 20

[1]Warfield, V. M. (2007). *Invitation to didactique*, p. 48. Philadelphia: Xlibris.

[2]Karplus, R., & Peterson, R. W. (1970). Intellectual development beyond elementary school II: Ratio, a survey. *School Science and Mathematics, 70*(9), 813–820.

[3] van Dooren, W., de Bock, D., Evers, M., & Verschaffel, L. (2009). Students' overuse of proportionality on missing-value problems: How numbers may change solutions. *Journal for Research in Mathematics Education, 40*(2), 187–211.

Chapter 21

Curves, Constructions, and Curved Surfaces

Our attention up to this point has been on shapes made with line segments, rays, or planar regions. But, as you know, straightness and flatness do not model many real-life objects well. In this chapter we first treat curves that do lie in the same flat surface, with the most attention given to circle-related vocabulary (including some angles) and constructions. Then we deal with several ideas associated with curved three-dimensional shapes. Finally, we present the Standards for Mathematical Practice and their description of mathematically proficient students.

21.1 Planar Curves and Constructions

Springs and the Slinky toy are examples of familiar curves in space, but our attention will first be on *planar curves*, curves that lie in a single flat surface. Draw almost any squiggly mark on a piece of paper, and you will have drawn a curve. (As you may know, in advanced mathematics it is sometimes con-venient to regard line segments and polygons as curves also.) Here are some important curves in mathematics; do you know names for them? (Notice that two are shown on coordinate systems, which are not part of the curves.)

Our focus will be on the circle, a shape you know well. Most of the associated vocabulary is familiar to you, but the following drawings may suggest one or two new ideas.

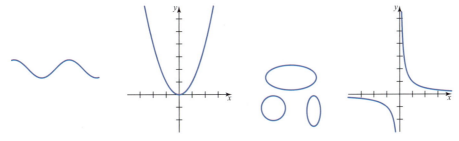

Circle, with center   Circular region   Radii (singular: radius)

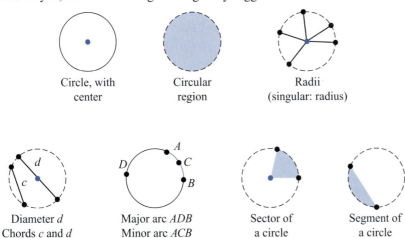

Diameter *d*   Major arc *ADB*   Sector of   Segment of
Chords *c* and *d*   Minor arc *ACB*   a circle   a circle

> *Think About ...*
> How are a chord and a diameter alike? How is a sector of a circle different from a
> segment of the circle?

 DISCUSSION 1   Properties of Circles

1. What are some properties associated with a circle? The properties might involve radii, diameters, arcs, symmetries, degrees, etc.

2. What are some quantities associated with a circle?

3. Are every two circles similar in the technical sense? Explain. ●

An angle with its vertex at the center of a circle cuts off (intercepts) a piece of the circle, called an **arc** of the circle. Sometimes, an arc is measured by the size of that angle, rather than by its actual length. For example, the two arcs cut off by the 54° angle below might each be described as arcs of 54°, even though they have different lengths. The actual *length* of a 54° arc will depend on the circle it is a part of. A common notation for the arc with endpoints A and B is $\overparen{AB}$. If the major arc is intended, a third point should be used: $\overparen{ADB}$.

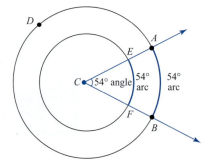

Notice that the number of degrees in major arc $\overparen{ADB}$ is 360° − 54°, or 306°.

An angle with its vertex at the center of a circle is a **central angle** of the circle, so a central angle and the arc it cuts off have the same number of degrees. Angle ACB has size 54°.

 ACTIVITY 1   A Conjecture About Inscribed Angles

An angle with its vertex on a circle and its sides along chords of the circle is called an **inscribed angle**. An inscribed angle intercepts an arc, which is the arc inside the opening of the angle.

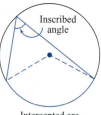

Intercepted arc

1. Measure the inscribed angle and the central angle for the intercepted arc. What relationship between the two measurements might apply? (You may have to extend the sides of the angles.)

2. In other circles, draw inscribed and central angles that intercept the same arc. Does the size of an inscribed angle always appear to be related to half the number of degrees in its intercepted arc? (Recall that an arc can be measured by the number of degrees in its related central angle.) ●

The Greeks considered the circle to be a "perfect" shape. Part of their admiration may have grown out of their knowledge of what can be accomplished with circles (drawn by a compass) and with lines (drawn with an unmarked ruler, or a straightedge). But other cultures have attached importance to the circle as well. For example, from the Navajo Indian viewpoint, "Everything an Indian does is in a circle, and that is because the Power of the World always works in circles, and everything tries to be round. . . ."[1]

Below and on the following pages are some examples of how straight lines and circles, drawn with a straightedge and a compass, can be used to undertake many geometric tasks without any need for measuring tools like a protractor or marked ruler.

The challenge:

Bisect ∠A below (i.e., "cut" it into two angles of the same size). You want to find the **angle bisector** of ∠A.

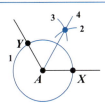

Step 1. Use A as the center and any radius to draw a circle.

Step 2. Draw arc with center X and a long enough radius.

Step 3. Draw arc with center Y and the same radius used in Step 2.

Step 4. Draw a ray through A and the point where the arcs intersect. That ray will be the bisector of the angle.

Think About …

Do you see the reflection symmetry in the construction of the angle bisector?

The challenge:

Copy angle B at point V on a straight piece.

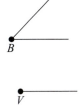

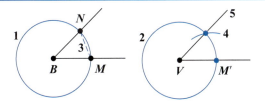

Step 1. Use B as the center and any radius to draw a circle.

Step 2. Use V as the center and the same radius as in Step 1 to draw a circle.

Step 3. Spread the compass to match the length of segment MN.

Step 4. Use that as the radius and M' as the center to draw an arc.

Step 5. Draw a ray through V and where the arc and the circle intersect. The resulting angle is a copy of ∠B.

The challenge:

Locate the midpoint of line segment *AB* below. (The construction also gives a line making right angles at the midpoint; hence, it is the **perpendicular bisector** of segment *AB*.)

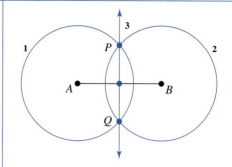

Step 1. Use *A* as the center and any radius longer than half the length of segment *AB* to draw a circle.

Step 2. Use *B* as the center and the same radius to draw a circle.

Step 3. Join the points where the circles intersect. That line intersects *AB* at its midpoint (and makes right angles with it).

Think About …

Do you see the reflection symmetry in the construction of the perpendicular bisector?

The challenge:

Locate the line that makes right angles with (is **perpendicular** to) line *CD* and goes through point *P*.

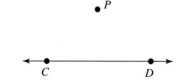

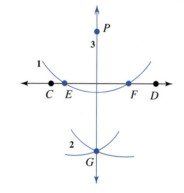

(The same steps work if point *P* is *on CD*.)

Step 1. Use *P* as the center and any radius long enough to give a circle cutting *CD* in two points *E* and *F* (you may have to extend *CD*).

Step 2. Using *E* and *F* as centers and the same radius for each new circle, draw arcs meeting at point *G*.

Step 3. Draw the line through *P* and *G*. It is perpendicular to *CD*.

Think About …

Do you see the reflection symmetry in the construction of the perpendicular?

Think About ...

Study the finished constructions for the angle bisector, the perpendicular bisector, and the perpendicular to a line. How do the symmetries give the desired results?

Notice Exercises 2 and 3 in this Grade 6 book. Try them.

Date _____ Time _____

LESSON 5·8 **Compass-and-Straightedge Constructions**

Use only a compass, a straightedge, and a sharp pencil. Use rulers and protractors only to check your work. Do not trace. Make your constructions on another sheet of paper. When you are satisfied with a construction, cut it out and tape it onto this page.

1. Copy this angle. When you are finished, check your work with a protractor.

2. Copy this quadrangle.

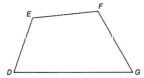

3. Construct a triangle that is the same shape as triangle *ABC* below but has sides twice as long.

192

Measurement and Geometry 2.0, 2.3

The final basic construction allows you to draw parallel lines.

The challenge:

Locate the line that passes through point *P* and is parallel to line *GH*.

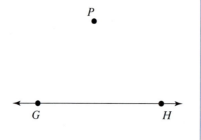

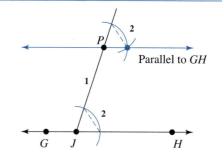

Parallel to *GH*

Step 1. From any point *J* on *GH*, draw ray *JP*.

Step 2. Copy one angle at *J* (e.g., ∠*PJH*) at point *P* so that one side is along the ray and in a corresponding position. The second side of this angle is the desired parallel.

When you glance at the above construction of the parallel lines, it looks as though the work at *J* has been slid up to *P* in the direction and distance going from *J* to *P*.

It is interesting that the straightedge-compass constructions do not actually involve making any measurements such as 5 cm or 48°. For example, if you bisect an angle that happens to be 80°, you will get an angle of 40°. Measuring with a ruler and protractor allows different—and natural—procedures.

 ACTIVITY 2   Ruler and Protractor Techniques

On separate paper, using your (metric) ruler and protractor (not a compass), draw a triangle that meets each set of specifications.

a. Its angle sizes are 65°, 75°, and 40°. (Why wasn't it necessary to mention the size of the third angle?)

b. The sides of the triangle making a 50° angle have lengths 5 cm and 8 cm.

c. The triangle has angles of 60° and 80° that share a side of length 6 cm.

Now compare your triangles with those of others. Are they all the same? ●

It is likely that in comparing your triangles in Activity 2 with those of others, you found that part (a) gives only triangles that appear to be similar, whereas parts (b) and (c) give triangles that are the same size and shape (i.e., **congruent** triangles) but perhaps flipped over. You may recognize part (b) as SAS and part (c) as ASA from secondary school.

TAKE-AWAY MESSAGE ... Circles, one type of planar curve, are surprisingly rich in related ideas and in the constructions they allow us to do. A ruler and a protractor alone allow us to make copies in some circumstances.

Learning Exercises for Section 21.1

1. Tell whether each statement is always true, sometimes true, or never true. Support your decisions.

a. A radius of a circle can have both endpoints on the circle.

b. A diameter of a circle is a chord of the circle.

continue

Notes

    c. A chord of a circle is a diameter of the circle.

    d. A radius of a circle has a length that is half the length of a diameter of the circle.

    e. Two circles with the same length for radius are exactly the same size.

    f. The distance between the center of a circle and a point on the circle is the same as the length of the radius.

    g. The bisector of an angle is a line of symmetry for the angle.

    h. The perpendicular bisector of a line segment is a line of symmetry for the line segment.

2. Draw a circle. Show in your drawing and label the following items: a chord that is not a diameter; perpendicular radii; a minor arc with endpoints Q and R; and a major arc with endpoints Q and R.

3. Show that each of these curved shapes will tessellate the plane.

    a.    **b.**

4. Describe the reflection symmetries and the rotational symmetries of each shape.

    a. a circle

    b. the curve below (it continues indefinitely in each direction)

5. For each angle size, sketch a circular region and a central angle of the given size, and tell what part of the whole circular region the sector of the angle is.

    a. 90°   **b.** 270°   **c.** 120°   **d.** 80°

    e. 180°   **f.** 45°   **g.** 135°   **h.** 60°

6. For each given circle and without measuring, find the number of degrees in each labeled angle. Arc PQ has 84°. (*Hint:* Activity 1)

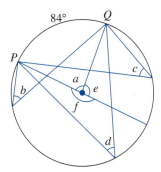 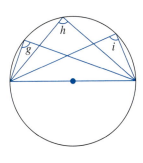

7. Give a partial justification for the likely conjecture about inscribed angles in Activity 1, using the special case shown below. (*Hints:* Angles in an isosceles triangle; exterior angle of triangle ABC)

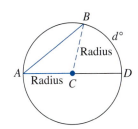

8. Using a compass and straightedge only, copy angle X and bisect it. Do the same for angle Y and angle Z.

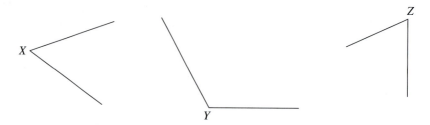

9. With a straightedge, sketch a scalene triangle. Then do the following:

  a. Using compass and straightedge only, find the midpoint of each side of the triangle.

  b. Draw segments joining each of the vertices with the midpoint of the opposite side. These segments are called the **medians** of the triangle. Does anything appear to be true?

10. In each diagram, copy the drawing and construct a line perpendicular to the given line and passing through the given point. Do not erase any key construction marks.

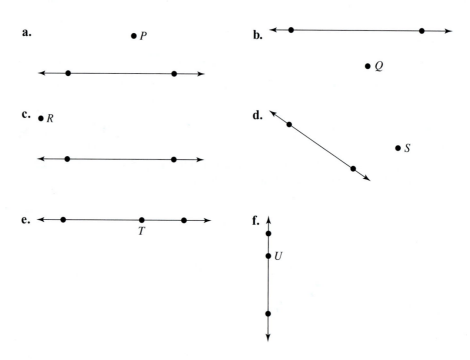

11. In each diagram, copy the drawing and construct a line that passes through the given point and is parallel to the given line. Do not erase any key construction marks.

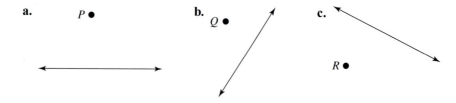

22. A line that touches a circle at exactly one point is called a **tangent to the circle**. A tangent is perpendicular to the radius drawn to that point. With a compass and straightedge, construct the tangents to circle C at points A and B.

Example of a tangent to a circle

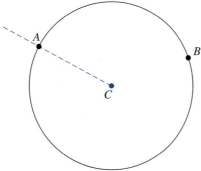

Supplementary Learning Exercises for Section 21.1

1. Name each object described for the circle with center C.

a. the region at a
b. the region at b
c. the angle at point C
d. the angle at point D
e. the piece of the circle at e

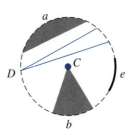

2. On the circle, sketch one of each of the following objects and label it:

a. a diameter (label it d)
b. a radius (label it r)
c. a sector of the circle (label it X)
d. a segment of the circle (label it Y)
e. a chord that is not a diameter (label it c)
f. a central angle (label it A)
g. an inscribed angle (label it B)
h. a (minor) arc $\overset{\frown}{PQ}$

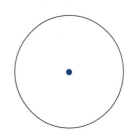

3. Give the size of each marked angle for the circles with centers C and D. Tell how you know. $\overset{\frown}{PQ}$ has $70°$. (*Hint:* Activity 1)

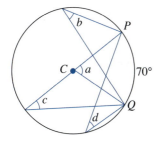

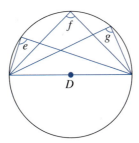

4. Draw line segment AB with a ruler, and pick any point C on the segment. Now pick any point D off the segment and draw ray CD.

a. Construct with a ruler and compass the bisector of $\angle ACD$ with ray CE. (Do not erase any key construction marks.)
b. Construct the angle bisector of $\angle BCD$ with ray CF. (Do not erase any key construction marks.)
c. Measure $\angle ECF$ with a protractor.

continue

d. Repeat parts (a–c) with different choices of points until you are ready to make a conjecture. Make the conjecture.

e. Give a line of reasoning to convince someone that your conjecture is true in general.

5. a. Construct (with a compass and ruler) the midpoints of the four sides of the rectangle below, and join them in order. (Do not erase any key construction marks.) Conjecture what type of quadrilateral the resulting shape is.

b. Give a line of reasoning showing that your conjecture in part (a) is true in general.

6. Construct with a compass and ruler a line of symmetry for a regular pentagon. (Do not erase any key construction marks.) How could you be certain that your line is indeed a line of symmetry?

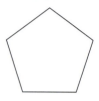

21.2 Curved Surfaces

Just as there are many planar curves, there also are many curved surfaces. Our focus will be on only three types: spheres, cylinders, and cones.

DISCUSSION 2   Perfectly Round

You are familiar with **spheres** (the surfaces of perfectly round balls, e.g.). What do you know about spheres? What do terms like center, radius, diameter, and **great circle** mean for a sphere?

A cylinder may be thought of this way (see the sketches in Figure 1). Trace around a given planar curve with a line, always keeping the line parallel to its earlier positions; then cut the resulting infinite surface with two parallel planes. The surface formed by the two planar regions and the part of the infinite surface between the two planes is a **cylinder**.

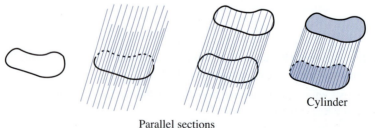

Parallel sections

Figure 1

The starting curve can be a circle, of course, and the tracing lines can make right angles with lines in the plane of the given curve. When both these conditions occur, we have the shape called the common **right circular cylinder**. For example, soda or soup cans are right circular cylinders. The distortion in two-dimensional sketches usually shows the circles as ellipses. Additional cross sections sometimes help show the curvature, as in Figure 2 shown below.

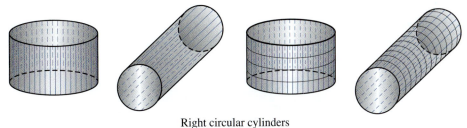

Right circular cylinders

Figure 2

As with prisms, the planar regions are called the **bases of the cylinder**, and the parallel bases are identical in size and shape (congruent). As with prisms, the line segments joining corresponding points on the two bases are not only parallel, but they also have the same length.

Think About …

Do you see that if the original "curve" were a polygon, the resulting "cylinder" would be a prism?

Cones can be described in a somewhat similar fashion. Again start with a curve and trace around the curve with lines, but this time have all the lines go through a fixed point. Cut through all the resulting surface with a plane. The surface between the fixed point and the cutting plane, along with the planar region, gives a **cone**. The fixed point is called the **vertex of the cone**, and the planar region is the **base of the cone**. If the original curve traced is a circle and all the segments joining the vertex to points on the base are the same length, then the cone is a **right circular cone**.

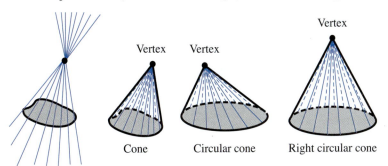

Cone   Circular cone   Right circular cone

Many objects in everyday use, if you can imagine the congruent bases, can be characterized as prisms and cylinders. You will probably begin to recognize objects, similar to those shown here, as prisms and cylinders or pyramids and cones.

Notes

Think About ...

In the description of a cone, what shape would you get if the original "curve" were a polygon?

 DISCUSSION 3   Shapes of Buildings

How houses are shaped is an interesting aspect of culture and of history. What shapes are houses in Western countries? What shapes are tepees, hogans, igloos, and yurts? Why are houses shaped as they are? ●

TAKE-AWAY MESSAGE ...   Spheres, cylinders, and cones are the basic three-dimensional curved surfaces and lead to other vocabulary useful in describing features of such curved surfaces.

Learning Exercises for Section 21.2

1. Tell whether each statement is always true, sometimes true, or never true. Support your decisions.

   a. A radius of a sphere can have both endpoints on the sphere.
   b. A radius of a sphere has a length that is half the length of a diameter of the sphere.
   c. A circular cone has reflection symmetry.
   d. A circular cylinder has rotational symmetry.
   e. Two spheres with radii of the same length are congruent.
   f. Two great circles of the same sphere have exactly the same size.
   g. Any two spheres are similar (in the technical sense).

2. Give the most informative technical description (e.g., right circular cone) for each description or drawing.

   a. a mailing tube   b. (the most common) lipstick containers
   c. a dunce cap   d. the sharpened end of a pencil
   e. the end of a sharp-pointed   f. (Assume circular
      carrot, cut at an angle   ends.)

3. Describe the reflection symmetries and the rotational symmetries of each shape.

   a. a right circular cylinder
   b. a right circular cone
   c. a sphere

4. a. Sketch a net for a right circular cylinder.
   b. Sketch a net for a right circular cone.
   c. Why do most flat maps of the world involve distortions of some regions of the world?

5. Where in the world can you find an example of the following objects?

   a. cylinder   **b.** cone
   c. sphere   **d.** great circle of a sphere

6. Sketch an example of each shape, if it is possible. If it is not possible, explain why.

   a. a circular cylinder that is not a right cylinder
   b. a cone that is not a circular cone
   c. a right circular cone
   d. a cylinder that has ovals as bases
   e. two nonintersecting great circles of a sphere
   f. half a sphere (a **hemisphere**), with a right circular cone having its vertex on the hemisphere and its base a great circular region

7. Prism is to pyramid as cylinder is to _____. Explain.

8. **a.** If the bottom layer of the cylinder shown holds $5\frac{2}{3}$ cups, how many cups would the whole cylinder hold? (What assumption did you make?)

   b. If 82.5 mL of water fills $\frac{3}{4}$ of another cylinder shaped like the one shown, how many milliliters of water will $\frac{3}{5}$ of the cylinder hold?

9. Sketch two spheres, one sphere with a radius twice as long as that of the other.

   a. How do the diameters of the two spheres compare?
   b. In your opinion, how do their volumes compare? Their surface areas? Explain.

10. **a.** Roll up a sheet of paper into a cylinder shape. If a plane cut through all the curved part, what shape(s) could the cross section be?
    b. Roll a sheet of paper into a conical shape. If a plane cut through all the curved part, what shape(s) could the cross section be?

11. Take a scrap sheet of paper and mark lines across the paper, parallel to the top and bottom of the sheet. Tape the top edge to the bottom edge. Assume that the open ends are covered.

    a. What shape do you have?
    b. Are the line segments you marked still parallel?
    c. Push the shape in (gently) to make a noncircular right cylinder.

12. *A curiosity* Cut off a long but narrow (about 1 to $1\frac{1}{2}$ inches) strip from a piece of scrap paper. Mark a line halfway between the longer edges.

    ```
    ┌ ─ ─ ─ ─ ─ ─ ─ ─ ─ ─ ─ ─ ─ ─ ┐
    └ ─ ─ ─ ─ ─ ─ ─ ─ ─ ─ ─ ─ ─ ─ ┘
    ```

    Twist one of the narrow ends 180° and tape it to the other end. The resulting curved surface is called a Moebius ("moy-bee-us") strip. Predict what you will get when you cut a Moebius strip along the line you marked. Then cut it to check your prediction.

Supplementary Learning Exercises for Section 21.2

1. Sketch an example of each shape. In your drawing, try to communicate the three-dimensional nature of the shape.

   a. a sphere and two great circles
   b. a circular cone in which the segments from the vertex to the base all have the same length

continue

    c. a circular cone in which the segments from the vertex to the base do *not* all have the same length

    d. a right cylinder that is not a circular cylinder

2. Look about you. What curved surfaces do you see?

3. Sketch a right circular cylinder with a hemisphere (half a sphere) at one base of the cylinder.

4. A cylindrical container is $\frac{3}{4}$ filled with $2\frac{1}{2}$ cups of water.

    a. How many cups would be required to fill the rest of the container?

    b. How many cups does half the container hold?

21.3 Issues for Learning: Standards for Mathematical Practice

The *Common Core State Standards* (CCSS)[2] consist of two intertwined parts: Standards for Content and Standards for Mathematical Practice. The Content Standards define grade-specific standards and give us an understanding of what concepts should receive instructional time at particular grade levels. The Standards for Mathematical Practices, however, describe the processes and proficiencies that students should develop for a sequence of topics and performances. They describe what the NCTM *Principles and Standards*[3] and some states describe as process standards, as well as what the National Research Council[4] describes as strands of mathematical proficiency. These are inclinations and habits of mind that we seek to develop in students as mathematicians at all grade levels. Because they describe varieties of expertise that mathematically proficient students exhibit, they are best understood by considering them along with examples from the mathematics of the content standards. The CCSS authors have provided full descriptions of these competencies at http://www.corestandards.org/assets/CCSSI_Math%20 Standards.pdf.

Mathematical Practices
1. Make sense of problems and persevere in solving them.
2. Reason abstractly and quantitatively.
3. Construct viable arguments and critique the reasoning of others.
4. Model with mathematics.
5. Use appropriate tools strategically.
6. Attend to precision.
7. Look for and make use of structure.
8. Look for and express regularity in repeated reasoning.

Through many of the earlier activities, you have been asked to reflect on properties, make conjectures about relationships regarding circles, inscribed angles, and triangles. As in other parts of the text, you were asked to give some justification for why some relationships might hold in general. Throughout the book, we have been encouraging you to make sense of aspects of mathematics that you may have been able to do procedurally, but may not have previously considered deeply. These practices are processes in which mathematicians routinely engage. We hope that as you teach, you will seek to develop mathematical proficiency in K–12 students.

Consider the following vignettes for a vision of what classrooms look like where such practices are supported.

Vignette 1. The teacher asks her fourth-grade class to solve 14×15. Rachel says that she and her partner Leah believe the product is 210. The teacher asks if everyone in the class agrees. Some students have gotten a different answer: 120.

Rachel: At first we thought the answer was 120, because 10 times 10 is 100 and 4 times 5 is 20. But then Leah said that this was not a reasonable answer because 12×10 is 120 and 14 is larger than 12 and 15 is larger than 10. The product should be larger than 120.

Teacher: What reasoning did you use to arrive at the answer 210?

Leah: Can I draw a picture and show you?

Teacher: That would be great.

Rachel and Leah go to the board and draw a rectangular area model to show the product of 14×15. Leah records the area of each rectangle.

	10	4
10	100	40
5	50	20

Rachel: The area of the largest rectangle is the product of 14×15. 14 is $10 + 4$. And 15 is 10 and 5. We add the areas of all the smaller rectangles to get the area of the rectangle that is 14×15. The 40 and 20 are easy to add. I borrow 10 from the 20 to build onto the 40 to make 50. Then I have 100 plus 50 + 50, which is again 100, then plus 10. So, 210.

The teacher records on the board:

$$100 + 50 + 40 + 20$$
$$100 + 50 + 40 + (10 + 10)$$
$$100 + 50 + (40 + 10) + 10$$
$$100 + 50 + 50 + 10$$

Teacher: We can write the 20 as $10 + 10$ and regroup the 10 with the 40. The 10 can be *associated* with 40. Why would we want to do that?

Jake: Adding 50 plus 50 is way simpler!

Mathematically proficient students begin by *making sense of problems*. Leah and Rachel began by estimating an answer and checked their answer to the problem against their estimate. Mathematically proficient students can check their answers by solving problems in more than one way, considering possible solution paths based on making sense rather than jumping to solutions.

Mathematically proficient students *construct viable arguments and critique the reasoning of others*. Rachel and Leah constructed an argument for the incorrectness of their initial estimate and their final correct reasoning. Mathematically proficient students *model with mathematics*. They modeled the problem using a rectangular area and used it to help them solve the problem and to communicate their reasoning. Rachel and Leah recognized the relationship between 40 and 50, sought to make the calculation easier (adding two 50s), and utilized the structure of the associative property of addition to make mental computations easier.

This illustrates that mathematically proficient students *look for and make use of structure.* As you reflect on these standards of practice, you should notice a distinction between these process standards and content standards. You may recognize some content standards in this vignette: understand concepts of area and relate area to multiplication and addition, multiply and divide within 100, use place-value understanding and properties of operations to perform multidigit arithmetic. In contrast, the fourth standard of mathematical practice, e.g., relates to choosing to model a multiplication problem with concepts of area to solve, as opposed to a content standard stating that they should be able to relate area to multiplication.

Vignette 2. In her fifth-grade classroom, a teacher presented the following problem:

Teacher: We are going to seat people with one chair on each side of a square table. We can put tables together, like so. When two tables are placed together, there is room for six chairs. When three tables are placed together, there is room for eight chairs. If we continue in the same manner, how many people can we seat with 6 tables side-by-side?

The students work for a time modeling the situation with square chips and drawings.

Deron: For 1, I got 4. For 2, I got 6. For 3, I got 8. For 4, I got 10. For 5, I got 12. For 6, I got 14. It's always two more than the one before. It's like every table that gets added adds a chair on the top and a chair on the bottom.

Teacher: Can you show us one way to record this? Deron, can you record this in a way that we can see the two seats that are added?

Deron writes:

No. of Tables	No. of Chairs
1	4
2	4 + 2
3	4 + 2 + 2
4	4 + 2 + 2 + 2
5	4 + 2 + 2 + 2 + 2
6	4 + 2 + 2 + 2 + 2 + 2

Teacher: What patterns and relationships do you see?
Deron: You could multiply instead of adding. Like, for the last two, $4 + 2 \times 4$ and $4 + 2 \times 5$.
Donte: There is always 4. But then every time you add a table, you add two more chairs.
Teacher: I like how you talk about how the number of tables is related to the number of chairs. That can be very powerful. Based on patterns you've discovered today, describe a relationship between tables and chairs. That relationship can help us answer questions like how many chairs I would be able to seat at 20 tables arranged in this way. Or 65 tables.

Students work for a time and Donte summarizes the relationship as follows: "For 20 tables, there are four seats and 19 times two seats, or for 65 tables there are four plus 64 times 2. It's like the first table has four chairs and the rest of the tables have two." Another student, Eva, uses her graphing calculator to plot points. Eva justifies her use of unconnected points by explaining that a connecting line would indicate there is a point at (1.5, 5) and this would mean that you could put five chairs at one-and-a-half tables, but one-and-a-half tables is not possible in this context.

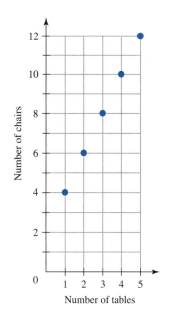

Mathematically proficient students *reason abstractly and quantitatively.* You may be familiar by now with the quantitative reasoning approach introduced in this book in sections on numbers and operations and algebraic reasoning. Here, we discussed that understanding a problem situation means understanding the quantities embedded in the situation and how they are related to one another. The students describe a relationship and *model with mathematics,* using tables, graphs, and words.

Mathematically proficient students *use appropriate tools strategically* and *attend to precision.* Initially, students used square tiles and drawings as tools to think about the patterns. They used the drawings and table to *look for and express regularity in reasoning.* Once they had discerned a pattern, Eva used her calculator, another tool, to plot points rather than a line. Eva and her classmates are able to interpret data in context, reasoning abstractly but contextualizing as needed to *make sense of the problem.*

It will take some time to understand the expertise we, as teachers, are seeking to develop in our students. In the meantime, we need to develop these processes and proficiencies in ourselves by making sense of problems, constructing viable arguments for our reasoning, using tools appropriately, and making sure we are precise in our use of mathematical terminology and notation.

21.4 Check Yourself

You should be able to work exercises like those assigned and to meet the following objectives:

1. Use the vocabulary associated with circles, spheres, cylinders, and cones. Some words should have already been familiar (center, radius, diameter, arc), but others may be less familiar to you (chord, major/minor arc, sector of a circle, segment of a circle, central angle, inscribed angle, hemisphere).

2. Use the relationship between the sizes of a central angle and an inscribed angle that intercept the same arc in the same circle.

3. Construct with a compass and straightedge, and with ruler and protractor: the bisector of a given angle, a copy of a given angle, the perpendicular bisector of a given line segment, the perpendicular to a given line through a given point, and the parallel to a given line through a given point.

4. Draw and recognize drawings of cylinders and cones, especially right ones.

5. Use whatever relationships have arisen during class and in the assigned exercises.

References for Chapter 21

[1]Ascher, M. (1991). *Ethnomathematics: A multicultural view of mathematical ideas*. Pacific Grove, CA: Brooks-Cole, p. 125.

[2]National Governors Association Center for Best Practices, Council of Chief State School Officers. (2010). *Common core state standards for mathematics*. Washington, DC: Author.

[3]National Council of Teachers of Mathematics. (2000). *Principles and standards for school mathematics*. Reston, VA: Author.

[4]National Research Council. (2001). *Adding it up: Helping children learn mathematics*. J. Kilpatrick, J. Swafford, & B. Findel (Eds.). Mathematics Learning Study Committee, Center for Education, Division of Behavioral and Social Sciences and Education. Washington, DC: National Academy Press.

Chapter 22

Transformation Geometry

You have already seen pieces of what is called *transformation geometry*. Your work with symmetry in Chapter 18 used movements, or transformations, such as reflections and rotations, with a focus on a particular shape. In Chapter 20 your work with similarities involved another type of transformation, the dilation. This chapter centers on the movements such as reflections and rotations that do not change the size or shape of a figure; these are often called **rigid motions**. Section 22.1 introduces the final two types of rigid motions, *translations* and *glide-reflections*. Sections 22.2, 22.3, and 22.4 treat more carefully the different types of rigid motions and introduce the idea of combining rigid motions. Section 22.5 extends some of the earlier work with tessellations, symmetry, and similarity. Section 22.6 examines visualization in the school curriculum.

22.1 Some Types of Rigid Motions

 ACTIVITY 1[1]   (Preliminary Homework) Let's Be Pick-y

Find all the different shapes that can be made from four toothpicks, subject to these two rules: (1) Toothpicks can touch only at endpoints, and (2) after the first toothpick, each toothpick either is perpendicular to the toothpick it touches or continues in the same straight line as the toothpick it touches. Record the different shapes you find, one shape to a box. Shapes are not different if they can be made to match. To get you started, two different shapes are shown.

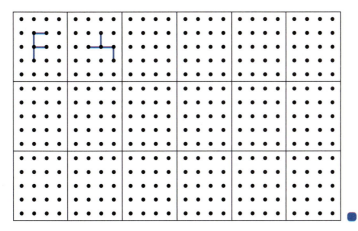

It is likely that you mentally or physically moved a four-toothpick shape to test whether it was different from one you already had, perhaps flipping it over. These mental or physical movements lead to a type of mathematics called **transformation geometry** and are called **rigid motions** (or **isometries**—*iso* means "same" and *metron* means "measure").

> A **rigid motion**, or **isometry**, is a movement that does not change lengths and angle sizes.

Think About …

What are some different types of rigid motions? What language would you use to describe how the shape shown at the start could end up at each of the other positions?

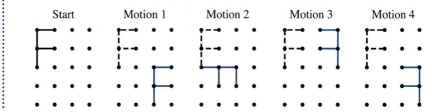

| Start | Motion 1 | Motion 2 | Motion 3 | Motion 4 |

Although you may have used the informal terms **slide**, **turn**, and **flip** or **mirror image** (terms that are sometimes used initially with elementary school children), the technical terms for the rigid motions involved in the Think About are **translation** (see Motion 1), **rotation about a point** (see Motion 2), and **reflection in a line** (see Motion 3). In a translation, each point moves in the same direction by the same distance, as the members in a marching band might. What about Motion 4? It looks as though the reflection in Motion 3 was further translated (slid) down. This type of combination of rigid motions in later work is called a **glide-reflection** and gives the fourth basic type of rigid motion.

 ACTIVITY 2   A Moving Experience

Pretend that the two-dimensional shapes below are on transparencies for an overhead projector. What shapes are the same? Which rigid motions, or isometries, help you decide?

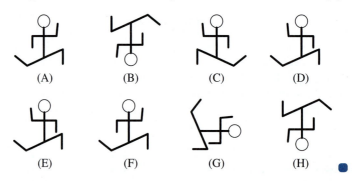

(A)   (B)   (C)   (D)

(E)   (F)   (G)   (H)

   So far, we have seen these types of isometries of the plane: *reflections* (shown in Activity 2, going from A to C), *rotations* (going from A to B), *translations* (going from A to F), and *glide-reflections* (going from C to F). If you started with shape A and imagined shape A being moved to shape F, you could call shape F the **image** of shape A under a translation: original → image. Because the shapes are identical, the words "original" and "image" help communicate which shape is the starting shape and which is the ending shape. Although we often focus on a particular shape, each rigid motion affects *every* point in the plane, not just the points in that shape.

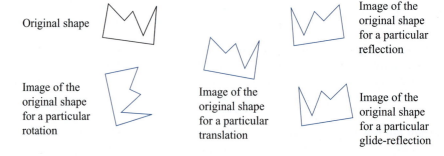

Original shape

Image of the original shape for a particular reflection

Image of the original shape for a particular rotation

Image of the original shape for a particular translation

Image of the original shape for a particular glide-reflection

What does the phrase "congruent figures" mean to you? Transformations allow a very general view of congruence for all figures, not just for triangles.

Two figures are **congruent** if one is the image of the other after some rigid motion.

This description allows shapes of any sort—segments, angles, polygons, curved regions, even three-dimensional figures—to be congruent. It also corresponds to a physical or a mental manipulation of one shape to see whether it matches the other shape exactly. A consequence of two figures being congruent is that any original part and its image part (referred to as *corresponding parts*, or *matching parts*, in high school) will always have the same measurements, if there is an appropriate measurement.

TAKE-AWAY MESSAGE . . . In the plane, there are four types of rigid motions (or isometries): reflections, rotations, translations, and glide-reflections.

This Grade 6 book uses the term *preimage* where we use *original*.

Wright Group, *Everyday Mathematics, Journal 1, Grade 6*, p. 178. Copyright © 2008, McGraw-Hill. Reproduced with permission of the McGraw-Hill Companies.

Learning Exercises for Section 22.1

1. Curricula in the middle grades can use these informal terms—*mirror image, slide, flip,* and *turn*—for the isometries. Which technical term goes with each informal term?

2. **a.** Try mental manipulation to tell which *single* type of rigid motion transforms shape *A* to each of the other shapes. It may help to draw a face or some mark on shape *A*. You may wish to check by tracing a copy of shape *A* and moving it to another shape.

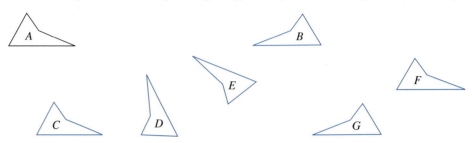

   b. Which of the shapes in part (a) are congruent? Explain.

3. What rigid motion is involved in each of the following situations? To retain the two-dimensional nature of the work in this section, assume that you are looking at a movie.

   a. a train moving along a straight track
   b. the motion of a fan blade
   c. a child sliding down a playground slide
   d. a clock hand moving
   e. a skateboarder skating in a circular bowl
   f. a doorknob moving
   g. one's two hands in crawling in a straight line

4. In parts (a–d) make a freehand sketch to show the image of the given triangle-rectangle shape for the stated isometry.

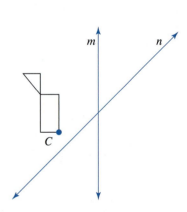

   a. a reflection in line *m*
   b. a reflection in line *n*
   c. a translation 6 cm to the right
   d. a rotation of 90° clockwise with center *C*
   e. Which of your shapes is congruent to the original shape? Explain.

5. The term *orientation* has a technical meaning that differs from everyday usage of the term. One can assign an orientation to a figure in this way: Pick any three noncollinear points on the figure—call them *P, Q, R*. Then the order *P–Q–R* assigns a clockwise or counterclockwise **orientation** to the figure (it may be helpful to think "*clock* orientation"). Different people, or different choices for the points, can assign a different orientation to the same figure. What is of value is knowing how the orientation of a figure is affected by each type of rigid motion. The orientation of the original and the orientation of its image might be the same, or they might be reversed. Make rough sketches or examine earlier ones to tell how the orientations of an original figure and its image are related for each type of rigid motion.

   a. translation   **b.** rotation   **c.** reflection   **d.** glide-reflection

6. Which of the isometries make sense for "moving" a three-dimensional shape?

7. Given four congruent isosceles right triangles, how many different shapes can you make, using all four triangles each time? Work mentally as much as possible, before you record and name the shapes you find. Which rigid motions did you use?

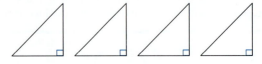

Supplementary Learning Exercises for Section 22.1

1. **a.** Give the technical names for the four types of rigid motions.
   b. Complete this definition: An isometry . . .

2. The shape shown to the right is to be rotated 83° clockwise with the center at point *P*. Can you say anything about length and angle measurements in the *image* of the shape, without performing the rotation? If so, tell what they are. If not, tell why.

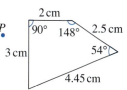

3. Name the single rigid motion that gives each lettered shape as the image of the original one. Explain how you decided.

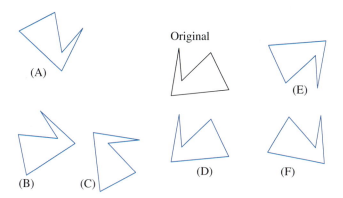

4. In Exercise 3 are the orientations (see Learning Exercise 5 in Section 22.1) the same for the original figure and each of its images?

5. In parts (a–d) make a freehand sketch to show the image of the given quadrilateral for the stated isometry.

   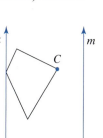

   a. a reflection in line *k*
   b. a reflection in line *m*
   c. a translation 3 cm straight down
   d. a rotation of 90° counterclockwise with center *C*
   e. Which of your images in parts (a–c) is congruent to the original quadrilateral? Tell how you know.

6. Using quadrilateral *A* as the original, tell what rigid motion gives each of the *B–H* congruent quadrilaterals as an image.

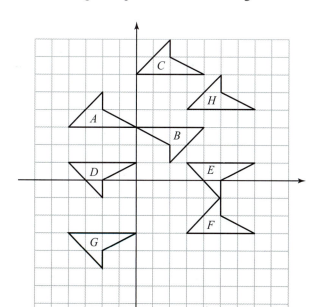

22.2 Finding Images for Rigid Motions

Earlier, your main task was to decide what rigid motion or isometry was involved when you were given a shape and its image. In this section your task is to find the image when you are given a shape and the type of rigid motion. With practice, drawing the image freehand can work out satisfactorily, or the structure provided by using grid paper can help in drawing images. In classrooms you may come across either computer software for finding the images of given figures for rigid motions or plastic devices, called MIRAs, for finding the images for reflections.

However, the paper-tracing methods of this section offer a low-tech (and therefore slower) means of finding images quite accurately and in a way that shows the motion involved. For making neater and more accurate drawings, use a ruler or the edge of a 3″ by 5″ card to draw line segments. Each paper-tracing method that follows uses two pieces of translucent paper. The thinnest paper you have can also work all right if you use a felt-tip pen.

First we illustrate a paper-tracing method for finding the image of a given figure for a particular translation. One way to describe a particular translation is to tell how far to move and in what direction. These two pieces of information can be communicated by a translation arrow, commonly called a **vector**. The length of the vector tells how far to translate, and the direction the vector is pointing toward tells the direction in which to translate.

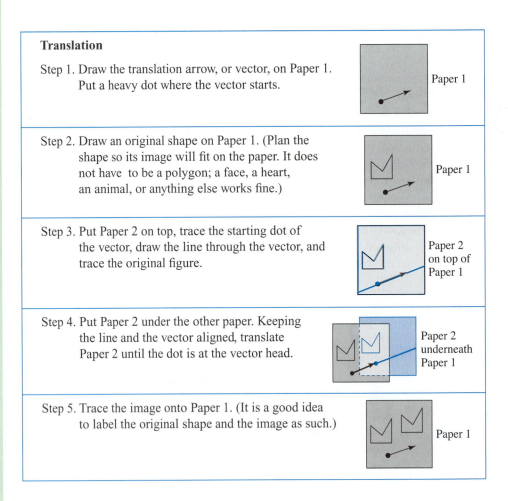

Translation

Step 1. Draw the translation arrow, or vector, on Paper 1.
Put a heavy dot where the vector starts.

Paper 1

Step 2. Draw an original shape on Paper 1. (Plan the shape so its image will fit on the paper. It does not have to be a polygon; a face, a heart, an animal, or anything else works fine.)

Paper 1

Step 3. Put Paper 2 on top, trace the starting dot of the vector, draw the line through the vector, and trace the original figure.

Paper 2 on top of Paper 1

Step 4. Put Paper 2 under the other paper. Keeping the line and the vector aligned, translate Paper 2 until the dot is at the vector head.

Paper 2 underneath Paper 1

Step 5. Trace the image onto Paper 1. (It is a good idea to label the original shape and the image as such.)

Paper 1

To find the image of a given shape for a reflection, you must know the line of reflection. Next, we illustrate a paper-tracing method for finding the reflection image of a shape.

Reflection

Step 1. Draw an original shape and the line of reflection on Paper 1. Put a heavy dot on the line of reflection for a reference point. (Plan so that the image will fit on the paper.)

Paper 1

Reference point on the line of reflection

Step 2. Put Paper 2 on top, and trace the figure, the line of reflection, and the reference point.

Paper 2 on top of Paper 1

Step 3. Flipping Paper 2 over and putting it underneath, align the lines of reflection and the reference points.

Paper 2 flipped over and underneath

Step 4. Trace the image onto Paper 1. Again, it is a good idea to label the original and its image.

Paper 1

To find the image of a given shape for a rotation, you must know (1) where the rotation is centered (the **center** of the rotation) and (2) an **angle** that shows the size and direction, clockwise or counterclockwise, of the rotation. We illustrate a paper-tracing method for accurately finding the rotation image of a given shape.

Rotation

Step 1. Draw an original shape on Paper 1. Pick a point for the center of the rotation, and draw the angle of the rotation so that its vertex is at this center. Plan ahead so the image will fit on Paper 1.

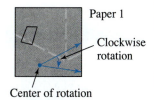

Paper 1

Clockwise rotation

Center of rotation

Step 2. With Paper 2 on top of Paper 1, trace the shape, the center of rotation, and the starting ray—the ray of the angle that would be rotated to give the second ray of the angle.

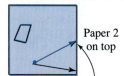

Paper 2 on top

Because rotation is clockwise, trace this ray and the shape.

continue

Step 3. Put Paper 2 underneath and align everything. With your pencil tip at the center of the rotation, turn the underneath paper in the proper clock direction until the starting ray is aligned with the second ray of the angle of rotation.

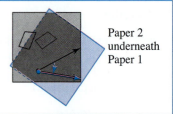

Paper 2 underneath Paper 1

Step 4. Trace the image onto Paper 1. (Again, it is a good idea to label the original and the image.)

Paper 1

ACTIVITY 3   Let's Move It

The aim of this activity is to make certain you can find images accurately using the illustrated methods (or perhaps some other methods your instructor may prefer). Each person in your group should choose one of the figures below, and then find its images for a translation, a reflection, and a rotation. (Do them separately because drawings can get cluttered quickly.) Choose your own vector (translation arrow), line of reflection, and center and angle of rotation.

You will practice the paper-tracing methods more in the Learning Exercises or on your own, as well as practice drawing images freehand and with dot paper. Finding the image for a glide-reflection is more involved; see Learning Exercise 5 in Section 22.4.

TAKE-AWAY MESSAGE . . .   Paper-tracing methods can help to find accurate images for translations, reflections, and rotations.

Learning Exercises for Section 22.2

1. Find the image for each indicated rigid motion using paper-tracing methods.
   a. the image of a shape of your choice for a translation with the vector shown at right
   b. the image of a shape of your choice for a rotation with a center of your choice and with a 90° clockwise angle
   c. the image of the following shape for a reflection in the line k:

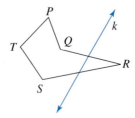

   d. the image of a shape of your choice for a rotation with a center on the shape and with a 90° counterclockwise angle

2. Copy and sketch *freehand*, as accurately as you can, the image of shape *G* for the rigid motion given in parts (a–c).

    a. a translation 5 cm east (label the image *H*)
    b. a 90° rotation clockwise, with center *X* (label the image *I*)
    c. a reflection in line *m* (label the image *J*)
    d. Check your freehand drawings.
    e. Which type of rigid motion is most difficult for you to visualize images?

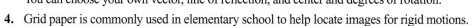

3. Choose another figure from Activity 3 and use tracing paper to find its images for a translation, a reflection, and a rotation. You can choose your own vector, line of reflection, and center and degrees of rotation.

4. Grid paper is commonly used in elementary school to help locate images for rigid motions.

    a. Find the image *S′* of the given pentagon *S* for a translation 4 units west.
    b. Find the image *S″* of the pentagon *S* for a rotation of 180°, with center *P*.
    c. Find the image *S‴* of the pentagon *S* for a reflection in line *m*.
    d. Find the image *S⁗* of the pentagon *S* for a reflection in line *n*.
    e. Find the image *S‴‴* of the pentagon *S* for a translation with vector *v*.

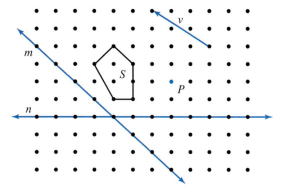

5. Find and label the image of the given quadrilateral shown below for each rigid motion described.

    a. image *A′* for a translation with vector *v*
    b. image *A″* for a translation with vector *w*
    c. image *A‴* for a reflection in line *j*
    d. image *A⁗* for a reflection in line *k* (shade it lightly)
    e. image *A‴‴* for a rotation of 90° counterclockwise, with center *Q*

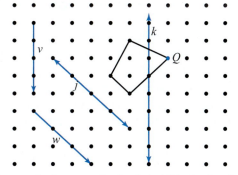

6. The regularity of isometric dot paper is also helpful in finding images. For each rigid motion described find the image of the flag shape *CAT* shown on the next page.

    a. translation with vector *v*
    b. reflection in line *m*
    c. rotation, 120° clockwise, center *C* [label it part (c)]
    d. rotation, 90° counterclockwise, center *T* [label it part (d)]
    e. rotation, 60° counterclockwise, center *A* [label it part (e)]

continue

Notes

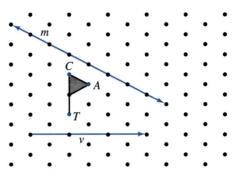

7. Copy the given quadrilateral *ABCD* on dot paper and sketch its image for a translation 6 spaces east. Label the image *A′B′C′D′*, where *A′* is the image of *A*, *B′* of *B*, etc. Draw the line segments joining *A* and *A′*, *B* and *B′*, *C* and *C′*, and *D* and *D′*. What do you notice? Does this make sense?

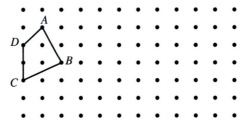

8. Using the same *ABCD* from Learning Exercise 7, sketch its image for a 90° clockwise rotation, using vertex *C* of the quadrilateral as the center of rotation. Label the image *A″B″C″D″*, where *A″* is the image of *A*, *B″* of *B*, etc.

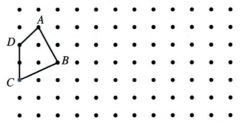

   a. Join each point *A*, *B*, and *D* to its image with a line segment. Are all these line segments the same length? Does this make sense for a rotation?
   b. Join each point *A*, *A″*, *B*, *B″*, *D*, and *D″* to the center *C* with a line segment. Which of these line segments have the same length? Does this make sense for a rotation?
   c. Examine the angles made by joining a point to the center *C* and then to the image of the point. Are all these angles the same size? How large is each of these angles? Does this make sense for a rotation?

9. Examine one of the reflections you made in previous exercises. Pick out three or four points on the original shape, and join each point to its image with a line segment. Are all these line segments the same length? What does appear to be true? Does this make sense for a reflection?

10. Which of the following vectors describe the same translation? Explain.

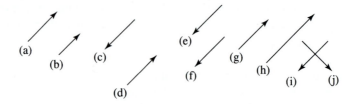

11. A rotation actually affects every point of the plane (imagine turning a sheet of stiff transparent plastic over the original). Which of the following rotations will have the same effect on every point of the plane? The center in each case is the same. Explain your decision.

    a. rotation of 90° clockwise   **b.** rotation of 90° counterclockwise
    c. rotation of 180° clockwise   **d.** rotation of 180° counterclockwise

12. Dot paper has some limitations when you wish to involve only points at the dots as vertices.

    a. Experiment with square dot paper to see which reflection lines allow the use of only points at the dots (being able to use the dots makes finding the images easier).
    b. Experiment with square dot paper to see which rotations allow the use of only points at the dots (which, again, makes finding the images easier).
    c. Experiment with isometric dot paper in ways similar to those in parts (a) and (b).

13. You and a friend are looking at a two-dimensional shape in a plane. Your friend is visualizing where the image of the shape would be for different rigid motions. For the questions that follow, what information would you need so that you could draw the image in the very same place your friend is visualizing it?

    a. Suppose that your friend is visualizing the image of the given shape for some translation. If you want to draw the image in exactly the spot your friend imagined it, what information about the translation would be essential?
    b. Suppose that your friend is visualizing the image of a given shape for some reflection. If you want to draw that image in exactly the spot your friend imagined it, what information about the reflection would be essential?
    c. Suppose that your friend is visualizing the image of a given shape for some rotation. If you want to find that image in exactly the spot your friend imagined it, what information about the rotation would be essential?

Supplementary Learning Exercises for Section 22.2

1. Copy the shape shown to the right and, using paper-tracing, find its image for each indicated rigid motion. Use separate paper for each part to avoid clutter.

    a. a reflection in a line of your choice
    b. a translation with a vector of your choice
    c. a clockwise rotation of 90° with a center of your choice
    d. a counterclockwise rotation of 90° with the same center as in part (c)

2. Sketch the image of *ABCD* for each of the following rigid motions:

    a. a translation with vector v to give $A'B'C'D'$
    b. a rotation of 90° clockwise, with center C, to give $A''B''C''D''$
    c. a reflection in line m to give $A'''B'''C'''D'''$
    d. a rotation of 180°, with center E, to give $A''''B''''C''''D''''$

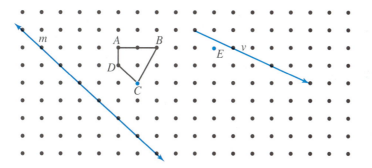

3. Using isometric dot paper, sketch the image of a quadrilateral *PQRS* for each of the following isometries:

    a. a reflection in a vertical line *x* of your choice
    b. a reflection in a horizontal line *y* of your choice
    c. a translation with a vector *w* of your choice
    d. a counterclockwise rotation of 90° with center *R*
    e. a clockwise rotation of 120° with a center *T* of your choice

4. Do these isometries give the same *final* location for every point?

    a. a 360° clockwise rotation with a center at point *X*
    b. a 360° counterclockwise rotation with a center at a different point *Y*
    c. a 0° rotation with a center at any point

5. Explain why the two vectors below describe the same translation.

22.3 A Closer Look at Some Rigid Motions

Some previous exercises should have suggested the following key relationships for some of the rigid motions. They are key relationships because, mathematically, knowing how a rigid motion affects each point is the most basic knowledge. In addition, these key relationships give a means of checking the results of rigid motions and of guiding the sketching of images. For someone who has difficulty with visualization, the key relationships give an alternative to strictly visual approaches. These key relationships also help in describing a rigid motion when only a shape and its image are given.

A common notation is to let *P′* represent the image of point *P*.

Type of rigid motion	Need to know	Key relationships
Reflection in line *m*	Line *m*	Line *m* is the perpendicular bisector of the line segment joining *P* and *P′*.
Rotation of *x*° with center *C*	Center *C*, *x*, and the clock direction	1. For each point *P*, the distance from the center *C* to *P* is the same as the distance from *C* to *P′*. 2. Every ∠*PCP′* has *x*°.
Translation	Distance and direction, or vector	All segments $\overline{PP'}$ have the same length and are parallel (the length and direction are summarized by the vector).

Glide-reflections receive more attention in Section 22.4.

 ACTIVITY 4   Practicing the Key Relationships

Using the original shape and its images shown on the next page, illustrate each of the key relationships given in the table above using one or more points. A sample point *P* and its image *P′* for the different rigid motions are shown.

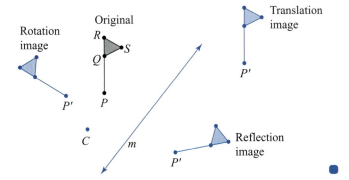

ACTIVITY 5 Constructing Images

With a compass and straightedge or tracing paper, construct the images of the given trian-gle below for the rigid motion described. (*Hint:* It may be a good idea to copy the pertinent parts for each on separate paper, because the drawing will get messy.)

a. a reflection in line m
b. a counterclockwise rotation of $x°$ with center C
c. a translation with vector v

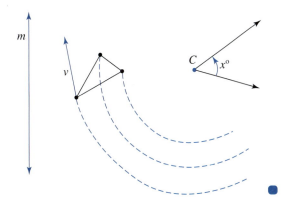

So far, the bulk of the work with rigid motions has involved finding the image, given a figure and information about a rigid motion. An important variation is this situation: Given a figure and its image, determine the information needed about the rigid motion involved. The key relationships help in describing a rigid motion *fully*: If the motion is a reflection, the key relationships help to find the line of reflection; if a rotation, the key relationships help to find the center and the angle; and if a translation, they help to find the vector.

> ### *Think About ...*
>
> Which rigid motions give Image 1 and Image 2 below? (Describe the rigid motion fully. For example, if it is a reflection, indicate the line of reflection.)

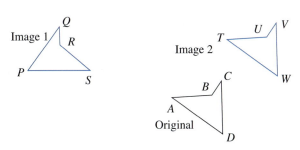

Notes

 ACTIVITY 6   Another Lost Rigid Motion

Which rigid motion gives the result shown below? Describe the rigid motion fully. For example, if it is a reflection, indicate the line of reflection.

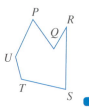

TAKE-AWAY MESSAGE . . .   Key relationships for the rigid motions give the underlying mathematical relationships for points and their images. The key relationships also can help in making freehand sketches or constructions of images of shapes. Furthermore, the key relationships can help to locate the full information about an unknown rigid motion.

Learning Exercises for Section 22.3

1. Draw a triangle with a ruler (and not too close to the edge of your paper). Then do the following:

    a. Use the key relationships to draw accurately the image of the triangle, for a translation 3.5 cm east. Write down how you used the key relationships. (*Hint*: Where are the images of the vertices of the triangle?)

    b. Use the key relationships to draw accurately the image of the triangle for a rotation of 90° counterclockwise, with the center at your choice of a point. Write down how you used the key relationships.

    c. With a new triangle and your choice for line *m*, use the key relationships to draw accurately the image of the triangle for the reflection in line *m*. Write down how you used the key relationships.

2. Trace the given shapes in each part. Then draw accurately the information needed for the motion described. (You may find it helpful to label points and their images.) Tell how you used the key relationships.

    a. The vector for the translation giving the following shape and its image. (Does it matter which shape is the original and which one is the image?)

Image

    b. The line of reflection for the reflection giving the following shape and its image. (Does it matter which shape is the original and which one is the image?)

          Image

continue

c. The angle and its clock direction for the rotation giving the following shape and its image. Measure the angle also. (Does it matter which shape is the original and which one is the image?)

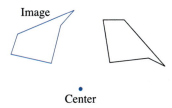

Image

•
Center

d. The line of reflection for the reflection of hexagon *RSTUVW* giving the image indicated.

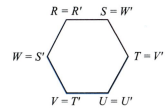

$R = R'$ $S = W'$

$W = S'$ $T = V'$

$V = T'$ $U = U'$

3. The shapes in parts (a–e) are different images of the given original shape. Trace the original and the image separately for each part to allow room for work. Identify the rigid motion involved, and then describe the rigid motion fully.

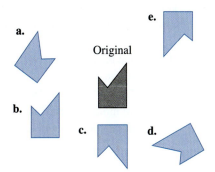

e.

a.

Original

b.

c.   d.

4. Notice that in Learning Exercise 2(d), point *R* is its own image. If the image of a point is the point itself, that point is called a **fixed point**. Tell where *all* the fixed points in the plane are, if there are any, for the following rigid motions:

   a. the reflection in some line *k*
   b. the translation 4 cm north
   c. the rotation with center *C* and angle 55° clockwise (*Hint*: There is at least one fixed point.)
   d. the rotation with center *Q* and angle 180°
   e. the rotation with center *M* and angle 360°

5. Copy the given figure and grid.

   a. Show the image of triangular region *ABC* for a 90° clockwise rotation with center *P*. Label it *A′B′C′*. (The key relationships may be helpful.)

continue

b. Now find the image of $A'B'C'$ (notice the primes) for a 90° *counter*clockwise rotation with center Q. Label it $A''B''C''$.

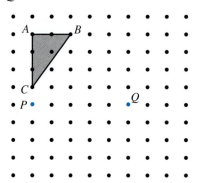

c. What single rigid motion will give $A''B''C''$ as the image of the original ABC? Describe it as completely as possible. (For example, if it is a reflection, what would be the line of reflection?)

6. a. If lines m and n are perpendicular, what is the image of line n for a reflection in line m?

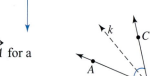

b. If k is the bisector of $\angle ABC$, what is the image of $\overrightarrow{BA}$ for a reflection in line k? The image of $\overrightarrow{BC}$?

Supplementary Learning Exercises for Section 22.3

1. On a blank sheet of paper, draw a triangle with a ruler (but not too close to any edge of the paper). Then use the key relationships to draw accurately the image of the triangle for each of the following rigid motions. Make clear how you used the key relationships.

   a. a translation 4 cm straight down the paper

   b. a rotation of 180°, with a center at one of the vertices of the triangle

   c. a reflection in some line that does not go through the triangle

In Exercises 2–5, identify the single rigid motion that is involved, and then locate the line of reflection (if reflection), center and degrees of rotation (if rotation), or vector (if translation).

2.

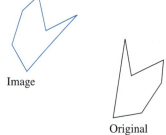

3.

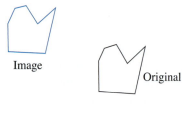

4.

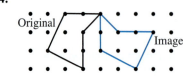

5. The drawing is on isometric dot paper.

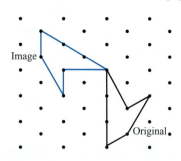

6. a. Find the image T' of the triangle T for a rotation of 120° counterclockwise, with center C, on the isometric dot paper. (*Hint:* Try to take advantage of the built-in 60° angles.)

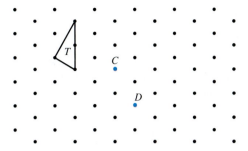

b. Now find the image T'' of the new triangle T' for a rotation of 120° *clockwise*, with center D.

c. What single rigid motion would give T'' as the image T?

22.4 Composition of Rigid Motions

Just as 8 and 2 can be *combined* to give another number—possibly 10, 6, 16, or 4, depending on whether the combining is done by adding, subtracting, multiplying, or dividing—rigid motions can be combined to give another rigid motion. The resulting isometry is called the **composition of rigid motions** (just as 16 is called the *product* of 8 and 2, e.g.).

The diagram in Figure 1 is an example of the composition of these two rigid motions: the translation, move 2 cm east, and the reflection in line k.

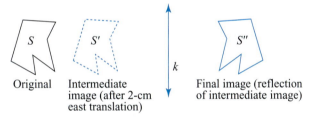

Original   Intermediate image (after 2-cm east translation) k Final image (reflection of intermediate image)

Figure 1

Although each of these two motions affects every point in the plane, let us focus on shape S. To arrive at the composition of these two rigid motions, first do the translation to get S'. The second motion, the reflection, then *starts* with S' as its original and gives S'' as the image of S'.

> The **composition** of two rigid motions is the *single* rigid motion that takes the original S to the final S''.

In this case, it looks as though a reflection in some new line m—a different line from line k—would give S'' as the image of S—i.e., the composition of the two given motions would be this single reflection. See Figure 2. (Notice that we use the original line k for the reflection and not its translation image, which is not even pictured here.)

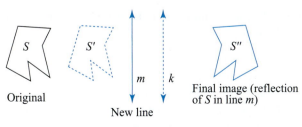

Original m k Final image (reflection of S in line m)

New line

Figure 2

Now let us look at how to express the composition of two rigid motions. Just as we write $8 + 2 = 10$ (or $8 - 2 = 6$, etc.), we could write

(reflection in line *k*) ∘ (translation 2 cm east) = (reflection in line *m*)

Notice the order in which the motions are written; this convention is important to observe when working with compositions.

In summary, the composition of two rigid motions is achieved by doing one rigid motion first, and then using the image from that motion as the *original* for the second rigid motion. Whatever motion would be the net effect of that combination on the original is a single motion, the composition of the two rigid motions. Using the symbol ∘ to represent this way of combining two rigid motions, we can write

(second motion) ∘ (first motion) = (the composition of the motions)

One helpful way of reading the symbol ∘ is to pronounce it as *after*. The order of recording the first and second motions may seem backward, but it is the conventional order in advanced mathematics.

Just as $287 + 95$ can be determined to be 382, a major question is how—or even whether—the composition of two rigid motions can be predicted from only information about the rigid motions. One composition that is of particular value occurs with the composition of a translation and a reflection. For theoretical reasons, the line of reflection must be *parallel* to the line of the vector of the translation, but the distance of the translation, or the length of the vector, can be any length. Suppose the translation is given by vector *v* in Figure 3, and the line of reflection is line *n* (notice that line *n* is parallel to the line the vector is on).

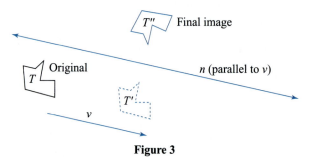

Figure 3

You should recognize that this type of composition has a name: a *glide-reflection*. Notice that the final image T'' is not the image of T for any single reflection, rotation, or translation.

> Whenever a translation and a reflection in a line parallel to the vector of the translation are combined, the resulting composition is a **glide-reflection.**

Glide-reflections are regarded as *single* rigid motions, even though they are defined in terms of two motions. This usage may seem strange, and it opens up the question of which other compositions will have special names. But there is a good reason for identifying glide-reflections. Adding just glide-reflections to the list of types of rigid motions makes the list *complete*. It will not be necessary to add any more types of rigid motions to the list.

> **Any composition of rigid motions can be described by a *single* reflection, rotation, translation, or glide-reflection.**

This last statement takes some time to sink in. It asserts, e.g., that the composition of 100 reflections in 100 different lines is the same as a *single* one of the rigid motions. It also assures us that, given any two copies of the same planar figure, one copy can be transformed into the other by a *single* motion.

ACTIVITY 7 Composition of Reflections in Two Parallel Lines

Carry out the composition (reflection in n) ∘ (reflection in m) and form a conjecture as to what type of rigid motion the composition of two reflections in parallel lines might always be.

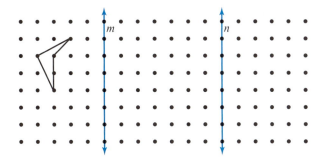

TAKE-AWAY MESSAGE . . . Composition of rigid motions, which is performed by following one motion by another, gives a sort of arithmetic for motions. One type of composition in particular, the glide-reflection, completes the list of rigid motions necessary to describe every composition of rigid motions of the plane. Certain compositions have predictable descriptions as single rigid motions.

Learning Exercises for Section 22.4

1. Make measurements on the example of composition given in Figure 2 to see if there is some relationship between line k, line m, and the 2-cm east translation.

2. Using your choice of original shape, find the composition of the two rigid motions described in each part. What single rigid motion does the composition seem to be?

   a. (translation, 4 cm east) ∘ (translation, 2.8 cm west)
   b. (translation, 3 cm northeast) ∘ (translation, 6 cm southwest)
   c. (reflection in line n) ∘ (reflection in line m), when lines n and m are parallel
   d. (reflection in line n) ∘ (reflection in line m), when lines n and m are not parallel
   e. (rotation, center P, $x°$ clockwise) ∘ (rotation, center P, $y°$ clockwise)
   f. (rotation, center Q, $a°$ clockwise) ∘ (rotation, center Q, $b°$ counterclockwise)
   g. (rotation, center R, 50° clockwise) ∘ (rotation, center S, 60° clockwise), where R and S are different points
   h. (rotation, center T, 40° clockwise) ∘ (reflection in line p), where line p goes through point T
   i. (rotation, center T, 40° clockwise) ∘ (reflection in line q), where line q does not go through point T

3. **a.** How does a glide-reflection affect the (clockwise or counterclockwise) orientation of a figure?
   b. Summarize how the different types of rigid motions affect the orientation of a figure.
   c. Does a glide-reflection give an image that is congruent to the original shape? Explain your thinking.

4. Using only orientation as a guide, what two types of rigid motions might each of the following possibly be? [*Hint*: See Learning Exercise 3(b).]

   a. the composition of a translation followed by a rotation
   b. the composition of a translation followed by two different reflections
   c. the composition of 4 different reflections in different lines
   d. the composition of 17 different reflections
   e. the composition of an even number of reflections; an odd number of reflections
   f. the composition of 3 different rotations, followed by 7 different translations, followed by 9 different glide-reflections

Notes

5. a. Copy and find the image of shape *PQRS* for the glide-reflection given by the translation with vector *w* and the reflection in line *m*. Use paper-tracing if you wish.

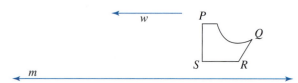

b. Label the (final) image of *P* as *P'* and of *Q* as *Q'*. With a ruler, draw the line segments joining *P* and *P'* and joining *Q* and *Q'*. How does the line of reflection seem to be related to these line segments? Check with other points and their images. This result is a key relationship for a point and its image for a glide-reflection.

6. a. In finding the composition of two rigid motions, does it matter in which order the rigid motions are done, in general? Explain.

b. With the translation and reflection that define a glide-reflection, does it matter in which order the motions are done? Your finding is the reason why glide-reflections are defined in such a particular way.

7. What single type of rigid motion gives each lettered shape as the image of the given one? Describe each of the rigid motions as fully as you can (e.g., if it is a reflection, what line is the line of reflection?). Explain how you decided.

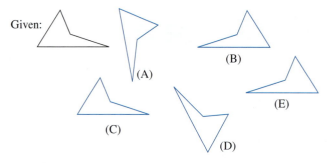

8. For each image of *QRST*, describe the rigid motion that gives the given image as indicated.

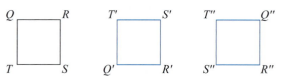

9. Trace the drawing and find two reflections so that their composition will give the same image as the original rigid motion.

a. This translation:

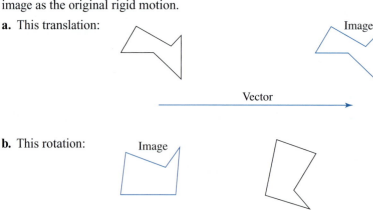

b. This rotation:

c. Find *other* pairs of reflecting lines for parts (a) and (b).

10. Make conjectures based on Learning Exercises 2(c), 2(d), and 9, and then gather more evidence.

11. Give an argument for this statement: Any rigid motion can be accomplished with at most three reflections. (*Hint*: See Learning Exercise 10.)

12. How many reflections, at minimum, might be needed to achieve the same effect as each composition described? Explain your reasoning.

   a. a composition of 17 reflections in different lines
   b. a composition of 24 reflections in different lines

13. Which rigid motion(s) could be used in describing each situation?

   a. footprints in the sand
   b. one's right hand in climbing a ladder
   c. one's two hands in climbing a ladder
   d. turning a microwave dish
   e. tuning in a different radio station
   f. adjusting a thermostat

14. Does a glide-reflection have any fixed points? That is, is there any point whose image under a glide-reflection is the original point?

15. a. What rigid motion is (reflection in line k) ∘ (reflection in line k)?
   b. What rigid motion is (translate 10 cm north) ∘ (translate 10 cm south)?

16. Experiment to find the single rigid motion equal to the composition of three reflections in parallel lines. Then describe that rigid motion in terms of the original lines (make some measurements on the original lines).

Supplementary Learning Exercises for Section 22.4

1. Name the four types of isometries.

2. In the composition, (reflection in line x) ∘ (translation 2 ft west), which rigid motion is done first?

3. Name the single type of rigid motion that each of these composition gives. Describe the rigid motion as fully as you can.

   a. (translation 4 in. south) ∘ (translation 6 in. north)
   b. (translation 6 in. north) ∘ (translation 4 in. south)
   c. (rotation, 58° clockwise, C) ∘ (rotation, 68° counterclockwise, C)
   d. (rotation 68° counterclockwise, C) ∘ (rotation 58° clockwise, C)

4. What single type of rigid motion gives each lettered shape as the image of the given one?

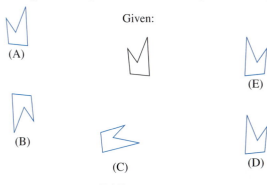

5. a. Line k and vector v are on parallel lines. Find the image of shape S for

   (reflection in line k) ∘ (translation with vector v).

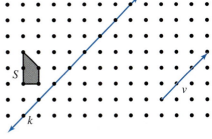

continue

b. What single type of rigid motion does the composition in part (a) give?

c. What would be different if you reversed the order and found the image for (translation with vector v) ∘ (reflection in line k)?

6. Explain why the following two rigid "motions" are considered to be the same. (*Hint:* How does each affect every point?)

- a rotation of $0°$ with any point as the center

- a translation of 0 cm in any direction

22.5 Transformations and Earlier Topics

Now that we have all the categories of rigid motions of the plane, we can review congruence, symmetry, tessellations, and similarity, discussing their relation to the types of transformations. Although many of the relationships carry over to three-dimensional geometry, we will work only with two dimensions here because most of our transformation geometry has been in two dimensions, with rigid motions (translations, rotations, reflections, glide-reflections) and with dilations.

The symmetries we studied earlier were either reflection symmetries or rotational symmetries, so a natural question is: *Are there figures that have translation or glide-reflection symmetries?*

DISCUSSION 1   A Long, Long Trail

Describe the reflections and rotations that are symmetries of each of the following figures. (Both patterns continue indefinitely.) Look for translations and/or glide-reflections that are also symmetries of the figures.

a.

b.

From Discussion 1 we see that each type of rigid motion is a candidate for being a symmetry of a figure. Symmetries of figures are special cases of transformations.

Tessellations also can be viewed with transformations (and congruence and symmetry) in mind.

DISCUSSION 2   What Isometries?

Keeping in mind that tessellations continue indefinitely, describe some rigid motions that give the same pattern for each of the tessellations suggested in the following two figures. Look for translations and glide-reflections as well as reflections and rotations.

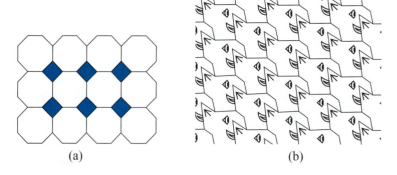

(a)   (b)

Rigid motions cannot themselves account for enlargements or reductions, of course. Dilations must be involved if there is a change in size. The terms *similarity* and *similar figures* allow the composition of a dilation and any rigid motion.

Notes

Two figures are **similar** if one is the image of the other under (1) a dilation or (2) the composition of a dilation and a rigid motion.

Rigidly moving an image after a size change does not negate the similarity; it just makes it more difficult to find the center of the size change. Fortunately, in most work with similar figures, the facts that all the ratios of pairs of corresponding lengths equal the scale factor and corresponding angles are the same size, are of greater importance than the location of the center of the size change.

Recall that for triangles, there is a simple test for similarity: Two triangles are similar if *two* angles in one have the same sizes as two angles in the other. How can that be? A moment's thought makes clear that the third pair of angles must also have the same size because the sizes of the three angles in every triangle add up to 180°. But how does having two pairs of angles the same size make certain that there is a size transformation? The following sketches give one way to convince yourself that there will be a size transformation relating two similar triangles that are not congruent. (Do you remember how to find the scale factor before you have the center?)

Two triangles with two pairs of angles having the same size:

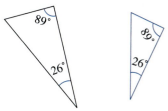

Rotate one triangle so the two sides of one angle are parallel to the sides of the corresponding angle in the other triangle.

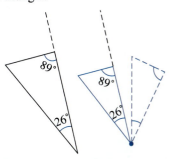

Join corresponding vertices to locate the center of the size transformation. (Can you find the scale factor another way now?)

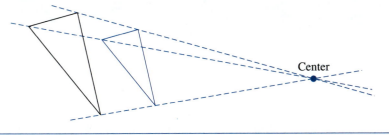

Center

Other situations might involve a reflection rather than a rotation.

 **DISCUSSION 3**   And in Space?

Can three-dimensional congruence, similarity, symmetry, and tessellations also be treated by a study of three-dimensional transformations? ●

TAKE-AWAY MESSAGE . . . The usual school topics of two-dimensional similarity and congruence, as well as symmetry and tessellations, can be treated in a very general fashion with transformations.

Learning Exercises for Section 22.5

1. Describe all the symmetries possible in each diagram. (An arrow means the pattern continues.)

   a.

   b.

2. Examine both statements given in each part. Is each statement true? Explain your decisions.

   a. If a shape has translation symmetry, then the shape is infinite.
      If a shape is infinite, then the shape has translation symmetry.
   b. If a shape has glide-reflection symmetry, then the shape is infinite.
      If a shape is infinite, then the shape has glide-reflection symmetry.

3. Which types of symmetry does each two-dimensional shape have?

   a. a line segment   b. a ray   c. a line   d. an angle

4. Design a figure that has translation symmetry. Does your figure also have glide-reflection symmetry?

5. Which transformations are symmetries for the tessellations given by the following (infinite) patterns?

   a. 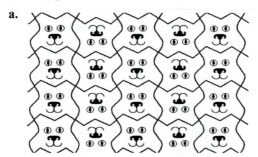   b.

   c.    d.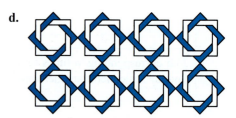

6. Information is given about angle sizes in the following four triangles. Which of the triangles, if any, are similar? Explain how you know.

   Triangle a: 65° and 32°   Triangle b: 65° and 35°
   Triangle c: 35° and 80°   Triangle d: 65° and 83°

7. Describe the image of a shape under the composition of a dilation with scale factor 1 and a rotation of $x°$, both with the same center.

8. Use the meaning of congruence from the transformation view to answer these.

   a. If shape 1 is congruent to shape 2, then is shape 2 congruent to shape 1? Explain.
   b. Is every shape congruent to itself? Explain.
   c. If shape 1 is congruent to shape 2 and shape 2 is congruent to shape 3, is shape 1 congruent to shape 3? Explain.

9. Use the meaning of similarity from the transformation view to answer these.
   a. If shape 1 is similar to shape 2, is shape 2 similar to shape 1? Explain.
   b. Is every shape similar to itself? Explain.
   c. If shape 1 is similar to shape 2 and shape 2 is similar to shape 3, is shape 1 similar to shape 3? Explain.

10. If it is possible, give an example of two triangles such that each pair of matching angles is the same size, but the triangles are not congruent. Must the two triangles be similar? Explain.

11. Discuss each diagram or description with transformations, symmetries, and tessellations in mind.

    a.

    b.

    c.

    d. any wallpaper design that is handy

12. Draw two quadrilaterals to show that having all four pairs of corresponding angles the same size is not enough to ensure that the quadrilaterals are similar.

13. **Pattern Blocks** (See the "Masters" if you do not have access to pattern blocks themselves.) Show patterns that would have different kinds of symmetry. (Don't forget translational and glide-reflection symmetries.)

Supplementary Learning Exercises for Section 22.5

1. Describe all the symmetries possible in each diagram below. Assume that all the patterns continue to the right and left indefinitely.

   a.
   b.
   c.

2. Sketch a new shape that has glide-reflection symmetry. Does your shape also have translation symmetry?

3. Describe the final image of a 3 cm by 4 cm rectangle under the composition of these three transformations: first, a rotation 90° clockwise with center C; then a size transformation with scale factor 5 and center C; and finally a glide-reflection from a 5-cm east translation and a reflection in a horizontal line.

4. Discuss the figure below with transformations, symmetries, and tessellations in mind.

Notes

22.6 Issues for Learning: Promoting Visualization in the Curriculum

People you encounter do not have equal amounts of innate ability in most areas—physical, social, artistic, and mental, e.g. Some have strengths where others might have weaknesses. So, it is no surprise that disparities with inborn visualization abilities occur in people. Some people have stronger visualization skills than others. To decide whether to turn right or left, many people have to turn a map so it is oriented in the direction they are facing. Others can read a map without needing to turn it, no matter how they themselves are oriented. Some people can interpret a two-dimensional drawing of a three-dimensional shape without difficulty; others struggle, at least at first.

For a time, a person's visualization abilities were regarded as unchangeable. But it is now recognized that, like most cognitive abilities, a person's visualization skills can be improved. Opportunities to improve visualization skills currently appear in most elementary (and beyond) school curricula, where they were once not a conscious concern but were given attention only by happenstance through topics in geometry.

These curricular efforts are welcome. Even though computers were conceived for numerical work, today computer graphics are encountered nearly everywhere. The production of such graphics and the processing of them perhaps led some to see visualization as an increasingly important part of our life: "Visual thinking is *the* thinking of the very near future" (p. 167).[2] Transformation geometry, with its visual experiences from rigid motions and size changes, gives teachers an opportunity to include or call attention to visualization in the curriculum.

Many of the activities you encountered in Chapter 17 on polyhedra called for visualization skills. For example, you were asked to look at nets in two dimensions and imagine how they could be folded so that they would result in three-dimensional shapes, such as a box. You may have noticed when doing those activities that students with stronger visualization skills found these activities easier to do than others. And, in this chapter, some of you may have found that visualizing what a transformation should look like was easier for you than for your classmates, which gave you an advantage during some of the activities and Learning Exercises.

Learning about and understanding transformations can begin in the early grades. Here is an example of how some researchers[3] introduced the fundamentals of transformation geometry to children in the primary grades, allowing these children to begin to practice visualization. These researchers found that the patterns which occur in quilting offered many opportunities for the study of geometric designs, particularly because they used transformations of designs, namely, slides, flips, and turns (see Learning Exercise 1 in Section 22.1). The researchers worked with four different second-grade teachers on a quilting unit in these teachers' mathematics classes. Students made quilt designs and formalized the processes they used to make them. They first used paper and crayons, next paper cut-outs, and then a computer program, to make *core squares*. Finally, they performed transformations on them. For example, the *side flip* below is obtained by a flip (reflection) over the vertical line down the middle of the core. A slide gives a pattern just like the core, only moved. Turns of 90°, 180°, and 270° could also be made.

Core   Side flip

By using two-sided paper cores, children could physically make these moves. They next had to create names for each motion so that they could talk to one another about them. Thus, the above flip through the vertical axis was called a LF (*left flip*), which could be made by grabbing the right edge of the core and flipping it over to the left. (What do you think an RF would be? Would it give a different result than a LF?) Developing a common notation so that the children could communicate about their designs was an important element of their work.

Chapter 23

Measurement Basics

Of all the occurrences of numbers in daily life, most of them will involve measurements, particularly if you include monetary values. If counts are also included as measurements (as they are in advanced mathematics), virtually all uses of numbers involve measurements. Many types of measurements share certain characteristics or key ideas. We identify these key ideas of measurement in Section 23.1. Although we have regarded length and angle size as familiar concepts in earlier chapters, we will also look at them from the viewpoint of key ideas of measurement in Section 23.2.

23.1 Key Ideas of Measurement

A given object has several characteristics, many of which we can quantify. For example, a particular woman has an age, a height, a shoe size, a blood pressure, a hair color, a certain amount of self-discipline, some degree of athleticism or degree of friendliness, and many other characteristics. Some characteristics are relatively easy to quantify and assign numerical values to, whereas others are not.

Finding the values of quantities associated with characteristics of objects or events often involves measurement. The term *measurement* may refer to the *process* of finding the value of a quantity, or it may refer to the *result* of that process. For instance, the measurement of the weight of an object might refer to how we measured the weight (e.g., with a pan balance) or to the resulting value (e.g., 18 pounds). This section covers four key ideas about the usual kinds of measurement, which are summarized in the Take-Away Message at the end of the section.

The first key idea is the process of measurement, particularly *direct measurement*. The **process of direct measurement** of a characteristic involves matching that characteristic of the object by using (often repeatedly) a **unit**, which is a different object having the same characteristic. The unknown weight of an object, e.g., might be determined on a pan balance by directly matching it with copies or partial copies of a known unit weight. The measurement (the result) is the number of units needed to match the measured object with respect to the characteristic being measured. The measurement is the value of the quantity.

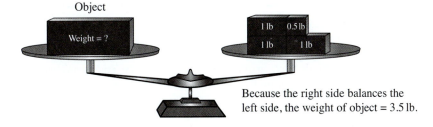

Because the right side balances the left side, the weight of object = 3.5 lb.

Measuring devices like the pan balance are often used for measuring weight, although one could roughly measure the unknown weight even more directly by hefting the object in one hand and then adding copies of the unit weight in the other hand until the weight in each hand felt the same. *Indirect measurements* of quantities involve mathematical or scientific principles, such as in determining a weight with the usual bathroom scale or a scale with a spring that compresses or stretches predictably.

The second key idea deals with the use of standard units.

DISCUSSION 1   Why Standard Units?

1. We usually use **standard units** when giving values of quantities. For example, the English system includes the ounce, pound, ton, inch, yard, mile, etc., and the metric system uses the gram, kilogram, metric ton, centimeter, meter, kilometer, etc. Why are standard units used?

2. Why are there so many different units for the same characteristic, even within one system of measurement? For example, the English system has the units ounce, pound, and ton for measuring weight.

3. Elementary school curricula often start measuring work with nonstandard units. For example, lengths might be measured with crayons or markers. Why? (Keep in mind the process of measurement.) ●

Most of the world uses the **metric system**, or **SI** (from *Le Système International d'Unités*). Thus, it is surprising that the United States, as a large industrial nation, has clung to the English system. Although the general public has not always responded favorably to governmental efforts to mandate the metric system, international trade is forcing us to be knowledgeable about, and to use, the metric system. Some of the largest U.S. industries have been the first to convert to the metric system from the English system.

DISCUSSION 2   Influences on the Choice of Standard Units

Why has the U.S. public resisted adopting wholeheartedly the metric system? Why do international trade and alliances influence our measurement system? ●

Scientists have long worked almost exclusively in metric units. Part of the reason is that virtually the entire world uses the metric system, but the main reason is how sensible and systematic the metric system is. A basic metric unit is carefully defined (for the sake of permanence and later reproducibility), and larger units and smaller subunits are related to each other in a consistent fashion (making the system easy to work in). In contrast, the English system's units are neither well defined (historically) nor consistently related. For example, the length unit *foot* possibly evolved from the lengths of human feet and is related to other length units in an inconsistent manner: 1 foot = 12 inches; 1 yard = 3 feet; 1 rod = 16.5 feet; 1 mile = 5280 feet; 1 furlong = $\frac{1}{8}$ mile; 1 fathom = 6 feet. Quick! How many rods are in a mile? A comparable question in the metric system is just a matter of adjusting a decimal point.

The basic SI unit for length (or its synonyms) is the **meter.** The meter is too long to show with a line segment here, but two subunits fit easily and illustrate a key feature of the metric system: units smaller and larger than the basic unit are multiples or submultiples of powers of 10.

1 decimeter (0.1 meter)

1 centimeter
(0.01 meter)

Furthermore, these subunits have names, such as *decimeter* and *centimeter*, and are consistently formed by putting a prefix on the word for the basic unit. The prefix *deci-* means 0.1, so *decimeter* means 0.1 meter. Similarly, *centi-* means 0.01, so *centimeter* means 0.01 meter. You have probably heard the word *kilometer*. The prefix *kilo-* means 1000, so *kilometer* means 1000 meters. Reversing your thinking, you can see that 10 deci-meters are in 1 meter, 100 centimeters are in 1 meter, and 1 meter is 0.001 kilometer.

An important advantage of the metric system is the consistent combination of the symbols for the basic units and the prefixes. In the case of the characteristic length, the use of the symbol m (for meter) along with a symbol for a prefix (such as d for *deci-*, c for *centi-*, etc.) allows us to report a length measurement quite concisely: e.g., 18 cm or 1.8 dm. Many of the metric system prefixes are given in Table 1 shown below; more are given in the Glossary. You should note that the symbols for the metric units do not end with periods (a convention that is also true for most of the unit abbreviations in the English system), and the same symbols are used for singular and plural. For example, m stands for meter and meters, and yd stands for yard and yards. In the English system, the exception to not using a period is the abbreviation for inch (in.).

Table 1 Metric Prefixes

Metric Prefix	Metric Symbol	Meaning of Prefix		Applied to Length
tera-	T	1 000 000 000 000	or 10^{12}	Tm
giga-	G	1 000 000 000	or 10^9	Gm
mega-	M	1 000 000	or 10^6	Mm
kilo-	k	1 000	or 10^3	km
hecto-	h	100	or 10^2	hm
deka-	da	10	or 10^1	dam
(no prefix)		(basic unit) 1	or 10^0	m
deci-	d	0.1	or 10^{-1}	dm
centi-	c	0.01	or 10^{-2}	cm
milli-	m	0.001	or 10^{-3}	mm
micro-	μ	0.000001	or 10^{-6}	μm
nano-	n	0.000000001	or 10^{-9}	nm

If you are new to the metric system, your first job will be to master the prefixes so you can apply them to the basic units for other characteristics. You may know a mnemonic for several of the metric prefixes. For example, the phrase "*k*ing *h*enry *da*nced, *d*rinking *c*hocolate *m*ilk" (although not using good capitalization) does give the most common metric prefixes. The table above shows some of the other metric prefixes.

Here is one final note on the metric system: Commas are not used in writing numbers. Where we write commas in multidigit numbers, SI would have us put spaces. So 13,438 m would be written as 13 438 m, and 0.84297 km would be written as 0.842 97 km. Some countries use the comma where we use the decimal point, so some agreement is necessary to avoid potential confusions in international trade and other communication. This SI recommendation is not observed in some elementary school textbooks published in the United States.

 ACTIVITY 1   It's All in the Unit

1. Measure the width of your desk or table in decimeters. Express that length in centimeters, millimeters, meters, and kilometers.

2. Measure the width of your desk or table in feet. Express that length in inches, yards, and miles.

3. In which system are conversions easier? Explain why. ●

The next discussion introduces the third key idea of measurement.

DISCUSSION 3 Are Measurements Exact?

Only one of these statements can be true. Determine which one is true, and explain why it is the only one.

a. "She weighed exactly 148 pounds."

b. "She is exactly 165 centimeters tall."

c. "It is exactly 93,000,000 miles to the sun."

d. "There were exactly 5 people at the meeting." ●

Because actual measurements (except for counts) are only approximations, a given value should not be read to imply that it is more nearly exact than is reasonable. For example, the value 38 pounds implies that the measuring was done only *to the nearest pound*. Thus, a reported "38 pounds" might describe a weight that is actually greater than or equal to $37\frac{1}{2}$ pounds and less than $38\frac{1}{2}$ pounds. Deducing that the value 38 pounds is exactly $38 \times 16 = 608$ ounces would be risky, even though 1 pound = 16 ounces. In actuality, the object could be as light as 600 ounces ($37\frac{1}{2} \times 16$) or as heavy as just under 616 ounces ($38\frac{1}{2} \times 16$).

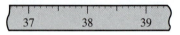

DISCUSSION 4 Interpreting a Measurement Given "to the Nearest"

Suppose a length is reported as $2\frac{3}{4}$ inches, to the nearest $\frac{1}{4}$ inch. Make a drawing of part of a ruler and argue that the length could be as short as $2\frac{5}{8}$ inches or as long as just under $2\frac{7}{8}$ inches. ●

Unfortunately, much school work with measurement treats the values as exact, so students can easily form the erroneous impression that the values *are* exact. But just as the ideal constructs of *line segment* and *angle* can be discussed, theoretically perfect measurements may also be discussed. For example, we can *imagine* a square with sides exactly 2 centimeters long, although drawing one even with a finely sharpened pencil is not possible. Likewise, we can *imagine* a line segment with a length of exactly 2 centimeters although we cannot actually produce one.

Which of these two ways would give a better measurement for the length of a board: measuring to the nearest inch or measuring to the nearest foot? In a direct measurement, choosing the smaller unit naturally allows the matching to be done more closely. So, usually a measurement with a smaller unit narrows the range of possible values for the measurement and lessens the error.

DISCUSSION 5 Measuring Piecemeal

How can you find the distance from Anyburg to Dayville? The area of the z-shaped region?

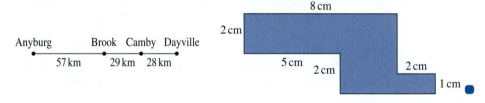

Your method in Discussion 5 likely involved the fourth key idea of measurement: You can measure an object by thinking of the object as cut into pieces, measuring each piece, and then adding those measurements.

TAKE-AWAY MESSAGE . . . Here is a summary of four key ideas of measurement.

1. (Direct) measurement of a characteristic of an object involves matching the object with a unit, or copies of a unit, having that characteristic. The matching should be based on the characteristic. The number of units, along with the unit, give the measurement, or the value, of the quantity associated with the characteristic.

2. Standard units are used because they are relatively permanent and enable communication over time and distance. Nonstandard units may be used in schools to allow a focus on the process of measurement.

3. Measurements are approximate. Smaller units give better approximations, although the numerical part of a reported value can have a bearing on the implied accuracy as well. (If we include counts as measurements, however, they can be exact.)

4. A quantity can be measured by thinking of the object being measured as "cut" into a (finite) number of pieces, measuring the quantity in each piece, and then adding up those measurements or values (sometimes called *additivity of measures*).

Learning Exercises for Section 23.1

1. Some quantities that are commonly measured include length, area, volume, weight, speed, time, temperature, intelligence, student achievement, force.

   Which of the above quantities are you measuring when you measure the following quantities?

   a. how far apart two people are
   b. how much soda is in a glass
   c. how warm is the lake water
   d. how heavy is a football
   e. how much room is left to draw on, on a piece of paper
   f. how fast you can run
   g. how long it takes to walk to school
   h. how large is a football field
   i. how much carpet is needed for a room
   j. how effective is a teacher

2. What are some possible units for the quantities listed in Learning Exercise 1?

3. What characteristics of a child could be measured with the given device?

   a. a thermometer   **b.** the child's math book   **c.** a stopwatch

4. For which quantities could each term or phrase be a unit? (A dictionary can be useful for some items.)

   a. mm of mercury
   b. number per square mile
   c. acre
   d. quire
   e. hectare
   f. cubit
   g. candela
   h. stone
   i. coulomb
   j. rating point (TV)
   k. decibel
   l. Scoville
   m. body mass index (BMI)
   n. section

5. Why would each item *not* be satisfactory as a unit for the given quantity?

   a. a rubber band, for length
   b. an ice cube, for weight
   c. a garage-sale item, for monetary value
   d. a person's judgment, for temperature
   e. a pinch of salt, for saltiness
   f. a sip, for amount of drink

6. Answer the following items about reports of measurements:

   a. Explain why this claim cannot be accurate: "The fish I caught was exactly 18 inches long!"
   b. "The fish weighed 123 pounds." What weights for the fish are possible and still have the claim be a true statement?

7. **a.** Show the shortest and longest lengths that could be reported as 7 units using this ruler.

   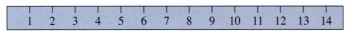

   b. Show the shortest and longest lengths that could be reported as $7\frac{1}{2}$ units using this ruler.

8. One kind of bathroom scale shows weights to the nearest half-pound.

   a. On consecutive days, a person's readings were 151.0 and 151.0 on the scale. Did the person weigh exactly the same amount each day? Explain.

continue

Notes

b. What weights would all give a reading of 162.0 on the scale?

c. What weights would all give a reading of 128.5 on the scale?

d. How much "off" could a reading of 116.0 be, on the scale? A reading of 223.5?

9. In parts (a–h), tell without calculating whether x is greater than 2000, equal to 2000, or less than 2000. Explain your thinking.

  a. x raisins weigh 2000 pounds   **b.** 2000 raisins weigh x pounds

  c. x yards = 2000 miles   **d.** 2000 yards = x miles

  e. x kilometers = 2000 meters   **f.** 2000 kilometers = x meters

  g. x tons = 2000 pounds   **h.** 2000 tons = x pounds

10. In parts (a–c), indicate which of the made-up units is larger. Tell how you know.

  a. 150 ags = 100 aps   **b.** 260 bas = 1.31×10^3 bos   **c.** 0.19 con = 0.095 cin

11. Tell how one of the four key ideas of measurement could be used in determining the weight of a wiggly puppy.

12. Give the number of unit square regions there are in each figure. Explain your reasoning, especially how you used key ideas of measurement.

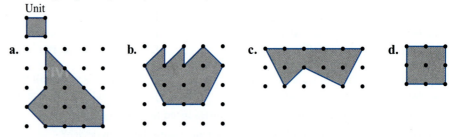

  e. Let the unit region be the region shown in part (d). What would be the values for the number of units in parts (a–c)? Explain your reasoning.

13. Sections of low picket fence come ready-made and cost $2.89 for a section 27 inches long. How much would it cost to put a fence along the borders of a flower patch shaped like the drawing below? Assume that pieces less than 6 inches long are too short to be useful. (What key idea or ideas of measurement did you use?)

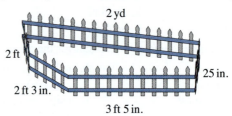

14. **a.** Add these lengths: $5\frac{3}{4}$ inches, $7\frac{7}{8}$ inches, and $3\frac{1}{2}$ inches.

  b. Make pieces of rulers (like the one below for $5\frac{3}{4}$ inches) for each measurement given in part (a). Then show where the shortest and the longest lengths that could be described as $5\frac{3}{4}$ inches (or $7\frac{7}{8}$ inches or $3\frac{1}{2}$ inches) would end.

  c. Add the *least* lengths that each of the measurements in part (a) could be describing.

  d. Add the *greatest* lengths that each of the measurements in part (a) could be describing.

  e. What is the range of values possible for the sum of the measurements in part (a)?

15. One teacher's gradebook computer program shows scores to the nearest whole point but keeps several decimal places in its memory. Use that information to explain why the gradebook program might give a sum of 241 for scores of 82, 73, and 87.

16. Think of the sizes of the units (think metric) to help you complete the following equalities of measurements:

    a. 76.3 cm = _____ m

    b. 2.7 m = _____ cm

    c. 19 mm = _____ cm

    d. 4.62 cm = _____ mm

    e. 0.62 km = _____ m

    f. 108 m = _____ km

    g. 8.7 dm = _____ cm

    h. 29 cm = _____ dm

    i. A person 155 cm tall is _____ meters tall.

17. For some purposes, 1 second (1 s) of time is too large a unit. Fast computers, e.g., need nanoseconds to describe times. Review the metric prefixes and complete the following:

    a. ___ ms = 1.3 s

    b. 143 ns = ___ s

    c. 500 ms = ___ s

    d. 2500 μs = ___ s

18. For historical reasons, the kilogram, rather than the gram, is the metric system's basic unit for mass. A raisin has a mass of about 1 gram (1 g). Complete the following conversions:

    a. 153.2 g = ___ kg

    b. 3.4 g = _____ mg

    c. 2.17 kg = _____ g

    d. 56 mg = _____ g

19. A map has the scale 1 000 000 : 1.

    a. In reality, how far apart in kilometers are two cities that are 8 cm apart on the map?

    b. Two locations that are actually 150 km apart should be how far apart on the map?

20. Sometimes, one can measure something by cutting it into an *infinite* number of pieces and then adding the measurements of all the pieces. What will the following infinite sum equal?

    $$\frac{1}{2} + \frac{1}{4} + \frac{1}{8} + \frac{1}{16} + \cdots = ?$$ (*Hint*: Use the diagrams for an idea.)

          etc.

21. When the French were designing the metric system in the late 1700s, they invited other countries to participate. England said that "reform of weights and measures was considered 'almost impracticable.'"[1] Why would anyone think that?

22. Decide whether each calculation is correct. Correct any calculation that is not correct. [*Recall*: 1 pound (lb) = 16 ounces (oz) and 1 gallon = 4 quarts.]

    a.
    $$\begin{array}{r} 4 \quad 1 \\ \cancel{5}.\,05 \text{ m} \\ -\ 3.\,52 \text{ m} \\ \hline 1.\,53 \text{ m} \end{array}$$

    b.
    $$\begin{array}{r} 4 \quad 1 \\ \cancel{5}:05 \text{ PM} \\ -\ 3:52 \text{ PM} \\ \hline 1:53 \end{array}$$

    c.
    $$\begin{array}{r} 5 \quad 1 \\ \cancel{6} \text{ lb } 3 \text{ oz} \\ -\ 2 \text{ lb } 8 \text{ oz} \\ \hline 3 \text{ lb } 5 \text{ oz} \end{array}$$

    d.
    $$\begin{array}{r} 5 \quad 1 \\ \cancel{6}.\,3 \text{ kg} \\ -\ 2.\,8 \text{ kg} \\ \hline 3.\,5 \text{ kg} \end{array}$$

    e.
    $$\begin{array}{r} 3 \quad 1 \\ 4 \text{ gal } \cancel{1} \text{ qt} \\ -\ \quad\quad 9 \text{ qt} \\ \hline 3 \text{ gal } 2 \text{ qt} \end{array}$$

    f.
    $$\begin{array}{r} 1 \quad 1 \\ 2 \text{ hr } 35 \text{ min} \\ +\ \quad 75 \text{ min} \\ \hline 3 \text{ hr } 10 \text{ min} \end{array}$$

23. The following are examples of nonmetric length units used in England at different times or in different settings:

    cubit, rod, ell, fathom, foot, furlong, hand, inch, knot, mile, pace, yard

    With those units in mind, describe one advantage of using the metric system approach to length units over the English system approach.

24. In earlier times in England, the following were units for measuring volume[1]:

2 mouthfuls	= 1 jigger		2 gallons	= 1 peck
2 jiggers	= 1 jackpot		2 pecks	= 1 half-bushel
2 jackpots	= 1 gill		2 half-bushels	= 1 bushel
2 gills	= 1 cup		2 bushels	= 1 cask
2 cups	= 1 pint		2 casks	= 1 barrel
2 pints	= 1 quart		2 barrels	= 1 hogshead
2 quarts	= 1 pottle		2 hogsheads	= 1 pipe
2 pottles	= 1 gallon		2 pipes	= 1 tun

  a. How many cups are in a gallon?

  b. Express each unit in terms of quarts.

  c. What feature of this system of volume units is somewhat like those of the metric system? What feature is different?

25. In 1893 the U.S. yard unit was defined to be equal to $\frac{3600}{3937}$ meter. Then in 1959 it was defined as 0.9144 meter. By how much did the new yard differ from the old yard?

26. ***Computer*** Two websites that focus on measurement are given. Browse the sites to see what they offer.

  a. U.S. Metric Association: http://lamar.ColoState.edu/~hillger/.

  b. National Institute of Standards and Technology (until 1988, the National Bureau of Standards): http://www.nist.gov/pml/wmd/metric/metric-pubs.cfm. Then Google "images for metric system" to sample what is available.

Supplementary Learning Exercises for Section 23.1

1. When you measure the width of a desk with the long side of a rectangular (not square) piece of cardboard, you get 7 units. If you measure the width with the *short* side of the piece of cardboard, would you get 7 units again, more than 7 units, or less than 7 units? Tell how you know.

2. You say to a friend on the telephone, "I drank 2 glasses of orange juice for breakfast." Your friend says, "Well, *I* drank 3 glasses." Why can't you tell who drank more orange juice?

3. Explain why each of the following statements cannot be completely true:

  a. "I measured, so I know I grew exactly 2 inches last year."

  b. "I weighed myself, and I'm excited! I lost exactly 2 pounds on my diet last week."

4. **a.** Show the shortest and longest lengths that could be reported as 3 units using the ruler below.

  b. Show the shortest and longest lengths that could be reported as $3\frac{1}{2}$ units using the ruler below.

5. In parts (a–d), tell without calculating whether x is greater than 100, equal to 100, or less than 100. Tell how you know.

  a. x m = 100 km   **b.** x km = 100 m

  c. x oz = 100 lb   **d.** x lb = 100 oz

6. These approximations are good for some purposes: 1 in. ≈ 2.5 cm; and 1 mi ≈ 1.6 km. Give an approximation for each of the following measurements:

  a. 10 cm ≈ _____ in.   **b.** 12 in. ≈ _____ cm

  c. 1 km ≈ _____ mi   **d.** 5 mi ≈ _____ km

1 cm

1 in.

7. Think of the sizes of the units (think metric) to help you complete the following equalities of measurements:

   a. 49 cm = _____ m
   b. 1.2 m = _____ cm
   c. 49 km = _____ m
   d. 49 m = _____ km
   e. 52 m = _____ dm
   f. 52 dm = _____ m

8. A unit for a piece of computer data is the *byte* (B). Computer users may see expressions such as the following. How many bytes does each represent? (See Table 1 on p. 519.)

   a. 512 MB
   b. 4 GB
   c. 2 TB

9. Calculate the following differences:

   a. 6 lb 5 oz − 2 lb 8 oz
   b. 7.3 kg − 3.6 kg
   c. 4 yd 2 ft 3 in. − 1 yd 2 ft 8 in.
   d. 5.28 m − 1.7 m
   e. 2 gal 2 qt − 3 qt
   f. from 2:45 PM to 5:30 PM (same day)

23.2 Length and Angle Size

Although we have treated length and angle size as familiar, they can also be looked at from the viewpoint of the key ideas of measurement. The basic classroom tools for measuring length and angle size are the ruler and the protractor. Children often use (simple) rulers as early as the first grade; they learn about protractors later, perhaps in Grade 4.

Length

We speak of the length of a piece of wire or a rectangle in two ways. The term **length** might refer to the quality or attribute we are focusing on, or it might refer to the measurement of that quality. The context usually makes clear which reference is intended.

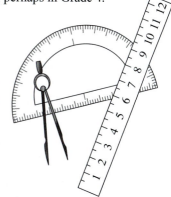

However, what might be puzzling to elementary school students is that there are several terms, all of which refer to the *length* attribute of different objects.

 DISCUSSION 6   Aliases for Length

"Height" and "width" are words that refer to lengths. What other terms refer to the same attribute as *length* does? ●

All the key ideas of measurement from Section 23.1 are useful in working with length. For example, early work in elementary grades often involves nonstandard units—paper clips, chains of plastic links, or pencils.

Figure 1

You should keep the approximate nature of measurements in mind. The heavy line segment shown in Figure 1 is about 4 "pencils" long, because four copies of the (same) pencil unit just about match the length of the line segment. If we measured to the nearest *half*-pencil, the measurement here would still be 4 pencils, and so we could say "4 pencils, to the nearest half-pencil" to communicate the greater precision.

The meter and the metric prefixes from the metric system give a variety of length units. If you have never established personal comparisons, or benchmarks, for some of the metric units, you should begin to do so. What does approximately 1 millimeter, 1 centimeter, 1 decimeter, 1 meter, and 1 kilometer mean to you? For many people, the width of a particular fingernail is a good benchmark for 1 cm.

1 centimeter
(0.01 meter)

1 decimeter (0.1 meter)

Estimating measurements is a very practical skill that is often neglected in school. Having benchmarks for units can help in estimating, say, the width of a room in meters or the perimeter of a piece of paper in decimeters.

The total length along a polygon is usually called its **perimeter** (*peri* means "around"; *metron* means "measure"). (Recall that a polygon is made up of just the line segments.)

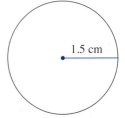

Perimeter = 10.2 cm

Think About …

What key idea of measurement is often involved in finding the perimeter of a polygon?

The length of a circle is usually called its **circumference** (*circum* means "around"; *ferre* means "to carry"). In everyday language, both *perimeter* and *circumference* may be used to refer to the sets of points, as well as the length measurement.

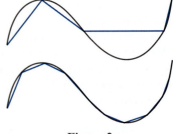

Circumference = 9.4 cm

Measuring the lengths of crooked curves presents difficulties. With short curves we can do the practical thing: Place a flexible measuring tape or a piece of string closely on the curve and then straighten out the tape or string and measure it. There are also small, wheeled devices that can be used to trace over a route on a map to find measurements of lengths. You may have seen an auto accident scene where a trundle wheel (a device made of a wheel) was used to measure distances.

One way to find lengths of curves that pays off is to approximate the curve by line segments and then observe what happens each time as smaller and smaller line segments are used. As all the segments become smaller and smaller, coming as close to zero length as you arbitrarily want, the sum of the lengths of all the line segments will get closer and closer to the (ideal) exact length of the curve. (Not only is this process plausible, it is also applied in the study of calculus.) In Figure 2, the shorter segments in the second drawing should give a better approximation of the length of the curve than the segments in the first drawing.

Figure 2

Angles and Angle Size

Let us turn our attention to angles and their measurement. As you may know, angles can be viewed in two different ways: (1) statically, as two rays or segments (the **sides** of the angle) starting from the same point (the **vertex**) or (2) more dynamically, as the result of a ray rotating from an initial position to a final position. The second way allows you to speak of measuring angles in a clockwise or counterclockwise direction.

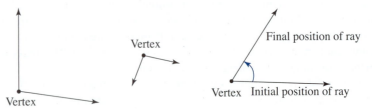

Which of the two angles shown to the right is larger? The size of an angle is determined by the opening, or the amount of rotation, or turn, involved in the angle. So the first angle has greater size. The visible length of the sides of the angle is irrelevant.

You already know some adjectives for angles: e.g., right, acute, obtuse, exterior, central, inscribed. Still other adjectives may be introduced in middle school. **Vertical angles** (or **opposite angles**) are formed when two lines cross; in the drawing shown here, a and c are a pair of vertical angles, as are b and d.

When two lines are crossed by a third line, eight angles are formed and can be grouped into four pairs of **corresponding angles,** angles that are located in corresponding positions, such as b and w, or d and y. Angles such as d and w in the drawing are called **alternate interior angles,** as are c and x.

The key ideas of measurement also apply to the measurement of angle size, of course. The standard unit used in the K–8 curriculum is the **degree** (°). Across the world, societies that were advanced enough to study astronomy, such as the Mesopotamians (in what is now Iraq) and the Mayans (in what is now Mexico and Central America), knew that there were about 365 days in a solar year, and they settled on 360 as a good number with which to define a unit for angle size. Hence, a full turn of a ray about its vertex goes through 360° (a circle has 360°). Choosing 360° for a full turn means that the degree is a fairly small unit. However, for angles used in ocean navigation or astronomy, even a degree is too large a unit. To obtain smaller units, each degree can be divided into 60 equal pieces, with each piece called a **minute** ('). Each minute can then be further divided into 60 **seconds** ("). (Notice that *minute* and *second* are the same words and relationships used with time, but here are about angle size.) The metric system recognizes the degree as a unit, but the more official SI unit is called the *radian* (about 57.3°). Radians do not usually appear until high school trigonometry, where assigning positive and negative signs to angle measurements to indicate counterclockwise and clockwise turns, respectively, also appears.

Because the degree is a relatively small unit, you could use a nonstandard unit larger than a degree, as illustrated in Figure 3, for introductory classroom work with angle-size measurement. In the drawing the size of angle X is approximately $4\frac{1}{2}$ units, and we could write $m(\angle X) = 4\frac{1}{2}$ *units,* where the letter m stands for the measurement of the angle. Indeed, rather than saying *size of the angle,* we could say *measure of the angle.*

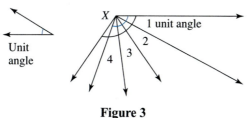

Figure 3

As you know, the usual classroom protractor shows 180°, the size of a half-turn. If you wish to measure an angle whose size is greater than 180°—e.g., the counterclockwise angle shown in the following sketch—you might think of the given angle as divided into two angles: the 180° angle and a smaller angle. Measuring the smaller angle with the protractor then enables you to tell the size of the whole angle, which is 209°.

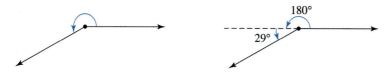

Think About …

Do you see another way to use the protractor to get the size of the counterclockwise angle?

Notes

You already know some important results about angle measurements. For example, the sizes of two supplementary angles add to 180°. And the sum of the sizes of the angles in any triangle is 180°. This last result leads to the more general result for n-gons: The sum of the sizes of the angles of an n-gon is $(n - 2)180°$. So, e.g., the sizes of the 8 angles in any octagon add up to $(8 - 2)180°$, or 1080°. [See the Grade 5 textbook page below for two methods of arriving at the $(n - 2)180°$ result.]

Think About …

How could you show that each angle in an *equiangular* octagon has 135°?

In this page from a Grade 5 textbook, how would you adjust Ignacio's method to get the correct sum?

Date _____   Time _____

LESSON 3·9   **Angles in Any Polygon**

1. Draw a line segment from vertex *A* of this octagon to each of the other vertices except *B* and *H*.

2. How many triangles did you divide the octagon into? _____

3. What is the sum of the angles in this octagon? _____

4. Ignacio said the sum of his octagon's angles is 1,440°. Below is the picture he drew to show how he found his answer. Explain Ignacio's mistake.

_____

_____

_____

_____

5. A 50-gon is a polygon with 50 sides. How could you find the sum of the angles

in a 50-gon?_____

_____

_____

Sum of the angles in a 50-gon = _____

Measurement and Geometry 2.0, **2.1**, **2.2**; Mathematical Reasoning 2.3, 3.0, 3.3   **89**

Wright Group, *Everyday Mathematics, Journal 1, Grade 5*, p. 89. Copyright © 2008, McGraw-Hill. Reproduced with permission of the McGraw-Hill Companies.

Several ideas about lengths or angle sizes can be involved in a single problem. For example, recall how corresponding lengths and angle sizes are related in congruent polygons and polyhedra or in similar polygons and polyhedra. With congruent polygons and polyhedra, corresponding lengths are equal, as are the sizes of corresponding angles. For similar polygons and polyhedra, corresponding angles are the same size, but every two corresponding sides give the same ratio (the scale factor) for their lengths.

EXAMPLE

The dashed line is a line of symmetry for heptagon *ABCDEFG*. Some angle sizes are given in the drawing. Find the sizes of the other angles of the heptagon. [*Recall*: The sum of the sizes of the angles of an *n*-gon is $(n-2)180°$.]

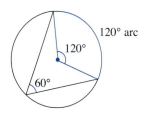

Solution

Because the dashed line is a line of symmetry, $\angle B$ is the same size as $\angle G$, or 140°. Similarly, $\angle F$ has size 135°. Because the segment *EDX* is straight and $\angle CDX$ has size 80°, $\angle D$ and its image $\angle E$ each has 100°. The only angle unaccounted for is $\angle A$. All the angles of the heptagon will total $(7-2)180°$, or 900°. The angles already determined total $140 + 140 + 135 + 100 + 100 + 135 = 750°$, which leaves 150° for $\angle A$. ●

Another result that you already know about angles (from Section 21.1) involves angles with their vertices at the center of the circle (central angles) or with their vertices on the circle (inscribed angles). A central angle cuts off an arc on the circle, and that arc can be measured by the same number of degrees as in the size of the central angle. The size of an inscribed angle cutting off the same arc is half the number of degrees in the arc or the central angle.

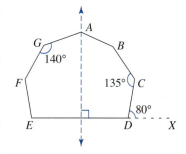

As with length, some ability at estimating angle size is often useful. Most people have a good mental picture of 90°, 180°, and 270° (if they have thought about the fact that $90° + 180° = 270°$). Thinking of a half or a third of 90° (a right angle) can help in estimating angle size.

TAKE-AWAY MESSAGE . . . The four key ideas of measurement from Section 23.1 all apply to length and angle size. There are many words that refer to the same characteristic as does length, including perimeter and circumference. Angle sizes are usually given in degrees, with 360° in a full turn and 180° in a half-turn. Several results about angle sizes enable us to find unknown measurements from known ones.

Learning Exercises for Section 23.2

1. Find the perimeter of each figure. (*Suggestion*: Use metric units.)
   a.

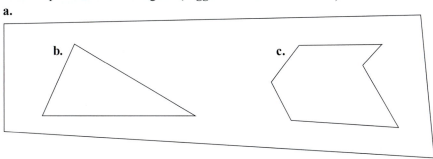

   d. Find the perimeter of each face of shape A and shape B from your kit of polyhedra.

2. Explain why the statement "My brand-new pencil is exactly $7\frac{1}{2}$ inches long" is not completely true.

3. Show a segment that would measure 6 cm to the nearest centimeter but 6.5 cm to the nearest half-centimeter. How many possibilities are there?

4. A student has a broken ruler with the first 3 inches missing. However, she has measured several segments by noting where on the ruler the two endpoints of each segment were. Find the lengths as she did, using the following information about where the two endpoints were. Can you determine them mentally?

   a. 4 inches and $7\frac{3}{4}$ inches   **b.** $5\frac{7}{8}$ inches and $9\frac{1}{4}$ inches   **c.** $6\frac{7}{16}$ inches and 10 inches

5. What is your personal benchmark for each length?

   a. 1 millimeter   **b.** 1 centimeter   **c.** 1 decimeter
   d. 1 meter   **e.** 1 kilometer

6. Estimate the length of each segment in metric units, and then measure each to check your estimating ability. (Notice that you can make up your own examples for further practice.)

   a. ——————————————————————————————
   b. ———————————————————————
   c. ——
   d. ————————
   e. the perimeter of the room you are in
   f. the height of the ceiling in the room you are in
   g. the length of a car you know
   h. the diameter of a quarter
   i. the height of this piece of paper
   j. the perimeter of this piece of paper

7. An imagined polygon has a perimeter of 24 cm, with each side having a length that is a whole number of centimeters. For each given polygon, what are all the possibilities for the lengths of its sides?

   a. a quadrilateral   **b.** a rectangle   **c.** a pentagon

8. If *n* identical square regions are placed so that touching regions share a whole side, what is an expression for the *maximum* perimeter possible? Use the side of the square as the unit of length.

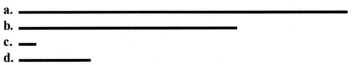

Sample trials

$n = 5$ $n = 5$ $n = 1$
Perimeter = 10   Perimeter = 12   Perimeter = 4

9. **a.** How does one find the shortest distance from a point to a line, as in the drawing to the left below? How does one find the distance between two parallel lines, as in the drawing to the right below?

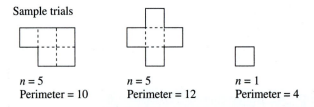

$P \bullet$

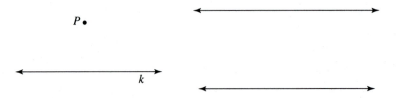

k

   b. For parallel lines, does it matter *where* you measure the distance between them?

10. What *length* can you measure to find out how many (horizontal) rows of squares are in the following shape?

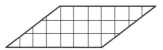

11. For the five figures below, the same-size circle is shown with an inside polygon and an outside polygon. Check that the perimeters of the outside polygons are getting smaller and the perimeters of the inside polygons are getting larger as you go from Figure A to Figure E.

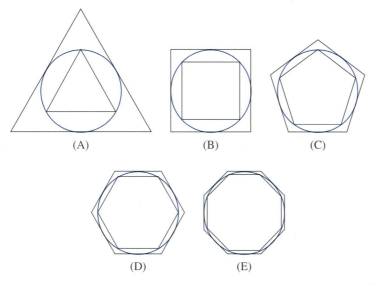

       (A)   (B)   (C)

       (D)   (E)

12. Complete each of the following measurement equalities:
    a. 1.2 km = ___ m   b. 5.3 m = ___ cm
    c. 62 mm = ___ cm   d. 3.25 dm = ___ cm
    e. We write "3.4 cm + 4.3 cm = 7.7 cm," but the sum could be as small as ___ cm.

13. Approximate the length of the curve first with segments 2 cm long, then with segments 1 cm long, and finally with segments 0.5 cm long. What would you estimate the length of the curve to be?

14. Estimate the size of each given angle. Then copy and measure to check your estimates. (*Note*: With some protractors, you may need to extend the sides of the angles. Also, as with line segments, you can make up your own angles for more estimation practice.)

  a.   b.   c.

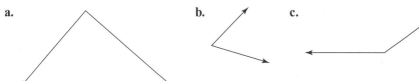

  d. Classify each of the angles in parts (a–c) as being acute, right, obtuse, or straight.

15. a. How many degrees must you turn a doorknob to open a door? Estimate and then check.
    b. A hippopotamus can open its mouth about 130°. Sketch an angle to show your estimate of such an angle, and then check by measuring.

16. How many degrees does the minute hand of a clock go through for each given time span?

    a. from 4:00 to 4:45   **b.** from noon to 12:35

    c. from 1:00 to 1:05   **d.** from 8:48 to 9:17

    e. from 2:00 to 3:00   **f.** from 2:00 to 3:30

    g. from 6:36 to 8:19   **h.** in a whole day

17. How many seconds are in a full 360° turn? One second is what part of a full turn?

18. How many degrees of longitude separate locations with the following degrees of longitude?

    a. Location P: 42° 16′ 23″ west longitude; location Q: 60° west longitude

    b. Location R: 115° 43″ west longitude; location S: 68° 32′ west longitude

    c. Location T: 34° 52′ west longitude; location U: 19° 48′ *east* longitude

19. Why are time zones roughly 15° of longitude wide?

20. The equatorial circle is about 25,000 miles long. What is the approximate length, in miles, of a 1° arc on the equatorial circle? A 1-minute arc? A 1-second arc?

21. Sketch angles with the following sizes by estimating: 30°, 45°, and 150°. Check your estimates by measuring your angles.

22. a. Give a justification that a pair of vertical angles will always be the same size (without relying on checking a particular example or examples); i.e., why must the ? angle in the drawing below have x°? (*Hint:* How is y related to the ? and x?)

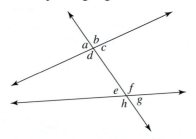

    b. Without using a protractor, give the sizes of all the angles with sizes not given in the drawing below.

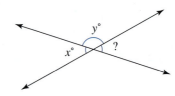

23. a. Give the four pairs of corresponding angles here.

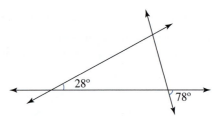

    b. Are the two angles in a corresponding pair always the same size?

    c. Suppose that the third line cuts two *parallel* lines. Experiment to see whether there is any relationship between the sizes of the two angles in a corresponding pair in this case. (It is easy to get parallel lines by using the lines on notebook paper.)

    d. Give two pairs of alternate interior angles in the previous drawing. How do they compare in size if the lines are parallel?

24. In the diagrams below, lines *m* and *n* are parallel. Give the sizes of the lettered angles.

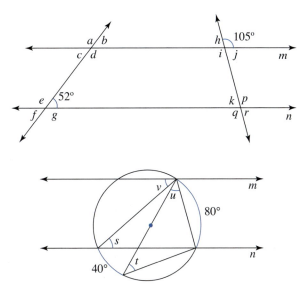

25. You know that the sizes of the angles of a triangle sum to 180°. Yet, direct measurements of the angles of a triangle may be "off," even when carefully measured.

    a. Give an argument for the sum being 180° based on using angles in parallel lines rather than direct measurement. Use the following diagram in stating your argument:

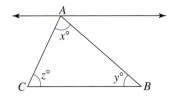

In parts (b–e), give the missing angle size(s) in the triangles described.

    b. Two angles have sizes 72° and 39°.
    c. A right triangle has an angle of 41°.
    d. A triangle has angles of 56° 39′ and 62° 43′.
    e. An isosceles triangle has an angle of 20°.   (There are two possibilities.)
    f. In every right triangle, what is the sum of the sizes of the acute angles?

26. In the following drawing, the triangle and the trapezoid are rotated 180° with the marked midpoints of sides as centers of the rotations. Give the measurements of angles (*a*) through (*g*), in terms of *w*, *x*, *y*, *z*.

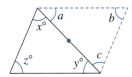 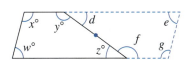

27. Recall that the sum of all the (interior) angles of an *n*-gon is $(n - 2)180°$.

    a. If the *n*-gon is regular or just equiangular, what is the size of each angle?
    b. Make a table showing the sizes of the angles in each equiangular *n*-gon from $n = 3$ to $n = 12$.
    c. As the number of sides of a regular *n*-gon grows larger and larger, what can you say about the size of each angle?
    d. The angle sum for some polygon is 4500°. How many sides does the polygon have?
    e. The angle sum for another polygon is 2520°. How many angles does the polygon have?

28. In a tessellation of the plane with regular polygonal regions, each vertex will involve the same arrangement of polygons, with the sizes of the angles at each vertex totaling 360°.

    a. Use the results from Learning Exercise 27(b) to argue that a tessellation involving exactly one kind of regular polygon is possible for only equilateral triangles, squares, and regular hexagons.

    b. A tessellation of the plane can involve more than one kind of regular polygon. Give some possibilities for such a tessellation, considering only the angles at a vertex. For example, because 90° + 135° + 135° = 360°, one square and two octagons could give a tessellation.

29. Recall from Section 17.5 that there are only five different kinds of convex regular polyhedra.

    a. Consider a vertex of a regular polyhedron. The sum of the angles with their vertices all at one vertex of the polyhedron must be less than 360°. Explain why.

    b. Argue that there are only five possible regular polyhedra.

    c. A semiregular polyhedron may involve more than one kind of regular polygon at each vertex, although each vertex must have the same arrangement. Give some possibilities for a semiregular polyhedron, considering only the angles at a vertex.

30. Skateboarders, skiers, or snowboarders may talk about "doing a 1080." What does that mean?

31. Three angles of quadrilateral STAR have sizes 63° 45′, 110° 25′, and 120.25°. Another quadrilateral is similar to STAR, with a scale factor 2.5. What are the sizes of the four angles in the second quadrilateral?

32. Tell whether the following statement is true or false, and explain your thinking: For different circles, arcs with the same number of degrees always have the same length.

33. Some teachers like to use tessellations to suggest some important geometric results. For example, fill in the measurements of the other angles that have the highlighted point as the vertex to get a justification that the sum of the sizes of the angles of a triangle is 180°, i.e., $x + y + z = 180$.

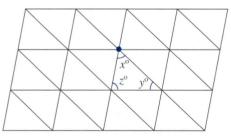

34. a. Any regular polygon can be fit inside some circle so that all the vertices of the polygon are on that circle. The center of the circle is the center of the regular polygon. Because the sides of a regular polygon are all equal, the arcs and the central angles they give must be equal. How large is each central angle for an equilateral triangle? A square? A regular pentagon? A regular hexagon? A regular n-gon?

    b. How could a teacher draw a regular heptagon with the help of a compass and a protractor?

35. In parts (a–c), find the number of degrees x in the exterior angle.

    a. 40° 80° $x°$

    b. 25° 30° $x°$

    c. $q°$ $p°$ $x°$

36. Just as angles in a plane are made when lines intersect, **dihedral angles** in space are made when two planes intersect. How might one measure a dihedral angle? For example, would it make sense to talk about a 90° dihedral angle? Explain your reasoning. (Giving your ideas, rather than looking up information, is of interest here.)

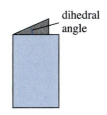

dihedral angle

37. Tell how you could get an angle of 94° if all you had was a wooden angle of 19° (and paper and pencil).

38. **a.** Although the 100-yard dash is common in the United States, the 100-meter dash is used in international track meets. One yard = 0.9144 meter. How do the 100-meter dash and the 100-yard dash compare? Give two answers, one for the additive comparison (difference) and one for the multiplicative comparison (ratio).
    b. Repeat part (a) for the mile (1760 yards) versus the 1500-meter run.

39. Complete each of the following measurement equalities. (Conversions for some English system length units are given in the Glossary; see **English system of measurement units**.)

    a. 4 miles = _____ yd   **b.** 1320 yd = _____ mi
    c. 1320 ft = _____ mi   **d.** 1.5 mi = _____ ft
    e. 12 yd = _____ in.   **f.** 84 in. = _____ yd

40. Technical words with everyday meanings or connotations can naturally be confusing to children (and others). Give different meanings or connotations for these.

    a. degree   **b.** yard   **c.** meter
    d. regular   **e.** right angle (vs. ____ angle)   **f.** plane

41. **a.** A hidden shape is congruent to quadrilateral *ABCD* shown in the following diagram. What do you know about specific measurements on the hidden shape?

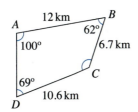

    b. Another hidden shape is similar to the given quadrilateral. What can you say about specific measurements on this hidden shape?

42. In the late 1700s, a Frenchman involved in the creation of the metric system thought the *grade*, a unit for angle size equal to one-hundredth of a right angle, would be useful.

    a. How many grades would be in a full circle?
    b. How many degrees would be in a centigrade?

43. You want to wrap the package at right with fancy ribbon as pictured. How many centimeters of ribbon do you need? Allow 25 cm for the bow.

44. **a.** With a compass and straightedge, construct a circle and a pair of perpendicular diameters for the circle. Join the endpoints of the diameters. What special quadrilateral do the endpoints appear to give? Can you justify that it *is* that shape?
    b. With a compass, draw a circle. Then, starting at some marked point, mark off points 1 radius apart around the circle. Join the points in order. What shape do these points appear to give?
    c. How might you construct a regular octagon and a regular 12-gon? [*Hint*: Use parts (a) and (b).]

45. Here is how to construct with compass and straightedge the two tangents to a circle from a point *outside* the circle: Use the segment joining the outside point P to the center C of the circle as the diameter of another circle. Where the two circles meet indicates how to draw the two tangents to the first circle. (The details of the construction are omitted in the drawing below.) Why does this method work?

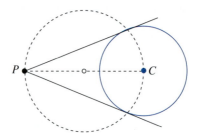

Supplementary Learning Exercises for Section 23.2

1. Estimate and then check by measuring with a ruler: How wide is your thumb and how long is your index finger, in centimeters?

2. Estimate the length of each segment in metric units, and then measure each to check your estimating powers.

   a. ─────────────────────

   b. ─────────────────────────────

   c. ───────────

3. We might measure and then write "$3\frac{1}{2}$ in. + $4\frac{1}{2}$ in. = 8 in.," but the sum could be as small as _____ in. or as large as _____ in.

4. Complete each of the following measurement equalities:

   a. 56 m = _____km   b. 5.6 m = _____cm
   c. 84 mm = _____cm   d. 84 mm = _____ m
   e. 75° = _____min   f. 75′ = _____ °

5. What is the perimeter of each of the following polygons?

   a. rectangle, length 18 in., width 1 ft, 3 in.
   b. rectangle with dimensions 1.63 m by 1.12 m
   c. rhombus with a side 3.57 cm
   d. kite with two sides 11.9 cm and 0.84 cm
   e. regular octagon with one side $4\frac{1}{3}$ yd

6. Estimate the size of each given angle. Then copy the angle and measure to check your estimate. With some protractors, you may need to extend the pictured sides of the angle.

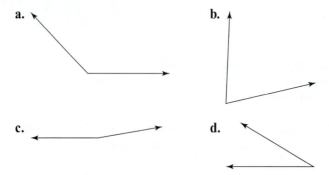

7. Without using a protractor, sketch angles with these sizes by estimating: 60°, 90°, and 120°. Check by measuring with a protractor.

8. Give the sizes of the missing angles in the polygons.

   a. triangle with two angles with sizes 62° and 49°
   b. triangle with two angles with sizes 39° 42′ and 100° 35′
   c. triangle with two angle sizes, 28° 30′ 45″ and 84° 42′ 27″
   d. quadrilateral with three angles of sizes 95°, 87°, and 9°
   e. parallelogram with one angle of size 75°
   f. rhombus with one angle of size 104°
   g. trapezoid with two angles on one of the parallel sides of sizes 110° and 83.5°
   h. equiangular octagon
   i. regular 15-gon
   j. isosceles triangle with largest angle 100°

9. Give the sizes of the lettered angles, if lines *x* and *y* are parallel.

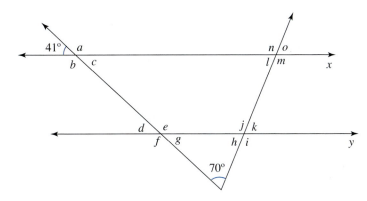

10. Give the missing sizes of angles and arcs (in degrees).

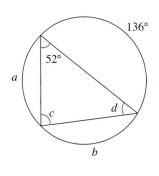

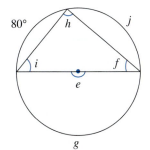

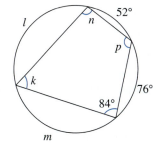

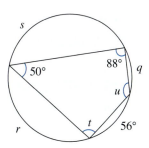

11. Which key idea of measurement did you use in Exercise 10?

12. Tell the number of sides of polygons with the following sums of angle sizes:

   a. 7200°   b. 4680°

23.3 Issues for Learning: Measurement of Length and Angle Size

As with geometry, the learning of measurement is not as satisfactory as one would hope for such a practical topic. Results on items from national and international tests give some indication of possible weak spots in the learning of measurement (though it is helpful to recognize that such broad results may well disguise the excellent performance of students in some individual classrooms). In one national testing,[2] given drawings of a clock, a ruler, a thermometer, and a bathroom scale, 28% of the fourth-graders could not identify which instruments best measure length, weight, and temperature. It is difficult to see how these students could have much understanding of measurement or how their schooling on measurement bore any relevance to life outside of school.

Some of the poor performance on tests is perhaps the result of how measurement is taught. Common concerns are (1) measurement is little more than vocabulary and filling in blanks in conversion exercises such as 4 feet = ___ inches, (2) measurement is taught without any actual measuring, and (3) the focus on measurement is on formulas without any emphasis on the foundations and meanings for the formulas.

DISCUSSION 7   When I Was Younger . . .

What are your memories of how you learned measurement in school? Did you have out-of-school experiences with measurement? ●

Neglect and limited foci may account for some of the poor test performance, but there are trouble spots in learning measurement that have become apparent through classroom research. For example, having children in a Grade 1–2 class[3] develop their own measuring tools by repeating a unit and then measuring with the tool revealed that, for measuring a segment like the one shown below, children had difficulty deciding whether the segment was $4\frac{1}{2}$ or $5\frac{1}{2}$ units long.

The research also made clear that counting units is not quite the same as thinking of ruler values in terms of a movement across a distance. Calling the line segments to be measured "Stuart Little trips" led to the latter understanding and even the insertion of the 0 on the homemade ruler.

Counting units

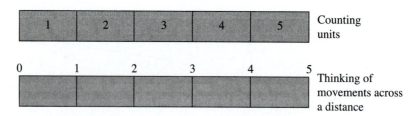

Thinking of movements across a distance

After a great deal of focus on the big ideas of measurement, children in an elementary school class were asked to summarize rules that were important for accurate measurement. Notice the big ideas of linear measurement that this child, whose work is shown on the next page, has come to understand.

Using your experiences over the past week, make a list of rules that you think are important to be followed for accurate measurement.

Rules for measurement: I think that one of the rule is that they should put ½ and ¼, and ¾ in the lines because if you put it in the spaces that means that you are counting the space but if you put the numbers in the lines you should already know what to call it. And you should have equal lines and if you would be very confused. And make sure that you are numbering right you're ruler because if you don't you will mestup big time like group #5! You should know were is the ending point and the starting point!

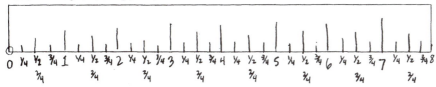

Create a ruler using those rules. Explain how the rules were translated into the creation of the ruler.

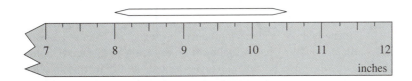

Children's early use of rulers can also be complicated by the many marks for fractions of units on the usual ruler, as performance on national test items shows. For example,[4] 29% of 9-year-olds in one national testing identified a ruler marking of $3\frac{3}{8}$ inches as $3\frac{1}{2}$ inches, as did 13% of the 13-year-olds. Only 2% of the younger group and 25% of the older group gave the correct reading of $3\frac{3}{8}$ inches. Items that involve a slight change from the standard use of a ruler often reveal a poor grasp of what the markings on a ruler mean. For example, in 2003 a fourth-grade item[2] from another test, shown below, gave these results: 42% thought the answer was 8 inches, 23% wrote $3\frac{1}{2}$, 14% wrote $10\frac{1}{2}$, and only 20% gave the correct answer.

What is the length of the toothpick in the figure above?

DISCUSSION 8   They Counted...

How do you believe the children who thought the answer was $3\frac{1}{2}$ inches were thinking? ●

The type of error students made on this example was also found in other research.[5] For example, a second-grade teacher broke off the first 2 inches of ruler, and then asked students how they could use this ruler to find the perimeter of a 5-inch by 7-inch card. One child measured the sides of the card as 7 inches and 9 inches and said the perimeter was 32 inches. Another student who had an unbroken ruler measured and said the perimeter was 24 inches. Some of the children in the class thought the first answer was correct, others thought the second was correct, causing quite a dilemma for the teacher.

These students had difficulties measuring with rulers, but they were at least learning to measure, which far too many students do not. On another national test,[6] fewer than half of the seventh-graders tested were able to find the perimeter of the two test items shown on the next page. Notice that not having the shape shown led to somewhat greater difficulty, even when less technical vocabulary was involved.

A. What is the perimeter of this rectangle? (46% correct)

B. What is the distance around a 4 by 7 rectangle? (37% correct)

Moreover, there are developmental considerations[7] in teaching ideas of measurement of just about any geometric quantity. For example, and perhaps surprisingly, a 5- or even 6-year-old often decides, "They would be the same," for the following item involving lengths:

> On which path, the dashed one or the zigzag one, would an ant walk farther, or would it be the same, going from *A* to *B*?

Similarly, for the line segments below, after agreeing that the two "sticks" in the drawing on the left are the same length, young children may say, after one stick is moved to the right, as in the second drawing, that the moved stick is longer. This *nonconservation of length* is common with young children.

Children also often have misleading intuitions about how angle size is measured. They often focus not on the amount of turn from one side of the angle to the other, but on the relative lengths shown for the sides or the distance between the endpoints of the segments shown for the sides.

Misleading intuitions

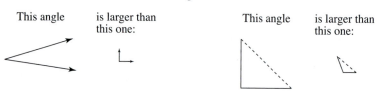

There is research evidence[8] of children's difficulties with conserving the size of angles. In one research study, children were shown an angle of 7.5° on a map, the amount that Amelia Earhart's navigation was off on her last flight. Some students thought that in "real life" that angle would be much bigger, "like 200°" (p. 293).

In this same study, another child drew the picture to the right to illustrate 60° and 90°, but was confused about which was larger because the arrow for the 60° angle was longer than that for the 90° angle.

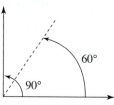

Still other students in this study were unsure about whether or not there could be curves in an angle and thought that the drawing shown below could be counted as an angle.

There are also difficulties with measuring angles because students do not know how to use a protractor properly, inasmuch as there is more than one line that could be selected as the terminal side of the angle.

Estimation of measurements is a practical skill that often receives little attention in the curriculum beyond looking for benchmarks for units of length. Estimations involving the measurement of length can take many forms, as the following tasks suggest[9]:

1. Estimate the length of this pencil in centimeters.

2. About how long is a pick-up truck in meters?

3. Get a meter stick and estimate the height of this building.

4. Which of those boards is about 4 feet long?

5. Think of something that is approximately 3 decimeters long.

Notice these characteristics in the examples:

- The object is present (tasks 1, 3, and 4) or not (task 2), or perhaps not named (task 5).

- The unit is present (task 3) or not (tasks 1, 2, 4, and 5).

- The estimated measurement is given (tasks 4 and 5) or not (tasks 1–3).

Estimations of angles could focus on larger than or smaller than 90°, then 45°, then 180°, and then 135°. One study with sixth-graders showed that estimating angle sizes before measuring them did give a better estimation performance than merely measuring the angles, and estimation transferred to other areas as well.

Making different choices for each characteristic can lead to different sorts of estimation tasks. Having the object and unit present and the estimated measurement given is, however, just a statement and not a task. (And not having any object, unit, or estimate is not a reasonable task.)

ACTIVITY 2 Different Estimation Tasks

Using different selections of the characteristics, make up two different estimation tasks, one involving length and one involving angles. ●

Studies[10] have shown that practice with the estimation of measurements does improve estimation performance, and so adults do estimate better than children do (although neither group does well).

TAKE-AWAY MESSAGE . . . Performance on wide-scale tests and results from research suggest that children's ability to deal with the concepts of length and angle is limited for a variety of possible reasons. Children often do not have a good foundation for the ideas of the quantities being measured, in this case length and angle, and they are sometimes not able to conserve length and angle size, thinking certain changes may affect size when they do not. These difficulties are compounded by the difficulties children have using rulers and protractors correctly. Estimation of length and angle size should be a part of the measurement curriculum.

23.4 Check Yourself

You should be able to work exercises like those assigned and to meet the following objectives:

1. Describe in general terms what the process of measurement involves.

2. State the four key ideas of measurement, and recognize when they have been used.

3. State the range of values that a given measurement covers. (For example, a length reported as 15.6 cm could be as short as 15.55 cm or almost as long as 15.65 cm.)

4. Recognize the "inverse" relationship between the size of the unit and the numerical part of the measurement.

5. Apply the four key ideas of measurement to length (or its special synonyms such as perimeter, height, circumference, etc.) and to angle size.

6. Use your personal benchmarks to estimate given lengths or given segments or lengths of named objects.

7. Estimate angle sizes and lengths, and use a protractor to measure given angles and a ruler marked in English or metric units to measure given segments.

8. Correctly use adjectives with angles (e.g., right angle, acute angle, obtuse angle, straight angle, supplementary angle, central angle, vertical angles, alternate interior angles, corresponding angles, exterior angle, inscribed angle, etc.), and deal with the relationships covered (e.g., angles with parallel lines, angles of a triangle, etc.).

9. Convert among metric units for length, convert among English units for length, and deal with angle sizes in degrees, minutes, and seconds.

10. Determine the total of the sizes of the angles in a polygon having a given number of sides, and the size of each angle in an equiangular or regular polygon having a given number of sides.

References for Chapter 23

[1]Klein, H. A. (1974). *The world of measurements*. New York: Simon & Schuster, p. 112.

[2]National Center for Education Statistics. (1999). http://nces.ed.gov/nationsreportcard/.

[3]Jaslow, L., & Vik, T. (2006). Using children's understanding of linear measurement to inform instruction. In S. Z. Smith & M. E. Smith (Eds.), *Teachers engaged in research: Inquiry into mathematics practice grades preK–2* (D. S. Mewborn, Series Ed.). Reston, VA: National Council of Teachers of Mathematics.

[4]Carpenter, T., Coburn, T. G., Reys, R. E., & Wilson, J. W. (1978). *Results from the first mathematics assessment of the National Assessment of Educational Progress*. Reston, VA: National Council of Teachers of Mathematics, pp. 95–96.

[5]Barrett, J. E., Jones, G., Thornton, C., & Dickson, S. (2003). Understanding children's developing strategies and concepts for length. In D. H. Clements & G. Bright (Eds.) *Learning and teaching measurement* (pp. 17–30). Reston, VA: National Council of Teachers of Mathematics.

[6]Kouba, V. L., Brown, C. A., Carpenter, T. P., Lindquist, M. M., Silver, E. A., & Swafford, J. O. (1988). Results of the fourth NAEP assessment of mathematics: Measurement, geometry, data interpretation, attitudes, and other topics. *Arithmetic Teacher, 35*(9), 10–16.

[7]Piaget, J., Inhelder, B., & Szeminska, A. (1960). *The child's conception of geometry*. New York: Harper & Row.

[8]Keiser, J. M. (2004). Struggles with developing the concept of angle: Comparing sixth-grade students' discourse to the history of the angle concept. *Mathematical Thinking and Learning, 6,* 285–306.

[9]Bright, G. W. (1976). Estimation as part of learning to measure. In D. Nelson & R. E. Reys (Eds.), *Measurement in school mathematics, 1976 Yearbook* (pp. 87–104). Reston, VA: National Council of Teachers of Mathematics.

[10]Sowder, J. (1992). Estimation and number sense. In D. Grouws (Ed.), *Handbook of research on mathematics teaching and learning* (pp. 371–389). New York: Macmillan.

ACTIVITY 2   Thinking INSIDE the Box

Using the smallest square region defined by the dots as the unit, find the area of the region enclosed by each of the following shaded polygons. (The dashed segments with the triangular regions are a hint that helps to avoid lots of estimating. Think about pieces of rectangular regions.)

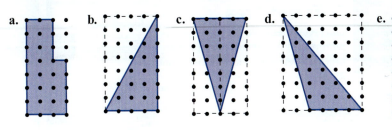

a.   b.   c.   d.   e.

If regions are congruent, the regions have equal areas. The "reverse" statement (called the **converse** of the original statement) is this statement: If regions have equal areas, then the regions are congruent. Examine the results for Activity 2 to show that the converse is not always true.

Three-dimensional shapes such as polyhedra have surfaces, so we can speak of the **surface area** of such shapes. For example, the six faces of a cube are square regions, so the surface area of the cube would be the sum of the areas of these six square regions.

 DISCUSSION 3   Finding Surface Areas

How would one find the surface area of three-dimensional shapes, such as polyhedra, from your kit of shapes? What key idea of measurement is involved? ●

EXAMPLE 1

Find the surface area of the given right rectangular prism.

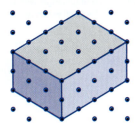

Solution

Right rectangular prisms are easy to work with, because the square regions (the units) that cover the faces are easily seen by drawing in the dashed lines, as shown to the right. The areas of the top and bottom faces are each 12 units; the areas of the two side faces are each 8 units; and the areas of the front and back faces are each 6 units. The surface area of the prism is then 52 units, i.e., 52 square regions. ●

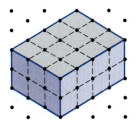

This Grade 4 page shows that figures with the same perimeter can have different areas.

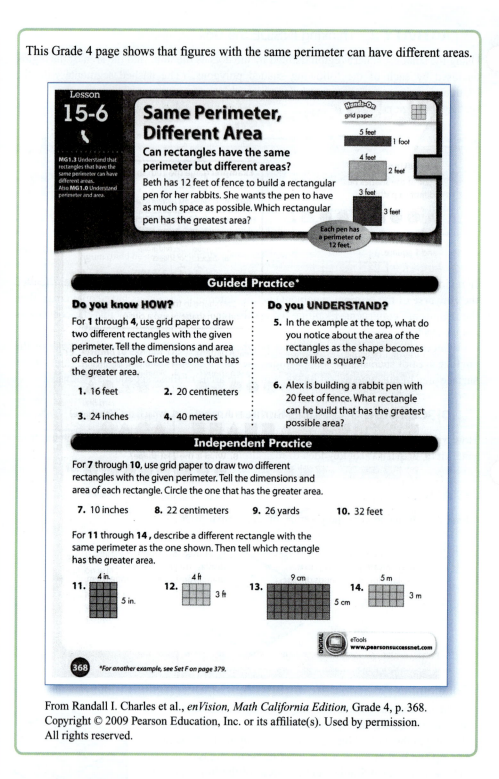

From Randall I. Charles et al., *enVision, Math California Edition,* Grade 4, p. 368.
Copyright © 2009 Pearson Education, Inc. or its affiliate(s). Used by permission.
All rights reserved.

Figures that are similar also have areas that are related. One way to think about how they are related is to focus on each square region in the area of the original shape and to examine the area of that square region in the enlarged (or shrunken) version having the scale factor *k*.

DISCUSSION 4 Relating Areas of Similar Shapes

The following diagram shows a square centimeter and its images for two different size changes. How are the areas of the original square centimeter and each image related? How can that relationship be predicted from knowing the scale factor?

Scale factor = ?

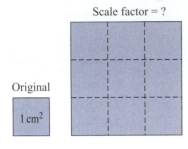

Original

1 cm²

Scale factor = ?

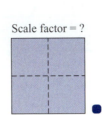

Because *every* square centimeter in an original shape gives k^2 square centimeters in a similar shape with scale factor k, the area of the new shape is always k^2 times the area of the original shape. If k is the scale factor, the relationship is expressed in equation form as follows:

$$\text{new area} = k^2 \times \text{original area}$$

EXAMPLE 2

A larger shape with scale factor 4 is similar to the octagon shown to the right. What are the area and perimeter of the larger shape?

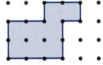

Solution

From a direct counting of the units, the given shape has an area of 8 square units and a perimeter of 14 length units. With the scale factor 4, the new area is $4^2 = 16$ times as large, or $16 \times 8 = 128$ square units, and the perimeter is 4 times as large, or 56 length units. (A drawing of the larger shape makes these results clear.) ●

TAKE-AWAY MESSAGE . . . Understanding area includes recognizing that *area of a region* means the number of area units required to cover the region. The key ideas of measurement all apply to area. Congruent shapes have the same areas, but not conversely: Shapes having the same area are not necessarily congruent. The surface area of a three-dimensional shape is the number of area units required to cover the shape, and it is often obtained by finding the areas of the individual surfaces that make up the shape and adding those results. The areas of similar shapes are related by the *square* of the scale factor: The area of the image is equal to the square of the scale factor times the area of the original.

Learning Exercises for Section 24.1

1. Find the areas of the shapes given on the next page by using pattern blocks (or pattern block cut-outs from the "Masters") as described below:

   Shape A in hexagonal regions, in trapezoidal regions, and in (wide) rhombus regions

   Shape B in hexagonal regions, in trapezoidal regions, and in (wide) rhombus regions

continue

Notes

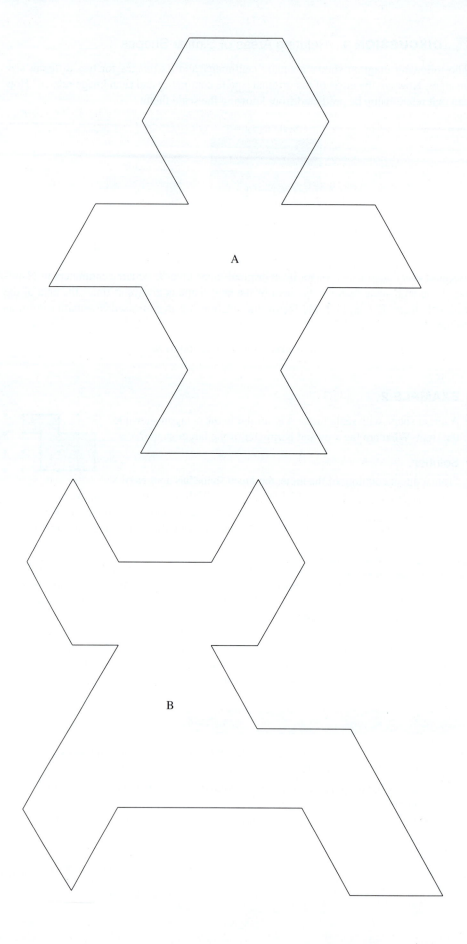

2. **a. Pattern Blocks** Invisible shape X has area 16 trapezoidal regions. What is its area in (wide) rhombus regions? Explain your reasoning.

   b. Invisible shape Y has area 21 trapezoidal regions. What is its area in (wide) rhombus regions? Explain your reasoning.

3. Some people use the size of a piece of toast as a benchmark for a square decimeter. Choose a benchmark for a square centimeter, a square meter, and an are.

4. Estimate each given area using an appropriate metric unit (think metric by using your benchmarks). Check your estimates when you can.

   a. the area of this page of this textbook
   b. the area of a postage stamp
   c. the area of the floor of the room you are in
   d. the area of a wall of the room you are in
   e. the area of a window in the room you are in
   f. the surface area of your body

5. Someone has been using the following nonstandard units in finding the areas of regions. For each given equality statement, tell whether x or 20 is greater, and explain.

| Unit P | Unit Q | Unit R |

   a. 20 P's = x Q's   **b.** 20 Q's = x R's
   c. 20 P's = x R's   **d.** 20 R's = x P's

   In parts (e–f), tell which of y or x is greater, and explain.

   e. y Q's = x R's
   f. x P's = y Q's
   g. If 24 P's = 7 Q's, how many P's make 1 Q? How many Q's make 1 P?
   h. If 8 Q's = 3 R's, how many Q's make 1 R? How many R's make 1 Q?
   i. If 3.7 blobs = 2.4 globs, then how many blobs make 1 glob? How many globs make 1 blob?

6. Al, Beth, and Cai are measuring the same regions, but they are using different units. (The squares indicated are all the same size.)

| Al's unit | Beth's unit | Cai's unit |

   For each person's measurements, give the measurements that the other two people will find for the regions.

   a. Al's measurement of a region is 16 of his units.
   b. Beth's measurement of a different region is $9\frac{1}{2}$ of her units.
   c. Cai's measurement of a third region is $6\frac{1}{3}$ of his units.

7. Make sketches to show how you could explain each of the following relationships:

   a. how many square inches are in a square foot and what part of a square foot a square inch is
   b. how many square feet are in a square yard and how many square yards make a square foot
   c. how many square yards are in a square mile

8. How many acres are in a square mile?

9. Complete, using your "metric sense" (your mental pictures), for the following units:

   a. 2.3 dm^2 = ___ cm^2 **b.** 45 dm^2 = ___ m^2
   c. 19.6 cm^2 = ___ dm^2 **d.** 0.04 m^2 = ___ cm^2

10. How long are the sides of a square region that has area 1 are (symbolized as 1 a)?

11. What is the approximate area enclosed by each shape?

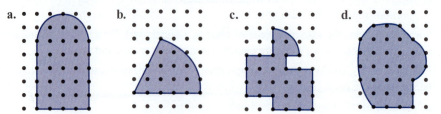

a.   b.   c.   d.

12. Determine the area enclosed by each polygon. Use the natural unit.

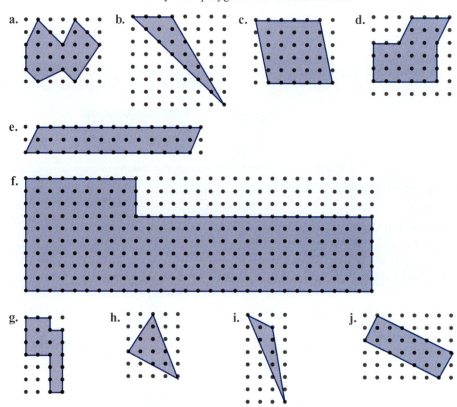

a.   b.   c.   d.

e.

f.

g.   h.   i.   j.

13. a. Describe an efficient way to find the surface area of a right regular octagonal prism.
   b. Describe an efficient way to find the surface area of a regular hexagonal pyramid.
   c. Describe an efficient way to find the surface area of a regular octahedron (see shape H in the kit of shapes).

14. Give the surface area of each polyhedron. Use the natural unit.

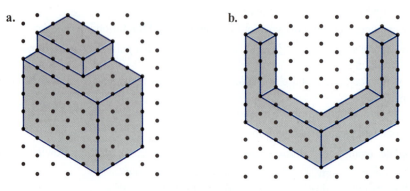

a.   b.

15. Draw a net for each shape to show the regions involved in finding the surface area.

   a. a right circular cylinder   **b.** a right circular cone

16. Is 4 square meters the same as a 4-meter square?

17. Many people confuse area and perimeter. How would you advise them, so that they do not mix up the ideas?

18. Rulers measure length, and protractors measure angle size. Is there a comparable school tool for measuring area? Mark a square centimeter grid on a transparent piece of plastic or on thin paper, and find the area of each given region.

a. 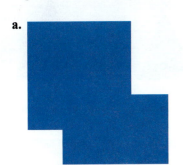   **b.**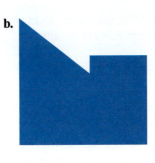

19. Give the area of each tangram piece for the situation described.

   a. The area of the whole square region is 1 area unit.
   b. The area of the whole square region is 47 area units.
   c. The area of the whole square region is $4\frac{2}{3}$ area units.

20. In the following diagram, suppose the shaded tessellation piece is 1 area unit. Give the area of the whole region pictured. What key idea of measurement did you use?

21. a. Ollie and Olivia plan to buy carpet to cover their basement floor. After measuring, they determine that the basement floor has an area of 15 yd². When they get to the carpet store, the salesperson asks for the area of the basement floor in square feet. Ollie reasons that because 1 yd is equal to 3 ft, 15 yd² will equal 45 ft². Is Ollie's reasoning correct? Explain, using a diagram to help in your explanation.

   b. Ollie and Olivia decide to tile their bathroom floor with ceramic tile from Italy. They know that the Italians use the metric system, so they cleverly measure the length and width of the floor in decimeters. The floor is 18 dm long and 15 dm wide. They want to be prepared for the salesperson's questions this time! What is the area of the bathroom floor in square decimeters? Square centimeters? Square meters? Explain your reasoning, using diagrams to help in your explanations.

22. Trace the given 8 by 8 square, and then cut on the solid line segments. Rearrange the four pieces to make a nonsquare rectangle. What are the length and width of the rectangle and its area? What was the area of the original square? What key idea appears to be violated here?

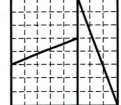

23. Two irregular shapes are similar with scale factor 5 involved. With sketches, show how the image of a square centimeter in the original shape looks in the larger shape.

24. **a.** Two polyhedra are similar with scale factor 7 involved. The surface area of the smaller polyhedron is 490 cm². What is the surface area of the larger polyhedron? Explain.
    b. Two polyhedra are similar with scale factor 8 involved. The surface area of the *larger* polyhedron is 640 cm². What is the surface area of the smaller polyhedron? Explain.
    c. Why is the ratio of the surface areas of these two pyramids likely *not* either 2.25 or $\frac{16}{9}$?

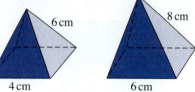

25. Two curved regions are similar, and they have areas of 500 cm² and 125 cm². What scale factor is involved? How do you know? (There are two possibilities.)

26. Without actually finding the measurements, describe how you might determine each given measurement of the shaded part of the figure at the right.

    a. the perimeter   **b.** the area

27. Use square-grid paper or squares to help you answer this question about shapes made with squares: For a given perimeter, will the area enclosed always be the same?

28. Is each statement correct? Correct any statement that needs correction.

    a. "The field was 6 square acres in area."
    b. "The area of a circle is 360 degrees."
    c. "It took 12 tattoo stickers to cover my whole face, so the perimeter of my face is 12 tattoo stickers."
    d. "The area of the rectangle is 20 decimeters."
    e. "One meter is 100 centimeters, so one square meter is 100 square centimeters."

29. Using the following units and *your answers for Learning Exercise 12,* give the areas of the figures in Learning Exercise 12:

    a. a right triangular region that is half the natural dot-paper unit, as shown at right (*Hint*: Do not count!)
    b. the region in part (j) of Learning Exercise 12

Supplementary Learning Exercises for Section 24.1

1. A book has 200 pages, each $8\frac{1}{2}$ in. by 11 in. What is the approximate total area of the pages in the book, *in square decimeters*? Use your metric sense rather than extensive calculations.

2. Three kings—Jong, Kong, and Long—have historically used different pieces of a standard regular hexagonal region as units for measuring area (see the drawings below).

King Jong's   King Kong's   King Long's
unit, J   unit, K   unit, L

Give each of the following area measurements as they would be expressed in the other two kingdoms. Show your work.

a. 30 J	**b.** 30 K	**c.** 30 L
d. 17 J	**e.** 17 K	**f.** 17 L
g. *m* J	**h.** *n* K	**i.** *p* L

3. How many square yards make a square mile? Write enough to make your reasoning clear.

4. How are a square millimeter and a square centimeter related in size? Tell how you know.

5. Using your metric sense for the units involved, complete each of the following measurement equations:

   a. $78 \text{ cm}^2 = $ _____ dm^2 **b.** $15 \text{ dm}^2 = $ _____ cm^2
   c. $1.4 \text{ m}^2 = $ _____ dm^2 **d.** $1.4 \text{ m}^2 = $ _____ cm^2
   e. $128 \text{ mm}^2 = $ _____ cm^2 **f.** $2.6 \text{ cm}^2 = $ _____ mm^2

6. Determine the area enclosed by each polygon. Use the natural dot-paper unit for area.

   a.   **b.**   **c.** 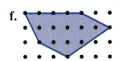   **d.**

   e. 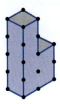   **f.** 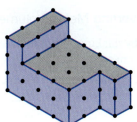   **g.**

7. Give the surface area of each polyhedron. Use the natural unit.

   a.   **b.**

8. Two regions are similar with a scale factor 1.2.

   a. If the area of the smaller region is 72 cm^2, what is the area of the larger region?
   b. If the area of the *larger* region is 72 cm^2, what is the area of the smaller region?

9. The areas of two similar regions are 9 in^2 and 81 in^2. What two scale factors could be involved?

10. A drawing of the school mascot is enlarged so that a 4-cm length in the original is 36 cm long in the enlargement.

    a. What is the ratio of the areas of the enlargement and the drawing?
    b. Without more information, why do you not know the area of either the drawing or the enlargement?

11. Using the following units and *your answers for Supplementary Learning Exercise 6,* give the areas of the figures in Supplementary Learning Exercise 6:

    a. a right triangular region that is half the natural dot-paper unit, as shown at right (*Hint:* Do not count!)
    b. the region in part (d) of Supplementary Learning Exercise 6

24.2 Volume

Any real object takes up space. Even a scrap of paper, a speck of dust, or a strand of cobweb takes up some space. The quantity of space occupied is called the **volume**. As with the terms length and area, the term *volume* is sometimes used for the attribute as well as the measurement or value of the attribute.

The topic of volume can appear in the curriculum as early as Grade 3, culminating in at least a few volume formulas in later grades. In this section, we treat only the main ideas of volume; we discuss the volume formulas in Chapter 25.

Boxes, cans, and other containers enclose an amount of space; we often refer to their **capacities** or to the volumes they enclose. The *material* actually making up a box involves a certain amount of space, so *literally,* the volume of the box would be the volume that the box would occupy if flattened out. Usually, *volume of the container,* however, refers to the capacity of the container rather than to the volume of the material that makes up the container.

The direct measurement of the volume of an object involves a matching of the object with a unit, an object having the volume attribute. Because every real object has volume, one could theoretically use any physical object as a unit for volume. The standard units, however, are cubic regions with the length of each edge of the cube having 1 length unit. For example, a cubic region with each edge 1 inch long is a standard unit called a **cubic inch** (abbreviated cu. in. although in^3 is increasingly common). Hence, volume basically involves counting the number of cubes required to match or fill a given shape.

Our attention will focus largely on those standard units in the metric system—e.g., cubic centimeters, cubic decimeters, and cubic meters. Because the cubes will have sides 1 centimeter, 1 decimeter, or 1 meter, you should mentally review your benchmarks for those length units.

Which is a cubic decimeter? Which is a cubic meter?

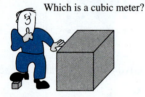

 DISCUSSION 5   Converting Metric Volume Units

Answer the following questions by thinking about the sizes of the units (and without using a formula). Explain your thinking in each case.

a. How many cubic centimeters does it take to fill a cubic decimeter box?
  That is, _____ cm^3 = 1 dm^3.   Vice versa, 1 cm^3 = _____ dm^3.

b. How many cubic decimeters does it take to fill a cubic meter box?
  That is, _____ dm^3 = 1 m^3.   Vice versa, 1 dm^3 = _____ m^3.

c. How many cubic centimeters does it take to fill a cubic meter box?
  That is, _____ cm^3 = 1 m^3.   Vice versa, 1 cm^3 = _____ m^3. ●

The cubic meter is too large a unit for many everyday purposes, so the metric system also recognizes a smaller unit, the **liter** (sometimes spelled litre), which is about 1.06 quarts, to measure volumes of liquids. You have likely seen bottles of water or soda that hold 1 or 2 liters, so you have some benchmark for the size of 1 liter. The symbol for liter is ℓ, or L. Because the lowercase printed letter l looks so much like the numeral 1, the value 1 l could be misread as eleven instead of 1 liter. We can apply all the metric prefixes to liter: e.g., 1 milliliter is 0.001 liter, or 1 mL = 0.001 L. Notice that although the liter and milliliter are units for volume, we do not say *cubic liter* or *cubic milliliter* because the words "liter" and "milliliter" already refer to units for volume.

The liter is related to the official cubic metric units for volume. Fortunately, the relationship is a nice one: **1 liter = 1 cubic decimeter.**

Notes

Think About ...

At one time a gallon of gasoline cost about $3.75 in the United States. A European visitor said to you, "That's interesting! I pay the equivalent of $1.90 per liter." Who (the visitor or you) pays more for gasoline?

 DISCUSSION 6 In the Hospital

Many medicines are measured in metric units. How are a milliliter and a cubic centimeter related in size? ●

Because the units cubic centimeter and milliliter are the same size, containers marked for measuring liquid volumes can use either unit. (As an aside, you may know that medical personnel most often write cc and say "see-see," rather than write cm^3 and say "cubic centimeter." They do so from long tradition rather than from any antipathy toward the official SI conventions.)

DISCUSSION 7 Volumes by Counting

What is the volume of each polyhedron? What key idea of measurement did you use? (Use the natural cubic region as the unit.)

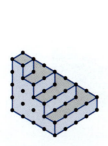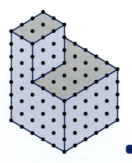

If two three-dimensional shapes happen to be congruent, then their volumes are equal, as are their surface areas. Recall that if two three-dimensional shapes are similar with scale factor k, then their volumes are also related in a predictable fashion although the two shapes may not be the same size:

$$\text{new volume} = k^3 \times \text{original volume}$$

In other words, the ratio, new volume to original volume, is the cube of the scale factor. One way to see this relationship is to think of each cubic region making up the original shape, and then think about what happens to each such cube under a size change. The length of each side of the original cube is now k times as long, so the image of the original cube will be a k by k by k cube. The following diagram shows examples with scale factor 2 and scale factor 3. Check to confirm that the images have 8 ($= 2^3$) and 27 ($= 3^3$) cubes, respectively. Hence, each cube in the original will give 2^3 and 3^3 cubes in the respective images. That is, the new volume will be 2^3 and 3^3 times as large as the original volume.

Original cube

Image with scale factor 2

Image with scale factor 3

EXAMPLE 3

Hidden shape X is congruent to the first polyhedron given in Discussion 7. Hidden shape Y is similar to that same polyhedron but larger with scale factor 6. What are the volumes of shapes X and Y?

Solution

Congruent shapes have the same volumes, so shape X has volume 30 cubic units. Volumes of similar shapes are related by the third power (cube) of the scale factor. Thus, the volume of shape Y is $6^3 \cdot 30$, or $216 \cdot 30 = 6480$ cubic units. ⬤

EXAMPLE 4

A larger prism is similar to a 2 by 3 by 2 right rectangular prism with scale factor 2. Sketch both prisms and compare their surface areas and volumes.

Solution

In the following sketch of the smaller prism, counting gives a surface area of 32 area units (square regions) and a volume of 12 volume units (cubical regions). Hence, the surface area of the larger prism should be $2^2 \cdot 32 = 128$ area units, and its volume should be $2^3 \cdot 12 = 96$ volume units. You should confirm the results for the larger prism by counting in the given sketch.

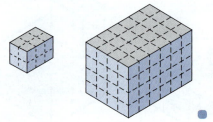

TAKE-AWAY MESSAGE . . .   Volume can be measured in the metric system using two types of units. One type is based on cubic regions with basic metric length units for edges. The other type is based on the liter, which is defined as 1 cubic decimeter. Here is a summary of all the relationships for similar space figures: Given a scale factor k, corresponding angles will have the same size, but new lengths will be k times as long as the corresponding original lengths, new areas will be k^2 times as large as corresponding areas in the original objects, and new volumes will be k^3 times as large as corresponding volumes in the original objects.

Learning Exercises for Section 24.2

1. Which attribute—length, angle size, area, or volume—could each of the following quantities involve? Give an appropriate metric unit in parts (a–m).

   a. the amount of water in a lake
   b. the amount of room on the lake for sailing
   c. the average depth of the lake
   d. the coastline of the lake
   e. the size of an oil slick on the lake
   f. the amount of water displaced by a ship
   g. the difference between sightings of two landmarks from a sailboat
   h. the amount of debris on the lake
   i. the amount of rope needed to moor a boat
   j. the amount of water that falls on the lake during a rainstorm

continue

k. the amount of water that evaporates from a lake
l. the part of the lake where swimming is allowed
m. the amount of temporary fencing needed to isolate the volleyball courts
n. how many identical advertising patches it would take to cover a sail

2. Find personal benchmarks for each given unit.

   a. 1 cm^3 **b.** 1 mL **c.** 1 dm^3 **d.** 1 L **e.** 1 m^3

3. What would be an appropriate metric unit for measuring the volume of each object?

   a. a ream of paper   **b.** a dose of medicine
   c. a gold nugget   **d.** a water tank
   e. a loaf of bread   **f.** an amount of liquid drunk daily

4. Estimate the volume of each object using metric units. (Check your estimates if it is feasible to do so. For example, some measuring cups have metric calibrations, and the labels for some objects give metric units.)

   a. a cup of water   **b.** a quart of milk
   c. a can of soda   **d.** the space of the room you're in

5. Design a net for a cube having 1 dm as the length of each edge. Would your net fit on the usual piece of paper?

6. For each equality, tell how x and y would be related ($x = y$, $x > y$, or $x < y$). Explain your reasoning.

   a. x spoonfuls $= y$ bathtubfuls   **b.** x truckloads $= y$ wheelbarrowfuls

7. Complete the following relationships and describe how you could derive the results (without using formulas):

   a. $1 \text{ dm}^3 = \underline{\quad} \text{ cm}^3$ **b.** $1 \text{ m}^3 = \underline{\quad} \text{ dm}^3$
   c. $1 \text{ L} = \underline{\quad} \text{ cm}^3$ **d.** $1 \text{ m}^3 = \underline{\quad} \text{ cm}^3$

8. Complete each equality.

   a. $1 \text{ cm}^3 = \underline{\quad} \text{ dm}^3$ **b.** $1 \text{ cm}^3 = \underline{\quad} \text{ m}^3$ **c.** $1 \text{ dm}^3 = \underline{\quad} \text{ m}^3$

9. Complete each statement.

   a. A decimeter is _____ as long as a centimeter, but a square decimeter is _____ as big as a square centimeter, and a cubic decimeter is _____ as large as a cubic centimeter.
   b. A meter is _____ as long as a centimeter, but a square meter is _____ as big as a square centimeter, and a cubic meter is _____ as large as a cubic centimeter.

10. Use your metric sense to complete each equality.

    a. $3.28 \text{ dm}^3 = \underline{\quad} \text{ cm}^3$ **b.** $3.28 \text{ dm} = \underline{\quad} \text{ cm}$
    c. $225.7 \text{ cm}^3 = \underline{\quad} \text{ dm}^3$ **d.** $225.7 \text{ cm}^2 = \underline{\quad} \text{ dm}^2$

11. **a.** Without using formulas, derive the relationship between a cubic inch and a cubic foot and the relationship between a cubic foot and a cubic yard. (*Hint*: Make a drawing.)

    b. One U.S. gallon is 231 cubic inches (a gallon is also 3.785 L). How many gallons are in a cubic foot?

12. A large measuring cup has $3\frac{2}{3}$ cups of water in it. A solid piece of metal is submerged in the cup, raising the water level to the $4\frac{1}{4}$ cup mark. What is the volume of the piece of metal? Which of the four key ideas of measurement (Section 23.1) is involved here?

13. How is area different from volume? Can anything have area without having volume? Can anything have volume without having area?

14. In parts (a–c), give the surface area and volume of the polyhedron. Use the natural units.

a. 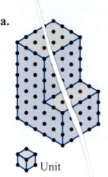 b. 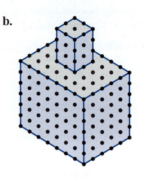 c.

Unit

d. If the small cube above were 1 centimeter on each edge, what is the volume of each polyhedron in parts (a–c), in milliliters?

e. Do any of the polyhedra in parts (a–c) have a volume greater than 1 liter?

15. Conveniently, 1 liter (or 1 cubic decimeter) of water has a mass equal to 1 kilogram, at 4°C and at standard air pressure. At that temperature and air pressure, what is the mass of 1 milliliter of water? One cubic centimeter of water? (Some plastic cubic centimeters used in classrooms are designed to preserve this relationship.)

16. Find a way of putting 4 cubes together face to face so that the volume is 4 cubic regions, but the surface area is 16 square regions. What is the surface area of most shapes made from 4 cubes?

17. Each of the sketches below represents the bottom story of a skyscraper and gives how many stories the skyscraper has. In each skyscraper, every story is just like its bottom story.

a. What is the volume of the bottom story of each skyscraper?

b. What is the area of the floor in the bottom story of each skyscraper?

c. Which skyscraper has the greater volume? Explain your reasoning.

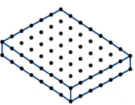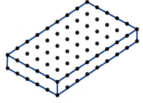

30 stories high   31 stories high

18. How many cubic regions would it take to fill the cylinder in the following sketch? What key idea(s) of measurement is (are) involved here?

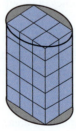

19. Cooking involves many volume units—e.g., teaspoon (tsp), tablespoon (Tbsp), cup (c), pint (pt), quart (qt), and gallon (gal). Using the relationships

$$1 \text{ Tbsp} = 3 \text{ tsp}, \quad 1 \text{ c} = 16 \text{ Tbsp}, \quad 1 \text{ pt} = 2 \text{ c}, \quad 1 \text{ qt} = 2 \text{ pt}, \quad 1 \text{ gal} = 4 \text{ qt}$$

complete the following conversions:

a. $4\frac{1}{2}$ Tbsp = _____ tsp

b. 5 tsp = _____ Tbsp

c. 8 Tbsp = _____ c

d. $\frac{2}{3}$ c = _____ Tbsp

e. 3 pt = _____ c

f. 1 gal = _____ pt or _____ c

g. $3\frac{1}{2}$ qt = _____ gal

h. $2\frac{1}{3}$ gal = _____ qt

i. $1\frac{1}{2}$ pt = _____ qt

j. $\frac{1}{3}$ qt = _____ pt

k. 1 c = _____ tsp

l. 3 Tbsp = _____ c

20. Consider the following diagrams:

a. Suppose that each shape is congruent to a hidden shape. Find the surface areas and the volumes of the hidden shapes. Tell how you know. Use the natural square and cubic regions as units.

b. Suppose now that each shape is similar to another hidden shape with scale factor 5. Find the surface areas and the volumes of the two hidden shapes. Tell how you know. Use the natural square and cubic regions as units. (A rough sketch of the image may be helpful.)

21. One polyhedron has surface area 5200 cm^2 and volume 24,000 cm^3. Give the surface area and the volume of the polyhedron's image for a dilation with the given scale factor. Write enough steps to make your thinking clear.

a. scale factor = 2

b. scale factor = $\frac{3}{4}$

c. scale factor = $3\frac{1}{5}$

d. scale factor = 110%

Supplementary Learning Exercises for Section 24.2

1. Complete each of the following to make true SI equations. Mentally picture each unit.

a. 2 dm^3 = _____ cm^3

b. 2 dm^3 = _____ mL

c. 49 mL = _____ cm^3

d. 1.3 m^3 = _____ dm^3

e. 4 m^3 = _____ cm^3

f. 1 m^3 = _____ L

2. Complete each of the following to make true English-system equations. Mentally picture each unit.

a. 4 ft^3 = _____ in^3

b. 27 ft^3 = _____ yd^3

c. 3 yd^3 = _____ ft^3

3. In each measurement equality, which of x and y is greater? Tell how you know.

a. x dm^3 = y m^3

b. x dm^3 = y cm^3

c. x ft^3 = y in^3

d. x ft^3 = y yd^3

4. Give the volume of each polyhedron in parts (a) and (b). Use the natural unit.

a. 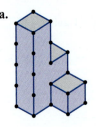   **b.**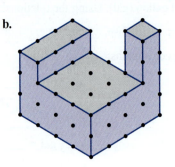

c. If the small cubes in parts (a) and (b) were 1 centimeter on each edge, what is the volume of each of the polyhedra, in milliliters?

d. If the small cubes in parts (a) and (b) were 1 foot on each edge, what is the volume of each of the polyhedra, in cubic feet? Does either have a volume greater than 1 cubic yard?

5. Child: "What is volume?" How would you reply?

6. Consider the following polyhedra:

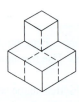 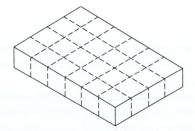

a. Suppose that each shape is congruent to a hidden shape. Give the surface area and the volume of each hidden shape. Use the natural square and cubic regions as units.

b. Now suppose that each shape is similar to a larger hidden shape with scale factor 6. Give the surface area and the volume of the hidden shapes. Use the natural square and cubic regions as units.

7. One right rectangular prism has surface area 856 cm^2 and volume 1.68 L. Give the surface area and the volume of the prism's image for a dilation with each given scale factor.

a. scale factor 2   **b.** scale factor $\frac{2}{3}$

c. scale factor 10   **d.** scale factor 10%

24.3 Issues for Learning: Measurement of Area and Volume

In Section 23.3 we wrote about some difficulties students have with learning measurement, in particular length and angle measurement, the focus of Chapter 23. In this section we continue to explore those difficulties by considering the concepts related to measuring area and volume. To begin to understand some of the difficulties with area measurement, consider this activity.

 ACTIVITY 3   Covering a Desktop

With a rectangular (not square) card like an index card, find out how many such rectangles would cover the top of your desk (no overlaps). Compare your result with that of others who have the same type of desk.[1] ●

Information from national tests indicates some of the types of difficulties students have with the concept of area. Consider the following two problems from a national test,[2] for which the performances by seventh-graders are given. The results show a disappointing ability to give the areas of rectangular regions. Only about half of the seventh-graders could answer each item correctly, even though the formula for the area of a rectangular region is usually encountered no later than Grade 4 and is routinely reviewed.

A. What is the area of this rectangle? (56% correct)

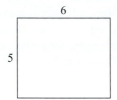

B. What is the area of this rectangle? (46% correct)

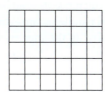

As is not unusual, the item closest to the concept of area, item B, is actually somewhat more difficult than item A (found by applying the formula). The most common error in determining the areas was to give the perimeters (and vice versa, giving areas for items seeking perimeters), illustrating confusion about the ideas of area and perimeter or about how they are calculated.

Middle school students have even more difficulty with measurement when a problem involves more than one step. Only about a third of U.S. eighth-graders were successful with the problem below.[3] (Because the question was a multiple-choice item with four options, you could expect 25% to be successful just from guessing.) Perhaps too much attention in classes to only straightforward, one-step problems leads students to expect a quick answer. Even though these results are given for middle schoolers, learning these concepts should have taken place in elementary school.

▶ A rectangular garden that is next to a building has a path around the other three sides, as shown.

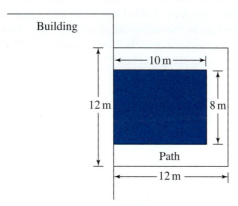

What is the area of the path?

A. 144 m²   **B.** 64 m²   **C.** 44 m²   **D.** 16 m² ◀

Research[4] has shown that developmental issues, as with length, are apparent with area ideas. Young children instructed to draw in squares of a given size to cover a rectangular region may draw several (rough) squares in the region but leave considerable gaps in the coverage. Later, they may proceed somewhat systematically, first drawing squares along the edges but then filling in the remaining region in a helter-skelter, often inaccurate, fashion. At some stage, they realize that a row or a column of squares can be repeated to cover a rectangular region, giving an efficient way of determining the total number of square regions required to cover the rectangular region. (The researchers also noted that grasping this row or column structure is very important to understanding the area model for multiplication.) This research suggests a good project for you. Ask some elementary school students, if you have access to any, to undertake this task of covering a region by drawing squares of the same size.

Another central concept for understanding measurement is that moving (rigidly) a shape does not change its measurements. For example, in Chapter 25 it is likely that in justifying the area formula for a parallelogram region, you will cut off a triangular region from one part of the parallelogram region and move it to another part of the parallelogram to obtain a rectangular region. If you were not aware that the area of the triangular region stays the same as you move it—and hence that the area of the final figure is the same as the area of the original parallelogram—this justification would not make sense. Related to this *conservation of area* is the idea of left-over, or complementary, area when a region is embedded in a larger region. For example, consider the following task:

▶ The drawings below show a top view of the same farm, with two different arrangements of the same barns (the dark regions). Which arrangement of barns gives more room for cows to roam over?

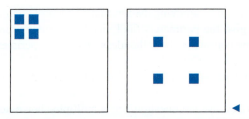

Until a certain age, people react perceptually and judge that there is more free room in the first drawing, not realizing that because the areas of the barns stay the same, the left-over, or complementary, area will be the same.[5] Again, if you have the opportunity to talk with a child 7–9 years old, you might try the task with him or her, using a sheet of paper for the farm and various numbers of blocks for the buildings.

Here is a variation of this task, in which children may give different answers to the two questions posed. Two identical grass fields have identically sized strawberry plots in them (shown in the first drawings in the following diagram). One strawberry plot is split and rearranged, as shown. The questions are, "Is there as much room for the strawberries now?" and "Is there as much room for the grass now?"

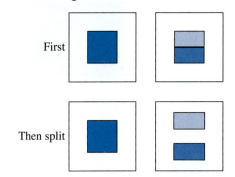

This task also involves good questions to pose to children ages 7–9. Their answers may prove interesting to you.

As you might suspect, children's concepts of volume also could be improved. Consider the following tasks:[5]

▶ Suppose that you have two identical, brand-new bars of modeling clay:

  a. Pretend that you roll one of the bars into a long, sausage shape, and that you roll up the other bar into a ball shape. Which shape—the sausage or the ball—will have more clay, or will they both have the same amount?

continue

Notes

b. Now pretend that you have a container of water and you are going to put a shape into the container so it is completely under water. Which shape—the sausage or the ball—would cause the water to rise higher, or would they both have the same effect? ◄

These ideas [*conservation of matter* in part (a) and *conservation of volume* in part (b)] take time to develop with children. For part (a), they might focus on the longer length of the sausage shape and regard the sausage as having more clay (or some might focus on the thinner aspect of the sausage and think that the ball has more clay). If you have drinking glasses of quite different dimensions, here is another conservation of volume task to try: Fill one container with liquid, pour the liquid into a different container, and ask whether the amount of liquid is the same. If the containers have quite different shapes, many children will focus on the height, say, of the liquid and judge that there is more liquid wherever it goes higher in the container. Clearly, a grasp of conservation of volume would be a prerequisite for understanding any argument involving moving parts of a three-dimensional region around. Try any of the preceding tasks with a primary school child if you have the opportunity.

Testing programs[6] also offer insight into students' understanding of volume. In a U.S. testing program, only 33% of fourth-graders could tell how many small cubes would make the larger cube, as in the drawing at the right, with 32% choosing 12 as the answer and 29% choosing 24. Research suggests that older children also have difficulty when attempting to give such counts from drawings of right rectangular prisms, because quite often they have difficulty "seeing" all the hidden cubes, or while counting somewhat systematically, unintentionally double-count. For example, for a 3 by 3 by 4 rectangular solid, a fifth-grader might count the 9 cubes in the front, say there are 9 more on the back (giving 18 so far), see the 12 cubes on the right, say there are 12 more on the left, and then arrive at a grand total of 42 cubes. Such a student is unaware that he is double-counting several cubes and missing others completely. Other children might avoid the double-counting but still miss counting all the cubes.

In working with similar figures with scale factor k, many students have difficulty in using the k^2 relationship between the areas (or the k^3 relationship between the volumes). Is the reason too much emphasis on formulas, without a good foundation in drawing shapes and their enlargements?

As with length and angle, students need opportunities to estimate area and volume. For example, ask children to estimate the area of the palms of their hands. They might first trace their hands on graph paper. Counting all the full squares inside the tracing is one way of obtaining an estimate. They can obtain a closer estimate by mentally combining parts of squares to make full squares, and keeping track of those used. The teacher might ask them to make an estimate of how many squares their palms would cover on another type of graph paper with larger squares. Or, using a box (preferably in the shape of a cube, but it need not be), ask how many such boxes would fill the classroom (or a closet). How might the box be used to determine a closer estimate? As noted in Section 23.3, studies have shown that practice with the estimation of measurements does improve estimation performance.

TAKE-AWAY MESSAGE . . . Both national tests and research have shown that many students do not understand the concept of area or of volume. Conservation of area and volume also plays a role. Children need both maturation and many opportunities in undertaking tasks that call for measuring area and volume. Learning formulas for area and volume will not make sense to students until they understand the concept of area as covering a region with square units and volume as filling a solid with cubic units. Children should also have experiences estimating area and volume.

Learning Exercises for Section 24.3

1. **a.** Why might so many third-graders have chosen 12 for the number of small cubes required to make the cube shown to the right? (*Hint:* Recall that many confuse a square with a cube.)
    b. How might a child have decided that the correct answer was 24?
    c. Where is this child's thinking incorrect? "Four in the front and four in the back; that's 8. Four on the right and four on the left; that's 8 more. 16 in all."

2. Consider the prism shown at the right.

    a. Explain how the fifth-grader was double counting in saying, "Nine in the front and 9 in the back; 18. Twelve on the right and 12 on the left; 24 more. 18 and 24 = 42."
    b. Which cubes is the fifth-grader overlooking?

24.4 Check Yourself

You should be able to work exercises like those assigned and to do the following:

1. Explain what *area* means, and describe surface area.

2. Apply the four key ideas of measurement (Section 23.1) to area and surface area.

3. Describe common area units (especially metric units), and convert among them, preferably by using your metric sense for area units and perhaps with the aid of a sketch.

4. Use the relationship between the areas or surface areas of two similar shapes.

5. Explain what *volume* means.

6. Apply the key ideas of measurement to volume.

7. Describe common volume units (especially metric units), and convert among them, preferably by using your metric sense and perhaps with the aid of a sketch.

8. Use the relationship between the volumes of two similar shapes.

References for Chapter 24

[1]Simon, M. A., & Blume, G. W. (August 1992). Understanding multiplicative structures: A study of prospective elementary teachers. In W. Geeslin & K. Graham (Eds.), *Proceedings of the 16th PME conference* (Vol. III, pp. 11–18). Durham, NH: University of New Hampshire.

[2]Kouba, V. L., Brown, C. A., Carpenter, T. P., Lindquist, M. M., Silver, E. A., & Swafford, J. O. (1988). Results of the fourth NAEP assessment of mathematics: Measurement, geometry, data interpretation, attitudes, and other topics. *Arithmetic Teacher*, 35(9), 10–16.

[3] *Trends in international mathematics and science study.* http://nces.ed.gov/timss/.

[4]Battista, M. T., Clements, D. H., Arnoff, J. A., Battista, K., & Borrow, C. V. A. (1998). Students' spatial structuring of 2D arrays of squares. *Journal for Research in Mathematics Education*, 29(5), 503–532.

[5]Piaget, J., Inhelder, B., & Szeminska, A. (1960). *The child's conception of geometry.* New York: Harper & Row.

[6]National Assessment of Educational Progress. http://www.nces.ed.gov/nationsreportcard/naepdata/.

Chapter 25

Counting Units Fast: Measurement Formulas

In our discussion of measurement, the emphasis so far has been on several key ideas of measurement and their application to the measurements of length, angle size, area, and volume. You may remember formulas for some of these measurements. Formulas are efficient, but too much emphasis on them can hide their purpose: They allow us to count units fast. In this chapter we *do* feature formulas, emphasizing where they come from and what the formulas tell us, as well as applying them.

25.1 Circumference, Area, and Surface Area Formulas

The basic idea of measurement is counting repetitions of a unit until an object is matched on some characteristic. This idea underlies every measurement formula. We are fortunate that for many shapes the counting can be accomplished rapidly and hence efficiently by using formulas. Our first formula involves a special kind of length.

Circumference

ACTIVITY 1 It's Easy as Pi

Draw several circles (or use several circular objects) and measure the circumference C and the diameter d of each circle with a tape measure or with a piece of string. Record the measurements in a table like the following one and look for a relationship (a formula) connecting C and d:

Diameter

Which circle	C	d	$C + d$	$C - d$	$C \times d$	$C \div d$

In Activity 1, you found that the values in one of the columns were always a little more than 3. Ideally, the values would have been the number we call π. The number π is an interesting number. You probably know that although π is certainly a real number, it cannot be written exactly as either a finite or a repeating decimal. Hence, π is an irrational number, and values such as 3.14 and $\frac{22}{7}$ only approximate π. The special symbol π is used to communicate the exact number it represents. You may not know that π has other uses besides being defined via circles—π comes up frequently in probability and statistics, e.g.

Activity 1 gives some *experimental* evidence that the ratio $C:d$ might always have the same value. Mathematicians would give an argument similar to this line of reasoning for circles in general: Any two circles are similar, so $C:C'$ will equal $d:d'$, or $\frac{C}{C'} = \frac{d}{d'}$. A little algebra then shows that $\frac{C}{d} = \frac{C'}{d'}$. That is, the ratio of the circumference of a circle to its diameter will always be the same. This ratio defines π.

For any circle with circumference C and diameter d,

$$\frac{C}{d} = \pi \qquad \text{or} \qquad C = \pi d$$

We will not give special attention to other formulas dealing with length. The reason is that the idea of perimeter for polygons, e.g., is better served by thinking of what perimeter means rather than learning perimeter formulas for particular polygons, such as rectangles or squares.

Area and Surface Area

Many of the formulas you may remember have to do with area. The emphasis here will be on how to *justify* the formulas, not just what the formulas are or how to use them. Rather than working through experiments that can lead to the formulas, as in Activity 1, most of the development of the formulas can be done deductively, as in the mathematician's argument stated above. But where should you start? In elementary school, the most common starting point is with the area of rectangular regions. For finding the areas of rectangles, you probably remember the formula *area* of a rectangular region equals *length* times *width*, or

$$A(\text{rectangular region}) = lw = bh$$

for rectangles with length l and width w, or base b and height h.

But how is it that units for length and width give us a count of the square units needed for *area*? That needs explanation.

 DISCUSSION 1 Multiplying Lengths Gives Area?

Give the dimensions of the following three rectangles. Find their areas. Why *does* multiplying the two dimensions of a rectangular region give its area? (Assume for each rectangle that the small interior squares have sides with lengths 1 length unit. Fractions may be involved.)

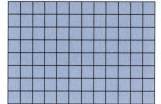 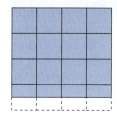 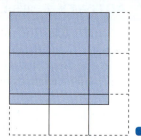

You are familiar with the hierarchical organization of quadrilaterals from Section 17.3, so it should be clear that the formula for the area of a rectangular region can be applied to a square region, where it takes the familiar form,

$$A(\text{square region}) = s^2.$$

(*Note*: This formula explains why s^2 is most often pronounced "s squared.")

The development from here on follows a path common in elementary school curricula. (There are other possible sequences in making sense of the area formulas; one sequence is mentioned in the Learning Exercises.) Now that rectangular regions can be handled by a formula, you can deal with parallelogram regions.

 DISCUSSION 2 Finding a Formula for A(parallelogram region)

How might one justify the formula A(parallelogram region) = bh, where b is the length of a base and h is the height for that base? The figures on the next page might give you an idea. (Why doesn't the rectangle formula work when using the lengths of the sides of the parallelogram?)

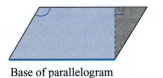

Base of parallelogram

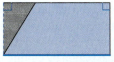

Base of parallelogram

According to the hierarchy of quadrilaterals, the formula for a parallelogram region should also apply to rectangular regions. Does it? Notice, however, that we are becoming quite removed from counting square regions. It is perhaps worth studying the argument for the parallelogram with the squares directly in view, as in the following drawings. The length of the base does indeed count the number of square regions in one row. The drawings also make clear why it is the *height* of the parallelogram that is important rather than the length of the other side; the height tells the number of rows of square regions there will be, whereas the length of the other side of the parallelogram does *not* tell the number of rows.

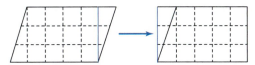

Think About …

Can *any* side of a parallelogram be used as the base in the $A = bh$ formula? What adjustment must you make?

With the formula

$$A(\textbf{parallelogram region}) = bh$$

for parallelogram regions in hand, you can next deal with triangular and trapezoidal regions.

ACTIVITY 2 Justifying the Area Formulas for Triangles and Trapezoids

1. Start with a triangle, as shown in the following drawing. The complete figure in the diagram suggests a way of deriving a formula for the area of the triangular region, using the formula for the area of a parallelogram region. (The blue point is the midpoint of a side of the triangle.) Your explanation should show that

$$A(\textbf{triangular region}) = \tfrac{1}{2}bh$$

    where b represents the length of the base of the triangle and h the height to that base.

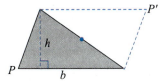

b = base of triangle
h = altitude or height of triangle

    (A height of a triangle is often called an **altitude**.)

2. Try the same method with a trapezoidal region to justify the following formula:

$$A(\textbf{trapezoidal region}) = \tfrac{1}{2}h(a + b)$$

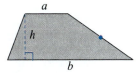

3. Check that the formula for trapezoidal regions also works for parallelogram, rectangle, rhombus, and square regions, as the hierarchy suggests it should.

Notes

Which polygons still need attention? Although we could give attention to other polygons, there is no absolute need to do so because we apply a key idea of measurement: *Every polygonal region can be cut into triangular regions.* Hence, in a pinch you can find the area of any polygonal region by cutting it into triangular regions, making several measurements of lengths, applying the A(triangular region) formula several times, and then adding the areas of all the triangular regions. Of course, if the region could be cut to give some parallelogram or rectangular regions, you could save some work by having fewer regions.

  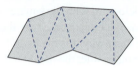

Pentagon   Concave quadrilateral   Octagon

It should be clear that the area formulas can be very useful in finding the surface areas of polyhedra because many of their faces will be regions for which you already have formulas. (See Example 1 below and the Grade 5 textbook page shown on page 570.)

EXAMPLE 1

Find the surface area of the triangular right prism shown at right (not drawn to scale).

Solution

Each of the two triangular bases has area

$$A = \frac{1}{2} \cdot 4 \cdot 5.8 = 11.6 \text{ cm}^2.$$

Because the prism is a right prism, the other faces are rectangular regions. The surface area is then 11.6 + 11.6 + 42 + 28 + 56, or 149.2 cm². ●

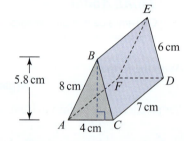

DISCUSSION 3   How Many Squares Make a Circle?

What are the areas of the five square regions—large and small—in the drawing shown here? Estimate the area of the circular region in the drawing. ●

The final area formula we will justify is that for a circle. As you may have guessed,

$$A(\text{circular region}) = \pi r^2$$

where r is the length of the radius of the circle. After your experience with the circumference of a circle in Activity 1, the appearance of π again should not be a complete surprise. Because a diameter of a circle is equal in length to two radii, the formula for the circumference $C = \pi d$ could be written $C = \pi(2r)$, or $C = 2\pi r$. This latter form is useful in deriving the formula for the area. The idea in finding the area formula for a circular region is to approximate the circular region with thinner and thinner, but congruent, isosceles triangular regions, and then to appeal to a basic idea from calculus. Notice that as more and more such triangles are fit inside the circle, the total of all their regions grows closer and closer to filling the whole circular region. By the time the regular 17-gon is used, one's eye can scarcely see the part of the circular region that is uncovered by all the isosceles triangular regions.

17-gon

Hence, it is very plausible that we can get arbitrarily close to the area of the circular region by taking enough of these identical, very thin isosceles triangles—i.e., by using a regular *n*-gon with *n* quite large. The other key is that if we use the unequal side of the isosceles triangle as base, then the height of the triangle for that base gets arbitrarily close to the radius of the circle, as the *n*-gon gets arbitrarily close to the circle itself (check the altitudes with the triangles in the circles above). The area enclosed by each of the *n* thin triangles will be

$\frac{1}{2}$ (length of side of *n*-gon) $\times$ (some length close to the radius)

For the following reasoning, let *some length close to the radius* be denoted by $\approx$ radius. To get close to the area of the circular region, we can add up the areas of all the *n* triangles:

A (circular region) $\approx A$ (all the *n* triangles)

or A (circular region) $\approx n \cdot \frac{1}{2} \cdot$ (length of side of *n*-gon) $\times$ ($\approx$ radius)

or A (circular region) $\approx \frac{1}{2} \cdot$ (*n* $\cdot$ length of side of *n*-gon) $\times$ ($\approx$ radius)

or A (circular region) $\approx \frac{1}{2}$ (perimeter of *n*-gon) $\times$ ($\approx$ radius)

Then, plausibly (or using the idea of *limit* from calculus), as the number of sides of the *n*-gon gets infinitely large, the perimeter of the *n*-gon "becomes" the circumference of the circle, or $2\pi r$, and the "$\approx$ radius" becomes the radius *r*. In symbols,

A(circular region) $= \frac{1}{2} (2\pi r)r$, or πrr, or

A(circular region) $= \pi r^2$

The more complicated nature of our derivation of the area formula for a circle may make clear why you have seen very few formulas for areas of regions involving curves or curved surfaces. We will cite one more area formula, without justifying it:

A(surface of a sphere) $= 4\pi r^2$

where *r* is the radius of the sphere. This formula is surprising because of the nonintuitive relationship that the surface area of the sphere is precisely 4 times the area of a circle having the same radius (which could be obtained on the sphere by a plane slicing through the center of the sphere).

You can find the surface area of a polyhedron by applying a key idea of measurement: Measure the area of each face and then add those measurements, as in Example 1 earlier.

The following table gives a summary of the area formulas and the one special formula for the length of a circle. Also useful for finding area is the idea of cutting a complicated region into pieces, and then adding the areas of the pieces to find the area of the whole region.

A(rectangular region) $= lw = bh$ A(square region) $= s^2$	
A(parallelogram region) $= bh$	
A(triangular region) $= \frac{1}{2}bh$	

continue

A(trapezoidal region) $= \frac{1}{2}h(a + b)$	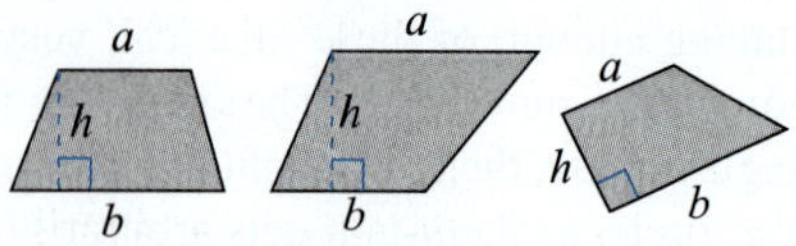
A(circular region) $= \pi r^2$ A(surface of a sphere) $= 4\pi r^2$ C(circle) $= 2\pi r = \pi d$	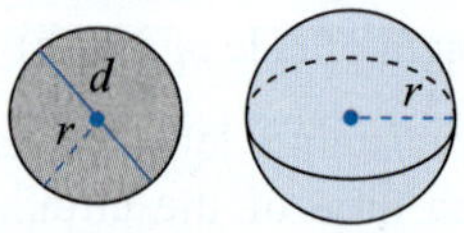
Surface area of a polyhedron $=$ sum of areas of its faces	(See Example 1 on p. 568.)

This page from a Grade 5 textbook explores surface area. Try Problem 2.

Date _____   Time _____

LESSON 11·7   Surface Area

The **surface area** of a box is the sum of the areas of all 6 sides (faces) of the box.

1. Your class will find the dimensions of a cardboard box.

   a. Fill in the dimensions on the figure below.

   b. Find the area of each side of the box. Then find the total surface area.

   Area of front = _____ in²

   Area of back = _____ in²

   Area of right side = _____ in²

   Area of left side = _____ in²

   Area of top = _____ in²

   Area of bottom = _____ in²

   Total surface area = _____ in²

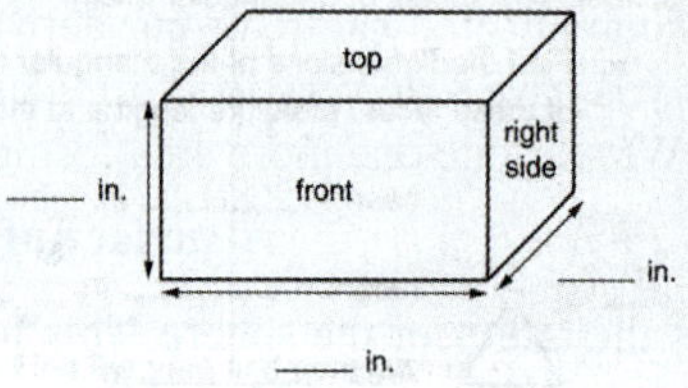

2. *Think:* How would you find the **area** of the metal used to manufacture a can?

   a. How would you find the area of the top or bottom of the can?

   _____

   _____

   b. How would you find the area of the curved surface between the top and bottom of the can?

   _____

   _____

   c. Choose a can. Find the total area of the metal used to manufacture the can. Remember to include a unit for each area.

   Area of top = _____

   Area of bottom = _____

   Area of curved side surface = _____

   Total surface area = _____

Measurement and Geometry 1.1, **1.3**, 1.4; Mathematical Reasoning 3.2   389

Wright Group, *Everyday Mathematics, Journal 2, Grade 5*, p. 389. Copyright © 2008, McGraw-Hill. Reproduced with permission of the McGraw-Hill Companies.

If you examine the area formulas, you will notice that each formula involves multiplying two length measurements. Why such a multiplication of lengths gives a count of area units is clearest with the rectangular regions, and then becomes increasingly obscure with other regions. But because we have good tools for measuring lengths, the formulas are convenient. It is noteworthy that the units behave like algebraic variables: cm × cm = cm², e.g. Scientists often check the units involved to see whether a particular formula is plausible.

💬 DISCUSSION 4   Check Your Units!

Just by considering the units, tell how you know that none of the following formulas for areas can be correct (d = the length of the diameter).

a. $A(\text{sphere}) = \dfrac{\pi d^3}{6}$ **b.** $A(\text{circle}) = \pi d$ **c.** $A(\text{circle}) = 2\pi r$ ●

TAKE-AWAY MESSAGE . . .   With so many formulas, it should be clear why the *concepts* of perimeter and area (and volume) get jumbled and even lost for many students, particularly if the formulas are presented without justification. Writing the entire formula in words helps one to remember it. It is important to remember what each formula gives: a count of the number of measurement units needed to match, cover, or fill an object.

Learning Exercises for Section 25.1

You will need a metric ruler for some of these Learning Exercises.

1. Organize all the formulas for geometric measurements, perhaps using index cards. Make clear what is being counted (length, angle size, area, or volume units). Include a sketch and highlight the parts that must be measured. Memorize any formulas you do not already know.

2. What is the perimeter of each given figure in centimeters? [*Suggestion*: Use metric for parts (c) and (d).]

   a. a circle with diameter 5 cm   **b.** a circle with radius 160 km

   c.   **d.**

3. The equator is about 25,000 miles long. Assuming that the equator is a circle, find the radius and diameter of our Earth. Use 3.14 for π.

4. The number π is approximately 3.1415926. (A mnemonic for the digits in π is using the number of letters in each word: May I have a large container of coffee?) Computers are sometimes tested by having them calculate many decimal places for π. (More than a billion decimal places have been calculated!)

   a. Sometimes $\frac{22}{7}$ is used as an approximation for π. About how far off is that value?

   b. Suppose one person uses 3.14 for π and another person uses 3.1415926 to calculate the circumference of a circle with radius 12 cm. By how much do their results differ?

continue

Notes

c. Suppose the two people in part (b) find the circumference of a 200-mile circular orbit for a satellite. By how much do their results differ? (The phrase "200-mile orbit" means 200 miles above our Earth. *Hint*: See Learning Exercise 3, or use as a rough approximation 8000 miles for the diameter of Earth.)

5. Which one of the following measurements is the *exact* circumference of a circle with diameter 5 cm? Explain.

    a. 15.7 cm   **b.** 15.708 cm   **c.** 5π cm   **d.** 15.7079630 cm

6. You would like to make a trundle wheel to use with grade school students. You want a wheel that will roll 1 meter in one full revolution. What should the diameter of the wheel be?

7. Suppose that a rope fits exactly around the equator. Then 40 feet more are added to the rope, and this longer rope is raised uniformly all around the equator. Which of the following could "walk" under the rope: a bacterium, an ant, a mouse, a 3-year-old child, and an adult human?

8. Answer each student whose question is given.

    a. *Student:* "It would be easier to just multiply the two sides of a parallelogram to get its area. Why do we have to find the height?"

    b. *Student:* "Where does the $\frac{1}{2}$ come from in $\frac{1}{2}bh$?"

9. Find the areas of the given regions. All units are centimeters.

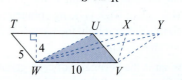

    a. parallelogram *PQRS* (at right)

    b. parallelogram: sides 9 and 16, with the 9 cm sides 8.5 cm apart

    c. rhombus: sides 12, height $4\frac{1}{2}$

    d. parallelogram *TUVW* (shown at right)

    e. triangles *WUV*, *WXV*, and *WYV* (shown at right)

10. Review how the formula for the area of a triangular region was derived. Rehearse how you would explain the formula, starting with the triangle at right.

11. The base of a triangle can be any side. For each side of the given triangles, draw the segments that show the height which corresponds to that side of that triangle. With compass and straightedge, *construct* one altitude for the first and third triangles.

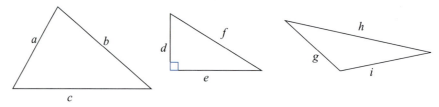

12. In finding the area of a triangle, does it matter which side you use for the base? Check your decision with the first triangle in Learning Exercise 11. (*Suggestion*: Use metric units.)

13. Count the unit squares in the following diagram to verify that the formula for the area of a triangle does give the correct number of units for the area:

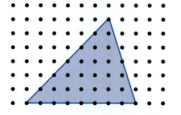

14. For the different choices of point *A*, which, if any, of the different triangles *ABC* below has the greatest area? The line that the *A*'s are on and the line that *B* and *C* are on are parallel. Explain. (*Hint*: Do *not* count squares.)

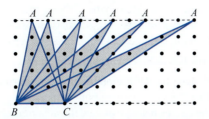

15. Does it matter where you measure the height of a trapezoid or a parallelogram in finding their areas? Explain.

16. Find the area of each trapezoidal region. (*Suggestion*: Use metric units.)

a.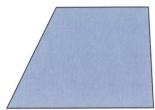

b.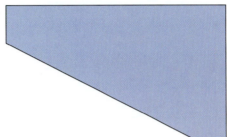

c. How can a trapezoidal region like the one in part (a) be cut into fourths with straight cuts?

17. a. What is the area of a trapezoid with height $7\frac{3}{4}$ inches and parallel sides with lengths $6\frac{1}{2}$ inches and $9\frac{1}{2}$ inches? (No decimals!)

b. What is the area of the trapezoid in part (a) if each of its dimensions is tripled?

18. i. Find the areas of the following shapes (a–h). Assume that lengths are in centimeters and that angles which look like right angles are right angles.

a. 2, 4, 6, 5

b. 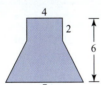 4, 2, 6, 7

c. 4, 7, 5, 6.3

d. 3.2, 3, 3.8, 4

e. 3

f. 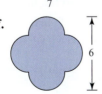 6

g. 2, 6

h. 7, 4

(The curves are congruent semicircles.)

ii. Suppose a company has to deal so frequently with shapes like the following shapes (i–l) that they would find it handy to have formulas for their areas. Find formulas for the company using just the variable measurements indicated.

i. *m*, *j*, *k*, *n*

j. *p*, *q*, *s*, *r*

k. *t*

l. *w*, *v*

(The curves are semicircles.)

iii. Find the perimeters for shapes (c–h, j–l). [Why are you not asked for the perimeters for shapes (a), (b), and (i)?]

19. a. A pizza shop charges $10.95 for a deluxe pizza with a 10-inch diameter. Is their price of $18.95 for a 14-inch-diameter deluxe pizza a better buy? (Assume that no pizza is wasted if you buy the larger size.)

　　b. Another pizza shop uses rectangular pans, either 8 inches by 10 inches or 8 inches by 14 inches. The smaller size costs $10.95. Is their price of $16.95 for the larger size a better buy? (Again, assume that no pizza is wasted.)

20. Each circle shown below has radius 2 cm.

　　a. What is the area of each circular region based on the formula? (Use 3.14 for π.)

　　b. What is the area of each grid square on both grids?

　　c. Using the grids in the circles below, estimate the area of each circle in cm². Which grid gives a better estimate?

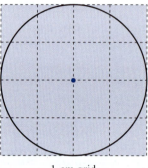　　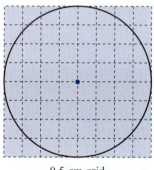

　　　　　　1-cm grid　　　　　　　　　　　0.5-cm grid

21. Sketch and find the areas of the sectors of each circle described below.

　　a. sector angle 120° and radius 12 cm

　　b. sector angle 54° and radius 9.6 cm

　　c. sector angle 288° and diameter 20 cm

　　d. In general, how could you find the area of a sector of a circle, knowing the radius and size of the central angle? Explain as though you were writing a book, describing your method.

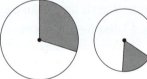

　　e. How could you find the area of a sector of a circle, knowing the radius and the actual *length* of its arc?

　　f. What is the perimeter of the sector in part (a), in centimeters? (Include the radii as part of the perimeter.)

　　g. What is the perimeter of the sector in part (c), in centimeters?

22. If n rows of n congruent circular regions are fit into a square, as shown in the drawings below, what percent of each square region is *not* covered by the circular regions?

　　96 cm　　　　　96 cm　　　　　96 cm

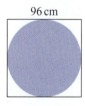　　　　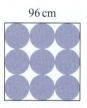　　etc.

23. a. What is the surface area of Earth in square miles? Use 8000 miles as the diameter of Earth.

　　b. Oceans cover about 70% of Earth's surface. How many square miles of *land* area are there?

　　c. What is the area of the Northern Hemisphere?

24. A store has two sizes of globes, 12" and 16" in diameter.

　　a. How do their surface areas compare?

　　b. What is the surface area of each globe?

25. Give the surface area of each right rectangular prism described below. (*Suggestion*: Making a rough sketch may help.)

    a. length 12 cm, width 8 cm, and height 10 cm
    b. height 1.2 m, depth 40 cm, and width 80 cm
    c. length $2\frac{1}{2}$ ft, width 3 ft, and height 8 in.
    d. length x cm, width y cm, and height z cm

26. A right prism is 24 cm high and has bases like the parallelogram shown at the right. What is the surface area of the prism?

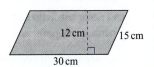
12 cm   15 cm
30 cm

27. Find an approximate area in square centimeters for the following region, using rectangular regions 0.5 cm wide (one rectangular region is sketched). How could you improve the approximation?

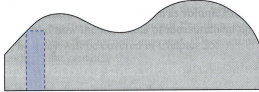

28. Explain how you know that each of the following area formulas, where w, x, y, and z are the lengths of the sides, cannot possibly be correct. (*Hint*: Think units.)

    a. $A(\text{quadrilateral region}) = w \cdot x \cdot y \cdot z$
    b. $A(\text{quadrilateral region}) = w + x + y + z$

29. A region with area 14.4 cm² undergoes a dilation with scale factor 125%.

    a. What is the new area?
    b. The new area is how many percent as large as the original area?
    c. The new area is how many percent larger than the original area?

30. Explain why a dilation with scale factor k gives an image of a rectangular region that has area k^2 times as much as the original rectangular region. Does the dilation affect the area of a triangular region by a factor of k^2 as well? Explain why or why not.

31. **a.** Justify this formula for the area of a kite: $A(\textbf{kite region}) = \frac{1}{2} dd'$, where d and d' are the lengths of the two diagonals of the kite. (Recall that the lines of the diagonals of a kite are perpendicular.)

    b. To which of these polygons does the formula in part (a) automatically apply: parallelogram, trapezoid, rectangle, rhombus, square? Explain your reasoning.

32. Drawing a polygon on square-grid dot paper so that every vertex of the polygon is at a dot is easy. Perhaps surprisingly, there is a way to predict the area of the polygon from just these two numbers: the number of dots on the polygon and the number of dots inside the polygon. Use figures on dot paper and organize your work, perhaps as suggested in the table below, in order to look for the relationship. Two examples are shown.

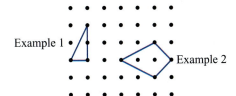

Example 1   Example 2

Number of points on the polygon	3	4	5	4	5	6	4	(etc.)
Number of points inside	0	0	0	1	1	1	2	
Area of polygon		1					3	

Example 1   Example 2

33. Our justification of the formula for the area of a triangular region involved creating a parallelogram region that was twice as large in area as the original triangular region. This led naturally to the formula $A = \frac{1}{2}(bh)$. (Note the use of parentheses to communicate that we took half the parallelogram's area.) It is interesting that algebraic variations of $\frac{1}{2} bh$ suggest other justifications. Explain how different approaches could have led to the following formulas. Note that the blue points shown are the midpoints of the line segments.

a. $A(\text{triangular region}) = \left(\frac{1}{2} b\right) \cdot h$ **b.** $A(\text{triangular region}) = \left(\frac{1}{2} h\right) \cdot b$

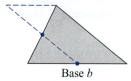

Base b 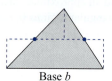

Base b

34. a. Review how the formula for the area of a trapezoidal region was derived. When finding a formula for the area of a trapezoidal region, you may have created a large parallelogram and taken half the area of its region, leading to $\frac{1}{2} [(a + b)h]$, or $\frac{(a + b)h}{2}$ for the area of the trapezoidal region. Algebraic variations of this expression suggest other possible derivations. For parts (b) and (c), explain how different approaches could have led to each expression by using the given diagram.

b. $A(\text{trapezoidal region}) = \left[\frac{1}{2} h\right] \cdot (a + b)$

c. $A(\text{trapezoidal region}) = \left[\frac{1}{2} (a + b)\right] \cdot h$
Hint: How long is the *midline*? How do you know? (The reasoning involved to answer this question may not be brief.)

35. Sketch a trapezoid $ABCD$, with $\overline{AB}$ and $\overline{CD}$ as the parallel sides. Imagine now that only points A and B move closer and closer to each other.
  a. What figure is $ABCD$ getting closer and closer to?
  b. Use the thinking in part (a) to show that the formula for the area of a trapezoid can give the formula for the area of a triangle.

36. An alternative development of the area formulas follows the sequence of steps in parts (a–c), starting from the familiar $A(\text{rectangular region}) = lw$, where l and w are the two dimensions of the rectangle.
  a. Derive the formula for the area of a right triangle (start with a right triangle, and relate it to some rectangle).
  b. Then derive the formula for the area of a general triangle, using part (a).
  c. Using the formula for the area of a triangular region, derive a formula for the area of a trapezoidal region.

37. Suppose that a sphere fits snugly into a right circular cylinder. How does the area of the sphere compare with the area of the cylinder without its two bases?

Supplementary Learning Exercises for Section 25.1

1. What does it mean to say, "Pi is an irrational number"?

2. Why is March 14 called "pi day" in many mathematics classrooms?

3. What does it mean to say that the formula for the area of a trapezoid can be applied to a rectangle or to a rhombus?

4. An official basketball hoop has an inside diameter of 18 in. A women's basketball has a circumference of 28.5 in. Could two basketballs go through the hoop at exactly the same time?

5. Find the perimeter of each of the following:

   a. circle with radius 6.2 m   (Use 3.14 for π.)

   b. circle with diameter 42 in.   (Use $3\frac{1}{7}$ for π.)

   c.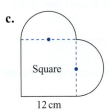
   Square
   12 cm

   d.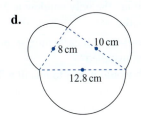
   8 cm
   10 cm
   12.8 cm

   e.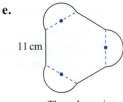
   11 cm
   The polygon is a regular hexagon.

6. What is the area enclosed by each of the shapes (a–d) in Exercise 5? [The triangle in part (d) is a right triangle.]

7. **a.** Show that the area of a circular region with radius 7 in. is about twice the area of a circular region with radius 5 in. even though 7 in. is not that much greater than 5 in.

   b. To make the result in part (a) believable, cut out a circle with radius 5 in. and trace it inside a circle with radius 7 in. but close to the circle.

8. Give the areas of the shaded portions of the following regions. Angles that look like right angles are right angles, and lines that look parallel *are* parallel.

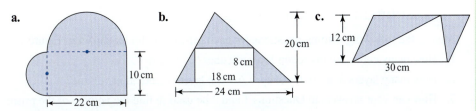

   a. 10 cm, 22 cm

   b. 8 cm, 18 cm, 24 cm

   c. 20 cm, 12 cm, 30 cm

9. Give the missing measurement in each part.

   a. A rectangular region has area 8.64 m² and one side 1.2 m. What is the other dimension of the rectangle?

   b. A triangular region has area 112 in² and base 14 in. What is the height of the triangle?

   c. A trapezoidal region has parallel sides of lengths 14.6 cm and 18.8 cm, and area 50.1 cm². What is the height of the trapezoid?

10. What is the surface area of each right rectangular prism with the given dimensions?

    a. 19.3 cm by 24 cm by 18.6 cm   **b.** 8 m by 4 m by 6 m

11. A right prism has as its base a parallelogram with one side 20 cm and the height to that side 8.4 cm. Another side of the parallelogram is 12 cm. The height of the prism is 9.6 cm. What is its surface area?

12. What is the surface area of each right circular cylinder described?

    a. diameter of base = 14 in., height of cylinder = 12 in.

    b. radius of base = 20 cm, height of cylinder = 8 cm

    c. area of base = 16π cm², height 7 cm

    d. area of base = about 254.47 cm², height 10 cm

13. A right cylinder has a base like the region shown in Supplementary Learning Exercise 8(a). If the cylinder is 5 cm tall, what is its surface area?

14. You have a closed box with surface area 236 in² and volume 240 in³. You need a larger box of the same shape in which all its lengths are 1.5 times as long as those in the original box. What are the surface area and volume of the larger box?

Notes

15. The area of a (perfect) ellipse is given by $A = \pi ab$, where a and b are the halves of the lines of symmetry of the ellipse, as in the sketch. What does the formula become as a and b change to the same value? Is that a surprise?

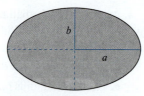

16. You and a friend want to redecorate a 12 ft by 18 ft room, with walls 8 ft high. The door to the room is 32 in. by 80 in., the only window is 3 ft by 4 ft, and the closet opening is 80 in. high by 5 ft wide.

   a. How much will it cost for wall-to-wall carpeting, if the carpet you want costs $15.60 per square yard when installed?
   b. Instead of carpet, how much would it cost to install a floor laminate at $2.99 per square foot, packaged in bundles that cover 13.1 square feet? Allow 5% extra for waste—useless left-over short pieces. (You would install the laminate.)
   c. How much will it cost to paint only the walls in the room (not inside the closet), if the paint you like costs $33 for a gallon and a gallon covers up to 400 square feet? (*Hint:* Make a drawing of each wall.)

25.2 Volume Formulas

Deriving formulas to find a volume through calculations instead of counting the cubic regions needed to match or to fill a three-dimensional object might seem like an impossible task. But, as you will see, we need only two formulas for the volumes of *all* prisms, cylinders, pyramids, and cones. Discussion 5 leads to one of those formulas.

DISCUSSION 5   Layer by Layer

Answer the following leading questions for each prism or cylinder shown in Figure 1:

1. How many cubes are in the top layer of cubes?
2. How many layers h are in the shape? Are the layers identical in volume?
3. How can your answers to Questions 1 and 2 be used to find the volume of the prism or cylinder?
4. What is the area B of the base? Numerically, how does the area of the base compare with the number of cubes in the top layer?
5. How can h and B give the volume of the prism or cylinder?

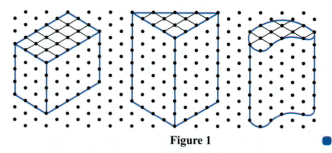

Figure 1

Discussion 5 should make apparent the formula V**(prism or cylinder)** $= Bh$, where B is the *area* of the base and h is the *height* of the prism or cylinder. (Notice the use of a capital B to remind us that the base involved here represents an area, not a length.) The type of base can lead to some other expression for B. For example, if the base is a rectangle with dimensions l and w, then $B = lw$, and so $V = Bh = (lw)h = lwh$. (You probably recognize this volume formula now if you had forgotten it.) Because the area of the base can be easily handled as it is encountered, using $V = Bh$ is good for *any* prism or cylinder. For example, if the shape is a circular cylinder with r as the radius for the base, then the $V = Bh$ formula could become $V = \pi r^2 h$. Most boxes in everyday living are shaped like prisms or cylinders, so this $V = Bh$ formula has wide applicability.

Or does it? Did you notice that all the shapes in Discussion 5 were *right* prisms or cylinders? Would the argument still apply to *oblique* prisms or cylinders? If you think of a tall stack of index cards, or crackers, or thin slices of bread, it should be apparent that changing the stack from a right prism/cylinder position to an oblique position does not change the volume of the stack (and leaves the height the same). The changed stack still involves the same cards, crackers, or bread as at the start, so the volume will be the same.

Another way to see that $V = Bh$ also applies to oblique prisms is to imitate the method you used for changing a parallelogram region into a rectangular region without changing the area. See the diagram in Figure 2. Can you imagine how to cut off part of an oblique pile of index cards, and move it to make a right prism shape? (Look at a pile of index cards arranged as an oblique prism.) You might have to cut twice, as with shape B from your kit of polyhedra.

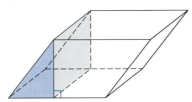 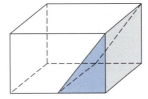

Figure 2

For any prism or cylinder, because each layer of cubes would be the same and numerically equal to the area B of the base and with the height h of the prism or cylinder telling the number of layers, the $V = Bh$ formula does indeed apply to any prism or cylinder, whether right or oblique.

EXAMPLE 2

In Example 1 in Section 25.1, we found the surface area of the triangular prism shown here. Now find its volume.

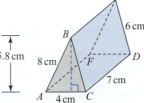

Solution

$V = Bh$ applies. In the earlier example, we found that the area of the base is $B = 11.6$ cm^2. The height for the triangular prism is $h = 7$ cm. So, $V = 11.6 \times 7 = 81.2$ cm^3. ●

 ACTIVITY 3 "Three Coins in the Fountain," or . . .

Get everyone's shape I from the polyhedra kits, and choose one person to be the assembler. (Others in the group can be advisers and assistant holders.) The task is to take two shape I's (or three, four, or however many it takes) to form a prism of some type. This activity will show how the volume of shape I is related to the volume of the prism you made:

$$V(\text{pyramid I}) = \underline{\quad} \cdot V(\text{related prism}) \qquad ●$$

Shape *I* is a special case, of course, and certainly the prism it gives is special. But the relationship holds for any pyramid, even though three copies of an arbitrary pyramid cannot be fit together to make a prism. As you might suspect, the same 1:3 relationship holds for the volume of a cone and the volume of the cylinder into which the cone would just fit.

Thus, the picture for prisms and cylinders **($V = Bh$)** and for pyramids and cones **($V = \frac{1}{3} Bh$)** is complete. The formulas could be applied to any appropriate shape—prism, cylinder, pyramid, or cone—with B perhaps being replaced by a suitable area formula for the base of the particular shape.

Here is one final formula for volume:

$$V(\text{sphere}) = \frac{4}{3} \pi r^3$$

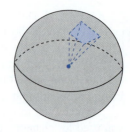

One possible justification is similar to that for the area of a circle in that it involves approximating the spherical region with pyramid regions. Imagine this reasoning: Each pyramid has its base vertices on the sphere, and its other vertex at the center of the sphere, as in this figure. There are many, many such pyramids, all identical and filling all of the spherical region except for small spaces close to the sphere. As the bases of the pyramids become smaller and smaller, the pyramids get closer and closer to filling the sphere completely.

The volume of the sphere is approximately the total of the volumes of all the pyramids. Each pyramid has volume $\frac{1}{3}Bh$, with h being very close to the radius r of the sphere as the number of pyramids increases greatly. In adding all these volumes, we can factor out the $\frac{1}{3}h$, leaving the sum of all the bases of the pyramids as the other factor. But by *taking the limit* (as they say in calculus), that sum becomes the surface area of the sphere, and the height h of each pyramid becomes the radius r of the sphere. So, the volume is $\frac{1}{3}r \cdot (4\pi r^2) = \frac{1}{3}(4\pi r^2) \cdot r$, or

$$V(\text{spherical region}) = \frac{4}{3}\pi r^3.$$

Just as with the area formulas, the volume formulas give the correct units. For example, if the unit for B is square centimeters and the unit for h is centimeters, then it makes sense that the volume formula gives cubic units: $\text{cm}^2 \cdot \text{cm} = \text{cm}^3$. This fact again illustrates the value of paying attention to the units involved.

A summary of the volume formulas follows. In addition, sometimes the key idea of imagining a complicated three-dimensional shape being cut into pieces that are easier to deal with, finding the volumes of the pieces, and then adding those results, is useful in finding the volume of the complicated shape.

For any prism or cylinder, $V = Bh$, where B is the area of the base and h is the height of the prism or cylinder

For any pyramid or cone, $V = \frac{1}{3}Bh$, where B is the area of the base and h is the height of the pyramid or cone

For a sphere with radius r, $V = \frac{4}{3}\pi r^3$

Problems can involve more than one formula, as the next example illustrates.

EXAMPLE 3

A custom-made solid iron support for an odd-shaped corner of a room is a right cylinder, 8 feet long with bases that are 135° sectors of circles with 4-inch radii. Only the curved part is to be coated with protective material. The iron weighs 490 pounds per cubic foot. (a) How much will the support weigh? (b) What is the area that is to be coated in square feet?

Solution

Making even a rough drawing is often helpful. This situation calls for the volume of the support, as well as the area of the curved part.

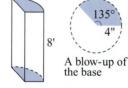

A blow-up of the base

a. $V = Bh$ applies; h is 8 feet, and B is the area of a 135° sector. That area will be $\frac{135}{360}$ or $\frac{3}{8}$ of the whole circle area: $B = \frac{3}{8}\pi 4^2 \approx 18.84$ square inches. The height h is given in feet, but 8' = 96". So, $V \approx 18.84 \times 96 \approx 1809$ cubic inches. Because $12^3 = 1728$, we know that the support has volume $1809 \div 1728$ cubic feet, or about 1.05 cubic feet. The support weighs $1.05 \times 490 \approx 515$ pounds.

b. In a net for the support, the curved part would look like a rectangular region ($A = lw$) with length 96" and width being $\frac{3}{8}$ of the circumference of a 4-in. radius circle (use

$C = 2\pi r$). So, this gives for the width $\frac{3}{8} \times 2\pi 4 \approx 9.42''$, and then the part to be coated has area 96×9.42, or about 904 square inches. The area to be coated would be $904 \div 144$, or about 6.3 square feet. ●

TAKE-AWAY MESSAGE . . . Three volume formulas and the key idea of cutting up a complicated shape allow us to determine easily the number of cubic units it would take to fill or match a great variety of shapes.

1. $V = Bh$ for any prism or cylinder with area of the base B and height h

2. $V = \frac{1}{3}Bh$ for any pyramid or cone with area of the base B and height h

3. $V = \frac{4}{3}\pi r^3$ for any sphere with radius r

Learning Exercises for Section 25.2

1. Would $V = Bh$ for each of the following figures? (B = area of the base, h = height of shape.) Explain.

a.   b.   c.

2. a. How many cubes, 1 cm on each edge, would fit into a right rectangular prism 7 cm by 10 cm by 12 cm?
   b. How many cubes, 2 cm on each edge, would fit into that same prism?
   c. What is the surface area of the prism in part (a)?

3. A plastics company makes many shapes with holes in them. One particular order calls for pieces 0.6 cm thick and shaped like the following shapes (a–c). For each shape, find the volume of plastic needed for an order calling for 1000 of each shape. Use metric units to make any measurements you need.

a.   b. 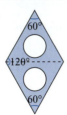 c.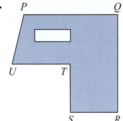

   d. What key ideas of measurement did you use in parts (a–c)?

4. We can make two circular cylinders without bases as follows: (1) An $8\frac{1}{2}$ inch by 11 inch piece of paper can be curved from top to bottom to give the circular cylinder without bases, and (2) the same-size paper can be curved from side to side to make the cylinder.

   a. How would the lateral areas of the two cylinders compare?
   b. How would the volumes of the two cylinders compare?
   c. If two shapes have equal surface areas, will they always have equal volumes? Give an example using three-dimensional shapes made from cubes.

5. a. A prism has bases that are 6 cm by 8 cm rectangular regions. The prism is 7 cm high. Sketch the prism. What is its volume?
   b. A rectangular pyramid fits exactly into the prism in part (a), with the base of the pyramid exactly on the base of the prism and with the other vertex of the pyramid on the other base of the prism. Sketch the pyramid inside the prism. What is the volume of the pyramid?

continue

Notes

c. What is the surface area of the prism in part (a), if it is a *right* prism? (*Hint*: Make a drawing to help keep track.)

6. a. The top face *PQR* of the given prism is shown at actual size to the right. What is the volume of the polyhedron? Measure in triangle *PQR* with metric units. (*Hint*: What additional measurement will you need to make for triangle *PQR*?)

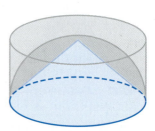

b. Point *S* in the top face of that same prism is joined to the vertices of the opposite face to give another polyhedron. What is its volume?

7. The Great Pyramid of Cheops, built around 4600 years ago, is a regular square pyramid 146.5 m tall and 230.4 m along each side of its base. What is its volume in cubic meters?

8. Plastic inflatable beach balls come in two sizes: 8-inch diameter when inflated, and 16-inch diameter when inflated.

a. How much air does each size beach ball hold when inflated?

b. How much will it cost for the plastic to make each beach ball, if the plastic costs 28¢ per square foot?

c. The ratio of 8 to 16 is 1 to 2. Why are the answers in parts (a) and (b) not related by a 1 to 2 ratio?

9. Food for farm animals might be stored in a silo. Silos are often shaped like right circular cylinders with a right circular cone on top. Suppose that one such silo has a diameter of 16 feet, a height of 12 feet for the cylinder, and an overall height of 15 feet.

a. Find the volume of the whole silo in cubic feet.

b. How many bushels does the cylindrical part of the silo hold? (A bushel is 2150.42 cubic inches.)

10. a. What are the surface area and volume of a sphere with radius 1 meter?

b. What is the volume of Earth in cubic miles? Use 8000 miles as the diameter of Earth.

c. What is the volume of our sun in cubic miles? Use 870,000 miles as the diameter of the sun.

d. How many Earths would make the volume of our sun?

11. Half a sphere (called a **hemisphere**) can fit exactly into a cylinder, as can a cone with its base at one base of the cylinder and its vertex on the other base.

a. Find the ratio of the volume of the hemisphere to that of the cone.

b. What is the ratio of the volume of the hemisphere to that of the cylinder? (Archimedes, an ancient Greek, knew these ratios.)

12. There are two bars of new clay, each shaped like a right rectangular prism with dimensions $\frac{1}{2}$", 2", and 8". One of the clay bars is rolled into a sphere, whereas the other bar retains its original shape.

a. Are the two final shapes congruent?

b. How do the volumes of the bar and the sphere compare?

c. What is the radius of the sphere?

13. a. A can of tennis balls often holds three tennis balls. What part of the can do the balls fill? (*Hint*: Let the radius of the can be *r*. What is then the height of the can?)

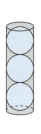

b. A sphere fits snugly inside a cube. What percent of the cube's volume is outside the sphere?

14. A cube and a sphere each encloses 1000 cm³. Which shape is taller, and by how much?

15. Many types of explosions depend on very fast burning, which in turn depends on the surface area exposed. Suppose a ball of gunpowder material is in the shape of a sphere with volume 36π cm³. Now suppose that this ball is broken into small spheres with diameters all equal to 0.1 mm. What is the ratio of the original surface area to the total surface area of all the smaller spheres?

16. In the 1990s an asteroid came quite close to Earth, and it will return in the future. At first the asteroid was believed to be $\frac{1}{3}$ mile in diameter, but later measurements indicated its diameter was $\frac{1}{10}$ mile.

    a. Assuming that the asteroid was spherical, find the volume of the $\frac{1}{3}$-mile-in-diameter version.

    b. What percent of the $\frac{1}{3}$-mile-in-diameter version is the $\frac{1}{10}$-mile version?

17. **a.** Consider shapes A, F, G, and I from your polyhedra kit. What would you measure to determine the surface area and volume of each shape?

    b. How might you find the volume of shape K?

18. Find the volume and surface area for each given object.

    a.

    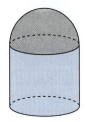

    Diameter = 6 cm
    Height of cylinder = 8 cm

    b. Radius of top = 5 cm

    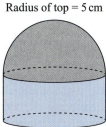

    Height of cylinder = 3 cm

    Give formulas for the volumes of the following, using only the measurements indicated by variables:

    c.

    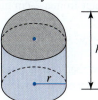

    Hemisphere on top of right circular cylinder

    d.

    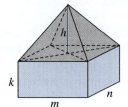

    The base is a rectangular region.

    e. Give formulas for the volume and surface area of a right circular cylinder that has height h and whose base has radius r.

19. The expression x^3 is most often pronounced "x cubed," instead of the more generic "x to the third power." Where does the "x cubed" come from?

20. **a.** Find the surface area of a cube with edges 8 cm long.

    b. Give a formula for the surface area of a cube with edge length equal to s cm.
    SA(cube) = _____

    c. Find the volume of a cube with edges 8 cm long.

    d. Give a formula for the volume of a cube with edge length equal to s cm.
    V(cube) = _____

    e. Can the surface area and the volume of a cube ever be the same? Can they ever be *numerically* the same?

21. Explain why a dilation with scale factor k gives an image of a right rectangular prism that has volume k^3 times as much as the volume of the original prism. Does the dilation affect the volume of a pyramid by a factor of k^3 as well? Explain why or why not.

22. **a.** Give a formula for the volume of a sphere in terms of the diameter d.

    b. Give a formula for the volume of a hemisphere in terms of the radius r.

continue

Notes

   c. Give a formula for the area of a sphere in terms of the diameter d.

   d. Give a formula for the area of a hemisphere in terms of the diameter d.

23. You want to make a dog house from a piece of plywood 4 feet by 8 feet. Design the largest dog house you can. (The plywood must cover the sides and the roof.)

24. Areas of rectangular regions and volumes of right rectangular prisms can give insight into some algebraically equivalent expressions. Using sketches and labeling subregions, find or verify equivalent expressions in the following:

   a. $(x + y)^2 =$ _____

   b. $(x + y)^3 = x^3 + 3x^2y + 3xy^2 + y^3$

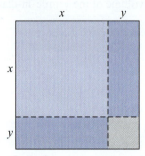

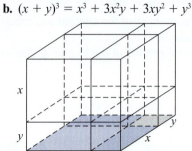

   c. $x(x + y) =$ _____
   (Make your own drawing.)

   d. $(x + 2)(x + 3) =$ _____
   (Make your own drawing.)

25. Here are examples of *Fermi problems*, named after Enrico Fermi, a physicist who was involved in early experiments with splitting the atom. In a Fermi problem, the idea is not to calculate carefully and extensively, but to use thoughtful estimates and minimal arithmetic (and probably scientific notation) as much as possible to arrive at an idea of the size of the answer.

   a. How many kernels of popped popcorn would fit into your mathematics classroom?

   b. What if the popcorn were not popped?

   c. How many toilet flushes could be made with all the water from a 2-cm rainfall on your city?

   d. How many dollar bills would it take to cover the surface of Earth?

   e. How many oranges would it take to fill Earth?

   f. If all the people in the world lay head-to-foot in a line, how many times would the line go around the equator (or what part of an equator if the line wouldn't go around once)?

Supplementary Learning Exercises for Section 25.2

Rough sketches are often helpful, even when not called for.

1. Sketch an oblique prism with curved regions as the bases. Indicate the measurements you would need to find the volume of your prism.

2. "I'm not sure. Are these formulas correct for spheres: $A = 4\pi r^3$ and $V = \frac{4}{3}\pi r^2$?"

   How would attention to length, area, and volume units help to recognize that the formulas are not correct?

3. Find the volume of each three-dimensional figure described.

   a. a right prism with height 7' and a square base whose side is 8'

   b. a right prism with lateral edges 5 m and an 8 m by 10 m rectangular base

   c. an oblique prism with height 15 cm and a rhomboidal base with sides 20 cm and height to one side 12 cm

   d. a pyramid 10 ft tall with base that is a 12' by 15' rectangular region

   e. a pyramid 18 in. tall with a trapezoidal base whose area is 36 in^2

   f. a pyramid with height 4 cm and a square base whose side is 5 cm

   g. a circular cylinder 45 cm high and base whose radius is 6 cm

   h. a long, thin circular cylinder 30 in. long and with diameter 3 in.

   i. How would your answers to parts (g) and (h) change, if the shapes were cones instead of cylinders?

4. Suppose that each of the shapes in Exercise 3(a), (c), and (h) is enlarged by a scale factor 4. What would the volumes of the enlargements be? (*Hint*: Do you remember how volumes of similar shapes are related?)

5. "You said that the volume of the 80 cm by 70 cm by 1.1 m box was 616,000, but I got only 6160." What did the student overlook (besides giving a unit)?

6. Find the volume and the surface area of each surface described.

   a. globe with radius 12 in.   **b.** spherical ball-bearing with diameter 2.5 cm
   c. enclosed hemisphere with radius 10 m
   d. enclosed hemisphere with diameter 30 ft
   e. enclosed hemisphere with radius r

7. The Pantheon in Rome is a building shaped like a hemisphere on top of a right circular cylinder. If the hemisphere were completed to make a full sphere, the sphere would touch the ground inside the cylinder. The diameter of the cylinder is about 140 feet. What is the volume enclosed by the Pantheon? What is its surface area (the hemisphere is the ceiling, and do not count the floor)?

8. **a.** A right rectangular prism 28 cm high has volume 7728 cm^3. One side of the base is 12 cm. How long is the other dimension of the base?
   b. A regular square pyramid is 15 cm tall and has volume 2420 cm^3. What is the *perimeter* of its base?
   c. A cone is 12 in. high and has volume 96 in^3. What is the area of its base?
   d. An enclosed hemisphere has volume $\frac{16\pi}{3}$ cm^3. What is its surface area?

9. What is the difference in the b and the B in $A = bh$ (for parallelograms) and in $V = Bh$ (for prisms and cylinders)?

10. **a.** Water supplies for cities are often measured in *acre-feet*, the volume of an acre of water to a depth of 1 foot. How many gallons are in 1 acre-foot? (*Recall*: 1 acre = 43,560 ft^2, and 1 gallon = 231 in^3.)
    b. In one community, the average daily usage of water per household (including outdoor use) is 400 gallons. About how many such households would use an acre-foot of water each day?
    c. Would the questions in parts (a) and (b) be answered more easily if all the measurements were in metric units?

25.3 Issues for Learning: What Measurement Is in the Curriculum?

Measurement topics in elementary school mathematics often focus on formulas rather than on the ideas involved. But measurement is much more than formulas! In Chapter 1 ("Part I, Reasoning About Numbers and Quantities"), we discussed different attributes and qualities of objects that can be quantified through measurement, such as Apgar scores used to measure the health of newly born babies, IQ tests to measure human intelligence, ways of quantifying and measuring the wealth of a nation, to name but a few. Understanding what it *means* to measure something, and finding and using appropriate units and instruments are all part of measurement.

The influences of the *Common Core State Standards for Mathematics*[1] (CCSS), as well as the earlier *Principles and Standards for School Mathematics*[2] (PSSM), no doubt will affect the elementary school curriculum in geometry (see Section 18.3) and mathematical practices (see Section 21.3) for some time. Both documents treat measurement in the curriculum. Indeed, the PSSM document notes that "the study of measurement is important in the mathematics curriculum from prekindergarten through high school because of the practicality and pervasiveness of measurement in so many aspects of everyday life" (p. 44). The following overview will provide you with the flavor of what to expect in textbooks for the topic of measurement.

Notes

Pre-K–2. Children should have experiences with length, time, and money, using both metric and English system units for length. They should understand that measuring requires repetitive use of an appropriate unit, such as a ruler to measure the length of their classroom. They should be able to make some comparisons and estimates of measures. Measurement language involving words such as "deep," "large," and "long" should become comfortable parts of their vocabulary. Students should be able to solve selected word problems dealing with measurements.

Grades 3–5. The use of standard units and tools such as rulers and protractors should continue. Students should understand length, area, volume, and angle measures and should use formulas for the area and perimeter of rectangles, as well as the volume of right rectangular prisms. They should have some practice at applying these to the surface areas of rectangular prisms and by cutting a more complicated shape into familiar pieces. Students should practice conversions within a system of units (e.g., changing a measurement given in centimeters to one in meters, or one given in feet to one in inches). They will work with coordinate systems and will continue to solve real-world and mathematical problems involving measurements.

Grades 6–8. Measurement topics now include developing formulas for the circumference and area of a circle and finding the areas of other regions through cutting them up (as in our Key Idea 4). Students should study surface areas and volumes of right prisms, pyramids, cylinders, cones, and spheres. Their study of congruence will involve rigid motions, and their study of similarity will include dilations. They should solve problems that involve scale factors, using ratio and proportion. Their study of angles will cover, e.g., supplementary angles, complementary angles, and vertical angles, as well as angle sums for polygons. They can apply the Pythagorean theorem (and its converse). Their continued work with coordinate systems will allow them to use the distance formula. Throughout, they will solve real-world and mathematical problems involving measurements.

Unfortunately, measurement has not received adequate attention in the curriculum, as evidenced by the performance on NAEP[3] measurement items. (Up to 20% of the NAEP mathematics items focus on measurement.) Until measurement begins to enjoy a greater emphasis in the curriculum, our students will continue to lack the understanding and skills needed in all walks of life.

25.4 Check Yourself

Formulas will offer an opportunity for your students to practice calculation (and estimation) in real contexts, particularly if they have to make length or angle measurements themselves. You should be able to do exercises like those assigned and to meet the following objectives:

1. Use the circumference, area, and volume formulas from this chapter. It is probably a good idea to list them all so that you can see whether any are similar and hence need special attention to keep straight. Knowing where the formulas come from and what the formulas tell you when they are used is essential. (See Appendix B for a summary)

2. Justify the area formulas using either the basic idea of area or formulas already established.

3. Show some appreciation of the strange but perfectly good number π.

4. Use a "layer" argument to justify the formula for the volume of any prism or cylinder, and know that the argument does not always apply to other shapes.

5. Use a key idea of measurement and your knowledge of formulas to give formulas for the perimeter, area, or volume of different sorts of shapes.

References for Chapter 25

[1]National Governors Association Center for Best Practices, Council of Chief State School Officers. (2010). *Common core state standards for mathematics.* Washington, DC: Author.

[2]National Council of Teachers of Mathematics. (2000). *Principles and standards for school mathematics.* Reston, VA: Author.

[3]Silver, E. A., & Kenney, P. A. (2000). *Results from the seventh mathematics assessment of the National Assessment of Educational Progress.* Reston, VA: National Council of Teachers of Mathematics.

Chapter 26

Special Topics in Measurement

The Pythagorean theorem is so important that it deserves its own section, Section 26.1. The other section in this chapter, Section 26.2, serves as a reminder that measurement is a pervasive process in life. That fact supports the view that measurement deserves a significant role in the school curriculum.

26.1 The Pythagorean Theorem

Relatively few results in elementary mathematics are associated with a name, perhaps because the original discoverer is not known. But in mathematics one result is always associated with Pythagoras: the Pythagorean theorem.

Right angles are common in many shapes encountered in the Western world, and the heights of triangles, trapezoids, parallelograms, prisms, pyramids, cylinders, and cones involve right angles. These right angles can often be associated with some right triangle.

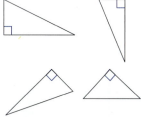

Right triangles come in all sizes and shapes. The sides of the triangle that give the right angle are sometimes called the **legs** of the right triangle, and the other side, the longest, is called the **hypotenuse**. It would seem that all we could rely on for *all* right triangles would be the one right angle and the fact that the hypotenuse is longer than either leg.

Fortunately, the lengths of the legs and the hypotenuse of a right triangle are related, although in a nonintuitive way. Perhaps that is the reason, along with its multitude of uses, why the relationship is so famous. The relationship is associated with Pythagoras (c. 570 BC), although the ancient Babylonians, Indians, and Chinese knew it long before him. The relationship is called the **Pythagorean theorem**.

> The **Pythagorean theorem**: In a right triangle having legs of lengths a and b and hypotenuse of length c,
>
> $$a^2 + b^2 = c^2$$

Interestingly, no well-developed algebraic notation existed at the time of Pythagoras for the square of a number. Where we write s^2 and think $s \cdot s$, the Greeks, not having that notation, thought of a geometric square! Hence, they interpreted the Pythagorean theorem geometrically, as follows.

Pythagoras's Version. For a right triangle, the sum of the areas of the square regions drawn on each leg is equal to the area of the square drawn on the hypotenuse. That is, area of Region I + area of Region II = area of Region III. The algebraic symbolism that we use for the square of a number, x^2, was not invented until the 1600s. We wonder what symbols people will use in the 2400s!

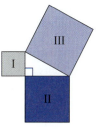

Notes

Here is some checking of the Pythagorean theorem. What is the answer to Problem 2?

Date   Time

LESSON 9·12   Verifying the Pythagorean Theorem   SRB 167

In a right triangle, the side opposite the right angle is called the **hypotenuse**. The other two sides are called the **legs of the triangle**.

hypotenuse c leg a b leg

Think about the following statement:
If a and b are the lengths of the legs of a right triangle and c is the length of the hypotenuse, then $a^2 + b^2 = c^2$.

This statement is known as the **Pythagorean theorem**.

1. To verify that the Pythagorean theorem is true, use a blank sheet of paper that has square corners. Draw diagonal lines to form 4 right triangles, one at each corner. Then measure the lengths of the legs and the hypotenuse of each right triangle, to the nearest millimeter. Record the lengths in the table below. Then complete the table.

Triangle	Leg (a)	Leg (b)	Hypotenuse (c)	$a^2 + b^2$	c^2
1					
2					
3					
4					

2. Compare ($a^2 + b^2$) to c^2 for each of the triangles you drew. Why might these two numbers be slightly different?

_____

_____

_____

3. Use the Pythagorean theorem to find c^2 for the triangle at the right. Then find the length c.

$c^2 = $ _____ units² c is about _____ units.

3 c 6

Algebra and Functions 1.0, 3.0, 3.2   363

Wright Group, *Everyday Mathematics, Journal 2, Grade 6*, p. 363. Copyright © 2008, McGraw-Hill. Reproduced with permission of the McGraw-Hill Companies.

The complete history of how the relationship was discovered and justified for *all* right triangles is lost, but since Pythagoras's time there have been more than 350 different mathematical justifications of it, including ones by Euclid, Leonardo da Vinci, and President James Garfield. Activity 1 shows an experimental way of verifying the relationship with the area interpretation, in a way that elementary students can understand.

ACTIVITY 1 Illustrating the Pythagorean Theorem with Areas

Notes

Trace the lettered regions in the given diagram, cut them out, and reassemble them to match the square region drawn on the hypotenuse.

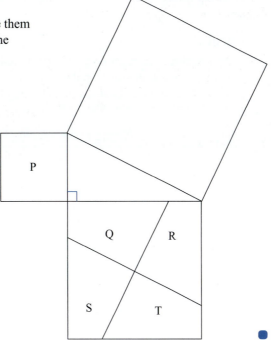

Following are some examples to give just a glimpse of the usefulness of the Pythagorean theorem. Knowing any two of the three lengths of the sides of a right triangle allows you to calculate the length of the third side.

EXAMPLE 1

Suppose a right triangle has one leg with length 5 cm and hypotenuse with length 13 cm. How long is the other leg?

Solution

Letting x represent the length of the other leg, the Pythagorean theorem asserts that $5^2 + x^2 = 13^2$. That is, $25 + x^2 = 169$, or $x^2 = 169 - 25 = 144$. So, $x = \pm\sqrt{144} = {}^+12$ or ${}^-12$ (${}^\pm 12$ or ±12 for short). The other leg is 12 cm long. (The negative square root, ${}^-12$, does not make sense because a negative length is not possible here.) ●

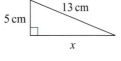

As you know, square roots of most whole numbers are not themselves whole numbers. For example, if the legs had lengths 3 cm and 8 cm, then $3^2 + 8^2 = h^2$ would give $9 + 64 = h^2$, or $h^2 = 73$, with $h = \pm\sqrt{73} \approx \pm8.5440037$. So, the length of the hypotenuse would be about 8.5 cm. (If the 3 cm and 8 cm were ideal measurements of ideal segments, the exact $\sqrt{73}$ cm would be appropriate.)

Although most calculators have a square-root key, finding a square root by trial and error can be fairly efficient (and emphasizes the meaning of square root). For example, suppose you want a decimal value for $\sqrt{119}$. Because $10^2 = 100$ and $20^2 = 400$, you immediately know that $\sqrt{119}$ is between 10 and 20. Because $11^2 = 121$, $\sqrt{119}$ must be slightly less than 11. Checking the value 10.9, we get $10.9^2 = 118.81$, so $\sqrt{119}$ is about 10.9.

> ⋮ *Think About ...*
> ⋮ How do you know that $\sqrt{784}$ is between 20 and 30? Between 25 and 30?

EXAMPLE 2

How is the height of an equilateral triangle related to its side?

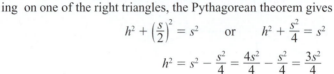

Solution

In many situations, we might know s and want to find h (see the drawing above at the right), rather than vice versa. The symmetry of the equilateral triangle gives two right triangles with legs h and $\frac{s}{2}$. By focusing on one of the right triangles, the Pythagorean theorem gives

$$h^2 + \left(\frac{s}{2}\right)^2 = s^2 \quad \text{or} \quad h^2 + \frac{s^2}{4} = s^2$$

$$h^2 = s^2 - \frac{s^2}{4} = \frac{4s^2}{4} - \frac{s^2}{4} = \frac{3s^2}{4}$$

Finally,

$$h = \pm\sqrt{\frac{3s^2}{4}} = \pm\frac{s\sqrt{3}}{2}$$

Thus, given just the length of a side of an equilateral triangle, we can find its height $\left(\sqrt{3} \approx 1.732\right)$. ●

EXAMPLE 3

A tower on level ground is to be supported by two wires of lengths 30 m and 40 m (after being fastened). The wires will be attached 24 m up the tower. How far from the foot of the tower will the wires hit the ground? (The planners are not sure there is enough room.)

Solution

Towers should meet level ground to make right angles, so the Pythagorean theorem is useful, as is a drawing. We get $x^2 + 24^2 = 30^2$, or $x^2 = 900 - 576$. So, $x^2 = 324$, and $x = 18$ (trial and error or calculator). Similarly, $y^2 + 24^2 = 40^2$, leading to $y = \pm\sqrt{1024}$, so $y = 32$. ●

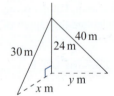

Because right angles are needed in finding heights for many area and volume problems, the Pythagorean theorem is often useful in such settings.

ACTIVITY 2   The Pythagorean Theorem Goes to Egypt

A regular square pyramid has a base with edges 10 cm long and with its other edges 12 cm long. Show that its volume and surface area are, respectively, 323.3 cm³ and 318 cm². (*Hint*: Find y first, then x.)

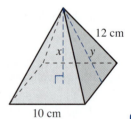

The Pythagorean theorem *begins* knowing that you have a right triangle and makes an assertion about the sides. The vice versa situation, called the **converse** in logic, also turns out to be true in this case. If you *begin* by knowing that $a^2 + b^2 = c^2$ for the sides a, b, and c of a triangle, then the triangle is a right triangle (the converse of the Pythagorean theorem). The ancient Egyptians knew this fact, and they used it to make right angles, as do carpenters and fence builders nowadays. For example, because $3^2 + 4^2 = 5^2$, finding points 3 feet and 4 feet from a corner so that the points are 5 feet apart will assure a right angle at the corner.

Think About …

Consider an isosceles right triangle with legs 1 unit long. Confirm that the hypotenuse has length $\sqrt{2}$ units.

You may know that $\sqrt{2}$ is irrational and thus cannot be written as the ratio of two integers: $\dfrac{\text{integer}}{\text{nonzero integer}}$. One legend says that the Pythagoreans had built a sort of "religion" around rational numbers, so when they found out that $\sqrt{2}$ was irrational even though it described a perfectly reasonable line segment, they were shocked and decided to keep it a secret. One of the group, however, told others. The legend says that the Pythagoreans took this person on a boat trip . . . and he did not come back.

TAKE-AWAY MESSAGE . . . The Pythagorean theorem is one of the very useful results for geometry. In a society where so many right angles occur, the theorem allows the calculation of many measurements. The theorem's converse, also true, allows one to create a right angle using only lengths.

Learning Exercises for Section 26.1

1. What does the Pythagorean theorem assert about each triangle?

   a.   **b.**   **c.**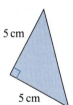

2. A common mistake with square roots is to think that $\sqrt{a^2 + b^2}$ is equal to $a + b$. Find a numerical example that shows this equality is incorrect.

3. Find the perimeter and area for each triangle pictured or described. Give an exact length and also an approximation, if appropriate.

   a.   **b.** 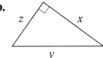   **c.**   **d.**

   e. right triangle with legs 1.6 m and 6.3 m
   f. right triangle with one leg 2.1 cm and hypotenuse 2.9 cm
   g. right triangle with legs 15 in. and 36 in.
   h. right triangle with one leg 4 m and hypotenuse 8.5 m

4. Find the length of a diagonal for each shape described.

   a. a square, 10 cm on each side
   b. a square, s cm on each side
   c. a rectangle, 10 cm by 24 cm
   d. a rectangle, m cm by n cm

5. Draw and find the length of each diagonal of each face of the following right rectangular prism and the *inside* diagonal of the prism, as shown. How long are the other *inside* diagonals? (*Hint:* Find the length of a diagonal of the base first.)

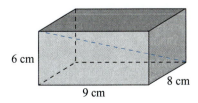

Notes

6. a. Give an expression for the length of an *inside* diagonal of a right rectangular prism that is x cm by y cm by z cm. (*Hint*: Learning Exercise 5.)

b. Will a 5-foot fishing rod fit inside a box that is 24 inches by 30 inches by 30 inches?

c. How long is an *inside* diagonal of a cube with edges of length e?

d. A 10-cm cube has all of its vertices on a sphere. What is the radius of the sphere?

7. The fire department needs to buy a ladder that will reach to the top of a two-story building—say, 22 feet. For safety's sake, when the full ladder is used, the distance the bottom of the ladder is from the building should be about $\frac{1}{4}$ the length of the ladder. The department can choose ladders that are 24', 28', and 32'. What size ladder should it buy? (*Hint*: If the ladder is x feet long, its bottom should be $\frac{x}{4}$ feet from the building.)

8. A plan calls for running a pipe from P to Q to R. The pipe costs \$5.89/m. If it is feasible to run the pipe directly from P to R, how much money would you save?

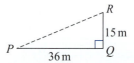

9. The following pyramid has a rectangular base, with several measurements and its other vertex given. What are the surface area and volume of the pyramid?

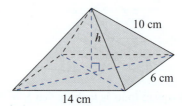

10. a. Find the areas of the semicircular regions drawn on the sides of the following right triangle, with $a = 6$ cm and $b = 8$ cm. How are the areas of the three regions related?

b. Use the variables a, b, and c to see in general whether the three areas are always related for a right triangle.

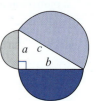

11. a. Show that the area enclosed by an equilateral triangle is given by $\frac{s^2\sqrt{3}}{4}$, where s is the length of a side of the triangle. (*Hint*: Example 2.)

b. Find the areas enclosed by the equilateral triangles drawn on the sides of the right triangle shown at the right, with $a = 5$ cm and $b = 12$ cm. How are the areas of the three regions related?

c. Use the variables a, b, and c to see in general whether the three areas are always related for a right triangle.

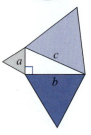

12. President Garfield's reasoning to justify that the Pythagorean theorem was based on the following drawing:

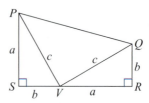

continue

Confirm that $a^2 + b^2 = c^2$ for the right triangle by doing the following steps:

a. Find the area of the whole shape by treating it as made up of three right triangular regions.

b. Then find the area of the whole shape by treating it as a trapezoid.

c. Compare the results from parts (a) and (b).

13. Triples of whole numbers such as the earlier 3–4–5 and 5–12–13 are called **Pythagorean triples** because they are positive integers that fit an $x^2 + y^2 = z^2$ relationship. Which of the following are Pythagorean triples?

a. 5, 6, and 7

b. 6, 8, and 10

c. 9, 12, and 15

d. 12, 16, and 20

e. If x, y, and z give a Pythagorean triple, will kx, ky, and kz also be a Pythagorean triple?

f. Use the idea in part (e) to generate 12 more Pythagorean triples.

14. a. Find the length of segment PQ on the given coordinate system.

b. Find the lengths of the other marked segments.

c. Point P has coordinate $(3, 5)$ and Q has coordinates $(1, 1)$. Give the coordinates of the other lettered points in the diagram. How can they be used to help find the lengths of the segments? For example, suppose you wanted to find the length of the segment joining the points with coordinates $(50, 30)$ and $(43, 6)$. Can it be done without extending the graph?

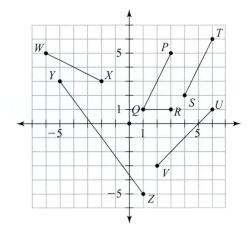

d. Your thinking for part (c) can be summarized compactly by the **distance formula:** $d = \sqrt{(x_1 - x_2)^2 + (y_1 - y_2)^2}$ gives the distance d between points (x_1, y_1) and (x_2, y_2). Verify that the distance formula gives the correct distances for some of your answers in parts (a) and (b).

15. Unlike the Pythagorean theorem, there are examples of situations in which an if-then statement is true, but its converse is not always true. In each of the following parts, decide whether both if-then statements are always true. If one is not, identify it and explain why it is not true.

a. If two rectangles are congruent, then they have equal areas.
If two rectangles have equal areas, then they are congruent.

b. If n^2 is a positive whole number, then n is a positive whole number.
If n is a positive whole number, then n^2 is a positive whole number.

c. If $x^2 = 4$, then $x = 2$.
If $x = 2$, then $x^2 = 4$.

d. If p is a factor of the product qr, then p is a factor of q or of r.
If p is a factor of q or of r, then p is a factor of the product qr.

16. Here is another ancient Greek result, called Hero's formula (or Heron's formula): The area of a triangle is given by $\sqrt{s(s-a)(s-b)(s-c)}$, where a, b, and c are the lengths of the sides of the triangle, and s is half the perimeter of the triangle.

   a. Apply the formula to a triangle with sides 15, 36, and 39.
   b. Verify using the converse of the Pythagorean theorem that the 15–36–39 triangle is a right triangle, and find its area using the triangle-area formula. Does the result agree with that of part (a)?

17. How many different lengths are possible on dot paper with 6 rows of dots and 6 dots in a row? The examples to the right show that there are 2 different lengths possible with 2 by 2 dots, and 5 different lengths possible with 3 by 3 dots. (In the drawing, two new ones overlap two of the old ones.)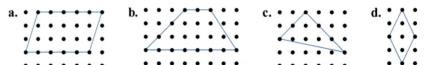

18. Give the perimeter and area of each polygon. Use the natural units.

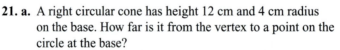

19. The following drawings are nets for polyhedra. Find the surface area and volume for each polyhedron. (*Hint*: Draw the polyhedron and label it with what you know.)

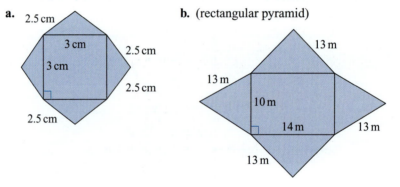

   a. 2.5 cm, 3 cm, 3 cm, 2.5 cm, 2.5 cm, 2.5 cm
   b. (rectangular pyramid) 13 m, 13 m, 10 m, 14 m, 13 m, 13 m

20. A large pile of sand has dimensions as in the figure at the right.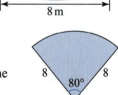

   a. Approximately what shape is the pile of sand?
   b. What is the volume of the pile of sand?
   c. How many children's sandboxes shaped like right square prisms 30 cm high with sides 1.5 m long will the sand fill?

21. a. A right circular cone has height 12 cm and 4 cm radius on the base. How far is it from the vertex to a point on the circle at the base?

   b. A piece of paper cut from an 8-inch circle and shaped like the sector to the right is cut out and rolled up to give a circular cone shape. What is the volume of the cone?

22. A new and larger triangle X is similar to the triangle shown below with the size transformation having scale factor 4. What is the area enclosed by triangle X?

29 cm

21 cm

23. An antenna on top of a building will be held steady by wires from P to A, B, C, and D, as pictured in the following diagram. How many feet of wire are needed? Assume that consecutive dots shown are 10 feet apart and allow 1 extra foot for each wire for tying the ends with a slight sag in the wires.

Notes

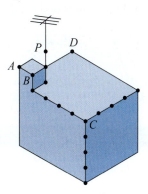

24. A segment of a circle is the region formed by a chord of a circle and the circle. (The shaded region in the drawing is a segment of the circle.) The area of a segment can be found for some angles, such as the 60° angle shown in the drawing, without trigonometry. Find the area of the segment shown for a circle with radius 10 cm. (*Hint*: Use a key idea of measurement.)

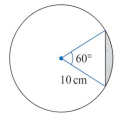

25. Give a geometric interpretation of the algebraic equation, $a^3 + b^3 = c^3$. Mathematicians conjectured for many years that there are no positive whole numbers a, b, and c that make the equation true, unlike the situation when the exponents all equal 2 (the Pythagorean triples).

26. A canal 4 m wide makes a 90° turn. Is it possible to make a bridge across the canal if you have only two boards, each 3.9 m long (and no nails)?[1]

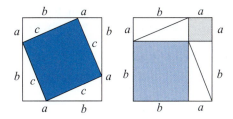

Boards, 3.9 m long

4 m wide

27. There are several congruent right triangles in the following squares. How can the drawings be used to justify the Pythagorean theorem?

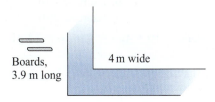

28. Find the missing angle sizes and lengths using the information in each sketch and your knowledge of triangles.

a.

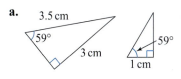

3.5 cm

59°

3 cm

59°

1 cm

b.

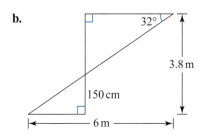

32°

3.8 m

150 cm

6 m

29. Starting with an isosceles right triangle with each leg 1 unit long, and building onto the hypotenuse continually as shown in the diagrams below, give a sequence like the one started. Find the lengths of the hypotenuses of the right triangles.

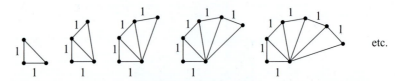

etc.

Supplementary Learning Exercises for Section 26.1

1. Find the *perimeter* of each right triangle described.

    a. legs 9 cm and 40 cm   **b.** legs 10 cm and 8 cm
    c. hypotenuse 50 m, one leg 14 m   **d.** hypotenuse 18 cm, one leg 10 cm

2. Find the *perimeter* of each shape.

    a. rectangle with diagonal 20 in. and one side 1 ft
    b. rhombus with diagonals 12 cm and 16 cm
    c. isosceles right triangle with one leg 7 cm
    d. square with diagonal 14.14 m
    e.

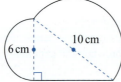

3. Give the area of each shape described in the Supplementary Learning Exercise.

    a. Exercise 2(a)   **b.** Exercise 2(b)
    c. Exercise 2(c)   **d.** Exercise 2(d)

4. Correct this student's work: $\sqrt{16 + 100} = 4 + 10 = 14$. How would you explain that it is incorrect?

5. In each part, the sides of a triangle have the given lengths. Which, if any, of the triangles is a right triangle? Tell how you know.

    a. 40 cm, 42 cm, and 58 cm   **b.** 9 in., 25 in., and 16 in.
    c. 28 cm, 1 m, and 96 cm   **d.** 1 ft, 3 ft, and $\sqrt{17}$ ft

6. A planning committee has two designs for a monument, with rectangular bases and the dimensions below. The interiors would be open. Find the volume of each design, and the outside surface area (not including the "floor"). (*Hint*: Yes, the triangles at the top are all right triangles.)

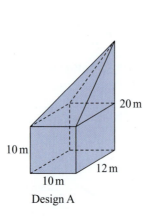

Design A

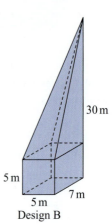

Design B

7. Find the surface area and volume of a tent (including its floor) with the dimensions given below.

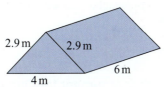

8. Find the distance between each of the following pairs of points on a coordinate system. Sketches may help.

    a. (5, 2) and (12, 2)   **b.** (3, 10) and (3, 5)   **c.** (a, b) and ($3a$, b)

    d. (c, d) and (c, $4d$)   **e.** (1, 5) and (4, 1) (*Hint:* Look for a right triangle.)

    f. (1, 1) and (6, 5)   **g.** (0, 14) and (5, 2)   **h.** (−1, 2) and (3, 5)

    i. (m, n) and (p, q)

9. Find the perimeter of polygon $ABCD$, where on a coordinate system point A is (3, 2), B is (7, 2), C is (10, 6), and D is (3, 9).

26.2 Some Other Kinds of Measurements

Working exclusively with measurements of geometric quantities can obscure the fact that all sorts of other quantities also are measured. Indeed, it is difficult to think of an everyday occurrence of a number that does not arise from a measurement, especially when counts and monetary cost are considered to be measurements.

Technical fields, of course, have many specialized measurements. But everyday life also uses a surprising number of measurements, even outside geometric quantities. You will recognize some of them in Discussion 1.

DISCUSSION 1 **From Rings to a Lottery**

What are units for values for the following quantities? (*Hint:* Look up carat and karat.)

a. the size of a diamond in a ring

b. the purity of the gold in a ring

c. the purity of the water you drink

d. the popularity of television shows

e. the amount of nutrition in a food

f. the healthiness of your gums

g. the chances of winning a lottery

In the elementary curriculum, working with actual measurements can not only teach something about quantitative aspects of many kinds of nonmathematical situations but also enable children to attach meanings to numbers. As a reminder, we repeat the key ideas about geometric measurements from Section 23.1. It will be of interest to find that they do not apply to all kinds of measurements.

Key Idea 1: The direct measurement of some characteristic in an object involves matching the object with a unit, or copies of a unit, having that characteristic. Many measurements are made indirectly using mechanical or scientific principles instead of direct comparisons.

Key Idea 2: Standard units are used because they ease communication and are relatively permanent in that they can be reproduced. The main system of standard units in the world is the metric system (SI).

Key Idea 3: Counts can be exact, but other measurements are approximate. Smaller units can give better approximations.

Key Idea 4: A quantity can sometimes be measured by thinking of the object as "cut" into a finite number of pieces, measuring the quantity in each piece, and then adding those measurements. (This one is the key idea that does not always apply.)

💬 DISCUSSION 2   Key-ing on Cereal

How might the key ideas of measurement be used with the quantity of calories in a particular kind of breakfast cereal? ●

Discussion 3 deals with standard units.

💬 DISCUSSION 3   Another Argument for One Simple System

You may have heard the following questions, which are given as "trick" questions:

a. Which is heavier, a pound of feathers or a pound of iron?

b. Which is heavier, a pound of feathers or a pound of gold? ●

If you were tricked by part (a) in Discussion 3 (the two are equally heavy), you are not alone. Many people, especially children, think about the *densities* of iron and feathers. You may also have been tricked in part (b): A pound of feathers is heavier than a pound of gold! Feathers (and iron) are measured with the *avoirdupois* ounce (28.34952 g), with 16 avoirdupois ounces in 1 avoirdupois pound (453.5924 g), whereas precious metals like gold are measured with the *troy* or *apothecaries'* ounce (31.10348 g), with 12 troy ounces in 1 troy pound (373.2417 g). The word "gallon" could also give trick questions. A *U.S. gallon* means something different from what a *Canadian gallon* does: 1 U.S. gallon = 231 cubic inches, but 1 Canadian (or Imperial) gallon = 277.42 cubic inches. Such complications are another good reason for having ONE worldwide standard system of units.

Angle size, length, area, and volume are, of course, part of every elementary school curriculum, as are time, temperature, and mass (or weight, although technically mass and weight are not the same idea). You know the basic standard SI units for length (the meter, m), time (the second, s), and mass (the kilogram, kg), and the basic units for area (the square meter, m^2) and volume (the cubic meter, m^3; also the liter, L). No doubt you are also somewhat familiar with the degree Celsius for temperature, which is a unit that is the same size as the official SI degree Kelvin unit.

As you know, the basic units can lead to units for other quantities, such as meters per second (m/s) for speed or the number of people per square kilometer for population density. Only a few such *per* units, used for quantities called **rates**, appear in the typical elementary school curriculum, perhaps because rates and ratios sometimes receive just computational attention in a few story problems or perhaps because they are conceptually more difficult. (Isn't that a reason why they should receive *more* attention?) For example, researchers and teachers have found that some children interpret a given speed like 50 miles per hour too narrowly. The children might think that asking how many miles one can go in a *half* hour at that speed does not make sense, because the speed is 50 miles per *hour*. Or, when asked to express the speed in terms of miles per century for a car that goes 50 miles per hour, one child said that would be impossible because the car would wear out before a century could pass.[2] Many children (and adults) have trouble viewing a rate as the *relationship between* the two quantities involved, much like the difficulty some children have in regarding a fraction as one number instead of the two numbers in the numerator and the denominator. Furthermore, measurements of rates do not follow all of the key ideas given earlier.

💬 DISCUSSION 4   Per-fection, Not Per-fiction

What might each of the following rates be measuring? What does each phrase or sentence mean?

a. 65 miles per hour

b. $16.50 per child

c. 7.2 grams of medicine X per liter

d. There was a return of 75% on the questionnaires we sent out. ●

A rate like $19.99 per square yard means $19.99 for every 1 square yard, or that the cost for each whole square yard is $19.99. One of the benefits of expressing rates in terms of one of the "per" units is that different rates can then be compared easily.

DISCUSSION 5   Who Went Fastest?

Who had the highest average speed?

   Al went 102 miles in 2 hours.

   Beth took 20 minutes to go 18 miles.

   Carla traveled 36 miles in $\frac{3}{4}$ hour.

   Dong went 180 miles in $4\frac{1}{2}$ hours. ●

DISCUSSION 6   Caution with Rates

a. A student has 112 honor points for her first 36 units, giving a grade point average (GPA) of 3.1. She then takes 6 units in summer school and gets a 3.5 GPA for the summer. Is her GPA on all her work now 3.3? Explain.

b. You go 100 miles in 2 hours, and then go 120 more miles in 3 more hours. What was your average speed for the whole trip? (*Hint*: It is not 45 mph.)

c. In summer school, a student earned a C in a 3-unit course and an A in a 1-unit course. She thought her average grade was B, or a GPA of 3.0, but a friend said that her GPA for the summer was 2.5. Who is correct? (The school operates on a system in which A is 4 honor points per unit, B is 3, etc.) ●

As the examples suggest, the *per* (meaning *for each* or *for every*) nature of rates means that finding a total rate based on unequal pieces of the *per* quantity, as the numbers of units in part (a) of Discussion 6 or the times in part (b) of Discussion 6, does not usually follow Key Idea 4. You must find the totals for both quantities involved in the *per* expression, rather than find the rate for each piece and then add them. Similarly, finding an average rate requires finding the totals for each quantity over the whole situation, not part by part. In contrast, if a rate is steady or the same all the time, then it can be measured or applied using any value of the *per* quantity. Or if, e.g., the numbers of hours spent for each distance had been the same in part (b) of Discussion 6 (say, each part of the trip took 2 hours), then the usual average of 50 mph and 60 mph, $(50 + 60) \div 2 = 55$, is correct. Another way to find the average speed is to look at 220 miles in 4 hours, which gives an average speed of 55 mph.

TAKE-AWAY MESSAGE . . .   It is not difficult to find reasons for the usefulness of a single set of standard units like the metric system, because people from so many different occupations or locations can measure quantities. Working with *per* quantities or rates requires some care, because they do not behave entirely like the geometric measurements studied.

Learning Exercises for Section 26.2

1. If you know someone from a culture quite different from that of the United States, ask her or him whether she or he used measurements units in the other culture that are not used in the United States (and, if so, tell what the units are used for).

2. Skim a newspaper for examples of different measurements. Tell what quantities are being measured.

3. Talk to someone with a major different from yours. What different sorts of measurements are made in that field?

4. Give the least and the greatest measurement possible for each reported measurement.

   a. 98 ms   **b.** 150 m, to the nearest meter   **c.** 1.42 L

5. Explain how Key Idea 4 might be applied in measuring the volume of a large amount of liquid.

6. Give a new example to show that Key Idea 4 is not reliable in rate situations.

7. *Dictionary* Many dictionaries include a table of weights and measures. Check such a table to see whether there are units with which you are not familiar.

8. These are all examples of nonmetric length units used in England at different times or in different settings: cubit, rod, ell, fathom, foot, furlong, hand, inch, knot, mile, nail, pace, and yard. Use these units to illustrate one advantage of the metric system approach to length units.

9. **a.** In your opinion, why has the United States been slow to embrace the metric system?
   b. Are there advantages/disadvantages in children's having to learn two sets of units—metric and English?

10. Answer each of the following questions:
    a. "I have taken 72 units, and my GPA is 3.1. If I take 15 units next term and get a 4.0, what will my GPA be then?"
    b. Nerita's record: For her first 30 units, 96 honor points; for her next 24 units, her GPA was 3.5; and last term her GPA for 12 units was 2.5. What is her overall GPA for those 66 units?
    c. Kien drove 175 miles in $3\frac{1}{2}$ hours, and then he ran into slow traffic and averaged only 25 mi/h for the next 75 miles. The final part of his trip took 2 hours for 100 miles. What was his average speed for the whole trip?
    d. A softball player batted .364 for her first 88 at-bats. Then she was slightly injured and hit only .250 for her next 40 at-bats. What was her overall batting average?

11. The body mass index (BMI) is one rough rate measurement of whether one is overweight. The BMI is defined by BMI $= \dfrac{\text{weight}}{\text{height}^2}$, where the weight is measured in kilograms and the height in meters. What are the BMIs of Alicia, Beth, and Carlita? *Approximations*: 1 kg = 2.2 lb, 1 m = 39.37 in.

    a. Alicia is 1.73 m tall and weighs 78 kg.
    b. Beth is 163 cm tall and weighs 145 lb.
    c. Carlita is 5' 2" tall and weighs 121 lb.
    d. Donna is 5' 7" tall. Her BMI is 28. How much does she weigh in kilograms? In pounds?
    e. Donna [see part (d)] wants her BMI to be under 25. How much weight does she have to lose?

12. Some curricula recommend using linked number lines to represent rate relationships. For example, questions about a medicine that uses 12 mL of water for every $\frac{3}{4}$ g of drug could involve the following drawing as an aid:

Use the drawing to help answer the following questions:

    a. How much water should be used with $\frac{1}{4}$ g of drug?
    b. How much water should be used with 1 g of drug?
    c. How much drug would be used with 24 mL of water?
    d. How much drug would be used with 20 mL of water?
    e. How much drug would be used with 6 mL of water?

13. Celsius temperatures (°C) and the common Fahrenheit (°F) temperatures are related as in the following diagram:

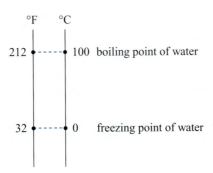

°F   °C

212 •----• 100   boiling point of water

32 •----• 0   freezing point of water

    a. From the diagram, which unit is larger, the Fahrenheit degree or the Celsius degree? How are the units related numerically? How do you know?

    b. From the diagram, 20°C is about what Fahrenheit temperature? How do you know?

    c. Use the diagram to determine the Fahrenheit temperature when it is ⁻40°C.

14. In each part, tell which of the speed units is larger. Explain your thinking.

    a. 1 km per s, or 1 m per s

    b. 1 km per s, or 1 km per h (h is the metric symbol for hour)

    c. 1 m per s, or 1 km per h

    d. 1 yd per s, or 1 mile per h

15. In the United States, the gasoline-miles relationship (*mileage*) is most often given in miles per gallon (fuel efficiency). Other countries might use something like gallons per mile (fuel consumption) if they used English units for volume and length.

    a. How are the two related? That is, how would 25 miles per gallon be expressed in terms of gallons per mile? Explain your thinking.

    b. Suppose you are visiting a metric country. How would 11 liters per 100 kilometers be expressed in terms of kilometers per liter? Explain your thinking.

16. An outstanding runner can run 100 meters in 10 seconds. What is the runner's speed in the given speed unit?

    a. in meters per second

    b. in kilometers per hour

    c. What does your answer for part (a) mean?

    d. What does your answer for part (b) mean?

17. An outstanding runner can run 100 yards in 10 seconds. What is the runner's speed in the given speed unit?

    a. in yards per second   **b.** in miles per hour

    c. Contrast your work for Learning Exercise 16(b) and for Learning Exercise 17(b). Which work was easier? Why?

18. **a.** Whose fudge was most expensive? Explain your thinking.

    ▶ Amy spent $13.98 for 2 pounds of fudge. Bea's 3.6 pounds of fudge cost $24.44. Conchita paid $1.60 for a quarter-pound of fudge. Danyell got $\frac{3}{8}$ pound of fudge for $2.70. ◀

    b. What was the average price per pound for the four kinds of fudge combined?

19. Answer the following questions:

    a. A teenager complains about not having enough clothes, so her parents say that she can buy 4 blouses, with an average cost of $25/blouse. The teenager sees $40 blouses and a $10 blouse on sale and calculates the average of $40 and $10. She buys 3 of the $40 blouses and a $10 blouse. Why were her parents unhappy?

continue

Notes

   b. At a lunch with four friends, you order a salad and drink for $9.99, but each friend orders a more elaborate meal costing $14.99. What is the average cost?

   c. For a large picnic, you buy 10 packages of chips for $2.59/bag and 2 bags of pretzels for $1.99/bag. Why is the average price per bag for the items *not* $2.29?

20. Here are two problems from a professional journal for teachers of mathematics in the middle grades.[3] What rates are involved? Solve the problems.

   a. ▶ Marlin's company sells a mineral supplement (for livestock) at $18.20 per 50-pound bag. If each cow is fed 10 ounces of this supplement per day, find the daily cost per cow. ◀

   b. ▶ A competing product sells for $16 per 50-pound bag but must be fed at the rate of 12 ounces per cow per day. Find the daily cost per cow. ◀

21. The smallest angle measured would cut off an arc 20 kilometers long if its vertex were placed at the center of a circle of radius 1500 light-years. What part of a degree is that angle? One report said that measuring this angle is comparable to seeing a single virus on Earth from the moon. (The speed of light is 3×10^8 m/s; a light-year is the distance light travels in a year.)

22. Water (including lakes and rivers) covers about 71% of Earth's surface. How many people per square mile would there be if all of Earth's population is spread out evenly over the land surface on Earth? Pretend that Earth is a perfect sphere.

23. Here are the nutrition facts from one kind of snack cracker:

Serving size 1 oz (28 g)
Servings per container 4.5
Calories 110

Calories from fat 30	**% daily value**
Total fat 3 g	5%
Saturated fat 0 g	0%
Cholesterol 0 mg	0%
Sodium 300 mg	3%
Total carbohydrate 19 g	6%
Dietary fiber 3 g	10%
Sugars 2 g	
Protein 4 g	
Vitamin A	0%
Vitamin C	0%
Calcium	16%
Iron	5%

   a. Find some rates that are included in the analysis.

   b. How many calories are in the whole box of snack crackers?

   c. How many milligrams of sodium are in the whole *daily value*, according to the analysis?

   d. Is it accurate to say that the crackers contain twice as much fiber as they contain fat? Explain.

   e. How many boxes of crackers would you have to eat to get your complete daily requirement of calcium?

24. One recipe for molasses squares calls for $\frac{1}{3}$ cup powdered sugar, $\frac{1}{3}$ cup molasses, $\frac{1}{8}$ teaspoon each of salt and soda, $\frac{7}{8}$ cup flour, and some other ingredients. This recipe gives 16 two-inch square bars.

   a. If a bakery used this recipe and wanted to make 300 molasses bars, what amounts of the given ingredients would be needed? How many square feet would the bars cover?

   b. A dieter decides that making 12 molasses bars would be acceptable. What amounts of the ingredients are needed for making 12 molasses bars?

   c. How are rates involved here?

25. A motorist traveled for 5 hours at an average speed of 55 miles per hour, and then went another 200 miles in $3\frac{1}{2}$ hours. What was the motorist's average speed for the whole trip?

26. What does this statement mean: "In our community, the average family has 2.6 children"?

27. a. If a hen and a half lay an egg and a half in a day and a half, how long will it take a dozen hens to lay a dozen eggs? (This is an old puzzler.)
  b. If 10 painters can paint 15 houses in 9 days, how many days will it take 20 painters to paint 20 houses?
  c. If 5 painters can paint 3 houses in 7 days, how many days will it take 8 painters to paint 10 houses?

Supplementary Learning Exercises for Section 26.2

1. a. In your opinion, what are the strong features of the metric system, compared with the English system?
  b. What are the disadvantages of the metric system, compared with the English system?
  c. If you were designing a system of measurement units, what features would you build in?

2. What is the rate of 50 mi/h in terms of the unit, mi/century? (Use 365 days for a year.)

3. You may use a dictionary to answer the following:
  a. How much would a 2-carat diamond weigh?
  b. What percent pure is 18-karat gold?

4. Why should it be expected that a gallon of gas will cost more in Canada than in the United States?

5. You want to carpet a 12 ft by 15 ft room. The carpet you like costs $12.99 per square yard, installed. About how much will the carpet cost?

6. Birthrates are often expressed in the number of births per 1000 population. Here are some data for U.S. births and birthrates (retrieved June 11, 2012, from http://www.infoplease.com).

Year	Number of births	Birthrate
1970	3,731,386	18.4
1980	3,612,258	15.9
1990	4,179,000	16.7
2000	4,058,814	14.7
2005	4,138,349	14.0
2009	4,131,019	13.8

According to these data, about what was the U.S. population in the following years?
  a. 2005   **b.** 1990   **c.** 1970

26.3 Check Yourself

You should be able to work exercises like those assigned and to meet the following objectives:

1. State the Pythagorean theorem and apply it to problems involving length, area, and volume (and right triangles, perhaps even ones not shown). The multitude of problems to which the Pythagorean theorem can be applied is one reason the theorem is so important.

2. Interpret the Pythagorean theorem and expressions such as x^2 and x^3 geometrically.

3. Deal with square roots.

4. Use the converse of the Pythagorean theorem to decide whether a triangle is a right triangle.

5. Tell why the metric system avoids certain difficulties that are involved in other sets of units.

6. Explain what a given rate means (e.g., 45 g/L).

7. Illustrate that Key Idea 4 does not always apply in certain rate situations, but show that you can deal with those situations.

References for Chapter 26

[1]Fomin, D., Genkin, S., & Itenberg, I. (1996). *Mathematical circles (Russian experience)* (M. Saul, Trans.). Providence, RI: American Mathematical Society. See Problem 22, p. 3.

[2.]Thompson, P. W. (1994). The development of the concept of speed and its relationship to concepts of rate. In G. Harel & J. Confrey (Eds.), *The development of multiplicative reasoning in the learning of mathematics* (pp. 179–234). Albany, NY: SUNY Press.

[3]Hilgart, F. (1996). Livestock production by the numbers: Taking the measure of things. *Mathematics Teaching in the Middle School, 1,* 712–717.

Part IV

Reasoning About Chance and Data

The study of chance originated in the seventeenth century as a means for winning at high-stakes gambling, and indeed, settings such as dice throwing are still used to teach probability. But in today's world, probability (the measurement of uncertainty) goes far beyond gambling situations. It undergirds much of the decision making that occurs in science and industry. Understanding the basics of probability is essential for making sense of many aspects of our daily lives, such as predicting the winner of a presidential race, investing in the stock market, considering the reliability information on a new car, or judging the effectiveness of a new drug.

Quantitative information permeates our culture—in sports, industry, budgeting, food, marketing, medicine, law, politics, and even the measures of educational quality. New technologies have aided in providing new and more powerful ways of collecting, analyzing, and interpreting data and making this information accessible and useful in decision making. Understanding how data are collected, analyzed, and interpreted helps us make sense of this vast array of information and is essential for statistical literacy. Probability plays a role in the development of statistical literacy, particularly in the decision-making process of interpreting data. The past two decades have seen enormous growth in the study of statistics and probability in the school curriculum.

The following problems provide a taste of the types of questions pursued in these chapters. Try working these problems. Save your answers to compare with later answers to these problems:

1. Assume there is a test for the HIV virus (which causes AIDS) that is 98% accurate. (That is, if someone has the HIV virus, the test will be positive 98% of the time, and if one does not have the virus, the test will be negative 98% of the time.) Assume further that 0.5% of the population actually has the HIV virus. Imagine that you have taken this test and your doctor somberly informs you that you've tested positive. How concerned should you be?

2. *The Monty Hall Problem.* On the *Let's Make a Deal* television show, the long-time host Monty Hall presented three doors to a contestant. Behind one door was the prize of the day and behind the other two doors were gag gifts. Contestants were asked to choose one door to open. Before the selected door was opened, Monty would open one of the other two doors to reveal a gag gift. The contestant was then asked whether he or she wanted to stay with the door chosen or switch to the other closed door. Would it matter?

3. Suppose you wanted to find the answer to the question, "Is there enough parking for students at this university?" What kind of data would help you answer this question? How would you go about collecting data to answer this question? (Think about who should be asked, how you would phrase the question, the number of people you would ask, and any other data that would be helpful.)

4. On March 26, 2012, a San Diego newspaper reported on a survey, undertaken by a reputable polling company, indicating that 56% of the local population (with a margin of 4%) favored an initiative for pension reform. The following day, a letter to the editor disagreed with the polling results, asserting that the poll surveyed only 611 people, a sample that was too small. How would you respond to the letter writer? What sample size would you be comfortable with? (San Diego has well over 1 million people.)

We first focus, in Chapters 27 and 28, on how to determine probabilities. Collecting, representing, and interpreting data are the topics of Chapters 29–31. Chapter 32 combines probability and statistics in the study of determining what are called confidence intervals. Chapter 33, the final chapter, returns to probability, with a look at expected value and at permutations and combinations.

Chapter 27

Quantifying Uncertainty

Chance and related terms such as *likelihood, odds,* and the more precise *probability* are all used to describe situations that involve uncertainty. We know that some events are more likely to occur than others. But what does "more likely" mean? What is the chance that a baby will be female? What is the likelihood of rain this week? What is the probability of a particular candidate winning an election? What are the odds for a particular horse winning a race?

Recall from Chapter 1 that a *quantity* is anything (an object, event, or quality thereof) that can be measured or counted and to which we can assign a value. In this chapter we explore ways of assigning values as measures of uncertainty, along with the basic vocabulary for *probability* (the usual mathematical term for chance and uncertainty). In particular, we discuss two basic methods of assigning numerical values to probabilities.

27.1 Understanding Chance Events

In this section we focus on understanding and quantifying **chance**. We first consider what is and what is not a *probabilistic situation.*

Many of us have experienced, at some time, events that were entirely unexpected. Consider the following coincidence:

> Rishad's sister, Shenae, rolled 10 sixes in a row while playing a board game. Rishad said, "Wow! I didn't think that could happen."

Sometimes events happen that we may think of as impossible. In this case, however, the event did happen, and so it is not a chance event. Suppose, though, that we reformulate the description of the event:

> Rishad's sister, Shenae, rolled 10 sixes in a row while playing a board game requiring the toss of one die. Rishad said, "If a billion people each rolled a die 10 times, I wonder what fraction of those people might get sixes on every roll?"

🖐 DISCUSSION 1   What's the Difference?

What is different about the two versions of Rishad's reaction? ●

There is an important difference between the original situation and its corresponding reformulation. The revised version does not refer to a specific event that has already happened on a specific occasion. *Rather, Rishad presumes that some process will be repeated.* Rishad's revised response moves the focus away from Shenae's accomplishment as an isolated occurrence and asks how common her accomplishment would be in a large number of similar situations. This revised response describes a probabilistic situation.

> A **probabilistic situation** is a situation in which we are interested in *the fraction of the number of repetitions* of a particular process that produces a particular result when repeated under identical circumstances a large number of times.

Notes

It often happens that people are thinking of a probabilistic situation but pose questions as if they were contemplating the outcome on a single occasion. Consider the following question and its rephrasings:

> *Question:* You toss two coins. What is the probability that you get two heads?
> *Rephrase 1:* Suppose a large number of people toss two coins. What fraction of the tosses will end up with two heads?
> *Rephrase 2:* Suppose one person tosses two coins a large number of times. What fraction of the tosses will end up with two heads?

Think About …
What is different in the two rephrasings?

Obtaining a six each of the 10 times Shenae rolled a die is highly unlikely, but over many such rolls it is possible to predict a pattern that can be expressed as a probability. Another way of looking at this phenomenon is to say that we cannot predict behavior in the short run, but when we undertake a process many times, there are patterns that help us make reliable predictions of what will occur in the future.

As we continue our discussion of probabilistic situations, keep in mind that we are *always* assuming that a process, such as tossing 10 dice or tossing 2 coins, is carried out a large number of times, even though this is not always stated.

We have already used some terms that need to be defined before we continue.

> A process undertaken a large number of times, together with the results, is often called an **experiment**. An **outcome** is a result of an experiment.

EXAMPLE 1

There are five balls in a bag. Two are red, labeled R_1 and R_2. Three are black, labeled B_1, B_2, and B_3. Suppose you draw out two balls, replace them, and draw again, doing so a large number of times. This process is an *experiment*. Suppose that the first draw is R_2 and B_1. This particular draw is an *outcome*. Suppose the second draw results in B_2 and B_3. The second draw is another *outcome*. ●

EXAMPLE 2

The rolling of a pair of dice a large number of times, noting the number of dots on top of each die, is a *probabilistic situation* because the rolling of the dice over and over again is done the same way. Carrying out this process of rolling dice many times and noting the numbers of dots on the top of the dice is an *experiment*. Each time the dice are rolled, some numbers of dots appear on top, say, 3 on one and 5 on the other. This *outcome* gives a sum of 8. ●

> An **event** is an outcome or a set of outcomes of a designated type. The **probability of an event** is the fraction of the times the event will occur when some process is repeated a large number of times. Fractions, decimals, and percentages are all used to express probabilities.

EXAMPLE 3

Look again at Example 1. Suppose you are interested in the probability that a draw of two balls would result in 2 black balls if the draw is undertaken many times. Obtaining 2 black balls would be an *event*. The *outcomes* that would be part of the event are draws of B_1B_2, B_1B_3, and B_2B_3. The outcomes that are *not* counted in the event are R_1R_2, R_1B_1, R_1B_2, R_1B_3, R_2B_1, R_2B_2, and R_2B_3. If the balls were the same size and weight, it would seem that each of these draws is equally likely. If you draw two balls a large number of times, say, 1000 times, you might expect that the fraction formed by the event's number of outcomes to the total number of outcomes would be $\frac{3}{10}$, i.e., the *probability* of drawing 2 black balls is $\frac{3}{10}$. ●

EXAMPLE 4

Continuing from Example 2, a roll of dice resulting in a 4 and a 6 is also an outcome, but this time the sum is 10. All possible outcomes that would give a sum of 10 give an *event.* If the dice are red and black, then a 4 on red and a 6 on black, a 5 on red and a 5 on black, and a 6 on red and a 4 on black would be the three possible outcomes in the event of obtaining a 10 on a roll of the dice. Now suppose that you undertake an experiment of tossing two dice 1000 times, and suppose that the event of the dots on top adding to 10 occurs 82 times. We would say that, based on this experiment, the *probability* of rolling a 10 is $\frac{82}{1000}$, or 0.082. ●

DISCUSSION 2   Practice Using These Terms

a. Give an example of an experiment involving a penny and a dime.

b. Describe an outcome of this experiment.

c. Describe a specific event for this experiment.

d. How can your experiment be considered a probabilistic situation?

e. How would you find the probability of the event you described in part (c)? ●

DISCUSSION 3   Describe the Repetition

How does the probability describe a situation in which a process is being repeated a large number of times?

1. The probability of having a boy is 51%.

2. What is the probability that it will rain tomorrow? There is a 20% chance of rain. ●

> *Think About …*
> What is different about the methods of assigning probabilities in Examples 3 and 4?

In the next section we will consider both of these methods.

TAKE-AWAY MESSAGE . . .   The probability of an event is the fraction of the number of times that the event will occur when some process is repeated a large number of times. The term *probability* is often meant when *chance, likelihood,* or *uncertainty* is used. People sometimes make statements that sound as though they are speaking of the outcome of a single occasion. However, in many cases the situation is probabilistic and people are actually interested in what happens when a process is repeated a large number of times because doing so leads to a more accurate determination of the actual probability.

Learning Exercises for Section 27.1

1. For each of the following statements, say whether or not the situation is probabilistic, and explain why or why not. If not, rephrase the statement so that it describes a probabilistic situation.

   a. the probability that the U.S. secretary of state is a woman

   b. the probability that a person will be able to read newspapers more intelligently after completing this course

   c. the probability that the soon-to-be-born baby of a pregnant Mrs. Johnson with two sons will be a girl

continue

Notes

    d. the probability that a woman's planned third child will be a girl, given the information that her first two children are boys

    e. the probability that it will snow tomorrow in this city

    f. the probability that Joe ate pizza yesterday

2. Because a probability is a *fraction* of a number of repetitions of some process, what is the least value a probability can have, as a percent? What is the greatest value it can have, as a percent? Explain your thinking.

3. Give examples that clarify the distinction between an uncertain situation and a probabilistic situation.

4. a. Describe an experiment involving a penny, a nickel, and a dime.

    b. Describe an outcome of this experiment.

    c. Suppose one specific event is getting exactly 2 heads. Which outcomes would be in this event? Which outcomes would not be in this event?

5. a. Describe an experiment involving a toss of 3 dice.

    b. Describe an outcome of this experiment.

    c. Describe an event related to the outcome of the experiment.

    d. List the outcomes that would be in the event, a sum of 5 for the dots showing on the dice.

6. a. Describe an experiment of the drawing of three cards from a deck of cards from which the jacks, queens, and kings have been removed. (*Note:* There are 52 cards in a deck— 13 cards are hearts, 13 cards are diamonds, 13 cards are clubs, and 13 cards are spades. A card with an ace counts as 1. Nine cards of each suit are marked 2 through 10. Ignore the jacks, queens, and kings, leaving 40 cards from which to draw.)

    b. Is drawing a 3 of hearts, a 4 of diamonds, and a 5 of spades an outcome or an event? Why?

    c. Is the drawing of three cards whose numbers add to 10 an outcome or an event? Why?

Supplementary Learning Exercises for Section 27.1

1. State whether or not the situation in each case is a probabilistic one, and tell why. If it is not a probabilistic situation, reformulate the situation so that it is a probabilistic one.

    a. the probability that you grew up in the same town in which you were born

    b. the probability that a twenty-year-old man graduated from college

    c. the probability that you went to the grocery store last Friday

    d. the probability that a five-card hand will be all black cards

2. In English there are many words that imply a probability. Match each given word with the most suitable value from this list:

0   0.03   0.5   0.85   1.0

    a. Sometimes   **b.** Never   **c.** Always   **d.** Rarely   **e.** Often

3. Wendy flipped a coin and rolled a six-sided die. She repeated this process 10 times and recorded the results.

    a. What are all the possible outcomes for this process?

    b. Wendy then counted up all the occurrences of a head with an odd number. Write the outcomes that would be included in that event.

    c. In your own words, describe the differences and similarities between outcomes and events.

4. If you buy a ticket for the lottery today, you are told that you have one chance in 4,500,000 of winning. You buy one ticket. Express this chance as a probability statement.

5. Suppose you draw 2 cards from a regular deck of 52 cards. Describe several outcomes. Describe several events. How do they differ?

27.2 Methods of Assigning Probabilities

Methods of assigning probabilities to events—determining what fraction of the time we should anticipate something to happen—rely on proportional reasoning. If we find that an event occurs $\frac{2}{3}$ of the time in a particular situation, then we use proportional reasoning to expect the event to occur approximately $\frac{2}{3}$ of the time for any number of times. So if the situation happened 300 times, we would expect the event to occur approximately 200 times.

There are two ways to assign probabilities: *theoretically* or *experimentally.* A probability that can be arrived at by knowledge based on a theory of what is likely to occur in a situation, such as when a fair coin is tossed, is **a theoretical probability**.

EXAMPLE 5

In the toss of one die, the possible outcomes are 1, 2, 3, 4, 5, and 6 dots on top. Each is equally likely. Thus, the theoretical probability of obtaining a 2 is $\frac{1}{6}$. If a coin is tossed, the theoretical probability of obtaining a head is $\frac{1}{2}$. With the spinner to the right, the theoretical probability of getting a 3 is $\frac{1}{4}$. The theoretical probability of getting a 2 is $\frac{1}{2}$. ●

An **experimental (or empirical) probability** is an application of the adage, "What has happened in the past will happen in the future." The experimental probability of an event is determined by undertaking a process a large number of times and noting the fraction of times the particular event occurs. Thus, an experimental probability value is determined by undertaking an experiment and observing what happens. Probabilities determined this way will vary, but the variation diminishes as the number of trials increases.

Determining a probability via an experiment is necessary when we have no prior knowledge of possible probabilities. The activities in this section will give you practice in assigning both experimental and theoretical probabilities.

ACTIVITY 1   A Tackful Experiment

We want to know the probability of a thumbtack landing point up if it is tossed from a paper cup.

a. Separate into groups. Within each group, toss a thumbtack onto the floor (or the top of a desk or table) 50 times. Keep track of how many times the thumbtack points up.

b. What would you say is the probability that a thumbtack tossed from a paper cup repeatedly will land with the point up?

c. Combine your data set with those of other groups. What would you now say is the probability that a thumbtack tossed from a paper cup repeatedly will land with the point up?

d. Which probability estimate do you trust more? Why? ●

Activity 1 is an example of assigning a probability experimentally (or empirically—i.e., through experience). You performed a process a large number of times, and then you used that information to formulate a probability statement. In the experiment, you noted the position of the thumbtack each time. That position is an outcome. The term *outcome* is used to mean one of the simplest results of an experiment. All of the outcomes together form what is called a *sample space,* so the sample space for this experiment has just two possible outcomes, *point up* and *point sideways.* You may have determined the probability of the event that a thumbtack lands with the point up by seeing what fraction of all the tosses the thumbtack landed point up. We estimate the probability of the event of a thumbtack landing point up by the fraction

$$\frac{\text{the number of times a thumbtack landed point up}}{\text{the total number of times the thumbtack was tossed}}$$

The terms *outcome, sample space,* and *event* are all important for understanding and determining both experimental and theoretical probabilities.

A **sample space** is the set of all possible outcomes of an experiment. If A represents an event, then $P(A)$ represents the probability of event A:

$$P(A) = \frac{\text{number of times } A \text{ happens}}{\text{total number of outcomes in the sample space}}$$

if all the outcomes are equally likely. Sometimes the probability of an event is called the **relative frequency** of the event.

EXAMPLE 6

The sample space consists of the two outcomes, point up and point sideways:

$$P(\text{point up}) = \frac{\text{number of times with point up}}{\text{total number of tosses}}$$

$$P(\text{point sideways}) = \frac{\text{number of times with point sideways}}{\text{total number of tosses}}$$

Note that $P(\text{point up}) + P(\text{point sideways}) = 1$. ●

EXAMPLE 7

Suppose a polling company called 300 people, selected at random. All were asked if they would vote for a new school bond. Of the 300 people interviewed, 123 said yes, 134 said no, and 43 were undecided. The *sample space* of all possible outcomes is yes, no, and undecided. An event Y might be getting an answer of yes. $P(Y) = \frac{123}{300}$. This fraction is also called the *relative frequency* of getting a yes response. ●

A diagram such as the one to the right (called a **Venn diagram**) is occasionally useful in thinking about sample spaces and their outcomes and events. You imagine that all of the outcomes are inside the sample-space box, with those favoring event A inside the circle labeled "Event A." Thus, if the sample space contains all the outcomes of tossing a thumbtack, A could represent the event that the point is up.

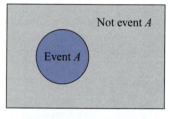

Similarly, the experiment described in Example 7 might give a Venn diagram such as the one to the right. As with all Venn diagrams, everything in the box is in the given sample space. That is, here the box represents all the people interviewed. In Venn diagrams, the sizes of the regions need not reflect the numbers of votes.

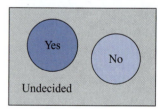

> ### Think About ...
> What is the sample space for randomly choosing a letter of the alphabet? For this experiment, make a Venn diagram that shows the event "chose a vowel." For this experiment, it is easy to be explicit about the outcomes.

When tossing thumbtacks in Activity 1, it may have occurred to you that, for efficiency's sake, you could toss more than one thumbtack at a time and note how many of them land point up and how many land point sideways. For example, your group could have tossed 10 thumbtacks at a time and gathered data more quickly by viewing the single toss as 10 repetitions of the basic one-thumbtack experiment. This approach is certainly possible, but because it literally is a different experiment from tossing just one thumbtack, it opens up other possibilities for descriptions of outcomes.

For a toss of 10 thumbtacks (with numbers 1 through 10 painted on for clarity), you could view the outcomes differently from before. If you use S to mean that a thumbtack

Notes

pointed sideways and U to mean it pointed up, a possible outcome could now be denoted by SSSUSUUSUS. Other outcomes are possible of course, such as USUUSUSSSU. The sample space would then consist of all possible lists with 10 S's and U's. From this set-up, an example of an event such as "all the outcomes in which 7 thumbtacks point up" makes sense.

Sometimes it is difficult to distinguish between outcomes and events. An event consists of one or more outcomes that are of particular interest, such as the event of getting a head and a tail on the toss of a penny and a dime. This event consists of two possible outcomes: a head on the penny and a tail on the dime, or a tail on the penny and a head on the dime.

But what if we toss only the dime, and the event is getting heads? Now the *only* outcome that satisfies this condition is getting heads when tossing the dime, so the outcome and the event are the same: getting heads on a toss of a dime. A clear understanding of what the sample space is can help to clarify the distinction between outcome and event.

DISCUSSION 4 Some Facts About Probabilities

1. Why is a probability limited to numbers between, and including, 0 and 1? When is it 0? When is it 1?

2. Suppose you calculate the probability of each possible outcome of an experiment, and add these probabilities. What should the sum be?

3. Suppose the probability of an event A is $\frac{2}{3}$. What is the probability of event A *not* occurring? ●

ACTIVITY 2 Balls in a Bag

One of your group will, without the others seeing, place five balls of two colors (or use other objects of two colors that cannot be distinguished by feel) into an opaque bag. Only that person knows what colors are in the bag. Others in the group draw a series of five balls, replacing each ball after drawing it and noting its color. The others then collectively make one prediction as to the number of each color of balls that are in the bag. The prediction is an *experimental* probability. How accurate is your prediction? ●

Consider the sample space for drawing one ball from a bag containing 2 red balls and 3 blue balls, replacing the ball, and then drawing again. This sample space consists of all possible outcomes and is made much easier by thinking of the balls as numbered, i.e., R_1, R_2, B_1, B_2, and B_3. Counting the possible outcomes requires that we list *all* possible pairings. Listing all outcomes should be done in a systematic, organized way to assure that all pairs are included and none are counted twice. The following shows a systematic listing:

R_1R_1	R_2R_1	B_1R_1	B_2R_1	B_3R_1
R_1R_2	R_2R_2	B_1R_2	B_2R_2	B_3R_2
R_1B_1	R_2B_1	B_1B_1	B_2B_1	B_3B_1
R_1B_2	R_2B_2	B_1B_2	B_2B_2	B_3B_2
R_1B_3	R_2B_3	B_1B_3	B_2B_3	B_3B_3

EXAMPLE 8

Suppose someone is interested in drawing 2 red balls. Drawing 2 red balls is then the event of interest. The four possible outcomes in this event are R_1R_1, R_1R_2, R_2R_1, and R_2R_2. The sample space has 25 possible outcomes in all, and it is reasonable to consider them equally likely. Thus, the theoretical probability of drawing 2 reds is $\frac{4}{25}$. ●

DISCUSSION 5   Things Change

In Activity 2, one ball was drawn from 2 red balls and 3 blue balls, and replaced. A second ball was then drawn. We call this **drawing with replacement.** Look back at the sample space in Activity 2. How would this sample space change if the first ball were *not* replaced? We call this **drawing without replacement.** What would now be the probability of drawing 2 reds, as in Example 8? ●

> ### Think About ...
> What patterns do you notice in the listed sample space of 25 outcomes for drawing two balls? If there had been 3 red balls and 3 blue balls, how would the listing change? If there had been only 2 blue balls and 2 red balls, how would the listing change?

It is even handy on occasion to think of an *impossible event*, such as getting 2 greens in the last experiment. The probability of an impossible event is, of course, equal to 0. Likewise, we can think of an event that is certain. For example, the event of *getting two balls that are either red or blue* will happen every time. Thus, this event is certain to happen and has probability 1.

> ### Think About ...
> What is the probability of NOT getting a red ball on the first draw and a blue ball on the second draw? What is the probability of getting either red or blue on the first draw? What is the probability of getting a red and then a green?

ACTIVITY 3   Heads Up

1. Suppose you toss a coin a large number of times. Predict what fraction of those times you would expect the coin to land heads up. Does your fraction give an experimental probability or a theoretical probability? Explain.

2. Try this experiment: Toss a coin 10 times and record for each toss what is on top. How many times did it land heads up? Combine your results with those from other people, and compare your prediction in Problem 1 about the probability of a coin landing heads up with the experimental probability of a coin landing heads up.

3. Look at your results in Problem 2. Did heads occur 3 or more times in a row (a *run* of 3 or more heads)? Did tails occur 3 or more times in a row (a run of 3 or more tails)? Such runs would not be unusual. ●

In the first problem in Activity 3, you probably made the assumption that coins are made in such a way that it is **equally likely** that a coin will land heads up as it is that the coin will land tails up (or at least it is *very* close to equally likely). Using this assumption, we can formulate a probability statement without actually performing the process a large number of times.

Recall that a theoretical probability is assigned based on determining the fraction of times an event will occur under ideal circumstances. Thus, the theoretical probability that a tossed fair coin will land heads up is $\frac{1}{2}$. You probably found in the second part of Activity 3 that your experimental probability was close to the theoretical probability, but you yourself may not have because you tossed the coin only 10 times.

> ### Think About ...
> For some event, when both its experimental probability and its theoretical probability are possible to determine, will the two probabilities be equal? Explain.

Return to the list of outcomes in the sample space for the experiment in which two balls were drawn (with replacement of the first ball) from a bag containing 2 red balls and 3 blue balls. If the balls were identical in every other way, then any one outcome is as likely as any other outcome in the list of 25 outcomes. The outcomes are *equally likely.* We can thus easily determine theoretical probabilities. Suppose we are interested in the event of drawing a red ball followed by a blue ball. This event occurs 6 out of 25 times in the sample space, so $P(\text{RB}) = \frac{6}{25}$.

Sometimes probabilities can be found either theoretically or experimentally. Other times, such as when tossing a thumbtack (for which there are not equally likely outcomes), finding the probability experimentally is necessary.

⋮ *Think About …*
⋮ Suppose the balls in the bag differ in size and weight. Do you think the outcomes are
⋮ equally likely? If not, how can you find *P*(RB)?

 ACTIVITY 4   Two Heads Up

1. If you toss a penny and a nickel, what is the probability of getting 2 heads? Answer this question experimentally.

2. To theoretically determine the probabilities in Problem 1, it is helpful to list the complete sample space.

Penny	Nickel
H	H
H	T
T	H
T	T

   Is each of the four outcomes equally likely? What is the probability of getting 2 heads? What is the probability of NOT getting 2 heads? How do you know?

3. Did your experimental probability from Problem 1 match the one from Problem 2? ●

The sample space in Activity 4 can give you information about other events. What is the probability of getting 1 head and 1 tail? (Notice from the list that this can happen in two ways.) What is the probability of getting no heads? Note that *P*(no heads) = 1 − *P*(one or more heads). More generally,

$$P(\text{not } A) = 1 - P(A),$$

for any event *A*.

 ACTIVITY 5   Give It a Spin

Spinners like those in board games can easily provide classroom experiments. Each experiment works the same way: Spin the arrow on the spinner and note where the arrow points. If the arrow lands on a line, spin again. (A partially straightened paper clip, with a pencil tip at one end, makes a serviceable spinner.)

Standard spinner   Paper clip spinner

   Suppose the spinner pictured here is used in an experiment of one spin.

1. What is the sample space for the experiment?

2. What angles are associated with the red, white, and blue outcomes?

3. What is the theoretical probability of each outcome?

4. What is *P*(not blue)?

5. How would you determine experimental probabilities for this experiment? ●

The term **odds** is used to compare an event with the corresponding nonevent. For example, we say the odds of drawing an ace from a regular deck of cards is 4 to 48, or 1 to 12. Or, if a brown bag held nine balls—3 red, 4 green, and 2 blue—we could say the odds of drawing a blue is 2 to 7. Note that a probability statement can be easily formulated from an odds statement. If the odds of drawing a blue are 2 to 7, then the probability of drawing a blue is $\frac{2}{2 + 7} = \frac{2}{9}$.

Notes

Statements about odds are not themselves probability statements, but they can give probability statements as follows. If the odds of Team A winning are given as "*a* to *b*," this means that

$$P(A \text{ winning}) = \frac{a}{a + b} \qquad \text{and} \qquad P(A \text{ losing}) = \frac{b}{a + b}$$

For example, if the odds in favor of Team A are "7 to 5," then the anticipated $P(A \text{ winning})$ is equal to $\frac{7}{12}$ and $P(A \text{ losing})$ is equal to $\frac{5}{12}$. The statement $P(A \text{ winning})$ is equal to $\frac{7}{12}$ means that if the game were played many, many times, you would expect Team A to win $\frac{7}{12}$ of the games. *Odds* can arise in other contexts. For example, a doctor might say, "The odds of recovery are 3 to 2," indicating that the probability of recovery is $\frac{3}{3+2} = \frac{3}{5}$.

Think About …

What does "The odds of recovery are 3 to 2" mean, if the doctor has in mind many, many cases?

Here is a page from a Grade 4 textbook. What could the children learn from Problem 2?

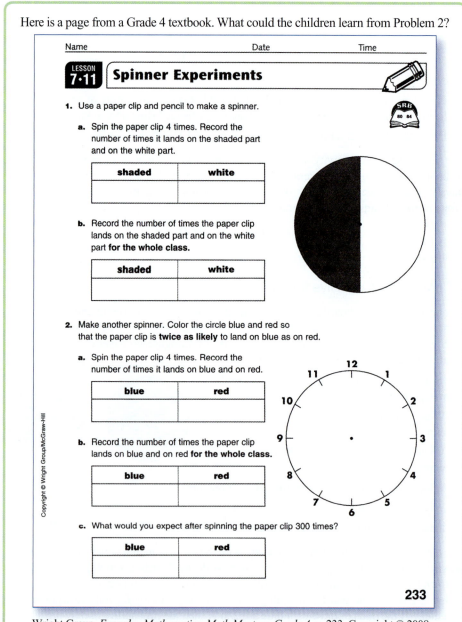

Wright Group, *Everyday Mathematics, Math Masters, Grade 4*, p. 233. Copyright © 2008, McGraw-Hill. Reproduced with permission of the McGraw-Hill Companies.

At this point you may wonder if all we do in probability is spin spinners, draw balls from a bag, or draw cards from a deck. In the next section we will see how we can use these same tools to simulate real-life probabilistic situations.

TAKE-AWAY MESSAGE . . Some vocabulary makes it easier to discuss a particular process and what happens: sample space, outcomes, events, equally likely outcomes. It may be possible to quantify the probability of an event in two ways: (1) by doing the experiment many times to get an experimental probability, or (2) by reasoning about the situation to get a theoretical probability. When an event is sure to happen, the probability of that event is 1. When an event cannot occur, the probability of that event is 0. If one knows the probability of a certain event A, then the probability of that event *not* occurring is P(not A), and P(not A) = 1 − P(A). The sum of the probabilities of all outcomes of an experiment is 1. Statements about *odds* can be changed to probability statements. If the odds in favor of event A are *a* to *b,* then the probability of A is $\frac{a}{a+b}$.

Learning Exercises for Section 27.2

1. Explain how each situation is similar to tossing a coin.

   a. predicting the sex of an unborn child
   b. guessing the answer to a true-false question
   c. picking the winner of a two-team game
   d. picking the winner of a two-person election

2. Explain how each situation may *not* be similar to tossing a coin.

   a. predicting the sex of an unborn child
   b. guessing the answer to a true-false question
   c. picking the winner of a two-team game
   d. picking the winner of a two-person election

3. Answer the question in each situation.

   a. A medical journal reports that if certain symptoms are present, the probability of having a particular disease is 90%. What does that percentage mean?
   b. A weather reporter says that the chances of rain are 80%. What does that percentage tell you?
   c. Suppose the weather reporter says that the chances of rain are 50%. Does that percentage mean that the reporter does not know whether it will rain or not?

4. Is it possible for the probability of some event to be 0? Explain.

5. Is it possible for the probability of some event to be 1? Explain.

6. Is it possible for the probability of some event to be 1.5? Explain.

7. Three students are arguing. Chad says, "I think the probability is 1 out of 2." Tien says, "No, it is 40 over 80." Falicia says, "It is 50%." What do you say to the students?

8. Suppose a couple is having difficulty choosing the name for their soon-to-be-born son. They agree that the first name should be Abraham, Benito, or Charles, and the middle name should be Aidan or Benjamin, but they cannot decide which to choose in either case. They decide to list all the possibilities and choose one first-name, middle-name combination at random. Is this a probabilistic situation? If so, what is the process that is being repeated?

9. Give the sample space for each spinner experiment shown on the next page, and give the theoretical probability for each outcome. How did you determine your theoretical probabilities? In this problem and others with drawings, assume that the angle sizes are as they appear—90°, 60°, 45°, etc. Recall that the sum of all the angles at the center of a circle must equal 360°.

continue

a. 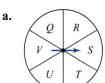 b.   c.   d.

e. Would your answers change if these were dartboards? (You are throwing blindfolded and the throw doesn't count unless you hit the board.)

10. How would you determine *experimental* probabilities for each outcome in the spinner experiments in Learning Exercise 9? Explain your reasoning.

11. Using the spinner in Learning Exercise 9(d) 12 times, is it possible to get region *J* 8 times? Explain.

12. Consider the spinner shown to the right in which the angle forming sector *X* is 120°. What is the theoretical probability of . . .

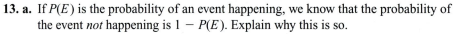

   a. getting *X*?   b. *not* getting *X*?
   c. getting *Y or X*?   d. getting *Y and Z* (simultaneously)?
   e. getting *Z*?   f. *not* getting *Z*?

13. a. If $P(E)$ is the probability of an event happening, we know that the probability of the event *not* happening is $1 - P(E)$. Explain why this is so.

   b. What is the probability of *not* drawing an ace from a deck of cards?
   c. What is the probability of a thumbtack *not* landing with the point up when tossed, based on your results from Activity 1?
   d. What is the probability of *not* getting 2 heads on the toss of a penny and a nickel? (Be careful with this one. *Hint:* What is the sample space?)
   e. What is the probability of *not* getting a 5 on the toss of one die?
   f. You have a key ring with 5 keys, one of which is the key to your car. What is the probability that if you blindly choose 1 key off the ring, it will *not* be the key to your car? What assumption(s) are you making?

14. One circular spinner is marked into four regions. Region *A* has an angle of 100° at the center of the circle, Region *B* has an angle of 20° at the center, and Regions *C* and *D* have equal angles at the center.

   a. Give the theoretical probability of each of the four outcomes with this spinner.
   b. Are any of the outcomes equally likely?
   c. What is the probability of *not* getting *B*?
   d. How would you determine experimental probabilities for this spinner?

15. Make sketches that describe the following spinners. Give the angle sizes of the regions.

   a. A spinner with five equally likely outcomes. (How could you do this accurately?)
   b. A spinner with five outcomes of which four outcomes are equally likely. The fifth outcome has a probability twice that of each of the others.
   c. A spinner with five outcomes of which three outcomes have the same probabilities and the fourth and fifth outcomes each have the same probability that is three times the probability of each of the first three outcomes.
   d. Two spinners that could be used to practice the basic multiplication facts from 0 × 0 through 9 × 9. (Is each fact equally likely to be practiced with your spinners?)
   e. Two spinners that could be used to practice the "bigger" basic multiplication facts, from 5 × 5 through 9 × 9.

16. a. List some situations in which the probability of a certain-to-happen event is 1.
   b. List some situations in which the probability of a certain-not-to-happen event is 0.

17. For each of the following situations, decide whether the situation calls for determining a probability experimentally or theoretically. If the probability is obtained experimentally, describe how you would determine it, and if the probability is theoretical, describe how you would find it.

   a. drawing a red M&M from a package of M&Ms
   b. getting a 2 on the throw of one die
   c. selecting a student who is from out of state on a particular college campus
   d. selecting a heart from a regular deck of cards

18. Consider the spinner to the right.

   a. How many times would you expect to get sector S in 1200 spins?
   b. How many times would you expect to get sector T in 1200 spins?
   c. Is it possible to get sector S 72 times in 100 spins? Explain.
   d. Is it possible to get sector T 24% of the time in 250 spins? Explain.
   e. What would you expect to get in 200 spins?
   f. You likely used proportional reasoning in parts (a–e). How is proportional reasoning being used?

19. In an experiment you are to draw one ball from a container (without looking, of course) and note the ball's color. You win if you draw a red ball. Each part below gives a description of the contents of two containers. For each part, which container provides the better chance of winning? If the chances are the same, say so. Explain your reasoning.

	Container 1		*Container 2*
a.	2 reds and 3 blues	or	6 reds and 7 blues
b.	3 reds and 5 blues	or	27 reds and 45 blues
c.	3 reds and 5 blues	or	16 reds and 27 blues
d.	3 reds and 5 blues	or	623 reds and 623 blues

20. What conditions (e.g., for the container, for the balls, how the drawing is done, etc.) are necessary to make reasonable the assumption that the outcomes of a draw-a-ball experiment are equally likely?

21. Suppose you hold only the 13 hearts from a regular deck of cards.

   a. You draw one card, then replace it, then draw again. What is the sample space for this experiment? What is the probability of drawing two face cards? (The jack, queen, and king are face cards.)
   b. This time, find the sample space if the first card is *not* replaced. Now what is the probability of drawing two face cards?

22. Your school released a message balloon as part of Heritage Day. The local news reports that the balloon will land somewhere within a 3-mile radius of where it was released. What is the probability that the balloon will land within the rectangular city park (0.5 mi × 0.25 mi) in that 3-mile radius? (Ignore weather that might influence its landing place.)

23. The probability situation in Exercise 22 is represented by an area model. How is this type of model different from and similar to a spinner model?

24. a. What probability statement says the same thing as "The odds for the Mammoths winning are 3 to 10"?
   b. What probability statement says the same thing as "The odds for the Dinosaurs losing are 6.7 to 1"?

25. a. If Team A is actually expected to win over Team B, how are x and y related in the statement, "The odds in favor of Team A *losing* are x to y"?
   b. If Team A and Team B play many, many times, what does the odds statement in part (a) mean?

26. Rephrase each of the following in terms of odds:

    a. The probability of making the field goal is $\frac{2}{3}$.

    b. The probability of drawing a red ball from the bag is $\frac{7}{12}$.

    c. The probability of not getting a winning card is $\frac{7}{30}$.

Supplementary Learning Exercises for Section 27.2

1. **a.** On the spinner at right, sectors A and D each have 120°, and sector C has a right angle. Find the probabilities for each outcome: arrow landing on A, B, C, and D.

    b. What does the value you found in part (a) for $P(B)$ mean?

2. Can a probability be negative? Why or why not?

3. **a.** On the given spinner, sector J has a right angle, and sectors K and L have the same angle size. Find the probabilities for the outcomes.

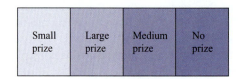

    b. What does the value you found in part (a) for $P(L)$ mean?

4. Draw the following spinners and describe how they can be drawn *accurately*:

    a. A spinner has four outcomes. Three outcomes have the same probability and the fourth outcome is half as likely.

    b. A spinner has three outcomes, each half as likely as the one before.

    c. A spinner has five outcomes. Outcomes A, B, and C have the same probability, and outcomes D and E have the same probability. The sum of the two different probabilities is $\frac{3}{8}$.

    d. A spinner has five outcomes. Outcomes A, B, and C have the same probability, and D and E have the same probability. The sum of the probabilities of A, B, and C is the same as the sum of the probabilities of D and E.

5. A school carnival has a game in which the blindfolded competitors throw a dart at a rectangular target. A prize is awarded according to what region the dart hits. If the dart misses the target, the thrower gets another try. Answer the following questions using this target. (Assume that all widths are the same and that the dart hits the target.)

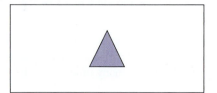

    a. What is the probability that a player receives a prize?

    b. What is the probability that a player wins the large prize?

    c. What is the probability that a player wins no prize?

    d. If the area of the whole target is 120 in^2, what is the area of each section?

    e. How does the area of each section relate to the probability of landing in this section?

6. Use the target area below (as in Supplementary Learning Exercise 5) to find the probabilities. The overall rectangle is 5 feet by 3 feet. The equilateral triangle is 1 foot on each side. What is the probability that the dart lands:

    a. within the equilateral triangle?

    b. within the rectangle but outside of the equilateral triangle?

7. Use the target to the right (as in Supplementary Learning Exercise 5) to find the indicated probabilities. The overall shape is a 3 unit by 3 unit square. The top and left sides are split evenly, and angles that look like right angles are right angles.

   a. What is $P(X)$? 　　**b.** What is $P(Y)$? 　　**c.** What is $P(Z)$?

8. Draw a rectangular area model for each situation.

   a. The probability of A is the same as that for B. The probability of C is twice that of each of the other two. A, B, and C are the only outcomes.
   b. The probability of each of A, B, C, and D is the same. The probability of E is one half that of the others. A, B, C, D, and E are the only outcomes.

9. For each situation would the probability best be found experimentally or theoretically? Explain how you would find it.

   a. the probability that a person chosen from this class is female
   b. the probability of getting a sum of 7 on the roll of two dice

10. Suppose the spinner here was spun 1800 times:

   a. How many times do you expect to spin an R?
   b. How many times do you expect to spin an S?
   c. How many times do you expect to spin a T?
   d. Is it possible to get the following frequencies?

R	S	T
705	425	670

11. Suppose a bag has an unknown number of blue and red marbles with more blue than red. Does the probability of drawing a red marble change if one of each color is added? If the probability changes, does it increase, decrease, or does it depend on the number of marbles to begin with?

12. Critique each probability question.

   a. An elementary school has 630 boys and 684 girls. What is the probability that the last student who walked out of the school yesterday was a boy?
   b. What is the probability that you have a dog?

13. Write each probability statement as a statement about odds.

   a. The probability for drawing a red queen from a standard deck of 52 cards is $\frac{1}{26}$.
   b. The probability of our chicken laying an egg with a double yolk is 0.125.
   c. The probability of Jim hitting the bull's eye is $\frac{2}{5}$.

14. Write each odds statement as a probability statement.

   a. The odds of Jon making a base hit are 3 to 4.
   b. The odds of a calico cat being male are 1 to 99,999.
   c. The odds of being put on hold when calling the doctor are 23 to 2.

15. There are 3 red balls, 1 yellow ball, and 2 blue balls in a brown bag.

   a. If you draw two in a row, with replacement, what is the probability of drawing two of the same color?
   b. How would the probability change if you were drawing without replacement?
   c. Answer the same questions, this time for drawing 2 yellow balls.

27.3 Simulating Probabilistic Situations

In the previous section we discussed ways of finding probabilities experimentally. We now undertake doing an experiment a large number of times and noting the outcomes.

 ACTIVITY 6 How Many Heads?

Suppose you want to find the probability of obtaining 3 heads and a tail when four coins are tossed. You could, of course, toss four coins a large number of times, and count the number of times that you obtain exactly 3 heads. The probability would be as follows:

$$P(3H) = \frac{\text{the number of times exactly three of the four coins show heads}}{\text{the total number of times four coins were tossed}}$$

In your group, toss four coins 10 times and record the number of heads each time. Combine your results with those of other groups to find the experimental probability of obtaining 3 heads on a toss of four coins. ●

Determining experimental probabilities can be very time-consuming when one actually carries out an experiment many, many times. Because numbers selected randomly are equally likely, we can use them to speed up this process. Using randomly selected numbers to find a probability in this manner is an example of a **simulation**.

> A **randomly selected** number or object from a set of numbers or objects is one that has a chance of being selected that is equal to that of any other number or object in the set.

 ACTIVITY 7   "Tossing" Four Coins Again

Once again, we will experimentally find the probability of obtaining 3 heads on a toss of four coins. This time we will simulate tossing four coins 100 times. To do so, compile a set of 100 four-digit numbers. There are many ways to do this. Your instructor will tell you whether to use the TI-73, Fathom, Excel, Illuminations, or a **table of randomly selected digits (TRSD)**. Each of these methods is explained in an appendix. See Appendix H (using the TI-73), I (using Fathom), J (using Excel), K (using the Illuminations website), or D (using a TRSD) and follow the given directions.

Once you have the set of 100 four-digit numbers, consider the digits in each number. An even digit can be used to represent heads, and an odd digit to represent tails. How many of the 100 four-digit numbers have three even digits and one odd digit? This number, call it x, can be used to find the probability of tossing 3 heads: $P(3H) = \frac{x}{100}$. How does this probability match the probability you found when you actually tossed coins in Activity 6? ●

There are many types of probabilistic situations in which a probability can be found with a simulation using a set of randomly selected numbers. For example, one can simulate drawing colored balls from a bag. Suppose we have a paper bag with 2 red balls and 3 blue balls. We want to know the probability of drawing a red ball on the first draw, replacing it, and then drawing a blue ball on the second draw. Keep in mind that the first ball is replaced—i.e., we are **drawing with replacement**.

Because there are five balls, a set of random numbers 1–5 can be used to simulate drawing with replacement. For example, 1 and 2 can represent red balls and 3, 4, and 5 can represent blue balls.

> *Think About …*
>
> How can this method also tell you the number of times you draw a blue ball followed by another blue ball? Why is it important to know that the first ball was replaced? (*Hint:* Think of outcomes such as 33, 44, 55.)

 ACTIVITY 8   Red and Blue

Return to the resource you used in Activity 7. Figure out a way to generate two-digit numbers, ignoring 0 and 6–9 for each digit. Let 1 and 2 represent red balls, and 3, 4, and 5 represent blue balls. Each of your two-digit numbers will represent a draw of two balls with replacement. For example, 24 would represent drawing a red ball (2), replacing it, and then drawing a blue ball (4).

1. Do 20 trials, and find the experimental probability of drawing a red ball and then a blue ball.
2. Combine your 20 trials with those of others, and from the combined data find the experimental probability of drawing a red and then a blue.
3. Use the same pairs of numbers to determine the probability of drawing 2 blues, first using just your data and then using the combined data from everyone. ●

In Problem 1 of Activity 8, would you have a different result if you did 100 trials instead of 20 trials? Which would you trust to be the better estimate of the probability? One conclusion you may have reached by now is that a large number of trials of a simulation will lead to a more accurate estimate of a probability.

Just as we used random numbers to simulate the drawing of balls from a bag, the drawing of the balls itself can be a simulation of some other event. What matters is that the objects (coins, balls, or other objects) are used to represent something else. For example, the toss of one coin may be used to simulate the unknown gender of a baby. Whether using some type of simulation or actually listing all elements in the sample space (which may not always be possible), a simulation is useful when it models some actual situation. For example, consider a lake having only bass (40%) and catfish (60%). Catching a bass and then a catfish can be simulated with a bag of 100 balls, 40 of which are one color and 60 of which are another color. Or you could consider 100 random two-digit numbers (00 through 99), where 00 through 39 could represent bass and 40 through 99 could represent catfish.

> ### *Think About …*
> Could the bass and catfish draws be represented using just five balls? How? Using one-digit numbers 0 through 9? How?

 ACTIVITY 9   Let's Spin

Explain how you could use the given spinner to simulate the drawing, with replacement, of two balls from a bag in which 2 balls are red, 3 are blue, and the balls are otherwise identical. (Assume all five sectors have the same central angle.)

Suppose you are a doctor with five patients who have the same bad disease. You have an expensive treatment that is known to cure the disease with probability $\frac{1}{4}$, or 25%. You are wondering what the probability of curing none of your five patients is, if you use the expensive treatment. How could this situation be simulated? Perhaps surprisingly, this situation can be simulated by drawing cards from a hat! Draw from a hat that has four identical cards except for the markings on the cards. One card is marked *cured* and three cards are marked *not cured*. Drawing these cards in repeated trials provides a model for the 25% cure rate.

To model the situation with the five patients, you would draw a card, note what it says, put it back in the hat, and then repeat this process four more times. After the fifth draw is finished, you would see whether any *cured* cards had been drawn—i.e., whether any "patients" were cured. Because probabilities should be interpreted *over the long run,* you would repeat this five-draws experiment many times to get an idea of the probability that you are concerned about—the probability that none of your patients would be cured.

> ### *Think About …*
> Why would you need to replace the card in this situation of the doctor with five patients?

You could simulate the same situation in other ways—e.g., with 1 green ball and 3 red balls in a paper bag. Drawing a green ball would represent a cure, and drawing a red ball would represent no cure. Thus, many real-life probabilistic situations can be modeled using simple tools such as cards, spinners, and balls in bags.

 DISCUSSION 6   Modeling Situations with an Experiment

Explain how each situation could be modeled by an experiment in which balls are drawn from a bag. Tell what each color would represent, how many colors of balls would be involved, and how you would proceed. How could each situation be modeled by drawing numbers from a hat?
a. predicting the sex of a child

continue

b. deciding on the chances of four consecutive successful space shuttle launches, if the probability of success each time is 99%

c. getting a 6 on a toss of an honest die

d. a beginner who is 10% accurate shooting an arrow at a balloon and hitting it

e. an expert who is 99% accurate shooting an arrow at a balloon and hitting it

f. predicting whether a germ will survive if it is treated with chemical X that kills 75% of the germs

g. the chances that 10 pieces of data sent back from space are all correct if the probability of each piece being correct is 0.9 ●

The processes in Discussion 6 depend on random draws of balls (or numbers, in the case of drawing a number from a hat). Hence, each ball has an equal chance of being drawn. For example, consider again drawing a ball from a bag of 2 red balls and 3 blue balls, *replacing* it, and then drawing again. The sample space for this experiment is made up of these equally likely outcomes:

R_1R_1	R_2R_1	B_1R_1	B_2R_1	B_3R_1
R_1R_2	R_2R_2	B_1R_2	B_2R_2	B_3R_2
R_1B_1	R_2B_1	B_1B_1	B_2B_1	B_3B_1
R_1B_2	R_2B_2	B_1B_2	B_2B_2	B_3B_2
R_1B_3	R_2B_3	B_1B_3	B_2B_3	B_3B_3

Recall that when outcomes are equally likely, it is possible to find the theoretical probability of an event by counting the outcomes associated with the event, and the probability of the event is then $\frac{\text{number of outcomes for an event}}{\text{number of outcomes in the sample space}}$. But what happens if the balls are *not replaced*? We next discuss drawing balls **without replacement**.

 DISCUSSION 7   Drawing Two Balls Without Replacement

We now change the situation so that we draw two balls from a bag containing 2 red balls and 3 blue balls, but we do *not* replace the first ball before drawing the second ball.

1. Write the sample space for this without-replacement situation. How is it different from the 25-outcome sample space above? (*Hint*: If a ball is not replaced, what balls are available for the next draw?)

2. What are some possible outcomes for this experiment?

3. What are some events that could be considered? ●

> *Think About ...*
>
> From Discussion 7, what patterns do you notice in your list of outcomes when drawing two balls without replacement? If there had been 3 red balls, how would the listing change? If there had been only 2 blue balls, how would the listing change?

Knowledge of probability is useful in understanding many types of situations. The problem in Activity 10 is just one example.

ACTIVITY 10   Free Throws

Tabatha is good at making free throws, and in the past she averaged making 2 out of every 3 free throws. At one game, she shot 5 free throws, and she missed every one! Her fans insist she must have been sick or hurt, or that something must have been wrong. They say it was not possible for her to miss all five shots. Are they correct? Is Tabatha's missing 5 free throws impossible? That is, is the probability of this event 0? ●

Here is a way to simulate this situation: Suppose we consider two-digit random numbers, from 01 to 99 (ignore 00). If the numbers are random, then $\frac{2}{3}$ of the time they should be numbers from 01 to 66, and $\frac{1}{3}$ of the time they should be numbers from 67 to 99. (Could we use 00–65 and 66–98, ignoring 99?) Randomly find five pairs of two-digit numbers. Let all numbers from 01 through 66 represent free throws made and all numbers from 67 through 99 free throws missed. Work in your groups to obtain 20 sets of five throws, combine them with others, and predict the probability of this event.

TAKE-AWAY MESSAGE . . . It may be surprising that, with clever choices of coding the numbers, a set of randomly selected numbers can be used to simulate so many situations. Different repetitions of the same simulation illustrate that different repetitions will most often give different, but close, experimental probabilities. In addition to random numbers, other entities such as balls drawn from a bag can be used to simulate a wide variety of events. In all cases, listing all the possible outcomes gives you the sample space and makes defining events and finding probabilities easier.

Learning Exercises for Section 27.3

1. Set up and complete a simulation of tossing a die 100 times. How do the probabilities of each outcome compare with the theoretical probabilities? Do the same simulation for 500 repetitions. What pattern do you notice as the number of repetitions increases?

2. **a.** Set up and complete a simulation to find the probability of getting green twice in a row if a spinner is on a circular region that is $\frac{1}{3}$ green, $\frac{1}{6}$ blue, $\frac{1}{3}$ red, and the rest yellow. The spinner is spun twice. Carry out the simulation 30 times and record the outcomes. (Your record should include colors.)

   Answer parts (b) through (e) based on the outcomes of your experiment in part (a), and tell whether your experimental values are close to the theoretical values.
   b. Which outcome is most likely? Why?
   c. What is the probability of getting a green on the first spin and a blue on the second?
   d. What is the probability of getting a green on one of the spins and a blue on the other? Why is this question different from the question in part (c)?
   e. What is the probability of not getting green twice in a row? (*Hint*: There is an easy way.)

3. Earlier in this section the following situation was simulated using cards: Suppose you are a doctor with five patients who have the same bad disease. You have an expensive treatment that is known to cure the disease with a probability of 25%. You are wondering what the probability of curing none of your five patients is, if you use the expensive treatment. Set up and complete a different simulation of this situation. (Because the question is about a group of five patients, use groups of size 5.) You decide how many times to do the simulation.

4. On one run of the free-throw simulation in Activity 10, the event YYYYN had a proportion 0.319. What does that mean?

5. Does a simulation using randomly generated numbers give theoretical or experimental probabilities? Explain.

6. **a.** Go to http://illuminations.nctm.org/, and click on Activities. In the options, type in Adjustable Spinner. Set the probabilities for the four colors in any way you want by moving the dots on the circle OR by moving the buttons for each color OR by setting the percents of each circle in the results frame. Set Number of spins to 1000, and click on Spin. You will see how the spinner works. *Write down* the numbers in the results frame. These are the relative sizes of the regions of the circles and show theoretical probabilities (but in percents). You can open five colors by adding a sector (click on $+1$). Once again, set the probabilities. Open the results frame to show the table at the bottom.
   b. Use this spinner activity to simulate a five-outcome experiment with unequally likely outcomes, and run the simulation 100 times. How do the experimental results compare with the theoretical ones in the table at the bottom of the screen?

continue

c. Repeat with a run of 1000 simulations. How do the experimental results compare with the theoretical ones?

d. Repeat with a run of 10,000 simulations. How do the experimental results compare with the theoretical ones?

Supplementary Learning Exercises for Section 27.3

1. a. Set up and carry out a simulation of spinning a spinner with a circular region that is $\frac{1}{5}$ red, $\frac{3}{10}$ green, $\frac{1}{5}$ yellow, and the rest purple. Perform the simulation 200 times.

   b. Which outcome occurred most? Is this different from what you expected?

   c. Find the experimental probabilities for each color. How do they compare to the theoretical probabilities?

   d. Based on the simulation, what is the probability of spinning yellow twice in a row?

   e. What is the probability of *not* getting 2 yellows in a row?

2. Set up a simulation for the following situation. A disease that affects trees has been reported near a farmer's grove. The farmer can use a treatment to prevent his trees from catching the disease. It is 65% effective in trees of the type the farmer owns. There are four trees on the perimeter that the farmer is most worried about. He wants to know the probability that these four trees will be safe if he gives them the treatment. Use your simulation with as many repetitions as necessary to predict this probability.

3. There are five Scrabble tiles in a bag. They are labeled G, O, N, I, and B. Set up and complete a simulation to find the probability that when the letters are pulled out without replacement, they will come in this order: B-I-N-G-O.

4. a. On a penalty kick, one soccer player makes a goal 35% of the time. Create a simulation for this situation.

   b. In a typical season this player kicks five penalty kicks. Use your simulation to model this 100 times.

   c. From your data, what are the chances that she will make a goal on each of the five penalty kicks?

   d. What is the probability that she makes a goal the first and last times, but not in between? (GMMMG = Goal, Miss, Miss, Miss, Goal.)

5. One study estimated that 59% of the adult population snore while sleeping. A family has four children. Design and run a simulation to find the probability that all four children will grow up to snore. What are the assumptions that you make in your simulation?

27.4 Issues for Learning: What Do Large-Scale Tests Tell Us About Probability Learning?

Children develop some basic notions of chance as they interact with the world. They begin to evaluate the likelihood of certain events, such as whether it will rain, and think about such events as impossible, unlikely, likely, or certain. Yet most of their thinking about probability is intuitive rather than based on evidence. In fact, it has long been known that even adults' ideas about probability (or chance or likelihood) are not reliable. Intuitions are often not correct. For example, when tossing a coin repeatedly, to think that a tails up is almost certain to occur after five consecutive heads up is common, not only with children, but also with adults (thus the label, the *gambler's fallacy*). And adults certainly know that a coin does not have a memory! Thus, it is not surprising to see these same intuitive ideas in children.

Due to the increasing importance of probability (and statistics) in a person's life and the poor performance of adults in situations where probability plays a role, more attention has been paid to probability (and statistics) in schooling over the last several years, thus offering the important opportunity to learn. As a result, the performance of U.S. students has exceeded the international average on some probability items in international testing programs.[1]

The National Assessment of Educational Progress (NAEP) is a national assessment of what American students know and can do in a number of important subjects. NAEP results

Notes

for the main assessments are based on carefully selected representative samples of students at Grades 4, 8, and 12. NAEP is called "The Nation's Report Card."[2] A few released test questions follow, which highlight some of the challenges students face in probabilistic thinking.

Students often do not appreciate the importance of knowing whether the outcomes are equally likely. They are willing to assign equal probabilities to any experiment that has, say, three outcomes. Spinners with unequal regions give one means of exposing the children to the fact that outcomes may not be equally likely. Experiments like tossing thumbtacks and noting whether the tip is up or down, or tossing a styrofoam cup and recording whether the cup ends up on its side, with its wide end down, or right side up, can also expose the students to outcomes that are not equally likely.

Consider the following released item for Grade 4 students from the 2011 test.[2] Only 12% of the fourth-graders got this item correct. The students are asked to choose between two spinners. It requires an explanation, which gives us insight into their thinking.

Item 1. Lori has a choice of two spinners. She wants the one that gives her a greater probability of landing on blue. Which spinner should she choose?

    ◯ Spinner A   ◯ Spinner B

Explain why the spinner you chose gives Lori the greater probability of landing on blue.

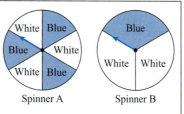

Spinner A   Spinner B

One student answered "Spinner B" and then gave the following reason:

> because on spinner B the blue is all together, but on spinner A she has a less chance because it is spread out.

Think About …

What difficulties does this student have understanding which spinner has the greater probability of landing on blue?

We do not have an explanation with the following item. Even though 66% of the fourth-graders were successful in choosing "1 out of 4" for Item 2 below, only 31% were successful in choosing the answer "3 out of 5" for Item 3, where the next-case view showed three possibilities. In general, a question in a "1 out of *n*" situation is much easier than one about a "several out of *n*" situation.

Item 2. The balls in this picture are placed in a box and a child picks one without looking. What is the probability that the ball picked will be the one with dots?

    A. 1 out of 4   **B.** 1 out of 3   **C.** 1 out of 2   **D.** 3 out of 4

Item 3. There are 3 fifth-graders and 2 sixth-graders on the swim team. Everyone's name is put in a hat and the captain is chosen by picking one name. What are the chances that the captain will be a fifth-grader?

    A. 1 out of 5   **B.** 1 out of 3   **C.** 3 out of 5   **D.** 2 out of 3

Think About …

What are some other possible reasons that performance on Item 3 was worse than that on Item 2? How would you redesign the two items to make them equally difficult?

The notion that probabilities are long-run results rather than next-case results may not be clear. This problem may be due partly to not attending to the fact that a probabilistic situation involves a process happening a large number of times. For example, we may say, "What is the probability that we get red if we spin the arrow on this spinner?" Although *we* understand this question to be about a fraction in the long run with many repetitions, *students* may interpret the question literally and focus on just the next toss. This thinking may give correct answers for some experiments but may lead to an incorrect understanding of probability. Doing an experiment many times when probability is first introduced may help establish the long-run nature of probability statements, with frequent reminders later on.

A more complicated probability involves a multistep experiment. In the following problem, more than 62% of the eighth-grade students were able to make an inference given the probability of a modified sample space in a multistep experiment.[2]

Item 4. Ken has a box that contains 12 marbles. The table below shows the number of marbles of each color that are in the box. Ken randomly selects 2 marbles from the box and keeps them. If Ken then randomly selects a third marble from the box, the probability that he will select a green marble is $\frac{2}{10}$.

Color	Yellow	Green	Orange	Blue
Number of Marbles	5	3	2	2

Which statement could be true about the first 2 marbles Ken selected?

A. One was yellow and one was green.   **B.** One was orange and one was yellow.

C. One was orange and one was blue.   **D.** Both were green.

E. Both were yellow.

TAKE-AWAY MESSAGE . . .   Children begin to form intuitive ideas about probability as they mature, but many of these ideas are incorrect and persist into adulthood. International and national tests provide valuable information about what children know and don't know about probability.

27.5 Check Yourself

Along with an ability to deal with exercises like those assigned and with experiences such as those during class, you should be able to meet the following objectives:

1. State and recognize the key features of a probabilistic situation, and rephrase the given wording of a nonprobabilistic situation to make it probabilistic.
2. Tell what the word "probability" means in the statement "The probability of that happening is about 30%." Express and interpret probability using the $P(A)$ notation.
3. Distinguish between experimental and theoretical probabilities, and describe how either one or both could be used to determine a probability in a given situation.
4. For a given task, use the following terms accurately: *outcome*, *sample space*, and *event*.
5. Explain what is meant by the term *random*.
6. Use these important results:
   - If event A is certain to happen, $P(A) = 1$.
   - If event A is impossible, $P(A) = 0$.
   - For any event A, $0 \le P(A) \le 1$, and $P(\text{not } A) = 1 - P(A)$.
7. Appreciate the usefulness of coin tossing, spinners, or drawing balls from a bag, and design an experiment to simulate some situation using one or more of those methods.
8. Find theoretical or experimental probabilities for a given event.
9. Design and carry out an experiment to simulate a situation using a TRSD (or Excel, Fathom, the TI-73, or Illuminations). Show how you determined the probabilities.
10. Distinguish between drawing with replacement and drawing without replacement.

References for Chapter 27

[1]See http://isc.bc.edu for test reports and, often, test items that have been released to the public.

[2]See http://nces.ed.gov/nationsreportcard/ for test reports and test items released to the public.

Determining More Complicated Probabilities

Notes

Finding probabilities of simple events, especially when outcomes are equally likely, is quite straightforward. But when we want to determine the probability of more complex or multistep experiments, we need some techniques, such as carefully constructed lists or tree diagrams, to help us. When events are considered together in some fashion to make a new event, we would like to have ways to relate the probability of this new event to the probabilities of the individual events. These topics are the focal points of this chapter.

28.1 Tree Diagrams and Lists for Multistep Experiments

One technique for analyzing a more complex or **multistep experiment** is to carefully list all the ways that the experiment could turn out, i.e., list the complete sample space, and then see which outcomes favor the event. We can determine the probability of the event by adding the probabilities of all the outcomes favoring that event. You did this for some events in Chapter 27. Often the tricky part of this technique is to find a method of listing all possible outcomes without forgetting or repeating any.

One way is to be systematic in listing the outcomes, as we did in the last chapter. Suppose we toss three coins. We might be modeling the genders for three unborn children or the win/loss possibilities for three games against opponents of equal abilities. What are the possible outcomes when tossing the three coins? It is tempting (but incorrect) to list them as 3 heads, 2 heads (and 1 tail), 1 head (and 2 tails), and no heads (3 tails), thus obtaining four possible outcomes. But the possible outcomes (and the actual probabilities of occurrences of 3 heads, 2 heads, 1 head, and no heads) become clearer if we say we are tossing a penny, a nickel, and a dime.

We need to write down the sample space of the toss of three coins systematically to assure that all possible outcomes are listed. In the second column of the following table, the first coin is listed four times as heads, then four times as tails. The second coin is then listed twice with heads, then twice with tails and then repeated, etc. There are eight possible outcomes, and the number of heads occurring each time is listed.

Outcome	Penny	Nickel	Dime	Number of Heads
1	H	H	H	3
2	H	H	T	2
3	H	T	H	2
4	H	T	T	1
5	T	H	H	2
6	T	H	T	1
7	T	T	H	1
8	T	T	T	0

If the coins are fair, all eight outcomes are equally likely. Each outcome has probability $\frac{1}{8}$. Notice that the number of heads in each outcome also provides information about the number of tails in the outcome. Thus, we know the probabilities for either the number of heads or the number of tails, as shown on the next page.

$$P(3H) = \frac{1}{8} = P(0T) \qquad\qquad P(1H) = \frac{1}{8} + \frac{1}{8} + \frac{1}{8} = \frac{3}{8} = P(2T)$$

$$P(2H) = \frac{1}{8} + \frac{1}{8} + \frac{1}{8} = \frac{3}{8} = P(1T) \qquad P(0H) = \frac{1}{8} = P(3T)$$

As you can see from the probabilities above, the four events 3H, 2H, 1H, and 0H are not equally likely. Even though counting the number of heads seemed to be a reasonable analysis of the toss-three-coins experiment, a closer look at the actual outcomes showed that there are eight possible outcomes, not four.

Think About ...

What is the sum of the four probabilities for the events 3H, 2H, 1H, and 0H? Why should you have expected to get a sum of 1 when adding these probabilities?

 DISCUSSION 1 Brownbagging

a. Suppose a brown paper bag contains 1 red ball and 2 green balls that are all identical except for color. For a first experiment you take out 1 ball, note its color, and then replace it. You then take out a second ball and note its color. Write a list that represents the sample space for the two draws. For example, let RR represent the outcome of drawing the red ball both times. (Be careful when denoting the green balls. Distinguish them as G_1 and G_2 when you list the elements in the sample space.) How many outcomes are in the sample space?

b. For a second experiment, suppose this time you draw from the bag twice and note the colors, but this time do not replace the first ball. Write a list to represent this sample space for the two draws, again keeping track of G_1 and G_2. How do the two sample spaces in parts (a) and (b) differ? ●

Discussion 1 is similar to some of the activities in the last chapter. In the first experiment when two balls were drawn, the probabilities for the second draw were in no way affected by the first draw, because all the balls were available at each draw. But in the second experiment when two balls were drawn, the probabilities for the second draw *were* affected by the first draw, because the first ball drawn was no longer available. Thus, it is important to keep in mind whether such an experiment is repeated **with replacement** or **without replacement**, which are terms used in the previous chapter.

Discussion 2 and its follow-up will suggest how, in a two-step experiment, the probability of a sequence of two individual outcomes can be related to the probabilities of the individual outcomes.

DISCUSSION 2 A Two-Step Spinner Experiment

Consider the two circular spinners that are illustrated to the right. The two-step experiment is to spin the arrows on both spinners and note the color you get on each spinner. We are interested in the probability of this outcome: green on Spinner 1 and green on Spinner 2.

Spinner 1   Spinner 2

a. Suppose that you do the experiment 1800 times. How many times do you expect to get green on Spinner 1? Why? How many times do you expect to get green on Spinner 2? Why?

b. Of those _____ spins having green on Spinner 1, how many do you expect to give green on Spinner 2 also? _____

A way of summarizing your answers to these last two questions is as follows:

Of 1800 repetitions of the experiment, it is reasonable to expect about _____ outcomes with green on both spinners, giving a probability of _____.

From this information, why is it a reasonable idea to say that the probability of the outcome green on both spinners is $\frac{300}{1800} = \frac{1}{6}$? ●

Using the probability notation discussed in the last chapter, we can write P(green on Spinner 1) to mean the probability of getting green on Spinner 1. The information inside the parentheses refers to the event you are interested in. Notice that because P(green on Spinner 1) $= \frac{1}{3}$ and P(green on Spinner 2) $= \frac{1}{2}$, their product is $\frac{1}{3} \times \frac{1}{2} = \frac{1}{6}$, which is equal to P(green on Spinner 1 *and* green on Spinner 2), as in part (b) of Discussion 2. In this situation, multiplying the individual probabilities gives the probability of both occurring. This fact will be useful with the next representation to be discussed, the tree diagram.

A way of organizing the information in order to determine the probabilities in cases such as the one in Discussion 2 is to use a **tree diagram**. Figure 1 shows a tree diagram for the two spinners. When a tree diagram has been set up properly, the probabilities along the branches can be multiplied to get the probability of each outcome in the sample space, as noted above. Hence, the tree diagram not only gives a systematic way to list all the outcomes, it also enables us to find the probabilities of all the outcomes. Tree diagrams can also be used to find the probabilities of more complicated events.

Spinner 1	Spinner 2	Outcome	Probability
R	R	RR	$\frac{1}{12}$
	B	RB	$\frac{1}{12}$
	G	RG	$\frac{1}{6}$
B	R	BR	$\frac{1}{12}$
	B	BB	$\frac{1}{12}$
	G	BG	$\frac{1}{6}$
G	R	GR	$\frac{1}{12}$
	B	GB	$\frac{1}{12}$
	G	GG	$\frac{1}{6}$

Figure 1 A tree diagram for the spinners in Discussion 2.

From this tree diagram we can find the probabilities of the outcomes to the whole experiment. This procedure is based on the multiplication result implied by Discussion 2. For example, what is the probability of getting green on Spinner 1 and red on Spinner 2? We see that this occurs only in the GR line, and P(GR) is the product of the probabilities on the two branches leading to GR, so $P(GR) = \frac{1}{3} \times \frac{1}{4} = \frac{1}{12}$. The probability of getting blue and then green is $\frac{1}{6}$, i.e., $P(BG) = \frac{1}{3} \times \frac{1}{2} = \frac{1}{6}$. Notice also that the sum of the probabilities of all nine outcomes is 1:

$$\frac{1}{12} + \frac{1}{12} + \frac{1}{6} + \frac{1}{12} + \frac{1}{12} + \frac{1}{6} + \frac{1}{12} + \frac{1}{12} + \frac{1}{6} = 1$$

As you may have suspected, the nine outcomes for this experiment are not equally likely. The diagram also helps us answer questions such as "What is the probability of getting one green and one red, regardless of the order?" This event can happen in two ways, RG or GR, so the probability of getting one green and one red is $\frac{1}{6} + \frac{1}{12}$, or $\frac{1}{4}$.

DISCUSSION 3   More Questions About the Spinner Experiment

Consider again the two spinners in Discussion 2. What is the probability of getting at least 1 green? Of getting no greens? Of getting at least 1 red? Of getting no reds? Of getting at least 1 blue? Of getting no blues? Of getting at most 1 green? ●

The outcomes of the spinner experiment are not equally likely. The reason is because the colors are not equally likely on the second spinner.

 **ACTIVITY 1**   Brownbagging Again

Here is a variation of the brownbagging experiment in Discussion 1. This time the bag contains 2 red balls and 3 green balls.

Experiment 1. Suppose you take out a ball, note its color, and then replace it. Draw a ball again, and note the color. We give two different tree diagrams that represent the sample space in which the first ball is replaced.

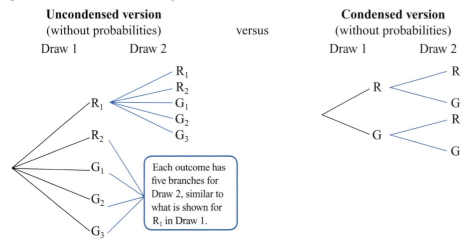

Uncondensed version
(without probabilities)   versus

Draw 1   Draw 2

Each outcome has five branches for Draw 2, similar to what is shown for R_1 in Draw 1.

Condensed version
(without probabilities)

Draw 1   Draw 2

a. On the uncondensed tree diagram, place the appropriate probabilities on each of the branches shown. What can you assume the probabilities would be for branches not showing?

   Notice that the uncondensed tree diagram gives 25 equally likely outcomes, each with a probability of $\frac{1}{25}$.

b. What numbers would go on the tree branches for the condensed tree diagram? For example, what is the probability of drawing a red ball, which is represented by the first branch? What is the probability of drawing a green ball? Once a ball is drawn and replaced, what are the probabilities for the second set of branches? The final probabilities should be $\frac{4}{25}, \frac{6}{25}, \frac{6}{25}$, and $\frac{9}{25}$. Compare your products along the branches in the condensed tree diagram with these answers.

Experiment 2. Suppose you take out a ball, and note its color. Then, without putting the first ball back (i.e., drawing without replacement), draw another ball.

   Below, we give a condensed tree diagram for Experiment 1 (drawing a ball and replacing it) and a condensed tree diagram for Experiment 2 (drawing a ball and not replacing it).

c. Finish the condensed tree diagrams for the two experiments. Label the branches to give the probabilities for the outcomes for each experiment. (Why do the probabilities change for the second draw in Experiment 2?)

d. Add the probabilities of all the outcomes for each experiment. Is the sum equal to 1 in each case?

Experiment 1 (with replacement)   **Experiment 2** (without replacement)

First draw	Second draw	Outcomes and probabilities		First draw	Second draw	Outcomes and probabilities

R
  R
       G
       R
  G
       G

R
  R
       G
       R
  G
       G

Think About …

Why is the brownbagging activity in Activity 1 considered to be a multistep experiment?

This Grade 6 page uses tree diagrams to find probabilities.

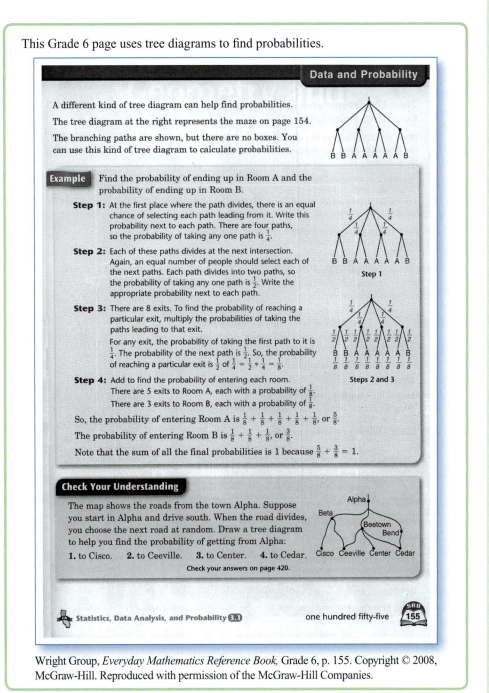

Data and Probability

A different kind of tree diagram can help find probabilities.

The tree diagram at the right represents the maze on page 154.

The branching paths are shown, but there are no boxes. You can use this kind of tree diagram to calculate probabilities.

B B A A A A B

Example Find the probability of ending up in Room A and the probability of ending up in Room B.

Step 1: At the first place where the path divides, there is an equal chance of selecting each path leading from it. Write this probability next to each path. There are four paths, so the probability of taking any one path is $\frac{1}{4}$.

Step 2: Each of these paths divides at the next intersection. Again, an equal number of people should select each of the next paths. Each path divides into two paths, so the probability of taking any one path is $\frac{1}{2}$. Write the appropriate probability next to each path.

Step 3: There are 8 exits. To find the probability of reaching a particular exit, multiply the probabilities of taking the paths leading to that exit.

For any exit, the probability of taking the first path to it is $\frac{1}{4}$. The probability of the next path is $\frac{1}{2}$. So, the probability of reaching a particular exit is $\frac{1}{2}$ of $\frac{1}{4} = \frac{1}{2} * \frac{1}{4} = \frac{1}{8}$.

Step 4: Add to find the probability of entering each room. There are 5 exits to Room A, each with a probability of $\frac{1}{8}$. There are 3 exits to Room B, each with a probability of $\frac{1}{8}$.

So, the probability of entering Room A is $\frac{1}{8} + \frac{1}{8} + \frac{1}{8} + \frac{1}{8} + \frac{1}{8}$, or $\frac{5}{8}$.

The probability of entering Room B is $\frac{1}{8} + \frac{1}{8} + \frac{1}{8}$, or $\frac{3}{8}$.

Note that the sum of all the final probabilities is 1 because $\frac{5}{8} + \frac{3}{8} = 1$.

Check Your Understanding

The map shows the roads from the town Alpha. Suppose you start in Alpha and drive south. When the road divides, you choose the next road at random. Draw a tree diagram to help you find the probability of getting from Alpha:

1. to Cisco.   **2.** to Ceeville.   **3.** to Center.   **4.** to Cedar.

Check your answers on page 420.

Statistics, Data Analysis, and Probability **3.1**   one hundred fifty-five   SRB **155**

Wright Group, *Everyday Mathematics Reference Book,* Grade 6, p. 155. Copyright © 2008, McGraw-Hill. Reproduced with permission of the McGraw-Hill Companies.

TAKE-AWAY MESSAGE . . .   Finding all of the outcomes of a multistep experiment requires care. One way is to make a systematic list. Another way is to make a tree diagram, which carries the added benefit of making it possible to find the probability of an outcome by multiplying the probabilities along the path to the outcome. Some tree diagrams can be shown in a condensed form. Experiments without replacement call for extra care in assigning probabilities.

Learning Exercises for Section 28.1

1. Suppose you plan to poll people by telephone about their opinions on a controversial issue. What method of calling would correspond to drawing with replacement? To drawing without replacement? Which method do you think is better? Why?

2. Make a tree diagram for this experiment: Toss a die and a coin. Note the number of dots on top of the die (1 through 6) and the face of the coin (H or T). Include the outcomes and their probabilities in your tree diagram.

3. From a box having 4 red balls and 6 green balls, you draw one ball, and then make a second draw. Make tree diagrams and give the outcomes and their probabilities for the following two versions of this experiment:

   a. if the first ball is replaced before the second draw.
   b. if the first ball is not replaced before the second draw.

4. The following table shows the beginning of the sample space for the throw of a pair of dice, with one die red (noted across the rows), and the other die white (noted down the columns). The experiment calls for recording the number showing on top of each die. Save your table for use in later Learning Exercises.

   a. Complete this table. Note that there will be a total of 36 outcomes in the sample space. Are all the outcomes equally likely?

Number on white die

	1	2	3	4	5	6
1	(1, 1)	(1, 2)	(1, 3)	(1, 4)	(1, 5)	(1, 6)
2	(2, 1)	(2, 2)	(2, 3)	(2, 4)	(2, 5)	(2, 6)
3						
4						
5						
6						

Number on red die: 3, 4, 5, 6

   b. Using the table, we could note that the probability of tossing a *sum* of 3 is $\frac{2}{36}$ because two of the equally likely 36 outcomes give a sum of 3: (1, 2) representing 1 on the red die and 2 on the white die and (2, 1) representing 2 on the red die and 1 on the white die. We can write this result as $P(\text{sum} = 3) = \frac{2}{36}$. Find the probabilities for all possible sums: $P(\text{sum} = 2)$, $P(\text{sum} = 3)$, . . . , $P(\text{sum} = 12)$.

   c. Would the same entries be useful for showing the outcomes if both dice were the same color?

5. A regular icosahedron is a regular polyhedron with 20 faces, so each of the digits, 0–9, can be written on two different faces. If you toss such a regular icosahedron twice, how many outcomes are there (assuming equally likely faces on top)? If you are using this die to practice addition, what is the probability that you will get 0 and 0 (thus practicing 0 + 0)? What is the probability that you will get a sum of 15? (Notice that this figure can be seen as three layers. The top and bottom layer each consists of five equilateral triangles that meet at a point. The middle layer consists of 10 equilateral triangles, each with an edge that meets a triangle in the top or bottom layers.)

6. **a.** In what possible ways can four (fair) coins fall when tossed? Find the probability of each outcome by listing the outcomes in the sample space and computing probabilities. Then find the sample space and compute all probabilities of the outcomes by making a tree diagram.

continue

b. Design and carry out a simulation of tossing four coins, with 48 repetitions. Compare the probabilities from the tree diagram to those you find through the simulation.

7. **a.** Make a tree diagram to determine the theoretical probability for this experiment: Spin the arrows (not shown) on each of the following three spinners, and note the color where the arrow lands on each spinner.

Spinner 1   Spinner 2   Spinner 3

   b. Give the sample space for the experiment and the probability of each outcome.

   c. What is the probability of getting at least 1 red?

   d. What is the probability of getting at least 1 black?

8. **a.** Give the sample space for the toss of two coins and a die.

   b. What is P(1 head and 1 tail and a number greater than 3)?

9. Suppose an unfair coin with $P(H) = 0.7$ is tossed twice. Make a tree diagram for the experiment, and determine the probability of each outcome. By how much do they differ from the probabilities of the outcomes with a fair coin?

10. Sometimes making a tree diagram for an experiment that involves many steps is not feasible. However, the idea behind the tree diagram can still be used. For example, imagine a tree diagram for tossing 10 different coins. Think of what the path through the imagined tree would look like to determine the probability of each of these outcomes.

   a. HHHHHHHHHH   **b.** HHTHTTTHHH   **c.** TTTTTHHHHH

   d. THTHTHTHTH   **e.** your choice of an outcome

Supplementary Learning Exercises for Section 28.1

1. Make tree diagrams and give all the outcomes and their probabilities for the following two versions of this experiment: A box has 6 red balls and 10 green balls. Draw one ball, and then make a second draw.

   a. *Version* 1: The first ball is replaced before the second ball is drawn.

   b. *Version* 2: The first ball is *not* replaced before the second ball is drawn.

2. **a.** Design a spinner to simulate the toss of a (dishonest) coin with probability of heads = 0.6.

   b. Make a tree diagram for (the simulation of) this experiment: Spin the spinner from part (a) twice, noting heads/tails for each spin. Include all the outcomes and their probabilities in your tree diagram.

   c. How do your probabilities from part (b) compare with those for an honest coin (probability of heads = 0.5)?

3. **a.** Give the sample space and the probability of each outcome for this experiment: Spin the arrow on each of the three spinners shown, and note the colors pointed to. (As always, if the spinner ends on a line, spin again.)

What is the probability of   **b.** getting all the same color?   **c.** getting 1 or more reds?

                                d. getting no reds?   **e.** getting exactly 1 red?

4. a. If you toss a fair coin 12 times, what is the probability of getting all tails?

   b. What does your answer for part (a) mean?

5. a. Make a tree diagram for spinning the given spinner twice. White and blue both have 60° angles.

   b. List the sample space.

   c. Find the probability for each outcome.

6. a. Make a table that shows the sample space for spinning the given spinner twice. (Each sector is the same size.)

   b. What is the probability of each outcome? Tell how you know.

7. a. Is spinning a spinner twice similar to drawing with replacement or without replacement?

   b. Make a new table for Supplementary Learning Exercise 6, showing the product for each pair spun. For example, 2, 3 would be 6.

   c. What is the probability of getting a product of 12?

   d. What is the probability that the product will be even?

8. Make a tree diagram for each given situation.

   a. A container has three balls numbered 1, 2, and 3. Draw one ball, record the number, and replace it. Then draw again and record its number.

   b. A container has three balls numbered 1, 2, and 3. Draw one ball, record the number, and do not replace it. Then draw another and record its number.

   c. A container has three balls numbered 1, 2, and 3. Draw one ball, record the number, and replace it. Then draw again, record it, and replace it. Then draw a third time and record the number.

   d. A container has three balls numbered 1, 2, and 3. Draw one ball, record the number, and do not replace it. Then draw another, record it, and do not replace it. Then draw a third time and record the number.

   e. Compare the four experiments in parts (a–d). How are the trees similar and how are they different?

9. a. Make a tree diagram for rolling a six-sided die followed by a four-sided die.

   b. Make a tree diagram for rolling a four-sided die followed by a six-sided die.

   c. How are the diagrams in parts (a) and (b) similar and different?

10. a. Make a tree diagram for spinning the following three spinners. The sections in Spinner A are all the same size, and the red section in Spinner B is 30°.

Spinner A   Spinner B   Spinner C

   b. List the sample space.   **c.** What is the probability of at least 1 purple?

   d. What is the probability of no red?

11. There are 35 numbered balls in a container. Five will be pulled out one at a time without replacement. Describe the tree diagram associated with this experiment.

12. The following fourth-grade question comes from a National Assessment of Educational Progress.[1] Only 25% of the students got the correct answer. What do you think is the most common wrong answer and why?

   ▶ In a bag of marbles, $\frac{1}{2}$ are red, $\frac{1}{4}$ are blue, $\frac{1}{6}$ are green, and $\frac{1}{12}$ are yellow. If a marble is taken from the bag without looking, is it most likely to be

   a. red   **b.** blue   **c.** green   **d.** yellow ◀

28.2 Probability of One Event *or* Another Event

In this section we discuss the probabilities of "or" combinations of events that have known probabilities, such as P(getting a sum = 7 *or* getting a sum = 11 when tossing two dice). In Section 28.3 we will say more about "and" combinations of two events.

The use of *or* in mathematics is not always the same as in everyday life. ***In mathematics, we use the word "or" in an inclusive sense. That is,* or *means "one or the other or both."***

Let us look again at the tree diagram for the two-step spinner experiment from Discussion 2 of the previous section. Recall that the outcome in the first spinner does not affect the outcome in the second spinner.

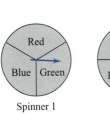

Spinner 1

Spinner 2

Spinner 1	Spinner 2	Outcome	Probability
	R	RR	$\frac{1}{12}$
R	B	RB	$\frac{1}{12}$
	G	RG	$\frac{1}{6}$
	R	BR	$\frac{1}{12}$
B	B	BB	$\frac{1}{12}$
	G	BG	$\frac{1}{6}$
	R	GR	$\frac{1}{12}$
G	B	GB	$\frac{1}{12}$
	G	GG	$\frac{1}{6}$

Suppose that we want to know the probability of getting a green on Spinner 1 *or* a green on Spinner 2. It is natural to expect that this probability can be obtained by adding the probabilities of the two events:

$$P(\text{green on Spinner 1}) + P(\text{green on Spinner 2}) = \frac{1}{3} + \frac{1}{2} = \frac{5}{6}$$

because green on Spinner 1 occurs $\frac{1}{3}$ of time and green on Spinner 2 occurs $\frac{1}{2}$ of the time. But if we look at the sample space and count the times that we get green on Spinner 1 OR green on Spinner 2, where OR means one or the other or *both,* we see that this result happens in the third, sixth, seventh, eighth, and ninth outcomes in the previous tree diagram: RG, BG, GR, GB, and GG. If we add up the probabilities of these outcomes, we have $\frac{1}{6} + \frac{1}{6} + \frac{1}{12} + \frac{1}{12} + \frac{1}{6}$, or $\frac{2}{3}$, and not the expected $\frac{5}{6}$.

Why did this discrepancy occur? By checking the tree diagram, we see that by taking all the cases where there is green on Spinner 1 (GR, GB, GG) along with all cases where there is green on Spinner 2 (RG, BG, GG), we used GG *twice.* Thus, adding the two probabilities, $P(\text{G on first}) + P(\text{G on second})$, gave us a value that is $\frac{1}{6}$ too much. This extra amount is due to the fact that $P(GG) = \frac{1}{6}$. We have to subtract the probability of green on Spinner 1 *and* green on Spinner 2 because GG appears twice. So $P(\text{G on first } or \text{ G on second}) = \frac{1}{3} + \frac{1}{2} - \frac{1}{6} = \frac{2}{3}$, which is the same result obtained by adding all the relevant cases in the tree diagram. We can generalize this situation as follows.

> **Addition Rule for Probability:**
>
> For events A and B,
>
> $$P(A \text{ or } B) = P(A) + P(B) - P(A \text{ and } B)$$
>
> Notice that the word "*and*" is used in the sense of *at the same time,* rather than addition.

A Venn diagram such as the one at right helps to make clear that the probabilities of any outcomes in A and simultaneously in B—the ones in the middle region—are considered *twice* in $P(A) + P(B)$, so subtracting $P(A \text{ and } B)$ once adjusts the sum properly.

Sample space

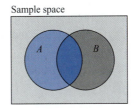

ACTIVITY 2   Using the Addition Rule for Probability

Use $P(A \text{ or } B) = P(A) + P(B) - P(A \text{ and } B)$ to find each given probability, and then check with the tree diagram for the two spinners in this section.

1. P(blue on Spinner 1 *or* green on Spinner 2) = _____

2. P(same color on both spinners *or* red on Spinner 1) = _____

3. P(blue on Spinner 1 *or* green on Spinner 1) = _____ ●

> *Think About ...*
>
> For Problem 3 of Activity 2, why wasn't it necessary to subtract something from $\frac{1}{3} + \frac{1}{3}$?

Thus far, one might think that the addition rule for probability applies only to spinners or other situations that have nothing to do with real life. Yet there are many situations in which the rules of probability are used for decision making. Consider the situation in Activity 3, where Karen is about to tell her employer when she will unlikely be able to work.

 ACTIVITY 3   Practicing the Addition Rule

Karen is signing up for a math class. She will be randomly assigned to one of 6 sections. Of the 6 sections offered, 3 sections are on Tuesday and Thursday and 3 sections are on Monday, Wednesday, and Friday. One of the Tuesday–Thursday sections and one of the Monday–Wednesday–Friday sections are evening classes. If she is assigned randomly to one of the sections, show that the probability that she will be in a Tuesday–Thursday section or an evening class is $\frac{2}{3}$. What do you think she should tell her employer? ●

TAKE-AWAY MESSAGE . . .   The addition rule can be used to find the probability of an *or* event in many instances. If the events do not have outcomes in common, the probability of one *or* another event happening is the sum of the two probabilities. If the two events do have outcomes in common, we must subtract the probability of the common outcomes so that no outcome is counted twice. This case can be stated as

$$P(A \text{ or } B) = P(A) + P(B) - P(A \text{ and } B)$$

When A and B have no common elements, $P(A \text{ and } B) = 0$.

Learning Exercises for Section 28.2

1. **a.** List the outcomes in the sample space for tossing a coin and a six-faced die. Do this in a manner that assures that *every* outcome is listed *exactly once*. A notation like P(H odd) refers to the probability of the event getting heads on the coin and an odd number on the die. Then find each probability.

    b. P(H3)   **c.** P(H3 *or* H4)   **d.** P(H even *or* H2)

2. **a.** List the elements in the sample space for one spin on a spinner partitioned into three equal parts: red, blue, and green, and one toss of a die. Then find each probability. (*Note:* R6 means red on the spinner and 6 dots on the die.)

    b. P(R6 *or* R2)   **c.** P(R *or* G)   **d.** P(R3 *or* G3)

3. **a.** Make a tree diagram for the sample space when drawing twice, with replacement, from a bag of three balls of which 2 are red and 1 is green. Find each probability.

    b. P(RR)   **c.** P(both the same color)
    d. P(RR *or* both the same color)   **e.** P(different colors *or* GG)

4. **a.** Make a tree diagram for the sample space when drawing twice, *without* replacement, from a bag of three balls of which 2 are red and 1 is green. Then find each probability.

    b. P(RR)   **c.** P(both the same color)
    d. P(RR *or* both the same color)   **e.** P(different colors *or* both green)

Notes

5. **a.** Make a tree diagram to find the sample space when spinning arrows on two spinners. Each spinner is divided into three equal sections: one section is red, the second is blue, and the third is green. Find each probability.

   b. *P*(RR)   **c.** *P*(both the same color)

   d. *P*(RR *or* both the same color)   **e.** *P*(1 red and 1 green *or* both green)

6. **a.** Make a tree diagram for spinning arrows (not shown) on the two spinners to the right. Find each probability. Colors on the spinners are red, yellow, blue, and green.

   b. *P*(RR)   **c.** *P*(both the same color)

   d. *P*(RR *or* both the same color)   **e.** *P*(1 red and 1 green *or* both green)

   f. *P*(red on at least one of the two spins *or* yellow on at least one of the two spins)

7. **a.** List the sample space for spinning arrows (not shown) on the three spinners at right. Find each probability.

   b. *P*(all the same color *or* red on one of the spinners)

   c. *P*(yellow on all three spinners *or* red on all three spinners)

   d. *P*(blue on at least one spinner *or* yellow on at least one spinner)

8. **a.** List the sample space for spinning arrows on the first spinner in Learning Exercise 7 and tossing a regular die. Find each probability.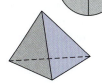

   b. *P*(Y2 *or* Y4)

   c. *P*(R3 *or* R odd)

   d. *P*(B2 *or* B odd)

9. **a.** A regular tetrahedron has four triangular faces, all equilateral triangles. The tetrahedron rests on one face, with the other three faces meeting at a vertex. (Think of a triangular pyramid.) The faces can be numbered 1 through 4.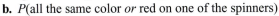
When a small regular tetrahedron is tossed, there is an equally likely chance of any of the four numbers falling face *down*. Write the sample space for the toss of a *pair* of tetrahedra. Find each probability for the tossing of two tetrahedra.

   b. *P*(odd number face down on both)

   c. *P*(even number face down on first tetrahedron *or* 4 face down on both)

   d. *P*(faces down sum to 8 *or* 4 face down on the second tetrahedron)

   e. *P*(faces down sum to 6 *or* faces down are both 3)

10. Using your sample space from Learning Exercise 4(a) in Section 28.1, find each probability for the toss of two dice.

   a. *P*(sum = 5 *or* sum = 6)

   b. *P*(sum = 14)

   c. *P*(sum = 9 or more)

   d. *P*(sum = 12 or less)

   e. *P*(sum < 11)

   f. *P*(sum at least 9)

   g. *P*(sum is not 7)

   h. *P*(sum is not 5)

   i. *P*(sum is at most 5)

   j. *P*(4 on red die *and* 6 on white die)

   k. *P*(4 on red die *or* 6 on white die)

11. Suppose you have a pair of tetrahedra. One is red on one face, yellow on two faces, and green on one face. The other is white and has faces marked 1, 2, 3, and 4.

a. Complete this table:

		Numbered Tetrahedron			
		1	2	3	4
Colored Tetrahedron	Red				
	Yellow				
	Yellow				
	Green				

b. If both tetrahedra are tossed, what is the probability of a red (facing down) and a 3 (facing down)? Of a yellow (facing down) and a number >1 on the other (facing down)? Of a green (facing down) and a number >4 (facing down) on the other? Of a yellow (facing down) on the colored one and a sum >2 of faces showing on the other?

12. Make an organized list of the possible gender combinations of four babies born to the same parents. Use it to answer the following questions:

a. What is the probability of having 2 girls and 2 boys?
b. What is the probability of having 4 girls?
c. What is the probability of the birth order being BBGG? Compare this with part (a).
d. What is the probability of having staggered births (i.e., a birth order such as GBGB)?

28.3 Probability of One Event *and* Another Event

The event *A* or *B* naturally includes any outcomes that *A* and *B* have in common. Hence, adding $P(A)$ and $P(B)$ counts twice the probabilities of those simultaneous outcomes in *both A* and *B*. This situation led to the addition rule for probability: $P(A \text{ or } B) = P(A) + P(B) - P(A \text{ and } B)$, where *and* means at the same time. The rule gives a sometimes welcome alternative to making an entire tree diagram. Occasionally, there is not enough information to make a tree diagram anyway, so the addition rule comes in handy.

We have noticed that at times $P(A \text{ and } B) = 0$. That is, event *A* and event *B* cannot happen simultaneously. This special case has a name.

> Two events are disjoint or mutually exclusive when it is impossible for them to happen simultaneously.

EXAMPLE 1

When tossing a pair of dice, one possible event is throwing a 2 on one of the dice. Another event might be throwing a sum of 11. These two events cannot both happen on one throw of the dice. Thus, they are disjoint. ●

EXAMPLE 2

In Activity 3 of Section 28.2, getting a Monday–Wednesday–Friday class would be disjoint from getting a Tuesday–Thursday class (unless Karen takes the same class twice, of course). ●

In the earlier experiment with spinners, suppose *C* is the event of getting 2 reds (with probability $\frac{1}{12}$), and *D* is the event of getting 2 greens (with probability $\frac{1}{6}$).

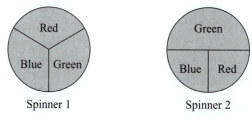

Spinner 1   Spinner 2

Spinner 1	Spinner 2	Outcome	Probability

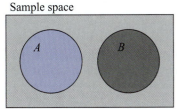

	R	RR	$\frac{1}{12}$
R	B	RB	$\frac{1}{12}$
	G	RG	$\frac{1}{6}$
	R	BR	$\frac{1}{12}$
B	B	BB	$\frac{1}{12}$
	G	BG	$\frac{1}{6}$
	R	GR	$\frac{1}{12}$
G	B	GB	$\frac{1}{12}$
	G	GG	$\frac{1}{6}$

Here $P(C \text{ and } D) = 0$, because there are no outcomes in which RR and GG can happen at the same time. This means that $P(C \text{ or } D) = P(C) + P(D) - 0$, or simply

$$P(C \text{ or } D) = P(C) + P(D) = \frac{1}{12} + \frac{1}{6} = \frac{1}{4}$$

for the disjoint events C and D.

Think About …

In the special case of *disjoint* events, you can just add the individual probabilities to find the probability of the *or* event. Why?

Disjoint events can be nicely represented with a Venn diagram, as shown below. The diagram makes clear that there are no outcomes in A and in B simultaneously, so $P(A \text{ and } B) = 0$ *when A and B are disjoint events* [and hence $P(A \text{ or } B) = P(A) + P(B)$ for disjoint events A and B].

Sample space

Events A and B are disjoint.

DISCUSSION 4   Are They Disjoint?

1. A coin is tossed and a die is cast. One event is getting heads on the coin toss. The second event is getting a 3 on the die cast. Are tossing a head *and* casting a 3 disjoint events? What is $P(H3)$? What is $P(H \text{ or } 3)$?

2. Two dice are cast. One event is getting a 6 on just one die. The other event is getting a sum of 11. Are these events disjoint? What is the probability of getting a 6 on just one die *and* getting a sum of 11? What is $P(6 \text{ on just one die or a sum of } 11)$?

3. Two dice are cast. One event is getting a 6 on just one die. The other event is getting a sum of 5. Are these events disjoint? What is the probability of getting a 6 on just one die *and* getting a sum of 5?

4. Give some examples of disjoint events. Remember, they cannot happen simultaneously. ●

During the time we have been analyzing more complex experiments, the idea of independent events has come up but has not yet been made explicit. To do so, we consider the following example: In the experiment with the two spinners shown in Discussion 2, the color from the first spin does not affect the color from the second spin; in such cases we say that getting a particular color on Spinner 1 and getting a specific color on Spinner 2 are *independent* of one another.

Two events are **independent** if the occurrence of one event does not change the probability of the other event. When two events are independent, the probability of both happening simultaneously is the product of the two probabilities. For *independent* events A and B,

$$P(A \text{ and } B) = P(A) \times P(B)$$

For example, when spinning the two spinners shown on the previous page, getting red on Spinner 1 and green on Spinner 2 are independent events. The probability of green on Spinner 2 is $\frac{1}{2}$, no matter what happens on Spinner 1. The idea of independence can apply more generally to events of all sorts.

The difference between *disjoint* events and *independent* events can be confusing at first. Consider the situations about the tossing of a pair of dice in the following examples.

EXAMPLE 3

Getting a 2 on one die and getting a sum of 11 are *disjoint* events. They *cannot* happen simultaneously. They are *not independent* events because a sum of 11 cannot be obtained if one die is a 2. ●

EXAMPLE 4

Getting a 5 on one die and getting a sum of 11 are *not disjoint* events. They *can* happen simultaneously, such as tossing a 5 and a 6. They can happen in the same outcome. But they are *not independent* events because knowing that there is a 5 on one die changes the probability of a sum of 11 from $\frac{1}{18}$ to $\frac{1}{6}$. Both $(5, 6)$ and $(6, 5)$ give a sum of 11, so the probability of a sum of 11 is $\frac{2}{36} = \frac{1}{18}$. But if a 5 has been tossed, then a 6 must be tossed on the other die to get a sum of 11, and that can happen with probability $\frac{1}{6}$. ●

EXAMPLE 5

Getting a 2 on one die and getting a 3 on the other die are *not disjoint* events because they can happen simultaneously. They are *independent* events because what turns up on the first die has no bearing on what turns up on the second die. ●

Two events with nonzero probabilities cannot be both disjoint and independent. If they are disjoint, $P(A \text{ and } B) = 0$. Yet, if the two events are independent, $P(A \text{ and } B) = P(A) \times P(B)$. Together, these two equations give $0 = P(A) \times P(B)$. So if $P(A) \neq 0$, then $P(B)$ must $= 0$. That is, if the two events are disjoint, then knowing the first event happened would change the probability of the second event (and in fact make it 0).

TAKE-AWAY MESSAGE . . . When two events cannot happen simultaneously, the events are said to be disjoint. When two events occur such that the first one has *no* effect on the second one, we say the events are independent. If events A and B are independent, $P(A \text{ and } B) = P(A) \times P(B)$. It is possible for two events to be (1) disjoint but not independent, (2) independent but not disjoint, and (3) not disjoint and not independent, but it is not possible for two events with nonzero probabilities to be both disjoint and independent.

Learning Exercises for Section 28.3

1. Give a new example for each situation described.

   a. disjoint events   **b.** independent events

2. Does the formula $P(A \text{ and } B) = P(A) \times P(B)$ always apply? Explain.

3. A skilled archer is shooting at a target. For each shot, her chances of hitting a bull's-eye are 98%.
    a. Make a tree diagram for her first two shots.
    b. If the archer has nerves of steel, is it reasonable to assume independence for the two shots? Did you assume independence in your tree diagram? How?
    c. What is the probability that the archer will hit the bull's-eye both times?
    d. What is the probability that the archer will hit the bull's-eye at least once?
    e. What is the probability that the archer will miss the bull's-eye both times?

4. Everyday usage of the word "or" is most often in an exclusive sense rather than the inclusive sense used in mathematics. Interpret each statement in both ways.
    a. "I think I will have an ice cream cone *or* a muffin."
    b. "I'm going to wear my tennis shoes *or* my sandals."

5. Suppose the weather forecast tells you that

    $P(\text{sun tomorrow}) = 0.7$ and $P(\text{good surfing tomorrow}) = 0.4$

    Find $P(\text{sun } and \text{ good surfing})$ and $P(\text{sun } or \text{ good surfing})$. Discuss the difference. What assumptions did you make?

6. Consider the following pairs of events. Which pairs are independent?
    a. drawing a number from a hat and replacing it; drawing again from the hat
    b. drawing a number from a hat and not replacing it; drawing again from the hat
    c. on a roll of two dice, rolling a sum of 4; rolling a double
    d. getting green on Spinner 1; getting green on Spinner 2
    e. shooting an arrow at a bull's-eye twice in a row
    f. your son waking up one morning with the flu; your daughter waking up the next morning with the flu
    g. graduating summa cum laude; acing final exams
    h. pulling a green M&M from a bag and eating it; then pulling a blue M&M from the same bag

7. Suppose you roll two tetrahedra, each with faces marked 1–4. Find each probability. (You may want to make a table of a sample space.)
    a. $P(\text{sum of 3 } and \text{ one 2})$ b. $P(\text{sum of 10})$
    c. $P(\text{sum of 3 } or \text{ exactly one 4})$ d. $P(\text{two 3s } and \text{ sum of 6})$

8. Suppose there are three crucial components on a space shuttle launch: rocket system, guidance system, and communications system. Suppose also that the probabilities that the systems function correctly are these: Rocket 0.95; Guidance 0.9; Communications 0.99. If one or more of the components does not work, the launch must be aborted. (We are not taking into account backup systems.)
    a. What is the probability that a launch will be aborted? What assumption must you make to determine this probability, based on the information you have? (*Hint for the calculation:* What is the probability that *all* the components work?)
    b. What does that probability in part (a) mean?

9. What alternate form does the formula for $P(A \text{ or } B)$ take in these special cases?
    a. If A and B are disjoint, then $P(A \text{ or } B) = $ _____.
    b. If A and B are independent, then $P(A \text{ or } B) = $ _____.
    c. If event B involves only some of the outcomes in event A, and no others, then $P(A \text{ or } B) = $ _____.

10. Here is an item from a national testing of eighth-graders:[2]

    ▶ A package of candies contained only 10 red candies, 10 blue candies, and 10 green candies. Bill shook up the package, opened it, and started taking out one candy at a time and eating it. The first 2 candies he took out and ate were blue. Bill thinks the probability of getting a blue candy on his third try is $\frac{10}{30}$ or $\frac{1}{3}$. Is Bill correct or incorrect? Explain your answer. ◀

    What would your answer be?

Supplementary Learning Exercises for Sections 28.2 and 28.3

1. **a.** List the sample space for rolling a four-sided die and flipping a coin. For parts (b–d), find the indicated probability.

    b. $P(2H)$ **c.** $P($an even *and* H$)$ **d.** $P(3)$

2. Find the following probabilities for spinning two spinners (not pictured): The first has four equal sections labeled red, green, blue, and yellow. In the second, red and green are the same size, and blue and yellow half that size.

    a. $P($same color both times$)$ **b.** $P(RY)$ (RY means the red first *and* the yellow second.)
    c. $P(YR)$ **d.** $P($they are the same color *or* exactly one is yellow$)$

3. **Experiment** Draw two balls from a bag containing 4 yellow, 2 blue, and 3 green balls, without replacement.

    a. List the sample space for the experiment.   **b.** $P($two different colors$)$
    c. $P($one ball yellow *and* the other ball blue$)$ **d.** $P(YB)$

4. Make a tree diagram for the following experiment (include the outcomes and their probabilities). Spin two spinners with the given degree amounts for the colored sectors:

    Spinner 1: red = 80°, blue = 80°, green = 160°, yellow = 40°
    Spinner 2: red = 180°, green = 60°, yellow = 120°

5. **a.** Show the entire sample space for drawing three balls from a bag containing 2 red balls, 1 green ball, and 1 yellow ball *without* replacement. (Use any listing method that you want.)

    b. $P(RRG)$ **c.** $P(GYR)$
    d. $P($at least two of the same color$)$ **e.** $P($all the same$)$

6. **a.** Show the entire sample space for drawing three balls from a bag containing 2 red balls, 1 green ball, and 1 yellow ball *with* replacement. (Use any listing method that you want.)

    b. $P(RRG)$ **c.** $P(GYR)$
    d. $P($at least two of the same color$)$ **e.** $P($all the same$)$

    f. How do your results differ from those of Supplementary Learning Exercise 5?

7. An eight-sided die consists of eight equilateral triangles numbered 1–8. Each set of four triangles meets at a point and then two sets of four are connected by the edges.

    a. Show the sample space of rolling two eight-sided dice by using a table.
    b. How many outcomes are there?
    c. Find $P($one is even *and* the other is odd$)$.
    d. Find $P($both are prime numbers$)$.

8. **a.** Using the information from Supplementary Learning Exercise 7, make a table showing the sums of the two rolls.

    b. Find $P($sum is 6$)$.
    c. Find $P($sum is 14$)$.
    d. What sum has the highest probability of happening?

9. Suppose that an experiment has exactly 40 equally likely outcomes. Furthermore, $P(X) = 0.35$ and event Y has 7 outcomes.

    a. If events X and Y share exactly 2 outcomes, find $P(X$ or $Y)$.
    b. If events X and Y are disjoint, find $P(X$ or $Y)$.
    c. If events X and Y are mutually exclusive, find $P(X$ or $Y)$.
    d. If events X and Y are independent, find $P(X$ and $Y)$.

10. What does it mean to say, "In mathematics the word 'or' is used in an inclusive sense"?

11. **Experiment** Toss a red die, a white die, and a green die, noting the number of dots on top of each die.

    a. How many outcomes are in the sample space?
    b. If the dice are honest, what is the probability of each outcome?

    Consider these events:

    X is the event of getting 3 dots on the red die and 4 dots on the white die.
    Y is the event of getting 6 dots on the green die.
    Z is the event of getting a sum of dots $= 7$.

    c. Find $P(X \text{ or } Y)$.   d. Find $P(X \text{ and } Y)$.
    e. Are events Y and Z disjoint?   f. Find $P(Y \text{ and } Z)$.

12. Which of the following situations give disjoint events? Explain your reasoning.

    a. drawing a heart and drawing a king from a standard deck of cards
    b. drawing a red card and drawing a club from a standard deck of cards
    c. rolling an even number and a prime number on a six-sided die
    d. owning a cat and owning a dog
    e. filing a state income tax form and a federal income tax form

13. Which of the following situations give independent events?

    a. drawing a heart on the first draw and a king on the second with no replacement
    b. hitting a base hit on two consecutive at-bats
    c. doing your homework and getting an A on the test
    d. a figure being a quadrilateral and a square

14. Consider the two spinners below and the natural experiment of spinning each spinner and noting the colors pointed to:

    a. Describe two events that are independent, or explain why it is not possible to do so.
    b. Describe two events that are not independent, or explain why it is not possible to do so.
    c. Describe two events that are disjoint, or explain why it is not possible to do so.
    d. Describe two events that are not disjoint, or explain why it is not possible to do so.

15. Suppose $P(A) = 0.2$ and $P(B) = 0.35$. If possible, describe a situation in which each probability can be true. If it is not possible, tell why.

    a. $P(A \text{ or } B) = 0.55$ b. $P(A \text{ or } B) = 0.35$ c. $P(A \text{ or } B) = 0.2$
    d. $P(A \text{ or } B) = 0.15$ e. $P(A \text{ or } B) = 0.25$ f. $P(A \text{ or } B) = 0.4$

28.4 Conditional Probability

Problem 1 in the introduction to Part IV (accuracy of HIV testing) is an example of a *conditional probability* problem, the kind we discuss in this section. Conditional probability allows the calculation of probabilities of events that are *not* independent, i.e., probabilities for which the occurrence of one event *does* affect the occurrence of the other event. We will solve Problem 1 in this section. But first we have to understand better the fundamental notions of conditional probability.

> The conditional probability of an event is the probability of an event that is contingent on another event's occurrence. A **contingency table** organizes data in a manner that allows us to determine the conditional probability of an event.

Data organized in a contingency table allow us to introduce conditional probability (and review some other recent ideas).

Notes

ACTIVITY 4   What Does a Contingency Table Tell Us?

Suppose a survey of 100 randomly selected state university students resulted in a sample of 60 male and 40 female students. Of the males, $\frac{2}{3}$ graduated from a high school in the state, whereas the remainder had high school diplomas received out-of-state. Of the females, $\frac{3}{4}$ were from in-state high schools. This information is represented in the following contingency table:

	In-state schools	Out-of-state schools	Totals
Male	40	20	60
Female	30	10	40
Totals	70	30	100

  a. Explain what each of the numbers in the table means.
  b. Is it possible to think about this as a probabilistic situation? If so, how?
  c. If you randomly select a state university student, what is the probability that the student selected is an in-state high school graduate, based on this information? That is, what is $P(I)$?
  d. What is the probability that the student both graduated in the state and is female? That is, what is $P(I \text{ and } F)$? Are I and F independent events? [*Hint:* Recall that if I and F are independent events, then $P(I \text{ and } F) = P(I) \times P(F)$.]
  e. What is $P(I \text{ or } F)$?
  f. What is the probability that the student graduated from an in-state high school, given that the student is female? ●

A common notation for a statement such as the question in Activity 4(f) is $P(I|F)$, read as "the probability of I, *given* F" or "the probability of I, on the condition that F has occurred." This is an example of a **conditional probability.** The probability of event I is now contingent on event F having occurred.

    When there is a given condition for an event, we restrict our attention to only a part of the original sample space. In $P(I|F)$ only the 40 females, of whom 30 are from in state, make up the restricted sample space. So $P(I|F) = 0.75$. (*Note: $I|F$* does not refer to division.)

> ### Think About ...
> What is the restricted sample space for $P(F|I)$? Determine $P(F|I)$ to show that order matters with conditional probabilities.

Conditional probabilities do not require contingency tables, however. For example, drawing balls from a bag, which simulates many probabilistic situations, can involve conditional probabilities. Suppose our paper bag contains 2 red balls and 3 blue balls (identical except for color) and the experiment is to draw a ball twice, noting the color each time. What is the probability of drawing a red on the second draw, $P(\text{R on second})$? If the ball is *replaced* after the first draw, then the probability of getting a red on the second draw is the same as the probability of getting a red on the first draw: Each probability is $\frac{2}{5}$. The draws are independent events.

    The tree diagram below represents the case in which the first ball is replaced. From the diagram we see that drawing red on both draws has probability $\frac{4}{25}$.

With Replacement

First draw	Second draw	Outcome	Probability
R ($\frac{2}{5}$)	R ($\frac{2}{5}$)	RR	$\frac{4}{25}$
	B ($\frac{3}{5}$)	RB	$\frac{6}{25}$
B ($\frac{3}{5}$)	R ($\frac{2}{5}$)	BR	$\frac{6}{25}$
	B ($\frac{3}{5}$)	BB	$\frac{9}{25}$

Notice that because the ball was replaced, drawing either color on the second draw was independent of what happened in the first draw, similar to what was true for the spinners when independence was first introduced.

However, what if the first drawn ball is *not replaced*? Now the possibilities for the second draw are obviously affected. The following tree diagram represents the situation in which the first ball was not replaced before drawing the second ball. Notice that the second set of branches shows the second outcome *given* that the first has occurred. Hence, the probabilities given with the second set of branches are *conditional* probabilities. The tree diagram takes into account the without-replacement aspect of this experiment. If a red ball is drawn on the first draw and not replaced, then the second draw is from four balls, of which only one ball is red. So $P(R|R) = \frac{1}{4}$ (in contrast to the $P(R|R) = \frac{2}{5}$ when the first ball is replaced).

Without Replacement

First draw	Second draw	Outcome	Probability

If a blue ball is drawn first, then of the four balls left, only two are red, so $P(R|B) = \frac{2}{4}$. Again, we get a value different from the $\frac{6}{25}$ when the first ball is replaced. Without replacement, what happens on the first draw obviously affects the probabilities for the second draw. The draws are not independent.

Because the probabilities have been adjusted, however, we can still multiply along the branches to give the probability of each outcome for the whole experiment. From the tree diagram,

$$P(R \text{ and } R) = P(R \text{ on first}) \times P(R \text{ on second}, \text{ given } R \text{ on first}) = \frac{2}{5} \times \frac{1}{4} = \frac{1}{10}$$

(without replacement). Recall that *with* replacement, this probability was $\frac{4}{25}$. So you have already used conditional probabilities in your earlier tree diagrams for no-replacement experiments when you adjusted the probabilities on the later branches.

Contingency tables and adjustments to tree diagrams allow us to answer many conditional probability questions. Contingency tables also allow an illustration of a more general computational procedure for conditional probability. For example, the contingency table in Activity 4 helped determine $P(\text{graduated from in state}, \text{ given that the person is female})$, or $P(I|F)$, by showing explicitly the reduced sample space: just the 40 females, with 30 of them in-state graduates. In other words, given that a person is one of the 40 females, the probability that that person is from in state is $\frac{30}{40}$. This statement is written $P(I|F) = \frac{30}{40}$.

Note also from the contingency table that the 30 is the number of students who were both in-state graduates *and* females, so $P(I \text{ and } F) = \frac{30}{100}$, and $P(F) = \frac{40}{100}$. Rewriting this statement gives

$$P(I|F) = \frac{30}{40} = \frac{\frac{30}{100}}{\frac{40}{100}} = \frac{P(I \text{ and } F)}{P(F)}$$

This form expresses the more general computational method for conditional probabilities:

$$P(A|B) = \frac{P(A \text{ and } B)}{P(B)}$$

We have been dealing with the notion that the probability of a second event may somehow be influenced by a first event happening. Think back about the manner in which we first encountered *independent* events. Events are independent if the occurrence of the first event does ***not*** affect the probability of the second event. That is, $P(A|B)$ is the same as $P(A)$ if A and B are independent events. Thus, the notion of conditional probability provides us with a natural way to regard independence of two events, A and B, using the earlier formula $P(A \text{ and } B) = P(A) \times P(B)$ version. That is,

$$P(A|B) = \frac{P(A \text{ and } B)}{P(B)} = \frac{P(A) \times P(B)}{P(B)} = P(A)$$

This result provides us with an alternative way of thinking about independence of events:

> **Events A and B are independent if $P(A|B) = P(A)$.**

If the occurrence of event A is not affected by the occurrence of event B, then the probability of A is not affected by the fact that B has occurred. That is, the probability of A *given B* is just the probability of A when A and B are independent.

DISCUSSION 5   A Declaration of Independence

If there had been 2000 red balls and 3000 blue balls to draw from (rather than 2 red and 3 blue), how different would be the probabilities for $P(R \text{ and } B)$ with replacement versus without replacement (in a two-draw setting)? Does it seem reasonable to "pretend" independence at times?

Unfortunately, rather than assuming independence in situations where events are not independent, too many times people do just the opposite: They assume two events to be dependent when they are independent, as is demonstrated in Discussion 6. ●

DISCUSSION 6   The Gambler's Fallacy

What assumption is incorrect in the following statements?

1. "It has come up heads 7 times in a row. I'm going to bet $20 it comes up tails the next toss."
2. "He has hit 10 shots in a row. He's bound to miss the next one."
3. "We have had 4 boys in a row. The chances that our next child will be a girl are almost certain." ●

We now have sufficient information to solve the HIV testing problem presented in the introduction to Part IV on page 605.

EXAMPLE 6

Assume there is a test for the HIV virus (the virus that causes AIDS) that is 98% accurate. (That is, if someone has the HIV virus, the test will be positive 98% of the time, and if one does not have the virus, the test will be negative 98% of the time.) Assume further that 0.5% of the population actually has the HIV virus. Imagine that you have taken this test and your doctor somberly informs you that you've tested positive. How concerned should you be?

Solution

Suppose we have a population of 10,000. We can use a contingency table to organize all the information we know. If 0.5% of the people have the virus, then 0.5% of 10,000 = 50 people have the virus. Thus, 9950 do not have the virus. The test is 98%

accurate, so of the 50 people who have the virus, 98% of them (or 49 people) will test positive, and 1 person will test negative. Of the 9950 people who do not have the virus, the test will be accurate for 98% of them, so 9751 people will test negative. That leaves 2% of the 9950, or 199, people who *mistakenly* test positive. Notice that we must take into account both *correct* and *incorrect* test results to make sense of these kinds of statements:

	Has the virus	Does not have virus	Totals
Positive	49	199	248
Negative	1	9751	9752
Totals	50	9950	10,000

Note that $P(\text{Pos}|\text{Virus}) = \dfrac{49}{50} = 0.98$. This means that given that a person *has* the virus, there is a 98% chance that the test will be positive for this person. However, we want to know what percent of the time a person who tests positive actually *has* the virus. That is, we are looking for $P(\text{Virus}|\text{Pos})$. Overall, $49 + 199 = 248$ test positive, but of those, only 49 have the HIV virus! Thus, 49 out of 248 people, or about *20% of those who test positive, actually have HIV.* ●

Think About …
Do the probabilities change with a population of a different size in the HIV testing problem?

TAKE-AWAY MESSAGE . .   When a condition is added to a probability statement, the restricted sample space gives a *conditional probability.* To express a conditional probability, we commonly use a notation such as $P(\text{event}|\text{condition})$. Tree diagrams can take conditional probabilities into account, and contingency tables provide a way of displaying the data so that conditional probabilities can be determined. The earlier idea of independent events A and B can be rephrased in terms of conditional probability: $P(A|B) = P(A)$.

Learning Exercises for Section 28.4

1. The following problem is from a Marilyn vos Savant column in *Parade* magazine (March 28, 1993). Is she right or wrong? Be sure you can defend your answer!

   ▶ Suppose it is assumed that about 5% of the general population uses drugs. You employ a test that is 95% accurate, which we'll say means that if the individual is a user, the test will be positive 95% of the time, and if the individual is a nonuser, the test will be negative 95% of the time. A person is selected at random and given the test. It's positive. What does such a result suggest? Would you conclude that the individual is highly likely to be a drug user? Her answer was "Given the conditions, once a person has tested positive, you may as well flip a coin to determine whether he or she is a drug user. The chances are only 50–50." ◀

2. **a.** A particular family has four children, and you know that at least one of the children is a girl. What is the probability that this family has exactly 2 girls and 2 boys? How is this different from not knowing that at least one of the children is a girl? (*Hint*: A tree diagram may help.)

   b. Suppose you are talking to a woman who has two children, and in the conversation she refers to one of her children as being a girl. What is the probability that she has 2 daughters? What is the probability that she has 2 daughters if the daughter mentioned is the older child?

3. Last year all families with six children in the small country of Candonia were surveyed. In 72 families the *exact order* of births of boys and girls was GBGBBG. What is your estimate of the number of families in which the *exact order* was GBBBBB? (Assume that births of boys and girls are equally likely in Candonia.)

4. Suppose you knew that a group of 100 persons surveyed contained 30% engineers and 70% lawyers. A certain person is selected at random from this group. Suppose the selected person is male, 45 years old, conservative, ambitious, and has no interest in political issues. Is it more likely that this person is a lawyer or an engineer?

5. This Learning Exercise is the second one from the introduction to Part IV. Try it again, and see if your answer is different from your earlier answer.

▶ *The Monty Hall Problem:* On the *Let's Make a Deal* television show, the long-time host Monty Hall presented three doors to a contestant. Behind one door was the prize of the day and behind the other two doors were gag gifts. Contestants were asked to choose one door to open. Before the selected door was opened, Monty would open one of the other two doors to reveal a gag gift. The contestant was then asked whether he or she wanted to stay with the door chosen or to switch to the other closed door. Would it matter? ◀

6. At the county fair, a clown is sitting at a table with three cards in front of her. She shows you that the first card is red on both sides, the second is white on both sides, and the third is red on one side and white on the other. She picks up the cards, shuffles them, hides them in a hat, and then draws out a card at random and lays it on the table, in such a manner that both of you can see only one side of the card. She says, "This card is red on the side we see. So it's either the red/red card or the red/white card. I'll bet you one dollar that the other side is red." Is this a safe bet for you to take?

7. Suppose a fair coin is tossed five times. Which outcome, if any, of the following has the greatest probability?

 a. HTHTH   **b.** HHHHH   **c.** TTTTH   **d.** THHTH

8. **a.** We learned that $P(A|B)$ is sometimes given by $P(A|B) = \dfrac{P(A \text{ and } B)}{P(B)}$. Use algebra to show that this definition, along with our earlier definition for independent events A and B, gives the $P(A|B) = P(A)$ relationship discussed in this section.

 b. If A and B are independent, is $P(A|B) = P(B|A)$?

9. A cab was involved in a hit-and-run accident at night. Two cab companies, Blue Cab and Green Cab, together own 1000 of the cabs in the city. Of these cabs, 85% are Green and 15% are Blue. A witness said that the cab in the accident was a Blue Cab. The witness was tested in similar conditions and made a correct identification 80% of the time. What is the probability that the cab in the accident was a Blue Cab, given the witness's statement?

10. Here are variations of Example 6 on HIV testing, to see which factors influence the level of concern.

 a. Suppose that the population is 30,000 rather than 10,000, with the same test accuracy (98%) and percent of HIV incidence the same (0.5%). What is the probability that a person who tests positive does have the HIV virus?

 b. Suppose there is a new test for the HIV virus that is 99% accurate (rather than the 98% in Example 6). Use a population of 20,000 but still with an HIV incidence of 0.5%. What is the probability that a person who tests positive does have the HIV virus?

 c. Suppose that the incidence of the HIV virus in one population is 3% (rather than the 0.5% of Example 6). Use a population of 10,000 and a test accuracy of 98%. What is the probability that a person who tests positive does have the HIV virus?

11. In a polling of 400 parents, 20% were high-income. Fifty high-income parents oppose vouchers. Vouchers are supported by 160 low-income parents.

 a. Complete the table, using the data above:

	High-income	Low-income	Totals
Support vouchers			
Oppose vouchers			
Totals			

continue

b. What is P(oppose vouchers|high-income)? (That is, what is the probability that a parent opposes vouchers, given that the parent has high income?)

c. P(high-income|oppose vouchers) = _____

d. What is the probability that a parent has high income OR opposes vouchers?

12. Gerd Gigerenzer, a famous psychologist of the Max Planck Institute in Berlin asked a group of physicians the following:

> to tell him the chance of a patient truly having a condition (breast cancer) when a test (a mammogram) that was 90 percent accurate at spotting those who had it and 93 percent accurate at spotting those who did not came back positive. He added one other important piece of information: that the condition affected about 0.8 percent of a population for the group . . . being tested. Of the twenty-four physicians to whom he gave this information, just two worked out correctly the chance of a patient really having the condition.[3] (pp. 106–107)

a. Find the probability of a false positive result, i.e., the probability that a patient without cancer tests positive.

b. Why do you suppose that so few physicians could find the probability that you found in part (a)?

Supplementary Learning Exercises for Section 28.4

1. Consider the following problem:

> ▶ A certain company employs 4800 people. Of these, 65% are women and 10% of all the employees have a college diploma. Of the men, 85% do not have a college diploma. How many female employees have college diplomas? ◀

Janine concluded that the answer is 312 women have a college diploma.

a. How did Janine arrive at this answer? What assumptions did she seem to make? What data did she use?

b. Was her answer correct? If not, what is the correct answer? (You may want to make a table to organize your data.)

2. In Supplementary Learning Exercise 1, 4% of the employees are in management.

a. Do we have enough data to determine how many women are in management? If so, how many?

b. Suppose 6.25% of the managers do not have a college diploma. What percent of the nonmanagement employees have a college diploma?

3. Sleep apnea is a condition that causes a person to stop breathing briefly while asleep. People with this condition suffer from various other problems as a result of poor-quality sleep. Snoring is often associated with sleep apnea. Use the following information to complete the following table for 1,000,000 people:

- 50% of people with apnea snore.
- 6.62% have sleep apnea.
- 59% of the population snore.

	Snore	Does not snore	Totals
Sleep apnea			
No sleep apnea			
Totals			1,000,000

4. Doctors estimate that 20% of the population actually have sleep apnea. Make another table that uses this statistic instead of the 6.62% above. What assumptions are you making in creating this new table?

5. Why is the "Gambler's Fallacy" (see p. 648) a fallacy?

6. The following question comes from a National Assessment of Educational Progress exam. On the examination, 49% of the students got this question correct.[4]

▶ The given table shows the gender and color of seven puppies.

Gender and Color of Puppies

	Male	Female
Black	1	2
Brown	1	3

If a puppy selected at random from the group is brown, what is the probability it is a male?

a. $\frac{1}{4}$ b. $\frac{2}{7}$ c. $\frac{1}{3}$ d. $\frac{1}{2}$ e. $\frac{2}{3}$ ◀

  a. What is *your* answer?
  b. How would you explain how to do this problem to a sixth-grade student?

7. Give two formulas that deal with independent events (pp. 642 and 648).

8. A survey about removing parking spaces along one business street to make additional lanes for traffic gave the following results, broken down by gender. Assume that future surveys give similar results.

	Males	Females
Favor removal	90	40
Oppose removal	30	60

  a. What is the probability that a respondent is male, given that the person favors removing the parking spaces?
  b. What is the probability that a respondent favors removing the parking spaces, given that the person is male?
  c. What is the probability that a respondent is male *or* the respondent opposes removing parking places?
  d. What is the probability that a respondent is male *and* the respondent opposes removing parking places?

9. Malta City has two major construction projects to consider: (1) Build a pedestrian bridge over a busy street and (2) build a new park. The budget will not allow both projects to be funded during the next budget year, so the City Council decided to put the choice to a vote. In a polling of 500 random citizens, 30% were parents of school children. Of these parents, 60% favored the park; 50% of the nonparent citizens favored the bridge. Assume that future votes give the same results. Make a contingency table to help find the following probabilities:

  a. P(favor bridge | parent)   **b.** P(parent | favor bridge)
  c. P(favor bridge *or* parent)   **d.** P(favor park *and* parent)

28.5 Issues for Learning: Research on the Learning of Probability

At one time there was not much research on young students' understanding of ideas pertaining to chance and uncertainty, but more information is now available[5–7] in addition to that from testing programs. Much of the research focuses on what children understand about chance. For example, children in one second-grade classroom[8] were surprised that, after many rolls of a pair of dice, the sum 7 "kept winning." Children were led by the teacher to investigate this phenomenon by making graphs of many rolls of the dice, and then discussing the different ways the numbers on a pair of dice could be displayed and why the sum 7 was the most likely.

In another research study,[9] third-grade children were provided with 16 lessons on probability topics. They were tested in the beginning and at the end of the experiment, and some of the children were interviewed. These researchers found that children at this age can make some progress in understanding probabilistic situations. This type of research has much to offer teachers, not only in terms of the research findings, but also in terms of the instructional tasks designed and used. For example, in one session children first discussed the following problem (p. 495), intended to help them begin to think about measuring uncertainty:

▶ Two basketball teams are playing tonight, the Red Hots and the Cyclones. The Red Hots have won 3 out of 7 games, and the Cyclones have won 6 out of 7 games. Who do you think will win tonight? ◀

The children were then given an activity sheet with a picture of a "race" between the colors red (for the Red Hots) and white (for the Cyclones), something like the following:

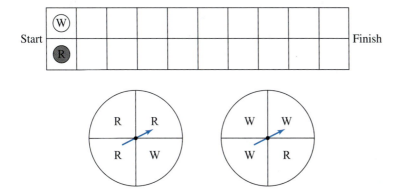

The task was for each of two players to pick one chip, either white or red, and put the chip in the start box. They would then spin one of the spinners, and the child with that color would move one space. Before playing, they were asked which spinner would be better for each of the two individuals, and why.

The children's responses to this activity illustrate growth in understanding sample space, the probability of an event, and comparing probabilities. Increasing levels of *understanding sample space* are illustrated by the children's responses when asked what colors would come up on the spinner: "Red. It's going to win." "Red and white." "Red and white, but they don't have the same chance." The increasing levels of *understanding of probability* are illustrated by these children's responses when asked why red has a better chance on the first spinner: "It's my favorite color." "There's more red." "Red. There's three pieces to one piece." The increasing levels of *comparing probabilities* are illustrated by their responses to being asked which spinner is best for the red chip: "It doesn't matter because I always spin red; it's my favorite color." "Spinner 1; it's got more red." "Spinner 1; there are three sections of red compared with one section of white."

This race task and the accompanying questions and responses show that the understanding of these concepts evolves over time and with instruction. The work also illustrates that instruction on probability concepts can begin early in elementary school.

The apparently developmental nature of children's understanding of proportionality[9] is one reason that more complicated probability situations do not come up until the upper elementary grades. Listing all the possible outcomes for a toss of three coins is not easy until children have some mental tool for a systematic approach (or instruction in a tool like a tree diagram). In considering the possibilities even when only two coins are tossed, the intuitive three-outcome view (both heads, both tails, mixed) can lead to an incorrect view that each outcome has probability $\frac{1}{3}$. These examples speak to the importance of a complete listing of all the outcomes to an experiment, as well as a clear description of exactly what the experiment involves.

Notes

> ### *Think About …*
> How would you explain to someone that the both-heads, both-tails, and one-head/one-tail outcomes for tossing two coins are not equally likely?

At a more basic level, children may find it difficult to deal with the ratio nature of a probability situation. For example,[10] consider the task in which students are asked which of two situations gives a better chance of drawing a marker with an X on it, as in the drawing below:

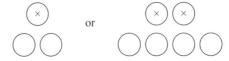

Children who do not have a ratio concept may focus on the X's ("winners") and choose the second situation (2:6) because there are two X's in it but only 1 X in the first situation. Others may focus on the non-X's ("losers") and choose the first situation because it has fewer losers than does the 2:6 one, or switch to that decision once the numbers of losers are pointed out. Still others might focus on the *surplus* losers (or winners). They may choose the 1:3 situation because, there is only one more loser than winners, whereas in the 2:6 situation, there are two more losers than winners. The researchers did not deal with a rearrangement such as the following to see whether that would influence the children's thinking:

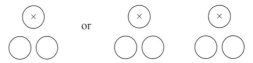

The same task with ratios such as 2:5 versus 6:13 seems to invite all sorts of (incorrect) reasoning among children who do not have proportional reasoning.

> ### *Think About …*
> How would you deal with a 2:5 as opposed to a 6:13 task (a) with drawings and (b) numerically?

Even though some understanding of ratio is essential for many probability concepts, students can do much work before having a fully developed ability to reason proportionally. In fact, the results of several studies suggest that topics such as conditional probability and independence are appropriate for middle school students. In these studies, students used their own invented strategies to make correct probability judgments, even in the absence of a complete understanding of rational numbers and numerical probabilities.[11]

    Other researchers have explored the manner in which probabilistic intuitions develop over time. The research is often aimed at determining what it is children know—i.e., what ideas about chance and uncertainty they have already formulated without instruction. This type of research can then be used to develop appropriate instruction. This research, too, is useful in the design of curricula for students, and for teachers to better understand their students' thinking. One such study[12] investigated the development of understanding of some basic ideas of probability over a large age span, with students in 5th, 7th, 9th, and 11th grades and some college students, none of whom had prior instruction on probability. Each of the problems given to these students was one for which a great deal of data showed there to be serious misconceptions. Here are two of the problems:

▶ In a lotto game, one has to choose 6 numbers from a total of 40. Vered has chosen 1, 2, 3, 4, 5, 6. Ruth has chosen 39, 1, 17, 33, 8, 27. Who has the greater chance of winning? ◀

▶ Dan dreams of becoming a doctor. He likes to help people. When he was in high school, he volunteered for the Red Cross organization. He accomplished his studies with high performance and served in the army as a medical attendant. After ending his army service, Dan registered at the university. Which seems to you to be more likely? Dan is a student of the medical school, OR Dan is a student. ◀

Think About ...

How would you answer these two questions?

The results were very interesting. For the first problem, where the correct answer is that both have the same chance of winning, the percents correct at the 5th, 7th, 9th, and 11th grades were 30, 45, 65, and 65. For the second problem, where being just a student is more probable than being a medical student, the percents correct were 15, 30, 20, and 60. Although we see some positive changes in these percents, the percents correct are still small even at the ninth-grade level. These results demonstrate the resilience of the misconceptions that children (and many adults) have about probability.

TAKE-AWAY MESSAGE . . . Research shows us that without instruction, misconceptions about probability sometimes continue into adulthood. With instruction, even students in the third grade can begin to reason probabilistically. Early opportunities to learn about probability may help to guide later intuitions, especially with more complicated situations. Hands-on experimentation may show children that outcomes can vary and that outcomes are not always equally likely. Research has shown how some children are reasoning when they give incorrect answers; perhaps carefully designed instruction can help to prevent incorrect reasoning. Proportional reasoning is the basis for a fuller understanding of probability.

28.6 Check Yourself

You should be able to work problems like those assigned and to meet the following objectives:

1. Draw a tree diagram for a given experiment, and determine the sample space and/or the probabilities of all the outcomes of the experiment. In some situations, you might use just the ideas of a tree diagram (as in calculating the probability of HTHHTTHTHH, in which case the tree diagram would be quite large).

2. Use the vocabulary of events knowledgeably: *disjoint* (or *mutually exclusive*) *events, independent events.*

3. Show your understanding of the mathematical use of *or* and *and,* particularly in these important results:

    a. $P(A \text{ or } B) = P(A) + P(B) - P(A \text{ and } B)$, for any events A and B;
    b. $P(A \text{ and } B) = P(A) \times P(B)$, for independent events A and B.

4. Use a contingency table and/or a tree diagram to answer conditional probability questions, use the $P(A|B)$ notation correctly, and compute by using $P(A|B) = \dfrac{P(A \text{ and } B)}{P(B)}$.

5. Express two different ways of thinking about independent events.

References for Chapter 28

[1]http://nces.ed.gov/nationsreportcard/ITMRLS/. (1992). Grade 4, Data Analysis, Statistics & Probability, "Probability of What Color Marble Being Drawn." Retrieved August 4, 2009.

[2]http://nces.ed.gov/nationsreportcard/pdf/studies/2007460_2.pdf. Downloaded October 2, 2012.

[3]Blastland, M., & Dilnot, A. (2009). *The numbers game.* New York: Gotham Books.

[4]http://nces.ed.gov/nationsreportcard/ITMRLS/. (2005). Grade 12, "Determine conditional probability from two-way table." Retrieved August 4, 2009.

[5]Shaughnessy, J. M. (1981). Misconceptions of probability: From systematics errors to systematic experiments and decisions. In A. P. Shulte & J. R. Smart (Eds.), *Teaching statistics and probability, 1981 Yearbook,* pp. 90–100. Reston, VA: National Council of Teachers of Mathematics.

[6]Shaughnessy, J. M. (1992). Research in probability and statistics: Reflections and directions. In D. A. Grouws (Ed.), *Handbook of research on mathematics teaching and learning,* pp. 465–511. New York: Macmillan.

[7]Shaughnessy, J. M. (2003). Research on students' understanding of probability. In J. Kilpatrick, W. G. Martin, & D. Schifter (Eds.), *A research companion to Principles and Standards for School Mathematics,* pp. 216–226. Reston, VA: National Council of Teachers of Mathematics.

Notes

[8]Conference Board of the Mathematical Sciences (2000). *The mathematical education of teachers.* Providence, RI: American Mathematical Society and Washington, DC: Mathematical Association of America.

[9]Jones, G. A., Langrall, C. W., Thornton, C. A., & Mogill, A. T. (1999). Students' probabilistic thinking in instruction. *Journal for Research in Mathematics Education, 30,* 487–519. The example here is adapted from the one that appears on pp. 495–496.

[10]Piaget, J., & Inhelder, B. (1975). *The origin of the idea of chance in children* (Trans. L. Leake, Jr., P. Burrell, & H. Fishbein). New York: Norton.

[11]Tarr, J. E., & Lannin, J. K. (2006). How can teachers build notions of conditional probability and independence? In G. A. Jones (Ed.), *Exploring probability in school: Challenges for teaching and learning*, pp. 233–259. Dordrecht, The Netherlands: Kluwer Academic Publishers.

[12]Fischbein, E., & Schnarch, D. (1997). The evolution with age of probabilistic intuitively based misconceptions. *Journal for Research in Mathematics Education, 28,* 96–105. The problems and information on results here were taken from p. 98.

Chapter 29

Introduction to Statistics and Sampling

Many times people make claims based on questionable evidence. For example, a newspaper might report that hormone therapy will decrease the likelihood of heart attacks in women. Later, another report will say that rather than decrease the risk, hormone therapy may increase the risk. Each claim is based on evidence that likely involved a selected group of women. The manner in which the selection of the group is made can cause an enormous difference in the outcome of a study. In this chapter we will consider ways of selecting a group to be studied so that a claim can be made with a reasonable chance of being true.

29.1 What Are Statistics?

Think back to the time you were choosing a college. Perhaps you always knew which college you would go to. Or perhaps the decision was already made for you. There is a chance, however, that your decision was not that easy or straightforward. Instead, you might have narrowed down your choices based on some characteristic such as reputation in your area of interest (for instance, colleges recognized for teacher preparation) or location (e.g., colleges very close to or very far from home). You then might have taken a closer look at other factors you considered important (such as the cost of attending for a year or starting salaries after graduation). If you did not experience this decision-making process yourself, you probably have friends or relatives who did.

As an example, take the case in which Miranda, a high school senior, narrowed the list of possible colleges to a few. She wanted to know the cost of attending each college for a year to help her choose a college to attend. Colleges usually provide this cost information by listing the separate costs of tuition, books, housing, food and miscellaneous personal items, and giving the total cost of attending for a year. Tuition cost generally does not vary from student to student, but books, housing, food, and personal costs usually do.

Suppose Miranda received a table containing the following information about costs for one year at University X:

Costs for One Year at University X

Tuition	$ 8310
Books	$ 645
Housing	$ 5835
Food	$ 5220
Personal	$ 3250
Total cost	$23,260

Think About ...

Where do the numbers in this table come from? For example, how was the cost of books determined? We know that not everyone spends the same amount of money on books in a year (nor does any individual spend the same amount from year to year).

Books, housing, food, and personal costs were probably determined by collecting and compiling information about the amounts students spent on these items in the previous year and computing an average for each item. Thus, we could say that these costs were determined statistically. Very simply put, the costs, item by item, of attending this college for one year are **statistics.** The cost of books is a **statistic.** The cost of housing is a statistic, etc.

> A **statistic** is a numerical value of a quantity used to describe data. The plural, **statistics,** is also used to mean the science of obtaining, organizing, describing, and analyzing data for the purpose of making decisions, as well as making predictions.

Notice that in the previous table, a specific number (a dollar amount) is reported for each item, but it is possible that no student will spend precisely $645 on books, $5835 on housing, $5220 on food, and $3250 on personal items. Although an individual student will probably not spend *exactly* $23,260 to attend this school for one year, these statistics still provide useful information about college costs.

In most instances, statistics are reliable and useful, as illustrated above. But statistics can also be unreliable or misleading. Just as it is possible to lie when giving an account of an event, it is possible to "lie" with statistics by inappropriately manipulating data, withholding information that is crucial to the interpretation of data, or presenting statistical information in a way that hides important information about the data. A person who understands fundamental ideas in statistics is more likely to recognize these unethical uses of data than a person who does not have this understanding. Statisticians and researchers generally follow professional standards for analyzing and presenting statistical information. In this and forthcoming chapters, we emphasize understanding and making sense of data and statistical information.

On the surface, statistics as a discipline appears to be a precise science that yields only one "right answer" to a question. One reason for this belief is that statistics is associated with mathematics, which is in turn associated with precision. Another reason is that a statistical analysis produces apparently precise numerical results. (For example, you've probably encountered statements such as "Families in that community have on average 2.6 children.") There are, however, some gray areas in interpreting statistics. As a rule of thumb, expect precision in some aspects of statistics, such as computing statistics or making honest graphs, but expect gray areas in aspects such as interpreting graphs and interpreting statistics.

💬 DISCUSSION 1   When Do We Use Statistics?

What are some other areas, besides the cost of college for one year, where one might use statistics to help in decision making? ●

You will be reminded in later chapters of the close relationship between probability and statistics. Much statistical decision making is based on probabilities. Even though probability is a powerful science, probability does not deal with absolutes. Instead, as you learned in Chapters 27 and 28, probability deals with chance and uncertainty. There is almost always an element of chance in making statistical decisions, and part of our work is to balance certainty and uncertainty in the study of statistics.

TAKE-AWAY MESSAGE . . .   *Statistics* are numbers that are used to represent quantitative data. *Statistics* also refers to the scholarly discipline dealing with these numbers of interest. Care must be taken to assure that the statistics used are not misleading but rather provide information useful in making a decision based on the data provided by statistics.

Learning Exercises for Section 29.1

1. "Statistics is the study of statistics." Why does that sentence make sense (and why is it grammatically correct)?

2. Give a possible statistic about each item:

   a. a football team
   b. a fourth-grade class
   c. one of your classes

3. Explain why each given conclusion is a misuse of statistics.

   a. "He hit 50 home runs last year, so I know he will hit 50 again this year."
   b. "I got 52% on the first quiz, so I know I will fail the course."
   c. "The newspaper reported that the unemployment rate is 8.2%. So out of the 40 students in my class, I can expect 3 of them to be unemployed."
   d. "In the last election, 64% voted not to increase taxes to cover the budget deficit. But at the company picnic, it seemed like almost everyone thought we needed the tax increase."

Supplementary Learning Exercises for Section 29.1

1. Tell whether each statement is a misuse of statistics.

   a. He received a B on the first test, so he will receive a B in the class.
   b. 50% of a group of 200 people are women, so 100 are women.
   c. Two years ago the stock increased by 2.5% and last year it increased 5%, so this year it will increase 10%.

2. Give a suitable statistic for each situation:

   a. computer usage among students
   b. dogs in a kennel
   c. crayons in a box containing 64

3. Give an example of a statistic from a newspaper article or a television report.

29.2 Sampling: The Why and the How

Finding a *representative sample* is extremely important in deriving statistics that can be used in making decisions. In this section we explore the constraints involved in sampling and some different ways of sampling.

 ACTIVITY 1   A Fifth-Grade Task

The problem below is adapted from one used by a researcher with a group of fifth-grade students.[1] The problem gives a situation in which samples are used because surveying everyone was not feasible. The students were asked to evaluate different sampling methods. In this activity, first decide how you would find an appropriate sample to address the problem, and then discuss and evaluate each of the nine sampling methods. Give reasons why each of the sampling methods is likely to provide valid, or invalid, statistics. (In this case, the numbers of yes and no responses are the statistics.)

▶ An elementary school with grades 1 through 6 has 100 students in each grade. A fifth-grade class is trying to raise some money to go on a field trip to Disneyland. They are considering several options to raise money and decide to do a survey to help them determine the best way to raise the most money. One option is to sell raffle tickets for a Wii U. How could they find out whether or not students were interested in buying a raffle ticket to win the game system? ◀

Nine different students each conducted a survey to estimate how many students in the school would buy a raffle ticket to win a Wii U. Each survey included 60 students, but the sampling method and the results are different for each survey. The nine surveys and their results are described as follows. (*Note:* In all cases, yes means that the student agreed to buy a raffle ticket.)

1. Raffi asked 60 friends. (75% yes, 25% no)

2. Marta got the names of all 600 students in the school, put them in a hat, and pulled out 60 of them. (35% yes, 65% no)

3. Spence had blond hair so he asked the first 60 students he found who also had blond hair. (55% yes, 45% no)

4. Jinfa asked 60 students at an after-school meeting of the Games Club. The Games Club met once a week and played different games—especially computerized ones. Anyone who was interested in games could join. (90% yes, 10% no)

5. Abby sent out a questionnaire to every student in the school and then used the first 60 that were returned to her. (50% yes, 50% no)

6. SuLin set up a booth outside the lunchroom, and anyone who wished could stop by and fill out her survey. To advertise her survey, she posted a sign that said WIN A Wii U. She stopped collecting surveys when 60 students had completed the survey. (100% yes)

7. Jazmine asked the first 60 students she found whose telephone number ended in a 3 because 3 is her favorite number. (25% yes, 75% no)

8. Dong wanted the same number of boys and girls and some students from each grade. So he asked 5 boys and 5 girls from each grade to get his total of 60 students. (30% yes, 70% no)

9. Paula didn't know many boys, so she decided to ask 60 girls. But she wanted to make sure she got some young girls and some older ones, so she asked 10 girls from each grade. (10% yes, 90% no) ●

Discussion 2 may help to crystallize some of your perceptions about the children's methods of sampling in Activity 1.

💬 DISCUSSION 2   Some Sampling Methods Are Better than Others

1. For each sample in Activity 1, why do you think the percentages came out the way they did?

2. What kinds of biases could show up in the students' samples?

3. Do you think the percentages would have changed if the sample size had changed? ●

The children in Activity 1 were each getting responses from only a portion of the total school population, but they actually wanted to know about the *whole* **population.** They were getting a **sample** of the total school population. It is important to understand the precise meanings of both words, "population" and "sample."

> A **population** is the entire group that is of interest. The group may be made up of items besides people: e.g., assembly line output, animals, trees, insects, and chemical yields. A **sample** is the part of the population that is actually used to collect data.

EXAMPLE 1

The blue dots represent a sample of size 10 from the population shown.

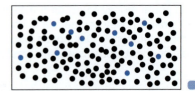

Notes

One might ask why we restrict ourselves to samples—why bother with a sample? Why not just use the entire population? There are many reasons for using a sample. First, it is not always possible to include all of the population. For example, suppose a company that makes lightbulbs wants to provide customers with an estimate of the number of hours each bulb will burn. By taking a sample of the bulbs and burning them until they burn out, the company can reach such an estimate. The company obviously does *not* want to burn out every bulb it makes to find out how long its bulbs will burn.

Second, gathering information is costly in terms of both time and money. If a political party wants information regarding people's feelings about whether or not a budget surplus should be spent on upgrading public schools and can get a very close estimate by polling 3000 people throughout the country, then it would be a waste of money and of people's time to poll 6000 people. Even the U.S. Census has not asked every person (the population) the same questions every year. Rather, until as recently as 2010, the Census Bureau used the responses from one out of every six addresses (a sample) answering a longer questionnaire to obtain information on the entire population. The census is mandated by the U.S. Constitution and occurs every 10 years.

Third, we also use samples instead of the entire population when we want timely results. If a legislator considering how to vote on a bill wants to know how people feel about increasing taxes to provide more schools and teachers, she cannot delay a decision until everyone in the state has been canvassed on this issue.

Fourth, and perhaps most important, the discipline of statistics allows us to interpret results from samples and to make assertions about the whole population. *A result from a relatively small, but carefully chosen, sample can give information about the whole population.* In this statement *relatively small* does not necessarily mean small in number, but rather it means far fewer than the whole population.

The distinction between sample and population is reflected in two other key terms.

A **sample statistic** is the result of a calculation (or count) based on data gathered from a sample. A calculation (or count) based on the population is called a **population parameter.**

Because we do not know the parameter of the population (unless the population is very small), we use a **sample statistic** as an estimate of the population.

EXAMPLE 2

In the previous example about testing lightbulbs, if the sample bulbs burned an average of 300 hours each, then the 300 hours is a *sample statistic*. We could use this value as an estimate of the (unknown) average number of hours *all* bulbs of the same type made by this company will actually burn. The (unknown) average number of hours of all the bulbs made by the company will actually burn is the *population parameter.* ●

EXAMPLE 3

Consider again the example about the legislator polling a sample of citizens to determine support for the increase in taxes. A result of 58% in support of the increase is a sample statistic. The legislator would then use that sample statistic to estimate the (unknown) population parameter as being (around) 58% of all the citizens in support of the increase. ●

In practice, if the population is relatively small and accessible, the data for the whole population might be collected, thus directly giving the population parameter. But for large populations, the population parameter is usually estimated by using the sample statistic. When it is otherwise difficult or impossible to find or calculate the population parameter, we estimate it using the sample statistic. *Sample* always refers to some subset of a *population.*

DISCUSSION 3   When to Use a Sample

In each given instance, tell whether you would estimate the population parameter by finding a sample statistic, or directly find the population parameter. Explain your answer.

1. The senior class in a school has 71 students, and you want to know how many of the seniors bought a class ring.

2. A pharmaceutical company is testing a new drug that it hopes will help AIDS victims, but side effects need to be known before seeking FDA approval.

3. The university (30,000 students) is having an election for student body president. You are in the running and want to know your chances of winning.

4. There is an outbreak of *E. coli* poisoning in town, and you want to know if all the victims ate something in common. ●

If we are to understand the degree to which a sample statistic reflects the population parameter, we need to be concerned about the issue of **sampling bias**.

> A sample is ***biased*** if the process of gathering the sample makes it likely that the sample will not reflect the population of interest.

> *Think About ...*
>
> In Activity 1, were any of the samples collected by the fifth-graders biased? If so, in what ways were they biased?

DISCUSSION 4   Different Types of Bias

A women's magazine wants to know what women in this country think about late-term abortions when the mother's life is in danger.

a. One editor suggests that readers be asked to respond to the question via e-mail. This type of sample is called a **self-selected** or **voluntary sample**. In what ways could this sample be biased? Would the sample of respondents be representative of women in the United States?

b. Another editor suggests that 10 people be sent to different cities to ask this question of the first 50 women who walk out of a large grocery store. This type of sample is called a **convenience sample**. Is this a better way to obtain an unbiased sample? Why or why not?

c. How would you suggest the sampling be done? ●

The sample most likely to be unbiased is one that is established in an entirely random manner. In the *American Heritage Dictionary,* the word *random* is defined as "having no specific pattern or purpose." But as you know from Section 27.2, in probability and statistics the term *random* is used to mean something more precise.

> A **random sample** is one in which *every member* of the given population has an equal chance of being selected for the sample.

One type of random sample is a **simple random sample (SRS)**, in which every possible sample of a particular size has an equal chance of being selected. There are various ways of obtaining a simple random sample, such as drawing names from a hat, using dice, using a spinner, or using a table of randomly selected digits.

DISCUSSION 5   What Should the Mayor Do?

A mayor wants to know what kind of support exists for building a new stadium with tax dollars. Her staff decides to take a random selection of telephone numbers from the phone book and poll the people who answer the phones.

a. Would this selection process yield a simple random sample of all the city's residents? Why or why not?

b. How could a simple random sample of the names in the phone book be generated? ●

There are times, however, when a random sample does not assure that subsets of different populations are represented. At such times we consider other types of sampling. For example, sometimes the population is made up of groups, or **strata,** and how the different strata respond might be of interest. A **stratified random sample** is often used when it is as important to find out how these groups differ on a particular quality or question as it is to find out how the entire population would respond. For example, pollsters often divide a population into groups by gender, age, or race, and then select random samples within each group. Dong (see Activity 1) stratified students by grade level to be sure that every grade was represented. If a personnel manager of a large business with several sites around the country wants to investigate whether or not promotions are being fairly given, she might stratify by gender, selecting a sample of females and a separate sample of males.

At times when random sampling is too difficult, **systematic sampling** is used. Systematic sampling is useful when the population is already organized in some way, but the manner of organization has nothing to do with the question under study. For example, if you want to find out how the residents of a college dorm feel about food service, you could interview the residents of every eighth room in the dorm.

Assuming that students are randomly assigned to dorm rooms, you could sample another way by randomly picking one floor and interviewing everyone on that floor. This last method involves what is called a **cluster sample.** A cluster is randomly selected, and all members in that cluster give data. Although individuals in a systematic sample or a cluster sample are not randomly selected as individuals, there is an assumption of a random ordering of the population being sampled, so the cluster sample should not be biased. Notice how this method is different from stratified sampling. A stratified random sample for the question about food service might be carried out by grouping the residents according to gender or according to vegetarians and nonvegetarians, for example.

There are, of course, times when a combination of sampling techniques is used. For example, exit polling (i.e., asking people whom they voted for as they leave the voting area) is often a mix of cluster sampling (because only some polling booths are selected) and convenience sampling.

TAKE-AWAY MESSAGE . . . A *sample* of a *population* allows one to gather information yielding a *sample statistic,* whereas collecting information from the entire population results in a *population parameter.* Many times it is necessary or sensible to gather data on a sample rather than on a population. It is important to realize that our ability to select a relatively small sample that will provide us with information about a much larger population is one of the strengths of the science of statistics. In order for a sample to be representative of the population, it must be selected with care. Whenever possible, a *simple random sample* (SRS) should be selected from which to gather data. If results from particular groups within the population are of interest, statisticians might use a *stratified sample.* When an SRS is difficult to obtain, statisticians sometimes resort to obtaining a *systematic sample* or a *cluster sample. Self-selected* or *voluntary samples* and *convenience samples* require extra precaution because of bias possibilities, but they are sometimes used.

Notes

Notes

In this sixth-grade lesson, describe John's, Tiffany's, and Angie's sampling methods and tell whether and why one method of sampling is better than the others.

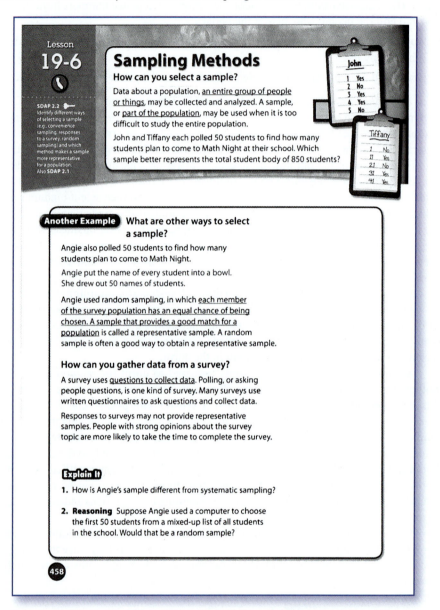

From Randall I. Charles et al., *enVision,* Math California Edition, Grade 6, p. 458. Copyright © 2009 Pearson Education, Inc. or its affiliate(s). Used by permission. All Rights Reserved.

Learning Exercises for Section 29.2

1. Suppose you are attending a university advertised as a school focusing on the liberal arts. Yet it seems that most of the people you meet are majoring in one of the sciences. What type of sampling would you use to find out whether students think of the university primarily as a liberal arts school or as a technical school? What is the population in this situation?

2. You are curious to know whether or not students at your university support gun control. What type of sampling might be reasonable, and why? What is the population in this situation?

3. A supermarket manager has received a complaint that the eggs from a certain distributor frequently are cracked. There are 60 cartons from the distributor on hand. He opens every fifth carton and checks the eggs for cracks. What type of sampling is he using? Is any other type of sampling advisable? What is the population in this situation?

4. A news organization might use exit polling by having its representatives stand outside preselected polling sites, randomly select people coming out, and ask them how they voted. Is this completely randomized sampling, stratified sampling, convenience sampling, or cluster sampling? Explain your answer.

5. The administration of a university wants to do a survey of alcohol use among students to see whether or not it is a problem. The administrators consider the following sampling methods to carry out the survey. Identify each sampling method and discuss any potential bias.

   a. They decide to select everyone in the fraternity houses.
   b. They divide up the student body by age: 17–20, 21–25, and older. They take a proportional sample from each age group. That is, 45% of the student body fall in the 17–20 interval, so they make sure that 45% of the sample are in that age group.
   c. They divide the student body based on whether the students reside on campus or off campus, and then they take a random sample of 50 from each group.
   d. They identify five major religious groups, take a random sample of 20 from each group, and then group together all the results.
   e. They hire someone to stand by the door to the student center and poll the first 100 students who enter.

6. Which kind of sampling plan was used in each of the following situations? Would the plan result in an unbiased sample?

   a. Complaints have been made about students not showering after physical education class at a high school. The administration is trying to decide whether to begin requiring showering. They divide the set of interested people into three groups: teachers, students, and parents. A random sample is selected from each group and asked to fill out a brief questionnaire about the need to shower after P.E. class.
   b. The managers of a large shopping mall are trying to decide whether to add another information booth to the concourse. They hire a student to stand by one door and ask the next 100 people who enter the mall whether they would like to see another information booth added to the mall and, if so, where.
   c. The managers of the new cafeteria on campus want to measure student satisfaction with the new facilities and food offerings. They decide to pass out questionnaires to students who go through the line between the times 8:15 and 8:30, 11:15 and 11:30, 2:15 and 2:30, and 5:15 and 5:30, and ask them to complete the questionnaires. They leave a box at the exit door for completed questionnaires.
   d. Three users of the Panther, a new car, reported problems with the trunk lid popping up while driving. So the car manufacturer test-auto drove the next 30 cars leaving the factory and found that the trunk lid did not pop open. It decided that the problem did not need to be addressed any further.
   e. A bookstore is contemplating putting in a coffee bar to attract customers. For three days (December 21–23), they ask each customer who buys a book whether he or she would like to see a coffee bar added to the store.

7. Give a new example of each situation described.

   a. a situation in which it would be better to poll the population rather than select a sample
   b. a situation in which a systematic sample would be a reasonable choice
   c. a situation in which a cluster sample would be easy to use yet still be reasonable

8. Is each given number a population parameter or a sample statistic? Explain.

   a. 27% of all residents of the United States older than 25 have a bachelor's degree or higher.
   b. 85% of college students polled said they experience stress when they take exams.
   c. In a fourth-grade class, more than half the students play musical instruments.

continue

Notes

Notes

d. A recent survey reported that shoppers spend an average of $32 on each trip to the mall.

e. The median selling price of all homes sold in Atlanta in one recent year was $197,000.

f. 3% of the runners in the marathon were from outside the United States.

9. The National Assessment of Education Progress tests mathematical knowledge in Grades 4, 8, and 12 across the nation. (Information about the test and test items that have been used may be found at http://nces.ed.gov/nationsreportcard/.) Here is one of the test items, first used in eighth grade. Answer it and justify your choice.

▶ A poll is being taken at Baker Junior High School to determine whether to change the school mascot. Which of the following would be the best place to find a sample of students to interview that would be most representative of the entire student body?

a. an algebra class   b. the cafeteria   c. the guidance office

d. a French class   e. the faculty room ◄

10. The *San Diego Union Tribune* chooses a question each week and asks readers to call in their responses. During the presidential campaign in 2008, readers were asked to predict who would win the Republican primary in New Hampshire. Of the 2831 responses, 78% chose Ron Paul, 8% John McCain, 6% Mike Huckabee, 5% Mitt Romney, 2% Rudy Giuliani, and 1% Fred Thompson. John McCain actually won the primary. What might have caused these numbers to be so different from those recorded in national polls?

11. Find an example in a newspaper, news magazine, or news program of a sample used to make predictions about the population. Bring it to class.

Supplementary Learning Exercises for Section 29.2

For the situations described in Supplementary Learning Exercises 1–4, tell what kind of sampling would be most appropriate to the situation. Explain your reasoning.

1. You want to know how many students drive to a certain university.

2. A principal of an elementary school is going to buy pizza for the students in the grade that brings in the most canned goods for the food drive. She wants to make sure that all the students will have toppings that they like.

3. The producer of a TV news show wants to know whether viewers will use the show's new website.

4. Owners of a milk company are concerned that the automatic gallon-measuring machine has become inaccurate. How can they determine whether this is an issue?

5. Describe a situation in which polling a population is better than sampling it. Explain why this would be appropriate.

6. What type of sampling is used in each situation?

a. A grocery store's registers give every tenth customer a telephone number to call to take a survey on customer service.

b. A university wants to find out whether students eat on campus. They asked students in English classes.

c. The police are interested in determining how many people are driving without a driver's license. They set up a roadblock at 6 PM and stop every tenth car until 10 PM.

d. A company obtains the e-mail addresses for all the students at a university. To find out what students think about the company's products, it sends each student a questionnaire about them.

e. To judge the popularity of a new product, a company includes a half-price coupon for it in the daily newspaper.

7. Is each given number a population parameter or a sample statistic? Tell how you know.

    a. Dogs are the favorite pet of 60% of Ms. Allen's fourth-graders.
    b. The newspaper reports that in the Jefferson School District 42% of the students have to repeat Algebra I.
    c. The newspaper surveyed parents in the Jefferson School District and found that 89% of them approved of the district schools.

29.3 Simulating Random Sampling

In different activities given in this chapter, you will want to generate a random sample of a population. Recall that a sample of a population is random if every member of the population has an equal chance of being selected. When statisticians want to find a random sample, they do not usually draw names from a hat, spin a spinner, or toss a die, although these are legitimate ways to sample randomly. To obtain a large sample, these methods would be very time-consuming. Instead, statisticians might use computer simulation software, a table of random numbers, or a computer or calculator with the capability of providing random numbers. In this section, though, you will do a simulation "by hand" to develop a sense of how the results of different trials can vary.

 ACTIVITY 2   Simulating the Estimation of a Fish Population

A fishing lake has only two kinds of fish: bass and catfish. Management wants to know what percent are bass and what percent are catfish.

1. How do two colors simulate this situation? What is the population? (Your instructor will supply the objects that represent the population of fish.)

2. Without looking, draw a sample of 10 fish, record the colors, and replace the fish. Is your sample a simple random sample? Explain.

3. Repeat several times. Does each sample have the same number of fish of each type?

4. Based on the evidence gathered, make a prediction about the percent of the fish that are bass and the percent that are catfish. ●

Although Activity 2 was intended to give a taste of simulating random samples, there is an important point illustrated: *Different samples, even simple random samples from the same population, can give different statistics.*

   The next activity uses random numbers to simulate in an efficient fashion many trials of a voting situation. Any method that produces random numbers can be used: a table of randomly selected digits (or TRSD), a calculator, or a computer program. (Appendix D provides a TRSD and directions for using it.)

 ACTIVITY 3   Is This a Good Way to Sample?

Consider the following problem:

▶ A newspaper reporter asked 5 people at random how each would vote on controversial legislation. Three of the 5 people said they would vote against it. Would this result be unusual if the voting population was evenly split on the legislation and the reporter's sampling procedure did not bias the results? ◀

You may think you already know the answer, but work through the procedure of finding random samples using random numbers. We must make selections so that each selection ends up being either for or against, when the total population is evenly split on the legislation. One way to do this would be to select a sequence of five random digits (for the 5 people) and let each even digit (0, 2, 4, 6, 8) represent *for* and each odd digit (1, 3, 5, 7, 9) represent *against*.

continue

Notes

Use your random numbers and record 5 votes. Keep track of the number of *fors* (0, 2, 4, 6, 8) and the number of *againsts* (1, 3, 5, 7, 9). Do this simulation 30 times.

   a. What fraction of your samples had 3 or more *againsts*? Is it an unusual split?

   b. Suppose now that the newspaper reporter asked 10 people for their opinions on this legislation. Would it be just as unusual to get at least 60% *againsts* when selecting 10 people at random from a population that is evenly split as it was when we selected 5 people at random? ●

Think About …

What do you think would happen in Activity 3 if you took a sample of 50? Would it be just as unusual to get at least 60% against?

As a rule of thumb, *larger samples give statistics that are usually closer to the population parameter than are the statistics from smaller samples.* But is a larger sample always that much better? This question is addressed in Chapter 32. For the time being, you should appreciate the speed with which random numbers can simulate random sampling.

Sometimes we are drawing samples from populations made up of groups that are of different sizes, as with the number of bass and the number of catfish in a particular lake. It is still possible to use random numbers to simulate the drawing of random samples. For example, if we wanted to draw samples of size 20 at random from a population having 56% in one group and 44% in another (such as 56% *for* and 44% *against*), then we could select *pairs* of digits from our random numbers. If a pair of digits is 01 through 56, then the pair represents a *for*. If a pair of digits is 57 or more (and letting the pair 00 represent 100), then it represents an *against*.

TAKE-AWAY MESSAGE . . . A random sample can be found in many ways, such as drawing names out of a hat. But for a large sample or a large number of samples, it is easier to use random numbers from a table, calculator, or computer program. Simulations show that a sample statistic can vary from sample to sample.

Learning Exercises for Section 29.3

1. Jonathan wanted to find a sample of a population that is made up of two groups, one group with 35% of the population and the other with 65%. He used pairs of digits from a random number table in which pairs 35 or less represented the first group and pairs of digits 36 or more represented the second group. But Maria chose to let numbers 34 or less represent the first group and 35 or more the second group. Could both their methods lead to an unbiased sample?

2. How would you use randomly generated numbers to find 30 random numbers from 1 to 500?

3. The following students volunteer for a committee to design a new curriculum in general education at a state university. The committee is to include only three students. Tell how you would choose three students at random by using random numbers.

Anders	Callgood	Halsord	Jones	Nguyen	Waterford
Aspen	Fuertes	Hunterlog	Lee	Smit	Winters
Bolchink	Gonzalez	Ingersoll	McLeod	Tubotchnik	Zbiek

4. Write down what you think would be a random sample of 50 numbers from 1 to 12. Next, use your favorite method (TRSD, calculator, or computer) to get 50 numbers ranging from 1 to 12. Did either sample of numbers have two of the same numbers next to one another? Three of the same numbers next to one another? Was the first set truly random?

5. **a.** Describe a simulation of estimating the fish population in Activity 2 by using random numbers for the drawing of samples of size 10. Assume that 25% of the fish are bass and the remainder are catfish.

continue

b. Use your simulation to get 5 samples. How do the percents in the samples compare with each other and also with those in the population?

6. A fitness club has 407 members. The manager wishes to know whether or not members are satisfied with the club offerings, including layout, classes, and equipment. He wants to interview 10 members, chosen at random. How could he use a simulation to select the 10 members?

7. A cosmetics company wants to know if its new suntan lotion formula will meet with approval from its customers. The company gives away samples at all of its 35 stores to customers who will agree to be surveyed by phone about the product. It collects 1786 phone numbers. Out of the 35 stores from which the company received more than 50 phone numbers, it plans to randomly select 5 stores and then randomly select 5 phone numbers from each of those stores. How could the company use simulations for both of these selections?

Supplementary Learning Exercises for Section 29.3

1. Describe how a random-digit table can be used to model rolling two six-sided dice.

2. **a.** Describe how a random-digit table can be used to model the sum of rolling two six-sided dice.
   b. Is this different from your answer in Supplementary Learning Exercise 1?
   c. In your simulation, what is the probability of getting a sum of 7?

3. Describe how a random-digit table can be used to model rolling two four-sided dice.

4. Five students are to be chosen from the following list:

   | Ron | Javier | Lucinda | Jimmy | Maria | Lam |
   | Mark | Karl | Jorge | Bonnie | Tranh | Teresa |

   a. Create a simulation to choose the five students at random.
   b. Use your simulation to get 10 samples.

5. **a.** Consider flipping a coin 100 times. Try to randomly write out 100 H's and T's without any help from a random generator.
   b. Use a random simulation of 100 flips. Record your results, in order.
   c. What is the longest number of heads in a row in each case?

29.4 Types of Data

Statistical data can be of different types. For example, movies might be described as comedies, dramas, adventures, westerns, or science fiction, but the number of minutes each takes might be of interest for other purposes. This section treats two main types of statistical data—categorical data and measurement data.

In the examples of sampling that we have discussed thus far, the surveys have asked questions such as "Are students interested in buying a raffle ticket for which a Wii game system is the prize?" The three possible answers (*yes, no,* or *maybe*) are not numbers. These responses represent different categories of answers, and so they give **categorical data**. Many times we create categories, such as classifying children in a school as receiving good grades in mathematics, average grades in mathematics, or poor grades in mathematics. We can even use numbers to identify the categories: 3 is good, 2 is average, and 1 is poor. But these numbers are just labels, and we do not expect to do any arithmetic with them. There is some order to the number labels; e.g., 3 is *better than* 2 or 1, but we really can't say how much better, and it would not make sense to say that a 3 is *three times as good* as a 1.

Other survey questions involve taking measurement of some quantity, such as the cost of attending University X for one year. These numbers are called **measurement data**, as opposed to categorical data. In such a case, we use a measurement scale in which the distance between units is constant. You are familiar with the idea of measuring things—a child's height, accuracy at answering basic multiplication facts, how long a lightbulb will last, a steak's weight, a car's value (its selling price), or a baby's temperature. If we measure height in inches, then the inch is our constant unit. If Jean is 44 inches tall, Ming is 48 inches tall, and Phil is 52 inches tall, then the difference between Ming's and Jean's heights is *the same as* the distance between Phil's and Ming's heights. If Mike is 72 inches tall and Paulie is 36 inches tall, then Mike's height is twice Paulie's height. We cannot make these types of comparisons with categorical data.

> Data that come from assigning objects or individuals into categories are called **categorical data**. This can be contrasted with **measurement data**, which can be measured, ordered, and operated on.

In Parts I–III of *Reconceptualizing Mathematics,* we called the things we measure *quantities.* In statistics, quantities are often called *variables.* A **statistical variable** is a property (of objects or people) on which we wish to collect data. But a statistical variable might also give categorical data, as when we ask what a person's favorite color is. So the use of *variable* is wider in statistics than it is in algebra.

When the data are obtained by measuring, the actual measurements are the values of the variable. When the data are categorical, the "values" of the variable are the different categories. The distinction between categorical and measurement variables is useful, as you will see in the next chapter, where you will study graphs of both categorical and measurement variables.

Measurement data sets are common in statistics, but deciding how to measure is not trivial. *Conceiving the object to be measured clearly enough to imagine a way to measure it is often the most difficult part of a statistical study of a new concept.* Curricula for Grades K–8 often have children decide what data should be collected that will answer the questions they may have suggested for a statistics project. Let us look at two more sophisticated examples to illustrate the difficulty of this decision: measuring "stuffness" and measuring parental tolerance of television violence.

How should we measure how much "stuff" something is made of? This isn't the same as how much something weighs, for anything weighs less at the top of Mount Everest or on the moon than it does at sea level, even though it contains just as much "stuff" in both locations. Scientists finally addressed this question by creating the concept of *mass.* Mass has served as an excellent measurement of the amount of "stuff."

How should we measure parents' tolerance of violence on television? This is a difficult question, yet it is illustrative of the types of information a researcher might want to study. We might determine parents' tolerance by giving them a rating scale and asking them to place their tolerance level in categories labeled from 0 to 6, where each number has a corresponding sentence describing what that number indicates. (Note that this scale would give categorical data, not measurement data.) Alternatively, we might measure parents' tolerance of violence by keeping track of the amount of TV time involving violent programming that parents knowingly allow their children to watch.

We use the last example to illustrate a point about measurement data. Notice that the two variables mentioned for measuring parents' tolerance of violence are quite different. The first variable has values 0 through 6 and is a categorical variable. The second variable might also have values 0 through 6, where each value gives the number of hours per day (rounded off to the nearest hour), and is a measurement variable. Yet the 0 to 6 values are conceptually different. In the second case, if we didn't agree to round off to the nearest whole numbers, we could find many values between 2 and 3. Also, we can make comparisons because the number 4 indicates 4 hours, which is twice as much as 2 hours. However, values with the rating-scale variable can be *only* whole number measurements. A rating of 4 is not twice as much as a rating of 2. The rating-scale numbers are separate, unconnected labels for categories, but the amount-of-time numbers represent an underlying continuum of numbers.

TAKE-AWAY MESSAGE . . . Statistical variables can give data of different types: categorical and measurement. Calculations can be performed on measurement data, but only counts can be performed with categorical data.

Learning Exercises for Section 29.4

1. Which type of data, categorical or measurement, is the response to each description?

    a. your favorite kind of ice cream   **b.** score on your last test
    c. kind of computer you prefer   **d.** amount of television you watched last night
    e. month in which you were born

2. A particular computer code uses 0 for a male and 1 for a female. Are the 0, 1 values categorical data or measurement data? Explain.

3. How could you define a way to measure the data for each given variable? Identify the type of data you would collect for each variable. If it is possible to use two ways of measuring a variable, one yielding categorical data and one yielding measurement data, describe both ways of measuring.

    a. attitude toward school
    b. willingness to work hard on a particular type of difficult task
    c. a dance performance
    d. safety of a particular crosswalk
    e. dental health
    f. pain intensity
    g. ability to play tennis

4. Explain how blood pressure can be considered measurement data. Is there a way to present blood pressure categorically?

Supplementary Learning Exercises for Section 29.4

1. Which type of data, categorical or measurement, is the response to each situation? Tell how you know.

    a. the order in which contestants finish a race   **b.** favorite movie
    c. amount of time spent studying   **d.** books read in a month
    e. children in a family

2. Give examples of measurement data for the following contexts. If possible, also give examples of categorical data.

    a. watching sports   **b.** swimming   **c.** shopping   **d.** singing contest

3. A survey has you check 5 for "strongly agree," 4 for "agree somewhat," 3 for "neutral," 2 for "disagree somewhat," and 1 for "strongly disagree." Are the values 1–5 categorical data or measurement data? Tell how you know.

29.5 Conducting a Survey

In this section you will begin a long-term project to conduct a survey of a sample of some population of people. There are several considerations that you will need to make. Of the following questions given, Questions 1–5 can be considered now with what you've learned in this chapter. Questions 6–8, which address how to represent and interpret the data you collect, should wait until after Chapter 30 and possibly Chapter 31.

Considerations include the following questions:

1. What information are you seeking?

2. What is the population you care about?

3. What type of sampling is reasonable, and how are you going to undertake it?

4. What size should your sample be?

5. What questions will you ask?

6. Once you have collected the data, how can you best represent them? (You should be able to justify your choice.)

continue

7. What reasonable interpretations can be made on the basis of the data?

8. How accurate are your results?

Answering Question 1 will depend on your own interests and will lead you to an answer for Question 2. The previous sections will help you with Questions 3 and 4. The example in this section will guide you in formulating your survey questions (Question 5). Information on how to answer Questions 6 and 7 will be provided in Chapters 30 and 31. To answer Question 8, try to provide a confidence interval (using the $\frac{1}{\sqrt{n}}$ rule in Chapter 32) if appropriate.

Your assignment, due at a date set by your instructor, is to conduct a survey and present and interpret the survey data. Here is an example of a survey study of the type you might undertake. We "walk through" Questions 1 through 5 with you.

EXAMPLE OF PREPARING FOR A SURVEY

1. *What information are you seeking?* (Express this information in the form of one or two research questions.)

   Do elementary school teachers approve of the use of calculators in their classrooms? If so, under what circumstances?

2. *What is the population you care about?*

   Elementary school teachers in the United States

3. *What type of sampling is reasonable, and how are you going to do it?*

   It is not possible for this survey to sample all elementary school teachers in the United States. Using only local teachers would be a biased sample. Therefore, the research question must be revised to make this assignment possible. We change the question to the following: Do elementary school teachers *in this locality* approve of the use of calculators in their classrooms? If they do, under what circumstances? The population we care about is also restricted to teachers within the school district. We next try to get as random a sample as possible of elementary school teachers in the local school district.

   Random sampling could require that we obtain the names of all elementary school teachers in the district, but this information is not readily available. If we want to sample teachers from a variety of types of schools and grade levels, we could use stratified sampling. Low- and middle-income levels of the school service area could be included, if that fits the locality. Some schools have Grade 6, whereas others do not.

   We limit our sample to teachers from Grades 1 through 5, and we divide the school district into five geographical areas such that each area represents a particular income level, if possible. We then list all elementary schools in each of the five regions, number the schools in each region, and use random numbers to choose one school from each region.

   Next we call that school and speak with the principal, asking the principal for permission to survey the teachers at that school with a written 1- to 2-minute questionnaire about calculator use. We deliver a set of questionnaires to the school and speak with the secretary about distributing them in mailboxes, placing our pick-up box nearby. We ask about a good time to pick up the questionnaires. If the principal refuses to allow us into the school, we choose another school in the same area at random, and we continue the process until we have 5 schools. Thus, our sample is stratified (by income level), but it also is a cluster sample (because we have taken all teachers at some schools).

4. *What size should your sample be?*

   We expect there to be, on average, 15 teachers at each school teaching Grades 1–5. This allows us to obtain answers from approximately 75 teachers. We expect them to be a representative sample of teachers in the city because the clusters are chosen randomly.

5. *What questions will you ask?*

    1. Do you allow your students to use calculators in math class?
        yes __ no __
        If <u>yes</u>, do you allow calculator use
    2. At any time?
    3. At restricted times? (If so, when?)
        If <u>no</u>, what are your reasons?
    4. Calculators are not available. yes __ no __
    5. Not allowed by _____.
    6. Parents are against it. yes __ no __
    7. I do not believe calculators should be allowed. yes __ no __

    (Of Answers 4–7, if more than one is yes, circle the most important reason.)
    Question 1 allows us to find a percent of teachers that allow calculators. The remaining questions help us interpret our results. ●

This example provides you with the type of questions you should ask. One matter of concern is whether or not teachers will complete and return the questionnaire. Many studies fail because there is an inadequate return rate. Teachers are very busy, so they may not bother responding if they are not convinced that it is worth their while. Thus, you may want to choose a question, population, and sample that will provide a better chance of completing this assignment. You should also promise to share your results with the teachers.

Assignment. With one or two others in your class, carry out a survey on a shared question, addressing the eight considerations given at the beginning of this section.

29.6 Issues for Learning: Sampling

You may be surprised to learn that some research on the idea of sampling has been carried out with children in elementary school. In a study[2] of the development of student understanding of sampling, students in the third, sixth, and ninth grades were asked, "If you were given a sample, what would you have?" Follow-up questions included the following:

1. a. Have you heard of the word "sample" before? Where? What does it mean?
   b. A newsperson on TV says, "In a research study on the weight of Tasmanian Grade 5 children, some researchers interviewed a sample of Grade 5 students in Tasmania." What does the word "sample" mean in this sentence?

2. a. Why do you think the researchers used a sample of Grade 5 children, instead of studying all the Grade 5 students in Tasmania?
   b. Do you think they used a sample of about 10 children? Why or why not? How many children should they choose for their sample? Why?
   c. How should they choose the children for their sample? Why?

    Here are some of the results. The third-graders were categorized as "small samplers" because they typically provided examples of samples, such as a food product, and described a sample as a small amount, an attempt, or a test, and they agreed to a sample size less than 15. When asked, "If you were given a sample, what would you have?" they would say such things as the following:

- "Something free."
- "A little packet of something."
- "You would be trying something."
- "At (the supermarket) . . . they cook something or get it from the shop, and they put it in a little container for you so that you can try it."
- "A blood sample is taking a little bit of blood."

In response to the questions of why the researchers did not test all Grade 5 students, the third-graders suggested that "they didn't have time to test all of them." The children used small numbers to indicate the number of students who should be tested: "Three to six" or "They could have used ten. [Why?] Because it's an even number and I like that number."

When asked how the researchers should choose a sample, common replies were "Go by random. [Why?] Because they're not really worried about what people they pick" or "Teacher might just choose people who've been working well or something." A slightly more advanced answer was "I would choose them in all shapes and sizes, some skinny, some fat. Then I'd compare them to another group and see what was the most average."

Some sixth-graders were also small samplers, but they gave answers similar to the third-graders whose answers were more advanced. The majority of them were categorized as "large samplers" who gave answers to questions about the weights of the Grade 5 students by saying, "[There are] about 10,000 children in Grade 5 in Tasmania and they took about, say, 50 of them and they found out, because 50 is a small portion of 10,000, they just found out the weights from there. [How choose?] Say, take a kid from each school. Take some—just pick a kid from random order. Look up on the computer; don't even know what the person looks like or anything. Pick that person." A more advanced answer to the question of how many Grade 5 children to test was "[Ten?] Probably some more, because if they only used 10, they could all be ones that weighed about the same, and there could be some that weighed less and weighed more in other places. [How many?] Probably about 100 or something. [How choose?] Just choose anybody; just close your eyes and pick them or something."

Only the ninth-graders exhibited real appreciation of the complexity of selecting an appropriate sample. The researchers suggest that teachers and curriculum planners be aware of the fact that students need help in making a transition from understanding a sample as something very small to understanding the variations that exist within any population. They also note how an understanding of these variations translates into the critical need for an appropriate sample size and for the lack of bias, before any reasonable conclusions can be drawn.

29.7 Check Yourself

This chapter contains a brief introduction to several statistical terms and to sampling. You should be able to work problems like those assigned and to meet the following objectives:

1. Distinguish between a population and a sample and between a (population) parameter and a (sample) statistic.

2. Give reasons why we use statistics from samples rather than from the whole population.

3. Understand the value of randomly selected samples and be able to simulate random samples.

4. Distinguish among several types of sampling: simple random, self-selected, convenience, stratified, systematic, and cluster sampling. Recognize which sampling types are likely to be valid and why others contain the strong possibility of bias.

5. Recognize and give examples of categorical data and measurement data.

References for Chapter 29

[1]Jacobs, V. (1997, March). Children's understanding of sampling in surveys. Paper presented at the American Educational Research Association Annual Meeting, Chicago.

[2]Watson, J. M., & Moritz, J. B. (2000). Developing concepts of sampling. *Journal for Research in Mathematics Education, 31,* 44–70. Quotes are from pp. 51–52, 56–58, and 60–61.

Chapter 30

Representing and Interpreting Data with One Variable

In this chapter you will learn how data can be used to make sense of everyday facts and events. At times, you might be interested in only one variable, such as test scores, heights of basketball players, or salaries of teachers. There are several ways to display such data. But simply displaying data often does not provide information sufficient to understand and interpret the data. Measuring the ways in which data cluster around some numbers and the ways in which the data are dispersed is also important. Having and using good data to make predictions can be extremely useful.

Representing and examining data can be made relatively easy with technology. Some calculators, such as the TI-73 from Texas Instruments, can give statistical calculations and representations. There are many applets available online that we use in this chapter to represent statistical information. (Applets are small computer programs designed to undertake limited, well-defined tasks.) Many other types of software may be available to you, such as Excel and Fathom. Excel is part of Microsoft Office and is available on most computers. Fathom, from Key Curriculum Press, is special software for statistical analysis and is available on some university computers. Appendices H–K provide instruction on how to use these tools for comparatively simple tasks. Your instructor may ask that you use a particular one of these tools. Additionally, you can find many useful statistical programs online by searching, e.g., with Google.

30.1 Representing Categorical Data

Graphical representations can be so informative that they often make their way into final reports, newspapers, and magazine articles. In this section we investigate some common graphical representations that are used to graph data with only one variable. We focus first on *categorical* data, i.e., data that can be separated into clearly defined categories and counted. We begin with a typical situation presented to elementary students.

▶ On Thursday morning, Jasmine polled the 27 students in her fourth-grade class to see how they traveled to school. She found that 7 of them took a bus, 8 of them traveled by car, and 12 walked. (She counted herself as one who walked to school.) ◀

This type of data is commonly presented in one of two ways: a **bar graph** (sometimes called a **bar chart**) or a **circle graph** (sometimes called a **pie chart**). Each type of graph is illustrated here. The vertical scale on this bar graph shows the actual counts, but many bar graphs give the percent of the total number instead.

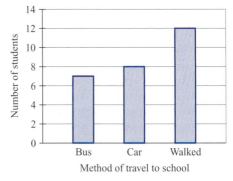

Figure 1 A bar graph showing Jasmine's transportation data.

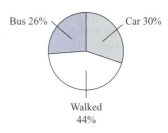

Figure 2 A circle graph showing Jasmine's transportation data.

DISCUSSION 1   Interpreting the Graphs

1. What questions does the bar graph answer? What questions does it leave unanswered?

2. Where did the numbers on the circle graph come from? What do they mean?

3. What questions does the circle graph answer? What questions does it leave unanswered? ●

Although nowadays most bar and circle graphs are made with technology, elementary students are often asked to make graphs without such tools. Making a graph can help one to understand other such graphs.

 ACTIVITY 1   Making a Bar Graph

If you know how many occurrences there are for each value of the variable, you can quickly construct a bar graph by hand using graph paper. Suppose all 420 students in Jasmine's school were asked about transportation. The results from their responses are as follows: 182 walked, 166 rode the bus, and 72 came by car. Draw an accurate bar graph using this information. Compare your graph for the entire school to the graph for Jasmine's class. (Take care in determining the scale you use for the bar chart.) ●

You no doubt noticed that with a large number of cases, deciding on a scale for the vertical axis becomes a necessity. The number of vertical spaces available or desirable, along with the maximum number to be represented, usually dictates the scale. For example, if the 182 walkers were to be represented by 10 vertical spaces, a scale of 20 walkers per space is reasonable. Should the bars touch one another? The usual conventions with categorical data are that the bars do not touch and that the data categories fall along the horizontal axis.

ACTIVITY 2   Making a Circle Graph

To generate our circle graph, we need to know the fraction or portion of the circular region of the graph that should be allocated to each value of our variable. For example, Jasmine found that 12 of her 27 classmates walked to school. That means that $\frac{12}{27}$ of her class are walkers, indicating that $\frac{12}{27}$ of the circle should be allocated to walkers. What fraction of the circle should be assigned to the bus riders? The car riders? Why?

Knowing that $\frac{12}{27}$ of the circle must be assigned to the walkers, we will figure out how to measure this amount. Because there are 360° in a circle, we can draw a sector or wedge of the circle having an angle size that is $\frac{12}{27}$ of 360°. That is, the angle size should be $\frac{12}{27} \times 360° = 160°$.

We can now draw our circle and use a protractor to measure out a sector with an angle of 160° (see the white part of the circle graph in Figure 2).

a. Find the angles for the other two modes of transportation (the bus and the car riders) and complete the circle graph.

b. Now make a circle graph for the whole-school situation. All 420 students in Jasmine's school were asked about their transportation modes: 182 walked, 166 rode the bus, and 72 came by car. ●

Because making a circle graph requires skill with fractions and measuring with a protractor, making a circle graph is an attractive assignment in the intermediate grades. However, *reading* the graph is a different skill.

A variation of a bar graph is the **pictograph,** in which pictures play the role of the bars. Figure 3 shows the transportation data for Jasmine's class in a pictograph. The "bars" are often vertical but may be horizontal, as in Figure 3 on the next page. Pictographs may also use labeled axes. Pictographs are common in news media because they offer an added visual appeal not present with bare-bones bar graphs.

Notes

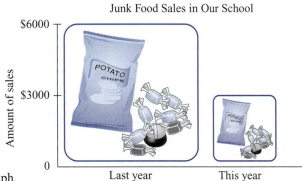

Method of Transportation to Jasmine's School
Walk (12)
Bus (7)
Car (8)

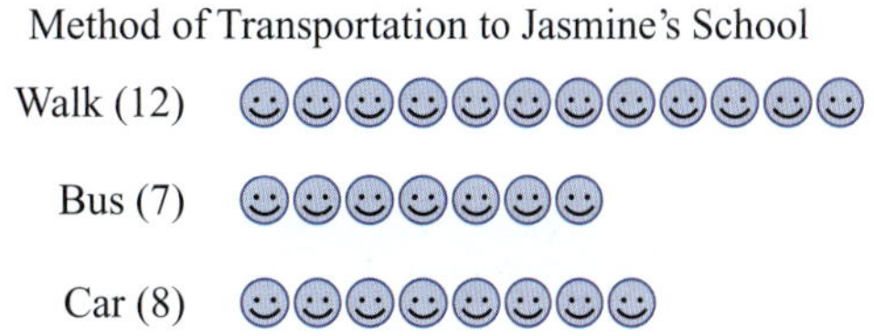

Figure 3 A pictograph for the transportation data.

One potential danger with pictographs in particular (and sometimes with bar graphs) is that the visual impression from the display can be deceptive. What do your eyes tell you about the relative amounts in the pictograph in Figure 4 below?

Junk Food Sales in Our School
$6000
$3000
0
Amount of sales
POTATO
CHIPS
Last year
This year

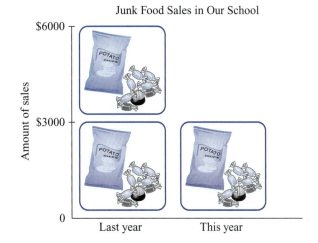

Figure 4 A deceptive pictograph.

Artistic demands may mean that the pictures should be similar (in the geometric sense), but then *both* dimensions in a two-dimensional drawing are changed. Hence, the area—what your eyes sense—is changed more than it should be. For example, in Figure 4 the smaller drawing should be only half as large as the larger drawing, because $3000 is half as large as $6000. But to our eyes it appears to be only one-fourth as large. Be alert for such distortions, especially for pictographs presented to argue a position. Figure 5 shows a more honest pictograph for the junk food data.

Junk Food Sales in Our School
$6000
$3000
0
Amount of sales
Last year
This year

Figure 5 A pictograph for the junk food data.

Think About ...

Why might a school administrator who was proud of the school's program against junk food prefer the graph in Figure 4 to the one in Figure 5?

TAKE-AWAY MESSAGE . . . Data consisting of counts are particularly well suited to making bar graphs (also called bar charts) and circle graphs (also called pie charts). Bar graphs are easy to make using graph paper. Circle graphs are more involved because the angle sizes that correspond to the data must be found. Each type of graph provides information not found in the other graph: The typical bar graph gives counts for the categories, and the typical circle graph gives percents for the categories. Pictographs, a variant of a bar graph, can give visual appeal but can leave incorrect impressions if not drawn with care.

Learning Exercises for Section 30.1

1. Refer to Jasmine's transportation data for her class, her bar graph, and her circle graph to answer the following questions:

   a. Is any information lost when Jasmine summarizes her data in a bar graph? When she constructs a circle graph?

   b. What are the advantages and disadvantages of a circle graph over a bar graph? Is one more effective in telling a story than the other?

2. In 2010 there were an estimated 82,724,222 students over 3 years of age enrolled in schools in the United States. Here is the breakdown of the data (from http://factfinder2.census.gov/):

Nursery and preschool	4,949,546
Kindergarten	4,182,694
Elementary school (Grades 1–8)	32,905,277
High school (Grades 9–12)	17,235,496
College or graduate school	23,451,209

   a. Make a circle graph using these data.

   b. Make a bar graph of these data. How will you mark your axes?

   c. Do any of these numbers surprise you? If so, which?

3. Make a bar graph of the following information given for the coastal water temperature (in degrees Fahrenheit) for Myrtle Beach, SC. (Place the months on the bottom axis.)

January	48.0	May	72.5	September	80.0
February	50.0	June	78.5	October	69.5
March	55.0	July	82.0	November	61.0
April	64.0	August	82.5	December	53.0

   Source: NODC Coastal Water Temperature Guides, August 8, 2012, http://www.nodc.noaa.gov/dsdt/cwtg/satl.html.

4. Using technology in some form, make a circle graph representing the following given information about transportation in Jasmine's class: 12 students walked to school, 7 took a bus, and 8 rode by car. Compare the resulting circle graph with the one you made in Activity 2, part (a).

5. Using technology in some form, make a circle graph for the transportation data given for Jasmine's whole school: 182 (walk), 166 (bus), and 72 (car). Compare it to the one you made for Activity 2, part (b). Use the two circle graphs for Learning Exercises 4 and 5 to see how the data for Jasmine's class compare to the data for the whole school.

6. The following is a problem from NAEP given to eight-graders.[1] Of the student responses, 79% were correct. What kinds of thinking are involved in answering the question?

   ▶ The given pie chart shows the portion of time Pat spent on homework in each subject last week. If Pat spent 2 hours on mathematics, about how many hours did Pat spend on homework altogether?

   a. 4   **b.** 8   **c.** 12   **d.** 16 ◀

   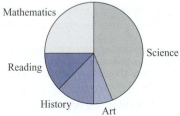

7. **a.** Consider the pictograph below for school enrollments in a rapidly growing district. How is it deceiving?

School Enrollments in Our District

Ten years ago   Now
(8040 children)   (16,420 children)

   b. Why might supporters of a large school bond find the graph in part (a) attractive?

   c. Sketch a less deceptive pictograph for the data in the graph in part (a).

8. **a.** An anti-tax group presents the graph below to argue against a tax increase. How is the graph deceptive?

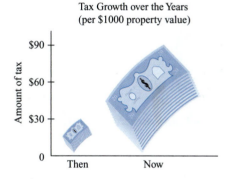

   b. Sketch a less deceptive pictograph for the data in the graph in part (a).

   c. Bar graphs may also be deceptive in the same way as the pictograph for part (a). What makes the following bar graph deceptive?

   d. Be alert when scales are altered. Sometimes changes in scale can leave an incorrect impression. What is exaggerated in the following bar graph by having the tax scale not start at $0?

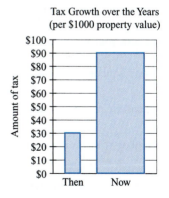

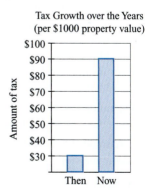

9. Two sixth-graders' circle graphs are shown on the next page. They were asked to choose a career, find out what the salary would be for that career, decide how they would spend the money, and illustrate this information on a circle graph. On a scale of 1–10, how would you grade each graph? Justify your decisions.

continue

Notes

a.

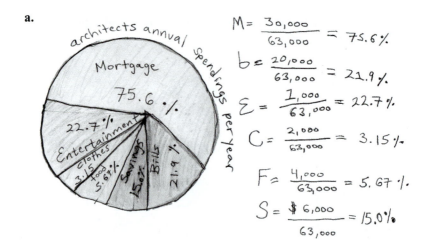

$$M = \frac{30,000}{63,000} = 75.6\%$$

$$b = \frac{20,000}{63,000} = 21.9\%$$

$$E = \frac{1,000}{63,000} = 22.7\%$$

$$C = \frac{2,000}{63,000} = 3.15\%$$

$$F = \frac{4,000}{63,000} = 5.67\%$$

$$S = \frac{\$6,000}{63,000} = 15.0\%$$

b.

Software Design Engineer: $79,000

$$Bills = \frac{28,440}{79,000}$$

$$Mortgage = \frac{23700}{79,000}$$

$$Food = \frac{1900}{79,000}$$

$$Clothes = \frac{7900}{79,000}$$

$$Savings/Extra = \frac{7900}{79,000}$$

$$Entertainment = \frac{3160}{79,000}$$

Supplementary Learning Exercises for Section 30.1

1. In the school that Jasmine's cousin attends, 90 children travel to school by bus, 45 children get to school by car, and 15 walk to school.

   a. Make a bar graph for those data by hand. Use percents on the vertical scale.
   b. Make a circle graph for those data by hand. Use a protractor for accuracy.
   c. Make a nondeceptive pictograph for those data. Let your basic symbol represent 15 children.
   d. Either Jasmine's school (Activity 1) or her cousin's school is in a rural area. Which one is? Give your reasoning.

2. Use technology in some form to make a bar graph and a circle graph for the data in Supplementary Learning Exercise 1.

3. Here are some estimated fertility rates (average number of births per woman) in 2008 for a few countries:

Afghanistan	6.58	Hong Kong	1.00	China (mainland)	1.77
Germany	1.41	Mexico	2.37	Philippines	3.00
United States	2.10				

   Source: The World Factbook, https://www.cia.gov/library/, May 2008.

a. Why would making a pictograph for the data be awkward?

b. Why is a circle graph not appropriate for these data?

c. Make a bar graph for the data.

d. What are some possible long-term implications for a low fertility rate such as Hong Kong's? For a high fertility rate such as Afghanistan's?

4. At a neighborhood block party residents compared the grocery stores at which they shop. The results are listed to the right.

Star Store	25
Global Grocer	37
McGee's Market	16
Kenny's Discounts	28

a. Make a bar graph of the data.

b. Make a circle graph of the data.

c. Which graph better shows the differences?

5. Make a circle graph for the following data: A high school competes in 10 different days of track events. Below are the total numbers of medals that Kensington High School won.

1st Place	12
2nd Place	28
3rd Place	13
4th Place	32

6. Make a circle graph for the following data: An employer polled her employees to find out how they traveled to work.

Drive alone	356
Carpool	123
Bus	160
Train	243
Walk/bike	22

7. The following item is adapted from a National Assessment of Educational Progress examination:[1] There are 1200 students enrolled in Adams Middle School. According to the graph at the right, how many of these students participate in sports?

a. 380   b. 456   c. 760   d. 820   e. 1162

54% of the students chose the right answer. What further questions of a different type could you ask about the situation?

Student Participation in Activities at Adams Middle School

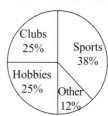

30.2 Representing and Interpreting Measurement Data

Bar graphs and circle graphs are excellent at providing information about categorical data for one variable. In this section we explore two ways of presenting measurement data for one variable: stem-and-leaf plots and histograms. Each type of representation provides different kinds of information. Continue to think about how different ways of displaying data fit various kinds of data.

Consider the following situation:

▶ Ms. Santos has collected data by asking her 23 students to estimate the length in feet of the classroom they were in. (It was 43 ft long.) These are the results of her poll: 35, 30, 40, 32, 46, 45, 54, 36, 45, 30, 55, 40, 40, 60, 40, 44, 42, 30, 24, 60, 42, 42, 40. ◀

In this situation the variable, which is length, is numerical, unlike the variable in the modes-of-transportation examples in Section 30.1. Unlike Jasmine's, Ms. Santos's data do not naturally fall into a few concise categories. Consequently, it would be inappropriate to attempt to represent her data using a bar graph or circle graph. To graph her *measurement* data, she needs another method.

One method that is sometimes used to represent small measurement data sets is a **stem-and-leaf plot.** The next activity will guide you through making this type of graph.

 **ACTIVITY 3**   How Long Is the Room?

1. First, make a vertical list of the *stems*. This list of stems usually consists of the left-most digits of the numbers in our data set. For example, because all the numbers in Professor Santos's data set fall between 24 and 60, our stem will consist of the numbers 2, 3, 4, 5, and 6, which stand for 20, 30, 40, 50, and 60, respectively. This list is usually written as follows:

$$
\begin{array}{c|}
2 \\
3 \\
4 \\
5 \\
6 \\
\end{array}
$$

Be sure to include all the stems that fall between the smallest and largest data values in the data set, even if there are no data values for that stem. For example, if there were no estimates in the 30s for Professor Santos's data, you would still include the stem, 3.

2. Once we have the stems listed, we are ready to add the *leaves*. For each value in our data set, we place the digit in the 1s place in the row that corresponds to the digit in the 10s place. For example, for the first value, 35, we place a 5 in the row that begins with the number 3:

$$
\begin{array}{c|l}
2 & \\
3 & 5 \\
4 & \\
5 & \\
6 & \\
\end{array}
$$

3. The next four values, 30, 40, 32, and 46, would appear as follows. These leaves should be aligned vertically so that the relative lengths of the rows will be obvious:

$$
\begin{array}{c|l}
2 & \\
3 & 5\ 0\ 2 \\
4 & 0\ 6 \\
5 & \\
6 & \\
\end{array}
$$

Finish this step by completing each row using the remaining values: 45, 54, 36, 45, 30, 55, 40, 40, 60, 40, 44, 42, 30, 24, 60, 42, 42, and 40.

4. The last step is to arrange each row in numerical order. The row beginning with the "3" has been rearranged for you. Complete this step by reordering the remaining rows:

$$
\begin{array}{c|l}
2 & \\
3 & 0\ 0\ 0\ 2\ 5\ 6 \\
4 & \\
5 & \\
6 & \\
\end{array}
$$

Think About …

Look back at the stem-and-leaf plot you just completed and think about what type of information it conveys. What do the relative lengths of the rows tell you?

Notes

Answer the questions posed here for a fifth-grade class. Can you think of another type of data set that would be interesting for fifth-graders to graph using a stem-and-leaf plot?

Name _____ Date _____ Time _____

STUDY LINK 6·3 | **Reading a Stem-and-Leaf Plot**

Use the information below to answer the questions.

SRB 118 119

Jamal was growing sunflowers. After eight weeks, he measured the height of his sunflowers in inches. He recorded the heights in the stem-and-leaf plot below.

1. How tall is the tallest sunflower? _____ in.

Which landmark is the height of the tallest flower? Circle its name.

minimum   mode

maximum   mean

2. How many sunflowers did Jamal measure? _____ sunflowers.

3. What is the mode for his measurements? _____ in.

4. Explain how to find the median for his measurements.

_____

_____

_____

_____

_____

Height of Sunflowers (inches)

Stems (10s)	Leaves (1s)
3	9 1
4	7 6 9 2 9
5	2 3 3 5 2 8 7 3
6	5 3 4
7	3

Practice

5. $62 * 53 =$ _____

6.
$$6,711 \\ -\ 4,140$$

7. $22\overline{)398} \rightarrow$ _____

8. $725 * 90 =$ _____

Statistics, Data Analysis, and Probability 1.0   **161**

You'll notice that by constructing the stem-and-leaf plot for Professor Santo's data set, we have inadvertently introduced categories—the 20s, the 30s, the 40s, the 50s, and the 60s, and we have grouped our data accordingly. In fact, the shape of the stem-and-leaf plot we have con-structed in Activity 3 is very similar to the shape of the graph in Figure 6 below.

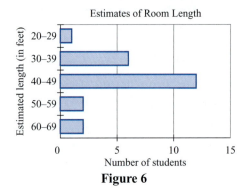

Figure 6

This graph in Figure 6 is merely a bar graph that has been turned on its side. By reorienting our graph as in Figure 7, we get back to the form of the bar graph we saw in Section 30.1.

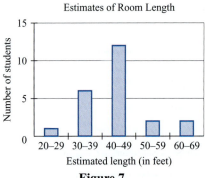

Figure 7

When dealing with measurement data, however, we often alter our bar graphs so that the bars fill up the horizontal scale (unless there are no values for a particular space). In this case, we draw the previous bar graph as in Figure 8.

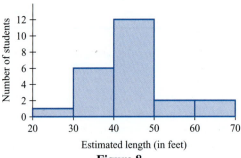

Figure 8

The modified bar chart in Figure 8 is called a **histogram.** *Histo* means cell, and you'll notice that the data have been put into "cells." The data in the 20 to 30 bar contain data greater than or equal to 20 and less than 30 ($20 \leq$ data value < 30), etc., imitating the stem-and-leaf plot categories. [Some books and software use a different convention to define the cells. For example, the cell could be defined by not including the 20 but including the 30 ($20 <$ data value ≤ 30). Thus, you should be careful when comparing histograms for the same data but created by different software.]

Think About …

Notice that there are no spaces between the bars in Figure 8. Why do you think this is? Why are there spaces between bars in a bar chart?

Some elementary school textbooks show histograms with a space between the bars, so it is not even a hard-and-fast rule that the bars touch. When the measurement data are all whole numbers, labeling for the cells such as the 20–29, 30–39, etc., is also common. The vertical scale here shows the counts, or frequencies, but most software can show the percent of all the cases on the vertical scale. Percents are handy if there are especially large numbers of cases.

The value of a histogram is that it gives an overall impression of the data. This impression is very valuable with large data sets, because the "forest is sometimes lost among all the trees." A histogram can give a sense of where the data values cluster, how they are spread out, and whether there are just a few extremely large or small values, called **outliers.**

Stem-and-leaf plots give a natural lead-in to histograms, but the resulting cell widths (or intervals) do not have to be 10 units. Other cell intervals are possible.

ACTIVITY 4   Making a Histogram Using Graph Paper

a. Using the data from Ms. Santos's class, choose intervals of 5 feet and make a histogram with cell intervals starting at 20, 25, 30, 35, 40, etc. How do these intervals compare with those in the histogram in Figure 8? Does the new histogram leave the same impression as the earlier one?

b. Next, make a histogram using a smaller interval, e.g., starting the cells with the even numbers, 20, 22, etc., through 68. Compare it with the histograms you made using larger intervals. Which gives a better idea of the overall pattern of the data? ●

The histograms resulting from the different cell widths illustrate the importance of choosing a good cell width so that the histogram gives a "feel" for the distribution of the data. There are various guidelines, such as having the number of cells be roughly the square root of the number of data values.[2] In a particular field of study, there is often a recommendation for the ratio of the lengths of the two axes. Neither of these guidelines usually arises in Grades K–8, however, but having a title and clearly labeled axes (with units mentioned) are naturally emphasized.

TAKE-AWAY MESSAGE . . . Stem-and-leaf plots can be useful in looking at measurement data. For small data sets, stem-and-leaf plots can be made quite easily with pencil and paper. If a stem-and-leaf plot is turned on its side, it looks something like a bar graph. A histogram can be viewed as a special type of bar graph in which the intervals are all the same width, and the bars usually touch to indicate that the entire interval is being represented by the bar over that interval. A histogram provides information on the shape of the distribution of the data. Changing the interval width can affect the shape of the histogram, but generally the overall shape remains somewhat the same.

Learning Exercises for Section 30.2

1. Refer back to Ms. Santos's estimation data, stem-and-leaf plot, and histogram to answer the following questions:

   a. Was any information lost when you made a stem-and-leaf plot of Ms. Santos's data? When you made a histogram of the data?

   b. What are the advantages and disadvantages of a histogram over a stem-and-leaf plot? Is one more effective in telling a story than the other?

Notes

2. In 2010 the state of Illinois had an estimated 4,752,853 households. Their incomes are reflected in this table (from the website http://factfinder2.census.gov/):

Income	Number of Households
Less than $10,000	337,453
$10,000 to $14,999	232,889
$15,000 to $24,999	522,814
$25,000 to $34,999	484,791
$35,000 to $49,999	655,894
$50,000 to $74,999	874,525
$75,000 to $99,999	598,860
$100,000 to $149,999	613,118
$150,000 to $199,999	223,384
$200,000 or more	209,125

    a. What is problematic in using a histogram to represent these data?

    b. Find a way to display the data in a graph or chart.

3. Which types of graphs (including stem-and-leaf) could be made with the following data sets? Justify your answers.

    a. the responses of 20 five-year-old children to the question "What is your favorite flavor of ice cream?"

    b. the heights of 20 five-year-old children

    c. the eye colors of students in your class

    d. the brands of toothpaste used by 500 dentists

    e. the growth each week of a plant

    f. the annual salaries of teachers in your school district

    g. the number of grams of chocolate in the 12 best-selling candy bars in the United States

    h. your weight measured every day for three months

For Learning Exercises 4–7, go to the Data Sets Folder and find the file called *60 Students,* or use the *60 Students* data printed in Appendix E. Cut and paste the data from the electronic form if you can (see the companion website at www.whfreeman.com/reconceptmath2e). Appendices H–K include information on different technologies.

4. Use data from the file *60 Students* to make the following graphs:

    a. a stem-and-leaf plot of hours worked per week

    b. a histogram of the GPAs of Republican students by using intervals of width two-tenths of a point per column (the first column should go from 0.6 to 0.8)

5. Use technology in some form to make a circle graph for political parties in the file *60 Students.* You will have to make counts for each political party.

6. Use technology in some form to make a histogram of hours worked per week based on *60 Students.*

7. Use technology in some form to make a histogram of hours of volunteer work based on *60 Students.*

8. Discuss the findings for Learning Exercises 6 and 7. From your own experience, do these findings make sense?

Supplementary Learning Exercises for Section 30.2

1. Here are the test scores for a class of 24 students:

75	81	78	69	81	69	87	91	97	81	62	78
54	77	83	85	84	93	79	86	82	85	81	73

    a. Make a stem-and-leaf plot for the test scores.
    b. Make a histogram for the test scores.
    c. What information is lost in the histogram that is contained in the stem-and-leaf plot?

2. Researchers at the Purdue University School of Veterinary Medicine deposited 25 female and 10 male fleas in the fur of a cat, in order to study the egg production of the flea. The following table gives the number of eggs produced by the fleas over 27 consecutive days:

Fleas Data Set

Day	Number of eggs	Day	Number of eggs	Day	Number of eggs
1	436	10	704	19	475
2	495	11	590	20	435
3	575	12	411	21	523
4	444	13	547	22	390
5	754	14	584	23	425
6	915	15	550	24	415
7	945	16	487	25	450
8	655	17	585	26	395
9	782	18	549	27	405

Source: Moore D., & McCabe G., *Introduction to the Practice of Statistics*, p. 27. New York: W. H. Freeman, 2009.

    a. Questions (i–iv) will help you decide on how the stem-and-leaf plot for the number of eggs should be drawn. How many rows will there be if

      i. the first digit is the stem?
      ii. the stem is split into two, one with leaves 0 to 4 and the other with leaves 5 to 9?
      iii. the first two digits are used as the stem?
      iv. Would the stem-and-leaf plot be more informative if the last digit was omitted?

    b. Draw a stem-and-leaf plot to illustrate the data.
    c. Describe the shape of the distribution of eggs and discuss any unusual features.
    d. What is the median of the data set? (Answer after Section 30.3.)

3. Which type(s) of graphs (including stem-and-leaf plots) could be made with the following data sets? Explain your reasoning.

    a. favorite class of 350 high school sophomores
    b. the total cost of one year's worth of textbooks for college freshmen at a certain college
    c. the name and number of copies sold of the 10 top bestsellers
    d. the weights of cats in a cat show
    e. brands of sneakers bought for the last four Summer Olympic Games
    f. the number of students enrolled in various dance classes

4. Make a stem-and-leaf plot for the following data:

22	37	30	62	32	26	31	27	26	27	30	26	29	24
26	81	42	29	33	35	45	49	39	26	37	42	41	39
33	30	74	33	49	38	61	21	41	30	24	34	60	24
61	26	35	34	34	26	47	41	27	39	30	34	38	29
35	31	41	38	28	27	31	37	45	24	34	33	28	37

Notes

5. Use technology to draw a circle graph showing the educational attainment of 18- to 64-year-olds in the United States:

Ages	Less than 9th grade	9th–12th grade, no diploma	High school diploma or equivalent	Some college, no Associate or 4-year degree	Associate degree	Bachelor's degree	Master's degree	Professional degree (such as DDS or JD)	Doctorate (such as PhD or EdD)
18–64	7701	17,431	56,635	37,266	16,282	34,928	11,869	2567	2010

Source: U.S. Census Bureau, *Current Population Survey: Annual Social and Economic Supplement, 2007.*
Data taken from http://www.census.gov/, May 2008.

30.3 Examining the Spread of Data

Summary information about a set of measurement data is often a more convenient way to convey information about the data than sharing the complete data set. Sometimes it is useful to know how spread out the data are. The **range** is the simplest statistic for measuring the spread. One type of graph, called a **box-and-whiskers graph** or sometimes a **box plot**, also shows how data are spread out. This section shows how to determine the range of a set of data and how to make and to interpret a box plot.

A simple way to gain some insight into the spread of a data set is to find the largest and smallest values of the data set, which can then be used to find the *range* of the data.

> The **range** of a data set is the difference between the largest and smallest values in the data set.

EXAMPLE 1

The range of the set of data 4, 12, 17, 7 is $17 - 4 = 13$.
The range of the scores 4, 4, 17, 17 is also $17 - 4 = 13$. ●

Knowing the range of the data gives us some idea of how the data are distributed. Often the range by itself does not convey as much information as we usually need, as Discussion 2 illustrates.

 DISCUSSION 2   Which Class Has Better Scores?

Consider the following situation:

▶ Ms. Kim and Ms. Jackson are concerned about the wide range in abilities of the students in their third-grade classes. Ms. Kim notes that on the school district third-grade mathematics test, her students scored between 22 and 98. Ms. Jackson's students scored between 49 and 100. ◀

a. Ms. Jackson claims that her students' test scores are more spread out than those of Ms. Kim's students. Could that be possible? Why or why not?

b. Which class would you say did better on the test, Ms. Kim's or Ms. Jackson's class? Why?

c. Suppose you found out that 40% of Ms. Jackson's students scored 70 or fewer points on the test, whereas 65% of Ms. Kim's class scored more than 75 points on the test. Would this information change your response to the previous question? ●

In the example of Ms. Kim's and Ms. Jackson's students' test scores, knowing the largest and smallest values of the data sets, and hence the range, gives us a very limited view of the data. It is often helpful to know the value for which a certain percentage of our data will fall below (or above). This is the idea behind **percentiles.** For example, in part (c) of Discussion 2, 70 would be the 40th percentile score for Ms. Jackson's class, because 40% of the students in Ms. Jackson's class scored at or below 70. In other words, we are separating the scores in Ms. Jackson's class into two groups. One group contains the scores less than or equal to 70, and the other group contains the scores greater than 70. Exactly 40% of the scores are in the first group.

> For a set of numbers, the ***n*th percentile** is the value p that separates the lowest *n percent* of the values from the rest. That is, *n percent* of the values are less than or equal to p.

(Sometimes the *n*th percentile p is defined as the percent of values below p. For very large data sets, which is the only time percentiles are usually presented, this difference in the definition does not really matter.)

EXAMPLE 2

The 60th percentile cuts off the lowest 60% of the data. For example, for the weights of newly born babies in Candonia, if the 60th percentile is 118 ounces, then 60% of the newly born babies weigh less than or equal to 118 ounces. ●

 ## DISCUSSION 3   When Is 40 Not 40?

In Ms. Jackson's class, what is the 40th percentile? Is the 40th percentile score for Ms. Jackson's class the same as the 40th percentile score for Ms. Kim's class? Why or why not? ●

> *Think About …*
>
> A teacher is talking to a parent whose child scored at the 65th percentile on a recent standardized test. The parent is concerned that the child is doing poorly. How could the teacher reply?

Of particular interest is the 50th percentile, also called the *median,* which separates the bottom half of the data from the top half of the data. Although the definition of percentile sounds precise, in truth it is not. For example, the 50th percentile p of some set of numbers should be a number such that 50% of the values are less than or equal to p. But for the set of values 2, 4, 5, 6, 8, 9, we could say that p is 5.2, or 5.8, or in fact any number between 5 or 6. Thus, we must have a more precise definition to ensure that there is just one median. Here is a procedure for finding the 50th percentile, the median.

> The **median,** i.e., the 50th percentile of a set of values, is the middle value in the ordered set of values. The median of an even number of values is the arithmetic average of the two middle numbers of the ordered set of values. The median of an odd number of values is the middle number in the ordered set of values.

EXAMPLE 3

a. Consider the set of numbers 2, 10, 5, 18, 7. We first order the numbers as 2, 5, 7, 10, 18. There is an odd number of values in the set, so the median is the middle number, 7.
b. Consider the set of numbers 2, 10, 5, 18, 7, 3. The ordered set is 2, 3, 5, 7, 10, 18. This set has an even number of values so we must find the average of the two middle numbers, 5 and 7. 5 + 7 = 12 and 12 ÷ 2 is 6, so 6 is the median of the set. Notice that the number 6 is not in the set. If the numbers had been 2, 3, 5, 5, 10, 18, the median would be 5, which *is* in the set. ●

Two additional percentiles are also important values. They are called *quartiles* and are especially helpful in providing insight into the distribution of data.

> The 25th percentile and 75th percentile scores are called **quartiles**. They are the scores we get by dividing up the data into four quarters (hence the name *quartile*). The 25th and 75th percentiles are often referred to as the **first** and **third quartiles,** respectively. (The median is the **second quartile**.) The first quartile can be thought of as the median of the bottom half of the values, and the third quartile can be thought of as the median of the top half of the values. The **interquartile range (IQR)** is the distance between the first and third quartile values.

The IQR gives a measure of the spread, although the individual percentiles do not.

Think About …

A data set of test scores has a first quartile score of 27, a median of 30, and a third quartile score of 36. What does each of the numbers 27, 30, and 36 mean? What is the interquartile range? What would the 50th percentile be?

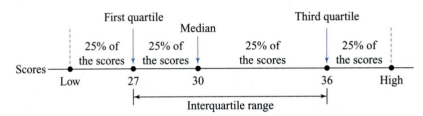

Finding the IQR gives the range of the middle 50% percent of all the scores, and hence the IQR is a measure of the spread of the data. In the case above, the IQR is $36 - 27 = 9$. Even though the first quartile is a single score, *first quartile* is often used to refer to all the scores in the lowest 25%, and similarly for the other quartile scores. The context usually makes clear whether *quartile* refers to the single value or to all the values in that quarter of the data.

Determining the quartile scores of a data set consists of finding the 25th, 50th, and 75th percentile scores (which are the first quartile, median, and third quartile, respectively). This process is illustrated in Activity 5.

ACTIVITY 5   Finding Quartiles

The following exam scores are from a class of 30 students:

65, 64, 46, 38, 58, 44, 65, 60, 50, 70, 55, 44, 68, 67, 66,
66, 81, 68, 51, 51, 75, 53, 47, 62, 59, 25, 78, 49, 77, 49

1. The first thing to do is order the data from smallest to largest. Making a stem-and-leaf plot can help order the data, as shown below.

$$
\begin{array}{r|l}
2 & 5 \\
3 & 8 \\
4 & 4\ 4\ 6\ 7\ 9\ 9 \\
5 & 0\ 1\ 1\ 3\ 5\ 8\ 9 \\
6 & 0\ 2\ 4\ 5\ 5\ 6\ 6\ 7\ 8\ 8 \\
7 & 0\ 5\ 7\ 8 \\
8 & 1 \\
\end{array}
$$

2. To find the median, we look for the value that is at the middle of our ordered data. If there is an odd number of data entries, there will be one data entry that has an equal number of data entries preceding and following it. If there is an even number of entries, we find the average of the two middle values. In this case, there are 30 scores, so we take the average of the 15th and 16th scores. That is, find the median. What is it?

3. Find the first and third quartile scores by finding the medians of the lower and upper halves of the data. Note that on either side of the median there are 15 scores.

4. Find the interquartile range for the data. ●

When presenting data, it is often helpful to give what is called the **five-number summary.** This summary consists of the smallest data value, the first quartile score, the median, the third quartile score, and the largest data value listed in order from smallest to largest. For the exam scores in Activity 5, the five-number summary is 25, 49, 59.5, 67, and 81.

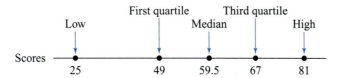

About half of our data values fall between the first and third quartile scores, so the smaller the interquartile range is, the closer these data values are together. Furthermore, statisticians often use the interquartile range to make judgments about what data values in a data set are extreme and thus warrant closer scrutiny, because extreme values might be errors or just not good representatives of the data set. Values that are either less than the first quartile score or greater than the third quartile score by more than one and a half times the length of the interquartile range are judged to be extreme and are labeled as *outliers*.

> An **outlier** is a data value such that the
>
> $$\text{data value} < \text{first quartile score} - 1\tfrac{1}{2} \times \text{IQR} \qquad or$$
>
> $$\text{data value} > \text{third quartile score} + 1\tfrac{1}{2} \times \text{IQR}$$
>
> where IQR is the interquartile range.

EXAMPLE 4

Suppose a data set has a first quartile value of 28, a median of 30, and a third quartile value of 36. Then the interquartile range, IQR, is $36 - 28 = 8$. Consider the following data value: $28 - 1\tfrac{1}{2} \times \text{IQR} = 28 - 12 = 16$. Thus, numbers less than 16 would be considered outliers. Next, consider $36 + 1\tfrac{1}{2} \times \text{IQR} = 36 + 12$ is 48. Numbers larger than 48 would also be considered outliers. ●

> *Think About …*
>
> Are there any outliers in the set of exam scores in Activity 5? Suppose the first score had been 20 instead of 25. Would 20 be an outlier? Why or why not? Which number(s) in the five-number summary for this revised set of scores would be affected? What is the possible significance of an outlier?

By using the interquartile range to make decisions about extreme data values, we are assuming that data values located in the center of the distribution are more reliable or more representative of the variable we are measuring than values located at the peripheries, and thus these data values should be given more weight in our analysis.

Notes

Outliers are not used in a five-number summary. If the first score had been 20 in the exam data, 20 would be an outlier. We would then use the next value, 38, as the first number in our five-number summary.

Quartile scores are often depicted graphically through the use of a **box-and-whiskers plot** or more simply a **box plot.** Box plots are made using the five-number summary. Figure 9 below shows a box plot of the data set consisting of the 30 exam scores with the five-number summary 25, 49, 59.5, 67, 81.

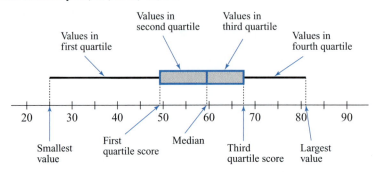

Figure 9 Box plot information, with IQR being the length of shaded box and no outliers.

The box plot takes a data set and graphically shows how the quartile scores break the data values into four parts. Because the box (shaded in Figure 9) is bounded by the first and third quartile scores, its length is the interquartile range. The two sections of the box contain the data values from the second and third quartiles of our ordered data set. The horizontal lines, or "whiskers," contain data values that are in the first and fourth quartiles. However, they may not include all the data values from these quartiles, because they extend only to include data values that are not outliers. The number of cases involved may or may not be displayed, depending on preference or the software used. For our test scores, $n = 30$ might be included.

Our data set in Activity 5 did not have any outliers, so the whiskers extend all the way to the smallest and largest values in our data set. You can verify that there are no outliers by using the box in Figure 9 to estimate how long one-and-a-half times the interquartile range would be, and then seeing if the smallest and largest values of the data set are below the first quartile score or above the third quartile score by more than this length. Software programs often display outliers as single points. (See Learning Exercises 6 and 7.)

DISCUSSION 4   Information from a Box Plot

a. What information does the box plot in Figure 9 convey about the data set it is representing? Think about specific questions the box plot might answer. What questions does it leave unanswered?

b. What does Figure 9 suggest about how the data are distributed, or spread out? Summarize in your own words what you think about the data distribution.

c. Because the whisker corresponding to the first quartile was longer than the whisker for the fourth quartile, Susan exclaimed, "There are a lot more scores below 49.5 than above 67.5." What is Susan likely to be thinking? How would you respond? ●

Although a quarter of all the cases is represented in each of the four pieces of a box plot, the length of the piece gives some information about the scores represented. A longer piece suggests that the scores may be spread out more than they would be in a shorter piece.

ACTIVITY 6   Making a Box Plot

Go to the data set *60 Students* in Appendix E.

a. Make a stem-and-leaf plot of GPAs for the female students.

b. Find the largest and smallest values.

c. Find the median.

d. Find the first and third quartiles.

e. Are there any outliers? If so, what are they?

f. Give the five-number summary of the data.

g. Make a box plot for this set of GPA data. ●

TAKE-AWAY MESSAGE . . . It is useful to know how spread out the data in a data set are. The range is one measure of spread. Percentiles also show how the data are spread. The *n*th percentile *p* of a set of data is the number for which *n*% of the values are less than or equal to *p*. (With small data sets, some percentiles are not meaningful.)

The 25th percentile of a data set is the value Q_1 for which 25% of the data are less than or equal to Q_1. Q_1 is called the first quartile. The 50th percentile is the median, and the 75th percentile is the third quartile, Q_3. The distance between the first and third quartiles is called the interquartile range. The smallest value, the first quartile, the median, the third quartile, and the largest value form a set of five numbers known as the five-number summary (ignoring outliers). These five values can be used to make a box plot—a graph that provides a visual summary of the information about a set of numbers.

Learning Exercises for Section 30.3

1. a. For which of the following data sets does it make sense to talk about quartile scores? Explain for each case why it does or doesn't make sense.

  i. the set of scores from a spelling test in a fifth-grade classroom
  ii. the pulse rates of 45 runners taken after a $2\frac{1}{2}$-mile run
  iii. the responses of 2000 college applicants to the question "What is your ethnicity?"
  iv. the colors of M&M's in a small bag of M&M's
  v. the number of M&M's in each of 28 small bags of M&M's

 b. Think of four more data sets for which quartile scores make sense.

 c. Think of four more data sets for which quartile scores do not make sense.

 d. What kind of data must we have in our data set so that we can use quartile scores?

2. Ms. Erickson, a principal in a local school, has just received the third-grade students' results from a standardized reading exam. When reviewing the total reading scores for the students in one reading group, she discovers that their scores fall between the 47th and 69th percentiles for all third-graders in the school district. Do you think the students in this group are below average, average, or above average readers? Why?

3. a. Find the five-number summary for the heights in centimeters of the 29 eleven-year-old girls given below.[3] (Try a stem-and-leaf plot to help you order the data.) Use the five-number summary to make a box plot. Are there any outliers?

    135, 146, 153, 154, 139, 131, 149, 137, 143, 147, 141, 136, 154, 151, 155,
    133, 149, 141, 164, 146, 149, 147, 152, 140, 143, 148, 149, 141, 137

 b. Sometimes it is useful to be able to make statements such as "The typical eleven-year-old girl's height usually falls between ____ cm and ____ cm." Do you think the first and third quartile scores from the data set in part (a) could be used as these boundaries? Why or why not?

4. Can you generate a data set with more than 10 data values such that the data values less than or equal to the median actually represent

 a. *exactly* 50% of the data?   **b.** *less than* 50% of the data?
 c. *more than* 50% of the data?

5. Can you generate a data set with more than 10 data values such that its box plot

    a. has no left whisker?   **b.** has no whiskers at all?   **c.** has an outlier?
    d. has a box made of only one rectangle, not two?   **e.** has no box?
    f. For those cases in parts (a–d) that are possible, explain why the data set you created generates a box plot with the desired characteristic. For those cases that are not possible, explain why.

6. Mr. Meyer gave a 100-point mathematics test to his fifth-grade class. He gave the tests to his aide to grade and record on the computer. Later, he generated a box plot of the scores to see how his class did. He got the following graph. Notice that the point on the left represents an outlier.

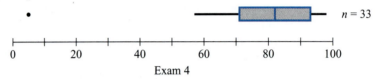

    a. How did Mr. Meyer's students do on the exam?
    b. About how many students scored below 70?
    c. What is the approximate value of the outlier?
    d. Mr. Meyer is shocked that someone received such a low score on the exam. When he checks the grade sheet on the computer, he discovers that the low score belongs to one of his best students in math. He wonders what happened. What do you think happened?

7. Below are two box plots that represent the distribution of the weights in pounds of 40 women and 20 men.

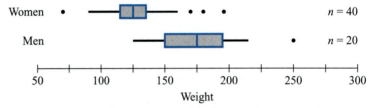

    a. List at least eight questions that can be answered by looking at the two box plots above, and provide answers for your questions. Try to avoid asking the same questions for each box plot—e.g., do not list both "What is the median weight of the males?" and "What is the median weight of the females?"
    b. List at least three questions that cannot be answered by examining these two box plots.
    c. What important points do you think the graph makes?
    d. If we assume that outliers on the right of the box-and-whiskers plots indicate overweight individuals, would it be correct to state that, in general, there are three times as many overweight women as men? Why or why not?

8. Suppose we grouped all the weights of the men and women from Learning Exercise 7 into a single data set and constructed a box plot. Draw a picture of what you think that plot would look like, and explain why it looks the way it does.

9. Examine the following histograms and explain why or why not each could be a histogram that best fits the weights data of the 20 males in Learning Exercise 7:

    a. 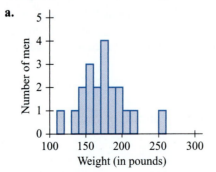   **b.**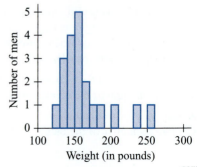

continue

c.

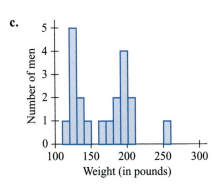

Weight (in pounds)

d.

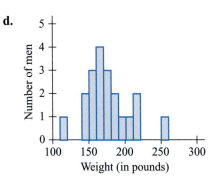

Weight (in pounds)

10. a. Complete the following table for *rules of thumb* (generally true statements) about the graphs we have studied:

	Form of data (measurement or categorical)	Number of variables	Shows actual data	Shows actual counts	Shows percent	Shows shape
Circle graph						
Histogram						
Stem-and-leaf plot						
Bar graph						
Box plot						

b. What do these five types of graphs have in common? Name at least two things. What are some of the differences among these types of graphs? Name at least three differences.

11. Go through the same steps as in Activity 6, but this time with the data on male students from *60 Students* given in Appendix E and in a file at the companion website, www.whfreeman.com/reconceptmath2e. Consider the two box plots together. Are there differences?

12. Go to the data file *World Statistics* in either the Data Sets Folder in the companion website at www.whfreeman.com/reconceptmath2e or Appendix E. Find a sensible way of making a stem-and-leaf plot for the data on land area for the countries given there. (*Hint:* Use a different unit for area.)

13. Open the data file *Life Expectancy at Birth* in either the Data Sets Folder in the companion website at www.whfreeman.com/reconceptmath2e or Appendix E. Make a box plot of the life expectancy at birth for females for the selected countries in 1999.

14. Use the data file *60 Students* in either the Data Sets Folder in the companion website at www.whfreeman.com/reconceptmath2e or Appendix E. Focus on the variable *hours worked per week*. Make a box plot and write a paragraph telling what you found.

15. Go to the data file *60 Students* and focus on the variable *hours of volunteer work per week*. Make a box plot and write a paragraph telling what you found.

16. Following is a problem from NAEP given to eighth-graders.[4] Seventy-three percent of the students gave correct answers. What is your answer?

▶ Gloria's diving scores from a recent competition are represented in the stem-and-leaf plot shown to the right. In this plot, 3|4 would be read as 3.4.

What was her lowest score for this competition?

5	2 5
6	1
7	7
8	0 2

a. 0.02   **b.** 1.0   **c.** 2.5   **d.** 5.2   **e.** 8.0 ◀

17. This is a problem that NAEP administered to students in Grade 8.[5] Fifty-one percent of these gave correct answers. What is your answer?

▶ The prices of gasoline in a certain region are \$1.41, \$1.36, \$1.57, and \$1.45. What is the median price per gallon for gasoline in this region?

a. \$1.41   **b.** \$1.43   **c.** \$1.44   **d.** \$1.45   **e.** \$1.47 ◀

Supplementary Learning Exercises for Section 30.3

See also Supplementary Learning Exercises 1(a–d) and 2(c–d) for Section 30.5.

1. For each of the following situations, would it make sense to give a five-number summary? Explain your reasoning.

    a. the responses of 92 fifth-graders to "What is your favorite kind of ice cream?"
    b. the heights of 92 fifth-graders, in inches
    c. the amount of milk drunk daily by each of 92 fifth-graders
    d. the responses of 92 fifth-graders to "What TV show do you always try to watch?"
    e. the responses of 92 fifth-graders to "How many minutes of TV do you watch most nights?"

2. Whose parents, if any, should be concerned about their child failing?

Arnold scored in the 54th percentile on a standardized reading test.
Bonnie's percentile score on a standardized reading test was 61.
Carl scored 52% on the semester reading test.

3. a. Tell all that you can about the following box plot of test scores:

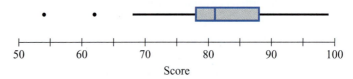

    b. What do you *not* know from this box plot?

4. Discuss the following statement in terms of statistical correctness:

All of the kids in our district will be scoring above the median on the district assessment exam.

5. One workday's worth of Old Faithful wait times in minutes is listed below in the order that they occurred:

51 82 58 81 49 92 50 88 62 93 56 89 51 79 58 82 52 88

    a. Draw a box plot showing the data.
    b. Is any information lost by drawing a box plot?

6. Create data sets with at least 10 elements that have the following:

    a. Minimum $= 14$ $Q_1 = 18$ $Q_3 = 22$ Maximum $= 40$
    b. Minimum $= 82$ Median $= 82$ $Q_3 = 94$ Maximum $= 98$
    c. $Q_1 = 19$ Median $= 27$ Maximum $= 75$ Interquartile range $= 40$

7. Draw a box plot for each set of data you created in Supplementary Learning Exercise 6.

8. Make a box plot for the set of data given in Supplementary Learning Exercise 4 (Section 30.2) on p. 687.

9. Match the histograms (a–c) to the most likely box plot (A–C).

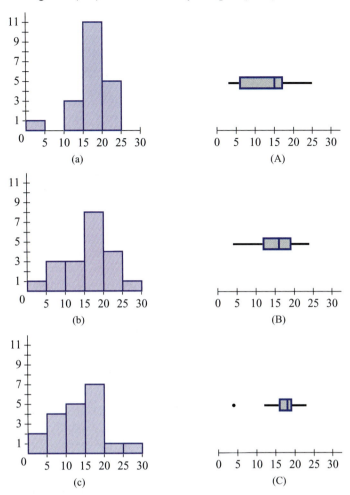

10. Find the five-number summaries for the following data sets:

   a. 73   58   84   86   72   71   95   86   73   74   77   72   88   91
   68   71   75   77   82   73   94   81

   b. 22   14   18   16   18   24   26   32   22   21   15   20   31   28
   26   24   17   19   20   15

11. Out of 485 students, Roger scored at the 78th percentile. How many students scored equal to, or lower than, Roger's score?

12. A group of 750 people take a test. How many scored between the 30th and 65th percentiles?

30.4 Measures of Center

Thus far we have encountered ways to talk about how data are spread out. Another way to investigate a data set is by selecting a numerical value that would best represent a *summary* score in the data set. When searching for a representative score in general, statisticians often consider three values: the *median,* the *mean,* and the *mode.* Until now in this chapter we have focused on the *median,* which is located at the middle of a data distribution.

The **mean** of a data set is found by computing the **arithmetic average** of the values in the set. Hence, the mean is the amount per person or per item. We sometimes refer to the mean as the *average*.

The **mode** of a data set is the data value that occurs most often in the data set. (Some data sets do not have a mode, and others have more than one mode.)

EXAMPLE 5

Consider the data set 3, 4, 4, 5, 6, 6, 7. The *mean* of this data set is

$$(3 + 4 + 4 + 5 + 6 + 6 + 7) \div 7 = 35 \div 7 = 5$$

The *median* is the middle number of the (ordered) data set and is also 5. However, this data set has more than one *mode* value. There are actually two modes, 4 and 6, and we would call the distribution of the data *bimodal*. ●

The median, mode, and mean are three representative values traditionally called *measures of center* or sometimes *measures of central tendency*. They provide anyone who is examining a data set with three choices concerning what value to choose as most representative for the data values in that set. (Unfortunately, these measures, particularly the mean and median, are sometimes confused in the media.)

Here is another example of finding and comparing the three measures of center. Consider the following situation:

▶ During office hours Professor Gonzales had 10 students individually come for help, and he spent the following number of minutes with each student: 9, 11, 12, 7, 7, 7, 13, 10, 5, 9. ◀

What are the median, mode, and mean for these data? We can quite easily find the median by first ordering the 10 numbers: 5, 7, 7, 7, 9, 9, 10, 11, 12, and 13, and then finding the middle value; the median is 9 (the average of 9 and 9). The mode is the most frequently occurring value, which is 7.

To find the mean, we need to find the average rate—i.e., the average number of minutes Professor Gonzales spent per student. As was shown in Example 5, to find the arithmetic average we add the data values and divide the sum by the total number of data values that were added together. The average number of minutes Professor Gonzales spent per student would be

$$\frac{9 + 11 + 12 + 7 + 7 + 7 + 13 + 10 + 5 + 9}{10} = \frac{90}{10} = 9$$

So, the mean is 9 minutes per student. But *what exactly* does this value tell you? You might respond that the mean is the number we get from adding up all the values and dividing by the number of addends. But this answer just describes the *procedure* for finding the average, not the *meaning* of the average. Alternatively, you might respond that the result means that, on average, Professor Gonzales spent 9 minutes with each student. This explanation, however, is also unsatisfactory because it doesn't convey much more meaning than saying that the average is 9 minutes per student. Suppose we let the 10 students meet with Professor Gonzales again, but this time we make sure that each student spends *exactly* 9 minutes, the value of the mean, with the professor.

First meetings with 10 students:

$$9 + 11 + 12 + 7 + 7 + 7 + 13 + 10 + 5 + 9 = 90 \text{ minutes}$$

Second meetings with 10 students:

$$9 + 9 + 9 + 9 + 9 + 9 + 9 + 9 + 9 + 9 = 90 \text{ minutes}$$

The total amount of time the professor spent in each group of meetings with the 10 students is the same: 90 minutes. Moreover, if we had chosen a time different from 9 minutes per student for the second set of meetings, there is no way that their total time would have been the same as the first total time. Only *9 minutes per student* for the second set of meetings would have resulted in the same total amount of time as the first set of meetings. Therefore, the average of 9 minutes per student is the *amount of time we could assign to each student and still have the total time spent with the professor remain the same.*

> Alternative explanation for the mean:
>
> The **mean** of a set of data from a single variable is the amount that each person or item would have if the total amount were shared equally among all persons or items.

In our example about meeting times, the mean can be thought of as a way of assuring a fair amount of time for each student without changing the total amount of time the professor would spend meeting with students. Viewed this way, the method we use to find the mean makes sense. To find this fair amount of time per student, we would have to find the total time and partition it equally into 10 parts, giving 9 minutes per student.

Think About ...

Why can only the mode (and neither the median nor the mean) be used with categorical data?

💬 DISCUSSION 5   Professor Childress's Office Hours

Professor Childress has five students individually come for help during office hours, and she spends the following number of minutes with them: 6, 4, 16, 13, 6.

a. What is the mean for the data set above? How can you interpret this mean? Is it reasonable?

b. What is the mode for the times Professor Childress spent with students? What does this number mean?

c. What is the median for the times Professor Childress spent with students? What does this number mean?

d. How are these three measures of center different? Which do you think is most representative of the time spent with each student? ●

One valuable feature about measures of center is that they allow us to compare data sets with unequal numbers of values. For example, it is difficult to tell just by looking at the times Professors Gonzales and Childress spent with their students which professor typically spent more time with each individual student. To address this question, we can compare measures of center—values that attempt to identify a typical, or representative, time spent with an individual student.

For example, by comparing the means of the two data sets, we see that both professors spent an average of 9 minutes with each student. Of course, any reasoning that involves comparing means is only as convincing as the ability of the mean to represent a typical time spent with an individual student. Some might argue that in this example, the medians or modes might better reflect the typical times spent by the professors with an individual student.

💬 DISCUSSION 6   Which Measure Is Better?

What is the median time spent with students by Professor Gonzalez and by Professor Childress? In your opinion, does the mean, median, or mode provide the most accurate reflection of the typical time the professors spent with an individual student? Explain why the measure of central tendency you chose is best. ●

It is interesting to notice how the median and mean are affected by extreme values. Consider the following situation:

Notes

▶ Mr. Sanders, the principal at Green Elementary School, was interested in how long it took for classes in each of six classrooms to clear the building during a fire drill. The six classes were timed during the drill and took 78, 97, 82, 69, 89, and 130 seconds to clear the building. Later, the teacher from the last classroom told him that she had transposed two digits; it actually took 103 seconds, not 130 seconds. ◀

Ordering the data with the inaccurate time, we get 69, 78, 82, 89, 97, 130.

Ordering the accurate data, we get 69, 78, 82, 89, 97, 103. The following table shows the mean and median (in seconds) for each data set:

	Inaccurate data	Accurate data
Mean	90.8	86.3
Median	85.5	85.5

For the inaccurate data set, notice how much greater the mean is affected by the extreme score of 130 seconds than is the median, which is not affected at all. Why is that? Because the mean uses *all* the numerical values in our data set, whereas the median uses at most only the middle two numerical values in the ordered data set, the mean is far more likely to be affected by extreme values. The mean uses more information about a data set, and for this and other reasons it is often preferred by statisticians over the median.

However, there are times when the data set is best represented by the median value instead of the mean. This superiority is especially true when we have extreme outliers that would greatly affect the value of the mean.

Think About …

Suppose that one year the median income of the high school graduating class of golfer Tiger Woods was $44,500 and the mean annual income for the class was $750,000. Why are the two numbers so different? Which one more closely represents the salaries of these graduates?

Think About …

Your cousin is planning to move to your city and buy a home there. Would she be better off knowing the median price or the mean price of homes in your city? Why?

Another aspect of working with the mean is that there are times when a *weighted average* is needed, such as in Activity 7.

 **ACTIVITY 7**   Finding the GPA

Consider the following story:

▶ Janice went to Viewpoint Community College for 3 semesters. She earned 42 credits and had a grade-point average (GPA) of 3.4. She then moved to State U for 5 semesters where she earned 80 credit hours before graduating and had a GPA of 2.8. What was her overall GPA when she graduated? She had to have an average GPA of 3 or better to enter a graduate program. Did she have that? ◀

1. What does "grade-point average" mean?
2. How many grade points did Janice earn at VCC?
3. How many grade points did Janice earn at SU?
4. What is the total number of grade points Janice accumulated?
5. What is the total number of credits Janice accumulated?
6. What can you do with the information from Questions 4 and 5?
7. Is the result in Question 6 the GPA that you predicted? ●

A common error made on this type of problem is to average the two grade-point averages given: 2.8 and 3.4. Doing so, however, does not reflect the fact that almost twice as many units were earned with the 2.8 grade-point average, and thus the final GPA should not be 3.1. We say that the two original GPAs are weighted differently, and we must find the *weighted average*, or *weighted mean.*

Think About ...

Suppose, in Activity 7, Janice had 42 credits rather than 80 credits at State U. How would Question 7 be changed?

TAKE-AWAY MESSAGE ... There are three ways to measure the "middle" of a data set: the mean, the median, and the mode. The mean is the arithmetic average of the data and can be interpreted as the amount per person or item if each received the same amount. In some cases, a weighted average must be calculated. If one value appears most frequently, it is called the mode of the data set. When values in a set of data are weighted differently, adjustments must be made in computing the average.

Learning Exercises for Section 30.4

1. Linda went to a craft sale and bought five items. She spent an average of $10 per item. Is it possible that

   a. exactly one of the items cost $10?
   b. none of the items cost $10?
   c. all the items cost different amounts?
   d. all the items cost the same?
   e. all but one of the items cost less than $10?
   f. all of the items cost more than $10?
   g. For each of the situations in parts (a–f) that you think is possible, generate a set of five prices that meet the condition. If you think the situation is not possible, state why.

2. Now suppose Linda had purchased six items with a mean price of $10. Respond to situations (a–f) in Learning Exercise 1.

3. Generate a data set having a mean that is less than the first quartile score.

4. Which measure do you think would be greater? Justify your answer.

   a. the mean or median price for a house in the United States
   b. the mean or median number of hours college students sleep per night
   c. the mean or median age of U.S. females at the time of their first marriage
   d. the mean or median age at death for U.S. males
   e. the mean or median number of children in U.S. families

5. The president of a local community college announced that the average age of students attending the school was 26 and that the median age was 19. Create a group of five students whose mean age is 26 and median age is 19.

6. In ancient Greece, architects and builders are said to have used the *golden rectangle,* a rectangle with a width-to-height ratio that is $1 : \frac{1}{2}(\sqrt{5} + 1)$, or 0.618.

   a. What does a width-to-height ratio of 0.618 mean?
   b. Would you expect every building whose construction was based on the golden rectangle to have a width-to-height ratio of exactly 0.618? Why or why not?

continue

c. The Native American Shoshoni used rectangles to decorate their leather goods. The following values are the width-to-height ratios of 20 rectangles taken from a sample of Shoshoni leather goods. Use measures of center to draw conclusions about whether or not the rectangles used by the Shoshoni to decorate leather goods were also modeled after the golden rectangle. Justify your conclusions. (The data are from Hand et al.[6])

0.693	0.662	0.690	0.606	0.570
0.749	0.672	0.628	0.609	0.844
0.654	0.615	0.668	0.601	0.576
0.670	0.606	0.611	0.553	0.933

d. Display the data from part (c) in a graph. Does your graph support or confound the conclusions you made in part (c)? Explain.

7. The following values are the silver contents (percent of weight) of Byzantine coins that date to the reign of Emperor Manuel I. The coins were minted during two different periods of his reign; the first coinage was minted early in his reign and the second coinage was minted late in his reign. Do you think that there was a difference between the silver content of coins minted during these two periods? Justify your response. State the assumptions you made to draw your conclusion. (The data are from Hand et al.[7])

Early coinage: 5.9, 6.8, 6.4, 7.0, 6.6, 7.7, 7.2, 6.9, 6.2
Late coinage: 5.3, 5.6, 5.5, 5.1, 6.2, 5.8, 5.8

8. "The class mean on the last examination was 79.8 points." What is one interpretation of the given statement (not how it was calculated)?

9. Which basketball team seems to be best, based on the following data? Explain your decision.

a. the Amazons, who have outscored their opponents by 184 points in 16 games
b. the Bears, who have outscored their opponents by 168 points in 14 games
c. the Cougars, who have outscored their opponents by 194 points in 17 games

10. Explain why the means, rather than the modes or medians, give a fairer way to compare the performances of two or more groups.

11. In a recent professional basketball game, each of the three leading scorers on the team scored an average of 21 points during the game. The remaining eight players scored an average of 6 points during the game. What was the team's average number of points scored per player?

12. Jackie had a grade point average of 3.3 at the community college she attended. She then transferred to a university and earned a grade point average of 3.7. Can you calculate her overall GPA?

13. Carla earned $64 a day for the first three days she was on the job, $86 a day for the next two days, and then $110 for each of the two weekend days, when she worked overtime. What was her average pay for the seven days?

14. Vanetta drove from Chicago to Indianapolis at an average speed of 50 miles per hour. She returned at an average speed of 60 miles per hour. What was her average speed for the trip?

15. Following is a problem from NAEP, given to eighth-graders.[8] Forty-five percent of the students gave the correct answer. What is your answer?

▶ The average weight of 50 prize-winning tomatoes is 2.36 pounds. What is the combined weight, in pounds, of these 50 tomatoes?

a. 0.0472　　　**b.** 11.8　　　**c.** 52.36　　　**d.** 59　　　**e.** 118 ◀

16. The following problem is another NAEP item.[8] There are four parts to the question. Try them. The percents of students getting the specified number of correct answers are as follows:

1 correct 19%   2 correct 16%   3 correct 18%   4 correct 38%

▶ Akira read from a book on Monday, Tuesday, and Wednesday. He read an average of 10 pages per day. Indicate whether each of the following is possible or not possible. ◀

Number of pages read

	Monday	Tuesday	Wednesday	Possible	Not possible
a.	4 pages	4 pages	2 pages		
b.	9 pages	10 pages	11 pages		
c.	5 pages	10 pages	15 pages		
d.	10 pages	15 pages	20 pages		

17. Fairway Clinic kept track one day of the ages of patients who came in for office visits. The following histogram shows the ages of patients who visited, broken down by decades. That is, the patients represented by the first bar were 10 and under.

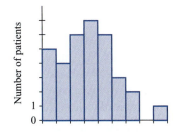

a. How many patients visited the clinic on this particular day?
b. Create a table that lists the intervals and frequencies.
c. Which is larger, the mean or the median? Can you tell from the graph? How?

18. Below are four histograms (a–d) and four box plots (A–D):

  i. Which box plot represents the data in each of the histograms?
  ii. In which data set(s) are the mean and median approximately equal?
  iii. In which data set(s) is the mean smaller than the median?

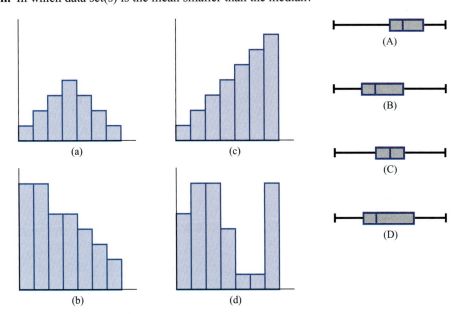

Supplementary Learning Exercises for Section 30.4

See also Supplementary Learning Exercises 1(e–h) and 2(a–b) for Section 30.5.

1. For each of the following contexts, state whether the mean, median, or mode is the most appropriate measure of center. Explain your answers.

    a. salaries of employees in a large company
    b. cost of gasoline
    c. types of car owned
    d. examination scores in a college class
    e. cost to buy a house in a particular region
    f. number of pets owned by children in a kindergarten class

2. a. Find the mean and median of the following data:

39	13	15	8	5	6	14	7	3	6	4	12

    b. Find the mean and median of the data in part (a) with the 39 removed. Explain why and how these new mean and median are different.

3. Consider the following heights of these students in a class:

5' 6"	5' 2"	4' 11"	4' 8"	4' 7"	5' 1"	4' 6"	4' 10"	5' 0"	4' 9"
5' 3"	4' 9"	4' 10"	4' 7"	5' 3"	4' 6"	4' 9"	4' 6"	4' 8"	5' 0"

    a. Converting the heights to feet only, find the mean.
    b. Converting the heights to inches only, find the mean.
    c. Did the values of the mean increase or decrease when going from feet to inches?
    d. Using appropriate technology, draw the histogram for heights in terms of feet and the histogram for heights in terms of inches. How do these histograms compare?

4. Whitney is baking cupcakes for a school bake sale. The first night she made three dozen. The next two nights she baked four dozen each night. The fourth night she made six dozen. What is the average number of cupcakes she made per night?

5. The mean weight of three cats is 14 pounds. Which of the following is possible?

    a. One of the cats weighed 20 pounds.
    b. None of the cats weighs more than 14 pounds.
    c. Each cat weighs less than 14 pounds.
    d. The cats weigh 14, 18 and 10 pounds.
    e. The cats weigh 13, 18, and 8 pounds.
    f. Two of the cats weigh less than 13 pounds each.

6. Javier earned $8.50 per hour for 32 hours and $12.75 per hour for 8 hours. What was his average hourly wage for those 40 hours?

30.5 Deviations from the Mean as Measures of Spread

We turn our attention again to measures of spread. The *range,* even though it is easy to determine just by subtracting the smallest value from the largest, is not used much by statisticians. The *interquartile range* provides one way of describing the spread of the data. Another way to measure spread is to consider the distance each data value is from the mean.

EXAMPLE 6

Consider the following situation about the number of pounds lost per month by three people on a weight-loss plan:

▶ Abe, Ben, and Carl joined a weight-loss program at the same time. Their monthly weight losses in pounds for 5 months are as follows:

         Abe: 3, 5, 6, 3, 3   Ben: 4, 4, 4, 4, 4   Carl: 10, 1, 0, 0, 9 ◀

Each person lost 20 pounds in 5 months, so the mean weight loss is the same: 4 pounds per month. In Abe's case, the median is 3, in Ben's case it is 4, and in Carl's case it is 1. (Remember to order the data before finding the median.) It appears that neither the mean nor the median by itself gives us a good picture of the weight lost by the three men. Ben's weight loss per month is consistent, whereas Carl's weight loss per month fluctuates so much that you would wonder about the efficacy of this weight-loss plan. ⬤

Is there some way we can show the differences in the spreads of the three data sets in Example 6? Suppose that in addition to computing the mean for each case, we measure the distances between the means and the data values. Adding these distances together for each case would give us some idea of how spread out the data are, because the greater the spread, the larger the distances between the data values and the mean would be. Thus, larger sums of distances from the mean would indicate a greater spread of data values.

Think About …
Of the three people in Example 6, who do you think would have the greatest sum of distances from the mean? Who would have the second greatest? Why?

Summing distances from the mean could be useful when comparing the spread of data sets having the same number of entries, as in Example 6. But, in general, to compare sets having possibly different numbers of entries, we compute the average distance from the mean per data entry, or in other words, the mean of the distances from the mean. We show how to do this next. Recall that in Example 6 the mean for each data set is 4.

Abe			**Ben**			**Carl**	
Values	**Distance from mean**		**Values**	**Distance from mean**		**Values**	**Distance from mean**
3	1		4	0		10	6
5	1		4	0		1	3
6	2		4	0		0	4
3	1		4	0		0	4
3	1		4	0		9	5
	Sum: 6			Sum: 0			Sum: 22
	Mean distance: $6 \div 5 = 1.2$			Mean distance: $0 \div 5 = 0$			Mean distance: $22 \div 5 = 4.4$

For these three tables, the sums of 6, 0, and 22 provide us with a good way of measuring the spread in each data set. Because the number of months is the same for each person, we divide by 5 (the number of months in each case), obtaining 1.2, 0, and 4.4 as the means of the distances between the month's weight loss and the overall mean weight loss for Abe, Ben, and Carl, respectively. We call the values 1.2 pounds, 0 pounds, and 4.4 pounds the **average deviations from the means** or **average deviations** (for short).

Note that average deviations correspond to our notion of which data sets are more spread out than others. For example, the average deviation for Carl's data is much higher than either Abe's or Ben's. This fact is consistent with our perception that Carl's data values are far more spread out than those for Abe or Ben. A side benefit of defining spread in this way is that we get a value of 0 for the average deviation if and only if all the data entries have the same value, as in Ben's case.

Suppose, however, that one person had been in the weight-loss plan longer than the others. Additional weight losses for one person would affect the sums but should not affect the way we measure spread.

Think About …
Suppose a fourth person was on a weight-loss program for 12 months and lost an average of 4 pounds per month, with an average deviation of 0.5. What types of numbers could you expect, month by month?

Notes

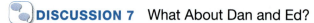

DISCUSSION 7   What About Dan and Ed?

1. Suppose another person, Dan, also had a mean weight loss of 4 pounds per month, with an average deviation of 0.8 pounds. How would the spread of Dan's values compare to those of Abe, Ben, and Carl?

2. Answer the same question in Problem 1 for yet another person, Ed, whose mean weight loss was 4 pounds per month, with an average deviation of 3.1 pounds. ●

Although the average deviations for the data sets we just used were easy to calculate and provided us with useful information about the spread of the data sets, there is another way of measuring spread that is somewhat more difficult to calculate, but statistically more useful, called the *standard deviation*. Computing this new measure is similar to but slightly more complex than finding the average deviation. Related to the standard deviation is a measure called the *variance*, which also captures the degree of spread of data values.

> The **variance** is found by summing the *squares* of the distances of the data values from the mean, and then dividing this sum by n, where n represents the number of data entries. The square root of the variance is called the **standard deviation**.

The standard deviation uses the same measurement units as the original variables, as does the average deviation, and thus is easier to interpret than the variance.

EXAMPLE 7

Consider again the data from Example 6 on weight loss by Abe, Ben, and Carl. To find the standard deviation of Abe's weight-loss data, we first determine the distances between the mean and each data value for Abe. (Recall that the mean was 4 pounds.) Then we square these values and sum them.

Weight loss value in pounds	Distance from mean	Squared distance from mean
3	1	1
5	1	1
6	2	4
3	1	1
3	1	1

The sum of the squared distances from the mean is $1 + 1 + 4 + 1 + 1 = 8$.

To find the *variance*, we divide the sum of the squared distances by 5, the number of data entries. This number, 1.6, is the variance. To find the standard deviation, take the square root of 1.6, which is about 1.26. Thus, the standard deviation is 1.26 pounds. Notice that the standard deviation in this case has the same units, pounds, as the data values do. In this case, the standard deviation, 1.26 pounds, is close to the average deviation found in Example 7, which was 1.2 pounds. ●

 ACTIVITY 8   What's the Standard Deviation?

Calculate the standard deviations for Ben's data and for Carl's data. How do the standard deviations for Abe's, Ben's, and Carl's data compare? Although the standard deviations vary from the average deviations, do they present a similar picture? ●

Notice that the standard deviation has the same desirable properties as the average deviation:

- Higher values for the standard deviation correspond to a greater amount of spread among scores in the data set.

- Any data set in which the data values are the same has a standard deviation of zero.

Thus, knowing the standard deviation gives us an indication of the spread, or the dispersion, of the data values. Statisticians use the standard deviation rather than the average deviation because the standard deviation has superior properties in more advanced work. (For technical reasons, in calculating the standard deviation of n values, they may divide by $n - 1$ rather than n, a practice that is reflected with some calculators, so different calculators can give slightly different values for the standard deviation.)

ACTIVITY 9   The Easy Way to Find a Standard Deviation

Calculating the standard deviation can be very time-consuming for large sets of values, so we usually let a computer or calculator do this work for us. Return to the exam scores in Activity 5 in Section 30.3. They were

65, 64, 46, 38, 58, 44, 65, 60, 50, 70, 55, 44, 68, 67, 66, 66, 81, 68, 51, 51, 75, 53, 47, 62, 59, 25, 78, 49, 77, and 49

a. Find the mean and the standard deviation of these scores, using a calculator or some other technology.

b. Change the 25 to 43 and again find the mean and standard deviation. How do this mean and standard deviation compare to those in part (a)? ●

Let us discuss both the numerical statistics and one visual representation for the same data.

DISCUSSION 8   Comparing Numbers and Graphs

In Activity 9 you found the mean and standard deviation for the exam scores. Figure 10 is a box plot for the same data. What do the mean, the standard deviation, and the box plot each tell you about the exam scores? Which data analysis method do you believe would give the most helpful information? Why?

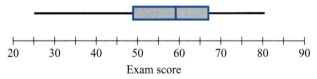

20   30   40   50   60   70   80   90
Exam score

Figure 10 A box plot with the five-number summary 25, 49, 59.5, 67, 81.   ●

  In the next section we will further explore the information that the mean and standard deviation often convey about the data sets from which they are derived. At this point, however, it is important that you keep in mind the two main ways we can talk about the spread of a data set. One way is through the five-number summary, depicted in a box plot. The other way to talk about the spread of a data set is through the standard deviation (which is favored by statisticians). When comparing similar data sets, we use this general rule: The larger the standard deviation, the greater the spread.

TAKE-AWAY MESSAGE . . . By themselves, the measures of center (mean, median, mode) do not give information about the spread of the data. The range, although easy to calculate, is not useful statistically. The average deviation provides an easy way to understand spread, based on the distance of values from the mean. However, the standard deviation is the preferred way of measuring the spread of the data and is the one most commonly used. The mean and standard deviation are more affected by extreme values than are the median and interquartile range. As we will see, the mean and standard deviation provide other important information about a data set.

Learning Exercises for Section **30.5**

Your instructor may recommend using technology of some sort to avoid lengthy calculations in Learning Exercises 3 and 4.

1. Find the average deviation for Professor Gonzales's and Professor Childress's time spent with students in Section 30.4. Do these numbers give more information about the time students spent with the professors than the means alone?

2. Answer the following questions about these four sets of data:

    Data set A: 69, 78, 66, 45, 58
    Data set B: 98, 78, 39, 55, 47
    Data set C: 67, 69, 72, 70, 71
    Data set D: 67, 69, 72, 70, 17

    a. Of sets A and B, which would you predict has the larger standard deviation? Why?
    b. Of sets C and D, which would you predict has the larger standard deviation? Why?
    c. Of the four sets, which would you predict has the largest standard deviation? Why?
    d. Of the four sets, which would you predict has the smallest standard deviation? Why?

3. A shoe factory is making a new shoe style to sell to women in a small country in the Pacific. To determine what sizes and how many of each size to manufacture, they collected a random sample of 50 women and measured their shoe sizes. Here are the results:

6	5	5.5	6.5	5.5	4.5	5	7	6	5.5	5.5	6.5	6
5.5	6	6	5	6.5	7	6	6	6.5	5	5.5	4.5	5
6.5	5.5	5.5	7	6.5	7.5	5.5	6	5.5	6.5	7.5	4.5	4
5.5	6	5	5	6.5	5	4	6	6	6	5.5		

    Calculate the mean and standard deviation of the sizes and interpret the results in terms of what relative quantities of which sizes the factory should make. Then find the five-number summary and decide which set of information gives you more assistance in making a decision for the shoe factory.

4. There were 100 points possible on a midterm exam in Mr. Kane's class. However, no student scored above 80, and he told them he would therefore grade them on the curve this way:

    Score 2 or more standard deviations above the mean—grade A
    Score between 1 and 2 standard deviations above the mean—grade B
    Score within 1 standard deviation above or below the mean—grade C
    Score between 1 and 2 standard deviations below the mean—grade D
    Score 2 or more standard deviations below the mean—grade F

    Here are their scores:

69	74	77	79	56	44	38	74	70	44	58	65	66	37	49
59	66	69	67	57	45	79	76	57	67	76	58	77	72	67
76	77	67	65	57	23	65	66	58	45	59	69	63	66	79

    How many students received a grade of A? B? C? D? F?

5. We have seen that outliers have a much stronger influence on the mean than on the median. How strongly do you think the standard deviation is influenced by outliers? To answer this question, calculate the standard deviation for Mr. Sanders's fire-drill data in Section 30.4. First calculate the standard deviation of the data in which the last value is 130 seconds. Then calculate the standard deviation with the last data value of 103 seconds. Explain why you think the standard deviation is or is not strongly affected by an outlier.

6. **a.** Find the mean and standard deviation of the grade point averages for females in the file *60 Student*s. (This file is in the Data Sets Folder in the companion website www.whfreeman.com/reconceptmath2e or Appendix E. Cut and paste from the electronic form if you can.) Using a calculator or other software is appropriate.
   b. Repeat part (a) for males. Compare the results.

7. Suppose you have already calculated the mean, the average deviation, and the standard deviation of the weights in pounds of 20 third-graders.

   a. If these weights were converted to kilograms and the mean and average deviation were once again calculated, would the second average deviation be (numerically) less than, greater than, or equal to the first average deviation? Explain your answer.
   b. If the standard deviation of these weights was calculated in kilograms, would it be (numerically) less than, greater than, or equal to the first standard deviation? Explain your answer.

8. Use this small data set to investigate parts (a) and (b):

   75, 80, 85, 90, 95

   a. How does adding or subtracting the same number from each data value affect the mean? The standard deviation?
   b. How does dividing each data value by the same number affect the mean? The standard deviation?

Supplementary Learning Exercises for Section 30.5

1. Consider the unordered scores 21, 5, 9, 9, 12, 8, 19, 16, 19, 13, 2, and 24.

   a. What is the range?
   b. What is the five-number summary?
   c. What is the interquartile range? Are there any outliers? Why or why not?
   d. Draw a box plot of the data.
   e. What is the mean of these data?
   f. What is the mode of these data?
   g. How do the measures of center compare?
   h. What is the average deviation of this set of data?
   i. What is the standard deviation of this set of data?
   j. How do the measures of spread compare?

2. The following table contains the averages of monthly rainfall over 30 years for New Orleans and San Diego:

Month	Jan	Feb	Mar	Apr	May	Jun	Jul	Aug	Sep	Oct	Nov	Dec
New Orleans	5.05	6.01	4.90	4.50	4.56	5.84	6.12	6.17	5.51	3.05	4.42	5.75
San Diego	1.80	1.53	1.77	0.79	0.19	0.07	0.02	0.10	0.24	0.37	1.45	1.57

   a. What is the average *yearly* rainfall for each city?
   b. What is the average *monthly* rainfall for each city?
   c. About what would you expect the standard deviation for the average monthly rainfall in each city to be?
   d. Compute the standard deviation for the average rainfall in each city, and compare those results with your predictions. Were you close?

30.6 Examining Distributions

Sometimes it is useful to consider the overall *shape* of the data when they are graphed as a histogram. For example, consider the histogram, shown on the next page, that represents the lengths of words from *The Foot Book* by Dr. Seuss.[9]

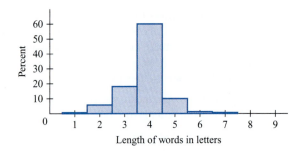

Length of words in letters

Think About …

What trends do you notice about the length-of-words data set from the histogram? How would you describe the distribution of these data to someone who couldn't see the histogram?

It is often helpful to draw a picture of the data by sketching a smooth curve that approximates a histogram of the data. Unlike the histogram with its jagged corners and potential for sharp, frequent spikes and dips, the smooth curve captures the general trends in the graph and portrays them in a way that is easily comprehensible to the reader.

Sketching this curve frequently amounts to nothing more than ignoring the sharp corners of the histogram and drawing the general shape of the graph. One way of doing this smoothing is to connect the midpoints of the tops of bars in the histogram with a smooth curve rather than lines, and then continue the curve down toward the horizontal axis. If the bars of the histogram are removed, the smooth curve representing the Dr. Seuss histogram might look like the following graph:

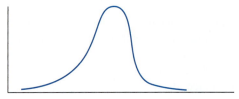

Sometimes deciding which trends (i.e. which bumps and dips) in the histogram are important in a data set is not so clear-cut. The difficulty in deciding which trends to report or emphasize comes from two sources. First, it is not always easy to determine just how much detail we should give to people who read our graphs. Should we emphasize every bump or dip in the histogram, or should we emphasize only the overall shape of the graph? If we provide more information, i.e., more rises and falls in our smooth curve, our readers are less likely to reach false conclusions about the data. Yet, if we provide too much detail, our readers may never be able to see any trends at all.

Common sense tells us that we must try to strike a balance between the two extremes. Striking a balance, however, often requires us to use our own judgment in deciding what level of detail is appropriate for anyone who will try to interpret our graphs.

The second source of difficulty comes up when the data we are analyzing represent only a sample of the total population we are trying to understand. It is most likely that the shape of the histogram of our sample will differ in some places from the histogram of our total population.

How can we be sure that the trends we are observing in the histogram of our sample are trends of the entire population and not just characteristics of the sample? In fact, we can never be sure unless we measure our total population. The larger the sample, however, the more likely it is that the total population also exhibits the same trend (ignoring the fact that a smaller sample might give good enough information for most purposes, as discussed in Chapter 29). Large differences between statistics for a large sample and for the total population are much less likely to occur than small differences. It is then to our advantage to include only large trends in our sketches of samples so that the sample curve is more likely to accurately reflect the shape of the graph of the total population.

✎ ACTIVITY 10 Delivery Dates and Due Dates

To illustrate these difficulties, consider the following histogram of the delivery dates of 42 pregnant women in relation to their due dates. A data score of 0 indicates that the woman delivered on her due date. A negative score means that the woman delivered that many days before her due date. A positive score means that the woman delivered that many days after her due date. Each bar has a width of seven days (one week). Thus, the bar over the zero represents the number of women who gave birth within three to four days of their due dates.

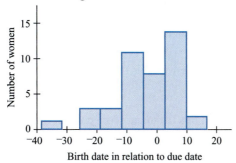

Draw a smooth curve to approximate the histogram above. Think about what trends your curve records and whether or not the trends are worth reporting. ●

Even if we are very precise in how we draw the curves that approximate histograms, doing so can obscure the similarities among data sets. For example, consider the three histograms shown below at right.

The first histogram[10] represents the heights in centimeters of 351 elderly women. The second histogram[11] represents the cholesterol readings for 320 male patients who show signs of narrowing of the arteries. The last histogram[12] represents the circumference in centimeters of the wrists of 252 males taken in conjunction with a test to determine body fat.

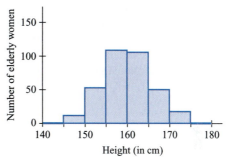

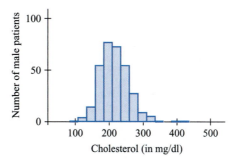

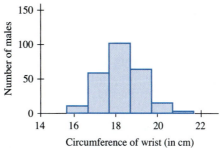

Think About ...

What are the differences in the shapes of these three histograms? How are they similar?

In the preceding three histograms, did you notice that most of the data points are located at the center of the distribution? The graphs also appear to be nearly *symmetrical*. In other words, the left and right halves of the graphs are almost mirror images of each other. Furthermore, all three graphs seem to have the given shape shown in a smooth-curve version.

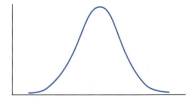

This kind of shape is a common one, and has been called the **bell curve** by statisticians due to its bell-like shape. The vertical scale is usually set so that 100% of the cases are under the curve—think of a histogram with very small widths in the bars.

> Data sets that generate a curve with this bell shape are often referred to as being **normally distributed.** The graph is called a **normal curve.**

The adverb *normally* is used in a technical sense and not in the everyday sense. There are many data sets, like the birth date in relation to the due date, that do not have a normal distribution. But there are many types of data sets that do have this kind of distribution for large samples: weights of people, heights of people, test scores on standardized tests, the life span of particular kinds of butterflies, and many other instances of measures dealing with the properties of living things. The data sets are also usually quite large.

Because normal distributions are so common, statisticians have studied them extensively and have discovered many important properties. For instance, they know that the *mean, median, and mode of a normally distributed data set have the same value*—the number on the horizontal axis that is at the center of the bell curve. They also know that for large, normally distributed data sets, 68% of the data values fall within one standard deviation of the mean; 95% fall within two standard deviations of the mean; and 99.7% fall within three standard deviations of the mean. The following graph illustrates this important property. (Note that 34% + 13.5% = 47.5%, which, when doubled, is 95%. Similarly, 95% + 2 · 2.35% = 99.7%.)

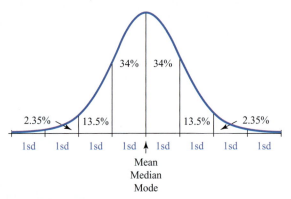

Figure 11 A normal distribution showing the measures of center and the percents in various parts of the distribution.

In the future, we will refer to this property of normal distributions as the **68–95–99.7 rule.** It is helpful to associate these percents with the graph in Figure 11. Notice that for normal distributions, knowing just the mean and the standard deviation gives you a great deal of information about what percents of the cases, and even how many cases, are in particular regions of the distribution.

Think About ...

If a normally distributed data set contained 2000 data points, how many of these points would you expect to differ from the mean by more than three standard deviations?

Suppose we consider the scores on three forms of a math placement test, each form given to a large group of students. The scores on each form are normally distributed. On Test Form 1 the mean is 66 and the standard deviation is 4; on Test Form 2 the mean is 66 and the standard deviation is 2; and on Test Form 3 the mean is 72 and the standard deviation is 6. If we graph all three sets of test scores on the same axis, we get the graph in Figure 12.

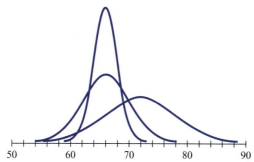

Figure 12 Distributions of scores on three test forms.

DISCUSSION 9   Understanding the Normal Distribution

a. How can you tell from the graphs of the test scores which test has the greatest standard deviation?

b. Which curve in Figure 12 represents which test form? How do you know?

c. Looking at the mean and standard deviation for Test Form 1, between what score values would you expect 99.7% of the test scores to fall? Does your prediction appear to agree with the graph?

d. Repeat the questions in part (c) for Test Forms 2 and 3.

e. Could you draw the bell-shaped curve of a normally distributed data set if you were given the mean, standard deviation, and the scale to use? If so, how? If not, why not? (What other information would you need to complete the graph?)

f. Suppose Maria took Test Form 1 and had a score of 70. Jannelle took Test Form 2 and also had a score of 70. Do you think one did better than the other? Why or why not? ●

Our first attempt at answering part (f) in Discussion 9 might be to ask which form of the test is more difficult. Because both students had scores of 70, if we knew which test form was more difficult, we would also know which student did better. But it is not clear which form is more difficult. Both test forms have a mean, or average, score of 66, which suggests they might be equally difficult. Does this mean that both students did equally well?

Instead of asking which test form is more difficult, we could ask which student scored higher *compared to other students* taking the same version of the exam. Maria's score of 70 on Test Form 1 is only one standard deviation (4 points) greater than the mean for Test Form 1. Because the scores for Test Form 1 are probably normally distributed, we can compute the percentage of students who scored a 70 or below on Test Form 1. From the graph of a normal distribution, we know that 50% of the students who took Test Form 1 received scores at or below the mean of 66. An additional 34% of the scores from Test Form 1 fall between the mean and one standard deviation above the mean. Thus, Maria scored higher than or the same as 84% of the students who took Test Form 1. (See Figure 13.)

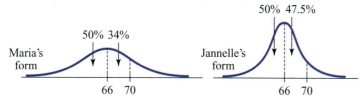

Figure 13 Maria's and Jannelle's scores of 70.

How did Jannelle do? The difference between her score of 70 and the mean of 66 on Test Form 2 is also 4 points—the same as Maria's. But the standard deviation for this test form is only 2 points. This means that Jannelle's score is two standard deviations above the mean for Test Form 2. So, Jannelle scored higher than the 50% of the students at or below the mean, plus the 34% who scored between the mean and one standard deviation above the mean, plus the 13.5% who scored between one and two standard deviations above from the mean. Thus, she did as well as or better than 97.5% of the students who took Test Form 2. (See Figure 13.) Clearly, Jannelle did better on her test (in comparison with the other students taking Test Form 2) than Maria did on hers (in comparison with the other students taking Test Form 1).

💬 DISCUSSION 10   Who Did Better?

1. Adrian had a score of 68 on Form 2 of the test. Julio had a score of 68 on Form 3 of the test. Who did better?

2. LeRoy had a score of 63 on Form 1 of the test and Brandon had a score of 65 on Form 3 of the test. Who did better? ●

Because of the nature of normal distributions, we can compare scores from two different normal distributions merely by comparing the number of standard deviations by which the scores differ from the means of their respective data sets. For example, in our comparison of Maria's and Jannelle's test scores, it was unnecessary to calculate the percent of students who scored at or below their scores to tell which girl did better. All we really needed to know was that Maria's score was one standard deviation above the mean and Jannelle's score was two standard deviations above the mean. From this comparison we could conclude that Jannelle did better.

Data values that have been converted into the number of standard deviations from the mean are called *z*-scores. For example, a data value *x* with a corresponding *z*-score of 1.5 indicates that the value *x* is 1.5 standard deviations above the mean.

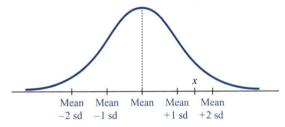

Mean −2 sd	Mean −1 sd	Mean	Mean +1 sd	Mean +2 sd

To calculate a *z*-score, all we do is subtract the mean from the original score, and divide this difference by the standard deviation.

> A **z-score** represents the number of standard deviations a data value *x* is from the mean. The *z*-score corresponding to the data value *x* from a normal distribution is
>
> $$z = \frac{x - \text{mean}}{\text{standard deviation}}$$

EXAMPLE 8

Recall that Maria's test score is 70 on Test Form 1 and Jannelle's test score is 70 on Test Form 2. What are Maria's and Jannelle's *z*-scores?

Solution

Because the standard deviation for Test Form 1 is 4 points, Maria's score of 70 gives the *z*-score $\frac{70-66}{4} = 1$. Because the standard deviation for Test Form 2 is 2 points, Jannelle's score of 70 gives the *z*-score $\frac{70-66}{2} = 2$, a much better *z*-score than Maria's. ●

DISCUSSION 11   Who Has the Better Score? *z*-Scores Can Tell Us

Refer to the three test forms given earlier: On Test Form 1 the mean is 66 and the standard deviation is 4; on Test Form 2 the mean is 66 and the standard deviation is 2; on Test Form 3 the mean is 72 and the standard deviation is 6.

a. Luis received a 76 on Test Form 3. What is his *z*-score? How did he do in comparison to Maria and Jannelle?

b. Jaime received a 72 on Test Form 3. What was his *z*-score? What does this indicate?

c. Anna's *z*-score on Test Form 1 was $^-1.5$. What does this mean? What was her test score?

d. What are the *z*-scores for Adrian, Julio, LeRoy, and Brandon (from Discussion 10)? Rank them according to their *z*-scores. ●

How do we know whether or not a distribution of values is normally distributed? Take a moment to make a close comparison between the bell-shaped curves suggested by the three histograms presented on p. 711 for the heights of 351 elderly women, the cholesterol readings for 320 male patients, and the circumference of wrist size of 252 males. We claimed that these sets of data were normally distributed. Did you notice that none of the histograms appeared to be perfectly bell-shaped? In all three graphs, there are slight differences between the left and right halves of the graph.

Based on these discrepancies, we could claim that none of these three data sets are normally distributed, because none of them are perfectly symmetrical. In fact, very few data sets precisely match the bell-shaped curve. If the differences are small, however, we can use our knowledge about the normal distribution to make statements about the data set we are analyzing and still make fairly accurate predictions. The larger the differences, though, the more careful we need to be about our claims.

DISCUSSION 12   Grading on a Curve?

Often students and teachers talk about grading on a curve, perhaps with the understanding that this type of grading would result in a grade distribution which is shaped like a normal distribution. But is the normal distribution a good model for grade distribution? Suppose that the scores on a test were normally distributed. What do you think of the following grading system for assigning grade values A through F?

- A: Students whose scores are more than two standard deviations above the mean
- B: Students whose scores are between one and two standard deviations above the mean
- C: Students whose scores are within one standard deviation of the mean
- D: Students whose scores are between one and two standard deviations below the mean
- F: Students whose scores are more than two standard deviations below the mean

a. What are the *z*-scores for each letter grade?

b. What percentage of the students would be assigned each letter grade? ●

TAKE-AWAY MESSAGE . . .   A histogram can be "smoothed out" to provide a curve that represents the data in a data set. Many large data sets that deal with natural characteristics (such as the height of males at age 15) will form a normal distribution, which is sometimes called a bell curve. In such a curve, 68% of the data are within one standard deviation of the mean, 95% are within two standard deviations of the mean, and 99.7% are within three standard deviations of the mean. We use *z*-scores to convert initial data values to numbers that allow us to compare values from different data sets having different normal (or near normal) distributions.

Notes

Learning Exercises for Section 30.6

Many of these problems will be much easier to solve if you use software or a calculator that calculates standard deviations.

1. **a.** Draw a curve you think represents the distribution of the lengths of words from a college textbook.
   b. Choose a fairly long paragraph from one of your college texts and count how many words of each length are in the paragraph. Then construct a histogram and draw a curve that describes the distribution of the lengths of words in your sample.
   c. Use your graph from part (b) to modify your original curve from part (a). What changes did you make and why? If your final curve looks different from the curve in part (b), explain why.

2. **a.** Construct a data set of 10 or more data entries such that the data appear to be normally distributed. How do you know that the data are approximately normally distributed?
   b. Construct a data set with 10 or more data entries that is clearly not normally distributed. Explain how you know that the data are not normally distributed.

3. What conclusions would you make about the students from three different classes whose math scores on a recent math test had the distributions shown in the following graphs?

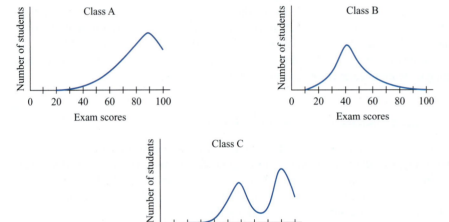

4. **a.** Does a negative z-score on a test indicate that the person lost points? Why or why not?
   b. What is the z-score for the mean?

5. Many data sets have a distribution that appears somewhat like a normal distribution, except that the graph on one side of the curve is longer than on the opposite side. The following graphs are examples of distributions that have uneven sides, and are called **skewed distributions:**

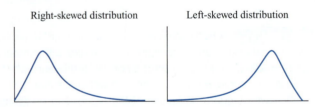

   a. Think of three sets of real-life data that might have a distribution that is skewed right. Explain why you think the data could be distributed that way.
   b. Think of three sets of real-life data that might have a distribution that is skewed left. Explain why you think the data could be distributed that way.

6. The data below represent the scores of 54 students on a vocabulary test:

> 14, 13, 16, 11, 11, 13, 11, 14, 14, 11, 19, 14, 13, 11, 14, 15, 14, 11,
> 13, 13, 11, 14, 12, 11, 13, 12, 11, 16, 12, 15, 15, 10, 11, 12, 10, 12,
> 11, 14, 13, 17, 11, 16, 16, 10, 12, 9, 12, 15, 10, 14, 13, 16, 13, 11

   a. Are the data normally distributed? Justify your answer.
   b. If the data are not normally distributed, do you think the data are close enough to being normally distributed that we can use many of the properties of normal distributions to describe the data set (such as the 68–95–99.7 rule)? Why or why not?

7. The 68–95–99.7 rule for normal distributions states that 95% of the data values in a normally distributed data set will be within two standard deviations of the mean. Generate numbers with a distribution that has fewer than 95% of the data values within two standard deviations of the mean. Can you generate a set that has many fewer than 95% of the data values within two standard deviations of the mean? How small can you make that percentage?

8. How can you tell from a graph that the median and mode have the same value as the mean in a normal distribution?

9. Draw pictures of at least two symmetric distributions that are clearly not normally distributed. Think of real-world situations that would yield data with these distributions.

10. Draw a picture of a symmetric distribution that has different values for the median and mode.

11. A fourteen-year-old boy is 66 inches tall. His six-year-old brother is 47 inches tall. The following table gives the means and standard deviations for the heights of six- and fourteen-year-old boys. Assume that the heights of boys at a particular age are normally distributed.

Age	Mean	Standard Deviation
6	45.8 inches	1.4
14	64.3 inches	2.8

   a. Which boy is taller for his age?
   b. If there are 11 other boys in the six-year-old's class at school, how many of them would you expect to be taller than he is?

12. Antonio's height is one standard deviation below the mean for boys his age. If the heights of boys are normally distributed, at what percentile is Antonio's height (what percent of the boys his age are as short or shorter than Antonio)?

13. Suppose a student had a z-score of $^-0.3$ on a standardized exam. Would you consider this a passing or failing score? Why?

14. Suppose you gave your students a test and converted their scores into z-scores. Could you tell from just the z-scores how well your class did on the test? Why or why not?

15. What are the advantages of converting raw scores to z-scores? What are the disadvantages of converting to z-scores? Is any information lost?

16. Consider the graphs of three normal distributions, P, Q, and R, on the same coordinate system:

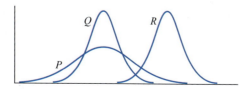

   a. Which distributions have the same mean? Tell how you know.
   b. Which distribution has the greatest standard deviation? Tell how you know.

Notes

Supplementary Learning Exercises for Section **30.6**

1. Graphs *A* and *B* show two normal distributions:

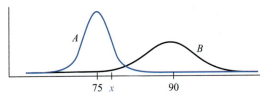

a. Which distribution has the greater mean? Tell how you know.
b. Which distribution has the greater median? Tell how you know.
c. Which distribution has the greater mode? Tell how you know.
d. Which distribution has the greater standard deviation? Tell how you know.
e. Which distribution has the greater variance? Tell how you know.
f. For which distribution does *x* have a greater *z*-score?

2. On the same test, Ann has a *z*-score of ⁻0.8 and Beth has a *z*-score of 0.8. Whose actual score is farther from the mean? Tell how you know.

3. a. Construct a set of data with at least 20 entries that is skewed right. Explain how you know that the data set is skewed right.
    b. Construct a set of data with at least 20 entries that is skewed left. Explain how you know that the data set is skewed left.

4. The following data are taken from Supplementary Learning Exercise 3 in Section 30.4. Parts (a) and (b) are repeated from that exercise.

5' 6"	5' 2"	4' 11"	4' 8"	4' 7"	5' 1"	4' 6"	4' 10"	5' 0"	4' 9"
5' 3"	4' 9"	4' 10"	4' 7"	5' 3"	4' 6"	4' 9"	4' 6"	4' 8"	5' 0"

a. Converting the numbers to feet only, find the mean and standard deviation.
b. Converting the numbers to inches only, find the mean and standard deviation.
c. For both parts (a) and (b), find the *z*-value for a student who is 5' 3".
d. Are the data normally distributed? Explain how you know.

5. If a data set is not normally distributed, does it make sense to discuss a *z*-value?

6. The mean for a particular test is 1200 and the standard deviation is 150. Find the *z*-value for each score:

    a. 1150   b. 1430   c. 1560   d. 1090   e. 850

7. Label the distribution shown in each graph with as many appropriate phrases as possible: symmetric, skewed right, skewed left, bimodal, normal, and nonnormal. (See Learning Exercise 5 in Section 30.6.)

a.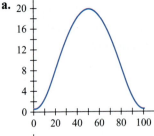

b.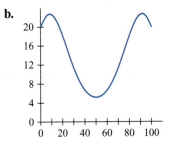

c.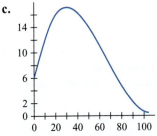

8. Draw a graph for each distribution described, if it is possible. If it is not possible, explain why not.

    a. a skewed distribution that has the same mean and median
    b. a normal distribution that has a different mean than median
    c. a symmetric distribution with a mean not equal to the median
    d. a symmetric distribution with a mean not equal to a mode

9. For the following distribution graph, draw lines marking the approximate mean, median, and mode:

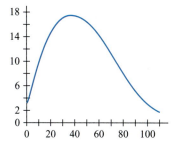

10. Suppose a distribution has a mean of 74, median of 68, and standard deviation of 5.4. What happens to each of these values if 15 is added to every score?

30.7 Issues for Learning: Understanding the Mean

Most children (and some adults) are puzzled by statements such as "The average number of persons per household is 3.2." This confusion is true even though these people know how to calculate an average. They understand average only as a procedure, and therefore they are puzzled about how to interpret "3.2 persons."

There are many ways to think about the mean, or the average, of a set of numbers. In one research study[13], children's understandings of average in Grades 4, 6, and 8 were studied. They had all been taught the procedure for finding the average, but they did not all understand average in the same way.

First, solve these two problems that children were asked to solve.

▶ **1.** The average price of nine bags of potato chips was $1.38. How could you place prices on the nine bags to make the average $1.38? Do not use $1.38 as one of the prices. ◀

▶ **2.** Consider the graph below that gives the amounts of allowance money. Imagine what the children's allowance would be if it were the "average" of the allowances shown on the graph.

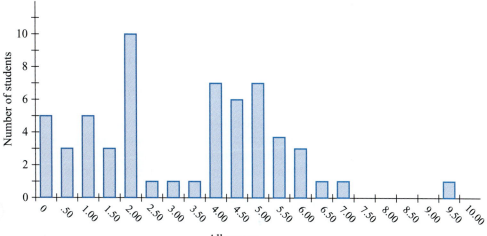

Notes

The manners in which students of different ages thought about these problems were quite surprising. Some thought of the average as the mode—the value with the highest frequency—and considered $2.00 to be the average allowance. These children did not yet view a set of numbers as an *entity*. Thinking about one number as representing the others was not possible for them.

Here is one response by a fourth-grader to Problem 1 about the average price of potato chips:

> OK, first, not all chips are the same, as you just told me, but the lowest chips I ever saw was $1.30, myself, so because the typical price was $1.38, I just put most of them at $1.38, just to make it typical, and highered the prices on a couple of them, just to make it realistic. (p. 27)

As you might expect, some considered the average to actually *be* the procedure for finding the average. These students did not think of the average as being representative of all the values and thus were not able to understand the problems. For example, on the potato chip problem, one fourth-grader said she would choose values where the "cents add to 38." She chose values such as $1.08 and $1.30, which added up to $2.38, so that the "average" would be $1.38. These students were concerned about the procedure rather than any meaning of average.

A third approach was to think of average as a "reasonable" number. These students based their answers on what they knew from everyday life, and they used a commonsense approach. On the allowances problem, one said that the allowance would depend on age and on what was reasonable. Thus, because the "typical allowance" was $1.50, no one should get $5.00.

Yet a fourth approach was to think about average as the midpoint. These students had a strong sense of "middle" even though they did not know what the median was. One said: "I'm trying to find what the number is that has about the same number on both sides" (p. 32). This strategy made it very difficult to work with nonsymmetrical distributions such as the allowance problem.

Finally, some students thought of average as a kind of "balance." This fifth approach at times led to thinking about average in terms of a pan balance, with equal weights on both sides. One student found the total price for the 9 bags of potato chips (9 × $1.38 = $12.42) and then tried to put prices on the bags that would total $12.42. (These same students found the allowance problem to be very difficult, however.)

A drawing may help to clarify this last student's method. Suppose that the mean for three values is 4. Then, to create values with that mean, one could try different sets of three values so that they would balance the three 4s. The values 2, 3, and 7 give one possibility. Note again that the mean can be thought of as the value that each would have, if the total amount (12 here) was divided into three equal amounts.

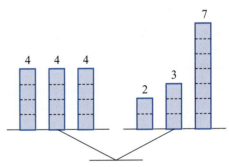

Students who used one of the last three approaches—(1) the mean as reasonable, (2) the mean as the middle, or (3) the mean as the basis for the total value—were beginning to understand the meaning of average, even though they were approaching this meaning in different ways. Children's use of these strategies moved from the first to the last as they matured and had more opportunities to work with the meaning of average.

So what does this research have to say about the teaching of averages? The researchers found that the curriculum could assist students in coming to a full understanding of average if it helped them first understand representativeness, i.e., that *one value can represent other values*. This notion is necessary to understand what the average, or the mean, is. The researchers also believed that the formal definition and procedure should come *after* discussions of values that could represent a set of values.

It can take many years for students to recognize mean as representative of a data set. Researchers believe that mean as "typical value" and mean as "fair-share" are intuitions on which the concept of mean can be built.[14] Average as fair-share connects well to the averaging algorithm. Elementary teachers sometimes enact a redistribution or leveling approach to get to the idea of the total items shared equally.

However, researchers also recognize that these conceptions of the mean are not as powerful for understanding that one value can represent many values. For developing this idea of the mean as "data reducer," students should spend time examining the shape of data and comparing two data sets.[14]

Learning Exercises for Section 30.7

1. The average price of 10 bags of potato chips was $2.24 each. Eight bags were all the same price and two bags cost $4.00 each. What was the price of each of the other eight bags?

2. Tim was given the following problem:

   ▶ Six bags of potato chips cost $1.60 each and another four bags cost $2.20 each. What is the average price of the potato chips? ◀

   a. Tim said the average price was $1.90. Karl said that the average price was $1.84. Who was correct?
   b. How did each boy obtain his answer?

3. Consider the fourth approach used by the students in the problem on allowances— thinking about the average as the midpoint. It appears that this child was trying to find the median allowance.

   a. What is the median allowance?
   b. What is the mean allowance?
   c. How do they compare?

4. Lili worked in a convenience store. Out of curiosity she kept track of the amount of money spent by shoppers during one hour, rounded to the closest dollar. Here is her tally:

   $7, $19, $12, $14, $12, $21, $16, $8, $19, $19, $17, $11, $13, $14, $19

   Suppose the children who worked on the allowance problem were also given this problem and asked to find the average amount spent on groceries. Consider the different ways they could answer. First make a bar chart of the data. On the horizontal axis label the bars as follows: below $10, between $10 and $20, between $20 and $30, etc. (Note that this is possible because no amount ended in 0.)

   a. A child who chose the mode in the allowance problem would say that the average spent here was what amount? Suppose the last amount was $12 rather than $19. What might the child say then?
   b. If a child thought that the average was actually the procedure rather than a number, what might that child say or do?
   c. If a child thought of the average as a "reasonable" number, what might that child say or do?
   d. If a child thought that the average was the midpoint, what might that child say or do?
   e. If a child thought of the average as a kind of "balance," what might that child say or do?

30.8 Check Yourself

In this chapter you learned how to represent data having one variable and how to interpret those representations. You know about different measures of center and different measures of spread, and how these measures assist in understanding data sets. And you have considered different shapes of distributions of data by studying and interpreting their graphs.

Notes

You should be able to work problems like those assigned and to meet the following objectives:

1. Make circle graphs, bar graphs, histograms, and stem-and-leaf plots for a variety of types of data, and know when to use each of these types of graphs.

2. Find the five-number summary of a set of numerical data and use that information to make a box-and-whiskers graph, i.e., a box plot. Interpret the data based on such a graph.

3. Find the mean and standard deviation of a set of data (using a calculator or computer) and interpret what these numbers tell you about the data.

4. Evaluate whether the mean, the median, or the mode best describes the center of data for a situation and know which of the three is most affected by outliers.

5. Tell whether a set of data, when graphed, appears to be normally distributed.

6. Understand and be able to use the 68–95–99.7 rule for normal distributions.

7. Find z-scores for data, and use them to explain differences in scores in different normally distributed data sets.

References for Chapter 30

[1]National Assessment of Educational Progress (NAEP). (2003). Item 8M6, No. 3, http://nces.ed.gov/nationsreportcard/.

[2]Moore, D. S. (1997). *Statistics: Concepts and controversies* (4th ed.). New York: W. H. Freeman.

[3]Hand, D. J., Daly, F., Lunn, A. D., McConway, K. J., & Ostrowski, E. A. (1994). *A handbook of small data sets*, New York: Chapman & Hall. This data set was taken from The Open University (1993). *MDST242 Statistics in Society, Unit C3: Is my child normal?* Milton Keynes, UK: The Open University, Figure 3.12.

[4]NAEP. (2003). Item 8M10, No. 8, http://nces.ed.gov/nationsreportcard/.

[5]NAEP. (2003). Item 8M12, No. 6, http://nces.ed.gov/nationsreportcard/.

[6]Hand, D. J., Daly, F., Lunn, A. D., McConway, K. J., & Ostrowski, E. A. (1994). *A handbook of small data sets*. New York: Chapman & Hall. This data set was taken from Dubois, C. (Ed.) (1970), *Lowie's selected papers in anthropology*, Berkeley: University of California Press.

[7]Hand, D. J., Daly, F., Lunn, A. D., McConway, K. J., & Ostrowski, E. A. (1994). *A handbook of small data sets*. New York: Chapman & Hall. This data set was taken from Hendy, M. F., & Charles, J. A. (1970), The production of techniques, silver content and circulation history of twelfth-century Byzantine Trachy, *Archaeometry, 12*, 13–21.

[8]NAEP. (2003). Item 8M7, Nos. 5 and 12, http://nces.ed.gov/nationsreportcard/.

[9]Geisel, Theodor Seuss (Pseudonym, Dr. Seuss). (1968). *The foot book*. New York: Random House.

[10]Hand, D. J., Daly, F., Lunn, A. D., McConway, K. J., & Ostrowski, E. A. (1994). *A handbook of small data sets*. New York: Chapman & Hall.

[11]Hand, D. J., Daly, F., Lunn, A. D., McConway, K. J., & Ostrowski, E. A. (1994). *A handbook of small data sets*. New York: Chapman & Hall. This data set was taken from Scott, D. W. (1992), *Multivariate density estimation*, Chichester, UK: John Wiley & Sons, Table B3, p. 275.

[12]Data for the last histogram were collected by K. W. Penrose, A. G. Nelson, and A. G. Fisher, FACSM, Human Performance Research Center, Brigham Young University, Provo, UT, and appeared on the StatLib website, http://lib.stat.cmu.edu.

[13]Mokros, J., & Russell, S. J. (1995). Children's concepts of average and representativeness. *Journal for Research in Mathematics Education, 26*, 20–39.

[14]Shaughnessy, J. M. (2007). Research on statistics learning and reasoning. In F. K. Lester (Ed.), *Second handbook of research on mathematics teaching and learning* (pp. 957–1009). Charlotte, NC: Information Age Publishing & National Council of Teachers of Mathematics.

Chapter 31

Dealing with Multiple Data Sets or with Multiple Variables

In Chapter 30 we examined graphical and numerical methods for presenting and interpreting data from a single variable. We now extend these methods to compare multiple data sets. We then consider sets of data with two variables and display them with scatter plots and line graphs. Analyzing such graphs must be done with care. Many mathematics textbooks include some of these topics in the upper elementary and middle grades.

Notes

31.1 Comparing Data Sets

Many times we collect data so that we can compare the scores of different groups on a single variable. Once we have collected data for different groups, we can compare group results by using many of the methods we previously learned about organizing and interpreting data, with some slight modifications. To illustrate how to compare group results, we once again return to the *60 Students* data file (see Appendix E). Suppose we want to compare the grade point averages (GPAs) of males and females in this group of students. One way is to compare and contrast the box plots for the two genders (see Figure 1). The main aim is to use the graphs to point us toward some possible conclusions about the data. (Note that any conclusions made here are not generalizable. They refer to this data set only.)

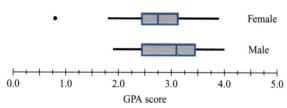

Figure 1 Box plots of GPAs for 60 students: female (top graph) and male (bottom graph).

Think About ...

What differences do you notice about the two box plots in Figure 1? What can you say about the medians? About the interquartile ranges? What do the slightly longer whiskers on the graph for female students mean? Can you make any predictions about the means and standard deviations for these two data sets?

Another way to analyze the two data sets is to compare histograms rather than box plots (see Figure 2). The two histograms shown on the next page compare the GPA data for the female students and the male students given in the *60 Students* data file.

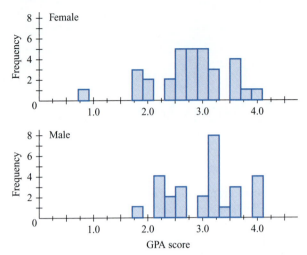

Figure 2 Histograms of GPAs for 60 students: female (top graph) and male (bottom graph).

Notice that the scales for both graphs are identical. The vertical axis scales give the counts (frequencies) for the students and the horizontal axis scales give the GPA scores.

DISCUSSION 1   GPAs

(You should keep the *60 Students* data file handy for this discussion.)

a. Using the histograms, make some comparison statements about GPAs for females and for males.

b. By how much do you think the overall mean GPA for females is affected by the one very low GPA? Suppose we were to delete that score and look at a histogram for GPAs for females again. How would the shape of this histogram compare with the first histogram for females?

c. Compare and discuss the information you receive from the box plots and histograms of male and female GPAs. ●

   The box plots and histograms allow us to see how scores are spread out in two different ways. The graphs themselves do not offer any explanation for the differences. To examine the differences, we need to look elsewhere. One possibility is that the very low GPA for one female may have affected the data. This possibility will be explored in Learning Exercise 2. Another possibility is that the number of hours worked per week may have affected grades. Suppose we consider the number of hours worked per week (over and above the number of hours spent on school work). Figure 3 at the top of the next page shows such a comparison for the two groups of students.

> *Think About …*
> From the graphs in Figure 3, can you tell which group, overall, works more hours each week?

It is often not easy to read the means from a histogram. If you thought the mean for the females was higher than that for the males, you were correct. The actual means for the number of hours worked per week are 16.0 for the females and 14.2 for the males. The standard deviations are 8.1 for the females and 8.8 for the males.

> *Think About …*
> How might the information about the mean number of hours worked per week explain the somewhat lower mean GPA for the females?

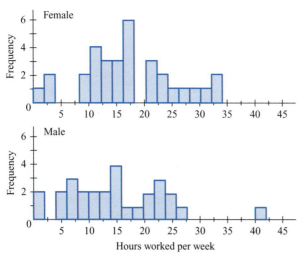

Figure 3 Histograms of hours worked per week: females (top graph) and males (bottom graph).

When comparing performances by different groups, it is important to think about other variables that might account for any apparent differences. Another caution is to recall that quite often you have statistics for samples but not for the whole population, and sample statistics can vary considerably from sample to sample, particularly for small samples.

ACTIVITY 1 Who's the Best Scorer?

At the conclusion of the first season of the Women's National Basketball Association, the three top scorers' totals for the 28 games of the regular season were as follows and are listed in order from Game 1 to Game 28:

Bolton-Holifield: 18, 27, 21, 20, 20, 23, 22, 25, 15, DNP, DNP, DNP, DNP, DNP, 16, 13, 18, 19, 22, 24, 34, 15, 34, 13, 13, 12, 12, 12 (DNP = Did not play)
Cooper: 25, 13, 20, 13, 11, 19, 13, 21, 16, 20, 13, 17, 30, 32, 44, 21, 34, 15, 30, 34, 17, 39, 21, 17, 17, 31, 15, 16
Leslie: 16, 22, 19, 20, 21, 19, 16, 22, 12, 7, 17, 10, 12, 18, 28, 17, 14, 23, 11, 14, 10, 17, 3, 23, 26, 13, 12, 10

a. Make either box plots or histograms to display these data.

b. How are the points-scored distributed for each player? What is the same about their distributions? What is different?

c. Generate a table or set of graphs that helps you determine which player should be selected as the top scorer. Justify your choice. Who is the second leading scorer? Justify your answer.

d. Generate a table or set of graphs that helps you determine which player is the most consistent scorer. Justify your choice. ●

Even though we can use side-by-side box plots or histograms to compare data for two groups, each graph gives information about a pattern in *one* variable. One of the disadvantages of box plots or histograms is that we cannot study patterns in two variables, such as the GPA and the hours worked per week for the same group of people, individual by individual. Another type of graph, a **scatter plot**, allows us to represent *both* the GPA *and* the hours worked per week for each student. The scatter plot in Figure 4 given on the next page shows this information for the 60 students. Each point represents an individual person. We could have used different symbols for the males and females, but rather than comparing these two groups, we focus just on representing two corresponding data values for *each* person.

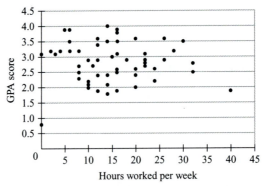

Figure 4 Scatter plot of GPA versus hours worked by each student in the *60 Students* data.

Think About …

Pick out a point in Figure 4. What does that data point tell you about that person?

Scatter plots contain a lot of information about the individuals, but it is difficult here to notice any overall patterns. We will consider scatter plots again in the next section, in which we discuss a graphical method for summarizing the data displayed in a scatter plot.

Scatter plots are extensively used when one might suspect a relationship between two variables. Data from the Trends in International Mathematics and Science Study (TIMSS)[1] present such a case. The data set provides mathematics and science scores for students in many countries. Are these two sets of scores related in some way? To find out, we again employ a scatter plot. The scatter plot can be constructed by thinking of our paired data as points on an *x-y* graph. We assign one variable to each axis and then graph the pairs of data.

Figure 5 shows a scatter plot in which the test values for mathematics and science for each country are represented by a point. For example, the U.S. performance on the tests is represented on the graph by the point (545, 565), because the average score of U.S. fourth-grade students was 545 on the mathematics test and 565 on the science test.

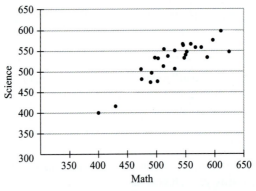

Figure 5 Scatter plot of average math scores versus science scores for fourth-grade students, using TIMSS data.

Notice that in the scatter plot above there is nothing special about our choice of assigning the mathematics scores to the *x*-axis and the science scores to the *y*-axis. The only time when this assignment matters is when we think that changes in one of the variables cause changes in the other. When this is the case, we always assign the *x*-axis to the variable we think is causing changes, often called the **independent variable.** This is a convention that has been agreed upon by statisticians and scientists to help them interpret graphs of variables that are suspected to have a causal relationship. For example, time is often an independent variable and thus is placed on the *x*-axis. The other variable can be called the **dependent variable**.

Notice also that we are not using the years of formal schooling or average age in the scatter plot, even though this information is available. However, separate analyses showed that there was little variation with these two variables taken into account, and thus it seems likely that they do not influence mathematics and science scores in the fourth grade. We will test this conjecture later.

When examining a scatter plot, what should we look for? Generally, we are interested in how the two variables are related. For example, suppose we want to know if TIMSS countries performed about the same in mathematics and science. In Figure 5, notice that the countries with high science test scores tended also to have high mathematics test scores— or, the other way round, that countries with high mathematics scores also had high science scores. When this pattern occurs, we say the two variables are **positively associated** (or more commonly, *positively correlated*). When an increase in one variable tends to happen in conjunction with a *decrease* in another variable, we say that these two variables are **negatively associated** (or *negatively correlated*). Consider the scatter plot in Figure 6. It plots locations across the United States according to their latitude and average low temperature in January[2] (computed by averaging the 31 daily low temperatures). A higher latitude reading is associated with a more northerly location. Notice that because temperature is affected by latitude, we place latitude on the *x*-axis as the independent variable.

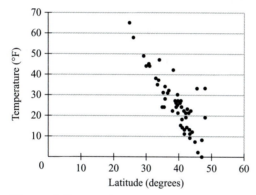

Figure 6 Scatter plot for temperature versus latitude in January.

As you might expect, higher latitudes tend to have lower January temperatures, resulting in a negative correlation between latitude and temperatures in January.

It is not always easy to tell whether two variables are positively or negatively correlated. For instance, a scatter plot of the average age of fourth-grade students and the average mathematics score for the countries in the TIMSS data set gives the following display (Figure 7).

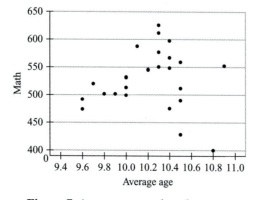

Figure 7 Average age and math scores.

You might expect that higher ages at Grade 4 would be associated with higher average test scores on the mathematics test. Does the given graph support this hypothesis? In fact, because of the scattering of the points, it is very difficult to conclude what type of relationship, if any, exists between these two variables; the graph seems to suggest that there may actually be no discernible overall relationship between age and performance on the mathematics test. However, consider the ages. Notice that there is a very small spread in the ages, from 9.6 to 11. Had the data involved ages from 6 to 12, the scatter plot might have looked very different.

This page from a middle-school mathematics textbook introduces the scatter plot. The scatter plot shown is based on data given earlier in the book.

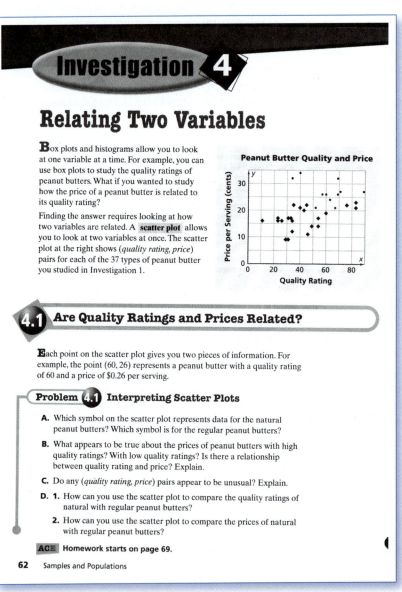

Investigation 4

Relating Two Variables

Box plots and histograms allow you to look at one variable at a time. For example, you can use box plots to study the quality ratings of peanut butters. What if you wanted to study how the price of a peanut butter is related to its quality rating?

Finding the answer requires looking at how two variables are related. A **scatter plot** allows you to look at two variables at once. The scatter plot at the right shows (*quality rating, price*) pairs for each of the 37 types of peanut butter you studied in Investigation 1.

Peanut Butter Quality and Price

4.1 Are Quality Ratings and Prices Related?

Each point on the scatter plot gives you two pieces of information. For example, the point (60, 26) represents a peanut butter with a quality rating of 60 and a price of $0.26 per serving.

Problem 4.1 Interpreting Scatter Plots

A. Which symbol on the scatter plot represents data for the natural peanut butters? Which symbol is for the regular peanut butters?

B. What appears to be true about the prices of peanut butters with high quality ratings? With low quality ratings? Is there a relationship between quality rating and price? Explain.

C. Do any (*quality rating, price*) pairs appear to be unusual? Explain.

D. **1.** How can you use the scatter plot to compare the quality ratings of natural with regular peanut butters?

   2. How can you use the scatter plot to compare the prices of natural with regular peanut butters?

ACE Homework starts on page 69.

62   Samples and Populations

Lappan, G., et al. *Connected Mathematics 2, Samples and Populations*, p. 62.
Copyright © 2009, Pearson. Reproduced with permission of Pearson Education, Inc.

> *Think About …*
>
> Can you think of three pairs of variables (other than those we considered from the TIMSS data set) that might be positively correlated? Negatively correlated? Where no discernible relationship seems to exist? Justify your selections.

> *Think About …*
>
> Would you expect the number of deaths from heat prostration to be positively correlated with the number of ice cream cones sold? Would that mean that ice cream consumption causes death from heat prostration? What other variable is relevant to both these variables?

Warning: A correlation between variables, whether positive or negative, does not necessarily mean that changes in one of the variables *caused* changes in the other.

EXAMPLE 1

We cannot conclude from the positive correlation between scores on the mathematics and science tests that success on the mathematics test caused success on the science test, or even that higher mathematics skills led to higher scores in science. Other factors, such as number of school days per year, number of hours spent on mathematics and science, extent of teacher training, or student work ethic (to name a few), might be responsible for this association. ●

EXAMPLE 2

The *San Diego Union-Tribune* (December 19, 2006) reported a study on men who did or did not shave daily. The researchers found that men who do not shave daily were three times more likely to suffer a stroke than men who shave daily. We were then told that the heightened risk had nothing to do with hair but rather with demographics. Hirsute males were more likely to smoke, be unmarried, and do manual labor. However, these factors did not fully account for the difference. Can you think of additional reasons for this strange finding? ●

> *Think About …*
>
> What do you think about the associations made in these two statements from *The Numbers Game* (p. 185)[3]?
>
> 1. People with bigger hands have better reading ability, so we should introduce hand-stretching exercises in schools.
> 2. In Scandinavia, storks are more likely to be seen on the rooftops where larger families live. Therefore, storks cause babies.

Scatter plots give one way of graphing two measurement variables, both measured for the same individuals. Another type of graph that relates two measurement variables is the **line graph**, sometimes called a **broken-line graph**. In effect, a line graph is based on a scatter plot with only one point for each value on the horizontal axis. The separate points are then joined by line segments to give the line graph. (Notice that with most scatter plots, you do *not* join every point.) Line graphs are often used to demonstrate change over time, especially when the data are collected only at isolated times. Time is almost always on the horizontal axis, and the scale should be uniformly spaced. (You may sometimes have to change the unit to make scale markings readable.) The line segments help the eye to see the overall trend. Figure 8 gives an example of a line graph, which represents data about the total U.S. population.

Year	Total U.S. population (in thousands)
1850	23,192
1900	75,995
1930	122,775
1950	150,697
1960	179,929

Year	Total U.S. population (in thousands)
1970	203,312
1980	226,456
1990	248,710
2000	281,422
2005	295,734

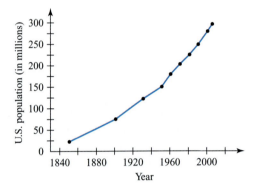

Figure 8 Line graph for U.S. population (in millions) over the years 1850–2005.

Think About …

Have the increases in population become steeper as time has passed? How might you account for the quite steep segments between 1950 and 1960, and between 1990 and 2005?

We must, of course, be careful in making and presenting graphs so that they do not distort data. Also, we must be careful in interpreting graphs. In the first graph[4] in Figure 9 below, the vertical axis begins at 0, and there is a great deal of white space.

Annual Insurance Employees, 1948–1993
"Insurance employees" is the percentage of full-time equivalent employees
who are working for insurance carriers. Allowing the data
to determine the scale ranges reveals interesting aspects of the data.

(a) A stable insurance industry

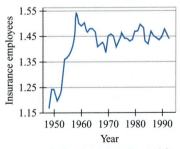

(b) The insurance industry workforce increased dramatically in the 1950s

Figure 9 Two graphs for the same data.

There appears to be little variability in the data in this first graph. The second graph uses numbers on the vertical axis that are bounded by the data, and now there appears to be more variability. The choice of scale and the inclusion of 0 on an axis often change the viewer's interpretation.

TAKE-AWAY MESSAGE . . . In this section we illustrated ways of comparing and analyzing different sets of data using box plots and histograms. When we have two measurement variables on each individual case, we can use a scatter plot. The overall layout of the dots on a scatter plot can indicate whether or not the sets of data appear to be related in some way, with language such as positively correlated or negatively correlated often used. However, even if a positive correlation, e.g., appears to exist, we cannot say that one variable caused the other. Line graphs give us a way of studying how two variables are related, especially when time is one of the variables.

Learning Exercises for Section 31.1

Data for Learning Exercises 1–4 can be found in the Data Sets folder at the companion website at www.whfreeman.com/reconceptmath2e, or in Appendix E.

1. Make box plots of the data on political party affiliation and the hours of volunteer work. The data are in the data files, in the folder called *60 Students,* or in Appendix E. (There is no file for "none" because the number of such cases is so small.) Can you see any differences in the three data sets: one for Democrats, one for Republicans, and one for Independents? Do the graphs lead you to make any conjectures?

2. In discussing the GPAs for males and for females in the *60 Students* data, a conjecture was that the one very low GPA for a single female might have seriously affected the statistics. The mean GPA for all 32 females is 2.74. Calculate the mean for the data on the 31 remaining females after the low GPA is dropped.

3. Three kindergarten classes were asked to choose their favorite book from a list of five books that had recently been read during story time in their classrooms. The five books were *Kat Kong* (K), *The Grouchy Ladybug* (L), *Ruby the Copy Cat* (R), *Stellaluna* (S), and *Tacky the Penguin* (T). Their choices follow:

Class	Students' Choices for Best Book
Ms. Wilson	L, S, L, L, K, K, T, K, S, L, T, T, K, K, T, L, K, K, R, K, T, T
Ms. Chen	R, S, T, T, T, K, T, R, R, K, T, K, L, K, K, S, L, S, K, K
Mr. Lopez	L, T, L, T, T, K, S, T, S, T, L, T, T, K, T, T, K, L, L, L, T, K, L

a. How do the three classes compare in their choices for best book? Support your conclusions with graphs and/or tables.
b. From the given data, which book would you say is best liked by kindergarten children?
c. Suggest possible reasons for differences in distributions among classes if such differences exist.

4. A science fair is coming up and your school has been invited to send one of its two science classes to participate in a science contest of general science knowledge. You have been asked to decide which science class to send. The science students recently took a test of general science knowledge to help you decide who should represent

your school in the contest. The test scores for each class are shown in the following two graphs:

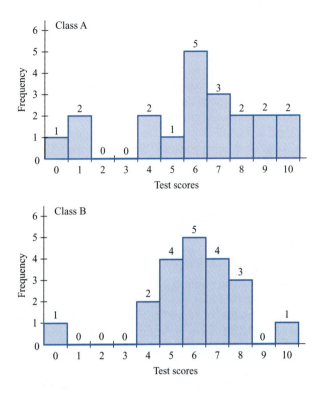

a. Based on the test scores, would you choose Class A or Class B to represent your school at the science contest? Why?

b. One student replied, "Class A and Class B have the same average test score (6 points) so I don't think it really matters. Toss a coin to decide." How would you respond to that suggestion?

5. The following box plots were adapted from a 2003 report of an assessment program in Australia:

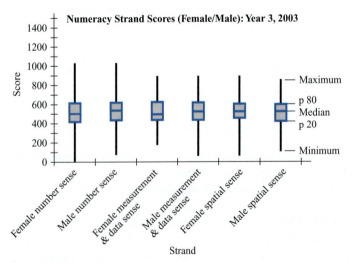

Figure 10 Range of performance for Year 3 female and Year 3 male students across the numeracy strands.

a. How does the form of the box plots differ from the ones thus far used in Chapters 30 and 31?

b. How do the males and females differ on the number sense strand? The measurement and data sense strand? The spatial sense strand?

6. The following data give the weights and cholesterol readings for 10 men. Make a scatter plot of the data:

Weight	153	175	200	178	253	142	193	240	173	204
Cholesterol	190	210	230	240	258	204	257	192	185	195

7. For each situation given, specify which types of graphs—circle graph, stem-and-leaf plot, bar graph, histogram, box plot, scatter plot, line graph—you think would best represent the situation. Explain why the graphs you selected would be relevant.

a. Easton has been offered a job in two different cities and wants to know which city has more affordable housing.

b. Vicki just finished grading her students' social studies and math tests. She is interested in getting a feel for how the class did on each test. Vicki also wants to see whether her students' scores in social studies are associated with their scores in mathematics.

c. Nathan notices that as the end of the school year approaches, his students seem to turn in fewer and fewer homework assignments. He is interested in investigating this trend.

d. Stella got caught in one too many April rainstorms and is beginning to wonder if this April is a particularly wet one for her city. She consults an almanac to get a list of the April rainfall totals for her city over the past 30 years.

e. Sanders, a marketing director, wishes to investigate the effectiveness of his advertising campaign. He consults his figures for the monthly costs of advertising and the company's monthly revenues over the past year.

8. The following graph is a scatter plot of the heights of 200 husbands and their wives (the couples were randomly selected). The heights are recorded in millimeters.[5]

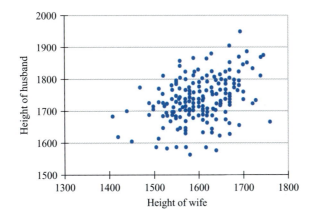

a. What does the point (1590, 1735) represent?

b. If someone claimed that the heights of husbands and wives are negatively associated, what would that mean?

continue

    c. If someone claimed that the heights of husbands and wives are positively associated, what would that mean?

    d. If someone claimed that the heights of husbands and wives are neither negatively nor positively associated, what would that mean?

    e. Based on the scatter plot, are the heights of husbands and wives positively associated, negatively associated, or neither?

    f. Notice that the scales on the axes do not start at the usual (0, 0). Does that give a distorted picture of the relationship?

9. Here are the monthly weights of a baby at birth and at the end of each of its first 11 months, in ounces: 120, 118, 123, 129, 137, 146, 152, 161, 168, 173, 179, 186. Make a line graph depicting these weights.

10. For each pair of variables given, tell whether you think that the two variables are positively associated, negatively associated, or not associated.

    a. a schoolchild's weight and the length of the child's hair

    b. the temperature of a child with an infection and the child's white-blood-cell count

    c. the time since a mortgage was started and the mortgage balance

    d. the average daily temperature and the sales of heavy coats

11. The following diagram gives another pair of line graphs,[6] this time interpreting growth over time of U.S. credit insurance. The data used are the same in the two graphs. Is one misleading? Why or why not?

Annual U.S. Credit Life Insurance in Force 1950–1989.
Different vertical scales give different impressions of the rate of growth over time.

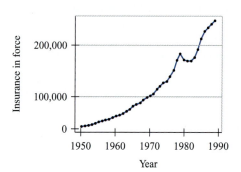

(a) U.S. credit life insurance market exploding.

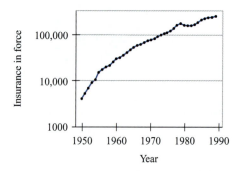

(b) U.S. credit life insurance market leveling off.

12. In 2009 there were approximately 3800 manatees in the waters of Florida. Unfortunately, many of them die due to accidents with watercraft. This data set[7] shows the number that died from 2000 through 2010. Make a line graph of the data.

Year	2000	2001	2002	2003	2004	2005	2006	2007	2008	2009	2010
Manatee Deaths by Watercraft	78	81	95	73	69	79	92	73	90	97	83

13. The following scatter plot and set of questions are taken from a book[8] written to help teachers prepare lessons on probability and data analysis for students in Grades 6 through 8. The scatterplot shows the population and the number of area codes for each state in the United States. Use the scatterplot to answer the following questions.

   a. How many states have five area codes? How did you determine the number?
   b. What does the point labeled "*A*" represent? How did you decide what it represented?
   c. The population of Canada is approximately 29,100,000. If a similar pattern exists there, how many area codes do you think Canada has? Explain your answer.
   d. The population of Great Britain is approximately 58,600,000. If Great Britain used the same system, how many area codes do you think you would need? Explain your answer.
   e. Suppose that a state has 14 area codes. What do you think the population of that state would be? Explain your answer.
   f. Describe the relationship between a state's population and the number of area codes assigned to it.

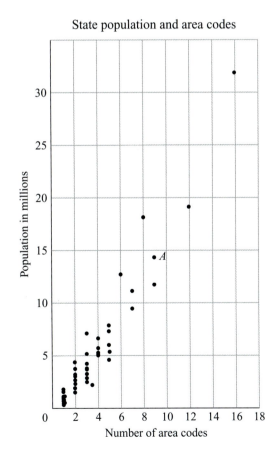

State population and area codes

14. Sixth-graders were asked to keep track of how they spent their time over 48 hours. Here are two students' choices for representing their data.

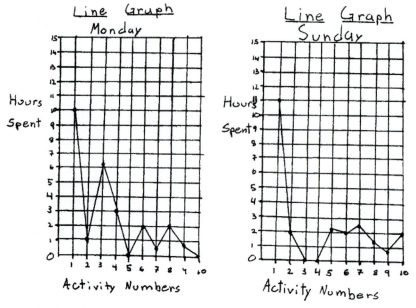

1. Sleeping
2. Skateboard
3. School
4. Sports
5. Church
6. Eating
7. TV
8. homework

9. Getting dressed
10. Shopping

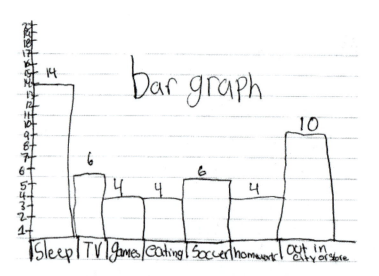

a. What do the dual line graphs enable you to see that a single bar graph does not?

b. What problems do you see with a choice of a line graph?

c. What representation would you choose? Why?

Supplementary Learning Exercises for Section 31.1

1. Smoking and Cancer

Smoke index: relative ranking of amount of smoking, with 100 being average
Cancer index: relative ranking of deaths due to lung cancer, with 100 being average

Occupation	Smoke Index	Cancer Index
Farmers, foresters, and fisherman	77	84
Miners and quarrymen	137	116
Gas, coke, and chemical makers	117	123
Glass and ceramics makers	94	128
Furnace, forge, foundry, and rolling mill workers	116	155
Electrical and electronics workers	102	101
Engineering and allied trades	111	118
Woodworkers	93	113
Leather workers	88	104
Textile workers	102	88
Clothing workers	91	104
Food, drink, and tobacco workers	104	129
Paper and printing workers	107	86
Makers of other products	112	96
Construction workers	113	144
Painters and decorators	110	139
Drivers of stationary engines, cranes, etc.	125	113
Laborers not included elsewhere	133	125
Transport and communications workers	115	146
Warehousemen, storekeepers, packers, and bottlers	105	115
Clerical workers	87	79
Sales workers	91	85
Service, sport, and recreation workers	100	120
Administrators and managers	76	60
Professionals, technical workers, and artists	66	51

Source: From Moore, D. S., & G. P. McCabe (1989). *Introduction to the Practice of Statistics.* New York: W. H. Freeman. Original source: *Occupational Mortality: The Registrar General's Decennial Supplement for England and Wales, 1970–1972.* London: Her Majesty's Stationery Office, 1978.

  a. Show the pairs of data from the table in a scatter plot.
  b. Are the data positively or negatively associated, or is there no association?
  c. What does your answer to part (b) tell you about the situation?

2. To use some computer programs to examine data statistically, the data must be entered according to positions on the page. The table below on the right describes how the data were entered and the table on the left is information about the data.

> NAME: Diamond Ring Pricing Using Linear
>   Regression
> TYPE: Random sample
> SIZE: 48 observations, 2 variables
>
> DESCRIPTIVE ABSTRACT
> This data set contains the prices of ladies' diamond rings and the carat size of their diamond stones. The rings are made with gold of 20-carat purity and are each mounted with a single diamond stone.
>
> SOURCE
> The source of the data is a full page advertisement placed in the February 29, 1992, *Straits Times* newspaper by a Singapore-based retailer of diamond jewelry.
>
> VARIABLE DESCRIPTIONS
> Column:
> **1** Size of diamond in carats (1 carat = 0.2 gram)
> **2** Price of ring in Singapore dollars

0.17	355	0.17	353
0.16	328	0.18	438
0.17	350	0.17	318
0.18	325	0.18	419
0.25	642	0.17	346
0.16	342	0.15	315
0.15	322	0.17	350
0.19	485	0.32	918
0.21	483	0.32	919
0.15	323	0.15	298
0.18	462	0.16	339
0.28	823	0.16	338
0.16	336	0.23	595
0.20	498	0.23	553
0.23	595	0.17	345
0.29	860	0.33	945
0.12	223	0.25	655
0.26	663	0.35	1086
0.25	750	0.18	443
0.27	720	0.25	678
0.18	468	0.25	675
0.16	345	0.15	287
0.17	352	0.26	693
0.16	332	0.15	316

a. To what does "Type" refer?

b. What are the two variables?

c. How much was the most expensive ring and what size was it?

d. Draw a scatter plot of the data using an appropriate scale. Technology may be helpful.

e. Are the variables positively or negatively associated? What does this mean in this context?[9]

3. Are the following variables positively associated, negatively associated, or have no association? Explain your reasoning.

a. length and age of a fish

b. the height of the surf and the number of people in the water

c. the price of gasoline and the average amount of coffee drunk daily

d. height and average number of miles driven each day

31.2 Lines of Best Fit and Correlation

The TIMSS data in Section 31.1 do seem to indicate that there is a relationship between mathematics scores and science scores. But how can we determine if there is a relationship worth noticing? This question has been answered by statisticians, who have a method for finding a *line of best fit* for a scatter plot and who have devised a numerical measure, the *correlation coefficient,* for the degree of association.

 **ACTIVITY 2**   Which Line Fits?

You noticed that there is some relationship between the math scores and the science scores on the TIMSS data. Consider again the scatter plot for science and math scores. On the given graph, draw a straight line that you think best fits the data.

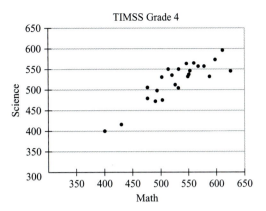

Figure 11 Scatter plot of average math versus science scores for the fourth grade using TIMSS data.

Next, compare your line with those of others in your group, and try to come to a resolution about which line is best and why. ●

From Activity 2, you might have concluded that it is very difficult to tell which straight line best captures the pattern in the data. Statisticians have investigated this problem, and they have developed a mathematical solution to find this line, which they call the *line of best fit* or *regression line*. Briefly, the method considers how close a particular line is to the points by measuring the vertical distances between the points and the line. The line that minimizes the sum of the squared vertical distances is the line of best fit. Calculators or software can do the necessary calculations quickly.

> The **line of best fit,** or **regression line,** is the line that is closest overall to the points in the scatter plot, and thus "best fits" the data.

Such a line is shown in the following scatter plot:

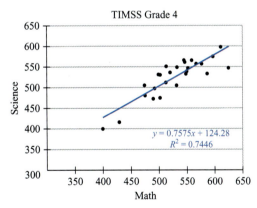

Figure 12 Line of best fit for the fourth-grade math and science scores.

In Figure 12, notice the equation in the display. By interpreting what the x and y represent here, we could say that the relationship between science scores S and math scores M is described by $S = 0.758M + 124.28$. If someone had a math score of 520, what would you predict that person's science score to be? If you calculated correctly in the equation, you could predict the science score to be about 518. Find 520 on the math line and move straight up to the regression line, then straight over to the science score. You should land in the neighborhood of 518. Thus, the line of best fit, or its equation, allows us to estimate values for one variable, given values for the other variable.

Think About ...

What is the slope of the regression line in the previous graph? What is the *y*-intercept? [*Hint*: Use the equation. Notice that the coordinate system shown does not start at (0, 0).]

You probably noticed that after the regression equation under the graph, the statement that $R^2 = 0.7446$ is given. R^2 is a technical measure of how well the line of best fit describes the data. We use this number to find the *correlation coefficient r*. (The switch to the lower-case *r* is for technical reasons.) We take the square root of 0.7446 to obtain a correlation coefficient for the math scores and the science scores. In this case, the correlation coefficient is 0.86. This number allows us to quantify the work we did earlier with correlation or association.

> The **correlation coefficient *r*** is a numerical measure of the association between the two variables. Its values will be between 1 and $^-$1, inclusive.

A perfect positive association would have the correlation coefficient 1 and a perfect negative association would have the correlation coefficient $^-$1.

DISCUSSION 2   What Do Correlation Coefficients Tell Us About the Data?

a. Consider the following seven scatter plots, each with the correlation coefficient *r* given. What changes in the patterns of the points on the scatter plots do you notice as the correlation coefficient goes from a number close to $^-$1 to a number close to 0? From a number close to 0 to a number close to 1?

b. For which graphs do you think the line of best fit would adequately capture a pattern of association?

c. What do you think a graph with a correlation coefficient of exactly 1 would look like? A graph with a correlation coefficient of exactly $^-$1?

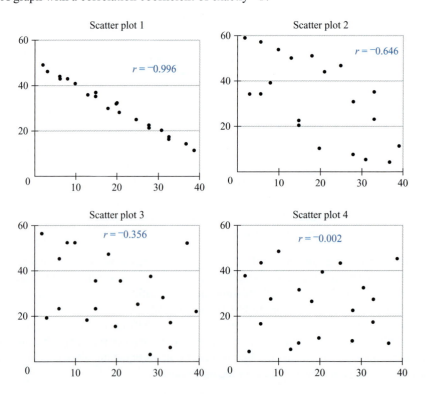

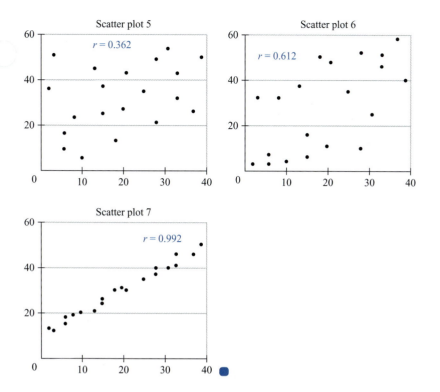

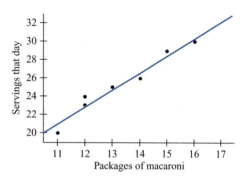

When the correlation between two variables is strong and is close to 1 or ⁻1, the line of best fit can be used to make predictions. Consider the case of a hospital cafeteria that keeps track of the number of packages of macaroni used in a day and the number of servings of macaroni served that day. Suppose these numbers are recorded for a week. We might have a graph with a line of best fit similar to the one in Figure 13.

Figure 13 Line of best fit for macaroni-servings data.

![pencil icon] **ACTIVITY 3** How Much Macaroni?

Using the line of best fit, make a prediction for how much macaroni should be used for 22 servings and for 27 servings. How many servings would you expect to get from $13\frac{1}{2}$ boxes? ●

Caution: Even though a line of best fit can always be calculated, it is not always appropriate to do so because there are times when two sets of data might be related but *not* in a linear fashion. For example, consider the scatter plot (at right) that shows the relationship between the number of people on the beach and the time of day. There is a relationship, with more people during the middle of the day and fewer people at night, but not a linear one.

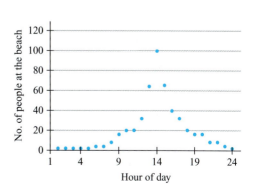

TAKE-AWAY MESSAGE . . .   In many situations with two measurement variables, the relationship between the two variables can be profitably analyzed by the line of best fit (or regression line). The line of best fit allows predictions for other values of the variables, either from the graph or from the equation for the line. The correlation coefficient, with values from 1 to ⁻1, gives a numerical measure of the degree of association between the two variables.

Learning Exercises for Section 31.2

1. Predict whether the correlation coefficient for each pair of the given variables is close to 1, ⁻1, or 0. Explain your answers.
    a. the age of a tree and the diameter of its trunk
    b. the temperature outside an office building and the temperature inside the building
    c. the amount of money a hot dog vendor charges for a hot dog and the number of hot dogs she sells each day
    d. the cost of a new car and its gas mileage
    e. the amount of time a student spends on a test and the student's score on the test (assume the student could take as long as he or she wanted to complete the test)
    f. the time a student spends studying for an exam and the score the student gets on the exam

2. The following table[9] contains the body and brain weights of 25 animals:

Species	Body Weight (kg)	Brain Weight (g)
Mountain beaver	1.35	8.1
Cow	465	423
Gray wolf	36.3	119.5
Goat	27.66	115
Guinea pig	1.04	5.5
Asian elephant	2547	4603
Donkey	187.1	419
Horse	521	655
Potar monkey	10	115
Cat	3.3	25.6
Giraffe	529	680
Gorilla	207	406
Human	62	1320
African elephant	6654	5712
Rhesus monkey	6.8	179
Kangaroo	35	56
Hamster	0.120	1
Mouse	0.023	0.4
Rabbit	2.5	12.1
Sheep	55.5	175
Jaguar	100	157
Chimpanzee	52.16	440
Rat	0.28	1.9
Mole	0.122	3
Pig	192	180

a. Here is a scatter plot of the data with a line of best fit:

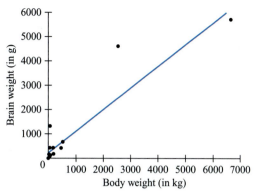

Brain weight (g) = 0.94 · Body weight (kg) + 191.22, $r^2 = 0.87$

Which three animals are denoted with points quite separate from the rest?

b. The correlation coefficient is $\sqrt{0.87} \approx 0.93$. What does this value tell you about brain weight and body weight?

c. Where is the human data point on this graph? Is the human species an outlier in some sense?

d. Suppose a new creature were discovered with a weight of 2000 kg. What could you expect the brain weight to be?

e. What does the correlation coefficient suggest about the association between body and brain weight? Does this surprise you? Explain.

3. After deleting information about the two species of elephant in the table given in Learning Exercise 2, the graphs look different:

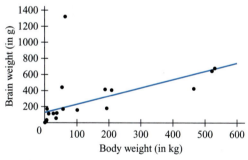

Brain weight (g) = 1.01 · Body weight (kg) + 130, $r^2 = 0.29$

a. What has changed in addition to the deletion of two animals? Do these changes make sense?

b. Is it safe to make predictions with this regression line? Why or why not?

c. How do you suppose the graph would look with the human species removed?

4. Go to http://www.cvgs.k12.va.us/, next DIG stats under Links, and then Graphical Analysis to download data on the Old Faithful geyser (Geyser Scatter Plot). You will find instructions for making a scatter plot of the data using Excel if you click on the blue Excel on the page. At this site, you will find many applications, together with data and data analysis done by high school students. Write a paragraph or page on what you found and did on this site.

5. Suppose that the equation for the line of best fit for some data is

$$y = 0.64x + 121, \quad \text{with } R^2 = 0.81$$

a. What is the correlation coefficient for the data?

b. What is the predicted y value for an x value of 60? Of 100?

continue

Notes

    c. What x value might give a y value of 249?

    d. In this situation, does the x-variable *cause* the y-variable?

6. The TIMMS Grade 4 data are shown with the variable Mathematics on the horizontal axis and the variable Science on the vertical axis.

    a. Make a new graph with Science on the horizontal axis and Mathematics on the vertical axis.

    b. The regression lines will look different, but take any one point and tell whether the meanings of the coordinates agree with those before the axes were changed.

    c. How do the correlations compare?

Supplementary Learning Exercises for Section 31.2

1. In a study of top female runners, researchers measured the stride rate for different speeds. The following table gives the average stride rate (steps per second) of these women versus the speed (ft/s):

Speed (ft/s)	15.86	16.88	17.5	18.62	19.97	21.06	22.11
Stride rate	3.05	3.12	3.17	3.25	3.36	3.46	3.55

Source: http://exploringdata.net/stride.htm.

    a. Draw a scatter plot of the data.

    b. The linear regression equation for these data is $y = 0.08x + 1.77$ with $R^2 = 0.998$. What does this R^2 tell you about the equation and the data?

    c. Make a table showing the predictions of the stride rate using the regression equation above.

Speed (ft/s)	15.86	16.88	17.5	18.62	19.97	21.06	22.11
Stride rate							

2. Tell whether each given scatter plot seems to have a correlation closer to 1, 0, or ⁻1. Explain your answer.

a.
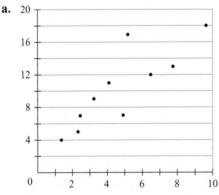

b.
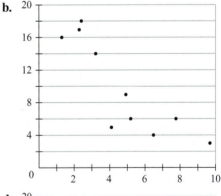

c.
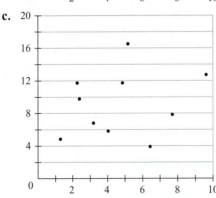

d.
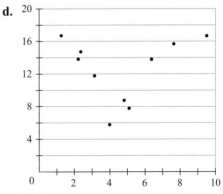

3. Using the graphs from Supplementary Learning Exercise 2, match the following equations and their correlations with their scatter plots:

a. Linear regression $(ax + b)$
regEQ$(x) = 0.335312x + 7.81062$
$r = 0.216177$
$R^2 = 0.046732$

b. Linear regression $(ax + b)$
regEQ$(x) = 0.204913x + 11.8287$
$r = 0.139031$
$R^2 = 0.01933$

c. Linear regression $(ax + b)$
regEQ$(x) = 1.51364x + 3.12535$
$r = 0.839144$
$R^2 = 0.704162$

d. Linear regression $(ax + b)$
regEQ$(x) = -1.87737x + 18.6987$
$r = -0.85194$
$R^2 = 0.725802$

4. Tell whether each given scatter plot seems to have a correlation closer to 1, 0, or ⁻1. Explain your answer.

a.

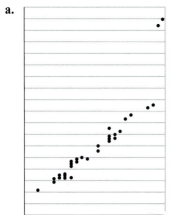

b.

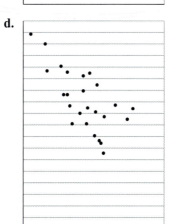

c.

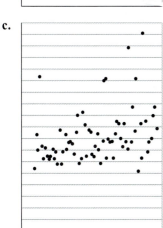

d.

5. Consider graph (d) from Supplementary Learning Exercise 2. How would the correlation change if we looked at only the x values from 0 to 4? How would it change if we looked at only the values from 4 to 10? How do these relate to the overall correlation of the graph?

31.3 Issues for Learning: More Than One Variable

Comparing two data sets is not as easy as it may first seem. The following problem[10] about the heights of fifth-graders was given to eighth-grade students:

▶ Jane and Sam had some data that showed the heights of professional basketball players in centimeters. They showed it to their classmates. Jan suggested that they make a display that shows the heights of the students in their class and the heights of the basketball players. Then they could answer the question: "Just how much taller are the basketball players than students in their class?" ◀

After the students had collected their data, they presented the following stem-and-leaf plot of the heights of fifth-graders, in centimeters:

```
12 |
13 | 8 8 8 9
14 | 1 2 4 7 7 7
15 | 0 0 1 1 1 1 2 2 2 2 3 3 5 6 6 7 8
16 |
17 | 1
```

The eighth-grade students were able to interpret this graph. For example, they could tell how many students were in the class, how many students were 152 centimeters tall, etc. These students were then given a stem-and-leaf plot of heights of basketball players, and they were also able to make interpretations of that graph, which indicated that they understood it. But then they were given the following stem-and-leaf plot that combined the two sets. The heights of basketball players appear in blue.

```
12 |
13 | 8 8 8 9
14 | 1 2 4 7 7 7
15 | 0 0 1 1 1 1 2 2 2 2 3 3 5 6 6 7 8
16 |
17 | 1
18 | 0 3 5
19 | 0 2 5 7 8 8
20 | 0 0 2 3 5 5 5 5 7
21 | 0 0 0 5
22 | 0
23 |
```

Although the eighth-grade students could make sense of each set of heights individually, they were unable to deal with both sets in a manner that showed they understood them. For example, they could describe a typical height for a fifth-grader, or a typical height for a basketball player, but they did not know how to compare these typical heights by finding the difference between them.

This example indicates that students should have practice making sense of more than one data set.

Think About …
How would you make sense of the data?

Children as early as fifth grade can make sense out of scatter plots and a simple form of regression lines. In one study[11] of fifth-graders, researchers wanted to know how students could learn science by investigating real situations. One situation was a study of water evaporation. Students worked in small groups and were encouraged to draw a picture to summarize and communicate the beginning stage of the experiment of comparing evaporation from jars of different shapes. They then used data tables to keep track of evaporation. A graph was needed to display the data. In this case, there were two sets of data to compare, and the best way of comparing them was a scatter plot. Finally, lines of best fit were drawn using a transparent ruler or a thread to move around until a line was found that best fit the data. In this case, there were two sets of data, so two lines of best fit were drawn. The graph invited questions to be asked, such as why the water in Jar 2 evaporated faster than the water in Jar 1. Thus, we see a sequence of activities that could be used for investigating any scientific question these children might explore: picture, table, graph, and question. The first three elements are demonstrated in this series of three figures.

FIGURE 4 Picturing the "Evaporation" experiment

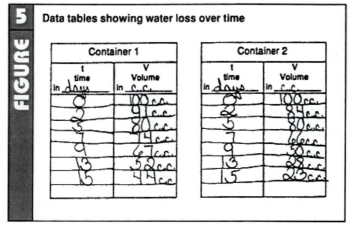

FIGURE 5 Data tables showing water loss over time

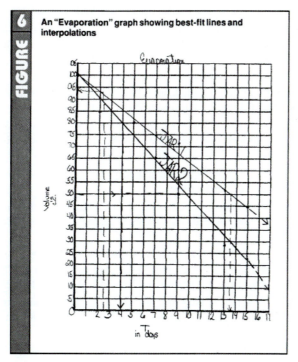

FIGURE 6 An "Evaporation" graph showing best-fit lines and interpolations

Notice here that these students were handling numbers that they themselves generated. They made sense of the numbers and knew what the tables and graphs represented. They were learning to think mathematically and scientifically.

31.4 Check Yourself

In this chapter you learned how to represent and compare more than one data set, how to work with data sets that have more than one variable for each person (or item), and how to interpret those representations. You should be able to work problems such as those assigned and to meet the following objectives:

1. Represent and compare two or more data sets using box plots, histograms, medians, and means.

2. On a coordinate system, make a scatter plot by graphing points that correspond to two variables within the same set, such as lengths and weights of 30 newborns, and then interpret the scatter plot.

3. Recognize scatter plots that represent positive relationships, negative relationships, or relationships that are neither positive nor negative.

4. Recognize scatter plots that represent strong, medium, or close-to-zero correlation coefficients.

5. Explain what information can be provided by a line of best fit and approximate where the line of best fit would fall on a scatter plot.

6. Use a given equation and R^2 value for a line of best fit to give values of one variable, given values for the other, and to determine the correlation coefficient.

7. Recognize that data sets that are strongly correlated do not necessarily show a cause-and-effect relationship, and give an example.

References for Chapter 31

[1]TIMSS data come from http://nces.ed.gov/timss/. This site is updated regularly.

[2]Hand, D. J., Daly, F., Lunn, A. D., McConway, K. J., & Ostrowski, E. A. (1994). *A handbook of small data sets*. New York: Chapman & Hall, pp. 208–209. The data were originally from Peixoto, J. L. (1990). A property of well-formulated polynomial regression models. *American Statistician, 44*, 26–30.

[3]Blastland, M., & Dilnot, A. (2009). *The numbers game*. New York: Gotham.

[4]Frees, E. W., & Miller, R. B. (1998). Designing effective graphs. *North American Actuarial Journal, 2*(2), 53–76. Page 55 appears here. Copyright 2009 by the Society of Actuaries, Schaumburg, IL. Reprinted with permission.

[5]Hand, D. J., Daly, F., Lunn, A. D., McConway, K. J., & Ostrowski, E. A. (1994). *A handbook of small data sets*. New York: Chapman & Hall, pp. 179–183. The data were originally from Marsh, C. (1988). *Exploring data*. Cambridge, UK: Policy Press, p. 315.

[6]Frees, E. W., & Miller, R. B. (1988). Designing effective graphs. *North American Actuarial Journal, 2*(2), 53–76. Page 57 appears here. Copyright 2009 by the Society of Actuaries, Schaumburg, Illinois. Reprinted with permission.

[7]See Florida Fish and Wildlife Conservation Commission report on probable cause of death for manatee. http://research.myfwc.com/manatees.

[8]Bright, G. W., Brewer, W., McClain, K., & Mooney, E. S. (2003). *Navigating through data analysis in grades 6–8*. Reston, VA: National Council of Teachers of Mathematics. Pages 100–101 are used here.

[9]Hand, D. J., Daly, F., Lunn, A. D., McConway, K. J., & Ostrowski, E. A. (1994). *A handbook of small data sets*. New York: Chapman & Hall, pp. 232–233. The data were originally from Jerison, H. J. (1973). *Evolution of the brain and intelligence*. New York: Academic Press.

[10]Bright, G. W., & Friel, S. N. (1998). Graphical representations: Helping students interpret data. In S. P. Lajoie (Ed.), *Reflections on statistics*, pp. 63–88 (particularly, pp. 80–81). Mahwah, NJ: Erlbaum. Redrawn here.

[11]Isaacs, A. C., & Kelso, C. R. (February 1996). Pictures, tables, graphs, and questions: Statistical processes. *Teaching Children Mathematics, 2*(6), 340–345.

Chapter 32

Variability in Samples

Recall that a **population parameter** is a measure of a characteristic of the whole population, whereas a **sample statistic** is a measure of a characteristic of the sample. In this chapter you learn to judge the validity of a claim and to draw conclusions based on a sample from the population. To judge the validity of a claim, we consider how likely the claim is to be true under certain conditions. To draw inferences from a sample, we consider how likely an observed sample statistic is after making assumptions about the sample's source. The emphasis in this chapter is on understanding how statistics are used to make decisions in settings involving some uncertainty. This understanding should help you become a smarter consumer and make educated decisions about statistical information you read and hear. *A fundamental and powerful idea that underlies this entire chapter is that a reasonably chosen sample is sufficient for making quite accurate predictions about the entire population from which the sample was drawn.*

32.1 Having Confidence in a Sample Statistic

By now you know that if you take more than one sample, you are likely to get different values for a statistic such as the mean. Is one value better than another? Sections 32.1 and 32.2 explore this question.

 DISCUSSION 1 What Would You Expect?

Suppose you know that 63% of the student body at State University is female. Suppose further that you work in the bursar's office, where students pay their registration fees.

a. Would you expect 6 of the next 10 people in line to be female? Is it possible that only 3 of the next 10 are female? Is it possible that none of the next 10 is female?

b. Would you expect exactly 63 of the next 100 students to be female? Would you be surprised if only 30 were female? Do you think it is possible that none of the next 100 students is female?

c. Explain how your previous work with probability helped you think about these questions ●

Most people who answer these questions say that they would not be overly surprised if the next 10 in line contained only 3 females, but would be somewhat more surprised if out of the next 100, only 30 were females. They would be very surprised if none of the next 10 students in line is female, and they do not think it is possible that none of the next 100 students is female. Where do these opinions come from? What is the basis for these intuitive notions? Are the notions correct?

Understanding probability leads us to reason that some statistical values for samples of a population are possible, while other values, if not impossible, are highly unlikely. But you may be bothered by the fact that it *is* possible, e.g., that only 3 of the next 10 students are female because the *sample statistic* 30% is not a good estimate of the *population parameter*, 63%. In other words, if there can be tremendous variability in sample statistics from the same population and if only one sample is used, then intuitively you might doubt that the resulting sample statistic will be a good estimate for the population parameter. Just how big a concern is this variability in sample statistics?

> ### *Think About ...*
> Does the variability in a sample statistic (from sample to sample) mean that you cannot trust sampling?

In this section you will find out how to react to sample variability. The conversation here is limited to samples in which there are two possible outcomes per item (or individual), such as *yes* or *no, for* or *against, female* or *male.* We are interested in the percent, or proportion, of each outcome within the sample. For example, if 25% say *yes,* then 75% say *no.*

If we already know the population parameter, then we would not need to draw samples to find estimates of this parameter. What would be the point of estimating a parameter whose true value we already know! Working with samples once the population parameter is known is *not* the way statisticians work. But taking this route, as we do here, is instructive *because it can lead to a better understanding of what sample statistics tell us.* The point of sampling is not to find estimates of something we already know. Rather, the reason we consider sample statistics here when we *do* know the population parameter is simply to come to a better understanding of how much credence we can place in a statistic drawn from *one* sample. We will assume for now that we are working with unbiased, random samples.

> ### *Think About ...*
> Why are unbiased, random samples important?

This section is intended to illustrate further the probabilistic nature of sample statistics. Just as we cannot be *sure* that tossing two dice will give us a sum less than 12, we know that our chances are fairly good that we will not get 2 sixes. In the same way, we cannot be absolutely certain of a population parameter unless we undertake a census of all members of the population, if that is even possible. But *sampling offers us a way to be **very close** to certain without the expense and time needed to check out every member of the population.*

🗨 DISCUSSION 2   Considering the Effect of Sample Size

A newspaper columnist drew a 10-person random sample of state voters and asked them whether they were for or against Proposition 223. Six of the 10 people said they were against the proposition.

a. Is the columnist safe in making a prediction that about 60% of the people in the state are against the proposition?

b. Suppose the random sample was 100 people, and 60% said they are against the proposition. Is she any safer in saying that 60% of the voters are against the proposition? ●

Suppose, *for instructional purposes only,* we already know that exactly 63% of the people in the state are against Proposition 223. Then we would ask: If 63% of the population are against Proposition 223, how likely is it that a randomly selected sample of 10 people from the population (admittedly a very small sample) will yield a percent against Proposition 223 that is "far away" from 63%? What is the chance that the sample will show, say, 40% against the proposition rather than about 63% against?

If a sample can provide an estimate of the population parameter that is far away from the actual population parameter, we have reason to be concerned, especially if the chances are great that such a discrepancy will happen. In Activity 1, we explore the likelihood that with sample size 10, the population parameter and the sample statistic are too far apart to consider the statistic a valid estimate for the parameter.

✏️ ACTIVITY 1   What 50 Samples of Size 10 Say

We demonstrate two ways of undertaking this activity. This one uses *Illuminations*, and Activity 2 uses *Fathom*. You need not do both.

To use *Illuminations* (http://illuminations.nctm.org), go to the menu item Activities and click on any one of the Grades. Look through the Activities until you find Adjustable Spinner, then click on it.

a. Set the number of sectors to 2, the number of spins to 10, and the theoretical probabilities to 63% (representing those against Proposition 223) and 37% (representing those for Proposition 223). Click on **Update.**

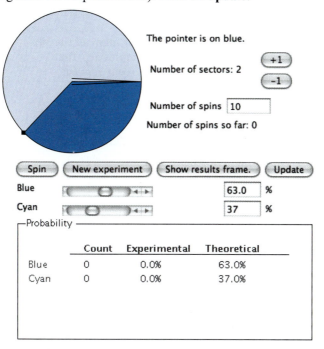

b. Click on **New experiment** and then **Spin** a few times. Each time, compare the **Experimental** and **Theoretical** probabilities, to get a feel for what is happening. Notice that on any one experiment, the experimental probabilities can vary a great deal from the theoretical probability.

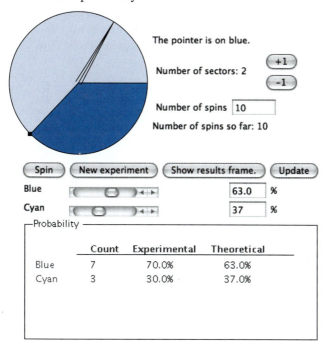

c. Suppose we spin 10 times, over and over for 50 trials, clicking on **New experiment** after each trial. In the following table, fill in the second column, which represents the experimental probabilities for the percent of people against Proposition 223 in each set of 10 spins. For example, the experimental probability for the trial illustrated above was 70%, so you would place a tick mark in the 70% row. After 50 rounds of spinning 10 times each, the 70% row may look like this:

70%	*### //*	7

Be sure to click on **New experiment** each time so that you get a new percentage that does not include the last trial. The numbers in the second column should add up to 50.

Percent in the sample that are *against* Proposition 223	How many of the 50 samples give that percent *against*
0%	
10%	
20%	
30%	
40%	
50%	
60%	
70%	
80%	
90%	
100%	

Study the results. What were the percentages you found? Did they cluster around 63%, the theoretical probability in this case? Why could none of the experimental probabilities be 59%? ●

 ACTIVITY 2   Once Again, What 50 Samples of Size 10 Say

The following display gives the results from a *Fathom* simulation of drawing 50 samples, each of size 10, from a population having 63% as the against rate. Each of the 50 balls represents one sample of size 10, with the percent against Proposition 223 for that sample shown underneath the ball.

O	O	O	O	O	O	O	O	O	O
40%	50%	80%	30%	60%	80%	70%	60%	40%	70%
70%	70%	90%	70%	60%	70%	50%	80%	40%	70%
80%	50%	70%	60%	50%	80%	50%	50%	80%	70%
40%	60%	70%	70%	80%	70%	50%	90%	60%	50%
80%	80%	60%	70%	70%	10%	70%	80%	70%	30%

a. What does the first ball and its percent mean?

b. Why are all the percents multiples of 10%?

c. Complete the following table, summarizing the findings for the percent against Proposition 223 for all 50 samples. (You may wish to make tally marks to the right of the table.)

Percent in the sample that are *against* Proposition 223	How many of the 50 samples give that percent *against*
0%	
10%	
20%	
30%	
40%	
50%	
60%	
70%	
80%	
90%	
100%	

Usually, there is a great variation in the percents when sample sizes are quite small, such as the sample size of 10 in the activity. ●

DISCUSSION 3   What Would Happen If . . . ?

Here is an extremely important question: As the sample size gets larger, what would happen to the spread of the results? Do you think you would be more or less likely to get 20% with a sample size of 10 or with a sample size of 100? Make predictions about what you would expect for sample size 1000, using 100 spins. Would you feel more confident of the result when the sample size is 500 compared to a sample size of 2000? Explain why or why not. ●

For each sample of size 100, we would obtain a sample statistic, just as we did for each sample of size 10 in Activity 2 (or 1). For example, suppose that in the fourteenth sample of size 100, forty-six (46%) say they are against Proposition 223. Once again, one sample may not give a very good estimate of the parameter, which is 63%, so getting 46% is worrisome. Rather than focusing on just *one* sample and its statistic, we focus instead on *what fraction of many samples* of 100 will give us 63% or more against the proposition— i.e., what fraction of these samples gives results that are a certain percent away from the actual parameter of 63%. Table 1 shows what happens when 1000 (rather than only 50) samples for *different sample sizes n* are found, using computer simulations.

DISCUSSION 4   What Does a Large Sample Size Buy?

Use Table 1 on the following page for this discussion.

a. In the column $n = 500$ what exactly does the 25 mean?

b. What is the sum of the numbers in each column? Why would you expect to get this sum?

c. What exactly does a sample size of 3000 give you that a sample size of 2000 does not?

d. Compare the results for a sample size of 1000 with those for sample sizes of 2000 and 3000.

e. How close to 63% are all the statistics from all the samples when the sample size is $n = 1000$?

f. Is your answer in part (e) close enough for most practical purposes? Why or why not? ●

Table 1 1000 Samples of Different Sizes from a Population of Which 63% Are Against Proposition 223 (the population parameter is 63%)

Percent that are *against* Prop 223, in a sample	Number of the 1000 samples of size *n* having the given percent *against* Prop 223 (from the first column)				
	Sample Size				
	n = 100	*n* = 500	*n* = 1000	*n* = 2000	*n* = 3000
48	1				
49	1				
50	1				
51	6				
52	6				
53	10				
54	14				
55	18	1			
56	29	3			
57	30	2			
58	55	6	2		
59	36	25	6	1	
60	72	73	50	12	6
61	76	105	116	64	41
62	84	189	226	229	226
63	85	179	240	335	421
64	77	163	204	273	251
65	99	135	99	73	52
66	68	74	50	13	3
67	68	26	6		
68	55	12	1		
69	34	5			
70	28	2			
71	14				
72	15				
73	8				
74	6				
75	0				
76	0				
77	1				
78	1				
79	1				
80	0				
81	0				
82	0				
83	1				

Table 1 is important because it shows that *although a sample statistic may vary from sample to sample, values closer to the population parameter are more probable than values farther from the population parameter. The sample statistics cluster around the population parameter.*

Notes

💬 DISCUSSION 5   Do You Really Have to Have a Huge Sample?

Table 1 represents the percent against Proposition 223 from 1000 simple random samples for different sizes *n*, where each sample is taken from the population of all citizens. Again, for the sake of seeing what happens, we assume that the population parameter is actually 63%.

a. What does Table 1 tell you about sample size? Notice that the population could have been very large.

b. What general patterns do you see in this table? What does your pattern tell you about the effect of sample size?

c. Complete this statement: "The larger the sample size, ..." ●

Another important point shown in Table 1 is that *as sample size increases, the more likely it is that a completely random, unbiased sample will give us a sample statistic that represents the population parameter quite well.* Here is a different way of thinking about this fact: With a sample size of 100, the chance is small, but not negligible, that we could have a sample statistic of either 53% or 74%, both of which may not be considered acceptable estimates when the population parameter is actually 63%. But when the sample size is 1000, nearly all the sample statistics are within 5 percentage points of the true parameter. Now the chances of getting a sample statistic of either 53% or 74% are minuscule.

To summarize: *The sample statistic of a larger sample will usually be closer to the population parameter than it would be for a smaller sample. However, **the sample need not be enormous** in order to get a good estimate of a population parameter.* Additionally, when we do *not* know the population parameter, which of course in reality we never do and must estimate it using a sample statistic, we can still be sure that with a large enough sample size, the sample statistic should be close to the population parameter.

So, in addition to the issue of *bias* discussed in Section 29.2, there is also the issue of *how many* individuals or items to sample to get a useful statistic without having an unnecessarily large sample size. Sample size affects the precision of the results of the sample—i.e., sample size affects how far off a sample statistic might be from the population parameter. A completely random sample of five is highly unlikely to give the precision you seek, no matter what the size of the population.

But what sample size would be sufficient? *Surprisingly enough, the size of the sample need not reflect the size of the population, so long as the population is much larger than the sample.*

Although increasing the size of a sample does increase the trustworthiness of the results, after a point the added precision of a larger sample size does not gain us enough additional certainty to justify the cost of obtaining data from a larger sample. For example, if the sample size is 3000 in Table 1, the percentage of votes *against* Proposition 223 in any one sample is within approximately 3 percentage points of the parameter. In most cases, this is close enough, and perhaps even closer than needed. Thus, to sample 5000 individuals would probably be a waste of time and money, even if the population were the entire country.

For surveys such as those we have been discussing, there is a rule of thumb that gives an idea of how far off our sample statistic might be—i.e., how precise the statistic is. For example, if we take a sample of 100, it is fairly safe (when we have reason to believe that the population proportion is fairly close to 50%) to predict a **margin of error** of $\frac{1}{\sqrt{100}}$, or $\frac{1}{10}$, or 10%. If we surveyed 100 people and asked whether they prefer Pepsi or Coke and found that 55% preferred Pepsi, then the interval from 55% − 10% to 55% + 10% is highly likely to contain the true population proportion. That is, the interval from 45% to 65% is highly likely to contain the population parameter.

> **A rule of thumb:** The expression $\frac{1}{\sqrt{n}}$ will give an estimate of the margin of error with a sample of size n if the population proportion is close to 50%.

Of course, this limited rule of thumb is just that—a quick estimate of the accuracy of a statistic based on a sample. There is a more standard and more robust way of determining how confident we can be that our sample percent represents the population proportion. The method involves some sophisticated computations that we will not undertake here. In the next section, we will look at an example that provides an intuitive idea of how to find what are called *confidence intervals*.

TAKE-AWAY MESSAGE . . .   If samples are unbiased, a statistic derived from a large sample is more likely to be closer to the corresponding parameter of the population than when the statistic is derived from a small sample. But just how large does the sample have to be? Surprisingly, the sample does not have to be extremely large, no matter what size the population. The sample size required will depend on how confident one wants to be about the sample statistic's proximity to the population parameter. In some circumstances, there is a rule of thumb for relating sample size to the exactness of the sample statistic.

Learning Exercises for Section 32.1

1. A polling company increased its sample size from 1000 people to 4000 people before a big election. Why would they do this?

2. Using Table 1 on p. 754, complete this table, which tells the number of samples with statistics that would be within a specific percent from the population parameter. Two entries in the table are provided for you.

1000 samples of size n	Number of samples in which the sample statistic is within this % of parameter:		
	≤10%	≤5%	≤2%
$n = 100$			421
$n = 500$			
$n = 1000$			
$n = 2000$			
$n = 3000$			991

3. According to Table 1 (remember that there is a total of 1000 samples), what is the likelihood of having a sample statistic of exactly 63% *against*

   a. if the sample size is 100?   b. if the sample size is 1000?
   c. if the sample size is 2000?   d. if the sample size is 3000?

4. A graduate student is planning her dissertation study. She plans to test a random sample of 20 fifth-grade students in a large school district on their attitudes toward violence in schools. Do you have any advice for her?

5. Which of the following samples would you want? Tell why.

   a. a sample with low bias and low precision
   b. a sample with high bias and low precision
   c. a sample with low bias and high precision
   d. a sample with high bias and high precision

6. Suppose we asked a random sample of 400 people if they preferred Coke or Pepsi and found that 54% of the people preferred Pepsi. What could we say about the true population proportion?

7. In a poll of 1631 people in a city of a million people, 53% said they would vote against funding a new library.

   a. Might funding for a new library still be approved? Tell how you know.
   b. Would you feel as you did in part (a) if only 100 people had been polled?
   c. Suppose that after polling 100 people, you found that 75% of the sample said they would vote against funding a new library. Do you think the measure will pass?

8. According to the U.S. Constitution, the federal government is obligated to take a census every 10 years. A census is intended to gather data from every person living in the United States rather than from a sample of the residents of the United States. Every household receives a form to complete.

   a. What do you think might be some of the problems involved in taking a census?
   b. In the 2000 census, it was estimated that 10 million people were missed. These people were primarily African Americans in inner cities and Latinos in Texas, New Mexico, and California. Why do you think these people were not counted?
   c. The undercount has serious political ramifications. What do you think they are?
   d. In the same census, it was estimated that about 6 million people were counted twice. Why do you think this double counting happened? Do you think there are also political ramifications involved in an overcount? Why or why not?
   e. Many people advocate an increased use of sampling in the census. What are the pros and cons of doing so?

9. When tossing three honest coins, is the likelihood of getting at least two heads smaller than, greater than, or equal to the likelihood of getting at least 200 heads when tossing 300 coins?

10. Imagine you have a huge jar of M&M's with many different colors of candies. We know that the manufacturer of M&M's puts in 40% browns. If you reach in and pull samples of 20 M&M's at a time, what do you think would be the likely interval for the numbers of browns you find in the samples? If you pull samples of 100 M&M's, what would be the likely interval for the number of browns?

11. You suspect that the population parameter is about 50%. What sample size would you use if you want the margin of error to be

   a. ±5%?   **b.** ±1%?   **c.** ±0.5%?

Supplementary Learning Exercises for Section 32.1

1. **a.** In the row that starts with 59 in Table 1 on p. 754, what do the numbers 36, 25, 6, and 1 mean?
   b. In the row that starts with 83 in Table 1, what does the 1 mean?

2. If there were a sample size of $n = 5000$ in Table 1 on p. 754, what do you think it would look like?

3. When can you use the $\frac{1}{\sqrt{n}}$ rule of thumb? What does the n refer to? What does the rule of thumb tell you?

4. Suppose 88% of a random sample of 1000 students said that they were in favor of longer library hours during final examinations. Why isn't the $\frac{1}{\sqrt{n}}$ rule of thumb applicable?

5. Is each statement true or false? Explain why any statement is false.

   a. With a larger and larger sample size, it is more probable that a random, unbiased sample will give a sample statistic that estimates the population parameter quite well.

   b. A sample statistic for a particular sample size will always be the same value from sample to sample.

   c. For a particular sample statistic, values closer to the population parameter are more probable than values farther from the population parameter.

6. Using data from Table 1 on p. 754, make a bar graph for the number of samples of each size below, showing the particular percents represented.

   a. $n = 1000$ (For 60%, there are 50 cases.)

   b. Does the bar graph suggest a normal curve?

32.2 Confidence Intervals

We can formalize some of the ideas discussed in Section 32.1 by using the terminology of confidence intervals. In this section we use examples and definitions to accomplish this task.

 DISCUSSION 6   Will the Incumbent Win?

Suppose you read a newspaper article that says, "A poll taken yesterday shows the incumbent has 55% of the vote, plus or minus 5 percentage points." What does this statement mean? ●

To understand the assertion made in Discussion 6, we must think about what would happen if several such polls were taken. Just as in probability statements (Section 27.1), we must think about repeating a process (such as taking a poll) a large number of times (even though in practice a poll involves just one sample).

So, suppose that 135 polls of 100 people each were taken over a brief period of time in Santa Maria, where the incumbent mayor, Luis Serrano, is running against Agnes Johnson. That is, 135 volunteers went out and each polled 100 randomly selected people. Table 2 below summarizes the results. One volunteer found that 44% of the sample favored Serrano, 2 volunteers found that in their samples 45% favored Serrano, etc.

Table 2 Frequencies of Percents for Serrano in 135 Samples

Percent	Frequency	Percent	Frequency	Percent	Frequency
44	1	52	11	60	6
45	2	53	10	61	3
46	0	54	12	62	4
47	4	55	13	63	1
48	3	56	11	64	2
49	6	57	9	65	1
50	9	58	8		Sum = 135
51	10	59	9		

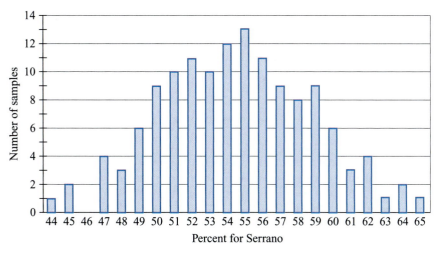

Figure 1 The distribution of percents for Serrano in 135 samples of 100.

Figure 1 shows a bar graph for this data set. All sample percents lie between, and include, 44% and 65%, but most sample percents appear to be in the interval from 50% to 60%. (Remember that the paper said the incumbent had 55% of the vote, plus or minus 5 percentage points—i.e., 50% through 60%.) But how sure can we be that this interval does include the true population parameter, the percent of the vote that Serrano will get?

Using the $\frac{1}{\sqrt{n}}$ rule of thumb (Section 32.1), each of the 135 sample percents gives an interval that has a good chance of including the population parameter. For example, here $\frac{1}{\sqrt{n}} = \frac{1}{\sqrt{100}} = 10\%$, so a sample having 54% in favor of Serrano gives rise to the interval 44%–64%, which has a strong chance of including the unknown population parameter. Furthermore, from Section 32.1 you know that with a large number of samples, their percents tend to cluster around the population parameter. Recall from Chapter 27 that the probability of an event A occurring is

$$P(A) = \frac{\text{number of times } A \text{ happens}}{\text{total number of outcomes}}$$

Over the 135 samples we have here, 108 of them have sample percents from 50% through 60% and each percent gives an interval that has a good chance of including the population parameter. Hence, P(including the population parameter with these intervals) $= \frac{108}{135} = 0.8$. That is, 80% of the sample intervals are likely to contain the population parameter. So, over many samples, there is an 80% probability that the 50% to 60% interval will contain the population parameter. This last sentence is an example of a *confidence statement*. The probability level, or *confidence level*, is 0.8, or 80%. The *confidence interval* is 50% to 60%.

> ### *Think About ...*
>
> 1. Confirm from Table 2 that the interval 55%, plus or minus 2%, covers the population parameter, with a probability of about 41%.
>
> 2. Instead of being 80% certain that the confidence interval 50% to 60% includes the population parameter, what would the confidence interval be if you wanted to be 95% certain that the interval included the population parameter? (*Hint:* 5% of the sample percents would be outside the interval.)

> The sample statistic and the margin of error enable us to find the **confidence interval**. A **confidence statement** tells the level of confidence, probabilistically speaking, that the confidence interval includes the population parameter.

EXAMPLE 1

A poll report may state that we can be 97% confident that a candidate will get 72% of the vote, ±5 percentage points. This is a confidence statement involving a 97% confidence level and a 5% margin of error, with a confidence interval of 67% to 77% (calculated from 72% ± 5%). The statement means that there is a 97% probability that the interval from 67% to 77% includes the population parameter. It also recognizes that there is a 3% probability that the interval does *not* include the population parameter. ●

In actuality, the confidence level is often not reported, in which case it is assumed to be 95% (if the source is reputable).

> ### *Think About ...*
>
> In a two-candidate race in the country of Timoco, a poll of a sample of residents reports that 52%, ±4%, of the residents would vote for Juan Gonzalez. What exactly does this statement mean? Can Mr. Gonzalez be quite certain of winning? Can Mr. Gonzalez be quite certain of winning if, instead, the report indicates with 95% confidence that 55%, ±4%, would vote for Mr. Gonzalez?

 ACTIVITY 3   Finding Confidence Intervals

Return to Table 1 on p. 754. For each sample size, tell what percent of the sample statistics had values within 10 percentage points of the (known) population parameter (63%). What percent of the samples had values within 5 percentage points of the population parameter? Within 3 percentage points? Within 1 percentage point? Give confidence statements in each case. ●

Using Table 1 again, let us apply the same reasoning in the case of polling people about Proposition 223. Suppose we take a sample of 100 and find the sample statistic. From Table 1, there are $1000 - 25 = 975$ values that lie within 10% of the (known) population parameter, i.e., greater than or equal to 53% and less than or equal to 73%, and $\frac{975}{1000} = 97.5\%$.

Thus, we can be 97.5% confident that the sample statistic is within 10 percentage points of the population parameter, so we would have a 97.5% confidence level with a confidence interval between (sample statistic − 10%) and (sample statistic + 10%). Similar reasoning would lead to the conclusion that we would be only about 78% confident that the sample statistic is within approximately 5 percentage points of the parameter. But if we have a sample of 1000, we can be about 98.5% certain that the sample statistic is within 3 percentage points of the population parameter. Thus, one way to obtain a smaller margin of error is to use a larger sample size. Practical considerations such as the added time and money to obtain a larger sample enter in, particularly because a moderate sample size gives quite good information.

The way that statisticians find confidence intervals in practice is different from the method we used in this section. In particular, statisticians do *not* take several different samples of the same size each. But by applying probability ideas, they can make such statements with some confidence. You know that probability statements do not offer certainty, and the same is true here. Consider the last Think About. Although we cannot say *absolutely* that, e.g., Gonzalez will get 50% or more of the votes, we can be confident enough to make related decisions.

Problem 4 in the introduction to Part IV on p. 605 addressed a poll undertaken to determine whether the public favored reform of the city pension system. The letter writer was not confident that the sample of size 611 people would be adequate to accurately predict the population parameter. Do you agree? You are asked to answer this question in Supplementary Learning Exercise 6.

TAKE-AWAY MESSAGE . . . A confidence statement tells us the level of confidence we may have that a particular confidence interval, determined by the sample statistic and a selected margin of error, contains the population parameter. Selecting the size of a sample should take into consideration both the goal of a close estimate and the time and financial costs of sampling. Choosing a sample larger than needed can be time-consuming and wasteful.

Learning Exercises for Section 32.2

1. The candidate for senator hired a firm to take a poll of people in his state to see if he had enough support to win the race against his only contender. The pollsters said they were 95% confident that he had the support of 54% of the voters in the state, $\pm 5\%$. Is there a chance he could lose the race? If he wanted to reduce the size of the confidence interval, what might he ask the pollsters to do? Would this solve his problem of uncertainty?

2. In 1936 Alf Landon and Franklin D. Roosevelt ran for the office of president of the United States. The *Literary Digest* had predicted the winner in several previous elections, and in 1936 the *Digest* predicted that Landon would win by a margin of 57% to 43% and would take 370 of 531 electoral votes. But we all know that Landon did *not* win. He received only 38% of the vote, and only 8 electoral votes! What happened? The magazine had mailed out 10 million sample ballots, of which 2.3 million were returned. Names and addresses were obtained from magazine subscriptions, car registrations, and telephone directories. Why was the *Digest* so mistaken? (*Hint:* How many people owned cars or telephones in 1936?)

3. In Chapter 27 you tossed a thumbtack 100 times and noted how many times it landed point up, and you used this number to state a probability that a thumbtack will land point up. Suppose the percent of times a thumbtack lands point up is normally distributed, with a standard deviation of 0.05. If you draw a normal curve for the percent of thumbtacks falling point up, what would be your 95% confidence interval for the parameter? Make a confidence statement about p, the percent of times a thumbtack will point up when tossed. (You will need your result from Activity 1 in Section 27.2.)

4. A newspaper reported that it had polled several voters and found 53% of the voters, with a margin of error of $\pm 2\%$, supported changing Social Security. As a reader of this newspaper and a person hoping someday to receive Social Security checks, what does this report mean to you?

5. What confidence interval would the $\frac{1}{\sqrt{n}}$ rule of thumb suggest if the observed proportion is 52% and the sample size is . . .

    a. 9?   **b.** 90?   **c.** 900?   **d.** 9000?

Supplementary Learning Exercises for Section 32.2

1. Stamfurth University has about 15,000 students. Its student council is considering raising student fees by $75 per student per semester, to be used to build an Olympic-sized pool. The contractor who had made a proposal for the work is anxious to know whether students will likely approve the new fee. The council members discuss the likelihood of the work being approved. Jeff says the entire student body needs to be

polled to be certain that the students will approve the work. Angie says that a sample of 100 should do, but Jan believes that a sample of 1000 is needed.

    a. What is the margin of error for Jeff's idea? Angie's and Jan's ideas?

    b. Suppose that a polling using Jeff's idea gives a result of 51% approval of the increase in student fees. Should the university proceed with the work? Explain your thinking.

    c. Suppose that a poll of 100 randomly selected students (Angie's idea) gives a 56% *yes* vote. Should the university proceed? Explain your thinking.

    d. Suppose that Jan's idea of using a sample of 1000 students gives a 54% *yes* vote. Should the university proceed? Explain your thinking.

2. A news report said that 52% of likely voters will vote for Candidate X, with a margin of error of 4%.

    a. Why should Candidate X not assume a win?

    b. What sample size might the poll have used?

3. A laboratory claimed that, based on a test of its new product for patients with a particular disease, 98%, plus or minus 2%, of such cases would be cured, with 95% confidence.

    a. What is the confidence interval for this claim?

    b. What is the level of confidence in this claim?

    c. Is it possible that someone with the disease will try the product and not be cured? Explain.

4. After asking 16 of her friends about two candidates for a campus office, Robin asserted, "I think that 56% will vote for Pat." What advice might you give Robin?

5. "Our confidence interval is 48%–54%, with confidence level 95%." What were the value of the sample statistic and the margin of error?

6. On March 26, 2012, the San Diego newspaper reported on a survey undertaken by a reputable polling company that indicated 56% of the public (with a margin of error of 4%) favored an initiative for pension reform. The following day, a letter to the editor disagreed with the polling results, stating that the poll surveyed only 611 people, a sample which was too small. Do you agree? If not, why not?

32.3 Issues for Learning: What Probability and Statistics Should Be in the Curriculum?

The chance and data topics you have studied in this book are part of the larger fields of, respectively, probability and statistics. Now that we have a state-led initiative known as the *Common Core State Standards (CCSS) for Mathematics*[1] that has been adopted (with modification) by 45 states, we can use it as a basis for discussing what could be included in the curriculum with regard to probability and statistics.

> *Think About …*
>
> What probability and statistics topics do you remember studying from your own Grades K–5 schooling? From your Grades 6–8 schooling?

Although probability and statistics topics have been promoted in Grades K–8 for some years, chapters devoted to them have sometimes been found toward the end of textbooks and not reached, or have been skipped because of a teacher's view that other topics are more important. But results from the National Assessments of Educational Progress[2] over a ten-year period suggest that more attention is now given to the topics, judging from the improvement in student performances at Grades 4 and 8.[3] The tests themselves include a noticeable number of items devoted to the general area of probability, statistics, and data analysis: About 10% of the items at Grade 4 and 15% at Grade 8 deal with these topics.[4]

So, your curriculum (and testing program) will almost certainly include attention to probability and statistics. The *Common Core State Standards* list *Statistics and Probability* as a separate domain at Grades 6 and 7. Before Grade 6, there are fundamental concepts and understandings that prepare students for the intense focus that follows in these grades. First, we'll discuss the expectations for Grades 6 and 7. This will help us think about what statistics and probability should be included in the curriculum for earlier grades.

The authors of the *CCSS* envision mathematically proficient students in Grades 6 and 7 as developing an understanding of mathematical variability. Such students should be able to summarize and describe distributions, which involves describing a distribution by its center, spread, and overall shape. As discussed in the Issues for Learning in Section 30.7, students need to understand, e.g., the mean as a "data reducer"—they need to see the mean as one value that is standing for many values. They should understand whether the mean is the appropriate measure of center depending on the context in which the data were gathered.

The authors describe students who use the measures of center and other measures of variability to compare two populations. The students should be able to use a sample of a population to draw inferences about the population. They also should be able to understand what constitutes appropriate and inappropriate sampling.

Students should investigate chance. They should understand that the probability of a chance event is a number between 0 and 1, that probability is determined by observing its frequency of occurrence over the long run. They use organized systematic thinking, tree diagrams, and tables to describe outcomes in a sample space. Probability ideas grow to include complementary events and mutually exclusive events. The students are expected to deal with more complicated experiments such as drawing twice from a container of cubes of two or more colors, and the related tree diagrams. Although the students should often carry out actual experiments to see the variation in outcomes, computer simulations can allow an experiment to be carried out a large number of times, very quickly.

This should sound familiar, as it is the work we did in the early chapters of Part IV. What are the activities in which elementary teachers should engage their students so they are ready for such sophisticated ideas?

Students should pose questions, gather relevant data, and represent the data with objects like cubes, pictures, tables, or graphs. They should give comments on what the overall data set shows. For example, in asking how many children have a pet, they could make a table showing tallies, or each child might put his or her name on a Post-It and then place it in the appropriate place to make a bar graph. They could then comment on the most common pet or discuss how many more children have one type than another type, or compare different representations, such as a table versus a bar graph. They should describe some familiar events as likely or unlikely.

Students should collect data for their questions, using observations, surveys, or experiments, and then learn to represent the data in a variety of ways such as tables, bar graphs, and line graphs. Students should be able to compare and contrast different data sets, or different representations of the same data set, and to draw conclusions that tentatively answer the original questions. Students should become familiar with measures of central tendency, particularly the median, of a set of numerical data. Probability vocabulary might include *certain, equally likely,* and *impossible.* By the time they approach Grade 6, students can understand that probabilities fall in the 0 to 1 range and are often expressed with fractions. The students should predict the probability of the outcomes of a simple experiment like spinning a spinner with regions of different colors, and then test the prediction by carrying out the experiment many times.

These statements illustrate the important role that topics in probability and statistics can play in the curriculum and their relevance to common media displays and data interpretations. Citizens will confront such displays and interpretations and should be able to evaluate the displays and react to the interpretations. You will notice that data displays receive much attention, because they can serve as a source for more sophisticated ideas from probability and statistics.

Notes

The American Statistical Association has also produced a document devoted to statistics in the pre-K–12 curriculum: *Guidelines for Assessment and Instruction in Statistics Education (GAISE).*[5] The authors advocate statistical literacy for everyone, and they begin by listing many of the ways in which statistics is used in the workplace and in our lives: in science, law, manufacturing, medicine, law enforcement, education, and public opinion polls, particularly in the political process. Statistical literacy is important for understanding and dealing with all these areas of our lives. Rather than providing topics for study by grade level, the report divides curricular topics into learning levels; each level builds on the concepts of previous levels and learners must begin at Level A. In Level A, e.g., students might undertake simple experiments for which data are collected in their own classroom; at Level B students would begin to understand the need for random selection for a sample survey, and at Level C they would design experiments that include randomization. For example, in Level A students might gather data on "What type of music is most popular among students in our class?"; at Level B students might want to compare different classes' favorite types of music; and at Level C students might design an experiment to determine "What type of music is most popular among students in our school?" (p. 16). The GAISE report, available online for an undetermined time, is a good resource for examples of questions and problems appropriate at each of the three levels.

Teachers also see a variety of statistics in their profession. School or district reports of test performances can include graphs or statistics. Research reports and articles in professional magazines will in all likelihood refer to means, standard deviations, and percentiles, as well as include graphs of various sorts. Understanding such topics will enhance a teacher's understanding of the reports and articles.

32.4 Check Yourself

In this chapter you learned that the size of a sample makes a difference in terms of the confidence you have that the statistic approximates the parameter. But you also learned that the sample size does not have to be extremely large to provide a good approximation of the parameter. An acceptable margin of error depends on your decision about the level of confidence needed.

You should be able to work problems such as those assigned and to meet the following objectives:

1. Use Table 1 to explain what is gained by increasing the sample size, but acknowledge what is lost by making the sample size too large.

2. Use, if appropriate, the $\frac{1}{\sqrt{n}}$ rule of thumb to determine a margin of error or a confidence interval, or to decide on a proper sample size for a desired margin of error.

3. Interpret or make a confidence statement, with a margin of error, about the results of a sample survey.

4. Interpret a newspaper report about the margin of error in a particular poll.

References for Chapter 32

[1]National Governors Association Center for Best Practices, Council of Chief State School Officers. (2010). *Common core state standards for mathematics.* Washington, DC: Author.

[2]See http://nces.ed.gov/nationsreportcard.

[3]Kloosterman, P., & Lester, F. K., Jr. (Eds.). (2004). *Results and interpretations of the 1990–2000 mathematics assessments of the National Assessment of Educational Progress.* Reston, VA: National Council of Teachers of Mathematics.

[4]Silver, E. A., & Kenney, P. A. (2000). *Results from the seventh mathematics assessment of the National Assessment of Educational Progress.* Reston, VA: National Council of Teachers of Mathematics.

[5]Franklin, C., Kader, G., Mewborn, D., Moreno, J., Peck, R., Perry, M., & Schaeffer, R. (2007). Guidelines for assessment and instruction in statistics education (GAISE) report. Alexandria, VA: American Statistical Association. Also at http://www.amstat.org/education/gaise/.

Chapter 33

Special Topics in Probability

This chapter introduces some topics in probability that have many applications. Expected value (Section 33.1) shows how to judge *payoffs* for probabilistic events that have numerical values associated with the outcomes. Section 33.2 treats events for which the order within an outcome is not important. The technical term for such outcomes is *combinations*.

Notes

33.1 Expected Value

You may have worked on the following exercise earlier (see the Learning Exercises in Section 28.4):

▶ At the county fair, a clown is sitting at a table with three cards in front of her. She shows you that the first card is red on both sides, the second is white on both sides, and the third is red on one side and white on the other. She picks them up, shuffles, hides them in a hat, then draws out a card at random and lays it on the table, in such a manner that both of you can see only one side of the card. She says, "This card is red on the side we see. So it's either the red/red card or the red/white card. I'll bet you one dollar that the other side is red." ◀

Is this a safe bet for you to take?

Recall that $P(W) = \frac{1}{3}$, so if you make the bet, it is certainly possible that you will win the dollar. This may not matter much to you, but if you were the *clown* and you played this game all day long, it would certainly matter. The clown is concerned about making money over the long run. She is concerned about the *expected value* in her situation. ***Expected value can arise whenever each outcome has a numerical value*** (in addition to its probability). The expected value gives the sum of the values (gained or lost), each weighted by the appropriate probability. It gives the average gain/loss per game or trial, for a large number of games or trials.

> The **expected value** of an experiment is the sum of the products obtained by multiplying the value associated with each possible outcome by its probability.

In the case of the clown situation, there are two possible outcomes, each with a dollar value. One outcome is that the other side is red, for which the clown gains \$1 with $P(R) = \frac{2}{3}$. The second outcome is that the other side is white, for which the clown loses \$1 (or ⁻\$1, from her viewpoint) with $P(W) = \frac{1}{3}$. So, the expected value is $\$1 \times \frac{2}{3} + (^-\$1) \times \frac{1}{3} = \frac{1}{3}$ of a dollar, or about 33¢. Thus, if she plays this game 500 times, she can expect to take home $500 \times \$\,0.33$, or about \$165.

Think About ...

From the viewpoint of the *player,* what is the expected value of the game? What does that value tell you, if you played the game many times, say, 100 times?

Let us consider another example.

EXAMPLE 1

One state lottery has a game called *Decco* in which a player chooses one card from each of the four suits in a regular deck of cards. A winning card is then drawn from each suit (from another deck of cards). For one match between a card selected in a suit and the winning card in that suit, $1 is awarded. For two matches, the prize is $5. For three matches, the prize is $50. And for four matches, the prize is $5000. It costs $1 to play the game. If a large number of people play this game, what can each person expect to win or lose, on average? (The lottery agency knows these probabilities: The probability of four matches is 0.000035, the probability of three matches is 0.00168, the probability of two matches is 0.0303, the probability of one match is 0.2420, and the probability of zero matches is 0.7260.)

Solution

We know the value of each outcome, so we can use the probabilities of the outcomes to find the expected value. Also, because it costs $1 to play the game, this amount, $^{-}$$1, affects every outcome. So, the expected value equals

$$(\$5000 \times 0.000035) + (\$50 \times 0.00168) + (\$5 \times 0.0303) + (\$1 \times 0.2420) + {}^{-}\$1$$

or $^{-}$34.75¢. Alternatively, because the cost of the game affects each outcome, the expected value could be calculated by subtracting the $1 cost from each of the awards, as follows:

$$(\$5000 - \$1) \cdot 0.000035 + (\$50 - \$1) \cdot 0.00168 + (\$5 - \$1) \cdot 0.0303$$
$$+ (\$1 - \$1) \cdot 0.2420 + (\$0 - \$1) \cdot 0.7260$$

The expected value again calculates to $^{-}$34.75¢.

The expected value of $^{-}$34.75¢ means that if a large number of people play this game once, the average loss per person would be about 35¢. From the lottery's viewpoint, that is an average gain of 35¢ per play for the lottery. (This is how the lottery makes money. Lotteries have been described as a tax on those who do not understand probability!)

With so many numbers involved, organizing the data in such a way as shown below is advisable. (Either method, or your own, can be used.)

Method 1

Outcome	Prize ($)	Probability	Prize × Probability
0 matches	0	0.726	0
1 match	1	0.242	0.242
2 matches	5	0.0303	0.1515
3 matches	50	0.00168	0.084
4 matches	5000	0.000035	0.175
(Cost to play)			($^{-}$1.00)
			Expected Value = $^{-}$0.3475

Method 2 (with the cost to play built into the value of each award)

Outcome	Value ($)	Probability	Value × Probability
0 matches	0 − 1	0.726	$^{-}$0.726
1 match	1 − 1	0.242	0.00
2 matches	5 − 1	0.0303	0.1212
3 matches	50 − 1	0.00168	0.0823
4 matches	5000 − 1	0.000035	0.1750
		Expected Value =	$^{-}$0.3475

Think About …

What is the ⁻1.00 in the last column in the first table? Which viewpoint, the lottery's or the player's, is this table using? How is the ⁻1.00 taken into account in the second table?

EXAMPLE 2

At a carnival, it costs 50¢ to spin the spinner shown at the right one time. Your winnings depend on where the spinner arrow ends up. (You spin again if it ends on a line.)

a. What is the expected value of the game, from the player's view? From the owner's view?

b. What do those expected values mean?

Win every time!

Solution (a–b)

The regions look as though their probabilities are $\frac{1}{2}$, $\frac{1}{3}$, and $\frac{1}{6}$, with respective payoffs (from the player's viewpoint) of 25¢, 50¢, and $1, but we must remember to take into account the 50¢ cost of the game.

Outcome	Probability	Prize ($)	Prize × Probability
Biggest region	$\frac{1}{2}$	$\frac{1}{4}$	$\$\frac{1}{8} = 12.5¢$
Medium region	$\frac{1}{3}$	$\frac{1}{2}$	$\$\frac{1}{6} \approx 16.7¢$
Smallest region	$\frac{1}{6}$	1	$\$\frac{1}{6} \approx 16.7¢$
(Cost to play)			⁻50¢
			Expected value $\approx$ ⁻4¢

We see that we can expect to lose about 4¢ per game over the long run. From the owner's viewpoint, she can expect to gain about 4¢ per game over the long run.

TAKE-AWAY MESSAGE . . . For many probabilistic situations, outcomes can have not only probabilities associated with them, but also other numerical values (monetary value, point value, etc.). The idea of expected value for such a situation then gives us an idea of the average payoff per repetition associated with repeating the situation over the long run.

Learning Exercises for Section 33.1

1. A board game involves tossing a regular die and moving the number of spaces indicated by the number of dots on the top of the die. If the die is tossed many times, what is the expected value of the number of spaces moved? What does that mean?

2. Another board game uses two dice, with the sum telling the number of spaces moved. What is the expected value of the number of spaces moved, for this game? What does that mean?

3. Suppose that a city sponsors a football bowl game every year. Over the years, city officials have come up with these figures: The probability of having teams that will draw a large crowd is 0.7, and the probability of having good weather is 0.85. They also know that concession sales depend on various combinations of these factors, as shown in the table on the next page. Find the expected value for concession sales. (Assume that the team attractiveness and quality of weather are independent events.)

continue

Conditions	Concessions revenue
Attractive teams, good weather	$480,000
Attractive teams, bad weather	$400,000
Unattractive teams, good weather	$320,000
Unattractive teams, bad weather	$200,000

4. You are in charge of the basketball toss at a school carnival. A participant gets one toss, winning a prize that costs you 49¢. You gather data that suggest that for the typical participant, P(making a basket) = 0.4. You would like to average a profit of at least 10¢ per participant. How much should you charge a participant?

5. You are a basketball coach. A good player on the opponent's team shoots 60% on 2-point shots from a certain lucky location. The player is, however, a poor free-throw shooter, making only 40% of her free-throws. You are trying to decide if it would be good strategy (if not good sportsmanship) to have your players foul the good player when she is about to shoot from her lucky location. Would this be a good strategy? (If fouled while shooting, the player gets two free-throws, each worth 1 point if made.)

6. A quiz show might have a contestant spin a wheel such as the following, with the contestant to answer a question if a money amount shows up:

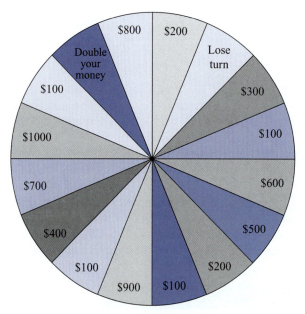

   a. What is the expected value for contestants who have $1000 already and spin (and answer the question correctly, if the spin allows a question)?
   b. What does your answer to part (a) mean?
   c. What is an expression for the expected value for contestants who have $x already and spin (and answer the question correctly)?

Supplementary Learning Exercises for Section 33.1

1. Suppose the six-sided die in Learning Exercise 1 in Section 33.1 on p. 767 is weighted so that the number 6 had a 50% chance of coming up. All the other numbers have an equal chance of coming up with regard to one another. What is the expected value of the number of spaces moved in the game with this die?

2. You are invited to play a game in which you roll two fair dice. If you roll a 7, you win $5.00. If you roll any other number, you lose the $1.00 it costs to play. What is the expected value for this game?

3. At the carnival it costs $1.00 to spin the spinner shown to the right one time. Your winnings depend on where the spinner lands. Should the owner expect to make money if the game is played many times? Explain.

4. For the past 60 days, a computer specialist recorded the number of calls that she received on a given day. She used this to compute the probabilities of receiving a given number of calls. For example, she computed a 30% chance of getting 3 calls on a given day. She is concerned about the expected value in her situation.

0 calls	1 call	2 calls	3 calls	4 calls	5 calls
0.05	0.15	0.25	0.30	0.10	0.15

    a. What is the expected value for the number of calls she receives?
    b. What does your answer in part (a) mean?

5. Your school's fund-raiser, the Fall Festival, is an event where dozens of booths are set up and manned by volunteers. At one booth, if the child wins the game, he or she wins a prize that costs 75 cents. You estimate that a child has a 30% chance of winning the prize. A person sells 25-cent tickets in one central place. It costs a certain number of tickets to play at each booth. What is the fewest number of tickets this booth should charge in order to make money?

33.2 Permutations and Combinations

Tree diagrams have been one of our most useful ways of analyzing more complicated experiments to find outcomes and probabilities. A tree diagram imposes an order on the outcomes, so we already have some background in working with ordered outcomes. This section explains how to count when order does *not* matter and how that can help determine probabilities in certain situations. We first review counting when order is taken into account.

> ⋮ *Think About ...*
> ⋮ Consider families that will have exactly four children. Which (if either) of the following has
> ⋮ the greater probability: having 4 boys in a row (BBBB), or having a boy, then 2 girls, and
> ⋮ finally another boy (BGGB)? (Assume that the probability of having a boy is the same as
> ⋮ the probability of having a girl.)

Most people think that $P(BBBB)$ should be less that $P(BGGB)$. But tracing each of BBBB and BGGB through a tree diagram gives the same result for each, $\left(\frac{1}{2}\right)^4$ or $\frac{1}{16}$.

> ⋮ *Think About ...*
> ⋮ Why is intuition so misleading here?

One reason why this probability question is not intuitive is that rather than interpret the situation with the *order* of the births in mind, people think of it as $P(4\ \text{boys})$ as opposed to $P(\text{a more balanced mix of boys and girls})$, without regard to the order of the births. For them, $P(4\ \text{boys}) < P(\text{some mix})$. And they would be correct, as you will soon see.

    When you think about it, a tree diagram deals with *ordered* selections. Something is done first, then something second, etc., so order is naturally imposed on the setting. But if order is not really an issue, then in the tree diagram, different branches may describe the same event.

> ⋮ *Think About ...*
> ⋮ BGGB gives one order in which a family could have four children, with 2 girls and 2 boys.
> ⋮ Find five other orders that give 2 girls and 2 boys. (If you cannot find five more, sketch a
> ⋮ tree diagram for the four births.)

Each of these six orders favors the event, 2 girls and 2 boys. So, there are *six* outcomes that favor that event, as opposed to the *one* outcome favoring BBBB.

Because each outcome has probability $\left(\frac{1}{2}\right)^4$ or $\frac{1}{16}$, $P(\text{BBBB}) = \frac{1}{16}$ is indeed less than

$$P(2 \text{ girls and } 2 \text{ boys}) = 6 \times \frac{1}{16} = \frac{3}{8}$$

even though $P(\text{BBBB})$ is exactly equal to the probability of any *one* particular birth order, such as BGGB.

A tree diagram for this setting does contain all of the six orders, so questions about $P(\text{have 2 girls and 2 boys})$ can be answered with the diagram. But doing so requires work that can get difficult when a tree diagram becomes large, as it would in using a tree diagram to find the probability of getting 8 or more heads in 10 tosses of a fair coin.

Permutations

Perhaps surprisingly, it is easier to see how to count *unordered* choices by first looking more closely at *ordered* choices. Ordered choices, called **permutations,** can be counted by means of the **fundamental counting principle**.

> **Fundamental Counting Principle:**
>
> If Act 1 can be performed in m ways, and Act 2 can be performed in n ways no matter how Act 1 turns out, then the sequence Act 1–Act 2 can be performed in $m \cdot n$ ways.

Here are some illustrations of the principle at work. Contrast using the fundamental counting principle to making a full tree diagram and counting all the routes.

EXAMPLE 3

Eight horses—Alabaster, Beauty, Candy, Doughty, Excellente, Friday, Great One, and High'n'Mighty—run a race. In how many ways can the first three finishers turn out?

Solution

Extend the fundamental counting principle to three "acts." Finishing first (Act 1) can happen in 8 ways. Then there are 7 ways in which finishing second (Act 2) can occur, and finally there are 6 ways in which third place (Act 3) can be filled. So, the first three finishers could occur in $8 \cdot 7 \cdot 6 = 336$ ways. ●

EXAMPLE 4

In how many ways can 10 tosses of a coin turn out?

Solution

Each of the 10 "acts" can occur in 2 ways (H or T), so there are $2^{10} = 1024$ different ways possible. ●

EXAMPLE 5

Given a list of five blanks, in how many different ways can A, B, and C be placed into three of the blanks, one letter to a blank? (Two blanks will be empty.)

Solution

As in the horse race, there are 5 choices of a blank for A, then 4 for B, and finally 3 for C. Thus, there are $5 \cdot 4 \cdot 3 = 60$ ways in which the three letters can be placed in the five blanks. Because A was mentioned first, you may wonder if this way of counting gives A too many of the choices. Write enough of the 60 ways, either in a systematic list or with a tree diagram, to convince yourself that the letters are treated fairly. ●

We can rephrase Example 3 using the word *permutation*: "How many permutations of 3 horses can be made from 8 horses?" Rephrasing Example 5 gives "How many permutations of 3 (blanks) can be made from 5 (blanks)?"

⋮ *Think About ...*
⋮ How would one calculate the number of permutations of 4 objects that can be made from
⋮ n objects?

The reasoning for the above Think About applies to the general situation and gives the following summary:

> There are $n(n - 1)(n - 2) \cdots$ **ordered selections (permutations)** of size r from n
> objects, where the product $n(n - 1)(n - 2) \cdots$ has r factors. We can use the common
> notation $_nP_r$ for the number of permutations of size r from n objects:
>
> $$_nP_r = n(n - 1)(n - 2) \cdots \quad \text{where } n(n - 1)(n - 2) \cdots \quad \text{has } r \text{ factors}$$

For example, $_6P_4 = 6 \cdot 5 \cdot 4 \cdot 3$ (4 factors).

⋮ *Think About ...*
⋮ What numerical expression describes the number of permutations of size 6 from
⋮ 6 objects, or $_6P_6$? What algebraic expression describes the number of permutations of
⋮ size n from n objects, or $_nP_n$?

Combinations

Order matters in spelling and numbers—RAT and TAR are different orders of the letters A, R, and T, and certainly have different meanings, as do 1234 and 4231. But many times order is not important, as in the "have 2 boys and 2 girls" situation earlier. The different orders BBGG, GBGB, GBBG (and others) all would describe the event "have 2 boys and 2 girls" even though they would be represented by separate paths through a tree diagram. Although we can use the fundamental counting principle to count ordered selections, it is not immediately clear that the number of selections when order does not matter can also be easily determined.

Fortunately, the number of ordered selections (permutations) can be exploited to tell the number of selections in which order does not matter. The technical term for such selections is *combinations*.

> A **combination** is a selection in which order does not matter. A common notation $_nC_r$
> is used for the number of selections of size r from n objects.

ABC, ACB, BAC, BCA, CAB, and CBA are six different permutations of the letters A, B, and C from the alphabet, but they represent just *one* combination of 3 letters from the alphabet. Notice that if we had started with the one combination ABC, we could have predicted the number of permutations of just those 3 letters by the earlier reasoning: $3 \cdot 2 \cdot 1 = 6$.

⋮ *Think About ...*
⋮ Choose one combination of 4 different letters from the alphabet. How many permutations
⋮ does this one combination give?

The same reasoning would apply to every combination of 4 different letters selected from the 26 letters in the alphabet. Each such combination gives $4 \cdot 3 \cdot 2 \cdot 1 = 24$ permutations. The total number of combinations of 4 letters chosen from the 26 letters (denoted $_{26}C_4$) can be determined from the total number of permutations in a way that builds on recognizing that (1) a given combination can give rise to several permutations and (2) all the permutations could be obtained by starting from the combinations and collecting all the permutations, combination by combination.

We give the reasoning as follows, but the calculations are not carried out so that the general form will be clearer: Because each combination of 4 letters of the alphabet gives $4 \cdot 3 \cdot 2 \cdot 1$ permutations, the result, $_{26}C_4 \cdot (4 \cdot 3 \cdot 2 \cdot 1)$, will yield the total number of *permutations* of 4 letters from 26 letters, which is $26 \cdot 25 \cdot 24 \cdot 23$. That is, $_{26}C_4 \cdot (4 \cdot 3 \cdot 2 \cdot 1) = 26 \cdot 25 \cdot 24 \cdot 23$, or

$$_{26}C_4 = \frac{26 \cdot 25 \cdot 24 \cdot 23}{4 \cdot 3 \cdot 2 \cdot 1}$$

Similarly but more generally, each combination of r objects gives rise to $r(r-1) \cdots 2 \cdot 1$ permutations of the r objects, so the total number of such combinations selected from n objects could also be calculated as in the example, using $n(n-1)(n-2) \cdots$ (with r factors) for the number of permutations.

That is, $_{26}C_4 \cdot r(r-1)(r-2) \cdots 2 \cdot 1 = n(n-1)(n-2) \cdots$ (until there are r factors). Hence, when we are concerned with the number of combinations, we would arrive at the following general formulation:

The number of **unordered selections (combinations)** of size r from n objects is given by

$$_nC_r = \frac{n(n-1)(n-2) \cdots \text{(until there are } r \text{ factors)}}{r(r-1) \cdots 2 \cdot 1}$$

Notice that the same number of factors, r, appears in both the numerator and denominator.

Some examples will show some calculations and then some applications to probability problems. Before looking at the solutions, ask yourself, "Is this about permutations or combinations?" One approach is to think of a specific selection, and then see whether changing the order in that selection would give a result that should be counted separately from the original one: For example, is CAB different from ABC in the situation? If the new order should be counted separately, then the problem deals with permutations. If the new one is the same selection as the original for the purposes of the problem, then the problem involves combinations.

EXAMPLE 6

In how many different ways could a committee of 5 people be chosen from a class of 30 students?

Solution

Order does not matter here, so we want to find $_{30}C_5$ or

$$\frac{30 \cdot 29 \cdot 28 \cdot 27 \cdot 26}{5 \cdot 4 \cdot 3 \cdot 2 \cdot 1} \quad \text{different ways}$$

Some simplification shows that this equals

$$\frac{29 \cdot 7 \cdot 27 \cdot 26}{1}, \quad \text{or} \quad 142{,}506 \text{ ways} \quad \bullet$$

EXAMPLE 7

If the first student chosen would be chair, the next student vice-chair, the third one secretary, and the fourth one treasurer, in how many different ways could a committee of 4 be chosen from a class of 30 students?

Solution

Example 7 deals with permutations, because order clearly matters. So, there are $30 \cdot 29 \cdot 28 \cdot 27$ or 657,720 ways to choose the committee chair, vice-chair, secretary, and treasurer. $\bullet$

EXAMPLE 8

In a family of four children with exactly two girls, in how many ways can the birth positions (first-born, second-born, etc.) for the two girls be assigned?

Solution

There are 2 birth positions to be chosen from the 4 births. Order does not matter because we are focusing on the girls (boys will automatically fill the other two birth orders). So, there are $\frac{4 \cdot 3}{2 \cdot 1}$, or 6, ways. Check this answer with your findings for the similar question given at the beginning of this section. Notice that the calculation tells only how many combinations there are, not what each of the combinations consists of. ●

Permutations and combinations are topics in this Grade 6 textbook. Try the exercises.

Independent Practice

For **9** through **16**, find the number of possible arrangements and state whether it is a permutation or combination.

9. Hugh and five friends are going to set a table for 6. How many possible seating arrangements are there?

10. There are 5 finalists in an art contest. In how many ways can the first, second, and third prizes be awarded?

11. Jose has 4 different shirts: red, blue, yellow, and white. He wants to pack 2 shirts for camp. How many possible pairs of 2 shirts can he pick?

12. For dinner, Jesse can choose 2 different side dishes from corn, potatoes, beans, rice, noodles, and peas. How many ways can he make his choices?

13. How many different ways can 5 cars be lined up in a showroom?

14. In how many possible ways can 9 children line up to go down the water slide?

15. Seven friends arrive one at a time for Victor's picnic. In how many possible orders can they arrive?

16. Angelo has 5 jackets. How many ways can he choose to give 2 of them to a clothing drive for charity?

Problem Solving

Use the illustration of the birds on the telephone wire for **17** and **18**.

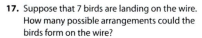

17. Suppose that 7 birds are landing on the wire. How many possible arrangements could the birds form on the wire?

    A 38   **B** 5,040   **C** 10,080   **D** 20,160

18. If 2 birds fly away, how many possible arrangements could the remaining birds form on the wire?

    A 38   **B** 64   **C** 120   **D** 40,320

19. Reasoning Which is less, the number of ways to choose 2 people from a group of 7 when order matters or when order doesn't matter?

20. Writing to Explain There are 24 ways to park x number of cars in order on the driveway. How many cars are there? Explain how you found your answer.

21. Think About the Process What numbers would you multiply to find the number of arrangements for 6 cows at a water trough? For 7 cows? Do you see a pattern? If so, describe it.

From Randall I. Charles et al., *enVision, Math California Edition, Grade 6*, p. 484.

Back to Probability

We now have all the pieces needed to handle more complicated probability problems. The following examples put together the pieces.

EXAMPLE 9

In a toss of 10 different honest coins, what is the probability of getting exactly 5 heads? (Guess first.)

Solution

Each (ordered) outcome has probability $\left(\frac{1}{2}\right)^{10}$. (Think of the tree diagram.) As with the birth orders, which particular coins of the 10 give heads is not of interest. What *is* of interest is all the different *combinations* of 5 coins that could give heads (with the other 5 coins giving tails). That number is $_{10}C_5 = 252$, so there are 252 outcomes favoring the event of getting exactly 5 heads in the 10 tosses. Because each one has probability $\left(\frac{1}{2}\right)^{10}$, the probability of exactly 5 heads in a toss of 10 coins is $252 \cdot \left(\frac{1}{2}\right)^{10} = \frac{252}{1024} = \frac{63}{256}$, or only about one-fourth, which is not at all the one-half that many would believe is the probability. ●

EXAMPLE 10

In 10 tosses of a "loaded" coin, with probability of heads = 0.7, what is the probability of getting exactly 5 heads?

Solution

The new feature here is how to handle the different probabilities, because the number of combinations would be the same as in Example 9: 252. If you think of tracing through a tree diagram, whenever you have an outcome with 5 H's and 5 T's, there must have been five 0.7s and five 0.3s on the 10 branches. That is, the probability of each path with 5 H's and 5 T's is $(0.7)^5 \cdot (0.3)^5$. With the loaded coin then, the probability of exactly 5 heads is $252 \cdot (0.7)^5 \cdot (0.3)^5$, or about 0.10. (Why is the answer for Example 10 less than that for Example 9?) ●

EXAMPLE 11

A new drug has probability 0.7 of curing a disease. What is the probability that at least 8 of 10 patients suffering from the disease will be cured if they use the drug?

Solution

This example is more complicated because there are three different probabilities to determine: P(exactly 8 cures), P(exactly 9 cures), and P(all 10 cures). The number of ways each of these events can happen must be determined, and then multiplied by the appropriate probability for each of the outcomes counted. The three events are mutually exclusive, so from the addition rule for probability, we can add those three individual results to answer the question.

For 8 cures, we get the probability

$$\frac{10 \cdot 9 \cdot 8 \cdot 7 \cdot 6 \cdot 5 \cdot 4 \cdot 3}{8 \cdot 7 \cdot 6 \cdot 5 \cdot 4 \cdot 3 \cdot 2 \cdot 1} \cdot (0.7)^8 \, (0.3)^2, \qquad \text{or} \qquad \text{about } 23.3\%$$

For 9 cures, we get

$$\frac{10 \cdot 9 \cdot 8 \cdot 7 \cdot 6 \cdot 5 \cdot 4 \cdot 3 \cdot 2}{9 \cdot 8 \cdot 7 \cdot 6 \cdot 5 \cdot 4 \cdot 3 \cdot 2 \cdot 1} \cdot (0.7)^9 \, (0.3)^1, \qquad \text{or} \qquad \text{about } 12.1\%$$

For 10 cures, we get

$$1 \cdot (0.7)^{10} \, (0.3)^0, \qquad \text{or} \qquad \text{about } 2.8\%$$

So, the probability of getting at least 8 cures would be about 23.3% + 12.1% + 2.8%, or 38.2%. ●

TAKE-AWAY MESSAGE . . . For a situation with complicated outcomes, think of a typical outcome. If changing the order within the outcome gives a different result for the event in question, then permutations are involved. If changing the order in the typical outcome does not give a different result, then combinations are involved. Although thinking about a tree diagram is helpful in determining the probability associated with each path through the tree, we can calculate the total *number* of permutations or the total *number* of combinations without drawing the tree diagram.

Notes

Learning Exercises for Section 33.2

1. Tell whether order is important or not important in each part.

   a. permutations   b. combinations
   c. assigning children to a seating chart
   d. making up a three-digit number by tossing an icosahedron (20 faces), with its faces marked with two copies of the ten digits 0–9
   e. picking four friends for a sleepover

2. a. List all the permutations of 2 letters, chosen from P, Q, R, S, and T.
   b. List all the combinations of 2 letters, chosen from P, Q, R, S, and T.

3. a. Explain how the fundamental counting principle enters in when counting the number of ways 5 children could be seated in a 5-chair row.
   b. Explain how the fundamental counting principle enters in when counting the number of ways 5 children could be seated in a 7-chair row (2 seats would be empty).
   c. Does part (a) involve permutations or combinations? Part (b)?

4. a. There are 7 different TV dinners in the freezer. In how many ways could Rosalie eat the TV dinners if she eats one dinner each workday night, Monday through Friday?
   b. A store stocks 7 different TV dinners. In how many ways could Rosalie pick 5 different TV dinners in that store?
   c. Explain how you decided what to do in parts (a) and (b).

5. In tossing a coin 10 times, how many different ways could

   a. exactly 2 heads appear?   b. exactly 8 tails appear?
   c. exactly 2 tails appear?   d. exactly 8 heads appear?
   e. Explain why the numerical answers for parts (a) and (d) are the same.

6. In a toss of 10 honest coins, what is the probability of getting

   a. exactly 3 heads?   b. exactly 2 heads?   c. at most 3 heads?
   d. Explain the meaning of the probabilities you found in parts (a–c).
      (*Hint:* Probabilities are about many, many trials.)

7. In families of four children, give the probabilities of having all boys, all girls, exactly 1 boy (and 3 girls), exactly 1 girl (and 3 boys), and 2 boys (and 2 girls). Explain the meaning of the probabilities you find.

8. Suppose that in a certain species of animal, P(male birth) $= \frac{2}{3}$. For litters of size 4, give the probabilities of all being male, of all being female, of having exactly one male, of having exactly one female, and having two of each sex.

9. In some card games with a deck of 52 cards, a "hand" is made up of 5 cards. How many different hands are possible?

10. a. If a bag contains 10 balls, how many different choices of 3 balls are possible?
    b. If the bag contains 6 red balls, how many different choices of 3 red balls are possible?
    c. If a bag contains 6 red balls and 4 blue balls, what is the probability that a draw of 3 balls will give all red balls? [*Hint:* See parts (a) and (b).]
    d. If a bag contains 4 blue balls, how many different choices of 2 blue balls are possible?
    e. If a bag contains 10 balls, how many different choices of 5 balls are possible?
    f. If a bag contains 6 red balls and 4 blue balls and if 5 balls are drawn, what is the probability that the draw will give 3 red balls and 2 blue balls? [*Hint:* Use parts (b), (d), and (e) as well as the fundamental counting principle.]

Notes

11. One lottery works this way: You pick 6 two-digit numbers, from 00 through 99. You are a big winner if all 6 of your choices match the 6 that come up in the lottery.

    a. What is the probability of being the big winner?
    b. If a big winner gets a million dollars and it costs $2 to enter the lottery, what is the expected value for playing the lottery?

12. One very skilled basketball player hits 85% of her free-throws. What is the probability that, in the long run, she will hit at least 8 of every 10 free-throws?

13. Numerical expressions such as $6 \cdot 5 \cdot 4 \cdot 3 \cdot 2 \cdot 1$ occur often enough in mathematics that there is a special notation for them. $6 \cdot 5 \cdot 4 \cdot 3 \cdot 2 \cdot 1$ is abbreviated as 6! (read as "6 factorial"). So, $6! = 6 \cdot 5 \cdot 4 \cdot 3 \cdot 2 \cdot 1 = 720$. Find the numerical values of the following:

    a. 7!   b. 10!   c. $(10 - 4)!$ d. $\dfrac{10!}{7!}$ e. $\dfrac{12!}{4!8!}$ f. $\dfrac{8!}{2!6!}$

14. Convince yourself that $_nP_r = \dfrac{n!}{(n - r)!}$.

15. Convince yourself that $_nC_r = \dfrac{n!}{(n - r)!r!}$.

16. Calculate the number of combinations of size x, given 5 objects, if x is

    a. 0   b. 1   c. 2   d. 3   e. 4   f. 5

17. a. The pattern of numbers that we have started at right is called **Pascal's triangle.** Can you generate the next two rows? What patterns do you notice in the triangle?

    b. Look at your answers to Learning Exercise 16 to see the relevance of Pascal's triangle to this section. What would the last row shown tell someone?

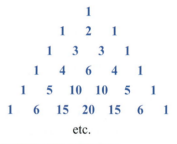

Supplementary Learning Exercises for Section 33.2

1. Telephone numbers (without the area code) are 7 digits. They don't begin with 0 or 1. How many different 7-digit numbers can be generated within a given area code?

2. Your classroom library has about 330 books. Are there enough codes to encode all your books with one of 4 letters (F for fiction, B for biography, S for science, H for history) and 2 digits? Explain.

3. There are 6 slots for CDs in my car. If I have 20 CDs, in how many different ways can I load the CD player?

4. The School Governance Committee must have 10 members. The committee consists of the principal or vice-principal, 1 staff person, 4 teachers, and 4 parents. Stanley School has 3 willing staff persons, 6 willing teachers, and 9 willing parents. In how many ways can the committee be formed?

5. Your sorority is forming a committee to consider what volunteer work can be done for the community. Your sorority has 13 sophomores, 15 juniors, and 9 seniors. How many different committees can be formed consisting of 1 sophomore, 2 juniors, and 1 senior?

6. If you were to draw a 5-card hand from a standard deck of 52 cards, what is the probability that it will have 3 face cards (from the 4 jacks, 4 queens, and 4 kings)?

7. License plates in one state consist of a digit followed by 3 letters and then 2 digits.

    a. How many different plates are possible if repetition is allowed on a given license plate?
    b. How many different plates are possible if no repetition is allowed on a given license plate?

8. Suppose a bag contains 8 different balls of the same size, weight, and texture.

    a. How many different choices of 4 balls are possible?

continue

b. If this bag of 8 balls contains 5 blue balls and 3 yellow balls, what is the probability that a draw of 3 balls will give all blue balls?

    c. If the bag contains 5 blue balls and 3 yellow balls and if 5 balls are drawn, what is the probability that the draw will give 3 blue and 2 yellow balls?

9. A bag contains 9 identical balls. Three are marked $1, four are marked $5, and two are marked $10. What is the probability that 2 balls which are pulled out (no replacement) will add up to at least $10?

33.3 Issues for Learning: Children Finding Permutations

In a long-term research study[1,2] of how children develop an understanding of mathematical proof, third-, fourth-, and fifth-graders were asked to build as many towers as possible, using four cubes of two different colors. One such tower might look like the one shown at right.

    The children were also asked to be able to convince their classmates that there were no duplicates and that all possible towers had been found.

 ACTIVITY 1   How Many Towers?

1. How many such towers are there? Draw them.

2. How can you show someone else that you have located all of them and that none of them are duplicates? ●

> *Think About ...*
>
> How did you begin this activity? Did you begin randomly by drawing every tower you could think of? Or did you organize your work in some way from the beginning?

At one point the researchers focused their attention on a pair of students, called Jackie and Meredith. The children, as you might expect, began to randomly build towers, while comparing each new one to previous ones. They began even naming some of their towers. For example,

The use of opposites turned out to be a good organizational tool for finding all the towers. After the children had built what they believed were all the possible towers, the teacher asked how they knew they had found all of them. The children then rearranged the towers as follows and offered their explanation to the teacher:

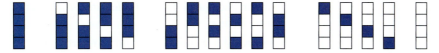

DISCUSSION 1   Is This All?

1. Describe the manner in which these students organized the towers.

2. How did they know they had found all of them?

3. How did they know there were no repetitions? ●

> *Think About ...*
>
> How does the fundamental counting principle apply to this problem?

Another child, Stephanie, also did this problem, but did not organize the towers so carefully. However, when she was asked to use five cubes instead of four, she realized that she had to organize her work in some way. She found groups of towers that she could describe. For example, she called the set of towers with exactly one blue in each of the five positions a *pattern*. She next looked at all the towers with two blue cubes in the first and second positions, then all the towers with blue cubes in the first, second, and third positions, towers with blue cubes in the first four positions, and then a tower with blue cubes in all five positions.

ACTIVITY 2   What Was Stephanie Thinking?

Make or draw towers in the manner used by Stephanie. Discuss with one another whether you think Stephanie has found all possible towers. If not, why not? ●

Stephanie knew she had found all towers with exactly one of a color and exactly four of a color, but continued to work on cases of exactly two of a color and exactly three of a color. After she concluded this task, she went on to six cubes of two colors, and at this point she made a table to be sure she had found all of them. For example, one column would be BBBBBB and another BBBBBW.

Over several months of work on the towers problems, Stephanie reached the point that she could use a "proof by cases" argument to show she had found all cases and none were repeated. Using this type of proof is quite sophisticated. "She persevered in trying to make sense of situations. . . . She was interested in reexamining and reconstructing earlier ideas, discussing her thinking with other classmates. She responded to teacher/researcher challenges to explain and to justify her thinking" (p. 212).

There were many other students involved in this study (conducted over several years) all solving a variety of problems involving permutations with the towers, pizza toppings, and clothing arrangements, for example. This research shows that children in elementary school can learn to think about such problems in ways that lead to finding all possible cases and being able to prove that they have found them all.

TAKE-AWAY MESSAGE . . .   Even young children, through exploration and encouragement, can understand and list permutations of objects by making drawings and carefully listing elements.

33.4 Check Yourself

Along with an ability to deal with exercises such as those assigned and your experiences in class, as appropriate, you should be able to meet the following objectives:

1. Define *expected value,* determine the expected value for a given situation, and explain what the result means.

2. State and apply the fundamental counting principle.

3. Distinguish between permutations and combinations, and calculate the number of permutations or combinations in a situation.

4. Solve given probability problems involving permutations and combinations.

References for Chapter 33

[1]Maher, C. A., & Martino, A. M. (1996). The development of the idea of mathematical proof: A 5-year case study. *Journal for Research in Mathematics Education, 27*(2), 194–214.

[2]Martino, A. M., & Maher, C. A. (1999). Teacher questioning to promote justification and generalization in mathematics: What research practice has taught us. *Journal of Mathematical Behavior, 18*(1), 53–78.

Video Clips Illustrating Children's Mathematical Thinking

The first six video clips associated with *Reconcepualizing Mathematics: Reasoning About Numbers and Quantities* were developed during a federally funded project at San Diego State University called *Integrating Mathematics and Pedagogy* (IMAP, REC 9979902). Video clip 7 was developed during another federally funded research endeavor called *Project Z: Mapping Developmental Trajectories of Students' Conceptions of Integers* (DRL-0918780). The others were developed for *Reconceptualizing Mathematics*. The clips are all from interviews of elementary school children. All of the teachers and students involved with the making of these videos have given their permission for the video clips to be used to help teachers better understand students' reasoning about, and understanding of numbers and operating on numbers. We ask that you, as a viewer of these clips, in turn respect these students and their teachers. Thus, if a student does not understand a topic, neither the student nor their teacher should be regarded in a negative fashion. There are many reasons why what you see in these clips could have come about, and you will be asked to explore some of these reasons. The purpose of each interview was to assess student knowledge, not to teach.

The video clips can be found at http://crmse.sdsu.edu/nickerson. They are ADA-compliant.

Although you may encounter some similar types of activities in a mathematics methods course, these videos are included with this mathematics content material because we have learned from experience that many prospective teachers come to view their own mathematics learning differently after watching these children. It will become clear to you that you will need to possess a deep understanding of mathematics if you are to successfully teach children such as those you see in the video clips. We hope that watching these children will motivate you to take seriously the study of mathematics at the elementary school level.

Video Clip 1: Strategies

In this clip, three first-graders are asked to add and subtract small numbers. Strategies children often use for adding, in a relative order of increasing sophistication, include (1) counting all, using manipulatives of some sort (in this case, small blocks); (2) starting with the first number and then counting on in some fashion, such as using fingers; (3) counting on from the larger addend rather than the first addend; and (4) compensating by using a known fact to produce a new fact. Counting with manipulatives such as in (1) also has different levels. Children sometimes try to count without moving the objects, and thus count too many or too few. To subtract, children use similar strategies, using manipulatives, counting on fingers, or using known facts to arrive at a new answer.

Questions for reflection and discussion

1. In the first interview, a girl is asked to take 5 objects away from 14 objects, and then 6 from 14. Describe the counting strategies she used. Were there any surprises here? If you were helping her, what problem would you give her next, and why?

2. The second girl is asked to add 7 to 4. How did she do this? Notice how she tapped her head when starting from 7. Her teacher taught students to "Put the big number in your head, then count on." Do you think this child actually undertook the process she describes here? Or was she just trying to help the interviewers understand how to do the problem?

3. How would you describe the strategy used by the boy in the third part of this clip? How did he decompose one of the numbers? What problem would you give him next to solve and why?

4. These children do not yet appear to know their basic addition and subtraction facts. Are any of them ready to learn them? As a teacher or parent, do you think that children should think about problems like these before being asked to develop instant recall of facts?

Notes

Video Clip 2: Javier

Javier, a fifth-grade, limited English proficient student, uses mental strategies to multiply 6 times 12 and 12 times 12. Before viewing the Javier video clip, tell how you could mentally find 6×12 and 12×12.

Questions for reflection and discussion

1. In each case, Javier appears to first do the problem mentally, then explains his reasoning to the interviewer. In each case, do you think his explanations matched his answers?

2. Were you surprised by his strategy for finding 12×12? Why or why not? How did Javier display good number sense?

3. Use the distributive property, as Javier did, to find 8×15 and 16×15.

4. Javier is learning English. Does he explain his thinking clearly to the interviewer?

Video Clip 3: Rachel

The purpose of this clip is to contrast the effectiveness of teaching only procedural rules with teaching for understanding. Doing so is difficult because one cannot select and video a "bad teacher" to be shown in this type of forum. We therefore asked a teacher who does teach for understanding to first teach a lesson (in this case, changing mixed numbers to improper fractions and vice versa) very procedurally, and then to teach the lesson about a month later in a way that focused on understanding. Rachel, a student in this class, is interviewed twice, once after each lesson.

Questions for reflection and discussion

1. In the first 50 seconds of the video clip, the teacher describes how she taught the first lesson. Do you think this type of teaching is typical of most teachers? How does this match up with your own introduction to work with fractions in elementary school?

2. Rachel is interviewed after this lesson. She is asked to change $\frac{9}{5}$ to a mixed number. She does not appear to be clear about what to do. Why did this happen, do you suppose?

3. The teacher is then interviewed about the second lesson she taught. What do you think she did differently?

4. Rachel is interviewed after the second lesson and asked to change $3\frac{3}{8}$ to an improper fraction. She begins by trying to apply the rule. What goes wrong here?

5. Rachel immediately tries to make a drawing to explain her reasoning. What happens then, and why? How does her drawing help her?

6. Many times teachers teach a lesson procedurally first because they have a lot of content to cover before standardized testing. They then go back and try to supply reasons for the procedures they've taught. Would you have any suggestions for these teachers about the advisability of this strategy of ordering lessons?

Video Clip 4: Ally

Ally is a fifth-grader whose teacher has been teaching fractions to her class. This clip is an excerpt from a much longer clip in which the interviewer first tried to diagnose what Ally knew about fractions. He next taught a lesson based on what was learned about Ally's knowledge of fractions, and finally he assessed what Ally had learned from the lesson. You are seeing only some of the first part of the interview with Ally followed by an interview with her teacher. Ally is a student at a school that has regularly had very high standardized text scores.

Notice how the interviewer gives Ally opportunities to work and talk about how she reasoned, without interrupting her or giving any indication about whether she was correct or not.

Ally is first asked to compare the following pairs of fractions by circling the larger in each pair:

$$\frac{1}{6} \text{ and } \frac{1}{3} \qquad 1 \text{ and } \frac{4}{3} \qquad \frac{3}{6} \text{ and } \frac{1}{2} \qquad \frac{1}{7} \text{ and } \frac{2}{7} \qquad \frac{3}{10} \text{ and } \frac{1}{2}$$

She is also asked to change $1\frac{1}{3}$ to an improper fraction and $\frac{13}{6}$ to a mixed number. Before viewing the video clip, do each of the above problems.

Questions for reflection and discussion

1. For each of the pairs of fractions Ally was asked to compare, describe how she reasoned about the pair.

2. What was the interviewer thinking when he chose a new pair of fractions for Ally to compare?

3. How well do you think Ally understands what a mixed number is? Why do you say this?

4. What did you learn about Ally from her teacher? Were you surprised by how much the teacher knew about Ally, and probably the other thirty-some students in the class?

5. Why did students in this teacher's class think multiplication and division of fractions were easier than adding and subtracting fractions?

6. If you were asked to tutor Ally, where would you begin, and why?

Video Clip 5: Felisha

During the summer after second grade, Felisha and three other students were given opportunities to explore basic fraction concepts over several days with a skilled teacher. In a postinterview, Felisha was asked to add two simple fractions, $\frac{3}{4}$ and $\frac{1}{2}$, even though the addition of fractions had not been taught during the prior sessions. She works silently for a few minutes before being asked about her reasoning.

Questions for reflection and discussion

1. Knowing that this child had not had any instruction on adding fractions, what knowledge do you think she did possess to undertake this work?

2. Was her explanation clear? Correct?

3. Were her drawings sufficient for solving this problem?

4. What problem would you pose next to Felisha? Why?

5. Do you think students can solve a lot of problems even though they've not been taught procedures for doing so? Why or why not?

Video Clip 6: Elliot

Elliot, a sixth-grade student, solves two division-of-fractions problems using his understanding of division. Before viewing this video clip, do the following division problems: $1 \div \frac{1}{3}$ and $1\frac{1}{2} \div \frac{1}{3}$.

Questions for reflection and discussion

1. Describe Elliot's reasoning for the first problem. Are his drawings correct? Are they the drawings you would use? Do you think Elliot has a good understanding of division? Why or why not?

2. Do you think Elliot has a good understanding of fractions? Why do you say that?

3. Is your answer the same as Elliot's for the second problem? What went wrong? (*Hint:* Attend to the last comment by Elliot. Was he thinking of the reference unit for the answer, or the reference unit 1? What was confusing here?)

4. How do you think you could help Elliot understand his error?

Video Clip 7: Violet

In this clip, a child is asked to solve problems involving integers. Violet is a second-grader. Second-graders have not had formal instruction on operations with integers.

Questions for reflection and discussion

Violet is asked to solve $^-3 -$ _____ $= 2$.

1. Describe the strategy that she used to reason about what number to add to $^-3$.

2. Knowing that she had not had any instruction on subtracting negative numbers, on what knowledge did she draw in her reasoning?

3. How did the number line support her thinking?

Video Clip 8: Definitions

In this clip, children of different ages are asked to give definitions. The interviewer asks follow-up questions to determine how the child classifies figures. In this clip, you also get a sense of a child's image of a concept even when he or she cannot clearly articulate a definition.

Questions for reflection and discussion

1. What do the children understand about the relationship among the different figures?

2. What nondefining attributes are included in their definitions?

3. Is it clear that they understand a geometric figure can be classified into more than one category?

Video Clip 9: Andreas

Andreas is a sixth-grader. He is asked to derive the formula for a triangle.

Questions for reflection and discussion

1. Are you surprised at what was and was not difficult for him?

2. When did you learn about where the formulas come from?

3. How important is it to understand the formulas we use and what meaning they have?

Video Clip 10: Mathematical Practices

These children are from a third-grade class. The teacher works hard at developing particular dispositions and habits of mind in her class. The mathematical practices are intertwined with but are distinct from knowledge of mathematics.

Question for reflection and discussion

1. In which of the mathematical practices do you see evidence of student engagement?

2. Choose one of the mathematical practices and reflect on how a teacher can support the development of this engagement when doing mathematics.

Video Clip 11: Measurement

The "Issues for Learning" section in this chapter describes some of the big ideas of measurement. This clip shows three children discussing measurement tasks that include a "broken ruler" task and measuring the same area with two different units.

Questions for reflection and discussion

1. Can you identify what big ideas they each understand?

2. What big ideas of measurement are still a challenge for them?

Video Clip 12: Nicholas

Nicholas is a third-grader. When he is asked to find the number of cubes that it will take to fill a box, he is actually finding the volume, though he doesn't use this terminology. Nicholas writes a mathematical expression for finding the volume of a particular right rectangular prism. He is asked to describe where the numbers come from and why that information tells us the number of cubes in the box.

Questions for reflection and discussion

1. What does he seem to understand?

2. What is difficult? Why do you think this is difficult for him?

3. Are there ways to ask the question of children that may support their seeing the connection between the formulas for volume and the three-dimensional objects they describe?

Summary of Formulas

Formula	Reasoning
Sum of angle sizes for n-gon $= (n-2)180°$	Split polygon with n sides into $(n-2)$ triangles.
$A(\text{rectangular region}) = lw$ Special case: $A(\text{square region}) = s^2$	w rows l square regions in each row
$A(\text{parallelogram region}) = bh$	
$A(\text{triangular region}) = \frac{1}{2}bh$	
$A(\text{trapezoid region}) = \frac{1}{2}h(a+b)$	
$A(\text{circular region}) = \pi r^2$ $C(\text{circle}) = 2\pi r = \pi d$	Area: Based on the total area of infinitely many triangles (Section 25.1) Circumference: Similarity (Section 25.1)
Surface area of a polyhedron $=$ sum of the areas of its faces	Key idea: Measure an object by cutting it into pieces, measuring each piece, and adding.
$V = Bh$ for any prism or cylinder with base area B and height h	Base area B numerically gives the number of cubes in one layer; h gives the number of layers.
$V = \frac{1}{3}Bh$ for any pyramid or cone with base area B and height h	Suggested by experiment; no general justification given in this book
$A(\text{surface of sphere}) = 4\pi r^2$ $V = \frac{4}{3}\pi r^3$ for any sphere	Area formula: No justification given in this book Volume formula: Based on the total volume of infinitely many pyramids (Section 25.2)

About the *Geometer's Sketchpad*® Lessons

The *Geometer's Sketchpad*® (GSP) is software by Key Curriculum Press (which can be found at http://www.keypress.com) that allows the drawing and manipulation of precise shapes. Prose can be added to a drawing, and both drawing and prose can be copied and pasted to word-processing software documents. The software can be a valuable resource for classroom demonstrations and the preparation of documents with geometry shapes.

We include a few optional online video lessons on using GSP. The lessons serve as an introduction to this powerful software program and perhaps a motivation to learn more about GSP. The lessons, developed by Dr. Janet Bowers, allow explorations of various shapes and the properties of those shapes. The lessons can be related to various sections of the chapters where these shapes and properties are learned. The lessons will not touch on all of GSP's features but should give you an idea of its capabilities. With replays of the lessons, you can learn how to make several kinds of drawings. You do not need to have the GSP software to view the lessons, but you do need the GSP software to practice on your own. The lessons can be found at http://crmse.sdsu.edu/nickerson. The lessons are ADA-compliant.

1. Special Triangles
2. Special Quadrilaterals
   a. Constructions by side lengths
   b. Constructions by diagonals
3. Transformation Geometry
4. Sum of Interior Angles

Each of these units contains the following features:

   a. Interactive java applets with reflection questions and answers
   b. Video tutorial showing you how to create the sketch using GSP
   c. Step-by-step construction directions that can be downloaded

Note: For students interested in furthering their knowledge of the *Geometer's Sketchpad* program, we suggest checking the professional development section of the Key Curriculum Press, including an online course that features further tutorials by the author of these materials, Janet Bowers, PhD.

Using the Table of Randomly Selected Digits (TRSD)

A Table of Randomly Selected Digits can be used to simulate probability situations, including random sampling. A TRSD is located at the end of this appendix. Please refer to that one for the following lessons. After that, you can use any table of randomly selected digits that you have.

1. Using the TRSD to simulate the toss of 4 coins

1. Locate the Table of Randomly Selected Digits in this appendix. This table was formed through a computer generation of random numbers, and then placed in sets of five digits to make it easier to read. The first column entries give row numbers, not randomly selected digits.

2. Turn to one of the pages in the table—it does not matter which one. Choose a place on the table by dropping your pencil and noting where the point is, or by simply looking up or closing your eyes and letting your finger drop onto the page.

3. Suppose your finger is in row 54 at the third set of five in the row—63186. Look only at the first four digits. Consider even digits to represent heads and odd digits to represent tails. 6318 would then represent the toss of 4 coins with 2 heads (represented by 6 and 8) and two tails (represented by 1 and 3). Make a five-column table with 0, 1, 2, 3, or 4 heads as the column headings. This random number (6318) would represent 2 heads, so place a tally mark in the column with 2 heads as the column heading. Then move down (or across) to the next set of digits, and again consider only the first four: 7808 would represent a toss resulting in 3 heads and 1 tail. Make a tally mark under 3 heads in your table. Continue this process. Each time this is done, make a mark in your table for the number of heads: 0, 1, 2, 3, or 4.

4. If several people work together, a large number of simulations of the tossing of 4 coins could be made in only a little time, if you share your numbers of heads with others in your group or in your class. Suppose each person simulates 20 tosses. If 20 of you do this simulation, then you have simulated tossing 4 coins 400 times. Suppose, over 400 times, you had 3 heads and 1 tail 89 times. Then your experimental probability of getting 3 heads and 1 tail on a toss of four coins is $\frac{89}{400}$ or about 22%.

2. Using five one-digit numbers to simulate the ball drawing in Section 27.3

In this lesson you simulate the drawing of one ball from a bag of 2 red balls and 3 blue balls, noting the color, replacing it, drawing a second ball, and noting its color.

1. Suppose you want to use only five digits in a simulation. To do so, let digits 1 and 2 represent red balls and 3, 4, 5 represent blue balls. Simply ignore the digits 0 and 6–9.

2. Starting, e.g., in the third set in row 40, we would read 42595 27977 32790 41917 as 4255232411, which translates into b r b b r b r b r r, or taken in consecutive pairs, they represent five simulations of our draw-two-balls experiment: br, bb, rb, rb, rr. If the event of interest is drawing red followed by drawing blue, this gives no (for the br), no (bb), yes (rb), yes (rb), and no (rr) for the five simulations. Start elsewhere to obtain additional simulations.

3. To find a random sample of 10 people from a population of 1000 people

1. Organize the 1000 people in some fashion, such as alphabetically. Then assign a three-digit number to each: 000 to the first person and 999 to the last (or 001 to the first person and 000 to the last).

2. Using the TRSD, choose any place in the table to begin. If the digits are arranged in sets of five, as in the TRSD that follows, ignore the fourth and fifth digits in each five-digit set. Go down a column, take the first three digits of each five-digit set for 10 sets.

3. Locate the individuals corresponding to each of the 10 three-digit numbers.

Notice that this sampling is similar to a situation in which tickets are given for some prize, perhaps a door prize at a party. Each ticket has two parts, and both have numbers. For each ticket given away, the other part is placed in a container. Thus, each person has been assigned a separate number. If there are 10 door prizes, the first 10 tickets taken from the container (in which the tickets have been thoroughly mixed and cannot be seen by the person selecting tickets) are associated with the 10 individuals who get prizes. Here the assignment of numbers is not based on an alphabetical listing but rather numbers are assigned by the order in which people come in the door, and the randomness develops when actually selecting the 10 numbers from a container.

Table of Randomly Selected Digits (TRSD)

1	00328	52694	37864	93224	74454	33886	17120	60932	39090	30674
2	05136	86903	95926	15771	14137	60923	53431	73737	94807	28815
3	68597	06896	48803	17910	23242	80947	11213	46626	41994	52952
4	16508	79646	33229	67582	44466	26119	42263	92222	36894	64348
5	55560	71569	01479	50838	01401	88489	78617	18785	18373	63431
6	78054	04561	17648	44578	24328	50951	20342	09327	22356	83965
7	94263	06048	23880	52193	18220	61567	75479	07541	89554	35294
8	79839	53271	98734	90449	25115	83661	57375	06879	88635	14464
9	27400	92094	82414	10697	82244	39683	50607	34439	47340	54524
10	22074	37659	72120	17121	74096	19442	67012	47905	01723	51894
11	13510	74328	03432	33897	25288	76819	68800	14987	16837	83429
12	06837	57432	66411	45577	36037	49384	28402	58229	95113	35294
13	73796	48766	55343	47984	81150	97408	66449	21678	28767	45622
14	79450	12243	67635	65136	28300	64656	13376	91884	20462	83225
15	07157	56219	48061	35955	44984	11451	81954	92823	45469	74357
16	16681	11972	02904	47489	19646	66585	59072	72368	55431	14384
17	49861	76253	02124	90355	07107	21633	79016	37457	93461	02636
18	07115	06152	48853	11337	67861	85780	73275	60249	32405	75483
19	98981	12028	49873	24083	45466	46227	24934	98524	34531	57160
20	16976	77713	43124	51125	24283	07367	77804	42972	50873	45392
21	44144	47146	44502	76636	65082	00154	34870	79978	88610	62278
22	61338	60404	58235	48140	12647	29989	47295	10621	39595	88988
23	23753	83397	68668	13454	61441	41226	07984	35666	02889	01286
24	02928	81245	33929	42222	91000	34699	06534	95955	01089	22128
25	79316	35342	18326	08735	11378	64982	10511	58252	18660	84592
26	84480	85717	54342	20628	08792	73587	20972	88047	38399	37733
27	02974	29874	18390	37192	91041	44234	50058	41625	26704	87527
28	32360	76575	71309	49530	55794	77803	26079	45151	55748	31066
29	33519	86479	36815	89900	34909	68591	55676	24541	38485	89951
30	13554	71227	42727	47318	34697	84231	56599	82546	11875	46183
31	40151	15156	73100	45755	82729	05132	22862	58245	46562	07497
32	51918	27296	04920	42450	23986	90834	08333	60073	98581	93100
33	14456	30581	64104	87668	12926	75034	35172	48588	04836	16888
34	97263	73552	83536	34733	43368	48072	38810	77930	77025	90136
35	43232	99440	22273	66259	51747	33965	98549	74348	00685	54153
36	05649	78633	78169	24966	16209	17728	28248	52596	66546	73614
37	03584	74328	02146	01357	51985	73958	64630	56757	32487	79447
38	20703	86561	41351	38059	07959	94954	67671	25201	23539	04925
39	49985	86109	07036	28695	37739	57596	79780	42643	20873	22679
40	24810	21716	42595	27977	32790	41917	16299	02555	17322	45460

continue

41	99520	65656	36410	12625	19648	93697	28112	10652	42235
42	62102	78228	98702	74293	59786	84054	16873	32793	84638
43	28776	14173	15452	82453	99591	11248	96008	19620	10672
44	30341	43795	36283	61785	12790	38549	18382	56866	07254
45	07335	34288	71837	40979	77215	50165	97525	16876	67902
46	63031	49913	21132	60661	81667	74420	05177	70525	56756
47	55227	08991	25807	75290	51036	46912	50137	19313	01054
48	07187	60009	75192	83242	45008	50832	44791	15005	86528
49	05578	79854	68733	56726	30931	96012	62957	29167	74916
50	06903	30507	31055	20622	58841	99640	94831	63019	31284
51	18885	12787	08182	82413	00304	08396	78964	27916	85172
52	94712	13085	30764	60275	82590	37636	47960	44288	73630
53	91733	89031	02448	70011	08342	53882	01936	35546	26526
54	35408	29675	63186	78080	46465	61655	42591	16026	22049
55	70347	61939	27336	92975	35678	16381	99724	14338	39027
56	03496	83975	45626	21813	48702	23651	06354	31410	00202
57	57546	45772	53082	42129	75946	89489	96562	65888	15154
58	89102	94034	36681	66499	15275	42090	87960	85226	26362
59	98643	19580	19467	08205	02933	34924	18592	13938	11455
60	95175	71806	30343	23956	95248	54688	20669	15852	11155
61	39051	64845	70678	51559	85434	73803	42769	60961	23117
62	46432	13908	13354	76460	41640	76001	26956	60228	21317
63	36605	18701	86180	00162	12305	69269	11991	87882	11119
64	21533	58679	08156	26606	25840	03851	51338	21161	51890
65	00157	31690	50321	32536	60370	17238	22151	73713	63289
66	63929	08404	59336	60290	29959	48480	48320	80443	55629
67	73663	55260	30418	60083	97982	09192	97978	67257	56478
68	68336	90747	51450	12909	09176	39197	77274	30751	26954
69	50821	73145	85981	42105	43973	88890	13109	71406	08502
70	35310	98983	37857	24024	66015	61904	84624	18102	10179
71	05813	53283	14926	15531	58898	61856	08999	03174	99705
72	17708	37063	19973	80923	71355	06252	31043	98657	33874
73	32944	23558	91868	58182	34669	10380	65135	15034	78889
74	73862	08659	68925	36053	98240	39044	79681	68779	01598
75	26478	87066	04550	07476	44506	23275	13929	17887	98828
76	62394	03161	70705	24091	20144	62002	42858	33828	53014
77	99875	80191	85743	05256	43429	42773	02235	77967	44038
78	00609	73923	25213	44171	27123	99542	23287	24658	93833
79	03866	23612	18808	41231	53141	21022	84255	92455	30274
80	96882	88567	41017	53993	90822	86724	38090	04896	12732

Data Sets in Printed Form

The data sets given in this appendix are also available at www.whfreeman.com/reconceptmath2e.

60 Students

Student	Gender	Pol Party	GPA	Career Aim	Hrs Work/Wk	Hrs Vol/Wk
1	M	Dem	3.2	Education	6	4
2	M	Dem	3.9	Engineering	5	4
3	F	Rep	3.1	Hospitality	3	0
4	F	Ind	2.9	Education	10	6
5	F	none	1.8	Counseling	14	6
6	F	Ind	2.5	Medicine	16	5
7	M	Rep	3.2	Law	4	2
8	M	Rep	2.6	Education	20	0
9	F	Rep	2.8	Prog/Tech	22	0
10	M	Rep	3.4	Education	12	0
11	M	Dem	3.1	Accounting	0	6
12	F	none	3.9	Business	16	4
13	F	Dem	2.0	Retail	10	3
14	F	Dem	2.4	Science	14	2
15	F	Dem	2.3	Education	8	0
16	M	Ind	2.2	Business	10	2
17	M	none	3.9	Engineering	6	0
18	F	Ind	3.2	Undecided	2	8
19	M	Dem	3.1	Education	16	4
20	M	Rep	2.2	Education	24	0
21	M	Dem	1.9	Other	40	0
22	F	Rep	0.8	Medicine	0	0
23	M	Rep	2.1	Journalism	14	8
24	M	Rep	4.0	Law	14	6
25	F	Rep	3.6	Education	12	4
26	F	Dem	2.9	Retail	10	0
27	F	Ind	2.7	Counseling	11	2
28	F	Dem	3.0	Accounting	15	0
29	F	Dem	2.6	Engineering	16	0
30	M	Dem	3.1	Engineering	22	0

continue

Student	Gender	Pol Party	GPA	Career Aim	Hrs Work/Wk	Hrs Vol/Wk
31	M	Dem	2.4	Education	20	2
32	F	Ind	2.6	Retail	24	2
33	M	Dem	2.9	Prog/Tech	22	0
34	M	Rep	3.6	Business	26	0
35	F	Rep	3.5	Medicine	30	0
36	F	Dem	2.5	Undecided	32	0
37	M	Dem	3.5	Art/Music	14	0
38	F	Ind	3.8	Law	16	4
39	M	Ind	4.0	Journalism	14	4
40	F	Ind	2.7	Retail	22	0
41	F	Dem	2.9	Accounting	26	2
42	M	Dem	2.4	Other	12	6
43	M	Rep	3.1	Art/Music	0	6
44	M	Rep	3.5	Other	6	2
45	M	Dem	2.5	Undecided	8	4
46	M	Rep	3.1	Education	22	2
47	F	Rep	3.2	Education	28	3
48	F	Rep	2.7	Engineering	8	8
49	F	Dem	2.8	Medicine	32	2
50	M	Dem	2.1	Art/Music	10	4
51	F	Rep	1.9	Prog/Tech	16	0
52	F	Dem	1.9	Business	12	0
53	F	Rep	2.0	Hospitality	20	2
54	F	Dem	3.6	Architecture	20	0
55	M	Dem	2.9	Education	18	0
56	M	Ind	3.2	Accounting	8	2
57	F	none	2.9	Law	12	2
58	F	Ind	2.6	Medicine	20	0
59	M	Rep	2.6	Education	24	0
60	F	Rep	3.5	Law	16	0

World Statistics

Countries	Land Area (1000 km^2)	Population in millions (in 2011)	Countries	Land Area (1000 km^2)	Population in millions (in 2011)
Russia	17,075	139	Thailand	513	67
Canada	9971		Spain	506	47
China + Taiwan	9567	1337	Sweden	450	

continue

Countries	Land Area (1000 km²)	Population in millions (in 2011)
United States	9364	313
Brazil	8547	203
Australia	7741	
India	3288	
Argentina	2780	
Kazakhstan	2717	
Sudan	2506	
Saudi Arabia	2150	
Mexico	1958	114
Iran	1633	78
Mongolia	1567	
Peru	1285	
Colombia	1139	
Pakistan	796	187
France	552	65

Countries	Land Area (1000 km²)	Population in millions (in 2011)
Japan	378	126
Germany	357	
Finland	338	
Vietnam	332	91
Malaysia	330	
Norway	324	
Italy	301	61
Philippines	300	102
New Zealand	271	
Indonesia	1919	246
Poland	313	38
UK	244	63
Korea (N + S)	221	72
Denmark	43	
Switzerland	41	

Source: http://www.infoplease.com/ipa/A0004379.html.

Life Expectancy at Birth

Country at birth	Males 1980	Males 1999	Males Change	Females 1980	Females 1999	Females Change
Australia	71.0	76.2	5.2	78.1	81.8	3.7
Austria	69.0	75.1	6.1	76.1	81.0	4.9
Belgium	70.0	74.4	4.4	76.8	80.8	4.0
Bulgaria	68.5	67.4	−1.1	73.9	74.9	1.0
Canada	71.7	76.3	4.6	78.9	81.7	2.8
Costa Rica	71.9	74.9	3.0	77.0	79.8	2.8
Cuba	72.2	73.3	1.1	n.a.	77.5	—
Czech Republic	66.8	71.4	4.6	73.9	78.2	4.3
Denmark	71.2	74.2	3.0	77.3	79.0	1.7
England and Wales	70.8	75.3	4.5	76.8	80.1	3.3
Finland	69.2	73.8	4.6	77.6	81.0	3.4
France	70.2	75.0	4.8	78.4	82.5	4.1
Germany	69.6	74.7	5.1	76.1	80.7	4.6

continue

Country at birth	Males 1980	Males 1999	Males Change	Females 1980	Females 1999	Females Change
Greece	72.2	75.5	3.3	76.8	80.6	3.8
Hungary	65.5	66.4	0.9	72.7	75.2	2.5
Ireland	70.1	73.9	3.8	75.6	79.1	3.5
Israel	72.2	76.9	4.7	75.8	80.6	4.8
Italy	70.6	75.6	5.0	77.4	82.3	4.9
Japan	73.4	77.1	3.7	78.8	84.0	5.2
Netherlands	72.5	75.3	2.8	79.2	80.5	1.3
New Zealand	70.0	75.7	5.7	76.3	80.8	4.5
Norway	72.3	75.6	3.3	79.2	81.1	1.9
Poland	66.0	68.2	2.2	74.4	77.2	2.8
Portugal	67.7	72.2	4.5	75.2	79.2	4.0
Puerto Rico	70.8	70.7	−0.1	76.9	79.8	2.9
Romania	66.6	67.1	0.5	71.9	74.2	2.3
Russian Federation	61.4	59.4	−2.0	73.0	72.0	−1.0
Scotland	69.0	72.7	3.7	75.2	78.1	2.9
Singapore	69.8	75.6	5.8	74.7	79.7	5.0
Slovakia	66.8	69.0	2.2	74.3	77.2	2.9
Spain	72.5	75.1	2.6	78.6	82.1	3.5
Sweden	72.8	77.1	4.3	78.8	81.9	3.1
Switzerland	72.8	76.8	4.0	79.6	82.5	2.9
United States	70.0	73.9	3.9	77.4	79.4	2.0

Source: Health, United States, 2004, Table 26. http://www.cdc.gov/nchs/hus04.ace.

60 Students Sorted by Gender

Student	Gender	Pol Party	GPA	Career Aim	Hrs Work/Wk	Hrs Vol/Wk
1	F	Rep	3.1	Hospitality	3	0
2	F	Ind	2.9	Education	10	6
3	F	none	1.8	Counseling	14	6
4	F	Ind	2.5	Medicine	16	5
5	F	Rep	2.8	Prog/Tech	22	0
6	F	none	3.9	Business	16	4
7	F	Dem	2.0	Retail	10	3
8	F	Dem	2.4	Science	14	2
9	F	Dem	2.3	Education	8	0
10	F	Ind	3.2	Undecided	2	8
11	F	Rep	0.8	Medicine	0	0
12	F	Rep	3.6	Education	12	4
13	F	Dem	2.9	Retail	10	0

continue

Student	Gender	Pol Party	GPA	Career Aim	Hrs Work/Wk	Hrs Vol/Wk
14	F	Ind	2.7	Counseling	11	2
15	F	Dem	3.0	Accounting	15	0
16	F	Dem	2.6	Engineering	16	0
17	F	Ind	2.6	Retail	24	2
18	F	Rep	3.5	Medicine	30	0
19	F	Dem	2.5	Undecided	32	0
20	F	Ind	3.8	Law	16	4
21	F	Ind	2.7	Retail	22	0
22	F	Dem	2.9	Accounting	26	2
23	F	Rep	3.2	Education	28	3
24	F	Rep	2.7	Engineering	8	8
25	F	Dem	2.8	Medicine	32	2
26	F	Rep	1.9	Prog/Tech	16	0
27	F	Dem	1.9	Business	12	0
28	F	Rep	2.0	Hospitality	20	2
29	F	Dem	3.6	Architecture	20	0
30	F	none	2.9	Law	12	2
31	F	Ind	2.6	Medicine	20	0
32	F	Rep	3.5	Law	16	0
33	M	Dem	3.2	Education	6	4
34	M	Dem	3.9	Engineering	5	4
35	M	Rep	3.2	Law	4	2
36	M	Rep	2.6	Education	20	0
37	M	Rep	3.4	Education	12	0
38	M	Dem	3.1	Accounting	0	6
39	M	Ind	2.2	Business	10	2
40	M	none	3.9	Engineering	6	0
41	M	Dem	3.1	Education	16	4
42	M	Rep	2.2	Education	24	0
43	M	Dem	1.9	Other	40	0
44	M	Rep	2.1	Journalism	14	8
45	M	Rep	4.0	Law	14	6
46	M	Dem	3.1	Engineering	22	0
47	M	Dem	2.4	Education	20	2
48	M	Dem	2.9	Prog/Tech	22	0
49	M	Rep	3.6	Business	26	0
50	M	Dem	3.5	Art/Music	14	0
51	M	Ind	4.0	Journalism	14	4
52	M	Dem	2.4	Other	12	6
53	M	Rep	3.1	Art/Music	0	6
54	M	Rep	3.5	Other	6	2

continue

Notes

Student	Gender	Pol Party	GPA	Career Aim	Hrs Work/Wk	Hrs Vol/Wk
55	M	Dem	2.5	Undecided	8	4
56	M	Dem	2.1	Art/Music	10	4
57	M	Dem	2.9	Education	18	0
58	M	Ind	3.2	Accounting	8	2
59	M	Rep	2.6	Education	24	0
60	M	Rep	3.1	Education	22	2

60 Students, Independent (Pol Party)

Gender	Pol Party	GPA	Career Aim	Hrs Work/Wk	Hrs Vol/Wk
F	Ind	2.9	Education	10	6
F	Ind	2.5	Medicine	16	5
F	Ind	3.2	Undecided	2	8
F	Ind	2.7	Counseling	11	2
F	Ind	2.6	Retail	24	2
F	Ind	3.8	Law	16	4
F	Ind	2.7	Retail	22	0
F	Ind	2.6	Medicine	20	0
M	Ind	2.2	Business	10	2
M	Ind	4.0	Journalism	14	4
M	Ind	3.2	Accounting	8	2

60 Students, Democrat (Pol Party)

Gender	Pol Party	GPA	Career Aim	Hrs Work/Wk	Hrs Vol/Wk
F	Dem	2.0	Retail	10	3
F	Dem	2.4	Science	14	2
F	Dem	2.3	Education	8	0
F	Dem	2.9	Retail	10	0
F	Dem	3.0	Accounting	15	0
F	Dem	2.6	Engineering	16	0
F	Dem	2.5	Undecided	32	0
F	Dem	2.9	Accounting	26	2
F	Dem	2.8	Medicine	32	2
F	Dem	1.9	Business	12	0
F	Dem	3.6	Architecture	20	0
M	Dem	3.2	Education	6	4

continue

Gender	Pol Party	GPA	Career Aim	Hrs Work/Wk	Hrs Vol/Wk
M	Dem	3.9	Engineering	5	4
M	Dem	3.1	Accounting	0	6
M	Dem	3.1	Education	16	4
M	Dem	1.9	Other	40	0
M	Dem	3.1	Engineering	22	0
M	Dem	2.4	Education	20	2
M	Dem	2.9	Prog/Tech	22	0
M	Dem	3.5	Art/Music	14	0
M	Dem	2.4	Other	12	6
M	Dem	2.5	Undecided	8	4
M	Dem	2.1	Art/Music	10	4
M	Dem	2.9	Education	18	0

60 Students, Republican (Pol Party)

Gender	Pol Party	GPA	Career Aim	Hrs Work/Wk	Hrs Vol/Wk
F	Rep	3.1	Hospitality	3	0
F	Rep	2.8	Prog/Tech	22	0
F	Rep	0.8	Medicine	0	0
F	Rep	3.6	Education	12	4
F	Rep	3.5	Medicine	30	0
F	Rep	3.2	Education	28	3
F	Rep	2.7	Engineering	8	8
F	Rep	1.9	Prog/Tech	16	0
F	Rep	2.0	Hospitality	20	2
F	Rep	3.5	Law	16	0
M	Rep	3.2	Law	4	2
M	Rep	2.6	Education	20	0
M	Rep	3.4	Education	12	0
M	Rep	2.2	Education	24	0
M	Rep	2.1	Journalism	14	8
M	Rep	4.0	Law	14	6
M	Rep	3.6	Business	26	0
M	Rep	3.1	Art/Music	0	6
M	Rep	3.5	Other	6	2
M	Rep	3.1	Education	22	2
M	Rep	2.6	Education	24	0

Note to students: The selected answers below give you some indication about whether you are on the right track. They are NOT intended to be the type of answers you would turn in to an instructor. ***Your complete answers should contain all the work toward obtaining an answer***, as described in the Message to Prospective and Practicing Teachers.

Here are a few examples of complete answers:

Section 1.2, Learning Exercise 2(a)

The quantities and their values are:

Regular price of CDs	$9.95
Regular price of tapes	$6.95
Discount this month	10%
Discounted price of CDs	90% of $9.95 = $8.96
Discounted price of tapes	90% of $6.95 = $6.26
New discount on 3 items	20%
Number of CDs bought	1
Number of tapes bought	1
Amount spent on CDs	$8.96
Amount spent on tapes	$6.26
Sales tax	6%
Amount spent	?

This drawing represents the problem

```
┌──────────┬──────────┬──────┐
│          │          │      │
│  $8.96   │  $6.26   │  6%  │
│          │          │      │
└──────────┴──────────┴──────┘
 ◄──────────────────────► 6% of $15.22 is $0.91
          $15.22
 ◄─────────────────────────────►
      $16.13 is the total spent.
```

Section 3.1, Learning Exercise 9

We need to find the weight of the sum of the medicine available from Companies A, B, and C. A diagram will help. Here is one possibility:

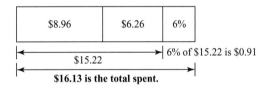

Company A's medicine is represented by a line marked 1.3 mg. Company C's medicine is represented next because it is easier to find—it is 0.9 mg less than A's medicine. Thus, Company C has 0.4 mg of medicine. The difference between Company B's medicine and Company C's is half of 1.3 mg, so it is 0.65 mg over and above Company C's medicine, which is 0.4 mg. Thus, Company B has 0.4 mg + 0.65 mg, which is 1.05 mg. The total medicine furnished by the companies is (1.3 + 0.4 + 1.05) mg, or 2.75 mg.

Section 5.4, Learning Exercise 2

$3 \times 10^4 \times 4 \times 10^6 = 12 \times 10^{10}$. This expression is not in scientific notation because $12 > 10$. In scientific notation this product would be expressed as 1.2×10^{11}.

Answers for Chapter 1

1.2 Quantitative Analysis

1. b.

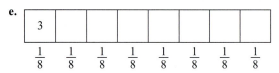

Age when sold $= 1998 - 1727 = 271$ years

Time period is $1998 - 1727 = 271$ years

e.

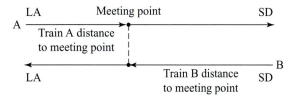

Each box represents 3 students. There are 8 boxes. Thus, there are 24 students.

2. a. See p. A-1.

3. This problem is an example of a word problem typically found in an algebra textbook that can be solved quite easily when all of the quantities to be used are listed and a drawing is made.

Quantities	Values
Speed of Train A	84 mph
Speed of Train B	92 mph
Distance traveled by Train A to meeting point	? miles
Distance traveled by Train B to meeting point	? miles
Total distance traveled by trains	132 miles
Time traveled by both trains, departure to meeting point	? hours

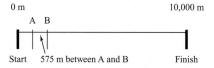

One solution: In 1 hour, the trains together will go $84 + 92 = 176$ miles. Since they need go only 132 miles, they will meet after $\frac{132}{176} = \frac{3}{4}$ hours. *A second solution* looks at how far each train goes in 1 minute: $\frac{84}{60}$ miles and $\frac{92}{60}$ miles, or $\frac{84}{60} + \frac{92}{60} = \frac{176}{60}$ miles together. Again, since they need to go 132 miles, the calculation $132 \div \frac{176}{60}$ will tell the number of minutes needed. Here is a *third solution*, the traditional algebraic one that also follows from a quantitative analysis. Note that the time traveled by both trains is the same, t hours.

Distance$_A$ + Distance$_B$ = 132 miles

$84t + 92t = 132$, so $176t = 132$ and

$t = 132 \div 176 = 0.75$ hour

Distance Train A travels in 0.75 hours is

   $0.75 \text{ h} \times 84 \text{ mph} = 63$ miles

Distance Train B travels in 0.75 hours is

   $0.75 \text{ h} \times 92 \text{ mph} = 69$ miles

Notice that algebra is not necessary for this problem, and need not be used in elementary school.

6. First, each quantity is listed, and then the value of each quantity is determined, if possible.

1. Distance (length) of the last part of the triathlon: 10,000 m
2. Distance between Aña and Bea at the start of the last part: 600 m
3. Distance between Aña and Bea during the last part: varies
4. Distance between Aña and Bea at the time they meet: 0 m
5. Distance between Aña and Bea at the end of the 10,000-m race: not known
6. Average speed (rate) at which Aña runs: 225 m/min
7. Average speed (rate) at which Bea runs: 200 m/min
8. Distance Aña runs in 1 minute: 225 m
9. Distance Bea runs in 1 minute: 200 m
10. Difference between the average speeds: Aña runs 25 m/min faster than Bea
11. Difference between the distance Aña runs in 1 minute and the distance Bea runs in 1 minute: 25 m
12. Time it takes Aña to run the last part of the triathlon: 10,000 m ÷ 225 m/min = 44.44 min
13. Time it takes Bea to run the last part of the triathlon: 10,000 m ÷ 200 m/min = 50 min
14. Time from when Aña begins the last running part to the time she passes Bea: not known
15. Time from when Bea begins the last running part to the time when Aña catches up: not known

Make a drawing (or drawings) to represent the problem and the relationships among the quantities. Consider the following diagram depicting the situation at the start of the running part of the triathlon. Aña and Bea are 600 m apart.

Now, consider the situation 1 minute later. Aña has run 225 m, Bea has run 200 m, and with a 600 m headstart, is at 800 m. So after 1 minute they are 575 m apart.

Using this process, I know they would be 550 m apart after 2 minutes, 525 m after 3 minutes, etc. Notice that the value of that distance between Aña and Bea *varies*, i.e., it is not 600 m except at the beginning. A critical value of the distance between Aña and Bea will occur when Aña overtakes Bea (if she does). At this time the distance would be 0.

Finally, solve the problem using the drawings. One way to think about it is this: In each minute of the race, Aña gains 25 m on Bea. Because Aña begins the race 600 m behind Bea, it will take Aña 600 m divided by 25 m/min, or 24 min, to catch up to Bea.

9. Speed of train leaving Moscow: 48 km/h
    Speed of train leaving Sverdlovsk: 54 km/h
    Time of travel: 12 h
    Distance traveled by first train: 48 km/h × 12 h = 576 km
    Distance traveled by second train: 54 km/h × 12 h = 648 km
    Distance between cities: 1822 km

Distance apart after 12 hours is

1822 km − (576 km + 648 km) = 598 km

1.3 Values of Quantities

2. **b.** miles per gallon

5. **a.** width of a fingernail   **b.** a paperclip   **c.** a quart
    d. width of a door   **e.** a little more than half a mile
    f. three cans of mushroom soup

    Of course, you may have others. You might want to compare your answer with those of others.

6. An inch is approximately 2.5 cm, a mile is approximately 1.6 km, a quart is approximately 1 liter.

1.4 Issues for Learning: Ways of Illustrating Story Problems

4.

Hat $19 (Sweater $38)

8.
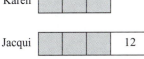

The 7 equally sized small boxes have 124 − 12 = 112 in them. Each small box has 112 ÷ 7 = 16. So, Karen has 48 stamps, Jacqui has 60, and Lynn has 16.

Answers for Chapter 2

2.1 Ways of Expressing Values of Quantities

2. There are 5 fingers on one hand, and 10 on both.

4. **a.** 2113

5. **a.** MMLXVI

6. **a.** 903

7. Twins, couple, dyad, brace, duet, duo, double, twosome, twice, bicompounds like bicycle, both, and dicompounds like dipolar are others.

2.2 Place Value

1. **a.** 35.7; 35   **g.** 234.7; 234   **l.** 2347; 2347

2. No. The ones place, not the decimal point, divides the number into two parts with similar names on both sides.

4. Three thousand two hundred; thirty-two hundred. They have the same value because they have the same number of hundreds: Three thousand is thirty hundreds.

6. **a.** and **b.** No regrouping into tens
    c. Regrouping (In this case, "borrowing" is not understood.)
    d. Digits are misplaced in quotient.
    e. No regrouping into tens. In all cases, the work indicates a lack of understanding of place value.

7. 1635 is exactly 1635 ones, is 163.5 tens, is 16.35 hundreds, is 1.635 thousands, . . . , is 16,350 tenths, is 163,500 hundredths.

9. To the right, yes. To the left, no. (For whole numbers.)

10. One thousand million

2.3 Bases Other Than Ten

3. **i.** 1000101_{two} **j.** 1000_{twelve}

4. 1, 10, 11, 100, 101, 110, 111, 1000, 1001, 1010, 1011, 1100, 1101, 1110, 1111, 10000, 10001, 10010, 10011, 10100

6. **a.** 43_{10} **b.** 34_{10}

9. **a.** <   **b.** =   **c.** >

12. 3 fives + 4 ones + $\frac{2}{5}$; 19.4_{ten}

14. **a.** $\frac{1}{4}$ in base twelve is the same amount as in base ten. $\frac{1}{4}$ is $\frac{3}{12}$ in base ten. $\frac{3}{12}$ in base twelve would be $\frac{3}{10}$, or 0.3_{twelve}.

17. **a.** 202_{seven} **b.** 400_{five}

18. **a.** 212   **b.** 78

19. Three longs and 4 small blocks; 3 flats and 4 longs

20. **a.** You should have 2 flats, 3 longs, and 4 small cubes, where the flat is five units by five units.
    b. You should have 2 flats, 3 longs, and 4 small cubes, where the flat is six units by six units.

22. You should have 2 flats, 3 longs, and 4 small cubes.

23. **a.** If the small cube is the unit, then 3542 could be represented with 3 large cubes, 5 flats, 4 longs, and 2 small cubes.
    b. 0.741 could only be represented if the large block is the unit. Then 0.741 would be represented with 7 flats, 4 longs, and 1 small cube.
    c. If the flat represents one unit, then 11.11 would be represented with 1 large cube, 1 flat, 1 long, and 1 small cube.

26. Bases larger than 4, otherwise the flats would have to be traded.

2.4 Operations in Different Bases

2. Block answers should yield:
    a. 1111_{five}

5. **a.** 101_{nine} **b.** 506_{seven}

7. **a.** 1030_{four} **b.** 413_{five}

8. Block or cut-out answers should yield:

   a. 121_{four} **b.** 132_{five}

9. Cut-outs for base 6 would consist of small squares, longs the length of 6 small squares, flats the size of 6 longs side by side.

Answers for Chapter 3

3.1 Additive Combinations and Comparisons

3. 50 students

4. C and D are combined, A and B are combined, A and D are compared, B and C are compared.

5. **a.** The point is that here there will be many answers. Any scores giving a difference between B's and B's opponents' scores = 34, with B's score greater, will solve the problem (but not uniquely).

6. Connie bought 13 Tootsie Rolls.

10. $\frac{1}{8}$

3.2 Ways of Thinking About Addition and Subtraction

1. **b.** Comparison

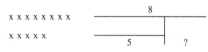

   c. Missing addend (5 + ? = 8)

2. **b.** Colleen had pennies in her piggy bank. She took out 17 and gave them to her brother. She counted the ones she had left and found she had 24 pennies left. How many pennies did she have in her piggy bank before she gave some to her brother?

4. **a.** The baseball team has 12 players and 9 were wearing gloves. How many players were not wearing gloves?

5. **c.** $m - p = x; m - x = p; p + x = m; x + p = m$

10. Mr. Lewis's students did not recognize the comparison problem as a problem calling for subtraction because they had only worked with take-away problems in class.

11. The drawings are quite easy to make.

   a. 8.25 ft   **b.** $\frac{6}{15}$ of the students   **c.** 6 cm

3.3 Children's Ways of Adding and Subtracting

2. *Case C*: 50 and 59 is: 50 and 50 is 100, and 9 more is 109, but take 2 away, so it's 107.

   Case D: 254 + 367: 2 plus 3 is 5 so it's 200 + 300, so it's somewhere in the 500s. Now add tens. They are 5 and 6. Start at 50, 60, 70, 80, 90, 100, you are in the 500s because of the 200 + 300 but now you're in the 600s because of the 50 and 60. But you've got one more ten. So 500 + 50 + 60, you'd have 610. Now add 4 more onto 10, which is 14. And add 7 more: 15, 16, 17, 18, 19, 20, 21. And that is 621.

3. and 4. *Case C*: (3) The student rounded the first number up to 40, then added the tens and the ones, and compensated for the rounding. (4) This method might be slightly more difficult because it would be easy to forget the need for compensating at the end.

Case D: (3) The student is adding from left to right: hundreds, then tens, then ones. So many steps and intermediate sums are hard to remember mentally, and so the student does make one error, then corrects it. (4) The method itself will not be difficult to remember because the direction from left to right is the way we read. However, the difficulty is remembering all the partial addends, so this might be easier if the solver could jot down partial sums along the way.

3.4 Ways of Thinking About Multiplication

1. **b.** Possibly . . .

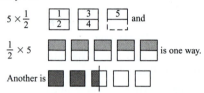

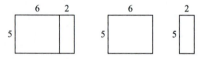

3. Not any fraction, but multiplying by a fraction less than 1. If the multiplier is a whole number or a fraction greater than 1, then "multiplication makes bigger" is correct.

5. 6 twelves, plus $\frac{1}{2}$ twelve—a repeated-addition multiplication as well as a part-of-an-amount multiplication.

7. **a.** 8 chicken breasts, $1\frac{1}{3}$ cups onion, $1\frac{1}{3}$ Tbsp olive oil, $2\frac{2}{3}$ cups apples, $1\frac{1}{3}$ Tbsp margarine, 2 cups apple juice, $3\frac{1}{3}$ Tbsp honey, $\frac{2}{3}$ tsp salt

9. **a.** *Property*: Multiplication is distributive over addition.

b. Note that no drawing is called for. A demonstration might be as follows:

$$3 \times (4 \times 5) = 3 \times 20 = 60$$
$$(3 \times 4) \times 5 = 12 \times 5 = 60$$

10. **a.**

R1	R2	R3	R4	R5	R6
W1	W2	W3	W4	W5	W6
B1	B2	B3	B4	B5	B6
G1	G2	G3	G4	G5	G6

Thus, the answer is 24.

b. 12

11. $2^8 = 256$, so there must be 8 ways to choose between two things to arrive at 256 different kinds of hamburgers.

13. $(3 \times 4 \times 3) \times 12486$, with the first multiplications from the fundamental counting principle, and the last from repeated addition. $449,496

15. **a.** 77.1   **b.** 0.76 (or just .76)   **c.** 17.85

17. 86,400 seconds in a day, from $24 \times (60 \times 60)$.

21. $4 \times 3 = n$, 12 different orders

22. $(20 \times 25) + (12 \times 18) = n$, 716 people

3.5 Ways of Thinking About Division

5. d. You can hike 3 kilometers in an hour on trails like the 4.5-km one. How long should you plan for going up the 4.5-km trail?

   e. Sample A runner prepares for a 10K run by running on Mount Azteca. How many times would it take her to run the 2.4-km trail to make 10 km?

6. a. In the testing, eighth-graders would give the actual results of the division calculation, $1500 \div 36 = 41\frac{2}{3}$, or 41 R 24, as the number of buses needed. But the company would realistically supply 42 buses.

8. a. *Repeated subtraction*: Draw a large rectangle for the 900 kg, and then mark off some smaller 12-kg pieces to get the idea across. The division tells how many there would be (75).

   b. *Sharing*: Use a rectangle for the 525 kg with several pretend 1-kg pieces shown, and then draw arrows to 3 empty spaces for the weeks to show the 1-kg pieces being distributed equally.

   c. *Sharing*: To find the average, you are answering the question of how many for one week drove the same number of miles each week. Draw a line segment labeled 5760. Divide it into 12 equal segments. Label the length of the small segments.

9. 8, 8, 8, 8, 8; $10^n a \div 10^n b = a \div b$

3.6 Children Find Products and Quotients

1. The first student's method is related to estimating because she first estimated the quotient to be between 2080 and 2240, and then found the remainder to obtain the exact answer.

3. *Advantages*: Attaches repeated subtraction to division conceptually; need not make best guesses (as in eventual algorithm); involves familiar subtraction rather than multiplication. Other advantages?

   Disadvantages: Slower; must understand a meaning for division (is this a disadvantage?); many subtractions so there is more possibility of error. Other disadvantages?

5. This procedure was invented by a very precocious child. Most first-graders would not have been able to do this. However, prospective teachers should recognize that they will sometimes have very bright students whose thinking is not always easily followed.

   To find $63 \div 9$:
   $60 \div 10 = 6$
   $6 + 3 = 9$
   $9 \div 9 = 1$
   $6 + 1 = 7$, so $63 \div 9 = 7$

3.7 Issues for Learning: Developing Number Sense

1. a. $135 + 98$, because each addend is greater than the corresponding addend in $114 + 92$.

3.

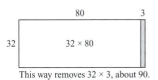

This way removes 32×3, about 90.

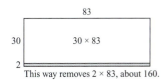

This way removes 2×83, about 160.

Thus, more is lost the second way, so the first way gives a closer estimate.

Or, if 32 is rounded to 30, and the 83 is not rounded, the answer is off by 2×83 or about 160. If 83 is rounded to 80 and the 32 is not rounded, the answer is off by 3×32 or about 90, so rounding the 83 yields a closer estimate. (The temptation is to say "round the 32 because only 2 is being dropped," but if 83 is rounded to 80, 3 is being dropped.")

4. a. None of the addends have 1 in the ones place, so the sum can't also.

6. a. 47 and 52 are "compatible," adding to about 100, and 36 and 69 are somewhat compatible, with a sum of about 100. Sum is about $200 + (20 \text{ or } 30)$: 220 or 230.

   d. $35,000 \div 50 = 700$

7. a. *Repeated subtraction*: There will be more 9.35s in 56 than there are 10s.

   Sharing: "Splitting" 56 into 10 amounts will give less than splitting it into fewer than 10 amounts.

8. 0.68×5 is trivial compared to 0.34×150, which is about a third of 150.

Answers for Chapter 4

4.1 Operating on Whole Numbers and Decimal Numbers

1. a. Let the flat be the unit, because a unit that can be cut into hundredths is needed. Then large cubes represent tens, longs represent tenths, and small cubes represent hundredths.

   Step 1: Lay out two rows: the first has 5 large cubes, 6 flats, and 2 longs. Directly below are 3 large cubes, 4 flats, 5 longs, and 2 small cubes.

   Step 2: Begin filling a third row. First bring down (directly below) the small cubes; there are 2 thousandths.

   Step 3: Bring down (to the third row) the longs. There are 7, representing 7 tenths.

   Step 4: Bring down the flats. There are 10. Trade for a large cube and place the cube in the next left column with the other cubes. The flats column is now empty, representing 0 ones.

   Step 5: Bring down all the large cubes. There are 8 plus the one from Step 4, for a total of 9, representing 9 tens.

   Step 6: Write down the number represented, from left to right: 9 cubes, 0 flats, 7 longs, and 2 small cubes, representing 90.72.

2. These involve just whole numbers, so the small cube can be the unit. It is instructive to act them out with base materials or toothpicks, noticing the trades.

   a. 1123_{five}

5. a. 500 (twice as many as for $4000 \div 16$)

   c. 125 (half as many as $4000 \div 16$)

   d. 125 (doubled dividend and divisor of c)

   e. 62.5 (doubled divisor of d)

   g. 1000   **h.** 10,000

8. a. Start with 3 flats, 9 longs, and 6 small cubes. Put 1 flat into each of 3 separate piles, then put 3 longs into each of those piles. Finally, put 2 small cubes into each pile, resulting in 1 flat, 3 longs, and 2 small cubes in each pile. The answer is 132. Trying a repeated subtraction is not practical because removing only 3 small cubes 132 times would involve extensive trading and a lot of time.

   b. Proceed much as in part (a), but the 2 flats will have to be traded for 20 longs (resulting in 25 longs). And after putting 6 longs into each of 4 piles, trade the remaining 1 long for 10 small cubes, resulting in 12 small cubes to be shared. There should be 6 longs and 3 small cubes in each of the four piles: 63.

   c. This time, after trading the flat for 10 longs, sharing, trading the remaining 2 longs for 20 small cubes (resulting in 27 small cubes), and sharing, there are 3 small cubes remaining. The answer is 46 R 3.

Answers for Chapter 5

5.1 Mental Computation

1. These are some possible ways:

   a. $43 \times (10 - 1) = 430 - 43$ (using the distributive property)
$= 430 - 30 - 13 = 387$

2. These are some possible ways:

   a. $365 + 35 + 40 = 400 + 40 = 440$ **b.** $758 - 30 = 728$

   c. $1000 + 431 = 1431$ **d.** $500 - 50 = 450$

   e. $124 \times \frac{100}{4} = 31 \times 100 = 3100$ **f.** $43 - 30 = 13$

3. a. $\frac{1}{4}$ of $60 = 15$ **b.** 7.8   **c.** 4.5

   d. $8 \times 10\%$ of $710 = 8 \times 71 = 8 \times 70 + 8 \times 1 = 560 + 8 = 568$ (or $4 \times 20\%$ of $710 = 4 \times 142 = 400 + 160 + 8 = 568$)

5.2 Computational Estimation

1. 0.76 is about $\frac{3}{4}$ and 62 is about 60, and $\frac{3}{4}$ of 60 is 45.

2. Now I want to be sure I have enough money, so I must be careful not to underestimate. I could say 80¢ $\times$ 60 or $48 would be close, and $50 (accounting for 2 additional tablets) is more than enough.

3. The second; the first would be less than 84.63, and the second would be more than 84.63.

6. Only some ways of estimating the answers are shown here.

   a. $50 \times 900 = 45,000$

   b. *A few ways*: $\frac{100}{4} \times 80 = 2000$ (so the estimate should be less than 2000)

$25 \times 76 \approx 25 \times 75 = 25 \times 25 \times 3 = 625 \times 3 = 1875$

or $25 \times 75 = 20 \times 75 + 5 \times 75 = 1500 + 375 = 1875$

or $25 \times 80 = 80 \times 25 = 8 \times 25 \times 10 = 2000$, so 25×76 is <u>a little less than 2000</u>,

or $25 \times 80 = 5 \times 5 \times 80 = 5 \times 400 = 2000$, so 25×76 is <u>a little less than 2000</u>,

or $20 \times 70 = 1400$ and $30 \times 80 = 2400$, so 25×76 is about in the middle of 1400 and 2400, which would be <u>about 1900</u>.

(There are more ways than what are shown here.)

7. a. The price is about $\frac{1}{4}$ off $50, or $12.50 off; about $37.50.

8. b. 32% is close to $\frac{1}{3}$, so an estimate is 4.

9. a. The student does not realize that rounding the $19 to $20 takes care of the 95¢, and that the answer should be less than $20 \times \$20 = \$400$. A better estimate would be just $20 \times \$20 = \$400$.

   b. The student does not take advantage of the fact that 25.3% is about $\frac{1}{4}$. A better estimate would be 30.

5.3 Estimating Values of Quantities

No answers are given for Section 5.3.

5.4 Using Scientific Notation for Estimating Values of Very Large and Very Small Quantities

1. a. 1.836×10^7 **b.** 2.04×10^2

2. See p. A-1.

3. a. 9.9×10^{18} **d.** 2.1×10^3

5. $314 \times 10^4 + 2.315 \times 10^4 = 316.315 \times 10^4 \approx 3.16 \times 10^6$; numbers must have the same power of 10 to be added.

7. Suppose n is 5. Then $2n$ is 10; $2^n = 32$; $n^2 = 25$; $10^n = 100,000$.

10. A gigabyte; 2^{30} is 1,073,741,824. See http://encyclopedia.thefreedictionary.com/gigabyte.

Answers for Chapter 6

6.1 Understanding the Meanings of $\frac{a}{b}$

2. d. *One possibility*: A rectangular region could be marked into 4 equal pieces, with none shaded.

   e. *One possibility*: 3 rectangular regions are shaded (showing that $3 = \frac{3}{1}$).

3. Be sure to contrast parts (a) and (b). In part (a), the unit is the whole piece of licorice whip. In part (b), the unit is the segment between any two consecutive whole numbers, and the labeling convention would have the $\frac{3}{4}$ in only one place. Mark half-way between 0 and 1, then half-way between $\frac{1}{2}$ and 1. The second mark is at $\frac{3}{4}$.

5. a. Two square regions of the same size, each marked into 2 equal pieces, with 3 of the pieces shaded.

   b. Start with 3 square regions; how you show the division can vary. Think of how 3 brownies could be shared by 2 people. One way would be to cut one brownie in half: $3 \div 2$. Each person gets $\frac{3}{2}$, or $1\frac{1}{2}$ brownies. Another way would be to cut each brownie in half; then each person would receive three halves.

6. c. Hexagon $= \frac{1}{2}$; trapezoid $= \frac{1}{4}$; blue rhombus $= \frac{1}{6}$;

triangle $= \frac{1}{12}$

9. Drawings should show that each of the following is the larger of two fractions.

  d. $\frac{12}{17}$ is larger, because something cut into 17 equal pieces gives larger pieces than something cut into 19 equal pieces does, and there are 12 pieces in each.

10. a. A is fair, B is fair (each piece is a quarter of the total rectangle), C is not fair, D is fair. (Can you see why? Recall the formula for the area of a triangle and apply here.) E is fair. F is not. (Try tracing and matching parts.) G and H are not fair, I is fair, J is not fair, K and L are both fair in that in each case the pieces are the same. However, in K, if $1.25 is charged for each of two pieces, then $5.00 is the cost of 8 pieces. Thus, 7 pieces would be less than $5.00—a good deal for the buyer but not for the Student Education Association. In L, the total number of pieces is more than 8, so the Student Association would make more money selling pieces of 2 rather than selling the whole cake.

  c. None

  d. That the pieces a fraction brings to mind should be the same size but not necessarily the same shape. Those considered not fair are those with pieces that are not the same size.

11. The shaded part of the second region shows $\frac{1}{4}$ (compare it to the equally sized first region). How the other markings are made is irrelevant with the first region as a guide. Without the first region as a guide, it would be just an "eyeball" estimate to say $\frac{1}{4}$.

15. b. $2\frac{1}{6}$ would be two of the whole box (all four pieces) in part (a), plus $\frac{1}{6}$ of another whole box.

16. a. Begin by partitioning the rectangle into 8 equally sized parts. (Why 8?) Then mark off *one* unit. Then find $\frac{3}{4}$ of that unit.

17. a. Mark the strip into 5 equal parts. Mark them 0, 0.25, 0.5, 0.75, 1, 1.25.

  b. If the 1.25 decimeter piece cost $1.60, then each of the 0.25 decimeter lengths should cost $1.60 ÷ 5, or 32¢, so 4 lengths (1 decimeter) would cost 4 × 32¢, or $1.28.

21. a. Discrete because. . . .

  b. Continuous because. . .

  c. Continuous because. . .

  d. Discrete (if asked for milk drunk, it would be continuous, but in this case it is discrete because. . . .)

22. a.

G	G	G	G		B	B

  b.

6.2 Equivalent (Equal) Fractions

2. a. One Way

Divide into thirds.   Then divide into fifths.

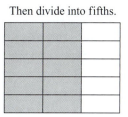

$\frac{2}{3}$ is the same as $\frac{10}{15}$

  b. and **c.** Sketches can be similarly drawn as the one for part (a).

4. $\frac{6}{10}$; GG GG GG BB BB shows the 6 girls in 3 of 5 equal-size groups.

5. a. $\frac{5}{8}$ **b.** $\frac{45}{32}$

6. a. $\frac{3}{4} > \frac{3}{5}$ because fourths are larger than fifths.

  b. $\frac{102}{101} > \frac{75}{76}$ because $\frac{102}{101} > 1$ and $\frac{75}{76} < 1$.

7. a. Show the 24 in 6 groups of 4 crayons each.

  b. Show the 24 in 4 groups of 6 crayons each.

  c. Show the 24 in 12 groups of 2 crayons each, and then group pairs of 2's to make 6 groups of 4 crayons each.

8. a. A set of 10 (or 20 or 30 or . . .) popsicle sticks will work.

9. The diagram involves units of two different sizes: 4 letters and 12 letters. The student meant, "Multiply both numerator and denominator by 3." Multiplying the fraction by 3 would certainly not give the same value.

10. Reducing usually refers to making something smaller, but equivalent fractions are equal to one another.

11. a. C   **b.** D

12. b. $\frac{1}{10} = 0.1 < 0.2 = \frac{1}{5} = \frac{3}{15} < 0.4$

14. d. $\frac{2}{5} > \frac{5}{16}$ **e.** $\frac{3}{4} > \frac{7}{11}$ **f.** $\frac{5}{14} > \frac{5}{16}$

6.3 Relating Fractions, Decimals, and Percents

2. a. $\frac{5}{8}$ **c.** $\frac{274}{3}$ **d.** $\frac{17}{10}$ **e.** $\frac{1}{1}$

  g. $\frac{53}{99}$ **h.** $\frac{53}{990}$ **i.** $4\frac{76}{99} = \frac{472}{99}$ **j.** $8\frac{94}{999} = \frac{8086}{999}$

3. a. Yes, infinitely many—can you find 10? How do you know $\frac{1}{2}$ works? Here are two others: $\frac{29}{60}, \frac{31}{60}$

  b. Yes, infinitely many—can you find 10?

4. a. One possibility is 0.45.

  c. One possibility is 1.35672.

5. $\frac{3}{17}, 0.21, \frac{1}{4}, 0.\overline{26}, \frac{11}{29}, \frac{11}{24}, 0.\overline{56}, \frac{2}{3}, \frac{12}{15}, \frac{5}{6}, 1.23$. Many can be ordered by using number sense. For example, $\frac{3}{17} < \frac{3}{15} = \frac{1}{5} = 0.2 < 0.21$. One more difficult ordering is $\frac{2}{3}, \frac{12}{15}, \frac{5}{6}$, but $\frac{12}{15} = \frac{4}{5}$, and ordering $\frac{2}{3}, \frac{4}{5}, \frac{5}{6}$ yields to number sense (compare to 1).

7. a. It could be 8.124973200000. . ., i.e., there are only 0s after the 2.

  b. It could be 8.1249732732732. . ., i.e., the 732 repeats.

  c. It could be 8.124973212112111211112 . . . (There is a pattern but it does not repeat.)

9. Complete the following table:

Decimal Form	Fraction	Percent
0.48	$\frac{48}{100}$	48
0.8	$\frac{80}{100}$ or $\frac{4}{5}$	80
4.57	$\frac{457}{100}$	457
0.001	$\frac{1}{1000}$	0.1

6.4 Estimating Fractional Values

5. A fraction is close to $\frac{1}{3}$ when the denominator is about three times as large as the numerator. A fraction is close to $\frac{2}{3}$ when the numerator multiplied by 3 is about the size of the denominator multiplied by 2. (For some fractions there could be more than one answer.)

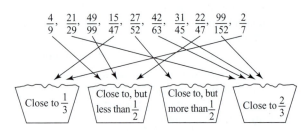

6. c. $\frac{3}{4}$ and $\frac{9}{10}$ are both close to 1, but $\frac{3}{4}$ is $\frac{1}{4}$ away from 1 and $\frac{9}{10}$ is $\frac{1}{10}$ away from 1. But $\frac{1}{10} < \frac{1}{4}$, so $\frac{9}{10}$ is closer to 1, and therefore $\frac{9}{10} > \frac{3}{4}$.

  d. $\frac{2}{5} < \frac{1}{2}$ but $\frac{3}{6} = \frac{1}{2}$, so $\frac{2}{5} < \frac{3}{6}$.

8. a. $\frac{2}{5}$ is less than $\frac{1}{2}$ and $\frac{3}{5}$ is greater than $\frac{1}{2}$, so $\frac{1}{2}$ is in between.

  b. *One way:* $\frac{8}{15} < \frac{9}{15} = \frac{3}{5}$ and $\frac{9}{12} = \frac{3}{4}$, so we need a fraction between $\frac{3}{5}$ and $\frac{3}{4}$. Rewriting as twentieths will give an answer. But symbolically at least, $\frac{3}{4\frac{1}{2}}$ looks as though it should work. $\frac{3}{4\frac{1}{2}} = \frac{3 \times 2}{4\frac{1}{2} \times 2} = \frac{6}{9} = \frac{2}{3}$, and $\frac{2}{3}$ is between the two.

10. a. $\frac{1}{8} = 0.125 = 12.5\%$ **b.** $\frac{1}{5} = 0.2 = 20\%$

  c. $\frac{1}{4} = 0.25 = 25\%$ **d.** $\frac{1}{3} = 0.3333. . . . = 33.\overline{3}\%$

  e. $\frac{2}{5} = 0.4 = 40\%$

  f. $\frac{3}{8} = 3 \times \frac{1}{8} = 3 \times 0.125 = 0.375 = 37.5\%$

11. a. Slightly less than $\frac{1}{4}$ **b.** Slightly less than 1

12. Here is just ONE possible way to estimate each.

  a. 10% of 800 + 5% of 800 is 80 + 40 = 120.

  b. 10% of 150 is 15. 90% is about 150 − 15, which is 135.

13. Sketch $\frac{4}{5}$, and then cut each fifth into two equal pieces.

17. a. About $\frac{1}{2} + 1 = 1\frac{1}{2}$ **b.** About $1\frac{1}{2} + 4 + 7 = 12\frac{1}{2}$

  c. About $15 - 5 = 10$ **d.** About $5 - 3 = 2$

  e. About $\frac{2}{3} - \frac{1}{3} = \frac{1}{3}$ **f.** About $4\frac{1}{2} + 6 - 2 = 8\frac{1}{2}$

Answers for Chapter 7

7.1 Adding and Subtracting Fractions

1. *One way:* $8 = 2^3$, $6 = 2 \times 3$, and $15 = 3 \times 5$. The denominator must have all of these factors, i.e., it must have at least 2^3, 3, and 5 as factors. $2^3 \times 3 \times 5 = 120$ is the smallest common denominator. Any other denominator must be a multiple of 120.

(For another way, see Exercise 16.)

3. c. $\frac{2}{3} - \frac{1}{2}$

$\frac{2}{3}$ is first shaded; $\frac{1}{2}$ is double-shaded. The portion of light shading is the difference: $\frac{1}{6}$ of the whole rectangle.

4. a. Let the yellow hexagon be the unit. Then $\frac{1}{2}$ is represented by 1 red trapezoid, and $\frac{1}{6}$ by a green triangle. Placed together, they can be replaced by 2 blue rhombuses, which represents $\frac{2}{3}$.

  e. Let 2 hexagons represent the unit. Now the trapezoid represents $\frac{1}{4}$. The number 4 would be represented by 8 hexagons. I need to take away $2\frac{1}{4}$. How much needs to be removed from 4 to reach $2\frac{1}{4}$? Replace 1 of the 8 hexagons with 2 trapezoids. Remove $2\frac{1}{4}$ (4 hexagons and a trapezoid), leaving 3 hexagons and a trapezoid, or 2 hexagons and 3 trapezoids, which represents $1\frac{3}{4}$.

7. He ate $\frac{1}{3} + \frac{1}{6} = \frac{1}{2}$ of a pizza. (*Note:* not $\frac{1}{6}$ of what was left, but $\frac{1}{6}$ of the whole pizza.)

9. *One way:* $\frac{2}{3} + \frac{1}{16} + \frac{1}{16} + \frac{1}{16}$, or $\frac{2}{3} + \frac{3}{16} = \frac{41}{48}$, so there will be $\frac{7}{48}$ of her estate left.

13.

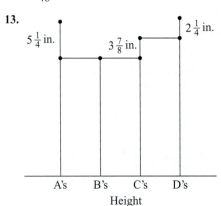

The diagram shows that D is $3\frac{7}{8} + 2\frac{1}{4} = 6\frac{1}{8}$ inches taller than B. So Donell is taller than Arnie, by $6\frac{1}{8} - 5\frac{1}{4} = \frac{7}{8}$ inches.

14. a. Caution: A common error is to lose track of the unit, which should be the same for the two fractions. Why is "Pam had $\frac{2}{3}$ of a cake. She ate half of it. How much cake did she eat?" incorrect (even though it does give the correct numerical answer)?

  b. and **c.** The mixed numbers encourage you to think of a measurement situation, using English system units.

15. a. $\frac{3}{4} + \frac{1}{4} = 1$, and $1 + \frac{5}{6} = 1\frac{5}{6}$

16. a. LCD = 40; $\frac{17 + 95 - 8}{40} = \frac{104}{40} = \frac{13}{5} = 2\frac{3}{5}$

17. a. Almost any example you try will show that subtraction is not commutative. Avoid subtracting a fraction that is equal to the other fraction, however, such as $\frac{12}{18} - \frac{18}{27}$, and $\frac{18}{27} - \frac{12}{18}$.

  b. So long as you avoid 0 as the third fraction, you should find a counterexample to $\left(\frac{a}{b} - \frac{c}{d}\right) - \frac{e}{f}$ being equal to $\frac{a}{b} - \left(\frac{c}{d} - \frac{e}{f}\right)$.

7.2 Multiplying by a Fraction

  3. *One way*: Cut each circular region into two equal pieces, giving 8 in all; then $\frac{5}{8}$ of the 4 circles would be 5 of the pieces. This is $\frac{5}{8}$ of 4, which is $2\frac{1}{2}$.

  A second way: Cut each circular region into 8 equal pieces; shade 5 pieces in each circle to get $\frac{5}{8}$ of the 4 circles. $\frac{5}{8}$ of 4 is $2\frac{1}{2}$ (or $\frac{20}{8}$ with the second method). To represent $4 \times \frac{5}{8}$, draw a region with 8 equal parts, shading 5 of them. Do this four times: $\frac{5}{8} + \frac{5}{8} + \frac{5}{8} + \frac{5}{8} = \frac{20}{8} = 2\frac{1}{2}$.

  5. a. 4   **b.** 2   **c.** 1   **d.** $\frac{1}{2}$

  7.

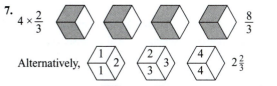

$4 \times \frac{2}{3}$ ... $\frac{8}{3}$

Alternatively, ... $2\frac{2}{3}$

$\frac{2}{3} \times 4$ ... $2\frac{2}{3}$

$\frac{2}{3}$ of the first 3 + $\frac{2}{3}$ of the last one

Alternatively, take $\frac{2}{3}$ of each hexagon and get the first row.

  8. $\frac{3}{4}$ of the rectangle is shaded. So, $\frac{2}{3}$ of $\frac{3}{4}$ of 1 is $\frac{1}{2}$.

  12. a. Make certain your drawing shows $\frac{2}{3}$ of the $\frac{3}{4}$ pie. The unit for the $\frac{2}{3}$ is the $\frac{3}{4}$ pie. The answer is $\frac{1}{2}$ of the pie, although your work may show $\frac{2}{4}$ or $\frac{6}{12}$ of the pie.

  13. a. Example: A recipe calls for $2\frac{3}{4}$ cups of sugar. Anh is making only $\frac{1}{4}$ of the recipe. How much sugar will she use?

14. b. No, "canceling" should be used only with multiplication or as a label for simplifying when using the $\frac{an}{bn} = \frac{a}{b}$ principle for equality of fractions and ratios.

15. a. $\frac{3}{8}$; here is a drawing to help you. The number of students in the class is irrelevant.

girls	boys		girls	boys
	F	Add lines		F
	F	to make		F
	F	equal		F
		pieces.		

  b. The same answer as in (a). Again, the number of students in the class is irrelevant.

18. b. $\frac{3}{14}$ $\frac{3}{9}$ $\frac{3}{7}$ $\frac{3}{4}$ $\frac{3}{2}$; because the denominators are descending, or because $\frac{3}{14}$ is close to 0, $\frac{3}{9}$ is $\frac{1}{3}$, $\frac{3}{7}$ is slightly less than $\frac{1}{2}$, $\frac{3}{4}$ is between $\frac{1}{2}$ and 1, and $\frac{3}{2} > 1$.

7.3 Dividing by a Fraction

  1. a. 16   **b.** 8   **c.** 4   **d.** 2   **e.** 1   **f.** $\frac{1}{2}$

  3. c. $24 \times 1\frac{1}{2}$ (or $24 \times \frac{3}{2}$) = 36   **d.** $7\frac{1}{2} \times 1\frac{1}{2}$ (or $7\frac{1}{2} \times \frac{3}{2}$) = $11\frac{1}{4}$

  e. $\frac{4}{5} \times 1\frac{1}{2}$ (or $\frac{4}{5} \times \frac{3}{2}$) = $1\frac{1}{5}$

  8. a. Using the hexagon as the unit, find $\frac{5}{6}$ of two hexagons. To do so, replace the two hexagons with 6 rhombuses. Then $\frac{5}{6}$ would be 5 rhombuses, which can be replaced with one hexagon and 2 rhombuses: $1\frac{2}{3}$.

  b. Let the hexagon be the unit. Replace it with 6 triangles. How many halves (how many trapezoids) can be used to match 5 triangles? 3 of the triangles would be one $\frac{1}{2}$. The other 2 triangles would be $\frac{2}{3}$ of another half, for a total of $1\frac{2}{3}$.

  Notice that parts (a) and (b) have the same answer. You do parts (c–f) in a similar fashion, noting equivalent answers.

  10. With 1 gallon, $\frac{3}{4}$ of the wall can be painted. So $\frac{3}{5}$ of a gallon of paint will cover $\frac{3}{5}$ of $\frac{3}{4}$ of the wall: $\frac{3}{5} \times \frac{3}{4} = \frac{9}{20}$

  12. You are explaining why $3\frac{1}{2}$ is the answer, referring to your sketch. (Where is the $3\frac{1}{2}$ in your sketch?)

  16. a. $\left[\frac{2}{3} \times (2 \times 9)\right] \div \frac{1}{2}$

  17. a. No—try almost any example.   **b.** No

7.4 Issues for Learning: Teaching Calculation with Fractions

  1. $\frac{1}{5}$ from $\frac{3}{4} \times \left[1 - \left(\frac{1}{3} + \frac{2}{5}\right)\right] = \frac{3}{4} \times \left[1 - \frac{11}{15}\right] = \frac{3}{4} \times \frac{4}{15}$

3. 5 full recipes, because $4 \div \frac{3}{4} = 5\frac{1}{3}$.

5. 333, because $75 \div 0.2 = 375$, but $200 \div 0.6$ is only $333\frac{1}{3}$.

7. $66\frac{2}{3}\%$; the excess amount was $\frac{1}{3} - \frac{1}{5} = \frac{2}{15}$. Comparing the $\frac{2}{15}$ to the planned $\frac{1}{5}$ gives $\frac{2}{15} \div \frac{1}{5} = \frac{10}{15} \approx 66\frac{2}{3}\%$.

Answers for Chapter 8

8.1 Quantitative Analysis of Multiplicative Situations

2. a. $4:3$ or $\frac{4}{3}$ **b.** $\frac{3}{4}$ **c.** $\frac{4}{3}$ **d.** $\frac{3}{7}$ **e.** $\frac{4}{7}$

8.2 Fractions in Multiplicative Comparisons

2. The first question, involving a multiplicative comparison, is answered by turning the ratio $9\frac{2}{7}:5$ into the fraction form, and simplifying. $9\frac{2}{7}:5 = \frac{9\frac{2}{7}}{5} = \frac{65}{35} = 1\frac{6}{7}$. Claudia ran $\frac{9\frac{2}{7}}{5} = \frac{65}{35} = 1\frac{6}{7}$ times as many laps as Juan. The second question involves another additive comparison and leads to Claudia's running $9\frac{2}{7} - 5 = 4\frac{2}{7}$ laps farther than Juan.

4. This is like a candy bar problem with an addition—the value (14 meters) for the whole amount. The given sentence translates into a $R:S = 3:4$ ratio, which gives $\frac{3}{7}$ as the R part of the whole ribbon. So the ribbon part is $\frac{3}{7} \times 14 = 6$ meters, and the strips part is 8 meters.

5. Your quantitative analysis should include the dogs' weights, the two multiplicative comparisons, the additive comparison, and the parts-whole relationships. A drawing is definitely helpful here, because there are two multiplicative comparisons and an additive comparison involved.

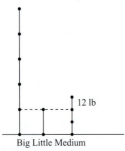

Big Little Medium

The other thirds of the medium dog's weight must also be 12 lb each, so the medium dog weighs 36 lb. The little dog's weight matches 2 of the 12-lb pieces, so the little dog weighs 24 lb. Finally, the big dog weighs 120 lb.

7. b. Cameron, $\frac{4}{5}$ of the bar; Don, $\frac{1}{5}$. Notice that there is no multiplicative comparison in part (b).

   c. Emily, $\frac{3}{4}$ of the bar; Fran, $\frac{1}{4}$

   d. Gay, $\frac{3}{10}$ of the bar; Haille, $\frac{6}{10}$ or $\frac{3}{5}$; Ida, $\frac{1}{10}$

   f. Mick and Nick, each $\frac{1}{4}$ of the bar; Ollie, $\frac{1}{6}$; Pete, $\frac{2}{6}$ or $\frac{1}{3}$

8. a. Al, 80¢ (*Note:* not $0.80¢); Babs, $1

   b. Cameron, $1.44; Don, 36¢

   c. Emily, $1.35; Fran, 45¢

   d. Gay, 54¢; Haille, $1.08; Ida, 18¢

   e. Judy, $1.20; Keisha, 40¢; Lannie, 20¢

   f. Mick, 45¢; Nick, 45¢; Ollie, 30¢; Pete, 60¢

10. Charity A, $16; Charity B, $8; Charity C, $64

Answers for Chapter 9

9.1 Ratio as a Measure

3. a. The usual way for most computers: the size of the tiny dots that make up the display (pixels)

   b. The number of dots of black per square inch

   c. The slope, just as you did with the ski slopes

4. Keep the angles the same, but multiply the length of each side by the same number to keep the ratio of new length:old length the same for every side. (The polygons are then said to be "similar," in the technical sense of geometry.)

9.2 Comparing Ratios

2. a. Stronger, because $\frac{8}{15} > \frac{1}{2}$. (How did you get the fractions?)

   b. Weaker than the intended recipe, since $\frac{9}{19} < \frac{1}{2}$ (number sense?), and weaker than the 8:15 one, since we already know it is stronger than the intended recipe.

   c. Many possibilities, so long as your ratios are slightly greater than 1:2—e.g., 11:20.

   d. Many possibilities, so long as your ratios are slightly less than 1:2— e.g., 11:24.

   e. Many possibilities, so long as your ratios are quite a bit greater than 1:2— e.g., 7:8.

   f. Many possibilities, so long as your ratios are quite a bit less than 1:2— e.g., 1:12

4. The first question translates into $10:4 = x:50$ (or $\frac{10}{4} = \frac{x}{50}$ or $\frac{5}{2} = \frac{x}{50}$), giving $x = 125$ Tbsp. Notice that another, less "mechanical" line of reasoning also makes sense: The cocoa requires $10 \div 4 = 2\frac{1}{2}$ Tbsp of cocoa per cup of milk. So for 50 cups, they need $50 \times 2\frac{1}{2}$, or 125, Tbsp. Then the ratio, $2:\frac{1}{8}$, for the number of Tbsp to the number of cups of cocoa, might lead either to $\frac{2}{\frac{1}{8}} = \frac{125}{x}$, or to the rate, 16 Tbsp per cup and $125 \div 16$. Either gives an answer of $7\frac{13}{16}$ cups of cocoa.

6. Try different values for the distances A and B can travel. Suppose B travels 10 miles in 2 hours, and A travels 12 miles in 3 hours; or 18 miles in 3 hours; or 24 miles in 3 hours. Would all of the numbers fit the situation? Now can you answer the question?

9. We need to determine B's rate of working. Using A's rate, A will make $9 \times 13 = 117$ parts in 9 hours, so B must have made $243 - 117 = 126$ parts in the 9 hours. B's rate, then, is $\frac{126}{9} = 14$ parts per hour, slightly faster than A's. So B would be more productive.

12. Mrs. Heath's class

15. There are different ways. (1) Replace the ratios with equal ratios having the same second entries — $(133 \times 115):(161 \times 115)$ and $(95 \times 161):(115 \times 161)$; and then compare 133×115 and 95×161. (2) Change the ratios to fraction form and then use methods that you earlier learned to determine whether or not two fractions are equal. Or (3) change the ratios to fractions and then to decimals (this way is probably easiest with a calculator). The fact that the ratios are equal may make one suspect that there might be an easier way. Simplify each ratio:

$$\frac{133}{161} = \frac{7 \times 19}{7 \times 23} = \frac{19}{23} \quad \text{and} \quad \frac{95}{115} = \frac{5 \times 19}{5 \times 23} = \frac{19}{23}$$

See Exercise 16 for a shortcut method.

16. **a.** $a:b = (ad):(bd) \left(\text{or} \frac{a}{b} = \frac{ad}{bd} \right)$ and

$c:d = (bc):(bd) \left(\text{or} \frac{c}{d} = \frac{bc}{bd} \right)$

Then the ratios (or fractions) are equal if $ad = bc$.

b. $x = 144$; $x = 39$

c. Not in general. Try it on, say, $\frac{1}{4} + \frac{1}{8}$ and $\frac{1}{2} \times \frac{1}{2}$.

17. Because $\frac{3}{4} + \frac{2}{4} = \frac{5}{4}$, which apparently would mean 5 hits in 4 at-bats, this way cannot be correct. Indeed, $3:4$ "+" $2:4$ here should equal $5:8$, 5 hits in 8 at-bats. (The quotation marks are a reminder that this does not represent the usual addition of numbers.) Critics say that ratios should be added this way: $a:b$ "+" $c:d = (a + c):(b + d)$.

21. **a.** The desired rate is $\frac{7 \text{ ugrads}}{2 \text{ grads}}$, because as the numbers of undergraduates and graduate students change, the ratio should stay the same.

b. $\frac{7 \text{ ugrads}}{2 \text{ grads}} = \frac{3.5 \text{ ugrads}}{1 \text{ grad}}$, or 3.5 undergrads per graduate student.

c. 3.5 undergrads for each graduate student, so if there are 100 graduate students, there should be 350 undergrads.

d. $\frac{7}{2} = \frac{x}{100}$. Multiplying both sides by 100 gives $x = 350$.

24. **a.** Andy drove faster, because he covered more miles in less time, so his miles per hour would be greater. (You may be helped by making up values for the quantities.)

9.3 Percents in Comparisons and Changes

2. Test 1: 75%; Test 2: 76%. The instructor would probably weight the tests the same, and give the average as $\frac{75\% + 76\%}{2} = 75.5\%$. But combining the points, you received 59 out of the 78 possible, or about 75.6%. If the scores were quite different, the two methods give noticeably different results (try the two methods on 10 out of 20 and 80 out of 100).

4. **a.** 58% **b.** 42%

7. $4.2 million is 90% of $4.67 million.

8. **a.** 25%

9. **a.** 25% (Notice the "larger than.")

10. 50% increase

12. About 10,780 ($10567 \div 0.98 \approx 10782.7$)

14. 100%

15. Pat's new salary will be $17.98 an hour, and her old salary was $14.50 an hour.

18. 24 points in the first half. Angie, Beth, and Carlita accounted for $\frac{1}{3} + \frac{3}{20} + \frac{1}{4} = \frac{11}{15}$ of the points (using fractions to avoid inaccuracies with the $\frac{1}{3}$). This means that the other 16 points represent $\frac{4}{15}$ of the total points, so $\frac{1}{15}$ of the total points would be 4 points, giving 60 for the total points scored. 40% of those points, or 24 points, were scored in the first half.

19. About $230. The sketch below may help, with a focus on the $\frac{5}{8}$ of her income being $450.

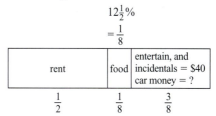

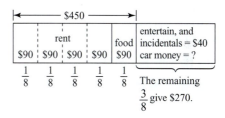

The remaining $\frac{3}{8}$ give $270.

22. **a.** $2\frac{2}{3}\%$ of all the students already qualify; $1\frac{1}{3}\%$ of all are newly eligible = 3600 students, so 1% of all = 2700 students and 100% of all = 270,000 students; 1200 students are newly eligible from urban schools, 900 from urban areas.

b. $11\% \approx \frac{1}{9}$, so about $9 \times 1500 = 13,500$ workers

26. About 9.1%, from $\frac{16}{176} \approx 9.09\%$. As in Learning Exercise 25, the same loss would apply to any gain of 10%, to return to the original starting value.

Answers for Chapter 10

10.1 Big Ideas About Signed Numbers

1. **a.** $^-75.2$, $^-22$, $^-3$, $^-1$, $^-\left(\frac{2}{3}\right)$, $\frac{3}{4}$, 1, $^+50$

2. **a.** 8 The additive inverse, or opposite, of $^-8$ is 8.

b. $^-0 = 0$. The "oppositing" across 0 would leave 0 where it is. More exactly, $0 + x = 0$ has $x = 0$ as the solution, so 0 behaves as its own additive inverse.

c. ^-a is not always negative. For example, if a is a negative number, its additive inverse, ^-a, is positive.

3. **a.** $^-9$ **b.** $^-9$

4. **a.** $^+1$, or 1 **b.** $^-4$ **c.** $^-4$

7. **a.** 6.25 would be one-quarter of the way between 6 and 7 on the number line.

b.

Same distance from zero as $\frac{3}{2}$ but in the opposite direction

8. a.

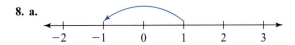

10.2 Children's Ways of Reasoning About Signed Numbers

1. a. A child starts by marking 1 on the number line and, also, the ending point of $^-3$. Since subtraction means to move to the left, the child counts 4 and places a 4 in the blank.

2. $^-1 + ^-4$

10.3 Other Models for Signed Numbers

1. a. 5 blue chips (or 5 more blue chips than gray)

3. a. $^+2$, or 2   **b.** $^-1$ **c.** $^-3$

4. a. Any negative number that does not have a terminating or repeating decimal. Examples include $-\sqrt{2}, -\sqrt{3}, -\sqrt{5}, -\sqrt{6}, \dots$, as well as special numbers like $-\pi$.

    b. Any negative number that can be written in the form
$$ -\frac{\text{whole number}}{\text{nonzero whole number}}.$$

10.4 Operations with Signed Numbers

1. a.

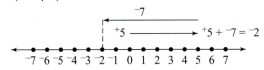

$^-4 + ^-2 = ^-6$

b. $^-4 + 2 = ^-2$

c. $^-4 - ^-2 = ^-2$

d. $^-4 - 2 = ^-6$

2. a. (Sample)

$$ ^+5 + ^-7 = ^-2 $$
$$ ^-7\ ^-6\ ^-5\ ^-4\ ^-3\ ^-2\ ^-1\ 0\ 1\ 2\ 3\ 4\ 5\ 6\ 7 $$

    c. $^+3$ **d.** $^-4$

4. a. $\frac{1}{3}$ **b.** $^-4\frac{4}{9}$

5. a. $^-3 + \ 5 = 2$ $2 - \ 5 = ^-3$
     $5 + ^-3 = 2$ $2 - ^-3 = \ 5$

    b. $3 + ^-5 = ^-2$ $^-2 - ^-5 = \ 3$
      $^-5 + \ 3 = ^-2$ $^-2 - \ 3 = ^-5$

    c. $^-32 + ^-29 = ^-61$ $^-61 - ^-29 = ^-32$
      $^-29 + ^-32 = ^-61$ $^-61 - ^-32 = ^-29$

9. a. $57{,}000 + 35{,}000 + ^-16{,}000 + ^-16{,}000 = n$. Company A earned $60,000 for the year.

12. a. High, 1.0 ft; low, 0 ft

    b. High, 1.1 ft; low, $^-0.3$ ft

13. a. $146.9°$ (from $135.9 - ^-11$; you may have seen the equivalent $135.9 + 11$ immediately)

    b. $225.16°$ **c.** $138.4°$ **d.** $189°$

    e. $215.4°$ **f.** $147.4°$

14. a.

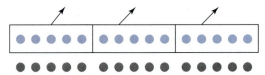

10.5 Multiplying and Dividing Signed Numbers

1. $\frac{^-2}{5}$ is the same as $^-2 \div 5$, which is negative, and can be written $^-\left(\frac{2}{5}\right)$.

$\frac{2}{^-5}$ is the same as $2 \div ^-5$, which is negative and can be written $^-\left(\frac{2}{5}\right)$.

Thus, all three expressions are naming the same number.

2. k. $\frac{^-1}{40}$ **l.** 50   **m.** 1   **n.** $^-1$

3. a. $3 \times ^-5 = ^-5 + ^-5 + ^-5 = ^-15$, so draw 3 groups, each with 5 gray chips.

    b. Start with 0 in the form of 15 blues and 15 grays. To find $^-3 \times 5$, subtract 5 blues three times, leaving 15 grays, or $^-15$.

6. a. $^+5 \times ^-4 = n\ (n = ^-20)$; she will have 20 fewer sheep than at present.

    d. $^-3 \times ^-30 = n\ (n = ^+90)$; he weighed 90 more pounds than at present.

8. a. No. The multiplicative inverse of most integers a is not an integer.

    b. 1 is its own multiplicative inverse, as is $^-1$.
      ($1 \cdot 1 = 1$ and $^-1 \cdot ^-1 = 1$.)

9. a. Multiplication is commutative.

    b. Addition is commutative.

    c. Multiplication is distributive over addition.

10.6 Number Systems

5. Be sure to set up addition and multiplication tables before doing this problem. When working with clock arithmetic on a 4-hour clock, the multiplicative inverse of 1 is 1, and of 3 is 3. But 2 does not have a multiplicative inverse.

7. Between every two real numbers, there is another real number. (Their average is one example.)

8. a. Not dense. There is not another integer between consecutive integers, like 1 and 2.

    b. Yes, dense. There is always a positive rational number between every two given positive rational numbers. For example, the average of two rational numbers is a rational number and is between the two.

9. Rewrite the given numbers with a large common denominator. For example, for part (a), use $\frac{3500}{2000}$ and $\frac{3600}{2000}$, and for part (b), $^-\left(\frac{2000}{3000}\right)$ and $^-\left(\frac{1860}{3000}\right)$. You can then find several rational numbers between the pairs.

10. a. Yes, the sum of any two even integers is an even integer. $2m + 2n = 2(m + n)$, and if m and n are integers, so is $(m + n)$.

  b. Yes, the sum of any two multiples of 3 is another multiple of 3.

  c. No, e.g., $3 + 5 = 8$ is not odd.

11. a. Yes, because addition is commutative.

  b. Yes, because addition is associative.

  c. Yes, because 0 is the additive identity.

  d. Yes, because the numbers are additive inverses.

  e. Yes, because the numbers are additive inverses.

  f. Yes, because addition is associative.

  g. Yes, rational numbers are closed.

  h. Yes, because addition is commutative and associative.

  i. Yes, because 0 is the additive identity.

  j. Yes, because the numbers are additive inverses.

Answers for Chapter 11

11.1 Factors and Multiples, Primes and Composites

1. a. 1, 5, and 25. There are no more because 25 is a perfect square of a prime number.

2. a. $k = 25 \times$ some whole number, or $k = 25m$, for some whole number m. A rectangular array would need 25 on one side and m on another side.

3. a. $216x = 2376$, or $x = 2376 \div 216$.

4. 21 squares could be in rectangular arrays only as 1 by 21, 21 by 1, 7 by 3, and 3 by 7. An array with 5 squares on one side would require 4.2 squares on the other side, but rectangular arrays allow only whole numbers on each side in number theory.

5. a. A number is a factor of itself.

  b. A number is a multiple of itself.

7. a. 13 is a factor of 39. True because there exists a whole number, namely 3, such that $13 \times 3 = 39$.

9. a. Yes, because for some whole numbers x and y, $m = kx$ and $n = my$, substituting the first into the second, $n = (kx)y = k(xy)$. The latter also shows that n is a multiple of k.

  b. Yes, because $m = kx$ and $n = ky$ for some whole numbers x and y and $m + n = kx + ky = k(x + y)$. So k is a factor of $m + n$.

  c. Suppose $k = 6$, $m = 12$, and $n = 15$. Now k is a factor of m but not of n. We have $m + n = 27$, and 6 is not a factor of 27. (This is a counterexample.)

11.

+	even	odd
even	even	odd
odd	odd	even

×	even	odd
even	even	even
odd	even	odd

  a. The sum of any number of even numbers is even.

  b. The sum of an even number of odd numbers is even (put them in pairs), so the sum of an odd number of odd numbers is odd.

13. Each of them has only 1 and itself as factors:

$$2, 3, 5, 7, 11, 13, 17, 19, 23, 29, 31, 37, 41, 43,$$
$$47, 53, 59, 61, 67, 71, 73, 79, 83, 89, 97$$

16. a. 1 and 829

19. c. Common factors of 18 and 24 are 1, 2, 3, 6. Common multiples of 6 and 18 are 18, 36, 54, etc. All multiples of 18 are also multiples of 6.

21. $28 = 1 + 2 + 4 + 7 + 14$

11.2 Prime Factorization

2. a.

Thus, the prime factors are 2, 3, and 17.

The prime factorization of 102 is $2 \times 3 \times 17$.

  b. $1827 = 3^2 \cdot 7 \cdot 29$

3. Different factor trees are possible, but these prime factorizations should be the same for the different factor trees (except possibly for the order of the factors):

  d. $10^4 = 2^4 \cdot 5^4$ **e.** $17{,}280 = 2^7 \cdot 3^3 \cdot 5$

  f. A factor tree shows all the prime factors, but most often it does not show all the possible composite factors of the number.

4. a. 2, 3, 7, and 11 are prime factors of $3 \times 7^3 \times 22$. (There are only these 4 prime factors. Any three of these is a correct answer.)

5. The prime factorization of a number gives all of the prime factors of the number. A prime factor of the number would be just one of those primes.

6. Think about the number 11^m, whose prime factorization has only 11 and powers of 11 as factors. Now think about the number 13^n, whose prime factorization has only 13 and powers of 13 as factors. The fundamental theorem of arithmetic says that prime factorizations are unique. Because 13 does not appear in the list of factors of 11^m, we know that 13 cannot be a factor of 11^m for any value of m. This says that 13^n cannot equal 11^m for any (nonzero) values of m and n.

8. a. $2^8 \cdot 7$ could not be a factor of m, since m does not have 7 in its prime factorization (which is unique and hence could not have another factorization with a 7 in it).

  b. Similarly, the part (b) number already has too many 2s to be a factor of m.

  c. The part (c) number is a factor of m, because all of its factors appear in the prime factorization of m.

10. a. 6   **b.** $3 \times 4 = 12$

  c. $45{,}000 = 2^3 \cdot 3^2 \cdot 5^4$ has 60 factors.

11. a. $19^4 \times 11^3 \times 2^5 \times n$ is a factor of $19^4 \times 11^4 \times 2^5$ if $n = 11$.

  b. $19^4 \times 22 \times 2^5 \times n$ cannot be a factor of $19^4 \times 11^4 \times 2^5$ because $19^4 \times 22 \times 2^5 \times n = 19^4 \times 11 \times 2^6 \times n$. There is one too many twos in $19^4 \times 11 \times 2^6 \times n$ to be a factor of $19^4 \times 11^4 \times 2^5$.

12. a. $19^4 \times 11^8 \times 2^5$. Yes, multiply q by 11^4.

  b. $19^4 \times 22^4 \times 2^5 \times 17 = 19^4 \times 11^4 \times 2^9 \times 17$. Yes, multiply q by $2^4 \times 17$.

14. a. 11   **e.** $(m + 1)(n + 1)(s + 1)$

15. $2^5 \cdot 5^9$ is one possibility.

16. $2^7 \cdot 11^2$ is one possibility. 2 can be replaced by other primes for other possibilities.

11.3 Divisibility Tests to Determine Whether a Number Is Prime

1. a. 2, 3, 4, 6, 8, 9 are (some of the) factors of 43,056. (5 and 10 are not factors.)

    b. 2, 3, 6, 9 are (some of the) factors of 700010154. (4, 5, 8, and 10 are not.)

2. a. The solution is not solely trial and error. If 2 and 3 but not 9 are factors, then the ones digit must be even and the sum of the digits must be a multiple of 3 but not of 9. If 4 is not a factor, then 4 is not a factor of the right-most two digits. By starting with the ones place, here is one result: 200,022. There are many others.

3. 3 must be a factor of the sum of the digits. The known digits give a sum of 29. So if the unknown digit is 1 or 4 or 7, then 3 will be a factor of the number. Similarly, if 9 is to be a factor, then 9 must divide the sum of the digits. With a current sum of 29, the only digit value that works for unknown digit is 7. (Why not 16?)

7. To be a multiple of 36, the number must be divisible by 4 and by 9. Thus, choose a two-digit number divisible by 4, say, 24, for the final two digits. To be divisible by 9, the sum of the digits must be divisible by 9, so annex a digit to 24 to obtain the number divisible by 9, say, 3. So 300,000,000,000,024 would be such a number. To obtain a number NOT divisible by 36, change the first digit to 2. Now the sum of the digits is not divisible by 9. Or, change the last two digits to form a number NOT divisible by 4, such as 23.

12. No, because in any two consecutive numbers, one is even so it has 2 as a factor.

13. b. 150 is divisible by 2, 3, 5, 6, 10, and 15, but not by 4, 8, 9, 12, or 18.

14. f. $1247 = 29 \times 43$ so 1247 is not prime. (Remember, all primes less than $\sqrt{1247}$ must be tested, i.e., for all primes equal to or less than 31 because $\sqrt{1247} \approx 35.3$.)

    g. 3816 is divisible by 2, so it is not prime. (In fact, $3816 = 2^3 \times 3^2 \times 53$.)

15. $4 = 2 + 2; 6 = 3 + 3; 8 = 3 + 5; 10 = 5 + 5; 12 = 5 + 7$ (you finish).

18. a, d, f, g, i. [Part (k)—121 and 22 have 11 as a factor in common; part (l)—39 and 169 both have the factor 13 in common.]

19. 125 and 243. (Note that $125 = 5^3$ and 243 is 3^5. There are many others.)

21. a. $128 + 494 + 381 = 1003 \to 4$

     $128 \to 11 \to 2; \quad 494 \to 17 \to 8; \quad 381 \to 12 \to 3;$
     $2 + 8 + 3 = 13 \to 4$

23. Here is one case. Let $p = 5$, and $n = 4$. Then by this theorem, $4^5 - 4$ should be divisible by 5; $4^5 = 4 \times 4 \times 4 \times 4 \times 4 = 1024$. When 4 is subtracted, the result is 1020, which is divisible by 5. You choose more to try.

24. Three-digit numbers less than 125 that are multiples of 5 are 100, 105, 110, 115, and 120. But 100, 110, and 120 are divisible by 2, so they can't work, which leaves 105 and 115. But 7 is a factor of 105, so that too must be discarded, which leaves only 115.

11.4 Greatest Common Factor, Least Common Multiple

1. a. 1, 2, 4, 8 (There are no more.)

    b. 1, 27, 49, 27×49 (There are more.)

2. Notice that there are an infinite number of multiples for each case. Here we list just four.

    b. $2 \times 27 \times 49, 27^2 \times 49, 27 \times 49^2, 27^2 \times 49^2, \ldots$

3. 60 years

5 and **6.** One can begin by factoring each of the numbers below.

$72 = 2^3 \cdot 3^2 \qquad 108 = 2^2 \cdot 3^3 \quad 144 = 2^4 \cdot 3^2 \qquad 150 = 2 \cdot 3 \cdot 5^2$
$350 = 2 \cdot 5^2 \cdot 7 \quad 567 = 3^4 \cdot 7 \quad 90 = 2 \cdot 3^2 \cdot 5 \quad 270 = 2 \cdot 3^3 \cdot 5$

    a. The LCM of 72, 108 is $2^3 \cdot 3^3 = 216$. The GCF of 72 and 108 is $2^2 \cdot 3^2 = 36$.

    b. The LCM of 144 and 150 is $2^4 \cdot 3^2 \cdot 5^2 = 3600$. The GCF of 144 and 150 is $2 \cdot 3 = 6$.

8. b. LCM is $37^6 \cdot 47^5 \cdot 67^6 \cdot 71$ and the GCF is $37^4 \cdot 47^5$.

9. 0 is always a common multiple of two numbers, but it is not useful as a denominator in adding or subtracting fractions, e.g.

10. Jogger A is at the start at 0 sec, 90 sec, 180 sec, etc.

Jogger B is at the start at 0 sec, 120 sec, 240 sec, etc.

The LCM of 90 and 120 is 360. So after 360 sec, or 6 minutes, they are again both at the start. Jogger A has gone around 4 times, and Jogger B has gone 3 times.

12. The gear with 12 teeth will rotate 5 times and the large one twice to get back to the original positions. Note that 60 is the LCM of 12 and 30: $12 \times 5 = 30 \times 2 = 60$.

14. Order 2 drums of base and 5 drums of color.

16. GCF(414,543) = 3, so \$3.

20. For example, $m = 2, n = 3; m = 4, n = 6; m = 6, n = 9; \ldots$

21. *Hint*: Think of "greatest common factor" in this order: first "factor," then "common factor," and finally "greatest common factor."

    a. $\dfrac{9}{10}$ **b.** $\dfrac{3}{4}$ **c.** $\dfrac{21}{25}$

23. a. $\dfrac{65}{72}$ **b.** $\dfrac{3}{8}$

25. a. $\dfrac{8}{9}$ **b.** $\dfrac{853}{4725}$

26. a. If x is a common factor of m and n, then x is also a factor of $m - n$.

    b. *One example*: If $m = 48$ and $n = 36$, then 6 is a factor of both; 6 is also a factor of $48 - 36 = 12$.

27. a. $\dfrac{15}{35} = \dfrac{3}{7}$ **b.** $\dfrac{28}{54} = \dfrac{2^2 \cdot 7}{2 \cdot 3^3} = \dfrac{14}{27}$

    c. $\dfrac{150}{350} = \dfrac{3}{7}$ **d.** $\dfrac{12}{144} = \dfrac{1}{12}$

28. b. $\dfrac{25}{72} + \dfrac{81}{567} = \dfrac{25}{2^3 \cdot 3^2} + \dfrac{81}{3^4 \cdot 7}$

$= \dfrac{25 \cdot 3^2 \cdot 7}{2^3 \cdot 3^2 \cdot 3^2 \cdot 7} + \dfrac{81 \cdot 2^3}{3^4 \cdot 7 \cdot 2^3} = \dfrac{1575 + 648}{2^3 \cdot 3^4 \cdot 7} = \dfrac{2223}{4536}$

Or, $\dfrac{25}{72} + \dfrac{81}{567} = \dfrac{25}{2^3 \cdot 3^2} + \dfrac{3^4}{3^4 \cdot 7} = \dfrac{25 \cdot 7}{2^3 \cdot 3^2 \cdot 7} + \dfrac{1 \cdot 2^3 \cdot 3^2}{7 \cdot 2^3 \cdot 3^2}$

$= \dfrac{175}{2^3 \cdot 3^2 \cdot 7} + \dfrac{72}{2^3 \cdot 3^2 \cdot 7} = \dfrac{175 + 72}{2^3 \cdot 3^2 \cdot 7} = \dfrac{247}{504}$

Note that $\dfrac{2223}{4536} = \dfrac{3^2 \cdot 13 \cdot 19}{3^4 \cdot 2^3 \cdot 7} = \dfrac{13 \cdot 19}{3^2 \cdot 2^3 \cdot 7} = \dfrac{247}{504}$. In the remaining parts, we will simplify fractions before adding or subtracting.

c. $\dfrac{36}{108} + \dfrac{41}{72} = \dfrac{2^2 \cdot 3^2}{2^2 \cdot 3^3} + \dfrac{41}{2^3 \cdot 3^2} = \dfrac{1}{3} + \dfrac{41}{2^3 \cdot 3^2}$

$= \dfrac{1 \cdot 2^3 \cdot 3}{2^3 \cdot 3^2} + \dfrac{41}{2^3 \cdot 3^2} = \dfrac{24 + 41}{2^3 \cdot 3^2} = \dfrac{65}{72}$

11.5 Issues for Learning: Understanding the Unique Factorization Theorem

1. k cannot be divided by 5, 11, or 17. Some students try to use divisibility rules to check for divisibility when this is not necessary.

3. a and b both have the same number of factors: $6 = (2 + 1)(1 + 1)$. A common error is to think that the larger number has more factors.

Answers for Chapter 12

12.1 Algebra as a Symbolic Language

1. a. 1800. $(ab) + (ac) = a(b + c)$, for every choice of numbers $a, b,$ and c. Distributive property (of multiplication over addition).

b. 100. $(ab) + (cb) = (a + c)b$, for every choice of numbers $a, b,$ and c. A variation of the distributive property (of multiplication over addition).

7. a. $x(x + 5)$ **b.** $(x \cdot 4)x$

10. a. $x + 2x + (x + 12) + (x - 3)$

b. $[x + 2x + (x + 12) + (x - 3)] \div 4$

11. a. $x + 10$ **b.** $x + y$ **c.** $x - 3$

d. $x - n$ **e.** $2x$

14. a. $2(x + 2 + x) = 4x + 4$ in.

15. The total of 3 fifties is *conceptually quite different* from the total of 50 threes, even though we know each will give the same total. The first is easier. We know from the previous section that this is an example of the remarkable relationship that always holds with the multiplication of two numbers: the commutative property of multiplication. This property is one that must be learned in school. It is not obvious to young children.

17. a. $3x + 2 = x + 10$

12.2 Algebra as Generalized Arithmetic

2. a. $5 \cdot 10^2 + 9 \cdot 10 + 8$
$- \ 3 \cdot 10^2 + 4 \cdot 10 + 7$
$\overline{ 2 \cdot 10^2 + 5 \cdot 10 + 1 = 251}$

b. $5x^2 + 9x + 8$
$- \ 3x^2 + 4x + 7$
$\overline{ 2x^2 + 5x + 1}$

When 598 and 347 are written out in long form, they are similar in structure to $5x^2 + 9x + 8$ and $3x^2 + 4x + 7$.

4. a. $\dfrac{12}{18} = \dfrac{2 \cdot 6}{3 \cdot 6} = \dfrac{2}{3}$

b. $\dfrac{4x^3}{7x^2} = \dfrac{4x \cdot x^2}{7 \cdot x^2} = \dfrac{4x}{7}$

In both cases, the numerators and denominators have common factors.

6. a. $\dfrac{5}{8} + \dfrac{3}{4} = \dfrac{5}{8} + \dfrac{3 \cdot 2}{4 \cdot 2} = \dfrac{5}{8} + \dfrac{6}{8} = \dfrac{11}{8}$ (or $1\tfrac{3}{8}$)

b. $\dfrac{3x}{(x + 2)(x + 3)} + \dfrac{2}{x + 2}$

$= \dfrac{3x}{(x + 2)(x + 3)} + \dfrac{2 \cdot (x + 3)}{(x + 2)(x + 3)}$

$= \dfrac{3x + 2 \cdot (x + 3)}{(x + 2)(x + 3)} = \dfrac{5x + 6}{(x + 2)(x + 3)}$

In both cases, one denominator is a multiple of another denominator, making a common denominator easy to find.

8. b. $\dfrac{3}{xy} - \dfrac{4x}{9n} = \dfrac{3 \cdot 9n}{xy \cdot 9n} - \dfrac{4x \cdot xy}{9n \cdot xy} = \dfrac{3 \cdot 9n}{9nxy} - \dfrac{4x^2y}{9nxy}$

$= \dfrac{27n - 4x^2 y}{9nxy}$

A common denominator is the product of the two denominators.

10. a. $\dfrac{5}{6} \div \dfrac{2}{3} = \dfrac{5}{6} \times \dfrac{3}{2} = \dfrac{15}{12} = \dfrac{5}{4}$ or $1\tfrac{1}{4}$

b. $\dfrac{x^2}{2y} \div \dfrac{xy}{3} = \dfrac{x^2}{2y} \cdot \dfrac{3}{xy} = \dfrac{3x^2}{2xy^2} = \dfrac{3x}{2y^2}$

In both cases, the quotient was found by inverting the divisor before multiplying.

11. a. $\ldots = x^2 + 2xy + y^2$

b. Describe the 8 pieces in the cube, and add them.

c. $\ldots = x^2 + xy$

d. $\ldots = x^2 + 2x + 3x + 6 = x^2 + 5x + 6$, with a drawing similar to that of part (a).

12.3 Numerical Patterns and Algebra

1. i. 2, 4, 6, 8; 4 is 30th entry

j. 2, 8, 5, 7; 8 (with 1, 4, 2, 8, 5, 7 as repeating numbers)

7. 2. The decimal has a repeating block, 1-4-2-8-5-7, six digits long. So from $99 \div 6 = 16$ R 3, there would be 16 full blocks, with 3 entries into the next block: 1-4-2.

8. a. 9 **c.** 1

d. The *ones* digits follow a repeating cycle of four digits: 7, 9, 3, 1. Find out how many full cycles there are in n, and look at the remainder by calculating $n \div 4$. The remainder tells you which number to go to in the next cycle. For example, 7^{778} will end in 9 because $778 \div 4 = 194$ R 2, and the second number in the (next) cycle is 9.

10. a. The first eight triangular numbers are 1, 3, 6, 10, 15, 21, 28, and 36, with the differences between two consecutive numbers increasing by 1 each time.

b. Yes, each bottom row increases by 1 from one triangle to the next.

11. a. The sum would be $\dfrac{n(n+1)}{2}$.

b. The successive sums give the triangular numbers, so the nth triangular number is $\dfrac{n(n+1)}{2}$.

12. a. Yes, the switch can be made.

b. 15 (If you keep getting 17, there is one crucial point at which you can avoid any moves in the wrong direction.)

13. Choice II. Examine powers of 2 (i.e., 2^{15} or 2^{20}).

12.4 Functions and Algebra

1. a. Yes, because each and every child has a first name.

b. Perhaps not, because more than one child might have the same last name, so if you say, "Smith," there might be two or more children who respond. If the last names of all the children are different, then the "last name → child" rule *would* give a function.

2. a. Yes, each element in the first set is associated with exactly one element in the second set.

c. Yes, each first entry in the ordered pairs is associated with exactly 1 second entry in the ordered pairs.

4. $f(x) = 100 - 22x$; $n = 10$

6. $output = 0.3 \times input - 7$; $n = 76$

11. a. $f(x) = x^2$; $n = 30$

b. $y = x^3$; $n = 9$

12. a. $f(n) = 4n + 1$

d. $f(n) = 3n + 4$

e. $f(n) = 5n + 6$

13. a. $n(n+1)$, or you may have seen $n^2 + n$.

d. $1 + 3 + 5 + 7 + \cdots + (2n - 1) = n^2$

16. a. 100°C, 0°C, 37°C, and about 21°C. (These values are worth memorizing.)

b. $F = \dfrac{9}{5}C + 32$

12.5 Algebra in Reasoning About Quantities

1. Dan, $7.80; Lincoln, $15.60; and Miguel, $23.40

5. 29 feet by $14\frac{1}{2}$ feet

6. First year, 11,000 tons; second year, 13,750 tons; and third year, 16,500 tons

9. $64,000,000 (Bestburg, this year); a diagram may be very helpful.

11. 6 hours

14. a. Monday, 15 min; Tuesday, 20 min; Wednesday, 40 min; Thursday, 30 min; and Friday, 15 min

Answers and Hints for Chapter 13

13.1 Using Graphs and Algebra to Show Quantitative Relationships

1. a. Samples: The amount of fuel in the tank decreases as the number of minutes increases. The amount of fuel leaked out increases as the number of minutes increases.

3. a. As the perimeter of a square increases, its area increases.

b. It could happen that the height of the triangle is decreasing in such a way that the product of the base and height (which is twice the area) remains constant. Or, it might happen that the height is constant, in which case the area increases as the base increases. Or, it might be the case that the height is decreasing in such a manner that the product of the base and height is decreasing, in which case the area decreases.

c. As the numerator gets larger, the value of the fraction increases.

5. Samples

a. Distance run—money raised; the money raised by Joella from her company is unaffected by the number of miles she runs.

b. Distance this year—distance last year; the distance run last year is the same as the distance run this year.

6. a. It decreases.

d. $G = 80 - 10T$

f. $T = 8$ minutes; when the tank is empty, $G = 0$, so $0 = 80 - 10T$, and $T = 8$.

h. The total of the fuel in the tank and the fuel leaked should equal 80 gallons. That equation is algebraically equivalent to the one in part (d).

8. c. Part (a) graph is straight; part (b), curved.

d. $p = 4s$; $A = s^2$

12. a. After the first year, there are $30 + (12 \times 5) = 90$ employees. So it will take 21 more months at the 10-per-month rate to reach 300 employees—33 months total.

b. A graph of the number of employees versus time in months should have two linear pieces: the first from (0, 30) to (12, 90), and the second from (12, 90) to (33, 300).

13.2 Understanding Slope: Making Connections Across Quantitative Situations, Graphs, and Algebraic Equations

2. a. The slope is $\dfrac{-10}{1}$, or $^-10$.

b. This slope tells us the amount of fuel the airplane is losing per minute.

c. The slope is $\dfrac{10}{1}$, or 10.

d. This slope tells me the amount of fuel that is leaking out per minute. (We could think of this as fuel being gained in the atmosphere.)

e. The slopes are alike except for the sign. That is, the first slope is negative, and the second is positive. The negative slope indicates the amount of fuel being lost by the plane per minute, whereas the positive slope indicates the amount of fuel being leaked into (gained in) the atmosphere per minute. In conventional graphs, a negative slope indicates that values from the vertical axis *decrease* as values from the horizontal axis increase.

4. d. Group 3 because . . .
(*Hint:* The heights tell only part of the story.)

5. d. Group 1 because . . .; the order of the candles is . . .

6. a. Cindy; she did not pick any more strawberries after 10 AM.

b. $19\frac{1}{5}$ quarts per hour

c. 9 AM; 28 quarts per hour

d. Barb; the slope of her line is the greatest of the three. Alternatively, Barb picks 28 quarts per hour, Annette picks $19\frac{1}{5}$ quarts per hour [part (b)], and Cindy picked 24 quarts per hour while she was picking.

9. a. Your table should have one column (or row) for the number of roses, and a second column (or row) for the total cost.

b. The total cost depends on the number of roses purchased.

c. If r = the number of roses purchased and C = the total cost (in $), then $C = 20 + 5r$.

d. Your graph should have the number of roses on the horizontal axis. It should include (1, 25) and have a slope = 5.

13.3 Linear Functions and Proportional Relationships

2. a. With P = the air pressure in pounds per square inch and h = height above sea level in feet, the slope is $\frac{12.1 - 14.7}{5280 - 0} \approx {}^{-}0.0005$ and $P = {}^{-}0.0005h + 14.7$. The relationship between P and h is not proportional.

b. As the distance below sea level increases by 33 feet, pressure increases 14.7 pounds per square inch. Not proportional.

4. a. 120 birds   **b.** $N = 4C$

7. a. For A, $p = 13h$ (where p is the number of parts and h is the number of hours).
For B, $p = 14h$ [Note that $\frac{243 - (9 \times 13)}{9}$ is 14. Or, $243 \div 9 = 27$, and $27 - 13 = 14$.]

b. Both expressions have the form $y = mx$ and so both are proportional relationships.

c. Both graphs start at (0, 0); the graph for A has slope 13, and the graph for B has slope 14. The slopes tell you how fast each worker works, 13 parts/h for A, 14 parts/h for B.

9. Only parts (b–d). In part (a), the equation is not of the correct form. In part (c), the variables are the number of crackers and the number of calories. In part (d), the variables are the number of tickets and the total dollar cost.

13.4 Nonlinear Functions

1. In the nonlinear cases and with enough points, your graphs should not be straight lines.

a. Nonlinear, because the equation cannot be written in the $y = mx + b$ form.

2. In the nonlinear cases and with enough points, your graphs should not be straight lines.

a. Nonlinear; $y = 5x(x - 2) = 5x^2 - 10x$, which cannot be written in the $y = mx + b$ form.

b. Linear; the given equation can be simplified to $y = \frac{-7}{2}$ or $y = 0x + \frac{-7}{2}$.

3. Samples

a. The number of gallons remaining in the tub decreases as the number of minutes the drain is open increases. If the outflow rate is steady, the relationship would be linear, perhaps of the form

(number of gal remaining) = (number of gal originally) − (outflow rate)x

b. The area increases as the length increases. $A = \left(\frac{1}{2}w\right)w$, or $A = \frac{1}{2}w^2$, a nonlinear equation.

c. The amount in your checking account will increase as the number of years increases. $A = 2500(1 + 0.0125)^y$, a nonlinear relationship between A and y. So after one year, you will have $2500(1 + 0.0125) = 2531.25$ dollars.

5. Each of the function rules is linear.

Answers for Chapter 14

14.1 Distance-Time and Position-Time Graphs

3. a. He has to travel 140 ft, at the rate of 5 ft/s, so it takes him 28 s to return home.

b.

t	0	5	10	15	20	25	30	35	40	43
d (from cave)	0	40	80	140	115	90	65	40	15	0

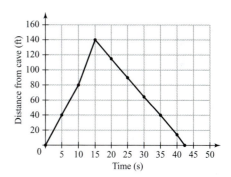

c.

t	0	5	10	15	20	25	30	35	40	43
total distance	0	40	80	140	165	190	215	240	265	280

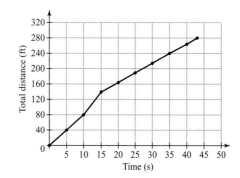

6. a.

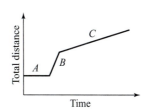

b. 0 ft/s

g. The corresponding slopes are the same in size (absolute value), with *E*'s slope being the negative of *B*'s slope.

7. $\dfrac{68 \times (365 \times 24 \times 60 \times 60 \times 1.86 \times 10^5)\ \text{miles}}{(2 \times 10^6)(365 \times 24)\ \text{hours}} \approx$
22,766 mi/h on average

8. Road 1 h, 15 min + canoe 36 min + jungle 16 h + path 3 h, 15 min = (in all) 21 h, 6 min. Was your explanation the same for each part?

14.2 Using Motion Detectors

1–2. a. The walker started 2 ft away from the motion detector, and then walked away from the motion detector at a constant speed for 6 s. (The speed was $\frac{2}{3}$ ft/s.)

b. The walker started 2 ft away from the motion detector, and then walked away from the motion detector at a constant speed for 3 s. The walker in part (b) walked faster than the walker in part (a). [Walker (b)'s speed was $\frac{4}{3}$ ft/s.]

4. a. Sample The walker started a short distance from the motion detector. He or she started walking away from the motion detector very slowly, and then gradually increased speed until he or she was walking very quickly.

14.3 Graphs of Speed Against Time

3.

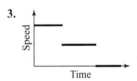

All three parts of the graph are straight horizontal line segments, each indicating constant speed. They differ by their positions on the vertical axis; the height signifies the speed in each of the cases. If you made this graph as a companion to that of the given graph, the times for the different pieces should match.

6. a. Sample Bart trotted evenly to the corner. There he saw that some friends had stopped ahead, so he sped up at a steady pace to catch them. When he did, they chatted for a little while. But then they realized they might be late to the game, so they decided to run as fast as they could.

9. a. C; the change in height per unit of horizontal distance is greater during C than in B.

b. G; during H, the speed of the roller coaster is about 0, but during G, the speed is positive although decreasing.

c. Very roughly. . .

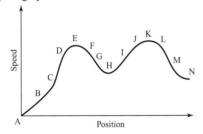

14.4 Interpreting Graphs

1. a. The fifth graph is most reasonable. The first part shows her speed increasing until she gets to a comfortable pace, at which she runs until she slows to a stop (the decrease of speed to 0). Then the increasing speed as she runs downhill fits the last part of the fifth graph.

b. Notice that the graph involves *distance traveled*.

c. Notice that some graphs are for distance-time and others speed-time.

3. *C*, because her speed is decreasing to 0.

8. The oven is turned on at *B* and heats up to 350°. It cools and warms up to keep the temperature around 350°.

Answers for Chapter 15

15.1 Finding Linear Equations

1. a. 2   **b.** $\dfrac{y-7}{x-4}$ or $\dfrac{y-17}{x-9}$

c. $y = 2x - 1$

2. a. $\dfrac{y-9}{x-5}$ **b.** $y = 6x - 21$

c. On a line, the slope is the same everywhere.

3. $y = 4x - 42$

7. $y = \frac{3}{4}x - 10$

11. $y = \frac{6}{7}x + \frac{33}{7}$

13. $y = \frac{^-5}{3}x + \frac{15}{2}$

17. With T = the Celsius temperature and h = the height in kilometers, $T = {}^-6.5h + 22.2$.

18. a. p: $\frac{^-3}{2}$ and q: $\frac{2}{3}$ **b.** $^-1$

19. b. Perpendicular; Slopes are $\frac{^-3}{4}$ and $\frac{4}{3}$.

20. a. Any equation giving a line with slope 7 but not passing through the point $(0, -4)$. For example, $y = 7x + 10$.

c. $y = {}^-\frac{1}{4}x + $ (your choice of a number)

21. 0, undefined. The product is not $^-1$. Any horizontal line and any vertical line are perpendicular, but the product of their slopes is not $^-1$.

22. a. $y = \frac{1}{5}x + 6\frac{2}{5}$ **b.** $y = {}^-5x + 22$

15.2 Solving Two Linear Equations in Two Variables

1. When the time is 2.5 hours, their distances will be the same, 7 miles.

2. $x = 3, y = 7$

6. $x = 1, y = {}^-5$

8. $x = 3, y = 3$

12. $x = 1, y = 5$

16. $x = 43, y = 31$

18. $x = 7\frac{1}{6}, y = 3$

21. a. Any linear equation with slope 3, but *y*-intercept not (0, ⁻10)

   b. Any linear equation obtained from the given one by multiplying both sides by the same nonzero number

22. a. Many possible answers, but (5, 7) should satisfy each one.

   b. Any linear equation with slope ⁻4, but a *y*-intercept different from (0, 6)

24. a. After $3\frac{2}{7}$ minutes, both elevators are at $164\frac{2}{7}$ feet.

   b. Passenger, 125 ft/min (headed down); freight, 50 ft/min

   c. Passenger, 575 ft; freight, 0 ft

26. a. After $1\frac{3}{8}$ minutes, both elevators are at 152.5 feet.

   b. Passenger, 180 ft/min; freight, 60 ft/min

27. a. The passenger elevator starts out above the freight elevator and keeps getting farther above (140 ft/min versus 25 ft/min), because both are headed up.

   b. The person likely found two equations and solved them, getting a negative value for the time. The negative value for time indicates that the two elevators would have been at the same height in the past *if* this trip had not started with time = 0.

28. Sketching the elevator trip on the graph indicates that the two elevators were at the same height during the passenger elevator's trip between 3 and 4 minutes. With (3, 240) and (4, 360) for the passenger elevator, and (1, 500) and (6, 200) for the freight elevator, we find that the elevators are at the same height ($333\frac{1}{3}$ ft, or 333 ft, 4 in.) after a trip time of $3\frac{7}{9}$ seconds.

29. The two elevators are at the same height at $12{:}03\frac{21}{44}$ and $12{:}08\frac{1}{4}$.

15.3 Different Approaches to Problems

   1. g. 125 ft

   h. Tony's graph: through (0, 50), with slope 10; Rita's graph: through (0, 0), slope 15

   7. 16 seconds, counting the head-start time

   8. $607\frac{1}{11}$ ft (including the 126-ft head start)

   9. 11 hours, 46 minutes

10. The plans give the same profit at sales of 200,000 CDs. Before that point, Option B gives more profit. After 200,000 CDs are sold, Option A yields more profit.

11. a. A: 0.05   B: 0.4

   b. Plan A costs less if you make more than 30 calls per month.

   c. Plan B costs less if you make fewer than 30 calls per month.

12. $200,000 in sales

15.4 Average Speed and Weighted Averages

   3. 12 mi/h (He has to cover the 18 miles in $1\frac{1}{2}$ hours.)

   5. Be careful—the number of credits at each place matters!

   6. 60 words per minute

   8. It is almost essential to draw a picture of the racetrack.

   a. Turtle will be ahead by 1.5 m.

   b. Rabbit will have forged ahead by 2 m. At the 20-m mark, when they head back, R will have traveled for 4 s, and T for $3\frac{1}{3}$ s. So on the way back, R will travel for 1.5 s and

thus go 15 m. T, on the other hand, will travel $5\frac{1}{2} - 3\frac{1}{3}$ s on the way back, or $2\frac{1}{6}$ s, and thus go 13 m. So R is ahead by 2 m after 5.5 s. When R heads back (after 4 s), T has been heading back for $\frac{2}{3}$ s, or 4 m. Because R catches up by 4 m each second (on the way back), it will take 1 more second for R to catch T, at the 10-m mark from either end, when both have traveled 5 s.

   c. It takes 4 + 2 = 6 s for R to finish. After 6 s, T has gone 36 m, 20 m over and 16 back. So T is 4 m behind when the race is over.

   d. The part (b) answers will change (R will be 4.5 m ahead; after the start, R was always ahead of T), but the part (c) answers will stay the same because it will take R the same amount of time to win.

10. a. $\frac{12}{60} = \frac{1}{5}$ of the way over

   b. 10 m/s (Rabbit ran for 36 seconds, total.)

   c. Turtle, because 3.173 m/s is a faster speed than the speed giving a tie

12. a. The graph coordinates are (0, 0), (0.5, 3), (1.5, 15). Speeds are 6 m/h and 12 m/h.

   b. 10 m/h. If Carly's speed did not vary, the graph would be a line from (0, 0) to (1.5, 15).

   c. 10 m/h

   d. 10 m/h

   e. Their averages are the same because they traveled the same distance in the same amount of time.

15.5 More About Functions

   1. b. If success could be assigned a value, then knowing the number of hours worked could predict the success value.

   c. For a given person, knowing his or her blood alcohol content could enable one to predict his or her reaction time.

   3. c. $2(x^2 + 3)$, or $2x^2 + 6$ **d.** 256   **e.** 676

   5. $y = \frac{1}{2}x^2$

   9. $f(x) = 5x - 3$

15.6 Issues for Learning: Topics in Algebra

   2. a. 0 (*Joke*: "All that work for nothing.")

   b. $45\frac{5}{8}$

   3. a. ⁻8   **b.** 55   **c.** ⁻$8\frac{7}{9}$ **d.** ⁻1

Answers for Chapter 16

16.1 Review of Polygon Vocabulary

   1. a. Scalene right triangle

   d. Rhombus

   e. Regular pentagon

   h. (Equiangular) hexagon

   i. Equilateral 12-gon, or concave equilateral 12-gon (see Learning Exercise 6 in this section). Using less-well-known prefixes, it is also an equilateral dodecagon.

   j. Rhombus

9. Possible (and mysterious) *Hint* 1: Add a column for twice the number of diagonals—can you relate that column to the column for the number of vertices? A pattern suggests an educated guess, but, as you will see, patterns cannot always be trusted! Hence, now that you have a conjecture, try to reason why it must be true. Better *Hint*, because it gives a general argument: Each vertex in an *n*-gon is joined to all but 3 of the vertices to give $n - 3$ diagonals at each vertex, but doing this at each of the *n* vertices will count each diagonal twice.

13. a. $\dfrac{(5 - 2)180}{5} = 108°$ **b.** $72°$

15. Each is true. Your reasoning?

20.

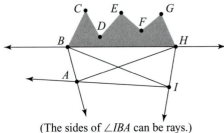

(The sides of $\angle IBA$ can be rays.)

21. There are many possibilities. Check to make sure the rays start at the correct points, and that *D* is the vertex of angle *FDE*. *TUV* should be shaded in, but not *PQRS*.

16.2 Organizing Shapes

1. The classification has each type of quadrilateral in a separate category, even though they share many characteristics besides having four sides. (The circles are a common way of showing the categories, and are not to be viewed as quadrilaterals themselves.)

2. a. AT   **b.** ST (Sketch a nonsquare rectangle.)
   c. ST (Sketch a parallelogram that has no right angles.)
   d. AT   **e.** ST   **f.** ST   **g.** ST

3. Samples

   a. *Shared*: Opposite sides parallel and equal in length; opposite angles the same size, etc. *Different*: Possible for angle sizes to differ, lengths of diagonal can differ, etc.

   b. *Shared*: Have pairs of adjacent sides the same length, diagonals make right angles, pairs of opposite angles have the same size, etc. *Different*: Possible for angle sizes to differ, possible for some sides to be different lengths, etc.

   c. *Shared*: Have 4 sides, 4 angles, 2 diagonals, angle sum is 360°, etc. *Different*: Trapezoids have parallel sides, quadrilaterals may not, etc.

4. Three are not possible.

6. a. That the parallel sides are horizontal, with one shorter than the other and above it

   b. (i) Perhaps the student thinks a parallelogram's angles cannot be right angles—that a parallelogram's sides must be "tilted."

     (ii) Perhaps the student thinks a rectangle must have unequal dimensions.

16.3 Triangles and Quadrilaterals

1. Kites and isosceles trapezoids do not fit easily into this Venn diagram.

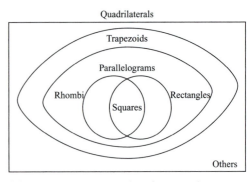

4. Stuck? Measure segments and angles—are there any possible relationships?

5. a. The angles can be placed next to each other so that the two outside sides appear to lie along a straight line, and their sum is therefore 180°.

   b. The new "placements" of the three angles again appear to lie along a straight line. The method does work with obtuse and right triangles, by folding down the vertex with the largest angle size.

16.4 Issues for Learning: Some Research on Two-Dimensional Shapes

4. Does your collection include rhombuses with 1, 2, or 4 right angles and with different orientations of the parallel sides?

Answers for Chapter 17

17.1 Shoeboxes Have Faces and Nets!

2. Did you start at the top and take advantage of the last cube drawn when adding later cubes?

3. b. 5 faces, 6 vertices, 9 edges

5. a.

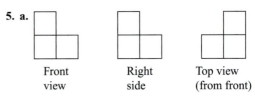

| Front view | Right side | Top view (from front) |

   b. Do you see *two* possible answers for the top view, depending on what might or might not be hidden in the back left corner?

   c. They are like views from the reverse (or mirror images).

6. a.

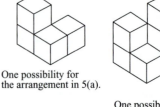

One possibility for the arrangement in 5(a).

One possibility for the arrangement in 5(b).

17.2 Introduction to Polyhedra

1. a. B, E (squares are special parallelograms), F

  e. H, J, K (and L and M, if used)

  f. K (What about C, E, F, and J? Some people regard a rectangle as a special isosceles trapezoid, as we will see later.)

  g. A, E, H (and L and M, if used)

  h. A, G

2. a. Lateral edges of a prism are equal in length and parallel to each other. The lateral faces of a prism are parallelogram regions (or special parallelogram regions, such as rectangular regions).

5. a. Unless you have a weird polyhedron such as a polyhedron with a square hole in it, the relationship will hold.

  b. 7

6. a. Right rectangular prism

  b. Right hexagonal prism (the common kind of pencil)

9. a. C, E, F

  b. Rectangular regions

  c. A and G

10. a. 14 (12 usually, but if the bases are. . . .)

17.3 Representing and Visualizing Polyhedra

1. a. Looking straight at the midpoint of an edge with the cube turned so the same amount seems to be above the edge as below the edge

  b. Looking from above and in front of an upper-right vertex from the right

  c. Looking straight on at one face of the cube

3. *Hint:* Add three hidden edges to the base and the two hidden lateral edges.

4. a. *Hint:* See Figure 1.

  b. *Hint:* Be sure that your lateral edges appear to be parallel and perpendicular to the edges of the base that they meet.

  c. *Hint:* Adjust the quadrilateral pyramid in Figure 8.

  d. *Hint:* Sketch the top base first, remembering that you are to look at the prism obliquely. Then sketch the lateral edges, keeping them parallel and of the same length. Finally, put in the edges of the bottom base, keeping in mind which edges are hidden.

  e. *Hint:* See Learning Exercise 3.

7. Nets (a) and (c) only [Did you notice that (b) and (d) are really the same net?]

8. e. Volume: 4 cube regions; surface area: 18 square regions

9. b. 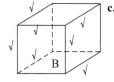 **c.**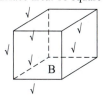

There might be other possibilities; compare yours with someone else's. This task is often difficult—what strategies did you use (e.g., focusing on edges that are *not* cut? Holding a cube as you imagined the cuts? Turning the net to make it easier to visualize where the base is? Working backwards from a cutout net to the cube?).

10. a. **b.**

  c. Same as part (b). (Do you see why?)

In each case, your version might be turned or even flipped from the answer given here.

12. *Hint:* There are 5 differently shaped tetrominoes.

13. a. There are 12 pentominoes. Did you find them all? How can you be sure?

  b. Each has area = 5 square regions, but the perimeters. . .

  c. No, e.g. . . . .

15. There are different possibilities for each part. One somewhat laborious way to check is to cut out your net and fold it. Another (recommended) way is to have a classmate or two look at yours.

17. a.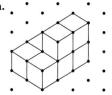

19. b.

17.4 Congruent Polyhedra

2. No—none can be moved to match another exactly.

3. The hidden shape will have the same-sized faces and (as a result) the same areas as the given shape, so 108 cm².

4. a. Yes. A rotation of 360° (or 0°) will make it match itself.

  b. Yes. Just reverse the motions that showed that P was congruent to Q.

  c. Yes. Do all the motions that show that R is congruent to S and then continue with the motions that show that S is congruent to T.

7. a.

17.5 Some Special Polyhedra

1. a. AT   **b.** ST   **c.** ST   **d.** AT

2. Knowing how many types (and what the types are) can enable you to ask, "Why are these the only ones?" and perhaps reveal something important about them. If only regular polyhedra can appear in some context for some theoretical reason, then you will know ahead of time what the possibilities are.

4. b. Try any quadrilateral prism that is not a cube.

c. Some of the triangles in your net should *not* be equilateral triangles.

5. a. The tetrahedron may be more easily seen with an actual cube, especially if the cube is transparent.

b. The other four vertices also give a regular tetrahedron.

9. a. Yes, because adjacent faces share edges, and all the edges of a given face are the same length.

b. Yes, because each face is the same type of regular polygonal region.

Answers for Chapter 18

18.1 Symmetry of Shapes in a Plane

1. Having reflection symmetry: A, B, C, D, E, H (two), I (two), M, O (two), T, U, V, W, X (two), Y

Having (nontrivial) rotational symmetry: HINOSXZ

4. a. 1 reflection symmetry only (Don't count a 360° rotation unless there are other rotational symmetries.)

b. 2 reflection symmetries; 2 rotational symmetries

c. 2 rotational symmetries (180° and 360°) [2 reflection symmetries if rhombus]

5. a. The bisector of the angle

b. Don't overlook the lines themselves.

c. There are actually infinitely many—do you see that?

e. 2 reflection symmetries (Do you also see 2 rotational symmetries?)

10. a. The segments must have the same length, so the triangle must be isosceles.

b. The two angles must have the same size.

c. *M* must be the midpoint of segment *BC*.

d. The two angles must have the same size.

14. a. Stuck? Try *rotational* symmetry.

b. The result should apply to special parallelograms, but it is not true for all kites.

18.2 Symmetry of Polyhedra

1. No, it is not possible. This right- versus left-handedness requires a reflection.

2. c. 5   **d.** 2

5. a. There are 2 planes giving reflection symmetries; they are. . . . There is 1 (nontrivial) rotational symmetry of 180°. (What is the axis?)

6. There are different possibilities. Compare with others. Do any require fewer than twice the original number of cubes?

7. *Hint*: The base must be a square.

10. Does each have the same final effect, regardless of any difference in motion along the way?

Answers for Chapter 19

19.1 Tessellating the Plane

1. What causes the difficulty?

4. If you have trouble, rotate the shape 180° about the midpoint of each side.

7. After experimenting a bit, you will believe, *Yes*.

8. c. The area of the larger triangle is 4 times the area of the smaller one. (Note that the *lengths* of the sides of the larger one are only twice the lengths of the sides of the smaller.)

19.2 Tessellating Space

1. Yes, there are many. . . .

2. Parts (b) and (d)

4. Volume is most often based on the number of cubical regions that fill a space; this space-filling by cubical regions is a tessellation.

Answers for Chapter 20

20.1 Similarity and Dilations in Planar Figures

2. a. Not similar. Even though the corresponding angles are all the same size, the given sides would involve different scale factors, 2 and $1\frac{6}{7}$: $12 = 2 \times 6$, but $13 = 1\frac{6}{7} \times 7$.

3–6. Usually, you can tell by looking at your result whether you have carried out the size transformation correctly; sides should be parallel to their images.

8. a. One pair of parallel sides have lengths 30.5 m; the other pair 12.2 m. The angle sizes are 115° and 65°.

9. a. $a = 50°$, $b = 60°$, $x = 7.2$ km, $y = 5\frac{1}{3}$, or $5.\overline{3}$ km if measurements are perfect.

d. $r = 75°$, $s = 75°$. So x and 28 correspond, as do y and 28.3, and 24 and 30. The 24 and 30 give the scale factor. $x = 22\frac{2}{5} = 22.4$ cm, $y = 35\frac{3}{8} = 35.375$ cm.

10. Yes, in both cases. If the center is on the figure, there are segments that overlap with rays, so it is visually different from the other cases.

12. A change in only the scale factor will result in a change in the size and location of the image.

14. a. Just one point!

b. After multiplying by ⁻2, measure in the opposite direction through the center.

15. Two are incorrect.

17. a. The new segment should be 3 copies of the original.

18. a. The first and third express the same relationship. (Your example?)

b. The first and third express the same relationship. (Your example?)

c. The first and third express the same relationship, and the second and fourth express the same relationship (but different from the first relationship).

19. a. 56 cm   **b.** 80 cm   **c.** 18 cm   **d.** 42 cm

22. Map scales are usually expressed as ratios or equations, so this one might be written 28 mi : 1 in. or 1 in. : 28 mi or 1 inch = 28 miles.

Many times the units are the same, so the scale here could be written 1,774,080 : 1 without any need to mention units. (There are 5280 feet in a mile and 12 inches in a foot.)

25. **a.** Here it is reasonable to have 1 cm = 30 million years. Then, starting with the Cambrian at 0 cm, Carboniferous would be at about 10.7 cm, Triassic at about 13.3 cm, Cretaceous at about 17.8 cm, Oligocene at 19 cm, Pleistocene at essentially the 20-cm mark.

  b. About 66.7 cm!

26. The original and the final image are similar; the scale factor is not 7, however.

28. **a.** Yes. All the angles in each triangle have 60°, so corresponding angles are the same size. And, even though for these triangles we need not check, every ratio of the lengths of corresponding sides is 7 : 12 (or 12 : 7).

  b. Yes. Again, the pairs of 60° angles assure that the triangles will be similar. How would you determine the scale factor for the chosen triangles?

  c. No. One triangle might have angles 90°, 45°, 45°; the other 90°, 30°, 60°.

  d. Yes. Every pair of corresponding angles has 90°, and because all the sides of each square have the same length, the equal ratios for similarity will be assured.

20.2 More About Similar Figures

3. The triangles in one pair are *not* similar.

4. **b.** A "tricky" part is seeing the correspondence after you have established similarity because the two triangles share an angle, giving a second pair of the same size along with the 114° angles. It may be helpful to redraw the triangles separately, with one triangle reoriented and the shared angles marked; $k = 5.6$ cm, and $n = 5.4 - 5.1 = 0.3$ cm.

  d. 30 ft

5. **a.** $2\frac{1}{4}$

  b. A rectangle, because size changes keep angles the same size

  c. 4 cm by $6\frac{2}{3}$ cm

  d. $26\frac{2}{3}$ cm^2 and 135 cm^2; the second is $\left(2\frac{1}{4}\right)^2$ times the first.

  e. Each could have been answered differently, because the 4 cm might have referred to the longer side and hence correspond to the 15-cm side rather than the 9-cm one.

20.3 Similarity in Space Figures

2. **b.** 1.5

3. Shapes (B) and (D) are similar, with scale factor 2 (or $\frac{1}{2}$, depending on your view of the *original*); shapes (B) and (E) are similar (allowing a reflection in a plane and a rotation as well as a size change), with scale factor $1\frac{1}{2}$ (or $\frac{2}{3}$); and (in the same way) shapes (D) and (E) are similar, with scale factor $\frac{3}{4}$ (or $1\frac{1}{3}$).

4. Shapes (B), (D), (F), and (G) are congruent (and therefore similar), and shapes (B), (D), (F), and (G) are similar to shape (A).

5. **b.** Because $36 = 12 \times 3$, $84 = 12 \times 7$, and $96 = 12 \times 8$ and all the corresponding angles are right angles, the prism in part (b) is similar to the given one. Alternatively, we could say that each of the ratios 36 : 3, 84 : 7, and 96 : 8 is equal to 12 : 1 or 12 (and the corresponding angles are all equal), so the prisms are similar.

  There are three other definite "Yes" and one "Well, probably, taking rounding into account."

  j. Yes. The scale factor is about 4.14.

6. **a.** 12 feet × 12 inches per foot = (scale factor) × 8 inches, so scale factor = 18. You might also get $\frac{1}{18}$ by using a different viewpoint as to the original figure, and the latter way is probably more common: "A $\frac{1}{18}$ scale model."

9. **c.** Yes, because. . .

  d. From the theory, the answer is $\left(1\frac{1}{2}\right)^2 = 2\frac{1}{4}$ times as large as the other.

  e. The ratio of the volumes is 27 : 8, but $\frac{27}{8} = \frac{3^3}{2^3} = \left(\frac{3}{2}\right)^3$, the cube of the scale factor.

10. **a.** Yes, with scale factor. . .

  b. Yes. The two scale factors are reciprocals, or multiplicative inverses.

  c. Yes, *X* and *Z* are similar. It is tricky to keep track of *new* and *original*, but using *Z* as the original and *X* as the final, the scale factor is (s.f. from *Z* to *Y*) × (s.f. from *Y* to *X*). Try a specific example to see how the product of the two scale factors enters in.

11. **c.** The ratio of the two should be the square of your scale factor.

  d. The ratio of the two should be the cube of your scale factor.

14. **a.** 8.8 cm by 13.2 cm by 22 cm

  b. 3 cm by 4.5 cm by 7.5 cm

18. **a.** Yes. All the angles are 90° and every ratio of the lengths of edges is 5 : 8.

  e. No. Although the ratios of the lengths of edges are all equal to 1 : 3, there is no assurance that the angles in the rhomboidal prism are 90°, as in the cube.

21. "Each cubic centimeter becomes a 4 cm by 4 cm by 4 cm cube in the enlargement, so each original cubic centimeter is now 4^3 cubic centimeters."

22. **a.** 25,600 cm^2

  b. 204,800 cm^3

25. The nets should be nets for cubes, of course, with sides/edges for the larger one twice the lengths for the smaller one.

26. Under the assumption that your shape is similar to that of the Statue of Liberty, your answers might be about 4 ft 6 in. for the statue's nose and 3 ft for its width of mouth.

Answers for Chapter 21

21.1 Planar Curves and Constructions

1. **a.** NT   **b.** AT   **c.** ST   **d.** AT

4. **a.** Any line through the center will give a reflection symmetry, and any size rotation with its center of rotation at the center of the circle will give a rotational symmetry.

b. There are many reflection symmetries (all the lines of reflection will be parallel) and many rotational symmetries (all 180°, but with different centers).

5. d. Because the whole circle has 360°, 80° would be $\frac{80}{360}$, or $\frac{2}{9}$, of the circumference.

e. $\frac{1}{2}$ **f.** $\frac{1}{8}$ **g.** $\frac{3}{8}$ **h.** $\frac{1}{6}$

6. a. 84°   **b.** 42°   **c.** 42°   **d.** 42°

e. 96°   **f.** 180°

7. The angle at A is the inscribed angle, intercepting the $d°$ arc. Triangle ABC is isosceles because of the radii, so the angles at A and B are equal. The size of the central angle $= 2 \times$ the size of the inscribed angle. Or, the size of the inscribed angle is half that of the central angle.

9. a. You can get the midpoint of a segment by constructing the perpendicular bisector of the segment.

11. Your eyes can usually tell you whether your work is all right.

12. a. Because the two distances are from the same radius, they are equal.

b. All the points should lie on the perpendicular bisector of segment AB.

13. a. *Hint*: 45 = half of 90

b. *Hint*: 135 = 90 + 45

c–d. *Hint*: $\frac{3}{4} = \frac{1}{2} + \frac{1}{4}$, and $\frac{1}{4}$ is half of a half.

14. a. Suppose a vandal or a malicious drunk came by.

b. Yes. Modify any regular polygon.

16. *Hint*: See Learning Exercise 5.

19. Using A as center, draw a circle with radius AB. Using B as center, draw a circle with radius AB. Either of the two points in which the circles intersect can be used for C.

21. a. Yes (SAS, or side-angle-side)

b. Not necessarily, but they are similar.

c. Yes (SSS, or side-side-side)

d. Yes, the third angle in GHI makes ASA (angle-side-angle) with Triangle 7.

21.2 Curved Surfaces

1. c. AT (but sometimes just 1)

d. AT (but sometimes just 1 nontrivial one)

e. AT   **f.** AT   **g.** AT

2. a. Right circular cylinder   **b.** Right circular cylinders

e. Oblique cone   **f.** Right circular cylinder

3. a. Using the line through the centers of the bases as axis, there are infinitely many rotational symmetries. With every line along a diameter of a cross section halfway down as axis, there are 180° and 360° rotational symmetries as well. Any plane perpendicular to the bases and passing through the centers of the bases will give a reflection symmetry, along with the plane parallel to the bases and halfway down the cylinder.

b. Do you see infinitely many rotational symmetries and infinitely many reflection symmetries? If not, reread the answer for part (a).

6. e. Not possible—every two great circles on a sphere will intersect.

8. b. 66 ml (*Hint*: How much will $\frac{1}{4}$ of the cylinder hold? $\frac{4}{4}$?)

9. a. The larger sphere has a diameter with a length twice the diameter of the smaller sphere.

10. a. Circles or ellipses, if the resulting cylinder is a circular cylinder

Answers for Chapter 22

22.1 Some Types of Rigid Motions

2. a. B reflection, C translation, D rotation, E rotation, F translation, G glide-reflection

b. Each is congruent to shape A, because it is the image of A for a rigid motion.

4. e. Each is congruent to the original, because each is the image of the original for some rigid motion.

5. a. Translations leave the orientation unchanged.

b. Rotations leave the orientation unchanged.

c. Reflections change the orientation.

7. There are at least 14. Decision making likely involves all the rigid motions.

22.2 Finding Images for Rigid Motions

1–2. Visual checks usually reveal whether images have been located correctly. Rotation images are the most difficult for many people.

4.
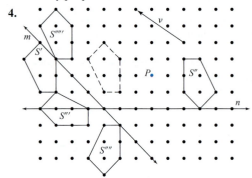

6.
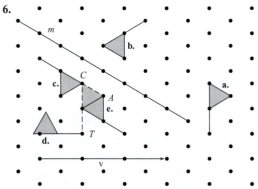

8. c. Each of the angles is 90°. This result makes sense for a rotation, because each point "travels" through the same angle.

11. Parts (c) and (d) will have the same effect on every point of the plane; turning 180° one clock direction will cause a shape to end up in exactly the same place as going 180° in the other clock direction.

22.3 A Closer Look at Some Rigid Motions

1. Again, your eyes usually tell you whether the result is all right, keeping in mind that most people have more trouble with rotations.

2. **a.** The vector should *end* at the figure to the left. Which shape is the original does matter, because the direction of the vector is opposite if the other figure is chosen as the original.

   c. The clock direction of the angle depends on which shape is the original.

4. **a.** Every point on line *k* is a fixed point.

   b. There are no fixed points.

   c–d. There is one fixed point, namely. . .

   e. Every point is a fixed point, for the 360° rotation.

5. **a–b.**

22.4 Composition of Rigid Motions

1. Lines *k* and *m* are perpendicular to the direction of the translation (and therefore are parallel) and are 1 cm apart. Note that 1 cm is one-half the 2 cm.

2. **a.** Translation, 1.2 cm east

   b. Translation, 3 cm southwest

   c. Translation. . . .

   d. Rotation. . . .

   e. Rotation. . . (unless $x = y$)

3. **a.** A glide-reflection changes the orientation of a figure.

   b. Translations and rotations do not change the orientation of a figure, but reflections and glide-reflections do.

4. **a.** A translation or a rotation (Each of the originals keeps the orientation the same, so their composition will also.)

   b. A translation or a rotation

   c. A translation or a rotation

5. **b.** The line of reflection appears to pass through the midpoints of the segments.

9. **a.** Your two lines of reflection should be perpendicular to the line of the vector, and half the vector's length apart.

   b. Your two lines of reflection should intersect at the center of the rotation; the angle they make will be half the size of the angle of the rotation.

   c. Perhaps surprisingly, there are many possibilities in each case.

11. Any rigid motion is one of these: reflection, translation, rotation, or glide-reflection. Each of translation/rotation can be achieved by the composition of two reflections

(Exercises 9–10), and a glide-reflection can be achieved by the composition of three reflections (two for the translation, plus the separate reflection of the glide-reflection).

12. **a.** 1 or 3 reflections

15. The rigid "motion" that leaves every point in the same place; this is legitimate, even though no net motion results.

22.5 Transformations and Earlier Topics

1. **a.** There are many translation symmetries, as well as many reflection symmetries. Describe several of the translation symmetries.

2. Only one of the statements is true in each part.

3. **a.** (Two, yes, two) reflection symmetries, (two) rotational symmetries (counting the 360° rotation)

   b. (Only one) reflection symmetry

   c. Reflection, rotational, translation, glide-reflection

   d. (One) reflection symmetry

6. *Hint:* What is the size of the third angle in each triangle?

8. **a.** Yes, just use the reverse of the rigid motion.

   b. Yes, use a 360° rotation.

9. **a.** Yes, . . .

   b. Yes, use a scale factor of 1.

   c. Yes, . . .

12. Two rectangles with dimensions not related by the same ratios

22.6 Issues for Learning: Promoting Visualization in the Curriculum

2. **a.**   **b.**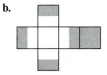

4. All three shapes are the same. It might be interesting to compare how others approached the exercise.

7. Here are two possible core squares. Are they different from each other? If you found another possible core square, are you certain it is different?

Answers for Chapter 23

23.1 Key Ideas of Measurement

1. **a.** Length

   b. Volume, although weight could conceivably be used

   c. Temperature

   d. Weight

   j. Your ideas, besides student achievement?

2. (Other correct answers are possible.)

   a. Yards, feet, inches, centimeters (!), millimeters (!!)

b. Milliliters, or ounces if weight is used

c. Degrees Fahrenheit or degrees Celsius

d. Grams or ounces

e. Square centimeters or square inches

f. Meters per second or yards per second

h. Square meters or square yards

j. Students' scores on tests or students' later success rate or. . .

3. a. Temperature, height

b. Height, weight, mathematics achievement

c. Running speed, walking speed, respiration rate—all of which involve measuring other quantities as well

4. a. Barometric pressure   **b.** Population density   **c.** Area

5. c. "One person's junk is another person's treasure," so the unit would have different values depending on who was using it.

d. You have perhaps read about perception experiments in which people's judgments of relative temperature were vastly influenced when one finger was in hot water, as compared to when the same finger was in cold water.

e. The actual amount in a *pinch* likely varies from person to person.

6. a. Real-life length measurements cannot be 100% exact.

7. a. Show segments as short as $6\frac{1}{2}$ units and up to $7\frac{1}{2}$ units.

9. a. $x > 2000$ because it would take more than 2000 raisins to weigh 2000 pounds.

b. $x < 2000$ because a raisin's weight is smaller than a pound.

c. $x > 2000$ because. . .

d. $x < 2000$ because. . .

10. a. The ap unit is larger than the ag, because it takes fewer aps to equal 150 ags.

11. Weigh yourself holding the puppy, and then. . .

12. a. $8\frac{1}{2}$ units, by counting repetitions of the unit; "cutting" the region into individual and part units and totaling them

e. $2\frac{1}{8}$ new units, because. . . ; 2 new units because. . . ; $1\frac{5}{8}$ new units because. . .

13. Probably $8 \times \$2.89 = \$23.12$, because some leftover pieces from a section would be too short to be useful elsewhere.

19. a. 80 km

b. 15 cm

20. 1

22. *Correct*: Only parts (a) and (d). *Hints*: The calculations are done as though the relationships are base-ten relationships, but 1 hour = 60 minutes, not 100; 1 pound = 16 ounces, not 10; 1 gallon = 4 quarts, not 10.

25. About $0.9144018 - 0.9144 \approx 0.0000018$ meter shorter

23.2 Length and Angle Size

4. Try doing these mentally, if you did them with paper and pencil.

a. $3\frac{3}{4}$ inches   **b.** $3\frac{3}{8}$ inches   **c.** $3\frac{9}{16}$ inches

6. h. Just under 2.5 cm, if you do not have a quarter handy

i. Almost 28 cm

j. About 99 cm, or about 1 m

7. Remember that the sides have to fit together to make a polygon: 1 cm, 1 cm, 1 cm, and 21 cm would be impossible. The complete answers are not given here.

a. Try to be systematic, possibly this way: 6-6-6-6; 5-6-6-7; 4-6-6-8; 3-6-6-9; 2-6-6-10; 1-6-6-11.

b. 1, 11, 1, 11 cm; 2, 10, 2, 10 cm; 3, 9, 3, 9 cm; 4, 8, 4, 8 cm; 5, 7, 5, 7 cm; 6, 6, 6, 6 cm. Notice how much easier the problem is with a rectangle, because the opposite sides must be the same length.

8. Experiment, gathering and organizing the data from several simpler, specific values for n (e.g., $n = 1$, $n = 2$, $n = 3$, etc.), and look for a pattern.

9. a. By measuring the distance of the line segment perpendicular to the line(s) from the point to the line or from one parallel line to the other

b. No, so long as the distance is measured along a perpendicular

11. Ancient Greeks used this idea: The perimeter (and area) of the circle should be between the perimeters (and areas) of the outer and inner polygons.

13. 4 and a little more, using a 2-cm segment to measure with; somewhat under 9, using a 1-cm segment; just under 18, using a 0.5-cm segment. The 0.5 cm fits very closely, so an estimate of just under 9 cm should be a good estimate.

18. a. $17° \ 43' \ 37''$ **b.** $46° \ 28' \ 43''$

19. *Hint*: $360 \div 15 = \ldots$

21. Remember to check your drawings by measuring with a protractor.

22. a. $y = 180 - x$, and $? = 180 - y = 180 - (180 - x) = 180 - 180 + x = x$. Thus, $? = x$.

23. a. c, g; a, e; d, h; b, f

b. No

c. Corresponding angles of parallel lines appear to be the same size.

d. c, e and d, f. If lines are parallel, the alternate interior angles are the same size.

25. a. *Hint*: What are the sizes of the other angles at vertex A, in terms of z and y?

27. a. Because the n angles total . . . and they are all equal in size, each is . . .

b. Your table should show these results: $3 \rightarrow 60°$; $4 \rightarrow 90°$; $5 \rightarrow 108°$; $6 \rightarrow 120°$; $7 \rightarrow 128\frac{4}{7}°$; $8 \rightarrow 135°$; $9 \rightarrow 140°$; $10 \rightarrow 144°$; $11 \rightarrow 147\frac{3}{11}°$; $12 \rightarrow 150°$. Notice that the angle sizes are getting larger. How do you know that they will never equal $180°$?

c. The sizes get larger and larger, but stay less than $180°$.

d. *Hint*: $(n - 2)180 = 4500$

29. a. If it were $360°$, it would be flat. (If it were more than $360°$, it would not be convex.)

b. What polygons give (same-sized) angles that will total less than $360°$ at a vertex? See Learning Exercise 27(b) and be sure to see whether there is more than one possibility for a given polygon.

c. *Hint*: See Learning Exercise 27(b) again.

30. The 1080 refers to the number of degrees in the turn the athlete can make when he or she is in air. So a 1080 refers to 3 full turns. Excellent athletes can do 720s and 900s as well.

35. a. 120°   **b.** 55°   **c.** $p + q$ degrees

36. The conventional way is to measure the angle formed by the two perpendiculars to the edge of the dihedral angle, at the same point on the edge, and with one in each plane making the dihedral angle.

37. *Hint*: Look at lots of multiples of 19.

38. b. The mile run is about 109 meters (or about 120 yd) longer than the 1500-m run. The ratio 1760 yd : 1500 m would most likely be given with the units the same. 1760 yd : 1640 yd = . . . , or 1609 : 1500.

42. a. 400

   b. 0.009° (This is not a temperature, even though you may know that the Celsius temperature scale was once called the centigrade scale.)

44. a. The endpoints give a square. Each angle intercepts half the circle and so has size 90°. Rotate 90° to see that the sides have the same length.

   b. A regular hexagon. Joining the points to the center of the circle shows 6 connected equilateral triangles, so each side of the hexagon has the same length, and the angle at each vertex is 120°.

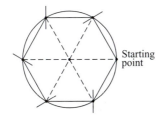

   c. *Octagon*: Bisect the central angles in part (a); 12-gon: Bisect the central angles in part (b).

Answers for Chapter 24

24.1 Area and Surface Area

2. a. 24 rhombus regions, because. . .

3. Here are possibilities, if you cannot think of any: nail of little finger for square centimeter, VW Beetle door for square meter, floor of an ordinary classroom (10 m by 10 m, or roughly 30⁺ ft by 30⁺ ft).

5. a. $20 > x$, because if the measurements are equal, then it will take more of the smaller units to give an area equal to x of the larger units (or, . . . it will take fewer of the larger units to make an area equal to 20 of the smaller units).

   b. $20 > x$, because. . .

7. a. Your sketch should confirm, e.g., that there are 144 square inches in 1 square foot. Hence, 1 square inch is $\frac{1}{144}$ of a square foot.

   b. . . . 9 sq ft in a square yard; $\frac{1}{9}$ sq yd makes a square foot

   c. *Hint*: There are 1760 yards in 1 mile.

9. a. 230   **b.** 0.45

11. a. ≈26 units   **b.** ≈13 units

12. a. 19 units

   b. *Hint*: "Surround" the triangle with a rectangle. 10.5 units

   c. "Cut off" a slice from the left end and put it on the right end. 25 units

   d. 21 units   **e.** 26 units

14. b. 82 units

16. No. The second one has area 16 square meters.

19. a. The different sizes are $\frac{1}{4}$ (regions I), $\frac{1}{8}$ (III, IV, V), and $\frac{1}{16}$ (regions II). Copying, cutting out the pieces, and moving them around should convince you.

   b. $11\frac{3}{4}$ area units (I), $5\frac{7}{8}$ area units (III, IV, V), and $2\frac{15}{16}$ area units (II)

21. a. Ollie is incorrect, because. . . [*Hint*: See Learning Exercise 7(b).]

22. The original 64 squares now seem to fill 65 squares! This strange result would violate the cutting-up key idea. The secret is that the apparent diagonal of the rectangle is really a very thin but long region, with area 1 square region.

24. a. 24 010 cm², because areas are related by the square of the scale factor. S.A.(larger) = $7^2 \cdot 490$.

   c. Not necessarily, because the pyramids are not similar ($4 : 6 \neq 6 : 8$).

27. Your work should show that figures with the same perimeter may not have the same area.

28. Each is incorrect, usually because of the inappropriateness of the unit for the quantity.

24.2 Volume

1. The units given here are possible, but others can be defended. The important thing is that the units be of the correct kind and not too large or small. Units like km or cm² for the area of a lake are either not appropriate (km is for length) or sensible (cm² is too small).

   a. Volume, km³ (possibly m³)   **b.** Area, km²

   c. Length, m or dam   **g.** Angle size, degrees

   h. Volume, m³   **i.** Length, m

2. If you are stuck, here are some ideas: a portion of your little finger (1 cm³, 1 mL), some plastic soda containers (1 dm³, 1 L), a box for a large washing machine (1 m³).

4. a. About 250 mL or 250 cm³   **b.** About a liter (slightly less)

   c. About $\frac{1}{3}$ liter   **d.** Perhaps about 300 m³

6. a. $x > y$ **b.** $x < y$

9. a. 10 times, 100 times, 1000 times

10. a. 3280   **b.** 32.8

11. a. Your drawings should suggest that 1 cubic foot = 1728 cubic inches; 1 cubic yard = 27 cubic feet; OR 1 cubic inch = $\frac{1}{1728}$ cubic feet; 1 cubic foot = $\frac{1}{27}$ cubic yard.

13. *Key to the second question*: Your answer to "Does *anything* include idealizations that exist just in the mind?"

14. a. Surface area = 150 square regions; volume = 108 cubic regions

  b. S.A. = 238 square regions; volume = 222 cubic regions

  e. No (1 L = 1000 mL or 1000 cm^3)

15. 1 gram, 1 gram

17. a. Count by rows. 42 cubic regions; 40 cubic regions

  b. 42 square regions; 40 square regions

  c. The first (30 stories, each 42 cubic regions = 1260 cubic regions vs. 31 stories, each 40 cubic regions = 1240 cubic regions)

20. a. Congruent shapes have the same measurements, so S.A. = 32 square regions and volume = 12 cubic regions for the first shape, and S.A. = 34 square regions, volume = 9 cubic regions for the second shape.

  b. Surface area of the hidden, larger version of the first shape = 800 square regions; volume = 1500 cubes.

21. a. 20 800 cm^2; 192 000 cm^3

  c. 53 248 cm^2; 786 432 cm^3

  d. 6292 cm^2; 31 944 cm^3

24.3 Issues for Learning: Measurement of Area and Volume

1. a. They might have counted the visible square regions.

  b. *One possibility*: Double the count as in part (a), for the hidden parts. Or, count the 4 in front, double for the back, giving 8. Do the same for the right and left sides, for another 8. Finally, get another 8 for the top 4, plus 4 for the bottom.

  c. The counts for the right and left are counting cubes that have been counted already.

2. a. For example, the 9 in the front and the 12 on the right both count the 3 cubes at the front right corner.

  b. The student is overlooking the cubes in the inner, middle columns.

Answers for Chapter 25

25.1 Circumference, Area, and Surface Area Formulas

2. a. 5π cm, or 15.7 cm

  c. The radius is about 3 cm, so $C = 6\pi$ cm, or 18.8 cm, or about 19 cm.

3. 3981 miles; 7962 miles (often referred to as 4000 miles and 8000 miles). *Note*: Earth is not a perfect sphere but is often treated as such.

4. a. $\frac{22}{7} \approx 3.1428571$, so . . .   **b.** 0.038 cm

6. 31.8 cm, or 32 cm

8. a. "Let's draw a parallelogram on graph paper. Check it out."

  b. "Remember how we put two triangles together to make a parallelogram? Then to find the area of one triangle, we had to take half the area of the parallelogram."

9. a. 12.16 cm^2 **b.** 76.5 cm^2

  e. 20 cm^2 each, even though their areas may look different

13. Each way should give an area of 24 square regions.

14. Each of the triangles has the same area, $7\frac{1}{2}$ units, even though the triangles may appear to the eye to have different areas. Because parallel lines are the same distance apart no matter where you measure it, the heights of all the triangles are the same (and the base is the same for each of them).

16. a. about 8.6 to 8.8 cm^2

18. i. Shape **a**: 15 cm^2, shape **b**: 30 cm^2, shape **c**: 29.75 cm^2

  Shape **d**: 12.08 cm^2, shape **e**: $4.5\pi + 9$, or about 23.14 cm^2

  Shape **f**: also 23.14 cm^2

  Shape **g**: $12 + 4.5\pi$, or about 26.14 cm^2

  Shape **h**: $28 + 2\pi$, or about 34.28 cm^2

19. a. What are the areas of the two pizzas? (Assume that they have the same thickness.) The larger pizza is cheaper by the square inch. If you were surprised by the answer, put the smaller pizza on top of the larger one and look at a slice of each with the same central angle.

20. a. 12.56 cm^2

  b. 1 cm^2 for first grid; 0.25 cm^2 for second

  c. Second grid gives better estimate

21. a. 48π or 150.8 cm^2

  b. 13.824π or 43.43 cm^2

  f. $24 + 8\pi$ or 49.13 cm

  g. $20 + 16\pi$ or 70.27 cm [for part (c)]

23. a. S.A. $\approx$ 200,000,000 or 2×10^8 sq miles.

  b. *One way*: The land area is about 30% of the total area, so find 30% of the S.A. from part (a): about 6×10^7 sq miles.

  c. 10^8 sq miles

24. a. Because the globes are similar, the ratio of their surface areas will be the square of the scale factor: $\left(\frac{16}{12}\right)^2 = \left(\frac{4}{3}\right)^2 = \frac{16}{9} \approx 1.78$. The larger globe has 1.78 times as much area as the smaller one does.

  b. Smaller: 144π sq in.; larger: 256π sq in. [Does this check with part (a)?]

26. 2880 cm^2

27. The estimate could be improved by using narrower rectangles.

31. a. See the right triangles? b. Rhombuses and squares, because. . .

32. If you get stuck, look at $\frac{\text{No. of dots on}}{2}$ + No. of dots inside, which is close to the correct number. The final result that does give the area is called Pick's formula.

25.2 Volume Formulas

1. In which ones are all the layers identical in volume?

3. Measurements can vary somewhat, and the production process may have changed measurements used in calculating the answers. If your answer is in the neighborhood of the listed answer, you were likely correct in your reasoning.

  b. With $B = 2.0$ cm^2 (or thereabouts—don't forget the holes), the volume of plastic needed for 1000 such pieces would be 1200 cm^3.

  d. The key idea about the measurement of the whole equaling the sum of the measurements of its parts

4. a. The same

   b. The top-bottom way gives a radius of $\frac{11}{2\pi}$ and a volume about 81.8 cubic inches, whereas the side-side way gives a radius of $\frac{8.5}{2\pi}$ and a volume about 63.2 cubic inches. So the top-bottom way gives 18.6 cubic inches more, or is about 129% as large as the side-side.

   c. *Hint*: Examine the surface area of a 1 by 1 by 5 right rectangular prism and find a shape made with 6 cubes having the same surface area.

6. Again, measurements can vary somewhat and the production process may alter the original dimensions, so if your answer is somewhat close to the one here, your measurements were probably correct.

   a. About 8.7 cm^3

8. a. 8″ ball, volume $= \frac{256\pi}{3} \approx 268.1$ in^3; 16″ ball, volume $= \frac{2048\pi}{3} \approx 2144.7$ in^3

   b. The areas are 64π in^2 and 256π in^2, so the costs of the plastic are about \$0.39 and \$1.56 (don't forget to change the square inches into square feet).

   c. Because the spheres are similar, the ratio of the volumes is $\left(\frac{1}{2}\right)^3$ or 1:8, and the ratio of the areas is $\left(\frac{1}{2}\right)^2$ or 1:4.

10. c. $\approx 3.45 \times 10^{17}$ cubic miles

   d. About 1,278,000, using the rounded answers in parts (b) and (c)

12. a–b. No, they are not congruent, but they have the same volume.

   c. $\sqrt[3]{\frac{6}{\pi}} \approx 1.24$, so about $1\frac{1}{4}$ inches

13. a. $\frac{2}{3}$ [Compare with Learning Exercise 11(b).]

   b. $1 - \frac{\pi}{6} \approx 47.6\%$

16. a. $\frac{4}{3}\pi\left(\frac{1}{6}\right)^3 \approx 0.019$ cubic miles

   b. $\frac{\text{later volume}}{\text{earlier volume}} \approx 0.027$, so the later volume is only 2.7% of the earlier volume, quite an error.

18. a. $V = 72\pi + 18\pi = 90\pi$, about 282.7 cm^3; S.A. $= 9\pi + (\pi 6)8 + 18\pi$, about 235.6 cm^2

   c. $V = \pi r^2(h - r) + \frac{2}{3}\pi r^3$

   d. $V = kmn + \frac{1}{3}mnh$

   e. $V = \pi r^2 h$, and S.A. $= 2\pi r^2 + 2\pi rh$.

20. a. 384 cm^2

   b. S.A.(cube) $= 6s^2$

   c. 512 cm^3

   d. V(cube) $= s^3$

   e. As measurements, the area and volume cannot be equal of course, because area and volume are completely different characteristics. The second question comes down to whether $6s^2 = s^3$ has any *numerical* solutions, and it does: $s = 6$ and the trivial (for here) $s = 0$.

21. Recall that the scale factor multiplies each length measurement, so in the formulas . . .

23. Without more restrictions on the shape of the dog house, there is no definite answer.

Answers for Chapter 26

26.1 The Pythagorean Theorem

2. Do not have either a or b equal to 0, and you will have a counterexample.

3. a. Perimeter $= 60$ cm, area $= 150$ cm^2

   b. $35 + \sqrt{175}$ cm, or 48.2 cm; $7.5\sqrt{175}$ cm^2, or 99.2 cm^2

5. 10 cm, $\sqrt{117} \approx 10.8$ cm, and $\sqrt{145} \approx 12$ cm are the lengths of the diagonals of the faces, and $\sqrt{181} \approx 13.5$ cm for each of the "inside" diagonals. How do you know that the triangle inside the prism is a right triangle?

6. a. $\sqrt{x^2 + y^2 + z^2}$

   b. No, but a 4-foot one would, just barely.

   c. $e\sqrt{3}$ units

   d. The diagonal of the cube is a diameter of the sphere, so. . . ($d = 10\sqrt{3}$ cm and $r = 5\sqrt{3} \approx 8.7$ cm)

9. Volume $= 28\sqrt{42} \approx 181$ cm^3; surface area $= 84 + 6\sqrt{91} + 14\sqrt{51} \approx 241.2$ cm^2

10. a. 4.5π cm^2, 8π cm^2, 12.5π cm^2. The sum of the areas on the legs equals the area on the hypotenuse, just as in the Pythagorean theorem!

12. How do you know that $\angle PVQ$ is a right angle? If you are stuck, it may be because you need $(a + b)(a + b) = a^2 + 2ab + b^2$.

13. Parts (b–c) give Pythagorean triples. Do you see how they are related to the triple 3, 4, 5?

15. One statement in each pair is false.

17. For the 6 rows with 6 dots in each row, 20 is not correct; there are only 19! (Which one is missing?) This problem is another example to show that patterns cannot be trusted 100%.

19. Suggestion: Use the Pythagorean theorem for the heights of the triangles and of the pyramid.

   a. A drawing helps to see that the altitude of each triangle is 2 cm ($1.5^2 + h^2 = 2.5^2$) and the height x for the pyramid is about 1.3 cm ($1.5^2 + x^2 = 2^2$). So, S.A. $= 3^2 + 4\left(\frac{1}{2} \cdot 3 \cdot 2\right) = 21$ cm^2, and $V = \frac{1}{3}Bh \approx \frac{1}{3} \cdot 3^2 \cdot 1.3 = 3.9$ cm^3.

20. a. Assume that the ground covered is a circular region, that the pile has lots of rotational symmetries, and that the pile will be shaped like a cone.

   b. 16π, or about 50.3 m^3

21. a. Make a drawing. Where is a right triangle?

24. Use a key idea of measurement and Learning Exercise 11(a).

26. *Hint*: Place one of the boards across the corner of the canal to make an isosceles right triangle.

28. Stuck? Does the Pythagorean theorem help?

26.2 Some Other Kinds of Measurements

5. If, say, your measuring container is smaller than a container of liquid, you can remove amounts with your measuring container, keeping a record of each amount and then adding these amounts when the large container is empty.

8. Rather than what appears to be a random collection of terms, the metric system uses prefixes with a basic term for all of its units of a particular type.

10. **c.** About 41 mph

  d. About 0.328

11. **d.** About 81 kg or 178 lb

  e. About 8.7 kg or 19 lb, at least

12. **a.** *One way*: Showing on the drug number line that $\frac{1}{4}$ is $\frac{1}{3}$ of $\frac{3}{4}$, so $\frac{1}{4}$ g should use $\frac{1}{3}$ of 12 mL, or 4 mL, on the water number line.

15. **a.** $\frac{1}{25}$ gallon per mile (The 25 miles per gallon means 25 miles per 1 gallon.)

  b. About 9.1 kilometers per liter

16. **a.** 10 m/s

  b. 36 km/h (if the runner could run that fast for an hour)

  c. The runner's rate is 10 meters for every second.

17. **a.** 10 yd/s   **b.** About 20.45 mi/h

  c. Metric (used in Learning Exercise 16) is easier, because the calculations are easier.

19. **a.** The total cost was $130, or $32.50/blouse.

  b. Your meal was roughly $10 of the total $70, so you could argue that you should pay only $10. Splitting the $70 evenly would mean you would pay $14.

21. $\dfrac{20}{3 \times 10^5 \text{ km/s} \cdot 3600 \text{ s/h} \cdot 24 \text{ h/day} \cdot 365 \text{ day/yr} \cdot 1500 \text{ yr}} \times 360° \approx$ $5.07 \times 10^{-13} = 0.000000000000507°$

24. **a.** $6\frac{1}{4}$ cups powdered sugar, $6\frac{1}{4}$ cups molasses, $2\frac{11}{32}$ (about $2\frac{1}{3}$) teaspoons each of salt and soda, $16\frac{13}{32}$ cups flour, . . . 1200 square inches, or $8\frac{1}{3}$ square feet

  c. For example, 16 bars per recipe, different amounts of ingredients per recipe or per 16 bars

25. About 55.9 miles per hour (475 miles traveled in 8.5 hours) (Why is the average larger than 55 mi/h?)

27. **b.** 6 days. There are different analyses possible. Here is one, using the compound unit, painter-day: It takes 6 painter-days to paint 1 house, so it will take 120 painter-days to paint 20 houses, which will take 20 painters 6 days.

  c. $14\frac{7}{12}$ days

Answers for Chapter 27

27.1 Understanding Chance Events

1. Remember that a probabilistic situation involves a certain process repeated a large number of times, *under the same circumstances.*

  a. Not probabilistic, because the secretary of state either is or is not a woman. *Reformulation*: The probability of a woman being among the next several U.S. secretaries of state. However, the fact that this choice is not made each

time under essentially the same circumstances means that it is not actually a probabilistic situation under any change in wording.

  b. Probabilistic (assuming that the class is not too small)

3. In an uncertain situation, you do not know what happened or is going to happen, or even whether the circumstances might differ in the future. In a probabilistic situation, some process will (in the future) be repeated in the same way a large number of times.

4. **a.** *Example of an experiment*: Actually tossing the coins and keeping track of the outcome.

  b. One outcome might be the penny showing heads, the nickel showing tails, and the dime showing heads.

  c. The event of getting exactly two heads can happen in three ways:

  The penny and nickel show heads and the dime shows tails.
  The penny shows heads, the nickel tails, and the dime heads.
  The penny shows tails, and the nickel and the dime show heads.

  ALL other outcomes would NOT be in this event.

27.2 Methods of Assigning Probabilities

1. **a.** The sex of a child is a two-outcome situation (male/female), as is the toss of a coin (heads/tails).

  b. The choice of answer is a two-outcome situation (true/false), as is the toss of a coin (heads/tails).

  c. The choice of winner is a two-outcome situation (Team 1/Team 2), as is the toss of a coin (heads/tails).

  d. The choice of winner is a two-outcome situation (Person 1/Person 2), as is the toss of a coin (heads/tails).

3. **a.** In a large number of similar patients with those symptoms, 90% had the disease.

  b. On a large number of days with the same conditions, it rained during 80% of the days.

4. Yes. A zero probability means that none of the outcomes are examples of the event—e.g., getting a 7 on a toss of a die.

5. Yes. A probability of one means that all of the outcomes are examples of the event—e.g., getting 12 or fewer on a toss of two dice.

9. **a.** The outcomes making up the sample space are $Q, R, S, T, U,$ and V. Each has theoretical probability $\frac{1}{6}$.

  b. The outcomes making up the sample space (and their probabilities) are $W(\frac{1}{2}), X(\frac{1}{4}), Y(\frac{1}{8})$, and $Z(\frac{1}{8})$.

10. You could spin the spinner 100 times (or any large number of times) and find the frequency for each outcome and divide by 100 (or the total number of spins). It is important that you be able to answer this question. If you cannot, review and contrast "experimental probability" and "theoretical probability."

11. Yes, it is possible to get 8 J's with 12 spins. With many, many spins, however, you would expect very close to half of them being J's. That is, you would expect the experimental probability to be close to the theoretical probability.

12. a. $\frac{1}{3}$　　**b.** $\frac{2}{3}$ (either from $\frac{1}{6} + \frac{1}{2}$, or from $1 - \frac{1}{3}$)

　　c. $\frac{5}{6}$　　**d.** 0　　　**e.** $\frac{1}{6}$　　　**f.** $\frac{5}{6}$

15. a. The spinner should be cut into five $360 \div 5 = 72°$ sectors.

　　b. Start with 6 equal sectors of $60°$ each, and omit one of the dividing radii.

17. a. Experimentally. Draw a sample of 20, say, and determine the fraction that is red.

　　b. Theoretically. Of the 6 outcomes, one outcome favors getting a 2, so $\frac{1}{6}$.

18. a. About 900 (How would you explain your reasoning?)

　　b. About 300

　　c. Yes, . . . (Your explanation is important.)

19. a. 6 reds and 7 blues, because in the long run you expect respective probabilities of $\frac{2}{5}$ and $\frac{6}{13}$, and $\frac{6}{13} > \frac{2}{5}$.

　　b. The chances are the same, because. . . .

　　c. 3 reds and 5 blues, because. . .

　　d. 623 reds and 623 blues, because. . .

26. a. The odds of making the field goal are 2 to 1.

27.3 Simulating Probabilistic Situations

1. The experimental probabilities should be close to the theoretical probabilities of $\frac{1}{6}$ for each outcome. A greater number of repetitions should give a value closer to the $\frac{1}{6}$ value.

2. a. Any appropriate simulation will do.

　　b. Green and red should show up the most, because their probabilities are greater than those for blue and yellow.

　　c. Theoretically, as you will find out later, the probability is $\frac{1}{18} \approx 0.0556$; the experimental results from the simulation for 1000 repetitions should be around that value.

　　d. In part (c), that the first spin gave the green was specified, whereas in part (d) . . .

　　e. Theoretically, $\frac{8}{9} \approx 0.89$, so experimental results for 1000 repetitions should be around that value. The easy way uses $P(\text{not green twice in a row}) = 1 - P(\text{green twice})$.

3. Results for curing none of the five should be around 0.237. It is possible that all five could be cured, but unlikely ($\frac{1}{1024} \approx$ 0.00098). With 10,000 repetitions, the results should be closer to the theoretical values given earlier in this answer. (One simulation might involve taking five random digits at a time and assigning the values 1 and 2 to cure, the values 3–8 to not-cure, and passing over any 9s or 0s.)

Answers for Chapter 28

28.1 Tree Diagrams and Lists for Multistep Experiments

2. Your tree diagram should show the 12 outcomes (why is it 12?); in a common notation, they are the following:
(1, H), (1, T), (2, H), (2, T), (3, H), (3, T), (4, H), (4, T),

(5, H), (5, T), (6, H), (6, T). Each has probability $\frac{1}{12}$, because we are assuming an honest die and an honest coin.

3. The condensed tree diagrams might look like these:

Experiment a	Experiment b
(with replacement)	(without replacement)

First draw	Second draw	Outcomes and prob.	First draw	Second draw	Outcomes and prob.

Experiment a:
- 0.4 → R → 0.4 → R RR 0.16
- R → 0.6 → G RG 0.24
- 0.6 → G → 0.4 → R GR 0.24
- G → 0.6 → G GG 0.36

Experiment b:
- 0.4 → R → 3/9 → R RR 2/15
- R → 6/9 → G RG 4/15
- 0.6 → G → 4/9 → R GR 4/15
- G → 5/9 → G GG 1/3

4. b. If the dice are "honest," there is no reason to expect one outcome to be more likely than another.

$P(\text{sum} = 2) = \frac{1}{36} = P(\text{sum} = 12); P(\text{sum} = 3) = \frac{1}{18};$

$P(\text{sum} = 4) = \frac{1}{12}; P(\text{sum} = 5) = \frac{1}{9}; P(\text{sum} = 6) = \frac{5}{36};$

$P(\text{sum} = 7) = \frac{1}{6}; P(\text{sum} = 8) = \frac{5}{36}; P(\text{sum} = 9) = \frac{1}{9};$

$P(\text{sum} = 10) = \frac{1}{12}; P(\text{sum} = 11) = \frac{1}{18}.$

What are $P(\text{sum} = 1)$ and $P(\text{sum} = 13)$?

Young children may focus on the numerical size of the number and often think that larger numbers or larger sums are more likely.

　c. Yes. The two dice are different dice, whether they are colored differently or not. Hence, 2 on die X, 3 on die Y is a different outcome from 2 on die Y, 3 on die X.

7. a. If we go from left to right in the tree, with the spinners in numerical order, the first level has two branches, each with probability 0.5. The next level has three branches, two labeled 0.25 for red and black and one 0.5 for white. The third level will have three branches, each with probability $\frac{1}{3}$. [Would a tree with a different order of spinners give different results in parts (b–d)?]

　b. Each of RRR, RRB, RRW, RBR, RBB, RBW, BRR, BRB, BRW, BBR, BBB, and BBW has probability $\frac{1}{24}$; each of RWR, RWB, RWW, BWR, BWB, and BWW has probability $\frac{1}{12}$.

8. a.
HH1	HH2	HH3	HH4	HH5	HH6
HT1	HT2	HT3	HT4	HT5	HT6
TH1	TH2	TH3	TH4	TH5	TH6
TT1	TT2	TT3	TT4	TT5	TT6

　b. $\frac{6}{24}$ or $\frac{1}{4}$

28.2 Probability of One Event *or* Another Event

1. a.
H1	H2	H3	H4	H5	H6
T1	T2	T3	T4	T5	T6

　b. $\frac{1}{12}$　　**c.** $\frac{2}{12}$　　**d.** $\frac{3}{12}$ or $\frac{1}{4}$

4. a.

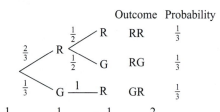

Outcome   Probability

R, RR	$\frac{1}{3}$
RG	$\frac{1}{3}$
GR	$\frac{1}{3}$

b. $\frac{1}{3}$ **c.** $\frac{1}{3}$ **d.** $\frac{1}{3}$ **e.** $\frac{2}{3}$

6. a.

Outcome   Probability

Outcome	Probability
BB	$\frac{1}{12}$
BR	$\frac{1}{12}$
BG	$\frac{1}{12}$
BY	$\frac{1}{4}$
RB	$\frac{1}{24}$
RR	$\frac{1}{24}$
RG	$\frac{1}{24}$
RY	$\frac{1}{8}$
YB	$\frac{1}{24}$
YR	$\frac{1}{24}$
YG	$\frac{1}{24}$
YY	$\frac{1}{8}$

b. $\frac{1}{24}$ **c.** $\frac{1}{12} + \frac{1}{24} + \frac{1}{8} = \frac{6}{24} = \frac{1}{4}$

7. a. BBR   BBY   BGR   BGY   BYR   BYY

RBR   RBY   RGR   RGY   RYR   RYY

GBR   GBY   GGR   GGY   GYR   GYY

YBR   YBY   YGR   YGY   YYR   YYY

Each outcome has probability $\frac{1}{24}$.

b. $P(\text{all one color}) = P(\text{YYY}) = \frac{1}{24}$; $P(\text{red on one spinner})$

$= \frac{1}{2}$ (interpreting "red on one spinner" to mean on *exactly* one). There is no overlap so the probability for part (b) is $\frac{1}{24} + \frac{1}{2} - 0 = \frac{13}{24}$. If you interpreted "red on one spinner" to mean "red on *at least* one spinner," your answer for part (b) would be $\frac{1}{24} + \frac{15}{24} - 0 = \frac{2}{3}$.

c. $P(\text{yellow on all three spinners}) = \frac{1}{24}$; $P(\text{red on all three}) = 0$

There is no overlap. So

$P(\text{yellow on all three or red on all three}) = \frac{1}{24}$

d. $P(\text{blue on at least one spin}) = \frac{12}{24}$ and

$P(\text{yellow on at least one}) = \frac{18}{24}$

$P(\text{blue on at least one AND yellow on at least one}) = \frac{8}{24}$

$P(\text{blue on at least one or yellow on at least one})$

$= \frac{12}{24} + \frac{18}{24} - \frac{8}{24} = \frac{22}{24}$

Note that this answer can be found also by counting all outcomes with at least one blue or one yellow. There are 22 such outcomes.

10. a. $\frac{9}{36}$ **b.** 0   **c.** $\frac{10}{36}$ **d.** 1

28.3 Probability of One Event *and* Another Event

1. See the narrative for the definitions. Try to come up with new examples.

2. No, it applies only when A and B are independent.

6. The independent events are in parts (a, d), and one other.

8. **a.** The probability that all three systems work is 0.84645, assuming. . . . Hence, the probability that at least one fails is . . .

   b. That probability means that with a very large number of launches, _____ will be aborted.

9. **a.** $P(A) + P(B)$ **b.** $P(A) + P(B) - P(A) \times P(B)$

   c. $P(A)$, because . . . (These are not worth memorizing.)

28.4 Conditional Probability

1. Suppose you have a population of 10,000. After you make the contingency table using the information you have (test is 95% accurate, 5% of the general population uses drugs), you should find that 950 of the 10,000 tested positive, and of those 475 (or 50%) do not use drugs. *Additional thought questions*: Would your conclusions be different if you had started with a population of, say, 5000? Would your conclusions be different if a larger portion of the population of, say, 8%, used drugs? *Reflection*: What if an Olympic athlete tested positive for marijuana use? Does this new understanding affect your perception of the situation?

2. **a.** One approach is to make a tree diagram to identify the sample space for 4 children in a family in order of birth (there are 2^4, or 16, possible outcomes). From there, you can determine the sample space for the event that at least 1 of the children is a girl (the outcome BBBB is eliminated), and with 6 outcomes having two of each gender, the probability of 2 girls and 2 boys is $\frac{6}{15} = \frac{2}{5}$. If you did not know that at least 1 child was a girl, the sample space would have all 16 outcomes, and the probability of 2 girls and 2 boys would be $\frac{6}{16} = \frac{3}{8}$. What assumptions have you made? (Think about independence, theoretical probabilities.)

   b. $P(\text{2 daughters}) = \frac{1}{3}$;

   $P(\text{2 daughters} \mid \text{older child is a daughter}) = \frac{1}{2}$.

   (List all the outcomes.)

4. Choosing from the pool *at random* gives a probability of 0.7 that the person is a lawyer and 0.3 that the person is an engineer.

6. The cards can be represented as R/R, R/W, and W/W. When the clown draws one card at random and you see a red side, you know it is not the W/W card. Thus, there are three sides left: 2 reds and 1 white. The other side of the card showing could be any one of those three sides, so the probability is not $\frac{1}{2}$. What is it?

8. a. Recall that if events A and B are independent,

then $P(A \text{ and } B) = P(A) \times P(B)$. Thus,

$P(A \mid B) = \dfrac{P(A \text{ and } B)}{P(B)} = \dfrac{P(A) \times P(B)}{P(B)}$, and the

$P(B)$ cancels out, leaving $P(A \mid B) = P(A)$.

b. No, unless $P(A) = P(B)$.

11. a.

	High-Income	Low-Income	
Support vouchers	30	160	190
Oppose vouchers	50	160	210
	80	320	400

b. $P(\text{oppose vouchers} \mid \text{high-income}) = \dfrac{50}{80} = 62.5\%$

c. $P(\text{high-income} \mid \text{oppose vouchers}) = \dfrac{50}{210} \approx 24\%$

d. $P(\text{high-income or opposes vouchers})$

$= \dfrac{80}{400} + \dfrac{210}{400} - \dfrac{50}{400} = \dfrac{240}{400} = 60\%$

Answers for Chapter 29

29.1 What Are Statistics?

1. The first word "statistics" refers to the discipline, the second refers to the numbers studied.

2. a. Samples: Percent of games won, average weight of players on the team, total number of points scored, total number of yards gained by passing, . . .

b. Samples: Average age, class average on last test, number of students in the class, number of siblings for the whole class, . . .

29.2 Sampling: The Why and the How

2. It would not be possible to select a completely random sample unless a list of all students could be obtained. If you think that almost all students eat on campus, or use the library, you could take a convenience sample at one of those places. Otherwise, you might try a stratified sample of dorm students and off-campus students (by selecting students from those who leave the parking lot). The population is the entire student body.

5. a. If the stereotype of fraternities (that excessive drinking is widespread) is true, this sample would be biased.

b. Stratified (random) sampling; The sample should be unbiased unless the samples from each group are not selected in some unbiased way (e.g., randomly).

c. If the two groups have about the same number of people, this sample should be all right. If not, the larger group would be underrepresented.

d. Although a stratified random sample, the choice of religious groups could bias the results about alcohol abuse. There is also the likelihood that the religious groups are of quite different sizes, so a very small group would be overrepresented and a very large group underrepresented.

e. Convenience sampling; There could be a bias for several reasons: Do students patronize the student center because alcohol is served there? What time of day would the polling be done? What if, say, the sample is drawn right after a large class populated mostly by a group likely to include alcohol abusers?

7. a. If an important decision that affects only a few people is to be made, all the people should be polled. For example, a family is deciding on a vacation locale. Or, a new drug has been tried on only 12 people.

b. Any assembly-line product, or particularly troublesome parts of an assembly line

c. If you want to know how people in a 15-block neighborhood feel about a new mall being planned for the neighborhood, you could randomly select three of the 15 blocks and then interview all the residents of those blocks. The information gathered in this way should not be biased in any particular way.

8. a. 27% refers to all people, and so it is a population parameter. (It may have been arrived at by a sample, but not necessarily. It might have been arrived at through the last census.)

b. 85% refers to the percent of students *polled* in a sample, so 85% is a sample statistic.

9. The cafeteria is most likely to contain a cross section of the student body. The algebra and French classes might represent just a segment of the student body, as might the students in the guidance office or the faculty room.

29.3 Simulating Random Sampling

2. Look at strings of three-digit numbers. For example, suppose you begin with the 4th row of the 4th column of the TRSD. Ignore numbers greater than 500. Moving down, take the first thirty numbers from 001 through 500. (Ignore 675, 508, 521, 904. The first number selected is 445. The second one is 106. The third and fourth are 171 and 338, etc.)

4. Human-made "random" samples usually are not random. For example, although genuinely random numbers will include consecutive repetitions of the same number, humans often avoid such repetitions.

29.4 Types of Data

1. a. Categorical   **b.** Measurement

2. Categorical data. The numbers are just codes for computer purposes.

Answers for Chapter 30

30.1 Representing Categorical Data

3. Typically, a graph has a title. What would be a reasonable title for this graph?

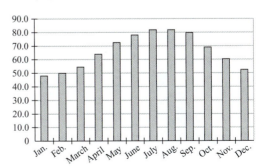

5. Here is most of the display. The color coding may not show up clearly here.

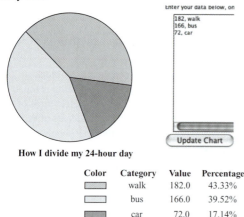

How I divide my 24-hour day

Color	Category	Value	Percentage
	walk	182.0	43.33%
	bus	166.0	39.52%
	car	72.0	17.14%

Sum of values = 420.0

Comparing the graphs for Learning Exercises 4 and 5 suggests that in Jasmine's class, a noticeably greater percent ride the bus and a smaller percent come by car than in the whole school.

7. a. Both length and width are doubled for the larger drawing, leaving the visual impression of being 4 times as large as the smaller drawing, when the actual enrollment is only twice as much.

   b. The graph leaves the impression that enrollment changes are much greater than they actually are.

30.2 Representing and Interpreting Measurement Data

3. e. Histogram, line graph, or stem-and-leaf plot

   f. Histogram, or stem-and-leaf plot

   g. Histogram (if the number of types of candy bars with grams in certain intervals is of concern) or stem-and-leaf plot; bar graph if focus is on grams versus type of candy bar

   h. Histogram, line graph, or stem-and-leaf plot. The first two would be easiest to make as the weights are made.

4. a. *Note:* Excel can order the data for you, if you are able to use it (go to Data in the menu heading at the top of the screen).

0	00023456668888
1	0000012222244444566666668
2	0000022222444668
3	022
4	0

6. Graphs may differ if an interval size $\neq 4$ is chosen.

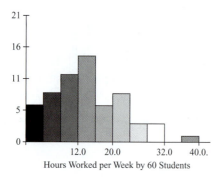

Hours Worked per Week by 60 Students

7. You may have chosen a different length for the cell width. With a maximum of 8, it is reasonable to have the cell interval of 1 hour. The Histogram Tool includes counts of values at the upper end of the interval in a particular cell.

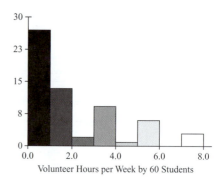

Volunteer Hours per Week by 60 Students

30.3 Examining the Spread of Data

1. a. i. Quartiles make sense because the data set is numerical and large enough for quartile scores to make sense.

   ii. Quartiles make sense because the data set has a large number of numerical data values.

   iii. Quartiles do not make sense because the data are nonnumerical (categorical).

   iv. Quartiles do not make sense because the data are nonnumerical (categorical).

   v. Quartiles make sense because the data set is measurement and has a large number of data values.

It does not make sense to talk about quartile scores for all numerical data sets. In fact, for small numerical data sets, quartile scores are often inappropriate. For example, finding all three quartile scores for a data set containing five numbers is not useful. Therefore, finding the quartile scores for the number of M&M's of each color in a bag is not very meaningful. However, finding the median of these numbers may be meaningful, suggesting a representative number of M&M's for each color.

   b. SAT tests, GPAs for a graduating class, GRE scores, e.g.

   c. Hair color of students in your class, e.g.

   d. A large number of data values that are numerical and can be ordered

3. a. low: 131; 1st quartile: 139.5; 2nd quartile (median): 146; third quartile: 150; high: 164

   b. You would still be leaving out the heights of 50% of the girls.

5. a. Has no left whisker: 5, 5, 5, 6, 6, 7, 7, 8, 8, 9, 9, e.g.

   b. Has no whiskers at all: 5, 5, 5, 6, 6, 7, 7, 7, 8, 8, 8, e.g.

   c. Has an outlier: 1, 7, 8, 9, 9, 10, 10, 11, 11, 13, 15.

   d. Has a box made of only one rectangle, not two: 7, 7, 7, 7, 7, 7, 7, 8, 8, 8.

   e. Has no box: 8, 8, 8, 8, 8, 8, 8, 8, 8, 8, 8.

   f. The following answers apply to the examples given in parts (a-d):

     a. The first quartile is the same as the minimum.

     b. The minimum is the same as the first quartile and the third quartile is the same as the maximum.

c. The IQR is $11 - 8 = 3$. $1.5 \times 3 = 4.5$. One value is more than 4.5 from 8.

d. The median is the same as the first quartile, but the third quartile is different. (This could be different. How?)

e. All the values are the same.

6. a. It is difficult to judge without knowing more about the nature of the test. But the scores look typical for a test out of 100.

b. The whisker for the first quartile scores does not show exactly where the scores are. The first quartile score itself does appear to be above 70. Thus, the 25% of the 32 scores, or 8, could almost all be in the low 70s, but all *could* be below 70. So we can say that at most 8 of the 32, plus the outlier for a total of 9, could be below 70, or if there is just one low score within the first quartile scores (with the others in the 70s), that score and the outlier would give only 2 scores below 70.

c. About 5

d. Perhaps the aide recorded this test score incorrectly on the computer, and the student actually scored a 95 rather than a 5. Mr. Meyer will probably go back and look at the test scores as they are recorded on each student's test!

8. Without more marks on the scores, the weights are just estimates. With 60 people involved, we will be looking for weights that give four intervals with about 15 people in each. About 10 women (W) weigh less than about 120, plus a few men (M) and more of the W to make 15 people, so the 1st quartile is about 122, say. Similarly, the median might be around 130, and the third quartile about 150. The low woman's weight is about 75 pounds, and the high man's weight is about 255 pounds. $1.5 \times (150 - 122) = 42$, so the 75 is an outlier and the low weight is 100; the 255 is also an outlier, so the high weight is about 215.

13. Here is most of the display.

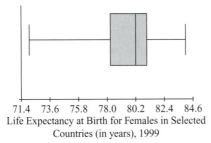

71.4 73.6 75.8 78.0 80.2 82.4 84.6
Life Expectancy at Birth for Females in Selected
Countries (in years), 1999

30.4 Measures of Center

1. a. Yes ($8, $9, $10, $11, $12, e.g.)

b. Yes ($8, $8, $9, $11, $14, e.g.)

c. Yes ($1, $2, $3, $6, $38, e.g.)

d. Yes (Is there more than one set of five prices that meets this condition?)

e. Yes [see answer to 1(c), e.g.]

f. No. If one thinks of the mean as a balancing point of the data set, then the mean would have to either be a value in the data set or a value that falls between the largest and smallest value in the data set.

4. a. The mean, because a large percentage of homes may range in price from $100,000–$500,000, but a small percentage of homes may have prices in the millions of dollars. The prices of these more expensive homes may be outliers, which tend to affect the mean (in this case, increase the mean) much more than they affect the median.

5. 18, 18, 19, 35, 40, e.g. Create your own set.

6. c. The mean of the ratio of the dimensions of the rectangles is 0.6605, the median is 0.641, and the mode is 0.606. However, without the outliers, the mean is 0.635 and the median is 0.622. It is likely that the Shoshonis also found the shapes of golden rectangles appealing, because the mean, median, and mode are fairly close to the ratio of 0.618034, particularly if outliers are disregarded.

d. The choice of graph is important. A histogram or box plot will show the spread.

8. A mean of 79.8 points can be interpreted as meaning that each person would have scored 79.8 points, if all the points for the class had been distributed equally.

9. The means are 11.5 points per game for the Amazons, 12 points per game for the Bears, and about 11.4 points per game for the Cougars. So one could argue that the Bears are best.

10. Because the mean gives a per-person or per-item value, it gives a way to avoid the complication from having groups of different sizes.

13. $83.43 per day

17. a. 34

c. The median, because of the relatively large number at the younger ages.

18. Histogram (a) goes with box plot (C); (b) with (B); (c) with (A); and (d) with (D).

30.5 Deviations from the Mean as Measures of Spread

2. a. B because of the one low value and/or one higher value

b. D because of the very low value at the end

c. Both B and D have a great deal of variation, so it is hard to tell.

d. C because there is very little variation among the numbers

4. The mean is 62.8, the standard deviation is 12.6. The score 1 deviation below the mean is 50.2, and the score 2 standard deviations below the mean is 37.6. The score 1 standard deviation above the mean is 75.4, and the score 2 standard deviations above the mean is 88. Therefore, the 2 students with scores below 37.6 ($62.8 - 25.2$) would receive an F, and the 6 students with scores between 37.6 and 50.2 would receive a D. The 28 students with scores between 50.2 and 75.4 would receive a C. The 9 students with scores between 75.4 and 88 would receive a B. A student with a score 88 or above would receive an A, but there were no such scores.

There would be 0 A's, 9 B's, 28 C's, 6 D's, and 2 F's.

6. a. Mean 2.74, standard deviation 0.65 (or possibly 0.66, depending on the technology)

b. Mean 2.99, standard deviation 0.60 (or possibly 0.62, depending on the technology). The mean for the males is somewhat higher, and their scores are slightly less spread out.

7. a. The second average deviation would be *numerically* less than the first average deviation because a kilogram is a bigger unit than a pound. The bigger the unit, the smaller the values of the weights. Because the values of the weights will be smaller, the value of the mean and all the other data values will also be *numerically* smaller. Thus, the second average deviation will also be numerically smaller. Taking units into account, the statistics are the same.

30.6 Examining Distributions

2. b. 2, 2, 2, 2, 3, 4, 5, 8, 10, 12, 12, 12, 12, 12, 12, e.g. This data set is not normally distributed because the mean is 7.333, the median is 8, and the mode is 12. Because the mode and the mean are relatively spread apart, and because most of the data values are not located relatively close to the mean, this data set is not approximately normally distributed. The data set gives two humps.

3. The mode in Class A was about 90; for Class B, 40; and for Class C, a 90. By and large, the students in Classes A and C outscored the students in Class B by a wide margin. The students in Class A tended to do better than those in Class C. In Class C the second most common score was 58, while there were only a small number of students who scored 70 or 80 on the exam. The teacher of Class C might wonder whether the students could be grouped into "studied" and "didn't study" categories. What other reasons might lead to such a "bimodal" distribution?

4. a. No, a negative z score just means that the score was below the mean.

   b. The z score that corresponds to the mean is 0.

6. a. Apparently not. Although the mean (12.87) and median (13) are about the same, the mode (11) differs. The shape of the histogram looks skewed right, rather than being symmetrical.

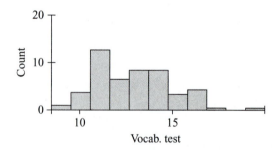

   b. The large number of occurrences of 11 makes it difficult to use the rule. With that exception, the rule might be applicable.

9. Symmetric distributions with two or more "humps" would not be normal. Such a two-humped ("bimodal") distribution might result from scores on a test for which several students studied but several others did not study.

15. *Advantages of converting raw scores to z scores:* It's easier to make comparisons across different normal distributions. *Disadvantages of converting to z scores* (i.e., is any information lost?): As in Exercise 14, you lose information about the data, such as the mean score on a test, or the mean height of American women, or some measure of how spread out the heights are in inches.

16. a. Distributions P and Q have the same mean. In graphs of normal distributions, the mean is the score corresponding to the highest point in the graph.

   b. Distribution P has the greatest standard deviation. Distributions Q and R appear to be equally spread out, whereas P is spread out more and that means its standard deviation is greater than that of the other two.

30.7 Issues for Learning: Understanding the Mean

1. $1.80

4. a. The child would probably pick 19, because there are more 19s than any other number. If the last number had been 12, there would be three 19s and three 12s, so it is hard to tell what the child might say—perhaps that there are two averages, $12 and $19.

   b. The child might say something such as "Well the lowest amount was $7 and the highest amount was like $21 so it would be like in the middle, maybe $14." (There might be, of course, several ways students might answer.)

   c. One possible answer among many might be "Yesterday I went to one of those places with my dad and he bought some milk and some cookies and some bread and a six-pack of Coke and he had to pay $9.00. So that was about average."

   d. "$14 is in the middle between $7 and $21 and there are about the same number of numbers between 7 and 14 as between 14 and 21."

   e. "I counted up the money spent and it was $221, and there were 15 amounts, so I think it would be $221 ÷ 15, which is about $15."

Answers for Chapter 31

31.1 Comparing Data Sets

3. a–b. Students may offer bar graphs or pie charts. You can also use the table below to make some comparisons. *Tacky the Penguin* (T) and *Kat Kong* (K) appear to be the overall favorites, although Mr. Lopez's class heavily favored *Tacky*. Students may offer the last bar graph that follows to support T for part (b).

Kindergarten books. txt

		Teacher			Row Summary
		Chen	Lopez	Wilson	
Favorite	K	7	4	8	19
	L	2	7	5	14
	R	3	0	1	4
	S	3	2	2	7
	T	5	10	6	21
Column Summary		20	23	22	65

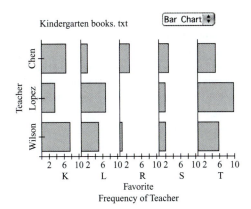

Kindergarten books. txt | Bar Chart ▼ |

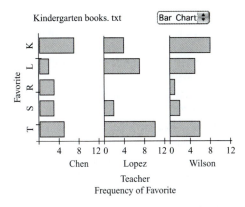

Kindergarten books. txt | Bar Chart ▼ |

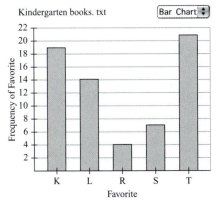

Kindergarten books. txt | Bar Chart ▼ |

c. A teacher's enthusiasm or some special contribution by a class member may have influenced the class' decisions.

6. Here is a sample. Cholesterol reading could be on the horizontal axis, but it is usually predicted from weight.

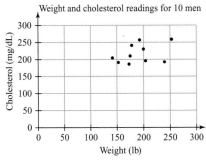

7. a. Box plots for the house prices in the two cities might give Easton enough information. A histogram, with its visual feedback, might be a good second choice.

b. A histogram or box plot of each set of test scores, because both may give her a sense of what a typical score might be and how spread out the test scores were. She could use a scatter plot of the students' two test scores to address the "associated?" question.

c. A scatter plot of the number of assignments turned in versus time would be useful in studying the trend.

d. Unless Stella sees a trend in the April rainfalls in recent years, either a histogram or a box plot, or even a stem-and-leaf plot, for the April rainfalls for the 30 years should be useful in judging whether or not this April is quite unusual.

e. Sanders would probably look at a scatter plot for the two types of costs. If the advertising campaign was in effect only part of the year, he might prefer separate costs-versus-month line graphs.

8. a. One of the couples consists of a woman who is 1590 mm and a man (her husband) who is 1735 mm.

b. Women who are taller tend to marry shorter men, while women who are smaller tend to be married to taller men.

c, e. They are generally positively associated. Women who are taller tend to be married to taller men, while women who are shorter tend to be married to shorter men.

d. The height of a wife would bear little relationship to the height of her husband. For example, knowing that a wife was short would give no information about the likely height of her husband.

f. Here, the general positive association is not distorted by the changes in scale.

10. a. No association

b. Positive association (As the first gets larger, so does the second.)

c. Negative association (As the first gets larger, the second gets smaller.)

d. Negative association

12. Here is a sample:

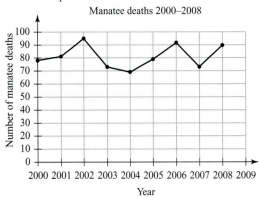

31.2 Lines of Best Fit and Correlation

1. d. Negative because the highest-priced cars often do not get good mileage. (An argument of "very close to zero" might be defended by pointing out that special features, like hydrogen fuel cars, are costly, but the correlation would not be -1.)

   e. Perhaps surprisingly, faculty experience suggests a negative correlation coefficient under unlimited time conditions (for exams intended for a set period of time). For speeded tests, with a strict time limit, the correlation coefficient is likely to be positive, although not $+1$.

   f. Positive, although probably not $+1$

2. a. Humans (62, 1320), Asian elephants (2547, 4603), and African elephants (6654, 5712)

   b. Such a high positive correlation coefficient means that as body weights increase, so do brain weights.

   c. Humans are at (62, 1320) and appear not to fit the line of best fit very well, so they can be regarded as outliers (note this is not the same sense as "outlier" for a box plot).

   d. Reading the graph will give an estimate, but the equation will give a more exact value:

   Brain weight = 0.94(2000) + 191.22
   = 2071.22 grams, about 2070 grams

   e. There is a strong positive association between the two weights. This is not surprising, because one would expect a larger animal to have a larger brain, in general.

3. a. Many points do not seem as close to the line as before. The equation is a little different, but the "outlier" status of humans is very obvious.

   b. The lower correlation coefficient, $\sqrt{0.29} \approx 0.54$, suggests a lower degree of association, so predictions may be off. (Including the outliers made the correlation seem stronger.)

   c. The regression line should fit the data better.

5. a. $\sqrt{0.81} = 0.9$

   b. About 159; about 185 (Why "about"?)

   c. 200

   d. No. Correlations do not automatically mean causation.

Answers for Chapter 32

32.1 Having Confidence in a Sample Statistic

3. a. $\frac{85}{1000}$, or 8.5%   **b.** $\frac{240}{1000}$, or 24%

   c. $\frac{335}{1000}$, or 33.5%   **d.** $\frac{421}{1000}$, or 42.1%

   Notice that even with a large sample, one cannot expect to "hit" the population parameter exactly.

4. A sample of size 20 cannot be expected to give a good confidence interval. Using the $\frac{1}{\sqrt{n}}$ rule of thumb, the

   $\pm \frac{1}{\sqrt{20}} \approx \pm 22\%$ gives a confidence interval covering 44%.

5. Low bias and high precision, because both conditions allow one to be more confident of the sample statistics.

8. a. Locating everyone, getting them to agree to be questioned, not counting anyone twice, training a large number of conscientious interviewers, developing appropriate questions, completing the census in a timely fashion, keeping track of all the data, etc.

   b. Homelessness, finding people, language difficulties, e.g.

   c. Have fewer representatives in Congress, do not get fair share of federal dollars that are distributed according to populations, e.g.

   d. Mobility (people moving during the census, e.g., or being at someone else's home when they were being interviewed), claiming more people in a residence than there were, e.g. Overcounts could result in the opposite effects as those in part (c): Having more representatives in Congress, getting more than the appropriate share of federal dollars, e.g.

   e. *Pro*: Much less expensive; if done carefully, it might even be more accurate.

   Con: People (especially Congress) might not trust the outcomes; tradition; . . .

10. With a sample of size 20, the number of browns could vary from, say, 4 to 12 (18% to 62%), or even more. With a sample of size 100, the number of browns should vary less percent-wise, say, from 30% to 50%.

32.2 Confidence Intervals

2. The mailings apparently did not give a representative sample, especially with a self-selected 23% returning the sample ballots (23% is not unusual in such a sampling). In 1936 those subscribing to magazines, or owning a car or telephone, may have been a richer part of the population and so gave a biased sample to start with.

4. With the margin of error in mind, the percent of the population supporting a change would be in the 51% to 55% interval. If the matter were to be voted on, there is a good chance that there would be some changes to Social Security. What those changes might be would affect your reaction.

5. a. $52\% \pm 33\frac{1}{3}\%$, or $18\frac{2}{3}\% - 85\frac{1}{3}\%$

   b. $52\% \pm 10.5\%$, or $41.5\% - 62.5\%$

Answers for Chapter 33

33.1 Expected Value

2. 7; this means . . .

4. Remember that from your viewpoint, when someone makes a basket, you lose 49¢. You would charge 30¢ or more. [Suppose you charge c cents. If the person makes a basket, you have gained c cents and lost 49¢. If the person does not make a basket, you have gained c cents. You want the expected value to be ≥ 10¢. The expected value is $(c - 49)0.4 + (c)0.6 = 0.4c - 19.6 + 0.6c = c - 19.6$. So if $c - 19.6 \geq 10$, then $c \geq 29.6$.]

6. a. Assuming that the $1000 does not count in determining expected value except to determine the value of "double your money," the expected value is

$$\frac{1}{16}(800 + 200 + \cdots + 100 + 1000) = \frac{7000}{16} = \$437.50$$

b. With a large number of spins, the contestant can expect to win an average of $437.50 per spin.

c. $\dfrac{6000 + x}{16}$ dollars

33.2 Permutations and Combinations

1. Order is . . .

   a. Important   **b.** Not important   **c.** Important

   d. Important   **e.** Not important

3. a. *Act 1*: Choose one of the 5 children for the first seat.
   Act 2: Choose one of the remaining 4 children for the second seat; . . .

   OR

   Act 1: Assign a first child to one of the 5 seats.
   Act 2: Assign a second child to one of the remaining 4 seats; . . .

b. *Act 1*: Assign a first child to one of the 7 seats.
   Act 2: Assign a second child to one of the remaining 6 seats; . . .

c. Both part (a) and part (b) involve permutations.

4. a. 2520

b. (Order does not matter.) 21

10. d. 6

   e. 252

   f. $\dfrac{20 \cdot 6}{252}$, or . . . (Why are the 20 and 6 multiplied?)

11. a. There are $\dfrac{100 \cdot 99 \cdot 98 \cdot 97 \cdot 96 \cdot 95}{6 \cdot 5 \cdot 4 \cdot 3 \cdot 2 \cdot 1} = 1{,}192{,}052{,}400$ combinations, choosing 6 numbers from 00 to 99. Only one of those will be the winner, so the probability of winning is $\dfrac{1}{1{,}192{,}052{,}400}$.

   b. $\dfrac{1}{1{,}192{,}052{,}400} \cdot 1{,}000{,}000 + \dfrac{1{,}192{,}052{,}399}{1{,}192{,}052{,}400} \cdot 0 \approx$ 0.0008 dollar, but the $2 cost means you can expect to lose all but a smidgen of your $2.

12. *Needed*: The probabilities of hitting exactly 8, exactly 9, and exactly 10 (and then add those).

$$45 \cdot 0.85^8 \cdot 0.15^2 + 10 \cdot 0.85^9 \cdot 0.15^1 + 1 \cdot 0.85^{10}$$
$$\approx 27.6\% + 34.7\% + 19.7\% \text{ or about } 82\%$$

13. a. 5040   **b.** 3,628,800   **c.** $6! = 720$

   d. 720. Note that the common factor 7! allows the fraction to be simplified.

   e. 495   **f.** 28

15. When $r = n$, $(n - r)! = 0!$, and what does that mean? So that the expressions make sense and still apply in cases involving 0!, 0! is defined to be $= 1$.

A number such as [1.1] at the end of an entry refers to Section 1.1, the section in which the entry first appears in different parts. Terms that are related to some noun or are adjectives may often be found with the associated noun. Some terms first appear in exercise lists, but are nonetheless important. The website http://nw.pima.edu/dmeeks/spandict/ currently gives English-to-Spanish translations of many mathematical terms.

68–95–99.7 rule—for a **normal distribution**, 68% of the scores are within one standard deviation of the mean; 95% are within two standard deviations of the mean; and 99.7% are within three standard deviations of the mean. [30.6]

abacus—a mechanical calculating device, still used in parts of the world. [2.2]

absolute value function—a function with function rule of the form $y = |x|$. For any real number x, the function rule returns x if $x \geq 0$ and the opposite of x if $x < 0$. [13.4]

absolute value of a number—on a number line, the distance of the number from 0; operationally, the number without a positive or negative sign. Denoted $|x|$. [10.4]

addends—numbers that are added; in $3 + 5 = 8$, 3 and 5 are addends. [3.2]

addition rule for probability—for events A and B, $P(A \text{ or } B) = P(A) + P(B) - P(A \text{ and } B)$. "Or" allows either *or both* events to occur. "And" is used in the sense of *at the same time*, rather than addition. [28.2]

addition/subtraction method—one method for finding the common solution to a system of equations. Contrast the **substitution method**. [15.2]

additive combination of quantities—the quantities are put together, either literally or conceptually, to form a new quantity. [3.1]

additive comparison of two quantities—the quantities are compared in a how-much-more or how-much-less sense. See **difference**. [3.1]

additive identity—that number (0) that, when added to any other number, gives a sum = the other number: $0 + a = a$ for all choices of a. [10.2]

additive inverse of a number—the number that, when added to the given number, gives a sum = 0. The numbers $^-2$ and 2 are additive inverses of each other, because $^-2 + 2 = 0$. Additive inverses are sometimes called opposites. [10.1]

algorithm—a systematic method for carrying out some process; e.g., the way you usually multiply 28 and 35 is an algorithm for multiplying whole numbers. [3.3]

altitude—see **height**. [25.1]

angle—(1) two rays from the same point; (2) a ray along with the result of its turning on its endpoint to a final position. The common point is its **vertex**, and the rays are its **sides**. [16.1, 23.2]

    acute angle—an angle smaller than a right angle (i.e., has size less than 90°). [16.1]

    adjacent angles—angles with a common side between them. [16.1]

    alternate interior angles—angles like those marked x and x' and also marked y and y' in the drawing on the next page. [23.2]

    central angle—an angle with its vertex at the center of a circle. [21.1]

    complementary angles—two angles having sizes that sum to 90°. [16.1]

    corresponding angles—(1) a pair of angles like those marked a in the drawing on the next page (also those marked b, those marked c, or those marked d in the drawing on the next page); (2) in congruent or similar shapes, angles that match. [23.2]

    dihedral angle—two half-planes from the same line. [23.2] (Learning Exercises there)

    exterior angle of a polygon—an angle formed by one side of a polygon with the extension of another side that passes through an endpoint. [16.1]

    face angle of a polyhedron—an angle of any polygonal face of the polyhedron. [16.3] (Learning Exercises there)

    inscribed angle—an angle with its vertex on a circle, and its sides chords of the circle. [21.1]

    interior angle of a polygon—an angle formed by two adjacent sides of the polygon. [16.1]

    obtuse angle—an angle having size between 90° and 180°. [16.1]

right angle—an angle of 90°. [16.1]

straight angle—an angle with its sides along a straight line (i.e., has size 180°). [16.1]

supplementary angles—two angles with sizes that sum to 180° (e.g., *x* and *y* in the diagram below). [16.1]

vertical angles—the nonadjacent angles formed when two lines cross (e.g., *a* and *d,* also *b* and *c,* in the diagram below). [23.2]

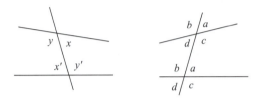

apex—in a pyramid, the vertex shared by all the lateral faces. [17.2]

arc of a circle—see **circle**. [21.1]

are—a metric system unit for **area** (100 square meters). [24.1]

area of a region—the number of square units that would be required to cover the region. The region could be a two-dimensional region, or it could refer to all the surfaces of a three-dimensional figure, in which case it is called the **surface area**. [24.1] Formulas for determining areas. [25.1]

arithmetic sequence—a pattern of numbers in which the same number, called the **common difference**, is added to (or subtracted from) any number in a list to get the next number in the list. [12.3]

array—a rectangular arrangement of objects. The **array model of multiplication** occurs in cases that can be modeled as a rectangle *n* units across and *m* units down. [3.4]

association between two variables—two variables are **positively associated** if a high value for one variable goes with a high value for the other variable, and low values go together in a similar way. The variables are **negatively associated** if an increase in one variable is accompanied by a decrease in the other variable. [31.1]

associative property
> **of addition**—$(a + b) + c = a + (b + c)$ for every choice of numbers *a*, *b*, and *c*;
> **of multiplication**—$(a \times b) \times c = a \times (b \times c)$ for every choice of numbers *a*, *b*, and *c*. Noun form: associativity of . . . [3.4]

average deviation from the mean (or **average deviation**)—the average of the (positive) distances of each score from the overall **mean**. [30.5]

axis—of a rotational symmetry of a three-dimensional shape—the line about which the shape is rotated. [18.2]

axis (plural: **axes**)—one of the number lines used in a **coordinate system**. [13.1]

balance display—an informal way to represent equations (or inequalities). [12.6]

bar graph or **bar chart**—a common way of graphing categorical data. Compare **histogram**. [30.1]

base of a prism or pyramid—see **prism** or **pyramid**. [17.2]

bell curve—the shape of the graph of a **normal distribution**; the graph is called a **normal curve**. [30.6]

benchmark for a number—a number that is close to the given number and is useful for the purpose at hand. [5]

benchmark for an amount—a familiar amount that gives a standard of comparison. [5.3]

bias—see **sampling bias**.

bisector
> **of a line segment**—a line that goes through the midpoint of the segment. [21.1 (construction of)]
> **of an angle**—a ray that cuts a given angle into two angles of the same size. [21.1 (construction of)]

box-and-whiskers plot or **box plot**—a graph summarizing a data set, quartile by quartile. [30.3]

capacity of a three-dimensional container—the **volume** that the container can hold. [24.2]

casting out nines—a technique for checking calculations with whole numbers. [11.3]

categorical data—data that come from assigning individuals or objects to categories, such as political party affiliation or auto manufacturer. [29.4]

center of a dilation—see **dilation.** [20.1]

center of a rotational symmetry of a shape—the point about which the shape is rotated. [18.1]

chance, chances, or **likelihood**—often used as an alternative to "probability." [27.1]

chord of a circle—see **circle**. Pronounced "cord." [21.1]

circle—the two-dimensional shape formed by all points at a given distance from a given point, called the **center** of the circle. Any segment from the center to a point of the circle is a **radius** (plural, *radii*) or the length of such a segment. An **arc** of a circle is the piece of the circle between two points of the circle; the shorter piece is a **minor arc** and the longer one, a **major arc**. A **chord** of a circle is any line segment with endpoints on the circle. A **diameter** of a circle is any chord through the center of the circle, or the length of such a chord. A **circular region** is the set of points inside a circle. A **sector** of a circle is the region formed by two radii and the circle (a pie shape). A **segment** of a circle is the region formed by a chord of the circle and the circle. [21.1] **Concentric circles** are circles with the same center. [21.1] (Learning Exercises there) The **circumference** of a circle is its length. [23.2]

circle graph or **pie chart**—a common way of graphing categorical data. [30.1]

circumference—the distance around a circle; the length of the circle. [13.1] (Learning Exercises there)

closed set—for a particular operation, a set is closed if every application of the operation gives an element of the set. For example, the set of whole numbers is closed under multiplication. [10.6]

closure property—see **closed set**. [10.6]

combination—the technical term in mathematics for a selection in which order does not matter (an **unordered selection**). AB and BA are the same combination. Compare **permutation**. [33.2]

common difference—see **arithmetic sequence**.

common factors of two (or more) numbers—each of the numbers will have its own factors. Common factors are those that are factors of both (or all the) numbers. [6.2]

common multiples of two (or more) numbers—each of the numbers will have its own multiples. Common multiples are those that are multiples of both (or all the) numbers. [7.1]

common ratio—see **geometric sequence**.

commutative property
of addition—$a + b = b + a$ for every choice of numbers a and b . . .
of multiplication—$a \times b = b \times a$ for every choice of numbers a and b. Noun form: commutativity of . . . [3.4]

comparison language—language that communicates (1) additive comparisons, as in *more/bigger/longer than* or *less than*, or (2) multiplicative comparisons, as in *times as much as*, or (3) both additive and multiplicative comparisons, as in *times bigger than* or *percent less than*. [20.1]

comparison subtraction situation—a situation in which an **additive comparison** is involved. [3.2]

compass—in mathematics, a tool for drawing circles. [21.1]

compatible numbers—a phrase sometimes used to describe numbers that give easy approximations or exact answers, e.g., in $62.1 + 24.3 + 40.9$, 62 and 41 are "compatible" because their sum is about 100; with multiplication, 4 and 25, or 5 and 20, e.g., are compatible because their product is 100. [3.7] (Learning Exercises there)

composite (number)—a whole number greater than 1 that is not a prime number. [11.1]

composition of rigid motions—the rigid motion that describes the net effect, from original shape to final shape, when one rigid motion is followed by another. [22.4]

concave shape—a shape such that there are some points that, when joined, give line segments that do not lie entirely inside the shape. (The shape is "dented in" at least once.) [16.1] (Learning Exercises there)

conditional probability—a probability that may or may not be influenced by the restriction due to some event (the condition). Notation: The conditional probability of A, given B, is denoted by $P(A|B)$ and read "Probability of A, given B." [28.4]

cone—the three-dimensional surface resulting when a fixed point is joined to every point of a two-dimensional curve, along with the region of the curve. The fixed point is the **vertex** (or apex) of the cone, and the two-dimensional region is the **base** of the cone. In a **right circular** cone, the base is a circular region and the line through the vertex and the center of the circular base is perpendicular to every diameter of the base. In **oblique circular** cones, such a line is not perpendicular to every diameter. [21.2]

confidence interval—the interval, sample statistic $\pm$ the margin of error, in which the population parameter is believed to be, in a **confidence statement**. [32.2]

confidence statement—the **level of confidence** (stated as a probability) that a **confidence interval** will contain the **population parameter**. [32.2]

congruent shapes—shapes for which some rigid motion gives one shape as the image of the other; this rigid motion assures that the shapes are exactly the same shape and the same size. [22.1, 17.4 (3D)]

conjecture—a tentative result, usually based on one or more examples or drawings; an educated guess. [16.3]

construction—in mathematics, the drawing of a geometric shape with the use of tools, most often a compass and straightedge. [21.1]

contingency table—a table showing outcomes for two (or more) events; useful in determining some conditional probabilities. [28.4]

continuous quantities—quantities that can be measured only by length, area, etc.; they are measured, not counted. Contrast **discrete quantities**. [3.2]

converse—the new statement obtained from a given if-then statement by interchanging the if and then clauses. [24.1]

convex shape—a shape such that all the segments joining points of the shape are on or inside the shape. [16.1] (Learning Exercises there)

coordinate system—in algebra, a means of showing relationships between variables. With two variables, two number lines, called **axes**, are usually placed at right angles with their zero points coinciding. Points in the system are located by using their positions relative to the axes. Each point is associated with an ordered pair of numbers, (x, y), called the **coordinates** of the point. [13.1]

correlation coefficient r—a numerical measure of the association between two **variables**, with values between (and including) 1 and ⁻1. [31.2]

counterexample—an example that shows that a given statement is not always true—e.g., 9 is a counterexample to "Every odd number is a prime number." [11.3, 16.3]

cross-multiplication algorithm—a shortcut algorithm for solving proportions with a missing value. [9.2] (Learning Exercise 16)

cube—a **polyhedron** made up of 6 square regions, also called a **regular hexahedron**. [17.1, 17.5]

cylinder—the three-dimensional surface formed by tracing around a two-dimensional curve with a segment of a fixed length, so that the segment stays parallel to all its positions, plus the region of the two-dimensional curve and the (parallel) two-dimensional region defined by the other endpoint of the segment. The two regions are the **bases** of the cylinder. A **circular** cylinder results if the two-dimensional curve is a circle, and is a **right circular** cylinder if the segment joining the centers of the two bases is perpendicular to every diameter of the bases, and an **oblique circular** cylinder otherwise. [21.2]

decagon—a **polygon** having 10 sides. [16.1]

degree (°)—for angle measurement, the size of an angle formed by a ray $\frac{1}{360}$ of a full turn from the other ray of the angle. Hence, there are 360 degrees in a full turn. A **minute** (′) for angle size is $\frac{1}{60}$ of a degree, and a **second** (″) is $\frac{1}{60}$ of a minute. [23.2]

denominator—the "bottom" number in a **fraction**. [6.1]

density of a set of numbers—given two numbers from the set, there is always a third number in the set that is between the given numbers. The set is said to be dense. [10.6]

dependent equations—two apparently different equations that are relatable by multiplication or division. Their graphs are identical. [15.2]

dependent variable—a variable with values that depend on the value of another variable; associated with the vertical axis. [13.2, 31.1]

diagonal of a polygon or **prism**—a line segment that joins two vertices of the **polygon** or **prism** but is not a side of the polygon or is not completely in a face of the prism. [16.1, 26.1] (Learning Exercises there)

diameter—the width of a circle, measured along a line through the center of the circle. [13.1, 21.1]

difference—in an **additive comparison**, the term applied to the quantity that tells how much more or less one is than the other. [3.1]

differencing method—a technique for suggesting the degree of a polynomial that describes the data in a table. [15.5]

digit—any of the simple symbols 0, 1, 2, 3, 4, 5, 6, 7, 8, 9. In base b, there would have to be symbols for $0, 1, 2, \ldots, b - 1$. [2.1]

dilation—a matching of the points of a plane (or space) such that the size of every angle is the same as in its image and such that the ratio of an image length to the original length is always the same value, called the **scale factor.** The **center** of a dilation is the point with which the matching is done. [20.1]

discrete quantities—quantities that are separate and nontouching; they can be counted. Contrast **continuous quantities**. [3.2]

distance formula—the distance d between points (x_1, y_1) and (x_2, y_2) is given by the formula $d = \sqrt{(x_1 - x_2)^2 + (y_1 - y_2)^2}$. [26.1] (Learning Exercises there)

distributive property—usually refers to the valuable distributive properties of multiplication over addition: $a \times (b + c) = (a \times b) + (a \times c)$, or sometimes $(b \div c) \times a = (b \times a) + (c \times a)$, for every choice of numbers a, b, and c. Noun form: distributivity. The latter form, with division replacing multiplication, is also a correct property ("right" distributivity of division over addition). Subtractions can replace the additions. [3.4]

dividend—in a division expression like $364 \div 15$, the 364 is the dividend. [3.5]

divisibility test—a short way of determining whether one number is a **divisor** (or a **factor**) of another. [11.3]

divisor—in a division expression like $5.64 \div 3$, the 3 is the divisor; in number theory, used as a synonym for factor. Cannot be 0. [3.5]

dodecahedron—a polyhedron having 12 faces; in a **regular dodecahedron** the faces are congruent regular pentagonal regions. [17.5]

drawing with/without replacement—generic terms for multistep experiments in which elements used ("drawn") at earlier steps are/are not replaced for later draws. [27.3, 28.1]

edge
    hidden edge—an edge in a drawing of a three-dimensional shape that cannot be seen from the viewpoint of the drawer; it is often indicated by dashed or lighter marks. [17.3]
    lateral edge—see **prism** and **pyramid**. [17.2]
    of a polyhedron—see **polyhedron**. [17.2]

empirical probability—synonym for **experimental probability**. [27.2]

English system of measurement units—a commonly used system in the United States, also called the **British** or **customary system**. A few of the units and their relationships follow:

Length units	*Area units*
1 mile (mi) = 5280 feet (ft)	Square inch (in^2), square foot (ft^2), etc.
$\quad$ = 1760 yards (yd)	1 acre (A) = 43,560 square feet
[1 nautical mile = 6076 feet]	
1 yard = 3 feet	*Volume units*
1 foot = 12 inches (in.)	Cubic inch (in^3), cubic foot (ft^3), etc.
1 fathom (fath) = 6 feet	1 gallon (gal) = 4 quarts (qt) = 8 pints (p or pt)
1 rod (rd) = 16.5 feet	1 (U.S.) gallon (gal) = 231 in^3
	1 (Imperial) gallon (gal) = 277.42 in^3

equal additions algorithm—a nonstandard algorithm for subtraction; taught in many other countries as their **standard algorithm** but not in the United States. [3.3]

equal fractions—fractions that name the same numerical value. Also called **equivalent fractions**. [6.2]

equal ratios—ratios that express the same **multiplicative comparison**. [9.2]

equally likely outcomes—**outcomes** that have the same probabilities. [27.2]

equiangular polygon—a polygon whose angles are all the same size. [16.1]

equilateral polygon—a polygon whose sides are all the same length. [16.1]

equivalent fractions—fractions that name the same numerical value. Also called **equal fractions**. [6.2]

Euler's formula for polyhedra—$V + F = E + 2$ (or any algebraic equivalent), where V is the number of vertices of a given polyhedron, F is the number of faces, and E is the number of edges. [17.2]

even number—a whole number that can be expressed as 2 times some whole number. The number 0 is an even number, as are 2, 4, 6, 8, ... [11.1] (Learning Exercises there)

event—an outcome or any collection of **outcomes**, usually described in words. Example: At least one head, on the toss of two coins, involves the outcomes HH, HT, and TH. A **certain** event is one that must happen (and so has probability = 1). An **impossible** event is one that cannot happen (and so has probability = 0). [27.1]

expected value of a process or experiment—the sum obtained by multiplying the value associated with each outcome by its probability and adding the products. [33.1]

experiment—a clearly specified process and its result. [27.1]

experimental (or **empirical**) **probability** of some event—a probability arrived at by repeating some process or experiment many times. The probability equals the ratio of the number of times the particular **event** occurs to the number of times the process is repeated and is often approximate. Also called the **relative frequency** of the event. Contrast **theoretical probability**. [27.2]

exponential function—a function with function rule of the form $y = b^x$, where b is a fixed number. [13.4]

face of a polyhedron—see **polyhedron**; **lateral face**—see **prism** and **pyramid**. [17.2]

factor, factor of—in multiplication, any of the numbers being multiplied—e.g., in 3.2×2.4, each of 3.2 and 2.4 is a factor; in number theory, a whole number that can appear as a factor in an expression yielding some whole number—e.g., 6 is a factor of 24 (because $6 \times 4 = 24$). [3.4]

factor tree—a diagram to arrive systematically at the **prime factorization of a number.** Contrast the different **tree diagram**. [11.2]

family of facts—for addition and subtraction, the set of four addition/subtraction equations that are related—e.g., $2 + 4 = 6$, $4 + 2 = 6$, $6 - 2 = 4$, and $6 - 4 = 2$ is a family of facts. For multiplication and division, it is the same idea: e.g., $3 \times 5 = 15$, $5 \times 3 = 15$, $15 \div 5 = 3$, and $15 \div 3 = 5$ is a family of facts. [3.2]

Fibonacci sequence—a pattern of numbers (most commonly, 1, 1, 2, 3, 5, 8, . . .) in which two consecutive numbers in a list are added to get the next number; named after the Italian mathematician Leonardo Fibonacci (c. 1175–1250). [12.3] (Learning Exercises there)

five-number summary—for a set of data, the five numbers (in order): smallest value, first **quartile** score, **median**, third quartile score, and largest value. [30.3]

fixed point—see **rigid motion**. [22.3] (Learning Exercises there)

flip—an informal name for a **reflection**. [22.1]

fraction—a **numeral** of the form $\frac{\text{number}}{\text{nonzero number}}$. A fraction could be signaling a part-whole relationship, a division, or a **ratio**. [6.1]

function—a correspondence that assigns each element of a first set to exactly one element of a second set. Both sets can be identical. The correspondence is often specified by a **function rule**—e.g., an equation. [12.4]

fundamental counting principle—if Act 1 can be performed in m ways, and Act 2 can be performed in n ways no matter how Act 1 turns out, then the sequence Act 1–Act 2 can be performed in mn ways. [3.4, 33.2]

Fundamental Theorem of Arithmetic—a statement asserting that a number (>1) is either a prime or has a unique prime factorization. Also called the **Unique Factorization Theorem**. Contrast the quite different **fundamental counting principle**. [11.2]

geometric sequence—a pattern of numbers obtained by multiplying a number in a list by the same number, called the **common ratio**, to get the next number in the list. [12.3]

glide-reflection—a type of rigid motion, the composition of a translation and a reflection in a line parallel to the vector of the translation. [22.1]

golden ratio—the value, approximately 1.618, that is approached by the ratios of consecutive numbers in a Fibonacci sequence. (Some texts give the reciprocal, 0.618, as the golden ratio.) [12.3] (Learning Exercises there)

great circle of a sphere—the largest circle possible on the surface of a sphere. Its radius and center are the same as those of the sphere. [21.2]

greatest common divisor (or **gcd** or **GCD**)—the number that is the greatest of all the common divisors of the given numbers. [6.2]

greatest common factor—see **greatest common divisor**. Also **gcf** or **GCF**. [6.2]

height
  of a prism or cylinder—a line segment from the plane of one base to the other base, so that it is perpendicular to every line in the other base that it meets; also, the length of the segment. [25.2]
  of a pyramid or cone—the line segment from the vertex of the pyramid or cone that makes a right angle with every line in the base that it meets; also, the length of that segment. [25.2]
  of a trapezoid or parallelogram—a line segment making right angles with each of two parallel sides; also, the length of that segment. [25.1]
  of a triangle—a line segment from a vertex of the triangle that makes a right angle with the opposite side; also, the length of that segment. Sometimes called the **altitude**. [25.1]

heptagon—a polygon having 7 sides. [16.1]

hexagon—a polygon having 6 sides. [16.1]

hexahedron—a **polyhedron** having 6 faces; a **regular hexahedron** is a cube. [17.5]

hierarchical classification—in the context of this book, a classification system in which shapes in a subcategory have the properties of the category (as well as other properties not shared with all shapes in the category). [16.2]

histogram—a bar graph for continuous data. [30.2]

hypotenuse of a right triangle—the side opposite the right angle. [26.1]

icosahedron—a **polyhedron** having 20 faces; in a **regular icosahedron** the faces are equilateral-triangular regions. [17.5]

image of a point or shape—the corresponding point or shape that a transformation gives for the point or shape. [22.1] Finding images for a size change [20.1]; finding images for rigid motions [22.2].

inconsistent equations—equations that have no common solution. Their graphs do not meet. [15.2]

independent events—two events A and B are independent if the occurrence of one event does not change the probability of the other event. Alternatively, $P(A|B) = P(A)$. For independent events A and B, $P(A \text{ and } B) = P(A) \times P(B)$. [28.3, 28.4]

independent variable—a variable used to predict another variable; associated with the horizontal axis. [13.2, 31.1]

inductive reasoning—arriving at a general statement by considering examples. [12.4, 16.2]

input (**input set**)—informal terms for an element of the first set (or the first set itself) in a function. [12.4]

integer—any of the numbers . . . , $^-3$, $^-2$, $^-1$, 0, 1, 2, 3, 4, . . . [10.1]

interquartile range (**IQR**)—the difference between the 25th percentile and 75th percentile (the **first** and **third quartile scores**) for a set of numerical data. [30.3]

irrational number—a number that cannot be written in the form $\frac{\text{integer}}{\text{nonzero integer}}$, i.e., a number that is not a **rational number**. Their decimals are **nonterminating** and do not **repeat**. [6.3]

isometric dot paper—paper with dots arranged in an equilateral triangle pattern, which is useful for one type of drawing of three-dimensional polyhedra. [17.1]

isometry—another name for a **rigid motion**. [22.1]

key words—individual words that sometimes, but not always, signal a particular operation—e.g., "left" often occurs in story problems in which subtraction gives the answer. Not recommended without paying attention to the whole context of the problem. [3.2] (Learning Exercises there)

kilometer—a metric system unit for measuring length or distance; equals 1000 **meters**; about 0.621 miles. [1.3]

kite—a quadrilateral with two consecutive sides having the same length and the other two sides also having the same length. [16.1, 16.2, 16.3]

lattice method for multiplication—a nonstandard **algorithm** for multiplication; occasionally appears in an elementary school textbook series. [4.1] (Learning Exercises there)

least common multiple or **lcm** or **LCM** (or **lcd** or **LCD**)—the number that is the least of all the common (nonzero) multiples of the given numbers (denominators). [7.1]

leg of a right triangle—either of the two sides making the right angle. [26.1]

length—the characteristic of one-dimensional shapes that is measured with a ruler; e.g., width, height, depth, thickness, perimeter, and circumference refer to the same characteristic. [23.2]

line—short for *straight line*, a straight path that continues forever in two directions. A **line segment** is the piece of a line between two points called endpoints. A **ray** is the piece of a line that starts at a point and continues forever in one direction.

line graph or **broken-line graph**—a way of representing two sets of measurement data. [31.1]

line of best fit or **regression line**—a statistically derived line that best summarizes the data in a **scatter plot**. [31.2]

linear equation—an equation whose graph is a straight line; $y = mx + b$ is a common linear equation. [13.2]

linear function—a function with function rule describable by a linear equation. [13.3]

lowest terms—see **simplest form**. [6.2]

margin of error—gives with the sample statistic an interval that will contain, with a certain probability, the population parameter. [32.1]

mean or **mean score**—the average rate for a data set, given per person or per item. If each person or item received the mean score, the total of these would be the same as the actual total. [30.4]

measurement data—numerical data that can be put in order and used with arithmetic operations. [29.4]

measurement division—see **repeated subtraction division settings**. [3.5]

measurement of a characteristic—(1) the process of direct measurement, in which the given object is compared to a unit with respect to the characteristic; or (2) the resulting number and unit (e.g., 3 inches). Standard units are most common, for purposes of communication and permanence (see, e.g., **metric system** and **SI**); key ideas of [23.1].

median of a triangle—a segment joining a vertex of the triangle to the midpoint of the opposite side. [21.1] (Learning Exercises there)

median value or **median**—in statistics, the middle value for a numerical data set, or the 50th **percentile**. [30.3, 30.4]

mental computation—calculation carried out mentally, most often in a nonstandard way. Useful in estimation but may also give exact answers. [5.1]

meter—the basic **metric system** unit of measurement of length or distance; about 39.37 inches or 1.09 yards. [1.3]

metric system—the most common international system of standard units for measurement; see **SI**. The basic unit for length is the meter (metre) [1.3]; see **metric prefixes** [23.1]; area units [24.1]; volume units [24.2].

minuend—the number from which another number is subtracted; in $18 - 6$, the minuend is the 18; the 6 is the **subtrahend**. [3.2]

mirror image—informal language for the image under a reflection. [22.1]

missing-addend situation—a situation that involves quantities that are related additively, but the addend for one of the quantities is missing; in $31 + n = 58$, n is a missing addend. Sometimes called a **missing-part** situation. [3.2]

missing-factor view of division—divisions that are based on multiplication settings where a factor is missing but the product is known. In $13 \times n = 182$, n is a missing factor and can be found by $182 \div 13$. [3.4]

mode—the data value that occurs most often in a data set. [30.4]

multibase blocks—blocks of different sizes that reflect the different **place values** for some numeration system. [2.3]

multiple of a number—any **product** involving the given number as a **factor**—e.g., 36 is a multiple of 12 (because $3 \times 12 = 36$). [3.4]

multiplicand—occasionally used in the United States to refer to the second **factor** in a multiplication expression. In 5×36, 36 is the multiplicand. [3.4]

multiplicative comparison of two quantities—the comparison of the two **quantities** by seeing how many times as large one of them is, compared to the other. See **ratio**. [8.1]

multiplicative inverse of a number—that number that, when multiplied by the given number, gives a product $= 1$—e.g., $\frac{11}{16}$ and $\frac{16}{11}$ are multiplicative inverses of each other, because $\frac{11}{16} \times \frac{16}{11} = 1$. Also called the **reciprocal**. 0 does not have a multiplicative inverse. [7.2] (Learning Exercises there)

multiplier—occasionally used in the United States to refer to the first **factor** in a multiplication expression. In 5×36, 5 is the multiplier. [3.4]

multistep situation or **experiment**—a process or experiment that involves more than one step or part. Examples: spinning the same spinner twice, or spinning two different spinners. [28.1]

negative integer—any of the numbers . . . , $^-3$, $^-2$, $^-1$. [10.1]

net for a three-dimensional shape—a two-dimensional pattern that gives the three-dimensional shape when folded up. [17.1]

n-gon—a polygon having n sides, where n is a whole number greater than 2. [16.1]

nonagon—a polygon having 9 sides. [16.1]

nonlinear function—a function whose graph is not a straight line. [13.4]

nonterminating decimal—a decimal that has infinitely many place values. [6.3]

normal distribution—informally, the way a **histogram** for many large data sets often looks: shaped like a bell. The data fit the **68–95–99.7 rule**. [30.6]

notations—symbols that represent geometric objects and are common and may even appear in elementary school mathematics textbooks. [16.1, 21.1]

  angles—named either by naming just the vertex, or if needed for clarity, as for angle D in the drawing below, naming a point on one side, the vertex, and then a point on the other side: $\angle D$ or $\angle CDE$.

  arcs of circles— $\overset{\frown}{ST}$, with S and T the endpoints of the arc; $\overset{\frown}{SUT}$ for clarity.

  line through D and E: $\overleftrightarrow{DE}$

line segment with endpoints at P and Q: $\overline{PQ}$

point—capital letters, A, B, etc.

polygon with vertices F, G, H, and I: *FGHI*

ray starting at C and going through D: $\overrightarrow{CD}$

line segments (and angles)—can also be indicated with small letters, which may also mean their lengths (and angle sizes). A book may also use a darkened-in dot for an endpoint to show it is definitely included (and then the open dot would mean the endpoint is not included).

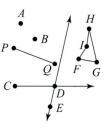

numeral—a word or symbol to communicate a number idea. [2.1]

numeration system—a system for naming or symbolizing numbers; two examples: our base-ten place-value system, Roman numerals. [2.1]

numerator—the "top" number in a **fraction**. [6.1]

octagon—a polygon having 8 sides. [16.1]

odd number—a whole number that can be expressed as 1 more than an even number. [10.6] (Learning Exercises there)

operator view of multiplication—a view that involves a fractional part of a quantity; also called **part-whole multiplication**. [3.4]

or—the mathematical use of "or" is inclusive and allows either or both. Example: Getting H on a first toss or H on a second toss would include HH as well as HT and TH. [28.2]

order of operations—a convention about which operation(s) should be done first when simplifying expressions involving addition, subtraction, multiplication, and/or division. This convention helps to avoid a lot of parentheses. [12.1]

orientation of a figure—a clock direction (clockwise or counterclockwise) assigned to a two-dimensional shape. [22.1] (Learning Exercises there)

outcome—one of the things that can happen when an experiment is carried out or thought about. Reserved for the simplest results. Example: A toss of two coins has these outcomes: HH, HT, TH, and TT. [27.1]

outlier—a score greater than the third quartile value + 1.5 times the **interquartile range**, or a score less than the first quartile value − 1.5 times the interquartile range. [30.3]

output (output set)—informal terms for an element of the second set (or the second set itself) in a function. [12.4]

parabola—the graph of a quadratic function. [13.4]

parallel lines or **parallel planes**—Lines in the same plane that never meet or planes that never meet. Notation: $x \parallel y$. Compare with **skew lines**.

parallelogram—a quadrilateral with both pairs of opposite sides parallel. To emphasize hierarchical concerns, *quadrilateral* can be replaced by *trapezoid*. [16.1, 16.3]

partitive division settings—see **sharing division settings**. [3.5]

part-whole—one meaning for a fraction. [6.1]

part-whole multiplication setting—a setting that involves a fractional part of a quantity; also called **operator view**. [3.4]

Pascal's triangle—the triangular arrangement of numbers, with each line obtainable by recording, for a fixed number of objects, the numbers of combinations of each possible size, 0, 1, 2, . . . [33.2] (Learning Exercises there)

pentagon—a polygon having 5 sides. [16.1]

pentomino—a connected two-dimensional shape made of 5 square regions with each square sharing at least one side with another square. [17.3] (Learning Exercises there)

percent—the number of hundredths in a ratio or a fraction when expressed as a decimal. Roughly, "out of one hundred" or "per hundred." [9.3]

percentile—the *n*th **percentile** is (sometimes roughly) the data value p for which $n\%$ of the data values are equal to or less than p. [30.3]

perfect number—positive integer that is equal to the sum of its factors excluding itself. [11.3]

perimeter of a polygon or closed curve—the distance along the polygon or closed curve. It is a type of length and must be kept distinct from the idea of area. [13.1, 23.2]

permutation—a selection or choice in which order is important (an **ordered selection**). AB and BA are different permutations. Compare **combination**. [33.2]

perpendicular bisector of a segment—a line that passes through the midpoint of the segment and makes right angles with the segment. [21.1 (construction of)]

perpendicular lines—lines that make right angles. [16.1, 18.2] (Learning Exercises there) Perpendicular to a line. [21.1 (construction of)] Notation: $x \perp y$ means lines x and y are perpendicular.

pi (π)—the number that expresses the ratio of the circumference of any circle to its diameter. Pi does not have an exact terminating or repeating decimal, so the small Greek letter π is used when the exact value is meant. Approximate values for π often used are 22/7 and 3.14, or more exactly but still approximately, 3.1415926536. [25.1]

pictograph—a bar graph or histogram in which the usual rectangles are replaced by symbols or pictures. [30.1]

pie chart—see **circle graph**.

place value—the numerical value associated with a particular position in a symbol—e.g., the place values in the base ten numeral 234 are 100 (for the position of the 2), 10 (for the position of the 3), and 1 (for the position of the 4). [2.2]

plane—a perfectly flat, endless surface. A **planar region** is made up of the points inside a closed two-dimensional (flat) surface. [16, 17]

polygon—a closed plane figure made up of line segments joined end to end without crossing over. The line segments are called the **sides** of the polygon, and the endpoints are the **vertices** of the polygon (singular—**vertex**). A **diagonal** is a line segment joining two vertices that are not consecutive. [16.1] There are many special polygons. One way of classifying some of them is given in a **hierarchy** (as in the following diagram). [16.3] Shapes lower in the diagram are special versions of any shape that they are connected to higher in the diagram—e.g., a rectangle is a special parallelogram or a special quadrilateral. A hierarchy recognizes that any fact known about all parallelograms, say, applies to any special parallelogram like a rhombus or rectangle or square.

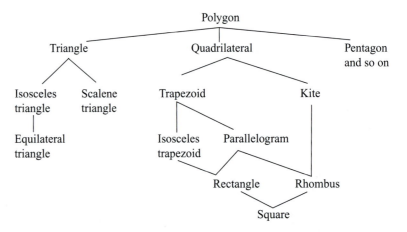

polygonal region—all of the points inside a polygon. The polygon is just the line segments. The distinction is possibly important because teachers often use a cardboard cutout to illustrate a polygon. The edges of the cutout show the polygon, but the cardboard itself actually shows a polygonal region. [16.1]

A triangle   A triangular region

polyhedron (plural, **polyhedra** or polyhedrons)—A closed three-dimensional surface made up of planar regions (flat pieces). [17.2] The planar regions are the **faces** of the polyhedron. The line segments where faces meet are called the **edges** of the polyhedron. The points at which edges meet are called the **vertices** (singular—**vertex**) of the polyhedron. A **diagonal** of a polyhedron is any line segment that is not a side or a diagonal of a face but joins two vertices. [26.1] (Learning Exercises there) **Pyramids** and **prisms** are special polyhedra. If all the faces of a polyhedron are **regular** polygons that are exactly alike and not arranged in some odd way, the polyhedron is called a **regular polyhedron**. [17.5]

polynomial in x—a sum of number multiples of whole-number powers of x. The addends are called **terms** of the polynomial. [12.2]

population—the total potential data source that is of interest (it may or may not involve people). Used to distinguish a portion of the complete data source (a **sample**) from the total data source (the population). A **population parameter** is a measure based on the entire population. [29.2] population parameter—see **population**. [29.2]

positive integer—any of the numbers 1, 2, 3, . . . , viewed as integers. [10.1]

prefixes—descriptive terms preceding a main term; prefixes can be of two types, general and metric.

General Prefixes (with some license)

	Latin	Greek		Latin	Greek
1	uni-	mono-	7	sept-	hept-
2	bi-	di-	8	oct-	oct-
3	tri-	tri-	9	non-	enne-
4	quad-, quadr-	tetr-	10	dec-	dek-
5	quint-	pent-	100	cent-	hect-
6	sex-	hex-	1000	mill-	kilo-
			-many	mult-	poly-

The Greek prefixes for 10, 100, 1000, etc. are used in the metric system for larger multiples of the base unit (e.g., dekameter, hectometer, kilometer), whereas the Latin prefixes for 10, 100, 1000, etc. are used for the fractional subunits (e.g., decimeter, centimeter, millimeter). See the following table:

Metric Prefixes

Prefix	Symbol	Meaning of Prefix	Applied to Length
yotta-	Y	10^{24}	Ym
zetta-	Z	10^{21}	Zm
exa-	E	10^{18}	Em
peta-	P	10^{15}	Pm
tera-	T	10^{12}	Tm
giga-	G	1 000 000 000 or 10^9	Gm
mega-	M	1 000 000 or 10^6	Mm
kilo-	k	1000 or 10^3	km
hecto-	h	100 or 10^2	hm
deka-	da	10 or 10^1	dam
no prefix		1 or 10^0	m
deci-	d	0.1 or 10^{-1}	dm
centi-	c	0.01 or 10^{-2}	cm
milli-	m	0.001 or 10^{-3}	mm
micro-	μ	0.000001 or 10^{-6}	μm
nano-	n	0.000000001 or 10^{-9}	nm
pico-	p	10^{-12}	pm
femto-	f	10^{-15}	fm
atto-	a	10^{-18}	am
zepto-	z	10^{-21}	zm
yocto-	y	10^{-24}	ym

prime (number)—a whole number that has exactly two factors. The numbers 0 and 1 are not prime numbers. [11.1]

prime factorization of a number—the expression of the number as the product of prime numbers. [11.2]

prism—a polyhedron having two faces (called **bases**) that are parallel and congruent (informally, *exactly alike in size and shape*) and whose other faces (called **lateral faces**) are parallelogram regions (or special parallelogram regions such as rectangular ones) formed by joining corresponding vertices of the bases. Some edges of a prism are on the two bases; the other edges are called **lateral edges**. [17.2] A prism is a **right prism** if the lateral edges (the ones not on the bases) are perpendicular to the edges at the bases, and is **oblique** otherwise. [17.2]

An oblique hexagonal prism   The two bases and   A right hexagonal prism
                              the lateral edges

probabilistic situation—a situation in which we are interested in the fraction of the number of repetitions of a particular process that produces a particular result when repeated under identical circumstances a large number of times. [27.1]

probability of an event—the fraction of the times that the event will occur when some process is repeated a large number of times. [27.1]

product—the result of a multiplication calculation. [3.4]

properties of operations—various relationships such as commutativity and associativity of addition that are important with numbers and algebraic expressions. [12.1]

proportion—an equality between two ratios. [9.2]

proportional relationship—the relationship between variable x and variable y is proportional if the ratio $y:x$ always has the same value. [13.3]

protractor—a tool for measuring angles. [16.1, Appendix G]

pyramid—a polyhedron with one face being any sort of polygonal region (often called the **base** of the pyramid) and with all the other faces being triangular regions with one vertex in common. [17.2] The edges not on the base are called **lateral edges**, and the nonbase faces are called **lateral faces**. If the base is shaped like a regular polygon and the lateral faces are all congruent, the pyramid is called a **regular pyramid** (see shapes A and G in the "Masters"). [17.2] (Learning Exercises there)

An oblique pentagonal   The base of the pyramid   The lateral edges of the
pyramid   (shaded)   pyramid in blue

Pythagorean theorem—the important relationship among the lengths of the three sides of a right triangle: The sum of the squares of the lengths of the **legs** equals the square of the length of the **hypotenuse**. [26.1]

Pythagorean triple—three nonzero whole numbers, a, b, and c, whose squares are related as expressed in the Pythagorean theorem: $a^2 + b^2 = c^2$. [26.1] (Learning Exercises there)

quadratic function—a function with function rule describable as a **polynomial** with 2 as the greatest power of the variable. [13.4]

quadrilateral—a polygon having 4 sides. [16.1]

qualitative graph—a coordinate graph without numbers on the axes; allows a study of relationships without concern for exact numerical values. [14.1]

quantitative analysis—determining the quantities in a situation and how they are related. Crucial for solving many story problems. [1.2]

quantitative structure—how the **quantities** in a situation are related. [1.2]

quantity—any measurable or countable aspect of any object, event, or idea. The **value of a quantity** is the measure or count of the quantity. The value of a variable quantity, or **variable**, is often represented by a letter (e.g., $x, y, h, d, \ldots$); the letter itself is often called a **variable**. [1.1, 12.5]

quartiles or **quartile values**—the 25th percentile, 50th percentile, and 75th percentile for a set of numerical data. [30.3]

quotient—the answer to a division calculation. [3.5]

quotitive division—see **repeated-subtraction division settings**. [3.5]

radius of a circle—see **circle**. [21.1]

random/randomly selected—a selection in which every item has an equal probability of being chosen. [27.3]

range—the difference between the largest and smallest values in a set of data. [30.3]

rate—a ratio that stays the same as the quantities that are involved vary. [9.2]

rate of change—the change in a variable for each change of 1 unit in a second variable. [13.2]

ratio of two quantities—the result of comparing the quantities multiplicatively (see **multiplicative comparison of two quantities**). Written $a:b$ or $\frac{a}{b}$. [8.1]

rational number—a number that can be written in the form $\frac{\text{integer}}{\text{nonzero integer}}$. In early work the integers are whole numbers. The decimal for a rational number is either terminating, or nonterminating but repeating. [6.3, 10.3]

ray—see **line**.

real number—any **rational** or **irrational** number. [6.3]

reciprocal of a number—a synonym for **multiplicative inverse of a number**. [7.2] (Learning Exercises there)

rectangle—a parallelogram with a right angle. (*Note*: This definition necessarily leads to all four angles being right angles.) [16.1, 16.3]

reflection in a line—a two-dimensional **rigid motion** in which a point on either side of the line has its image as though the line were a mirror. See also **symmetry**. [22.1, 22.2]

reflection symmetry with respect to a plane (for a three-dimensional shape)—a reflection in a plane that gives the original shape as the image. Also called mirror-image symmetry or plane symmetry. [18.2]

regression line—see **line of best fit**. [31.2]

regrouping algorithm for subtraction—the standard algorithm for subtraction in the United States; "borrowing" is sometimes used but not recommended. [3.3]

regular polygon—a polygon that is both **equiangular** and **equilateral**. [16.1]

regular polyhedron—a polyhedron whose faces are all the same regular polygonal regions and have the same arrangement at each vertex. Also called **Platonic solids**. [17.5]

relative frequency of an event—sometimes used as a synonym for **experimental probability**. [27.2]

relatively prime numbers—numbers that have only 1 as a common factor. [11.3]

remainder—in a take-away subtraction setting, the amount left over; in a division calculation like $17 \div 3$, the amount left over when no more 3s can be subtracted: $17 \div 3 = 5$ R 2. [3.2, 3.5]

repeated-addition multiplication settings—settings in which quantities with the same values are combined additively; abstractly, a short way to describe calculations in which the same addend is repeated—e.g., $5 + 5 + 5 + 5 + 5 + 5$ can be described by 6×5 (U.S. convention on order of the 6 and 5). [3.4]

repeated-subtraction division settings—settings in which same-valued quantities are taken away from a quantity. Sometimes called the **measurement** or the **quotitive** view of division. [3.4]

repeating decimal—a **nonterminating** decimal that has a block of digits that repeat endlessly, as in 4.13535353535 . . . (forever). That number may be abbreviated by $4.1\overline{35}$. [6.3]

representation—a way of communicating or thinking about something; in geometry, a drawing, a model, a net, a word, or even an equation can be a representation. [17.3]

rhombus—a **kite** that is also a **parallelogram**. (*Note*: This definition necessarily leads to all four sides having the same length.) [16.1, 16.3]

rigid motion—a matching of the points in the plane (or space) with points, so that original lengths are the same as lengths in the **images**. [22] A **fixed point** for a rigid motion is a point that is its own image. [22.3] (Learning Exercises there)

rotation about a point—a two-dimensional rigid motion in which the plane is turned about a point, called the **center** of the rotation. See also **symmetry**. [22.1]

rotational symmetry with respect to a point (for a two-dimensional shape) or **with respect to a line** (for a three-dimensional shape)—a rotation about the point, called the **center**, or about a line, called the **axis**, such that the image is the same as the original shape. [18.1, 18.2]

ruler—in mathematics, a device marked for measuring length, most often with straight edges. Useful in constructions, even without the marks. [21.1]

Russian peasant algorithm—a nonstandard algorithm for multiplying whole numbers. [4.1] (Learning Exercises there)

sample—a portion of a **population**. In a **simple random sample** (**SRS**), each member of the population has an equal probability of being selected for the sample, a good way to limit **sampling bias**. A **stratified random sample** is a random sample with each of different groups (or **strata**) represented. A **cluster sample** may be used when the cluster is believed not to represent any bias. In **systematic sampling**, the sample is chosen in a systematic manner, such as every fifth person. A **convenience sample** is a sample chosen just because of availability. A **self-selected** or **voluntary sample** is a sample based on those willing to respond. [29.2]

sample space for a process or experiment—the set of all the possible **outcomes** for the process or experiment. Example: A toss of a nickel and a dime has this sample space: HH, HT, TH, and TT. [27.2]

sampling bias—possible influences that mean the sample may not reflect the entire population but only a special part of the population. [29.2]

scaffolding algorithm—a nonstandard algorithm for division that may be used as a lead-in to the **standard algorithm.** [3.6]

scale factor—the common ratio of the image length to the original length in similar shapes. [20.1 (2D), 20.3 (3D)]

scatter plot—a coordinate graph of data in which each point gives numerical values for two variables. [31.1]

scientific notation—the expression of a number in the form $a \times 10^b$, where $1 \leq a < 10$ and b is an integer. [5.4]

sector of a circle—see **circle**. [21.1]

segment of a circle—see **circle**. [21.1]

sharing division settings—settings in which a quantity is put into a designated number of equal-valued amounts. Sometimes called **partitive division**. [3.5]

SI—The metric system of standard units. SI is short for *Système International d'Unités*, the International System of Units. The strengths of the system are that subunits and larger units are related to a basic unit by a power of 10 and that the basic units are carefully defined for reproducibility (in most cases). The table for metric prefixes (under **prefixes**) gives the subunits and larger units and applies them to the meter (symbol: m), the basic unit for length. [23.1]

signed numbers—positive and negative numbers. Called signed numbers because of the + or − sign, as in $^+4$ or $^-12$. [10.1]

similar shapes—shapes that are related by a dilation, possibly along with a rigid motion of some sort. Noun—**similarity**. [20.1 (2D), 20.3 (3D)]

simplest form (fractions)—a **fraction** in which the **numerator** and **denominator** have no common factors other than 1. Preferred to the common "reduced" form. [6.2]

simulation—a representation of something, perhaps by coding with random numbers, spinners, or colored balls, e.g. [27.3]

size change or **size transformation**— see **dilation**. [20.1]

skew lines—lines in space that never meet but are not parallel (e.g., the bottom edge of the front wall, and the top edge of the side wall). Contrast with **parallel lines**, which are in the same plane.

skewed distributions—if the histogram for a data set is not symmetrical and has a "tail" to one side, the data set and distribution are said to be "skewed left" if the tail is to the left, and "skewed right" if the tail is to the right. [30.6] (Learning Exercises there)

slide—an informal name for a **translation**. [22.1]

slope of a line—the rate of change for the two variables involved; the slope is calculated by $\dfrac{\text{change in } y\text{-values}}{\text{change in } x\text{-values}}$ for two points (x, y) on the line and is commonly denoted as $\dfrac{\text{rise}}{\text{run}}$. [13.2]

speed—the ratio of the distance traveled to the time elapsed. Speed is a **rate** often given as miles per hour or feet per second, e.g. [13]

sphere—the three-dimensional set of points at a fixed distance from a fixed point, called the **center** of the sphere. [21.2] A **hemisphere** is half a sphere. [21.2] (Learning Exercises there)

square—a rectangle that is also a rhombus. (*Note*: This definition necessarily leads to all four angles being right angles and all four sides having the same length.) [16.1, 16.3]

standard algorithm—one of the usual methods for adding, subtracting, multiplying, or dividing numbers; may vary from country to country. [3.3]

standard deviation—the (positive) square root of the **variance**. [30.5]

standard units—in measurement, an accepted system of units, such as the **metric system**. [23.1]

statistic—a numerical value of a quantity used to describe data. **Statistics** as a title refers to the science of obtaining, organizing, describing, and analyzing data, often for the purpose of making decisions or predictions. A sample statistic is based on a **sample**. [29.1, 29.2]

statistical variable—an attribute being focused on as a source of data. With **categorical data**, the attribute involves only different categories, such as those that would result from naming children's "favorite pie." With **measurement data**, the attribute is a quantity that gives values which can be ordered on a numerical scale with equal units. ("Variable" in algebra usually refers to a measurement variable.) [29.4]

stem-and-leaf plot—a way of organizing measurement data that can easily be changed to a **bar graph** or a **line graph**. [30.2]

straightedge—a tool for drawing line segments. Usually a ruler but the measuring marks are not to be used. [21.1]

substitution method—one approach to finding the common solution to a system of equations. Contrast the **addition/subtraction method**. [15.2]

subtraction of integers—a formal definition is $a - b = a + {}^-b$, for every choice of integers a and b. More basic meanings for subtraction often apply. [10.2]

subtrahend—the number being subtracted in an expression like $18 - 6$; 6 is the subtrahend; 18 is the **minuend**. [3.2]

sum—the result of an addition. [3.1]

surface area—see **area**. [24.1]

symmetry of a (two-dimensional or three-dimensional) shape—a rigid motion that gives an image that is the same shape as the original shape. The type of rigid motion can be suggested by an adjective: reflection symmetry, line symmetry or mirror-image symmetry, or rotational symmetry. [18.1, 18.2]

table of randomly selected digits—a table of digits that have been selected without bias. Abbreviated **TRSD**. [27.3, Appendix D]

take-away subtraction situation—a situation in which a part of an amount is removed from the whole amount. (The whole amount can also be removed.) [3.2]

tangent to a circle—a line that has exactly one point in common with the circle. A tangent is perpendicular to the radius to that point. [21.1] (Learning Exercises there)

tangram—a type of puzzle, typically made up of pieces cut in a certain way from a square region. [16.1] (Learning Exercises there)

terms of a polynomial—the expressions added in a polynomial. For example, $5x^2$, ${}^-3x$, and 7 are the terms of $5x^2 - 3x + 7$ (because $5x^2 - 3x + 7 = 5x^2 + {}^-3x + 7$). [12.2]

terms of a sequence—the individual entries in a list. [12.3]

tessellation of a plane (or space)—a covering of a plane (or space) with a limited number of types of regions. In a **regular** tessellation of the plane, all the regions are one type of regular polygonal region. [19.1 (2D), 19.2 (3D)]

tetrahedron—a polyhedron having 4 faces; a **regular tetrahedron** has faces that are equilateral-triangular regions. [17.5]

tetromino—a connected two-dimensional shape made of 4 square regions with each square sharing at least one side with another square. [17.3] (Learning Exercises there)

theoretical probability—a probability arrived at by judging what is likely to occur in a situation by using knowledge of the situation, without actually trying the situation. Contrast **experimental probability**. [27.2]

transformation geometry—a term for a school geometry topic that features **rigid motions** and **size changes**. [22]

translation—a type of rigid motion in which the image of a point is the point that is a fixed distance in a fixed direction from the original point. [22.1]

trapezoid—a quadrilateral with at least one pair of opposite sides parallel. Many books use *exactly one pair*, but the definition here allows the trapezoid family to relate to other special quadrilaterals. To retain that value without causing other difficulties, the **isosceles trapezoid** is defined awkwardly, as a trapezoid in which both angles adjacent to one of the parallel sides are the same size. Under either definition, the two sides of an isosceles trapezoid that may not be parallel have the same length. [16.1] (Learning Exercises there), [16.3]

tree diagram—in counting, a diagram for systematically showing settings that fall under the **fundamental counting principle**. Contrast **factor tree**. [3.4]

tree diagram—in probability, one systematic way to determine and record the outcomes to an experiment, especially a multistep situation. [28.1]

triangle—a polygon having 3 sides. In an **acute triangle**, all three angles are acute; a **right triangle** has a right angle, and an **obtuse triangle** has an obtuse angle. The lengths of the sides of a **scalene triangle** are all different, but an **isosceles triangle** has at least two sides with the same length, and an **equilateral triangle** has the same length for all three sides. See also **polygon**. [16.1]

triangular numbers—the numbers 1, 3, 6, 10, 15, . . . , $\dfrac{n(n+1)}{2}$, . . ., so called because they give counts of dots that can be arranged in a triangle. The arrangement of 10 bowling pins gives the fourth triangular number ($n = 4$). [16.1] (Learning Exercises there)

TRSD—see **table of randomly selected digits**.

turn—an informal name for **rotation**. [22.1]

Unique Factorization Theorem—a statement asserting that a number (>1) is either a prime or has a unique prime factorization. Also called the **Fundamental Theorem of Arithmetic**. [11.2]

unit—in measurements, the size of the comparison object, e.g., mile; see **standard units**. (With fractions, what is equal to 1.) [1.1, 6.1, 23.1]

unit fraction—a fraction with numerator 1, such as $\frac{1}{4}$. [6.1]

unit ratio—a **ratio** in which the second entry is 1, as in 7 to 1. [9.2]

value of a quantity—the measure of the **quantity** or the number of items counted; a value of a quantity involves a number and a unit—e.g., 3 pints (for the quantity, the amount of milk in the jug). [1.1]

variable—see **quantity**.

variance—in statistics, the average of the squares of the differences between each score and the overall **mean**. [30.5]

vector—the direction and distance associated with a given **translation**, often shown by an arrow of the correct length and pointed in the correct direction; in general, a quantity that may have two or more subquantities. [22.2]

Venn diagram—a geometric representation of a sample space, usually with a rectangular region containing circular regions representing events. [27.2]

vertex of a polygon—see **polygon**; of a polyhedron—see **polyhedron**; of an angle—see **angle**.

volume of a closed three-dimensional region—The number of three-dimensional units that could fit inside the region, or that could be used to make an exact model of the region. The units are usually cubic regions (e.g., cm^3). The metric system also uses the liter (litre). [24.2]; [25 (formulas for determining volumes)]

weighted average—an average recognizing that different items can have values which should be given different weights (e.g., an A in a 1-unit course versus an A in a 3-unit course, in calculating grade-point average). [15.4]

whole number—any of the numbers 0, 1, 2, 3, 4, etc.

z-score—a derived score representing the number of **standard deviations** a given score is from the **mean**. It is most useful with scores from a **normal distribution**. [30.6]

Index

Masters for Base Materials, Pattern Blocks, Dot Paper, and Nets

The following pages contain masters for these materials:

1. **Two-dimensional materials for homework in bases two, four, five** (two sheets), and **ten**

2. **Pattern Block patterns**   Most of the shapes are quite useful with fraction work, but the square and the narrow rhombus do not have easy rational relationships to the other four (they are good for geometry, however). The 3D blocks have one-inch sides.

   a. (Regular) hexagons (colored yellow in a common 3D version)
   b. (Isosceles) trapezoids (red)
   c. Wide rhombuses (blue)
   d. (Equilateral) triangles (green)
   e. Squares (orange)
   f. Narrow rhombuses (natural wood)

3. **Dot Paper**   Two pages of each of the isometric dot paper and the square dot paper are included. You may wish to photocopy a page when you need dot paper.

4. **Nets**   Patterns A through O give 3D shapes when they are cut out and taped.

SOME BASE
TWO PIECES

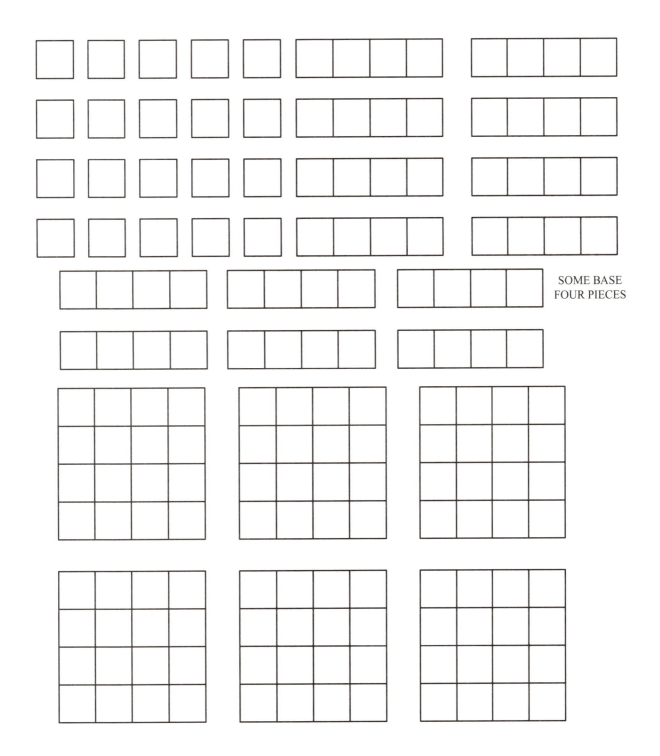

SOME BASE
FOUR PIECES

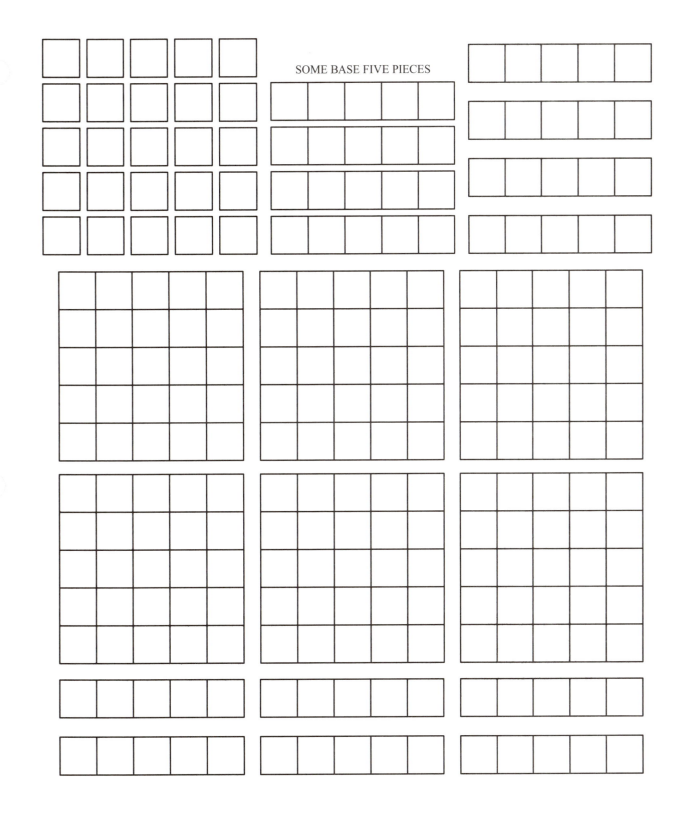

SOME BASE FIVE PIECES

SOME BASE FIVE PIECES

SOME BASE TEN PIECES
(not based on centimeters)

BASE TEN FLATS

Isometric Dot Paper

Square Dot Paper

In commercially available Pattern Blocks, the hexagonal regions are usually yellow and the trapezoidal regions are usually red.

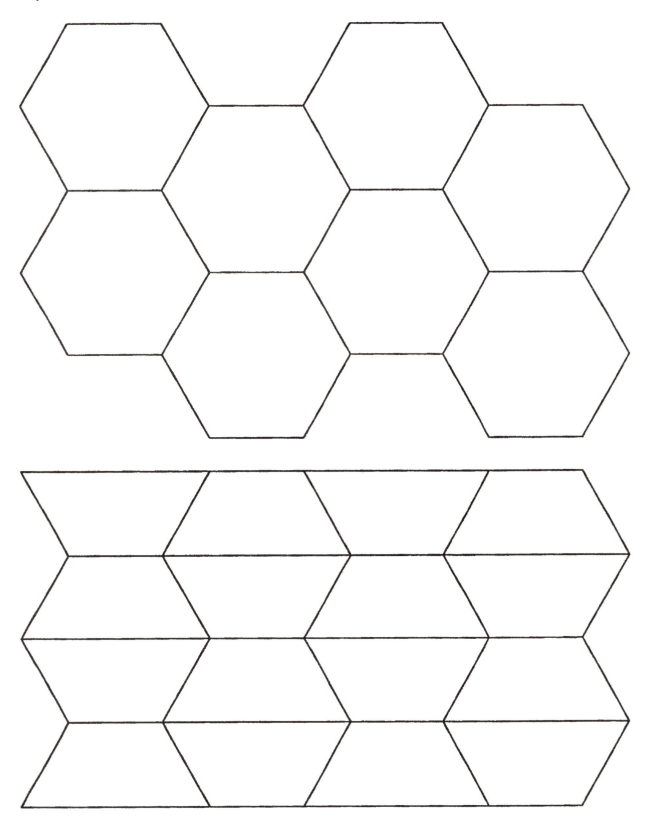

In commercially available Pattern Blocks, the triangular regions are usually green and the square regions are usually orange.

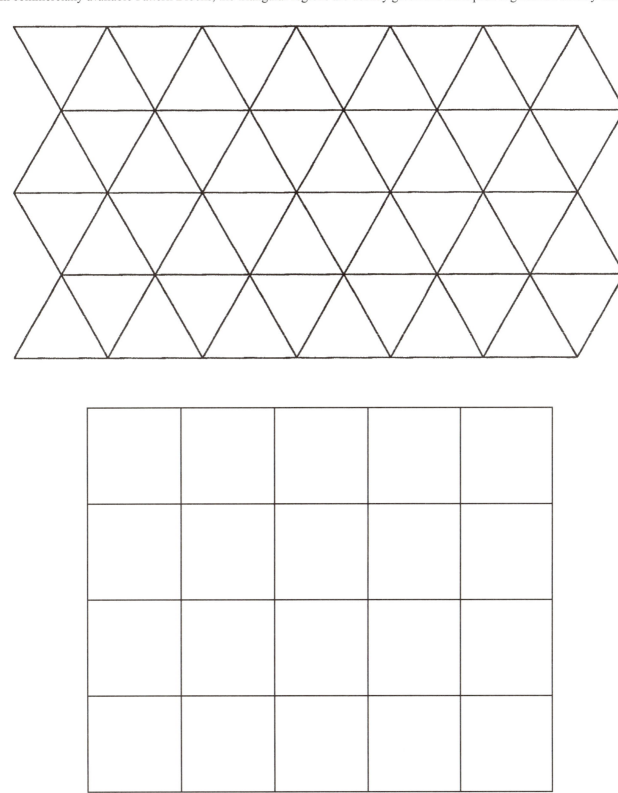

In commercially available Pattern Blocks, the wide rhomboidal regions are usually blue and the narrow rhomboidal regions are usually natural wood color.

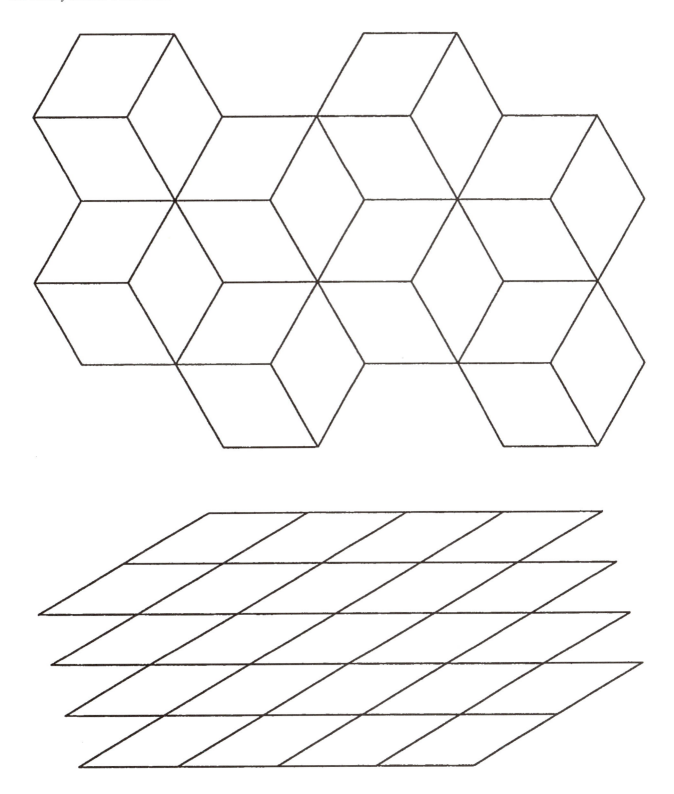

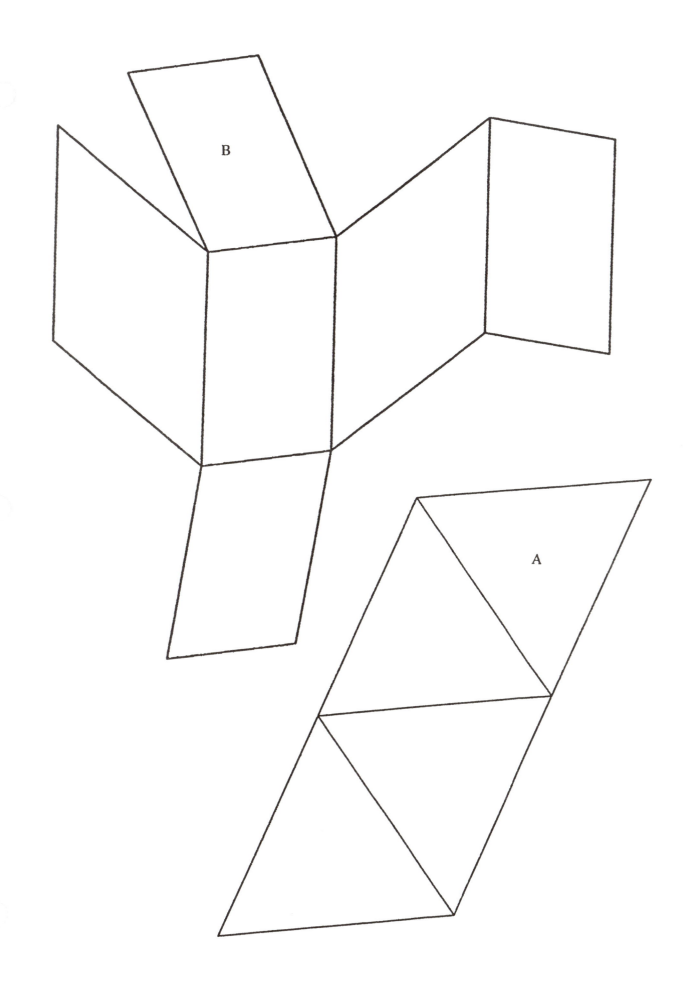

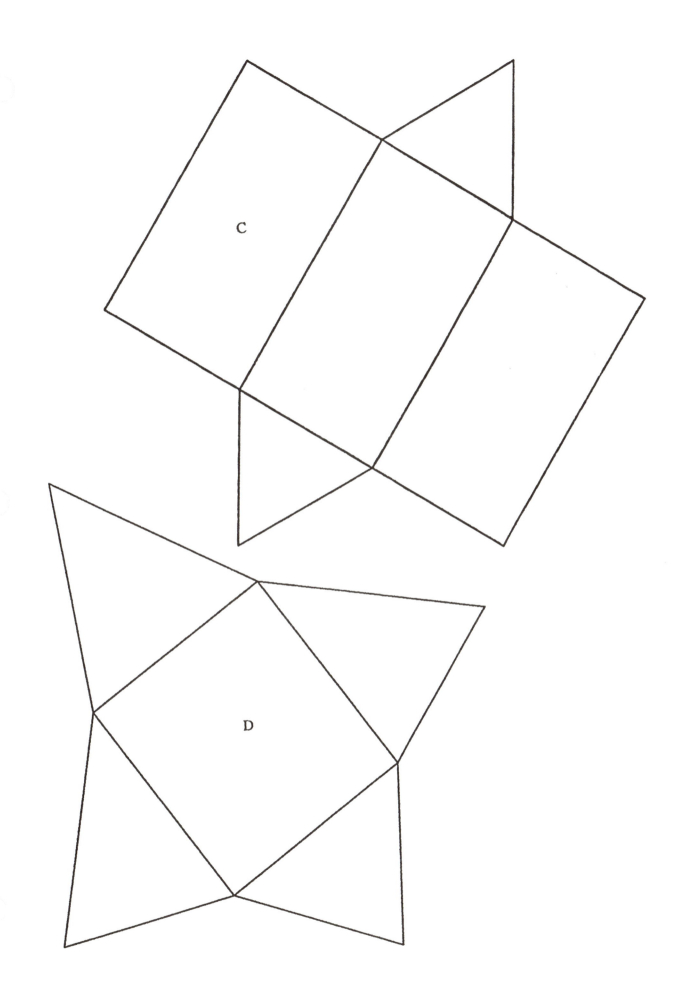

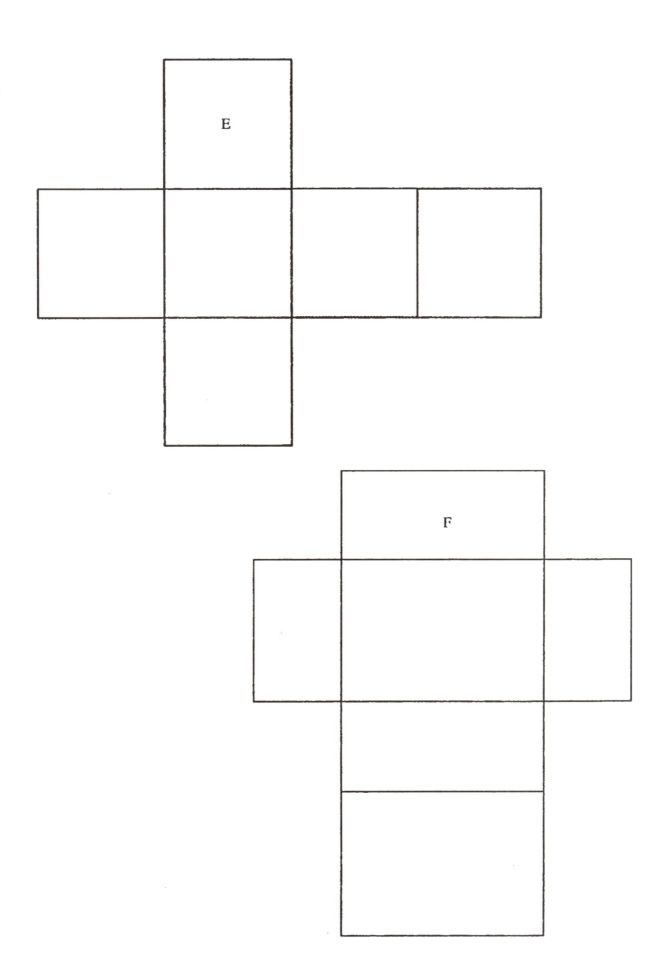

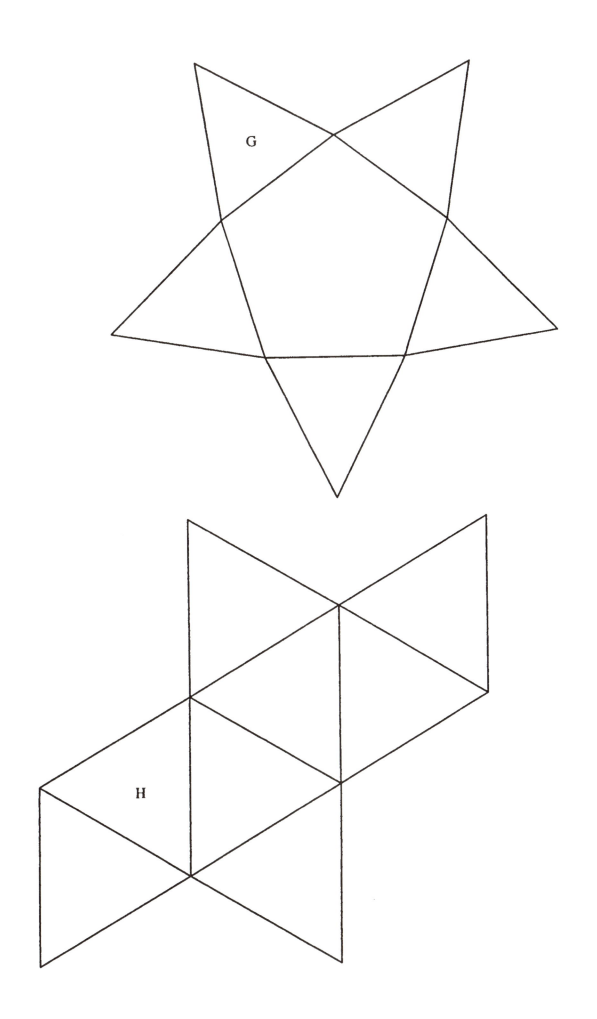

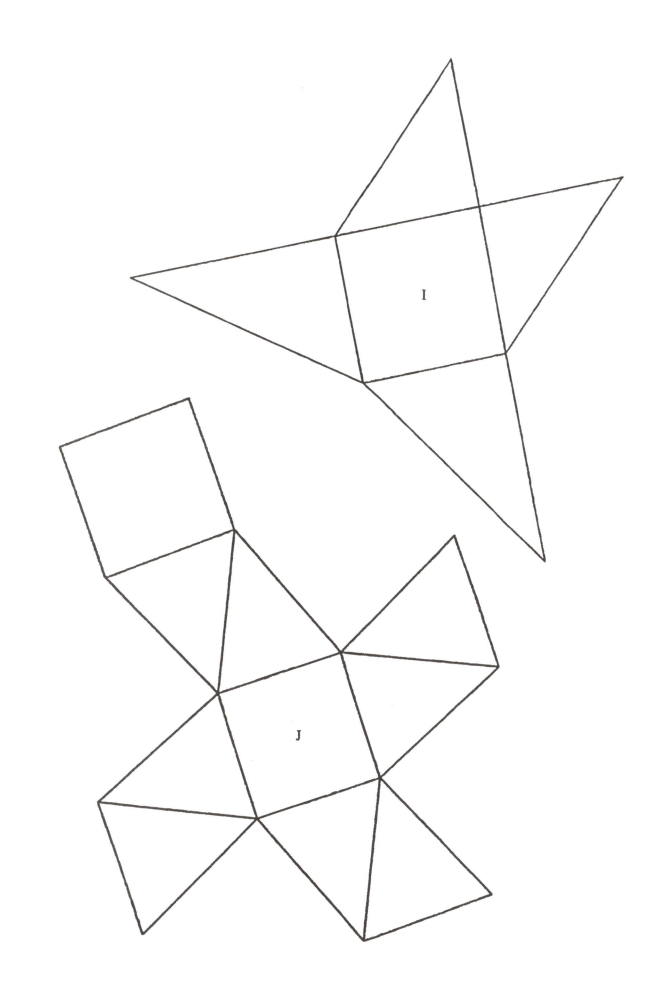

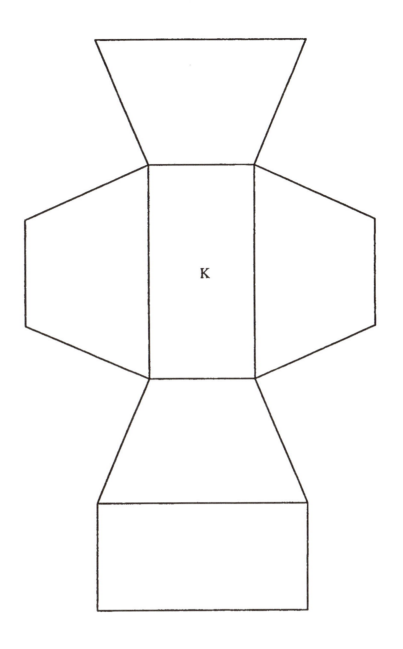

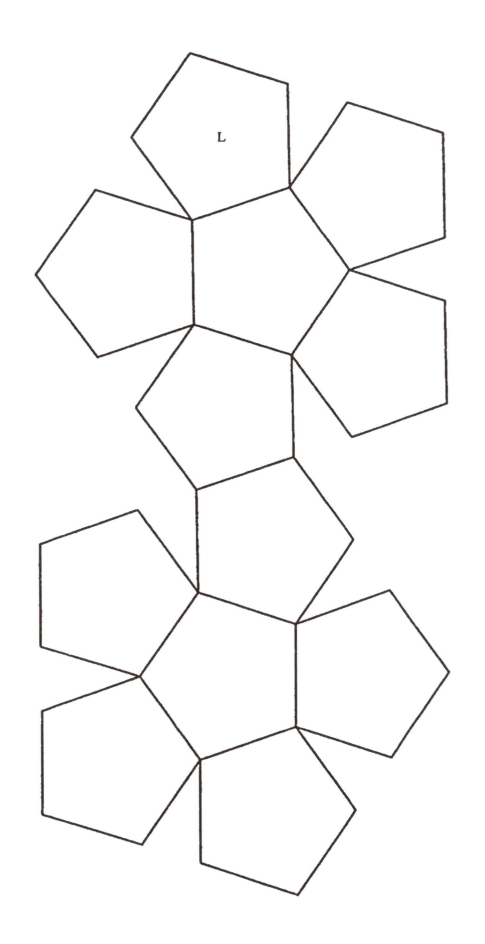

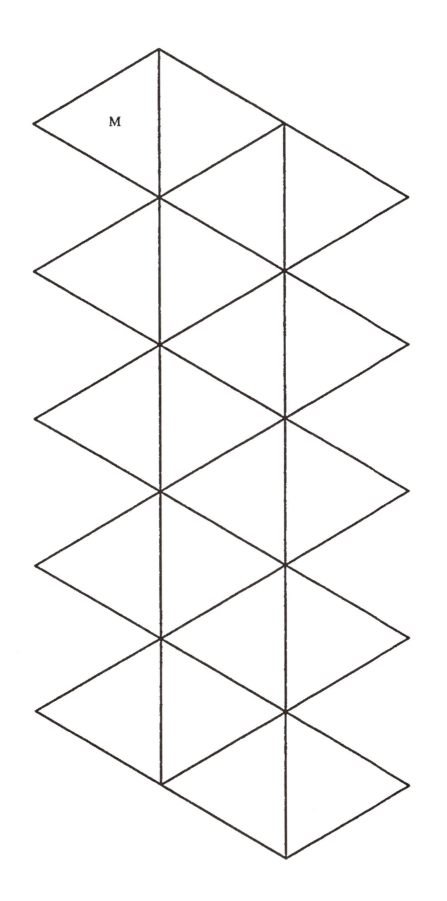

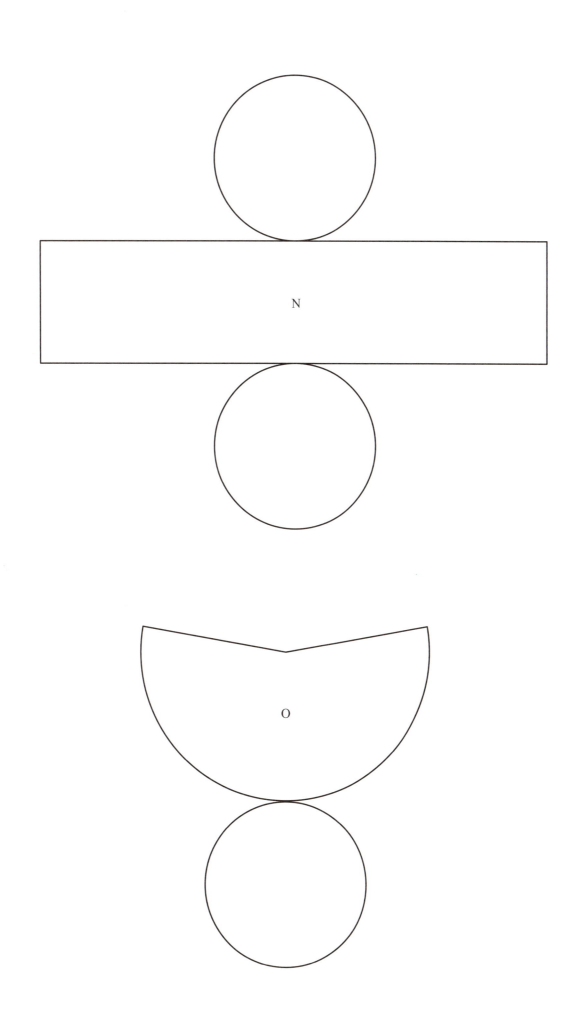